T0181653

Wastewater to Water

Makarand M. Ghangrekar

Wastewater to Water

Principles, Technologies and Engineering Design

 Springer

Makarand M. Ghangrekar
Department of Civil Engineering
Indian Institute of Technology Kharagpur
Kharagpur, West Bengal, India

ISBN 978-981-19-4050-7   ISBN 978-981-19-4048-4   (eBook)
https://doi.org/10.1007/978-981-19-4048-4

This Springer imprint is published by the registered company Springer Nature Singapore Pte Ltd.
The registered company address is: 152 Beach Road, #21-01/04 Gateway East, Singapore 189721, Singapore

Preface

Water supply systems and wastewater collection systems are among the most important infrastructure supports required for any society. This subject is imperative in the domain of Civil Engineering, Environmental Engineering and allied specializations. This book is written as a textbook, which encompasses all the necessary components involved in wastewater collection, treatment, reuse or disposal. Thus, this textbook will be beneficial for the undergraduate students of Civil Engineering and Chemical Engineering, and also for the postgraduate students of Environmental Engineering, Environmental Science and other associated interdisciplinary courses. Further, the design examples provided for designing the collection systems and wastewater treatment systems will be advantageous for practicing engineers involved in wastewater and sanitary engineering and design of industrial effluent treatment plants. To understand the topic covered in this book, a fundamental knowledge of intermediate-level science and mathematics is desired. Other than this, there are no specific prerequisites for understanding the topics covered. All other necessary background knowledge required is adequately described in every chapter.

This textbook provides exposure to the reader to what is wastewater, the different types of pollutants that can be present in it, and the emerging contaminants of concern. The necessity and requirements of collection systems have been described, theory of estimating the quantity of sewage generation and stormwater surface runoff and hydraulic design of the sewers are also elucidated. Fundamentals of reactor engineering are also well explained, which are not covered in many of the wastewater engineering textbooks. This will assist students to understand reactor engineering in a better and easier way and thus it will put them at ease in grasping the design of different reactors, which are covered later in this book. Further, the topics like self-purification of natural streams, physical unit operations and chemical unit processes used in wastewater treatment are described in details covering fundamental principles involved along with solved examples.

This book also expounds on the fundamentals of biological wastewater treatment, which will be of great assistance to the students, not having a strong background in the subject of biology, to better apprehend the design of biological wastewater treatment systems. Aerobic, anaerobic and hybrid biological wastewater treatment systems for the removal of organic matter and nutrients are described with adequate details to make the students ready for designing wastewater treatment plants comprising of these treatment systems. Design procedures for popularly used biological treatment systems, such as activated sludge process, sequencing batch reactor, moving bed biofilm reactor, trickling filters, bio-towers, up-flow anaerobic sludge blanket reactor, rotating-disk biological contactor, hanging sponge reactor, oxidation ponds, facultative ponds, etc., are described with adequate theoretical depth for a new person to easily understand these. The design examples provided will also be beneficial for the professionals working in Wastewater Engineering. A separate chapter is dedicated to the management of sludge generated from effluent treatment plants. The solved examples provided and the end of the chapter exercises will also be a useful resource material for setting up question papers.

Tertiary wastewater treatment systems including advanced oxidation processes, disinfection, and emerging technologies required to be adopted for facilitating reuse of the treated water are elaborated with design examples, which will be additional strength for this book as many of these technologies have not been covered in other textbooks. A separate chapter expounding on onsite sanitation solutions has also been provided, with details on the microbial fuel cell-based onsite toilet waste treatment system with the capability to generate electricity, which is a unique feature of this book. The last chapter provides another unique individuality to this textbook and will educate the readers about economic analysis and cost comparison of the different technologies typically used in wastewater treatment, with examples considering typical Indian market prices. Once the concepts of evaluating life cycle costing are picked up by the readers, then these can be extended to costing for any country or region to facilitate comparison and selection of the most appropriate technology.

Thus, this book is written with the specific intent to completely educate undergraduate and postgraduate students in wastewater engineering and to aid working professionals in the field of Environmental Engineering. It typically covers the syllabus of Wastewater Engineering course offered by most of the Asian universities and even universities from other continents. The fundamentals, basic principles, related mechanisms and physical, chemical and biological aspects involved are elaborated for easy understanding of the students. However, for expanding knowledge on any of the topics, the readers may go through the key references provided at the end of each chapter. Thus, an attempt has been made to develop a basic subject understanding in wastewater engineering for the readers covering all possible topics and technologies that are popularly used in practice as well as describing the forthcoming technologies. After studying this book, the reader is expected to gain proficiency and develop the capability to design a treatment plant for any characteristics of wastewater to meet the desired treated water quality norms for disposal or reuse for various purposes.

Kharagpur, India   Makarand M. Ghangrekar

Acknowledgement

This book is dedicated to my family and all the students and colleagues who encouraged me to write a textbook and extended endless support in this endeavor. The author is deeply indebted to his teachers for making him capable to write the textbook 'Wastewater to Water: Principles, Technologies and Engineering Design'. The experience gained by the author, while undertaking teaching and research in this subject for nearly 30 years and handling industrial consultancy on effluent treatment plant design/troubleshooting of reactor operation, and designing or vetting of sewage treatment plant designs, developed an in-depth subject understanding of this topic while getting exposed to various technologies used in practice and developing an understanding of the detailed engineering involved in this subject. The author deeply acknowledges the assistance received from the Research Scholars working with him and their knowledge contribution in some of the chapters in this book. The chapter-wise contributions made by the Research Scholars in this book are: Shraddha Yadav—Self-purification of natural streams; Rishabh Raj and Akash Tripathy—Physicochemical operations and processes for wastewater treatment; Swati Das and Santosh Kumar—Fundamentals and principles of biological wastewater treatment; Dr. Neethu—Aerobic wastewater treatment; Dr. Sovik Das—Aerobic wastewater treatment systems and Emerging technologies; Dr. Pritha Chatterjee—Hybrid aerobic wastewater treatment system and Anaerobic wastewater treatment; Shreeniwas Sathe—Biological processes for nutrient removal and also Tertiary wastewater treatment systems; Arun Kumar—Sludge management; Dibyojyoty Nath—Natural systems; Monali Priyadarshini—Advanced oxidation processes; Dr. Indrajit Chakraborty—Disinfection and Economics of Wastewater treatment; Azhan Ahmad—Emerging technologies; Dr. Indrasis Das—Onsite sanitation and Dr. Gourav Dhar Bhowmick for his contribution in membrane processes covered in this book. Mahendra Mulkalwar provided valuable inputs on market prices for writing the economics of the wastewater treatment systems. Without backup support from these scholars, the author would not have taken up this task of book writing. Also, Dr. Atul Shinde and Manikanta Doki extended technical/editing assistance as and when required and the sons of the author, Onkar and Atharva, extended help during example problem-solving for the successful completion of this textbook.

The author sincerely thanks the publisher for providing the opportunity for writing a book on the topic of his professional interest and staff of the publishing house for extending excellent cooperation during this process of preparation of the manuscript, and further editing, typesetting and publishing. The author is grateful to the authority and stakeholders of IIT Kharagpur, India for the encouragement and support received during the entire process of book writing.

Makarand M. Ghangrekar,
Ph.D., MASCE, MEASA, FNAE

Contents

About the Author

Makarand M. Ghangrekar is a Professor of Environmental Engineering in the Department of Civil Engineering, Institute Chair Professor, Head, PK Sinha Centre for Bioenergy and Renewables and former Head of the School, Environmental Science and Engineering at Indian Institute of Technology Kharagpur, India. He is a Civil and Environmental Engineer and received his Ph.D. in Environmental Science and Engineering from Indian Institute of Technology Bombay, India. He had been visiting Scientist at Ben Gurion University, Israel and University of Newcastle upon Tyne, UK under Marie Curie fellowship by European Union and had stint as faculty of various capacities in renowned engineering colleges and research institutes. He has been involved in teaching of Wastewater Engineering, Wastewater Management, and Industrial Water Pollution Control courses for undergraduate and post-graduate levels for more than 26 years and the experience he gained during teaching and handling research and consultancy projects in this subject has been poured into this textbook for benefit of the budding engineers. He has provided effluent treatment plant designs for several industries in India and abroad and sewage treatment plant design/ vetting to many Government agencies. His basic research interests are in the subject areas of water and wastewater treatment systems, biological wastewater treatment, anaerobic wastewater treatment, and bio-electrochemical processes for wastewater treatment and value-added product recovery. During his professional career of 28 years, he has guided 25 Ph.D. Research Scholars and 52 Master students' projects. He has contributed 248 research papers in journals of international repute, and 52 book chapters. His research work

has been presented in more than 250 conferences in India
and abroad. He has also delivered invited lectures in the
various reputed universities globally. He is recognized worldwide in scientific community for his research contribution
in the development of bio-electrochemical processes and his
research group stands among the top three research labs in
the world in terms of scientific publications. The very first
of its kind microbial fuel cell-based onsite real human waste
treatment system coined as 'Bioelectric toilet' was developed
by him, which received extensive publicity in both electronic
and print media.

Abbreviations

AC	Annualized cost
ADP	Adenosine diphosphate
AF	Anaerobic filter
AMX	Amoxicillin
AnRBC	Anaerobic rotating biological contactor
AnSBR	Anaerobic sequencing batch reactor
AO7	Acid orange 7
AOPs	Advanced oxidation processes
ASP	Activated sludge process
ATP	Adenosine triphosphate
AUB	Acetate utilizing bacteria
BAF	Biological activated filter or Biological aerobic filter
BAT	Best available technologies
BCM	Billion cubic meters
BDD	Boron doped diamond
BEPs	Bio-electrochemical processes
BOD	Biochemical oxygen demand
BP-4	Benzophenone-4
BPA	Bisphenol A
BPM	Black plastic media
BP-S	Bipolar series
CB	Conduction bond
CBZ	Carbamazepine
CD	Current density
CE	Coulombic efficiency
cfu	*Colony-forming unit*
CI	Cast iron
CIP	Ciprofloxacin
CNT	Carbon-based nanotube
COD	Chemical oxygen demand

CPC	Compound parabolic collector
CPCB	Central Pollution Control Board
CPHEEO	Central Public Health and Environmental Engineering Organisation
CSTR	Continuous-flow stirred tank reactor
CWs	Constructed wetlands
DAF	Dissolved air flotation
DBD	Dielectric barrier discharge
DBPs	Disinfection by-products
DCF	Diclofenac
DDE	Dichloro-diphenyl-dichloroethylene
DDT	Dichloro-diphenyl-trichloroethane
DEHP	Di-Ethylhexyl Phthalate
DHS	Downflow hanging sponge
DINP	Di-Isononyl Phthalate
DLVO	Derjaguin Landau Verwey Overbeek
DNA	Deoxyribonucleic acid
DO	Dissolved oxygen
DOM	Dissolved organic matter
DRDO	Defence Research and Development Organisation
DS	Desalination
DSA	Dimensionally stable electrodes
DWC	Double walled corrugated
DWF	Dry weather flow
E. Coli	*Escherichia coliforms*
EA-ASP	Extended aeration activated sludge process
EBPR	Enhanced biological phosphorus removal
EC	Electrocoagulation
ECs	Emerging contaminants
ED	Electrodialysis
EDCs	Endocrine disrupting compounds
EDI	Energy dissipating inlet
EE	Electrical energy
EO	Electro-oxidation
EPR	Electron paramagnetic resonance
EPS	Extracellular polymeric substances
ESR	Electron spin resonance
F/M	Food to microorganisms ratio
FBBR	Fluidized bed bio-reactor
FB-SMAR	Fluidized-bed submerged media anaerobic reactor
FOG	Fat, oil, and grease

FRP	Fibreglass-reinforced plastic
FS	Flat sheet
FSL	Full supply level
FSW-CWs	Free surface water constructed wetlands
GAC	Granular activated carbon
GLS	Gas-liquid-solid
GRP	Glass-fibre reinforced plastic
GT	Gas transfer
HAA	Haloacetic acid
HC	Hydrodynamic cavitation
HDPE	High-density polyethylene
HEX	Hexafilter
HF	Hollow fibre
HOA	Hydrogen oxidizing acetotrophs
HOM	Hydrogen oxidizing methanogens
HP	Horsepower
HPLC	High-performance liquid chromatography
HRAP	High-rate algal pond
HRT	Hydraulic retention time
HSSF-CWs	Horizontal sub-surface flow constructed wetlands
HUB	Hydrogen utilizing bacteria
HVAC	Heating, ventilation, and cooling
INR	Indian Rupees
IUCN	International Union for Conservation of Nature
LCC	Life cycle costing
LED	Light-emitting diode
LP	Low pressure
LPCD	Litre per capita per day
MBBR	Moving bed biofilm reactor
MBR	Membrane bioreactor
MCC	Microbial carbon-capture cell
MCE	Mineralization current efficiency
MCRT	Mean cell residence time
MDC	Microbial desalination cell
ME	Membrane extraction
MEC	Microbial electrolysis cell
MF	Microfiltration
MFC	Microbial fuel cell
MFFs	Media filling fractions
MLD	Million litres per day
MLSS	Mixed liquor suspended solids

MLVSS	Mixed liquor volatile suspended solids
MMO	Mixed metal electrodes
MoEFCC	Ministry of Environment, Forest and Climate Change
MP	Medium pressure
MPN	Most probable number
MP-P	Monopolar parallel
MP-S	Monopolar series
MTH	Malathion
NED dihydrochloride	N-(1-naphthyl)-ethylene-diamine dihydrochloride
NER	Normalized energy recovery
NF	Nanofiltration
NHE	Normal hydrogen electrode
NHOA	Non-hydrogen-oxidizing acetotrophs
NOM	Natural organic matter
NPSH	Net positive suction head
NRB	Nitrate reducing bacteria
NSAID	Nonsteroidal anti-inflammatory drugs
NTU	Nephelometric turbidity unit
OCV	Open circuit voltage
OHPA	Obligate hydrogen-producing acetogens
OLR	Organic loading rate
OV	Operating voltage
PAH	Polycyclic aromatic hydrocarbons
PBSA	2-Phenylbenzimidazole-5-sulphonic acid
PCBs	Polychlorinated biphenyls
PCC	Plain cement concrete
PCPs	Personal care products
PE	Population equivalent
PEC	Proton exchange membrane
PES	Poly-ethyl-sulphone
PF	Plate and frame
PMCs	Persistent and mobile chemicals
PP	Polypropylene
PPCPs	Pharmaceutical and personal care products
PR	Process recovery
PST	Primary sedimentation tank
PV	Pervaporation
PVC	Polyvinyl chloride
PVDF	Polyvinylidene difluoride
PW	Present worth
PWF	Present worth factor

RAS	Return activated sludge
RBC	Rotating biological contactor
RBTS	Reed bed treatment system
RCC	Reinforced cement concrete
RCS	Reactive chlorine species
RNA	Ribonucleic acid
RO	Reverse osmosis
ROS	Reactive oxygen species
rpm	Revolution per minute
RTD	Residence time distribution
SAF	Submerged aerobic filter
SB	Sponge biocarriers
SBR	Sequencing batch reactor
SB-SMAR	Static-bed submerged media anaerobic reactor
sCOD	Soluble chemical oxygen demand
SDS	Sodium dodecyl sulphate
SEC	Specific energy consumption
SF-CWs	Surface flow constructed wetlands
SHRI	Sludge Hygienization Research Radiator
SLR	Solid loading rate
SMA	Specific methanogenic activity
SMP	Soluble microbial products
SMX	Sulfamethoxazole
SOD	Sediment oxygen demand
SOR	Surface overflow rate
SPD	Sustainable power density
SRB	Sulphate reducing bacteria
SRT	Solid retention time
SS	Suspended solids
SSF-CWs	Sub-surface flow constructed wetlands
SST	Secondary sedimentation tank
SSV	Settled sludge volume
STPs	Sewage treatment plants
SVI	Sludge volume index
SW	Spiral wound
SWMM	Storm water management model
TB	Tubular
TC	Total cost
TCF	Trichlorfon
tCOD	Total chemical oxygen demand
TDS	Total dissolved solids

TF	Trickling filter
TFN	Thin film nanocomposite
THC	Total hydrocarbon
THM	Trihalomethane
ThOD	Theoretical oxygen demand
TKN	Total Kjeldahl Nitrogen
TMP	Transmembrane pressure
TOC	Total organic carbon
TP	Total phosphorus
TPA	Terephthalic acid
TS	Total solids
TSS	Total suspended solids
UAnSB reactor	Up-flow anammox sludge bed reactor
UASB	Up-flow anaerobic sludge blanket
UF	Ultrafiltration
UNESCO	United Nations Educational, Scientific and Cultural Organization
UPVC	Unplasticized polyvinyl chloride
UPWF	Uniform present worth factor
US	Ultrasounds or Ultrasonication
USEPA	United States Environmental Protection Agency
UV	Ultraviolet
VB	Valance bond
VFA	Volatile fatty acid
VOCs	Volatile organic compounds
VS	Volatile solids
VSSF-CWs	Vertical sub-surface flow constructed wetlands
WHO	World Health Organization
WT	Water treatment
WWT	Wastewater treatment
WWTPs	Wastewater treatment plants

Introduction

1

1.1 Background

The ever-increasing growth of modern civilization and the consequential pollution load that it puts on the bio-geosphere is alarming (Kovalakova et al., 2020). Rapid urbanization not only has led to the increase in generation of liquid and solid waste from cities and urban communities, however, such urbanization also has considerably depleted the natural water resources. Additionally, intense water usage, to cater to the demands of urbanization, has led to a severe water crisis scenario globally. Global water demand is presently estimated as 4600 billion m^3 per year and it is projected to rise to 6600 billion m^3 per year by 2050 (Boretti & Rosa, 2019).

The countries with high population density will face more severe water crisis as the disparity between availability and supply would be highest in such places. For example, in a report by Government of India, in June 2018, India witnessed the worst water crisis in the recorded history of the country (Shah & Narain, 2019). By 2030, it is projected that India's water demand will be twice the available supply, implying severe water scarcity for hundreds of millions of people. As per the report of National Commission for Integrated Water Resource Development, the per day water requirement by 2050 in high use scenario is likely to be about 1,180 BCM (billion cubic meter); whereas, the present-day availability is 695 BCM. The total availability of possible water in the country (1,137 BCM) is also slightly lower than this projected demand (NITI Aayog, 2018). Although such estimates are often conservative in nature, however the figures do not fail to impress upon the awful need of alternative water sources.

In the nineteenth century, technology innovation, industrialization and urbanization has occurred without properly addressing the environmental ethics. However, growing concern of fresh water scarcity and severe pollution in existing water bodies had compelled the then engineers to have pragmatic thinking to address the environmental issues. In 1870s, it was realized for the first time, that the improvement in human health could be brought about by improvements in the sanitation infrastructure. Land application of sewage that time was not explored as an option and sewage was discharged in rivers in the pretext that the river offers sufficient dilution to keep the biochemical oxygen demand (BOD) under control. Later in 1914, when it was realized in England, by Arden and Lockett, that if the sewage is aerated in the settling tank for a few hours it led to reduction in oxygen demand and this technology was named as activated sludge process (ASP) (Kiely, 1998), which is still one of the most popularly used aerobic wastewater treatment processes for sewage and industrial wastewaters.

Around the same time, in many cities of America, piped water supply was adopted by supplying water abstracted mainly from surface water sources, such as rivers, lakes and reservoirs. The sewage generated was collected from a separate network of pipes and after removal of visible suspended impurities it was discharged back to the water bodies. However, this practice resulted in the typhoid epidemic in many American cities that time and the necessity for further treatment of sewage before discharging to the water body was realized. Thus, necessity of appropriate sewage treatment has been realized for protection of human health.

In light of this situation, the regulatory bodies and policy framing authorities around the globe are focussing on framing more stringent norms for wastewater treatment, focussing not only on reducing pollution of water bodies but also on encouraging reuse of treated water so as to reduce stress on fresh water. The Environmental Protection Agency of United States of America (USEPA) came up with a nation-wide water reuse action plan, the first version of which was floated in public domain on February 2020 (EPA, 2020). Asian countries like India and Singapore are also emphasizing on the reuse of treated wastewater for different purposes. The Central Public Health and Environmental Engineering Organization (CPHEEO), and the Central Pollution Control Board (CPCB) are the two regulatory bodies of the Government of India that have laid down guidelines for onsite and offsite reuse of treated sewage and industrial wastewaters (CPHEEO, 2013; Schellenberg et al., 2020).

In addition, water recycling projects, such as NEWater executed by the Singapore Government (Bai et al., 2020) and reuse of treated wastewater in agricultural fields in Israel, indicate that there is a silent yet powerful paradigm shift in the water usage practices globally. Although, the lessons learnt from the experience of different countries that facilitated water reuse express that this shift can only be successful through rigorous trial and error and it is the longer chosen path towards water sustainability, as such steps require implementation of energy intensive and often high-cost treatment options as well as huge infrastructural support, to produce assured quality treated water for reuse (Tal, 2006).

In previous century, many of the water supply and sewerage projects were approved considering the need of the project for the society without proper evaluation of the environmental impacts. However, now for any engineering developmental projects, environmental impact assessment is made a mandatory part of the proposal, which should address adequately the positive/negative impacts that such projects are likely to cause to the environment. No longer any project can be pushed forward only based on the economic benefit or societal need without properly evaluating its environmental impacts and properly addressing environmental ethics for long term protection of ecology, environment and people. The sewerage projects are not exception to this and environmental impacts of such projects need to be evaluated. This is particularly very important while taking a final decision on disposal of the treated sewage and to estimate the possible environmental impacts that such disposal will have. However, many of the engineering educational courses are still lacking in teaching environmental ethics of such projects up to the mark, which need to be urgently addressed to make future engineers aware of these concerns and to avoid environmental ill effects of any such engineering project.

1.2   Water and Wastewater

Water is a renewable resource present on the earth, which is naturally recycled in the hydrological cycle, with surface waters having short residence time as compared to the groundwater, where the groundwater is having long residence time. This hydrological cycle renews the water and provide continuous source of water. In this hydrological cycle evaporation is more pronounced from ocean (approximately 7 times) as compared to that from land surface. This water vapour rises into the atmosphere and the lower temperature aloft causes condensation of it and then it precipitates mostly in the form of rain and sometimes as snow. The precipitation that is received on land, depending on properties of catchment area and soil cover, some fraction forms surface run-off, some infiltrate in soil and some fraction is returned to the atmosphere via evaporation.

Out of total quantity of water on the earth, 96.5% is available as saline water in ocean and remaining 3.5% is distributed as ground water (both fresh and saline put together as 1.69%), polar ice (1.7%), lakes (0.013%), rivers (0.0002%), soil moisture (0.0012%), and remaining as marshes, biological water and atmospheric water (Kiely, 1998). Out of this total water available the fresh water reserves are only 2.5%, majorly distributed as ground water of about 0.76%, lakes and rivers contributing 0.0072% and remaining 1.7% as polar ice. Thus, out of total quantity of freshwater reserves polar ice represents 68.6%, groundwater about 30.1%, lakes 0.26% and marshes and rivers about 0.03 and 0.006%, respectively.

For fulfilment of our need for water, traditionally the human being is dependent on this scanty freshwater reserves. The urban civilization, agriculture and industrialization traditionally was dependent on surface and ground water. With the commencement of industrialization, intensification of agriculture and increasing human population, the water demand has increased dramatically, resulting in stress on the fresh water reserves and due to limited availability of fresh water in the recent time, engineers are also exploring possibilities of utilizing sea water by adopting certain desalination technologies or membrane processes, like reverse osmosis, to convert this saline water into fresh water to facilitate use in domestic and industrial applications.

The variety of water sources explored for urban and rural water supply systems may include groundwater, surface water, spring water, desalinated seawater, brackish water, and reclaimed water (Meron et al., 2020). For example, Spain, USA, Israel, Singapore, and Saudi Arabia are heavily relying on water reuse and desalination (IWA, 2014; Tal, 2006), while in Denmark up to 2010, groundwater was the only water supply source. In Sweden and Germany, in the same year, only 22% and 61%, respectively, of the supplied water came from groundwater, the remaining quantity was coming from surface water, spring water, and artificially recharged groundwater (IWA, 2014).

The importance of water quality as a factor constraining water use has often gone unacknowledged in the analyses of water scarcity. Though water is renewable, however freshwater resources are finite and per capita availability of fresh water is reducing every year. It is estimated that the potential water availability per capita (m^3/year), which was 17,532 in 1960 and 10,058 in 1980, will reduce to 5,717 in the year 2025 (Sophocleous, 2004). Here it is important to understand that a country is said to be water stressed when annual water supplies drop below 1,700 m^3 per person. According to this definition many countries or regions in the countries are water stressed as on today. Hence, judicious use of available water resources and avoiding pollution of the water body is utmost necessary to ensure adequate quantity, with necessary quality of the water available for future generations. It is necessary to get convinced that though the water is to a large extent a recyclable resource, however protecting its quality to make it usable is our responsibility. The urban water cycle that exists today is depicted in the Fig. 1.1. The linkage between the sources of water, wastewater generation by stake holders, the role of the water and wastewater treatment plants and disposal/reuse of the treated water is depicted in the Fig. 1.1, formulating urban water cycle.

The water is said to be pristine when it is fresh, clean and free from any pollution. Any changes in the natural quality of water due to manmade activities or due to receipt of hydrological storm runoff carrying lot of sediments load or any geogenic activity implies pollution of it. Natural events such as floods, mud flows, hurricanes can also lead to deterioration of water quality. However, the man-made activities, such as discharge of untreated sewage, industrial effluents, uncontrolled recreational activities and receiving urban storm water in water bodies, can lead to sever water pollution. Pollution of water bodies means directly or indirectly introduction of substances that will result in harmful effects to aquatic life, pose hazards to human and animal health, bring hindrance to aquatic activities and bring impairment of water quality to make it unsuitable for agriculture, domestic, industrial as well as recreational usages. Disposal of hot water, even if free from other pollutants, will also lead to water pollution as it will alter the spectrum of life otherwise supported by the water body and bring change in the dissolved oxygen available in the water.

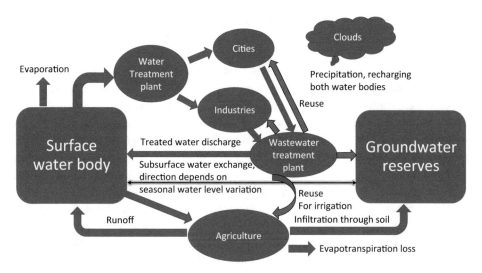

Fig. 1.1 Water cycle that exists nearby urban areas. Green arrows should be encouraged more after ensuring proper quality standards to facilitate treated wastewater reuse for sustainable urban water management

Land–water linkage also indirectly influences the water quality and affect the aquatic life in the water bodies. Land use for urbanization, industrialization, agriculture, landfill sites and other recreational activities all influence the quality of water in nearby water body and hence aquatic life supported by it. For example, intensive application of fertilizers in agriculture land will lead to increase in concentration of these nutrients in the storm water runoff from these agricultural fields, leading to increased concentration of nutrients in the adjoining water body. Similarly, leachate from landfill site, if reaches the water body, will add recalcitrant compounds and heavy metals, apart from a range of organic contaminants to the water body. Construction of hydraulic structures, such as dams, water diversion structures within and between catchments and canals, can also have a bearing on the change in the water quantity as well as quality.

For any of the water use there are specified quality requirements defined in terms of physical, chemical and bacteriological quality requirement to ensure the water is suitable and safe for that specified purpose. If water does not meet these quality requirements, it is deemed to be of poor quality and unacceptable for that particular use and hence require quality manipulation to make it suitable for that specific use, thus demanding treatment prior to use of it. Depending on a particular use, the required water quality is decided, which is actually some value judgement on the constituents present in the water. Hence, defining a water quality suitable for all uses is difficult or rather not practicable.

For domestic water supply the higher standards of the water quality are essential; whereas, the water quality requirement for irrigation is not that stringent and for industry water supply the quality requirement varies from product to product. Agriculture, by far the largest consumer of water, also suffers when water supply becomes saline due to over-exploitation of groundwater. When water authority supply water in urban area, water quality requirement for potable purpose is targeted and accordingly the raw water from source is being treated to meet this potable water quality. Though the quality parameters vary from country to country, however for most of the important parameters the acceptable values recommended by various authorities are mostly comparable.

For distributing water for urban area to meet domestic water demand, depending on the seasonal variation in the water quality at source, it is necessary to treat this raw water collected from the source. This raw water source is typically located some distance upstream from the city so as to tap better quality water. The treatment plant is generally located near the city and the treated water, after ensuring the quality requirements for potable use, is distributed to the city through water distribution system, which is huge network of pipelines. Upon use of this water, for various purposes in the city, the wastewater generated is collected through another network of pipes, called as sewerage. Generally, this sewage collected through conduit is taken to a downstream location away from the city, and having lower elevation to explore maximum gravity flow, where it is treated and disposed off. The treated sewage may be discharged either back to the water body or utilized for irrigation or for industrial water supply, if industries are around.

While treating this sewage the pollutants level present in sewage is brought down to meet the stipulated discharge norms defined by the regulating agency. Therefore, when this treated sewage is discharged to water body, depending on the dilution that is offered by the water present in the water body some level of pollutant concentration will still be there. Though naturally water quality may get replenished particularly for oxidizable organic pollutants, however due to continuous discharge of this partially treated sewage in the same water body the water quality might not get naturally replenished to the desired level. For the downstream urban area if this water is to be taken as raw water, either it will be unfit in quality or this will add additional burden on the treatment of water to make it potable and hence escalate the cost of treatment. This aspect is very important in managing urban water supply and sewage and the engineers involved must do holistic planning rather than providing solution for a city in isolation.

After use for one purpose, the resulting water can be potentially reused for satisfying water demand for some other activity by providing necessary affordable treatment to it, if required. Domestic, industrial and agricultural activities produce large amount of used water containing a wide range of pollutants in it. Though the natural waterways offer readily available conduits for these waste streams, this practice leads to pollution of the ground water as well as surface water bodies, making them unfit for downstream use. Hence, these waste streams should be properly treated before they are getting discharged in the water bodies. In spite of providing treatment to meet the regulatory norms, pollutants in considerably reduced concentrations are still getting discharged in the water bodies; hence, making it imperative to treat water before using it to fulfil domestic and industrial needs as per the quality requirements. Treatment before use is required to ensure that the unwanted constituents are removed to meet the quality requirement for specified use, which may be domestic water supply, industrial water supply, irrigation water requirement and aesthetic enhancement of the landscape of the city and accordingly the quality requirement will vary. However, after use for any of the purpose the treatment of wastewater generated is necessary to meet the regulatory norms for discharge/disposal of this treated effluent in the environment.

Water quality degradation in water bodies interferes with vital and legitimate use of water at local, regional or international levels. Setting quality criteria for water and defining standards are therefore necessary to ensure that the appropriate quality water is made available for a particular purpose. These regulations are being used as administrative means to manage and maintain quality of water for the users. These water quality standards are set by international agency, such as World Health Organization (WHO), European Union, or country specific federal agencies or state specified standards. The objective of imposition of such standards is the protection of the end user, such as humans, animals or industries. These standards are thus set taking into cognizance the different possible uses of that water. These standards generally cover the limits for pollutants/substances likely to remain present in water, such as suspended solids, dissolved organic matter, pathogens, nitrogen, salinity, trace elements, etc., and some of the countries are even coming up with the standards for micropollutants or emerging contaminants.

1.3 Environmental Legislation Related to Water

Most of the countries have their properly laid down quality standards for controlling the quality of the treated sewage discharged to the environment, thus in a way regulating the pollution likely to be caused by this treated sewage discharge. For member countries, under urban wastewater treatment regulation, 2002, European Union (Sl. No. 254/2001), regulates the concentration of the pollutants to be present in treated sewage. According to these regulations BOD_5 at 20 °C of wastewater discharged to surface water bodies should not exceed 25 mg/L, chemical oxygen demand (COD) not to exceed 125 mg/L and total suspended solids not to exceed 35 mg/L. Depending upon the population size, the phosphorous and nitrogen are regulated as 2 and 15 mg/L, respectively, for population between 10,000 and 100,000 and for cities having population more than 100 thousand these values are 1 mg/L and 10 mg/L, respectively.

In 2016, the European Union (Official Journal of European Union, L 152, 09-06-2017) further provides guidelines for identifying best available technologies (BAT) for common wastewater management systems. Which states that this BAT is not applicable where average annual BOD level in the effluent from biological wastewater treatment plant is ≤20 mg/L. It is recommended that either BAT for TOC or COD will be applicable, with TOC being preferred, with the yearly average values in the range of 10–33 and 30–100 mg/L, respectively. The lower values are to be achieved when few tributary wastewater streams containing mostly easily biodegradable organic matter is being treated; whereas, the upper end of the range may be up to 100 mg/L for TOC and up to 300 mg/L for COD, both yearly averages, provided the removal efficiency is $\geq 90\%$ as a yearly average and the BOD level in effluent is ≤ 20 mg/L. The total suspended solids (TSS) concentration is stipulated as 5–35 mg/L, considering that the lower range can be easily achieved by sand filtration, micro/ultra-filtration or membrane bioreactors.

For United Sates the Environmental Protection Agency (EPA) recommends BOD limit of 30 mg/L and TSS limit of 30 mg/L for 30 days average values. For BOD determination using nitrification suppression the limit is 25 mg/L (EPA, 2010). In Environmental (Protection) Amendment rules, 2017 (GSR 1265(E)), Ministry of Environment, Forest and Climate Change notification, Government of India, the applicable quality criteria for India for discharge of sewage treatment plant effluent into water bodies as well as for land disposal/application are defined for metro cities, state capitals and other cities separately. These metro cities include Mumbai, Delhi, Kolkata, Chennai, Bengaluru, Hyderabad, Ahmedabad and Pune. Some of the state capitals are given some relaxation, such as capital of north-eastern states, Jammu and Kashmir and Union territories. For these metro cities and selected state capitals the limit for BOD is 20 mg/L and for TSS the limit is 50 mg/L. Whereas, for any city in the country the limit for faecal coliform count is less than 1000 most probable number (MPN) per 100 mL. For other cities and some of the state capitals and Union territories the BOD is relaxed up to 30 mg/L and TSS up to 100 mg/L.

However, the National Green Tribunal of India has recommended the BOD limit for mega-cities (population more than 10 million) and metro-cities (population more than a million) to be 10 mg/L, limit for COD to be 50 mg/L and desirable limit for faecal coliform to be 100 MPN per 100 mL and permissible limit 230 MPN per 100 mL. There is also a limit recommended for total nitrogen and phosphorous as 10 mg/L and 1 mg/L, respectively, for these mega- and metro-cities. Some relaxation is recommended for other smaller cities and for disposal in deep marine outfall. Similarly, the quality criteria for regulating the presence of pollutants in the treated sewage before it is discharged to environment is set by other countries and for most of the nations these limits are in the similar range as stated above (Appendix 1.1).

1.4 Necessity and Essential Requirements of Wastewater Treatment System

The water body which is considered to be suitable for public water supply should be reasonably free from pollution. There are guidelines laid by different nations for defining the quality of raw water, which can be acceptable for public water supply; however, this water is treated further in water treatment plant to ensure that the quality of water meets potable water quality standards for that country. Generally, for the treatment of raw water collected from impounded reservoir the conventional water treatment plant comprises of cascade aeration for removal of dissolved gases, such as ammonia and hydrogen sulphide, and imparting freshness to the water by solubilizing more oxygen in water. This is followed by coagulation, flocculation and sedimentation for removal of majority of the colloidal solids. The water after clarification is treated further in rapid sand filters to remove the smaller flocs to produce treated water with turbidity typically close to 1 NTU (Nephelometric turbidity unit) and in any case less than 5 NTU. For such turbidity free water, the disinfection will be effective, which follows the sand filter. This disinfected water after ensuring some minimum residual dose of disinfectant, typical 0.2 mg/L of free residual chlorine, is considered to be safe and distributed for domestic consumption.

Thus, it is to be understood that the water treatment plants have capability to remove only suspended solids, typically colloidal particles, and killing pathogenic microorganisms present in the raw water. Thus, the raw water having soluble impurities higher than potable limits, cannot be accepted as a source water for public water supply as conventional water treatment plants are not having capabilities to remove such soluble organic or inorganic matters. Rather such compounds present will react with chlorine and may form further harmful and toxic disinfection products. Thus, one must remember that there is conceptual difference between water treatment plant and sewage treatment plant, since the purpose and scope of both are entirely different.

The daily activities of human beings produce both liquid and solid wastes. The water that is supplied by the water authority or water extracted from the private water sources, after it is being used for domestic activities, such as cloth washing, bathing, utensils washing, and flushing toilets, gets fouled and forms the liquid waste stream generated from the houses. The sources of wastewater generation can be defined as a combination of the water used for carrying the waste generated from residences, institutions, and commercial and industrial establishments. Depending upon the type of sewerage system used, this wastewater generated from households, when transported through sewers, may get mixed with groundwater, surface water and storm water, thus increasing its quantity further.

If the untreated wastewater is allowed to accumulate or it is discharged in water bodies, it will lead to highly unhygienic conditions. The organic matter present in the wastewater will undergo decomposition with production of malodorous gases. If the wastewater is discharged without treatment in the water body, this will result in the depletion of dissolved oxygen (DO) from the water bodies. Due to depletion of DO, the survival of aquatic life will become difficult, finally leading to anaerobic conditions in the receiving waters. The nutrients present in the wastewater can stimulate the growth of aquatic plants, leading to problems like eutrophication.

In addition to organic matter and nutrients, the untreated domestic wastewater usually contains numerous pathogenic or disease-causing microorganisms, that dwell in the human intestinal tract. The wastewater also contains inorganic suspended solids such as gritty materials, silt and clay. The continuous deposition of this inorganic materials may reduce the capacity of water body considerably over a period. Though, total solids content, both soluble and suspended solids put together, in sewage is typically less than 1% and remaining more than 99% is water, still this quantity of solids present in sewage pose considerable nuisance, if released untreated in environment. Hence, it becomes necessary to treat the sewage before it is disposed either in waterbody or on land. Since sewage contains more than 99% water, the fundamental laws of fluid mechanics for design of sewerage infrastructure are applicable for sewage, of course with some change in properties/coefficients as compared to freshwater.

The nature of treatment given to the sewage at treatment plant is way different than what is there in water treatment plant producing potable water, as discussed previously. Sewage treatment plant is designed to remove floating solids and settleable suspended solids during primary treatment and colloidal and soluble organic matter removal in secondary treatment. Depending on the effluent standards applicable a nitrogen removal from secondary treated effluent may be facilitated and for disposal of this treated sewage in water bodies, likely to be used as source of potable water, this secondary treated effluent is disinfected. However, the xenobiotic compounds such as personal care products, antibiotics and other medicines, which are generally present, though at very low concentrations, will not be considerably removed during primary and secondary treatment of sewage and may form disinfection by-products during chlorination of secondary treated effluents. Formation of these disinfection by-products might be more harmful than the basic compounds present in secondary treated effluent at low concentration. Hence, proper care in selection of disinfectant need to be exercised when this disinfected water is discharged in water bodies, which are to be used as source water for public water supply.

1.5   Objectives of Sewage Collection, Treatment and Disposal

The objective of sewage collection, treatment and disposal is to ensure that sewage discharged from communities is properly collected, transported, and effectively treated to the required degree. Thus, this treated sewage when discharged will not to cause threat to human health or unacceptable damage to the natural environment and it is safely disposed off without causing any health or environmental problems. Thus, an efficient sewerage scheme should have the following goals:

- To provide a good sanitary environmental condition of the city ensuring protection of public health.
- To efficiently transport, treat and dispose-off all liquid wastes generated from the community to prevent any favourable conditions for mosquito breeding, habitat for flies or bacteria growing.

- To treat the sewage, as per requirement, so as not to endanger the water body or groundwater or land to get polluted where it is finally disposed-off. Thus, it should ensure protection of the receiving environment from degradation or contamination.
- In water stress areas, the sewage is to be treated to meet the desired quality norms to encourage reuse of treated sewage for various purposes in the locality.

1.6   Wastewater Treatment

The efficient collection, treatment and safe disposal of wastewater generated from urban area is necessary. This will facilitate protection of environment and environmental conservation, because the wastewater collected from cities and towns must ultimately be returned to receiving water body or to the land or shall be reused to fulfil certain needs, such as industrial reuse, horticulture applications, aesthetic enhancement, etc. The sewage treatment plants constructed near the end of nineteenth century were designed to remove suspended matter alone by the principal of simple gravity settling as provided in primary treatment. It soon became apparent that this primary treatment alone was insufficient to protect the water quality of the receiving water body. This was mainly realized due to the presence of organic material, in colloidal and dissolved form, in the sewage after primary settling. Thus, in the beginning of twentieth century several treatment systems, called secondary treatment, were developed with the objective of removal of soluble and colloidal organic matter from primary treated sewage. For this secondary treatment, biological methods are generally used. Aerobic biological treatment processes were popularly used as a secondary treatment, and are still the first choice.

In the second half of twentieth century, it became clear that the discharge of effluents from even the most efficient secondary treatment plant could lead to the deterioration of the quality of receiving water body. This could be attributed partly to the discharge of ammonia in the effluent. In the receiving water body, this secondary treated effluent exerts an oxygen demand for the biological oxidation of ammonia to nitrate, a process called nitrification. However, even when nitrification is carried out at the treatment plant itself, the discharge of effluent can still be harmful to the water quality due to introduction of nitrogen in the form of nitrate and phosphorus as phosphate. The tolerance limits of nitrates for the water when used as raw water for public water supplies and bathing ghats is 50 mg/L as NO_3^- (WHO, 2017).

The availability of nitrogen and phosphorous tends to cause an excessive growth of aquatic life; notably, autotrophic organisms such as algae, that can use carbon dioxide rather than organic matter as a source of carbon for cell synthesis. Thus, explosive growth of autotrophic biomass can occur when nitrogen and phosphorus are abundantly available. Although, this biomass will produce photosynthetic oxygen in the water during daytime, however after sunset it will consume oxygen; hence, the DO concentration will decrease and might reach to the levels that is too low to sustain the life of other (macro)organisms. This phenomenon of eutrophication has led to the necessity of tertiary treatment systems, in which nitrogen and/or phosphorus are necessary to be removed, along with solids and organic materials.

Once the minimum effluent quality has been specified, for maximum allowable concentrations of solids (both suspended and dissolved), organic matter, nutrients, and pathogens, the objective of the treatment is to attain reliably these set standards. The role of design engineer is to develop a treatment scheme that will address adequately all the pollutants under consideration and guarantee the technical feasibility of the scheme, taking into consideration other factors, such as construction and maintenance costs, the availability of construction materials and equipment, as well as specialized skilled personals for operation and maintenance of the treatment plant.

The primary treatment of sewage consists of screens (for removal of floating matter), grit chamber (for removal of inorganic suspended solids) and primary sedimentation tank (for removal residual settleable solids, which are mostly organic). Skimming tanks may be used for removal of oils; however, in conventional sewage treatment plant no separate skimming tank is used and oil removal is achieved by collecting the scum in primary sedimentation tank. This primary treatment alone will not produce an effluent with an acceptable residual organic matter concentration. Hence, invariably biological treatment method(s) is(are) used to effect secondary treatment for removal of organic material, which is present in primary treated effluent in colloidal and soluble forms. In biological treatment systems, the organic matter oxidation is catalysed by bacteria mediated metabolism. Depending upon the requirement of the final effluent quality, tertiary treatment and/or pathogen removal may be included.

Majority of wastewater treatment plants are using aerobic biological wastewater treatment for the removal of organic matter. The popularly used aerobic processes are the activated sludge process (ASP), oxidation ditch, trickling filter, aerated lagoons, sequencing batch reactor (SBR) and moving bed biofilm reactor (MBBR). Stabilization ponds are also being used, wherever adequate land is available, which utilizes both aerobic and anaerobic metabolism. Since last three decades, due to increase in power cost in almost all countries and subsequent increase in operation cost of aerobic processes, more attention is being paid for the use of anaerobic treatment systems for the treatment of wastewaters including sewage. High-rate anaerobic processes, such as up-flow anaerobic sludge blanket (UASB) reactor and anaerobic filters, are also used for secondary sewage treatment at many places.

After this secondary treatment, depending on the mode of disposal, suitable tertiary treatment may be given for killing pathogens, nutrient removal, suspended solids removal, etc. Generally secondary treatment followed by disinfection will meet the effluent standards for disposal into water bodies. However, in presence of organic matter, though with much reduced concentration, disinfection by chlorination will lead to formation of disinfection by-products, many of which are carcinogenic. When the treated sewage is disposed off on land for irrigation, the level of disinfection required will depend on the type of secondary treatment and type of crops cultivated with restricted or unrestricted public access.

1.7   Emerging Concerns for Sewage Treatment

Generally domestic sewage does not contain inorganic matter or refractory organic compounds at highly toxic concentrations. However, due to increased use of personal care products and cosmetics these compounds are also found to be present in the sewage. As recommended by WHO, three representative endocrines disrupting compounds may be considered as benchmarks, for assessing their occurrence and treatment efficacy, wherever necessary, with values of 0.1 μg/L for bisphenol A, 0.3 μg/L for nonylphenol and 1 ng/L for beta-estradiol. In addition, unconsumed pharmaceuticals, such as antibiotics, are also reported to be present in sewage, though at very low concentrations in the range of mostly μg/L to maximum reaching up to a mg/L. The worldwide annual per capita consumption of pharmaceuticals is 15 g and it is three to ten times higher in developed countries (50–150 g) (Zhang et al., 2008). From a review of consumption and excretion patterns of 17 pharmaceuticals and 2 hormones for the Spanish population of 42 million in 2003, it was concluded that the annual consumption ranged from 12.0 kg for 17α-ethinylestradiol to 276.1 tons for ibuprofen (Carballa et al., 2008). These persistent and mobile chemicals (PMCs) are generally released into the aquatic environment by non-point sources including drugs excreted or disposed to the domestic sewerage system or through landfills leachate (Arikan et al., 2008; Nikolaou et al., 2007), effluents from hospitals and runoff from animal husbandry and aquaculture sites (Lin et al., 2008).

The antimicrobial resistance caused by various PMCs is leading to huge adverse health, financial and social costs, and it is of critical importance to limit the spread of antimicrobial resistance (Broom & Doron, 2020; McGettigan et al., 2019; Sethuvel et al., 2019). It is thus critical to understand what are the pharmaceuticals and even personal care products that are being consumed. For example, triclosan, an anti-bacterial agent, which is widely used in many personal care products, represents an environmental risk due to development of concomitant resistance to clinically important antimicrobials (Karmakar et al., 2019; Lu et al., 2018). These emerging contaminants are persistent and can have adverse effect on metabolism even at such low concentrations. In urban area, depending upon the type of industries, which are allowed to discharge liquid effluents into the public sewers and the dilution that is being offered by sewage, the municipal wastewater may have these inorganic substances or toxic organic compounds with the concentration more than the discharge limits stipulated by the authorities. Certain compounds, such as sulphates, metals such as chromium, etc., if presents at higher concentrations, may disturb the secondary treatment of the sewage.

Conventional wastewater treatment plants (WWTPs) are mainly designed for the treatment of biodegradable organic matter and suspended solids; hence, these pharmaceuticals and personal care products are not efficiently removed and are present in the effluent of these WWTPs in the range of ng L^{-1} to μg L^{-1}, thus entering the natural ecosystems (Yang et al., 2011). For instance, the influent and effluent concentrations of one of the widely used pharmaceuticals Ciprofloxacin (CIP) in a WWTPs were measured to be 5451 ng L^{-1} and 919 ng L^{-1}, respectively (Mohapatra et al., 2016). On the other hand, the influent and effluent concentration of Metorpolol, a beta blocker, in the same WWTP was 7743 ng L^{-1} and 4735 ng L^{-1}, respectively (Table 1.1).

The concentration of other pharmaceuticals detected by different researchers in the influent and effluent of WWTPs are presented in Table 1.1. These contaminants can

Table 1.1 Concentration of pharmaceuticals detected at the inlet and outlet of wastewater treatment plants globally

Pharmaceuticals	Influent (ng L^{-1})	Effluent (ng L^{-1})	Reference
Antibiotics			
Acetaminophen	134,320	58	Mohapatra et al. (2016)
Azithromycin	2500	1300	Guerra et al. (2014)
Clarithromycin	8000	7000	Guerra et al. (2014)
Levofloxacin	2539	523	Mohapatra et al. (2016)
Sulfamethoxazole	2900	490	Mohapatra et al. (2016)
Antifungal agents			
Triclosan	2900	490	Guerra et al. (2014)
Trimethoprim	1620	894	Mohapatra et al. (2016)
Anti-inflammatory			
Codeine	5700	3300	Guerra et al. (2014)
Diclofenac	109	104	Mohapatra et al. (2016)
Ibuprofen	8600	470	Guerra et al. (2014)
Naproxen	5033	166	Behera et al. (2011)
Lipid regulators			
Clofibric acid	65	6	Behera et al. (2011)
Gemfibrozil	1040	654	Mohapatra et al. (2016)
Beta blockers			
Atenolol	2600	1030	Mohapatra et al. (2016)
Metoprolol	6	3	Behera et al. (2011)
Propranolol	1962	615	Tran et al. (2018)

Modified from Azhan et al. (2021)

adversely affect the environment because of their persisting toxicity in the water body; hence, need to be removed by application of effective advanced wastewater treatment technologies to make water suitable for use or protect water body from accumulation of these compounds. Hence, in near future the sewage treatment plants necessarily should have such tertiary treatment capabilities to effectively remove these emerging contaminants to ensure that the treated sewage is safe for discharge in water body or onsite reuse.

1.8 Commonly Used Terminologies and Definitions

Industrial wastewater: It is the wastewater generated from the industrial processes and commercial areas referred as industrial effluent. Some of the industrial effluents will have similar types of pollutants present as compared to sewage, however at varying concentrations, such as food processing industrial effluents. Whereas, some of the industries will produce effluents which may contain objectionable organic and/or inorganic compounds that may or may not be amenable to conventional sewage treatment processes and may require alteration or additional unit operation(s)/process(es) for removal of these compounds, which are either not present in sewage or present at very high concentrations as compared to sewage.

Municipal sewage: It is the liquid waste originating from the domestic uses of water. It includes sullage, discharge from toilets, urinals, wastewater generated from commercial establishments, institutions, industrial establishments and also the groundwater and storm water that may enter into the sewers. Its decomposition produces large quantities of malodorous gases, and it contains numerous pathogenic or disease producing microorganisms, along with high concentration of organic matter and suspended solids. Depending upon the effluent characteristics received from industries and commercial establishment the characteristics of municipal sewage may be different than sanitary sewage and may contain some pollutants at higher concentrations and some of the pollutants which are not present in sanitary sewage.

Night Soil: It is a term used to indicate the human and animal excreta entering in sewers through water carriage system.

Sanitary sewage: Sewage originated from the residential buildings comes under this category. This is very foul in nature. It is the wastewater generated from the lavatory, basins, urinals and water closets of residential buildings, office building, theatre and other institutions. It is also referred as domestic wastewater. Apart from presence of organic matter in suspended and soluble form, presence of pathogens is a bigger concern for this waste stream. Sewage also contains sufficient nitrogen and phosphorous to support biological wastewater treatment process. However, when untreated sewage is discharged to stagnant water body, presence of these nutrients may lead to eutrophication.

Sewage treatment plant: It is a facility designed to receive the wastewater from domestic, commercial and industrial sources located within the city, which are allowed to discharge liquid effluents in sewers. It is designed to remove materials present in the wastewater that damage water quality and compromise public health and safety when discharged into water body or land. It is combination of unit operations and unit processes arranged in particular order to effectively treat the sewage to desirable quality standards as defined by regulating authority for safe disposal of treated effluent.

Sewer: It is an underground conduit or pipe through which sewage is conveyed to a point of treatment plant or discharge or disposal. There are three types of sewer systems that are commonly used for sewage collection. Separate sewers are those which carry the house hold and industrial wastewaters only. In separate system, storm water drains are separately provided, which carry rain water from the roofs and street surfaces towards the point of disposal. Combined sewers are those which carry both sewage and storm water together in the same conduit. Whereas, in case of partially separate system the sewers carry part of the storm runoff from roof and paved court yards and rest of the runoff is carried in separate storm water drains.

House sewer (or drain) is used to discharge the sewage from a building to a lateral sewer located nearby street. *Lateral sewers* collect sewage from the house sewers. *Branch sewer* or *submain sewer* is a sewer which receives sewage from a relatively small area and this branch sewer discharge it to main sewer. *Main sewer* or *trunk sewer* is a sewer that receives sewage from many tributary branches and submain sewers, serving as an outlet for a large territory. The main sewer transport sewage to the treatment plant. *Depressed sewer* is a section of sewer constructed lower than adjacent sections to pass beneath an obstacle or obstruction. It runs full under the force of gravity and at greater than atmospheric pressure. The sewage enters and leaves the depressed sewer at atmospheric pressure. *Intercepting sewer* is a sewer laid transversely to main sewer system to intercept the dry weather flow of sewage and additional surface and storm water as may be desirable. An intercepting sewer is usually a large sewer, flowing parallel to a natural drainage channel, into which a number of mains sewers discharge sewage. *Outfall sewer* receives entire sewage from the collection system and finally it is discharged to a common point. *Relief sewer* or *overflow sewer* is used to carry the flow in excess of the capacity of an existing sewer.

Sewerage: The term sewerage refers to the infrastructure which includes device, equipment and appurtenances for efficient collection, transportation and pumping of sewage, however sewage treatment plants (STPs) were generally excluded from sewerage in the past and STPs were considered as separate entity. However, with increased acceptability of the decentralized sewage treatment system now sewage treatment facility is also considered as part of sewerage scheme. Basically, it is a water carriage system designed and constructed for collecting and carrying of sewage through sewers to the point where adequate treatment will be provided to meet the effluent quality norms suitable for disposal as per the desired mode.

Stormwater: It indicates the rain water surface run-off generated from the locality.

Subsoil water: Groundwater that enters into the sewers through cracks and leakages in sewers or joints is called subsoil water.

Sullage: This refers to the wastewater generated from bathrooms, kitchens, washing places and wash basins without much faecal contamination. Composition of this waste does not involve higher concentration of organic matter and it is less polluted water as compared to sewage and it is also referred as *grey water*.

Wastewater: The term wastewater includes presence of both organic and inorganic constituents, in soluble or suspended form, and mineral content in water at concentrations higher than that generally found in fresh water. Generally, the organic portion of the wastewater undergoes biological decompositions and the mineral matter may react with other constituents present in water to form dissolved solids. Inorganic matter, like nitrogen compounds, can also be removed employing biological treatment, whereas for higher degree of removal of other inorganic compounds physico-chemical treatment methods are required.

Questions

1.1. What are consequences of raw sewage discharge to the aquatic environment?

1.2. When a country is considered to be water stressed? What are the ways to reduce this stress?

1.3. Describe how pathogens enter in the water body and adverse effect caused due to their presence.

1.4. Detailed the adverse effects of pharmaceuticals and antibiotics generally found in the secondary treated sewage that is released into natural water body.

1.5. Describe different nomenclature used for sewers employed in sewerage system.

1.6. Describe pristine water.

1.7. Prepare a note on the freshwater availability. How the stress on the freshwater availability can be reduced?

1.8. Write a note on the environmental legislation related to discharge of treated effluent.

1.9. Discuss the typical quality standards adopted for discharge of treated sewage in water bodies in few countries.

1.10. For removal of which pollutants, the conventional sewage treatment plant is designed?

1.11. Discuss the difference between water treatment plant producing potable water and sewage treatment plant producing treated effluent safe for disposal in water body or on land.

1.12. Write a short note on presence of emerging contaminants in sewage.

References

Arikan, O. A., Rice, C., & Codling, E. (2008). Occurrence of antibiotics and hormones in a major agricultural watershed. *Desalination, 226*, 121–133.

Azhan, A., Priyadarshini, M., Das, S., & Ghangrekar, M. M. (2021). Electrocoagulation as an efficacious technology for the treatment of wastewater containing active pharmaceutical compounds: A review. *Separation Science and Technology*. https://doi.org/10.1080/01496395.2021.1972011

Bai, Y., Shan, F., Zhu, Y., Xu, J., Wu, Y., Luo, X., Wu, Y., Hu, H., & Zhang, B. (2020). Long-term performance and economic evaluation of full-scale MF and RO process—A case study of the Changi NEWater Project Phase 2 in Singapore. *Water Cycle, 1*, 128–135. https://doi.org/10.1016/j.watcyc.2020.09.001

Behera, S. K., et al. (2011). Occurrence and removal of antibiotics, hormones and several other pharmaceuticals in wastewater treatment plants of the largest industrial city of Korea. *Science of the Total Environment, 409*(20), 4351–4360. https://doi.org/10.1016/j.scitotenv.2011.07.015

Boretti, A., & Rosa, L. (2019). Reassessing the projections of the world water development report. *npj Clean Water, 2*, 1–6. https://doi.org/10.1038/s41545-019-0039-9

Broom, A., & Doron, A. (2020). Antimicrobial resistance, politics, and practice in India. *Qualitative Health Research, 30*(11), 1684–1696.

Carballa, M., Francisco, O., & Lema, J. (2008). Comparison of predicted and measured concentrations of selected pharmaceuticals, fragrances and hormones in Spanish sewage. *Chemosphere, 72*, 1118–1123.

CPHEEO. (2013). *Manual on sewerage and sewage treatment systems—2013*. Central Public Health & Environmental Engineering Organisation (CPHEEO), Government of India.

EPA. (2010). *National pollutant discharge elimination system (NPDES), Chapter 5, Technology based effluent limitations*. Environmental Protection Agency.

EPA. (2020). *National water reuse action plan improving the security, sustainability, and resilience of our nation's water resources collaborative implementation (Version 1)*. Environmental Protection Agency.

Guerra, P., et al. (2014). Occurrence and fate of antibiotic, analgesic/anti-inflammatory, and antifungal compounds in five wastewater treatment processes. *Science of the Total Environment, 473–474*, 235–243. https://doi.org/10.1016/j.scitotenv.2013.12.008

IWA. (2014). *International statistics for water services*. International Water Association.

Karmakar, S., Abraham, T. J., Kumar, S., Kumar, S., Shukla, S. P., Roy, U., & Kumar, K. (2019). Triclosan exposure induces varying extent of reversible antimicrobial resistance in Aeromonas hydrophila and Edwardsiella tarda. *Ecotoxicology and Environmental Safety, 180*, 309–316.

Kiely, G. (1998). *Environmental engineering*. Irwin/McGraw Hill.

Kovalakova, P., Cizmas, L., McDonald, T. J., Marsalek, B., Feng, M., & Sharma, V. K. (2020). Occurrence and toxicity of antibiotics in the aquatic environment: A review. *Chemosphere, 251*, 126351. https://doi.org/10.1016/j.chemosphere.2020.126351

Lin, A. Y. C., Yu, T. H., & Lin, C. F. (2008). Pharmaceutical contamination in residential, industrial, and agricultural waste streams: Risk to aqueous environments in Taiwan. *Chemosphere, 2008*(74), 131–141.

Lu, J., Jin, M., Nguyen, S. H., Mao, L. K., Li, J., Coin, L. J. M., Yuan, Z. G., & Guo, J. H. (2018). Non-antibiotic antimicrobial triclosan induces multiple antibiotic resistance through genetic mutation. *Environment International, 118*, 257–265.

McGettigan, P., Roderick, P., Kadam, A., & Pollock, A. (2019). Threats to global antimicrobial resistance control: Centrally approved and unapproved antibiotic formulations sold in India. *British Journal of Clinical Pharmacology, 85*(1), 59–70.

Meron, N., Blass, V., & Thoma, G. (2020). A national-level LCA of a water supply system in a Mediterranean semi-arid climate—Israel as a case study. *The International Journal of Life Cycle Assessment, 25*, 133–1144.

Mohapatra, S., Huang, C. H., Mukherji, S., & Padhye, L. P. (2016). Occurrence and fate of pharmaceuticals in WWTPs in India and comparison with a similar study in the United States. *Chemosphere, 159*, 526–535. https://doi.org/10.1016/j.chemosphere.2016.06.047

Nikolaou, A., Meric, S., & Fatta, D. (2007). Occurrence patterns of pharmaceuticals in water and wastewater environments. *Analytical and Bioanalytical Chemistry, 387*, 1225–1234.

NITI Aayog. (2018). *Composite water management index: A tool for water management.* NITI Aayog, Government of India. http://hdl.handle.net/2451/42272

Schellenberg, T., Subramanian, V., Ganeshan, G., Tompkins, D., & Pradeep, R. (2020). Wastewater discharge standards in the evolving context of urban sustainability—The case of India. *Frontiers in Environmental Science, 8*, 30. https://doi.org/10.3389/fenvs.2020.00030

Sethuvel, D. P. M., Anandan, S., Michael, J. S., Murugan, D., Neeravi, A., Verghese, V. P., Walia, K., & Veeraraghavan, B. (2019). Virulence gene profiles of Shigella species isolated from stool specimens in India: Its association with clinical manifestation and antimicrobial resistance. *Pathogens and Global Health, 113*(4), 173–179.

Shah, S. H., & Narain, V. (2019). Re-framing India's "water crisis": An institutions and entitlements perspective. *Geoforum, 101*, 76–79. https://doi.org/10.1016/j.geoforum.2019.02.030

Sophocleous, M. (2004). Global and regional water availability and demand: Prospects for the future. *Natural Resources Research, 13*(2), 61–75.

Tal, A. (2006). Seeking sustainability: Israel's evolving water management strategy. *Science, 313*(5790), 1081–1084. https://doi.org/10.1126/science.1126011

Tran, N. H., Reinhard, M., & Gin, K. Y. H. (2018). Occurrence and fate of emerging contaminants in municipal wastewater treatment plants from different geographical regions—a review. *Water Research, 133*, 182–207. https://doi.org/10.1016/j.watres.2017.12.029

WHO. (2017). *Guidelines for drinking-water quality* (4th ed., incorporating the first addendum). World Health Organization.

Yang, X., Flowers, R. C., Weinberg, H. S., & Singer, P. C. (2011). Occurrence and removal of pharmaceuticals and personal care products (PPCPs) in an advanced wastewater reclamation plant. *Water Research, 45*, 5218–5228. https://doi.org/10.1016/j.watres.2011.07.026

Zhang, Y., Geißen, S. U., & Gal, C. (2008). Carbamazepine and diclofenac: Removal in wastewater treatment plants and occurrence in water bodies. *Chemosphere, 73*, 1151–1161.

Quantity Estimation of Sanitary Sewage and Storm Water

2

2.1 Sewage and Storm Water Collection

In any urban area it is necessary to provide a collection system for the efficient and effective collection of the sewage generated from that locality and convey it to a point where treatment will be given to this collected wastewater to facilitate reuse of it or safe disposal of it either in water body or on land. The objective of providing this collection system is to ensure that the sullage, sewage or excreta discharged from the community is properly collected, transported and treated to the required level to facilitate safe reuse or disposal without causing any adverse health or environmental impacts.

Apart from provision of this sewerage scheme, it is necessary to provide proper collection system for the surface run-off generated during the storms. In absence of such storm water collection system, stagnation of this water will lead to the development of unhygienic condition in that area, due to decomposition of organic matter, such as leaf litter washed away along with surface runoff from the locality, and it will lead to the spreading of foul odour and place for breeding flies. Depending upon the annual distribution of the rainfall, consideration of few other engineering factors and economic assessment, either a combined system may be provided for a city, which carries both sewage and storm water together in the single drain, or a separate system may be provided for collection of these two separately, i.e., sewers (closed conduits) for collection of sewage and separate storm water drains (closed or open drains) for collecting storm water.

The design of this collection system is influenced by topographical and other engineering considerations. The final decision of the type of system and final design will depend on the design period adopted, stage wise population aimed to be served with expected quantity of sewage flow and variation in the flow anticipated. The hourly fluctuation in sewage flow again depends on the extent of area and size of population served, with lesser fluctuation expected with increase in population served and vice versa. Topography, ground slope, availability of site for locating treatment plant and intermediate pumping stations, as per requirement, and the mode of final disposal will also play important role.

If mode of final disposal of treated sewage is in a water body, available hydraulic gradient considering high flood level in the water body need to be taken into consideration. If final disposal is for irrigation, availability of irrigation lands and readiness of people to accept treated sewage for irrigation and modalities of charges need to be worked out. Though in the past, mostly centralized sewage treatment plant was provided for most of the cities, possibility of decentralized sewage treatment, and availability of land for installing such decentralized sewage treatment plants in the city need to be explored. In addition, the mode of utilization of treated effluent from these decentralized sewage treatment plants need to be explored to encourage reuse of treated sewage within the city itself, such as for sprinkling lawns, gardens and playgrounds, aesthetic enhancement, roadside horticulture, cooling water in commercial establishments, etc.

While designing the sewerage system, care must be exercised that it should not have adverse social and environmental impacts. In order to achieve this, the location of the discharge point for disposal of final treated effluent in the river should be away from any intake well for water supply or the intake structure should be properly protected by construction of barrage to avoid contamination of water used as a source for water supply; especially this is very important during low flow conditions in the river. The applicable effluent discharge quality standards must be ensured, particularly with respect to BOD, microbial quality, nitrate and presence of any toxic and persistence substances to protect water quality downstream of the point of discharge, mainly if there are bathing ghats or river bank recreational activities. It should be ensured that the adequate dilution is being offered by the water flowing in the stream to the incoming treated sewage so as not to affect aquatic life forms present in that river.

Proper precautions need to be taken to avoid possibility of any ground water pollution either due to selection of pond system as a treatment technology or use of treated effluent for irrigation. The choice of technology to be selected for treatment of sewage also depends on availability of land, power supply and requirement of stage wise design of the plant in phased manner. For coastal cities the treated sewage should be discharged sufficiently offshore so as not to adversely affect the water quality of beaches. Land acquisition for installation of intermediate pumping stations could be crucial issue due to odour problem nearby such stations because of pumping of raw sewage and decomposition of organic matter in the sump. This odour problem will be critical if sewage is hold in a sump for longer time. Among the few selected options, a most optimum and feasible solution is considered for implementation, which will have lesser overall cost, i.e., capital and operating costs.

2.2 Types of Collection System

2.2.1 Background

For safe disposal of the sewage and storm water generated from a locality, efficient collection, conveyance, adequate treatment and proper disposal of storm water and treated sewage is necessary. To achieve this, following conditions should be satisfied.

- While getting transported through collection system and after treatment wherever sewage is disposed off, it should not pollute the drinking water source, either surface water bodies or groundwater, or water bodies that are used for bathing or recreational purposes.
- During conveyance of untreated sewage, it should not be exposed so as to have access to human being or animals and should not give unsightly appearance or odour nuisance, and should not become a place for breeding flies.
- It should not cause harmful effect to the public health and adversely affect the receiving environment.

The collection system is meant for collection of the sewage generated from individual houses, commercial establishments, institution and small-scale industries and transporting it to a common point where it can be treated as per the needs before disposal. In the past, wastewater generated from water closets was collected by conservancy methods and other liquid waste was transported through open drain to finally join natural drains. Since, carts were used to carry the excreta, it was not hygienic method for transportation to the disposal point. Hence, now the collection and conveyance of sewage is done in a water carriage system, where it is transported in closed conduit using water as a medium of transport.

2.2.2 Types of Sewerage System

The sewage and storm water can be collected in a single drain or it can be collected separately depending on the frequency of rainfall in that area, hydrological features of the area, mode of disposal of treated sewage and few other engineering constraints, such as availability of space for laying two separate lines and cost of the system. Based on these considerations the sewerage system can be mainly of three types, viz., combined system, separate system and partially separate system. To address problem related to a particular locality at times vacuum or pressurized sewer system can also be used.

Combined system: In combined system along with domestic sewage the run-off resulting from storms is also carried through a same conduit of sewerage system. In countries

where rainfall is more or less uniformly distributed over a year such system may be beneficial. Whereas in the countries receiving seasonal rainfall, restricted for only few days in
a year, actual rainy days are very few and for such cities the combined system will face
the problem of maintaining self-cleansing velocity in the sewers during dry season, as the
sewage flow will be far lower as compared to the combined design discharge. The design
discharge here is the summation of maximum sewage flow plus a storm water run-off
resulting from a rain with designed frequency.

Separate system: In separate system, separate pipes/drains are used; one carrying
sewage and other carrying storm water run-off. The sewage collected from the city is
transported to a point where it is treated adequately, to meet the desired quality standards,
before it is discharged into the water body or used for irrigation. Whereas, the storm
water collected can be directly discharged into the water body since the storm run-off is
not as foul/polluted as sewage and no treatment is generally provided. However, developed countries are even trying to treat this to protect water quality in the receiving water
bodies. The separate system is advantageous and expected to be economical for bigger
towns. However, economics of the system cannot be generalized and detailed economic
analysis is essential before reaching to the conclusion on which system is economical for
that specific city.

Partially separate system: In this partially separate system part of the storm water,
especially collected from roofs and paved courtyards of the buildings, is admitted in
the same drain along with sewage collected from residences and institutions, etc. The
storm water runoff from the other places is collected separately using separate storm
water drains. This system reduces the house plumbing work and proves to be beneficial.
Also, due to addition of some fraction of storm water, because of increased discharge the
flushing of sewers will occur due to development of self-cleansing velocity within the
sewers.

Vacuum type system: The vacuum sewer collects sewage from multiple sources and
conveys it to the sewage treatment plant (STP); however, it has limited capacity to elevate
sewage up to 9.0 m. The sewage generated from the buildings is collected in a sewage
pit, which has level control device to get vacuum pump activated once sufficient quantity of sewage is collected in the pit. From this pit it is lifted to higher elevation sewer
or to the sewage treatment plant. Possibility needs to be explored to take advantage of
ground slope available to minimize the suction head. Each sewage collection pit is fitted
with a pneumatic pressure-controlled vacuum valve, which automatically opens after a
predetermined volume of sewage has entered the sump. When the vacuum valve closes,
atmospheric pressure is restored inside the pit. Overall, the sewer lines are so installed
that the vacuum created at the central station is maintained throughout the valve pits in
network served by that vacuum station. A disadvantage of this system is the need of
uninterrupted power supply to the vacuum pumps installed at central station and hence
limiting application of this system for high profile multistorey residential buildings and
not for the public sewer systems.

Pressurized sewers: This system need to be used for serving the houses in low laying area. Typically, in this system sewage from houses and establishments in that locality is collected in a common sump, from where using submersible pump it is lifted and injected in a sewer located at higher elevation on the shoulder of the road. This system is suitable for laying sewers in an area with undulating terrain, rocky strata and high groundwater table level, as pressurized sewers can be laid close to the ground and anchored well. In addition, there will not be infiltration of ground water as sewers are laid at shallow depth from ground and any leakage from sewer can be quickly detected and repaired. Also, this will require essentially smaller diameter pipes and above all construction of deep manholes will be avoided. It is desirable to provide self-cutting pumps to avoid clogging of the pumps and to ensure long term operation without involving frequent maintenance. A screen ahead of the sump is essential and separate oil and grease trap can be provided, if high oil and grease concentration is expected. Continuous power supply is essential for this system, which could be a drawback of this system at a place where power is not available continuously.

2.2.3 Advantages and Disadvantages of Combined System

In an area where rainfall is spread more or less uniform throughout a year, there is no additional need of flushing of sewers, as self-cleansing velocity will be developed in sewers due to higher discharge because of addition of storm water along with sewage. This system will also support only one set of pipes required for house plumbing, thus reducing cost of house plumbing. In congested localities, it is easy to lay only one pipe rather than two pipes as required in other systems, favouring combined system. However, combined system is not suitable for the area where rainfall is restricted to few days in a year, because in this case the dry weather flow will be small due to which self-cleansing velocity may not be developed in sewers, resulting in silting in the sewers. In addition, in case of combined system large wastewater flow is required to be treated at sewage treatment plant, before disposal; hence, resulting in higher capital and operating cost of the treatment plant. In addition, the capacity of the intermediate pumping stations, wherever required, will be large making this system uneconomical in terms of capital as well as operating cost. Apart from this, the overflow of combined sewers during rains will spoil public hygiene as it contains organic matters and pathogens.

2.2.4 Advantages and Disadvantages of Separate System

In separate system, where two separate drains are used, the sewage flows in separate pipe, hence the quantity of wastewater required to be treated at sewage treatment plant is small, resulting in economy of treatment. This system might result less costly as only sanitary sewage is transported in closed conduit and storm water can be collected and conveyed through open drains taking advantage of natural slopes. Hence, intermediate pumping stations and pumping of treated wastewater for final disposal may only be required for sewage and not essential for storm water, thus making this system economical due to reduced flow required to be handled for pumping as compared to combined system. However, due to handling sewage and storm water flow separately in separate pipes, the self-cleansing velocity may not develop at certain locations in sewers and hence flushing of sewers may be required. This is particularly applicable for the early stages of laterals, where discharge received in the sewers is very less to develop self-cleansing velocity. Since, this system requires laying two sets of pipes, it may be difficult to execute in the congested area. In addition, this system will require maintenance of two sets of pipelines and appurtenances, hence increasing maintenance cost of the system.

2.2.5 Advantages and Disadvantages of Partially Separate System

In partially separate system, economical and reasonable size sewers are required since part of the storm water is admitted to the sewers. The work of house plumbing is reduced in this system as rain water from roofs, sullage from bathrooms and kitchen, etc., are combined with discharge from water closets, hence requiring single plumbing in the house. Since some fraction of storm water is admitted to the sewers this will increase discharge and help in developing self-cleansing velocity and hence flushing of sewers may not be required. However, due to addition of some fraction of storm water it will increase the capacity and operating cost of intermediate pumping stations, wherever required, as well as final pumping of treated effluent for disposal as compared to the separate system. This system will also demand larger capacity STP as compared to separate system. Also, in dry weather condition the self-cleansing velocity may not develop in the sewers at certain locations demanding flushing of the sewers.

2.3 Considerations for Selecting Type of Sewerage System

Certain aspects are evaluated and considered before finalizing the type of collection system to be recommended for a city under consideration. The separate system requires laying of two sets of conduits; whereas, in combined system only one bigger size conduit is required. Laying of these two separate sets of conduits may be difficult in the congested localities. In combined system sewers are liable for silting during non-monsoon season, hence they are required to be laid at steeper gradients to maintain self-cleansing velocity. Laying sewers with steeper gradient will require more numbers of intermediate pumping stations in the collection system, particularly when a topography is like a flat terrain. This will make the system costly due to increased capital cost of construction for more numbers of pumping stations and also it will increase operation and maintenance cost of these pumping stations. Also, in case of combined system large capacity pumping stations are required to safely handle the flow that is likely to be generated during highest design storm considered.

In case of separate system pumping is only required for sewage; whereas, pumping can be avoided for storm water drains, as these are not very deep and normally laid along the natural slopes. In case of combined system, large quantity of wastewater is required to be treated before discharge; hence, large capacity treatment plant is required, thus increasing capital cost for construction of higher capacity sewage treatment plant and enhancing operating cost of such plants. Whereas, in case of separate system, only sewage is being treated before it is discharged into natural water body or used for irrigation and no treatment is generally given to the rainwater collected before it is discharge in to natural water body, thus keeping the capacity of sewage treatment plant under control and reducing capital expenditure as well as operating costs.

While finalizing the choice of sewerage scheme, increment in future population and accordingly increment in water demand due to commercial, recreational and industrial expansion should be taken into account so as to permit timely expansion of the sewerage facilities. However, provision for future should not be very excessive of the actual discharge estimated for early years so as to avoid deposition in the sewers. Based on site conditions, the economy of the system needs to be evaluated and the one offering least cost is to be selected for implementation.

2.4   Design Period

The duration of time up to which the capacity of a sewer or any other appurtenances will be adequate to serve its function is called as the design period. Thus, the future period for which the provision is made in designing the capacities of the various components of the sewerage scheme is referred as the design period. The design period for a particular component of the sewerage scheme depends upon the following:

- Ease and difficulty in expansion/augmentation including consideration of their location,
- Amount and availability of investment, rate of interest and inflation,
- Anticipated rate of population growth, including shifts in communities, industries and commercial expansions expected,
- Hydraulic design constraints of the systems designed, and
- Life of the material/structure and equipment employed.

The laterals and submains are generally designed for peak flow likely to be developed from that area upon full development as per the city master plan. Trunk sewers, interceptors and outfall sewers are difficult and uneconomical to enlarge frequently, hence for these sewers longer design period is considered. However, though such trunk sewer is designed for longer time period, it is also necessary to acquire or reserve the right of way for future expansion of larger size trunk sewer, as per requirement to meet the future needs of handling increased sewage discharge. The sewerage scheme in general is designed for thirty years, after accounting the duration required for design and construction, which is typically two to five years. Following design period can be considered for different components of sewerage scheme.

1. Laterals less than 15 cm diameter: Full development.
2. Trunk or main sewers: 30 to 40 years.
3. Treatment Units: 15 to 20 years to facilitate future expansion in phase manner
4. Pumping Station : 30 years, since cost of the civil works will be economical for full design
5. Pumping machinery: 10 to 15 years, since life is up to 15 years.
6. Effluent disposal/utilization: 30 years, this provision in capacity will be economical in initial stage itself.

2.5   Estimation of Per Capita Sewage Generation

Sanitary sewage is the used water for various purposes by the residents in the city that is drained into sewers. During use of potable water supplied by the water authority in kitchen, bathrooms, lavatories, toilets and cloth washing, substantial amounts of pollutants will be released in this wastewater generated from households. The sewage collected from the municipal area consists of wastewater generated from the residences, commercial centres, institutions, recreational activities and permissible industrial wastewaters discharged into sewers. Some quantity of groundwater, depending on sewer invert level and water table, and some fraction of storm water from the area during storm will also get admitted in sewers thereby increasing the discharge; hence, some allowance is generally made while deciding the design discharge for this quantity. Before designing the sewer, it is necessary to know the design discharge, i.e., quantity of sewage, which will flow in it after completion of the project.

Accurate estimation of sewage discharge is necessary for hydraulic design of the sewers to decide required diameter of the sewers. Estimation of the sewage flow is based on the per capita water demand by the population at present and based on this the projected value of population at the end of the design period considering increase in population as well as change in per capita water demand at the end of design period. Far lower estimation of sewage flow than reality will soon lead to inadequate sewer size after commissioning of the scheme and the sewers may not remain adequate to serve for the entire design period. Likewise, very high discharge estimation will lead to larger sewer size affecting economy of the sewerage scheme. Whereas, very less discharge actually flowing in the sewer than design discharge may lead to deposition of settleable solids in the sewers, since not meeting the criteria of the self-cleansing velocity. Actual measurement of the discharge is not possible if the sewers do not exist; and where the capacity of the existing sewers is inadequate and need to be increased, still actual accurate present discharge measurement is not possible due to limited capacity of existing sewers leading to overflows in the existing system. Since sewers are design to serve for some more future years, engineering skills and experience have to be used to accurately estimate the sewage discharge.

2.5.1 Sources of Sanitary Sewage

The flow that occurs in sewers in separate sewerage system or the flow that occurs during non-rainy days in combined system is referred as 'dry weather flow'. This flow indicates the flow of sanitary sewage. This depends upon the rate of water supply, type of area served, economic conditions of the people, weather conditions and infiltration of groundwater in the sewers, if sewers are laid below groundwater table. The water that is supplied by water authority when used by the residents of the city will lead to the wastewater generation. Quantity of the sewage generation is little less than the quantity of water supply due to some fraction of water lost during distribution as well as some fraction being used for the purpose, which will not generate any wastewater, e.g., water used for gardening, sprinkling lawns, air coolers, etc. The quantity of sewage generation is typically estimated based on the per capita water demand keeping following points in mind.

- Water supplied by the agency for domestic usage, after desired uses it is discharged into sewers as sewage.
- Water is supplied to the industries for various industrial processes by the agency. Depending on the type of pollutants present in such industrial wastewaters the industry may be asked to provide some treatment or without any treatment may be allowed to discharge in sewers. Thus, some quantity of this water supplied to the industries after usage in different industrial applications is discharged as wastewater into sanitary sewers.
- The water authority also supply water to the various public places, such as schools, theatres, hotels, restaurants, hospitals and commercial complexes. Considerable fraction of this water after desired use joins the sewers as wastewater.
- Residents may draw water from private source, such as open well or tube well, to fulfil domestic water demand. After uses, this wastewater is discharged in to the sewers.
- Similarly, in addition to the water that is supplied by the water authority, industries and commercial establishments might also be allowed to utilize ground water or surface

water for fulfilling various demands. Fraction of this water is converted in to wastewater in different industrial processes or used for public utilities within the industry, thus generating wastewater. This is ultimately discharged in to sewers from the industries, which are allowed to discharge their wastewater in public sewers.

- Infiltration of groundwater into sewers through leaky joints and cracks in sewers and manholes is unavoidable when ground water table is above sewer invert level.
- Entrance of rainwater in sewers will occur during rainy days in sewers through faulty joints or cracks or through manhole covers.

2.5.2 Evaluation of Sewage Discharge

Reasonable accurate estimation of sewage discharge is necessary; otherwise, sewers may prove inadequate resulting in overflow much before the end of design period, or may prove too large in diameter to make the system uneconomical and hydraulically inefficient leading to deposition of suspended solids in the sewers. Hence, before designing the sewerage system it is important to estimate the discharge/quantity of the sewage, which will flow in it after completion of the project and at the end of design period. Considering the different sources of fulfilling water demand by public and industries, purpose of water use and likely possibility of groundwater entry in sewers, apart from accounted water that is supplied by water authority major fraction of which will be converted to wastewater, following addition and subtractions are considered while estimating the sewage quantity.

a. *Addition due to unaccounted private water supplies*: Individual households, institutions, industries and commercial establishments may be using water supply from private wells, tube wells, etc. This will lead to the wastewater generation more than the water quantity supplied by water authority. Part of this water extracted from private source, after desired uses, is converted into wastewater and ultimately discharged into sewers. This quantity can only be estimated by actual field observations.

b. *Addition due to infiltration*: Due to groundwater seepage in to sewers through defective joints or cracks formed in the sewers and manholes, additional water quantity than water supplied by water authority will result. This quantity of entry of ground water in sewer depends upon the relative level of the water table with respect to the sewer invert level. If water table is well below the sewer invert level, the infiltration can occur only during rain, when water is moving down through soil. If sewer invert level is below water table, infiltration of ground water will occur in sewer. Quantity of this water entering in sewers depends upon the permeability of the soil and it is very difficult to estimate exactly; however, allowance for ground water infiltration under worst conditions should be made while estimating sewage design discharge.

This amount of infiltration may range from 0.009 to 0.926 m^3/km.day per cm of pipe diameter (Lee and Lin, 2007). This quantity of infiltration may be considered to be 3 to 5% of the peak hourly flow or approximately 10% of the average daily flow of sewage. In India, depending on geological stratum and workmanship of laying sewers the Ministry of Urban Development recommends allowance of minimum 500 to maximum 5000 L/km.day. Also, an allowance of groundwater entry per manhole of minimum 250 L/day per manhole to maximum 500 L/day per manhole is recommended (CPHEEO, 1993). During rains, storm water drainage may also infiltrate into sewers. This inflow is difficult to estimate and generally no extra provision is made for this quantity. This extra quantity can be taken care of by extra empty space (free board) left at the top in the sewers, which are designed for running $2/3$ to $3/4$ full at maximum design discharge.

c. **Subtraction due to water losses**: During water distribution the water quantity that is lost through leakage in the water distribution system itself and through faulty house connections doesn't reach to the consumers; hence, this fraction lost will not appear as a sewage.

d. **Subtraction due to water not entering the sewerage system**: Some purpose for which the water is used may not generate any wastewater, e.g., boiler feed water, water sprinkled over the roads, streets, lawns, and gardens, water consumed in industrial product, water used in air coolers in arid regions, etc. Hence, there is no wastewater generation for utilization of water for such purposes.

2.5.3   Estimating Net Quantity of Sewage

Considering the above addition and subtraction over the accounted quantity of water that is supplied by water authority, the net quantity of sewage generation can be estimated as per Eq. 2.1.

$$
\begin{aligned}
\text{Net quantity of sewage generation} = &\text{ Accounted quantity of water supplied by the water works}+ \\
&\text{Addition due to unaccounted private water supplies}+ \\
&\text{Addition due to infiltration}- \\
&\text{Subtraction due to water losses}- \\
&\text{Subtraction due to water not entering the sewerage system}
\end{aligned}
\tag{2.1}
$$

Generally, 75–80% of the accounted water supplied by water authority is considered as quantity of sewage produced; however, this is applicable when the authority supply water to entire population. In fully developed areas the net quantity of sewage generation may reach up to 90% of the water supply. In developing countries, where sewers are provided in peri-urban areas of city, however the water supply is not there and people are using private source, then the net quantity of sewage generation may be more than the quantity of water supplied. Similarly, in arid regions the quantity of sewage generated may be as low as 40% of the water supply. The sewers should be generally designed considering minimum quantity of 100 LPCD (litre per capita per day) for the population to be served at the end of design period.

2.5.4 Sewage Quantity from Commercial Establishments and Institutions

The industries and commercial establishments often have source of water other than the water supplied by city water authority and may discharge their wastewater into the sanitary sewers. The quantity of sewage generation from such establishments can be estimated from the water consumption expected as detailed in Table 2.1 (CPHEEO, 2013).

2.6 Variation in Sewage Flow

Variation occurs in the flow of sewage over annual average daily flow due to seasonal variation in water demand. Fluctuation in sewage flow also occurs from hour to hour in any given day. The typical hourly variation in the sewage flow is shown in the Fig. 2.1. If the flow is monitored near its origin, the peak flow will be more pronounced. If the sewage has to travel a long distance the peak will reduce because of the time required in collecting sufficient quantity of sewage required to fill the sewers and time required in travelling from different distances within the sewer network to reach to the point where it is gauged. As sewage flow in sewer lines, more and more sewage will get mixed with it

Table 2.1 Institutional water demand

Sr. No	Type of institution/commercial establishment	Water demand, L
1	Hospitals including laundry Beds less than 100 Beds more than 100	 340 450
2	Hostels, boarding schools, colleges	135
3	Airports, train and bus stations, sea ports Per duty staff Per passenger	 70 15
4	Restaurants, per seat	70
5	Offices, day schools, day college, per person	45
6	Factories, per duty staff	45
7	Cinema halls and theaters	15

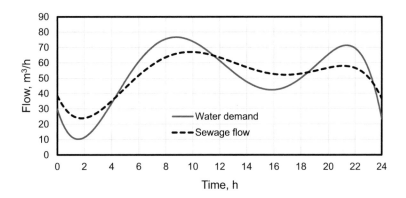

Fig. 2.1 Typical hourly variations in sewage flow as compared to water demand

due to continuous increase in the tributary area being served by the sewer line. This leads to reduction in the fluctuations in the sewage flow and the lag period goes on decreasing.

This magnitude of variation in the sewage flow rate varies from place to place and it is very difficult to predict. For smaller township this variation will be more pronounced due to lesser length of sewers and hence lower travel time required before the sewage reaches to the main sewer; however, for large cities this variation will be less pronounced due to larger extent of area served involving longer length of sewers. Considering the hourly variation, the maximum hourly flow shall be considered as two times the annual average hourly flow. Further this maximum hourly flow will be affected due to change in water demand per season and it might get amplified by a factor of 1.5 for accounting these seasonal variations. Thus, the maximum hourly flow variation will become three times the annual average hourly flow. For cities serving population less than 50,000 the peak factor may be more than 2.5 and generally peak factor value up to 3.0 is considered in design; whereas, for city population more than 750,000 the peak factor will be about 2 (CPHEEO, 2013). For cities with larger population a peak factor may get dampen further and a value of 1.5 is considered in design. Therefore, for outfall sewer of large cities the peak flow can be considered as 1.5 times the annual average daily flow and even same is considered for design of the treatment facility. This peak factor also depends on the population density, topography of the area and hours of water supply and it may be better to estimate it for individual case.

The minimum flow developed in sewers is also important to ensure self-cleansing velocity in the sewer during this less flow to avoid silting in sewers. This flow will be generated in the sewers during late night/ very early morning hours. The effect of this flow is more pronounced on lateral sewers than the main sewer. While designing the sewers, they must be checked for this minimum velocity generated. Considering the seasonal variation, the minimum hourly flow will be about $2/3$ of annual average flow and considering the hourly variation in a day the minimum hourly flow variation will be $1/2$ of minimum hourly flow estimated above considering seasonal variation. Thus, minimum hourly flow will be $1/3$ of the annual average hourly flow. In designing sewers, minimum flow may be considered as $1/3$ to $1/2$ of the average flow. The overall variation between the maximum and minimum flow is more in the laterals and less in the main or trunk sewers. This ratio of peak flow to average flow may be more than 6 for laterals and about 1.5 to 3 in case of main sewers depending on the total population served.

2.7 Population Forecasting

Design of water supply and sanitation scheme is based on the forecasted population of a particular town for the selected design period. Any underestimation in this population forecasting will make the system inadequate for the intended purpose; similarly, overestimation of population will make the system costly and uneconomical. Over the years change in the population of the city will occur and the system designed should take into account of the population at the end of the design period so that the capacity will not fall short till the end of design period. The factors, like increase in population due to births, decrease due to deaths, increase/decrease due to migration, and increase due to annexation of adjoining villages in the city limits will affect the change in population of a city. The net effect of the births and deaths on population is termed as natural increase when births are more or natural decrease if deaths are more than births. From the census population records, the present and past population records for the city under consideration can be obtained. After collecting these population numbers, the population at the end of design period is forecasted, considering the factors discussed above. Using various methods based on mathematical correlation of the past population growth pattern followed by the city or considering the growth pattern of population of cities already grown in the past under similar conditions, the population for the city under consideration is forecasted.

2.7.1 Arithmetical Increase Method

This method is suitable for large and old city which have already undergone considerable development and are represented by well settled and established communities. When this method is used for small, average or comparatively new cities, it will give lower population estimate than the actual population number that might reach at the end of design period. In this method the average increase in population per decade is evaluated from the past census records. This average increment in population is added to the present population to find out the population of the next decade. Thus, in this method it is assumed that the decadal increase of population of a city under consideration is considered as constant. Hence, $dP/dt = C$, i.e., rate of change of population with respect to time is considered as a constant. Therefore, population after nth decade will be estimated as per Eq. 2.2.

$$P_n = P + n.C \qquad (2.2)$$

Table 2.2 Estimation of increment in population, incremental increase and geometric increase in population

Year	Population	Increment	Geometrical increase rate of growth	Incremental increase
1971	85,654	–		
1981	100,567	14,913	14,913/85,654 = 0.174	–
1991	118,155	17,588	17,588/100,567 = 0.175	2,675
2001	139,635	21,480	21,480/118,155 = 0.182	3,892
2011	165,347	25,712	25,712/139,635 = 0.184	4,232
2021	197,226	31,879	31,879/165,347 = 0.193	6,167
	Average	(111,572/5) = 22,314		(16,966/4) = 4242

where, P_n is the population after 'n' decades and 'P' is present population of the city.

Example: 2.1 Predict the population for a town for the year 2031, 2041, and 2051 using arithmetical increase method from the following past census population data.

Year	1971	1981	1991	2001	2011	2021
Population	85,654	100,567	118,155	139,635	165,347	197,226

Solution

The estimation of increment in population each decade is presented in Table 2.2 (Column 3). From this increment observed for each decade, average increment is estimated by dividing the total of increment by 5 to get a value of 22,314. Thus, here it is considered that the population of the city will increase for the future with the constant increment of 22,314 people per decade. Using this average increment value, the population of the future decades can be forecasted using Eq. 2.2 as stated below:

Population forecast for year 2031 is, $P_{2031} = 197{,}226 + 22{,}314 \times 1 = 219{,}540$

Similarly, $P_{2041} = 197{,}226 + 22{,}314 \times 2 = 241{,}854$

$P_{2051} = 197{,}226 + 22{,}314 \times 3 = 264{,}168$.

2.7.2   Geometrical Increase Method or Geometrical Progression Method

In this method the percentage increase in population from decade to decade is assumed to remain constant and the city population in future will increase with this constant percentage increase. Geometric mean increase is used to find out the future increment in population. This method gives higher values of population forecasting and hence this method should be used for a new industrial town at the early stage of development for only few initial decades. Thus, using geometrical increase method the population at the end of nth decade 'P_n' can be estimated using Eq. 2.3.

$$P_n = P(1 + I_G/100)^n \qquad (2.3)$$

where, I_G is the geometric mean (%); P is the present population; and n is number of decades.

Example: 2.2 As per the past population census data given in Example 2.1, predict the population for the year 2031, 2041, and 2051 using geometrical progression method.

Solution
The rate of geometrical increase is estimated as per the Table 2.2, Column 4. From these values of geometrical increase observed for each decade, mean geometrical increase is estimated as:

Geometric mean $I_G = (0.174 \times 0.175 \times 0.182 \times 0.184 \times 0.193)^{1/5}$.

$= 0.1815$ i.e., 18.15%

Therefore, population forecasting for the year 2031 will be:

$P_{2031} = 197,226 \times (1 + 0.1815)^1 = 233,023$.

Similarly for year 2041 and 2051 the forecasted population can be estimated as:

$P_{2041} = 197,226 \times (1 + 0.1815)^2 = 275,316$.

$P_{2051} = 197,226 \times (1 + 0.1815)^3 = 325,286$.

2.7.3 Incremental Increase Method

This method is modification of arithmetical increase method and gives higher population estimation than arithmetical increase method and lower population estimation than geometrical increase method. It is suitable for an average size growing city under normal condition, where the population growth rate is found to be in increasing order. While adopting this method the average incremental increase in population is also considered along with average increase in population for forecasting the future population. For each decade, the incremental increase is determined from the past population census data and the average value is added to the present population along with the average rate of increase in population. Thus, population P_n after n^{th} decade is estimated as per Eq. 2.4.

$$P_n = P + nX + \frac{n(n+1)}{2} Y \qquad (2.4)$$

where, P_n is the population after n^{th} decade; X is an average increase in population; and Y is the incremental increase in population.

Example: 2.3 As per the past population census data provided in Example 2.1, predict the population for the year 2031, 2041, and 2051 using incremental increase method.

Solution

The increment in population per decade is estimated (Column 3, Table 2.2) and from this increment values, incremental increase in population per decade is obtained. These values are provided in Column 5 in Table 2.2. Using these average values of increment and incremental increase the future population of the city shall be forecasted as stated below.

Population in year 2031 is, $P_{2031} = 197{,}226 + (22{,}314 \times 1) + [(1\ (1+1))/2] \times 4{,}242$ $= 223{,}782$.

For year 2041, $P_{2041} = 197{,}226 + (22{,}314 \times 2) + [(2\ (2+1)/2)] \times 4{,}242 = 254{,}580$.

For year 2051, $P_{2051} = 197{,}226 + (22{,}314 \times 3) + [(3\ (3+1)/2)] \times 4{,}242 = 289{,}620$.

2.7.4   Decreasing Rate of Growth Method

If population growth of completely grown old city is plotted, the curve will follow like 'S' shape, i.e., early growth of population takes place at increasing rate coming to steady growth in middle; whereas, the later population growth of the city happens at the decreasing rate. For such nearly fully developed city the rate of increase in population will keep decreasing as the population will be nearing to the saturation population. For forecasting population for such cities decreasing rate of growth method will be best suitable, wherein the average decrease in the rate of growth is estimated. The latest percentage increase in population is then modified by deducting the decrease in rate of growth for each decade.

Example: 2.4 Predict the population for a town for the year 2031, 2041 and 2051 using decreasing rate of growth method from the following past census population data.

Year	1971	1981	1991	2001	2011	2021
Population	85,654	100,567	118,155	137,635	158,347	180,220

Solution

The increment in population and percentage increase in population are worked out as explained earlier and the values are presented in the Table 2.3. The decrease in percentage increase in population is then estimated as given in Column 5 of the Table 2.3. Using this

Table 2.3 Solution for decrease rate of growth method of population forecasting

Year	Population	Increment	Geometrical increase rate of growth	Percentage increase	Decrease in percentage increase
1971	85,654	–			
1981	100,567	14,913	14,913/85,654 = 0.174	17.4	–
1991	118,155	17,588	17,588/100,567 = 0.175	17.5	+ 0.1
2001	137,635	19,480	19,480/118,155 = 0.165	16.5	−1.0
2011	158,347	20,712	20,712/137,635 = 0.151	15.1	−1.4
2021	180,220	21,873	21,873/158,347 = 0.138	13.8	−1.3

average percentage decrease the latest percentage increase in population is corrected to forecast population for each future decade. Considering the last three decades decreasing values of the percentage increase in population the average decrease in percentage increase is equal to $(1.0 + 1.4 + 1.3)/3 = 1.233$. Hence the population for the future decade can be forecasted as explained below.

For population estimation for 2031, the net percentage increase in population will be $13.8 - 1.233 = 12.567$. Therefore, population for 2031 will be:

$P_{2031} = 180,220 + (12.567/100) \times 180,220 = 202,868$.

Similarly for year 2041, $P_{2041} = 202,868 + ((12.567 - 1.233)/100) \times 202,868 = 225,861$.

Population for year 2051, $P_{2051} = 225,861 + ((11.334 - 1.233)/100) \times 225,861 = 248,675$.

2.7.5   Graphical Method

In simple graphical method, the population records of last few decades are plotted to a suitable scale on graph (Fig. 2.2). The population curve is smoothly extended for getting future population. This extension should be done carefully and it requires proper experience and judgment. Thus, for the population data given in Example 2.1, the forecasted population for the year 2031, 2041, and 2051 will be 231,353; 270,219; and 313,275, respectively. Notice that the incremental increase method predicts lesser population numbers for 2041 and 2051, whereas the geometrical increase method predicts higher population than predicted by this graphical method. Similarly, the arithmetical increase method predicts lesser population numbers. Hence, the best way of applying this method is to extend the curve by comparing with population curve of some other already grown similar cities having the similar growth condition, as explained later.

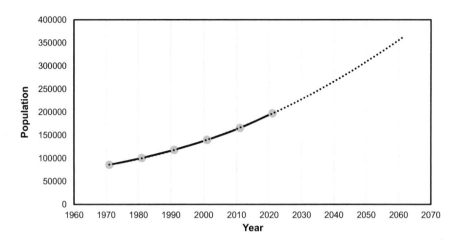

Fig. 2.2 Simple graphical method of population forecasting

2.7.6 Comparative Graphical Method

In this method the census populations of cities which have already developed under similar conditions are plotted. The curve of past population of the city under consideration is plotted on the same graph. The curve is extended carefully by comparing with the population curve of some similar cities having undergone population growth under similar conditions. The advantage of this method is that the future population of a city can be predicted from the present population even in the absence of some of the past census records. This method is further elaborated in Example 2.5.

Example: 2.5 The census population records of a city X in the past years 1990, 2000, 2010 and 2020 were 32,000; 38,000; 43,000 and 50,000, respectively. The cities A, B, C and D are considered to be developed in the past under similar conditions as city X is expected to grow. It is required to forecast the population of the city X for the years 2030 and 2040. The population data of cities A, B, C and D of different decades are given below:

  City A: 52,000; 62,000; 72,000 and 87,000 in 1980, 1990, 2000 and 2010, respectively.
  City B: 50,000; 58,000; 69,000 and 76,000 in 1981, 1990, 2001 and 2008, respectively.
  City C: 50,000; 56,500; 64,000 and 70,000 in 1984, 1990, 2000 and 2008, respectively.
  City D: 50,000; 54,000; 58,000 and 62,000 in 1981, 1993, 2002 and 2009, respectively.

Solution
Population curves for the cities A, B, C, D and X are plotted (Fig. 2.3). Then a mean curve is also plotted by dotted line as shown in the figure. The population curve X is extended beyond 50,000 population matching with the dotted mean curve. From the curve, the population forecast for city X is 58,000 and 68,000 in year 2030 and 2040, respectively.

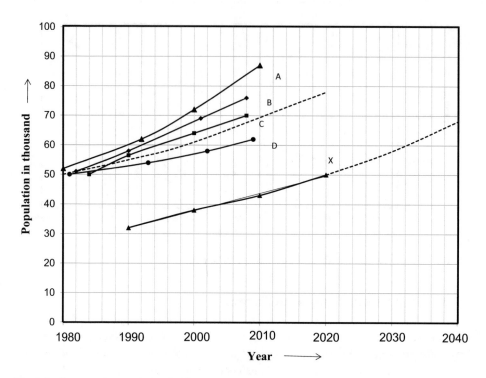

Fig. 2.3 Comparative graph method for population forecasting

2.7.7 Master Plan Method

The development of big cities and metropolitan cities is generally planned and regulated by local bodies according to master plan prepared for the city. This master plan is prepared for next 25 to 30 years stating how the city will grow and amenities will be developed. According to the master plan the city is divided into various zones, such as residential, commercial, institutional, industrial, parks and gardens, etc. Thus, the expansion of the city is regulated by the by-laws issued by the local authority. The population densities are fixed based on the assigned usage for these various zones in the master plan. From this zone wise population density, the total population and hence, the water demand and wastewater generation for that zone can be worked out. Adopting this method, after summation of population numbers for all zones it is very easy to access precisely the design population.

2.7.8 Logistic Curve Method

This method is used when the growth rate of population due to births, deaths and migrations takes place under normal situation and it is not subjected to any extraordinary changes like war, earth quake, epidemic or any natural disaster. Under this normal situation the population follows the growth curve characteristics of living things within limited space and economic opportunity. If the population of a city is plotted with respect to time, the curve so obtained under normal condition will be S-shaped curve and it is known as logistic curve (Fig. 2.4). This method of population forecasting is applicable for already developed large cities.

In Fig. 2.4, the curve shows an early growth from point A to B at an increasing rate, i.e., demonstrating geometric growth or log growth, where rate of change of population $\frac{dP}{dt} \propto P$, the transitional middle curve from B to D follows linear trend and hence arithmetic increase method is applicable for population forecasting in this region, thus $\frac{dP}{dt}$ = constant. For later population growth from point D onwards, the rate of change of population is proportional to the difference between saturation population and existing population, thus $\frac{dP}{dt} \propto (P_s - P)$. A mathematical solution for this logistic curve from A to E can be represented by first order equation, as given in Eq. 2.5.

$$\log_e\left(\frac{Ps - P}{P}\right) - \log_e\left(\frac{Ps - P_0}{P_0}\right) = -K.P_S.t \tag{2.5}$$

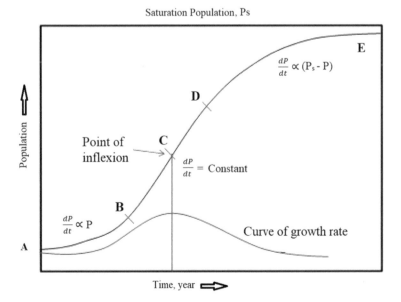

Fig. 2.4 Logistic curve for population growth

where, P is a population at any time t from the origin, i.e., from point A; P_s is saturation population for the city; P_0 is the population of the city at the start point A; K is constant and t is time in years.

The Eq. 2.5 can be rearranged as Eq. 2.6.

$$\log_e\left(\frac{Ps - P}{P}\right)\left(\frac{P_0}{Ps - P_0}\right) = -K.P_s.t \tag{2.6}$$

After solving Eq. 2.6 and expressing for P, i.e., population at any time t, the Eq. 2.7 can be obtained.

$$P = \frac{P_s}{1 + \frac{Ps-P_0}{P_0}\log_e^{-1}(-K.P_s.t)} \tag{2.7}$$

Substituting $\frac{Ps-P_0}{P_0} = m$ (a constant) and $(- K.P_s) = n$ (another constant), the Eq. 2.7 will become Eq. 2.8.

$$P = \frac{P_s}{1 + m\log_e^{-1}(n.t)} \tag{2.8}$$

This Eq. 2.8 is the required equation of the logistic curve, which will be used for predicting population. If only three pairs of characteristic population values P_0, P_1, P_2 at time $t_0 = 0$, t_1 and $t_2 = 2t_1$, respectively, extending over the past record are chosen, the saturation population P_s and constant 'm' and 'n' can be estimated by using Eqs. 2.9, 2.10 and 2.11, respectively.

$$P_s = \frac{2P_0 P_1 P_2 - P_1^2(P_0 + P_2)}{P_0 P_2 - P_1^2} \tag{2.9}$$

$$m = \frac{Ps - P_0}{P_0} \tag{2.10}$$

$$n = \frac{2.3}{t_1}\log_{10}\left(\frac{P_0(P_s - P_1)}{P_1(P_s - P_0)}\right) \tag{2.11}$$

Example: 2.6 Forecast population of a city for year 2031 using logistic curve method. The past population of a city in three consecutive decades, i.e., in year 2001, 2011 and 2021 was 36,300; 76,400 and 118,700, respectively. Determine (a) the saturation population, (b) the equation of logistic curve, and (c) the expected population in 2031.

Solution

The past population data for the city are:

$P_0 = 36,300$ at $t_0 = 0$; $P_1 = 76,400$ at $t_1 = 10$ years; and $P_2 = 118,700$ at $t_2 = 20$ years

Estimating the saturation population using Eq. 2.9.

$$P_s = \frac{2P_0 P_1 P_2 - P_1^2(P_0 + P_2)}{P_0 P_2 - P_1^2}$$

$$= \frac{2 \times 36,300 \times 76,400 \times 118,700 - 76,400 \times 76400 \times (36,300 + 118,700)}{36,300 \times 118,700 - 76,400 \times 76,400}$$

$$= 161,203$$

$$m = \frac{P_s - P_0}{P_0} = \frac{161,203 - 36,300}{36,300} = 3.44$$

$$n = \frac{2.3}{t_1} \log_{10} \frac{P_0(P_s - P_1)}{P_1(P_s - P_0)}$$

$$= \frac{2.3}{10} \log_{10}\left(\frac{36,300(161,203 - 76,400)}{76,400(161,203 - 36,300)}\right) = -0.113$$

Population for year 2031

$$P = \frac{P_s}{1 + m\log_e^{-1}(n.t)} = \frac{161,203}{1 + 3.44 \times \log_e^{-1}(-0.113 \times 30)}$$

$$= \frac{161,203}{1 + 3.44 \times 0.0337} = 144,456$$

Hence, forecasted population for the year 2031 will be 144,456 persons.

2.8   Estimation of Quantity of Sanitary Sewage

For a city under consideration the population of the city is projected at the end of design period considering the appropriate method of population forecasting. Wherever master plan is prepared for the town planning or any other relevant agency the decision regarding the design population should be taken giving due consideration to population figures reported in master plan. Also, it should be clear that, though the population of total city is being projected, the population growth will not be uniform throughout the city and there will be variation in zone or ward wise population growth. This information will also be required for designing the sewers in these zones.

Once design population is finalized the data on present per capita water demand is obtained. It is necessary to have a clarity that this per capita water demand will change with time as the city grows, due to occurrence of more commercial and recreational activities with growth of city and upliftment in the standard of living of people over the period of time. Hence, possible increment in per capita water demand at the end of design period should be decided and using this water demand and forecasted population the quantity of sewage generation can be estimated. If the municipality is allowing industries to discharge wastewater into sanitary sewers this additional quantity need to be estimated based on local survey to finalized the municipal sewage discharge. Based on the size of the city appropriate peak factor should be considered to finalize the design discharge.

Example: 2.7   For a city for which population forecasting is done in Example 2.3, determine design discharge for designing outfall sewer with design period of 30 years. The rate of water supply at present is 150 LPCD.

Solution

The present rate of per capita water supply is likely to increase at the end of design period. Considering this increment in per capita water demand to be 20%, the per capita demand at the end of design period will be 180 LPCD. The forecasted population at the end of design period, i.e., 2051 is 289,620. Hence, considering sewage generation of 80% of the water supply the average sewage flow will be:

$289,620 \times 180 \times 10^{-3} \times 0.8 = 41,705.28$ m^3/day $= 0.483$ m^3/sec.

Considering peak factor of 2.25 the design discharge will be 1.09 m^3/sec.

2.9   Estimation of Storm Water Runoff

During precipitation the fraction of it that flows over ground surface forms the storm water runoff. This storm runoff depends on many factors, as discussed later. This flow can be estimated either using rational method or empirical formula, which are catchment specific. Storm water collection system is not generally design for the peak flow that will occur by the rarest rain with once in 10 years or more return period; however, it is designed to provide enough capacity of the drains to avoid frequent flooding of the area. When rain exceeds the design value some flooding will occur, which causes inconvenience; however, it is permitted to keep the cost of storm water collection system under control to make it economical.

This frequency of permissible flooding may vary from place to place depending on the importance of the area. Commercial and industrial areas and areas of strategic importance are subjected to less frequent flooding and accordingly capacity of storm water drains is finalized so as to restrict occurrence of flooding in this area with frequency of once in 2 years. The central and comparatively high-priced areas in the city are subjected to once in two years flooding and the peripheral areas may be subjected to flooding once in a year and accordingly the frequency of rain is considered in design (CPHEEO, 2019). The CPHEEO has further provided the detailed guideline for considering the design return period for different cities (CPHEEO, 2019), these values are provided in Table 2.4. Though, this storm water runoff is expected to be considerably less polluted as compared

Table 2.4 Recommended design return period for various types of urban catchments

Sr. No.	Urban catchment	Return period	
		Mega cities	Other cities
1	Central business, commercial, industrial, and urban residential core area	Once in 5 years	Once in 2 years
2	Urban residential peripheral area	Once in 2 years	Once in 1 year
3	Open space, parks and landscape	Once in 6 months	Once in 6 months
4	Airports and other critical infrastructure such as railway stations, power stations	Once in 100 years	Once in 50 years

Source CPHEEO, 2019

to sanitary sewage; however, it is necessary not to allow it to remain stagnant in a locality for long, otherwise decomposition of littered organic matter present in this will start and it will lead to unhealthy conditions. The problem will be more aggravated when it is getting mixed with sanitary sewage due to presence of pathogens in addition to organic matter.

2.9.1 Factors Affecting the Quantity of Stormwater

The surface run-off resulting after precipitation contributes to the stormwater. This estimated quantity of stormwater reaching to the storm sewers or drains is very large as compared to the design discharge estimated for sanitary sewage. The factors affecting the quantity of stormwater flow are as stated below:

i. Area of the catchment,
ii. Slope and shape of the catchment area,
iii. Porosity of the soil,
iv. Obstruction in the flow of water as trees, fields, gardens, etc.
v. Initial state of catchment area with respect to wetness,
vi. Intensity and duration of rainfall,
vii. Atmospheric temperature and humidity,
viii. Number and size of ditches present in the area.

Larger the catchment area served larger is the storm water discharge expected. Shape of the catchment also affects the storm water discharge. The regular fan shape catchments are likely to generated more discharge than the long ribbon shape catchment due to increased time of concentration in the later. Steeper slope of catchment will reduce the fraction of precipitation seeping in to the soil and results in more storm water runoff; whereas, flat topography will provide more opportunity for the rain water to seep in to the soil due to very less velocity of flow and will generate lesser storm water runoff. Permeable soil strata will result in lesser quantity of the surface runoff during storm.

Similarly, obstruction in the surface flow, such as trees, fields, gardens, will reduce the flow velocity and permit more infiltration of storm water in the soil and hence, generate lesser surface runoff as compared to areas free from such obstructions for all other conditions remaining same. When the catchment is completely dry more fraction of storm water will be retained by soil to fulfil the soil moisture content; hence, generating lesser surface runoff during storm as compared to condition when soil moisture is already saturated. Similarly, a dry and hot weather condition will support more evaporation loss and generate lesser surface runoff. Also, when there are unfilled ponds and ditches present in the catchment area, the surface runoff from the catchment will not be maximum unless these water bodies get filled; hence, this will also reduce the storm water runoff unless these ditches get filled.

2.9.2 Measurement of Rainfall

The rainfall intensity could be measured by using rain gauges and recording the amount of rain falling in unit time. The rainfall intensity is usually expressed as mm/h or cm/h. The rain gauges used can be manual recording type or automatic recording type rain gauges. Meteorological department makes the everyday rainfall data available, however only from this data the storm runoff cannot be estimated unless the intensity of rainfall is known. This data on intensity of rainfall can be obtained from auto recording rain gauges. It is necessary to collect past 20 to 25 years of the data on intensity of rainfall in that particular area. If such data is not available then data of town having similar climatic conditions can be considered for designing the storm water drains.

2.9.3 Methods for Estimation of Quantity of Storm Water

The maximum surface runoff likely to be generated from a catchment area under the rainfall with designed frequency need to be estimated for designing the drain section. This discharge will be typically maximum when the entire catchment area will start contributing, i.e., for the rain with duration equal to the time of concentration. Hence, for computing the design flow for storm water drains the intensity of rainfall for a duration equal to the time of concentration is the one to be considered. Once this intensity of rainfall is finalized the rainfall runoff can be estimated using either (a) Rational method, or (b) Empirical formulae method. Rational method of runoff estimation is most popularly used. In both these methods, the quantity of storm water is considered as function of intensity of rainfall, coefficient of runoff and area of the catchment.

2.9.3.1 Time of Concentration

The time duration after which the entire catchment area will start contributing to the surface runoff is called as the time of concentration. In other words, it is duration required for the raindrop that is fallen on the most remote part of the catchment to reach to the point where the discharge is being measured, while this drop is flowing along the ground to enter the sewers and then through available sewers, if any. Based on the rainfall intensity and duration pattern following inferences can be drawn:

i. The rainfall with duration smaller than the time of concentration will not produce maximum discharge of surface run-off, as at no point of time entire catchment will contribute for surface runoff.
ii. The runoff shall not be maximum even for the rain having duration more than the time of concentration. This is because with longer the duration of rainfall, the intensity expected is lesser.
iii. The storm water runoff shall be considered to be maximum when the duration of rainfall is equal to the time of concentration (t_c) and it is referred as 'critical rainfall duration'.

The critical rainfall duration or time of concentration is equal to summation of inlet time and time of travel as given by Eq. 2.12 (Kang et al., 2008) (Fig. 2.5).

$$\text{Time of concentration } (t_c) = \text{Inlet time } (t_i) +$$
$$\text{time of travel } (t_t) \text{ from point P to Q} \qquad (2.12)$$

Inlet Time: The time required for the rain in falling on the most remote point of the tributary area to flow across the ground surface along the natural drains or gutters

Fig. 2.5 Estimation of time of concentration for a given catchment

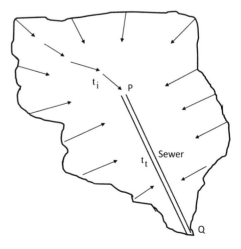

up to inlet of sewer is called inlet time (Fig. 2.5). The inlet time 't_i' can be estimated for a particular catchment using relationships similar to Eq. 2.13, once the values of these constants are determined from the past rainfall records. These coefficients will have different values for different catchments.

$$t_i = [A. L^n / H]^m \text{ or } t_i = C\left(\frac{L}{S}I^2\right)^{1/3} \tag{2.13}$$

where, t_i is the inlet time, minute; L is the length of overland flow in kilometre from critical point to mouth of drain; H is the total fall of level from the critical point to mouth of drain in meter; A, n and m are constants, specific for a catchment. In second equation C is coefficient; L is in m; S is slope of ground surface; and I is intensity of rainfall, mm/h (Lee and Lin, 2007).

Time of Travel: The time required by the storm water to flow in the drain channel from the mouth to the point under consideration or the point of concentration is called as time of travel. This time of Travel (t_t) can be calculated knowing length of drain divided by the velocity of the flow in the drain.

2.9.3.2 Runoff Coefficient

The total precipitation when falling on any area is disseminated as infiltration, evaporation, storage in ponds or reservoir and remaining as surface runoff. Thus, some fraction of the total rainfall received will result in surface runoff. This coefficient which is to be multiplied with the quantity of total rainfall to determine the quantity of surface runoff that will reach the storm water drains is known as 'runoff coefficient' or 'imperviousness coefficient'. The runoff coefficient depends upon the porosity of soil cover, wetness and type of ground cover. In a given catchment the area may have combinations of different types of ground cover having different runoff coefficients for each one of them depending on the imperviousness. The typical runoff coefficients for the different type of ground cover are provided in the Table 2.5. Thus, the overall runoff coefficient for the catchment area can be estimated using Eq. 2.14.

$$\text{Overall runoff coefficient, } C = \frac{A_1.C_1 + A_2.C_2 + \ldots + A_n.C_n}{A_1 + A_2 + \ldots + A_n} \tag{2.14}$$

where, A_1, A_2, ...A_n are types of area cover with corresponding runoff coefficients of C_1, C_2, ...C_n.

2.9.3.3 Rational Method

The runoff generated during the storm is proportional to the intensity of the rain for a duration equal to time of concentration for the rainfall frequency considered. In addition to the intensity of the rainfall, the area of catchment and overall runoff coefficient will decide the total discharge resulting from the rain. Thus, the storm water surface discharge

Table 2.5 Runoff coefficient for different types of ground cover in a catchment

Type of Cover	Coefficient of runoff
Business areas	0.70–0.90
Water tight roofs	0.70–0.95
Apartment areas	0.50–0.70
Single family area	0.30–0.50
Parks, playgrounds, lawns	0.10–0.25
Forests and woody area	0.10–0.20
Paved streets with good quality asphalt, stones, bricks or blocks	0.80–0.90
Water bound macadam roads	0.25–0.60
Gravel surfaced walkways	0.15–0.30

(Q_s) can be estimated by using rational method as per Eq. 2.15 (Chen et al., 2004) or Eq. 2.16 (Yang et al., 2000).

$$Q_s = \frac{C.I.A}{360} \qquad (2.15)$$

where, Q_s is the quantity of storm water runoff, m^3/sec; C is the overall coefficient of runoff; I is the intensity of rainfall (mm/h) for the duration equal to time of concentration for a decided frequency; and A is the catchment area in hectares.

$$\text{Or } Q_s = 0.278C.I.A \qquad (2.16)$$

where, Q_s is the storm water runoff discharge, m^3/sec; I is the intensity of rainfall, mm/h; and A is area of catchment, km^2.

2.9.3.4 Empirical Formulae

There are various empirical formulae, which are used for determination of runoff from very large catchment area. Various empirical relationships are established based on the past rainfall records for the specific catchment conditions suiting a particular region. If sufficient past rainfall records are available, such functional relationship can be established between the dependable variables, such as rainfall intensity, area, runoff coefficient and overall slope of the catchment. Such empirical formula can then be used for prediction of storm water runoff for that particular catchment. This empirical formula is catchment specific and formula developed for one catchment with particular values of the coefficients cannot be used for other, unless the conditions of the catchment and weather conditions are similar. These Empirical equations could have following forms as given in Eqs. 2.17 and 2.18:

$$Q = \frac{C.I.A}{m}\sqrt[n]{S/A} \qquad (2.17)$$

$$Q = \frac{C.A^n}{m} \qquad (2.18)$$

where, S is the slope of the area in meter per thousand meters; A is the drainage area in km^2 or hectares; m and n are constants/coefficients specific for that catchment representing other dependable variables, which are not considered in the empirical relation in proportion to the area and slope of the catchment.

For estimation of storm water runoff EPA (2015) recommends storm water management model (SWMM). This is a distributed model that allows a study area to be subdivided into any number of irregularly shaped sub-catchments to best capture the effect that spatial variability in topography, drainage pathways, land cover, and soil characteristics have on runoff generation. Generation of storm water runoff is computed on a sub-catchment basis using Eq. 2.19.

$$Q = \frac{1.49}{n}S^{1/2}R_x^{2/3}A_x \qquad (2.19)$$

where, n is the surface roughness coefficient which will have typical values of 0.012 for smooth asphalt pavement, 0.014 for tar and sand pavement, 0.017 for concrete pavement, 0.022 for business land use, 0.035 for semi-business and industrial land use, and 0.04 for dense residential area. S is the apparent or average slope of the sub-catchment (m/m), A_x is the area across the sub-catchment's width through which the runoff flows (m^2), and R_x is the hydraulic radius associated with this area (m).

Various empirical formulae proposed by American engineers are given in Eqs. 2.20, 2.21, 2.22 (Sarvaas, 1924) and Eq. 2.23 (Chow, 1962).

Burkli-Ziegler Formula

$$Q = ARC\sqrt[4]{\frac{S}{A}} \qquad (2.20)$$

McMath Formula

$$Q = ARC\sqrt[5]{\frac{S}{A}} \qquad (2.21)$$

Colonel Dickens Formula

$$Q = 825\sqrt[4]{M^3} \qquad (2.22)$$

where (for Eq. 2.20 through Eq. 2.22), A is drainage area (acres); Q is discharge reaching the sewers (cubic feet per second); M is the area of catchment (square miles); R is the average rate of rainfall during the heaviest rainfall (inches per hour); S is general fall of the area per 1000 m; and C is a constant, which may vary from 0.9 in the Burkli-Ziegler formula to that of to 0.3 for McMath.

   Modified Myers Formula

$$Q = 100p\sqrt{M} \tag{2.23}$$

where, Q is discharge (cubic feet per second); M is drainage area (square miles); p is numerical percentage rating on the Myer's scale. The Myer's scale is a device which furnishes a standard by which the flood flow characteristics in different streams can be roughly compared.

   Chamier (1898) claimed a formula (Eq. 2.24) by taking a special account of the duration of the rainfall in investigating the sufficiency of waterways, he obtained reasonably close results between computed discharges by the formula and actual measurements.

$$Q = 640RC\sqrt[4]{M^3} \tag{2.24}$$

where, Q is the maximum discharge at the outlet (cubic feet per second); R is the average rate of greatest rainfall for such duration as it will allow the flood water flowing to the outlet from the farthest extremity of the catchment area (inches per hour); C is co-efficient of surface discharge, giving the proportion of rainfall that may be expected to flow-off the surface; and M is catchment area (square miles).

2.9.3.5   Flood Frequency Analysis

Carter (1961) developed an empirical equation (Eq. 2.25) relating the mean annual flood to the lag time, drainage area and percent of impervious cover to determine the effect of urbanization on the mean annual flood in the vicinity of Washington, D.C (Espey & Winslow, 1975).

$$Q = 223KA^{0.85}T^{-0.45} \tag{2.25}$$

where, Q is mean annual flood and it is equivalent to the flood having a recurrence interval of 2.33 years (cubic feet per second); A is drainage area (square miles); T is lag time (h); K is adjustment factor based upon the degree of imperviousness of the area. The factor K is expressed as given by Eq. (2.26):

$$K = \frac{0.30 + 0.0045I}{0.30} \tag{2.26}$$

where, I is the percent of impervious cover.

2.9.4   Empirical Formulae for Rainfall Intensities

It is observed that generally the rains having longer duration are of lesser intensity and the rain with the shorter duration is having higher probability of more intensity, with the exception of extreme weather conditions faced in the recent times in many countries. Empirical relationships between rainfall intensity and duration can be established based on long term past rainfall experience for that region. The longer the past records are available, the better dependable will be the forecast done by such empirical relation. Thus, it is evident that the rainfall intensity is inversely proportional to the duration of rainfall and the typical relationship is presented in the Fig. 2.6. In general, the empirical relationships will have any of the forms given in Eq. 2.27 or Eq. 2.28. Under typical Indian conditions, for example the intensity of rainfall generally considered in design is usually in the range 12 mm/h to 20 mm/h.

$$I = \frac{a}{t + b} \tag{2.27}$$

$$\text{or } I = \frac{b}{t^n} \tag{2.28}$$

where, a, b, and n are constants; I is the intensity of rainfall (mm/h or cm/h); and t is the duration of rainfall in minutes. In Eq. 2.27 the value of 'a' could be in the range of 700 to 3000 and 'b' around 10 to 20, depending on the duration of the rainfall and other environmental conditions in that area.

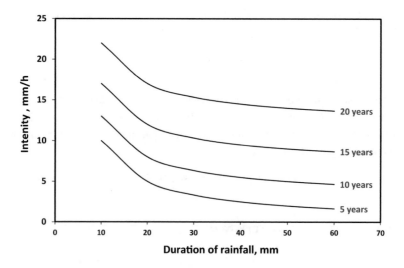

Fig. 2.6 Duration of rainfall and intensity relationship for different frequency of rainfall

Example: 2.8 Determine a designed discharge for a combined sewer system. The total population of the town is 40,000 and rate of water supply is 150 LPCD. The catchment area is 80 hectares and the average coefficient of runoff is estimated to be 0.56. The critical rainfall duration is 33 min and the rainfall and duration has the relationship of $I = 960/(t + 15)$, mm/h and t is in min.

Solution

Estimation of sewage discharge.

Consider 80% of the water supply will result in sewage generation, thus:

Quantity of sanitary sewage $= 40,000 \times 150 \times 0.80 \times 10^{-3} = 4800$ m^3/day $= 0.0556$ m^3/sec.

Considering peak factor of 2.5, the design discharge for sanitary sewage $= 0.0556 \times 2.5$

$= 0.139$ m^3/sec

Estimation of storm water discharge.

Intensity of rainfall, $I = 960/(t + 15)$.

Therefore, $I = 960/(33 + 15) = 20$ mm/h.

Hence, storm water runoff, $Q = C.I.A/360 = 0.56 \times 20 \times 80/(360) = 2.489$ m^3/sec.

Therefore, design discharge for combined sewer system $= 2.489 + 0.139 = 2.628$ m^3/sec.

Example: 2.9 A catchment area is extended to 350 hectares having the surface cover as given below:

Type of cover	Coefficient of runoff	Percentage
Roofs	0.90	20
Pavements and yards	0.80	15
Lawns and gardens	0.15	22
Asphalt surface roads	0.80	14
Open ground	0.10	17
Macadamized roads	0.40	12

Estimate the overall coefficient of surface runoff and the design discharge of storm water runoff, if intensity of rainfall is 20 mm/h for rain with duration equal to time of concentration. If population density in the area is 330 persons per hectare and rate of water supply is 135 LPCD, estimate design discharge for separate system, partially separate system, and combined system.

Solution

Estimation of overall runoff coefficient.

Overall runoff coefficient $C = [A_1.C_1 + A_2.C_2 + ... + A_n.C_n] / [A_1 + A_2 + ... + A_n]$.

$$= \frac{(0.20 \times 0.90 + 0.15 \times 0.80 + 0.22 \times 0.15 + 0.14 \times 0.8 + 0.17 \times 0.1 + 0.12 \times 0.4)}{0.20 + 0.15 + 0.22 + 0.14 + 0.17 + 0.12}$$

$= 0.51$.

Estimating quantity of storm water.

$Q = C.I.A/360$

$= 0.51 \times 20 \times 350/360 = 9.917$ m^3/sec.

Thus, design discharge for storm water drains of separate system will be 9.917 m^3/sec.

Estimation of sewage discharge.

Assuming 80% of the water supply will result in sewage generation.

Quantity of sanitary sewage $= 330 \times 350 \times 135 \times 0.80 \times 10^{-3} = 12,474$ m^3/day $= 0.144$ m^3/sec.

Considering peak factor of 2.25, the design discharge for sanitary sewers of separate system $= 0.144 \times 2.25 = 0.325$ m^3/sec.

The design discharge for combined system $= 0.325 + 9.917 = 10.242$ m^3/sec.

Estimation of discharge for partially separate system.

Storm water discharge falling on roofs and paved courtyards will be admitted to the sanitary sewers. Hence, estimating this quantity of storm water as below:

Average coefficient of runoff $= (0.90 \times 0.2 \times 350 + 0.80 \times 0.15 \times 350) / (0.35 \times 350)$ $= 0.857$.

Storm water discharge admitted to sanitary sewers $= 0.857 \times 20 \times 0.35 \times 350 / 360$. $= 5.832$ m^3/sec.

Therefore, design discharge in the sanitary sewer of partially separate system:

$= 5.832 + 0.325 = 6.157$ m^3/sec,

The design discharge for storm water drains of partially separate system will be.

$= 9.917 - 5.832 = 4.085$ m^3/sec.

Example: 2.10 For a small town having total population of 45,000 and total area of 48 hectares, determine the design discharge for the combined system. The rate of water supply is 170 LPCD and 75% of it is resulting in generation of sewage. The peak factor for sewage discharge will be 2.5. Average impermeability coefficient of this area is 0.46 and time of concentration is 32 min. The rainfall intensity is given by the relation $I = 150/t^{0.625}$, where I is mm/h and t is in min.

Solution

Estimating sewage flow.

$= 45,000 \times 170 \times 0.75 \times 10^{-3} = 5737.5$ m^3/day $= 0.0664$ m^3/sec.

Hence design discharge of sanitary sewage $= 0.0664 \times 2.5 = 0.166$ m^3/sec.

Estimating quantity of storm water runoff.

Intensity of designed rain $= 150/t^{0.625} = 150/32^{0.625} = 17.19$ mm/h.

Hence, storm water runoff $Q = CIA/360 = 0.46 \times 17.19 \times 48/360 = 1.054$ m^3/sec.

Therefore, design discharge for combined sewer system $= 1.054 + 0.166 = 1.220$ m^3/sec.

Example: 2.11 A catchment consists of 4.8 km^2 of agricultural area with runoff coefficient $C = 0.22$, a woody area of 3.2 km^2 with $C = 0.15$ and 2.2 km^2 of grass cover with $C = 0.30$ and single housing residential area including roads of 1.5 km^2 with $C = 0.60$. The water course is 2.8 km in length having a fall of 22 m. The intensity–duration–frequency relation for this area is expressed by the relation:

$$I = \frac{72T^{0.22}}{(t_c + 10)^{0.4}}$$

where I is in mm/h, T (return period, frequency of rainfall) is in years, t_c is in minutes. The time of concentration (t_c) for this catchment is given by the equation $t_c = 0.00033 \times L^{0.68} \times S^{-0.38}$, where t_c is in h, Length (L) is in m and S is slope. Determine the peak runoff for 20 years return period using rational method.

Solution
Total area $= 4.8 + 3.2 + 2.2 + 1.5 = 11.7$ km^2.

Slope of water course: $S = \Delta H/L = 22/2800$.

Time of concentration $t_c = 0.00033 \times L^{0.68} \times S^{-0.38} = 0.00033 \times 2800^{0.68} \times (22/2800)^{-0.38}$.

$= 0.46$ h $= 27.58$ min.

Intensity, $I = \frac{72T^{0.22}}{(t_c+10)^{0.4}} = \frac{72 \times 20^{0.22}}{(27.58+10)^{0.4}} = 32.63 \; mm/h$.

Overall runoff coefficient $= (4.8 \times 0.22 + 3.2 \times 0.15 + 2.2 \times 0.3 + 1.5 \times 0.6)/(4.8 + 3.2 + 2.2 + 1.5)$.

$= 3.096/11.7 = 0.2646$.

Design discharge $Q = CIA = 0.2646 \times 32.63 \times 10^{-3} \times 1/3600 \times 11.7 \times 10^6 = 28.06$ m^3/sec.

Example: 2.12 A watershed of total area of 18 hectares comprises of upstream non-residential area of 10 hectares and remaining downstream residential area of 8 hectares having existing storm water drain of 1.8 km in length. The time of concentrations for non-residential and residential area are 22 min and 20 min and overall runoff coefficients are 0.36 and 0.64, respectively. For rainfall frequency of once in 5 years the intensity duration has a relation $I = 82 \times e^{-0.028 \times t_c}$, where t_c is time of concentration in min and I is mm/h. The velocity of the flow in existing storm water drain is 0.90 m/sec. Estimate the design flow for storm water drain for non-residential area of catchment contributing the storm flow at the entry of the existing sewer and at the end of the 1.8 km of existing length of the drain when whole catchment is contributing.

Solution

Rainfall intensity, I, for non-residential area $= 82 \times e^{-0.028 \times t_c} = 82 \times e^{-0.028 \times 22} = 44.29\, mm/h$.

Rainfall intensity, I, for residential area $= 82 \times e^{-0.028 \times t_c} = 82 \times e^{-0.028 \times 20} = 46.84\, mm/h$.

Time of concentration = inlet time + time of travel.

Time of travel = 1800/0.90 = 2000 sec = 33.33 min.

Hence, time of concentration for whole catchment = 22 + 33.33 = 55.33 min.

Rainfall intensity, I, for total area $= 82 \times e^{-0.028 \times t_c} = 82 \times e^{-0.028 \times 55.33} = 17.42\, mm/h$.

Design discharge for non-residential part of the catchment = $0.36 \times 44.29 \times 10 /360$.
= 0.443 m^3/sec.

Discharge from only residential part of the catchment = $0.64 \times 46.84 \times 8 /360 = 0.666$ m^3/sec.

Overall runoff coefficient = $(0.36 \times 10 + 0.64 \times 8) / 18 = 0.484$.

Hence, the storm runoff for entire catchment = $(0.484 \times 17.42 \times 18) /360 = 0.422$ m^3/sec.

Hence, the design discharge for storm water drain at the end of existing drain will be highest discharged generated among these, i.e., 0.666 m^3/sec.

Questions

2.1. How design discharge for sanitary sewage is estimated?

2.2. What is dry weather flow? How much variation in this flow is expected to arrive at maximum and minimum flow?

2.3. Describe variation in sewage flow. Design of which component of sewerage scheme will be affected due to this flow variations?

2.4. What is design period? What are the parameters deciding this?

2.5. Describe evaluation of per capita sewage discharge.

2.6. Describe the factors affecting storm water runoff from a catchment.

2.7. Explain rational and empirical methods of storm water runoff estimation.

2.8. What is time of concentration and critical rainfall duration? Describe importance of this in determining design discharge.

2.9. Provide the comparative merits and demerits of the separate and combined system of sewerage.

2.10. A 1200 m long storm sewer collects water from a catchment area of 45 hectare, where 35% of area is covered by roof (C = 0.9); 20% of area is covered by pavements (C = 0.8) and 45% of area is open land (C = 0.13). Determine the average C, time of concentration, intensity of rainfall and diameter of storm water sewer if it is running full, considering following: (i) the time of entry = 3.0 min, (ii) velocity of flow = 1.25 m/sec, (iii) for storm water drain, n = 0.013 and slope = 0.001, (iv) intensity of rainfall, cm/h $= 75/(t_c + 5)$. (Answer: C $= 0.5335$; $t_c = 19$ min; I $= 3.125$ cm/h; D $= 1.473$ m)

2.11. Estimate the quantity of sewage to be considered for designing the sewage treatment plant for a city with past census population as given in the table below. Use suitable population forecasting method using appropriate design period. The per capita domestic water demand of 135 LPCD should be considered. In addition to this about 35 LPCD water is supplied to meet commercial and institutional water demand.

Year	1971	1981	1991	2001	2011
Population	350,000	385,000	430,000	490,000	580,000

(Answer: By incremental increase method P2031 $= 749,999$; Sewage flow $= 1.18$ m^3/sec; design flow $= 1.77$ m^3/sec)

2.12. What are various types of sewerage systems used? Describe merits of these systems.

2.13. The population data for a town is given below. Find out the population in the year 2031, 2041 and 2051 by (a) arithmetical (b) geometric (c) incremental increase methods.

Year	1981	1991	2001	2011	2021
Population	82,000	1,18,000	1,60,000	2,08,000	2,42,000

(Answer: Arithmetic increase method 282000, 322000, 362000; geometric increase method 314600, 408980, 531674; incremental increase method 281333, 319999, 357998)

2.14. In two past consecutive decades the population of a town was 40,000; 100,000 and 130,000. Determine: (a) saturation population; (b) equation for logistic curve; (c) expected population in next decade. (Answer: Saturation population 137500; m $=$ 2.4375; n $= -0.187$; forecasted population 136283)

2.15. Explain different methods of population forecasting covering best applicability of each method at a stage of the city growth.

References

Carter, R. W. (1961). Magnitude and frequency of floods in suburban areas, in Short papers in the geologic and hydrologic sciences: U.S. Geol. Survey Prof. Paper 424–B, p. B9–B11.

Chamier, G. (1898). Capacities required for culverts and flood openings. *Transactions of the Institute of Civil Engineers, 134*, 313–323.

Chen, H., Chen, R. H., Yu, F. C., Chen, W. S., & Hung, J. J. (2004). The inspection of the triggering mechanism for a hazardous mudflow in an urbanized territory. *Environmental Geology, 45*(7), 899–906.

Chow, V. T. (1962). Hydrological determination of waterway areas for the design of drainage struc-
    tures in small drainage basin. University of Illinois Bulletin, Vol. 59, No. 65.
CPHEEO. (1993). Manual on Sewerage and Sewage Treatment (2nd Ed.). Central Public Health and
    Environmental Engineering Organization, Ministry of Urban Development, Government of India,
    New Delhi, India.
CPHEEO (2013). Manual on Sewerage and Sewage Treatment, Part A, Engineering (3rd Ed.).
    Central Public Health and Environmental Engineering Organization, Ministry of Urban Devel-
    opment, Government of India, New Delhi, India.
CPHEEO (2019). Manual on Storm Water Drainage System. Central Public Health and Environment
    Engineering Organization (CPHEEO), Ministry of Housing and Urban Affairs, Government of
    India. http://mohua.gov.in/upload/uploadfiles/files/Volume%20I%20Engineering(3).pdf
EPA (2015). Storm Water Management Model Reference Manual Volume I Hydrology (Revised).
    Office of Research and Development, Water Supply and Water Resources Division, Environ-
    mental Protection Agency, United States of America. https://nepis.epa.gov/Exe/ZyPDF.cgi/P10
    0NYRA.PDF?Dockey=P100NYRA.PDF
Espey, W. H., Winslow, D. E. (1975). Quantity aspects of urban stormwater runoff. Short
    Course Proceedings of Application of Stormwater Management Models, University of Mas-
    sachusetts/Amherst, pp. 83–137.
Kang, J. H., Kayhanian, M., & Stenstrom, M. K. (2008). Predicting the existence of stormwater first
    flush from the time of concentration. *Water Research, 42*(1–2), 220–228.
Lee C.C. and Lin S.D. (2007). Handbook of Environmental Engineering Calculations (2nd Ed.). New
    York, McGraw-Hill.
Sarvaas, J. (1924). *Storm water discharge in city areas.* Paper & Discussion, Proceedings of Victorian
    Institute of Engineers, University of Melbourne, Australia, pp. 173–188.
Yang, X., Zhou, Q., & Melville, M. (2000). An integrated drainage network analysis system for agri-
    cultural drainage management. Part 2: the application. *Agricultural Water Management, 45*(1),
    87–100.

Sewer Materials, Shapes, Patterns of Collection System and Appurtenances

3

3.1 Shapes of Open Drain Section

For conveyance of the less polluted water such as sullage and rain water, i.e., wastewater from kitchens, bathrooms and for collecting storm water from courtyards, roads and open grounds, the open drain section can be used. However, open drain section is not recommended for conveying wastewater generated from toilets as it is highly polluted. These open drains are laid on both sides of the road along the boundary wall of buildings for collecting this less polluted water. Though, some municipalities prefer these types of open drain due to low maintenance cost this is less hygienic as the drains are open to atmosphere. Hence, it should be preferred for carrying only storm water runoff and not for sullage. These drains require periodic cleaning as deposition of solids will occur due to non-development of self-cleansing velocity. They are designed to carry the discharge by gravity and accordingly bed slope needs to be provided to ensure self-cleansing velocity.

For delivering the functional requirements these drains should have smooth inside surface with proper plastering from inside. The grade at which these drains are laid should offer a self-cleansing velocity even when carrying dry weather flow. The drain section should have sufficient carrying capacity with adequate free board considering the maximum discharge in the drain. The materials used for construction should be non-corrosive with sufficient resistance to abrasion and the section should be structurally safe and stable. The dimensions of the section selected should be adequate enough, so as to facilitate easy cleaning of these drains.

The typical section used in practice for open drains are: (i) trapezoidal, (ii) rectangular, (iii) semi-circular, (iv) U-shaped and (v) V shaped section (Fig. 3.1). The rectangular section is suitable for handling large discharge as self-cleansing velocity will not develop in this section while handling smaller discharges, which will lead to settling of suspended solids. The trapezoidal section is better in this regard as cross sectional area is lesser

© The Author(s), under exclusive license to Springer Nature Singapore Pte Ltd. 2022   59
M. Ghangrekar, *Wastewater to Water*, https://doi.org/10.1007/978-981-19-4048-4_3

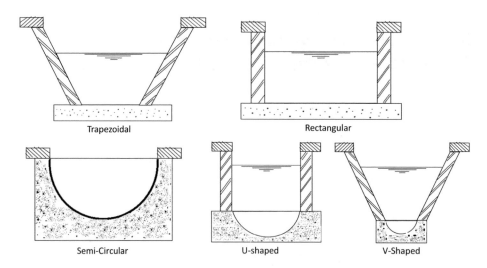

Fig. 3.1 Shapes of the section commonly used for open drains

at lesser depth; however, this section is also suitable for handling moderate and high discharge. These sections are constructed by laying plain cement concrete bed over which sides are constructed with stone or brick masonry and surface is plastered. For deeper drains the sides can also be made with reinforced cement concrete. Large open drains are generally having rectangular or trapezoidal cross-section.

Semi-circular section is mostly used in practice for handling smaller discharge. This section is stable and facilitates easy cleaning. This can be constructed using half rounded stone-ware pipes. These are not suitable for handling large discharges. For handling large discharge, the semi-circular section can be converted to U-shape section by raising the sides using stone or brick masonry and plastered inside. The base can be made with stoneware half rounded pipe or concrete itself. For handling larger flow variation V-shape section is better, since it will develop self-cleansing velocity at lower flow due to less cross-section of flow. The bottom section can be made with stone-ware and sides above that can be made with masonry or concrete or whole section could be made up of concrete.

Depending upon the estimated flow to be carried the drain section shall be designed. The depth of the open drain should preferably be kept lesser than 1.5 m; however, for area receiving higher rainfall the drain section could be up to 3.0 m deep. The velocity developed in the drain should be self-cleansing; however, the velocity at maximum discharge should not cross the limiting velocity, so as to protect the section from erosion. Depending on the material used for construction of the section, the limiting velocity could have value of 1.5 m/sec for stone pitched drains and 2–2.5 m/sec for cement concrete lined sewer.

3.2   Shapes of Closed Sewers

Sewers are generally circular pipes laid below ground level, slopping continuously towards the outfall end. Circular sewer section is ideal from bearing the load of back-fill material and live load, if any, on the sewers laid under the public roads. Minimum diameter of circular sewer used should be 150 mm and for larger cities serving population more than one hundred thousand this minimum diameter can be 200 mm. For house sewer pipe connecting to public sewer the minimum diameter should be 100 mm or higher depending on the number of people served.

Sewers are designed to flow partially full under gravity. Shapes other than circular are also used to better handle variation in flow rates or handling higher discharge considering convenience of construction of large sewers in situ. Other than circular shape, the shapes which are generally used for sewers are (Fig. 3.2 a through i): (i) Standard Egg-shaped sewer; (ii) New or Modified egg-shaped sewer; (iii) Horse shoe shaped sewer; (iv) Parabolic shaped sewer; (v) Semi-elliptical section; (vi) Rectangular section; (vii) U-shaped section; (viii) Semi-circular shaped sewer; and (ix) Basket handled shape sewer.

Standard egg-shaped sewers, also called as ovoid shaped sewer, and new or modified egg-shaped sewers are used in combined sewers or for sewers where high flow variation is expected. This shape helps in maintaining velocity of flow more or less stable with variation in the depth of flow. Thus, these sewers can generate self-cleansing velocity even during dry weather flow. The egg-shaped section is superior for both load transmission as well as maintaining velocity at minimum flows, thus having ability to flush out sediments at the bottom smaller cross section of flow when peak flow reaches. These are normally made up of reinforced cement concrete, either cast in situ or pre-cast.

Horse shoe shaped sewers, with shape similar to horse shoe, and semi-circular sections are used for large sewers handling larger discharges, such as trunk and outfall sewers. These are generally cast in situ made with concrete. Rectangular or trapezoidal sections are used for conveying large flow of storm water. These shapes are easy to construct, however will have poor hydraulic properties as hydraulic radius increases when flow increases. U-shaped section is used for larger sewers and especially in open cuts. Other sections of the sewers have become absolute due to difficulty in construction on site and non-availability of these shapes readily in market. Basket handle section is used for large combined sewer and it is constructed in situ. The purpose of cunette (a small cross section dug in the bottom of a much larger channel) is to carry dry weather flow. Such cunette can also be used in rectangular or trapezoidal section to improve ability of these shapes in handling small dry weather flow discharges in case of combined sewer.

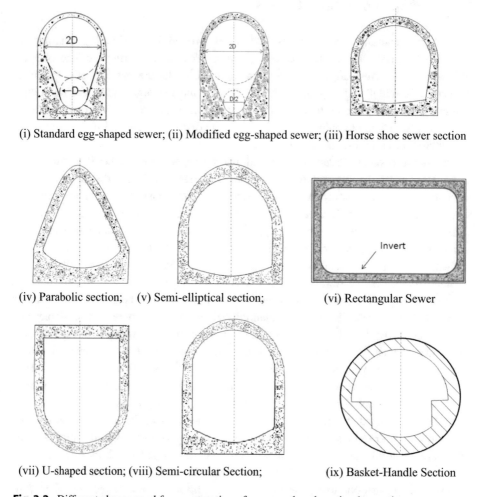

(i) Standard egg-shaped sewer; (ii) Modified egg-shaped sewer; (iii) Horse shoe sewer section

(iv) Parabolic section;   (v) Semi-elliptical section;   (vi) Rectangular Sewer

(vii) U-shaped section; (viii) Semi-circular Section;   (ix) Basket-Handle Section

Fig. 3.2 Different shapes used for construction of sewers other than circular section

3.3   Materials Used for Construction of Sewers

As compared to storm runoff the sanitary sewage is highly polluted wastewater, which will undergo decomposition of organic matter while the sewage it getting transported through sewers. Typically, under anaerobic conditions hydrogen sulphide will be liberated leading to corrosion inside the sewers. In addition, the raw sewage contains sand and grit matters, which will lead to physical abrasion of the inside surface of the sewers. Considering these factors, the sewer material selection should be finalized. Single material might not meet all the conditions that may be encountered in sewer design; hence, selection of material should be made considering the requirements and different materials could be selected for parts of a single project. The factors that will influence selection of materials for sewers are characteristics of wastewater, availability of pipes in the sizes required including materials for joining pipes, ease of handling and transport, procedure for laying and jointing, water imperviousness, physical strength, resistance to corrosion, resistance to abrasion, durability, cost of the pipes and cost of handling and installation.

3.3.1 Important Factors to be Considered for Selecting Material for Sewer

The factors to be considered while selecting material for making sewer pipes are discussed below in details.

a. *Resistance to corrosion*: Sewer carries wastewater that releases corrosive gases, such as H_2S. In contact with moisture this hydrogen sulphide can be converted into sulphuric acid and can lead to the corrosion of sewer pipe material from inside. Hence, selection of corrosion resistance material is must for ensuring long life of sewer.

b. *Resistance to abrasion*: Sewage contains considerable concentration of suspended solids, part of which are inorganic solids, such as sand or grit. These particles while moving at high velocity can cause erosion of interior surface and hence lead to wear and tear of sewer pipe internally. Eventually with time, this abrasion will reduce thickness of pipe and thus, reduce hydraulic efficiency of the sewer by making the interior surface rough.

c. *Hydraulically efficient material*: The sewer material should offer a smooth interior surface to have less frictional coefficient to make sewer hydraulically efficient. Also, the material selected should have sufficient resistance to abrasion so as to maintain this smoothness for longer duration of operation, thus the hydraulic carrying capacity should not drop down as the sewer becomes old.

d. *Strength and durability*: The sewer pipe material should have sufficient strength to withstand all the forces that are likely to occur on them. Sewers are subjected to external loads of backfill material and traffic load, if any. In normal course, sewers are not subjected to internal pressure of water. To withstand external load safely without failure, sufficient wall thickness of pipe or reinforcement in case of concrete pipes is essential. In addition, the material selected should be durable and should have sufficient resistance against natural weathering action to ensure longer life of the pipe.

e. *Imperviousness*: The material selected for sewer should not allow seepage of sewage from sewer to surrounding, thus the material selected for pipe should be impervious. If such seepage occurs it will pollute the nearby area and ground water, hence imperviousness property of the material is important.

f. **Weight of the material**: The material selected for sewer should have less specific weight, thus to make pipe light in weight. The lightweight pipes are easy for handling, transport, laying and jointing, offering advantage while construction of sewers.
g. **Economy and cost**: Sewer material should be less costly to make the sewerage scheme economical and affordable.

3.3.2 Materials for Sewers

3.3.2.1 Brick Sewers

Bricks can be used for construction of large size combined sewer or particularly for storm water drains. This brick sewer is plastered from outside with sulphate resistance cement to avoid entry of tree roots and groundwater through brick joints. These sewers are lined from inside with stone ware or ceramic block to make them smooth and hydraulically efficient. Lining also makes the sewer resistant to corrosion. Brick sewers shall have cement concrete or stoneware material for invert and a 12.5 mm thick cement plaster with neat finish for the remaining surface to maintain it smooth. Due to comparatively higher cost, larger space requirement, slower progress of constriction of sewers, brick is now used only for sewer construction in special cases. Failures of this sewer may occur mainly due to the disintegration of the bricks or the mortar joints.

3.3.2.2 Asbestos Cement Sewers

Asbestos cement is manufactured from a mixture of asbestos fibres, silica and cement. Asbestos fibres are thoroughly mixed with cement to act as reinforcement. These pipes are available in size of 80–1000 mm internal diameter and length up to 4.0 m. These pipes can be easily assembled without skilled labour with the help of special coupling, called as 'ring tie coupling' or simplex joint. These pipes and joints are adequately resistant to external electrolytic corrosion under most natural soil conditions and the joints are flexible to permit 12° deflection for curved laying. These pipes are generally used for vertical transport of sewage to ground in building plumbing, for transport of rainwater from roofs in multistorey buildings, and for transport of less foul sullage, i.e., wastewater from kitchen and bathroom.

These pipes are light in weight as compared to concrete and stoneware sewers; hence, these are easy to carry and transport. They can be easily cut to required length and assemble without skilled labour. The interior surface of the pipe is smooth (Manning's n = 0.011); hence, can make hydraulically efficient sewer. However, asbestos pipes are structurally not very strong and cannot withstand high external load. These pipes are also susceptible to corrosion by sulphuric acid. When bacteria produce H_2S, in presence of water, H_2SO_4 can be formed leading to corrosion of pipe material from inside the sewer.

3.3.2.3 Plain Cement Concrete (PCC) or Reinforced Cement Concrete (RCC) Sewers

The concrete sewers can be made for any desired strength and could be strong in tension as well as compression and can be cast in situ or precast pipes. These pipes offer resistant to erosion and abrasion. This material can be economical for medium and large size sewer sections. These pipes are available in wide range of sizes and the trench can be opened and backfilled rapidly during maintenance of sewers. However, these pipes can get corroded and pitted by the action of H_2SO_4. Hence, the carrying capacity of the pipe reduces with time because of corrosion, if proper protection is not given. At higher velocities of flow the pipes are susceptible to erosion by sewage containing silt and grit. These shall be manufactured with sulphate resistant cement and with high alumina coating on the inside at the manufacturers works itself. The concrete sewers can be protected internally by vitrified clay linings. With protection lining they are used for almost all the branch and main sewers.

Plain cement concrete (1: 1.5: 3) pipes are available up to 0.45 m diameter and reinforcement cement pipes are available up to 1.8 m diameter. However, plain cement concrete pipes are used in sewer systems on a limited scale only and generally reinforced concrete pipes are preferred. These pipes can be cast in situ or precast pipes. Cast-in-situ reinforced concrete sewers are constructed where they are more economical, or when non-standard sections or a special shape is required or when the headroom and working space are limited. All formwork for concrete sewers shall be unyielding and tight and shall produce a smooth sewer interior surface. Precast pipes are better in quality than the cast in situ pipes. The reinforcement in these pipes can be different, such as single cage reinforced pipes, used for internal pressure less than 0.8 m; double cage reinforced pipes used for both internal and external pressure greater than 0.8 m; elliptical cage reinforced pipes used for larger diameter sewers subjected to external pressure; and Hume pipes with steel shells coated with concrete from inside and outside. Nominal longitudinal reinforcement of 0.25% is provided in these pipes.

3.3.2.4 Vitrified Clay or Stoneware Sewers

Vitrified clay or stoneware pipes are used for house connections as well as lateral sewers. Size of this pipe is available from 10 cm up to 30 cm internal diameter with a length of 0.9 m. These pipes are rarely manufactured for diameter greater than 60 cm and length greater than 1.2 m. These are joined by bell and spigot flexible compression joints. Specially process clay is moulded to the shape of pipe and spigot and bell end are made and these are air dried for about two weeks. Then these are burned at about 1100–1200 °C for a period of about 10 days. During last stage of backing, glazing is done by spraying sodium chloride by chemical reaction between sodium and melted silica to obtained smooth, impervious and hard surface.

These pipes are resistant to corrosion from most of the acids, hence fit for carrying polluted water such as sewage. Interior surface of this pipe is smooth and it is hydraulically efficient offering adequate resistance to abrasion by grit matter. These pipes are highly impervious and reasonably strong in compression. A minimum crushing strength of 1,600 kg/m length of pipe is generally adopted for all sizes of pipes manufactured (IS: 4127-1983). These pipes are durable and economical for small diameter sewers. The pipe material is impervious and does not absorb water more than 5% of its own weight, when immersed in water for 24 h. However, these pipes are heavy, bulky and brittle; hence, difficult to transport and need careful handling. These pipes cannot be used as pressure pipes, because they are weak in tension. In addition, due to small individual pipe length these require large number of joints and skilled labour for jointing the pipes.

3.3.2.5 Cast Iron Sewers

These pipes are strong and capable to withstand greater tensile, compressive, as well as bending stresses; however, these are costly. These pipes are available in diameters varying from 80 to 1050 mm and 3.66 to 5.5 m in length. These are covered with protective coatings from inside and outside to impart corrosion resistance. A variety of joints are available for cast iron pipes, including socket, spigot, and flanged joints. Cast iron pipes are used for outfall sewers, rising mains of pumping stations, and depressed section of inverted siphons, where pipes are running under pressure. These are also suitable for sewers under heavy traffic load, such as sewers below railways and highways. They are suitable for laying sewers on carried over piers in case of low-lying areas. They offer 100% leak proof sewer line to avoid any possibility of groundwater contamination due to leakage of sewage. Though, cast iron pipes are having corrosion resistance in most natural soils; however, these are less resistant to acid corrosion or carrying highly septic sewage; hence, generally lined from inside with cement concrete, coal tar paint, epoxy, etc. Whenever necessary, this pipe can be laid with deflection of $2°-5°$, from straight alignment, depending on type of joints used.

3.3.2.6 Steel Pipes

These are used under the situations such as rising main sewers, under water crossing, bridge crossing, necessary connections for pumping stations, laying of sewers over self-supporting spans to cross low-lying area, railway crossings, etc. Steel pipes can withstand high internal pressure, impact loads and vibrations much better than CI pipes. These pipes are more ductile to withstand water hammer pressure better, however these pipes cannot withstand high external load and may collapse when negative pressure is developed in sewer. They are susceptible to corrosion and are generally not used for partially flowing sewers. They are protected internally and externally against the action of corrosion. These pipes shall be coated inside by high alumina cement mortar or polyurea and outside they are coated with epoxy paint to protect against corrosion.

3.3.2.7 Ductile Iron Pipes

Ductile iron pipes can also be used for conveying the sewage efficiently and it is a better pipe material than cast iron, however it is costly material. These pipes are having better capability to withstand water hammer. The predominant wall material is ductile iron, a spheroidized graphite cast iron normally prepared using the centrifugal cast process. These pipes are available in diameter ranging from 80 to 1000 mm and length from 5.5 up to 6 m. Internally these pipes are coated with cement mortar lining or any other polyethylene or poly wrap or plastic bagging/sleeve lining to inhibit corrosion from the wastewater being conveyed. Various types of external coatings are used to inhibit external corrosion of pipe from the environment.

Ductile iron is still believed to be stronger and more fracture resistant material having properties like impact resistance, high wear and tear resistance, high tensile strength and ductility. However, like most ferrous materials it is susceptible to corrosion. These pipes are strong offering both inner and outer smooth surfaces free from lumps, cracks, blisters and scars and typically 30% lighter in weight as compared to CI pipes. A typical life expectancy of thicker walled pipe could be up to 75 years, however with the current thinner walled ductile pipe the life could be about 20 years in highly corrosive soils without adopting any corrosion control measures like cathodic protection.

3.3.2.8 Plastic Sewers (PVC Pipes)

Plastic pipes are used for internal drainage works in houses. These pipes are available in sizes ranging from 75 to 315 mm external diameter and can be used in drainage works. They offer smooth internal surface. The additional advantages they offer are resistant to corrosion, light weight of pipe, economical in laying, jointing and maintenance, the pipe is tough and rigid, and ease in fabrication and transport of these pipes. The material of normal polyvinyl chloride (PVC) pipes becomes brittle over the period of operation when exposed to solar radiation or heating and cooling and can lead to deformation in pipes and cracks formation; hence, unplasticized polyvinyl chloride (UPVC) material should be preferred for sewers carrying sewage and storm water.

3.3.2.9 High Density Polyethylene Pipes

High density polyethylene (HDPE) pipes are not brittle like AC pipes and other pipes and hence hard fall during loading, unloading and handling shall not cause any damage to these pipes. They can be joined by welding or can be jointed with detachable joints up to 630 mm diameter (IS: 4984-1995). These are popularly used for conveyance of industrial wastewaters. They offer all the advantages as offered by PVC pipes; however, PVC pipes offer very little flexibility and normally considered rigid; whereas, HDPE pipes are flexible hence best suited for laying in hilly and uneven terrain. Flexibility allows simple handling and installation of HDPE pipes. Because of low density, these pipes are very light in weight, making them easy to handle and reducing transportation and installation cost. HDPE pipes are non-corrosive and offer very smooth inside surface due to which head loss is minimal and also this material resist scale formation. These pipes could be manufactured with smooth external surface or non-smooth external surface, known as double walled corrugated (DWC) pipes. These HDPE pipes are available with internal diameter ranging from 75 to 1,000 mm and length up to 6 m.

3.3.2.10 Glass Fibre Reinforced Plastic Pipes

This martial is widely used where corrosion resistant pipes are required. Glass-fibre reinforced plastic (GRP) can be used as a lining material for conventional pipes to protect from internal or external corrosion. It is made from the composite matrix of glass fibre, polyester resin and fillers. These pipes have better strength, durability, high tensile strength, low density and high corrosion resistance. These are generally manufactured up to 3.0 m diameter and length of 6, 9 and 12 m (IS: 12709-1994). Glass reinforced plastic pipes represent the ideal solution for transport of any kind of water, chemicals, effluent and highly septic wastewater like sewage, because they combine the advantages of corrosion resistance with a mechanical strength that can be compared with the steel pipes. Typical properties that result in advantages in GRP pipes use can be summarized as follows:

- Light weight of pipes that allows for ease of laying and transport.
- Possibility of nesting of different diameters of pipes, thus allowing additional saving in transport cost.
- Length of pipe is larger than other pipe materials, hence require lesser number of joints.
- Easy installation procedures due to the kind of mechanical bell and spigot joint.
- Corrosion resistance material, hence no protections such as coating, painting or cathodic protection are necessary.
- Smooth internal surface that minimizes the head loss and avoids the formation of deposits.
- High mechanical resistance due to the glass reinforcement.
- The material is absolute impermeable hence no leakage.
- Very long expected life of the material.

3.3.2.11 Pitch Impregnated Fibre Pipes

Pitch fibre pipes are cheaper, lightweight and easier to handle material made of wood cellulose impregnated with inert coal tar pitch of about 70% by weight. The early method of manufacturing these pipes was to roll up sheets of pitch impregnated paper onto a spindle until the desired wall thickness had been obtained. This form of manufacture was crude compared with the present-day technique, which produce a homogeneous wall of cellulose and asbestos pitch fibre completely impregnated with hard coal tar pitch. They are typically manufactured with 50–225 mm nominal diameter and length varying from

1.5 to 3.0 m and can be connected with each other in any weather condition with push fit joints without the use of jointing compound or rubber ring joints (IS: 11925-1986).

Due to longer length, the cost of jointing, handling and laying is reduced for this pipe. They can be cut to required length on the site. These pipes have shown their durability in service. These pipes are flexible, resistant to heat, freezing and thawing (CPHEEO, 2013). They also offer corrosion resistance and these pipes are unaffected by acids and other chemicals attack and can withstand pH range of 0.5–12.5. Aggressive soil conditions, even those with an artificially high sulphate content, created by the use of clinker, ash or slag fills, are not detrimental to these pipes. These pipes are recommended for septic tank connections, house sewers, down pipes, storm drains, and industrial waste drainage.

3.4   Corrosion of Sewers

While sewage is flowing through the sewers, organic material is likely to get deposited in the sewers, particularly during the hours of low flow in the sewers. This settled organic matter will undergo degradation due to presence of bacteria. Since, majorly anoxic conditions are prevailing in the sewers and sulphate is also present; sulphur reducing heterotrophic bacteria will utilize this organic matter as carbon source and reduce sulphate to produce sulphide. Also, anaerobic fermentation of settled organic matter will occur, producing volatile fatty acids and being water soluble, these produced acids will reduce pH of the sewage. This combination of reduced pH and production of hydrogen sulphide, which will be released in the free head space present in the sewers, produce corrosive environment inside the sewer. This hydrogen sulphide will get dissolved in the moisture condensed on the sewer crown portion. At such locations, whatever oxygen is available will be utilized by *Thiobacillus* bacteria for oxidizing hydrogen sulphide to sulphuric acid. For the sewers made with concrete, steel or iron, this acid formation will solubilize the material and lead to severe destruction of the crown and failure of the sewer. The problem is aggravated under warmer climate and when the sewers are laid with flatter gradient offering more retention time to offer extensive damage to the sewers.

Corrosion of sewers can be avoided by providing proper corrosion resistant lining to the sewers and having forced ventilation. Forced ventilation can reduce the crown condensation of hydrogen sulphide, by removing it from the headspace of sewer and releasing it to the atmosphere. Forced ventilation will also provide sufficient oxygen for reducing production of hydrogen sulphide and volatile fatty acids. Sewers flowing full, such as outfall sewer, face less problem of sewer corrosion. For lateral sewers the vitrified clay material for sewer has demonstrated long service life offering resistance to corrosion. Plastic pipes, such as high-density polyethylene and glass fibre reinforced pipes, are also expected to give satisfactory service life. For larger size sewers, where concrete is commonly used material, the sewers should be casted with integral lining or lined with plastic, clay tiles or asphaltic compounds to give protection against corrosion.

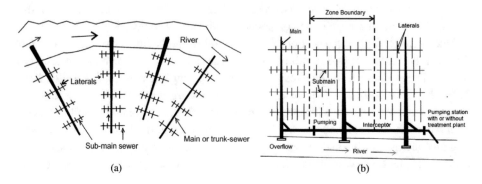

Fig. 3.3 **a** Perpendicular pattern and **b** interceptor pattern of collection system

3.5   Patterns of Collection System

The network of sewers consists of house sewers discharging the sewage to laterals. The lateral discharges sewage into branch sewers or sub-mains and sub-mains discharge it into main sewer or trunk sewer. The trunk sewer carries sewage to the common point where adequate treatment is given to the sewage and then it is discharged. The patterns of collection system depend upon: the location and methods of treatment and options for disposal of treated sewage, the topographical and hydrological features of the area, type of sewerage system employed, and extent of area to be served. Various geometrical patterns that can be adopted for collection systems as per the suitability are perpendicular pattern, interceptor pattern, radial pattern, fan pattern, and zone pattern.

3.5.1   Perpendicular Pattern

In this pattern sewers carrying storm water are laid perpendicular to the direction of flow of the water body, such as river, to maintain shortest possible path to discharge storm water in natural water body (Fig. 3.3a). It is suitable for separate system and partially separate system for storm water drains. This pattern is not suitable for combined system, because sewage treatment plants are required to be installed at many places; otherwise, it will pollute the water body in which the sewage is getting discharged.

3.5.2   Interceptor Pattern

As the city develops this pattern brings improvement over the perpendicular patter by tapping the polluted storm water drains or sewers by providing common interceptor to take wastewater to a point where it will be treated before getting discharged in to the

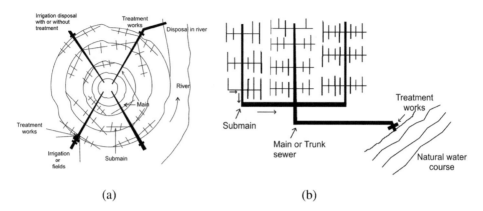

Fig. 3.4 **a** Radial pattern of collection system, and **b** fan pattern of collection system

water body. Thus, perpendicular sewers are intercepted with large size interceptor sewer (Fig. 3.3b). Interceptor sewer carries sewage to a common point, where it can be disposed off after providing proper treatment. Overflow arrangement should be provided for handling excessive flow during storms. This will allow surplus flow than the capacity of interceptor to be discharged in natural water course.

3.5.3   Radial Pattern

The radial pattern takes the advantage of natural slope available in case the city has hilly terrain, sloping outwards from the centre. In this pattern sewers are laid radially outwards from the centre; hence, this pattern is called as radial pattern (Fig. 3.4a). Since, a greater number of disposal options are essential for this pattern and water body will not be available all around the city, this pattern is suitable where land disposal of treated sewage for agriculture application is acceptable or industrial reuse of treated sewage could be a viable option. Hence, multiple disposal options requirement is the drawback in this pattern.

3.5.4   Fan Pattern

This is suitable pattern for a city situated at one side of the natural water body, such as river. The sewers are laid in a way so that the entire sewage flows to a common point, where one treatment plant is located (Fig. 3.4b). Number of converging main sewers and sub-mains are used in this pattern forming a fan shape; hence it is named as fan pattern. Single treatment plant requirement is the advantage of this pattern. However, the drawback of this pattern is that larger diameter pipe is required for main sewer near to

Fig. 3.5 Zone pattern of
collection system

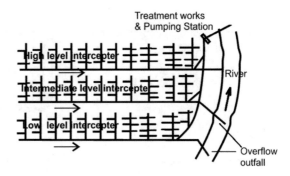

the treatment plant as entire city sewage is collected at a common point. In addition, with new development of the city the load on existing treatment plant will increase and also with increase in quantity of sewage generation due to increment in population beyond design capacity the submains and mains will required to be replaced with higher capacity. Hence, augmenting capacity of existing sewers will be difficult due to replacement of mains and sub-mains with larger diameter sewer section.

3.5.5 Zone Pattern

For serving very large cities, where pattern with single interceptor will be overloaded, the city is divided in different zones and an interceptor sewer is provided for each zone collecting the entire sewage generated from that zone. More numbers of interceptors are provided in this pattern (Fig. 3.5). This pattern is more suitable for sloping area due to requirement of a lesser number of pumping station than flat areas in which case a greater number of intermediate pumping stations will be essential to control the maximum depth at which the sewers will be required to be laid. If the final disposal of sewage is in water body it is recommended to have large size receiving interceptor sewer collecting the sewage from these interceptors to take it to a common point where it will be treated before it is discharged in water body. Whereas, for storm water drains, treatment will generally be not required and these interceptors provided in different zones can discharge the storm water directly into water body.

3.6 Construction of Sewers

3.6.1 Laying of Sewer Pipes

Sewers are generally laid starting from their outfall ends towards their starting points. With this style of laying sewers, the advantage of utilization of the tail sewers even during the initial periods of its construction is possible. The complete detail drawings should be available before execution of the sewer construction. This working drawing should have details of alignment of sewers, size of sewer, location of manhole and other appurtenances, invert level at each junction, type of bedding and details of the trench to be excavated with width and depth. It is common practice, to first locate the points where manholes are required to be constructed, ensuring stipulated spacing between the manholes, as per drawing, i.e., L-section of sewer, and then laying the sewer pipe straight between the two manholes. The positions of the manholes are marked on ground and they are numbered. The construction work of the manhole can be taken up then as per the drawing ensuring invert level in the manhole, as per design. The sewers can be then laid between two adjacent manholes.

The central line of the sewer and edge line of the sewer trench are marked on the ground and an offset line is also marked parallel to the central line at suitable distance, about half the trench width plus 0.6 m. This line can be drawn by fixing the wooden pegs at 15 m intervals and this offset line will be used for finding out centre line of the sewer simply by offsetting. Maintaining alignment and necessary gradient is very important in sewer construction. The trench of suitable width is excavated between the two manholes and the sewer is laid between them. Depending on the nature of soil and depth of excavation a shoring support may be required for excavation in the loose soil or sandy soil strata. Proper fencing protection should be provided to the excavated trench portion, till it is backfilled and road is restored, to guide the adjoining traffic by putting appropriate sign boards.

The trench is excavated and the bottom bed is properly dressed. The minimum width of the trench should be 75 cm and for larger sewer diameter the width of the trench shall be outer diameter of pipe plus 30–40 cm, so as to get sufficient space within the trench for jointing the sewer pipes. At location where sewer invert level is below ground water table or water-logged area or near sea coast, arrangement for continuous dewatering of the trench during excavation and laying of sewers is required to be provided. The trench is excavated up to a level of the bottom embedding concrete or up to the invert level of the sewer pipe plus pipe thickness if no embedding concrete is provided. The designed invert levels and desired slope as per the longitudinal section of the sewer should be precisely transferred to the trench bottom.

The trench bedding is provided for making surface uniform with desired gradient and hard enough for laying and resting sewers. In case of hard soil strata, the sewer pipes may be laid directly on the soil bedding. This is known as impermissible bedding. In case of ordinary bedding on hard surfaces the bottom of the bedding is dressed in a way to fit the quadrant of the pipe. Thus, the bottom portion of the trench is excavated to confirm the shape of the pipe itself. For weaker soil strata a layer of compacted strata made-up of stone chips and gravel may be used, which is known as granular bedding or first-class bedding.

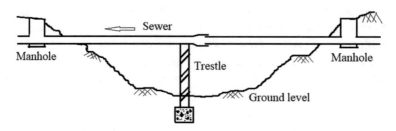

Fig. 3.6 Trestle supports for passing the sewer line over depressed area

In case loose or weaker soil strata when strength to the bedding is imparted by laying cement concrete, it is named as cradle bedding. The concentre bedding of 100–450 mm is provided depending on the pipe diameter. For imparting additional strength, wherever required in case of softer grounds, the sewer may be completely encased with concrete with top concrete cover of minimum 75 mm and bedding support of minimum 150 mm. While sewer line crossing the low laying area (depression) a trestle support needs to be provided near each joint in the sewer (Fig. 3.6).

After bedding concrete is laid in required alignment and levels. The sewer pipes are then lowered down into the trench either manually or for bigger pipe diameters with the help of machines. The sewer pipe lengths are usually laid from the lowest point with their sockets facing up the gradient, on desired bedding. Thus, the spigot end of new pipe can be easily inserted in the socket end of the already laid pipe. In case of collar joint both the pipes are placed matching levels. The grade and alignment are checked and proper joining material is applied in place. Further excavation and bedding preparation is then carried out for laying the sewer pipes between the next consecutive manholes. Thus, the process is continued till the entire sewers are laid out.

3.6.2   Testing of Sewers

Test for straightness of alignment: The sewers laid between the two manholes should be tested for straightness of alignment. This is done by placing a mirror at one end of the sewer line and a lamp at the other end. If the pipe line is straight, full circle of light will be observed. If the image is crescent then the alignment is adjusted, and joints are fixed again properly.

Test for leakage or water test: The sewers are tested before backfilling after giving sufficient time for the joints to set for no leakage. This leakage might occur either through improper joints or through cracks in sewer pipes. For this sewer pipe sections are tested between the manhole to manhole under a test pressure of about 1.5 m water head. To carry this, the downstream end of the sewer is plugged and water is filled in the manhole at upper end. The depth of water in manhole is maintained at about 1.5 m above the sewer invert. The sewer line is physically inspected for about half an hour and the leaking joints, if any, are identified and properly repaired.

3.6.3   Backfilling of Trench

After the sewer line has been laid and tested, the trenches are back filled using the excavated material. The earth should be laid equally on both side of the pipe with layer of 15 cm thickness. Each layer should be properly watered and rammed (compacted) to achieve consolidation of the filling material. The top of the trench surface is made similar to the original road surface.

3.7   Sewer Appurtenances

The structures, which are constructed at suitable intervals along the sewerage system to help its efficient operation and maintenance, are called as sewer appurtenances. These include: (i) Manholes, (ii) Drop manholes, (iii) Lamp holes, (iv) Clean-outs, (v) Street inlets called Gullies, (vi) Catch basins, (vii) Flushing tanks, (viii) Oil and grease trap, (ix) Inverted siphons, and (x) Storm water regulators. As per the requirement presence of few of these is essential for efficient operation of the sewerage system.

3.7.1   Manholes

The manhole is masonry or RCC chamber constructed at suitable intervals along the sewer lines, for providing access into sewer. Thus, the manhole helps in inspection, cleaning and maintenance of sewers, as and when required. These are provided at every change in alignment, bends, junction, change of gradient or change of diameter of the sewers. The sewer line between the two manholes is generally laid straight with even gradient. For straight sewer line manholes are provided at regular interval depending upon the diameter of the sewer. The spacing of manhole is recommended in IS 1742-1983. For sewer up to 0.3 m diameter or sewers which cannot be entered for cleaning or inspection the maximum spacing between the manholes recommended is 30 m, and 300 m spacing is recommended for sewer pipe greater than 2.0 m diameter (Table 3.1). A spacing allowance of 100 m per 1 m diameter of sewer is a general thumb rule in case of very large sewers (CPHEEO, 1993). The internal dimensions required for the manholes are provided in Table 3.2 (CPHEEO, 1993). The minimum width of the manhole should not be less than internal diameter of the sewer pipe plus 150 mm benching on both the sides.

Table. 3.1 Spacing of manholes for sewers with different diameter

Pipe diameter	Spacing
Less than 0.3 m	30 m
0.4–0.9	45–90 m
0.9–1.5 m	90–150 m
1.5–2.0 m	150–200 m
Greater than 2.0 m	300 m

Table. 3.2 Minimum internal dimensions for manhole chambers for different depth of sewers

Depth of sewer	Internal dimensions
0.9 m or less depth	0.90 m × 0.80 m
For depth between 0.9 and 2.5 m	1.20 m × 0.90 m, 1.2 m dia. for circular
For depth above 2.5 and up to 9.0 m	For circular chamber 1.5 m dia.
For depth above 9.0 and up to 14.0 m	For circular chamber 1.8 m dia.

3.7.1.1 Classification of Manholes

Depending upon the depth the manholes they can be classified as: (a) shallow manholes, (b) normal manholes, and (c) deep Manholes. These are described below.

Shallow Manholes: These manholes have a depth of 0.7–0.9 m and these are constructed at the start of the branch sewer or at a place not subjected to heavy traffic conditions (Fig. 3.7). These are provided with light cover at top and these are also called as inspection chamber.

Normal Manholes: These manholes are 1.5 m deep with internal dimensions of 1.0 m × 1.0 m square or rectangular with 1.2 m × 0.9 m (Fig. 3.8). These are provided with heavy cover at its top as these may be expected to support the traffic load.

Deep Manholes: These manholes are more than 1.5 m deep. The section of such manhole is not provided uniform throughout (Fig. 3.9). The size in upper portion is reduced

Fig. 3.7 Shallow manhole

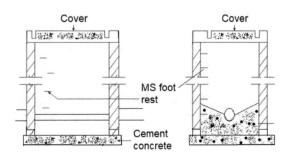

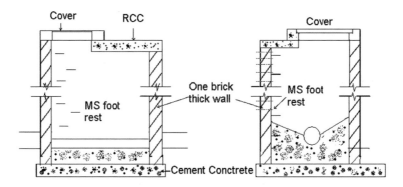

Fig. 3.8 Rectangular manhole for depth 0.9–1.5 m

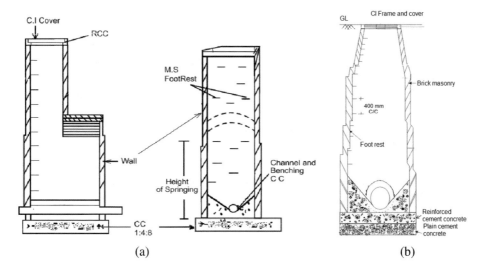

Fig. 3.9 **a** Rectangular and **b** circular deep manhole cross section

by providing an offset to reduce the material requirement for construction. Steps are provided in such manholes for descending into the manhole. These are provided with heavy cover at its top to support the anticipated traffic load.

3.7.1.2 Other Types of Manholes

Straight—through manholes: This is the simplest type of manhole, which is built on a straight run of sewer without any side junctions. While connecting sewers of different diameters, the soffit or crown level of the two sewers should be the same, except where special conditions require otherwise.

Junction Manholes: These types of manholes are constructed at every junction of two or more sewers, and on the curved portion of the sewers, with curved portion situated within the manhole. This manhole can be constructed with the shape other than rectangular to suit the curve requirement and achieve economy. The soffit of the smaller sewer at junction should not be lower than the soffit level of the larger sewer. Gradient of smaller sewer may be altered from the previous manhole to reduce the difference of invert at the point of junction to a convenient amount.

Side entrance Manholes: Where it is difficult to obtain direct vertical access to the larger sewer from the top ground level due to some obstruction, such as other pipe lines of water or gas, the access shaft should be constructed at the nearest convenient position off the line of sewer, and connected to the manhole chamber by a lateral passage. Floor of the side entrance passage should fall at about 1 in 30 towards the sewer and this passage should enter the chamber not lower than the soffit level of the sewer. Necessary steps or a ladder with safety chain or removable handrail should be provided to reach the benching from the side entrance above the soffit.

Drop Manholes: When a sewer connects with another sewer with the difference in level between invert level of branch sewer and water line in the main sewer at maximum discharge greater than 0.6 m, a drop manhole is built either with vertical or nearly vertical drop pipe from higher sewer to the lower one (Fig. 3.10). The drop manhole is also required in the same sewer line in sloping ground, when drop of more than 0.6 m is required to control the gradient and to satisfy the limiting velocity, i.e., non-scouring velocity. The drop pipe may be outside the shaft and encased in concrete or supported on brackets inside the shaft. If the drop pipe is outside the shaft, a continuation of the sewer should be built through the shaft wall to form a rodding and inspection eye, provided with half blank flange (Fig. 3.10). When the drop pipe is inside the shaft, it should be of cast iron and provided with adequate arrangements for rodding and with water cushion of 150 mm depth at the end. The diameter of the drop pipe should be at least equal to incoming pipe.

Scraper (service) type manhole: All sewers having diameter above 450 mm should be provided with a scraper type manhole at intervals of 110–120 m. This manhole should have clear opening of 1.2 m × 0.9 m at the top to facilitate lowering of buckets.

Flushing Manholes: In flat ground for branch sewers, when it is not possible to obtain self-cleansing velocity at all flows, due to very little flow, it is necessary to incorporate flushing device. This can be achieved by making grooves at intervals of 45–50 m in the main drains in which wooden planks are inserted and water is allowed to head up. Upon removal of the planks after heading up of the sewage, the sewage will rush with high velocity facilitating cleaning of the sewers. Alternatively, such flushing can be carried out by using water from overhead water tank through pipes and flushing hydrants or through fire hydrants or tankers and hose. Flushing manholes are provided at the head of the sewers where self-cleansing velocity is difficult to maintain. With short duration storage or addition of external water, sufficient velocity shall be imparted in the sewer to wash

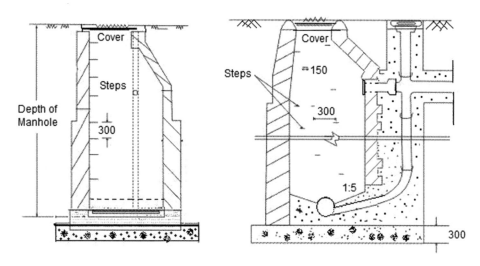

Fig. 3.10 Drop manhole

Fig. 3.11 Inverted siphon

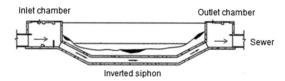

away the deposited solids. In case of heavy chocking in sewers, care should be exercised to ensure that there is no possibility of back flow of sewage into the water supply mains.

3.7.2 Inverted Siphons

A depressed sewer, known as inverted siphon, is a sewer that runs full under gravity flow at a pressure above atmosphere in the sewer. Inverted siphons are used to pass under obstacles, such as buried pipes, subways, etc. (Fig. 3.11). This terminology 'siphon' is a misnomer, as there is no siphon action in this depressed sewer. As the inverted siphon requires considerable attention for maintenance, it should be used only when other means of passing an obstacle in line of the sewer are impracticable.

3.7.3 Stormwater Inlets

Storm water inlets are provided to admit the surface runoff generated during storms to the sewers. These are classified in three major groups, such as curb inlets, gutter inlets,

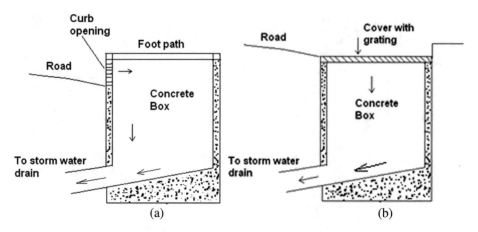

Fig. 3.12 **a** Curb inlet and **b** gutter inlet

and combined inlets. They can be either depressed or flush with respect to the elevation of the pavement surface. The structure of the inlet is constructed with brickwork or stone masonry with cast iron grating at the opening confirming to IS: 5961-1970. At the location where traffic load is not expected, fabricated steel grating can be used. The clear opening shall not be more than 25 mm. The connecting pipe from the street inlet to the sewer should be minimum of 200 mm diameter and laid with sufficient slope. A maximum spacing of 30 m is recommended between the inlets, which depends upon the road surface, size and type of inlet and rainfall pattern in that area.

Curb Inlet: These are vertical opening on the road curbs through which stormwater flow enters the stormwater drains. These are preferred where heavy traffic is anticipated (Fig. 3.12a).

Gutter Inlets: These are horizontal openings in the gutter which is covered by one or more grating through which stormwater is admitted (Fig. 3.12b).

Combined Inlets: In this, the curb and gutter inlet both are provided to act as a single unit. The gutter inlet is normally placed right in front of the curb inlets.

3.7.4   Catch Basins

Catch basins are provided to stop entry of the heavy debris present in stormwater into the sewers. However, their use is discouraged these days because of the nuisance due to mosquito breeding apart from posing substantial maintenance problems. At the bottom of the basin space is provided for the accumulation of impurities. Perforated cover is provided at the top of the basin to admit rain water into the basin. A hood is provided to prevent escape of sewer gas (Fig. 3.13).

Fig. 3.13 Catch basin

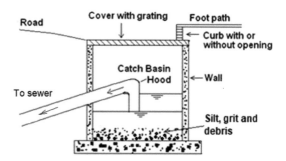

Fig. 3.14 Clean-out

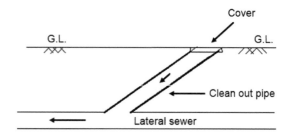

3.7.5   Clean-Outs

Clean-out is a pipe, which is connected to the underground sewer. The other end of the clean-out pipe is brought up to ground level and a cover is placed at ground level (Fig. 3.14). A clean-out is generally provided at the upper end of lateral sewers in place of manholes. During blockage of pipe, the cover is taken out and water is forced through the clean-out pipe to lateral sewers to remove any obstacles in the sewer line. For heavier obstacles, flexible rod may be inserted through the clean-out pipe and moved forward and backward to remove such obstacle.

3.7.6   Flow Regulator or Overflow Device

These are provided for preventing overloading of sewers, pumping stations, treatment plants or disposal arrangement, by diverting the excess flow over the design capacity of sewer to a relief sewer. The overflow device may be side flow or leaping weirs according to the position of the weir, siphon spillways or float actuated gates and valves.

Side flow weir: It is constructed along one or both sides of the combined sewer and it delivers the excess flow during storm period to a relief sewer or natural drainage course (Fig. 3.15). The crest of the weir is set at an elevation corresponding to the desired depth of flow in the sewer. The weir length must be sufficiently long for effective regulation of the flow.

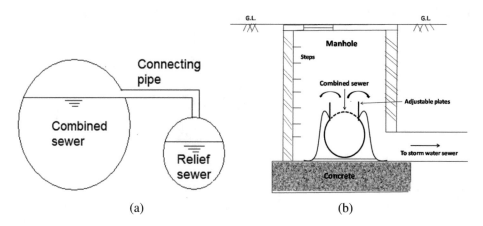

Fig. 3.15 **a** Side flow weir **b** overflow weir adjustable plate arrangement

Leaping weir: The leaping weir is used by providing a gap or opening at the invert of a combined sewer. The leaping weir is formed by a gap at the invert of a sewer through which the dry weather flow will fall and over which a major portion of the entire storm will leap. This leaping weir has an advantage of operating as regulator without involving moving parts. The grit material present in the sewage gets concentrated in the lower flow channel in this case. From practical consideration, it is desirable to have moving crests to make the opening adjustable. When the sewage discharge in combined sewer is small, the flow falls directly into the intercepting sewer through the opening. However, when the discharge exceeds a certain limit, the excess flow leaps or jumps across the weir and it is carried to natural stream or river. This arrangement is shown in the Fig. 3.16b.

Float actuated gates and valves: The excess flow in the sewer can also be regulated by providing an automatic mechanical regulator. These are actuated by the float according to

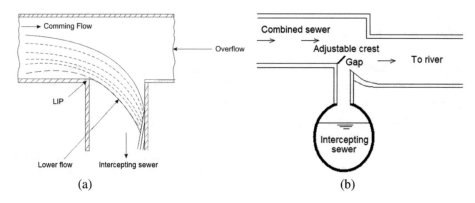

Fig. 3.16 Leaping weir **a** without adjustable crest and **b** adjustable crest type

Fig. 3.17 Siphon spillway

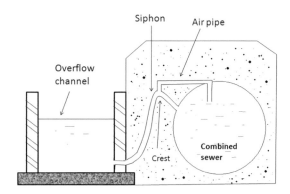

the water level in the sump interconnected with the sewers. Since, moving part is involved in this, regular maintenance of this regulator is necessary.

Siphon spillway: The siphon arrangement of diverting excess sewage from the combined sewer is most effective because it works on the principle of siphon action and it is operated automatically. The overflow channel is connected to the combined sewer through the siphon as shown in Fig. 3.17. An air pipe is provided at the crest level of siphon to activate the siphon, when water will reach in the combined sewer at stipulated level.

3.7.7   Flap Gates and Flood Gates

Flap gates or backwater gates are installed at sewer outlets to prevent back flow of water during high tide, or at high flood level in the receiving stream. These gates can be rectangular or circular in shape and made up of wooden planks or metal alloy sheets. These gates should be designed such that the flap should get open at a very small head difference. Adequate storage in outfall sewer is also necessary to prevent back flow into the system due to the closure of these gates at the time of high tides, if pumping is to be avoided.

3.7.8   Sewer Ventilators

Ventilation to the sewer is necessary to make provision for the escape of gases to take care of the exigencies of full flow and to keep the sewage as fresh as possible. In case of stormwater, this can be done by providing ventilating manhole covers with grating. In modern sewerage system, provision of ventilators is not essential due to elimination of intercepting traps in the house connections, which is allowing ventilation.

Fig. 3.18 Lamp hole

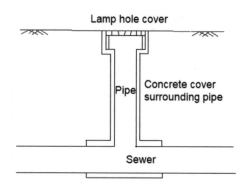

3.7.9 Lamp Hole

It is an opening or hole constructed in a sewer for a purpose of lowering the lamp inside it. It could be made with stoneware or concrete pipe, which is connected to sewer line through a T-junction as shown in the Fig. 3.18. The pipe is provided with concrete cover to make it stable. Manhole cover of sufficient strength should be provided at ground level to take the load of anticipated traffic. As and when required an electric lamp can be inserted in the lamp hole and the light of lamp is observed from adjacent manholes. If the sewer length is unobstructed, the light of lamp will be seen. Lamp hole is constructed when construction of manhole is difficult. In present practice as far as possible the use of lamp hole is avoided. This lamp hole can also be used for flushing the sewers. If the top cover is perforated it will also help in providing ventilation to the sewer, such lamp hole is known as fresh air inlet.

Questions

3.1. What should be properties of the material expected to be used for sewer construction?
3.2. Write a note on different materials used for sewer construction.
3.3. With schematic describe various shapes used for sewer section.
3.4. What are the advantages and drawback of the circular section of sewer?
3.5. Describe different patterns used for sewerage collection system.
3.6. Write a short note on testing of the sewers.
3.7. Discuss occurrence of corrosion in sewers. What measures are adopted for protecting the sewers against corrosion?
3.8. What are the different types of bedding used while laying the sewer line?
3.9. Describe in detail how construction of sewer line is executed.
3.10. Write short notes on laying of sewer pipes. What are the tests conducted on the sewers after laying before backfilling the trench?
3.11 Prepare note on sewer maintenance.
3.12 Define sewer appurtenances. What are the appurtenances used in sewerage network?

3.13. Describe different types of manholes used in sewerage collection system.

3.14. When drop manhole is required to be provided in sewers?

3.15. Describe different types of storm water inlets used in storm water collection system.

3.16. Why flow regulator device is used in sewers? Describe different types of regulators generally used.

3.17. Describe junction manhole, side entrance manhole and through manhole.

References

CPHEEO (1993). *Manual on Sewerage and Sewage Treatment* (2nd ed.). Central Public Health and Environmental Engineering Organization, Ministry of Urban Development, Government of India, New Delhi, India.

CPHEEO. (2013). *Manual on sewerage and sewage treatment, Part A, Engineering* (3rd ed.). Central Public Health and Environmental Engineering Organization, Ministry of Urban Development, Government of India, New Delhi, India.

IS: 11925-1986. *Specifications for pitch impregnated fibre pipes and fittings for drainage purpose.* Indian Standards, Bureau of Indian Standards, Manak Bhavan, New Delhi, India.

IS: 12709-1994. *Glass fibre reinforced plastics (GRP) pipes, joints and fittings for use for Potable water supply—Specification.* Indian Standard first revision 2009, Bureau of Indian Standards, Manak Bhavan, New Delhi, India.

IS: 1742-1983. *Code of practice for building drainage.* Indian Standard reaffirmed in 2002, Bureau of Indian Standards, Manak Bhavan, New Delhi, India.

IS: 4127-1983. *Code of practice for laying of glazed stoneware pipes, CED 24: Public Health Engineering.* Indian Standards, Bureau of Indian Standards, Manak Bhavan, New Delhi, India.

IS: 4984-1995. *High density polyethylene pipe for water supply-specification.* Indian Standard, 4th revision, Bureau of Indian Standards, Manak Bhavan, New Delhi, India.

IS: 5961-1970. *Specification for cast iron gratings for drainage purposes.* Indian Standard reaffirmed in 2003, Bureau of Indian Standards, Manak Bhavan, New Delhi, India.

Hydraulic Design of Sewerage System

4

4.1 Hydraulic Design of Sewers

4.1.1 General Consideration

Generally, sewers are laid at continuous falling gradient towards the outfall point with circular or any other suitable pipe cross section. Storm water drains are constructed separately, as surface drains at suitable gradient with either rectangular or trapezoidal section. Sewers are designed to carry the maximum quantity of sanitary sewage likely to be produced from the contributing area to the particular sewer. Storm water drains are designed to carry the maximum estimated storm runoff produced by rain of design frequency and of duration equal to the time of concentration, which is likely to produce maximum discharge by the contributing catchment area. The sewers with diameter less than 2100 mm are generally precast section and these are laid in straight alignment between the two manholes. Larger sections with cast-in-situ may be laid with some curved alignment in plan.

The most common location for sanitary sewers in small roads is along the centre of the carriage way, so that single sewer can serve houses on both sides of the road with equal length of house connections both sides. On wider roads the sewers are laid on both sides of the road adjacent to curb or below footpath. Though, sewers can also be laid on backside of the property to have single sewer line serving parallel row of houses; however, there may be difficulty in access to the sewers for maintenance in this case. The flow in the sewer is said to be steady, if the rate of discharge at any point remains constant, and it is said to be unsteady flow when discharge varies with time. When velocity and depth remain constant throughout the sewer it is steady open channel flow and said to be uniform flow; whereas, when depth and/or velocity are changing it is considered as non-uniform flow. For simplifying the hydraulic design, the sewer flow is assumed to be

M. Ghangrekar, *Wastewater to Water*, https://doi.org/10.1007/978-981-19-4048-4_4

steady flow condition. However, for rising main of sewage pumping station and other places, where is it pressure flow, turbulent flow condition prevails.

4.1.2   Requirements of Design and Planning of Sewerage System

The sewerage scheme is designed to collect entire sewage effectively and efficiently transport it from the houses to the point of treatment and disposal. To achieve this the sewer section provided should be adequate in size to avoid overflow and related possible health hazards. Hence, for evaluating proper diameter of the sewer, correct estimation of sewage discharge is necessary. The flow velocity inside the sewers should neither be so large as to require heavy excavation and high lift pumping, nor should be so small leading to the deposition of solids in the sewers.

The sewers should be laid at least 2–3 m deep from ground level to be able to collect sewage generated even from basement, if there is any, by gravity flow. Shallow sewers can be laid minimum 0.4 m depth below ground. For sewers with live traffic load above, a minimum depth of 1 m below ground level is considered safe. However, the depth at which sewers are laid should be such that it is able to receive sewage flow by gravity from the adjacent houses. The sewage in sewer should flow under gravity with 0.5–0.8 full at the designed discharge, i.e., at the maximum estimated discharge supposed to flow in sewer. The minimum diameter of the sewer used is 150 mm and minimum diameter for house drainage is 100 mm. Under no circumstances the sewer diameter should be reduced downstream. The sewage collected is conveyed to the point usually located in low-lying area of the city, where the treatment plant is located.

4.1.3   Difference Between Water Supply and Sewerage Networks

The sewerage in any city is a huge network of connected sewers running in 100s of kilometres depending upon the city area. Though, water supply network is also similar huge network, however there are major differences between the water distribution network and sewerage collection system as presented in the Table 4.1.

4.1.4   Provision of Freeboard in Sewers

Sanitary sewers: Sewers with diameter less than 0.4 m are generally designed to run half-full at maximum discharge, and sewers with diameter greater than 0.4 m shall be designed to flow $2/3$ to $3/4$ full at maximum discharge. This free space left in the sewer provides factor of safety to counteract against the following factors:

Table 4.1 Comparison between the water distribution network and sewerage collection system

Water supply pipes	Sewer pipes
• It carries potable water free from any suspended impurity and microorganisms • Velocity higher than self-cleansing is not essential, because suspended solids are not present • It carries water under pressure. Hence, the pipe can be laid up and down the hills and the valleys within certain limits • These pipes are flowing full under pressure	• It carries polluted water containing organic and inorganic solids, which may settle in the pipes. It can cause corrosion of the pipe material. It can also cause abrasion of the interior surface of the sewer due to friction from sandy matter • To avoid deposition of solids in the sewers, self-cleansing velocity is necessary to maintain at all possible variation in discharge • It carries sewage under gravity; hence, it is required to be laid at a continuous falling gradient in the downward direction towards outfall • Sewers are design to run partial full at design discharge. This extra space ensures non-pressure gravity flow in sewer. This will minimize leakage from the sewer, from the faulty joints or crack, if any

(a) It will ensure that sewers will never flow full eliminating development of pressure flow inside the sewer.

(b) Safeguard against any lower estimation of the quantity of wastewater to be collected at the end of design period due to private water supply by public and industries discharging effluent to the sewers.

(c) Large scale infiltration of storm water through underground cracks or open joints in the sewers can be accommodated.

(d) Unforeseen increase in population or water consumption and the consequent increase in sewage production, resulting in increased discharge, that can also be handled.

Storm water drains: Storm water drains are provided with nominal freeboard, above their designed full supply line, since the overflow from storm water drains is not much harmful. A minimum free board of 0.3 m is generally provided in case of storm water drains.

4.2   Hydraulic Formulae for Estimating Flow Velocities

Sewers of any shape are hydraulically designed as open channel flow, except in the case of inverted siphons and discharge lines of pumping stations, where the flow is pressure flow. Generally, Manning's formula is used for design of sewers as an open channel flow, and Darcy-Weisbach frictional head loss formula and Hazen-William's formula are used for flow full closed conduit or pressure flow design. These and other equations used for determining velocity of flow are described below.

4.2.1 Chezy's Formula

The velocity v in m/sec for the flow in the open channel can be given by using Eq. 4.1 as per Chezy's formula (Herschy et al., 1998).

$$v = C\ R^{1/2}\ S^{1/2} \tag{4.1}$$

where, C is Chezy's constant, non-dimensional; R is hydraulic radius, m; and S is the slope of hydraulic grade line, non-dimensional.

4.2.2 Manning's Formula

This is the most commonly used formula for designing sewers. The velocity of flow through sewers can be determined using Manning's formula as per Eq. 4.2 (Chow, 1955).

$$v = \frac{1}{n} R^{2/3} S^{1/2} \tag{4.2}$$

where, v is the velocity of flow in the sewer, m/sec; $R\ (= a/p)$ is the hydraulic mean depth of flow, m; a is the cross-sectional area of flow, m^2; p is wetted perimeter, m; n is the Manning's rugosity coefficient, which depends upon the type of the channel surface, i.e., smoothness of internal surface offered by the pipe material and lies between 0.011 and 0.015. For brick sewer it could be 0.017 and 0.03 for stone facing sewers; and S is the hydraulic gradient, equal to invert slope for uniform flows.

The value of n will vary with the pipe material and condition of the pipe, whether it is in good condition or fair condition. For example, for good condition and fair condition cement concrete pipe with collar joint, the n will have value of 0.013 and 0.015, respectively; similarly, these values for good condition and fair condition salt glazed stone ware pipe are 0.012 and 0.015, respectively. For RCC pipes with socket and spigot joint the design value for n is 0.011. For cast iron unlined pipe, the value for n is 0.013; whereas, with cement mortar lining it will be 0.011. The asbestos and plastic pipes are also having value of n as 0.011, since these pipes offer smooth interior surface (CPHEEO, 2019; MSMA, 2012).

4.2.3 Hazen-Williams Formula

For pipe flowing full and for the pumping mains the turbulent flow conditions are considered to ensure that the suspended matter does not settle in the pipes. Hazen-William's equation can be used for hydraulic design of such sewers as per Eq. 4.3 (Liou, 1998).

$$V = 0.849\, C\, R^{0.63}\, S^{0.54} \qquad (4.3)$$

where, V is the velocity of flow, m/sec; C is Hazen-William's coefficient; R and S are hydraulic mean depth and hydraulic gradient, respectively, as defined earlier.

For circular sewer section flowing full, the equation will be represented by Eq. 4.4.

$$V = 0.3545\, C\, D^{0.63}\, S^{0.54} \qquad (4.4)$$

Equation for estimating discharge will be given by Eq. 4.5 for flow full condition of the pipe.

$$Q = 0.2784\, C\, D^{2.63}\, S^{0.54} \qquad (4.5)$$

where, Q is the discharge in m³/s.

The Hazen-Williams coefficient 'C' varies with life of the pipe and obviously it has high value when the pipe is new and as the pipe become old the value become lower due to erosion of interior surface of sewer pipe. For example, for RCC new pipe the value for C is 150 and the value recommended for design is 120, as the pipe interior may become rough with time. The design values of 'C' for asbestos cement pipe, plastic pipes, cast iron pipe, and steel pipe lined with cement are 120, 120, 100, and 120, respectively (CPHEEO, 2019). Hazen-William's equation and Darcy-Weisbach equations can give comparable results at moderately high Reynold's number, if proper value for C is selected. The common errors associated with Hazen-William's equation and related nomographs are incorrect selection of C, using nominal diameter of pipe rather than actual diameter and use of this equation under the circumstances where it is not best applicable, such as low Reynold's number and fluid other than water.

The value of the friction coefficient, such as Manning's 'n' and Hazen-William's 'C', depends on the pipe material, Reynold's number, size and shape of the pipe and depth of flow. These coefficients are usually considered independent of depth of flow, velocity of flow and viscosity; whereas to represent the friction conditions these coefficients must be dependent on relative roughness of pipe material and Reynolds Number.

Example: 4.1 If the cement concrete sewer having diameter of 200 mm is flowing full under gravity and laid with gradient of 4 in 1000, determine the velocity generated in sewer using different hydraulic formulae. Manning's n = 0.011; Chezy's C = 60; Hazen-William C = 120.

Solution

Hydraulic mean depth = a/P = $\frac{\pi D^2}{\frac{4}{\pi D}}$ = D/4 = 200/4 = 50 mm = 0.05 m; S = 4/1000 = 1/250

Chezy's formula

Using C = 60, velocity $v = C\ R^{1/2}\ S^{1/2}$ = $60 \times 0.05^{1/2} \times (\frac{1}{250})^{1/2}$ = 0.85 m/sec

Manning's formula

Using Manning's n = 0.011, velocity v = $\frac{1}{n}R^{2/3}S^{1/2}$ = $\frac{1}{0.011}0.05^{2/3}\left(\frac{1}{250}\right)^{1/2}$ = 0.78 m/sec

Hazen-William formula

Using C = 120, velocity $V = 0.849\ C\ R^{0.63}\ S^{0.54}$ = $0.849 \times 120 \times (0.05)^{0.63}\ (1/250)^{0.54}$ = 0.783 m/sec

4.2.4 Velocity of Flow in Sewers

4.2.4.1 Minimum Velocity: Self-cleansing Velocity

With hourly variation in the flow expected in the sewer the depth of flow will vary and with increase in depth of flow the velocity will increase in the sewer. This self-cleansing velocity is exerted by the scouring action of the flowing sewage. While designing the sever the objective is to ensure that the velocity developed in sewer will not allow settling of suspended solids in the sewer. The velocity that would not permit the solids to settle down and even scour the deposited particles of a given size is referred as a self-cleansing velocity. It is important to ensure that this minimum velocity, self-cleansing velocity, should at least develop once in a day so as not to allow any deposition in the sewers. Otherwise, if such deposition takes place, it will obstruct free flow causing further deposition and finally leading to the blocking of the sewer section. For adequate transport of the suspended solids present in the sewage, without allowing settling in the sewer, this minimum velocity or self-cleansing velocity can be estimated using Eq. 4.6 as proposed by Shield (Dias & Matos, 2001).

$$Vs = \frac{1}{n}R^{1/6}\sqrt{K(Ss-1)d'} \tag{4.6}$$

Also, the slope at which the sewers should be laid so as to have self-cleansing will be given by the Eq. 4.7.

$$S = \frac{K}{R}(Ss - 1)d' \tag{4.7}$$

where, Vs is the self-cleansing velocity, m/s; n is the Manning's rugosity coefficient, R is hydraulic mean depth, m; K is the factor whose value depends on the shape and cohesiveness of the solids in suspension, non-dimensional; Ss is the specific gravity of the suspended solids, non-dimensional; and d' is the particle size, m.

Based on the Shield's expression, Camp derived the expression for circular pipe flowing half full or completely full, where $R = D/4$; so that this velocity can be estimated without knowing the diameter of the pipe to simplify the design by fixing velocity first. Now, the Manning's equation can be rearranged as $R^{2/3}/n = Vs/S^{1/2}$. From Darcy-Weisbach equation (Liou, 1998), Eq. 4.8, the value of S can be substituted in this Manning's equation to obtain Eq. 4.9.

$$S = H/L = fV^2/2gD \tag{4.8}$$

$$\text{Hence, } R^{2/3}/n = \sqrt{\frac{2gD}{f}} \tag{4.9}$$

Substituting, this in Eq. 4.6 and substituting $R = D/4$, the expression will have form as given in Eq. 4.10.

$$V_s = \sqrt{\frac{8K}{f'}(Ss - 1)g \cdot d'} \tag{4.10}$$

where, K is the constant, having values for clean inorganic solids $= 0.04$ and for organic sticky solids $= 0.06$; f' = Darcy-Weisbach friction factor (for sewers $= 0.03$); g is gravity acceleration; d' is the size of suspended particles, m; and Ss is the specific gravity of the solids in suspension. The Ss will have value of 2.65 for inorganic solids such as sandy particles and for organic solids it may vary in the wide range of 1.03–1.6 and generally representative value of around 1.2 is considered.

Example: 4.2 Using Shield's equation determine the self-cleansing velocity and the gradient required for the sewer with diameter of 250 mm for adequately transporting sand particles of 1 mm size and specific gravity of 2.65. In addition, estimate the self-cleansing velocity and gradient required for carrying organic solids of 5 mm size and specific gravity of 1.2. Consider, $K = 0.04$ for sand and 0.06 for organic solids and $n = 0.013$.

Solution

The velocity Vs for the sandy particles will be estimated using Eq. 4.6 and substituting the values as stated below:

$$Vs = \frac{1}{0.013}\left(\frac{0.25}{4}\right)^{1/6}\sqrt{0.04(2.65-1)\times 0.001} = 0.393 \text{ m/sec}$$

Similarly, the velocity Vs required for organic solids will be:

$$Vs = \frac{1}{0.013}\left(\frac{0.25}{4}\right)^{1/6}\sqrt{0.06(1.2-1)\times 0.005} = 0.375 \text{ m/sec}$$

The gradient S shall be estimated using Eq. 4.7 and substituting appropriate values.

$$S = \frac{0.04}{0.25/4}(2.65-1)\times 0.001 = 1.056 \times 10^{-3} = 1 \text{ m in } 947 \text{ m}$$

The gradient required for transport of organic solids without allowing settling will be:

$$S = \frac{0.06}{0.25/4}(1.2-1)\times 0.005 = 9.6 \times 10^{-4} = 1 \text{ m in } 1042 \text{ m}$$

Answer: Since lower invert slope and velocity is required for the transport of lighter organic solids the self-cleansing velocity required for sandy particle, i.e., 0.393 m/sec and gradient of 1 m in 947 m need to be provided, which will also ensure no settling for organic solids.

For adequate transport of the suspended solids present in sewage, i.e., sand up to 1 mm diameter with specific gravity 2.65 and organic particles up to 5 mm diameter with specific gravity of 1.2, it is necessary that a minimum velocity of about 0.40 m/sec should be maintained in the sewers even at minimum flow. For grit bearing wastewater the velocity of 0.6 m/sec is considered to be safe at present peak flow, which may reach to 0.8 m/sec at design peak flow. Designing the sewer at the velocity of about 0.8–1.2 m/sec for the design peak flow is normally being followed.

The gradient required also varies with the discharge that sewer is handling. The gradients required for small sewers are recommended to be 6 in 1000 for carrying peak discharge of 2 L/sec, 2 in 1000 for peak discharge of 10 L/sec and 1 in 1000 when the peak discharge is 30 L/sec. This will ensure a velocity of 0.6 m/sec at peak flow in early years. Smaller diameter sewers are required to be laid at steeper gradient, e.g., for 0.15 m diameter sewer the gradient required may be around 1 in 150; whereas, for 0.30 m diameter the gradient required may be around 1 in 300 or so. During early years since the present flow is lesser than the design flow, it is necessary that the velocity of 0.6 m/sec is developed at least once in a day, when maximum discharge is reached, to ensure no deposition of solids in sewers occurs during the early years of operation.

While finalizing the sizes and gradients of the sewers, they must be checked for the minimum velocity that would be generated at minimum discharge, i.e., about 1/3 of the average discharge. For designing the sewers, the flow velocity at full depth is generally kept at about 0.8 m/sec or so. Since, sewers are generally designed to flow at $^1/_2$ to $^3/_4$ full at maximum discharge, the velocity at 'designed discharge' (i.e., $^1/_2$ to $^3/_4$ full) will

Sewer material	Limiting velocity, m/sec
Vitrified tiles	4.5–5.0
Cast iron sewer	3.5–4.5
Stone ware sewer	3.0–4.5
Cement concrete	2.5–3.0
Brick lined sewer	1.5–2.5

Table 4.2 Limiting velocity or non-scouring velocity for different sewer material

even be more than 0.8 m/sec. Thus, the minimum velocity generated in sewers will help in the following ways:

- Adequate transportation of suspended solids,
- Keeping the sewer size under control; and
- Preventing the organic matter present in sewage from decomposition by moving it faster, thereby preventing evolution of foul gases.

4.2.4.2 Maximum Velocity or Non-scouring Velocity

Due to the continuous abrasion caused by suspended solids present in sewage, the interior surface of the sewer pipe gets scored over time of service. The scoring will be pronounced at higher velocities of flow than what can be tolerated by the pipe materials. This wear and tear of the sewer pipes will reduce the life span of the pipe and their hydraulic carrying capacity, due to increase in roughness of interior surface. In order to avoid this, it is necessary to limit the maximum velocity that will be produced in sewer pipe at any time. This limiting or non-scouring velocity mainly depends upon the material used for construction of sewer. Harder the sewer material, offering resistance to abrasion, more will be the value for limiting velocity. The limiting velocity for different sewer material is provided in Table 4.2. The problem of maximum or non-scouring velocity is severe in case of laying sewers in hilly areas, where ground slope is very steep. To overcome this and to keep maximum velocity in the sewer under control drop manholes are constructed at suitable intervals along the length of the sewer, so as not to lay sewers at steeper gradients.

4.2.4.3 Effect of Variation of Flow on Velocities in a Sewer

Due to diurnal variation in sewage flow, the discharge flowing through sewer will vary considerably for time of hour in a day. Hence, there occur variation in a depth of flow and thus, variation in hydraulic mean depth (R). Due to change in this R, there will be changes in the flow velocity, since it is proportional to $(R)^{2/3}$. Hence, it is necessary to check the sewer flow for minimum velocity of about 0.40 m/sec at the time of minimum flow (i.e., about $1/3$ of average flow) and the velocity of about 0.8–1.2 m/sec should be developed at a time of average flow of design discharge. The velocity should also be checked for maximum velocity, i.e., non-scouring velocity at the maximum discharge.

To make the economical design of sewerage scheme, for flat terrain the sewers are designed to achieve self-cleansing velocity at maximum discharge, which will occur once in a day. This will permit flatter gradient for sewers and sewers will be cleaned once this discharge is reached in a day. At upper stretches of the laterals, as they flow only partially full even at design discharge, due to limitation on the minimum diameter of the sewer to be used and additional limitations of gradient with which sewers have to be laid to avoid deeper excavation, adequate self-cleansing velocity may not be achieved. In this case a flushing arrangement need to be provided for flushing the sewers to wash away the deposition or remove the deposition using suction pumps.

For mild slopping ground, the condition of developing self-cleansing velocity can be achieved at average flow to make the design economical. Whereas, in case of hilly areas, sewers shall be designed for achieving self-cleansing velocity at minimum discharge; however, the design must be checked for non-scouring velocity at maximum discharge. Since, in this case following natural ground slopes if the sewers are laid, it will cross the limiting velocity. Hence, drop manholes are constructed in case of hilly terrain at suitable interval to keep the gradient of the sewer under control so as not to cross the limiting velocity.

Example: 4.3 Design a sewer to carry a maximum discharge of 400 L/sec running half-full while carrying maximum discharge. A good condition sewer pipe is used, made with salt glazed stoneware having Manning's rugosity coefficient of 0.012 and it is laid with the gradient of 1 in 900.

Solution
Discharge, $Q = A.V$; $R = A/P$ and for half-full sewer, $R = D/4$
    Solving,

$$0.40 = \left(\pi D^2/8\right)(1/n)R^{2/3}S^{1/2} = \left(\pi D^2/8\right)(1/0.012)(D/4)^{2/3}(1/900)^{1/2}$$

Solving for D, we get D = 0.973 m and velocity of flow is 0.54 m/sec.

Comment: This problem was possible to solve since the sewer was flowing half-full. If the pipe is flowing partially full, other than half full, it is not this easy to solve the equations and it becomes time consuming. Hence, to have quick design of sewers the hydraulic elements are converted to the ratio of actual depth of flow to the full diameter of the sewer, which makes it easy to design the sewer as explained later.

4.3 Hydraulic Characteristics of Circular Sewer Flowing Partially Full

Circular sewer section is the most commonly used section due to advantages that it offers, such as these pipes can be easily manufactured, this shape offers greatest hydraulic mean depth when flowing half-full or flowing full. For a given area this section offers less perimeter, hence section is economical due to least material quantities required. The sewer is design to flow partially full and this section offers advantage of developing velocity of flow more than that developed under flow full condition when sewer is flowing between half-full to flow full. Below half-full, due to reduction in depth of flow, the velocity as well as discharge reduces considerably.

To understand the variation in the hydraulic elements with variation in the depth of flow in the sewer let us consider the circular sewer section flowing partially full as shown in Fig. 4.1. Where, the diameter of the sewer is 'D' and depth of partial-flow is 'd'. Hence, while the sewer is flowing full cross-sectional area of flow 'A' will be $\pi D^2/4$, wetted perimeter 'P' will be 'πD' and hydraulic mean depth 'R' will be '$D/4$'. Now, while the sewer is flowing partially full let 'α' be the subtended angle of the water surface made with the center of circle. Using basics of trigonometry, the hydraulic properties at the partial depth of flow can be established.

The depth at partial flow shall be obtained from Eq. 4.11

$$d = \left[\frac{D}{2} - \frac{D}{2} \cos\left(\frac{\alpha}{2}\right) \right] \tag{4.11}$$

Hence, proportionate depth d/D can be obtained using Eq. 4.12.

$$\frac{d}{D} = \frac{1}{2}\left[1 - \cos\left(\frac{\alpha}{2}\right) \right] \tag{4.12}$$

Similarly, the proportional area a/A can be expressed as per Eq. 4.13.

Fig. 4.1 Section of a circular sewer running partially full

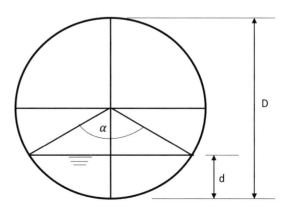

$$\frac{a}{A} = \left[\frac{\alpha}{360} - \frac{\sin \alpha}{2\pi} \right] \tag{4.13}$$

The proportionate perimeter p/P will be represented by Eq. 4.14.

$$\frac{p}{P} = \frac{\alpha}{360} \tag{4.14}$$

Proportionate hydraulic mean depth r/R shall be obtained using Eq. 4.15.

$$\frac{r}{R} = \left[1 - \frac{360 \sin \alpha}{2\pi \alpha} \right] \tag{4.15}$$

Proportionate velocity v/V will be represented by Eq. 4.16, when value of manning's rugosity coefficient 'n' is considered to be varying with depth and if it is considered to be constant then proportionate v/V will be given by Eq. 4.17.

$$\frac{v}{V} = \frac{N}{n} \frac{r^{2/3}}{R^{2/3}} \tag{4.16}$$

$$\frac{v}{V} = \frac{r^{2/3}}{R^{2/3}} \tag{4.17}$$

Similarly, the proportional discharge q/Q will be given by Eq. 4.18 if n is considered variable with depth and if it is considered constant then by Eq. 4.19.

$$\frac{q}{Q} = \frac{a}{A} \frac{N}{n} \frac{r^{2/3}}{R^{2/3}} \tag{4.18}$$

$$\frac{q}{Q} = \frac{a}{A} \frac{r^{2/3}}{R^{2/3}} \tag{4.19}$$

In all above equations (Eq. 4.11 through Eq. 4.15) except 'α' all other things are constant (Fig. 4.1). Hence, for different values of 'α', all the proportionate elements can be easily estimated. Values of these hydraulic elements can be obtained from the proportionate graph prepared for different values of d/D (Fig. 4.2). The value of Manning's n can be considered constant for all depths. In reality, it varies with the depth of flow as tractive resistance is affected by depth of flow and wetted perimeter, hence it can be considered as variable with depth and accordingly the hydraulic elements values can be read from the graphs for different depth ratio of flow. The value of N can vary from $n = 1.22$ N when the depth of flow is 0.1D, 1.28 N for 0.3D and 1.23 N for 0.5D, to n equal to N when sewer flowing full (WPCF, 1960).

From the hydraulic element proportional plot (Fig. 4.2) it is evident that the velocities in partially flowing circular sewer sections can exceed than those in full section and it is maximum at d/D of 0.8. Similarly, the discharge obtained is not maximum at flow full condition, however it is maximum when the depth is about 0.95 times the full depth.

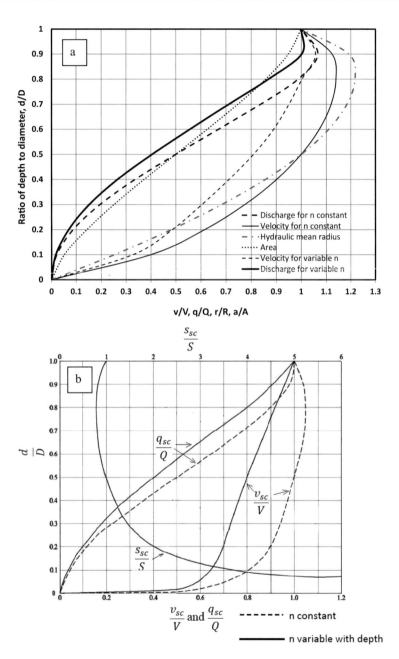

Fig. 4.2 a Proportionate graph for circular sewer section, and **b** for same self-cleansing velocity at all depth (modified from CPHEEO Manual, 1993)

Table 4.3 Hydraulic elements of circular sewer section for different d/D

d/D	a/A	p/P	r/R	n is constant for all depth, n/N = 1		N is variable with depth		
				v/V	q/Q	n/N	v/V	q/Q
0.1	0.052	0.205	0.254	0.401	0.021	1.220	0.329	0.017
0.2	0.143	0.295	0.482	0.615	0.088	1.266	0.486	0.070
0.3	0.252	0.369	0.684	0.776	0.196	1.282	0.605	0.153
0.4	0.373	0.436	0.857	0.902	0.337	1.266	0.713	0.266
0.5	0.500	0.500	1.000	1.000	0.500	1.235	0.810	0.405
0.6	0.626	0.564	1.110	1.072	0.671	1.205	0.890	0.557
0.7	0.748	0.631	1.185	1.120	0.838	1.177	0.952	0.712
0.8	0.858	0.705	1.217	1.140	0.988	1.136	1.003	0.869
0.9	0.949	0.795	1.192	1.124	1.066	1.064	1.057	1.002

Since it is difficult to read accurately the values from the graphs given in Fig. 4.2, these values are presented in Table 4.3 for reading accurate values to be used in practice for different d/D. The intermediate values can be estimated doing interpolations.

The sewers flowing with depths between 50 and 80% full need not have to be placed on steeper gradients to be as self-cleansing as sewers flowing full. For example, when variation of n with depth of flow is neglected, the flow velocity is maximum when depth of flow is $0.8D$ and discharge is maximum when depth of flow is $0.95D$. The reason is that velocity and discharge are function of tractive force intensity, which depends upon friction coefficient as well as flow velocity generated by gradient of the sewer.

For designing laterals and sub-mains, sometimes they are designed to flow half-full at design discharge. The rationale behind this assumption is that it provides adequate factor of safety against the peak flows, however designing sewer this way is not justified for design of mains and outfall sewers. The nomograms developed for Manning's equation can also become handy to decide the diameter of the pipe (CPHEEO, 2013).

For a typical design of sewers, where the flow variation is estimated for which sewer has to be designed to find depth of flow and velocity developed in sewer, using basic fundamentals of trigonometry and designing the sewers will be time consuming, hence the proportional values of the hydraulic element, as presented in Table 4.3 or Fig. 4.2, will be handy to readily do the design of the sewer section. This will be more convincing while going through the solved examples.

Example: 4.4 A branch sewer 100 m long having diameter of 800 mm is laid with a slope of 0.002. This sewer is discharging the sewage in main sewer with invert level of this sewer at the junction manhole at 92.300 m. If the discharge in the branch sewer is 760 m^3/h, what will be depth of flow and velocity in the sewer if it is freely discharging in main sewer without having any surcharge. What will be water level in upstream manhole? Consider $n = 0.013$ and constant with depth. If the water level in the receiving main sewer is at the crown level of this branch sewer, then what will be water level in the upstream manhole?

Solution

Sewer is laid with slope of 0.002, i.e., 1 in 500, hence invert level at the entry of branch sewer in upstream manhole should be at least $92.300 + (100/500) = 92.500$ m

Now, using Manning's formula (Eq. 4.2), the velocity when the sewer is flowing full is estimated

$$V = \frac{1}{n} R^{2/3} S^{1/2} = \frac{1}{0.013} 0.2^{2/3} \times 0.002^{1/2} = 1.177 \text{ m/sec}$$

$$\text{Discharge } Q = A.V = \left(\pi D^2 / 4\right) \times 1.177 = 0.591 \text{ m}^3/\text{sec}$$

Actual flow in sewer is 760 m^3/h $= 0.211$ m^3/sec; therefore $q/Q = 0.211/0.591 = 0.357$

For q/Q of 0.357, the $d/D = 0.412$, and for this d/D the $v/V = 0.914$ (these values can be obtained from Table 4.3 or reading value from nomogram, Fig. 4.2).

Therefore, depth of flow while carrying 0.211 m^3/sec is $0.412 \times 0.8 = 0.33$

Similarly, velocity in the sewer $= 0.914 \times 1.177 = 1.076$ m/sec

For free delivery of the sewage in the main sewer without any surcharge, the invert level difference of this receiving sewer and main sewer shall be more than 0.33 m

Considering the invert level difference of 0.350 m, thus invert level at entry of the branch sewer in upstream manhole will be $92.500 + 0.350 = 92.850$ m

Hence, the water level in upstream manhole $= 92.500 + 0.35 + 0.412 \times 0.8 = 93.180$ m (**Answer**)

When the water level in the receiving main sewer is at the crown level of this branch sewer then this sewer will flow full.

For discharge of 760 m^3/h $(= 0.211$ m^3/sec), the velocity generated in the sewer can be estimated using $Q = V.A$

Thus, velocity $= \frac{0.211}{\frac{\pi}{4} 0.8^2} = 0.42$ m/sec

In this case head loss in this sewer will be:

Frictional head loss, $H_f = flv^2/2gD = (0.03 \times 100 \times 0.42^2)/(2 \times 9.81 \times 0.8) = 0.034$ m

Additional entry and exit loss will be $KV^2/2 g$ and considering value of 0.5 and 1 for K for entry and exit, respectively, the head loss will be:

Entry loss $= 0.5 \times 0.42^2/(2 \times 9.81) = 0.0044$ m and exit loss will be 0.009 m, hence total loss $= 0.0135$ m

Hence, water surface in the upstream manhole will be

$= 92.500 + 0.800 + 0.034 + 0.013 = 93.347$ m (**Answer**)

4.3.1 Designing Sewer to Be Same Self-Cleansing at Partial Depth as Full Depth

The sewers can be design to exert same self-cleansing effect at partial flow as it will have when flowing full. To achieve this the required ratios of s_{sc}/S, v_{sc}/V, q_{sc}/Q can be estimated considering the tractive force exerted at flow full condition (T), which should be same for the partial flow (τ). Subscript 'sc' is used to denote self-cleansing at partial depth of flow equivalent to that obtained in full flow condition. Consider a layer of sediment of unit length, unit width and thickness is deposited at the invert of the sewer (Fig. 4.3). The sewer is laid with slope of θ degree with horizontal. Now, the drag force or the intensity of tractive force (τ) exerted by the flowing water on a channel is given by Eq. 4.20.

$$\tau = \gamma_w \cdot R \cdot S \qquad (4.20)$$

where, γ_w is the unit weight of water; R is hydraulic mean depth; and S is the slope of the invert of the sewer per unit length.

With the assumption that the quantity of tractive force intensity exerted on the sediment at full flow and partial flow implies equality of cleansing, i.e., for sewers to be same self-cleansing at partial depth as what it will be when flowing full, hence equating these tractive force ($\tau = T$) as per Eq. 4.21 and solving we can get Eq. 4.22, representing relation between ratio of slope and hydraulic mean gradient so as to have same self-cleansing effect at partial depth of flow.

$$\gamma_w \cdot r \cdot s_{sc} = \gamma_w \cdot R \cdot S \qquad (4.21)$$

Hence, $s_{sc} = (R/r)S$

$$\text{Or} \quad \frac{s_{sc}}{S} = \frac{R}{r} \qquad (4.22)$$

Therefore, from Manning's equation one can get ratio of velocity for partial depth to full depth as per Eq. 4.23.

$$\frac{v_{sc}}{V} = \frac{N}{n}\left(\frac{r}{R}\right)^{2/3}\left(\frac{s_{sc}}{S}\right)^{1/2} \qquad (4.23)$$

Or, by substituting $r/R = S/s_{sc}$, the ratio of velocity can be obtained as Eq. 4.24.

Fig. 4.3 A sediment particle moving on the sewer invert laid at θ degree with horizonal

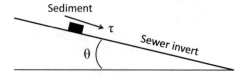

$$\frac{v_s}{V} = \frac{N}{n}\left(\frac{r}{R}\right)^{1/6}$$ (4.24)

Thus, the ratio for discharge can be represented by Eq. 4.25.

$$\frac{q_{sc}}{Q} = \frac{N}{n}\frac{a}{A}\left(\frac{r}{R}\right)^{1/6}$$ (4.25)

The variation of v_{sc}/V and q_{sc}/Q with d/D is presented in the Fig. 4.2b. The variation of s_{sc}/S with d/D is also presented in the Fig. 4.2b. From the shape of s_{sc}/S curve it is evident that for sewer flowing more than half-full the gradient required to maintain same self-cleansing effect as flowing full may be marginally less, whereas at the depth of 0.2D the gradient required is double than half-full flow to have same self-cleansing effect as flow full condition. For flow occupying lower depth than 0.2D, more steeper gradient is required.

4.3.2   Solved Examples on Design of Sewers

Example: 4.5 A sewer having diameter of 400 mm is to be designed to flow at 0.3 depth on a suitable grade ensuring a degree of self-cleansing equivalent to that obtained at full depth at a velocity of 0.8 m/sec. Find the required gradient, associated velocity and discharge at this 0.3 depth. Consider Manning's rugosity coefficient, $n = 0.013$ and variation of it with depth can be neglected.

Solution
Manning's formula for partial depth

$$v = \frac{1}{n}r^{2/3}s^{1/2}$$

For full depth

$$V = \frac{1}{0.013}0.1^{2/3}S^{1/2} = 0.80$$

Therefore, using $V = 0.80$ m/sec, $N = n = 0.013$ and $R = D/4 = 100$ mm and solving for S

$$S = 0.0023 \text{ i.e., } 1 \text{ in } 429.14$$

This is the gradient required for full depth for velocity of 0.8 m/sec.

$$Q = A.V = (\pi/4)(0.4)^2 \times 0.80 = 0.101 \text{ m}^3/\text{sec}$$

At depth $d = 0.3D$, (i.e., for $d/D = 0.3$) we have $a/A = 0.252$ and $r/R = 0.684$ (for neglecting variation of n).

Now for the sewer to be same self-cleansing at 0.3 depth as it will be at full depth, the gradient (s_{sc}) required can be worked out using relation $s_{sc} = (R/r)S$.

Therefore, $s_{sc} = S/0.684$

$= 0.0023/0.684 = 0.00363$ (i.e., 1 in 297.4)

Hence, the sewer is required to be laid with gradient of 1 in 297.4, so as to have same self-cleansing effect exerted as it will have while flowing full at velocity of 0.8 m/sec.

Now, the velocity v_{sc} generated at this gradient is given by Eq. 4.24.

$$v_{sc} = V \frac{N}{n} \left(\frac{r}{R}\right)^{1/6}$$
$$= 0.8 \times (0.684)^{1/6}$$
$$= 0.751 \, \text{m/sec} \, (\textbf{Answer})$$

The discharge q_{sc} relation is given by Eq. 4.25.

$$q_{sc} = Q \frac{N}{n} \frac{a}{A} \left(\frac{r}{R}\right)^{1/6}$$
$$q_{sc} = 0.101 \times 1 \times (0.252) \times (0.684)^{1/6}$$
$$= 0.024 \, \text{m}^3/\text{sec} \, (\textbf{Answer})$$

Example: 4.6 A combined sewer is to be designed to serve an area of 40 km^2 with an average population density of 140 persons/hectare. The average rate of water supply is 170 L/capita.day with sewage generation of 80% of water supplied. The peaking factor for sewage flow is twice that of the average sewage flow. The rainfall intensity of design rain of 15 mm per day shall be considered for discharge estimation. The average runoff coefficient for this catchment can be considered as 0.44. What will be the design discharge for the sewer? Find the diameter of the sewer if running full at maximum discharge with velocity of 0.9 m/sec. If this sever has to flow at 0.8 depth at maximum discharge determine the diameter required and velocity of flow if laid with gradient of 1 in 1800. For $d/D = 0.8$, $q/Q = 0.988$, $v/V = 1.14$. Consider constant $n = 0.012$.

Solution

Total population of the area = population density × area = $140 \times 40 \times 10^2 = 560,000$ persons

Average sewage flow = $170 \times 0.80 \times 560,000$ L/day = 76.16×10^6 L/day = 0.88 m^3/sec

Hence, maximum sewage flow = 2 × average sewage flow = $2 \times 0.88 = 1.763$ m^3/sec

Now, storm water runoff = $0.44 \times (15/1000) \times 40 \times 10^6 \times [1/(24 \times 60 \times 60)] = 3.056$ m^3/sec

Total flow of the combined sewer = sewage flow + storm flow = 1.763 + 3.056 = 4.819 m^3/sec

Hence, the flow carrying capacity of the sewer required is 4.819 m^3/sec

Therefore, diameter of the sewer required while flowing full at the velocity of 0.9 m/sec can be calculated as:

$$\pi/4(D)^2 \times 0.90 = 4.819 \, \text{m}^3/\text{sec}$$

Hence, D = 2.612 m (**Answer**)

Estimating diameter of the sewer if flowing 0.8 depth

If the sewer has to flow 0.8 depth at this discharge (q) of 4.819 m^3/sec, from the proportionate values for d/D of 0.8, the q/Q = 0.988, hence discharge (Q) when the sewer is flowing full will be:

$$Q = 4.819/0.988 = 4.877 \, \text{m}^3/\text{sec}$$

Using Manning's formula

$$4.877 = \left(\pi D^2/4\right) \times \frac{1}{n}\left(\frac{D}{4}\right)^{2/3} s^{1/2}$$

In above equation all values are known except D, hence solving with S of 1 in 1800, D = 2.178 m (**Answer**)

Similarly, velocity V while flowing full will be 4.877/($\pi D^2/4$) = 1.31 m/sec

Hence velocity at partial flow of 0.8 depth = 1.31 × 1.14 = 1.49 m/sec (**Answer**)

Example: 4.7 Find the minimum velocity and gradient required to transport coarse sand through a sewer of 500 mm diameter flowing full. The properties of the solids to be kept in suspension are: sand particles of 1.0 mm diameter and specific gravity of 2.65 and organic matter of 5 mm average size with specific gravity of 1.2. The friction factor for the sewer material may be assumed 0.03 and Manning's rugosity coefficient of 0.012. Consider K = 0.04 for inorganic solids and 0.06 for organic solids.

Solution

Minimum velocity i.e., self-cleansing velocity required for inorganic sand particles:

$$v_s = \sqrt{\frac{8K}{f'}(S_s - 1)gd'}$$

$$v_s = \sqrt{\frac{8 \times 0.04}{0.03}(2.65 - 1) \times 9.81 \times 0.001}$$

$$= 0.4155 \, \text{m/sec say } 0.42 \, \text{m/sec}$$

Similarly, for organic solids this velocity will be 0.396 m/sec

Therefore, the minimum velocity in sewer to keep both the types of solids in suspension is 0.42 m/sec

Now, diameter of the sewer $D = 0.5$ m

Hydraulic mean depth $= D/4 = 0.5/4 = 0.125$ m

Using Manning's formula:

$V = (1/n) R^{2/3} S^{1/2}$

$0.42 = (1/0.012) \times (0.125)^{2/3} \times s^{1/2}$

Hence, $S = 1/2460$

Therefore, gradient of the sewer required is 1 in 2460.

Example: 4.8 Design a sewer running 0.7 times full at designed discharge for a town provided with the separate sewerage system. The town is having a total population of 72,000 persons. The water supply per capita to the town is 160 L/day. The manning's $n = 0.013$ for the pipe material and permissible slope is 1 in 800. Variation of n with depth may be neglected. Check for minimum and maximum velocity assuming minimum flow equal to $1/3$ of average flow and maximum flow as 3 times the average. For $d/D = 0.7$, $q/Q = 0.838$, $v/V = 1.12$; also, for $q/Q = 0.094$, $d/D = 0.206$ and $v/V = 0.625$.

Solution

Average water supplied $= 72,000 \times 160 \times [1/(24 \times 60 \times 60 \times 1000)] = 0.133$ m^3/sec

Considering average sewage production per day as 80% of water supply

Hence, average sewage discharge $= 0.133 \times 0.8 = 0.107$ m^3/sec

Maximum sewage discharge $= 3 \times 0.107 = 0.32$ m^3/sec

Now for $d/D = 0.7$, $q/Q = 0.838$, $v/V = 1.12$

Therefore, $Q = 0.32/0.838 = 0.382$ m^3/sec

$$Q = \frac{1}{n} \frac{\pi D^2}{4} \left(\frac{D}{4}\right)^{2/3} S^{1/2}$$

Now,

$$0.382 = \frac{1}{0.013} \frac{\pi D^2}{4} \left(\frac{D}{4}\right)^{2/3} \left(\frac{1}{800}\right)^{1/2}$$

Therefore, D $= 0.742$ m

$V = Q/A = 0.382/(\pi \times 0.742^2/4) = 0.884$ m/sec

Now, for $d/D = 0.7$ the $v/V = 1.12$

Therefore $v = 1.12 \times 0.884 = 0.99$ m/sec

This velocity is less than limiting velocity and it is self-cleansing velocity, hence it is fine.

Check for minimum velocity.

Now $q_{min} = 0.107/3 = 0.036$ m^3/sec.

$q_{min}/Q = 0.036/0.382 = 0.094$.

From proportional chart, for $q/Q = 0.094$, $d/D = 0.206$ and $v/V = 0.625$.

Therefore, the velocity at minimum flow $= 0.625 \times 0.884 = 0.55$ m/sec

This velocity is self-cleansing velocity; hence it is fine.

$d_{min} = 0.206 \times 0.742 = 0.153$ m

Comment: If the velocity at minimum flow is not satisfactory, increase the slope or try with reduction in depth of flow at maximum discharge or reduction in diameter of the sewer.

Example: 4.9 The minimum and maximum flows estimated for design of cement concrete sewer are 5,185 and 15,200 m^3/day, respectively. If this trunk sewer has to flow 0.8 times full at maximum discharge and should have self-cleansing effect at all flows as offered by velocity of 0.9 m/sec while flowing full. Determine diameter and slope of this trunk sewer. What will be minimum depth of flow and velocity? Consider n = 0.012.

Solution

Minimum flow = 5,185 m^3/day = 0.06 m^3/sec and maximum flow = 15,200 m^3/day = 0.176 m^3/sec

This sewer is flowing 0.8 full at maximum discharge.

From the proportionate value (Table 4.3) for d/D of 0.8, $q/Q = 0.988$

Hence, discharge Q for flow full section = 0.176/0.988 = 0.178 m^3/sec

Therefore, for flow full condition $Q = 0.178$ m^3/sec and $V = 0.9$ m/sec

C/s area of sewer = $Q/V = 0.198$ m^2, hence diameter $D = 0.502$ m

From Manning's formula for full depth, the required slope will be

$$V = \frac{1}{n} R^{2/3} S^{1/2} = 0.9 = \frac{1}{0.012} \left(\frac{0.502}{4} \right)^{2/3} (S)^{1/2}$$

Solving for S, the $S = 0.00185$, i.e., 1 in 539 m

This is the slope required for full depth of flow to generate velocity of 0.9 m/sec.

Now $Q_{min}/Q = 0.06/0.178 = 0.337$

For q/Q of 0.337, $d/D = 0.4$, $a/A = 0.373$, $r/R = 0.857$ (Table 4.3, neglecting variation of n).

Now for the sewer to be same self-cleansing at 0.4 depth as it will be at full depth, the gradient (s_{sc}) required can be worked out using relation $s_{sc} = (R/r)S$

Therefore, $s_{sc} = S/0.857 = 0.00185/0.857 = 0.00216$ i.e., 1 in 461.92 so as to have same self-cleansing effect exerted while flowing full at velocity of 0.9 m/sec.

Now, the velocity v_{sc} generated at this gradient is given by Eq. 4.24.

$$v_{SC} = V \frac{N}{n} \left(\frac{r}{R} \right)^{1/6}$$

$= 0.9 \times (0.857)^{1/6} = 0.877$ m/sec, which is self-cleansing as it is more than 0.6 m/sec

Answer: The diameter required for the sewer $= 0.502$ m; gradient required $= 1$ in 461.92; depth at minimum discharge $= 0.4D = 0.4 \times 0.502 = 0.20$ m; and velocity at minimum discharge $= 0.877$ m/sec.

4.4   Design of Storm Water Drains for Separate System

The storm water drains are designed to avoid flooding in the area and carry the runoff generated from the rain of designed frequency. Topography with minimum ponding and storage will result in higher peak flows, which will require larger structure and this will also lead to the potential of flooding in the downstream area. The flow at any time in the storm water collection system is not only dependent on the rainfall intensity but also depends on the system designed for storm water collection upstream, which is contributing to the discharge. Hence, instead of considering only the local area, while designing storm water collection drains, cumulative effect of urban area should be considered. Due to local area design, such storm water collection system will lead to frequent flooding due to increased runoff received from upstream and decrease in groundwater recharge because of decreased in retention time due to surface paving in urban area, which is increasing the runoff coefficient.

The minor system or component involved in design of storm water collection system consist of ditches, canals and sewers designed to accommodate a storm of moderately short occurrence interval of 2–5 years depending upon the importance of the locality. With this frequent flooding of the area can be avoided due to occurrence of the frequently occurring moderate storms. The major system involves path followed by the runoff, exceeding the capacity of the minor system. Thus, the component of the major system includes streets, adjoining lands, drainage rights of way for surface flow, and any natural drainage channels existing in that locality. These major systems should be capable of holding the flow resulting from the rain of lower frequency, such as once in 20 years or so. Ideally, the storm water collection system should be designed involving measures for reducing the quantity of runoff. This might require consensus from building and zoning regulatory authority to adopt measures for reducing surface runoff. Some of the measures that could be adopted are suggested below.

- In case of separate system, the storm water from roofs should be discharged on grassed surfaces rather than pavements to increase retention time and increased opportunity for infiltration. This will reduce the runoff as well as magnitude of peak.
- Porous pavements or interlocking paving blocks or gravel surfacing should be encouraged surrounding the buildings and parking lots to permit infiltration. Due to rough surface the velocity of the flow will reduce, thus helping to diminish the peak flow and encouraging infiltration to recharge ground water.

- Rather than the grid system of collection, aligning the drains with contour grading to produce more circuitous flow path that will dampen the peak flow due to more retention in the collection system itself. If the flow channels are made pervious this will also result in reducing total discharge.
- Increasing the surface detention by provision of detention ponds and infiltration basins or infiltration trenches.

While designing the minor systems, in storm water drains velocities provided are generally higher than sanitary sewers, since relatively coarser solids are expected to effectively transport in these drains. Hence, velocities at flow full section are generally kept at 0.75–2.5 m/sec. The maximum allowed velocity is decided by the limiting velocity of material selected for construction of drain.

Storm water drains are designed for full design flow likely to be generated from the rain of design frequency. Though, these drains may not always receive this full discharge and solid deposition will occur at low flow, however the periodic larger runoff events can aid in scouring of accumulated debris. If the natural slope does not permit maintaining self-cleansing velocity in the section, arrangement for manual or mechanical cleaning of drain section shall be provided.

Storm water is collected from streets into the link drains, which in turn discharge into main drains of open type. The main drain finally discharges the water into open water body. As far as possible gravity discharge is preferred, however when it is not possible, pumping need to be employed. While designing, the alignment of link drains, major drains and sources of disposal are properly planned on contour maps. The maximum discharge expected in the drains is estimated using rational formula. The longitudinal sections of the drains are prepared keeping in view the full supply level (FSL) so that at no place it should go above the natural ground surface level along the length of the drain. After deciding the FSL line, the bed line is fixed (i.e., depth of drain) based on following consideration.

a. The bed level should not go below the bed level of source into which storm water is to be discharged.
b. The depth in open drain should preferably be kept less than a man height.
c. The depth is sometimes also decided based on available width in the congested areas.
d. The drain section should be economical and velocities generated should be non-silting and non-scouring in nature.

The drain section is finally designed using Manning's formula. Adequate free board (Table 4.4) is provided over the design water depth at maximum discharge.

Example: 4.10 Design unlined trapezoidal section storm water drain collecting surface runoff from a catchment area of 40 hectares. The inlet time is 13 min and time of flow in the drain in upstream sewer is 17 min. The average runoff coefficient for the catchment is 0.45.

Table 4.4 Minimum free board for storm water drain

Drain size	Free board
Up to 300 mm bed width	100 mm
Beyond 300 mm and up to 900 mm bed width	150 mm
Beyond 900 mm and up to 1500 mm bed width	300 mm

Source CPHEEO (2019)

The slope available for drains is 1 in 1800. The permissible maximum velocity through the drain is 0.80 m/sec. Consider Manning's coefficient n = 0.022. Rainfall intensity and time relation is $1570/(t + 24)$, mm/h, where 't' is in min. Also consider for the drains having discharge up to 15 cumecs the relation of depth of flow 'd' to the bottom width 'w' as $d = 0.5.w^{0.5}$.

Solution

Estimating storm water runoff
Time of concentration = travel time + inlet time = 13 + 17 = 30 min
  Intensity of rainfall = 1570/(30 + 24) = 29.07 mm/h
  Using rational formula, runoff $Q = CIA/360 = 0.45 \times 29.07 \times 40/360 = 1.4535$ m^3/sec
  Design of drain section
  Keeping the water depth to 0.80 m and side slope of 1:1 in cutting the drains with free board of 0.3 m
  Hence for depth of 0.80 m, the bottom width will be:
  $0.80 = 0.5 \times w^{0.5}$
  Hence, w = 2.56 m, thus, provide 2.6 m as bottom width, hence with 1:1 slope, the top width will be $2.6 + 2 \times 0.80 = 4.2$ m.
  Now, c/s area of the drain = $0.8 \times (4.2 + 2.6)/2 = 2.72$ m^2
  Wetted perimeter = $w + 2\sqrt{2}\, d = 2.6 + 2\sqrt{2} \times 0.8 = 4.863$ m
  Hence, $R = A/P = 2.72/4.863 = 0.56$ m
  Now, $Q = A \times V = A \times \frac{1}{n}R^{2/3}S^{1/2} = 2.72 \times 1/0.022 \times 0.56^{2/3} \times (1/1800)^{0.5}$
  Hence $Q = 1.98$ m^3/sec, however the design discharge is only 1.4535 m^3/sec
  Hence, restricting the width to 1.9 m
  $A = (1.9 + 3.5)0.8/2 = 2.16$ m^2
  $P = 1.9 + 2\sqrt{2} \times 0.8 = 4.163$ m
  $R = 2.16/4.163 = 0.519$ m
  Hence, Q = $2.16 \times 1/0.022 \times 0.519^{2/3} \times (1/1800)^{0.5} = 1.494$ m^3/sec (hence it is fine)
  Now $V = Q/A = 1.494/2.16 = 0.692$ m/sec, which is less than 0.80 m/sec, hence it is fine and the design is satisfactory.
  The top width of the channel including free board of 0.3 m will be $1.9 + 1.1 + 1.1 = 4.1$ m
  Hence, final dimension of the trapezoidal drain section is 1.9 m bottom width, total depth of 1.1 m (including 0.3 m free board) and tope width of 4.1 m.

4.5 Sewage and Stormwater Pumping Stations

4.5.1 Introduction

There are certain locations, where due to natural slope of the ground it is possible to convey sewage by gravity to a central (or decentralized) treatment facility or stormwater is conveyed up to the disposal point entirely by gravity. Whereas, in case of large extent of area being served with flat ground, localities at lower elevation or widely undulating topography it may be essential to employ intermediate pumping station in the collection system for conveyance of sewage to the treatment plant. Sewage and stormwater are required to be lifted up from a lower level to a higher level at various places in a sewerages system. Pumping of sewage is also generally required at the sewage treatment plant. Availability of land, scope for future expansion, type of sump and the pump used, and aesthetic are the important consideration for designing the sewage or storm water pumping station.

There is difference between pumping of sewage and fresh water, due to polluted nature of the wastewater containing suspended solids and floating solids, which may clog the pumps. The dissolved organic and inorganic matter present in the sewage may chemically react with the pump and pipe material and can cause corrosion. The pathogens present in the sewage may pose health hazard to the workers. Settling of organic matter in the sump well and subsequent decomposition will lead to spreading of foul odour in the pumping station, requiring proper design to avoid such deposition of solids. In addition, variation of sewage flow with time makes it a challenging task to handle variable inlet flow and accordingly operate pumps with different capacities.

Pumping is often required for (i) untreated domestic wastewater in collection system, (ii) combined domestic wastewater and stormwater runoff in collection system, (iii) stormwater runoff in separate system, (iv) sludge at a wastewater treatment plant, (v) treated domestic wastewater at treatment plant for disposal, and (vi) recycling treated water or mixed liquor at treatment plants. Each pumping application has specific purpose and requires specific design and pump selection considerations. At sewage treatment plant, pumping is also required for extraction of grit from grit chamber and pumping may also be essential for conveying separated grease and floating solids to disposal facility.

Selection of proper location for the sewage and storm water pumping station is necessary to ensure that the area to be served can adequately drained. The site selection for the pumping station is important and the area selected should never get flooded. Possible future development and growth and accordingly increase in capacity in future should be given due consideration while finalizing the location. Location of the pumping station should be finalized considering the future expansion and expected increase in the sewage flow. There need to be enough space in the pumping station to replace low-capacity pump with higher capacities as per the need in future. The site selected should not get flooded during the storms and easily accessible in all seasons. In case of storm water pumping station, the location should be such that for the rain with runoff exceeding the capacity of the pumping station the impoundment should not cause any damage to the pumping station.

Initial sewage flow received at the pumping station is too less as compared to design flow at the end of design period. This should be kept in mind while selecting capacity of the pumps, so as to avoid too frequent on and off of pumps or holding sewage for longer time to avoid decomposition of organic matter in the sump. The capacity of the pumping station is based on the present and future sewage flow. Generally, design period up to 15 years is considered for pumps. The civil structure and the pipelines shall be adequate to serve for the design period of 30 years.

Generally pumping station should contain at least three pumping units of such capacity to handle the maximum sewage flow if the largest unit is out of service. The pumps provided in the pumping station at sewage treatment plant should be selected to provide as uniform a flow as possible to the treatment plant. All pumping stations should have an alarm system to signal power or pump failure and every effort should be made to prevent or minimize overflow. Flow measuring device, such as venturi meter, shall be provided at the pumping station. In all cases raw-sewage pumps should be protected by screens or racks unless special devices such as self-cutting grinder pumps are provided. Housing for electric motors should be made above ground and in dry wells electric motors should be provided protection against flooding. The ventilation in dry well should be adequate, preferably of forced air type, and accessibility for repairs and replacements should be ensured.

4.5.2 Types of Pumps

The pumps used in the sewerage system or at sewage treatment plant for pumping of sewage, sewage sludge, grit matter, etc. could be of following types depending on the suitability:

a. Radial-flow centrifugal pumps
b. Axial-flow and mixed-flow centrifugal pumps
c. Reciprocating pistons or plunger pumps
d. Diaphragm pumps
e. Rotary screw pumps
f. Pneumatic ejectors
g. Air-lift pumps.

Other pumps and pumping devices are available, but their use in wastewater conveyance and treatment is not that popular.

Radial-flow centrifugal pumps: These pumps consist of two parts, casing and impeller. Impeller of the pump rotates at high speed inside the casing. Sewage is drawn from the suction pipe into the pump and curved rotating vanes of impeller throw it up through outlet pipe because of centrifugal force. Radial-flow pumps throw the liquid entering at the centre of the impeller out into a spiral volute or casing. The impellers of all centrifugal pumps can be closed, semi open, or open depending on the application. The impeller of the centrifugal pump has vanes, which are either open or have both front and back shrouds. Open impeller type pumps have no shrouds and are more suitable for sewage pumping, because suspended solids and floating matter present in the sewage can be easily pumped without clogging. Whereas, semi-open pumps have only back shroud. These pumps can have a horizontal or vertical design. These pumps are commonly used for any capacity and head. These pumps have low specific speed up to 4200.

Axial- flow centrifugal pumps: Axial-flow design of centrifugal pump can handle large flow capacities but only with reduced discharge heads. They are constructed vertically. The vertical pumps have positive submergence of the impeller. These are used for pumping large sewage flow, more than 2,000 m^3/h and head up to 9.0 m. These pumps have relatively high specific speed of 8000–16,000. The water/wastewater enters in this pump axially and the head is developed by the propelling action of the impeller vanes. These axial pumps would have the open impeller to avoid clogging.

Mixed flow pumps: Mixed flow pumps develop heads by combination of centrifugal action and the lift of the impeller vane on the liquid. This pump has single impeller. The flow enters the pump axially and discharges in an axial and radial direction into volute type casing. The specific speed of the pump varies from 4200 to 9000. These pumps are used for medium pumping head ranging from 8 to 15 m. This mixed flow pumps, particularly of high specific speed will be generally semi-open impeller type.

Most water and wastewater can be pumped with centrifugal pumps; however, they should not be used for the following:

- Pumping viscous industrial liquids or sludges, where the efficiency of centrifugal pumps is very low. For such applications, positive displacement pumps are used.
- These are not suitable for pumping of low flow against high head, except for deep-well applications. Large numbers of impellers required is a disadvantage for the centrifugal pump for pumping wastewater against high heads.

The rotational speed of impeller affects the capacity, efficiency, and extent of cavitation. Cavitation can be a problem even if the suction lift is kept within permissible limits and it should be checked. Centrifugal pumps are classified on the basis of their specific speed (*Ns*) at the point of maximum efficiency. The specific speed of the pump is defined as speed of the impeller in revolution per minute (rpm) such that it would deliver discharge of 1 m^3/min against a head of 1.0 m and it is determined using the following equation (Zhang et al., 2017):

$$Ns = \frac{3.65n\sqrt{Q}}{H^{0.75}} \qquad\qquad (4.26)$$

where, Ns is the specific speed; Q is the flow, m^3/min; H is the head, m; and n is speed, rpm.

For more suction lift, the pumps with low specific speed are suitable than the pumps with high specific speed. The axial flow pumps with high specific speed require positive suction head and some minimum submergence for trouble free operation and they will not work with any suction lift. It is advisable to avoid suction lift for the centrifugal pumps. Hence, pumps are typically installed either to work submerged in the wet well or installed in the dry well at such a level that the impeller will be below the level of the water in the wet well.

Positive displacement pumps: These pumps include reciprocating piston, plunger, and diaphragm pumps. Almost all reciprocating pumps used in water and wastewater engineering are metering or power pumps. A piston or plunger is reciprocated in a cylinder, which is driven forward and backward by a crankshaft connected to an outside driving unit. Adjusting metering pump flow involves merely changing the length and number of piston strokes. A diaphragm pump is similar to a reciprocating piston or plunger; however, instead of a piston, it contains a flexible diaphragm that oscillates as the crankshaft rotates. Plunger and diaphragm pumps feed stipulated volume of chemicals (acids or caustics for pH adjustment) to a water or wastewater stream. These are not suitable for sewage pumping because solids and polythene pieces and rugs present in the sewage may clog them. These pumps have high initial cost and very low efficiency.

Rotary screw pumps: In rotary screw pump, a motor rotates a vane screw or rubber stator on a shaft to lift or feed sludge or solid waste material to a higher level or to the inlet of another pump. These are used in the square grit chamber for removal of grit, which is fairly free from organic matter.

Air pumps: These are pneumatic ejectors and airlifts pumps. In pneumatic ejector, wastewater flows into a receiver pot and an air pressure system blows the liquid to a treatment process at a higher elevation. The air system can use plant air (or steam), a pneumatic pressure tank, or an air compressor. The air pump system has no moving parts in contact with the wastewater; thus, no clogging of impeller is involved. Ejectors are normally maintenance free and offer longer service life. Airlift pumps consist of an updraft tube, an air-line, and an air compressor or blower. Airlifts blow air at the bottom of a submerged updraft tube. While traveling upward the air bubbles expand, thus reducing density and pressure within the tube. Higher flows can be lifted for short distances by this pump. Airlifts are used in wastewater treatment to transfer mixed liquors or slurries from one reactor to another. These pumps are having very low efficiency and they can lift the sewage for small head.

4.5.2.1 Efficiencies of Pumps

Efficiencies of the pumps range from 85% for large capacity centrifugal pumps (radial-flow and axial-flow and mixed-flow centrifugal pumps) to less than 50% for many smaller units. For reciprocating pistons or plunger pumps efficiency is about 30% onward depending on horsepower and number of cylinders. For diaphragm pumps, efficiency is about 30%; and for rotary screw type pump, pneumatic ejectors type pump and air-lift pump it is below 25%.

4.5.2.2 Materials for Construction of Pumps

For pumping of water using radial-flow centrifugal pumps and axial-flow and mixed-flow centrifugal pumps normally bronze impellers, bronze or steel bearings, carbon steel or stainless shafts, and cast-iron housing is used. For domestic wastewater pumping using radial-flow centrifugal pumps and axial-flow and mixed-flow centrifugal pumps similar material is used except that they are made from cast iron or stainless-steel impellers. For industrial wastewater and chemical feeders using radial-flow centrifugal pump or reciprocating piston pump or plunger type pump, a variety of materials are used depending on the requirement of corrosion resistance. In diaphragm pumps the diaphragm is usually made of rubber. Rotary screw type, pneumatic ejectors type and air-lift pumps normally have steel components.

4.5.3 Pumping System Design

To choose the proper pump, the environmental engineer must know the capacity, head requirements, and characteristics of the wastewater to be pumped. This section addresses the capacity and head requirements for the selected pump.

4.5.3.1 Capacity of Sump and Pumps

To compute capacity of pump average system flow rate is being assessed and variation in the flow rate is estimated. For example, when pumping sewage from a sewerage system, the pump must handle peak flows roughly two to three times the average flow, depending on population size being served. Variation in summer and winter and future needs also dictate the capacity of the pumps. Population increase trends and past flow rates should also be considered in this evaluation. The capacity of the pumping station should be so determined that the pump of minimum duty should also run for at least 5 min. The smallest capacity pump should be capable to pump sewage from wet well and discharge it by maintaining self-cleansing velocity of about 0.6 m/sec in the delivery main (rising main).

The sump should have enough capacity to prevent pump motors from overheating due to extensive cycling and the capacity should be small enough to accommodate cycling times that will reduce septicity and odour problem due to long duration of holdup of sewage. In addition, the capacity of the well should be such that with any combination of inflow and pumping, the cycle of operation for each pump will not be less than 5 min and the maximum detention time in the wet well will not exceed 30 min at average flow (CPHEEO, 2013). Alternately, the wet well capacity can also be decided so that it contain sufficient volume of wastewater to permit the pump to run for at least 2 min and restart not more than once in 5 min (Steel & McGhee, 1979). The pump running time (T_r) and the filling time (T_f) of the sump can be estimated using Eq. 4.27 and Eq. 4.28, respectively (Lee & Lin, 2007).

$$T_r = \frac{V}{D - Q} \tag{4.27}$$

$$T_f = \frac{V}{Q} \tag{4.28}$$

where, T_r is pump running time, min; T_f is the filling time of sump with pump off, min; V is the wastewater storage volume of the sump, m^3; D is discharge at peak flow, m^3/sec; Q is the daily average inflow, m^3/sec.

Thus, the total cycle time (T) can be computed using Eq. 4.29.

$$T = T_r + T_f \tag{4.29}$$

Typically, submersible pumps can have 4 to 10 cycles per hour and preferably it can be kept below 6 cycles per hour for individual pump. If pumps selected have capacity equal to the peak flow rate, in this case the volume required for the sump can be computed as per Eq. 4.30 (WEF, 1993).

$$V = T \times Q/4 \tag{4.30}$$

where, V is the volume required for the sump, m^3; T is the pump cycle time, min; and Q is sewage peak flow, m^3/min.

The capacity of the pumps installed should meet the peak flow rate with about 100% standby. Two or more number of pumps should be provided. The capacity and number of pumps for larger pumping station are so selected that variation in the flow rate can be adjusted by throttling the delivery valve or manipulating speed of the pump without starting or stopping the pumps too frequently. The general practice is to provide three pumping sets in small stations consisting of one pump of capacity equal to dry weather flow (DWF), second pump with capacity of two times DWF and third pump of capacity three times DWF. For larger pumping stations five pump sets shall be provided with capacities of 2 units of 0.5 DWF, 2 units of one DWF and one large pump with capacity of three times DWF. The suction and delivery sides shall not be less than 100 mm for sewage pumping to avoid clogging.

Example: 4.11 A sewage pumping station is to be designed for an academic residential campus having total residential population of 30,000 and non-residential 4,000 persons working in various academic and commercial units housed in the campus. If the water demand for residential is estimated to be 150 L per capita per day and for non-residential staff as 70 L per capita per day, what will be average quantity of sewage generation considering 75% of the water supplied generates sewage. Consider the peak factor of 3.5 for estimating peak flow and average flow to minimum flow ratio is 5. Design the wet well for the pumping station.

Solution
Estimating the average quantity of sewage generation per day

$$= (30,000 \times 150 + 4,000 \times 70) \times 10^{-3} \times 0.75 = 3585 \text{ m}^3/\text{day} = 2.49 \text{ m}^3/\text{min}$$

Hence, peak sewage flow $= 2.49 \times 3.5 = 8.715 \text{ m}^3/\text{min}$.
The minimum flow $= 2.49/5 = 0.498 \text{ m}^3/\text{min}$
Design of wet well
Estimating volume of the wet well required for 2 min run time at peak flow

$$= 8.715 \times 2 = 17.43 \text{ m}^3$$

Estimate the volume required for total cycle time of 5 min at average flow using Eq. 4.29.

$$T = \frac{V}{D-Q} + \frac{V}{Q} = \frac{V}{8.715 - 2.49} + \frac{V}{2.49} = 5 \text{ min}$$

Hence, $V = 8.893 \text{ m}^3$
Since the volume of the wet well required for the pump running time of 2 min is more than the volume required for the cycle time, providing volume of the wet well as 17.43 m^3.
Estimating actual pumping cycle

$$T = \frac{17.43}{8.715 - 2.49} + \frac{17.43}{2.49} = 9.8 \text{ min}$$

If working depth of the wet well provided is 2.0 m then diameter of the wet well required is 3.33 m. Thus, total depth of wet well will be 0.5 m (provided from well bottom to lowest water level to ensure minimum submergence) + 2.0 m + 0.6 m free board, thus total depth = 3.1 m below incoming sewer invert. In case of excessive excavation, the free board from the invert of incoming sewer can be reduced.

Alternately, if the pumps provided are both of the peak flow handling capacity and each pump operated for 4 cycles per hour, then two pumps will operate for 8 cycles per hour. The time between each start will be 60/8 = 7.5 min. Using Eq. 4.30, the volume required for the wet well will be:

$$V = T \times Q/4 = 7.5 \times 8.715/4 = 16.34 \, \text{m}^3$$

For working depth of 2 m the diameter required will be 3.23 m and depth will be 3.1 m. However, wet well volume of 17.43 m^3 is recommended so as to have pump run time of 2 min.

4.5.3.2 Head Requirement of Pump

Semi-open or open impeller type centrifugal pumps are most commonly used for the pumping of sewage. For pumping of slurries and sludge reciprocating pumps are widely used. Head describes the pressure in terms of lift. The discharge head on a pump is a sum of the following contributing factors:

(a) *Static head* (h_d): The difference in elevation through which the wastewater must be lifted, i.e., the lowest water level in the wet well and the highest point on the discharge side.

(b) *Friction head* (h_f): The resistance to flow is caused by friction in the pipes, valves, and bends, which need to be overcome and this add on to the static head of pumping. Entrance and transition losses shall also be considered as applicable. The loss of head in friction in the pipes shall be estimated from the Darcy-Weisbach equation, $h_f = fLv^2/(2gD)$ (Liou, 1998).

(c) *Velocity Head* (h_v): The head required to impart energy into a fluid to induce velocity. Normally this head is quite small, however to maintain self-cleansing velocity of minimum 0.6 m/sec this head needs to be ensured. This is estimated as $v^2/2\,g$.

(d) *Pressure Head* (h_p): The pressure differential head that the pump must develop to deliver water on the delivery side under higher pressure, if there is any residual water head above the discharge pipe. The pressure on water in sump well is usually atmospheric pressure, whereas when pumping into sewers there would be potential head at the point of delivery, against which the pump has to deliver. Thus, this is the difference between pressures on the liquid in the wet well and at the point of delivery.

Thus, total Head (H) of pumping shall be expressed by the Eq. 4.31.

$$H = h_d + h_f + h_v \pm h_p \tag{4.31}$$

The dynamic head on the pump can also be estimated using Bernoulli's equation as per Eq. 4.32 (Shojaeefard et al., 2012).

$$H = \frac{P_d}{\gamma} + \frac{V_d^2}{2g} + z_d - \frac{P_s}{\gamma} + \frac{V_s^2}{2g} + z_s \tag{4.32}$$

where, H is the dynamic head required for pumping, m; γ is the specific weight of water, N/m^3; g is the gravity acceleration, m/sec^2; P_d and P_s are gauge pressure at delivery and suction, N/m^2, respectively; similarly, z_d and z_s are the elevation at discharge and suction end, m.

4.5.3.3 Suction Lift

The amount of suction lift that can be handled must be carefully computed for the type of pump selected. It is limited by the barometric pressure, which depends on elevation and temperature, the vapor pressure that also depends on temperature, friction and entrance losses at the suction side, and the net positive suction head (NPSH). The NPSH factor depends on the shape of the impeller and it is obtained from the pump manufacturer.

4.5.3.4 Power Required for Pump

The power required to drive the pump for lifting and delivering sewage is a function of flow Q and the total head H and it can be calculated using Eq. 4.33 (Jun et al., 2016).

$$P = 0.163 \times Q \times H \tag{4.33}$$

where P is water power, kW; Q is the discharge m^3/min; H is the total head, m.

The power output of a pump is the work done to pump the wastewater to a higher elevation per unit time. Thus, the theoretical power output (kW) or horse power is a function of discharge and total head of pumping, which can be assessed as per Eq. 4.34 (Jun et al., 2016) or Eq. 4.35.

$$P \text{(in Watt)} = \gamma QH \tag{4.34}$$

where, Q is in m^3/sec, H is in m, and P (in HP) = P (in Watt)/746 is the conversion of W to HP, (where, 1 HP = 746 W, conversely 1 kW = 1.341 HP).

The actual power required for pumping is determined taking in to account the efficiency of the pump and driving unit and it is popularly referred as brake horsepower (BHP). The Eq. 4.35 can be used for determining the brake horsepower.

$$\text{BHP} = (w \cdot Q \cdot H)/(75 \cdot \eta_p \cdot \eta_m) \tag{4.35}$$

where, Q is the discharge (m^3/sec); H is the head of water (m); w is the density of water (kg/m^3); η_p is the efficiency of the pump; and η_m is the efficiency of the driving motor.

Alternately, the power required for pumping in kW is given by Eq. 4.36.

$$P = \rho^* g * H^* Q/(1000 * \eta) \tag{4.36}$$

where, ρ is the density of water, kg/m^3; g is gravity acceleration, m/sec^2; H is the head of pumping, m; P is power, kW; η is overall efficiency of pump and driving motor.

The efficiency of the pump is defined as the ratio of the power output ($P = \gamma QH$) to the input power of the pump. Similarly, the motor efficiency is defined as the ratio of the power applied to the pump by the motor to the power input to the motor. The overall efficiency of the pump set is the product of these two, i.e., ratio of power output of pump to the power input to the motor.

Example: 4.12 Estimate the water power for a pump delivering a wastewater discharge of 1.5 m^3/min against a total system head of 15 m.

Solution

$$P = 0.163\,QH = 0.163 \times 1.5 \times 15 = 3.668\ kW$$

Estimating in horse power
1 HP = 550 Ft.Lb/sec = 550 × 0.305 m/ft × 1 × 4.448 N/(lb.sec)
= 746 N m/sec (Joule/sec = Watt)
Therefore, 1 HP = 746 W or 0.746 kW
Hence, 3.668 kW = 4.92 HP

4.5.4   Types of Pumping Stations

Wide variety of arrangements can be made at the pumping stations depending on size and applications. The pumping-station configurations can be classified as: wet well/dry well, wet well only with submersible pumps, and wet well only with non-submersible pumps.

Wet well and dry well: In wet well and separate dry well configuration, two pits (wells) are made, one to hold the wastewater (wet well) and other (dry well) to house the pumps and appurtenances (Fig. 4.4a, b). This is required for wastewater that cannot be primed or conveyed long distances under suction heads. This option is typically used to pump large volumes of raw sewage, where uninterrupted flow is critical and wastewater solids could clog suction piping. While construction costs of this type may be higher and a heating, ventilation, and cooling (HVAC) system is necessary due to installation well below ground; however, ease of operation and maintenance is the advantage of this arrangement as pumps can be accessed any time for the maintenance. This configuration is best for operation and maintenance activities because operators can see and touch the equipment.

Wet well with non-submersible pumps: In this configuration, wet well holds the wastewater. The pumps are installed above the water level in wet well (Fig. 4.4c). This option is used in areas where the wastewater can be "pulled" through suction piping, e.g., treated or finished water or where shutdowns or failures would not be immediately critical, e.g., a

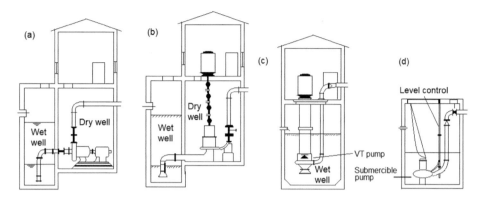

Fig. 4.4 Pumping stations **a** pumping station with horizontal pumps installed in dry well, **b** pumping station with vertical pump in dry well, **c** pumping station with vertical pumps in wet well, and **d** wet well with submersible sewage pump

package sewage treatment plant's raw wastewater lift stations, equalization of secondary treated wastewater, etc.

Wet well with submersible pumps: In this configuration, wet well holds both the pumps and the wastewater being pumped. The pump impeller is submerged or nearly submerged in the wastewater (Fig. 4.4d). Additional piping is not required in this type to convey the wastewater to the impeller. This option is common worldwide, and the submersible centrifugal pumps can be installed and operated cost-effectively. When vertical (VT) pumps are installed the driving motor is mounted on the floor above the ceiling of the wet well to provide easy access for maintenance.

In selecting the best design for an application, the designer should consider the following aspects:

- Many gases are formed by domestic wastewater, including CH_4 that is flammable. When pumps or other equipment are located in rooms below ground level, the possibility of explosion or gas build-up exists, and ventilation is extremely important. Particularly care should be taken for not allowing H_2S to accumulate, which could be fatal to the maintenance personal if increased beyond 1 ppm concentration.
- When wastewater is pumped at high velocities or through long lines, the water hammering caused by pumped water could be a problem. Valves and piping should be designed to withstand these pressure waves. Even pumps that discharge to the atmosphere should use check valves to cushion this surge.
- Coarse bar screens (opening 50–100 mm) shall be provided ahead of pumping station when centrifugal pumps are installed.
- The dry-well design is most preferred. The pumping station must be able to adjust the variation of incoming wastewater flow. The smallest capacity pump should be able to

pump from the wet well and discharge at a self-cleansing velocity of about 0.6 m/sec. Pumping stations typically include at least two pumps and a basic wet-well level control system. One pump is considered a "standby" pump, although the controls typically cycle the pumps in rotation during normal flows so they receive equal wear.

Example: 4.13 Design a raw sewage pumping station for residential complex. A per capita water demand of a complex is 170 LPCD having total population of 35,000 persons. The sewage generated from this residential complex is required to be lifted for 5.4 m of static head from invert of the sewer and 180 m distance. Consider loss of head in bends, valves and exit of 0.5 m. Determine (a) size of the sump well, (b) power required for the pump, and (c) diameter of the rising main. Assume any other practically implementable data, if required.

Solution

Estimation of sewage flow
Considering sewage generation equal to 80% of the water supply
  Average sewage flow $= 35,000 \times 170 \times 0.8 \times 10^{-3} = 4760$ m^3/d $= 0.055$ m^3/sec $= 3.3$ m^3/min.
  Peak sewage flow, considering peak factor of 3 $= 0.165$ m^3/sec $= 9.9$ m^3/min
  Considering velocity of 0.6 m/sec in rising main at average flow,
  Hence, area required is Q/V $= 3.3/(0.6 \times 60) = 0.092$ m^2, therefore diameter required

$$D = \sqrt{\frac{0.092 \times 4}{\pi}} = 0.342 \, \text{m}$$

  Provide diamter of 0.34 m, hence actual velocity during average flow $= 0.055 \times 4/(\pi \times 0.34^2) = 0.606$ m/sec and at peak flow the velocity will be 1.818 m/sec.
  Design of sump well
  Design the sump for minimum time of 2 min for maximum capacity pump to run continuously.
  Quantity of sewage $= 0.165 \times 60 \times 2 = 19.8$ m^3
  Estimate the volume required for total cycle time of 5 min at average flow using Eq. 4.29.

$$T = \frac{V}{D-Q} + \frac{V}{Q} = \frac{V}{9.9-3.3} + \frac{V}{3.3} = 5 \, \text{min}$$

  Solving for V, it is equal to 11 m^3
  If two pumps, each of peak flow capacity, are provided then considering each pump will operate for 4 cycles per hour, thus two pump making 8 cycles per hour with on and off frequency of 7.5 min

$$V = T \times Q/4 = 7.5 \times 9.9/4 = 18.56 \, \text{m}^3$$

Hence, the volume of the sump required is higher of this, i.e., required for pump running for min 2 min $= 19.8$ m^3

Quantity of sewage in rising main $= (\pi D^2) \times L/4 = \pi \times 0.34^2 \times 180/4 = 16.34$ m^3

Net storage capacity of the sump $= 19.8 + 16.34 = 36.14$ m^3

Provide a wet well sump unit with effective water depth of 2.0 m. Hence the surface area required for the sump $= 19.8/2 = 9.9$ m^2.

Provide circular shaped sump well having surface area of 9.9 m^2 and depth of 2.0 m, hence diameter required will be 3.551 m. Actual depth of sump will be below invert level of the incoming sewer $= 2.0 + 0.3 + 0.5 = 2.3$ m.

Check for detention time of sewage in the sump at average inflow $=$ volume/flow $= 19.8/(3.3) = 6$ min (less than 30 min, hence acceptable)

Check for minimum duration of pumping

If pump with the maximum discharge of 0.165 m^3/sec (peak flow) is operated,

The maximum duration of storage at average flow $= 6$ min.

Volume of sewage collected at average flow $= 0.055 \times 60 \times 6 = 19.8$ m^3.

Hence duration of pumping for maximum capacity pump $= 19.8/(0.165 \times 60) = 2$ min.

Hence, for lower capacity pump the continuous duration of operation will be more than 2 min, which is greater than minimum operation duration recommended.

$$\text{Actual cycel time} = \frac{V}{D-Q} + \frac{V}{Q} = \frac{19.8}{9.9 - 3.3} + \frac{19.8}{3.3} = 9 \, \text{min}$$

This is greater than 5 min, hence design in fine.

Calculating power of pump

For pump with maximum discharge capacity equal to peak flow

Considering friction factor of 0.033, the frictional head loss $= h_f = fLv^2/(2gD)$

$$= 0.033 \times 180 \times (1.818)^2/(2 \times 9.81 \times 0.34)$$

$$= 2.943 \, \text{m}$$

Velocity head $= v^2/2\,g = (1.818)^2/(2 \times 9.81) = 0.168$ m

Total head of pumping $= 5.4 + 2.3 + 2.943 + 0.5 + 0.168 = 11.311$ m

Considering efficiency of pump $= 65\%$ and efficiency of motor $= 75\%$; hence kW of motor required for highest capacity pump (to be able to pump peak flow), using Eq. 4.36:

$$= 9800 \times 0.165 \times 11.311/(1000 \times 0.65 \times 0.75) = 37.52 \, \text{kW}$$

Provide 3 pumps, one with 37.52 kW (say 40 kW) to handle peak flow alone and other two pumps of capacity to handle of 1 DWF (head of 8.546 m, 9.45 kW for DWF, say 10 kW) and for the third pump of 2 DWF capacity which will handle flow of 0.11 m^3/sec (velocity of 1.212 m/sec, frictional head loss of 1.308 m, velocity head of 0.075 m, thus total head of 9.583 m) the power required will be 21.19 kW.

Questions

4.1. A 900 m long storm sewer collects water from a catchment area of 60 hectares, where 30% area is covered by roof (C = 0.9), 20% area by pavements (C = 0.8) and 50% area is covered by open plots (C = 0.20). Determine the diameter of storm water drain. Assume the time of entry = 3 min; velocity at full flow in sewer = 1.45 m/sec; gradient of sewer = 0.001, and roughness coefficient = 0.013. The intensity of rainfall, cm/h = 75/(t + 5), where t is in min.

4.2. Explain the importance of considering minimum and maximum velocity while designing the sewers.

4.3. Describe 'Self-cleansing velocity' and 'limiting velocity'.

4.4. Explain important consideration while finalizing alignment and bed line of storm water drain.

4.5. Find the gradient required in sewer of 0.5 m diameter to maintain self-cleansing velocity at flow full condition.

4.6. What is a difference between pumping of sewage and pumping of fresh water?

4.7. Define the requirements of pumping station in sewerage scheme.

4.8. Describe specific speed of the centrifugal pumps.

4.9. Discuss pumps used in sewage pumping station. How capacity and power required for pumps are worked out at pumping stations?

4.10. Based on what consideration the capacity of the sump well is decided?

4.11. Describe the criteria for selection of site for pumping station. What are the facilities/ accessories required in the pumping station?

4.12. Describe different types of pumping stations and the types of pumps used in each.

4.13. A per capita sewage generation of a township is estimated to be 160 LPCD having total population of 52,000 persons. The sewage generated from this town is required to be lifted for 10 m of static head and 180 m distance. Consider loss of head in bends and valves of 0.4 m. Determine (a) size of the sump well, (b) horse power required for the pump, and (c) diameter of the rising main. Consider friction factor of 0.033, efficiency of pump = 65% and efficiency of motor = 75%. Assume any other additional suitable data, if required.

4.14. A storm water drain 460 m long receives at its head storm water from an area of 20 hectares. The time of entry for the area is 5 min, time of travel is 17 min and the overall coefficient of runoff for the catchment is 0.5. The fall available is 5.15 m. Design the sewer considering ¾ full at design discharge. What will be velocity of flow? Is it self-cleansing? Consider, manning n = 0.013; intensity of rainfall, cm/h = 75/(t + 10); d/D = 0.75, q/Q = 0.94, v/V = 1.08. (Answer: I = 23.4 mm/h; q = 0.65 m³/sec; Q = 0.69 m³/sec; D = 0.61 m; v = 2.54 m/sec (self cleansing)

4.15. The drainage area of a sector of town is 18 hectares. The classification of surface is as follows:

% of total area	Type of surface	Coefficient of runoff
20	Pavement	0.85
20	Roof cover	0.80
15	Unpaved area	0.20
30	Gardens	0.20
15	Bushes	0.15

If the time of concentration is 33 min, find the maximum run off using Rainfall intensity, cm/h $= 900/(t_c + 60)$, where t_c is in min. What will be required diameter of the sewer if it is flowing full at this discharge? If the sewer is running 3/4th full at maximum discharge what will be diameter of the sewer required? Consider, invert slope of 1 in 450, $n = 0.013$, and for $d/D = 0.75$, $q/Q = 0.94$, $v/V = 1.08$. (Answer: For flow full condition D $= 1.27$ m and for d/D $= 0.75$ at maximum discharge D $= 1.30$ m)

4.16. Comments on the following with respect to sewage pumping station: (i) capacity of wet well, (ii) number of pumps and standby capacity.

4.17. A town has a population of 100,000 persons with per capita water supply of 170 LPCD. Design a main sewer running 0.7 full at maximum discharge. Manning's n constant at all depth $= 0.013$. Sewer is laid with the slope of 1 in 500. Consider peak factor of 3. Also calculate velocity of the flow and comment on the velocity. For $d/D = 0.7$; $v/V = 1.120$; $q/Q = 0.838$.

4.18. Explain the factors that define the range of velocities of flow in sewers. What are the approximate values of these velocities?

4.19. Calculate the velocity of flow and corresponding discharge in a sewer of circular section having diameter equal to 1.0 m and laid at a gradient of 1 in 500. The sewer runs 0.6 full at maximum discharge. Use Manning's formula $n = 0.012$. For $d/D = 0.6$; $v/V = 1.072$; $q/Q = 0.671$. (Answer: Q $= 1.16$ m^3/sec; V $= 1.48$ m/sec; q $= 0.78$ m^3/sec; v $= 1.59$ m/sec)

4.20. A 30 cm diameter sewer having an invert slope of 1 in 150 was flowing full. What would be the velocity of flow and discharge? Consider $n = 0.013$. Is the velocity self-cleansing? What would be the velocity and discharge when the sewer is running 0.8 times its full depth? For $d/D = 0.8$, the proportionate velocity $(v/V) = 1.14$ and proportionate discharge $(q/Q) = 1.067$. (Answer: Q $= 0.079$ m^3/sec; V $= 1.12$ m/sec (Self cleansing); q $= 0.084$ m^3/sec; v $= 1.28$ m/sec)

References

Chow, V. T. (1955). A note on the Manning formula. *Eos, Transactions American Geophysical Union, 36*(4), 688–688.

CPHEEO. (2013). *Manual on sewerage and sewage treatment, part A, engineering*, 3rd edn. Central Public Health and Environmental Engineering Organization, Ministry of Urban Development, Government of India, New Delhi, India.

CPHEEO. (2019). *Manual on storm water drainage systems, volume-I, part A: Engineering design.* Central Public Health and Environmental Engineering Organisation, Ministry of Housing and Urban Affairs, Government of India.

Dias, S. P., & Matos, J. S. (2001). Small diameter gravity sewers: Self-cleansing conditions and aspects of wastewater quality. *Water Science and Technology, 43*(5), 111–118.

Herschy, R. W., Rumney, G. R., Oliver, J. E., Mosley, M. P., & Holland, P. G. (1998). Chézy formula. *Encyclopedia of Hydrology and Water Resources,* 121–121.

Jun, M., Srivastava, R., Jeong, J., Lee, J., & Kim, M. (2016). Simple recycling of copper by the synergistic exploitation of industrial wastes: A step towards sustainability. *Green Chemistry, 18,* 3823–3834.

Lee, C. C., & Lin, S. D. (2007). *Handbook of environmental engineering calculations* (2nd ed.). McGraw-Hill.

Liou, C. P. (1998). Limitations and proper use of the Hazen-Williams equation. *Journal of Hydraulic Engineering, 124*(9), 951–954.

MSMA. (2012). Urban stormwater management manual for Malaysia. Department of Irrigation and Drainage, Government of Malaysia.

Shojaeefard, M. H., Tahani, M., Ehghaghi, M. B., Fallahian, M. A., & Beglari, M. (2012). Numerical study of the effects of some geometric characteristics of a centrifugal pump impeller that pumps a viscous fluid. *Computers & Fluids, 60,* 61–70.

Steel, E. W., & McGhee, T. J. (1979). *Water supply and sewerage* (5th ed.). McGraw-Hill.

WEF. (1993). Design of wastewater and stormwater pumping stations. Water Environment Federation, Alexandria, Virginia, USA.

WPCF. (1960). Manual of practice no. 9, Water Pollution Control Federation. Washington DC, USA.

Zhang, W., Yu, Z., & Zhu, B. (2017). Influence of tip clearance on pressure fluctuation in low specific speed mixed-flow pump passage. *Energies, 10*(2), 148.

Classification and Quantification of Major Water Pollutants

5

5.1 Classification of Water Pollutants

When polluted waste streams are discharged in water bodies, they will add many unwanted ingredients in water and might lead to the increase in concentrations of these constituents beyond acceptable limits for aquatic life or making the water quality unfit for human consumption or even for the recreation. Depending on the adverse effects caused upon release of a particular pollutant in the water body the various types of water pollutants can be classified into major categories, such as: (1) organic pollutants, (2) pathogens, (3) nutrients and agriculture runoff, (4) suspended solids and sediments (organic and inorganic), (5) soluble inorganic pollutants (salts, metals, and chemicals), (6) thermal pollution, (7) radioactive pollutants, (8) nano-particles and (9) emerging contaminants.

5.1.1 Organic Pollutants

Organic pollutants can be further divided into following categories:

(a) *Oxygen demanding organic matters*: Wastewaters containing high concentration of oxidizable organic matters, when discharged in the water body, will exert biochemical oxygen demand (BOD) and reduce dissolve oxygen (DO) available in the water. Domestic and municipal wastewaters, wastewater from food processing industries, dairies, canning industries, slaughter houses, paper and pulp mills, tanneries, breweries, distilleries, etc. have considerable concentration of biodegradable organic compounds present either in suspended, colloidal or in dissolved form, which will undergo aerobic biodegradation by microbial activity in presence of adequate DO in the receiving water body. Thus, the DO available in the water body will be consumed

M. Ghangrekar, *Wastewater to Water*, https://doi.org/10.1007/978-981-19-4048-4_5

for aerobic oxidation of organic matter present in the wastewater. This depletion of DO will be a serious problem, adversely affecting the aquatic life, particularly if DO falls below 4.0 mg/L.

This decrease of DO is an index of pollution. When the DO level falls below 3.0 mg/L the hardy fish, Indian carp, channel catfish, guppy, and eel will also face difficulty in survival (Colt, 2006) and when DO falls below 0.5 mg/L aerobic bacteria will also face difficulty in continuing aerobic respiration and the condition in the water body will become anaerobic imparting a blackish colour to the water. Upon development of such dark colour, solar light will not be able to penetrate in the entire water; thus, adversely affecting the photosynthetic oxygen yield and hence, under such conditions it will require longer stretch of river for naturally replenishing the water quality, due the combine action of deoxygenation and simultaneous reaeration. Thus, very long stretch of the river will get affected once anaerobic condition develops, making the water unsuitable for water supply as well as even for recreation. Under anaerobic condition, hydrogen sulphide and ammonia will be released and it will impart foul smell to the water.

(b) **Synthetic organic compounds**: Synthetic organic compounds are also likely to enter the aquatic system through various manmade activities, such as effluents released from production units of these compounds, spillage during transportation, and their usage in different domestic, agricultural, commercial and industrial applications. These compounds include an array of synthetic pesticides based on organochlorine or organo-phosphorous compounds, synthetic herbicides, synthetic detergents, food additives, pharmaceuticals, insecticides, paints, synthetic fibres, plastics, solvents and volatile organic compounds (VOCs) (Barrera-Diaz, 2009).

Petroleum and its derivatives, chlorinated and phenolic pesticides and industrial wastes containing synthetic dyes, polyaromatic hydrocarbons, pharmaceutical active compounds are endocrine disruptors (Semrany et al., 2012). Most of these compounds are toxic and biorefractory organics, i.e., they are resistant to microbial degradation. Even concentration of some of these in traces may make water unfit for different uses. Detergents can form foam and volatile substances may cause explosion in sewers. Polychlorinated biphenyls (PCBs) are used in the industries since 1930s, which are complex mixtures of chloro-biphenyls. Being fat soluble, PCBs move readily through the environment and within the tissues or cells. Once introduced into environment, these compounds are exceedingly persistent and their stability to chemical reagents is also high.

The pesticide dichloro-diphenyl-trichloroethane (DDT) is a prohibited insecticide in developed countries; however, it is still used in some developing or underdeveloped countries. It is an organic compound synthesized from petrochemicals and chlorine. One of the commonly found breakdown product of it in water bodies is DDE (dichloro-diphenyl-dichloroethylene) that is very persistent and fat soluble. The DDT is known to reduce photosynthesis in phytoplankton. In some of the estuaries

DDT level has been found to be around 0.01–0.2 µg/g of body weight in aquatic organisms (Myers et al., 1980). At higher levels (2–5 µg/g of body weight) it may lead to mortality in shrimp, crab and fish. The biocides Malathion, Parathion and Dipteres are nerve poisons, which are used to control insect pests on crops. These are very toxic to fish, though less than DDT and are not retained but rather slowly inactivated and excreted.

The detection of these compounds even at ng/L magnitude in water is now possible, because of improvements in analytical methods. The presence of these compounds even at very low concentration in drinking water is a very serious issue. Adverse effects of these emerging contaminants, on flora, fauna, humans and biological wastewater treatment process performance, are emphasized in literature (Semrany, 2012). In addition to their potential toxicity, these bio-refractory compounds can also cause taste and odour problems in water (Barrera-Díaz et al., 2009). Few thousands of such compounds are being used in different products and we have knowledge of adverse effects of only very few of these compounds. It is necessary to develop deeper understanding of ill effects of most of these persistent compounds and decide their acceptable threshold limit in water so as to protect aquatic life and to curb entry of these compounds in food chain at concentration beyond acceptance.

(c) *Oil*: Oil is a natural product which results from the plant remains fossilized over millions of years, under marine conditions. It is a complex mixture of hydrocarbons and it can be degradable under bacterial action, however the biodegradation rate is different for different oils and tar is considered to be one of the slowest to degrade. Oil enters in to the water bodies through oil spills, leak from oil pipes and effluents released from production units and refineries. Being lighter than water it spreads over the surface of water, breaking the contact of water with air, hence resulting in reduction of DO. This pollutant is also responsible for endangering water birds and coastal plants due to coating of oils and adversely affecting the normal activities. It also results in reduction of light transmission through surface waters, thereby reducing the photosynthetic activity of the aquatic plants. Oil includes polycyclic aromatic hydrocarbons (PAH), some of which are known to be carcinogenic (Haritash & Kaushik, 2009). Immiscible components of oil become emulsified and disperse in water, which will demand addition of the emulsion breaking agent for removal of this fraction of oil to minimize toxicity of it, if any.

5.1.2 Pathogens

The pathogenic microorganisms enter into water body through sewage discharge as a major source or through the effluents from industries like slaughterhouses. Viruses, bacteria and protozoa are of faecal origin, many of which are pathogenic, and these can cause water borne diseases, such as cholera, typhoid, dysentery, polio and infectious hepatitis in human. Pathogens, such as viruses, are much smaller than bacteria and protozoa. Even in developed countries accidental releases from sewage treatment facilities as well as urban and farm runoff contribute harmful pathogens to waterbodies. Thousands of people across the United States are sickened every year by Legionnaires' disease, a severe form of pneumonia caused from water sources like cooling towers and piped water.

Water contamination has a long presence in history of human civilization, with descriptions in the *Sushruta Samshita* about water-borne diseases, resembling to cholera, in an Indian text written in Sanskrit as early as 500–400 B.C. (Colwell, 1996; Pandey et al., 2014). Although in recent times, cholera infections have not been reported in developed countries, mainly due to improved sanitation, millions of people each year are still getting infected by *Vibrio cholera* in developing countries (Nelson et al., 2009). The World Health Organization reports about 3–5 million cholera cases and 10,000–120,000 deaths, mainly in developing countries, due to cholera every year. Over time, cholera has caused millions of deaths in developing as well as developed countries (Colwell, 1996).

Indicator organisms, such as faecal coliforms and *Escherichia coli*, are used commonly to estimate the levels of pathogens in water body, i.e., to quantify the potential water-borne pathogens footprints of water resources. For decades, public health officials/scientists are evaluating the water quality by enumerating faecal coliforms and *E. coli* levels in rivers, lakes, estuaries, and coastal waters. Infectious diseases caused by pathogens are the third leading cause of death in the United States, and it is the leading cause of death in the world (Binder et al., 1999). In India, regular monitoring of water quality in the rivers and wells in the country revealed that the total coliform counts in some places are far exceeding the desired level in water to be fit for human consumption (CPCB, 2013).

5.1.3 Nutrients

The agriculture run-off, liquid effluent from fertilizer industry and sewage contains substantial concentration of nutrients like nitrogen (N) and phosphorous (P). These wastewaters provide nutrients to the plants and may stimulate the growth of algae and other aquatic weeds in the receiving waters, which may gradually change the status of sluggish water body, like lake, from oligotrophic to a eutrophic system. Thus, the value of the water body is degraded. In long run, water body reduces DO, leads to eutrophication and ends up as a dead pool of water due to continuous deposition of dead cells of algae, microorganisms and their decomposition products. People swimming in the eutrophic waters containing blue-green algae can have skin and eye irritation, gastroenteritis and vomiting. High nitrogen levels in the water supply, causes a potential risk, especially to infants. This is when the methaemoglobin results in a decrease in the oxygen carrying capacity of the blood (blue baby disease) as nitrate ions in the blood readily oxidize ferrous ions in the haemoglobin.

In freshwater systems, eutrophication is a process whereby water bodies receive excess inorganic nutrients, especially N and P, which stimulate excessive growth of plants and algae. Under excess availability of nutrients blue green algae, diatoms like *Asteriorella* and *Fragilaria* and Cyanobacteria like *Anabaena* dominate in the water body. This excessive growth will lead to considerable reduction in light penetration due to increased turbidity of the water, which will lead to the death of microphytes because of shading effect. Eutrophication can happen naturally in the normal succession of some freshwater ecosystems. However, when the nutrient enrichment is due to the activities of humans, sometimes referred to as "cultural eutrophication", the rate of this natural process is significantly intensified. Two major nutrients, N and P, occur in streams at excessive concentrations in various forms as ions or dissolved in solution. Aquatic plants convert dissolved inorganic forms of nitrogen (nitrate, nitrite, and ammonium) and phosphorus (orthophosphate) into organic or particulate forms for use in higher trophic production. The main effects caused by eutrophication can be summarized as follows:

- Species diversity decreases and the dominant biota changes
- Increase in plant and animal biomass occurs in the water body
- Increase in turbidity of water
- Rate of sedimentation increases, shortening the lifespan of the lake, and
- Anoxic conditions may develop in the waterbody.

5.1.4   Suspended Solids and Sediments

These comprise of silt, sand and minerals eroded from land entering in the waterbody mainly through the surface runoff during rainy season and through municipal sewage discharge. This can lead to the siltation, which will reduce storage capacity of reservoirs. Presence of suspended solids can block the sunlight penetration in the water, which is required for the photosynthesis by bottom vegetation. Deposition of the solids in the quiescent stretches of the stream or ocean bottom can impair the normal aquatic life and affect the diversity of the aquatic ecosystem. If the deposited solids are organic in nature, they will undergo decomposition leading to the development of anaerobic conditions. Finer suspended solids, such as silt and coal dust, may injure the gills of fish and cause asphyxiation.

Another great concern and worries the scientists are having recently is the presence of microplastics in water bodies. A microplastic is any piece of plastic smaller than 5 mm in size. The life cycle of micro-plastics is still uncertain, however most scientific experts estimate it to be greater than 450 years. The reason for the long-life span is because plastic is a relatively new product in the environment. Hence, the bacteria have not developed enough capability to break down the molecular structure of plastic.

The impact of microplastics has been studied in a variety of shellfish as they are very much exposed to microplastics. The exposure has been found to impact their reproductive systems, even affecting their offspring. Eating these micro-plastics also leads to their off-spring 'growing up' to be smaller and less robust. Through the fish source the microplastic can enter human food chain. Presence of microplastic has been observed even in table salt. Though the exact effect of microplastics on human health has not been analysed yet completely; however, ability of microplastics to carry metals, like lead and cadmium, certainly will induce toxicity in the body. If human consume food and water contaminated with micro-plastics, the stomach can absorb chemicals that are released by these micro-plastics and transport them into blood. Hence, controlling presence of microplastic in water bodies is of greater concern to protect aquatic organisms as well as human being, since the study so far has emphasized on severe adverse effects of microplastic on fish.

5.1.5 Soluble Inorganic Pollutants

Apart from the organic matter discharged in the water body through sewage and industrial wastewaters, higher concentration of heavy metals and other inorganic dissolved pollutants also contaminate the water. Inorganic compounds containing nitrogen and phosphorous compounds can lead to the excessive autotrophic algal growth. Some of these compounds are non-biodegradable and persist in the environment. These pollutants include mineral acids, inorganic salts, trace elements, metals, metals compounds, complexes of metals with organic compounds, cyanides, sulphates, etc. Accumulation of these heavy metals can have adverse effect on aquatic flora and fauna and may constitute a public health problem, where contaminated organisms are used for food. Metals in high concentration can be toxic to biota, e.g., Hg, Cu, Cd, Pb, As, and Se. Copper greater than 0.1 mg/L is toxic to microorganisms.

Long-term exposure to cadmium may cause kidney damage, may cause skeletal damage and it has also been identified as a carcinogen (Semrany, 2012). Acute mercury exposure may be responsible for lung damage. Chronic poisoning is characterized by neurological and psychological symptoms, such as tremor, changes in personality, restlessness, anxiety, sleep disturbance, and depression. Metallic mercury may cause kidney damage. Metallic mercury is an allergen, which may cause contact eczema, and mercury from amalgam fillings may give rise to oral lichen (McParland & Warnakulasuriya, 2012). The symptoms of acute lead poisoning are headache, irritability, abdominal pain, and various symptoms related to the nervous system. Inorganic arsenic is acutely toxic, and intake of large quantities leads to gastrointestinal symptoms, severe disturbances of the cardiovascular and central nervous systems, and eventually death. In survivors, bone marrow depression, hemolysis, hepatomegaly, melanosis, polyneuropathy, and encephalopathy may be observed. Ingestion of inorganic arsenic may induce peripheral vascular disease, which in its extreme form might lead to gangrenous changes (Semrany, 2012).

5.1.6 Thermal Pollution

Thermal pollution is defined as an abrupt increase or decrease in temperature of a natural water body by human influence. The waterbody may be ocean, lake, river, or pond. This normally occurs when industries or any other activity take water from a natural resource and put it back with an altered temperature. Considerable thermal pollution results due to discharge of hot water from thermal power plants, nuclear power plants, and industries, where water is used as a coolant. As a result of hot water discharge, the temperature of water body increases. Rise in temperature reduces the DO content of the water, affecting adversely the aquatic life.

The effects of thermal pollution are diverse, however the thermal pollution damages water ecosystems and reduces/alters aquatic animal populations. Plant species, algae, bacteria, and multi-celled animals all respond differently to significant temperature changes. Organisms that cannot adapt to the changed temperature can die of various causes or can be forced out of the raised temperature area. Reproductive problems can further reduce the diversity of life in the polluted area. This alters the spectrum of organisms, which can adopt to live at that changed temperature and DO levels. This process can also wipe away streamside vegetation, which constantly depends on levels of oxygen and temperature. When organic matter is also present, the bacterial action increases due to rise in temperature; hence, resulting in rapid decrease of DO. The discharge of hot water leads to the thermal stratification in the water body, where hot water being lighter will remain on the top (Ghangrekar & Chatterjee, 2018).

5.1.7 Radioactive Pollutants

Radioactive contamination, also known as radiological contamination, is the deposition of or the presence of radioactive substances on surfaces or within solids, liquids, or gases (including the human body), where their presence is unintended or undesirable. Such contamination presents a health hazard because of the radioactive decay of the contaminants, which emit harmful ionizing radiations, such as alpha particles or beta particles, gamma rays or neutrons. The degree of hazard is determined by concentration of the contaminants, the energy and type of the radiation being emitted, and the proximity of the contamination to organs of the body (Ghangrekar & Chatterjee, 2018).

Radioactive materials originate from mining and processing of ores, use of radioactive isotopes in research, agriculture, medical and industrial activities, such as I-131, P-32, Co-60, Ca-45, S-35, C-14, etc. Accidental radioactive discharge from nuclear power plants and nuclear reactors, e.g., Sr-90, Cesium Cs-137, Plutonium Pu-248, Uranium-238, Uranium-235, and uses and testing of nuclear weapons release these radioactive isotopes in environment. The primary harm from these nuclear wastes comes from radiation, which can cause health problems in humans and other living organisms and degrade the quality of surrounding air, water, and soil. These isotopes are toxic to the life forms; they accumulate in the bones, teeth and can cause serious disorders. The safe concentration for lifetime consumption is 1×10^{-7} microcuries per ml (Barrera-Díaz et al., 2009).

Alpha (α) radiation comprises of emission of two protons and two neutrons from the nucleus; whereas beta (β) particles are electrons that are emitted from an unstable nucleus, which is a result of unprompted transformation of neutron into proton and electron. The Gamma (γ) rays have no mass or charge and are simply a form of electromagnetic radiation that travels with the speed of light. Exposure to all these forms of radiation is dangerous to living beings. The low exposure level can cause somatic and/or genetic damage. Sometimes the exposure may result in adverse effects like higher risk of cancer, leukaemia, sterility, and reduction in life span. Genetic damage by increasing the mutation rate in chromosomes and genes affects the future generations.

5.1.8 Nanoparticle Pollution

With recent advancement of nano-technology and its vast applications in various fields, these nanoparticles are also appearing in the effluents generated from its processing units or industries utilizing these for various products or liquid waste from laboratories, where research on nanoparticles is being undertaken. Nanotechnology and nanomaterial development are experiencing unprecedented expansion in this twenty-first century due to their unique physical and chemical properties, which include enhanced catalytic, antimicrobial, and oxidative properties. Nanoscale materials are being used in different applications, such as water purification, electronics, biomedical, pharmaceutical, cosmetic, energy and environmental, catalysis, material applications and many other engineering processes. Due to multifaceted immense potential of this technology, there has been a worldwide increase in nanotechnology research and development (Nowack & Bucheli, 2007).

The unique properties these materials possess at nanoscale is the main reason of nanotechnology being getting so much attention. Materials that may seem inert at millimetre or micron scale may have very different properties at nanoscale. Below the 100 nm size threshold, both the surface area-to-mass ratio and the proportion of the total number of atoms at the surface of a structure are large enough that surface properties become important. This leads to an altered chemical reactivity due to substantially altered catalytic properties as well as different thermal and electrical conductivities and tensile strength. At nano-scale, quantum effects may begin to apply themselves, changing optical, electrical, and magnetic behaviours of the material (O'Brien & Cummins, 2008). Another peculiarity of nanoparticles is that their properties can be significantly altered by changing the surface characteristics, e.g., engineering surfaces of nanoparticles with proteins and polymers, adding/subtracting atoms to alter the band-gap (range of electron energy levels) of a quantum dot or coating the surface of particles to reduce agglomeration.

Nanoparticles are having many application areas depending on their surface characteristics, e.g., reacting to certain wavelengths in the case of semiconductor quantum dots for use as sensors or specific area targeting for drug delivery. As the use of nano-functionalized products increases, the potential for environmental exposure and contamination is also increasing (O'Brien & Cummins, 2011). Nanoparticle pollution is also called "invisible pollution" and considered to be the most difficult pollution being managed and controlled. Long-term exposure to nanoparticles may be expected to cause serious damage to the human's respiratory tract, lung diseases, heart diseases, and premature death (Gao et al., 2015).

Still very little is understood about the adverse effect of these nano-particles once released in aquatic environment and effect on animal and human health. Developing this knowledge about ill effects is utmost essential for protection of health and aquatic ecosystem. There are no common principles for protection of the personnel working in the waste industry on how to handle various forms of nano-wastes. Currently, the quantities or concentrations of nanomaterials in waste streams or in the environment remain unmonitored (Klaine et al., 2008). Apparently, it is assumed that the current quantities are low; however, with rapid introduction of new nanoproducts into the market, and discovery of new nanomaterials of unknown impacts to the environment, this scenario is likely to change dramatically in near future.

There is a paucity of toxicity data and its relationship to the physico-chemical properties of nanomaterials. The complexity of managing nano-wastes is also due to the property of dynamic transformation of the same. It is highly essential to develop universal principles and technologies of managing these wastes urgently. A case-by-case approach being adopted presently may prove uneconomical, laborious, and even impractical considering the number and types of nanomaterials with their different sizes and properties being released in environment (Ghangrekar & Chatterjee, 2018). Further efforts are highly necessary to set standards for nanomaterials detection, characterization, and treatment. Even the ill-effects and the dynamics of characteristics of nanomaterials are not very well understood, demanding further systematic approach to study adverse effects of nanomaterials, which are getting released in the environment.

5.1.9 Emerging Contaminants

Emerging contaminants (ECs) are those chemical compounds about which we have very limited knowledge of their ill effects and no published health standards are available. However, these ECs are characterized or recognized by verifiable threats to the environment or human health. Varieties of ECs mainly of anthropogenic origin are identified in the recent times and these have captured a attention of scientific community due to their harmful impacts on both environment as well as on human health even at very low concentrations (Lei et al., 2015). These are relatively recently detected pollutants, and hence, a gap exist between the available knowledge and their fate, effect, behaviour and identification as well as treatment or removal technologies (Gogoi et al., 2018).

Due to the rapid growth of urbanization, transportation, industry and agriculture, release of these chemicals is increasing, which can have hazardous harmful effects on the environment (Rosenfeld & Feng, 2011). Hence, these contaminants are of emerging concern. The ECs may be of different origins, such as industrial, agricultural, laboratory, hospital wastewater, municipal (domestic), man-made nanomaterial, gasoline additives, etc., which are generated from the products, services and lifestyle prevalent in the modern society (Kroon et al., 2020). These ECs can be categorized in four broad groups, namely pharmaceutical and personal care products (PPCPs), pesticides and herbicides, dyes, and industrial chemicals (Table 5.1). The presence of ECs in the environment is alarming owing to their capability of eco-toxicity. The route through which these ECs find their way in the environment is majorly via the effluent discharged from industry, domestic wastewater and agricultural runoff.

The conventional wastewater treatment plants are not capable of removing ECs; hence, cannot prevent the release of ECs into water bodies when secondary treated effluents are released. Technologies best suitable for removal of these ECs and quantification of harmful impacts of these chemical compounds on aquatic life and human health are not yet concisely documented (da Silva Vilar et al., 2021). Advanced oxidation processes are being used for removal of some of these ECs; however, standardization of design for effective removal is yet to be achieved. Reverse osmosis (RO), nano-filtration (NF), and adsorption can also be used for removal of ECs; however, there are limitations associated with these methods in terms of high capital and operating costs.

5.2   Quantification of Major Pollutants

5.2.1   Collection and Preservation of Samples

Before doing analysis of the water or wastewater for a particular parameter quantification, it is necessary to ensure that the sample collected is representative and appropriate protocol is followed for collection and preservation of the sample, so as to make analysis results true representative. In general, the objective of sampling is to collect a small volume of water/wastewater sample that is enough in volume to be conveniently transported to and handled in the laboratory. This small volume sample should be accurately collected taking all necessary precautions, so as to represent the larger volume of water/ wastewater being sampled. The relative proportions or concentrations of all pertinent components must be same in the sample as that in the main water/wastewater being sampled. In addition, the samples must be handled with care so as no significant changes in composition occur before the tests are being performed. It is not possible to generalize the terms of detailed

Table 5.1 Classification and sources of ECs

Types of ECs		Sub-types of ECs	Representative ECs	Major sources[#]
PPCPs	Personal care products	Sunscreens	2-Phenylbenzimidazole -5-sulphonic acid (PBSA), Benzophenone-4 (BP-4), 4-methylbenzylidene camphor	Cleaning effluent from all sector, personal care products industrial effluent, domestic wastewater
		Fragrances (musk compounds)	Tonalide, Galaxolide, Musk xylene, Musk ketone	
		Cleaning products	Soap, Handwash, Detergents, Phenyl	
		Beauty products	Deodorants, Lotions, Facial cream, Talcum powder	
	Pharmaceuticals	Nonsteroidal anti-inflammatory drugs (NSAID)	Ibuprofen, Naproxen, Diclofenac, Indomethacin, Celecoxib, Mefenamic acid	Domestic wastewater effluent, pharmaceutical industrial effluent, hospital effluent, aqua-culture effluent and livestock farms effluent
		Beta-blockers	Metoprolol succinate, Acebutolol, Metoprolol tartrate	
		Antibiotics,	Penicillin, Azithromycin, Tetracylines	
		Antidepressant	Citalopram, Fluvoxamine, Paroxetine,	
		Hormones	Anastrozole, Premarin	
		Lipid regulators	Ezetimibe, Colesevelam, Implitapide, Niacin	
		Anticonvulsants	Acetazolamide, Carbamazepine	
Dyes		Natural	Jack fruits, Hina, Turmeric, Onion	Domestic effluent, textile industrial effluent, laboratory effluent
		Artificial	Fast green, Oil red O, Orange G, Picric acid, Eosin Y, Safranin	

(continued)

Table 5.1 (continued)

Types of ECs	Sub-types of ECs	Representative ECs	Major sources[#]
Industrial chemicals	Plasticizer	Di-Ethylhexyl Phthalate (DEHP), Di-Isononyl Phthalate (DINP), Bisphenol A (BPA)	Industrial, domestic and laboratory wastewater effluent
	Flame retarder	Mineral wool, Asbestos cement, Gypsum boards	
Pesticides	Molluscicides	Metaldehyde, Metal salts, Methiocarb, Acetylcholinesterase inhibitors	Agricultural runoff, aquaculture effluent, domestic wastewater, livestock farms effluent
	Fungicide	Calcium polysulfide, Cyproconazole, Carbanilate fungicides	
	Insecticides	Aldrin, Chlordane, pyrethroids, Chlordecone, DDT	

[#] (Ghangrekar et al., 2021; Gita et al., 2017; Mahmood et al., 2016; Rout et al., 2021)

procedures for collection of samples for all analysis, because of the varied purposes and procedures of the tests for which the sample is being collected.

5.2.1.1 General Precautions

Care should be taken to collect truly representative sample of existing conditions and to handle sample in such a way that it does not deteriorate or become contaminated before it reaches to laboratory for analysis. It is advised to rinse the sample container with the same water being sampled for 2.0 to three times before filling the container. For some sources grab sample may be representative, however for some source representative sample can only be obtained by making composites of samples, which have been collected over a period of time or at many different sampling points at the same time. Sometimes, it will be more informative to analyse multiple samples separately collected at different time instead of one composite sample, so as to understand variation in that particular parameter with time.

Make a record of every sample collected, and identify every bottle, preferably by attaching a tag or label. The record should have sufficient information to provide positive identification of the sample at later time, as well as name of the person who collected the sample, date and time of sampling, exact location, water temperature, and any other data that may be appropriate for the purpose of sample correlation, such as weather condition, water level, stream flow, location, etc.

Before samples are collected from the distribution systems, flush the lines sufficiently to ensure that the sample is representative of the supply, taking into account the diameter and length of the pipe to be flushed and the velocity of flow. Collect samples from wells only after the well has been pumped sufficiently to ensure that the sample represents the groundwater that feeds the well. Sometimes it will be necessary to pump at a specified rate to achieve a characteristic drawdown, if this determines the zones from which the well is supplied. When samples are collected from a river or stream, the water characteristics may vary with depth, stream flow, and the distance from shore and from one bank to other. An integrated sample shall be taken in this case from top to bottom in the middle of the stream in such a way that the sample is made composite according to the flow. If only grab or catch sample is to be collected, it is best to take it in the middle of the stream and at mid depth. Lakes and reservoirs are subject to considerable variations from normal causes, such as seasonal stratification, rainfall, runoff, and wind. The choice of location, depth, and frequency of sampling will depend on local conditions and the purpose of the sampling. In general, the sampling procedure should take account of both of the tests or analyses to be performed and the purpose for which the results are being generated.

5.2.1.2   Types of Samples

Grab or Catch Samples: This sample is collected at a particular time and place and it can represent only the composition of the source at that time and place of sampling. When source water/ wastewater quality is known to be fairly constant in composition over the considerable period of time or over substantial distance in all directions, under such circumstances the grab sample may be considered as a representative. The examples of sampling from such sources could be water supply, some surface waters, final effluent from wastewater treatment plant, and rarely influent wastewater to the treatment plant.

When variation in characteristics and flow at source is expected with time, grab samples collected at suitable time intervals and analysed separately can be of great value in documenting the extent of these variations. The sampling interval can be chosen based on the frequency with which the changes are expected. When composition of the source varies with place rather than time, the samples collected from appropriate locations, with less emphasis on timing, may provide the most meaningful information. Maximum care must be taken while collecting sludge sample to collect representative sample, though there is no definite procedure/guideline for collection of sludge sample.

Composite Samples: The composite sample is the mixture of grab samples collected at the same sampling point at different time in a day. It is also sometimes referred as time composite, when it is necessary to distinguish this type of samples from others. These time composite samples are most useful for observing average concentrations that are to be used, e.g., in estimating the organic loading or the efficiency of a wastewater treatment unit. Composite samples can save the laboratory resources, efforts and expenses by analysing single representative sample rather than analysing number of samples and then averaging the results. For this purpose, a composite sample representing 24-h period is considered standard for most parameter determinations. Under certain circumstances, a composite sample representing one shift or a complete cycle of a process operation could be preferable.

For determination of particular water/wastewater characteristic, which is subjected to the significant and unavoidable changes on storage, composite samples cannot be used, e.g., analyses for dissolved gases, residual chlorine, temperature, and pH. In such case, it is necessary to perform such analysis on individual samples as soon as possible after collection and preferably at sampling point. Time composite samples should be used only for determination of characteristics that cannot change under existing conditions of sample collection and preservation.

For time composite sampling, collect individual portion of samples in a wide mouth bottle of minimum 120 mL capacity. Collect these portions at pre-decided time interval and mix at the end of the sampling period or combine in a single bottle as collected. If preservatives are to be used, add them initially to the sample bottle, so that all portions of the composite are preserved as soon as collected. It is desirable, and often absolutely essential, to combine the individual samples in volume proportional to the flow. A final volume of 2–3 L is sufficient for sewage and industrial wastewaters.

Integrated Samples: The water quality information required may sometimes be represented best by analysing mixture of grab samples collected from different points at the same time. Such mixtures are called as integrated samples. For example, while sampling water from river or stream that varies in composition across its width and depth, integrated samples are representative. In such case a mixture of samples representing various points in the cross-section, in proportion to their relative flows, may be useful for evaluating average composition. The integrated samples are also useful for deciding combined treatment of various separate wastewater streams generated from an industry.

Under certain situations, local variations in the values of a particular parameter are more important and the samples are required to examine separately rather than integrated samples. For example, in case of lake, while monitoring a parameter like DO, where characteristics of water varies significantly with both depth and horizontal location, neither total nor average figures are especially significant. The preparation of integrated samples usually requires special equipment to collect a sample from a known depth, without contamination by the overlying water. Prior knowledge about the volume, movement, and composition of the various parts of the water to be sampled is necessary.

5.2.1.3 Quantity of Sample

In general, 2–3 L sample volume is sufficient for general chemical analysis. For certain special determinations, larger samples may be necessary. Do not use the same sample for chemical, bacteriological, and microscopic examinations, since the methods of collection, handling and preservation are different. Hence, separate samples should be collected for that specific analysis following the suitable procedure for collection.

5.2.1.4 Preservation of Samples

Preservation techniques can bring the chemical and/or biological changes that can inevitably continue after the sample is collected from the source. However, certain changes may occur in the chemical structure of the constituents present in water sample that are function of physical conditions. Some parameters to be determined may get affected due to storage of samples with or without reduction in temperature. Some cations are subjected to be lost by adsorption on, or ion exchange with, the walls of glass containers. These include aluminium, cadmium, chromium, copper, iron, lead, manganese, silver, and zinc, which are best collected in separate clean bottle and acidified with concentrated hydrochloric or nitric acid to a pH below 2.0 in order to minimize precipitation and adsorption on the walls of the container.

The characteristics, such as, temperature, pH, concentration of dissolved gases may change even during very short duration of storage. Hence, make the determination of temperature, pH and dissolved gases in the field itself while sample is collected. With changes in the pH-alkalinity-carbon dioxide balance, calcium carbonate may precipitate and cause a decrease in the values for calcium and for total hardness. Microbiological activity may be responsible for bringing changes in the nitrate-nitrite-ammonia content, for decrease in phenols and in BOD, or for the reduction of sulphate to sulphide. Residual chlorine may be reduced to chloride. Sulphide, sulphite, ferrous iron, iodide, and cyanide may be lost through oxidation. Colour, odour, and turbidity may increase or decrease. Hence, it is impossible to prescribe absolute procedure for the prevention of all possible characteristics of the sample. To a large extent, the dependability of water analyses must rest on the experience and judgment of the analyst.

5.2.1.5 Time Interval Between Wastewater Sample Collection and Analysis

In general, the shorter the time that elapses between collection of samples and analyses, the more reliable will be the analytical results. For certain parameters and physical values, immediate analysis in the field is necessary to obtain reliable analysis, because the sample composition may change before it reaches to the laboratory. The time that may be allowed to elapse between collection of samples and analyses depends on the characteristics of the sample, the particular analysis that is intended to be carried out, and condition of storage. Changes caused by growth of organisms are greatly retarded by keeping the sample in the dark and at low temperature until the analysis.

5.2.1.6 Wastewater Sample Preservation Methods

Preservation of samples is difficult because almost all preservatives interfere with some of the other analytical tests. Analysis of the wastewater sample immediately after collection is always advisable. Sample storage at low temperature (4 °C) is perhaps the best way to preserve most of the qualities of the samples until the next day. Use of chemical preservatives is only encouraged when such chemicals will not interfere with the analysis being made. The chemical preservative should be added in the sample bottle initially so that all portions of the sample will be preserved as soon as collected. No single method of preservation is entirely satisfactory for all parameters, hence choose the preservative with due regard to the analysis to be made.

Sample preservation methods are relatively limited and are intended generally to retard biological action, retard hydrolysis of chemical compounds and complexes, and reduce volatility of constituents. The preservation methods like pH control, chemical addition, refrigeration, and freezing are generally adopted. For determination of parameters, such as acidity and alkalinity, the sample can be stored up to 24 h in refrigerator. The BOD determination should be carried out preferably within 6 h of sample collection and sample should be preserved in refrigerator for that duration. For COD determination, the sample should be analysed as soon as possible. In case the sample required storage for few hours, add concentrated sulfuric acid to bring pH below 2.0. Turbidity determination should be done on the same day and sample can be stored in dark up to 24 h. The dissolved oxygen fixing should be carried out at site as per the procedure described for determination of DO. Analysis of temperature and pH should be carried out immediately after the sample collection.

5.2.1.7 Quality Control in Chemical Analysis

It is important to maintain stipulated quality of analysis in laboratory so that the results obtained will represent the true quality of the water being analysed and the results are reproducible. This is achieved by employing trained and experienced manpower for analysis, using good physical and equipment facility, properly calibrating the instruments, and using certified reagents and standards. Internal quality control programme should be carried out by routine analysis of control sample at least once in a day on which unknown samples are being analysed. In addition, confirm the ability of laboratory to produce acceptable results by analyses of few reference samples from external agency once or twice a year.

5.2.1.8 Expression of Results

Analytical results are usually expressed in milligram per litre (mg/L). The term parts per million (ppm), which is weight to weight ratio, is equivalent to mg/L considering one litre of water weighs 1 kg. For concentration lower than 1 mg/L, it is convenient to express the results in terms of micrograms per litre (μg/L). The term 'percent' is preferred when the concentration is greater than 10,000 mg/L, one percent being equivalent to 10 g/L. Results can also be expressed in milliequivalents per litre (meq/L). It is obtained by dividing mg/L of the element or ion by its equivalent weight in grams. There are various other units to express the analytical results depending upon the suitability and application. For analysis, such as turbidity and colour, reference is made to some arbitrary standards and the results are expressed in comparison to that standard.

5.2.2 Measurement of pH

Measurement of pH is one of the most important tests used in water chemistry. It is a measure of the acid–base equilibrium achieved by various dissolved compounds. In most natural waters, it is controlled by the carbon dioxide–bicarbonate–carbonate equilibrium system. In every phase of water and wastewater treatment processes, i.e., in acid base neutralization, water softening, precipitation, coagulation, disinfection and corrosion control, the pH has an independent role. At a given temperature, the intensity of the acidic or basic character of a solution is indicated by pH or hydrogen ion activity. By definition it is the negative logarithm of the hydrogen ion activity (H^+) as expressed by Eq. 5.1.

$$pH = -\log_{10}\left(\text{activity of } H^+\right) \tag{5.1}$$

In dilute solution, the hydrogen ion activity is approximately equal to the concentration of hydrogen ion. Pure water is very slightly ionized and at equilibrium the ionic product $Kw = [H^+][OH^-] = 1.01 \times 10^{-14}$ at 25 °C and $[H^+] = [OH^-] = 1.005 \times 10^{-7}$. Thus, a logarithmic form is given by Eq. 5.2.

$$\left(-\log_{10}[H^+]\right) + \left(-\log_{10}[OH^-]\right) = \log[Kw] \tag{5.2}$$

Or,

$$pH + pOH = pKw \tag{5.3}$$

Form the above equilibrium it is clear that the pH for an aqueous solution lies between 0 and 14. The pH of most of raw water sources lies within the range of 6.5–8.5. The slightly basic pH is due to the presence of bicarbonates and carbonates of the alkali and alkaline earth salts. Since, the pH is defined operationally on a potentiometric scale, the measuring instrument is also calibrated potentiometrically with an indicating (glass) electrode and a reference electrode using standards buffers having assigned values. The reference electrode consists of a half-cell that provides a standard electrode potential. Generally, calomel or silver-chloride electrodes are used as a reference electrode. Several types of glass electrodes are available. The glass electrode consists essentially of a very thick-walled glass bulb, made of low melting point glass of high electrical conductivity, blown at the end of a glass tube. This bulb contains an electrode which has a constant potential, e.g., a platinum wire inserted in a solution of hydrochloric acid saturated with quin-hydron. The bulb is placed in the liquid for which the pH is to be determined.

It is necessary to calibrate the electrode system against standard buffer solution of known pH. Buffer tablets having pH of 4.0, 7.0 and 9.2 are available. Because buffer solution may deteriorate in quality with time as a result of contamination, prepare it fresh for calibration work. Before use, remove electrodes from storage solutions as recommended by the manufacturers and rinse it with distilled water. Dry electrode by gently blotting with a soft tissue paper. Calibrate the instrument with electrodes immersed in a buffer solution within 2 pH units of sample pH. Remove electrodes from buffer, rinse thoroughly and blot dry. Immerse in a second buffer below pH of 10.0, approximately 3 pH units different from the first, the reading should be within 0.1 units for the pH of the second buffer. For sample analysis establish equilibrium between electrodes and sample by stirring sample to ensure homogeneity and measure the pH.

5.2.3   Measurement of Acidity and Alkalinity

5.2.3.1   Measurement of Acidity

Acidity of water is its quantitative capacity to react with a strong base to a designated pH. The measured value may vary significantly with the end-point pH used in the determination. Acidity is a measure of an aggregate property of water and can be interpreted in terms of specific substances only when the chemical composition of the sample is known. Strong mineral acids, weak acids such as carbonic and acetic, and hydrolysing salts, such as iron or aluminium sulphates may contribute to the measured acidity according to the method of determination. Acids contribute to corrosiveness and influence chemical reaction rates, chemical speciation, and biological processes. The measurement also reflects a change in the quality of the source water.

In the titration of a single acidic species, as in the standardization of reagents, the most accurate end point is obtained from the inflection point of a titration curve. The inflection point is the pH at which curvature changes from convex to concave or vice versa. Because accurate identification of inflection points may be difficult or impossible in buffered or complex mixtures, the titration in such cases is carried to an arbitrary end-point pH based on practical considerations. For routine control of titrations or rapid preliminary estimates of acidity, the colour change of an indicator may be used for the end point. Samples of industrial wastes, acid mine drainage, or other solutions that contain appreciable amounts of hydrolysable metal ions, such as iron, aluminium, or manganese, are treated with hydrogen peroxide to ensure oxidation of any reduced forms of polyvalent cations, and boiled to hasten hydrolysis. Acidity results may be highly variable if this procedure is not followed exactly.

Ideally the end point of the acidity titration should correspond to the stoichiometric equivalence point for neutralization of acids present. The pH at the equivalence point will depend on the sample, the choice among multiple inflection points, and the intended use of the data. Dissolved carbon dioxide (CO_2) usually is the major acidic component of unpolluted surface waters. Handle samples from such sources carefully to minimize the loss of dissolved gases. In a sample containing only carbon dioxide-bicarbonates-carbonates, titration to pH of 8.3 at 25 °C corresponds to stoichiometric neutralization of carbonic acid to bicarbonate. Because the colour change of phenolphthalein indicator is close to pH of 8.3, this value generally is accepted as a standard end point for titration of total acidity, including CO_2 and most weak acids. Metacresol purple also has an end point at pH of 8.3 and gives a sharper colour change.

For more complex mixtures or buffered solutions, selection of an inflection point may be subjective. Consequently, use fixed end points pH of 3.7 and pH of 8.3 for standard acidity determinations via a potentiometric titration in wastewaters and natural waters, where the simple carbonate equilibrium discussed above cannot be assumed. Bromophenol blue has a sharp colour change at its end point of 3.7. The resulting titrations are identified, traditionally, as "methyl orange acidity" (pH of 3.7) and "phenolphthalein" or total acidity (at pH of 8.3) regardless of the actual method of measurement.

Interferences: Dissolved gases contributing to acidity or alkalinity, such as CO_2, hydrogen sulphide, or ammonia, may be lost or gained during sampling, storage, or titration. Minimize such effects by titrating to the end point promptly after opening sample container, avoiding vigorous shaking or mixing, protecting sample from the atmosphere during titration, and letting sample become no warmer than it was at collection.

In the potentiometric titration, oily matter, suspended solids, precipitates, or other waste matters may coat the glass electrode and cause a sluggish response. Difficulty from this source is likely to be revealed in an erratic titration curve. Do not remove interferences from sample because they may contribute to its acidity. Briefly pause between titrant additions to let electrode come to equilibrium or clean the electrode occasionally.

In samples containing oxidizable or hydrolysable ions, such as ferrous or ferric iron, aluminium, and manganese, the reaction rates at room temperature may be slow enough to cause drifting end points. Do not use indicator titration with coloured or turbid sample that may obscure the colour change at the end point. Residual free available chlorine in the sample may bleach the indicator. Eliminate this source of interference by adding one drop of 0.1 M sodium thiosulfate ($Na_2 S_2 O_3$).

Selection of procedure: Determine sample acidity from the volume of standard alkali required to titrate a portion to a pH of 8.3 (phenolphthalein acidity) or pH 3.7 (methyl orange acidity of wastewaters and grossly polluted waters). Titrate at room temperature using a properly calibrated pH meter, electrically operated titrator, or colour indicators.

Sample size: The range of acidities found in wastewaters is so large that a single sample size and normality of base used as titrant cannot be specified. Use a sufficiently large volume of titrant (20 mL or more from a 50-mL burette) to obtain relatively good volumetric precision while keeping sample volume sufficiently small to permit sharp end points. For samples having acidities less than about 1000 mg/L as calcium carbonate ($CaCO_3$), select a volume with less than 50 mg $CaCO_3$ equivalent acidity and titrate with 0.02 N sodium hydroxide (NaOH). For acidities greater than about 1000 mg as $CaCO_3$/L, use a portion containing acidity equivalent to less than 250 mg $CaCO_3$ and titrate with 0.1 N NaOH. If necessary, make a preliminary titration to determine optimum sample size and/or normality of titrant. After titration the acidity of the water/wastewater sample can be expressed as per Eq. 5.4.

$$\text{Acidity, as mg } CaCO_3/L = \frac{(A \times B) - (C \times D) \times 50000}{mL \; of \; sample} \qquad (5.4)$$

where, A is mL of NaOH titrant used; B is the normality of NaOH; C is the mL of H_2SO_4; and D is the normality of H_2SO_4, if acid is added for preserving samples.

5.2.3.2 Measurement of Alkalinity

Alkalinity is acid- neutralizing capacity of water. The measured value may vary significantly with the end-point pH used. Alkalinity is a measure of an aggregate of all the titratable bases present in water and can be interpreted in terms of specific substances only when the chemical composition of the sample is known. Alkalinity is significant in many uses and treatment of natural water and wastewaters. The alkalinity of surface water is primarily a function of carbonate, bicarbonate, and hydroxide content, and it is taken as an indication of the concentration of these constituents. The measured values also may include contributions from borates, phosphates, silicates, or other bases, if these are present. Alkalinity in excess of alkaline earth metal concentrations is significant in determining the suitability of water for irrigation. Alkalinity measurement is used in the interpretation and control of water and wastewater treatment processes. Raw domestic sewage has an alkalinity more or less similar to that of the water supply.

When alkalinity is due to entirely carbonate or bicarbonate, the pH at the equivalence point of the titration is determined by the concentration of CO_2 at that stage. The CO_2 concentration depends, in turn, on the total carbonate species originally present and any losses that might have occurred during titration. Traditionally alkalinity is measured by titration to pH of 8.3 irrespective of the coloured indicator, if any, used in the determination. The sharp end-point colour changes produced by Metacresol purple (pH of 8.3) and Bromocresol green or Methyl Orange (pH of 4.5) make these indicators suitable for the alkalinity titration.

Interferences: Soaps, oily matter, suspended solids, or precipitates may coat the glass electrode and cause a sluggish response. Allow additional time between titrant additions to let electrode come to equilibrium or clean the electrodes occasionally. Do not filter, dilute, concentrate, or alter sample.

Selection of procedure: Determine sample alkalinity from volume of standard acid required to titrate a portion to a designated pH at room temperature. Report alkalinity less than 20 mg $CaCO_3/L$ only if it has been determined by the low-alkalinity method. Colour indicators may be used for routine and control titrations in the absence of interfering colour and turbidity and for preliminary titrations to select sample size and strength of titrant. Appropriate sample size of 25–50 mL for titration is generally adequate. For low alkalinity method, titrate a 200-mL sample with 0.02 N H_2SO_4 from a 10-mL burette. Standard sulfuric acid or hydrochloric acid, 0.02 N, is used for titration. The results obtained from the phenolphthalein (pH of 8.3) and total alkalinity (pH of 4.5) determinations offer a means for stoichiometric classification of the three principal forms of alkalinity present in water sample. The classification attributes the entire alkalinity to bicarbonate, carbonate, and hydroxide, and assumes the absence of other (weak) inorganic or organic acids, such as silicic, phosphoric, and boric acids.

To the selected volume of sample add Phenolphthalein or Metacresol purple indicator and titrate till pH of 8.3 to measure phenolphthalein alkalinity. Add methyl orange or Bromocresol green indicator and titrate till pH of 4.5 to measure total alkalinity. As the end point is approached make smaller additions of acid and make sure that pH equilibrium is reached before adding more titrant. The alkalinity can be estimated as per the Eq. 5.5.

$$\text{Alkalinity, mg CaCO}_3/\text{L} = \frac{A \times N \times 50000}{mL\ of\ sample} \tag{5.5}$$

where, A is mL standard acid used and N is the normality of standard acid.

5.2.4 Measurement of Solids Present in Wastewaters

Solids refer to the suspended or dissolved matters present in water or wastewater. Presence of solids may affect water quality adversely in a multiple way. Water with high dissolved solids generally is consider inferior for potability, since it may induce an unfavourable physiological reaction in the consumers. For this reason, a limit of 500 mg dissolved solids/L is desirable for drinking waters. Highly mineralized water is also unsuitable for many industrial applications. Water having high concentration of suspended solids (SS) may be aesthetically unsatisfactory for purposes, such as bathing and recreation. Solids analyses are important in the control of biological and physical wastewater treatment processes and for assessing compliance with wastewater effluent standards set by regulatory agency.

"Total solids" (TS) is the term used to quantify the residue left in the crucible after evaporation of a water/wastewater sample and its subsequent drying in an oven at a defined temperature. Total solids include total suspended solids (TSS), the portion of total solids retained by a filter, and total dissolved solids (TDS), the portion of solids that passes through the filter. The type of filter holder, the pore size, porosity, area, and thickness of the filter and the physical nature, particle size, and amount of material deposited on the filter are the principal factors affecting separation of suspended solids from dissolved solids.

Fixed solids are defined as the residue of total, suspended, or dissolved solids after ignition for a specified time at a specified temperature. The weight loss on ignition is called volatile solids (VS). Determinations of fixed and volatile solids do not distinguish precisely between inorganic and organic matter because the loss on ignition is not confined to be associated with only organic matter. It includes losses due to decomposition or volatilization of some mineral salts. Better characterization of organic matter can be made by such tests as total organic carbon, BOD, and COD. Settable solids are the class of solids settling out of suspension within a defined period.

The temperature at which the water/wastewater samples are dried has an important bearing on result. Residues dried at 103–105 °C will retain water of crystallization and also some mechanically occluded water. Loss of CO_2 will result in conversion of bicarbonate to carbonate. Loss of organic matter by volatilization will be very negligible at this temperature. Since, removal of occluded water is marginal at this temperature, attainment of constant weight may be very slow while drying at 103–105 °C.

Solids (or wastewater containing solids) dried at 180 ± 2 °C will lose almost all mechanically occluded water. Some water of crystallization may remain, especially if sulphates are present. Partial loss of organic matter might occur due to volatilization at this elevated temperature of drying. Loss of CO_2 will result from conversion of bicarbonates to carbonates. Some chloride and nitrate salts may also be lost. In general, evaporating and drying water samples at 180 °C yields values for dissolved solids closer to those obtained through summation of individually determined mineral species than the dissolved solids values secured through drying at the lower temperature. Results for solids (or wastewaters) having high oil or grease may be questionable because of the difficulty of drying to constant weight in a practical time. Some special purpose analysis may demand deviation from the stated procedures to include an unusual constituent with the measured solids.

5.2.4.1 Determination of Total Solids Dried at 103–105 °C

A well-mixed wastewater sample is poured and evaporated in a pre-weighed empty dish and dried to constant weight in an oven at 103–105 °C. The increase in weight over that of the empty dish represents the total solids. The results may not represent the weight of actual dissolved solids and suspended solids in wastewater samples. Highly mineralized water with a significant concentration of calcium, magnesium, chloride, and/or sulphate may be hygroscopic and require prolonged drying, proper desiccation, and rapid weighing. Disperse visible floating oil and grease with a blender before withdrawing a sample portion for analysis. Limit sample volume so as no more than 200 mg residue will remain on dish, since excessive residue on the dish may form a water-trapping crust.

If volatile solids are to be measured ignite clean evaporating dish/crucible at 550 ± 50 °C for 1 h in a muffle furnace and take empty weight of the dish/crucible. If only total solids need to be measured then heat clean dish at 103–105 °C for 1 h. Choose a sample volume that will yield a residue between 2.5 and 200 mg. Transfer a measured volume of well-mixed sample to pre-weighed dish and evaporate to dryness on a steam bath or in a drying oven. Dry evaporated sample for at least 1 h in an oven at 103 to 105 °C, cool dish in desiccator to balance temperature, and weigh. Repeat cycle of drying, cooling, desiccating, and weighing until a constant weight is obtained. The total solids present in that wastewater sample is expressed as per the Eq. 5.6.

$$\text{Total solids, mg/L} = \frac{(A - B) \times 1000}{Sample\ volume,\ mL} \qquad (5.6)$$

where, A is the weight of dried residue + dish, mg; and B is the weight of empty dish, mg.

5.2.4.2   Determination of Total Dissolved Solids Dried at 180 °C

A well-mixed water/wastewater sample is filtered through a standard glass fibre filter, and the filtrate is evaporated to complete dryness in a weighed dish and dried to constant weight at 180 °C. The increase in dish weight represents the total dissolved solids present in the water/wastewater sample. The results may not agree with the theoretical value for solids calculated from chemical analysis of sample. The filtrate from the total suspended solids determination may be used for determination of total dissolved solids. Water samples containing high concentration of calcium, magnesium, chloride, and/or sulphate content may be hygroscopic and require prolonged drying period, proper desiccation, and rapid weighing. Samples having high bicarbonate require careful and possibly prolonged drying at 180 °C to ensure complete conversion of bicarbonate to carbonate.

If volatile solids are also to be measured in that case ignite the cleaned evaporating dish at 550 ± 50 °C for 1 h in a muffle furnace. However, if only TDS are to be measured, heat the clean dish at 180 ± 2 °C for 1 h in an oven. Store dish in desiccator until needed. Weigh immediately before use. Choose sample volume so as to yield between 2.5 and 200 mg of dried residue. Filter measured volume of well mixed water sample through glass-fibre filter disk, 22–125 mm diameter, ≤ 2 μm pore size and wash with three successive 10 mL volume of distilled water, allowing complete drainage between washings, and continue suction for about 3 min after filtration is complete. Transfer filtrate to a weighed evaporating dish and evaporate to dryness in an oven. Dry for at least 1 h in an oven at 180 ± 2 °C. Cool in a desiccator to bring to room temperature and weigh. Repeat the cycle of drying, cooling, desiccating, and weighing until a constant weight is obtained. The TDS present in the sample (mg/L) can be estimated using same equation as Eq. 5.6.

5.2.4.3   Determination of Total Suspended Solids Dried at 103–105 °C

A well-mixed sample is filtered through a pre-weighed glass-fibre filter paper, 22–125 mm diameter, ≤ 2 μm pore size and the residue retained on the filter is dried to a constant weight at 103–105 °C. The increment in weight of the filter represents the total suspended solids retained on it. Exclude large floating particles or any agglomerates of nonhomogeneous materials present in the wastewater sample, unless their inclusion is must in accuracy of the final result. For samples with high dissolved solids thoroughly wash the filter to ensure removal of the dissolved material. Prolonged filtration times resulting from filter clogging may produce high result owing to excessive solids capture on the clogged filter.

Filter a measured volume of well mixed sample through the pre-weighed glass fibre filter. Wash with three successive 10 mL volumes of distilled water, allowing complete drainage between washings and continue suction for about 3 min after filtration is complete. Carefully remove filter from filtration apparatus and place it in pre-weighed crucible. Dry the crucible with filter for at least 1 h at 103–105 °C in an oven, cool in a desiccator to room temperature and weigh. Repeat the cycle of drying, cooling, desiccating, and weighing until a constant weight is obtained. The SS present in the water sample shall be estimated using Eq. 5.7.

$$\text{Total suspended solids, mg/L} = \frac{(A - B) \times 1000}{Sample\ volume,\ mL} \tag{5.7}$$

where, A is the weight of the crucible + filter paper + dried residue, mg; and B is the empty weight of crucible + dry filter paper before filtration, mg.

5.2.4.4 Determination of Fixed and Volatile Solids Ignited at 550 °C

The residue from total solids, total dissolved solids or total suspended solids determination after drying at respective lower temperature for these solids measurement, is ignited to constant weight at 550 ± 50 °C. The remaining solids after ignition represent the fixed total, fixed dissolved, or fixed suspended solids, respectively, while the weight lost upon ignition is the volatile solids. Volatile suspended solids determination is useful in control of wastewater treatment plant operation as it offers an approximation of the amount of bacterial biomass estimation in the activated sludge. Errors in the volatile solids determination may occur by loss of volatile matter during drying. Determination of low concentrations of volatile solids in the presence of high fixed solids concentrations may be subjected to considerable error. In such cases, measure the suspected volatile components by total organic carbon.

Ignite residue produced during total solids, total dissolved solids or total suspended solids determination to constant weight in a muffle furnace at a temperature of 550 ± 50 °C. Ensure the furnace is up to the temperature before inserting sample. About 15–20 min ignition is adequate. Let the crucible cool partially in air until most of the heat has been dissipated. Transfer it to a desiccator for final cooling to room temperature in a dry atmosphere. Weigh crucible as soon as it has cooled to room temperature. Repeat cycle of igniting, cooling, desiccating, and weighing until a constant weight. The volatile suspended solids present shall be estimated using Eq. 5.8 and fixed solids are estimated as per Eq. 5.9.

$$\text{Volatile suspended solids, mg/L} = \frac{(A - B) \times 1000}{Sample\ volume,\ mL} \tag{5.8}$$

$$\text{Fixed suspended solids, mg/L} = \frac{(B - C) \times 1000}{Sample\ volume,\ mL} \tag{5.9}$$

where, A is weight of residue + ashless filter paper + crucible before ignition, mg; B is the weight of residue + crucible and filter after ignition, mg; and C is the weight of empty crucible plus filter paper, mg. Similar expressions shall be used for total volatile and total fixed solids and volatile dissolved solids and fixed dissolved solids determination, substituting respective weights in the above expressions.

5.2.4.5 Determination of Settleable Solids

Settleable solids present in surface waters and wastewaters can be determined and reported on either as a volumetric basis (mL/L) or a weight basis (mg/L). The volumetric test requires only an Imhoff cone; whereas, the gravimetric test requires other apparatus and a glass vessel with a minimum diameter of 9 cm and volume of 1 L. For volumetric measurement fill an Imhoff cone to the 1 L mark with a uniformly mixed sample. Allow the settleable solids to settle for 45 min. Gently stir sides of cone with a rod or by spinning, allow settling for 15 min more, and record volume of settleable solids in the cone as millilitres per litre. Where a separation of settleable and floating solids occurs, do not estimate the floating material as a part of settleable matter.

For gravimetric measurement determine total suspended solids of well-mixed sample first. Pour a well-mixed sample into a glass vessel of not less than 9 cm diameter having volume not less than 1 L, say a measuring cylinder. Allow it to stand quiescent for 1 h and without disturbing the settled or floating solids, siphon about 250 mL from centre of container at a point halfway between the surface of the settled material and the super-natant liquid surface. Determine total suspended solids (mg/L) of this supernatant liquor representing the non-settleable solids. Thus, difference between the total suspended solids (mg/L) and non-settleable solids (mg/L) will give settleable solids (mg/L).

5.2.4.6 Determination Sludge Volume Index

Sludge volume index (SVI) is the volume in mL occupied by 1 g of a sludge after 30 min of settling. The SVI is used to monitor settling characteristics of activated sludge and other biological suspension. Although SVI is not supported theoretically, experience has shown it to be useful parameter in routine process control. The suspended solids concentration of a well-mixed sludge sample is first determined. For this sludge sample a settled sludge volume (SSV) after 30 min of settling is determined in a measuring cylinder of 1 L volume. From this SSV the SVI can be estimated as per Eq. 5.10.

$$SVI\left(\frac{mL}{g \ of \ SS}\right) = \frac{Settled \ sludge \ volume \ (\text{mL/L}) \times 1000}{Suspended \ solids \ (\text{mg/L})} \tag{5.10}$$

5.2.5 Determination of Dissolved Oxygen (DO)

The solubility of atmospheric oxygen in fresh water ranges from 14.6 mg/L at 0 °C to about 7.0 mg/L at 35 °C under normal atmospheric pressure. Since it is poorly soluble gas its solubility directly varies with the partial pressure at any given temperature. Analysis of DO is one of the important tests in wastewater engineering. It is necessary to know DO levels to keep a check on stream pollution. The DO test is the basis of BOD test, which is an important parameter to evaluate pollution potential of wastes. Maintaining minimum level of DO is necessary for all aerobic biological wastewater treatment processes.

Dissolved oxygen present in the sample oxidizes the divalent manganese to its higher valency, which precipitates as a brown hydrated oxide after addition of NaOH and KI. Upon acidification, manganese reverts to divalent state and liberates iodine from KI equivalent to DO content in the sample. The liberated iodine is titrated against $Na_2S_2O_3$ (N/80) using starch as an indicator. Ferrous ion, ferric ion, nitrite, microbial mass and high suspended solids concentration are the main sources of interference. Modifications in the estimation procedure to reduce these interferences is necessary (APHA, 2017).

The manganese sulphate solution is prepared by dissolving 480 g of tetrahydrate manganous sulphate diluted to 1000 mL. Filter the solution if necessary. This solution should not give colour with starch when added to an acidified solution of KI. The Alkali iodide-azide reagent is prepared by dissolving 500 g of NaOH and 150 g of KI and dilute to 1000 mL. In this, 10 g of NaN_3 dissolved in 40 mL distilled water is added. This solution should not give colour with starch solution when diluted and acidified.

Stock solution of sodium thiosulphate (0.1 N) is prepared by dissolving 24.82 g of $Na_2S_2O_3.5H_2O$ in boiled cooled distilled water and diluted to 1000 mL. This stock solution is preserved by adding chloroform (5 mL/L). Form this stock solution a standard sodium thiosulphate, 0.025 N, is prepared by diluting 250 ml of stock $Na_2S_2O_3$ solution to 1000 mL with freshly boiled and cooled distilled water. Preserve by adding chloroform (5 ml/L). For precision, this solution will have to be standardized against standard of dichromate solution for each set of titrations.

A freshly collected wastewater/water sample is filled in a 300 mL BOD bottle. Add 2 mL of $MnSO_4$ solution followed by addition of 2 mL of NaOH + KI + NaN_3 solution. The tip of the pipette should be lowered below the liquid level while adding these reagents and put the stopper immediately. Mix it well by inverting the bottle 2–3 times and allow the precipitate to settle leaving 150 mL clear supernatant. Add 2 mL of concentrated H_2SO_4 and mix the solution well till precipitate goes into solution. Take 203 mL in a conical flask and titrate against $Na_2S_2O_3$ using starch as an indicator. When 2 mL $MnSO_4$ followed by 2 ml NaOH + KI + NaN_3 is added to the sample, 4.0 ml of original sample is lost. This 203 mL that is taken for titration will correspond to 200 ml of original sample. Each mL of 0.025 N $Na_2S_2O_3$ represents 0.2 mg of O_2, thus using Eq. 5.11 the DO present in the sample shall be estimated.

$$D.O. \text{ in mg/L} = \frac{0.2 \times 1000 \times mL\ of\ Thiosulphate}{200} \qquad (5.11)$$

The method outlined is known as Winkler Modification and also as Alsterberg Azide Modification. The reagents NaOH + KI + NaN_3 are used in this method to eliminate interference caused by nitrite and due to higher concentration of ferric ion. When the samples contain ferrous ion Rideal Stewart modification protocol is adopted. In this 0.7 mL conc. H_2SO_4 is added followed by 1 mL of 0.63% $KMnO_4$ addition immediately after collection of samples in the BOD bottle itself. If ferric ions are present in large concentration, add 1 mL of 40% KF solution. Remove excess $KMnO_4$ using potassium oxalate just sufficient to neutralize $KMnO_4$ as excess oxalate give negative error.

While determining DO of activated sludge mixed liquor containing biological flocs having high demand for O_2, copper sulphate sulfamic acid flocculation modification is followed. In this method, samples are fixed by adding 10 mL copper sulphate sulfamic acid reagent to 100 mL of the sample. This reagent is prepared by adding 32 g sulfamic acid to 475 mL distilled water plus 50 g of $CuSO_4$ in 500 mL of distilled water and 25 mL acetic acid. To overcome interference of complex sulphur compound, such as wastewater from sulphite pulp industry, the sample is pre-treated with alkaline hypochlorite solution to convert polythionates to sulphates and free sulphur. Excess of hypochlorite is destroyed by addition of KI and sodium sulphite. However, this test is difficult to perform accurately.

The DO from water and wastewater sample can also be determined using standard make DO meter with DO probe. Calibrate the instrument to be used before measurement as per the instruction given by the manufacturer and then the probe can be inserted in the water sample, with a care of not entrapping air bubble near the probe sensor while inserting it in the sample. The DO meter will display the stable reading after reaching equilibrium after few seconds.

5.2.6 Determination of Biochemical Oxygen Demand

Biochemical oxygen demand (BOD) is defined as the amount of oxygen required by aerobic microorganisms for stabilizing biologically decomposable organic matters present in a wastewater under aerobic conditions. The BOD test is widely performed to determine: (i) biodegradable organic matter concentration present in the wastewaters, (ii) to assess the degree of pollution in lakes and streams at any time and their self-purification capacity, and (iii) to monitor the efficiency of wastewater treatment methods.

Since this test involves measurement of O_2 consumed by bacteria, while stabilizing organic matter under aerobic conditions, it is necessary to provide favourable conditions of nutrient availability, near neutral pH, absence of any microbial growth inhibiting substances and stable and favourable temperature. Because of the low solubility of O_2 in water, strong wastes are always diluted to ensure that the demand does not increase the available dissolved oxygen. A mixed group of microorganisms should be present in the sample and if such microorganisms are not present in the sample, then it should be seeded artificially. External seeding is essential for most of the industrial wastewaters, whereas for determination of BOD for sanitary sewage seeding is not required and unseeded BOD test is performed. Temperature is typically controlled at 20 °C. The test is conducted for 5 days of incubation and for biodegradable wastewater about 70% of the oxygen demand is satisfied during this period.

Since DO estimation is the basis of BOD test, sources of interference in BOD test are the same as in the DO test. In addition, lack of nutrients in dilution water, lack of an acclimated seed microorganisms and presence of heavy metals or other toxic materials, such as residual chlorine, are major sources of interference in this test. Aerate the required volume of distilled water by bubbling compressed air for overnight to attain DO close to saturation. Try to maintain the temperature near 20 °C. Add 1 ml each of phosphate buffer, magnesium sulphate, calcium chloride, and ferric chloride solutions for each litre of dilution water, and mix well. In the case of the wastewaters, which are not expected to have sufficient bacterial population, add a seed to the dilution water. Generally, 2 mL of settled sewage is considered sufficient for 1000 mL of dilution water. For determination of BOD for complex wastewater generated from industry other than food processing industry, ensure that the seed used is acclimated to that wastewater and in this case using sewage as a seed might not be appropriate.

Neutralize the wastewater sample to a pH of around 7.0, if it is highly alkaline or acidic. The sample should be free from residual chlorine. If it contains residual chlorine, remove it by using Na_2SO_3 solution. Samples having high DO content, i.e., DO of 9 mg/L due to either algal growth or some other reason, reduce the DO content by aerating and agitating the samples. Make several dilutions of the prepared sample so as to obtain about 50% depletion of DO in dilution water but not less than 2 mg/L and the residual dissolved oxygen after 5 days of incubation should not be less than 1 mg/L.

Siphon the dilution prepared as above in four labelled BOD bottles as demonstrated and put stopper immediately. Keep one bottle for determination of the initial DO and incubate three bottles at 20 °C for 5 days. In India, incubation for three days at temperature of 27 °C is acceptable representing BOD_3 at 27 °C. Ensure that the bottles have a water seal. Since when sewage is used as seed and added as about 2 mL per litre of the dilution water the organic matter present in this 2 mL of sewage will also exert the oxygen demand, hence this need to be subtracted from the oxygen demand exerted in five days of incubation in the BOD bottle. To estimate the oxygen demand exerted by this dilution water, prepare a blank in duplicate by siphoning seeded dilution water (without sample) to measure the O_2 consumption in dilution water. Fix the bottles kept for immediate DO determination and blank by adding 2 ml of $MnSO_4$ followed by 2 ml NaOH + KI + NaN_3 as described in the estimation of DO. Determine DO in the sample and in the blank on initial day and after 5 days. Alternately the DO can also be determined using DO probe.

The BOD present is the sample is then estimated using Eq. 5.12.

$$BOD, \frac{mg}{L} = \frac{(DO_i - DO_f) - (1-p)(B_i - B_f)}{p} \tag{5.12}$$

where, DO_i and DO_f are the initial and final DO in the sample bottle on 0th day and after 5 days of incubation, mg/L; B_i and B_f are the initial DO and final DO in the blank; p is the dilution fraction (=volume of sample (mL)/total volume of BOD bottle, i.e., 300 mL).

The DO depletion of blank is multiplied by the factor $(1 - p)$, because only that much volume of the blank is present in the sample bottle and remaining is the sample itself.

5.2.7 Determination of Chemical Oxygen Demand

The chemical oxygen demand (COD) test determines the oxygen required for chemical oxidation of organic matter present in the wastewater with the help of strong chemical oxidant. The test can be employed for the same purpose as the BOD test taking into account limitations of the test, which lies in its inability to differentiate between the biologically oxidisable and non-biodegradable organic/inorganic matter that will also get oxidize during COD determination. However, COD determination has an advantage over BOD determination, as in former the result can be obtained within 4 h as compared to 5 days (or 3 days) required for BOD test. Further, the COD test is relatively easy, gives reproducible results and it is not affected by interference as in the case of BOD test, which lacks in precision.

In COD test, the organic matter present in the wastewater sample gets oxidized completely by $K_2Cr_2O_7$ in the presence of H_2SO_4 and silver sulphate as catalyst to produce $CO_2 + H_2O$. The excess $K_2Cr_2O_7$ remaining after reaction is titrated with $Fe(NH_4)_2(SO_4)_2$. For ensuring complete oxidation of organic matter, it is necessary to see that volume of H_2SO_4 equals the volume of sample plus dichromate solution. The dichromate consumed gives oxygen required for oxidation of organic matter. The dichromate remaining can also be measured using spectrophotometric observation after proper calibration.

Fatty acids, straight chain aliphatic compounds, chlorides, nitrates and iron are the main interfering radicals. The interference caused by chlorides can be eliminated by the addition of $HgSO_4$ to the sample prior to addition of other reagents. About 0.4 g of $HgSO_4$ is adequate to complex 40 mg Cl^- ion in the form of purely ionized $HgCl_2$. Addition of Ag_2SO_4 to concentrated H_2SO_4 (22 g/9 lb) as a catalyst stimulates the oxidation of straight chain aliphatic and aromatic compounds. The NO_2 exerts COD of 1.14 mg/mg NO_2. Sulphamic acid in the amount 10 mg/mg of NO_2 may be added to $K_2Cr_2O_7$ solution to avoid interference caused by NO_2.

While performing the test with open reflux method, place 0.4 g $HgSO_4$ in a reflux flask. Add 20 mL of appropriately diluted sample with distilled water as per the need and mix well. Add pumice stone or glass beads followed by 10 mL of standard 0.25 N, $K_2Cr_2O_7$. Add slowly 30 mL of H_2SO_4 containing Ag_2SO_4 by mixing thoroughly. This slow addition along with swirling will prevent fatty acids to escape out due to high temperature. Mix well and if the colour turns to green either take fresh sample with more dilution or add more $K_2Cr_2O_7$ and acid. Connect the flask to condenser arrangement. Mix the contents before heating, improper mixing will result in bumping and sample may be blown out. Reflux for minimum 2 h and allow to get cool and then wash down the condenser with distilled water. Collect a minimum of 150 mL from reflux flask and titrate excess $K_2Cr_2O_7$ with 0.1 N $Fe(NH_4)_2(SO_4)_2$, using ferroin indicator. Sharp colour change from blue green to wine red indicates end point of the titration. Reflux blank in the same manner using distilled water instead of sample. The COD present in the sample can be estimated using Eq. 5.13.

$$COD, mg/L = \frac{(A - B) \times N \times 8000}{mL \, of \, sample} \qquad (5.13)$$

where, A is mL of $Fe(NH_4)_2(SO_4)_2$ required for blank; B is the mL of $Fe(NH_4)_2(SO_4)_2$ required for sample; N is the normality of $Fe(NH_4)_2(SO_4)_2 = 0.1$ N.

The same procedure can be followed in closed reflux COD determination in Teflon capped vials of about 10 mL capacity by appropriately selecting sample volume and dichromate volume and digesting for 2 h in COD digester. After digestion for 2 h, allow vials to cool at room temperature and the absorbance can be observed in the spectrophotometer at wavelength of 600 nm. Preparing and using the standard calibration curve for potassium dichromate solution, the observed absorbance for the sample shall be converted to the COD value as concentration in mg/L.

5.2.8 Determination of Total Organic Carbon

The organic carbon present in water and wastewater is comprised of different types of organic compounds. The total organic carbon (TOC) is a more appropriate and direct indication of total organic content. For TOC determination of wide variety of water samples high-temperature combustion method is widely used. Sample is homogenised if it contains insoluble matter or solids. Inorganic carbon is removed from sample before analysis by adding phosphoric acid or sulfuric acid to reduce pH to 2.0 or less followed by purging purified CO_2-free any gas for about 10 min. The sample is taken out using syringe, and introduced in to the standard TOC analyser and response is recorded. Analyse the corrected analyser response of standards and samples. Formulate a standard curve of corrected analyser response vs. TOC concentration. Deduct procedural blank from each sample analyser response and compare to standard curve to determine the TOC content (APHA-AWWA-WEF, 2017).

5.2.9 Determination of Nitrogen

Nitrogen is present in wastewater as organic or inorganic compounds either under reduced or oxidised forms (Roig et al., 2001). With the enforcement of the regulation on nitrogen content in the treated wastewaters, different forms of nitrogen (nitrate, nitrite, ammoniacal, Kjeldahl) are required to be determined that is present in the wastewaters. In general, about 15–50 mg/L of nitrogen is present in sanitary sewage. Around 40% of the total nitrogen is present in the form of organic nitrogen and the remaining is present as ammonia (Reed, 1985).

5.2.9.1 Ammoniacal Nitrogen

For determination of ammonia nitrogen, the sample pH is adjusted to 9.5 using 6 N NaOH solution to reduce hydrolysis of cyanates and compounds of organic nitrogen. Ammonia free distilled water with volume of 500 and 20 mL borate buffer is added to a distillation flask along with a few glass beads or boiling chips. Distillation is continued in boric acid solution until no traces of ammonia are seen. Replace this distillation flask with the sample distillation flask. For preparation of sample distillation flask, take 500 mL of dechlorinated water/wastewater sample. Add 25 mL of borate buffer and adjust the pH to 9.5 using 6 N NaOH solution. Distillation is carried out at the rate of 6–10 mL/min with the tip of delivery tube immersed in receiving boric acid solution. Collect distillate in a 500 mL Erlenmeyer flask containing 50 mL indicating boric acid solution. A minimum of 200 mL of distillate is collected and diluted to 500 mL.

For ammonia determination by titration select sample volumes as 250, 100, 50 and 25 mL for 5–10, 10–20, 20–50 and 50–100 mg/L of ammonia nitrogen concentration present in the sample, respectively. For sludge and sediment sample collect approximately 1 g dry weight of sample in crucible. Wash sample into a 500 mL Kjeldahl flask with water and dilute it to 250 mL. Distillate it in a boric acid solution as absorbent and add a piece of paraffin wax to distillation flask and collect only 100 mL distillate in case of solid samples for which the ammonia is to be determined.

Titrate collected 200 mL distillate diluted to 500 mL with standard 0.02 N H_2SO_4 titrant in presence of mixed indicator, containing methyl red and methylene blue, until colour turns pale lavender. Carry out the same procedure for a blank. The ammoniacal nitrogen from the sample can be estimated using Eq. 5.14 or Eq. 5.15.

$$1. \text{ Liquid sample : } NH_3\text{-N} = \frac{(A - B) \times 280}{mL\ Sample} \text{mg/L} \tag{5.14}$$

$$2. \text{ Sludge or sediment sample : } NH_3\text{-N} = \frac{(A - B) \times 280}{g\ dry\ wt\ sample} \text{mg/kg} \tag{5.15}$$

where, A is the volume of H_2SO_4 titrated for sample, mL and B is the volume of H_2SO_4 titrated for blank, mL.

5.2.9.2 Nitrite Nitrogen

Nitrite (NO^{2-}) is determined by associating diazotized sulfanilamide with N-(1-naphthyl)-ethylene-diamine dihydrochloride (NED dihydrochloride) through intermediate formation of a reddish-purple azo dye at pH of 2.0–2.5. Chemical incompatibility makes it questionable to coexist NO_2^-, free chlorine, and nitrogen trichloride (NCl_3). The NCl_3 yields a false red colour after addition of colour reagent. The ions Sb^{3+}, Au^{3+}, Bi^{3+}, Fe^{3+}, Pb^{2+}, Hg^{2+}, and Ag^+ hinder the precipitation. Cupric ion lowers the detection due to catalytic decomposition of the diazonium salt. Presence of coloured ions can modify the colour. Filter out any suspended solids if present in the sample.

It is recommended to use fresh sample for determination as bacteria start converting NO_2^- to NO_3^- or NH_3. In special circumstances store the sample at 4 °C or freeze at –20 °C if required preservation for the duration of 1–2 days. Acid should not be used for preservation of the sample for nitrite determination. Filter 50 mL of wastewater sample through 0.45 μm pore size membrane filter paper, correct the pH between 5.0 and 9.0 with 1 N HCl or NH_4OH followed by addition of 2 mL of colour reagent to sample and mix well. After 10 min or before 2 h, measure absorbance at 543 nm using spectrophotometer. Use light path as 1, 5 and 10 cm for 2–25, 2–6 and less than 2 μg/L NO_2^--N concentrations, respectively. Plot absorbance of standards against NO_2^--N concentration and formulate a standard curve. Estimate the concentration of sample from curve.

5.2.9.3 Nitrate Nitrogen

Due to the maximum chances of presence of interfering constituents, determination of nitrate (NO_3^-) becomes difficult. Also, dissolved organic matter and other solutes absorb UV light, along with NO_3^-, hence it may not be feasible to determine NO_3^- at only one wavelength in spectrophotometer. Measuring the UV absorbance at 220 nm for NO_3^- determination can be done; however, since soluble organic matter can also absorb UV light at this wavelength, a second measurement can be made at 275 nm to correct the NO_3^- value (APHA-AWWA-WEF, 2017). At 275 nm the dissolved organic matter cannot absorb the UV light.

The UV spectra for NO_3^- and NO_2^- are similar, however NO_3^- concentration is usually much higher in polluted water than NO_2^- concentrations. At wavelengths <210 nm, a weak absorption of bicarbonate occurs, however it does not affect the NO_3^- second-derivative signal. Presence of Iron (Fe) along with copper (Cu) extremely interfere the test at concentration of 20 mg/L. The UV absorbance spectrum reaches to a peak of 203 nm when NO_3^--N concentration is less than 3 mg/L. As the concentration increases the peak slowly moves to 207 nm and then the peak moves from wavelength 224 nm to higher for second-derivative. The slit width of spectrophotometer may affect the profile of the UV and second-derivative spectra.

Prepare 0.5, 1.0, 1.5, 2.0, 2.5 mg/L up to 7 mg/L of NO_3–N standards to use as a calibration standard. Prepare filtered sample water/ wastewater and take 50 mL of it and 1 mL of 1 N HCl solution and mixed well. Take sample in cuvette, measure absorbance at 220 nm against distilled water sample for which the absorbance is set to zero to obtain NO_3^- absorbance reading. Take absorbance reading at 275 nm to determine interference due to presence of dissolved organic matter. Subtract the absorbance reading obtained at 275 nm twice from the absorbance reading obtained for samples and standards at 220 nm to get absorbance due to NO_3^-. If the correction value deviates more than 10% from the absorbance obtained at 220 nm, this test is not recommended for such samples. Construct standard curve for absorbance of NO_3^- versus known NO_3^--N concentration of standards. Thus, using corrected sample absorbance the nitrate concentration present can be estimated from the graph.

5.2.9.4 Total Kjeldahl Nitrogen

The sum of ammonia nitrogen and organic nitrogen is represented as "Kjeldahl nitrogen". In Kjeldahl method the tri-negative state of nitrogen is determined. Organic nitrogen includes proteins, peptide, urea and many other synthetic organic compounds. The total Kjeldahl nitrogen (TKN) is unable to determine the different forms of nitrogen like nitrite, nitrate, nitroso, nitro, nitrile, semi-carbazone, oxime, azine, azide, hydrazone and azo (APHA-AWWA-WEF, 2017). Fresh sample should be used for more accurate results. If analysis is not possible on fresh sample, then acidify the sample with concentrated H_2SO_4 at pH of 1.5–2.0 for Kjeldahl digestion and store at 4 °C. Prohibit the use of $HgCl_2$ as it interferes with the removal of ammonia.

Nitrate produces N_2O, which leads to negative interference if present in surplus of 10 mg/L. Also, positive interference due to nitrate is possible if satisfactory organic matter in a low state of oxidation is present in the sample. Presence of large amount of salt or inorganic solids may lead to surge the digestion temperature about 400 °C, at this point pyrolytic nitrogen loss happens. In order to prevent this condition, maintain the balance of acid-salt by adding more H_2SO_4. This can be avoided by adding 10 mL conc. H_2SO_4/3 g COD in digestion flask. In Kjeldahl digestion CO_2 and H_2O releases during digestion of organic matter by H_2SO_4.

Select sample size as 500, 250, 100, 50 and 25 mL for 0–1, 1–10, 10–20, 20–50 and 50–100 mg/L of organic nitrogen concentration expected in the sample, respectively, in 800-mL Kjeldahl flask. Adjust the sample pH to 7.0. Dilute the sample to 300 mL, if required. Add 25 mL borate buffer followed by 6 N NaOH till the pH of 9.5 is achieved. Add few boiling chips or glass beads and boil this 300 mL sample. If ammonia nitrogen is also to be determined distil this in boric acid solution and follow the procedure as described earlier for ammonia nitrogen determination using Eq. 5.14. The residue in the distillation flask will have organic nitrogen present in it.

Cool Kjeldahl flask and carefully add 50 mL digestion reagent (consisting of 6.7 g of K_2SO_4, 0.365 g of $CuSO_4$ and 6.7 mL of conc. H_2SO_4) to distillation flask. Add few glass beads and heat it under close hood with exhaust arrangement to remove acid fumes. Boil till the profuse white fumes are observed and volume comes down to 25–50 mL. Continue digestion for next 30 min till sample becomes pale green. Cool it and dilute to 300 mL with distilled water and mix well. Add 50 mL sodium hydroxide-thiosulfate reagent in very careful manner to form an alkaline layer at flask bottom. Ensure complete mixing and connect the flask to a steamed-out distillation apparatus. The pH of the solution should exceed 11.0.

Collect 200 mL distillate in 50 mL of boric acid solution with the tip of condenser below level of boric acid absorbent solution and maintain the temperature in condenser below 29 °C. Titrate with standard 0.02 N H_2SO_4 titrant until colour turns pale lavender. Carry a reagent blank and standards through same procedure. The organic nitrogen can be estimated using Eq. 5.14. The summation of ammonia nitrogen and organic nitrogen will give TKN present in the sample.

5.2.10 Determination of Phosphorus

In water and wastewater phosphorus is present as phosphates. It is found in particulates or detritus or in the body of aquatic organisms. Phosphate is classified as (a) orthophosphate, (b) condensed phosphates, and (c) organically bound phosphates. Phosphate in water body comes through pollution from sewage, industrial waste and application of fertiliser or from natural contact with minerals (Maiti, 2001). Phosphorus of all the forms get converted to inorganic forms (orthophosphate) after digesting the sample. Determination of phosphorus is performed in two steps. In first step conversion of the phosphorus to dissolved orthophosphate is carried out and in second step this dissolved orthophosphate is determined following colorimetric method.

If the sample is heated, positive interference is produced due to arsenate and silica, if present. Negative interferences are caused by presence of arsenate, fluoride, thorium, bismuth, sulphide, thiosulfate, thiocyanate, or excess molybdate. Ferrous ion induces blue colour but does not interfere with the results if less than 100 mg/L. Oxidation with bromine water may remove interference caused by sulphide.

Take a 50 mL of water sample in a Kjeldahl flask and 1 mL of concentrated H_2SO_4 is added followed by addition of 5 mL of concentrated HNO_3. Digest it to a volume of 10 mL and continue until solution becomes colourless. Cool down the sample and add 20 mL of distilled water. Add a drop of phenolphthalein indicator and start adding 1 N NaOH solution slowly until the faint pink colour appears. The solution is filtered, if particulate material or turbidity is present, and transferred to 100 mL volumetric flask to make it up to 100 mL with distilled water. From this, take 20 mL of sample and add 4.0 mL of ammonium molybdate reagent followed by 10 drops of stannous chloride reagent. After 10 min, however before 12 min, measure the absorbance in spectrophotometer at 690 nm and compare with a calibration curve, using distilled water as blank. The phosphate can be estimated using Eq. 5.16.

$$\text{Phosphorous, P (mg/L)} = \frac{mg\ P\ (in\ approximately\ 104.5\ mL\ final\ volume)}{mL\ sample} \times 1000$$

$$(5.16)$$

5.2.11 Determination of Most Probable Number for Identification of Coliforms

For testing, if the water is free from bacterial contamination, the multiple tube fermentation technique can be adopted to identify presence of coliform, which is used as an indicator microorganism. The coliform group consists of several genera of bacteria belonging to the family of Enterobacteriaceae. These are facultative anaerobic, non-spore forming, Gram-negative rod-shaped bacteria having ability to ferment lactose. These bacteria can be identified by using multiple tube fermentation technique or by membrane filter (MF) technique. In case of multiple tube fermentation, the results of the multiple tubes are expressed in terms of most probable number (MPN). This number represents probability of number of coliforms present in the sample.

The determination of the MPN is based on statistics and the results are associated with the frequency of occurrence of a series of positive results. Generally, five or ten test tubes with effective dilutions are used to find the growth of bacteria. The accuracy of test increases as per the increase in the test tubes with effective dilution. Suspend 35.6 g of Lauryl Tryptose Broth in 1000 mL distilled water, dissolve the medium completely. Use three groups of the series of test tubes of 5 or 10 in numbers. Add 9 mL of broth to each test tubes containing inverted Durham tubes for the detection of fermentation gas. Sterilise all the tubes in autoclave at 121 °C for 15 min.

Cool down all the test tubes in laminar flow chamber under UV light. Transfer 1 mL of sample to each of 5 tubes of one group of broth, transfer 0.1 mL of sample to 5 tubes of next group of broth and transfer 0.01 mL of sample to 5 tubes of remaining last group of broth tubes. Incubate the tubes at 35 °C for 24 h. If no gas or acid reaction (evident by shades of yellow colour) is evident after 24 h, continue until 48 h and re-examine. After incubation, observe the fermentation gas in Durham's tube. Record the number of positive results from each set and compare with the standard chart to give presumptive coliform count per 100 mL (Gunasekaran, 2007). For the combination of tubes not appearing in the standard chart the MPN can be estimated using Eq. 5.17 (Thomas, 1942).

$$\text{MPN per 100 mL} = \frac{\text{Number of positive tubes} \times 100}{\sqrt{(\text{mL of sample in negative tubes}) \times (\text{mL of sample in all tubes})}}$$

(5.17)

While, this protocol is followed for determination of coliform, similar procedure can be followed by selecting suitable test media, if available, for other microorganisms.

5.3   BOD Model and Relation with COD and TOC

5.3.1   Estimation of Organic Matter Content from Wastewater

As explained earlier the organic matter present in the water/wastewater can be analysed in laboratory by determining BOD, COD or by determining TOC. The relevance of these tests and relation between them are discussed below.

5.3.2   Biochemical Oxygen Demand

The BOD test is based on the premise that the biodegradable organic matter contained in a water sample will be oxidized to CO_2 and H_2O by the catalytic action of microorganisms using molecular oxygen as terminal electron acceptor. For example, the general overall oxidation reaction for glucose can be expressed as per Eq. 5.18.

$$C_6H_{12}O_6 + 6O_2 \rightarrow 6CO_2 + 6H_2O \qquad (5.18)$$

The theoretical oxygen demand for glucose per gram of carbon can be estimated from the Eq. 5.18.

$$\text{Oxygen demand} = (\text{Gram of oxygen used})/(\text{Gram of carbon oxidized})$$
$$= 192/72 = 2.67 \text{ g/g of carbon}$$

Thus, glucose will have a theoretical oxygen demand of 2.67 g per g of carbon present, which could be best represented as TOC, i.e., COD to TOC ratio for glucose is 2.67. Similarly, oxygen demand per gram of glucose will be $192/180 = 1.067$ g of O_2.

The actual BOD that will be exerted during test will be much less than theoretical oxygen demand due to incorporation of some of the carbon into newly synthesized bacterial cells. The test is performed under the conditions similar to those that occur in natural water to measure indirectly the amount of biodegradable organic matter present in the water in terms of oxygen demand. A water sample is inoculated with bacteria (about 2 mL of sewage per litre of dilution water) that consume the biodegradable organic matter present in the sample to obtain energy for their life processes. The organisms also utilize oxygen in the process of consuming the organic matter, the process is called as 'aerobic' oxidation. This oxygen consumption will be proportional to the concentration of organic matter present in the wastewater sample. More is the organic matter concentration more is the amount of oxygen that will be utilized for oxidation.

Thus, the BOD test is the indirect measurement of concentration of biodegradable organic matters present in wastewater in terms of the oxygen requirement to convert them into stable end product. Not all of the organic matters are biodegradable within the stipulated incubation period and the actual test procedure lacks in precision due to different inoculum seeds used and many folds dilution required. However, it is still the most widely used laboratory test to quantify organic matter because of the direct conceptual relationship between BOD and oxygen depletion in receiving water bodies. This BOD test is required to be performed for the following listed purposes.

- To determine amount of oxygen required for biochemical stabilization of organic matter present in that wastewater.
- To determine suitability of biological treatment method, depending on the COD/BOD ratio, and deciding the sizing of the treatment unit. The wastewater is considered to be most suitable for biological treatment when the COD/BOD ratio is less than 2. Whereas, for wastewater having COD/BOD ratio up to 6, acclimation of the inoculum is necessary for effective biological treatment. For wastewater having higher COD/BOD ratio than six, biological treatment may not be a very suitable option for this wastewater.
- The BOD can also be used for monitoring efficiency of the treatment process, irrespective of whether the technology adopted in treatment plant is aerobic or anaerobic.
- To ensure compliance with wastewater discharge permits set by the regulatory agency, as for most of the wastewaters final effluent BOD is being regulated by the agency.

During aerobic oxidation of organic matter present in the wastewater sample, the organic matter will be converted into stable end product, CO_2. Depending on the other constituents present in the wastewater other oxidized products, such as sulphate (SO_4), orthophosphate (PO_4) and nitrate (NO_3) can also be formed.

The simple representation of carbonaceous BOD can be explained as stated in Eq. 5.19.

$$\text{Organic matter} + O_2 \xrightarrow{\text{Aerobic microorganisms}} CO_2 + H_2O + \text{New Cells} + \text{Stable products}$$

$$(5.19)$$

This reaction will continue till sufficient DO is available in the water. When the DO is not available, condition in water body will become anaerobic and fermentative reduction process will occur. The reaction under anaerobic conditions is stated in the Eq. 5.20.

$$\text{Organic matter} \xrightarrow{\text{Anaerobic microorganisms}} CO_2 + CH_4 + \text{New Cells} + \text{Other products} (NH_3, H_2S)$$

$$(5.20)$$

5.3.2.1 The BOD Test

Biochemical oxidation is slow process and theoretically takes an infinite time to go to completion, i.e., to achieve maximum oxidation of organic matter. Even for easily biodegradable organic matter after incubation for prolonged duration the BOD value determined will be slightly lesser than the COD of the same sample, due to accumulation of end product of bacterial metabolism and leftover bacterial cell mass in the BOD bottle, which itself is organic matter. During the first few days the rate of oxygen depletion is rapid because of the high concentration of organic matter present in the wastewater sample. Whereas, as the concentration of organic matter decreases with incubation time, the rate of oxygen consumption will decrease. Also, initially the concentration of easily biodegradable organic matter will be more and as the time proceeds fraction of this component will deplete faster as compared to total organic matter reduction. Hence, during initial days of BOD test the rate of BOD exerted will be more as compared to later days.

During the latter part of the BOD curve, oxygen consumption is mostly associated with the decay of the microorganisms that grew during the early part of the test. The oxygen consumption profile will typically follow the pattern as shown in Fig. 5.1. For sewage, within 20 days period of incubation, the oxidation of carbonaceous organic matter will be about 95–99% complete, and in the first five days, the typical period used for BOD determination, 60–70% oxidation of organic matter is complete. The 20 °C temperature used is an average temperature value typically for slow moving streams in temperate climate. Since biochemical reaction rates are temperature dependent, different results would be obtained at different temperatures of incubation.

The biochemical oxygen demand is generally represented as BOD_5 at 20 °C, which states the amount of oxygen consumed for biochemical oxidation of organic matter for first five days of incubation at temperature of 20 °C. Under Indian conditions, the BOD values determined for 3 days of incubation at temperature of 27 °C are also acceptable. The actual BOD of the unseeded sample can be estimated using Eq. 5.21. Whereas, for determination of BOD for a seeded BOD test Eq. 5.12 (as stated earlier) shall be used.

$$\text{The 5-day BOD of wastewater sample} = \frac{DO_i - DO_f}{p} \qquad (5.21)$$

where, DO_i and DO_f are the initial and final DO of diluted wastewater sample and p is the dilution fraction equal to volume of wastewater sample to the total volume of the BOD bottle, i.e., 300 mL.

The total volume of the BOD bottle used for test is usually 300 mL. The dilution water (distilled water) is aerated for sufficient time to correct DO close to the saturation value. Nutrients and buffer solutions are added to the dilution water to provide nutrient for

Fig. 5.1 Variation in DO profile and BOD with duration of incubation during BOD test

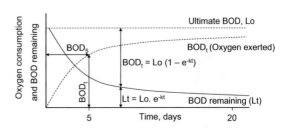

microbial growth and to maintain pH near neutral, as explained earlier. Sufficient amount of seed is added to the dilution water to ensure adequate concentration of microbial population to carry out the biodegradation of organic matter present in sample. Since, the saturation value of DO for water at 20 °C is only 9.1 mg/L, it is usually necessary to dilute the wastewater samples to keep final DO level, at the end of incubation period of about 1.5 mg/L or higher. Hence, according to BOD values expected for the wastewater sample appropriate dilution should be carried out. In addition, to have accuracy in the BOD test result, the DO drop during five days of incubation should also be about 1.5–2.0 mg/L, and accordingly the final dilution should be decided.

Example: 5.1: Dilution requirement for BOD test: For a wastewater sample the expected BOD_5 value is about 300 mg/L. The initial DO of prepared dilution water is 7.8 mg/L. Determine the dilution requirement for performing BOD test.

Solution

$BOD \sim 300$ mg/L and $DO_i = 7.8$ mg/L,

To ensure that minimum DO of about 1.5 mg/L will remain in BOD bottle after five days of incubation, the amount of DO available for satisfying BOD in the dilution water is $(7.8–1.5) = 6.3$ mg/L.

Hence, the dilution required will be $= 300 / 6.3 = 47.62$

Thus, dilution recommended for this BOD test is 50–60 times.

Example: 5.2. BOD determination: While performing the BOD test, a blank BOD bottle containing only seeded dilution water has its DO level dropped by 1.2 mg/L during a 5-day incubation. A sample BOD bottle was filled with 6 mL of wastewater and the rest seeded dilution water, which experienced a DO drop of 5.8 mg/L during the same time period. What would be BOD_5 of the wastewater?

Solution

Dilution fraction $p = 6/300 = 0.02$.

Therefore, $BOD_5 = [5.8 – 1.2 (1 – (0.02))]/(0.02) = 231.2$ mg/L~231 mg/L (**Answer**)

5.3.2.2 BOD Model

The rate at which the oxygen is consumed is considered to be directly proportional to the concentration of biodegradable organic matter remaining at any time in the BOD bottle. Thus, in accordance with first order reaction kinetics the BOD reaction can be formulated as explained in Eq. 5.22.

$$\frac{dL_t}{dt} = -K L_t \qquad (5.22)$$

where, L_t is the amount of first order BOD remaining in wastewater at any time t, and K is the first order BOD reaction rate constant, time^{-1}.

Integrating with time $t = 0$ to t

$$\int_0^t dL_t = -KL_t.dt$$

i.e.,

$$\left[\log L_t\right]_0^t = -K.t$$

Solving, the Eq. 5.23 or Eq. 5.24 can be obtained, depending on the log type selected, for estimating the amount of BOD remaining at time 't' (Fig. 5.1).

$$L_t = L_0.e^{-K.t} \tag{5.23}$$

Or,

$$L_t = L_0.10^{-K.t} \tag{5.24}$$

where L_0 or BOD_u at time t = 0, is the ultimate first stage BOD initially present in the sample.

The relation between BOD reaction rate constant K with base e (in case of natural log) and K with base 10 (for log to the base 10) can be given by Eq. 5.25.

$$K(\text{base } 10) = K(\text{base } e)/2.303 \tag{5.25}$$

Thus, the amount of BOD that has been exerted (amount of oxygen that is consumed) at any time 't' is given by Eq. 5.26.

$$BOD_t = L_0 - L_t = L_0\left(1 - e^{-K.t}\right) \tag{5.26}$$

Therefore for the five-day BOD, the Eq. 5.26 will become:

$$BOD_5 = L_0 - L_5 = L_0\left(1 - e^{-5K}\right)$$

For polluted water and wastewater, a typical value of K (base e, 20 °C) is 0.23 per day and K (base 10, 20 °C) is 0.10 per day. These values vary widely for the wastewaters depending upon the organic matter constituents present and their biodegradability in the range from 0.05 to 0.3 per day for K (base 10) and 0.23–0.7 for K (base e).

The ultimate BOD (L_0) is defined as the maximum amount of BOD that can be exerted by that wastewater. It is difficult to assign exact time to achieve this ultimate BOD, and theoretically it will be achieved when infinite time of incubation is used. From the practical point of view, it can be said that when the BOD curve is approximately horizontal (Fig. 5.1) the ultimate BOD has been achieved. The time required to achieve this ultimate BOD depends upon the characteristics of the wastewater, i.e., chemical composition of the organic matter present in the wastewater and biodegradable properties of the organic compounds present and also on temperature of incubation. At higher temperature, for same concentration and nature of organic matter the ultimate BOD will be achieved in shorter time as compared to incubation at lower temperatures. At lower temperature of incubation, it will require more time to reach ultimate BOD. The ultimate BOD best expresses the concentration of biodegradable organic matter based on the total oxygen required to oxidize it. However, the ultimate BOD value does not indicate how rapidly oxygen will be depleted in the receiving water. Oxygen depletion is related to both the ultimate BOD and the BOD rate constant (K).

The BOD reaction rate constant (K) depends on the following parameters:

i.   Nature of the organic matter present in the wastewater
ii. Ability of microorganisms in the system to utilize the waste, and
iii. The temperature.

Nature of the waste: Thousands of organic compounds exist in aquatic environment with different chemical composition in nature. All organic compounds will not have same degradation rate. Simple sugar and starch are rapidly degradable matters and hence, they will therefore represent a high value of BOD rate constant. Cellulose degrades much more slowly and human hairs are not degradable during five-days duration of incubation in BOD test or during biological treatment of wastewaters with few hours of retention time in a reactor. For complex wastewaters, like sewage, the BOD rate constant depends upon the relative proportions of the various organic components present in it. The BOD rate constant is high for the raw sewage (K (base e) = 0.35–0.7 per day) and low for the treated sewage (K (base e) = 0.15–0.23 per day), owing to the fact that, during wastewater treatment the easily biodegradable organic matter will get more completely removed than the less biodegradable fraction of organic matters. Hence, in the treated wastewater, relative proportion of the less biodegradable organic matter will be higher, giving lower BOD rate constant.

Ability of organisms to utilize waste: Every microorganism is limited in its ability to oxidize the organic compounds. Many organic matters can only be oxidized by a particular group of microorganisms. In natural environment, where the water course is receiving particular organic compound, the microbiota that have capability to degrade such organic matter will grow in the predominance. However, the culture used during BOD test might have very small fraction of such organisms, which can degrade that particular organic compound present in the wastewater. As a result, the BOD value for limited incubation duration and the rate constant determined based on this laboratory test would be

lower than that in the natural water environment, where the wastewater is regularly discharged. Hence, the BOD test should be conducted with the microbial seed, which has
been acclimated to that wastewater, so that the rate constant determined in the laboratory
can be compared to that in the natural water body, where such wastewater is being continuously discharged. The BOD determined using acclimated seed culture will provide
correct estimation of oxygen required while treating the wastewater in aerobic biological
reactor.

Temperature: Biochemical reactions are temperature dependent and activity of the
microorganism increases with the increase in temperature up to certain value, and activity
decreases with drop in temperature. Since, the oxygen utilization in BOD test is caused
by microbial metabolism, the rate of utilization is similarly affected by the temperature.
The BOD is determined usually at a standard temperature of 20 °C. However, the water
temperature may vary from place to place for the same river; hence, the BOD rate constant is required to be adjusted to that temperature of receiving water using relationship
given by Eq. 5.27.

$$K_T = K_{20}\,\theta^{(T-20)} \tag{5.27}$$

where, T is the actual temperature, °C; K_T is the BOD reaction rate constant at the
temperature T, day^{-1}; K_{20} is the BOD reaction rate constant at 20 °C, day^{-1}; θ is the
temperature coefficient. This θ has a value of 1.056 in general and 1.047 for temperature
greater than 20 °C. This is because increase in reaction rate is higher, when the temperature increases from 10 to 20 °C as compared to when the temperature is increased from
20 to 30 °C.

Example: 5.3 A water flowing in the river, after receiving treated wastewater, has a temperature of 17 °C. The first order BOD reaction rate constant determined in the laboratory for
this mixed water had a value of 0.18 per day. Determine the fraction of maximum oxygen
consumption that will occur in first five days?

Solution

Determine the BOD rate constant at the river water temperature of 17 °C.

$$K_{17} = K_{20}(1.056)^{(T-20)} = 0.18(1.056)^{(17-20)} = 0.153 \text{ per day}$$

Using this K_{17}, find out the fraction of maximum oxygen consumption in five days:

$$BOD_5 = L_0\left(1 - e^{-0.153\times5}\right)$$

Therefore, $BOD_5 / L_0 = 0.535$
Thus, 53.5% of the ultimate oxygen demand will be satisfied in the five days.

Example: 5.4 The BOD of a sewage incubated for three days at 27 °C has been found to be 180 mg/L. What will be the five-day 20 °C BOD? Consider $K = 0.12$ (base 10) at 20 °C, and $\theta = 1.047$.

Solution

BOD at 27 °C $= 180$ mg/L

$\quad K_{20} = 0.12$ per day

$\quad$ Now

$$K_{27} = K_{20}\,\theta^{(T-20)}$$
$$K_{27} = 0.12(1.047)^{27-20} = 0.166 \text{ per day}$$

Now, $BOD_t = L_0\,(1 - 10^{-kt})$

$$180 = L_0\left(1 - 10^{-0.166 \times 3}\right)$$

Hence, $L_0 = 263.8$ mg/L

This is ultimate BOD, which is independent of incubation temperature.

Now, BOD_5 at 20 °C can be estimated as given below using $K = 0.12$ at 20 °C.

BOD_5 at 20 °C $= Lo\,(1 - 10^{-kt}) = 263.8\,(1 - 10^{-0.12 \times 5}) = 197.5$ mg/L ~ 198 mg/L

Thus, five-day BOD at incubation temperature of 20 °C for this wastewater will be 198 mg/L.

Example: 5.5 The final DO in an unseeded sample of diluted sewage having an initial DO of 8.3 mg/L is measured to be 2.8 mg/L after 5 days of incubation. The dilution fraction is 0.03 and reaction rate constant $K = 0.23$ day^{-1}. Determine (a) BOD_5 and BOD_u of this wastewater. What will be BOD remaining after 5 days.

Solution

(a) Oxygen demand for first 5 days

$$BOD_5 = (DO_i - DO_f)/p = (8.3 - 2.8)/0.03 = 183.33 \sim 183 \text{ mg/L}$$

(b) Ultimate BOD of this wastewater

$$BOD_u = L_0 = BOD_t/\left(1 - e^{-kt}\right) = 183.33/\left(1 - e^{-0.23 \times 5}\right) = 268.3 \sim 268 \text{ mg/L}$$

(c) After 5 days, 183 mg/L of oxygen demand out of total 268 mg/L will be satisfied. Hence, the remaining oxygen demand will be 268 – 183 = 85 mg/L.

Example: 5.6 Determine the three-day BOD and ultimate first stage BOD for a wastewater for which the BOD_5 at 20 °C is 230 mg/L. The reaction rate constant K (base e) $= 0.23$ per day. What will be BOD_3 at 27 °C? Consider $\theta = 1.047$.

Solution

Ultimate BOD, $BOD_u = L_0 = BOD_t/(1 - e^{-Kt}) = 230 / (1 - e^{-0.23 \times 5}) = 336.6$ mg/L
  Therefore, three-day BOD of this wastewater will be:

$$BOD_3 \text{ at } 20\,°C = L_0 - L_3 = L_0\left(1 - e^{-Kt}\right) = 336.6\left(1 - e^{-0.23 \times 3}\right) = 167.8 \text{ mg/L} \sim 168 \text{ mg/L}$$

Now, $K_{27} = K_{20}\, \theta^{(T-20)}$

$$K_{27} = 0.23(1.047)^{27-20} = 0.317 \text{ per day}$$

Hence, BOD_3 at 27 °C will be:

$$= L_0\left(1 - e^{-Kt}\right) = 336.6\left(1 - e^{-0.317 \times 3}\right) = 206.6 \sim 207 \text{ mg/L}$$

Note BOD determined at 20 °C for five days incubation and at 27 °C at three days incubation are having different values and are not same.

5.3.2.3 Interpretation of the BOD Test Results

While interpreting the BOD test results for industrial wastewaters, following factors must be considered:

- The seed used for performing BOD test must be acclimated to the wastewater, thus the lag period required for acclimation is eliminated.
- The first order BOD reaction rate constant should be established based on long term BOD tests on both raw wastewater and treated effluents. The rate constants for untreated and treated wastewater are not same for most of the wastewaters. The BOD reaction rate constant is having higher value for untreated wastewater and lower value for treated wastewater. For example, for a raw sewage the BOD reaction rate constant is about 0.15–0.3 and that for treated sewage it is around 0.05–0.15 (base 10). The value of K (base e) for raw wastewaters varies in the range 0.35–0.7 and that for treated wastewaters it will be 0.12–0.23. Hence, direct comparison of BOD for untreated and treated wastewater may not be valid.
- In case of acidic wastewater, all samples must be neutralized before putting into BOD test bottles.
- For wastewaters, where organic matter is present in suspended form, interpretation of the BOD test result is difficult due to lag time involved in hydrolysis of suspended organic matter before actual oxidation of it starts during the test. Hence, for such

wastewater correlation between BOD and COD is difficult. The BOD reaction rate coefficient affected significantly by the size of the particles in wastewater. For example, for settleable solids with size more than 100 μm, the rate constant will have value of 0.08 per day (base e); whereas, for colloidal organic matter in the range of 0.1–1.0 μm the rate constant will be having value of about 0.22 per day (base e) and for dissolved organic matter the rate constant (base e) will have value up to 0.39 per day, respectively.

5.3.2.4 Nitrification in BOD Test

During the hydrolysis of proteins, if present in the wastewaters, non-carbonaceous matter, such as ammonia is produced. In addition, when the living microbes die, from excreta waste, and from nitrogen containing organic compounds, the nitrogen tied to organic molecule is converted to ammonia by bacterial and fungal action. Under aerobic conditions, this ammonia will be converted to nitrate, in two step reactions (Eqs. 5.28 and 5.29) called as nitrification as per the reactions presented below,

$$2NH_3 + 3O_2 \xrightarrow{Nitrosomonas} 2NO_2^- + 2H^+ + 2H_2O \tag{5.28}$$

$$2NO_2^- + O_2 \xrightarrow{Nitrobacter} 2NO_3^- \tag{5.29}$$

Thus, the organic matter containing nitrogen present in wastewater will have oxygen requirement even for nitrification. The oxygen demand associated with the oxidation of ammonia to nitrate is referred as the nitrogenous BOD. Due to slow growth rate of nitrifying bacteria, this BOD demand normally occurs from 6 to 10 days (Fig. 5.2). This is one of the reasons to restrict the incubation duration of 5 days for BOD determination to eliminate oxygen demand for nitrification and to find out only carbonaceous oxygen demand. Incidentally, the five-day period was chosen for the BOD test, because the Themes River requires five days to flow from its origin to join the sea. Since, the interest was for protecting river water quality to understand how much BOD will be exerted form the wastewater being released in the river.

Fig. 5.2 Nitrification that can occur during BOD test performed for longer duration of incubation than 5 days

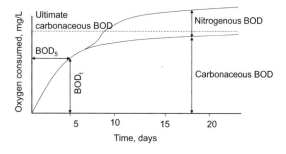

5.3.3   Other Measures of Oxygen Demand

5.3.3.1 Chemical Oxygen Demand

The chemical oxygen demand of the wastewater can be determined by performing COD test as explained earlier. During this COD determination total oxidizable content of the waste is oxidized by dichromate under acidic condition. In this test to determine the oxygen requirement of the wastewater, strong oxidizing agent 'potassium dichromate' is used under acidic environment by addition of sulphuric acid to accelerate the rate of reactions. In addition, the reflux flasks (or closed reflux vials), used for the test, are heated to 150 °C for 2 h in presence of silver sulphate as catalyst. In presence of silver sulphate catalyst, the recovery of most of the organic compounds is greater than 92%. Thus, COD test measures virtually all the oxidizable organic compounds, whether biodegradable or not, except some aromatic compounds, which resists dichromate oxidation. The results are more reproducible for COD test and less than 2% deviation is expected in the analysis.

The COD value of a wastewater is proportional to BOD only for readily soluble organic matter in dissolved form, e.g., sugars. No correlation between BOD and COD can be expressed when the organic matter is present in suspended form (under such situation filtered samples should be used) and for complex wastewater containing refractory organic substances. For readily biodegradable wastewaters, such as dairy effluent the $BOD_u = 0.92 \times COD$. This correlation between BOD and COD for sewage is presented in the Fig. 5.3 (Haandel & Lettinga, 1994).

The COD is faster determination, however does not give any idea about the nature of organic matter whether biodegradable or biorefractory. Hence, determination of BOD is essential for the wastewater to know biodegradable organic matter fraction present in it. The BOD is not very useful test for routine treatment plant control due to long incubation period required and getting results only after 3 or 5 days. Thus, it is important to develop correlation between BOD and COD (or TOC) based on long term experience for that wastewater, so that COD (or TOC) can be used as an alternate parameter for routine analysis and control of the treatment plant. Once COD values are known, the BOD can be estimated using correlation.

Fig. 5.3 Correlation between BOD and COD for sewage at 20 °C incubation and K (base 10) = 0.10 per day. The COD_b represents the COD of biodegradable fraction of organic matter and BODu representing ultimate BOD

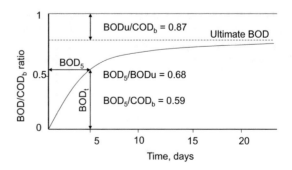

5.3.3.2 Theoretical Oxygen Demand

Theoretical oxygen demand (ThOD) for the wastewater can be estimated as oxygen required for oxidizing the organic matter to end products, if chemical composition of the wastewater is known as in case of industrial wastewaters. However, for wastewater like municipal sewage estimation of ThOD is not possible due to difficulties in establishing chemical composition of numerous organic molecules present in it with different concentrations. The ThOD can be estimated based on the oxidation equation for each of the constituent organic matter present in the wastewater. For most of the organic compounds (except aromatics resisting dichromate oxidation) COD is equal to ThOD. For example, for glucose, the ThOD can be worked out as using Eq. 5.18 (as provided earlier).

$$C_6H_{12}O_6 + 6O_2 \rightarrow 6CO_2 + 6H_2O \tag{5.18}$$

Therefore, $ThOD = (6\,M_{o2})/(M_{C6H12O6}) = (6 \times 16 \times 2)/(12 \times 6 + 1 \times 12 + 6 \times 16) = 1.067$ g per g of glucose.

TOC is related to COD through carbon–oxygen balance. From Eq. 5.31, the ratio of COD/TOC can be estimated.

$$COD/TOC = (6\,M_{o2})/(6\,M_C) = 2.67$$

Depending on the organic matter in question COD/TOC ratio may vary from zero (for organic matter resistant to dichromate oxidation) to 5.33 for methane (most reduced organic compound). Since, organic matter undergoes changes during biological oxidation, COD/TOC and BOD_5/TOC, BOD_5/COD values will change during the phases of treatment. For example, for wastewater suitable for biological wastewater treatment, for untreated wastewater BOD_5/COD ratio is about 0.3–0.8, whereas this ratio after primary treatment and secondary treatment becomes 0.4–0.6 and 0.1–0.3, respectively. Similarly, the BOD_5/TOC ratio for such raw wastewater could be in the range of 1.2–2.0, which will change to 0.8–1.2 and 0.2–0.5 after primary and secondary treatment, respectively.

Example: 5.7: Correlation between BOD, COD and TOC: Determine BOD/COD, BOD_5/TOC ratios for bacterial cells ($C_5H_7NO_2$). Consider BOD first order reaction rate constant to be 0.23 day^{-1} (base e).

Solution

Estimation of COD

$C_5H_7NO_2 + 5O_2 \rightarrow 5CO_2 + NH_3 + 2H_2O$

Hence COD per milligram of cells $= 5 \times 32/(12 \times 5 + 7 \times 1 + 14 \times 1 + 32) = 160/113$
$= 1.42$ mg COD/mg of cells

Determination of BOD₅

BOD_5/BOD_u

$$= \left(1 - e^{-K.t}\right) = \left(1 - e^{-0.23 \times 5}\right) = 0.68$$

$$BOD_5 = 0.68 \times 1.42 \times 0.9 \text{ mg } O_2/\text{mg } C_5H_7NO_2 \left(\text{considering } BOD_u = 0.9 \text{ COD}\right)$$
$$= 0.87 \text{ mg } BOD_5/\text{mg } C_5H_7NO_2$$

Estimating TOC of the cells

$$TOC = \frac{12 \times 5}{113} = 0.53 \text{ mg } TOC/\text{mg } C_5H_7NO_2$$

Therefore,

$$\frac{BOD_5}{COD} = \frac{0.87}{1.42} = 0.613$$

and

$$\frac{BOD_5}{TOC} = \frac{0.87}{0.53} = 1.64$$
$$\frac{TOC}{COD} = \frac{0.53}{1.42} = 0.37$$

Example: 5.8: Correlation between BOD, COD and TOC: For a wastewater containing 180 mg/L of ethylene glycol ($C_2H_6O_2$), 120 mg/L of phenol (C_6H_6O), 60 mg/L of sulphide (S^{2-}) and 120 mg/L of anthraquinone-2-sulfonic acid sodium salt (ASS, $C_{14}H_7NaO_5S$), which is non-biodegradable, estimate (a) COD and TOC, (b) BOD₅ if the K_e is 0.23/day, (c) If after treatment, the BOD₅ is 20 mg/L, estimate the COD. (d) Estimate BOD/COD and COD/TOC ratio for raw and treated wastewater, if 90% of ethylene glycol and phenol are being removed during the treatment. For treated wastewater $K_e = 0.12$/day and consider $BOD_u = 0.90$ COD for wastewater.

Solution

(a) *Estimation of COD*

Ethylene glycol

$$C_2H_6O_2 + 2.5 O_2 \rightarrow 2CO_2 + 3H_2O$$
$$COD = \frac{2.5 \times 32}{12 \times 2 + 1 \times 6 + 32} \times 180\frac{\text{mg}}{\text{L}} = 232 \text{ mg/L}$$

Phenol

$$C_6H_6O + 7O_2 \rightarrow 6CO_2 + 3H_2O$$

$$COD = \frac{7 \times 32}{12 \times 6 + 1 \times 6 + 16} \times 120\frac{mg}{L} = 286 \text{ mg/L}$$

Sulphide

$$S^{2-} + 2O_2 \rightarrow SO_4^{2-}$$

$$COD = \frac{2 \times 32}{32} \times 60\frac{mg}{L} = 120 \text{ mg/L}$$

Anthraquinone-2-sulfonic acid sodium salt (Kiwi et al., 1993)

$$C_{14}H_7NaO_5S + 15O_2 \rightarrow 14CO_2 + 3H_2O + NaHSO_4$$

$$COD = \frac{15 \times 32}{14 \times 12 + 7 \times 1 + 23 + 16 \times 5 + 32} \times 120\frac{mg}{L} = 186 \text{ mg/L}$$

Hence, total COD $= 232 + 286 + 120 + 186 = 824$ mg/L

(b) *Estimate of TOC*

Ethylene glycol

$$\frac{2 \times 12}{62} \times 180\frac{mg}{L} = 70 \text{ mg/L}$$

Phenol

$$\frac{12 \times 6}{94} \times 120\frac{mg}{L} = 92 \text{ mg/L}$$

Anthraquinone-2-sulfonic acid sodium salt

$$\frac{14 \times 12}{310} \times 120\frac{mg}{L} = 65 \text{ mg/L}$$

Total TOC is $70 + 92 + 65 = 227$ mg/L.

(c) *Estimation of BOD$_5$*

The ultimate BOD can be estimated as $BOD_u = 0.90COD$
    ASS being non-biodegradable, in BOD estimation it is excluded

$$BOD_u = (232 + 286 + 120) \times 0.90 \text{ mg/L} = 574 \text{ mg/L}$$

$$BOD_5 = BOD_u(1 - e^{-kt}) = 574\left(1 - e^{-0.23 \times 5}\right)$$

$$= 392 \text{ mg/L}$$

Hence, BOD_5 of the untreated wastewater will be 392 mg/L.

(d) The BOD_u of the effluent can be estimated from BOD_5 of 20 mg/L as stated below:

$$\frac{20}{1 - e^{-5 \times 0.12}} = 44.3 \text{ mg/L}$$

The COD for this BOD_u is 44.3/0.90 = 49 mg/L
Hence effluent COD = 186 mg/L + 49 mg/L + residual by-products

$$= \text{slightly greater than 235 mg/L.}$$

(e) *Estimating BOD/COD and COD/TOC ratio for raw and treated wastewater*

Ratio	Untreated wastewater	Treated wastewater
BOD/COD	392/824 = 0.476	20 / 235 = 0.085
COD/TOC	824/227 = 3.63	235 / 81.2 = 2.89

Considering 90% removal of ethylene glycol and phenol and no removal of ASS, the TOC for the effluent can be estimated to be:
Ethylene glycol:

$$\frac{2 \times 12}{62} \times 180 \times 0.1 = 7.0 \text{ mg/L}$$

Phenol:

$$\frac{12 \times 6}{94} \times 120 \times 0.1 = 9.2 \text{ mg/L}$$

Anthraquinone-2-sulfonic acid sodium salt

$$\frac{14 \times 12}{310} \times 120 = 65 \text{ mg/L}$$

Total effluent TOC is 7.0 + 9.2 + 65 = 81.2 mg/L.
Thus, effluent TOC will be slightly more than 81.2 mg/L, depending on the microbial cell concentration present in the effluent, which will also add to the TOC.

Questions

5.1 How man made persistent organic pollutants are adversely affecting the aquatic environment?

5.2 What is micro-plastic? What are the adverse effects of microplastic pollution?

5.3 Prepare a detailed report on the occurrence of microplastics in lakes, rivers and ocean. Throw some light on the magnitude of this problem to be faced by human beings for years to come.

5.4 What are the adverse effects and causes of thermal pollution?

5.5 Write brief note on pollution due to nanomaterials.

5.6 Write a short note on collection and preservation of the samples.

5.7 Explain grab sample, composite sample and integrated sample.

5.8 Explain objectives of conducting BOD test.

5.9 Explain BOD reaction rate constant and describe parameters on which it is dependent.

5.10 Draw a curve for BOD exerted and remaining with respect to time for biodegradable wastewater and derive mathematical expression for both.

5.11 Why only about 60% BOD is satisfied during BOD test determination, whereas during actual wastewater treatment in aerobic process more than 90% of BOD can be removed during 5 to 6 h of retention time in biological reactor?

5.12 BOD of a sewage incubated for 3 days at 27 °C was measured to be 110 mg/L. Calculate BOD_5 at 20 °C. Consider $k = 0.23$ per day (base e) and temperature coefficient $= 1.047$. (Answer: $L_0 = 179$ mg/; BOD_5 at 20 °C $= 122$ mg/L)

5.13 Describe nitrification during BOD test.

5.14 Explain correlation between BOD, BOD_u and COD for sewage.

5.15 How COD/TOC and BOD/TOC ratio will change with stages of wastewater treatment?

5.16 Describe the limitations of BOD test, how size of biodegradable particles will affect the rate of reaction?

5.17 Describe how acidity and alkalinity of the wastewater sample is determined.

5.18 What are the different forms in which solids are present in the wastewater? Provide short description on determination of each of these.

5.19 A sample of wastewater has 4 day 20 °C BOD value of 75% of the final. Find the BOD reaction rate constant per day. If the ultimate BOD of this wastewater is 300 mg/L, what will be BOD_3 at 27 °C? Consider temperature coefficient $= 1.047$.

5.20 Following observations were made during BOD determination in the lab:
   (i) Initial dissolved oxygen in blank $= 6.3$ mg/L and in sample it is 6.0 mg/L.
   (ii) Final DO after 3 days of incubation in blank $= 5.7$ mg/L and in sample $= 1.8$ mg/L.

   The sample was 5% diluted and incubated at temperature of 27 °C for three days. Determine BOD_3 at 27 °C and BOD_5 at 20 °C, considering BOD reaction rate constant of 0.23 per day (base e), temperature coefficient $= 1.047$.

References

APHA-AWWA-WEF. (2017). *Standard methods for the examination of water and wastewater* (23rd ed.). American Public Health Association (APHA), American Water Works Association (AWWA), Water Environment Federation (WEF), Washington DC.

Barrera-Díaz, C., Linares-Hernández, I., Roa-Morales, G., Bilyeu, B., & Balderas-Hernández, P. (2009). Removal of biorefractory compounds in industrial wastewater by chemical and electrochemical pretreatments. *Industrial and Engineering Chemistry Research, 48*(3), 1253–1258.

Binder, S., Levitt, A. M., Sacks, J. J., & Hughes, J. M. (1999). Emerging infectious diseases: Public health issues for the 21st Century. *Science, 4*(5418), 1311–1313.

Colt, J. (2006). Water quality requirements for reuse systems. *Aquacultural Engineering, 34*, 143–156.

Colwell, R. R. (1996). Global climate and infectious disease: The Cholera Paradigm. *Science, 4*(5295), 2025–2031.

CPCB. (2013). *Performance evaluation of sewage treatment plants under NRCD*. Central Pollution Control Board. Government of India. https://cpcb.nic.in

da Silva Vilar, D., Torres, N. H., Bharagava, R. N., Bilal, M., Iqbal, H. M. N., Salazar-Banda, G. R., Eguiluz, K. I. B., & Ferreira, L. F. R. (2021). Emerging contaminants in environment: Occurrence, toxicity, and management strategies with emphasis on microbial remediation and advanced oxidation processes. *Microbe Mediated Remediation Environmental Contaminants* https://doi.org/10.1016/B978-0-12-821199-1.00001-8

Gao, Y., Yang, T., & Jin, J. (2015). Nanoparticle pollution and associated increasing potential risks on environment and human health: A case study of China. *Environmental Science and Pollution Research, 22*(23), 19297–19306.

Ghangrekar, M. M., & Chatterjee, P. (2018). Water pollutants classification and its effects on environment. In R. Das (Ed.), *Carbon nanotubes for clean water, carbon nanostructures* (pp. 11–26). Springer International Publishing AG, part of Springer Nature 2018.

Ghangrekar, M. M., Kumar, S., & Chakraborty, I. (2021). Environmental Impacts and necessity of removal of emerging contaminants to facilitate safe reuse of treated municipal wastewaters. In R. Yadava (Ed.), *Chapter 6 in environmental degradation: Challenges and strategies for mitigation*. Springer, Singapore.

Gita, S., Hussan, A., & Choudhury, T. G. (2017). Impact of textile dyes waste on aquatic environments and its treatment. *Environment and Ecology, 35*, 2349–2353.

Gogoi, A., Mazumder, P., Tyagi, V. K., Tushara Chaminda, G. G., An, A. K., & Kumar, M. (2018). Occurrence and fate of emerging contaminants in water environment: A review. *Groundwater for Sustainable Development, 6*, 169–180. https://doi.org/10.1016/j.gsd.2017.12.009

Gunasekaran, P. (2007). *Laboratory manual in microbiology*. New Age International. New Delhi, India.

Haritash, A. K., & Kaushik, C. P. (2009). Biodegradation aspects of polycyclic aromatic hydrocarbons (PAHs): A review. *Journal of Hazardous Materials, 169*(1–3), 1–15.

Kiwi, J., Pulgann, C., Pennger, P., & Gratzel, M. (1993). Beneficial effects of homogeneous photo-Fenton pre-treatment upon the biodegradation of anthraquinone sulfonate in wastewater treatment. *Applied Catalysis B Environmental, 3*, 85–99.

Klaine, S. J., Alvarez, P. J. J., Batley, G. E., Fernandes, T. F., Handy, R. D., & Lyon, D. Y. (2008). Nanomaterials in the environment: Behavior, fate, bioavailability, and effects. *Environmental Toxicology and Chemistry, 27*(9), 1825–1851.

Kroon, F. J., Berry, K. L. E., Brinkman, D. L., Kookana, R., Leusch, F. D. L., Melvin, S. D., Neale, P. A., Negri, A. P., Puotinen, M., Tsang, J. J., van de Merwe, J. P., & Williams, M. (2020). Sources, presence and potential effects of contaminants of emerging concern in the marine environments of the Great Barrier Reef and Torres Strait. *Australia Science of the Total Environment, 719*, 135140. https://doi.org/10.1016/j.scitotenv.2019.135140

Lei, M., Zhang, L., Lei, J., Zong, L., Li, J., Wu, Z., & Wang, Z. (2015). Overview of emerging contaminants and associated human health effects. *Biomed Research International*, 1–12. https://doi.org/10.1155/2015/404796

Maiti, S. K. (2001). *Handbook of methods in environmental studies: vol 1: Water and wastewater analysis*. ABD Publishers. India.

Mahmood, I., Imadi, S. R., Shazadi, K., Gul, A., & Hakeem, K. R. (2016). Effects of pesticides on environment. In *Plant, soil and microbes* (pp. 253–269). Springer International Publishing, Cham. https://doi.org/10.1007/978-3-319-27455-3_13

McParland, H., & Warnakulasuriya, S. (2012). Oral lichenoid contact lesions to mercury and dental amalgam—A review. *Journal of Biomedicine and Biotechnology, 2012*, 589569.

Myers, A. A., Southgate, T., & Cross, T. F. (1980). Distinguishing the effects of oil pollution from natural cyclical phenomena on the biota of Bantry Bay, Ireland. *Marine Pollution Bulletin, 11*, 204–207.

Nelson, E. J., Harris, J. B., Glenn, M. J., Calderwood, S. B., & Camilli, A. (2009). Cholera transmission: The host, pathogen and bacteriophage dynamic. *Nature Reviews Microbiology, 4*(10), 693–702.

Nowack, B., & Bucheli, T. D. (2007). Occurrence, behavior and effects of nanoparticles in the environment. *Environmental Pollution, 150*(1), 5–22.

O'Brien, N., & Cummins, E. (2008). Recent developments in nanotechnology and risk assessment strategies for addressing public and environmental health concerns. *Human and Ecological Risk Assessment, 14*(3), 568–592.

O'Brien, N. J., & Cummins, E. J. (2011). A risk assessment framework for assessing metallic nanomaterials of environmental concern: Aquatic exposure and behaviour. *Risk Analysis, 31*(5), 706–726.

Pandey, P. K., Philip, H., Kass, M. L., Soupir, S. B., & Singh, V. P. (2014). Contamination of water resources by pathogenic bacteria. *AMB Express, 4*, 51. https://doi.org/10.1186/s13568-014-0051-x

Reed, S. C. (1985). Nitrogen removal in wastewater stabilization ponds. *Journal (water Pollution Control Federation), 57*(1), 39–45.

Roig, B., Pouly, F., Gonzalez, C., & Thomas, O. (2001). Alternative method for the measurement of ammonium nitrogen in wastewater. *Analytica Chimica Acta, 437*(1), 145–149.

Rosenfeld, P. E., & Feng, L. G. H. (2011). Emerging contaminants. In *Risks of hazardous wastes*. Elsevier, pp. 215–222. https://doi.org/10.1016/B978-1-4377-7842-7.00016-7

Rout, P. R., Zhang, T. C., Bhunia, P., & Surampalli, R. Y. (2021). Treatment technologies for emerging contaminants in wastewater treatment plants: A review. *Science of the Total Environment, 753*, 141990. https://doi.org/10.1016/j.scitotenv.2020.141990

Semrany, S., Favier, L., Djelal, H., Taha, S., & Amrane, A. (2012). Bioaugmentation: Possible solution in the treatment of bio-refractory organic compounds (Bio-ROCs). *Biochemical Engineering Journal, 69*, 75–86.

Thomas, H. A. (1942). Bacterial densities from fermentation tube tests. *Journal of American Water Works Association, 34*(4), 572–576.

Fundamentals of Reactor Engineering and Overview of Sewage Treatment

6

6.1　Background

The wastewater treatment plant comprises of unit operations and unit processes, where the concept of mass balance, hydraulic regime in the reactor, and reactor engineering are important to ensure the desired efficiency of the pollutant removal. In unit operations, the treatment or removal of the pollutant occurs by the action of the physical forces. Whereas in the unit processes, the treatment is brought about predominantly by the chemical or biochemical reactions. A given wastewater treatment plant utilizes a number of unit operations and unit processes to achieve the desired level of treatment for the wastewater, typically to meet the discharge limit or to satisfy quality criteria for reuse. This collective wastewater treatment scheme is called as flow scheme, flow sheet, flow schematic, process diagram, or process train. For a given type of wastewater, many such process schematics can be developed, comprising of unit operations and processes, to achieve desired degree of treatment. However, the most desired process train would be the one, which will give reliable performance at the least capital as well as operating cost.

Basic considerations for developing the treatment flow sheet include: (a) characteristics of the wastewater to be treated and degree of treatment required; (b) regulatory discharge quality norms as set by the agency; (c) preference and expertise of the designers and availability of the equipment; (d) level of expertise available with the local plant operators; (e) plant economics; (f) topography of the site, land availability, and hydraulic requirement, if any; (g) available infrastructure on site; and (h) minimum environmental impacts and maximum environmental benefits.

For unit operation, involving physical forces, the concept of mass balance and hydraulic regime within the reactor are important in modelling the performance of such operations. For biochemical and chemical processes, apart from mass balance and hydraulics in the reactor, reaction rate determines the sizing required for the reactor and hence, these parameters play an important role in modelling the reactor performance. Thus, this chapter is intended to introduce the type of reactors used in wastewater engineering and describing process analysis for benefit of the reader to develop clear understanding of this subject. In addition, an overview of the sewage treatment plant is provided to introduce the readers with different reactors used for complete treatment of sewage to meet regulatory quality norms for safe discharge of treated effluent.

6.2   Wastewater Treatment Classification and Plant Analysis

The degree of treatment required can be determined by comparing the influent wastewater characteristics to the required effluent characteristics, adhering to the regulations set by a controlling agency. A number of different treatment alternatives can be developed to achieve the treated wastewater quality requirements.

6.2.1   Classification of Treatment Methods

The individual treatment unit selected in the flow scheme is usually classified as one of the following:

- Physical unit operation
- Chemical unit process
- Biological unit process.

Physical unit operation: Treatment method in which the application of physical forces predominates for removal of pollutant(s) is known as physical unit operation. Most of these methods are based on physical forces, e.g., screening, mixing, flocculation, sedimentation, flotation, and filtration.

Chemical unit process: Treatment method in which removal or conversion of pollutants is brought by the addition of chemicals or by other chemical reactions is known as chemical unit process, e.g., precipitation, gas transfer, adsorption, and disinfection.

Biological unit process: Treatment method in which the removal of contaminants is brought about by biochemical activity is known as biological unit process. This is primarily used to remove biodegradable organic substances from the wastewater, either present in colloidal or dissolved form. In this biological unit process, organic matter is converted into gases that can escape into the atmosphere and into bacterial cells, which can be further removed by settling. Biological treatment can also be used for nitrogen, phosphorous and sulphate removal from the wastewater.

Different treatment methods used in wastewater treatment plants are classified in three different categories as stated below:

- **Primary treatment**: This refers to physical unit operations employed for the removal of floating matters and settleable suspended solids from the wastewater.
- **Secondary treatment**: This refers to chemical and biological unit processes. Biological processes are mostly used for the removal of colloidal and soluble organic matters from the wastewater. Chemical or biological unit processes can also be used for the removal of nutrients and other soluble impurities present in the wastewater by precipitation or chemical oxidation.
- **Tertiary treatment**: Refers to any one or combination of two or all three, i.e., physical unit operations and chemical or biological unit processes, used after secondary treatment to offer polishing treatment to the wastewater to make it suitable for disposal or reuse.

6.2.2 Elements of Plant Analysis and Design

The important terminologies used in the analysis and design of treatment plants are stated below (CPHEEO, 2013).

Reactor: Reactor refers to the vessel or containment structure along with all of its appurtenances, in which the unit operation or process takes place. Although, unit operation or unit process used may be a natural phenomenon, however they are accelerated, initiated or controlled by altering environment in the reactor. To achieve control over the rate of reaction, appurtenances are fitted to the reactor as per the requirement.

Flow sheet: It is the graphical representation of a particular orderly combination of unit operations and unit processes used in a treatment plant to achieve desired level of treatment so as to produce final treated water suitable in quality for reuse or disposal, as per the case.

Process loading criteria: The criteria used as the basis for sizing the individual unit operation or unit process is known as process loading criteria or process design criteria.

Solid balance: It is determined by identifying the quantities of solids entering and leaving each unit operation or unit process. This solid balance is important to understand performance of the reactor.

Hydraulic profile: This is used to identify the elevation of free surface of wastewater as it flows through various treatment units. Finalizing hydraulic grade line for reactors in sequence is important to ensure gravity flow from one reactor to other.

Plant layout: It is the spatial arrangement, i.e., in plan, of the physical facilities of the treatment plant identified in the flow sheet. While making this spatial arrangement due care is taken to accommodate future augmentation of the treatment plant.

6.3   Order of Reaction

The reactions occurring during wastewater treatment are slow and hence, kinetic considerations are important for design. The general equation used for relating the rate of change
of concentration with respect to time can be expressed as per Eq. 6.1.

$$dS/dt = K \cdot S^n \tag{6.1}$$

where, S is the concentration of the reacting substance, K is the reaction rate constant per
unit time, t is the time elapsed and n denotes the order of the reaction ($n = 1$ for first
order reaction, $n = 2$ for second order reaction, and so on).

The value of K depends on the environmental conditions in the reactor, such as (a)
temperature, (b) availability of nutrients and growth factors, (c) presence of toxicity, if
any and (d) presence of catalysts. Zero order reactions ($n = 0$) are independent of the
substance concentration and hence their rate (dS/dt) is constant. Certain catalytic reactions
occur in this way and sometimes even biological reaction may follow zero order reaction
kinetics.

In first order reactions, the rate of change of concentration of substance is proportional
to the concentration of that substance in the reactor. This concentration of the substance
and rate will diminish with respect to time. Decomposition of single substance exhibits the
true first order reaction kinetics. Biological stabilization of organic matter in batch reactor
is a typical example of a pseudo-first order reaction. The rate of reaction is proportional
to the concentration of a single item, organic matter in this case, provided the other
parameters controlling the reactions are favourable. If the substrate concentration (organic
matter) is maintained constant within the narrow range (as in the case of continuous flow
completely mixed reactors), then the rate of reaction is practically constant and thus, it is
like pseudo-zero order type of reaction. Some biological treatment systems behave in this
manner.

There are various complex processes whose overall rate is approximately first order in
nature. With a complex substrate present in sewage or industrial wastewater, overall reaction rate may appear like a first order reaction, although the individual substrate among
the several may exhibit the zero-order reaction kinetics. This is because, the rate of reaction may be higher initially due to higher utilization of easily biodegradable substrate,
but rate will eventually slow down with respect to time due to more complex substrates
left in the reactor. Sometimes the attached growth biological process is also modelled as
second order reaction, e.g., super rate trickling filter.

6.4 Types of Reactors Used in Wastewater Treatment

In a reactor, the reactions that are undergoing could be heat liberating (exothermic) and/or heat absorbing (endothermic). Operating temperature is important in the design of reactor as it will affect the reaction rate. However, in most of the reactors used in wastewater engineering, the temperature change due to reactions is usually negligible. The reactor may also be classified according to the number of phases involved in the reactions, such as liquid, solid and gas. The reaction is known to be homogeneous, when it occurs only in single phase, and when the reaction occurs at least in the presence of two phases it is referred as heterogenous. To accelerate the rate of reaction sometimes catalyst is used. This catalyst used is neither reactant nor product and it can either accelerate or hinder a rate of reaction. Accordingly, the reactions occurring in a reactor may be classified as catalytic or non-catalytic reaction.

Wastewater treatment is generally achieved in reactors in which physical, chemical or biological changes occur to facilitate treatment to the wastewater by removing targeted pollutant(s). The types of reactors used in the wastewater treatment shall be classified as (a) batch reactor, (b) plug-flow or tubular-flow reactor, (c) completely mixed or continuous-flow stirred tank reactor (CSTR), (d) arbitrary-flow reactor, (e) packed bed reactor, and (f) fluidized bed reactor. These reactor types are briefly discussed below.

(a) *Batch reactor*: These reactors are operated as fill and draw type. In this type of reactor, the wastewater flow is not continuous in the reactor, i.e., no flow enters or leaves the reactor, when reaction cycle is on. The reactors are operated in batch mode with fill time, reaction time, and withdrawal time. For example, BOD bottles incubated during the test and sequencing batch reactor (SBR). The reactor content may be completely mixed to ensure that no temperature or concentration gradient exists within the reactor. All the elements in the reactor, under batch mode of operation, are exposed to treatment for the same period of time for which the wastewater is held in the reactor. Hence, performance of the batch reactor is comparable with ideal plug flow reactor.

(b) *Plug-flow (tubular flow) reactor*: In this reactor, the fluid particles pass through the tank and are discharged in the same sequence in which they enter the tank. The particles remain in the tank for a time equal to theoretical detention time. There is no overtaking, falling behind, intermixing or dispersion of the particles. Longitudinal dispersion is considered as minimal, which can typically occur in high length to width ratio of the tanks or in a pipe flow if the velocity is not excessive. Such condition is also considered to occur when there are several CSTRs operated in series. The examples of plug flow reactor used in wastewater treatment are a rectangular grit chamber and aeration tank of conventional activated sludge process with high length to width ratio.

(c) *Continuous-flow stirred tank (completely-mixed) reactor*: In this reactor, the incoming wastewater volume is dispersed immediately throughout the tank as it enters the tank.

Thus, the content in the reactor is perfectly homogeneous at all points in the reactor. This can be achieved in square (or slightly rectangular) and circular shape of the tank (plan view) that are uniformly mixed. The particles leave the tank in proportion to their statistical population. The concentration of pollutant in the effluent leaving the reactor is the same as that present in the reactor.

(d) *Arbitrary flow*: In this reactor, any degree of partial mixing between plug flow and complete mixing condition exists. Each element of the incoming flow resides in the reactor for different duration of time. It is also called as intermixing or dispersed flow and it lies between ideal plug flow and ideal completely mixed reactor. This flow condition can be used in practice to describe the flow conditions in most of the reactors. The ideal plug-flow and CSTR are the limiting cases and actual flow regimes will range in a broad spectrum between these ideal models. These intermediate cases represented by arbitrary flow could be plug-flow with dispersion and intermediate mixed flow.

(e) *Packed bed reactor*: They are filled with some packing medium, such as rock, slag, ceramic or synthetic plastic media. These packed bed reactors can be anaerobic filter, when completely filled and no air is supplied, or aerobic (trickling filter), when flow is intermittent or submerged aerobic filter, when compressed air is supplied from the bottom. The media offering surface for biofilm attachment is stationary and the wastewater moves through the voids present in the packed bed provided in the reactor.

(f) *Fluidized bed reactor*: This reactor is similar to packed bed reactor except packing medium is expanded by the upward movement of fluid (or air) than resting on each other as in case of a fixed bed reactor. The porosity or degree of fluidization can be controlled by controlling flow rate of fluid (wastewater or air). Recently, the moving bed reactor concept is also being encouraged in the wastewater treatment by selecting media material, which has specific gravity close to that of wastewater. Thus, this media can freely move in the reactor and does not require high up-flow velocities to keep media in fluidized state as required in case of fluidized bed reactor.

6.5   Flow Patterns of Reactors

The flow pattern in the reactors depends on mixing conditions developed in them. This mixing in turn depends upon the shape of the reactor, energy spent per unit volume of the wastewater present in the reactor, the size and scale of the reactor, up-flow velocity of the liquid, rate of biogas generation (in case of an anaerobic reactor) or the rate of air supplied (in case of an aerobic reactor), etc. Flow pattern affects the time of exposure to the treatment and substrate distribution in the reactor. Depending upon the flow pattern the reactors can be classified as:

(a) Batch reactors,
(b) Ideal plug flow reactors,
(c) Ideal completely-mixed reactors,
(d) Non-ideal, dispersed flow reactors, and
(e) Series or parallel combinations of the reactors.

The hydraulic regime in the reactor can be defined with respect to the 'Dispersion number', Eq. 6.2, which characterizes mixing condition in the reactor (Arceivala and Asolekar, 2007).

$$\text{Dispersion Number} = D/UL \tag{6.2}$$

where, D is the axial or longitudinal dispersion coefficient, L^2/t; U is the mean flow velocity along the reactor, L/t; L is the length of axial travel path, L.

For ideal plug flow $D/UL = 0$, since, dispersion is zero by definition. However, in practice $D/UL \leq 0.2$ indicates the regime is approaching ideal plug flow conditions. The D/UL value of 3 to 4.0 or greater indicates that reactor is approaching completely mixed conditions.

The residence time distribution (RTD) of the fluid elements that compose the effluent, characterizes each flow model. Tracer study on the reactor may be carried out to make it possible to determine the residence time distribution curve for the reactor. The tracer input given may be pulse (slug release) or step inputs (continuous release of tracer). This is described in details for each type of the reactor in this section.

It is interesting to observe that facultative aerated lagoons and oxidation ponds vary widely in their mixing conditions, practically from lower to higher value of D/UL, and one must give consideration to this aspect while designing by deciding in advance what mixing regime is desired. Aeration tank of activated sludge process and Carousel type oxidation ditch exhibit well-mixed conditions and hence, can be designed as the completely-mixed reactor (Arceivala and Asolekar, 2007). Temperature variation from 12 to 20 °C has very little influence on the dispersion conditions in the reactor. However, cross wind has significant effect and might encourage short-circuiting in the larger reactor, like oxidation pond, hence wide variation in the dispersion number determined at different time might occur.

6.5.1 Plug Flow Reactor

In a long-tubular channel, for a plug flow reactor of volume V, let a wastewater flow of Q is applied, thus the mean residence time, i.e., theoretical residence time, t_o, will be V/Q. If a pulse or slug of dye with mass 'M' is released in the influent, it will move into the plug flow reactor and form a band or a piston that will move through the reactor.

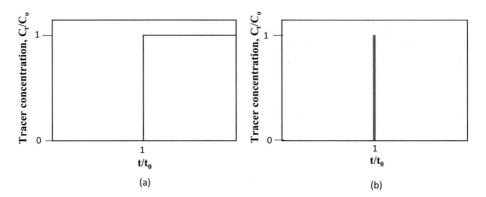

Fig. 6.1 Tracer curve for the plug flow reactor **a** with continuous release of dye and **b** for slug release of dye

Eventually this band (piston) will reach to the other end of the reactor and will appear in the effluent of the reactor. The RTD curve is a plot of C_t/C_0 versus t/t_0; where, t is the time after release of the dye, C_t is the dye concentration in the effluent of the reactor after time 't', and C_0 is the concentration of the dye released.

The RTD curve for the continuous release of the dye is presented in the Fig. 6.1a, whereas for slug release of dye (tracer) in the plug flow reactor RTD curve is shown in Fig. 6.1b. In ideal plug flow reactor, the dye released will make its appearance in the effluent after a time 't' from the release of the dye, which will be equal to the theoretical residence time of the plug flow reactor. Thus, t/t_0 in this case will be equal to one.

6.5.2 Non-ideal Plug Flow Reactor

In case of non-ideal plug flow reactor, where longitudinal mixing or plug flow with dispersion is occurring, the mixing will be intermediate between completely mixed reactor and plug flow reactor. In such a reactor, when a slug dose of dye is added, it will not appear at the outlet end of the reactor as shown in Fig. 6.1b as slug, rather due to dispersion and longitudinal mixing, the base of the curve will be widened as presented in Fig. 6.2. In this case, the average residence time of the tracer in the reactor will be lesser than the theoretical residence time. The exact nature of the curve depends on the type of reactor, its geometrical shape, size, mixing intensity, dispersion coefficient, etc., and based on the tracer experiment on that reactor, the shape of the curve can be obtained.

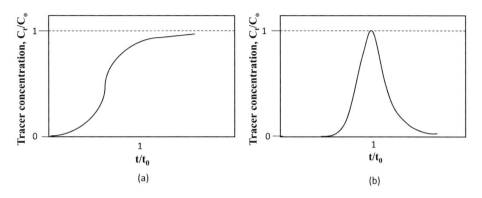

Fig. 6.2 Tracer curve for the non-ideal plug flow reactor **a** with continuous release of dye and **b** for slug release of dye

6.5.3 Completely Mixed Reactor

In a completely mixed reactor with volume V receiving wastewater with flow rate of Q, when a slug (pulse) dose of dye of mass S is released, it will get immediately mixed in the total volume V of the reactor to get a uniform concentration of the dye as C_0. This concentration C_0 will be equal to S/V immediately after release of the dye in the reactor, i.e., the time 't' elapsed is considered to be zero. As the wastewater flow continues in the reactor, this administered slug dose of the dye with initial concentration of C_0 in the reactor will start getting diluted making the concentration C_t with elapse of time t. After some time, the dye will be completely washed out from the reactor and C_t will become zero. The curves for the tracer input with continuous dye addition and slug dose of dye addition are presented in the Fig. 6.3a, b, respectively.

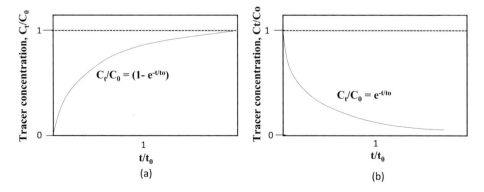

Fig. 6.3 Tracer curve for the completely mixed reactor **a** with continuous release of dye and **b** for slug release of dye

A mass balance equation in generalized form can be written for this dye addition as presented in Eq. 6.3.

$$\text{Accumulation} = \text{Inflow} - \text{Decrease due to reaction} - \text{Outflow} \qquad (6.3)$$

If the dye addition is done as a slug dose, then after time $t = 0$, there is no input dye addition in the completely mixed reactor. Also, if there is no reaction that is occurring in the reactor for the degradation of dye, then Eq. 6.3 can be expressed as Eq. 6.4.

$$dC_t \times V = 0 - 0 - C_t \times Q \times dt \qquad (6.4)$$

Rearranging the Eq. 6.4 will give Eq. 6.5.

$$\int_{C_0}^{C_t} \frac{dC_t}{C_t} = \frac{Q}{V} \int_0^t dt \qquad (6.5)$$

Solving the integration to give Eq. 6.6, where $V/Q = t_0$, i.e., the mean time of residence in the reactor. This Eq. 6.6 is the equation for the tracer curve for a slug dye released in the completely mixed reactor (Fig. 6.3b).

$$\frac{C_t}{C_0} = e^{-\frac{Q}{V}t} = e^{\frac{-t}{t_0}}. \qquad (6.6)$$

6.6   Concept of Mass Balance

To understand the changes occurring in the reactor during any reactions that are taking place, the fundamental approach of mass balance is adopted, which is based on the premise that for non-nuclear reactions, the mass can neither be created nor be destroyed and it can only be converted from one form to another. Thus, mass that accumulates in the reactor will be equal to the mass of inflow to the system minus the mass leaving the system and minus the mass converted during the reactions. This statement can be stated as per the Eq. 6.3, as stated earlier.

In case of unit operation, where there is no reaction that occurs in the reactor, the decrease due to reaction is absent and mass of pollutant accumulated in unit operation will be difference between inflow and outflow. In general analysis of mass balance, consider a reactor as shown in Fig. 6.4, with system boundary as shown by the dotted line in the figure. Consider, Q is the influent flow rate with reactant concentration of C_0, V is the volume of the reactor, which is equal to the liquid volume, C_t is the concentration of the pollutant (reactant) and being completely mixed reactor consider this concentration will leave the reactor with flow rate of Q. Thus, for this reactor Eq. 6.3 can be expressed as per Eq. 6.7, where r is considered as the rate of reaction of the pollutant, mass/volume time.

Fig. 6.4 Description of the reactor considered for mass balance analysis

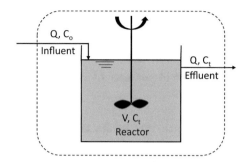

$$dC_t \times V = Q \times C_0 \times dt - V \times r \times dt - C_t \times Q \times dt \qquad (6.7)$$

Rearranging the terms following Eq. 6.8 can be obtained.

$$\frac{dC_t}{dt} V = Q{\cdot}C_0 - V \cdot r - Q{\cdot}C_t \qquad (6.8)$$

If the rate of reaction is considered as the first order reaction, then $r = KC_t$. Also, dC_t/dt represents the change in the pollutant concentration with time due to the reaction(s) taking place in the reactor. Substituting value for r, the Eq. 6.9 can be obtained.

$$\frac{dC_t}{dt} V = Q{\cdot}C_0 - V \cdot KC_t - Q \cdot C_t \qquad (6.9)$$

Diving by V, the Eq. 6.10 shall be obtained.

$$\frac{dC_t}{dt} = \frac{Q}{V}C_0 - KC_t - \frac{Q}{V}C_t \qquad (6.10)$$

Since, most of the unit processes are designed for steady state, the rate of change of pollutant concentration will be zero once the reactor reaches the steady state, after allowing for sufficient start-up time. Also, V/Q represents the hydraulic retention time (θ). Thus, substituting value of V/Q and dC_t/dt equal to zero in Eq. 6.10 and solving, the Eq. 6.11 can be obtained. From this equation, the concentration of pollutant in the effluent of the reactor can be predicted.

$$C_t = \frac{C_o}{1 + K\theta} \qquad (6.11)$$

Example: 6.1 In a completely mixed biological wastewater treatment process the influent BOD to the reactor is 180 mg/L. If the reaction rate $K = 0.5$ per hour and the desired effluent BOD concentration is 20 mg/L, find out the retention time required.

Solution

$C_0 = 180$ mg/L; $C_t = 20$ mg/L; and $K = 0.5$ h^{-1}, using Eq. 6.11 and solving for θ:

$$20 = \frac{180}{1 + 0.5\theta}$$

Solving, $\theta = 16$ h

Hence, retention time of 16 h is required to get 20 mg/L effluent BOD concentration.

6.6.1 Analysis of Batch Reactor

Consider the batch reactor shown in the Fig. 6.5, where it is filled and operated for reactions to occur with time, hence C_t will change with time with initial value of C_0. The reactor is operated as completely mixed system, hence at any given time the concentration of the reactant will be uniform in the reactor, which will of course be changing with time. Since, in Eq. 6.3, the inflow and outflow term will be zero, being batch operation, the mass balance equation will become Eq. 6.12.

$$\text{Accumulation} = -\text{Decrease due to reaction} \qquad (6.12)$$

Considering the first order rate of reaction, with reaction rate constant K, the Eq. 6.12 can be written as Eq. 6.13 by considering the accumulation equal to $V \cdot dC_t/dt$.

$$V \cdot dC_t/dt = -K C_t V \qquad (6.13)$$

Rearranging the Eq. 6.13 for integration, Eq. 6.14 can be obtained.

$$\int_{C_0}^{C_t} \frac{dC_t}{C_t} = -K \int_0^{\theta} dt \qquad (6.14)$$

Integrating within the limits for t equal to zero to θ, the concentration of reactant will change from C_0 to C_t and Eq. 6.15 can be obtained.

Fig. 6.5 Batch rector considered for mass balance

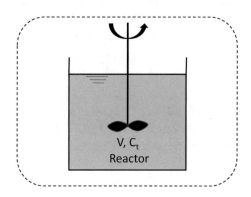

V, C_t
Reactor

$$C_t = C_0 e^{-K\theta} \tag{6.15}$$

Example: 6.2 If first order reaction rate constant for the batch reactor is 0.4 per hour, what will be the percentage removal of the pollutant that will occur during the reaction time of 6 h? If initial concentration of the pollutant is 120 mg/L, what will be the effluent concentration?

Solution
From Eq. 6.15, substituting $\theta = 6$ and $K = 0.4$ h^{-1}, the C_t/C_0 can be obtained.
    Thus, $C_t/C_0 = e^{-0.4 \times 6} = 0.0907$
    Removal efficiency $= (1 - C_t/C_o) \times 100 = 90.93\%$
    Effluent concentration of pollutant $= C_t = 0.0907 \times C_0 = 0.0907 \times 120 = 10.88$ mg/L.

6.6.2 Analysis of Completely Mixed Reactor

Consider the continuously mixed reactor where wastewater with the pollutant (reactant) is being admitted continuously and the treated wastewater is leaving the reactor continuously, as shown in Fig. 6.4 earlier. Due to completely mixed nature of the reactor, the pollutant entering the reactor with incoming flow Q will be instantaneously mixed in the reactor and hence, concentration of pollutant in the effluent will be same as the concentration of it present in the reactor.

    For such completely mixed (CSTR) reactor, considering the first order rate of reaction with rate constant K, the mass balance Eq. 6.3 can be written using variables associated with this CSTR to represent Eq. 6.9 and solution to this equation can be obtained as per Eq. 6.11, as stated earlier. For first order reaction, the residence time required for the CSTR will be more as compared to batch or plug flow reactor for achieving the same degree of treatment. This is illustrated later in the solved examples. The efficiency of the CSTR reactor can be improved by using CSTRs in series rather than using single large CSTR of volume equal to total volume of CSTRs used in series.

6.6.3 Analysis of Plug Flow Reactor

A plug flow reactor (Fig. 6.6) is operated under steady state condition; thus, the flow and influent concentration of pollutant are considered constant and also effluent pollutant concentration would reach the stable concentration. Considering the ideal plug flow conditions prevails in the reactor, thus there is no overtaking or falling back of fluid elements and negligible dispersion occurring in the reactor. The concentration of pollutant entering the reactor is C_0 and as the wastewater moves through the reactor, the concentration 'C_t' will keep changing and finally when the effluent is leaving the reactor the pollutant concentration will be C_e. Due to gradient in the concentration along the length of the reactor,

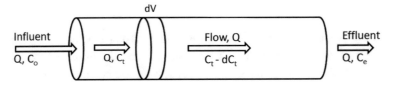

Fig. 6.6 Plug flow reactor considered in mass balance analysis

for analysis consider a small volume 'dV' of the reactor (Fig. 6.6). The mass balance equation (Eq. 6.3) at steady can be written as per Eq. 6.16.

$$0 = \text{Inflow} - \text{Outflow} - \text{Decrease because of reaction} \tag{6.16}$$

Or

$$\text{Inflow} = \text{Outflow} + \text{Decrease because of reaction}$$

The material balance for the differential reactor of volume dV and rate of reaction r can be written referring to the Fig. 6.6 as per Eq. 6.17.

$$QC_t = Q(C_t - dC_t) + r \cdot dV \tag{6.17}$$

Thus,

$$QdC_t = r \cdot dV \tag{6.18}$$

Equation 6.18 represents the general equation for design of plug flow reactor. Considering the first order rate of reaction, then $r = -KC_t$. Using this, the Eq. 6.18 becomes Eq. 6.19 and after rearranging for integration, it will yield Eq. 6.20. Solving Eq. 6.20 will result in Eq. 6.21.

$$QdC_t = -KC_t \cdot dV \tag{6.19}$$

$$\int_{C_o}^{C_e} \frac{dc_t}{C_t} = -\frac{K}{Q} \int_{0}^{V} dV \tag{6.20}$$

$$ln\frac{C_e}{C_o} = -K\frac{V}{Q} \tag{6.21}$$

whereas, V/Q is the hydraulic retention time θ. Substituting it in Eq. 6.21 and solving we can get the final Eq. 6.22.

$$C_e = C_0 \cdot e^{-K \cdot \theta} \tag{6.22}$$

Example: 6.3 A plug flow aeration tank of activated sludge process is treating wastewater having a BOD of 160 mg/L at a hydraulic retention time of 5 h. The rate of reaction is considered to be first order with value of reaction rate constant as 0.4 per hour. Find out effluent concentration that can be obtained at this retention time of 5 h.

Solution

Using Eq. 6.21 and substituting $C_0 = 160$ mg/L, $K = 0.4$ h^{-1}, and $\theta = 5$ h, the solution will be obtained.

$$C_e = 160 \cdot e^{-0.4 \times 5} = 21.65 \text{ mg/L} \sim 22 \text{ mg/L}$$

Hence, effluent BOD concentration of this plug flow reactor will be 22 mg/L.

6.6.4 Analysis of CSTR Reactors in Series

When multiple CSTRs are arranged in the series with either same HRT in each reactor or different HRTs, then the performance of this reactor should be obtained considering performance of the each of the CSTR involved in series. The problem become further complex, if the CSTRs involved are having different HRTs. For a group of CSTRs working in series with same HRT for each reactor, the solution can be obtained using Eq. 6.11. For example, if there are two CSTRs operated in series and initial concentration of pollutant to be removed is C_0 and the concentration of pollutant in the effluent of first and second CSTR is C_1 and C_2, then from Eq. 6.11 we get following Eq. 6.23 and Eq. 6.24, respectively, for predicting this effluent pollutant concentrations from these CSTRs working in series.

$$C_1 = \frac{C_o}{1 + K\theta} \tag{6.23}$$

and

$$C_2 = \frac{C_1}{1 + K\theta} \tag{6.24}$$

Substituting value of C_1 from Eq. 6.23 in Eq. 6.24, the Eq. 6.25 will be obtained, which is the result of combining the performance of both the CSTRs to obtain the effluent concentration of second CSTR, C_2, which is the final effluent concentration here in this case.

$$C_2 = \frac{C_0}{(1 + K\theta) \times (1 + K\theta)} = \frac{C_0}{(1 + K\theta)^2} \tag{6.25}$$

Thus, generalizing the Eq. 6.25 for n number of CSTRs in series with each having equal residence time, the Eq. 6.26 will be obtained.

$$C_n = \frac{C_o}{(1 + K\theta)^n} \tag{6.26}$$

where, C_n is the effluent concentration of the pollutant from the nth CSTR in series, which is the final concentration; K is the rate constant; θ is the hydraulic retention time, here in this case it is same for all CSTRs involved in series.

The solution for CSTR reactors in series can be obtained graphically as well for number of reactors involved in series. For this, a graph of reaction rate 'r' versus concentration is plotted. Knowing the residence time in each CSTR, a line with the slope of $-(1/\theta)$ is drawn from the C_0. The point at which this will intersect to the plot of 'r' versus concentration, it will represent reaction rate of first CSTR and a vertical line from the intersection point to x-axis will give the effluent concentration for first CSTR (Reynolds and Richards, 1996). Similar exercise is repeated for the subsequent CSTRs to estimate the effluent pollutant concentrations for them. This can be further understood from the material balance equation, which can be written as Eq. 6.27 for steady state conditions.

$$\text{Inflow} = \text{Outflow} + \text{Decrease due to reaction} \tag{6.27}$$

Considering the number of CSTRs in series operation with variable declaration as indicated in Fig. 6.4, the Eq. 6.27 can be expressed as Eq. 6.28.

$$Q \cdot C_0 = Q \cdot C_1 + r_1 \cdot V \tag{6.28}$$

where, $-r$ could be represented as KC_t^n, with n as order of reaction. Dividing by Q and rearranging, we get Eq. 6.29.

$$C_1 = C_0 - r_1 \cdot V/Q = C_0 - r_1 \cdot \theta \tag{6.29}$$

Since, $V/Q = \theta$

Now from the graph (refer Fig. 6.7 of Example 6.5), the line drawn from the point $(C_0, 0)$ with a slope of $-(1/\theta)$ will intersect the rate curve at r_1 and vertical line from this point to axis will provide the abscissa, representing the effluent concentration of the pollutant from the first CSTR as C_1. Thus, for the slope line we can write the Eq. 6.30.

$$\text{Slope} = -\frac{1}{\theta} = \frac{-(r_1 - 0)}{C_1 - C_0} \tag{6.30}$$

Solving above equation for C_1, we will get the same equation as represented in Eq. 6.29. The Eq. 6.30 can be generalized for the nth number of CSTRs in series as per Eq. 6.31 (Reynolds and Richards, 1996).

$$-\frac{1}{\theta} = \frac{-(r_n)}{C_n - C_{n-1}} \tag{6.31}$$

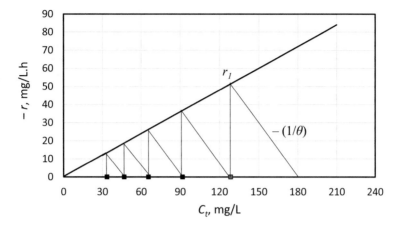

Fig. 6.7 Graphical solution for five CSTRs operated in series

Example: 6.4 A pollutant P with the initial concentration of 180 mg/L is intended to be removed from the series of CSTR with all having equal residence time of 1 h. The reaction rate constant has a value of 0.4 per hour. Estimate the final effluent concentration of P and concentration after each stage of CSTR. If the effluent limit of this pollutant P is 20 mg/L, are these five CSTRs in series adequate to meet this effluent limit? If not, how many CSTRs of retention time of 1 h in each will be required?

Solution
Using the generalized Eq. 6.26 and using $C_0 = 180$ mg/L, $K = 0.4$ h^{-1}, $\theta = 1$ h and $n = 5$ the C_5 can be predicted.

$$C_5 = \frac{C_0}{(1 + K\theta)^n} = \frac{180}{(1 + 0.4 \times 1)^5} = 33.47 \text{ mg/L}.$$

This effluent concentration is not meeting the discharge limit. Hence, for determining the number of CSTRs required by solving the Eq. 6.26 for n and using $C_n = 20$, $C_0 = 180$ mg/L, $K = 0.4$ h^{-1}, and $\theta = 1$ h.

$$20 = \frac{180}{(1 + 0.4 \times 1)^n} \quad \text{Solving for } n, \text{ the } n = 6.53,$$

Hence, 7 CSTRs in series will be required to meet the discharge standard of less than or equal to 20 mg/L of P. The effluent concentration of P in each stage of CSTR can be estimated using value of n as 1, 2, 3, and 4 in Eq. 6.26. These effluent concentrations at each stage will be $C_1 = 128.57$ mg/L, $C_2 = 91.84$ mg/L, $C_3 = 65.60$ mg/L, and $C_4 = 46.86$ mg/L.

Example: 6.5 A pollutant P with the initial concentration of 180 mg/L is intended to be removed from the series of CSTR with all five having equal residence time of 1 h. The reaction rate constant has a value of 0.4 per hour. Estimate the final effluent concentration of P and concentration after each stage of CSTR using graphical method. Consider the first order reaction $-r = KC_t$. Also, compare the performance of this CSTR with plug-flow reactor of similar residence time.

Solution

For $-r = KC_t$, the reaction rates at the different concentrations for $K = 0.4$ h^{-1} are given below:

C_t, mg/L	30	60	90	120	150	180	210
$-r$, mg/L h	12	24	36	48	60	72	84

Plotting the graph of $-r$ versus C_t, as represented in Fig. 6.7. Now, the slope of $-(1/\theta) = -1$

By drawing a line with the slope of -1 from point $(180, 0)$ and then dropping a vertical line from the point where it is intersecting the plot of $-r$, which will provide a value of effluent concentration of pollutant from the first CSTR, i.e., 128 mg/L. Similarly, other values can be obtained as 92, 65, 47 and 33 mg/L, for subsequent CSTRs.

The performance of the plug-flow reactor with retention time of 5 h is given by Eq. 6.22.

$$C_e = C_o \cdot e^{-K \cdot \theta} = 180 \times e^{-0.4 \times 5} = 24.36 \, \text{mg/L}$$

Thus, it can be inferred that when 5 CSTRs in series are able to reduce the pollutant concentration to 33 mg/L from influent concentration of 180 mg/L, the plug-flow reactor with same retention time as total of CSTRs will be able to reduce it to about 24 mg/L; thus, demonstrating better efficiency.

6.6.5 Analysis of Dispersed Plug Flow Reactor

The flow regime occurring in this reactor is in between the ideal plug-flow and the ideal completely mixed reactor due to dispersion and longitudinal mixing, and hence, it is considered as non-ideal regime. Wehner and Wilhelm (1956) developed an equation for such dispersed plug-flow reactor as presented in Eq. 6.32.

$$\frac{C_t}{C_0} = \frac{4i \, e^{\frac{UL}{2D}}}{(1+i)^2 e^{i\frac{UL}{2D}} - (1-i)^2 e^{-i\frac{UL}{2D}}} \tag{6.32}$$

where, U is the velocity in the axial direction, m/sec; L is the length of the reactor, m; D is the longitudinal dispersion coefficient, m^2/sec; $i = [1 + 4K\theta(D/UL)]^{0.5}$, which is dimensionless; K is the rate constant, per unit time; and θ is the hydraulic residence time, sec.

Depending on the intensity of the mixing and the turbulence present in the reactor, the dispersion model will range from plug flow at one extreme to completely mixed at other. The required reactor volume for these dispersed plug flow reactors will be between plug-flow reactor and completely mixed reactor. The longitudinal dispersion coefficient, D, is related to the mixing condition prevailing in the reactor and it measures the degree of axial dispersion in the reactor. Thus, as explained earlier, for the dispersion number D/UL tending to zero indicates negligible dispersion and plug flow reactor configuration. Whereas, when this dispersion number tends to infinity, it indicates considerable dispersion and hence, completely mixed reactor configuration.

For a conventional activated sludge process, which is designed as a plug flow reactor, the value of dispersion number ranges between zero and 0.2. For dispersed activated sludge process, the value of the dispersion number will range from 0.2 to 4.0 depending on the degree of mixing prevailing in the reactor. Whereas, for completely mixed activated sludge process provided with mechanical aeration, the dispersion number will have value greater than 4 and for oxidation ponds, it is generally in between 0.1 and 2.0 (Arceivala and Asolekar, 2007). This dispersion number for a particular reactor can be determined by performing the tracer studies following the procedure described in literature (Levenspiel, 1972).

For first order reaction rate kinetics, the retention time required for the batch reactor and plug flow reactor is same for same percentage of pollutant removal. Whereas, for same first order reaction rate, the hydraulic retention time required for the dispersed plug flow reactor will be higher and for completely mixed reactor, it will be further higher for the same degree of treatment as compared to plug flow reactor. Even though the residence time required for completely mixed reactor is higher, due to other advantages of the CSTR, they are preferred for wastewater treatment. The advantages that the CSTR offers are better ability to handle fluctuations in the incoming wastewater characteristics, including any marginal toxicity, due to dilution that this reactor offers. Since in CSTR, incoming wastewater is getting mixed with the larger volume of the wastewater present in the reactor basin; hence, minimizing the concentration shocks and favouring more or less stable reactor performance.

6.7   Overview of Sewage Treatment Plant

6.7.1   Characteristics of Municipal Wastewater

Characterization of wastewater is essential for suggesting an effective and economical treatment. It helps in the choice of treatment methods and deciding the extent of treatment, assessing the beneficial uses of treated wastewater and utilizing the self-purification capacity of natural water bodies in a planned and controlled manner. While analysis of wastewater for a particular case is advisable, for which the treatment plant design is being proposed, data from the other neighbouring cities may be utilized during initial stage of planning.

Domestic sewage comprises of spent water from kitchen, bathroom, lavatory, etc. Factors that contribute to variations in characteristics of the domestic sewage are daily per capita water consumption, quality of water supply and the type (intermittent or continuous), condition and extent of sewerage system, weather conditions in that locality and habits of the people. Municipal sewage, which contains domestic sewage, wastewater generated from commercial establishments and institutions, and industrial wastewaters, may differ in characteristics from place to place depending upon the type of industries and commercial establishments discharging wastewaters to the sewers. In this case it becomes necessary to characterise the wastewater generated from that city for designing the treatment plant.

Temperature of the sewage varies from place to place depending on the season and location of the place. In general, under South Asian conditions the temperature of the raw sewage is observed to be varying between 15 and 35 °C at various places in different seasons. Temperature lower than 10 °C is common in cities located at higher altitude and latitudes, particularly in winter seasons. The pH of the fresh sewage is slightly more than the water supplied to the community. Decomposition of organic matter may lower the pH of sewage with time, while the presence of industrial wastewater may cause extreme fluctuations. Generally, the pH of raw sewage is in the range 5.5 to 8.0. Fresh domestic sewage has a slightly soapy and cloudy appearance. With time the sewage becomes stale, darkening in colour and liberate pronounced obnoxious smell due to microbial activity that releases gases, like ammonia and hydrogen sulphide.

Though sewage generally contains less than 0.5% solids (typically around 0.1–0.2%), the rest being water, still the nuisance caused by the solids is a matter of concern, as these solids are highly degradable and therefore need proper disposal. These solids may be classified into dissolved solids, suspended solids and volatile suspended solids. Knowledge of the volatile or organic fraction of solid, which decomposes, becomes necessary as this constitutes the load on biological treatment units. Estimation of suspended solids, both organic and inorganic, gives a general idea of the load on sedimentation and grit removal facility. Dissolved inorganic fraction is to be considered when sewage is used for irrigation or any other reuse is planned. In general, the suspended solids present in the sewage range between 300 and 700 mg/L and total dissolved solids (TDS) concentration, depending on the TDS present in the water supplied, is in the range of 200–500 mg/L. The concentrations of these solids may be higher in case of municipal wastewaters.

The BOD observed for raw sewage is generally in the range of 100–400 mg/L, whereas for cities from developed countries higher than 400 mg/L of BOD values are reported. In general, the COD of raw sewage at various places is reported to be in the range 200–700 mg/L. It is not possible to establish a general relationship between the experimental five-day BOD and the ultimate BOD for all wastewaters and it is wastewater specific. For sewage with first order BOD reaction rate constant $K = 0.23$ day^{-1} at 20 °C, the BOD$_5$ is 0.68 times of the ultimate BOD, and ultimate BOD is 87% of the COD. Thus, the COD/BOD ratio for the sewage is around 1.7.

Generally, the domestic sewage contains sufficient nitrogen and phosphorous to take care of the needs of the biological treatment. Nitrogen content in the untreated sewage is observed to be in the range of 20–60 mg/L measured as TKN. The concentration of PO$_4$ in raw sewage is generally observed in the range of 5–10 mg/L. Presence of these nutrients in treated sewage is a matter of concern when the treated effluent is to be reused. Presence of pathogenic microorganisms is another serious concern for discharge of sewage in the water bodies as these can lead to waterborne disease. In terms of indicator organism *Escherichia coliforms* (*E. Coli*) number in raw sewage, it is in the range of 10^5–10^7 per 100 mL.

Some heavy metals and compounds such as chromium, copper, cyanide, which are toxic may find their way into municipal sewage through industrial discharges. Knowing concentration of these compounds is important if the sewage is to be treated by biological treatment methods or disposed off in stream or on land. With the use of several chemical compounds in personal care products, such as shampoos, soaps, cleansers, skin ointments, etc., as well as consuming antibiotics, the presence of emerging contaminants in the sewage is a growing concern. Though these compounds are presently at very low concentrations in sewage, however they can still have adverse effect on uses of such water as discussed earlier. In general, all these compounds are within toxic limits in sanitary sewage; however, with receipt of industrial wastewater discharges they may cross the limits in municipal wastewaters. For many of these emerging contaminants there is no discharge limit that is defined in many countries.

6.7.2 Sewage Treatment Flow Sheet

The sewage treatment plant flow sheet involves selection of an appropriate combination of unit operations and unit processes to achieve a desired degree of pollutant removal to achieve the set effluent quality standards. The selection of unit operations and processes primarily depends on the characteristics of the untreated sewage and the required level of contaminants permitted in the treated effluent. The design of process flow sheet is important in overall design of sewage treatment plant and requires thorough understanding of the functioning of the treatment units. It calls for optimization of wastewater treatment system coupled with stage wise optimal design of individual unit operation/process to offer a minimal cost design solution for treatment.

The major pollutants to be removed from domestic sewage are both organic and inorganic suspended solids (SS), biodegradable organic matter present in colloidal and soluble forms, and pathogens. Concentration of BOD, SS and coliforms present in the final effluent are being considered as the performance indicator for sewage treatment plants. In addition, reduction of nitrogen and phosphorous could also be required to meet the specified effluent quality standards in many countries.

The conventional flow sheet of sewage treatment plant consists of unit operations such as screen, grit chamber, and primary sedimentation tank (PST), followed by unit process of aerobic biological treatment, such as activated sludge process (ASP) or trickling filter. The sludges removed from primary and secondary sedimentation tanks are digested anaerobically followed by drying of anaerobically digested sludge on sand drying beds. This process flow sheet is presented in Fig. 6.8.

It is possible to replace the aerobic secondary biological wastewater treatment, activated sludge process or trickling filter, by low-cost treatment systems such as oxidation ditch, aerated lagoon or waste stabilization ponds. In such secondary treatment options, the primary sedimentation tank can be eliminated. In case of waste stabilization ponds, even grit chamber is generally not provided and after screens the sewage is pumped to this pond, which will offer both primary as well as secondary treatment. Also, anaerobic digestion of the sludge is not required in this option. Some of the process flow sheets are shown in Fig. 6.9.

With the better understanding of microbiology and biochemistry of anaerobic treatment, it is feasible to treat dilute organic wastewater, such as domestic wastewater, directly

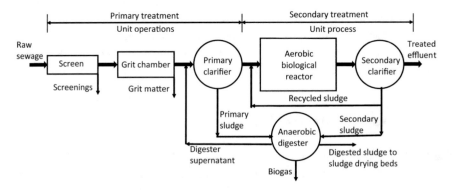

Fig. 6.8 Flow-sheet of conventional sewage treatment plant

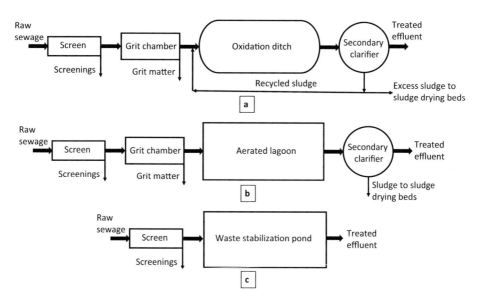

Fig. 6.9 Sewage treatment flow sheet using **a** oxidation ditch, **b** aerated lagoon, and **c** waste stabilization pond

using anaerobic treatment. Anaerobic reactors, such as up-flow anaerobic sludge blanket (UASB) reactor, fluidized-bed submerged media anaerobic reactor (FB-SMAR) and anaerobic filter (AF) or static-bed SMAR (SB-SMAR) and anaerobic rotating biological contactor (AnRBC) are being used for treatment of sewage. Out of these UASB reactor and anaerobic filter are popular reactor configuration used in South Asian, African and South American countries and even in few other countries. Since, UASB reactor has internal settler hence external secondary sedimentation tank will not be required in this case. It is generally reported that BOD_5 removal efficiencies of UASB reactor may range from 70 to 80% and COD removal efficiency from 60 to 70%, while treating low strength municipal wastewater. Consequently, post treatment will generally be required to achieve the prescribed effluent standards. Generally, oxidation ponds or moving bed biofilm reactor (MBBR) or activated sludge process are being used for the post treatment of anaerobic reactor effluent. In case of post treatment in oxidation pond, secondary clarifier is not required after UASB reactor or after oxidation pond. The process flow sheet utilizing anaerobic reactor is depicted in Fig. 6.10.

Several other treatment flow schemes can be developed and are being used in practice other than the schemes described. In recent times the secondary treatment options, such as MBBR, sequencing batch reactor (SBR), extended aeration activated sludge process, and membrane bioreactor (MBR) are also gaining popularity for effective treatment of the sewage. While using these reactor configurations PST is eliminated in most places. Since, effluent total nitrogen and/or nitrate concentration are also being regulated in many

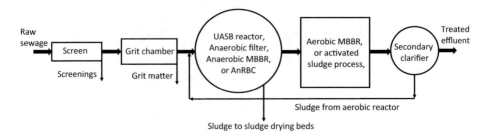

Fig. 6.10 Process flow sheet employing anaerobic treatment system

countries, after secondary biological treatment for removal of organic matter biological nitrification and denitrification is used. The reactors used for nitrification (aerobic) and denitrification (anoxic) could be either suspended growth process or attached growth process. By maintaining anoxic zone in SBR, separate nitrification and denitrification will not require and it is capable of organic matter removal as well as nitrogen removal. Also, in case of RBCs, which is typically used in stages (three to five) in last stages nitrification and denitrification can occur.

After effective removal of solids, organic matter and nitrogen as per the effluent quality standards, it is necessary to carry out disinfection particularly when treated sewage is to be discharged into water bodies that are likely to be used for drinking water source downstream or even recreation. Though chlorination is being used at most places for disinfection, it has serious drawback of formation of disinfection by-products, which will have adverse effect on aquatic life and even for human consumption. Hence, other alternative disinfections, such as ozonation and ultraviolet irradiation, are also being explored and these are finding increased acceptance in sewage treatment plants.

Questions

6.1 A completely mixed reactor is treating wastewater with flow rate of 6 L/sec having influent pollutant concentration of 180 mg/L. Considering the first order reaction rate constant of 0.35 per hour, determine the required retention time to meet the effluent pollutant concentration of 20 mg/L. What will be the volume required for the plug flow reactor if the same effluent concentration is to be achieved?

6.2 Define the different types of reactors used in wastewater treatment.

6.3 Prepare short note on the dispersion number and how it defines the nature of the reactor.

6.4 It is proposed to use five CSTRs with equal residence time in series for treatment of wastewater having initial pollutant concentration of 200 mg/L and flow rate of 20 m^3/h. Following the first order reaction rate, determine the residence time and volume required for each CSTR. The final treatment efficiency expected is 90% and the first order reaction rate constant is 0.4 per hour. What will be total volume of the reactor required? What will be volume required for single CSTR to achieve the same 90% of the pollutant removal?

6.5 Describe primary, secondary and tertiary treatment of wastewater.

6.6 Describe unit operation and unit processes giving examples.

6.7 Describe process flow sheet of sewage treatment plant utilizing UASB reactor as first stage secondary treatment.

6.8 With help of schematics describe the sewage treatment plant flow scheme utilizing (a) SBR, (b) MBR, and (c) MBBR.

References

Arceivala, S. J., & Asolekar, S. R. (2007). Wastewater treatment for pollution control and reuse. Tata McGraw-Hill Education.

CPHEEO. (2013). Manual on sewerage and sewage treatment, Part A, engineering (3rd ed.). Central Public Health and Environmental Engineering Organization, Ministry of Urban Development, Government of India.

Levenspiel, O. (1972). Chemical reaction engineering (2nd ed.). Wiley.

Reynolds, T. D., & Richards, P. A. (1996). Unit operations and processes in environmental engineering (2nd ed.). PWS Publishing Company.

Wehner, J. F., & Wilhelm, R. H. (1956). Boundary conditions of flow reactor. Chemical Engineering Science, 6, 89.

Self-purification of Natural Streams

7

7.1 Self-purification

Natural streams are vital and irresistible freshwater systems for the sustenance of eco-life on earth. The anthropogenic sources (industries and domestic households) often discharge treated or partially treated effluent (wastewater) into the natural depressions, lakes, fish-ponds, natural streams, sea or the oceans. Deterioration in the stream water quality and oxygen content occurs due to mixing of wastewater containing low dissolved oxygen, high suspended solids, organic compounds, toxic compounds, etc. (Bronfman, 1992). The condition of a polluted stream basically relies on the balance between the available and influx oxygen resources and biochemical oxygen demand (BOD) of the organic pollutants present in the wastewater that is being discharged in the river at any time.

The BOD for the degradation of organics by microbial species is intimately associated with the depletion in the dissolved oxygen (DO) content in the natural water body. The fate of DO is an important criterion, which needs to be maintained to avoid anoxic condition in the natural water systems. The threshold quantity of DO require for the survival of fish and aquatic animals is 4 mg/L (Patel & Vashi, 2015). A minimum of about 2.0 mg/L of DO is essential to support higher life forms. A number of factors affect the amount of DO available in a river. The oxygen demanding wastes deplete DO, plants add DO during day time, however remove it at night and respiration of organisms reduces DO. In summer, rising temperature reduces solubility of oxygen, while lower flows reduce the rate at which oxygen enters the water from atmosphere.

The rivers or natural streams have the capability to purify the external pollutants and regain its natural condition. The self-purification process of any natural stream can be observed in the four zones, namely degradation zone, active decomposition zone, recovery zone, and clean zone (Fig. 7.1).

© The Author(s), under exclusive license to Springer Nature Singapore Pte Ltd. 2022   207
M. Ghangrekar, *Wastewater to Water*, https://doi.org/10.1007/978-981-19-4048-4_7

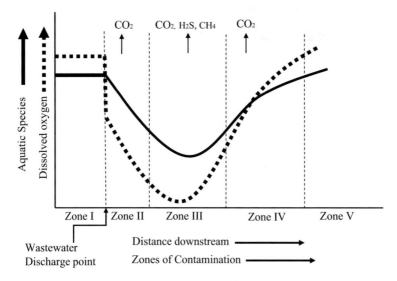

Fig. 7.1 Variation of DO and aquatic species in zones of contamination after a polluted water containing oxidizable organic matter is discharged into the river. Zone I and V: clear water zone; Zone-II: degradation zone; Zone-III: active decomposition zone; Zone-IV: recovery zone

(i) **Degradation zone** (Zone II): It is the zone of strong pollution, which is characterised by dark colour of water, high turbidity, floating solids, reduced DO nearly up to 40% of saturation DO, i.e., 3–4 mg/L for the natural stream at the temperature around 30 °C. This reduction in DO inhibits the sustenance of aquatic life and cause death of fish. This condition also favours sludge deposition, increased carbon dioxide concentration, and support faster de-oxygenation and slower re-oxygenation. It extends downstream up to a definite length from the wastewater receiving location. The degradation zone is typically unfavourable for algae and fish; however, certain sewage fungi and bottom worms can survive under such conditions.

(ii) **Active decomposition zone** (Zone III): It is the most polluted zone in the process of self-purification of the river. This zone is characterised by gray colour of water, reduction in DO even up to zero, production of gases due to anaerobic condition, such as methane, hydrogen sulphide, carbon dioxide, ammonia, etc. There is also formation of scum layer due to presence of sludge particles and other finer solids, because of the buoyancy imparted by evolved gases under anaerobic condition. The rise in DO content is also observed in the same zone. Fish and algae are typically unable to survive in this zone; however, the presence of protozoa, fungi and larvae of sewage fly is observed in this zone.

(iii) **Recovery zone** (Zone IV): It is the zone of better re-aeration and characterised by increasing DO (more than 40% of saturation DO) and presence of protozoa, algae, tolerant fish, snails, mussels, etc. In this zone, the water starts getting clearer and reduction in turbidity is evident.

(iv) **Clear water zone** (Zone I and V): This zone is characterised by the DO content up to saturation concentration or close to this value. The river water is clear with no bottom sludge settlement, minimal turbidity, and presence of ordinary fish, like game fish and other aquatic species is evident.

7.2   Factors Affecting Self-purification

The dynamic nature of self-purification in natural streams is exhibited by mechanism of intra and inter natural factors, i.e., sunlight, dilution, sedimentation, oxidation and reduction. Thus, the process also includes transportation and transformation of pollutant, adsorption by sediments, and sedimentation through terrigenous and biogenic suspension. Characterization on the basis of physical, chemical and biological attributes represents different stages of self-purification. However, the intensity of self-purification varies with the pollution load and on the other natural factors as described below.

(i) **Dilution and Dispersion**: The decrease in concentration of organic pollutants upon discharge into large volume of natural streams/river occurs due to factors like dilution and dispersion. It minimizes the potential trouble of pollution due to discharge of wastewater into river because of dilution offered by water present in the river. Due to dilution, sufficient amount of DO could be still available in the river water to prevent it falling below the critical value.

(ii) **Sedimentation**: The settlement of solids present in wastewater is termed as sedimentation. These settled solids will undergo decomposition at the bed of river, if biodegradable. The convective currents of river stream affect the settlement potential of solids, as strong flow currents oppose sedimentation and weak flow currents favour sedimentation.

(iii) **Biochemical oxidation**: The organic matter present in the wastewater gets oxidized as the wastewater is discharged into river and the oxidization by microbial species results in the depletion of DO. The occurrence of deficiency in DO is compensated by the replenishment from atmospheric oxygen and photosynthetic oxygen release. The oxidation process continues till the complete stabilisation of organic matter by action of microorganism is achieved and it is one of the major factors responsible for self-purification of rivers. The rate of oxidation of organic matter depends on the type of wastewater being discharged in the river, depending upon the chemical composition of the organic matter present. This oxidation rate is faster at higher temperature and slower at lower temperature.

(iv) **Reduction**: The reduction process, in absence of DO, results in hydrolysis of complex organic matter either biologically or chemically into liquids, gases or simpler matter by anaerobic bacteria at the bottom depth of the waterbody. The simpler end products are easy to oxidize and ultimately mineralized.

(v) **Sunlight**: The sunlight plays crucial role in photosynthetic activity, where autotrophs derive energy from sunlight for food preparation. In presence of sunlight algae utilize the carbon dioxide and release oxygen, which replenishes the oxygen content of the natural stream. Sunlight helps in better oxidation of organic matter and subsequently enhances the self-purification capability.

(vi) **Temperature**: Temperature influences the saturation DO content of river water as higher the temperature, lower will be the DO content and vice-versa; since solubility of any gas in water reduces with rise in temperature. The acceleration in the rate of biological and chemical activity on rise in temperature, results in rapid decline of DO and sometimes lead to anaerobic condition in case of heavy organic pollution load being released in the river. Since, the activity of microorganisms is more at the higher temperature, hence the biodegradation of organic matter will be faster and self-purification will take less time at warmer temperature than in winter. However, if anaerobic condition develops, it will affect longer stretch of the river downstream the point of discharge of wastewater.

(vii) **Turbulent motion**: The turbulence helps in better re-oxygenation from the atmospheric air and helps in maintaining the aerobic conditions in the river. However, turbulence may also have detrimental effect by retarding the growth of algae (useful for reaeration) due to increase in turbidity. High velocities through vertical cross-section cause more turbulence and reaeration to avoid vertical stratification of pollutants.

(viii) **Rate of reaeration**: The rate at which oxygen re-enters is a reaeration rate and it greatly influences the self-purification capacity of any river. The higher the rate of reaeration, faster will be the mechanism of self-purification. Adeney and Becker (1921), proposed Eqs. 7.1 and 7.2, expressing the rate of reaeration is directly proportional to the saturation deficit at particular time. These equations are formulated for the quiescent water condition; however, this law of solution can also be valid for condition of many natural streams.

$$q = (100 - q_1) \cdot \left(1 - e^{-f\left(\frac{a}{v}\right)t}\right) \tag{7.1}$$

$$dq/dt = (100 - q_1) \cdot f\left(\frac{a}{v}\right) \cdot e^{-f\left(\frac{a}{v}\right)t} \tag{7.2}$$

where q is the quantity of gas dissolved in percentage saturation; q_1 is initial concentration of soluble gas; f is the escape coefficient of gas from the liquid per unit area and volume; a is surface area; t is exposure time; v is volume of liquid; dq/dt

is the rate of reaeration; and $100 - q_1$ represents saturation deficit. The experimental values of the escape coefficient 'f' per unit area and volume generally range from 0.34 (for tap water) to 0.61 (for seawater).

(ix) **Neoteric Factors**: Apart from conventional factors, the self-purification of natural streams is affected by some neoteric (not so conventional) oxygen demands, such as sediment oxygen demand (SOD). This SOD represents the oxygen required by the benthic organisms for biological respiration and chemical oxidation of reduced substances at the interface of an organic sediment and overlying water. The periphyton or the microbes which proliferate to underwater substances are the significant source of SOD (Butts and Evans, 1978). However, attached filamentous algae, slime bacteria, such as *Sphaerotilua* and *Leptomitus*, may also represent an oxygen demand. The production of oxygen by photosynthesis and respiration by bottom and suspended agents, such as algae, bacteria or aquatic macrophytes, changes the overall oxygen balance in any stream, as discussed earlier. Biological oxidation of inorganic nitrogen by different chemolithotropic bacteria, i.e., ammonia oxidizers and nitrite oxidizers, also results in the depletion of DO for the process called as nitrification. The presence of unoxidized ammonia and nitrite in stream water is also toxic to fish even at lower concentration (Horne and Goldman, 1994; Schurr and Ruchti, 1975).

7.3   Streeter-Phelps Oxygen Sag Analysis

Generally, the unpolluted natural stream tends to hold maximum quantity of DO (i.e., close to saturation DO) at the existing temperature and atmospheric pressure condition, which varies between 14.6 mg/L (at 0 °C) and 7.6 mg/L (at 30 °C). In a highly polluted stream, the increased rate of oxygen depletion and decreased rate of oxygen replenishment result in overall less availability of DO in water. The increased rate of oxygen depletion is followed by increased rate of re-oxygenation until the equilibrium is achieved. When the rate of reaeration increases the rate of deoxygenation, the gradual recovery of DO starts in the river water.

(i) **Re-oxygenation rate**: The rate at which atmospheric oxygen gets dissolved in the river water or any other water body in order to compensate the declining oxygen is called as a re-oxygenation rate. A curve plotted between the rate at which oxygen is supplied and the time of flow shows the re-oxygenation curve (Fig. 7.2a). It depends on the depth of water bodies, such as for shallow ponds/lakes/rivers the rate of re-oxygenation is more; the condition of water bodies, i.e., more re-oxygenation in running condition as compared to quiescent; and the temperature of water. The

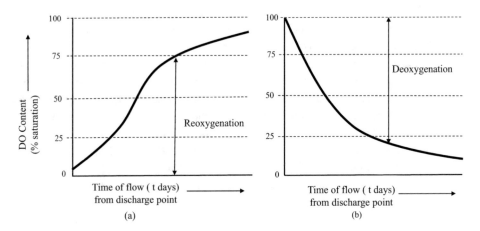

Fig. 7.2 **a** Reoxygenation curve and **b** deoxygenation curve

re-oxygenation rate is lower at higher temperature due to reduction in saturation con-centration of oxygen. The DO deficit is also driving factor for re-oxygenation. The greater DO deficit causes more re-oxygenation since oxygen solubility rate is pro-portional to the deficit, i.e., difference between saturation concentration and existing concentration.

(ii) *De-oxygenation rate*: When a wastewater containing biodegradable organic matter is discharged in the river, the amount of DO in river water reduces with time due to consumption for oxidation of organic matter. The quantity of unoxidized organic matter remaining at the given time and temperature of water decides the rate of de-oxygenation, which coincides with rate of BOD exertion. A curve plotted between the DO depletion and time of flow from point of pollution represents the de-oxygenation curve (Fig. 7.2b) (Streeter & Phelps, 1925).

7.4   Mathematical Representation of Oxygen Sag Curve

The mathematical analysis of biochemical oxygen demand (deoxygenation) and DO replenishment (reoxygenation) is proposed for the first time by H. S. Streeter and E. B. Phelps in 1925, popularly known as Streeter- Phelps Model. The difference in the sat-uration DO and actual DO content represents the oxygen deficit at that particular time 't' in the river water (Eq. 7.3). The oxygen deficit should ideally be close to zero for clean river. This DO deficit present in the river water varies with the rate of re-oxygenation and de-oxygenation.

$$\text{Oxygen deficit(D)} = \text{DO}_{\text{saturation}} - \text{DO}_{\text{actual}} \qquad (7.3)$$

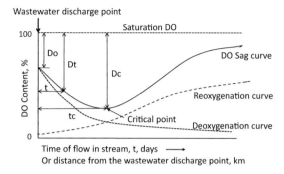

Fig. 7.3 Dissolved oxygen deficit curve for the natural stream polluted with discharge of wastewater and point of critical DO deficit

The analysis of oxygen sag curve can be easily done by superimposing the rates of deoxygenation and reoxygenation. The superimposition of both the de-oxygenation and re-oxygenation curves gives the resultant deficit of oxygen at a given instant of flow and the curve is known as oxygen sag curve or oxygen deficit curve as represented in Fig. 7.3.

The amount of oxygen deficit (D_t) (in percent DO saturation) and the oxygen balance, i.e., $100 - D_t$, at particular time of flow can be obtained using this curve. As the de-oxygenation rate increases, the sag in oxygen deficit curve also increases indicating increase in oxygen deficit. The net change in DO deficit rate (Eq. 7.4) can be obtained by combining the rate of de-oxygenation and re-oxygenation as detailed below.

Net rate of oxygen deficit = rate of de-oxygenation – rate of re-oxygenation

$$\frac{dD_t}{dt} = K'L_t - R'D_t \tag{7.4}$$

where K' is the coefficient of de-oxygenation for the river water mixed with wastewater (base e), same as BOD reaction rate constant (K); R' is the coefficient of re-oxygenation for the river water (base e); t is time in days; L_t is the first stage BOD remaining at time t (in mg/L); and D_t is the deficit in dissolved oxygen at time t (in mg/L).

By substituting, $L_t = L_o \cdot e^{-K' \cdot t}$, where L_o is ultimate first stage BOD, the Eq. 7.4 is modified to Eq. 7.5 and rearranging will result in Eq. 7.6.

$$\frac{dD_t}{dt} = K'L_o \cdot e^{-K't} - R'D_t \tag{7.5}$$

$$\frac{dD_t}{dt} + R'D_t = K'L_o \cdot e^{-K't} \tag{7.6}$$

After solving the first order differential Eq. 7.6 from the known boundary conditions the solution will be represented by Eq. 7.7.

$$D_t = \frac{K' \cdot L_o}{R' - K'}\left[e^{-K' \cdot t} - e^{-R' \cdot t}\right] + \left[D_o \cdot e^{-R' \cdot t}\right] \tag{7.7}$$

The above mathematical equation, if expressed for log to the base 10, is popularly known as Streeter-Phelps equation (Eq. 7.8).

$$D_t = \frac{K \cdot L_o}{R - K}\left[10^{-K \cdot t} - 10^{-R \cdot t}\right] + \left[D_o \cdot 10^{-R \cdot t}\right] \qquad (7.8)$$

where D_t is the deficit in DO after time t (mg/L); L_o is the ultimate first stage BOD (mg/L); D_o is initial deficit of DO at the point of wastewater discharge (mg/L); R is the reoxygenation coefficient (base 10). The K is coefficient of de-oxygenation (base 10), which is practically equal to the first order BOD reaction rate constant determined in laboratory and it varies with temperature. The value of K for other temperatures can be estimated using Eq. 7.9.

$$K_{(T)} = K_{(20)}[1.047]^{T-20} \qquad (7.9)$$

Under field conditions, the re-oxygenation coefficient varies with velocity and depth of stream and correlation of it is expressed by Eq. 7.10.

$$R_{(20)} = \frac{3.9 \, v^{0.5}}{y^{1.5}} \qquad (7.10)$$

where v is average velocity of stream (m/sec); and y is the average depth of stream in m. The value of R also varies with temperature and for temperature other than 20 °C it can be worked out using Eq. 7.11.

$$R_{(T)} = R_{(20)}[1.016]^{T-20} \qquad (7.11)$$

The values of re-oxygenation constant (day^{-1}) (at temperature 20 °C) vary according to the size of river, flow velocity and type of natural streams as indicated in Table 7.1.

At a particular time, both the re-oxygenation and de-oxygenation rate becomes algebraically equal and it corresponds to the **critical point**. After the attainment of critical point, the oxygen deficit decreases with time and ultimately becomes close to zero. The time at which the lowest DO (critical DO deficit) occurs is the critical time (T_c) and it is obtained by equating the differentiated Eq. 7.8 to zero; and which is empirically equal to Eq. 7.12.

$$T_c = \frac{1}{R - K}\log\left[\left\{\frac{KL_o - RD_o + KD_o}{KL_o}\right\}\frac{R}{K}\right] \qquad (7.12)$$

The distance from the point of release of wastewater at which critical deficit occurs is S_c and it is obtained from the Eq. 7.13.

$$S_c = T_c \cdot v \qquad (7.13)$$

Table 7.1 Values of re-oxygenation rate constant (base e) and self-purification constant for diverse water-bodies at 20 °C

Natural waterbody	Re-oxygenation constant (at 20 °C) (R', day^{-1})	Self-purification constant ($f = \frac{R}{K}$)
For waterfalls and rapid flows	Greater than 1.15	Greater than 5.0
For swift streams	0.69–1.15	3.0–5.0
For large streams (normal velocity)	0.46–0.69	2.0–3.0
For large streams (low velocity)	0.35–0.46	1.5–2.0
For large lakes and sluggish water bodies	0.23–0.35	1.0–1.5
For small ponds	0.1–0.23	0.5–1.0

Modified from Ain et al. (2019) and Makariem (2004)

where v is the mean flow velocity of river after receiving the discharge of wastewater.

The maximum deficit in oxygen or critical DO deficit ($D_{critical}$) is given by Eq. 7.14

$$D_{critical} = \frac{K}{R} L_0 10^{-K \cdot Tc} \tag{7.14}$$

For natural log Eq. 7.14 can be expressed as Eq. 7.15, which is obtained by putting $dDt/dt = 0$ in Eq. 7.6.

$$D_{critical} = \frac{K'}{R'} L_0 e^{-K' \cdot Tc} \tag{7.15}$$

The constant $\frac{R}{K} = \frac{R'}{K'} = f$ is termed as constant of self-purification and its value range from 0.5 to 5.0 for natural streams; however, in case of waterfalls and rapid flowing streams, it is above 5.0 (Table 7.1). The T_c in terms of self-purifying constant (f) can be expressed as Eq. 7.16 after substituting $R/K = f$ in Eq. 7.12.

$$T_c = \frac{1}{K(f-1)} \log\left[\left\{1 - (f-1)\frac{D_o}{L_o}\right\} f\right] \tag{7.16}$$

Critical DO deficit in term of self-purification constant is given by Eq. 7.17.

$$D_{critical} = \frac{L_o}{f} [10]^{-K T_c} \tag{7.17}$$

Taking log on both sides of Eq. 7.17, the Eq. 7.18 can be obtained.

$$\log D_{critical} = \log \frac{L_o}{f} - KT_c \qquad (7.18)$$

substituting the value of T_c from Eq. 7.16 in Eq. 7.18, it will result in Eq. 7.19.
Therefore,

$$\log D_{critical} = \log \frac{L_o}{f} - \frac{K}{K(f-1)} \log\left[f\left\{ 1 - (f-1)\frac{D_o}{L_o} \right\} \right] \qquad (7.19)$$

On rearranging the Eq. 7.19, a simplified form as Eq. 7.20 can be obtained.

$$\left\{ \frac{L_o}{D_{critical} \cdot f} \right\}^{f-1} = f\left[1 - (f-1)\frac{D_o}{L_o} \right] \qquad (7.20)$$

Above equations are used to identify the influence of wastewater discharge at the particular time and distance at the downstream side of the discharge point on the river. It helps in determining the degree of effluent treatment or the quantity of dilution required in the discharged wastewater to maintain the adequate DO for supporting aquatic life in the natural stream. The Streeter-Phelps equation can also be used as water quality modelling tool to describe the variation of DO with the released BOD because of the oxidizable organic matter discharged in any waterbody and such analysis helps in maintaining the sufficient DO in the water body by regulating the BOD of the wastewater being discharged.

Example: 7.1 The re-oxygenation constant for a large river is 0.44 per day. The DO of 8.4 mg/L and ultimate oxygen demand of 20 mg/L was found for the mixture (river water + wastewater) after receiving wastewater from an outfall sewer. Downstream of the wastewater discharge point the stream is flowing with the velocity of 0.75 m/sec and de-oxygenation rate constant (base 10) is found to be 0.12 per day. Estimate the amount of dissolved oxygen at a distance of 50 km downstream from the point of discharge of the wastewater? Consider saturation DO value of 9.16 mg/L.

Solution
Given, $R = 0.44$ per day; $K = 0.12$ per day; velocity of stream = 0.75 m/sec.

Saturation DO = 9.16 mg/L; actual DO = 8.4 mg/L.

Initial deficit $(D_o) = 9.16 - 8.4$ mg/L = 0.76 mg/L.

Time required for 50 km = distance downstream/velocity of flow in large stream.

t = $\frac{50 \times 10^3}{0.75}$ = 66666.67 sec = 0.7716 days.

DO deficit after time t,

$$D_t = \frac{K \cdot L_o}{R - K}\left[10^{-K \cdot t} - 10^{-R \cdot t} \right] + \left[D_o \cdot 10^{-R \cdot t} \right]$$

$$D_t = \frac{0.12 \times 20}{0.44 - 0.12}\left[10^{-0.12 \times 0.7716} - 10^{-0.44 \times 0.7716}\right] + \left[0.76 \times 10^{-0.44 \times 0.7716}\right]$$

$D_t = 2.98$ mg/L, this is DO deficit at 50 km from the point of wastewater discharge. Hence, DO at 50 km downstream $= 9.16 - 2.98 = 6.18$ mg/L (**Answer**).

Example: 7.2 A natural stream having initial DO of 8.4 mg/L receives wastewater discharge of 7 m³/sec. The stream is having total discharge of 100 m³/sec. Upon receipt of the wastewater the stream is flowing with a velocity of 0.3 m/sec. The BOD_5 of wastewater getting discharged is 30 mg/L and BOD_5 of natural stream is 1.1 mg/L. The constant for de-oxygenation is having a value of 0.1 per day (base 10). Find the critical DO deficit and time and distance at which this critical deficit will occur at downstream of point of wastewater discharge in the stream. Consider purifying coefficient (f) of stream as 2.0, DO of the wastewater as 0.6 mg/L, and saturation DO value of 8.7 mg/L.

Solution
The initial DO of natural stream $= 8.4$ mg/L.
  $K = 0.1$ per day.
  DO of mix (wastewater + stream) at t = 0, i.e., at point of discharge $= \frac{8.4 \times 100 + 0.6 \times 7}{100 + 7}$.
  DO of mix $= 7.89$ mg/L.
  Initial oxygen deficit (D_o) $= 8.7 - 7.89 = 0.81$ mg/L.
  Also, $BOD_5 = \frac{L_sQ_s + L_RQ_R}{Q_S + Q_R} = \frac{30 \times 7 + 1.1 \times 100}{7 + 100} = 2.99$ mg/L.
  BOD_5 of the mix $= Y_5 = 2.99$ mg/L.
  $Y_5 = L_O\left(1 - 10^{-K \times 5}\right)$.
  Hence, the ultimate BOD of the mix 'L_O' $= 2.99/\left(1 - 10^{-0.1 \times 5}\right) = 4.37$ mg/L.
  Now using the equation

$$\left\{\frac{L_o}{D_{critical} \cdot f}\right\}^{f-1} = f\left[1 - (f-1)\frac{D_o}{L_o}\right] = \left\{\frac{4.37}{D_{critical} \times 2}\right\}^{2-1} = 2\left[1 - (2-1)\frac{0.81}{4.37}\right]$$

$D_{critical} = 1.34$ mg/L (**Answer**).
  Now using the equation

$$T_c = \frac{1}{K(f-1)}\log\left[\left\{1 - \frac{(f-1)D_O}{L_O}\right\}f\right]$$

$$T_c = \frac{1}{0.1(2-1)}\log\left[\left\{1 - \frac{(2-1)0.81}{4.37}\right\}2\right] = 2.12 \text{ days (\textbf{Answer})}$$

Distance (S_c) $=$ velocity of stream $\times$ travel time $= 0.3$ m/sec $\times 2.12 \times 24 \times 60 \times 60$ sec $\times 10^{-3}$
                      $= 54.95$ km (**Answer**)

Example: 7.3 An industrial area disposes its effluent at the rate of 0.18 m³/sec into a stream with flow rate of 1.20 m³/sec at a point A. At a location fairly upstream of A, the stream has dissolved oxygen of 8.6 mg/L, and industrial effluent has dissolved oxygen of 1.4 mg/L. The temperatures of industrial effluent and stream water are 27 °C and 22 °C, respectively, and 5-day BOD of 25 mg/L and 3 mg/L, respectively. The de-oxygenation constant has value of 0.20 per day (base e) and re-oxygenation constant of the mixture is 0.40 per day (base e). Approximate saturation concentration of DO, at 22 °C and 23 °C is 8.73 mg/L and 8.56 mg/L, respectively. Determine (a) temperature of mix, (b) self-purification coefficient at the temperature of mixture, (c) time, distance and magnitude of critical dissolved oxygen deficit. Consider temperature coefficient for de-oxygenation and re-oxygenation as 1.047 and 1.016, respectively, and velocity of flow after mixing at downstream of point A as 0.9 m/sec.

Solution

De-oxygenation constant (K') at 20 °C (base e) = 0.2 per day.
   K at 20°C (base 10) = 0.2/2.3 = 0.434 × 0.2 = 0.087 per day.
   Re-oxygenation constant (R') at 20 °C (base e) = 0.4 per day.
   R at 20 °C (base 10) = 0.4/2.3 = 0.434 × 0.4 = 0.174 per day.

Estimation of DO, BOD and Temperature of mixture

$$DO \text{ of mixture} = (DO \text{ of wastewater} \times Q_S + DO \text{ of river} \times Q_R)/(Q_S + Q_R)$$
$$= (1.4 \times 0.18 + 8.6 \times 1.2)/(0.18 + 1.2) = 7.66 \text{ mg/L}$$

$$BOD_5 \text{ of mixture} (\text{at } 20\,°C) = (BOD \text{ of wastewater} \times Q_S + BOD \text{ of river} \times Q_R)/(Q_S + Q_R)$$
$$= (25 \times 0.18 + 3 \times 1.2)/(0.18 + 1.2) = 5.87 \text{ mg/L}$$

$$\text{Temperature of mixture} = (\text{wastewater temp.} \times Q_S + \text{river water temp.} \times Q_R)/(Q_S + Q_R)$$
$$= (27 \times 0.18 + 22 \times 1.2)/(0.18 + 1.2) = 22.65\,°C \,(\textbf{Answer})$$

Ultimate BOD of mixture = $BOD_5/(1 - 10^{-K \cdot t}) = 5.87/(1 - 10^{-0.087 \times 5}) = 9.28$ mg/L.
   K and R at temperature of 22.65 °C
   K at 22.65 °C = $K_{(20°C)} [1.047]^{T-20} = 0.087 \times [1.047]^{22.65-20} = 0.098$ per day.
   Similarly, $R_{(22.65°C)} = R_{(20°C)} [1.016]^{T-20} = 0.174 \times [1.016]^{22.65-20} = 0.181$ per day.
   Hence, self-purification coefficient (f) at 22.65 °C = 0.181/0.098 = 1.847 (**Answer**).

Oxygen deficit of the mixture
DO of the mixture = 7.66 mg/L.
   Saturation concentration of DO at 22.65 °C shall be obtained by interpolation of given saturation concentration at 22 and 23 °C, which will be = 8.62 mg/L.
   Initial DO Deficit = 8.62 − 7.66 = 0.96 mg/L.
   Time after which critical deficit occurs is T_c, given by

$$T_c = \frac{1}{K(f-1)} \log\left[\left\{1 - \frac{(f-1)D_O}{L_O}\right\}f\right]$$

$$T_c = \frac{1}{0.098(1.847-1)} \log\left[\left\{1 - \frac{(1.847-1)0.96}{9.28}\right\}1.847\right]$$

$$T_c = 2.73 \text{ days}$$

Distance at which critical oxygen deficit occurs,
$= \text{velocity} \times \text{time} = 0.9 \times 2.73 \times 24 \times 3600 \times 10^{-3} = 212.28 \text{ km}.$
Value of critical deficit ($D_{critical}$) is given by

$$\left\{\frac{L_o}{D_{critical} \cdot f}\right\}^{f-1} = f\left[1 - (f-1)\frac{D_o}{L_o}\right]$$

$$\left\{\frac{9.28}{D_{critical} \times 1.847}\right\}^{1.847-1} = 1.847\left[1 - (1.847-1)\frac{0.96}{9.28}\right]$$

$D_{critical} = 2.71$ mg/L (**Answer**).

Example: 7.4 A sewage generated from a city after secondary treatment is getting released in a river as per the details provided in the Table 7.2. Determine DO deficit profile for 100 km downstream of the point of discharge of wastewater. Velocity of flow downstream of point of wastewater discharge is 0.45 m/sec. The reaeration rate constant R' is 0.4 per day and deoxygenation rate constant K' has a value of 0.23 per day (base e).

Solution
River discharge $= 1.15$ m^3/s, sewage discharge $= 0.26$ m^3/sec.
  BOD of mix $= \frac{(1.15 \times 2.4 + 0.26 \times 28)}{(1.15 + 0.26)} = 7.12$ mg/L.
  DO of mix $= \frac{(1.15 \times 8.3 + 0.26 \times 1.2)}{(1.15 + 0.26)} = 6.99$ mg/L.
  Temperature of mix $= \frac{(1.15 \times 24 + 0.26 \times 26)}{(1.15 + 0.26)} = 24.37$ °C.
  Saturation value of DO at 24.37 °C is 8.36 mg/L.

Table 7.2 Characteristics of river and wastewater released in the river	River	Secondary treated effluent from STP
	Flow rate $= 1.15$ m^3/sec	Flow rate $= 0.26$ m^3/sec
	5-day BOD at 20 °C $=$ 2.4 mg/L	5-day BOD at 20 °C $= 28$ mg/L
	Temperature 24 °C	Temperature 26 °C
	DO $= 8.3$ mg/L	DO $= 1.2$ mg/L

Estimating ultimate BOD

$L_t = L_0 \left(1 - e^{-K \times t}\right).$

$7.12 = L_0 \left(1 - e^{-0.23 \times 5}\right).$

Hence, $L_0 = 10.42$ mg/L.

Initial DO deficit $(D_0) = 8.36 - 6.99 = 1.37$ mg/L.

Deoxygenation and reoxygenation coefficients estimation at 24.37 °C temperature

$K_T = K_{20} \left(\theta\right)^{T-20}.$

Hence, $K'_{24.37} = 0.23 \, (1.047)^{24.37-20} = 0.281$ day^{-1}.

$R_T = R_{20} \left(\theta\right)^{T-20}.$

Hence, $R'_{24.37} = 0.40 \, (1.016)^{24.37-20} = 0.429$ day^{-1}

Critical time $T_c = \dfrac{1}{R' - K'} \log_e \dfrac{R'}{K'} \left(1 - \dfrac{D_0 \times \left(R' - K'\right)}{K' \times L_o}\right)$

$\quad = \dfrac{1}{0.429 - 0.281} \log_e \dfrac{0.429}{0.281} \left(1 - \dfrac{1.37 \times (0.429 - 0.281)}{0.281 \times 10.42}\right) = 2.374$ days

Critical DO deficit, $D_{cxitical} = \dfrac{K'}{R'} L_0 e^{-K' \cdot Tc}$

$\quad = \dfrac{0.281}{0.429} 10.42 e^{-0.281 \times 2.374} = 3.50$ mg/L

Distance at which it occurs $= X =$ velocity $\times$ time

$= (0.45 \text{ m/s}) \times (2.374 \times 24 \times 60 \times 60 \text{sec})$

$= 92,301 \text{ m} = 92.30 \text{ km}$

For profiling the DO deficit estimate the DO deficit at every 20 km downstream of the discharge point. The time required for river flow to reach at 20 km distance,

$$t_{20 \, km} = \dfrac{(20 \times 1000)}{(0.45 \times 24 \times 3600)} = 0.514 \text{ day}$$

The DO deficit at 20 km can be estimated as below:

$$Dt = \dfrac{K'Lo}{R' - K'} \left[e^{-K't} - e^{-R't}\right] + Do \cdot e^{-R't}$$

where $K' = 0.281$ d^{-1}, $R' = 0.429$ d^{-1}, $Do = 1.37$ mg/L and $Lo = 10.42$ mg/L and $t = 0.514$ day. Substituting these values in the equation the DO deficit at 20 km $= 2.35$ mg/L.

Similarly, DO deficit at 40 km (i.e., $t = 1.029$ days) $= 2.97$ mg/L.

DO deficit at 60 km (i.e., $t = 1.543$ days) $= 3.33$ mg/L.

DO deficit at 80 km (i.e., $t = 2.058$ days) $= 3.48$ mg/L.

and DO deficit at 100 km (i.e., $t = 2.572$ days) $= 3.49$ mg/L.

Table 7.3 provides the DO deficit and DO available in the river 100 km downstream of the discharge of secondary treated sewage.

Table 7.3 The DO deficit at different points along length of river

Distance in km	Time in days	DO deficit, mg/L	DO, mg/L = (8.36 − DO deficit)
0	0	1.37	6.99
20	0.514	2.35	6.01
40	1.029	2.97	5.39
60	1.543	3.33	5.03
80	2.058	3.48	4.88
92.30	2.374	3.50	4.86
100	2.572	3.49	4.87

It is to be noted that even after traveling 100 km downstream the point of wastewater discharge, the river water DO has not replenished due to receipt of this wastewater, however the critical DO in the river water is not falling below 4 mg/L, hence expected to support the aquatic life in the river water.

Questions

7.1. What are the factors responsible for self-purification of natural streams and what steps should be taken to control stream pollution?

7.2. Explain the zones of contamination in the polluted stream and elaborate how the existence of aquatic life is affected by DO variation?

7.3. Discuss the impact of de-oxygenation and re-oxygenation process in the natural stream with the help of curves?

7.4. A sugar mill discharges 2000 m^3 per day of wastewater into a stream with typical flow rate of 0.22 m^3/sec. The DO concentration of discharged wastewater and stream water are 0.8 mg/L and 8.24 mg/L, respectively. The deoxygenation rate constant has a value of 0.22 per day (base e) and the reoxygenation constant is 0.5 per day (base e). Determine the following with the use of data given:
   (a) Critical time (T_c), critical deficit ($D_{critical}$) and the distance at which it occurs
   (b) DO deficit at $t = 0.5T_c$ and $t = 2T_c$
   (c) DO concentration at 500 km downstream from the point of wastewater discharge

Parameter	Wastewater	Surface water
5-day BOD at 20 °C (mg/L)	30	0.8
Temperature (°C)	28	25
Velocity (m/sec)	–	0.48

Saturation value of DO at 25.28°C as 8.20 mg/L

7.5. A semi-treated effluent having flow rate of 700 L/sec, BOD of 55 mg/L, DO of
      1.8 mg/L and temperature of 25 °C is discharged into a natural stream having flow
      rate of 11 m^3/sec, BOD of 3.0 mg/L, DO of 8.2 mg/L and temperature of 22 °C. The
      downstream velocity of the natural stream is 0.186 m/sec. Determine the combined
      discharge, BOD, DO and temperature of the mixture (natural stream + effluent). Also,
      estimate the critical DO, its location and DO concentration at 100 km downstream
      side from the discharge point. The deoxygenation rate constant has a value of 0.12 per
      day at 20°C and reoxygenation constant is 0.5 per day (both expressed with logarithm
      having base 10). Consider saturation DO at combined temperature as 8.16 mg/L.
7.6. A wastewater having BOD of 27 mg/L is being discharged with flowrate of 0.25 m^3/sec
      through outfall sewer in to a river having BOD of 1.5 mg/L with flowrate of 4 m^3/sec.
      The stream is developing a mean velocity of 0.17 m/sec at downstream of discharge
      point. The DO of river water at upstream of discharge is 7.8 mg/L and for wastewater
      it is 1 mg/L. Considering the deoxygenation rate constant of 0.2 per day and reaeration
      rate constant of 0.4 per day (both expressed for base e) for the temperature of the mix
      flow, determine the time and distance for the critical DO deficit, and the minimum DO
      that will reach in the river water. Consider saturation DO after mixing as 8.16 mg/L.
7.7. A dairy effluent is discharged into a river and ultimate BOD of the mixture is equal to
      18 mg/L. The initial oxygen deficit is 2 mg/L. The reoxygenation and deoxygenation
      rate constants for the effluent-river water mixture are 0.5 per day (base e) and 0.25
      per day (base e), respectively. Calculate the time after which the DO deficit becomes
      the 20% of the initial DO deficit.
7.8. Explain variation of DO and aquatic species in zones of contamination after a polluted
      water containing oxidizable organic matter is discharged into the river.
7.9. Describe the factors that will affect the deoxygenation and reoxygenation in natural
      streams.
7.10. Write a note on Streeter-Phelps oxygen sag analysis.

Answers

7.4.
   (a) Critical time (T_c) = 2.05 days; critical deficit ($D_{critical}$) = 1.56 mg/L; and the
       distance at which it occurs = 85.01 km
   (b) DO deficit at t = 0.5T_c and t = 2T_c will be 1.395 mg/L and 1.269 mg/L, respectively.
   (c) DO concentration at 500 km downstream from the point of wastewater discharge
       will be 8.01 mg/L.
7.5. Combined BOD = 6.11 mg/L; Combined DO = 7.82 mg/L; Combined temperature =
      22.18 °C; Critical DO = 8.16 − 1.37 = 6.79 mg/L; Distance of critical DO deficit =
      22.3 km; DO concentration at 100 km downstream = 7.74 mg/L.
7.6. The time and distance for the critical DO deficit = 2.6 days and 38.189 km, respectively,
      and the minimum DO that will reach in the river water = 6.75 mg/L.

7.7. Time after which the DO deficit becomes the 20% of the initial DO deficit $= 15.18$ days.

References

Ain, C., Rudiyanti, S., Haeruddin, & Sari, H. P. (2019). Purification capacity and oxygen sag in Sringin river, Semarang. *International Journal of Applied Environmental Sciences, 14*, 1–16. ISSN 0973-6077.

Adeney, W. E., & Becker, H. G. (1921). The determination of the rate of solution of atmospheric nitrogen and oxygen by water. *Philosophical Magazine Series 6, 42*(247), 87–96. https://doi.org/10.1080/14786442108633734

Bronfman, A. M. (1992). Black Sea environment, self-purification in the context of the problems of anthropogenic ecology of the sea. *GeoJournal, 27*(2), 141–148. https://doi.org/10.2307/41145452. http://www.jstor.org/stable/41145452

Butts, T. A., & Evans, R. L. (1978). *Sediment oxygen demand studies of selected Northeastern Illinois streams*. Circular 129, State of Illinois, Department of registration and education. https://www.isws.illinois.edu/pubdoc/C/ISWSC-129.pdf

Horne, A. J., & Goldman, C. R. (1994). *Limnology* (2nd ed., pp. 576). McGraw-Hill, Inc. https://sswm.info/sites/default/files/reference_attachments/HORNE%20and%20GOLDMAN%201994%20Understanding%20Lake%20Ecology.pdf. https://www.enr.gov.nt.ca/sites/enr/files/dissolved_oxygen.pdf

Makariem, N. (2004). *Guidelines on stipulation of accommodating capacity of load of water pollution in water sources* (Decree of the State Minister for Environmental Affairs No. 110/2003 dated June 27, 2003). http://extwprlegs1.fao.org/docs/pdf/ins48777.pdf

Patel, H., & Vashi, R. T. (2015). Chapter 2—*Characterization and treatment of textile wastewater* (pp. 21–71). Elsevier. ISBN 9780128023266. https://doi.org/10.1016/B978-0-12-802326-6.00002-2

Schurr, J. M., & Ruchti, J. (1975). Kinetics of oxygen exchange, photosynthesis, and respiration in rivers determined from time-delayed correlations between sunlight and dissolved oxygen. *Schweiz. z. Hydrologie, 37*(1), 144–174. https://doi.org/10.1007/BF02505184

Streeter, H. W., & Phelps, E. B. (1925). *A study of the pollution and natural purification of the Ohio river*. United States Public Health Bulletin No. 146. United States Public Health Service. https://udspace.udel.edu/bitstream/handle/19716/1590/C%26EE148.pdf?sequence=2&isAllowed=y

Unit Operations and Chemical Unit Processes 8

8.1 Introduction

The expression 'physico-chemical' treatment of wastewater is a generic term used for different physical operations as well as chemical processes, which are primarily used for removal of suspended, colloidal, and soluble contaminants, such as grits, suspended organic solids, hardness, heavy metals, etc., from wastewater (Fig. 8.1). The physical treatment deals with the application of physical phenomena, such as gravity settling, adsorption, sieving and screening to remove the intended pollutants from wastewater and thus it is often referred to as 'unit operation' as they rely only on physical movement of contaminants without altering their original nature. Screens, primary clarifiers, secondary settling tanks, gravity filters, etc. are some of the commonly used physical unit operations in a typical wastewater treatment plant.

On the other hand, in chemical processes different chemicals are added in the wastewater to be treated to bring about chemical changes in the nature of contaminants or aqueous medium to facilitate separation of the pollutant under consideration so as to produce treated effluent free from this pollutant and hence, they are termed as 'unit processes'. Coagulation-flocculation, water-softening, disinfection, chemical precipitation, etc. are some of the examples of frequently employed unit processes. Principally dissimilar unit operations and unit processes are often used in combination during treatment of wastewater to attain higher contaminant removal efficiency and better effluent quality. For instance, a coagulation-flocculation unit is always followed by a clarifier to separate coagulated flocs form the wastewater. Together, the physical and chemical units constitute a vital component of water and wastewater treatment plants and are imperative for achieving the desired effluent quality. Due to their complementary nature and interdependency, the physical and chemical treatment units are frequently clubbed together and referred to as physico-chemical treatment of wastewater.

© The Author(s), under exclusive license to Springer Nature Singapore Pte Ltd. 2022   225
M. Ghangrekar, *Wastewater to Water*, https://doi.org/10.1007/978-981-19-4048-4_8

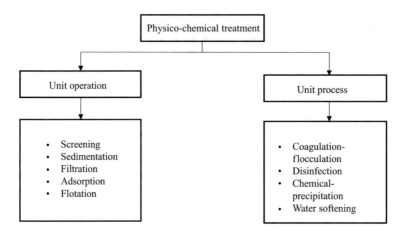

Fig. 8.1 Different physico-chemical options adopted for wastewater treatment

In conventional sewage treatment plant screen, grit chamber and primary sedimentation tank (PST) are used for offering primary treatment for removal of floating matter, grit particles, and settleable organic solids, respectively. Generally, oil and grease concentration in municipal wastewater is very less and separate skimming tank is not provided. In this case, the oil and grease are removed by arresting the scum by provision of baffle ahead of the effluent weir in PST. However, when oil and grease concentration is high, typically greater than 50 mg/L, separate skimming tank or other alternate technology need to be employed for removal of it. Other unit operations and processes, other than screen, grit chamber and PST, are not used as part of primary treatment in conventional sewage treatment plant (as described in Chap. 6). However, depending upon the characteristics of the municipal wastewater and industrial effluent involving some additional unit operation/process may be necessary, particularly to improve efficiency of the follow-up treatment unit. For example, for improving efficiency of suspended solids removal, coagulation, flocculation, and sedimentation is often used in industrial wastewater treatment plants. Similarly, when treated sewage or industrial wastewater is to be reused for cooling water, demineralization may be required as tertiary treatment to bring TDS under control.

8.2 Screens

Screening is the first unit operation carried out in a wastewater treatment plant and it consists of placing inclined parallel bars fixed at a certain distance apart in the channel (Fig. 8.2). Screens are installed ahead of pumping stations as the first step as well as bar screen is installed as first unit operation in the wastewater treatment plant. The primary role of the screen is to remove floating materials and coarser particles from the influent

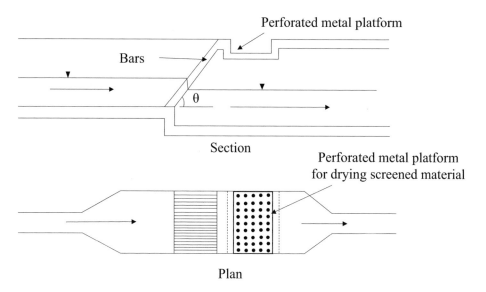

Fig. 8.2 Schematic diagram of a screen channel

raw sewage. The main idea of removing these materials is to protect pipes, pumps, valves, and other equipment from floating materials and increase the efficiency and life of these equipment.

The screening system consists of parallel bars, rods, wire mesh and openings that may be of any shape (normally rectangular and circular). During the construction of the screen, these points must be considered: degree of screening required, proper removal and disposal of screened out material, and odour potential. The cross-sectional area of the screen chamber is always constructed greater (around 2–3 times) than the arriving sewer line and the length of screen channel should be adequate to avoid eddies current formation around the screen so as not to allow these materials to get dislodged from the screen opening.

The screen chamber comprises of a rectangular conduit and the floor of the channel is kept slightly lower (7–15 cm) than the invert of an influent sewer. Channel bed slope is designed to attain self-cleaning velocity in the channel, which would avoid the deposition of grits and other heavier suspended solids. The entrance of the screen chamber should have a smooth transition and divergence to minimize the head loss. An effluent structure of the screen chamber also must have a uniform convergence and arrangement to separate the effluent from each bar rack should be provided as per requirement (Fig. 8.3).

Screens are designed to carry peak flow and a minimum of two bar screens must be provided with the arrangement to divert the flow to meet the maintenance requirement. Each of these screens should be designed to handle the peak flow. Classification of the screen can be done based upon the opening size as coarse (bar screen), medium and fine

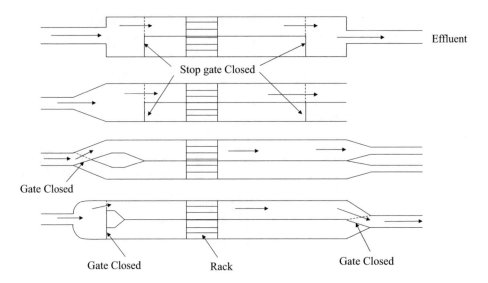

Fig. 8.3 Double chamber bar screen and possible flow arrangements that can be used

screens and based on cleaning operation, it can be grouped into manually cleaned screen and mechanical cleaned screen.

8.2.1 Classification of Screens

Screens can also be classified as fixed screen, where the bar rack is stationary. The screen could be band screen consisting of an endless perforated band that passes over upper and lower sprocket. The screen can be drum screen or strainer, which consists of rotating cylinder having screen covering the circumferential area of the drum. The screen can also be designed as steps.

Sometimes grinder or comminutor is used in conjunction with coarse screens to cutdown or grind the coarser floating materials. Comminutor utilize cutting teeth (or shredding device) on a rotating or oscillating drum that passes through stationary combs (or disks). Hence, while the wastewater is passing through this arrangement of thin opening with cutting edge, the objects of large size are shredded to the size of 6–10 mm. Provision of bye pass to this device should always be made. Such comminutors are recommended for small sewage treatment plants of up to 1 MLD (CPHEEO, 2013).

8.2.1.1 Coarse Screen

The coarse screen, also known as bar screen or trash rack, works more like a protective device and it is installed as the first unit operation in wastewater treatment plants. Bar screen and coarse woven-wire screen are the most common type used for removal of

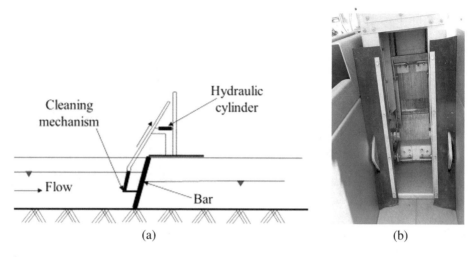

Fig. 8.4 Reciprocating rack screen **a** schematic diagram, and **b** photograph of the screen (adopted with permission from Kusters Water catalogue, Spartanburg, SC—USA)

coarser floating matters from municipal wastewaters. Bar screens are also used ahead of the raw sewage pumping station with a relatively large clear opening of 75–150 mm to protect the pumps from clogging. The bar screen used at sewage treatment plant can have a clear opening of 20–50 mm and it can be cleaned manually or by mechanical units. Screens are set at an angle of 30°–60° with the horizontal for manually cleaned racks and with angle of 60°–85° for mechanically cleaned screen. This inclined position increases the effective submerged opening area of the screen, thus helping to keep head loss under control.

8.2.1.2 Reciprocating Rake Screen

In reciprocating rake screen (Fig. 8.4), rakes travel up to the base of the screen, engross the bars and pull the screening to the top of screen from where screenings are removed. The advantages of the reciprocating rake can be listed as it can handle large objects, low operating and maintenance cost, no submerged moving parts, effective raking, and efficient discharge of screened materials. The relatively high price due to use of stainless steel, long cleaning cycle and more headroom requirement can be listed as some of the drawbacks of this screening.

8.2.1.3 Chain-Driven Screen

Chain-driven screens (Fig. 8.5) are mechanically cleaned bar screens and these are classified based on whether the screen is racked from the upstream or downstream sides and whether the rack returns to the bottom of the chamber from upstream or downstream (as front clean and back return screen or front clean and front return screen, etc.). This screen

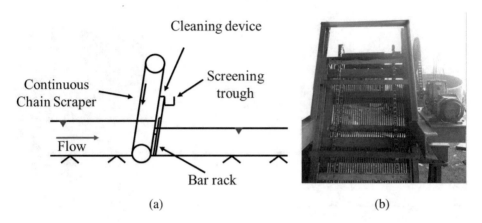

Fig. 8.5 Chain driven screen **a** schematic diagram, **b** photograph of chain driven screen (adopted with permission from Envirospec, Kolkata, India catalogue)

mechanism has a short cleaning cycle. However, the main disadvantage of this mechanism is that it has submerged moving part that requires dewatering of screening channel before any maintenance.

8.2.1.4 Continuous Belt Screen

It is mechanically cleaned coarse screen and considered as a new development in the screening system. Continuous belt screen (Fig. 8.6) is a self-cleaning continuous screening belt that can remove fine and coarse solids. Large numbers of racks (screening element) are attached with drive chains and number of these racks depends upon the depth of channel. This screen has submerged sprocket and screen opening range from 0.5 to 30 mm. Advantages of this screen is the maintenance can be done above the wastewater operating level and this unit has no problem of jamming. However, replacement of screen is time consuming process and operation of this screen is expensive.

Fig. 8.6 Continuous belt screen

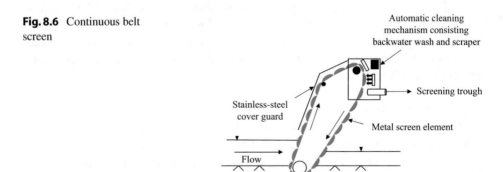

Fig. 8.7 Rotatory drum screen

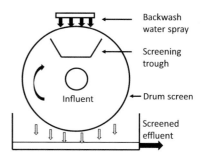

8.2.1.5 Fine Screen

Fine screens have a clear opening generally ranging from 0.035 to 6 mm, and they are cleaned mechanically. Fine screens are mechanically cleaned screens made with perforated plates, woven wire cloths or placing very closely spaced bars with clear openings of less than 20 mm. Clear spacing less than 6 mm is typical opening adopted in practice for fine screens. Fine screen is not advisable to be used in sewage treatment plant as first stage screen due to clogging problems; however, a fine screen can work as a pre-treatment unit for industrial wastewater. In case of sewage treatment these are typically used after coarse screening. Fine screens with finer openings (less than 1 mm) can also be used after secondary treatment to increase the effluent quality by reducing concentration of suspended solids. The fine screens can be a static wedge-wire type, step type, centrifugal screen, or drum-type screen. Stationary screens are cleaned by moving brushes, rakes, or teeth and movable screens are cleaned continuously during operation preferably by water jet.

8.2.1.6 Rotatory Drum Screen

It consists of a rotating cylinder with a screen covering the circumference surface area of drum (Fig. 8.7). The wastewater enters the drum axially and leaves radially out from the screen material provided at the circumference. The solids deposited at the interior surface of the drum are removed by a jet of water from the top and discharged into a trough that is placed below the jet. For tertiary treatment, the micro-strainers used are having a very fine size screens and are used to polish secondary treated effluent or remove algae from the effluent of stabilization ponds. Opening size of 1–5 mm and 0.25–2.5 mm is generally used for primary treatment and opening size of 6–40 μm is used for polishing treatment of secondary effluents. The primary disadvantage of the rotatory drum screen is high head loss compared to the bar screen.

8.2.1.7 Band Screen

Band screen is used to screen out fine materials with perforated band being attached to a drive chain, which passes over the upper and lower sprocket and it acts as a screening mechanism. The flow pattern is through the front to the back sides of the panel (Fig. 8.8).

Fig. 8.8 Band screen

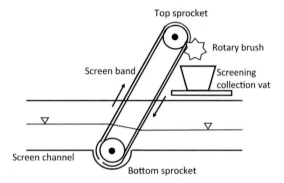

Brushes and water jet are used to remove the material retained in the screen. Generally, Band screens are used for the primary treatment of industrial effluent. Opening size of 0.8–2.5 mm is generally provided in this screen.

8.2.1.8 Step Screen

The design of step screen may contain two step-shaped sets of perforated metal plates, in which the first (bottom) is movable and the second is fixed (Fig. 8.9). The mobile plate travels in a vertical motion to carry the filtered solids to the next step and then after draining the water the screening material is discharged to the trash bin. Step screen is self-cleaning device with opening from 1 to 6 mm.

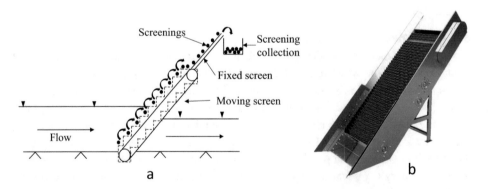

Fig. 8.9 Step screen **a** schematic and **b** photograph (adopted with permission from Mr. Dhaman Anand Koka, CTO, M/s Fundametal Technologies, Puducherry, India)

8.2.2 Specifications of Bar Screen

There is a difference in velocities of flow between the channel upstream of the screen and the wastewater moving through the screen. This difference affects the operation of screen. Lower velocity through the screen opening will result in more screening material collected on screen. However, lower velocity than self-cleansing in the screen channel will lead to the deposition of grit and other settleable solids in the approach channel at upstream of the screen. Hence, the approach velocity in the screen channel should not fall below the self-cleaning velocity. The suggested approach velocity for the bar screen is 0.6–0.75 m/sec for wastewater containing grit particles. This velocity should be attained by managing the bed slope of the channel. The length of the screen chamber should be adequate for not allowing eddies to form at the screen section due to expansion of the flow at inlet and contraction of flow at outlet, if any.

Recommended velocity through the bar screen opening for hand cleaned bar screens is about 0.3 m/sec at average flow and the maximum recommended velocity shall be about 0.80–1.2 m/sec at maximum flow. At peak flow, if the velocity through the screen is kept in the range of 0.6–1.2 m/sec it will give satisfactory performance results. If the velocity through the screen at peak flow exceeds above 1.5 m/sec it will lead to the excessive turbulence at the screen section, thus allowing the floating solids to pass through the screen opening.

Due to placement of the bar rack in the passage of flow in screen channel, this obstruction will lead to the head loss. Head loss through the screen should be controlled so that it should not cause excessive built-up of the water level upstream of the screen; hence, to avoid the influent sewer to operate under pressure due to flow full conditions. This head loss in screen is a function of approach velocity in the screen channel and the velocity through the opening between the bars. Using Bernoulli's equation this head loss through the screen can be estimated (Fig. 8.10). Applying Bernoulli's theorem on the upstream and downstream of the screen, Eq. 8.1 can be obtained.

$$d_1 + \frac{V_1^2}{2g} = d_2 + \frac{V_2^2}{2g} + \delta d \qquad (8.1)$$

Thus, δd, the head that is lost, can be expressed as Eq. 8.2.

Fig. 8.10 Wastewater profile in the bar screen channel

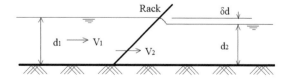

$$\delta d = d_1 - d_2 = \frac{V_2^2 - V_1^2}{2gC^2} \tag{8.2}$$

where d_1 and d_2 are upstream and downstream depth of the water with respect to rack, respectively; V_1 and V_2 are velocity upstream and downstream of screen, respectively; g is acceleration due to gravity; and C is the discharge coefficient having typical value of 0.84 (Lee & Lin, 2007), hence for C^2 the value is 0.7. Thus, the equation for head loss (h) through the screen can be written as Eq. 8.3.

$$h = \frac{1}{0.7} \frac{V_2^2 - V_1^2}{2g} \tag{8.3}$$

Kirchmer (1926) defined the head-loss through the bar screen as per Eq. 8.4.

$$h = \beta \left(\frac{W}{b}\right)^{\frac{4}{3}} h_V \cdot \sin \theta \tag{8.4}$$

where h represents head-loss (m), β is bar shape factor (with a value of 1.79 for circular bar, 2.42 for sharp edge rectangular bar, 1.83 for rectangular bar with semi-circular upstream, 1.67 for rectangular bars with both upstream and downstream faces as semi-circular), W is the width of bar facing the flow (m), b represents the clear spacing between the bars (m), h_V ($= V^2/2g$) is the velocity head of flow approaching the bar (m), V represents the geometric mean of approach velocity (m/sec), g is gravity acceleration (m/sec^2) and θ is the inclination angle of bar rack with the horizontal.

For fine screens or the screen having rack other than parallel bars the head-loss through the clean or partially clogged bar can be estimated using Eq. 8.5, which is obtained by simplifying Eq. 8.3.

$$h = 0.0729 \left(V^2 - v^2\right) \tag{8.5}$$

where h represents head loss (m), V is velocity through the screen opening (m/sec) and v is the velocity of flow before the screen, i.e., in screen channel upstream of screen section (m/sec).

Head-loss through the fine screen can also be estimated as per Eq. 8.6, as per the discharge equation through orifice.

$$h = \frac{1}{2g \cdot C_d} \left(\frac{Q}{A}\right)^2 \tag{8.6}$$

where C_d represents the coefficient of discharge (typically 0.6 for cleaned racks), Q represents discharge through the screen (m^3/s) and A represents the effective open submerged area (m^2) and g is the gravity acceleration.

Generally, minimum head loss of 0.15 m is maintained in course screen. However, for the clogged hand-cleaned screen, the maximum head loss should not exceed 0.3 m. For the mechanically cleaned screen, the manufacturer specify head loss and it can vary from 0.15 to 0.6 m. The head loss for clean water through the clean screen is generally less. However, while wastewater flows through the screen the head loss will be typically more, and it depends on the nature of the floating solids present and cleaning frequency adopted in the screen. In addition, while wastewater being continuously flowing through these bars, over the period of operation sewage fungi will grow on these bars affecting the net area of opening and coefficient of discharge. This will lead to more head loss than that estimated for clean bars. Hence, minimum allowance of 0.15 m is recommended even if the estimated value is lesser. The head loss for manually cleaned screen is estimated when the screen is half clogged and for mechanically cleaned screens, which are continuously getting cleaned, manufacturer specifications are referred.

The angle of inclination of the bar rack with horizontal for manually cleaned screen is kept between 30° and 60°; whereas for the mechanically cleaned screen, angle varies from 45° to 85°. Since, the manually cleaned screen will be cleaned after some time interval, hence to ensure that sufficient area of opening is available for wastewater to pass through the screen even after retention of floating solids the rack is placed at smaller angle with horizontal. Whereas, for mechanically cleaned screen, since the cleaning is done continuously the clogging of opening will not be much, hence in this case the rack can be placed more vertical.

For a mechanically cleaned bar, the clear spacing between bars could be between 10 and 75 mm. For a manually cleaned bar screen, the clear spacing between the bars could vary from 20 to 50 mm. Based on the screen's position, the clear spacing between bars should be changed to meet the requirement. A course screen with clear spacing of 20–30 mm followed by fine screen with clear spacing between the bar of 6 and 12 mm is often used in sewage treatment plant. For industrial wastewater treatment, the spacing of the bars could be anything between 6 and 20 mm as per the requirement. A bar screen with an opening of 75–150 mm is used before the pumping station of raw sewage.

The width of the bar facing the wastewater flow can be 5–15 mm and depth of bar can vary from 25 to 75 mm. These bars shall be welded with the support plate from the downstream side to avoid deformation of alignment of bars due to hydrostatic pressure and impact of floating matters. Hence, to ensure that the bars provided are having sufficient moment of inertia to withstand these forces the bar with size less than 5 mm × 25 mm is not recommended to be used.

8.2.3 Quantities and Disposal of Screening Material

The quantity of screening material collected in the wastewater treatment facility can vary based on the geographical location, sewer system (separate or combined), type of rack or screen installed and with selection of spacing between the bars or opening size provided. Generally, the amount of screening material collected varies from 3.5 to 35 L per 1000 m^3

of wastewater being treated (typically 15 L per 1000 m^3, i.e., 1 ML of wastewater treated). This value can increase up to 225 L per 1000 m^3 of wastewater treated in case of a combined sewerage system during storms (Metcalf & Eddy, 2003). The screening material collected generally contains dry solids of 10–20% with the bulk density varying from 640 to 1120 kg/m^3 (WEF and ASCE, 1991). The screening material has a moisture content of 80–90% and it also contains organic solids, which can start decomposing, causing foul smell and nuisance. In case of large wastewater treatment plants, the screening material collected can be disposed of either by burning in an incineration plant or by dumping at a sanitary landfill. For small wastewater treatment plant, screenings may be disposed off by burial on the plant site.

Example: 8.1 If the head loss measured though the clean screen is 32 mm and approach velocity of flow in the screen channel is 0.6 m/sec, estimate the velocity through the clear screen opening. What will be the ratio of vertical projected clear width of opening to the gross width for this bar rack, if placed at an angle of 45° with horizontal? What will be head loss if the screen opening is half clogged? Use Eq. 8.3 for head loss calculation.

Solution
From Eq. 8.3,

$$0.032 = \frac{1}{0.7} \frac{V_2^2 - 0.6^2}{2 \times 9.81}$$

Solving for V_2, we have $V_2 = 0.894$ m/sec, i.e., the velocity through the clear screen opening.

Now for estimating the submerged inclined opening area of the screen using continuity equation, where A_1 and A_2 are the cross-sectional area of flow in channel and submerged area of screen opening, respectively.

$$Q = A_1 \times V_1 = A_2 \times V_2$$

$$A_1/A_2 = V_2/V_1 = 0.894/0.6 = 1.490$$

A_2 is the inclined submerged area of opening, hence vertical projected opening area will be $A_2 \times \sin\theta$. Hence the ratio of area will be:

$$A_1/A_2 \sin\theta = 1.49/\sin\theta = 2.107$$

Since depth of flow is same the ratio of vertical projected clear width of opening to the gross width = 1/2.107 = 0.474 (**Answer**).

Now if the screen is half clogged then velocity through the screen opening will be doubled, i.e., $0.894 \times 2 = 1.788$ m/sec, hence head loss will be:

$$= \frac{1}{0.7} \frac{1.788^2 - 0.6^2}{2 \times 9.81} = 0.206\,\text{m}$$

Thus, head loss of 0.206 m will occur when this screen is half-clogged.

Example: 8.2 Design the bar screen chamber for minimum sewage flow of 8 MLD, average sewage flow of 20 MLD and maximum sewage flow of 45 MLD. Provide size of the bar as 8 mm × 40 mm and clear spacing between bars as 30 mm.

Solution

The average flow of 20 MLD is equal to 0.231 m^3/sec. Similarly, the minimum flow of 8 MLD is 0.092 m^3/sec and maximum flow of 45 MLD is 0.52 m^3/sec.

As per recommendations, provide size of the bar as 8 mm × 40 mm with an inclination angle of 45° with horizontal and clear spacing between bars as 30 mm. Considering maximum velocity through screen as 0.3 m/sec at average flow and 0.80 m/sec at maximum flow.

Step 1: Calculate the submerged area of screen opening required.

$$Submerged\ area = \frac{flow}{velocity\ through\ channel}$$

Net submerged area of opening at average flow = average flow/velocity through screen opening

$$= (0.231\,\text{m}^3/\text{sec})/(0.3\,\text{m/sec}) = 0.77\,\text{m}^2$$

Net submerged area at maximum flow = (0.52 m^3/sec)/(0.80 m/sec) = 0.65 m^2.

As the area requirement at average flow is more, hence provide the net submerged area of screen opening 0.77 m^2.

Step 2: Calculate the submerged vertical cross section area of screen.

If *n* number of bars are used in the screen, then ratio of width of opening to the gross width is

$$= [\{(n+1) \times \text{clear opening}\}/\{(n+1) \times \text{clear opening} + \text{width of bar} \times n\}]$$

Considering provision of 30 bars, then the ratio of width of opening to gross width will be

$$= [\{(30+1) \times 30\}/\{(30+1) \times 30 + (8 \times 30)\}] = 0.794$$

Hence, gross submerged area of screen = 0.77/0.794 = 0.97 m^2.
Submerged vertical cross section area = gross area × sin θ

$$= 0.97 \times \sin 45 = 0.686\,\text{m}^2$$

This is the submerged vertical cross-sectional area of the screen channel.

Now, velocity of flow in chamber $= 0.231/0.686 = 0.337$ m/sec.

This velocity is less than the self-cleaning velocity of 0.42 m/sec. Hence change the inclination angle to 30° with horizontal.

Submerged vertical cross section area $= 0.97 \times \sin 30 = 0.485$ m^2.

Velocity of flow in chamber $= 0.231/0.485 = 0.48$ m/sec (>0.42 m/sec).

Step 3: Calculation of size of the channel.

Gross width of screen chamber $= (30 + 1) \times 30 + 8 \times 30 = 1170$ mm $= 1.17$ m.

Hence, wastewater depth at average flow $= 0.485/1.17 = 0.414$ m.

Provide free board of 0.3 m.

Therefore, total depth of screen chamber required $= 0.414 + 0.3 = 0.714$ say 0.75 m.

Size of screen channel $= 1.17$ m $\times 0.75$ m.

Step 4: Bed slope of calculation.

$$V = \frac{1}{n} \cdot R^{\frac{2}{3}} \cdot S^{\frac{1}{2}}$$

$$R = A/P = (1.17 \times 0.414)/(2 \times 0.414 + 1.17) = 0.2424$$

$$S = \left[(0.48 \times 0.013)/(0.2424)^{2/3}\right]^2 = 0.000257$$

i.e., 1 in 3882, hence bed slope of channel is 1 in 3882 m.

Step 5: For estimation of head loss for 50% clogged screen, use clear spacing of 15 mm instead of 30 mm.

$$h = \beta \left(\frac{W}{b}\right)^{\frac{4}{3}} h_V \cdot \sin\theta = 2.42 \left(\frac{8}{15}\right)^{\frac{4}{3}} \frac{0.48^2}{2 \times 9.81} \cdot \sin 30$$

$$= 0.00614 \text{ m} = 6.14 \text{ mm} \ (<150 \text{ mm})$$

Thus, provide 150 mm head loss at the screen section.

Step 6: Check for adequacy of the depth of the channel at maximum flow i.e., 0.52 m^3/sec considering the depth of flow upstream of screen as D.

Hence R $= 1.17$D/(1.17 + 2D)

$$Q = 0.52 = A \times v = 1.17D \times (1/0.013) \times (1.17D/(1.17 + 2D))^{2/3}(1/3882)^{1/2}$$

Solving for D, we have depth of flow as 0.755 m, hence considering free board of 0.3 m the depth of channel required is 1.055 m, thus provide a screen channel total depth of 1.1 m.

Velocity of flow in approach channel at maximum flow $= 0.52/(1.17 \times 0.755) = 0.59$ m/sec, nearly close to 0.6 m/sec, hence it is acceptable or else the width of the channel can be reduced to increased velocity.

Similarly head loss at maximum flow when screen is half clogged will be

$$h = 2.42 \left(\frac{8}{15}\right)^{\frac{4}{3}} \frac{0.59^2}{2 \times 9.81} \cdot \sin 30 = 0.0093 \text{ m} = 9.3 \text{ mm}$$

This is less than 150 mm head loss provided; thus, design is acceptable.

8.3 Theory of Sedimentation

Sewage is a complex mixture of polluting organic and inorganic matters in the used water. The solids in the sewage can be present in the form of dissolved, colloidal, pseudo colloidal and suspended solids. Fraction of the suspended solids could get settled under action of gravity when favourable condition is provided. Sedimentation is the process of removing the suspended solids from the wastewater and most of these suspended solids in water are heavier than the water and tend to settle down in stagnant water under the action of gravity. The suspended solids remain in suspension due to the turbulence in moving water. Nevertheless, as soon as a basin is provided to reduce the turbulence, these suspended solids start settling down.

These settled solids are called as sludge and this sludge is removed from bottom of settling tank as per suitability and treated separately for improving dewater capability. Generally, 30–40% of pollution load in terms of biochemical oxygen demand from sewage can be reduced by primary sedimentation. The velocity at which solids settles is called settling velocity and it is dependent on factors, like density and viscosity of wastewater; shape, size, and specific gravity of settling particles; and surface nature of the solids. In sewage treatment plant, sedimentation occurs in grit chamber, primary sedimentation tank (PST), secondary sedimentation tank (SST), and in sludge thickeners.

8.3.1 Classification of Settling

The settling can be classified into four different types based on particles tendency to interact with each other and their movement on basis of concentration in settling tank. These four types of settling are described in details to understand clearly the difference between them.

8.3.1.1 Discrete Particle Settling or Type I Settling

The Type I settling occurs when inert solids at low concentration settle in the wastewater as a discrete particle without interfering settling of other particles. This operation of settling in wastewater with low solids concentration happens with particles settling as an individual entity by gravity forces without interacting with neighbouring particles. There is no apparent flocculation or interaction between the particles while they are settling.

This type of settling occurs in the grit chamber for settling of sandy particles as well as during plain sedimentation of surface waters. Stokes and Newton's law are applicable in this type of settling as per settling conditions. Stable terminal settling velocity is gained by particles under gravity force once the particle starts movement in a constant acceleration field, which is opposed by the drag force until it equals the impelling force. Once the equilibrium is reached between gravity acceleration impelling force and drag force, then onward the settling occurs at a constant velocity, referred as a 'terminal settling velocity'.

The settling of discrete and non-flocculating particles can be explained mathematically via the law of sedimentation formulated by Sir Isaac Newton. This law computes the terminal settling velocity by equating the effective weight of the particle with the drag force (as per Eq. 8.7). Drag force oppose the settling of particle, while effective weight of particle accelerates the settling in starting phase, till the terminal settling velocity is achieved, at which both downward and upward forces become equal.

$$Drag\ force = Total\ weight - Buoyancy \tag{8.7}$$

$$Drag\ force = \frac{C_d A \rho_w v_p^2}{2} \tag{8.8}$$

$$Total\ weight = \frac{4\pi r^3 \gamma_s}{3} \tag{8.9}$$

$$Buoyancy = \frac{4\pi r^3 \gamma_w}{3} \tag{8.10}$$

where C_d is a drag coefficient; A represents the cross sectional or projected area of particle in direction of motion (m^2), ρ_w is the density of water (kg/m^3), v_p is settling velocity of particle (m/sec), r represents the radius of particle (m), γ_s is the unit weight of solid particle (kg/m^3) and γ_w represent the unit weight of water (kg/m^3).

Solving Eq. 8.7 and placing the corresponding values for spherical particle yields Newton's law (Eq. 8.11)

$$v_p^2 = \frac{4g(S_s - 1)d_p}{3C_d} \tag{8.11}$$

where v_p represents terminal velocity of particle, S_s is specific gravity of the particle, d_p is the diameter of the particle, g is gravity acceleration and C_d is a drag coefficient as defined earlier.

The numerical value of drag coefficient depends on whether the flow regime around the particle is laminar or turbulent. Thus, the drag coefficient (C_d) of a particle is the function of Reynolds number (N_R). Three distinct types of flow or region can be classified depending upon the Reynolds number: (a) laminar flow ($N_R < 1$), where viscous forces are more important than the inertia forces; (b) transitional flow ($1 < N_R < 1000$), where inertia and viscous forces are of equal importance; and (c) turbulent flow ($N_R > 1000$), where inertia forces are most important (Dey et al., 2019). The shape of the particle also affects the drag coefficient; however, for calculation purpose, it can be assumed that the particle is spherical. The drag coefficient can be estimated using Eq. 8.12 and Reynold's number using Eq. 8.13.

$$C_d = \frac{24}{N_R} + \frac{3}{\sqrt{N_R}} + 0.34 \tag{8.12}$$

$$N_R = \frac{\rho_w \cdot v_p \cdot d_p}{\mu} = \frac{v_p \cdot d_p}{v} \tag{8.13}$$

where ρ_w represents the density of medium where particle is settling (kg/m^3), μ is the dynamic viscosity of water (N sec/m^2), v is kinematic viscosity of water (m^2/sec), v_p represents settling velocity of particle (m/sec) and d_p is diameter of particle (m).

During settling in the laminar region, force of viscosity is predominant and the first term of Eq. 8.12 is considered to calculate drag coefficient. For the turbulent region, inertia force is dominant, and the first two terms of Eq. 8.12 become zero or reduced to the minimum value. In the transition region, all the terms of Eq. 8.12 are used to calculate drag coefficient.

$$\text{For laminar flow } (N_R < 1), C_d = \frac{24}{N_R} \text{ and } v_p = \frac{(S_s - 1) \cdot g \cdot \rho_w d_p^2}{18\mu}$$

For Turbulent flow ($N_R > 1000$), the drag coefficient for spherical particle

$$C_d = 0.4 \text{ and } v_p = \sqrt{3.33g(S_s - 1)d_p}$$

$$\text{For Transition flow } (1 < N_R < 1000), C_d = \frac{24}{N_R} + \frac{3}{\sqrt{N_R}} + 0.34$$

Sir George Gabrial Stokes solved Eq. 8.11 for the spherical particles falling under laminar condition ($N_R < 1$) and derived the Stokes' formula for terminal settling velocity (v_s) as expressed in Eq. 8.14. A settling in a dilute suspension in water and wastewater follows Stokes law when the viscous forces are dominant.

$$v_s = \frac{g(\rho_s - \rho_w)d_p^2}{18\mu} \tag{8.14}$$

For transition flow ($1 < N_R < 1000$), the settling velocity of the particle can be determined by trial and error or by performing multiple iteration by first estimating the settling velocity using Stokes law (8.14) and then iterating in Newton's equation (Eq. 8.11) till value of the settling velocity converges.

8.3.1.2 Ideal Sedimentation Basin

In 1936, Camp (1936) designed an ideal settling basin for the removal of discrete particles. He proposed four different regions in settling tank as inlet zone, settling zone, outlet zone and sludge settled zone. Following points are assumed during the design and experiment: Type I settling of particle occurs, quiescent condition prevails in settling zone, even distribution of flow occurs throughout the cross-section of settling chamber, uniform distribution of particles is present throughout the depth and no scoring (i.e., resuspension) of the particles occurs after settlement. The wastewater containing uniformly distributed sediments enters the settling tank with a uniform velocity (V), discharge (Q) and dimensions of the tank are length-L, breadth-B and settling height-H. The flow velocity V can be calculated as per Eq. 8.15.

$$V = \frac{Q}{A_c} \tag{8.15}$$

where V represents the horizonal flow through velocity of wastewater (m/sec), Q is the flow rate of wastewater (m^3/sec) and A_c represents the cross-sectional area of the tank (B × H) through which wastewater flows (m^2).

A particle introduced at the topmost inlet point will be removed from the wastewater if it has a settling velocity (v_s) good enough to reach at the bottom of the tank during the detention time (Fig. 8.11). Detention time (t_0) can be defined as the mean time period for which the wastewater remains in the settling tank. The relation between settling velocity and detention time can be written as per Eq. 8.16.

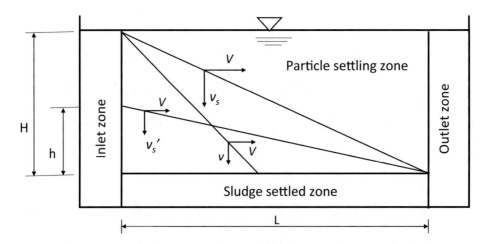

Fig. 8.11 An ideal discrete particle settling in a rectangular sedimentation tank

$$v_s = \frac{H}{t_0} \tag{8.16}$$

where v_s represents the settling velocity of a particle (m/sec), H is representing settling depth, and t_0 is the detention time of wastewater in the settling basin.

Detention time can also be computed as per Eq. 8.17.

$$t_0 = \frac{L \cdot B \cdot H}{Q} \tag{8.17}$$

where L represent the length of settling zone (m), B is the width of tank (m) and H represent the height of settling zone (m).

Solving Eqs. 8.16 and 8.17 the Eq. 8.18 shall be obtained.

$$v_s = \frac{H}{\frac{L \cdot B \cdot H}{Q}} = \frac{Q}{L \cdot B} = \frac{Q}{A_s} \tag{8.18}$$

where A_s = surface area of the tank (m^2).

It can be referred that the particles with settling velocities v, which is equal or greater than v_s, will settle down and such particles will be completely removed from the wastewater. This quantity of settling velocity can also be called as surface overflow rate (SOR) or surface loading rate and it has unit m^3/m^2 h or m^3/m^2 day. The SOR can be understood as the settling velocity of a suspended particle, which is introduced at the highest point of the settling chamber and it will reach at the lowermost point of the tank near the exit using full detention time. Camp idea of overflow velocity is considered as an innovative postulation, as it not only removed the depth of tank from design formulation but also it showed that detention time (t_0) has no effect on particle removal (Camp, 1936).

The particle with a settling velocity (v_s') lower than the surface overflow rate will not get entirely removed (Fig. 8.11) and removal of such particles will be partial. Thus, the fraction of the particles removed (P) can be estimated as per Eq. 8.19.

$$P = \frac{v_s'}{v_s} \tag{8.19}$$

As per assumption, the particles are uniformly distributed over the entire depth of tank with quiescent condition prevailing in the settling tank. From the trajectory of particles (Fig. 8.11) with constant horizontal flow velocity (V) and different settling velocity, it can be referred that both particles, i.e., the particle with settling velocity v_s and smaller particle with settling velocity v_s', will move downward within the given detention time and travel the depth H and h, respectively (Eq. 8.20). Also, from the geometric consideration or after solving Eqs. 8.16 and 8.19, the expression (Eq. 8.21) for estimating the fraction removal of particles entering at different heights of the inlet can be obtained. Thus, for the particles having settling velocity v_s', which is less than v_s, the partial removal of such particles will be equal to h/H or v_s'/v_s.

$$t_0 = \frac{v_s}{H} = \frac{v_s'}{h} \tag{8.20}$$

$$P = \frac{h/t_0}{H/t_0} = \frac{h}{H} \quad \left(\text{also, can be expressed as } P = \frac{v_s'}{v_s} \right) \tag{8.21}$$

Thus, all particles with settling velocity v_s' that enter the settling tank at the depth of h or lower will be removed. Since with the assumption that all the particles are uniformly distributed throughout the depth of the settling tank at inlet, the fractional removal (P) criterion is illustrated in Eq. 8.21.

Large variation in particle size exists in wastewater admitted to the settling tank. Hence for determining overall removal efficiency of the settling tank, an experimental analysis will be required. For this a batch settling test is performed in a column and samples are collected at suitable time interval and at different depths of the column and solid concentrations are determined. Upon analysis of this test results the data for construction of a plot as shown in Fig. 8.12 will be available, which is a cumulative settling velocity distribution. The fraction of the total particles removed for a designed velocity of v_s, i.e., overflow rate adopted in design will be given by Eq. 8.22.

$$\text{Fraction removed} = (1 - P_o) + \frac{1}{v_s} \int_0^{P_o} v dP \tag{8.22}$$

where $(1 - P_o)$ is the fraction of particles with settling velocity v greater than v_s, and second term represent the fraction of the particles with velocity v less than v_s.

As per the calculations, it shows that the depth of the tank doesn't have any effect on the efficiency of the tank. The efficiency of the settling will change with the change in the surface overflow rate. In application of this law for designing settling tank, design factors must be compensated to allow the wind effect, short circuiting, dissipation of energy at inlet, upward draw at outlet, difference of density and temperature between incoming sewage and wastewater already present in settling tank and its content, and turbulence created by the movement of sludge removal equipment. Though, settling is

Fig. 8.12 A settling velocity distribution curve, where P represents the weight fraction of the particles with settling velocity less than design settling velocity v_s

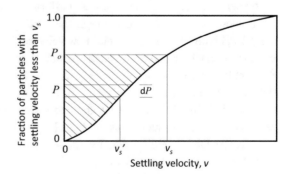

said to be the surface phenomenon, and surface area provided governs the efficiency, for above mentioned factors some minimum depth of clear settling zone is required to be provided to ensure discrete particle settling. Due to provision of this clear depth for settling, volume of the settling tank and hence detention time of the tank will also become the design parameter.

8.3.1.3 Flocculent Settling

Flocculant settling is the Type II settling in which settling of flocculent particles occurs under dilute suspension. Depending on the nature of the particles even in relatively dilute concentration in wastewater some particles do not act as discrete particle rather they tend to coalesce with each other and to form particle aggregate (flocs). Thus, due to flocculation they increase their size while settling and settle at faster velocity. Flocculent settling occurs when the particles form the flocs and settle fast due to an increase in size. Degree of flocculation depends on many factors, like range of particle size, depth of tank, contact time, velocity gradient, concentration of particles and overflow rate. This type of settling occurs in PST and upper portion of secondary settling tanks due to lower concentration of settling flocs.

Due to continuous change in shape and size, the Stokes equation is not applicable in flocculent settling. There are two main mechanisms related to flocculent settling: (a) difference in settling velocity of particle helps the faster settling particle to overtake the particle with slow settling velocity and coalesce with them, and (b) velocity gradients within wastewater that cause particles in region of a higher velocity to overtake those in adjacent stream paths moving at slower velocities. A batch settling test is usually required to evaluate the settling characteristics of flocculent settling particles.

Settling column test: This test is conducted to evaluate the applicable residence time and surface overflow rate for designing the sedimentation tank. Flocculation is a complex process and it depends upon the opportunity for contact, depth of the basin, overflow rate, velocity gradients in the system, size and shape of particles, surface properties and concentration of settling particles. The combined effect of all these variables can only be evaluated by performing laboratory experiments. The settling column is constructed with the same height as the proposed depth of the settling tank with a sampling port provided at equal intervals along the height of column (Fig. 8.13). The internal diameter of the settling column should be at least 14 cm to minimize the wall effect. During the test, the temperature should be maintained uniform to avoid convection currents and uniform distribution of solids should be maintained throughout the depth of the column at the beginning of the test. The duration of the test is assumed to be equivalent to the settling time proposed for the tank and samples are withdrawn from the ports and examined for suspended solids concentration. The percentage removal is plotted against time and depth. The curve of equal percentage removal is drawn. The fraction of particles removed can be estimated using the Eq. 8.23.

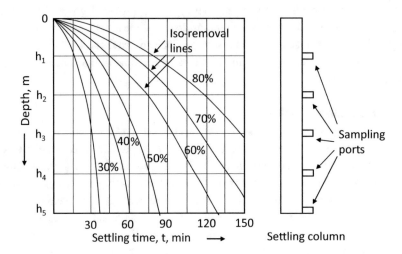

Fig. 8.13 Settling column and iso-percent removal lines obtained after performing settling test for the analysis of flocculent settling

$$R = \sum_{h=1}^{n} \left(\frac{\Delta h_n}{H} \right) \left(\frac{R_n + R_{n+1}}{2} \right) \qquad (8.23)$$

where R represents the total suspended solids removal in percentage, n is the number of iso-percent removal curves, Δh_n is the distance between the isopercent removal curves, i.e., ordinate (m), H represents total height of settling column (m), R_n represents the iso-percent removal curve number n, R_{n+1} represents iso-percent removal curve number n + 1.

The wastewater containing suspended solids should be admitted to the column in such a way that uniform distribution of solid particles is present throughout the depth of a settling column. Settling should takes place under quiescent conditions. At pre-decided time intervals, samples are withdrawn from the ports and analysed for suspended solids. Percentage removal of solids is analysed for each sample collected and it is plotted as a number (%) against time and depth. The curve of equal percentage removal is drawn between the plotted points.

While using the laboratory experimental results for designing the sedimentation tank, variation in performance under actual field condition should be taken into account. The efficiency of sedimentation tank, with respect to suspended solids or BOD removal, is affected by the following:

- Eddy currents formed by the inertia of incoming fluid,
- Wind induced turbulence created at the water surface of the uncovered tanks,
- Thermal convection currents,

- Cold or warm water causing the formation of density currents that allow cold water to move along the bottom of the basin, and
- Thermal stratification in hot climates.

Because of the above reasons the removal efficiency of the tank and detention time has correlation $R = t/(a + b \cdot t)$, where 'a' and 'b' are empirical constants, 'R' is expected removal efficiency, and 't' is nominal detention time. To account for these non-optimum conditions encountered in the field, due to continuous entry of wastewater and exit of it at outlet end of the sedimentation tank, and due to ripples formed on the surface of the water because of wind action, etc., the settling velocity (surface overflow rate) obtained from the laboratory column test is often multiplied by a factor of 0.65–0.85, and the detention time is multiplied by a factor of 1.25–1.50. This will give adequate treatment efficiency in the field conditions as obtained under laboratory test.

Example: 8.3 The settling test was performed in the laboratory using settling column of total height of 2.5 m. Four numbers of sampling ports were provided to the column at the equal interval of 0.5 m from bottom and fifth one at bottom. Samples were collected from these ports at every 30 min and from the results the iso-removal lines are obtained as plotted in the Fig. 8.14. Determine the overall removal of solids after 1.0 h of settling.

Solution

Percentage removal

$$= \frac{\Delta h_1}{h_5} \times \frac{(R_1 + R_2)}{2} + \frac{\Delta h_2}{h_5} \times \frac{(R_2 + R_3)}{2} + \frac{\Delta h_3}{h_5} \times \frac{(R_3 + R_4)}{2} + \frac{\Delta h_4}{h_5} \times \frac{(R_4 + R_5)}{2}$$

Fig. 8.14 Results of the laboratory settling column test representing iso-removal lines

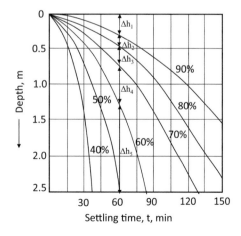

$$+ \frac{\Delta h_5}{h_5} \times \frac{(R_4 + R_5)}{2}$$

For iso-percentage removal curves shown in the Fig. 8.14, after 60 min of settling the values of $\Delta h_1 = 0.313$ m; $\Delta h_2 = 0.156$ m; $\Delta h_3 = 0.281$ m; $\Delta h_4 = 0.531$ m; $\Delta h_5 = 1.219$ m; and h_5 is 2.5 m. Substituting values in the above equation, following values will be obtained.

$$\frac{\Delta h_1}{h_5} \times \frac{(R_1 + R_2)}{2} = 0.313(100 + 90)/(2.5 \times 2) = 11.894\%$$

$$\frac{\Delta h_2}{h_5} \times \frac{(R_2 + R_3)}{2} = 0.156(90 + 80)/(2.5 \times 2) = 5.304\%$$

$$\frac{\Delta h_3}{h_5} \times \frac{(R_3 + R_4)}{2} = 0.281(80 + 70)/(2.5 \times 2) = 8.430\%$$

$$\frac{\Delta h_4}{h_5} \times \frac{(R_4 + R_5)}{2} = 0.531(70 + 60)/(2.5 \times 2) = 13.806\%$$

$$\frac{\Delta h_5}{h_5} \times \frac{(R_5 + R_6)}{2} = 1.219(60 + 50)/(2.5 \times 2) = 26.818\%$$

Therefore, total removal under quiescent settling condition is 66.25%. To achieve this removal the detention time recommended for the actual settling tank is $1 \times 1.5 = 1.5$ h.

8.3.1.4 Hindered and Compression Settling

The hindered settling or Type III settling or zone settling occurs with an intermediate concentration of the particles, where the particles are close to each other and interparticle forces hinder the settling of neighbouring particles. Thus, in hindered settling the adhesion or cohesion forces between suspended particles are adequate to disrupt the settling of adjoining particles. Their positions concerning to each other remain the same and they settle as a unit due to these attraction forces, because of this phenomenon it is also called zone settling. This condition arrives in system that contain reasonable concentration of suspended particles and the particle are so close to each other that all settle at a constant velocity forming a sludge blanket. At the top of the settling particles blanket, there will be distinct solid–liquid interface between the settling particles zone and clear supernatant water on top. This type of settling happens at intermediate depth in a secondary settling tank generally provided with biological treatment facilities, lime-softening sedimentation, and sludge thickeners.

As the hindered settling continues, a compressed layer of particles begins to appear at lower part of settling tank and simultaneously it starts forming a structure where particles have close physical contact. Water below this compressed layer containing high concentration of solids is displaced upward through the interstices between this compacting sludge zone. Due to this movement, hindered settling zone contains a gradient in solids concentration, which is increasing from surface to compression settling zone. This type of settling transpires in the lower part of sludge thickening facilities.

Compression or Type IV settling is the settling of the particles at a high concentration, such that the particles are touching each other and further downward movement is only possible with compression of the particles beneath. This Type IV settling occurs in the sludge thickener and at lower parts of the secondary sedimentation tank. Though discrete and flocculent particles both can exhibit the hindered and compression settling, however it is more common for flocculating particles. Classical laws of sedimentation are not applicable for hindered and compression settling. Hence, a column settling test is required to be performed to determine the settling characteristics and settling velocity.

Analysis of hindered settling (Type 3): When a high concentration of suspended solids is present in a wastewater, both hindered settling (Type III) and compression settling (Type IV) will usually occur at the lower depth of the sedimentation tank in addition to discrete and flocculent settling occurring at top portion. This condition will typically occur when the mixed liquor from the activated sludge process is admitted to the secondary clarifier for settling of sludge mass. The settling phenomenon that occurs is illustrated in Fig. 8.15, when a concentrated suspension of particles, with uniform concentration throughout, is placed in cylinder.

Due to high concentration of particles in the hindered settling zone, the liquid tends to move up through the interstices of contacting particles. As a result, the particles settle as a zone or 'blanket', maintaining the same relative position with respect to each other. As the solids settle as a zone, a relatively clear layer of water is produced above the particles in the settling column. The rate of settling in the hindered settling region is a function of concentration of solids and their characteristics.

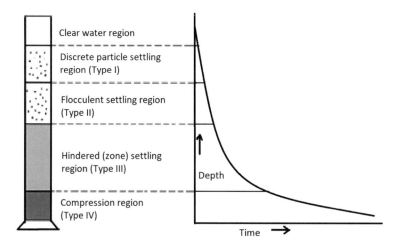

Fig. 8.15 Illustration of settling regions for mixed liquor from activated sludge process

As settling continues, a compressed layer of particles begins to form at the bottom of the cylinder in the compression settling region. Thus, in hindered settling region a gradation in solids concentration exists from solid-water interface of settling region to that found in the compression settling region. Due to variability of nature of solids and concentration, settling test is necessary to determine the settling characteristics. Based on the data of column settling test performed in laboratory, two different approaches as: (a) single batch test result and (b) solid flux method can be used to determine settling area for sedimentation tank. The second approach of solid flux method involves series of settling tests at different suspended solids concentrations.

Area requirement based on single batch test result: The final surface overflow rate selected for design of sedimentation tanks is based on: (a) the area necessary for clarification, (b) the area required for thickening, and (c) the rate of sludge withdrawal, which will decide the actual overflow from the sedimentation tank. Since the area required to support free settling of particles in the settling region is usually lesser than the area required for thickening, the rate of free settling is rarely the controlling factor. In case of activated sludge process, where light, fluffy bulking floc particles are present, it is conceivable that the free or flocculent settling velocity of these particles could control the design. However, this is exceptional case and generally the area required for thickening will decide the area required for clarifier.

For a column of height H_o having initially a uniform solid concentration of C_o, the position of interface as the time elapses is given in Fig. 8.16. The rate at which interface subsides is equal to the slope of the curve at that point of time. The area required for thickening can be estimated using Eq. 8.24.

$$A = \frac{Q \cdot t_u}{H_o} \tag{8.24}$$

Fig. 8.16 Analysis of the single batch settling test result

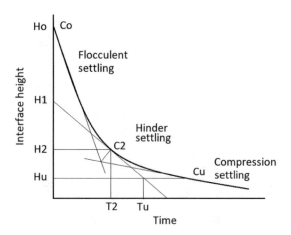

where A is the area required for sludge thickening, m^2; Q is the wastewater flow rate in the tank, m^3/sec; H_o is the initial height of interface in the settling column, m; and t_u is the time required to reach the desired underflow concentration, sec.

The critical concentration controlling the sludge handling capability of the tank will occur at a depth H_2 (Fig. 8.16), where concentration is C_2. This concentration 'C_2' is determined by extending straight line portion of the hyperbola and bisecting angle of intersection to get the point 'C_2' on the settling curve (Metcalf & Eddy, 2003).

The time 't_u' can be determined by adopting following steps. Construct a horizontal line at the depth 'H_u' that corresponds to the depth at which the solids are at the required targeted underflow concentration 'C_u'. The value of H_u can be determined using Eq. 8.25.

$$H_u = \frac{C_0 \cdot H_o}{C_u} \tag{8.25}$$

Draw a tangent to a settling curve at point C_2. From intersection of this tangent line and horizontal line from 'H_u' draw a vertical line. This vertical line will determine 't_u'. With this value of 't_u' the area required for thickening is computed using relation given in Eq. 8.24.

The area required for clarification can be determined from the straight-line portion of the settling curve at initial settling stage of flocculent region. The larger of the two areas is the controlling value. Although 'C_u' in settling test will occur at much longer time, due to continuous withdrawal from the bottom of tank this time may not reach in settling tank, hence 't_u' is approximated from the tangent.

Example: 8.4 Calculate the area required for the secondary sedimentation tank of activated sludge process. In a settling cylinder of 2.0 m height the settling test was performed, and the settling curve as shown in Fig. 8.17 was obtained with initial mixed liquor suspended solids concentration, C_o, of 4000 mg/L. Determine the area to yield a thickened sludge concentration C_u of 18 g/L with a wastewater inflow of 1000 m^3/day to the sedimentation tank. In addition, determine the solids loading rate and the overflow rate of the secondary sedimentation tank.

Solution
Determination of the area required for thickening

$$H_u = \frac{C_0 \cdot H_o}{C_u}$$

$$H_u = \frac{4000 \times 2.0}{18,000} = 0.44 \text{ m}$$

Drawing a horizontal line at height $H_u = 0.44$ m (Fig. 8.17). Construct a tangent to the settling curve at point C_2, the midpoint of the region between hindered and compression

Fig. 8.17 Results of the batch
settling test showing profile of
the sludge blanket with time

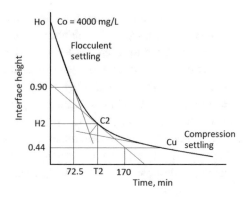

settling. From the intersection of the tangent at C_2 and horizontal line at the height H_u
($= 0.44$ m) drop a vertical line to get t_u, which is equal to 170 min.

Hence, the required area for achieving concentration of 18 g/L of sludge is (Eq. 8.24):

$$A = \frac{Qt_u}{H_o} = \frac{1000}{24 \times 60} \times \frac{170}{2.0} = 59.03 \text{ m}^2$$

This area should be adequate for clarification also; hence, estimate the area required
for clarification as illustrated below.

Determine subsidence velocity 'v' from the beginning of the straight-line settling
portion of the curve, by considering velocity of particles present at the interface.

$$v = \frac{2.0 - 0.9}{72.5} \times 60 = 0.91 \text{ m/h}$$

Determination of overflow rate

The overflow rate is proportional to the liquid volume above the sludge zone, since the
sludge is drained from bottom and only the supernatant water will overflow from the
settling tank.

$$Q = 1000 \times \frac{2.0 - 0.44}{2.0} = 780 \text{ m}^3/\text{day}$$

Determination of area required for clarification

The area required for clarification is obtained by dividing overflow rate by the calculated
settling velocity.

$$A = \frac{Q}{v} = \frac{780}{0.91} \times \frac{1}{24} = 35.71 \text{ m}^2$$

Thus, the controlling requirement is the thickening area, which is 59.03 m², since it is exceeding the area required for clarification.

Determination of the solids loading rate

$$\text{Solids loading rate}\left(\text{kg/m}^2\,\text{day}\right) = \frac{1000\left(\frac{\text{m}^3}{\text{day}}\right) \times 4000\left(\text{g/m}^3\right)}{10^3\left(\frac{\text{g}}{\text{kg}}\right) \times 59.03} = 67.76\ \text{kg/m}^2\,\text{day}$$

Determination of hydraulic loading rate

$$\text{Hydraulic loading rate} = \frac{780}{59.03} = 13.21\ \text{m}^3/\text{m}^2\,\text{day}$$

8.4 Grit Chamber

The grit chamber is provided after the screen and before primary sedimentation tank as a second unit operation of primary treatment for removal of sand and grit matters from the wastewater. Grit chamber is an essential part of the municipal wastewater treatment plant. Generally, provision of grit chamber is not required in industrial wastewater treatment plants. It is designed to remove the inorganic particles with specific gravity of around 2.65 and diameter of smallest particle to be removed as 0.20 mm. In practice, sometimes grit chamber is designed for removal of smallest particle of size 0.15 mm. Grit particles consist of sand, gravel, cinder, or any heavy particle with a specific gravity higher than that of organic matter. The location and design of grit chamber are important aspects to protect the pumps and other equipment from abrasion and wear. Absence of grit chamber in the sewage treatment plant will lead to increased maintenance of equipment, grit deposition in pipelines, channels, and conduits and it will also reduce the cleaning interval of digester because of higher accumulation these inert solids in digested sludge.

The conventional horizontal flow rectangular grit chamber is designed for peak flow and flow through velocity is maintained constant throughout the grit chamber under all variations in the wastewater inflow. Separate removal of suspended inorganic solids in grit chamber and settleable suspended organic solids in primary sedimentation tank is necessary because of different nature and mode of further handling, treatment, and disposal of these solids. Grit can be disposed off after washing. This washing is required for the collected grit matter for removal of higher size organic solids settled along with grit particles and for removal of organic solids attached on these sandy particles. Whereas, the suspended solids settled in primary sedimentation tank, being organic matter, require further treatment before disposal.

8.4.1   Classification of Grit Chamber

Grit chambers are classified into four general types as: horizontal flow rectangular grit chamber, square grit chamber, aerated grit chamber and vortex flow grit chamber. In rectangular horizontal flow type grit chamber, the inflow passes through the narrow channel in a horizontal direction by maintaining nearly plug-flow condition. Velocity is maintained constant in a rectangular grit chamber for varying flow by providing a velocity control weir at the effluent end. The square grit chamber has a square shape in plan and continuous grit removal arrangement is provided. The aerated type grit chamber consists of a mechanism to supply air at the bottom of the chamber that causes a spiral roll pattern of wastewater flow in the grit chamber. The vortex system consists of a cylindrical chamber or conical chamber and flow enters this chamber tangentially from top, thus creating the vortex flow, wherein gravitational and centrifugal forces help to capture grit, separating the wastewater free from grit.

8.4.2   Constant Velocity Horizontal Flow Rectangular Grit Chamber

Settling of grit particles in horizontal flow rectangular grit chamber follows discrete particle settling, i.e., Type I settling. The theoretical concept of horizontal basin is based upon Camp's ideal settling basin (1946), as discussed earlier under the theory of sedimentation. The basic assumption is that horizontal flow velocity is maintained uniform over the variation in the depth of flow in the grit chamber and grit particle settles at a constant settling velocity. Horizontal flow grit chamber is an enlarged channel, which can be divided into four separate zones focusing on different functions (Fig. 8.18), such as inlet zone, outlet zone, settling zone, and sludge zone. The purpose of inlet zone is to distribute the incoming wastewater uniformly throughout the cross-section of the chamber. The separate outlet zone is provided than settling zone so that because of the contraction of the wastewater, while exiting the grit chamber through narrow opening of weir, any turbulence created should not disturb settling of the grit particle. The settling zone is provided for free settling of the grit particles following discrete settling and sludge zone collects the settled grit till periodically it is removed from the chamber using grab bucket arrangement or grit pump.

The grit chamber is generally designed for settlement of smallest grit particle with diameter of 0.2 mm so that the particles having size equal to or higher than 0.2 mm will be completely removed because of exhibiting higher settling velocity. In addition, the horizontal velocity through the grit chamber should be adequate for resuspending the organic solids, if settled; whereas, the grit particles settled in the chamber should not be scored. If settling velocity of any particle (v_s) is greater than surface overflow rate (settling velocity of smallest particle) of the grit chamber (v_o), then all such particles will

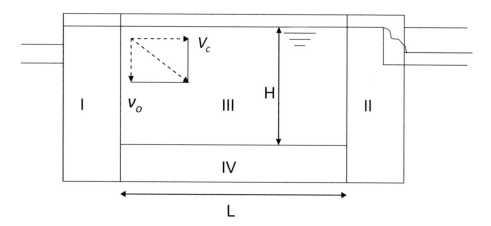

Fig. 8.18 Different zone of grit chamber. Zone I is inlet zone; zone II is outlet zone; zone III represents clear settling zone; and zone IV representing sludge zone

settle. Otherwise, for solid particles having settling velocity lesser than v_o, they will only be partially removed, as explained earlier.

Required detention time for grit chamber shall be estimated using Eq. 8.26:

$$Detention~Time~(t) = \frac{length~of~settling~zone~(L)}{Horizontal~Veocity~(v)} = \frac{Height~of~settling~zone~(H)}{Settling~Velocity~(v_o)} \quad (8.26)$$

Rearranging the terms in the Eq. 8.26, we can get relation $L/H = v/v_o$. Where L is the length of settling zone in the grit chamber and H is the settling depth. Knowing the settling velocity of the smallest particle to be removed and critical horizontal velocity (V_c), once the depth of grit chamber is decided, the length required for the grit chamber shall be estimated using this relation $L/H = V_c/v_o$.

To prevent the particle's scouring from grit settled zone, the horizontal velocity should be less than critical scouring velocity (V_c). This scouring velocity can be estimated as per Eq. 8.27.

$$V_c = \sqrt{\left[\frac{8\beta}{f}g(S_s - 1)D\right]} \quad (8.27)$$

where β is constant and value depends upon the grit material (0.04 for uni-granular sand and 0.06 for non-granular sticky gritty matter), f is the Darcy-Weisbach friction factor (0.03), g is the gravitational acceleration, m/sec^2, S_s represents specific gravity of grit particle (typically 2.65), and D denotes the diameter of the grit particle, m.

The critical velocity for scouring of a sand particle with a specific gravity of 2.65 and size of 0.20 mm is 0.228 m/sec. The grit chamber is generally designed for horizontal velocity slightly lesser than this value so that already settled grit particles are not being scoured and re-lifted in suspension. Also, at minimum flow the horizontal velocity generated in grit chamber should not be so low that it will allow even organic solids to get settled in the grit chamber, hence maintaining constant velocity with variation in the depth of flow is necessary for proper functioning of the grit chamber.

Design guidelines for rectangular horizonal flow grit chamber: Rectangular grit chamber is the oldest type of grit chamber used in sewage treatment plants, which is made of a long narrow channel (Fig. 8.19). This is designed to maintain a velocity near the critical scouring velocity with plug flow condition prevailing for the least mixing and to provide adequate time for grit particles to get settle. Higher length to width ratio of the chamber is provided for ensuring plug flow condition. A minimum allowance of about twice the maximum depth or 20–50% of the theoretical length of the chamber should be provided for the inlet and outlet zones. The width of the grit chamber is kept between 1 and 1.5 m. The flow is kept shallow with a recommended freeboard of a minimum of 0.3 m and a grit accumulation space of about 0.25 m. A detention time of 30–60 sec is recommended for the grit chamber. In a large treatment plant, two or more grit chambers can operate parallelly to meet the flow requirements. Grit settled at bottom of horizontal flow grit chamber is removed by scraper-conveyor mechanism or chain driven grit conveyor buckets and plough. In small plants, manual removal of settled grit is also an feasible option.

Example: 8.5 Design a rectangular grit chamber for treatment of sewage generated from a population of 70,000 persons with an average per capita water consumption of 130 LPCD.

Solution

Estimation of sewage flow
Considering 80% of the water supply contributes to sewage generation.

Amount of average sewage generated $= 130 \times 70,000 \times 0.8 \times 10^{-3} = 7280$ m^3/day $= 0.084$ m^3/sec.

Considering the peak flow factor of 2.5, the design flow will be.

Maximum sewage flow $= 2.5 \times 0.084 = 0.21$ m^3/sec.

The velocity of flow should be less than the scouring velocity (0.228 m/sec) and recommended detention time is between 30 and 60 sec. Hence, considering the horizontal velocity of 0.2 m/sec with a detention time of 45 sec for design.

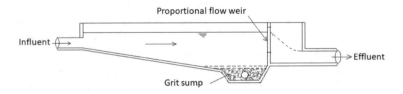

Fig. 8.19 Elevation cross section of horizontal flow rectangular grit chamber

Calculating dimensions of grit chamber
Now, length of chamber = horizontal velocity × detention time

$$= 0.2 \text{ m/sec} \times 45 \text{ sec} = 9 \text{ m}$$

Volume of chamber = flow × detention time

$$= 0.21 \text{ m}^3/\text{sec} \times 45 \text{ sec} = 9.45 \text{ m}^3$$

Cross sectional area of flow = Volume/length
$$= 9.45 \text{ m}^3/9 \text{ m} = 1.05 \text{ m}^2$$

Generally, the width of grit chamber is kept between 1 and 1.5 m. Hence providing the width of 1.5 m.

Now, depth of wastewater in grit chamber = 0.7 m.

After providing free board and grit accumulation zone,

Final depth of the grit chamber = 0.7 + 0.3 + 0.25 = 1.25 m.

Provide additional 25% length for inlet and outlet zone,

Hence, final length = 1.25 × 9 = 11.25 m.

Thus, the final dimension of the grit chamber will be 11.25 m (length) × 1.5 m (width) × 1.25 m (depth).

Example: 8.6 Design a horizontal flow grit chamber with rectangular cross section for treating an average and maximum sewage flow of 7.28 MLD and 18.2 MLD, respectively. At location of sewage treatment plant, a maximum temperature of 34 °C is observed for sewage during summer and that in winter the minimum temperature is 15 °C. Design the grit chamber satisfying the velocity ratio with ratio of depth and length.

Solution
The settling velocity of the grit particle will be minimum at lower temperature, i.e., 15 °C. At this temperature kinematic viscosity is 1.14×10^{-2} cm^2/sec.

In *first trial* assume Reynolds number 'R' less than or equal to 1, hence using Stokes' law for estimating the settling velocity.

$$Vs = \frac{g}{18}\left[\frac{S-1}{v}\right]D^2$$

$$Vs = \frac{981}{18}\left[\frac{2.65-1}{1.14x10^{-2}}\right]0.02^2$$

Thus, $Vs = 3.15$ cm/sec.

Reynolds Number $R = v \cdot D/v = 3.15 \times 0.02/1.14 \times 10^{-2} = 5.53 >$ which is greater than 1.

Therefore, Vs is not equal to 3.15 cm/sec because the equation for Vs is valid only for $R < 1$. Using $Vs = 3.15$ cm/sec, calculate R and C_D and then again Vs till it converges. Proceed with estimation of C_D and using Newton's equation estimate the settling velocity for few trials.

Thus, in subsequent trial

$$Vs = 2.4 \text{ cm/sec}$$

$$R = 2.4 \times 0.02/\left(1.14 \times 10^{-2}\right) = 4.21$$

$$C_D = \frac{24}{4.21} + \frac{3}{\sqrt{4.21}} + 0.34 = 7.50$$

From Newton's equation

$$Vs = \sqrt{\left[\frac{4}{3}\frac{981}{7.50}(2.65 - 1)0.02\right]}$$

Thus, $Vs = 2.4$ cm/sec (2074 m/day), which is same as the value used for determining Reynold's number, hence the solution of Vs is converged.

Estimation of critical horizonal velocity
Now for $\beta = 0.06$, f = 0.03, and D = 0.02 cm

$$Vc = \sqrt{\left[\frac{8\beta}{f}g(S - 1)D\right]}$$

$$Vc = \sqrt{\left[\frac{8 \times 0.06}{0.03}981(2.65 - 1)0.02\right]} = 22.76 \text{ cm/sec}$$

Designing the grit chamber
Now Q = 18.2 MLD = 0.2106 m³/sec.

Adopt horizonal velocity of 0.22 m/sec.

Therefore, C/S area of grit chamber A = Q/V = 0.2106/0.22 = 0.958 m².

If width of 1.2 m is provided, the depth required = 0.7979~0.8 m

Provide total depth = 0.8 + 0.3(free board) + 0.25(space for grit accumulation)

$$= 1.35 \text{ m}$$

Now $Vo/Vc = H/L = 2.4/22$.

Therefore, theoretical length $L = 22 \times 0.8/2.4 = 7.33$ m.

Provide 3 m extra length for inlet and outlet zones.

Therefore, total length $= 3 + 7.33 = 10.33$ m say 10.35 m.

Total working volume $= 0.8 \times 10.35 \times 1.2 = 9.936$ m^3.

Hence, overall detention time $= 9.936/0.2106 = 47.2$ sec (which is within 30–60 sec).

The final dimension of the grit chamber is 10.35 m (length) $\times$ 1.2 m (width) $\times$ 1.35 m (depth).

8.4.2.1 Velocity Control Device for Horizonal Flow Rectangular Grit Chamber

The grit chamber is designed to handle the maximum dry weather flow. Even, when two grit chambers are provided, each one is designed to handle peak flow. However, variation in incoming sewage flow can cause varying velocity in the grit chamber, instigating turbulence and scoring of settled solids at high velocities in the grit chamber or low velocities may permit even suspended organic solids to get settled in grit chamber. Hence for proper functioning, the velocity should not be allowed to change in the grit chamber despite of the variation in the flow.

This requirement can be met by employing a velocity control device, such as a proportional flow weir or Parshall flume at the outlet end of the grit chamber. The size and shape of the opening between the weir plates of proportional flow weir are provided such that the discharge is directly proportional to liquid depth in the grit chamber. Thus, constant velocity is maintained within the permissible variation (5–10%) even at varying discharges. A suitable method to control the velocity between recommended range of 0.15–0.30 m/sec through the grit chamber is achieved by constructing a control section at the outlet portion of channel such as flow weir, Sutro weir (it is half of proportional flow weir cut symmetrically along the vertical central axis) and Parshall flume. Operating experience of grit chamber has shown that if the horizontal flow through velocity in the grit chamber is maintained in the range of 0.23–0.3 m/sec it gives satisfactory performance.

8.4.2.2 Proportional Flow Weir

Proportional flow weir is the combination of a weir and an orifice, which maintains a constant velocity in a chamber by varying the cross-sectional area of flow through the weir so that the depth is proportional to the flow. The sharp edges generated by the curve at the bottom-most part of grit chamber are truncated on both the side because such sharp opening will not contribute to the flow due to entrapment of solids leading to blockage. These edges are trimmed from the channel wall at a minimum distance of 75 mm (Fig. 8.20), and height of the vertical edge (a) is 25–35 mm. To compensate for this loss in area, the weir's edge is lowered by $a/3$ than the theoretical level. Weir is constructed from 0.1 to 0.3 m above the bottom of the chamber to give space for grit storage and mechanical grit clearing. Weir should be situated at an elevation as to give a

Fig. 8.20 Proportional flow
weir used for controlling
velocity in the grit chamber

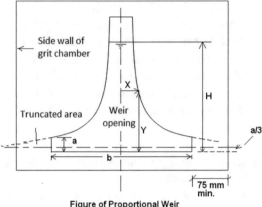

Figure of Proportional Weir

free fall into outlet channel and each grit channel is provided with a separate control weir. Head loss should be between 0.06 and 0.6 m (CPHEEO, 2013) and in the mechanized units, the free fall over the exit weir should not be less than 0.2 m during peak flow. For grit chamber provided with proportional flow weir the rectangular cross section is ideal.

The discharge through a proportional weir is calculated by using the following Eq. 8.28.

$$Q = C \cdot b\sqrt{2ab} \cdot \left[H - \frac{a}{3}\right] \qquad (8.28)$$

where Q is flow rate (m³/sec), C is constant with the value of 0.61 for symmetrical sharp-edged weir, a is height of the vertical edge (Fig. 8.20), which is in the range of 25–35 mm, b represents the width of the weir, H denotes the height of water above the crest of the weir.

The equation of the curve forming on the edge of the weir is given by Eq. 8.29.

$$x = \frac{b}{2}\left[1 - \frac{2}{\pi}\tan^{-1}\sqrt{\left(\frac{y}{a} - 1\right)}\right] \qquad (8.29)$$

where x represents the width of the weir at the liquid surface and y is depth of liquid at the weir.

8.4.2.3 Parshall Flume

Parshall flume was developed in 1926 by Ralph M. Parshall at Colorado State University, while working on the canal system as a flow measuring device. Parshall flume is a common choice for grit chambers, which is usually installed at the end of the chamber as a velocity control device (Fig. 8.21). Parshall flume has advantages like minimum head loss, it acts as discharge measuring device, self-cleansing device, works in submerged conditions, and has negligible problem of clogging. One Parshall flume can serve 2–3 grit chambers for discharge measurement.

Parshall flume consists of three parts: converging section, throat section and diverging section (Fig. 8.22). The floor of the throat section is inclined downwards and for diverging section it is inclined upwards. The dimensions of all the parts are fixed to achieve the transition of flow from subcritical ($F_R < 1$) to supercritical ($F_R > 1$) flow and this transition is known as free flow condition. Free flow condition is obtained by narrowing the throat width (W) and by incorporating a drop in floor level of the throat. In free flow, the wastewater depth at downstream (h_d) is minimum and only upstream depth (h_u) needs to be measured in order to obtain the discharge value.

This geometry creates a critical depth to occur near the beginning of the throat section and generate a backwater curve. By measuring the depth of water in this backwater curve, h_u, the discharge can be estimated (Fig. 8.22). The ideal section of the grit chamber constructed with Parshall flume as velocity control device is parabola, however combination of rectangular and trapezoidal cross section is preferred in practice due to difficulty in constructing the parabolic section. This combination of trapezoidal and rectangular cross

Fig. 8.21 Grit chamber with Parshall flume as a system for controlling velocity

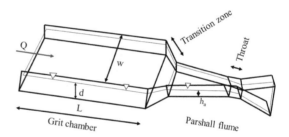

Fig. 8.22 Parshall flume

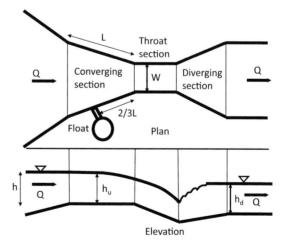

Fig. 8.23 Proposed
cross-sectional area instead of
the ideal parabolic
cross-section area of the grit
chamber provided with
Parshall flume

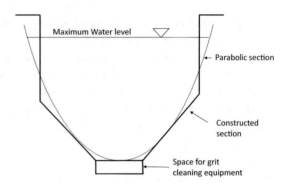

section is so selected that it approximates the parabolic cross-sectional area as shown in
Fig. 8.23.

Sometimes, the transition of flow from subcritical to supercritical is not achieved due to
different reasons (change in velocity, deposition of solids, etc.) and this failure is termed
as submergence condition. During the submergence condition, h_d increases up to a significant head and can lead to the backwater effect, thus giving rise to the submergence of
the upstream infrastructure. In case of submerged flow even downstream depth of flow is
required to be measured (Fig. 8.24). However, the manufacturer provides tolerance limit
or submergence limit up to which the flume should works sufficiently. Submergence limit
(S_t) is the ratio of secondary point head (h_d) to the primary point head (h_u) (Fig. 8.24)
and ranges from roughly 50 to 80% depending upon the size of throat. For estimating the
discharge Eqs. 8.30 or 8.31 can be used. In Eq. 8.30, for the throat width of 0.3–2.4 m
the value of C is 2.26 for SI units.

$$Q = CWh_u^{1.522\, W^{0.26}} \tag{8.30}$$

Or

$$Q = C \times h_u^n \tag{8.31}$$

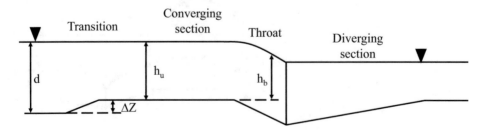

Fig. 8.24 Parshall flume showing submergence condition

Table 8.1 Standard dimension of throat width

Flow range (ML/day)	0–5	5–30	30–45	45–170	170–250	250–350	350–500	500–700	700–850	850–1400
Throat width (mm)	75	150	225	300	450	600	900	1200	1500	2400

where Q is the rate of flow through Parshall fume (m³/sec), C represents the discharge coefficient depending on the throat width (provided by the manufacturer), h_u is primary head point measured at the convergence section, n is discharge exponent depending upon the throat width (provided by the manufacturer; normally-1.5).

Measurement of flow through a Parshall flume is calculated by an empirical formula provided by the manufacturer, and the design and discharge characteristics related to Parshall flume has been standardized (Table 8.1) in ASTM D1941-91 (2013), ISO-9826, 1992 and CPHEEO (1993). Moreover, EPA (1995) has also provided an empirical relation for the discharge calculation as stated in Eq. 8.30.

8.4.2.4 Parabolic Grit Channel

A constant flow through velocity in a channel is important to ensure settling of grit particles and not organic solids and to maintain plug flow conditions in a tank. To achieve this constant flow, the shape of the cross-sectional area of flow needs to be changed with the change in discharge. Let's assume that the horizontal velocity is constant in the channel and the channel's width (x) and depth (y) are varying to maintain the constant velocity (Eq. 8.32; Fig. 8.25).

$$V_c = Q / \int_{y=0}^{y=y} x\,dy \tag{8.32}$$

Also, the discharge Q can be expressed as Eq. 8.33 and solving will result in Eq. 8.34.

$$Q = C_1 \times \int_{y=0}^{y=y} x\,dy \tag{8.33}$$

Fig. 8.25 Cross-section of grit chamber with varying width and depth

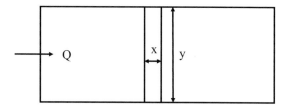

$$Q = C_1 \, x \, y \tag{8.34}$$

Discharge through the velocity control device is calculated by the Eq. 8.35.

$$Q = C_2 y^n \tag{8.35}$$

By equating Eqs. 8.34 and 8.35, the Eq. 8.36 will be obtained.

$$C_1 \, x \, y = C_2 y^n \tag{8.36}$$

Let's assume that the velocity control device is a flume, so the value of n will be 1.5.

$$C_1 \, x \, y = C_2 y^{1.5} \tag{8.37}$$

$$y = C_3 x^2 \tag{8.38}$$

The general equation $x^2 = 4py$ (Fig. 8.26) represents the parabola. Therefore, it can be said that whenever the Parshall flume is used as a flow control device, the grit channel should be constructed as a parabola to maintain the constant horizontal velocity. However, for easy construction of the channel, a trapezoidal section is used instead of a parabolic section (Fig. 8.23). Area of the parabola section is calculated as per Eq. 8.39 and discharge using Eq. 8.40.

$$A \ (parabola) = \frac{8}{3}\sqrt{p} \times y^{\frac{3}{2}} \tag{8.39}$$

$$Q \ (through \ parabolic \ channel) = v_c \times A \ (parabolic) \tag{8.40}$$

Fig. 8.26 Cross section of parabolic grit chamber

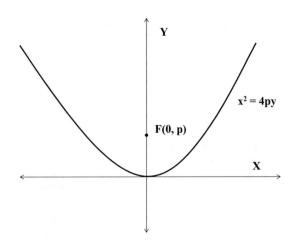

$$v_c \times A = C h_u^n \tag{8.41}$$

$$v_c \times \frac{8}{3} \sqrt{p} \times y^{\frac{3}{2}} = C h_u^n \tag{8.42}$$

During venturi flow, the depth of flow in grit chamber is equal to h_u and horizontal flow velocity (v_c) is taken as 0.3 m/sec (camp's velocity). Now, p can be calculated by using Eq. 8.42; moreover, by knowing p for a section (Fig. 8.26), the designer can calculate the different values of x and y co-ordinate.

Example: 8.7 Design a parabolic grit chamber with Parshall flume to treat a maximum sewage flow of 100,000 m³/day. Consider $n = 1.5$, $C = 1.8$, width of throat section $W = 762$ mm.

Solution
Flow to be treated $Q_{max} = 100,000$ m³/day $= 1.157$ m³/sec.
  Now $W = 762$ mm, $C = 1.80$, and $n = 1.555$.

Calculation of h_u using Eq. 8.31

$$Q_1 = C \cdot h_u^n$$

$$H_u = 0.75 \text{ m}$$

Calculation of p using Eq. 8.42

$$Q = v_c \times \frac{8}{3} \sqrt{p} \times y^{\frac{3}{2}}$$

$$v_c = 0.3 \text{ m/sec}$$

$$y = h_u = 0.75 \text{ m, hence solving to get } p = 4.96 \text{ m}$$

Calculation of x

$$x^2 = 4py$$

$$x = \sqrt{4 \times 4.96 \times 0.75}$$

$$x = 3.86 \text{ m}$$

Thus, using the equation $x = \sqrt{4 \times 4.96 \times y}$ the coordinates of the parabolic section of channel can be obtained.

8.4.3   Square Grit Chamber

The horizontal flow rectangular grit chamber faces a problem of sedimentation of organic matter along with grit particles, requiring external washing of the grit before disposal. This problem of external washing of grit can be eliminated by providing square shape of the grit chamber rather than long rectangular channel. This shape will also facilitate compact design of sewage treatment plant. Square grit chamber (Fig. 8.27) is a relatively recent design as compared to a rectangular grit channel. In square grit chamber, screened sewage is distributed throughout the cross-section by a series of vanes from one side of square, and the wastewater flows in a straight line maintaining non-ideal plug flow conditions and exit from the weir provided on the opposite side.

Square grit chamber is designed considering the surface overflow rate (800–1400 m^3/m^2 day). The design is made for at least 95% removal efficiency for grit particles of diameter 0.15 mm at peak flow (Metcalf & Eddy, 2003). The horizontal flow velocity in this square grit chamber is kept at 0.3 m/sec at peak flow. Thus, at lower flow some organic solids will get settled due to low velocity generated. Settled grit is collected in a sump or a collection chamber, located at the side of a tank, by rotating rakes provided

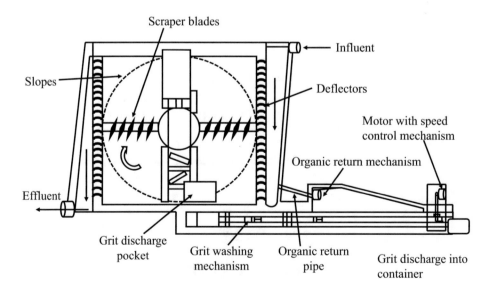

Fig. 8.27 Plan view of the square grit chamber

with scraper. The grit is moved and washed from the sump by reciprocating rake mechanism (grit washing mechanism) or inclined screw pump mechanism used for grit pumping from the grit chamber. Pushing the grit matter over the rake results in adequate abrasion to bring organic solids in suspension and washing of grit occurs. Washing of grit results in cleaner and drier grit free from any organic solids that might have settled along. In this grit screw or reciprocating pump, arrangement is so made that the organic solids after cleaning of settled grit particles are returned to the grit chamber (Fig. 8.27); whereas, the processed grit is sent for disposal.

It is recommended to provide at least two grit chambers for ensuring continuous operation of the treatment plant in case of cleaning and maintenance work. Grit washing mechanism is the main advantage of square grit chamber as it considerably reduces the cost of external grit washing. Square grit chamber also reduces the overall footprint of the treatment plant due to its compact design, requiring lesser space for construction, when compared to a conventional grit chamber.

Example: 8.8 Design a square shaped grit chamber for the peak sewage flow rate of 85 MLD. Consider the grit size to be removed as 0.15 mm with a specific gravity of 2.65. Consider the kinematic viscosity of water as 1.14×10^{-6} m^2 sec^{-1}.

Solution

Estimation of settling velocity
Applying Stokes law for settling velocity (v_s):

$$v_s = \frac{g}{18}\left(\frac{S-1}{\vartheta}\right)d^2$$

where g, is the acceleration due to gravity (9.81 m sec^{-2}); S is the specific gravity of the particle; υ, is the kinematic viscosity of water, which is given as 1.14×10^{-6} m^2 sec^{-1} in this case, and d is the diameter of the particle in m.

Arranging all the values in equation,

$$v_s = \frac{9.81}{18} \times \left(\frac{2.65 - 1}{1.14 \times 10^{-6}}\right) \times \left(0.15 \times 10^{-3}\right)^2$$

$$v_s = \frac{0.3642}{20.52} = 0.0177 \text{ m/sec} = 17.74 \text{ mm/sec}$$

Checking for Reynold's no. (R_e) for consistency of estimated settling velocity.

$$R_e = \frac{v_s D}{\vartheta} = \frac{0.0177 \times 0.15 \times 10^{-3}}{1.14 \times 10^{-6}} = 2.32$$

As $R_e > 1$, hence Stokes law is not valid.

Iteration 1

Calculating C_D using $R_e = 2.32$.

$$C_D = \frac{24}{R} + \frac{3}{\sqrt{R}} + 0.34 = \frac{24}{2.32} + \frac{3}{\sqrt{2.32}} + 0.34 = 10.344 + 1.96 + 0.34 = 12.644$$

Recalculating v_s with the estimated C_D,

$$v_s = \sqrt{\frac{4 \times g \times (S - 1) \times d}{3 \times C_D}} = \sqrt{\frac{4 \times 9.81 \times (2.65 - 1) \times 0.15 \times 10^{-3}}{3 \times 12.644}}$$

$$= \sqrt{\frac{0.0097119}{37.932}} = 0.01600 \text{ m/sec}$$

Iteration 2

Recalculating R_e with this revised v_s

$$R_e = \frac{0.01600 \times 0.15 \times 10^{-3}}{1.14 \times 10^{-6}} = 2.105$$

Estimating C_D with the revised R_e,

$$C_D = \frac{24}{2.105} + \frac{3}{\sqrt{2.105}} + 0.34 = 11.40 + 2.06 + 0.34 = 13.8$$

Calculating v_s with

$$v_s = \sqrt{\frac{4 \times g \times (S - 1) \times d}{3 \times C_D}} = \sqrt{\frac{4 \times 9.81 \times (2.65 - 1) \times 0.15 \times 10^{-3}}{3 \times 13.8}}$$

$$= \sqrt{\frac{0.0097119}{41.4}} = 0.01536 \text{ m/sec}$$

Conducting similar iteration few more times, at end of iteration number 5, the settling velocity will be 0.0148 m/sec. Hence, conduct subsequent iteration 6 as stated below.

Iteration 6

Recalculating R_e with this revised v_s

$$R_e = \frac{0.01489 \times 0.15 \times 10^{-3}}{1.14 \times 10^{-6}} = 1.96$$

Estimating C_D with the revised R_e,

$$C_D = \frac{24}{1.96} + \frac{3}{\sqrt{1.96}} + 0.34 = 12.244 + 2.142 + 0.34 = 14.726$$

Calculating v_s with

$$v_s = \sqrt{\frac{4 \times g \times (S-1) \times d}{3 \times C_D}} = \sqrt{\frac{4 \times 9.81 \times (2.65-1) \times 0.15 \times 10^{-3}}{3 \times 14.726}}$$

$$= \sqrt{\frac{0.0097119}{44.178}} = 0.01482 \text{ m/sec}$$

The value of v_s converged, thus indicating that the final value of v_s is 0.0148 m/sec.

Calculation of design surface overflow rate
Assuming a performance efficiency (η) of 75% and a n value of 1/8 corresponding to a very good performance, design surface overflow rate is given as follows (CPHEEO, 2013):

$$\eta = 1 - \left[1 + nV_s / \left(\frac{Q}{A}\right)\right]^{\frac{1}{n}}$$

Rearranging,

$$\frac{Q}{A} = \frac{nV_s}{\left((1-\eta)^{-n} - 1\right)} = \frac{\frac{1}{8} \times 0.0148}{(1-0.75)^{-\frac{1}{8}} - 1} = \frac{0.00185}{0.189} = 0.00977 \text{ m}^3/\text{m}^2 \text{ sec}$$

$$= 844.13 \text{ m}^3/\text{m}^2 \cdot \text{day}$$

Determination of grit chamber dimensions
Surface area, $A = Q/\text{SOR} = 85{,}000/844.13 = 100.69 \text{ m}^2$.
  Hence, the sides of the grit chamber will be $= \sqrt{100.69} = 10.03$ m.
  Say a size of 10.05 m is adopted for the grit chamber.
  Provide a retention time of 60 sec.
  Flow, $Q = \frac{85000}{3600 \times 24} = 0.9838$ in m^3 sec^{-1}.
  Volume required $= 0.9838 \times 60 = 59$ m^3.
  Hence, depth required $= \frac{59}{10.05 \times 10.05} = 0.584$ m
  Check for horizontal velocity.
  Horizontal flow through velocity $= \frac{0.9838}{0.584 \times 10.05} = 0.168$ m/sec.
  Critical velocity (Eq. 8.27) $= Vc = \sqrt{\left[\frac{8 \times 0.06}{0.03} 9.81(2.65-1)0.00015\right]} = 0.197$ m/sec.
  Thus, the horizonal velocity generated at peak flow of 0.168 m/sec is less than the critical velocity of 0.197 m/sec, hence the grit particles settled will not be resuspended.
  Providing a freeboard of 0.3 m and a grit accumulation depth of 0.25 m, the total depth of the grit chamber will be 1.134, thus provide a depth of 1.15 m. Provide two grit chambers of this dimensions.

Hence, the final dimensions of the square grit chamber will be 10.05 m $\times$ 10.05 m $\times$ 1.15 m. Provide two grit chambers of this dimensions.

8.4.4   Aerated Grit Chamber

Aerated grit chamber consists of a rectangular tank with an air pumping device. Air pumping device pumps air with the average flow of 0.2–0.5 m^3/m min. The screened influent sewage comes in contact with the pumped air, and a spiral flow pattern is generated in this aerated grit chamber (Fig. 8.28). This spiral flow induces agitation and mixing and hence any organic solids attached to the grit particles will be brought in suspension, thus favouring settling of only grit particles. The size of the smallest grit particle (usually 0.20 mm) that will be removed depends upon the velocity of this spiral roll. Appropriate velocity of the roll can assist in achieving the required overall removal of grit particles, which can be tweaked by adjusting the airflow quantity. Wastewater passes through the grit chamber in a spiral flow and tends to make two to three passes (roll) across the tank's bottom at maximum flow and the number of rolls tend to increase with the decrease in flow. The velocity of roll is the function of shape of the chamber (Eq. 8.43) and rate of air diffusion, and the number of rolls in tank is estimated as per Eq. 8.44.

$$\frac{\Delta L}{\pi D} = \frac{v_H}{v_T} \tag{8.43}$$

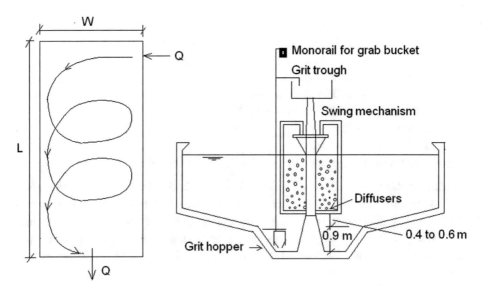

Fig. 8.28   Aerated grit chamber showing **a** helical flow pattern and **b** cross-section of the chamber

$$n = \frac{L}{\Delta L} \tag{8.44}$$

where ΔL is the length of one spiral roll, D is depth of water in the chamber, v_H is the velocity component of the spiral flow in axial direction ($Q/A_{(cross\ section)}$), v_T is the velocity component of the spiral flow tangential to the spiral circulation path in the plane of grit chamber's cross section (0.3 m/sec), n is number of rolls in the chamber, L is length of the grit chamber.

The advantages of aerated grit chamber are enlisted as follows: pre-aeration reduces septic conditions; head loss can be minimized; and this aerated grit chamber can give constant removal efficiency of grit over a wide range of flow. However, higher power consumption and technical assistance during maintenance limits the application of aerated grit chambers. A minimum of two grit chambers are installed to meet the flow demand during maintenance or during the malfunctioning of one unit and covered aeration tank is generally recommended to avoid the release of volatile organic compounds (VOCs) during aeration.

Grit removal mechanism consists of grit bucket, chain bucket conveyor, screw conveyors, jet pumps, airlift pump, or tubular conveyors. The grit is collected in the grit hopper (the collection chamber) and pumped out for direct disposal as aerated grit chamber eliminates the necessity of external grit washing as the settled grit will be well washed and free from organic solids.

Design recommendations for aerated grit chamber: The air diffusers are situated at around 0.45–0.6 m from the floor of the grit chamber (Fig. 8.28), while the grit hoppers are placed at 0.9 m depth with sharp sloping sides under the air diffusers (Metcalf & Eddy, 2003). There is always a chance of an increase in the volume of sewage because of the pumped air, which should be considered appropriately during the design of aerated grit chambers.

- Depth of wastewater in the grit chamber: 2–5 m,
- Length of the aerated grit chamber: 7.5–20 m,
- Width to depth ratio: 1:1–5:1,
- Width of the grit chamber: 2.5–7 m,
- Detention time during peak flow: 2–5 min with typical value of 3 min
- Air supply: 0.3 m^3/min m of length of the grit chamber.

Example: 8.9 Design an aerated grit chamber for treatment of sewage with average flow rate of 0.8 m^3/sec and peak factor of 2.5.

Solution

Peak sewage flow $= 0.8 \times 2.5 = 2$ m^3/sec.

Determination of volume of the grit chamber
Consider a detention time of 3 min.

$$V = Q \times \textit{detention time}$$

$$V = 2 \times 3 \times 60 = 360 \text{ m}^3$$

Construct two aerated grit chambers for flexibility in cleaning and maintenance.

$$\text{Volume of each chamber} = 180 \text{ m}^3$$

Calculation of dimensions of grit chamber
Assume, width to depth ratio = 1.5:1 and depth = 3 m

$$\text{Hence, width} = 3 \times 1.5 = 4.5 \text{ m}$$
$$\text{Length} = 180/(4.5 \times 3) = 13.33 \text{ m}$$

Provide 20% extra length at inlet and outlet section. Hence, the dimension of the tanks will be 16 m $\times$ 4.5 m $\times$ 3 m.

Estimating air supply requirement

$$\text{Quantity of air} = 0.3 \times 16 = 4.8 \text{ m}^3/\text{min}$$

Check for surface overflow rate

$$\text{SOR} = 2/(2 \times 13.33 \times 4.5) = 0.017 \text{ m/sec} = 1.7 \text{ cm/sec}$$

Which is less than 2.4 cm/sec, that is settling velocity of smallest grit particle, hence design is safe.

8.4.5   Vortex Type Grit Chamber

Vortex is a region of fluid in which the flow revolves around an axis and this axis can be straight or curved. Vortex type grit chamber achieves this vortex flow by two different methods. In first method, wastewater enters the chamber tangentially and a constant flow velocity is maintained by the turbine, while the pitch blade aids in the separation of organic particles from the grit (Fig. 8.29). In the chamber, grit particles follow a toroidal flow pattern due to the rotation of propeller and while moving in downward direction settles into the hopper (grit collection chamber) from where an airlift pump or grit pump lift the settled particles to the sump for further disposal. The effluent exits the vortex grit

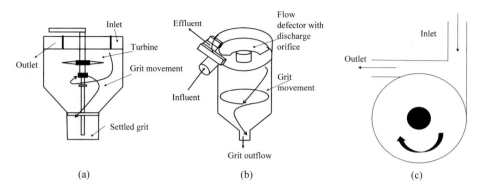

Fig. 8.29 Vortex type grit chamber, **a** with turbine, **b** without turbine, **c** typical arrangement in plan

chamber in the direction perpendicular to the flow path direction from top, hence grit particles will not be able to appear in the effluent due to inertia of toroidal flow path.

In second type of vortex grit chamber, the vortex flow is achieved by the wastewater entering tangentially at the top of the tank and existing from the effluent pipe located at the centre (Fig. 8.29). Inertia force plays a significant role in removing the grit particles in vortex grit chamber and the grit particles remain in the chamber while the grit-free wastewater moves out after treatment.

Head loss in vortex type grit chamber depends on the size of the particle to be removed and it increases as the size of particle decreases. In a vortex type grit chamber, the smaller sized particles need higher tangential flow velocity to get separated from wastewater stream, resulting in higher head loss during the flow. Moreover, removal of finer grit particles is more energy intensive due to excess centrifugal force requirement. These grit particles settle in the tank by gravity, while the organic particles attached to grit are also brought in suspension as the grit particle moves along the wall causing abrasion and thus removal of any organic matter attached with grit. Belt conveyer mechanism is preferred to remove settled grit from the vortex grit chamber.

Vortex type grit chamber has many advantages like high energy efficiency, minimum head loss, small footprint, no submerged bearing, and constant removal efficiency in a wide range of flow. The requirement of dewatering mechanism, need of grit washing and rags on turbine blade, if used, are still the major drawbacks of these systems. For designing the vortex grit chamber, the detention time of 20–30 sec (at average flow) is generally adopted. The height of vortex tank is kept at 2.7–4.5 m and diameter of tank may vary from 0.9 to 7.2 m depending on the sewage flow to be handled.

8.4.6   Sludge De-gritting

Sludge de-gritting is practiced in treatment plants, where separate grit chambers are not provided and grit particles are allowed to settle along with organic suspended particles in the PST. Grit removal is achieved by pumping the settled sludge from PST to cyclone de-gritter, which acts as a cyclone separator and the organic fraction of primary sludge is separated from the grit by the action of vortex. Cost analysis is to be performed before deciding whether the grit chamber is preferred instead of sludge de-gritter.

8.4.7   Collection and Disposal of Grit

The quantity of grit collected depends upon: (a) the type of sewerage system, (b) the geography of the drainage area, (c) sewer conditions, (d) variety of industries present in the area and (e) the type of soil. Around 0.004–0.2 m^3 of grit per million litres of wastewater treated could be collected from a municipal wastewater treatment plant. While designing the grit handling facility a typical value of 0.015 m^3/ML of wastewater treated is to be considered. The moisture content of the grit ranges from 13 to 65% and volatile content from 1 to 56% depending on effectiveness of the suspended organic solids not being allowed to get settled in the grit chamber. The specific gravity of grit particles ranges from 1.3 to 2.7 with bulk density of around 1600 kg/m^3 (Metcalf & Eddy, 2003).

The grit is disposed after cleaning for removal of organic solids and clean grit is characterized as grit free from organic particles with the absence of odour. Washing the grit is one method to obtain clean grit. Grit washing machine must be included whenever there is a possibility of suspended organic solids to get settled along with the grit. After washing the grit, for removing organic solids, it is subjected to disposal by dumping or burying or by landfilling. However, disposal of grit must be carried out after meeting the environmental regulations.

8.5   Skimming Tank

Fat, oil, and grease (FOG) are organic pollutants, whose defining characteristics are low water affinity and very low biodegradability. The FOG are present either as free particles or are coalesced with other suspended solids, while their concentration is reported between 50 and 150 mg/L in domestic wastewaters (Pintor et al., 2016). Discharge of FOG in the environment can cause damage to the aquatic lifeform as a thin sheet of FOG formed on the water surface will cut-off air water contact and reduce the oxygen solubility, and it also reduces light penetration into the waterbody. Moreover, FOG also affects the traditional wastewater treatment plant by inhibiting the biological activity in the activated sludge process, clogging the pipes and pumps, leading to the corrosion in pipes and damaging the probes of different monitoring equipment.

The removal of FOG is achieved by using gravity separation, dissolved air floatation, adsorption, electro-coagulation, and biological removal methods. However, the use of gravity separation methods like grease traps, skimming tanks, and aerated skimming tanks predominate in wastewater treatment plants. The typical FOG limit for the effluent is 10–20 mg/L (EPA, 1995); ideally, this limit can be achieved by using grease traps at the source prior to the discharge in the sewerage system. Grease traps are small concrete tanks with scum baffle and it is designed for a detention time of 3–5 min with a flow-through velocity of 2–6 m/h (EPA, 1995). These grease traps can retain up to 80% of influent FOG; however, due to poor maintenance and negligence the FOG can enter the sewerage system.

Skimming tank: A skimming tank is a rectangular or circular chamber designed for a detention time of 1–15 min (EPA, 1995). During the detention period, FOG rise and get collected on the surface while the wastewater flows out from deep outlets. A deep outlet helps to remove any solids that are settled during operation. The surface area of the tank depends upon the minimum rising velocity of FOGs and an inlet is designed to maintain uniform horizontal flow through velocity. These simple long and shallow tanks are provided for small treatment plants in industries. Instead of providing separate skimming tank, in conventional sewage treatment plant these FOG are removed in PST by providing extra baffle ahead of the effluent weir.

Aerated skimming tank: Aerated skimming tank is an elongated rectangular trough-shaped tank with a relatively large surface area (Fig. 8.30). The aerated skimming tank uses both aided and induced floatation systems to remove FOG, and air is pumped in the form of bubbles to promote the floatation and separation of FOG from wastewater. The wastewater passes through the turbulence created by rising bubbles and FOG rise upward with the air bubbles. The scum collected at the surface of wastewater is pushed into the stilling zone located at both the sides of the tank by a scum removal mechanism and wastewater free from FOG leaves the tank. The recommended detention time for an aerated skimming tank is 3 min with airflow of 180 L/m^3 at peak flow (EPA, 1996). The efficiency can be increased by passing chlorine gas along with compressed air, as chlorine gas can destroy the colloidal protective effect of protein, which clutches the grease in coalesce form. Usually, FOG recovered from the stilling chamber cannot be reused in industries and thus disposal by burning or burial is the only option left. However, recent investigations have shown that the collected FOG can be used for the generation of biodiesel and biogas.

Skimming tank in hot climate: In a hot climate like India, oil and grease do not coagulate and congeal quickly, and concentration of these pollutants in municipal wastewater is also low to cause any significant disturbance to other treatment processes. Thus, skimming tanks are not provided in India at sewage treatment plants. However, onsite removal of oil and grease using different traps are used for treatment of industrial wastewater before

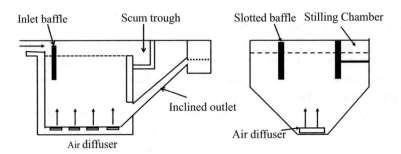

Fig. 8.30 Skimming tank used for removal of oil and grease

entering the sewer, which can save time and resources compared to the treatment of these pollutants in the wastewater treatment plants. In municipal sewage treatment plant, the small concentration of oil and grease present in the sewage can be removed by provision of baffle ahead of the effluent weir of primary sedimentation tank.

8.6   Primary Sedimentation Tank

Sewage reaching the treatment plant contains a considerable concentration of suspended solids of which about 70% are organic in nature. These suspended organic solids contribute to the biological oxygen demand of wastewater. Moreover, suspended particles can interfere with biological processes and would increase the load on secondary treatment units, enhance oxygen demand for aerobic processes, thus increasing overall size and operating cost of the secondary treatment. Therefore, removal of suspended organic particles becomes an essential step in wastewater treatment plants, particularly when some of the aerobic treatment processes are used for biological secondary treatment.

Traditionally, a primary sedimentation tank (PST) or primary clarifier is used to remove suspended organic solids by employing gravity settling operation. In PST, flocculant or type II settling is predominant during the removal of suspended solids; however, discrete or type I settling can also be observed. During settling operations, 50–70% of total suspended solids are removed; furthermore, 30–40% reduction in BOD and 25–75% decrease in bacterial count is also achieved (EPA, 1996). Due to the flocculant settling, surface overflow rate is an important parameter in achieving the desired efficiency of settling operation.

The suspended organic particles settle at the bottom of the sedimentation tank forming concentrated slurry, which is also referred to as primary sludge or simply sludge. Primary sludge is biologically active as it is mainly comprised of putrescible solids and prolonged detention period can cause generation of odorous gases due to the anaerobic degradation of organic particles present in the sludge. In addition to odour nuisance, the gas bubbles released from sludge rises upwards that can cause resuspension of sludge solids as well as it can hinder with the ongoing settling process occurring in the settling zone of the PST. Thus, deciding the adequate detention time with a suitable overflow rate is imperative in maintaining the optimum removal efficiency of the sedimentation unit.

The sludge is removed periodically from the bottom of the PST to avoid the biological anaerobic digestion of sludge in the PST. Specific gravity of this primary sludge ranges from 1.03 to 1.6. Settling tank can also serve the purpose of removing scum from the influent wastewater, removal of leftover inert smaller size inorganic particles that could not be removed in the grit chamber, and collection of return liquor from sludge dewatering unit as well as digester supernatant.

The primary clarifier is designed as either a rectangular or circular tank with an overflow rate of 30–50 m^3/m^2 day, while maintaining the detention time of 2 h. Choosing the clarifier configuration depends on the space requirement, availability of technical manpower, and cost of the project. Moreover, in large sewage treatment plants a minimum of two sedimentation tanks are constructed to meet the requirements during maintenance period and in case of any technical breakdown of the PST.

In continuous flow type tank, the velocity of influent sewage is reduced up to a predefined limit (flow-through velocity). The flow enters the tank horizontally or radially depending on the geometrical configuration (rectangular or circular, respectively) with a designed overflow rate for the removal of required particles. Inlet and outlet (weir, orifice, baffle, etc.) controls are used to regulate the flow velocity throughout the tank and sludge removal mechanism is used to remove the sludge periodically. The continuous flow through sedimentation tank is capable of handling large design flows, easier for operation and maintenance, and has reduced workforce requirement.

8.6.1   Type of Sedimentation Tank

8.6.1.1   Rectangular Primary Sedimentation Tank

It consists of a long rectangular tank with an inlet and outlet device and a sludge/scum removal mechanism (Fig. 8.31). Wastewater enters the tank through an inlet pipe and settling tank consists of multiple discharge ports to evenly distribute the flow to the entire width of the tank with a maximum horizontal distance of 2 m and it is responsible for uniformly spreading the influent wastewater in the tank. Inlet devices are designed to truncate the high inlet velocity, while minimizing the turbulence, eddy currents and short-circuiting in the clarifier for maintaining nearly laminar flow throughout the sedimentation basin. Settled sludge is collected into sludge hopper provided generally near the inlet end of the rectangular tank. However, based on the location of the sludge hopper the PSTs are classified as Gould type I when location of the sludge hopper is near the inlet end and Gould type II in which the sludge hopper is located at the middle of clarifier. Hopper

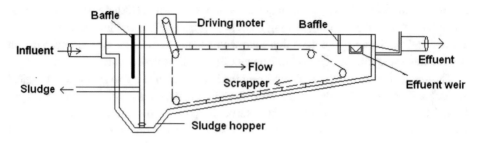

Fig. 8.31 Rectangular sedimentation tank

is always constructed at a suitable slope to avoid the accumulation of sludge on the side walls as fresh sludge tends to be sticky in nature.

Inlet design: Inlets in a PST are designed to maintain the inlet port velocity in the range of 0.075–0.150 m/sec and to minimize density currents to promote flocculation of suspended solids. Typical PST inlet consists of simple weir, submerged weir, orifice, and baffles. Simple weir is effective in distributing the flow throughout the width of the settling basin; however, sometimes weir can generate a vertical flow velocity component in the sludge hopper resulting in the resuspension of collected sludge.

Baffles can be solid wall (target baffle), perforated and figure type; each of this has its own function in the tank. Target baffle and figure baffle are placed 0.6–0.9 m in front of the inlet port and their function is to dissipate the kinetic energy of incoming flow, while perforated baffles are placed throughout the tank at a submergence depth of 0.45–0.60 m below the water level to reduce the density current while promoting flocculation. Generally, submerged weir in combination with the baffles are used as inlet for enhancing the performance of PSTs. Moreover, US EPA has recommended that the baffles must be extended up to minimum half of the tank depth, while the diameter of slots in case of perforated baffle should be between 50 and 100 mm. The scum baffle ahead of the outlet point is provided to obstruct the flow of wastewater to trap scum and prevent any resuspended solids from escaping along with the effluent. Scum baffles are extended above the water level up to 300 mm with provision of scum collection sump or hopper at one side of the tank with manual or mechanical scrapping of the scum towards this sump. Distance between the weir and scum baffle should not be less than 0.6 m.

Outlet design: Sometimes near the end wall of the tank, a strong upward current is generated and it is responsible for the resuspension of sludge particles. Hence, while designing a PST the weir loading rate is kept under control to minimize the turbulence at the exit of the tank. To minimize the effect of hydraulic pattern and to reduce the solids carry-over into the effluent channel, a combination of V-notch weir with effluent launder is constructed. Launders are always built with the V-notch weir and these launders can be placed longitudinal or lateral.

Sludge and scum removal: The common sludge collection systems usually provided in a PST are chain and flight scraper, reciprocating rack, traveling bridge, and floating type sludge collector. Generally, chain and flight scrapers are used for sludge removal and they consist of a pair of conveyor chains with attached scraper flights. Conveyor chains are made of metals and scrapers are built using fiberglass. Sludge is pushed into a hopper by the scraper. After pushing the sludge into the hopper, the flight scraper moves to the top and travels up to its original position, the conveyor chain controls this cyclic motion. This return of flight at the top of the surface works as a scum (oil, grease, etc.) removal device, while returning to its original position. Usually, the sludge hopper is placed at the inlet end of the clarifier, and a trapezoidal shape hopper is constructed with a side slope of 1.2:1–2:1 (Metcalf & Eddy, 2003).

The bridge type mechanism is supported on the sidewall and travels up and down along the length using rail or rubber wheel and the scrapers are suspended from the bridge. It can also be modified to collect scum during its return travel. The scraper pushes the sludge into the hopper from where the sludge is pumped to the sump for further treatment. High construction and maintenance cost limits its use in treatment plants.

In large sedimentation tanks, the hopper will be a transverse trough with a cross collector mechanism to push the solid to a side, where the pump and sludge take-off conduits are located. Cross collector can be chain and flight scraper or screw-type collector. Scum is collected into scum trough and the return flight scraper pushes the scum into it. The scum is pumped out of trough and is send for further treatment.

Permissible horizontal velocity of wastewater through the PST: To avoid resuspension (scouring) of settled particles, horizontal velocities through the PST should be kept sufficiently low. Following equation (Eq. 8.27, as given previously) by Camp can be used to calculate the critical velocity, Vc, which is the horizontal velocity that will just produce scour (m/sec).

$$Vc = \sqrt{\left[\frac{8\beta}{f} g (S-1)D \right]} \tag{8.27}$$

where β is constant equal to 0.04 for uni-granular sand and 0.06 for non-uniform sticky material; f is Darcy–Weisbach friction factor (0.02–0.03); g is gravity acceleration; S is specific gravity of the particle to be removed (1.2–1.6); and D is diameter of the particle, m.

For organic particle with size of 0.1 mm and specific gravity of 1.25 this velocity will be about 0.063 m/sec. Hence, horizontal flow through velocity through the PST during peak flow should be lesser than this value to avoid resuspension of the organic settled solids. Hence, flow through velocity of 1 cm/sec at average flow is generally used for design of PST so as the settled organic solids should not get resuspended due to higher flow through velocity of wastewater in the tank.

8.6.1.2 Circular Tank

A circular type primary clarifier (Fig. 8.32) is also equipped with an inlet and outlet device, and a sludge/scum removal mechanism. The theoretical flow pattern in a circular tank is radial, wherein wastewater enters the tank either through central feed system or by periphery feed system. Primary clarifiers are commonly designed with a central inlet and a peripheral outlet system. Circular tanks are preferred over rectangular tanks due to less maintenance and construction costs. However, for places where multiple tanks are required, circular tanks become a costly because of their larger footprint and due to the requirement of more piping facilities.

Inlet arrangement: Wastewater is pumped to the centre of the tank through an inlet pipe, which may be either a submerged horizontal pipe or an inverted siphon pipe laid beneath the tank floor. The inlet pipe opens into a central hollow pillar or well and this hollow pillar is broadened to provide enough number of inlet distribution ports. The distribution ports are located at the depth of 0.3–0.6 m below the water surface. The baffle provided forms the central hollow well consists of energy dissipating inlet (EDI), which helps to reduce the inflow velocity as wastewater enters the well from the ports of inlet pipe. The EDI provides continuous distribution of flow throughout the well and promotes flocculation in the tank. The diameter of EDI can vary from 10 to 20% of tank diameter with a detention time of 8–10 sec. Sewage enters from the central hollow pillar or well into the settling tank with the velocity of 0.30 m/sec to 0.45 m/sec at average flow and velocity should not increase beyond 0.75 m/sec at maximum flow to avoid turbulence and resuspension of solids. Inlet baffle depth extends up to 2 m below the water surface to reduce the short-circuiting (Metcalf & Eddy, 2003).

Outlet design: As an outlet device, a peripheral v-notch weir device is constructed, which allow overflow of settled wastewater into an effluent trough/launder. A baffle is provided ahead of the weir, with a spacing of about 0.3 m, which helps to retain the floating solids. This scum baffle is provided to prevent the scum from flowing out along with

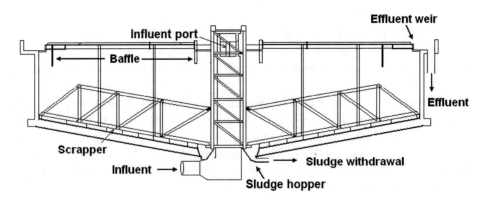

Fig. 8.32 Sectional elevation of circular sedimentation tank

the effluent. The design of the scum baffle is similar as that described for a rectangular sedimentation tank.

Sludge and scum removal: A revolving sludge scraper mechanism with radial arms and ploughs are used for the circular sedimentation tank and it is fixed just above the floor level with a rotation speed of about two per hour. Ploughs are used to push the sludge towards the central hopper, after which the collected sludge is transported from the central hopper to a sludge sump for further treatment. The bottom of the sedimentation tank is sloped (1 vertical to 12 horizontal) like an inverted cone to facilitate sliding of settled sludge towards sludge hopper. For a small tank (diameter<9 m), the revolving bridge with a sludge scraper is spanned across the circular tank while being supported on beams. The revolving bridge for a large tank is supported on the wall of the circular tank on one side and on a pillar at the centre, and a walkway is constructed on the top of the tank for the ease of maintenance. Scum removing blades are connected to the revolving sludge removal mechanism and positioned at the surface of the wastewater. The blade pushes the scum to a scum trough. The central well can also be provided with a scum skimmer if required.

8.6.1.3 Multilevel Clarifiers

Construction of multilevel or stacked clarifiers started around 1960s in Japan in order to minimize the land requirement in the construction of wastewater treatment plants. The operation of stacked clarifiers is the same as of a rectangular sedimentation tank in terms of flow patterns, and settling and removal of particles. In a multilevel clarifier, two tanks are constructed one above the other, and each clarifier is fed separately, resulting in parallel flow-through tanks. Better control of odour, less surface area, low volatile organic carbon emission, less pumping requirement are some of the advantages of multilevel clarifiers.

8.6.2 Factors Affecting Performance of Sedimentation Tank

Short-circuiting: Short-circuiting is a phenomenon where a substantial amount of wastewater crosses the tank without being held for the required detention time. Tracer study or plotting of residence time distribution curve helps to determine the extent of short-circuiting. In field situations, displacement efficiency is computed, which is the ratio of actual detention time to the theoretical detention time. Using a suitable inlet and outlet device or construction of a narrow sedimentation channel can avoid short-circuiting in PST.

Density current: Due to the difference in density between influent and wastewater present in the clarifier at the bottom, density current may be formed. Density current will lead to a change in actual flow through velocity with a short-circuiting effect (Fig. 8.33a). Inlet and outlet structures are helpful to control the phenomena of density current up

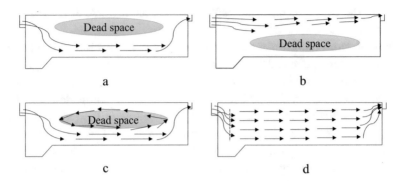

Fig. 8.33 Different flow pattern **a** short-circuiting due to density current and solid concentration, **b** short-circuiting due to temperature difference when the incoming wastewater is warmer than the temperature of the wastewater in the settling tank or when the incoming wastewater is having less solids concentration than the wastewater present in the tank, **c** short-circuiting due to cross wind flow, **d** ideal flow with suitable inlet device (Modified from Metcalf & Eddy, 2003)

to some extent by distributing the flow uniformly throughout the basin or by breaking the feed stream and dissipating the energy of flow (Fig. 8.33d). Plate and tube settlers, intermediate diffuser wall, and redistribution baffles are used to overcome this problem.

Temperature differentials: The difference of even one degree Celsius in temperature of influent and previously present wastewater can cause short-circuiting (Fig. 8.33b). Warm water rises to the surface and reaches the effluent launders in a fraction of detention time. Suitable inlet devices are adequate to neutralize the effect of temperature differentials short-circuiting.

Wind effect: Wind force has a significant effect on performance of open to air PSTs. The high-velocity wind pushes the wastewater flow and creates a surface current moving in the direction of the wind. Surface wind force in opposite direction of flow force the liquid at the surface to move backwards and creates an underflow current in the direction of flow, resulting in the formation of a circulating current in the sedimentation basin (Fig. 8.33c). Circulating current can lead to short-circuiting of the influent in PST in addition to the scouring of settled solids. Covering the top of a large size sedimentation tank may reduce the effect of wind and formation of surface currents.

Solid concentration: The difference in solids concentration between influent and already present wastewater in the sedimentation tank can cause the short circuiting. This effect is like as discussed above in density current flow and hence leading to short-circuiting. The only difference being the cause of current formation, which is the concentration gradient between incoming wastewater and wastewater present in sedimentation tank.

Equipment movement: The equipment and their movement rate may affect the per-formance of the clarifier. Sludge collection mechanisms like chain and flight of scrapers, hydraulic vacuum units move through the wastewater of the tank to scrap the settled sludge, and the high movement rate of these devices can lead to the formation of under-current and short-circuiting, which can have an adverse impact on the ongoing settling operation.

8.6.3 Solids Removal Efficiency of Sedimentation Tanks

Many factors influence the removal of suspended organic solids though settling operation in PSTs, such as thermal stratification in a hot climate, thermal convection current, wind turbulence on the wastewater surface, eddy current formed by the inertia of incoming fluid, and density current formation as discussed in the previous section. These factors make the derivation of any mathematical relationship for estimating the removal of suspended organic solids in PST complicated and cumbersome. The possibility of predicting the settling velocity and detention time by column settling test is also questionable. However, for the laboratory settling column test results a multiplying a factor of 0.65–0.85 is recommended for settling velocity and a factor of 1.25–1.50 for the detention time to accommodate these factors.

Empirical relationship has been developed between organic suspended solids removal and detention time by observing the operation of actual PST. Different researchers have proposed various models for establishing a suitable empirical relationship between removal efficiency and other operational parameters of the sedimentation tank. Crites and Tchobanoglous (1998) have proposed the empirical equation (Eq. 8.45) between the TSS and BOD removal efficiency with detention time.

$$R = \frac{t}{a + b \cdot t} \tag{8.45}$$

where R represents the expected removal efficiency, t is nominal detention time (h), a and b are empirical constants. For BOD removal at 20 °C, 'a' is 0.018 and 'b' is 0.020 and in case of TSS removal, value of 'a' is 0.0075 and for 'b' it is 0.014.

Tebbutt and Christoulas (1975) have proposed the empirical model (Eq. 8.46) between removal efficiency of TSS and overflow rate based of pilot-scale plant located at the University of Birmingham; however, this model has been tested and evaluated at different sewage treatment plants and satisfactory results were found.

$$E_s = a \times e^{-\frac{b}{S_i} - c \times q} \tag{8.46}$$

$$a = 1.71 - 0.03T \tag{8.47}$$

$$b = 683.6 - 21.13T \tag{8.48}$$

$$E_c = 0.733E_s - 0.08 \qquad (8.49)$$

where E_s represents the suspended solid removal efficiency, %; a and b are empirical constants depending on temperature; T is the temperature of wastewater, °C; S_i represents the influent concentration of suspended solids, mg/L; q denotes the surface overflow rate, m³/m² day; c is constant (0.0035 m/day); and E_c is COD removal efficiency, %.

8.6.4 Design Recommendations for PST

Primary sedimentation tanks can be circular or rectangular tanks designed using average dry weather flow and checked for peak flow condition. The numbers of tanks to be provided are determined by limitation of tank size. Two tanks in parallel are normally preferred in large sewage treatment plants to facilitate maintenance of any tank. The diameter of circular tank may range from 3 to 60 m (up to 45 m typical) and it is governed by structural requirements of the trusses, which supports scrapper in case of mechanically cleaned tank. Though rectangular tank with length up to 90 m has been used in the past, however usually length more than 40–45 m is not preferred due to increase in self weight of the scrapping mechanism and more power required to drive it. Width of the tank is governed by the size of the scrappers available for mechanically cleaned tank.

The depth of mechanically cleaned tank should be as shallow as possible, with minimum of 2.15 m. In addition, 0.25 m for sludge settling zone and 0.3–0.5 m as free board is provided. The average depth of the tank used in practice is about 3.5 m and the range adopted is 3–5 m. The floor of the tank is provided with slope 6–16% (8–12% typical) for circular tank and 2–8% for rectangular tanks. The scrappers are attached to rotating arms in case of circular tanks and to endless continuous chain in case of rectangular tanks. In rectangular tanks, the solids are collected in the sludge hoppers at the influent end, and this collected sludge is withdrawn at fixed time intervals. In case of circular tank, the sludge hopper is provided at the centre surrounding the central feed inlet pipe. The scrapper velocity of 0.6–1.2 m/min (0.9 m/min typical) is used in case of rectangular tank and flight speed of 0.02–0.05 rpm (0.03 typical) is used in case pf circular tanks.

Surface overflow rate: For designing the sedimentation tank for municipal wastewater treatment, the surface overflow rate ranges from 30 to 50 m³/m² day at an average flow and 80–120 m³/m² day at peak flow without activated sludge recycle. When the activated sludge is recycled into the PST, the overflow rate ranges from 24 to 32 m³/m² day for average flow and 48–70 m³/m² day for peak flow. The performance of the sedimentation is mainly dependent on the surface overflow rate adopted.

Weir loading: Theoretically, weir loading is the quantity of water passing over per meter length of weir per day and it has minimal effect on the efficiency of PST. However, the weir loading is considered during the design of sedimentation tanks and typically values ranging from 125 to 500 m³/m² day are adopted (generally less than 185 m³/m² day).

Higher weir loading can be considered when PST is followed by a secondary treatment and weir loading rate up to 300 m^3/m^2 day is used during peak flow conditions.

Detention time: The detention time at average flow for a PST is generally kept in the range of 1 h to a maximum of 2.5 h; however, the cold climate has an increasing effect on liquid viscosity, causing decreasing in settling velocity. This decrease in settling velocity affects the removal efficiency and longer detention time is required to achieve the required removal efficiency. The factor of safety to be adopted in required detention time with variation in temperature can be estimated using Eq. 8.50 (WPCF, 1985), where temperature is in °C. Thus, this factor should be multiplied to the normal retention time to get satisfactory performance of sedimentation tank even in winter.

$$Factor\ of\ safety = 1.82 \times e^{-0.03 \times Temperature} \tag{8.50}$$

Example: 8.10 Design a primary sedimentation tank for treatment of municipal wastewater with an average flow of 12 MLD and a peak factor of 2.0.

Solution

Estimation of required surface area
As per the recommendation, the surface overflow rate of a PST is 40 m^3/m^2 day at an average flow and up to 120 m^3/m^2 day for peak flow.

$$\text{Hence, the surface area required} = \text{flow/surface overflow rate}$$
$$= 12000/40 = 300 \text{ m}^2 \text{ at average flow}$$

Check surface overflow rate at maximum flow.
Surface overflow rate at maximum flow = $12,000 \times 2.0/300 = 80$ m^3/m^2 day (<120 m^3/m^2 day).

Calculating dimensions of the tank
Provide width of the sedimentation tank as 9 m.
Theoretical length of the tank = 300/9 = 33.33 m.
Length = 33.33 + 0.70 (inlet allowance) + 3.00 (outlet allowance) ~37 m.

Calculation of flow through velocity
As per the recommendation, provide detention time of PST to be 1.5 h.

$$\text{Depth of tank} = \text{surface loading rate} \times \text{detention time}$$
$$= 40 \times 1.5/24 = 2.5 \text{ m}$$

Volume of tank = 2.5 m $\times$ 37 m $\times$ 9 m

Flow through velocity at average flow = 12000((2.5 × 9 × 24 × 60 × 60) = 0.006 m/sec

= 0.62 cm/sec(<1 cm/sec as required at average flow)

Flow through velocity at peak flow = 12,000 × 2.0/(2.5 × 9 × 24 × 60 × 60) = 0.012 m/sec.

This velocity is less than the scoring velocity of 0.06 m/sec.

Total depth = 2.5 + 0.5 (free board) + 0.25 (sludge zone) = 3.25 m.

Verifying the weir loading rate
Permissible weir loading is 185 m^3/m day.

Length of weir required = 12,000/185 = 64.86 m.

Provide three effluent collection channels with each having width of 0.5 m spaced at 0.5 m from each other and final weir at the end wall of the settling tank, thus the length of the weir provided will be 9 × 7 = 63 m. Plus provide weir on side wall of the settling tank offering additional 2 m, thus total weir length provided will be 65 m.

8.7   Secondary Sedimentation Tank

The secondary sedimentation tank (SST) is provided after the secondary biological wastewater treatment to facilitate the sedimentation of the microbial cells produced during biological oxidation of organic matter. If these cells produced are not removed, complete treatment will not occur as these biological cells will represent about 30–60% of the organic matter present in untreated wastewater in the effluent of aerobic treatment facility. Depending on the type of reactor used, particularly for the suspended growth process, fraction of this settled cells is returned to the biological reactor and remaining sludge mass is wasted as excess sludge for further treatment.

In SST while the biological sludge mass is getting settled in the tank it will follow clarification and then Type III settling (hindered settling) and compaction settling, if targeted. Generally separate sludge thickener is used in large sewage treatment plant for further concentrating the sludge settled in SST. The design guidelines for secondary sedimentation tank for different biological wastewater treatment processes are presented in the Table 8.2 (CPHEEO, 2013).

Weir loading rate for the secondary sedimentation tank is kept less than or equal to 185 m^3/m day. Other guidelines for the dimensions of the tank are like as described earlier in case of primary sedimentation tank.

Example: 8.11 Design secondary sedimentation tank for treatment of 11 MLD of treated effluent coming from aeration tank of conventional activated sludge process. The MLSS in aeration tank is 3300 mg/L and peak flow factor is 2.0.

Table 8.2 Design parameters for secondary sedimentation tank

Parameter	Overflow rate, m^3/m^2 day		Solid loading rate, kg/m^2 day		Depth, m	Detention time, h
	Average	Peak	Average	Peak		
SST for trickling filter	15–25	40–50	70–120	190	2.5–3.5	1.5–2.0
SST for activated sludge process	15–25	40–50	70–140	210	3.5–4.5	1.5–2.0
SST for extended aeration activated sludge process	8–15	25–35	25–120	170	3.5–4.5	1.5–2.0

Solution

Adopt surface loading rate of 20 m^3/m^2 day at average flow.

Therefore, surface area required = 11,000/20 = 550 m^2.

The surface overflow rate at peak flow = 22,000/550 = 40 m^3/m^2 day.

This is well within the limit of 40–50 m^3/m^2 day, hence area is acceptable.

Check for solid loading rate

At average flow solid loading rate = $11{,}000 \times 3300 \times 10^{-3}/550$ = 66 kg SS/m^2 day.

Hence, within the limit of 70 – 140 kg SS/m^2 day.

At peak flow solid loading rate = $22{,}000 \times 3300 \times 10^{-3}/550$ = 122 kg of SS/m^2 day.

This is less than permissible limit of 210 kg of SS/m^2 day.

Diameter of the tank for 550 m^2 area = 26.47 m.

Provide detention time of 2 h, hence volume = $11{,}000 \times 2/24$ = 916.67 m^3.

Hence depth of the tank = 916.67/550 = 1.67 m.

Provide depth of 2.0 m + 0.3 m for sludge accumulation and 0.4 m free board. Hence total depth = 2.7 m.

Check for weir loading

Weir loading = $11{,}000/(\pi * D)$ = 132.35 m^3/m day,

It is less than 185 m^3/m day, hence design is fine.

8.8 Plate Settler and Tube Settler

Sedimentation tank and its design is evolving from standard rectangular tank to other modified settlers and the requirement of industries, such as reliable performance with lesser footprint, have influenced this development. Inclined settler also has an industrial origin; however, the theories related to compact inclined plate settler have been in academic discussions for a long time. The concept started by Hazen (1904) as horizontal tray settlers and Camp (1946, 1953) provided critical research required for its development. The first-ever inclined settler was used in Sweden in 1955. Since then, inclined settler has been used in primary, secondary, or tertiary treatment as a replacement of traditional settling tank due to their capability to overcome problems like short-circuiting, wind circulation, and shock loading.

Inclined settler (Fig. 8.34) is a shallow settling device, typically consisting of bundles of incline plates or tubes of various geometries. This arrangement helps to increase the effective surface area of the tank by a factor of 8–10, which improves the settling as settling depends more on effective settling area rather than detention time. These plates are fixed at an angle between 45° and 60° (generally 55°) with the horizontal that helps in self-cleaning of settlers and maximizes the use of the plate surface area. As the solids accumulate over the plates, gravitational force exceeds the shear resistance and the accumulated mass slides down. If the inclination of plate angle is lower than 45°, accumulation of solids between the plates affects the operation as it can lead to turbulence between the plates and require frequent cleaning for removal of these accumulated solids. However, the inclination angle up to 7° has been used by providing the periodic back-flushing.

Decreasing the spacing between plates can also increase the surface area; however, below a certain spacing, it can lead to turbulence between plates, causing resuspension of solids. It is recommended to maintain the Reynold's number lower than 800 between the plates to avoid resuspension of solids with a nominal spacing of 50 mm between the plates. The inclined length of the plate should be between 1 and 2 m and the surface overflow rate is used in the range of 20–40 m^3/m^2 day (Metcalf & Eddy, 2003). Inclined settlers also reduce the detention time as the solid settling particles have to travel less distance before falling on to the surface and each pair of plate works as a separate settling chamber, whose depth is only about 50 mm. The settler can achieve the same efficiency of BOD and TSS removal under 15 min of retention time for an equal surface overflow rate compared to the traditional settling tank.

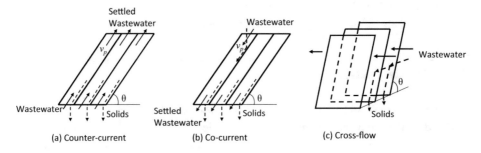

Fig. 8.34 Schematic diagram of plate settlers: **a** counter-current, **b** co-current, and **c** cross flow

The main disadvantage of plate and tube settlers is the frequent accumulation of solids in inclined plates and pipes, which adversely affects the flow distribution in the settling unit and causes odour problems due to the degradation of accumulated sludge solids. To overcome these issues flushing of sludge on a day-to-day basis is recommended.

Inclined settlers are constructed for one of the three ways (Fig. 8.34) concerning the direction of flow and corresponding direction of settling of solids as (a) counter-current, (b) co-current, and (c) cross flow. Theoretical analysis and development of flow geometry for these plate settlers were proposed by Yao (1970).

8.8.1 Counter-Current Plate Settler

In the counter-current plate settler, the wastewater is pumped to the settling module in an upward direction from bottom of the plates. The wastewater flows against the gravity between the channel formed by inclined plates and exits from the outlet located at the top of the tank. The solid particles tend to settle on the surface of the lower inclined plate. These accumulated solids slide to the bottom of the tank by gravity and this sludge is removed periodically for further processing. Counter-current design is most common in the industry as they are less expensive to install and operate.

In up-flow or counter-current settling, the time (t) for the particle to settle vertical distance between two inclined plates is given by Eq. 8.51 and length of surface (L) required to be provided for this detention time is given by Eq. 8.52.

$$t = \frac{d}{v_s \cos \theta} \tag{8.51}$$

$$L = \frac{d\left(v_p - v_s \sin \theta\right)}{v_s \cos \theta} \tag{8.52}$$

where d is the perpendicular distance between the two plates (m), v_s is settling velocity of the particle, v_p represents flow velocity between the parallel plates (m/sec), θ is the inclination angle of the plate (Fig. 8.35).

The particles with higher settling velocity than v_s will settle in the plate settler and the relation between v_s and v_p can be written as per Eq. 8.53 (AWWA, 1999). All solids with settling velocity greater than v_s will be completely removed.

$$\frac{v_s}{v_p} = \frac{d/\cos \theta}{L + d \tan \theta} \tag{8.53}$$

In the case of inclined settler, where many plates are used, the flow velocity (v_p) is calculated using Eq. 8.54.

$$v_p = \frac{Q}{(n-1)wd} \tag{8.54}$$

Fig. 8.35 Velocity vector
within the plate and geometric
similarity for counter-current
settling

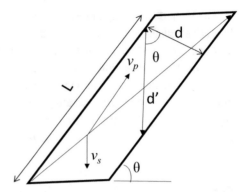

where Q is flow rate (m³/sec), n represents the total number of the plates and, w is the
width of plate (m).

8.8.2   Co-current Plate Settler

For down-flow (co-current settling), wastewater is fed from the top of the tank and moves
in the downward direction and the solid particles also settle in the downward direction
by gravity. The time for the particle to settle between two inclined surfaces is the same
as stated in Eq. 8.51. However, the relation between settling velocity and flow through
velocity between plates can be summarised as in Eq. 8.55.

$$\frac{v_s}{v_p} = \frac{d/\cos\theta}{L - d \times \tan\theta} \tag{8.55}$$

8.8.3   Cross-Flow Plate Settler

The wastewater flows in the horizontal direction, while settled solids move in the downward direction. In the cross-flow reactor, resuspension of solids is usually less of a
problem. The development of cross-flow reactors is predominant in Japan. Settling time
is the same as counter-current settling (Eq. 8.51) and the relationship between v_s and v_p
can be written as (Eq. 8.56):

$$\frac{v_s}{v_p} = \frac{d/\cos\theta}{w} \tag{8.56}$$

8.8.4 Lamella Clarifier

There are number of proprietary lamella clarifier designs available as per the manufacturer. Inclined plates used may be based on hexagonal or rectangular tubes or special geometrical proprietary shape. Some possible design characteristics of lamella clarifier include:

- A tube or plate spacing of 50 mm,
- Tube or plate used are of length 1–2 m,
- Plates are pitched between 45° and 70° with horizontal to allow for self-cleaning, lower pitch angle than this will require arrangement for backwashing,
- Typical loading rates are 5–10 m/h,
- Typical retention time of 60–120 min is used for design.

Lamella clarifiers (Fig. 8.36) can handle a maximum feed water concentration of 10,000 mg/L of grease and 3,000 mg/L of suspended solids. The expected separation efficiency for a typical unit is 90–99% removal for free oils and greases under standard operation conditions. For emulsified oils and greases the removal efficiency could be 20–40%, when no chemical amendment is practiced; whereas, when emulsion breaking chemical agents are added the efficiency could improve to 50–99%. This is capable to produce treated water with a turbidity of around 1–2 NTU.

The surface loading rate (also known as surface overflow rate or surface settling rate) for a lamella clarifier falls between 10 and 25 m/h. For these settling rates, the retention time in the clarifier is low, at around 20 min or less, with operating capacities tending to range from 1–3 m^3/m^2 h of projected surface area.

Other design details: The specific hydraulic loading on plates should be limited to 2.9 m/h to ensure laminar flow is maintained between plates. The plates should be inclined at a 50–70° angle with the horizontal to allow for self-cleansing. This results in the projected plate area of the lamella clarifier taking up approximately 50% of the space of a conventional clarifier without plates. Typical spacing between plates used is 50 mm,

Fig. 8.36 Lamella clarifier

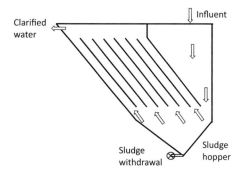

though plates can be spaced in the range of 50–80 mm apart, given that the particles greater than 50 mm in size have been removed in pre-treatment stages. Depending on the scale of the system, total plate lengths can vary and most plates used have a length of 1–2 m; however, the plate length should allow for 125 mm free top water level, with about 1.5 m of space left below the plates at the bottom of the clarifier for collection of sludge. The material used for plates should be made of stainless steel, except for situations in which the system has been dosed with chlorine to prevent algal growth. In these circumstances, the plates may be plastic or plastic coated. The feed wastewater should be introduced at least 20% below the base of the plate to prevent disturbance of the settling zones at the base of the plates.

Example: 8.12 A counter-current flow plate settler has plates with the dimensions of 3.5 m × 1.5 m and inclination of plate is 60°. Surface over flow rate adopted is 25 m/day. Calculate the smallest size of particle with specific gravity of 1.05 that will be removed with 100% efficiency. Also estimate the time required for this particle to settle between the plates and velocity of flow between the plates. Consider kinematic viscosity of wastewater as 1.2×10^{-6} m^2/sec.

Solution
Assume the perpendicular spacing between the plates as 50 mm.
$V_0 = 25$ m/day $= 2.89 \times 10^{(-4)}$ m/sec.
$L = 3.5$ m, W $= 1.5$ m.

Calculation of diameter of the particle
Let's assume Stoke's law is valid, the kinematic viscosity of wastewater is 1.2×10^{-6} m^2/sec and specific gravity of particle is 1.05.

$$V_s = \frac{g(SG_s - SG_w)D^2}{18\,v}$$

$$2.89 \times 10^{-4} = \frac{9.81(1.05 - 1)D^2}{18 \times 1.2 \times 10^{-6}}$$

$D = 0.11$ mm.
Since Reynold's number for this is $v \cdot d/v = 0.026$, Stoke's law is applicable and hence smallest size of particles completely removed will be 0.11 mm.

Calculation of detention time and flow between the plates

$$t = \frac{d/\cos\theta}{v_s}$$

$$t = \frac{\frac{0.05}{\cos 60°}}{2.89 \times 10^{-4}}$$

$t = 346$ s $= 5.8$ min.

Hence, time required for this particle to settle between the plates is 5.8 min.

$$\frac{v_s}{v_p} = \frac{d/\cos\theta}{L + d\tan\theta}$$

$$v_p = \frac{2.89 \times 10^{-4} \times (1.5 + 0.05\ \tan\ 60°)}{0.05/\cos\ 60°}$$

$v_p = 0.0046$ m/sec $= 16.5$ m/h is the velocity of flow between the plates.

Example: 8.13 Design a plate settler for a secondary treatment of industrial wastewater having flow rate of 1.5 MLD. The dimensions of a plates be used are 1.3 m $\times$ 3.0 m and these are placed at an inclined angle of 55° with horizontal.

Solution

Considering a SOR of 20 m^3/m^2 day, area of plate settler $= \frac{1500}{20} = 75$ m^2.

The dimensions of a plates are 1.3 m $\times$ 3.0 m and it is placed at an inclined angle of 55° with horizontal.

Projected plan area of each plate $= 1.3 \times 3 \times \cos 55° = 2.237$ m^2.

Number of plates required $= \frac{75}{2.237} = 33.5$ (Provide 34 plates).

Hence, a total number of 34 plates having dimension of 1.3 m $\times$ 3 m and inclined at 55° to the horizontal are required with a spacing of 50 mm between them.

8.9 Equalization

Sewage generation varies throughout the day or year depending upon the factors like geographical location, seasonal variation of weather, diurnal variation, rate of water supply, etc. The wastewater quantity generation and characteristics also vary in the industry depending on the type of product and processes used in the industry and in commercial sectors depending on operational hours. During peak hours, the generation of municipal wastewater can be 2.8 times of average hourly generation and during the rainy season, flow can increase exponentially. However, the treatment facilities are designed for a constant design flow rate. Hence, a holding tank is used to collect the incoming wastewater to avoid shock loading on the follow-up wastewater treatment system. This holding tank is called as flow equalization tank as it controls the hydraulic and even the resulting organic shock loading, thus ensuing the higher efficiency of biological and physical treatment.

Flow equalization is not essential for large capacity sewage treatment plants due to the negligible effect of varying flow. However, for treatment facilities of smaller size or industrial treatment plants, where wastewater quantity and characteristics can vary significantly with time, an equalization basin becomes a necessity to maintain the uniform outflow rate, to operate follow up treatment units with uniform loading that is free from surges. As per the recommendations, construction of an equalization basin is advisable when the peak factor exceeds three by a wide margin.

The benefits of flow equalization tanks are to bring under control the fluctuation in organic matter concentration, maintain more or less stable pH, and regulate the concentration of toxic compounds in the wastewater stream, if there are any. These benefits lead to a stable performance of biological wastewater treatment with enhanced quality of activated sludge and the effluent. Moreover, the disadvantages of equalization tank are additional capital cost, requirement of land area, and additional operational and maintenance requirements.

8.9.1 Location and Types of Equalization Basin

The optimum location of the equalization basin depends on the kind of treatment required for the influent wastewater, area availability, sewage collection system, and influent wastewater characteristics. In standard practice, an equalization tank is provided after the primary treatment and before biological treatment. The use of an equalization basin before primary treatment can cause settlement of the influent settleable solids in the equalization basin. Such settlement will lead to the release of foul odour due to the degradation of settled solids; hence, sufficient mixing and aeration should be provided in equalization basin to prevent the settling of solids. This additional requirement of aeration and mixing increases the cost of treatment plant. Moreover, the primary treatment facilities don't require constant pH or organic loading for efficient treatment, hence placing equalization basin after PST is ideal location.

Equalization basins can be classified into two types: in-line and off-line equalization tanks. The in-line equalization tank receives all the flow coming from the primary treatment, whereas the off-line equalization tank receives only the surplus flow. The pumping cost is less in an off-line equalization tank; nevertheless, constant concentration of pollutants is best maintained in an in-line equalization tank.

8.9.2 Volume Requirement for the Equalization Tank

The volume requirement of the equalization tank is estimated by plotting a cumulative mass inflow and outflow volume, in which these are plotted against time, i.e., for 24 h in case of municipal wastewater and complete cycle of operation in case of industrial wastewaters (Fig. 8.37). Since the outflow from the equalization basin will be pumped at the constant rate, the cumulative outflow comes as a straight line starting from origin till end of the equalization cycle time; while the cumulative inflow volume mass comes as a curve with one or multiple peaks. For calculation of required volume for equalization,

Fig. 8.37 Cumulative mass volume curves to estimate volume required for equalization

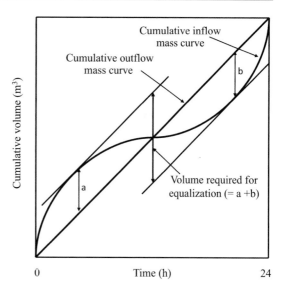

from the maximum cumulative surplus and maximum cumulative deficit peak point, in case of multiple peaks, tangents parallel to the cumulative outflow mass curve are drawn. The required volume for equalization is equal to the vertical distance (ordinate) between these tangents.

Analytical method of calculating the volume of tank is illustrated in Example 8.14, in this method all the data are tabulated for one whole cycle and values for cumulative surplus or deficit are obtained based on the constant cumulative outflow pumping capacity of the equalization basin. The summation of maximum cumulative deficit and maximum cumulative surplus gives the required volume for equalization. In practice, the volume of the equalization tank is kept 10–20% more than this theoretical value estimated to meet the following demands: (a) to accommodate concentrated stream for dilution, (b) to avoid complete drawdown of basin to protect the floating aerator from damage, and (c) to meet any unseen changes in the wastewater inflow.

8.9.3 Basin Geometry and Construction

Basin geometry depends upon the type of equalization basin and its purpose. If a tank is used for flow and mass loading, a continuous flow stirred tank reactor with proper inlet and outlet device may be preferred. Proper mixing equipment and cleaning facility should be installed for better operation. Basin can be constructed using concrete, steel, or earthen material. In case of availability of land, earthen type of construction is economical with proper lining on the walls and floor to make them impervious and constructed with multiple inlet and outlet ports with suitable interior side slope of 3 horizontal to 1 vertical.

A synthetic liner or compacted clay liner is recommended to avoid the ground water contamination by the infiltration of wastewater in the earthen equalization tank. Type and thickness of liner depends upon the land availability, topography, soil property and ground water level.

Minimum water depth in the equalization basin must be between 1.5 and 2 m and this minimum wastewater depth is maintained to protect the aerator and other machines. Moreover, floating aerator is provided with concrete pad to minimise soil erosion at the tank bottom if constructed with soil excavation and embankment. In the case of high groundwater level, embankment failure of tank is a real threat and it is required to be included in tank design to provide pumping and drainage of groundwater. Some operational appurtenances like facilities to clean the basin walls from attachment of solids and grease, high-water take off for the removal of floating materials and foam, and odour control facilities are also provided for easy operation and maintenance.

8.9.4 Mixing Requirement

Mixing mechanism helps to avoid the settling of suspended solids in the equalization basin. Mixers with adequate power should be installed to facilitate mixing. Mechanical mixers with capacity of 0.004–0.008 kW/m^3 of the basin volume will ensure adequate mixing. Most commonly, submerged or surface aerators are used. In case of installing diffused air mixing an air requirement of about 3.74 m^3/m^3 is recommended to ensure that all solids are kept in suspension. Aeration in the tank helps to maintain the aerobic medium, avoid septic conditions to develop, apart from providing mixing. An automatically controlled flow regulating device is also required at places where gravity discharge from the basin is used. A flow measuring device is compulsory at the outlet point to monitor equalized flow.

Example: 8.14 Estimate the volume required for an equalization basin for the flow variation data given below:

Time	8	9	10	11	12	1	2	3	4	5	6	7
Flow, m^3	28	28	87	110	98	60	36	34	32	48	68	84
Time	8	9	10	11	12	1	2	3	4	5	6	7
Flow, m^3	96	80	36	24	20	18	13	13	18	24	25	26

Solution

From the flow data given above estimate the total flow generated per day by summation of each hourly flow. Thus, the total flow per day = 1104 m^3.

Hence, average pumping rate will be = 1104/24 = 46 m^3/h.

From this estimate the cumulative inflow and cumulative pumping as presented in Table 8.3.

From the Table 8.3, the maximum cumulative deficit is 36 m^3 and maximum cumulative surplus is 243 m^3.

Hence, the capacity required for equalization = 243 + 36 = 279 m^3.

Providing 20% extra capacity to have minimum storage volume of wastewater in the equalization basin to protect mechanical floating aerators, if used, and to provide some

Table 8.3 Estimation of cumulative deficit and surplus for equalization volume requirement

Given date		Cumulative inflow	Cumulative pumping	Cumulative surplus	Cumulative deficit
Time	m^3	m^3	m^3	m^3	m^3
8	28	28	46		18
9	28	56	92		**36**
10	87	143	138	5	
11	110	253	184	69	
12	96	349	230	119	
1	60	409	276	133	
2	36	445	322	123	
3	34	479	368	111	
4	32	511	414	97	
5	48	559	460	99	
6	68	627	506	121	
7	84	711	552	159	
8	96	807	598	209	
9	80	887	644	**243**	
10	36	923	690	233	
11	24	947	736	211	
12	20	967	782	185	
1	18	985	828	157	
2	13	998	874	124	
3	13	1011	920	91	
4	18	1029	966	63	
5	24	1053	1012	41	
6	25	1078	1058	20	
7	26	1104	1104	0	

dilution without allowing tank to go empty. Hence, volume recommended for equalization basin $= 335$ m^3.

Provide power @ 0.004 kW/m^3 for mechanical mixer, hence the power requirement is 1.34 kW.

Provide liquid depth of 3 m, hence the area required will be 111.67 m^2. Hence, provide a circular shape equalization basin with diameter of 11.9 m. The total depth of the basin will be $3.0 + 0.5$ (free board) $= 3.5$ m.

8.9.5 Design of Equalization Basin for Concentration Dampening

The equalization basin can also be designed to keep the maximum concentration of BOD or COD or any other pollutant under control in the effluent coming out of the equalization basin. In case of nearly constant wastewater inflow with variation in the wastewater characteristics a normal statistical distribution of wastewater composite analyses can provide the required equalization retention time (Eq. 8.57)

$$t = \frac{\Delta t \left(Si^2 \right)}{2 \left(Se^2 \right)} \tag{8.57}$$

where Δt is the time interval over which wastewater samples are composited, h; t is the equalization detention time, h; Si^2 is variance of the influent wastewater concentration (square of standard deviation); Se^2 is variance of the effluent concentration at a specified probability (e.g., 97.5 or 99%).

When completely mixed basin is used for wastewater treatment, e.g., ASP or aerated lagoon, this volume of secondary treatment system can be considered as a part of equalization volume. For example, if completely mix aeration tank provided is having retention time of 5 h, and retention time required for equalization is 10 h, then equalization basin shall be provided with retention time of 5 h.

Equalizing both volume and strength of the wastewater: For the wastewater when both the flow and strength vary randomly the equalization requirements can be worked out from material balance as per Eq. 8.58 (Eckenfelder, 2000).

$$CiQT + C_0V = C_2QT + C_2V \tag{8.58}$$

where Ci is the concentration of pollutant entering the equalization basin over the sampling time interval T; Q is average flow rate over sampling interval (T); C_0 is concentration in the equalization basin at start of sampling interval; V is volume of equalization basin; and C_2 is concentration leaving the equalization basin at end of sampling interval.

It is assumed that the effluent pollutant concentration is constant during the time interval selected. Hence, effluent concentration after each time interval can be obtained using Eq. 8.59, or deciding the acceptable variation in concentration the retention time required

(V/Q) for equalization can be estimated.

$$C_2 = \frac{Ci\,T + Co\,V/Q}{T + V/Q} \tag{8.59}$$

8.10   Neutralization

Neutralization is a process to neutralize acidic or alkaline wastewaters for their effective treatment or for meeting the discharge norms. Neutralization can also be required as a pre-treatment for some industrial wastewaters to correct the pH, as highly acidic or alkaline pH of wastewater can cause damage to the plant equipment and could also hamper the biological or chemical processes occurring in different treatment units. These methods are selected based on the cost associated with the neutralizing agent and equipment requirements.

Methods adopted for neutralization of acidic wastewater are:

- mixing alkaline wastewater (if available),
- use of limestone,
- use of lime slurries,
- use of soda ash,
- use of caustic soda.

Methods generally used for the treatment of alkaline wastewater are:

- addition of sulfuric acid,
- blowing waste boiler flue gas (14% CO_2),
- mixing acidic wastewater (if available).

Limestone chips: Limestone reacts with the acidity and produces calcium salt as a product. Reaction occurs on the surface of limestone chips, hence high surface area with adequate contact time is required for wastewater to be neutralized. The disadvantage of using limestone can be listed as (a) occurrence of unintentional reactions, (b) requirement of high surface area of limestones, (c) calcium salts deposition on the limestone surface making it non-reactive, (d) increase in the TDS and SS of the wastewater, and (e) no control on the process if there is variation in the characteristics of the wastewater to be neutralized. Hence, the lime stone bed neutralization is effective for the wastewater having reasonably constant acidity. Thus, this is not effective for wastewater where flow and concentration vary with time. In addition, the H_2SO_4 concentration in wastewater should be limited up to 0.6% to avoid coating of limestone with nonreactive $CaSO_4$ and excessive CO_2 evaluation, which will limit neutralization.

Lime slurry: Lime is delivered as a dry product and it needs to be crushed into fine powder and mixed into water to form a slurry. Lime slurries exhibit a high reaction rate compared to limestone bed; however, due to low solubility in wastewater, it is difficult to handle the lime slurry. Slurry also leads to the generation of sludge that need to be dewatered and disposed. The reactions of lime slurry with wastewater can be accelerated by providing heating and agitation. The reaction can be complete within 5–10 min. Quicklime (CaO) is generally used with 8–15% of lime slurry for neutralization of acidic wastewater. Sodium hydroxide, sodium carbonate, and calcium bicarbonate are widely used in small treatment plants for neutralization either completely or in combination with lime, while lime-based chemicals are predominant in large-scale treatment plants. Lime-based chemicals are cheap but less convenient due to slower reaction rate and variation in quality of lime received with season.

Flue gas: Several industries produce flue gas by the burning of fuel for energy and heat requirement. Blowing the flue gas through alkaline wastewater could offer an economical method of neutralization. The flue gas containing 14% of CO_2 can be admitted either gas bubbled in wastewater or spray tower with counter-current can be used. Carbon dioxide reacts with wastewater to neutralize the alkalinity present. The flue gas is to be pre-filtered to remove the sulphur dioxide, hydrogen sulphide and other unburned carbon particles before pumping into the wastewater.

Design recommendations: Neutralization tank should be constructed with an anti-corrosive material or a protective lining shall be provided to make it corrosion resistant. The batch mode of operation is usually adopted for small-scale treatment plants with flow less than 350 m^3/day. For neutralizing a larger volume of wastewater, continuous flow system is generally preferred. For a successful design of the neutralization, information on influent wastewater characteristics and laboratory experimental results are necessary. Suitable pH probe should be installed in the neutralization tank to measure real time pH of the wastewater.

Neutralization of strong acidic or alkaline wastewater is a highly nonlinear process, particularly when close to neutral. Hence, stepwise addition of chemical (acid/base) in two to three steps is preferred for better control of the process (Fig. 8.38). The two-step process is predominant in wastewater treatment plants. In first stage the pH of the acidic wastewater can be corrected up to 6.0–6.5 and remaining pH correction in second stage shall be done with sodium or calcium bicarbonate. Thus, any extra addition of bicarbonate will not shoot up the pH to alkaline range and better control over the process can be exercised. Airflow rate of 0.3–0.9 m^3/m^2 min at a depth of 2.7 m is used for air mixing or mechanical mixer with capacity of 0.04–0.08 kW/m^3 shall be used to ensure adequate mixing of the acid/alkali added. Since, small volume of chemical is added to larger wastewater volume, mixing is very important.

Example: 8.15 An acidic wastewater generated from para-boil rice processing unit with flow rate of 12 m^3/h is to be neutralized to pH of 7.0 using lime to make it suitable for

Fig. 8.38 Typical titration curve for strong acidic wastewater neutralization

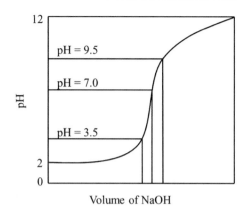

further biological treatment. The titration for this wastewater in laboratory showed total lime requirement of 1780 mg/L. A two-stage control neutralization system is proposed to be used with first stage requiring 1600 mg/L and the second stage 180 mg/L. Design the neutralization tanks.

Solution

Average lime dosage in first stage $= 12$ m^3/h $\times 1600$ mg/L $\times 10^{-3} \times 24 = 460.8$ kg/day.

Average lime dosage in second stage $= 12 \times 180 \times 10^{-3} \times 24 = 51.84$ kg/day.

With this dosage each basin should be designed with detention time of 7.5 min.

Volume $= 12 \times 7.5/60 = 1.5$ m^3.

Use two tanks with total height 1.3 m and liquid depth of 1.0 m and diameter of 1.40 m.

To ensure proper mixing the power required is 0.04–0.08 kW/m^3 say 0.06 kW/m^3.

Hence, provide 100 W of mixer in each reaction tank. Also provide one or two standard wall baffles, 180° apart, 1/12–1/20 of the width of the tank diameter, i.e., 10 cm.

8.11   Dissolved Air Flotation

Flotation is a unit operation in which finer suspended solids or oil and grease are removed from the wastewater by making them to float on the free water surface by imparting buoyancy to these finer solids by rising air bubbles throughout the water column. In dissolved air flotation (DAF) systems, the wastewater is pressurized at a high pressure (3.5–5 atm) in the presence of ample amount of compressed air to attain saturation air solubility at that pressure. The saturation of air in the wastewater is achieved by mixing compressed air with wastewater under pressure for about three to five minutes in a retention tank. The pressurized wastewater-air mixture is then depressurized to atmospheric pressure in the flotation unit causing release of microbubbles of air from solution (10–100 μm) through-out entire volume of wastewater. The sludge flocs, SS, and oil globules are floated by these minute air bubbles released, which make them to float upward towards the surface of water and get concentrated in the form of scum. As the air bubble travels through the water medium, it gets adhered to the suspended particles or enmeshed into the solid matrix of flocs present in the wastewater. Thus, the average density of these solid-air agglomer-ates becomes less than that of water; hence, due to buoyancy these particles start to float on the surface of the water forming a layer of scum. This scum layer is then removed using skimming mechanism and the treated effluent is passed for further treatment.

Design parameters for dissolved air flotation unit: The design of DAF unit is governed by the type and nature of solid particles to be removed. There are different factors that affect the efficiency of DAF systems, such as concentration of suspended solids, quantity of air released, particle-rise velocity and the solids loading rate. Nonetheless, air to solid ratio is the most critical parameter affecting the performance of DAF unit.

Air to solid (A/S) ratio: The ratio of volume of air released for flotation to the mass of suspended solids to be floated is the most critical parameter governing the performance of DAF systems as inadequate quantity of air will result in the partial or incomplete flotation of solids and no significant improvement will be noticed by the application of excessive air. The required value of *A/S* depends upon the type and concentration of the suspension; hence, it must be determined experimentally using a device known as "laboratory flotation cell".

Saturation of air in water is directly proportional to pressure and inversely proportional to temperature of water (Henry's Laws). In case of wastewater the relationship between pressure and solubility, the slope of the curve varies as compared to unpolluted water depending on the constituents present in the wastewater. The quantity of air that will be theoretically released from solution when pressure is reduced to atmospheric pressure can be estimated as per Eq. 8.60.

$$A = s_a \frac{P}{Pa} - s_a \tag{8.60}$$

where A is air released at atmospheric pressure per unit volume at 100% saturation (cm^3/L); s_a is the air saturation at atmospheric pressure, cm^3/L; P is absolute pressure, kPa; Pa is atmospheric pressure, kPa.

The actual quantity of air released will depend upon turbulent mixing conditions at the point of pressure reduction and on degree of saturation obtained in the pressurizing system. Solubility of air is less in wastewater as compared to that in water; hence a correction factor 'f' may have to be applied as given in Eq. 8.61. The retention tank will generally achieve about 50–90% of the saturation value.

$$A = s_a \left(\frac{f \cdot P}{Pa} - 1 \right) \tag{8.61}$$

where f is the fraction of saturation in retention tank.

Release of sufficient air bubbles in floatation unit to float all solids is important. Insufficient quantity of air release will result only in partial floatation of solids and excessive air will result in no improvement, however increase in operating cost. The performance of DAF can be related to an air/solids (A/S) ratio, i.e., mass of air released per mass of solids (SS) present in the influent wastewater as given in Eq. 8.62.

$$\frac{A}{S} = \frac{s_a}{S_i} \frac{R}{Q} \left[\frac{f \cdot P}{Pa} - 1 \right] \tag{8.62}$$

where Q is wastewater flow; R is pressurized recycle; Si is influent oil and/or SS concentration, and A/S is the air to solids ratio required.

In a system, without pressurized recycling, where all the wastewater flow is pressurized for solubilizing air, the A/S ratio for any operating pressure can be determined using the following relationship (Eq. 8.63).

$$\frac{A}{S} = \frac{1.3 s_a ((f \times P) - 1)}{S_i} \tag{8.63}$$

The numerical value of 1.3 represents weight of air in milligrams of one mL of air under higher than atmospheric pressure. The above relationship (Eq. 8.63) holds good for DAF systems, where treated effluent is not recycled, and entire inflow is pressurized (Fig. 8.39). The corresponding equation for DAF systems with effluent recycling (Fig. 8.40), where only the recycled flow is pressurized is given as (Eq. 8.64).

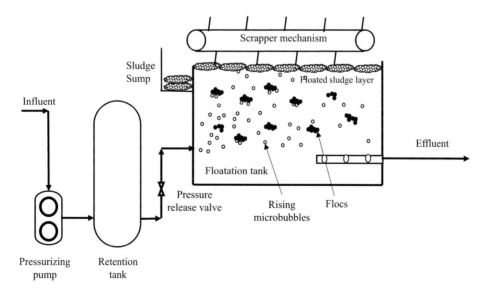

Fig. 8.39 Dissolved air flotation unit without recycling

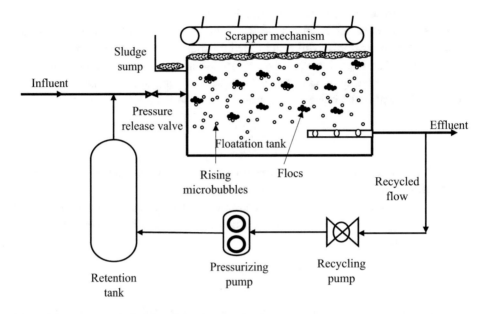

Fig. 8.40 Dissolved air flotation unit with recycling

$$\frac{A}{S} = \frac{1.3s_a(f \times P - 1)R}{S_iQ} \tag{8.64}$$

where R is pressurised recycled flow (m³/day); P is absolute pressure, atm ($P = (p + 101.35)/101.35$), p is the gage pressure, kPa; and Q is influent wastewater flow, m³/day.

Hydraulic loading rate and retention time: Sufficient retention time should be provided in floatation unit to ensure proper contact between air bubbles and suspended particles occurs. With increase in hydraulic retention time (HRT), suspended solids concentration in the effluent will reduce, whereas the floating sludge layer will become more concentrated as the particles are continuously transported to the scum layer by moving air bubbles. Moreover, if hydraulic loading rate is too high, turbulence can occur during the flow of water disrupting the formation of stable sludge/scum layer at the surface, while unnecessary reduction in hydraulic loading can render the design ineffective in achieving desired removal efficiency. Typical design value of hydraulic loading rate for a DAF unit should be in the range of 30–120 m³/m² day. Additionally, providing 3–5 min of detention time in the retention tank is crucial for optimum performance of DAF unit. It ensures that the influent is uniformly supersaturated with air before it enters the flotation unit.

Solid loading rate: Solid loading rate is denoted as the mass of suspended solids present in the influent per hour per unit surface area of the tank. The typical solid loading rate is in the range of 2–5 kg/m² h. Solid loading rates can be enhanced by adding polymers and coagulants; however, solid loading rate greater than 10 kg/m^2 h is discouraged due to operational difficulties at such high solids loading rates.

Example: 8.16 Design a dissolved air floatation unit for treatment of a wastewater with flow rate of 0.35 m^3/min to be operated at a maximum temperature of 30 °C. The wastewater contains non-emulsified oil and non-settleable suspended solids with total concentration of 110 mg/L. It is intended to reduce the oil to less than 15 mg/L using this DAF. Laboratory studies showed that the alum dose requirement is 30 mg/L. For effluent oil and grease concentration of 15 mg/L, the optimum A/S ratio of 0.05 was found and the required surface loading rate was 0.08 m^3/min m^2. The absolute pressure of 3.5 atm was used for pressurizing the recycled flow to dissolve air during laboratory test. The alum sludge production is 0.64 g per g of alum used. The sludge scum concentrated has solids content of 2.5%. At 30 °C the weight solubility of air is 20.9 mg/L. The value of f shall be considered as 0.5. Design the DAF with and without recycling system. Estimate the area required for floatation unit and quantity of sludge generated.

Solution

Design without recycling
Estimate the required pressure

$$\frac{A}{S} = \frac{1.3s_a((f \times P) - 1)}{S_i}$$

$$0.05 = \frac{1.3 \times 20.9 \times ((0.5 \times P) - 1)}{110}$$

Hence, absolute pressure P = 2.40 atm (relative pressure of (243.73 − 101.35) = 142.38 kPa).

Surface area required for floatation unit, A = 0.35/0.08 = 4.375 m^2.

Design with recycled system
Determining required recycle rate

$$\frac{A}{S} = \frac{1.3s_a(f \times P - 1)R}{S_iQ}$$

$$0.05 = \frac{1.3 \times 20.9 \times ((0.5 \times 3.5) - 1)R}{110 \times 0.35 \times 60 \times 24}$$

Hence, $R = 138.29$ m^3/day.
The required surface area will be

$$A = \frac{Q+R}{loading} = \frac{0.35 + \left(\frac{138.29}{60 \times 24}\right)}{0.08} = 5.58 \text{ m}^2$$

Estimating sludge quantities generated

$$\text{Oil and solids sludge} = (110 - 15)\text{mg/L} \times 0.35\frac{\text{m}^3}{\text{min}} \times 1440\frac{\text{min}}{\text{d}} \times 10^{-3}$$

$$= 47.88 \text{ kg/day}$$

$$\text{Alum sludge} = 0.64 \times 30\frac{\text{mg}}{\text{L}}\text{alum} \times 0.35\frac{\text{m}^3}{\text{min}} \times 1440\frac{\text{min}}{\text{d}} \times 10^{-3} = 9.68 \text{ kg/day}$$

$$\text{Total sludge} = 47.88 + 9.68 = 57.56 \text{ kg/day}$$

Since sludge contains 2.5% solids, i.e., 25 g/L (= 25 kg/m^3), hence the total sludge volume generated per day = 57.56/25 = 2.3 m^3/day

8.12   Coagulation

Wastewater contains large variety of solids of varying properties, shapes and sizes, which are either present in dissolved or suspended form. Primary sedimentation plays a vital role in the removal of substantial quantity of settleable suspended particles present in the sewage, making the settled effluent suitable for further treatment in follow-up secondary treatment processes. The sedimentation efficiency of primary clarifiers is greatly affected by the size and nature of the settling solids. The relatively larger solid particles (size >1 μm) settle via gravity settling during plain sedimentation. Plain sedimentation of municipal sewage and many of the industrial wastewaters containing large fraction of colloidal particles (0.01–1 μm), such as those coming from coal washing units, petroleum refineries, etc.; which are inefficient for gravity settling as colloidal particles tend to remain in suspension even after providing longer HRT. Different types of solids generally encountered in wastewaters and their removal method is elaborated in Table 8.4.

Poor settling characteristics of colloidal particles is mainly because of their repulsive nature caused by the presence of similar surface charge, which is typically negative in most cases. These colloidal particles undergo random motion in bulk solution, which is referred to as Brownian motion, however destabilizing effect of such random movement of particle is negligible as the negative surface charge prevent colloidal particles from agglomeration. Hence, certain chemical agents known as coagulants and flocculant aids are supplemented to the wastewater to destabilize these colloidal particles by neutralizing the surface charge to form larger flocs so that these large size flocs can exhibit higher settling velocity to facilitate their removal in sedimentation tank.

Table 8.4 Types of solid particles present in wastewaters and removal mechanism

Type of solid particle	Typical size range	Removal process
Large floating matter	>6 mm	Coarse screens
Small floating matter	2–6 mm	Fine screens
Grit (sand, silt etc.)	>0.2 mm	Discrete gravity settling in grit chamber
Suspended organic solids	>0.1 mm	Gravity settling in sedimentation tank
Colloids	0.001–1 µm	Coagulation, flocculation and sedimentation
Dissolved solids	<0.001 µm	Biological treatment, membrane process

Thus, the main objective of coagulation and flocculation processes is to increase the size of particles so that they can settle in a reasonable time frame. During coagulation, the repulsive negative charge is neutralized by the addition of suitable coagulants along with external agitation and then these fine size particles are allowed to aggregate and form flocs of larger size during flocculation. These large size flocs are removed by employing sedimentation or for industrial wastewaters sometimes alternate technology of dissolved air flotation is used.

8.12.1 Properties of Colloidal Solid Particles

Undissolved solids in wastewater can be broadly categorized as settleable suspended solids and colloids or colloidal solids, though it is difficult to draw exact border for separating both these types of solids. Settleable suspended solids are those which can be readily removed from wastewater using gravity settling in reasonable amount of time; whereas, colloidal solids cannot be removed in reasonable time period and require addition of settling aids in the form of coagulants. During the processes of coagulation and flocculation, the factors responsible for keeping colloidal solids in stable suspension are suppressed using certain chemical agents. Hence, it is important to understand the characteristics of colloids, which keep them in suspension and thus it would assist in improving the efficacy of coagulation and flocculation.

Colloidal solids in water can be classified as hydrophilic or hydrophobic colloidal solids depending upon their affinity for water. Due to presence of the water-soluble groups on the surface of colloidal particles, hydrophilic colloids have affinity for water. Amino, carboxyl, hydroxyl and sulphonic are some of the examples of water-soluble groups present on the surface of hydrophilic colloids. These groups promote the hydration and cause a water surrounding layer or film around the colloid particle, which is called as water of hydration or bound water or water hull. Colloids of organic matter and protein and their degradation products are generally hydrophilic in nature. Hydrophobic colloids have no affinity or very little affinity, if any, for water. Hence, in aqueous medium they are not having significant water surrounding them (water of hydration). Inorganic colloids, such as clay particles, are hydrophobic in nature. Destabilization of hydrophilic colloids is difficult and require more coagulant dose as compared to hydrophobic colloids.

8.12.2 Surface Charge of Colloids

The most important property of colloidal particles responsible for keeping them in suspension for extended time period without allowing them to aggregate is their surface charge. The development of surface charge depends upon the composition of the medium (wastewater in this context) and nature of colloids. For majority of colloidal particles present in wastewaters the surface charge is negative. The surface charge is developed through four principal ways as described below.

a. *Isomorphous replacement*: During isomorphous replacement, ions in the lattice structure of colloids are replaced with similar ions with lower valency from the solution. Such phenomenon is generally observed with clays and other soil particles. For instance, Al^{+3} ion replacing Si^{+4} ion in a solid SiO_4 crystal makes the entire lattice negatively charged as Al^{+3} ion has one less valence electron than Si^{+4}, which creates the charge imbalance as one oxygen atom remains unpaired when Al^{+3} ion enters in the crystal lattice.
b. *Preferential or selective adsorption*: Inert particles, such as silica, oil droplets, etc., allow anions, like hydroxyl ion, humic acid, dye, etc., to attach on their surface and hence, acquire a net negative charge on them.
c. *Ionization*: Dissociation of compound, such as hydroxyl group, carboxyl group, amino group, etc., occurs depending on the pH of the wastewater due to gain or loss of protons. For instance, silica having hydroxyl group on in its exterior surface attains different charge depending on the pH of the solution (Eq. 8.65).

$$- Si - OH_2^+ \leftrightarrow Si - OH \leftrightarrow Si - OH^-$$
$$pH \ll 2.0 \qquad pH = 2.0 \qquad pH \gg 2.0 \qquad (8.65)$$

d. *Structural imperfections*: In mineral crystal, charge can develop due to imperfect formation of crystals or due to the cleavage of bonds at the edge of the crystal.

8.12.3 Electrical Double Layer

A colloidal dispersion in solution doesn't carry a net charge due to the formation of electrical double layer. This double layer consists of two layers, a strongly held fixed

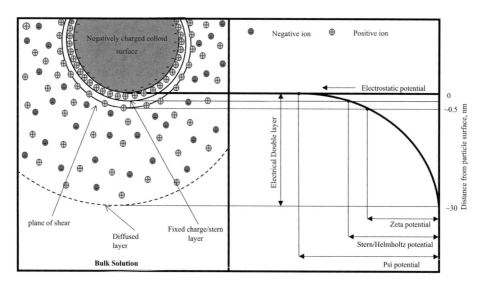

Fig. 8.41 Surface charge on a negatively charged colloidal particle in water

inner layer known as 'Stern layer' or 'Helmholtz layer' and a loosely held outer diffuse layer (Fig. 8.41). The inner layer is thinner in comparison to outer diffuse layer with a stretch of about 0.5 nm. The inner layer comprises of positively charged counter ions adsorbed over the negatively charged surface of the colloidal particle via electrostatic and Van der Waals forces of attraction, which are strong enough to resist the thermal agitation. The diffuse layer lies beyond the Stern layer and has sparsely distributed ions. Depending upon the nature of the solution, the diffuse layer can extend up to 30 nm.

8.12.4 Zeta Potential and Psi Potential

When electric current is passed through an electrolyte solution containing charged particles, the particles will be attracted towards one of the electrodes depending upon its surface charge. This movement is referred to as electrophoresis. During the movement, particles drag cloud of ions along with some portion of water and the electrical potential between this cloud surface (also referred to as plane of shear as shown in Fig. 8.41) and the bulk solution is called as zeta potential (ζ); whereas the potential drop between surface of the colloid and body of the bulk solution is referred to as psi (ψ) potential. In a zeta potential range of ± 0.5 mV coagulation will be effective.

Zeta potential can be calculated using Eq. 8.66 as:

$$\zeta = \frac{4\pi q d}{\varepsilon}$$ (8.66)

where ζ is zeta potential (mV); q is the charge per unit area; d is thickness of the layer surrounding the shear surface of particle through which the charge is effective and ε is dielectric constant of the liquid medium.

8.12.5 Particle Stability

Colloidal solutions in which particles remain in suspension without aggregating under natural condition are referred to as stable suspension. The electrical double layer surrounding the particle creates an energy barrier preventing them from agglomerating over a long period of time. The stability of the colloidal particle in wastewater is mainly due the force of electrostatic repulsion, which counterbalances Van der Waals forces of attraction between them and it is also referred to as electrostatic stability. This electrostatic repulsion between similar charge colloidal particles does not allow particles to come close enough for Van der Waals forces of attraction to come in to effect to favour flocculation. In other words, when the sum of repulsion forces is higher than net attractive force, the colloidal suspension is considered to be stable suspension. However, stability of a suspension can also be affected by concentration of colloids, pH of suspension and addition of external destabilizing compounds, such as coagulant, which can remove or reduce stability of the colloidal solution.

8.12.6 Theory of Coagulation

Coagulation is a chemical process in which destabilization of colloidal particles is carried out by addition of suitable coagulants to the water or wastewater, such as alum or hydrated aluminium sulphate $(Al_2(SO_4)_3.18H_2O)$, ferric chloride $(FeCl_3)$, etc. Destabilization of colloids leads to the increment in the size of agglomerated particles, which ultimately favours settling of these from the water or wastewater. The increase in size occurs due to particle collision, which happens when the charge on the colloids is neutralized, so as to allow particles to come sufficiently close enough for Van der Waals force to dominate; thus, leading to agglomeration. Series of complex reactions occur during coagulation process along with numerous side reactions owing to the presence of different substances in wastewater. Hence, the determination of exact mechanism of coagulation process becomes very difficult. However, coagulation action is assumed to take place by following four mechanisms as demonstrated in Fig. 8.42.

a. *Electrical double layer contraction*: Electrostatic repulsion between colloidal particles is significantly reduced if the thickness of electrical double layer is reduced. It allows the particle to come in contact with each other and remain attached due to Van der Waals forces of attraction. More is the reduction in the thickness of double layer

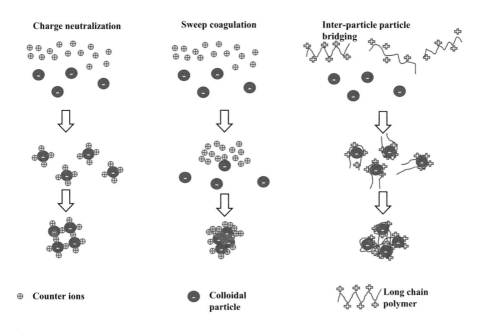

Fig. 8.42 Different mechanisms involved in the process of coagulation

higher will be the tendency of the particles to aggregate. The reduction in thickness of double layer is achieved by addition of potential determining ions or counterions (ions with unequal activities, which dissolves preferentially in a solution), electrolytes and hydrolysed metal salts.

The effectiveness of counter ion depends upon its charge and it can be assessed using Schulze-Hardy rule, which states that the critical coagulant concentration varies with the ionic valence z of the counter-ion as z^{-n}; where value of n ranges from two to six and can be predicted using the theory of Derjaguin, Landau, Verwey, and Overbeek (DLVO) (Metcalfe & Healy, 1990; Trefalt et al., 2020). For instance, ionic concentrations of Na^+, Ca^{2+}/Fe^{2+} and Al^{3+} required to coagulate fine suspended and colloidal particles are usually in the ratio of $1:1/2^6:1/3^6$ according to DLVO model (Davis, 2010). The potential determining ions interact with colloidal particles and reduce their surface charge. The magnitude to which surface charge of colloids is neutralized depends upon the concentration of counterions added to the bulk solution. Excessive dosage can reverse the particle surface charge and nature of electrical double layer imparting stability to these colloids. However, use of potential determining ion is discarded in wastewater treatment as huge concentration of ions are required to be added to bring about sufficient reduction in thickness of the double layer.

On the other hand, electrolytes reduce the thickness of the diffused electrical layer and thereby decreasing the zeta potential. Electrolytes are seldom used in wastewater treatment as main coagulants due to the requirement of exceedingly large concentration to meet the requisite demand; however, they are often employed along with the main coagulant for enhancing coagulation efficiency. Hydrolysed metal salts, such as aluminium sulphate (alum), ferrous sulphate, ferric chloride, etc., are the most prominent coagulants employed in the treatment of wastewater owing to their low cost and removal of wide variety of impurities including organic matter, inorganic particles as well as reducing pathogenic microbes present in the wastewater. Ferric coagulants are more preferred due to wider working pH range as compared to near neutral pH requirement for alum. At times, natural polyelectrolytes such as cellulose derivatives and synthetic polyelectrolytes like polymerised high molecular weight monomers are also applied along with hydrolysed metal salts to enhance the coagulation efficiency.

b. *Adsorption and charge neutralization*: In this, the reversely charged intermediates, having opposite charge to that of colloids present in wastewater, are formed from the dissociation of coagulants in water that tries to adhere on to the surface of colloids and thus neutralizing the surface charge on them. When inorganic coagulants, such as alum or ferric chloride, are added in the wastewater, they ionize immediately. For instance, ferric chloride is commonly used as a coagulant for the treatment of wastewater as it dissociates in water to form chloride (Cl^-) anions and ferric (Fe^{3+}) cations. Further, the ferric ions react spontaneously with water to form a number of aqua-metallic ions and release proton (Eq. 8.67 through Eq. 8.69).

$$Fe^{3+} + H_2O \rightarrow FeOH^{2+} \qquad (8.67)$$

$$Fe^{3+} + 2H_2O \rightarrow Fe(OH_2)^+ + 2H^+ \qquad (8.68)$$

$$Fe^{3+} + 3H_2O \rightarrow Fe(OH)_3 + 3H^+ \qquad (8.69)$$

These aqua-metallic ions become part of the ion cloud surrounding the colloidal particle. Due to the high surface affinity, these ions get readily attached to the surface of colloid, thus neutralizing its surface charge (Shammas, 2005). This leads to disintegration of electrical double layer surrounding the colloidal particle. In absence of any electrostatic force/potential the ion cloud dissipates, which allows particle to particle contact without any hinderance. Again, as discussed earlier excessive dosage of coagulant can restabilize the colloidal particle due to charge reversal and redevelopment of electrical double layer.

c. *Adsorption and interparticle bridging*: It is assumed that large molecules are formed via dissociation of inorganic coagulants and synthetic polymers. These molecules have rather large surface area, which acts as an adsorption site for colloidal particles. The adherence of particle is assisted by hydrogen bonding, electrostatic forces, dipole interaction and Van der Waals forces of attraction. These large molecules entrap many colloidal particles and then agglomerate; thus, acting as bridge between colloidal particles to form larger mesh like floc, which settles relatively quickly producing a denser sludge.

d. *Sweep coagulation*: When added to water or wastewater, alum and ferric coagulants react with alkalinity present in water or wastewater to produce sticky and gelatinous products, such as $Al(OH)_3$, $Fe(OH)_3$, etc. Being heavier than water these gelatinous products settle readily and swiftly. While these molecules settle, the colloidal particle present in the medium gets adhered and entrapped on their sticky surface; thus, sweeping away the colloidal particle from the suspension. Hence it is referred to as sweep coagulation.

8.12.7 Types of Coagulant

In general, water and wastewater treatment processes employ two categories of coagulants as (a) polyvalent metallic salts and (b) polymers. Coagulant selection and dosing are essential component for the effective coagulation of the colloidal particles in wastewater. A good coagulant should have the following properties.

i. Should produce high charge species in water/wastewater that carries effective coagulation, reducing the net total suspended solid (TSS) concentration and turbidity.
ii. Should not produce any toxic intermediates, which renders the effluent unfit for discharge or reuse.
iii. Should have no or less solubility for a pH range in which water and wastewater treatment is carried out.

8.12.7.1 Metallic Salts

As discussed previously, trivalent salts of aluminium and iron are most frequently used coagulants in water and wastewater treatment plants. There are many inorganic metallic coagulants available in the market, some of the most widely used inorganic metallic coagulants are discussed here along with their applicability, working range of pH, advantages and disadvantages (Table 8.5).

Alum ($Al_2(SO_4)_3 \cdot xH_2O$): Alum or hydrated aluminium sulphate is the most commonly used coagulant for raw water treatment with chemical formula being $Al_2(SO_4)_3 \cdot xH_2O$, where x is usually 18. At pH of 4.0, about 51.3 mg/L of Al^{3+} remains in solution; whereas at pH of 9.0, about 10.8 mg/L of Al^{3+} is in solution and it is least soluble at pH of approximately 7.0. Aluminium sulphate ionizes in water to produce sulphate anions (SO_4^{2-}) and aluminium cations (Al^{3+}). The Al^{3+} cations react with water spontaneously to form different aqua-metallic ions of aluminium along with release of hydrogen

Table 8.5 Different types of inorganic coagulants generally used for wastewater treatment

Coagulant type	Chemicals	Coagulation pH range	Advantages	Disadvantages	References
Inorganic metallic salts	Alum ($Al_2(SO_4)_3.18H_2O$); Aluminium Chloride ($AlCl_3$)	5.5–7.5	Stable, availability in solid and liquid form, easier handling and application, higher colour and turbidity removal than ferric coagulants at comparatively lower dosage	High alkalinity consumption, higher coagulant residuals in finished water leading to increase in total dissolved solids concentration, sulphate and chloride ions tend to increase corrosivity; possibly increased risk of Alzheimer's disease upon long term consumption of water	Berube and Dorea (2008), Brandt et al. (2016), Chow et al. (2008, 2009), Reiber et al. (1995), Shin et al. (2008), Sillanpää and Matilainen (2014), Smith and Kamal (2009), Uyak and Toroz (2007)
	Ferric chloride ($FeCl_3$), Ferric sulphate ($Fe_2(SO_4)_3$)	4.0–9.0	Better removal of natural organic matters (NOM) than aluminium salts, denser floc formation leads to relatively quicker settling, readily available	High alkalinity demand, sludge formed is corrosive in nature; requires additional chemical for stabilization and corrosion control of finished water	Brandt et al. (2016), Budd et al. (2004), Rigobello et al. (2011), Sharp et al. (2006), Sillanpää and Matilainen (2014), Uyak and Toroz (2007), Uyguner et al. (2007), Zhao et al. (2011)

(continued)

Table 8.5 (continued)

Coagulant type	Chemicals	Coagulation pH range	Advantages	Disadvantages	References
Inorganic polymers	Polyaluminium chloride ($Al_x(OH)_yCl_z$)	6.0–9.0	Lower alkalinity requirement, lesser susceptibility to temperature variations and better NOM removal than alum salts, low dosage, reduced sludge generation, and decreased residual coagulants in treated water	Not effective in removing highly hydrophobic and higher molar mass NOM; relatively stable Al species are formed which doesn't hydrolyse further during coagulation; coagulant efficiency is dependent upon hydrolysis product and species speciation	Brandt et al. (2016), Cheng et al. (2008), Dongsheng et al. (2006), Sillanpää and Matilainen (2014), Trinh and Kang (2011), Yan et al. (2009), Zhan et al. (2011)
	Polymeric ferric sulphate $[Fe_2(OH)_n(SO_4)_{(6-n)/2}]_m$	>4.5	Works in broader pH range, lower sensitivity to temperature, lesser dosage and residual iron in treated water, and comparatively less corrosive than ferric chloride	Polymeric species formed are affected by hydrolysis condition; limited application as coagulant is still in developmental stage	Brandt et al. (2016), Lei et al. (2009), Sillanpää and Matilainen (2014), Zouboulis et al. (2008)

ion (Eq. 8.70 through Eq. 8.72). Furthermore, when alum is added to wastewater (or water) containing alkalinity in the form of calcium and magnesium bicarbonate, it produces sticky and gelatinous precipitate of aluminium hydroxide ($Al(OH)_3$) (Eqs. 8.73 or 8.74), which attracts fine suspended impurities over its surface and gets settled out readily in the follow up sedimentation tank.

$$Al^{3+} + H_2O \rightarrow Al(OH)^{2+} + H^+ \tag{8.70}$$

$$Al^{3+} + 2H_2O \rightarrow Al(OH)_2^+ + 2H^+ \tag{8.71}$$

$$Al^{3+} + 3H_2O \rightarrow Al(OH)_3 \downarrow + 3H^+ \tag{8.72}$$

$$Al_2(SO_4)_3 \cdot 18H_2O + 3Ca(HCO_3)_2 \rightarrow 2Al(OH)_3 + 3CaSO_4 + 6CO_2 + 18H_2O \tag{8.73}$$

$$Al_2(SO_4)_3 \cdot 18H_2O + 3Ca(OH)_2 \rightarrow 2Al(OH)_3 + 3CaSO_4 + 18H_2O \tag{8.74}$$

It is evident from the reaction shown in Eq. 8.73 that one mole of alum reacts with three moles of bicarbonate alkalinity to produce two mole of sticky gelatinous precipitate of $Al(OH)_3$. Hence, alum is also referred to as 1:2 coagulant. Floc charge is positive below pH of 7.6 and negative above 8.2 and between these pH limits floc charge is mixed. Also, the product calcium sulphate ($CaSO_4$) imparts hardness to the water; whereas, CO_2 increases acidity of the water that is being treated. If alkalinity is not present in the water naturally, it shall be added using lime ($Ca(OH)_2$) or soda ash (Na_2CO_3) to facilitate the formation of gelatinous precipitate of $Al(OH)_3$ in wastewater.

Copperas/hydrated ferrous sulphate ($FeSO_4 \cdot 7H_2O$): Copperas follows a similar removal mechanism as that of alum. It reacts with alkalinity present in water to produce sticky gelatinous floc of ferric hydroxide ($Fe(OH)_3$) (Eq. 8.75 through Eq. 8.77), which adsorbs fine suspended and colloidal particles from the suspension on its surface. Consequently, larger and denser flocs with enhanced settleability characteristics are formed in the wastewater.

$$FeSO_4 \cdot 7H_2O + Ca(HCO_3)_2 \rightarrow Fe(HCO_3)_2 + CaSO_4 + 7H_2O \tag{8.75}$$

$$Fe(HCO_3)_2 + 2Ca(OH)_2 \rightarrow Fe(OH)_2 + 2CaCO_3 + 2H_2O \tag{8.76}$$

$$Fe(OH)_2 + \frac{1}{4}O_2 + \frac{1}{2}H_2O \rightarrow Fe(OH)_3 \downarrow \tag{8.77}$$

Chlorinated copperas ($Fe_2(SO_4)_3 + FeCl_3$): It is a preferred coagulant for wastewater treatment due to cheaper and wide working range of pH. Chlorinated copperas is produced by the addition of chlorine in copperas (Eqs. 8.78 to 8.80).

$$6FeSO_4 \cdot 7H_2O + 3Cl_2 \rightarrow 2Fe_2(SO_4)_3 + 2FeCl_3 + 7H_2O \qquad (8.78)$$

$$Fe_2(SO_4)_3 + 3Ca(OH)_2 \rightarrow 2Fe(OH)_3 \downarrow + 3CaSO_4 \qquad (8.79)$$

$$2FeCl_3 + 3Ca(OH)_2 \rightarrow 2Fe(OH)_3 \downarrow + 3CaCl_2 \qquad (8.80)$$

Insoluble hydrous ferric oxide is produced over a wide pH range of 3.0–13.0. The floc charge is positive in acid range and negative in alkaline range of pH. The charge is mixed over the pH range of 6.5–8.0.

Sodium aluminate ($Na_2 \cdot Al_2O_4$): Sodium aluminate is an alkaline compound used as coagulant. It reacts in water, which does not have natural alkalinity, with Ca or Mg salts present in water to form sticky gel like precipitates of calcium or magnesium aluminate (Eq. 8.81). This is an expensive coagulant and it is preferred to treat water with inadequate natural alkalinity. When natural alkalinity is present it reacts with alkalinity as per Eq. 8.82 to form $CaAl_2O_4$.

$$Na_2 \cdot Al_2O_4 + Ca/Mg \begin{cases} Cl_2^- \\ SO_4^{2-} \\ CO_3^{2-} \\ HCO_3^- \end{cases} \rightarrow Ca/Mg\ Al_2O_4 \downarrow + 2Na \begin{cases} Cl^- \\ SO_4^{2-} \\ CO_3^{2-} \\ HCO_3^- \end{cases} \qquad (8.81)$$

$$Na_2 \cdot Al_2O_4 + Ca/Mg(HCO_3)_2 \rightarrow Ca/Mg\ Al_2O_4 \downarrow + Na_2CO_3 + CO_2 + H_2O \quad (8.82)$$

8.12.7.2 Polymers

Sometimes, long-chain molecules of ionizable synthetic organic polymers are used as primary coagulant. These polymers comprise of ionizable groups, such as amino, sulfonic, carboxyl, etc., and depending upon the charge they carry, they can be categorized into cationic (containing +ve groups), anionic (containing –ve groups) and ampholytic (containing both +ve and –ve groups). Generally, cationic polymers, such as polydiallyl dimethyl ammonium chloride (poly-DADMAC) and epichlorohydrin dimethylamine (epi-DMA), are used in conjugation with metallic salts for coagulation of wastewater with very high suspended solids concentration, because they produce positively charged counterions (as colloids are negatively charged). In addition, due to their ability to reduce metals salts coagulant requirement, it leads to the significant decrease in sludge production.

8.12.8 Governing Parameters for Coagulation

8.12.8.1 Alkalinity

Alkalinity is defined as the concentration of all the ions present in water those are capable of neutralizing hydrogen ion (H^+). In other words, it is the capability of water to neutralize an acid. Alkalinity in wastewater is usually due to presence of anions, such as carbonate ions (CO_3^-), bicarbonate ions (HCO_3^-), hydroxyl ions (OH^-) or due to presence of dissolved gases like CO_2 and organic matter. Alkalinity is crucial for the formation of sticky gelatinous metallic hydroxide precipitates, which is essential for an effective coagulation process. If the wastewater does not contain sufficient alkalinity, it is required to be added externally to facilitate the formation of gelatinous precipitates responsible for carrying out sweep coagulation. In absence of alkalinity, when aluminium and iron coagulants are used, the pH of the effluent decreases due to formation of acids as illustrated in Eqs. 8.83 and 8.84. When alkalinity is present in the wastewater formation of acid is prevented and hence, pH of water does not change drastically.

$$Al_2(SO_4)_3 \cdot 18H_2O \leftrightarrow 2Al(OH)_3 + 3H_2SO_4 + 12H_2O \tag{8.83}$$

$$FeCl_3 + 3H_2O \leftrightarrow Fe(OH)_3 + 3HCl \tag{8.84}$$

8.12.8.2 The pH

Maintaining pH of water/wastewater during coagulation is of prime importance. It plays a vital role in the production of precipitates formed by trivalent metallic coagulants, which is responsible for coagulation. Variation in pH of wastewater may lead to the lesser production of these precipitates due to more solubility of metals at acidic pH; thus, seriously hampering the coagulation process. Whereas, higher pH can cause redispersion of coagulated particle in the wastewater. The size and nature of coagulated particles is also affected by the pH, which in turn affect their settling rate in the wastewater. For inorganic coagulants aluminium and iron, pH plays a vital role in governing the hydrolysis and polymerization reaction. Consequently, the pH acts as determining factor for type of ions and hydrolysis products formed in the solution along with their charge density.

Aluminium is readily soluble when pH of medium is below 4.5 and it is generally present in the form of Al^{3+} species such as $[Al(H_2O)_6]^{3+}$. As the pH rises solubility of aluminium also declines due to formation of less soluble species, such as $[Al_{13}(OH)_{32}]^{7+}$. In the pH range of 6.0–9.5, aluminium tends to dissociate into floc forming hydroxides such as $Al(OH)_3$ and $Al(SO_4)OH$, with minimum solubility of Al^{+3} ion being 2.7×10^{-5} mg/L at pH of 6.0 (Sawyer et al., 2003). Therefore, alum is most effective in the pH range of 5.5–7.5, having optimum coagulating pH of 6.0. However, when pH of the medium is greater than 9.5, soluble aluminate ion will be the dominant species, thereby drastically reducing the coagulation efficiency of aluminium coagulants.

On the other hand, solubility of iron in water depends upon the oxidation state in which iron exists. Ferrous ions (Fe^{2+}) are completely soluble in water irrespective of pH. On the contrary, ferric ion (Fe^{3+}) is soluble in water at pH of 3.5 or less and insoluble if pH is greater than 3.5 and attains minimum solubility of 5.6×10^{-5} mg/L at a pH of 4.2, thus making working range of iron-based coagulant much broader than that of aluminium coagulants (Sawyer et al., 2003). The optimum pH range of coagulation is specific to coagulant used and type of wastewater being treated (Table 8.5). Thus, the pH of the wastewater should be monitored and controlled carefully to produce the optimum result for coagulation.

8.12.8.3 Rapid Mixing of Coagulant

Addition of coagulant in wastewater is accompanied by rapid mixing, which not only ensures complete mixing of coagulant but also provides a minimum energy referred as 'threshold energy' to the coagulating precipitates, which is required for neutralizing and overcoming the protective negative charge on the surface of the colloidal particles present in the wastewater. The mixing of coagulant in wastewater can be achieved using baffles in the mixing basins or via mechanical mixers. When mechanical mixers are employed, intensity of mixing is represented using a term referred to as 'temporal mean velocity gradient' denoted by G, which is estimated using Eq. 8.85. The value of G during rapid mixing is usually kept around 300 sec^{-1} for optimum coagulation.

$$G = \sqrt{\frac{P}{\mu V}} \qquad (8.85)$$

where P is power applied to shaft (W); μ is dynamic viscosity of the wastewater (N sec/m^2); and V is the volume of the tank (in m^3).

Flash mixer is a mechanical type of rapid mixing device used to impart high intensity mixing of coagulant to the wastewater. In flash mixer, typically a motor driven high speed impeller fitted on vertical shaft is provided in a rectangular basin to achieve requisite G value for ensuring optimum coagulation (Fig. 8.43). Circular tank can also be used with provision of baffle to break the vortex motion formation to ensure mixing. The retention time of wastewater in flash mixing tank is a key factor affecting the efficiency of coagulation process. Insufficient mixing will not ensure homogenous concentration of coagulant throughout the wastewater volume leading to inadequate coagulation. On the other hand, too large detention time would require larger basin volume incurring unnecessary financial burden. In usual practise, retention time of flash mixer is kept in the range of 30 sec to 2 min with a typical design value of 60 sec.

8.12.8.4 Coagulant Dosage

The optimum dosage of coagulant required depends upon the type of coagulant being used and the desired degree of treatment. If the dose of coagulant is less than optimum,

Fig. 8.43 Flash mixer with impeller arrangement

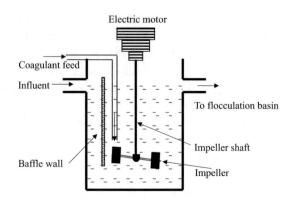

desired removal efficiency will not be achieved due to lesser formation of coagulating precipitates. Whereas, if the dosage of coagulant is higher than optimum, redispersion of flocs can occur as colloids are restabilized leading to the increase in turbidity of the treated water. The optimum dosage of coagulant is determined experimentally for a particular application using 'Jar test'. In a typical jar test six identical glass beakers/jars are taken and filled with equal volume (about 500 mL) of water/wastewater (Fig. 8.44a). The pH of the raw water/wastewater is checked and adjusted in the working range of coagulant that is being used for the test and in actual treatment of the raw water/wastewater.

While performing jar test a varying dosage of coagulant solution is added to each jar filled with water or wastewater. The coagulant solution is typically prepared by mixing 10 g of coagulant in 1 L of distilled water; thus 1 mL of solution contains 10 mg of coagulant. After coagulant dose addition in the jars, rapid mixing (impeller speed of about 100 rpm) is provided for one minute and for next 30 min the impeller speed is reduced to 30–40 rpm to support slow mixing for flocculation. After this the mixing is stopped and flocs are allowed to settle for next 30 min in the jars. After settling turbidity of supernatant water in each jar is determined using nephelometer. A graph (Fig. 8.44b) is plotted between residual turbidity (y-axis) and coagulant dosage (x-axis). Dose corresponding to lowest residual turbidity represents optimum coagulant dose for the particular sample of raw water or wastewater.

In cases, where optimum working range of pH for coagulant is also to be determined along with optimum dosage, two jar tests are conducted for the determination of optimum pH and optimum dosage separately. At first, the test is carried out by varying pH of wastewater samples in all jars at constant coagulant dosage and then determining the turbidity of the supernatant followed by the variation of coagulant dosage at optimum pH determined in the previous step.

Example: 8.17 A coagulation assisted sedimentation is proposed to treat a textile industry effluent having flow of 1000 m³/day. After performing jar test, it has been determined that 250 mg/L of pure ferric chloride ($FeCl_3$) is required for treating this wastewater. The influent

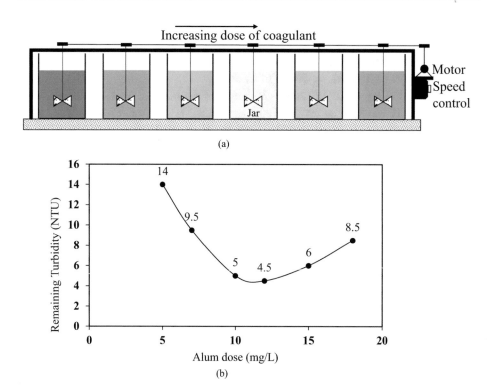

(a)

(b)

Fig. 8.44 Jar test **a** schematic diagram of a typical jar test apparatus and **b** plot for a result of a jar test performed for different alum dose at constant pH

suspended solids concentration is measured to be 550 mg/L, whereas the effluent suspended solids concentration required is 50 mg/L. Solids content of sludge is 1% (by weight) and average specific gravity of sludge solids is 1.2. Calculate the volume of sludge to be disposed per day from the sedimentation unit. If influent has total alkalinity of 1600 mg/L as $CaCO_3$. Calculate if the external alkalinity is required to be added per year, if 85% pure calcium oxide is available.

Solution

Ferric chloride reacts with alkalinity present in wastewater to produce ferric hydroxide precipitate as shown in the Eq. 8.86.

$$2FeCl_3 + 3Ca(HCO_3)_2 \leftrightarrow 2Fe(OH)_3 \downarrow +3CaCl_2 + 6CO_2 \tag{8.86}$$

Calculation of annual requirement of ferric chloride and calcium oxide

2 mol of $FeCl_3$ requires 3 mol of $Ca(HCO_3)_2$ to produce 2 mol of $Fe(OH)_3$ precipitate.

2×162.2 mg $FeCl_3$ requires 3×162 mg of $Ca(HCO_3)_2$ to produce 2×106.9 mg of $Fe(OH)_3$ precipitate.

1 milli-equivalent of $Ca(HCO_3)_2$ = 1 milli-equivalent of $CaCO_3$.

i.e., 81 mg of $Ca(HCO_3)_2$ = 50 mg of $CaCO_3$.

Therefore, 162 mg of $Ca(HCO_3)_2$ = 100 mg of $CaCO_3$.

2×162.2 mg $FeCl_3$ requires 3×100 mg of $CaCO_3$ to produce 2×106.9 mg of $Fe(OH)_3$ precipitate.

1 mg $FeCl_3$ requires $(3 \times 100/(2 \times 162.2))$ mg of $CaCO_3$ to produce $(2 \times 106.9/(2 \times 162.2))$ mg of $Fe(OH)_3$ precipitate

Therefore, 250 mg $FeCl_3$ requires $(3 \times 100/(2 \times 162.2)) \times 250$ mg of $CaCO_3$ to produce $(2 \times 106.9/(2 \times 162.2)) \times 250$ mg of $Fe(OH)_3$ precipitate.

Alkalinity reqd. as $CaCO_3 = (3 \times 100/(2 \times 162.2)) \times 250$ mg/L = 231.2 mg/L.

Alkalinity present in wastewater as $CaCO_3$ = 1600 mg/L, hence no external addition of alkalinity is required to support the coagulation.

Annual requirement of $FeCl_3 = (250 \times 1000 \times 365) \times 10^{-3}$ = 91,250 kg per year.

Calculation of sludge produced

Influent suspended solid (SS) concentration = 550 mg/L.

Effluent SS concentration = 50 mg/L.

SS removed = 550 − 50 = 500 mg/L.

SS removed per day = $(500 \times 1000 \times 10^{-3})$ kg = 500 kg.

$Fe(OH)_3$ precipitate produced per day = $(2 \times 106.9/(2 \times 162.2)) \times 250 \times 1000 \times 10^{-3}$ = 164.77 kg.

Total solid in sludge per day = 500 + 164.77 = 664.77 kg/day.

Solid content of sludge = 1% by weight, i.e., 10 g/L = 10 kg/m^3.

Volume of sludge = 664.77/10 = 66.477 m^3/day.

Example: 8.18 During a jar test experiment, alum was used as coagulant for treating a wastewater sample with no alkalinity. It was observed that flocs were not able to form due to lowering of pH of the water sample. If the amount of alum added was 10 mg/L, calculate the amount of sodium hydroxide required in the form of caustic soda powder with 50% purity in mg/L to neutralize the excess acid to bring the pH of wastewater sample to pH of 7.0.

Solution

$$Al_2(SO_4)_3 \cdot 18H_2O + 3Ca(HCO_3)_2 \rightarrow 2Al(OH)_3 + 3CaSO_4 + 18H_2O + 6CO_2$$
$$\quad\ \text{1 mole} \qquad\qquad \text{3 mole} \quad \text{2 mole} \qquad\qquad\qquad\qquad \text{6 mole}$$

666 mg of $Al_2(SO_4)_3 \cdot 18H_2O$ reacts with $3Ca(HCO_3)_2$.

If alkalinity is not present in water following reaction occurs. Sulphuric acid formed as one of the products lowering the pH of water as shown in Eq. 8.87.

$$Al_2(SO_4)_3 \cdot 18H_2O \rightarrow 2Al(OH)_3.3H_2O + 3H_2SO_4 + 6H_2O \qquad (8.87)$$

$$H_2SO_4 \leftrightarrow 2H^+ + SO_4^{2-} \qquad (8.88)$$

1 mol of $Al_2(SO_4)_3 \cdot 18H_2O$ produces 3 mol of H_2SO_4 as per Eq. 8.87.

1 mol of H_2SO_4 yields 2 mol of H^+ ion as per Eq. 8.88.

Hence, 1 mol of $Al_2(SO_4)_3 \cdot 18H_2O$ yields 6 mol of H^+ ion.

i.e., 666 g of $Al_2(SO_4)_3 \cdot 18H_2O$ yields 6 g of H^+ ion.

10 mg/L of $Al_2(SO_4)_3 \cdot 18H_2O$ yields $(6/666) \times 10$ mg/L of H^+ ion = 0.09 mg/L of H^+ ion = 9×10^{-5} mol/L

pH = $-\log[H^+]$ = $-\log(9 \times 10^{-5})$ = 4.05.

Clearly, the pH is out of range for coagulation with alum, which is in the range of 5.5–7.5

Hence, to neutralize sulphuric acid to bring the pH to 7.0 (i.e., to neutralize all of sulphuric acid formed) NaOH is required to be added (Eq. 8.89)

$$H_2SO_4 + 2NaOH \leftrightarrow Na_2SO_4 + 2H_2O \qquad (8.89)$$

Therefore, for each mole of sulphuric acid, 2 mol of NaOH is required for neutralization.

i.e., 98 g H_2SO_4 requires (2×40) g of NaOH.

Also, 1 mol alum produces 3 mol of H_2SO_4.

i.e., 666 g alum produces (3×98) g H_2SO_4.

Hence, 10 mg/L alum will produce = $(3 \times 98/666) \times 10$ mg/L of H_2SO_4 = 4.41 mg/L of H_2SO_4.

Hence, 4.41 mg/L H_2SO_4 will require $(2 \times 40/98) \times 4.41$ mg/L of NaOH = 3.6 mg/L of NaOH.

Therefore, Total amount of caustic soda powder required (with 50% purity) = 3.6/0.5 = 7.2 mg/L of NaOH.

8.13 Flocculation

Flocculation is the process, where coagulated or neutralized particles are brought together in contact of each other causing them to agglomerate and form larger flocs, which can be readily settled out through sedimentation. To assist and increase the opportunity for the particles to collide and come in contact, slow mixing is used in the flocculation tank.

8.13.1 Mechanism of Flocculation

During coagulation, addition of coagulant and rapid mixing of wastewater leads to destabilization of colloidal particles and growth of primary flocs (small sized flocs formed
initially due to destabilization of colloids); however, primary flocs and residual particles
are further allowed to come in contact with each other and progressively agglomerate
into a larger, and denser floc during flocculation operation for enhancing settling rate of
particles to attain better removal efficiency. The formation of larger flocs is assumed to
be occurring in two stages, namely perikinetic flocculation and orthokinetic flocculation,
which are discussed in the following section.

8.13.1.1 Perikinetic Flocculation

The phenomena in which interparticle collisions occur as a result of Brownian motion
(Fig. 8.45a) is termed as perikinetic flocculation (O'Melia, 1970; Overbeek, 1952). Such
type of flocculation commences immediately after destabilization of colloidal particles
and it is driven naturally by diffusion of colloidal particles in suspension due to thermal
agitation. Perikinetic flocculation lasts only for short duration and ceases to occur once
the agglomerated floc size reaches the limiting value beyond which Brownian motion
has negligible effect on floc formation. Typically, aggregation of particles smaller than
0.1 μm diameter occurs during perikinetic flocculation leading to the formation of flocs
of size ranging from 1 to 100 μm (Davis, 2010). Perikinetic flocculation is sometimes
also referred to as 'microscale flocculation' as only smaller size flocs are generated in
this stage. Since, effect of Brownian motion to support this perikinetic flocculation is
negligible, an artificial mixing is used to support this in treatment plants.

   The rate of change in concentration of colloidal particles in suspension with respect to
time due to perikinetic flocculation, J_B, can estimated using Eq. 8.90 as (O'Melia, 1972):

$$J_B = \frac{dc}{dt} = -\frac{4\eta kTc^2}{3\mu} \tag{8.90}$$

where c represents the total concentration of colloidal particles in suspension at time
t; η stands for collision efficiency factor representing the fraction of the total number
of collision successful in producing flocs; k is Boltzmann's constant; T is the absolute

(a)   (b)

Laminar flow   Turbulent flow

Fig. 8.45 Different types of flocculation **a** perikinetic flocculation and **b** orthokinetic flocculation

temperature; and μ denotes the viscosity of the fluid. The negative symbol in Eq. 8.90 signifies decrease in concentration of colloids with time due to their agglomeration in the suspension.

8.13.1.2 Orthokinetic Flocculation

The term orthokinetic flocculation (Fig. 8.45b) is defined as the phenomena, in which interparticle contacts are produced by laminar or turbulent fluid motion (Yusa, 1977). Flocculation of relatively larger particles (size greater than 1 μm) occurs due to advection or bulk motion of fluid, such as stirring and hence, it is also referred to as 'macroscale flocculation'. The principal parameters governing the rate and extent of orthokinetic flocculation in wastewater is applied velocity gradient and duration of flocculation. Velocity gradient is induced in flocculation tank via slow mixing using mechanical mixers or by providing baffles along the basin, which develops relative velocity between particles in suspension, thereby augmenting the opportunity to contact and agglomerate.

For a given wastewater, higher velocity gradient increases the number of contacts between the particles leading to greater quantity of floc formation; however, ultimate floc size gets reduced at higher velocity gradient as the breakdown of aggregated particles occurs due to the turbulence and shearing action of liquid. On the other hand, more time will be required to reach optimum floc size at lower velocity gradient owing to the subsequent reduction in number of contacts between the particles. Hence, for a given velocity gradient size of ultimate floc formed will not grow further beyond a limiting flocculation period.

The rate of reduction in concentration of colloidal particles with time, J_G, can be expressed using Eq. 8.91 (O'Melia, 1972):

$$J_G = \frac{dc}{dt} = -\frac{2\eta \Gamma d^3 c^2}{3} \tag{8.91}$$

where diameter of the colloidal particles is denoted as d and the velocity gradient of the laminar flow is symbolised using Γ. Negative symbol is used before the Eq. 8.91 in order to account for the decrease in concentration of the particles due their aggregation with time.

8.13.2 Factors Affecting Flocculation

a. *Concentration of colloidal particles in wastewater*: Higher the concentration of colloidal particles in the wastewater, more is the opportunity for them to come in contact with each other and agglomerate. On the other hand, dilute wastewater with low turbidity shows less tendency for flocculation as the colloidal particles are fewer in number and farther apart from each other, which reduces their opportunity to come in contact with each other and agglomerate.

b. ***Slow mixing***: During flocculation, the slow mixing is induced in the wastewater to enhance the opportunity of colloidal particles to come in contact with each other. This is in contrast with coagulation process, where violent agitation is induced via flash mixing of wastewater. Higher mixing intensity in flocculation can induce shear effect on flocs, which can cause them to disintegrate and disperse in the water again. Hence, controlled mixing is vital in making the flocculation process efficient and effective. Generally, the value of temporal mean velocity gradient in the range of 10–75 sec^{-1} is used for flocculator. Mixing in flocculation tanks can either be caused by using mixing basins or via application of mechanical mixers.

    i. *Mixing basins*: In these basins mixing effect is induced by creating turbulence in the wastewater as it flows across the basin. The turbulence is produced by the sudden change in direction of flow of water using baffle walls (Fig. 8.46). Generally, two types of mixing basins are used, namely 'around and end type' and 'over and under type'. Flow of water takes place in horizontal direction in around and end type of mixing basin, whereas in the latter it is based on upward and downward flow of water.

    ii. *Mechanical mixers*: As the name suggests, mechanical devices such as radial and axial flow impellers with shaft and paddle arrangement are used for inducing slow mixing in the flocculation tank. These mechanical mixers can either be mounted horizontally or vertically depending upon the design and geometry of the flocculation tank and paddle configuration as represented in Fig. 8.47a, b.

c. *Intensity of mixing:* For measuring mixing intensity a parameter referred to as 'temporal mean velocity gradient' (G) with unit of sec^{-1} is used in coagulation and flocculation process. It signifies the relative velocity between two particles located at a particular distance from each other in a medium (Fig. 8.48) and it can be calculated using Eq. 8.92.

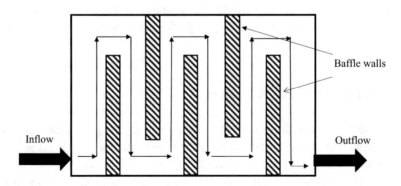

Inflow

Outflow

Baffle walls

Fig. 8.46   Mixing basin with baffle walls

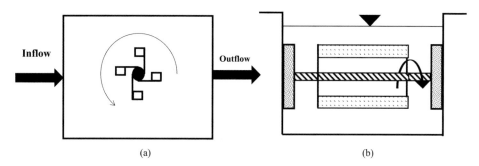

Fig. 8.47 Mechanical mixers **a** plan view of horizontal mixing (vertical shaft) type, **b** elevation view of vertical mixing (horizontal shaft) type

Fig. 8.48 Two colloidal particles separated by a distance x and moving with different velocities v_1 and v_2, respectively

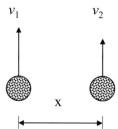

$$G = \frac{v_1 - v_2}{x} \tag{8.92}$$

where v_1 and v_2 are velocities of colloidal particles 1 and 2 having distance 'x' between them.

Determining the exact distance between two colloidal particles is difficult and hence, more mechanical approach is used while estimating the value of G, using Eq. 8.93, particularly when mechanical mixers are employed during flocculation (Fig. 8.47).

$$G = \sqrt{\frac{P}{\mu V}} \tag{8.93}$$

where P is the power applied to shaft in watts; μ is dynamic viscosity of the wastewater in N-sec/m^2; and V is volume of the tank in m^3.

Power (P) required to drive a shaft (Fig. 8.49) can be estimated using Eq. 8.94 as:

$$P = F_D \times v_r \tag{8.94}$$

Fig. 8.49 Shaft and paddle
arrangement on a mechanical
mixer

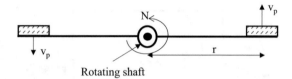

Rotating shaft

where F_D is the drag force acting on the paddle surface, N; v_r is the relative velocity
of moving paddle (m/sec) with respect to the flowing wastewater and it is calculated as
difference between v_p and v_w i.e., $v_p - v_w$; v_p represents the velocity of paddle mounted
on the shaft of the mixer; and v_w is the velocity of flow of water in the flocculation tank,
m/sec.

If angular velocity or rotational speed of shaft is known, then v_p can be calculated
using Eq. 8.95.

$$v_p = \omega r = \frac{2\pi N}{60} r \qquad (8.95)$$

where ω is the angular velocity of shaft in rad/sec; N represents rotational speed of shaft
in rpm; and r is the distance between the centre of the paddle and shaft.

Since velocity of water is much lower than that of velocity of paddle, for practical
purpose it is safe to assume that v_r is equal to v_p. The drag force acting on a paddle can
be estimated using Eq. 8.96.

$$F_D = (1/2) C_D \rho_X A_D v_p^2 \qquad (8.96)$$

where F_D is the drag force acting on the paddle, N; C_D is the dimensionless coefficient
of drag; ρ_w is density of water, kg/m^3; A_p is total area of paddle over which the drag
force is acting, m^2; and v_p is the velocity of paddle, m/sec.

Substituting value of F_D in Eq. 8.94, power required for driving the shaft can be
obtained using Eq. 8.97.

$$P = C_D \rho_w A_p v_p^3 / 2 \qquad (8.97)$$

Substituting value of P in Eq. 8.93, the Eq. 8.98 can be obtained for estimating value
of G.

$$G = \sqrt{\frac{1}{2}\left(\frac{C_D \rho_w A_p v_p^3}{\mu V}\right)} \qquad (8.98)$$

d. *Mixing time*: Adequate mixing time is imperative to an effective coagulation and flocculation process. In usual wastewater treatment practices, mixing time required for
   flocculation ranges from 20 to 30 min to ensure proper floc formation. Hence, most of

the flocculation tanks are designed for a hydraulic retention time (t) keeping in mind the requisite mixing time criteria. A parameter 'Gt' (product of velocity gradient and hydraulic retention time) representing 'conjugation opportunity' is often used as an important design criterion for flocculation tank. It affects the size and density of floc; thus, requires careful attention during hydraulic design of flocculation tank to promote growth of readily settleable flocs.

The parameter G directly affects the density of the flocs; whereas detention period 't' is responsible for determining the size of the floc formed. It has been observed that greater G value with shorter detention time favours the formation of denser small flocs, while lower G value with large detention time tend to produce larger lighter flocs. Aim of flocculation is to generate denser and larger flocs as they can be settled swiftly. Hence, it is advantageous to install multiple mixers along the length of a flocculation tank with varied value of G (Fig. 8.50). Mixer near the inlet is operated under higher value of G and the mixer near the outlet is operated at lowest value of G. This promotes faster mixing along with formation of smaller denser flocs near inlet. As wastewater flows through the flocculator, the mixing intensity is tapered and the mixers at outlet is operated at least value of G (with or without higher t), which generates larger and denser floc in the flocculation tank. A continuous horizontal flow of wastewater is maintained for facilitating such agglomeration in flocculator, thereby improving the removal efficiency and performance.

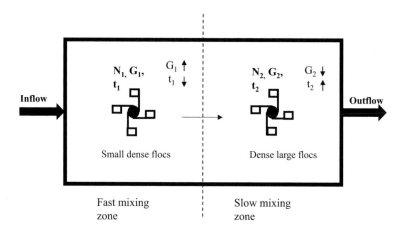

Fig. 8.50 Multiple mixer arrangement in flocculator with varied value of G and t

8.13.3 Design Considerations

The flocculators are designed for peripheral velocity of the paddle blades in the range of 0.1–0.9 m/sec. The relative velocity with respect to adjoining water near the external tip of the paddle is generally ¾ of the peripheral velocity of the paddles, which is maintained below 1.0 m/sec to prevent excessive velocity gradient for avoiding disintegration of flocs (Peavy et al., 1985). The total paddle blade area should be 15–20% as per Droste and Gehr (2018) and 10–25% (CPHEEO, 1999) of the flocculation basin cross sectional area. The 10% minimum area of blades will ensure sufficient mixing and this area of blade should not be more than 25% to avoid excessive rotational flow.

The depth of the flocculator is generally adopted as 3–4.5 m when mechanical mixers are installed; whereas, for baffled flocculators depth of 1.5–3 times distance between the baffles is suggested with a minimum distance of 0.45 m between the baffles (CPHEEO, 1999). Hydraulic retention time in flocculator is kept between 10 and 30 min (20 min typical). The temporal velocity gradient G is kept between 10 and 75 sec^{-1}. For tapered flocculator of three zones, in first zone the G value will be 60–70 sec^{-1}, in the middle zone 30–40 sec^{-1} and in the third zone it can be 10–20 sec^{-1} to support large size floc formation.

Though, the actual value of product of G and retention time t required for a particular wastewater can be worked out experimentally. The flocculators are usually designed to maintain Gt in the range of 10^4–10^5 (Droste & Gehr, 2018). Moreover, manual on Water Supply and Treatment, CPHEEO (1999) recommends Gt value of $(2–6) \times 10^4$ for aluminium coagulants and $(1–1.5) \times 10^5$ for ferric coagulants.

Example: 8.19 A flocculation basin is to be designed to handle 60 MLD of wastewater flow. The basin should have four compartments, each equipped with a vertical mechanical mixer with shaft and paddle arrangement having motor and brake efficiency of 90% and 75%, respectively. The shafts are to be fixed horizontally at mid-depth along the width of the tank and has gradually varying G to provide tapered flocculation. Kinematic viscosity of the water is 1.31×10^{-6} m^2/sec. Calculate:

i.   Dimension of flocculator
ii. Power required for driving each mixer
iii. The rpm of paddle wheel.

Provide two flocculators operating in parallel and two paddle blades per paddle wheel with one on either arm. Take coefficient of drag, C_D as 1.8; Depth/width of flocculator = 1.

Solution

(i) Since the flocculation basin should have four compartments with varying G values, provide: $G_1 = 60$ sec$^-$ in first compartment; $G_2 = 45$ sec$^-$ in second compartment; $G_3 = 30$ sec$^-$ in third compartment; and $G_4 = 15$ sec$^-$ in fourth compartment

$$G_{avg} = \frac{60 + 45 + 30 + 15}{4} = 37.5 \text{ sec}^-$$

Assume, $Gt = 6 \times 10^4$

Therefore, $t = \frac{6 \times 10^4}{37.5} = 26.67$ min.

Provide detention time of flocculator as 27 min (it is in range of 10–30 min, hence acceptable)

Volume of flocculator $= \frac{60 \times 10^6}{24 \times 60 \times 10^3} \text{m}^3/\text{min} \times 27\text{min} = 1125 \text{ m}^3$.

Two flocculators has to be provided, hence volume of each tank $= \frac{1125}{2} = 562.50 \text{ m}^3$.

Given, Depth/width = 1, i.e., d/w = 1

Providing, d = 4.5 m, hence including 0.5 m free board, total depth will be 5 m and width will be 4.5 m.

Hence, length of flocculation tank, $L = \frac{562.50}{4.5 \times 4.5} = 27.78$ m = 28 m (*provided*).

Four compartments are to be provide, assuming equal dimension of each compartment

Length of each compartment, $l = \frac{28}{4} = 7$ m.

Hence, dimension of each compartment = 4.5 m × 4.5 m × 7.0 m

(ii) Now, $G = \sqrt{\frac{P}{\mu V}}$ hence, $P = \mu V G^2$

Power requirement for each compartment P_i can be computed as, $P_i = \mu V_i G_i^2$

Where, $V_i =$ volume of compartment i and G_i mean temporal velocity gradient of mechanical mixer in compartment i; i = 1–4

$$P_1 = \mu V_1 G_1^2 = \frac{\left(1.31 \times 10^{-6} \times 10^3\right)\frac{\text{kg}}{\text{m s}} \times (4.5 \times 4.5 \times 7)\text{m}^3 \times (60 \text{ sec}^-)^2}{0.9 \times 0.75} = 990.36 \text{ W}$$

Similarly,

$$P_2 = \mu V_2 G_2^2 = \frac{\left(1.31 \times 10^{-6} \times 10^3\right)\frac{\text{kg}}{\text{m s}} \times (4.5 \times 4.5 \times 7)\text{m}^3 \times (45 \text{ sec}^-)^2}{0.9 \times 0.75} = 557.08 \text{ W}$$

$$P_3 = \mu V_3 G_3^2 = \frac{\left(1.31 \times 10^{-6} \times 10^3\right)\frac{\text{kg}}{\text{m s}} \times (4.5 \times 4.5 \times 7)\text{m}^3 \times (30 \text{ sec}^-)^2}{0.9 \times 0.75} = 247.59 \text{ W}$$

$$P_4 = \mu V_4 G_4^2 = \frac{\left(1.31 \times 10^{-6} \times 10^3\right)\frac{\text{kg}}{\text{m s}} \times (4.5 \times 4.5 \times 7)\text{m}^3 \times (15 \text{ sec}^-)^2}{0.9 \times 0.75} = 61.90 \text{ W}$$

However, considering the efficiency of motor and brake efficiency of 90% and 75%, respectively, the power actually transferred to flocculator for compartment one = 668.49 W. Similarly for compartment 2–4 the actual power transferred for flocculation will be 376.03 W, 167.12 W, and 41.78 W, respectively.

(iii) Providing a clear gap of 3.2 m between the paddle of two adjacent mixers and a clear gap of 1.6 m between tip of the paddle and the flocculator inlet and outlet walls, as well as 0.35 m clearance from the basin floor and side walls to avoid any turbulence and breakage of floc.

The working cross-sectional (c/s) area of each compartment = 4.5 m × 4.5 m = 20.25 m^2

Paddle blade area should be greater than 10 % of c/s are of compartment and less than 25% c/s of the area of the compartment as per design requirements.

i.e., paddle blade area must be greater than of 2.025 m^2 and less than 5.062 m^2.

Providing mechanical mixer with paddle width of 0.4 m and length 3.8 m located on either arm making a diameter of 3.8 m (Fig. 8.51). Thus, distance between the centre line of the paddles will be 3.4 m.

Therefore, area of paddle blades per mixer, Ap = (0.4 m × 3.8 m) × 2 = 3.04 m^2

This shall be between 2.025 m^2 and 5.062 m^2, hence acceptable.

Now, $P = \tfrac{1}{2}\, C_D\, \rho_w\, A_p\, v_p{}^3$

Thus,

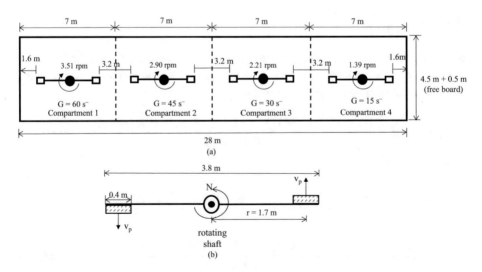

Fig. 8.51 Flocculation unit **a** different compartments of flocculator with tapered G value of mechanical mixers **b** rotating paddle and shaft arrangement

$$v_p = \left(\frac{2P}{A_p \rho_w C_D}\right)^{\frac{1}{3}}$$

$C_D =$ dimensionless coefficient of drag $= 1.8$ (given)
$\rho_w =$ density of water $= 1000$ kg/m^3
$A_p =$ total area of paddle over which the drag force is acting $= (0.4 \times 3.8) \times 2 = 3.04$ m^2

For first mixer

$$v_{p1} = \left(\frac{2 \times 668.49}{3.04 \times 1000 \times 1.8}\right)^{\frac{1}{3}} = 0.625 \frac{m}{sec}$$

$$v_{p1} = \frac{2\pi r N_{p1}}{60} \text{ hence } N_{p1} = \frac{60 v_{p1}}{2\pi r} = 60 \times \frac{0.625}{2\pi\left(\frac{3.4}{2}\right)} = 3.51 \text{ rpm}$$

Similarly, for second mixer

$$v_{p2} = \left(\frac{2 \times 376.03}{3.04 \times 1000 \times 1.8}\right)^{\frac{1}{3}} = 0.516 \frac{m}{sec}$$

$$v_{p2} = \frac{2\pi r N_{p2}}{60}, \text{ hence } N_{p2} = \frac{60 v_{p2}}{2\pi r} = 60 \times \frac{0.516}{2\pi\left(\frac{3.4}{2}\right)} = 2.90 \text{ rpm}$$

For third mixer,

$$v_{p3} = \left(\frac{2 \times 167.12}{3.04 \times 1000 \times 1.8}\right)^{\frac{1}{3}} = 0.394 \frac{m}{sec}$$

$$v_{p3} = \frac{2\pi r N_{p3}}{60} \text{ hence } N_{p3} = \frac{60 v_{p3}}{2\pi r} = 60 \times \frac{0.394}{2\pi\left(\frac{3.4}{2}\right)} = 2.21 \text{ rpm}$$

For fourth mixer,

$$v_{p4} = \left(\frac{2 \times 41.78}{3.04 \times 1000 \times 1.8}\right)^{\frac{1}{3}} = 0.248 \frac{m}{sec}$$

$$v_{p4} = \frac{2\pi r N_{p4}}{60} \text{ hence } N_{p4} = \frac{60 v_{p4}}{2\pi r} = 60 \times \frac{0.248}{2\pi\left(\frac{3.4}{2}\right)} = 1.39 \text{ rpm}$$

8.14   High-Rate Flocculation

Advancement in technology and better understanding of the treatment processes have not only improved the efficiency of treatment units but also brought down their detention time significantly without compromising with the effluent quality. Reduction in detention time and increase in the loading rates goes hand in hand with each other, where decrease in the former causes the later to increase. Designing a process as high rate becomes even more crucial when dealing with large flow rates, where time and related costs are of paramount importance. The aim of high-rate flocculation is to enhance the rate at which flocs are formed subsequently reducing the detention period and increasing the loading rate on the flocculation tank. Different techniques that are employed for high-rate flocculation are ballasted flocculation, dense solid flocculation, etc. High-rate flocculation units are combined with high-rate settlers, such as Lamella plate settlers, to achieve high-rate clarification of solids.

8.14.1   Ballasted Flocculation

In this type of flocculator, ballasting agent in the form of silica micro-sand of size in range of 0.1–0.3 mm (having large specific surface area and specific gravity of 2.65) is added to the wastewater along with metallic salts such as alum or ferric chloride. The entire process can be divided into three major stages. The first stage being a rapid mixing stage where coagulant is added. The second stage is a maturation stage or flocculation stage, where micro-sand is fed to the system and clarification stage is third in which usually lamella settler is employed for settling of flocs. This entire process can be visualized as depicted in Fig. 8.52. Destabilized suspended solids get adhered to the surface of this micro-sand thereby forming a dense floc referred to as ballasted floc, which settles rapidly. Sometimes polymers are also added in the wastewater to assist in the ballasting, i.e., bonding of micro-sand and destabilized particles. The flocs are then settled in the following clarifier units. Settled sludge is fed to cyclone separators to detach micro-sand from the flocs, from where the micro-sand is recycled back to the flocculation basin, while the remaining solids may be subjected to further treatment before they are disposed.

Ballasted flocculation can be readily employed in water treatment facilities to produce high quality treated water achieving 90–99% turbidity removal and in wastewater treatment for the efficacious removal of suspended and colloidal particles. However, ballasted flocculation system doesn't work well for the influent predominantly containing soluble contaminants. Critical parameters affecting the efficiency of ballasted flocculation are selection of coagulant and polymer dose, concentration of ballasting agent, mixing intensity and hydraulic loading rate. The typical liquid retention time in ballasted flocculation varies from 10 to 15 min, thus significantly minimising the footprint of treatment units.

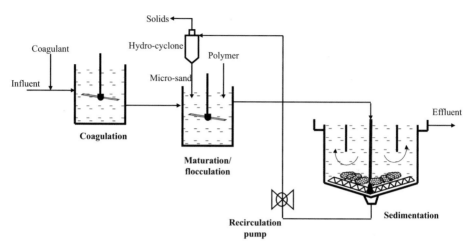

Fig. 8.52 Schematic diagram of a ballasted flocculation process

Typical design parameters values adopted in ballasted flocculation for wastewater applications are as follows (Ghanem et al., 2007; Young & Edwards, 2000; Young & Edwards, 2003):

- HRT = 2 min (mixing stage) and 2–6 min (maturation stage), 5 min (settling stage)
- Surface overflow rate = 50–120 m/h
- G_{mixing} = 600 sec^{-1}
- $G_{maturation}$ = 50–150 sec^{-1}
- Coagulant dosage = 50–150 mg/L
- Polymer dose = 0.5–10 mg/L (1 mg/L typically)
- Micro-sand dose = 1–10 g/L.

8.14.2 High Density Sludge Process

High density sludge (HDS) process (Fig. 8.53) employs recycling of chemically conditioned sludge to accelerate the rate of flocculation and it is usually employed for acidic industrial influent containing dissolved metals as primary pollutants and contaminated runoffs, such as acid mine drainage (AMD). The process begins with feeding of wastewater to a rapid mixer tank, where the influent is mixed with a mixture of sludge and lime for about 5 min to neutralize the acidic influent. The sludge lime mixture is produced in lime-sludge mix tank with the addition of lime (neutralizing agent) and coagulant (such as ferric sulphate, ferric chloride, etc.) to the recycled sludge coming out from clarifier.

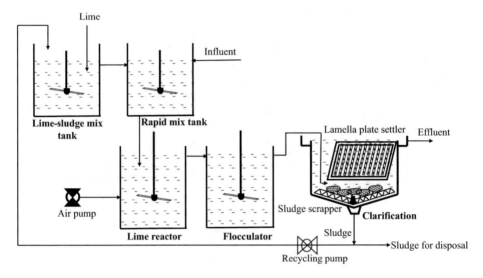

Fig. 8.53 Simplified representation of high-density sludge flocculation and clarification

The aim of neutralizing agent is to bring the pH of influent to a point at which metal of concern become insoluble (pH usually in range of 9.0–9.5) (Murdock et al., 1994). The resulting slurry from rapid mixing tank then cascades into a lime reactor in which complete precipitation of metal into hydroxide occurs at the elevated pH. Typical detention time provided in lime reactor is about 30 min for ensuring complete precipitation. Additionally, provision for pumping of air to lime reactor may also be installed to assist in precipitation of dissolved iron from the influent (Aubé, 1999).

The wastewater is then mixed with flocculants like polymers in a flocculators, which promotes bonding of chemically conditioned solids and suspended particles present in the wastewater. Moreover, lighter oily contaminants like grease and finer solids are also separated out during the flocculation operation. The entire process accelerates the rate of flocculation forming dense and homogeneous flocs. The wastewater is then fed to a lamella plate separator, where flocculated solids are removed, and the supernatant is processed for further treatment. A part of settled sludge is recycled back and rest is disposed off from the clarifier. Sludge-lime mixing prior to neutralization is a vital component of HDS process. The mixing of sludge and calcium hydroxide causes precipitation on the surface of flocs present in the sludge, thereby increasing their size and density, and thus accelerating the coagulation-flocculation process.

Questions

8.1. What is screening and comment on quantity of screening material collected and disposal options for these materials?

8.2. Summarise the design recommendation for bar screen.

8.3. Describe the fine screen and different types of fine screens used?

8.4. Comment on the coarse screen and its types?

8.5. Describe the grit chamber and its use in wastewater treatment plants?

8.6. Describe surface overflow rate and how performance efficiency of grit chamber is related to it.

8.7. Write short notes on the following: (a) design recommendations of rectangular grit chamber, (b) vortex flow grit chamber, (c) aerated grit chamber, (d) disposal of grit, (e) skimming tank and its importance, (f) ideal settling basin, and (g) neutralization.

8.8. Design an aerated grit chamber for treatment of sewage generated from population of 100,000 with a rate of water supply of 210 LPCD.

8.9. Describe the different types of settling involved in wastewater treatment.

8.10. Write about Stock's law and derive it.

8.11. Describe the factors affecting the performance of primary sedimentation tank and expected BOD and SS removal from it.

8.12. Summarise the recommendations for design of primary sedimentation tank.

8.13. Write about inclined plate settler and its advantages and disadvantages.

8.14. Describe the equalization tank and its use in sewage treatment plant.

8.15. Find the head loss through a clogged bar screen with following details: clogging is 50%, approach velocity is 0.5 m/sec, velocity through the clear rack is 0.8 m/sec, and flow coefficient for clogged bar screen is 0.7.

8.16. Design a circular primary sedimentation tank for primary treatment of sewage with flow rate of 15 MLD.

8.17. How the volume of equalization basin will be estimated for nearly constant wastewater flow and to keep COD concentration of wastewater under control?

8.18. Design an equalization basin for a constant out flow from the basin for further biological treatment of industrial wastewater.

Time	8–11	11–14	14–17	17–20	20–23	23–2	2–5	5–8
Flow, m^3	22.3	43.2	16.8	41.1	39.6	11.1	11.1	8.1

8.19. What is difference between coagulation and flocculation?

8.20. What are different mechanisms through which coagulation occurs?

8.21. What are the different coagulants that can be used in treatment of wastewater?

8.22. Describe parameters governing performance of coagulation.

8.23. Describe factors affecting performance of flocculators.

8.24. Why alkalinity is required for effective coagulation using alum?

8.25. Explain the parameter G and Gt used in coagulation and flocculation.

8.26. Determine the alkalinity required to be added per day in form of $CaCO_3$ and volume of sludge produced per day in an effluent treatment plant with a design flow of 30.2

MLD having suspended solids concentration of 500 mg/L subjected to coagulation-flocculation unit and alum is used as only coagulant with a dosing of 60 mg/L. Assume, wastewater has no alkalinity and all suspended solids are removed. Consider solids content in the sludge as 2.5% (by weight).

8.27. Design a flocculation tank having a design flow rate of 40 MLD and detention time of 30 min. Consider dynamic viscosity of the water as 1.31×10^{-3} N-sec/m^2 and C_D = 1.8. Provide three flocculators with average G value of 30 sec^{-1}. The paddle blade to be installed have width of 0.4 m. Calculate (i) dimension of flocculator; (ii) power requirement of each mixer (provide 1.5 times of calculated power to account for any power loss); (iii) the rpm of paddle wheel.

8.28. Design a plate settler of concurrent flow for wastewater inflow of 0.05 m^3/sec. Consider kinematic viscosity of wastewater as 1.2×10^{-6} m^2/sec and specific gravity of particle is 1.05. Consider size of the plates 1.2 m × 3 m and plate angle of 55°. What will be velocity through the plates?

8.29. Settling column test is performed and column has three ports at 0.5 m height each. Sampling is done at every half an hour and result are plotted in Fig. 8.54. Determine the percent removal of solids after one hour of settling. ($\Delta h_1 = 0.30$, $\Delta h_2 = 0.20$, $\Delta h_3 = 1.5$)

Answers

8.8. Considering peak factor of 2.5, depth of 2 m, width of 4 m, dimensions of grit chamber 13.2 m (L) × 4 m (W) × 2.5 m (D); air requirement = 4 m^3/min.

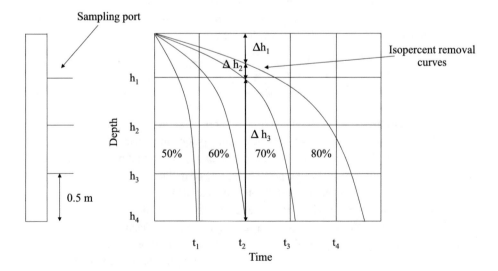

Fig. 8.54   Result of settling column test

8.15. Head loss = 16.8 cm

8.16. Two sedimentation tanks with diameter of 15.5 m and total depth of 3.25 m.

8.18. Volume required for equalization = 44.04 m^3; provide volume of 53 m^3; power for mixer = 250 W.

8.26. Alkalinity required as $CaCO_3$ = 815.4 kg/day; sludge production = 620.96 m^3.

8.27. Provide two flocculator each of volume 416.67 m^3; dimensions 4.5 m × 4.5 m × 7 m; paddles with dimension of 3.8 m length and 0.4 m width; rpm required will be 2.4, 1.98, 1.48.

8.28. 105 plates required. Velocity through the plates = 26.64 m/h.

8.29. 77.25%.

References

Aubé, B. C. (1999). Innovative modification to high density sludge process. In *Proceedings for Sudbury II* (pp. 13–17), Sadburry, Canada.

American Water Works Association. (1999). *Water quality and treatment* (5th ed.). American water works association, Denver, USA.

ASTM Designation: D1941-91. (2013). *Standard test method for open channel flow measurement of water with the parshall flume*, ASTM International, West Conshohocken, PA, U.S.A.

Berube, D., & Dorea, C. (2008). Optimizing alum coagulation for turbidity, organics, and residual Al reductions. *Water Science and Technology: Water Supply, 8*(5), 505–511.

Brandt, M. J., Johnson, K. M., Elphinston, A. J., & Ratnayaka, D. D. (2016). *Twort's water supply*. Butterworth-Heinemann.

Budd, G. C., Hess, A. F., Shorney-Darby, H., Neemann, J. J., Spencer, C. M., Bellamy, J. D., & Hargette, P. H. (2004). Coagulation applications for new treatment goals. *Journal-American Water Works Association, 96*(2), 102–113.

Camp, T. R. (1936). A study of the rational design of settling tanks. *Sewage Works Journal, 8*(5), 742–758.

Camp, T. R. (1946). Sedimentation and the design of settling tanks. *Transactions of the American Society of Civil Engineers, 111*(1), 895–936.

Camp, T. R. (1953). Sedimentation basin design. *Sewage and Industrial Wastes, 25*(1), 1.

Cheng, W. P., Chi, F. H., Li, C. C., & Yu, R. F. (2008). A study on the removal of organic substances from low-turbidity and low-alkalinity water with metal-polysilicate coagulants. *Colloids and Surfaces A: Physicochemical and Engineering Aspects, 312*(2–3), 238–244.

Chow, C. W., Fabris, R., Leeuwen, J. V., Wang, D., & Drikas, M. (2008). Assessing natural organic matter treatability using high performance size exclusion chromatography. *Environmental Science & Technology, 42*(17), 6683–6689.

Chow, C. W., van Leeuwen, J. A., Fabris, R., & Drikas, M. (2009). Optimised coagulation using aluminium sulfate for the removal of dissolved organic carbon. *Desalination, 245*(1–3), 120–134.

CPHEEO (1993). *Manual on Sewerage and Sewage Treatment* (2nd ed.)., Central Public Health and Environmental Engineering Organization, Ministry of Urban Development, Government of India, New Delhi, India

CPHEEO. (1999). *Manual on water supply and treatment*. CPHEEO, Ministry of Urban Development. Government of India, New Delhi, India

CPHEEO. (2013). *Manual on sewerage and sewage treatment system.* Central Pollution Health & Environmental Engineering Association, Ministry of Housing and Urban Affairs. Government of India, New Delhi, India.

Crites, R., & Tchobanoglous, G. (1998). *Small and decentralized wastewater management system.* McGraw-Hill.

Davis, M. L. (2010). *Water and wastewater engineering: Design principles and practice.* McGraw-Hill Education.

Dey, S., Ali, S. Z., & Padhi, E. (2019). Terminal fall velocity: The legacy of Stokes from the perspective of fluvial hydraulics. *Proceedings of the Royal Society A: Mathematical, Physical and Engineering Sciences, 475*(2228), 20190277.

Dongsheng, W., Hong, L., Chunhua, L., & Hongxiao, T. (2006). Removal of humic acid by coagulation with nano-Al13. *Water Science and Technology: Water Supply, 6*(1), 59–67.

Droste, R. L., & Gehr, R. L. (2018). *Theory and practice of water and wastewater treatment.* Wiley.

Eckenfelder, W. W. (2000). *Industrial water pollution control.* McGraw-Hill International.

EPA. (1995). *Waste water treatment manuals: Preliminary treatment.* Environmental Protection Agency.

EPA. (1996). *Waste water treatment manuals: Primary, secondary and tertiary treatment.* Environmental Protection Agency, Ireland.

Ghanem, A. V., Young, J. C., & Edwards, F. G. (2007). Mechanisms of ballasted floc formation. *Journal of Environmental Engineering, 133*(3), 271–277.

Hazen, A. (1904). On sedimentation. *Transactions of the American Society of Civil Engineers, 53*(2), 45–71.

ISO-9826. (1992). International Organization for Standardization. (1992). *Measurement of liquid flow in open channels—Parshall and SANIIRI Flumes* (ISO Standard No. 9826:1992).

Kirschmer, O. (1926). *Untersuchungen über den Gefällsverlust an Rechen* [Investigation of head losses at racks]. Hydraulisches Institut Mitteilung (1), Technische Hochschule, München, Germany.

Lee, C. C., & Lin, S. D. (2007). *Handbook of environmental engineering calculations.* McGraw-Hill Education.

Lei, G., Ma, J., Guan, X., Song, A., & Cui, Y. (2009). Effect of basicity on coagulation performance of polyferric chloride applied in eutrophicated raw water. *Desalination, 247*(1–3), 518–529.

Metcalfe, I., & Healy, T. (1990). Charge-regulation modelling of the Schulze-Hardy rule and related coagulation effects. *Faraday Discussions of the Chemical Society, 90*, 335–344.

Metcalf & Eddy Inc., Tchobanoglous, G., Burton, F. L., Tsuchihashi, R., & Stensel, H. D. (2003). *Wastewater engineering: Treatment and resource recovery* (4th ed.). McGraw-Hill Professional.

Murdock, D., Fox, J., & Bensley, J. (1994). Treatment of acid mine drainage by the high density sludge process. In: *Proceedings of the International Land Reclamation and Mine Drainage Conference and Third International Conference on the Abatement of Acid Drainage*, Pittsburgh, April (Vol. 1, pp. 241–249).

O'Melia, C. R. (1972). Coagulation and flocculation. In *Physico-chemical process for water quality control* (pp. 61–109). Wiley Interscience.

Overbeek, J. T. G. (1952). Kinetics of flocculation. *Colloid Science, 1*, 278–301.

Peavy, H. S., Rowe, D. R., & Tchobanoglous, G. (1985). *Environmental engineering.* McGraw-Hill.

Pintor, A. M. A., Vilar, V. J. P., Botelho, C. M. S., & Boaventura, R. A. R. (2016). Oil and grease removal from wastewaters: Sorption treatment as an alternative to state-of-the-art technologies. A critical review. *Chemical Engineering Journal, 297*, 229–255.

Reiber, S., Kukull, W., & Standish-Lee, P. (1995). Drinking water aluminum and bioavailability. *Journal of the American Water Works Association, 87*(5), 86–100.

Rigobello, E. S., Dantas, A. D. B., Di Bernardo, L., & Vieira, E. M. (2011). Influence of the apparent molecular size of aquatic humic substances on colour removal by coagulation and filtration. *Environmental Technology, 32*(15), 1767–1777.

Sawyer, C. N., McCarty, P. L., & Parkin, G. E. (2003). *Chemistry for environmental engineering* (5th ed.). McGraw-Hill.

Shammas, N. K. (2005). Coagulation and flocculation. In *Handbook of environmental engineering, Volume 3: Physicochemical treatment processes* (pp. 103–139). The Humana Press Inc., Lenox.

Sharp, E., Jarvis, P., Parsons, S., & Jefferson, B. (2006). Impact of fractional character on the coagulation of NOM. *Colloids and Surfaces A: Physicochemical and Engineering Aspects, 286*(1–3), 104–111.

Shin, J., Spinette, R., & O'melia, C. (2008). Stoichiometry of coagulation revisited. *Environmental Science & Technology, 42*(7), 2582–2589.

Sillanpää, M., & Matilainen, A. (2014). NOM removal by coagulation. In *Natural organic matter in water: Characterization and treatment methods* (Vol. 55). Butterworth-Heinemann. Mikkeli, Finland.

Smith, E., & Kamal, Y. (2009). Optimizing treatment for reduction of disinfection by-product (DBP) formation. *Water Science and Technology: Water Supply, 9*(2), 191–198.

Tebbutt, T. H. Y., & Christoulas, D. G. (1975). Performance relationships for primary sedimentation. *Water Research, 9*(4), 347–356.

Trefalt, G., Szilágyi, I., & Borkovec, M. (2020). Schulze-Hardy rule revisited. *Colloid and Polymer Science, 298*(8), 961–967.

Trinh, T. K., & Kang, L. S. (2011). Response surface methodological approach to optimize the coagulation–flocculation process in drinking water treatment. *Chemical Engineering Research and Design, 89*(7), 1126–1135.

Uyak, V., & Toroz, I. (2007). Disinfection by-product precursors reduction by various coagulation techniques in Istanbul water supplies. *Journal of Hazardous Materials, 141*(1), 320–328.

Uyguner, C., Bekbolet, M., & Selcuk, H. d. (2007). A comparative approach to the application of a physico-chemical and advanced oxidation combined system to natural water samples. *Separation Science and Technology, 42*(7), 1405–1419.

Water Environment Federation and the American Society of Civil Engineers. (1991). *Design of municipal wastewater treatment plants*. Water Environment Federation, Alexandria, VA, U.S.A.

Water Pollution Control Federation. (1985). *Clarifier Design: WPCF Manual Practice FD-10*. Water Pollution Control Federation, Alexandria, VA, USA.

Yan, M., Liu, H., Wang, D., Ni, J., & Qu, J. (2009). Natural organic matter removal by coagulation: Effect of kinetics and hydraulic power. *Water Science and Technology: Water Supply, 9*(1), 21–30.

Yao, K. M. (1970). Theoretical study of high-rate sedimentation. *Journal-Water Pollution Control Federation, 42*(2), 218–228.

Young, J. C., & Edwards, F. G. (2000). Fundamentals of ballasted flocculation reactions. *Proceedings of the Water Environment Federation, 2000*(14), 56–80.

Young, J. C., & Edwards, F. G. (2003). James C. Young, Findlay G. Edwards. *Water Environment Research, 75*(3), 263–272.

Yusa, M. (1977). Mechanisms of pelleting flocculation. *International Journal of Mineral Processing, 4*(4), 293–305.

Zhan, X., Gao, B., Wang, Y., & Yue, Q. (2011). Influence of velocity gradient on aluminum and iron floc property for NOM removal from low organic matter surface water by coagulation. *Chemical Engineering Journal, 166*(1), 116–121.

Zhao, Y., Gao, B., Shon, H., Cao, B., & Kim, J.-H. (2011). Coagulation characteristics of titanium (Ti) salt coagulant compared with aluminum (Al) and iron (Fe) salts. *Journal of Hazardous Materials, 185*(2–3), 1536–1542.

Zouboulis, A., Moussas, P., & Vasilakou, F. (2008). Polyferric sulphate: Preparation, characterisation and application in coagulation experiments. *Journal of Hazardous Materials, 155*(3), 459–468.

Fundamentals of Biological Wastewater Treatment

9

9.1   Background

The treatment of contaminated water is indispensable to maintain both the biotic and abiotic ecology that supports human life on the earth. However, the discharge of domestic, municipal and industrial effluents containing both organic and inorganic constituents into rivers, ponds and other aquatic environments causes water pollution (Singh et al., 2020). Therefore, treatment of these wastewaters before being discharged into natural water bodies is highly recommended to alleviate the ever-increasing degradation of natural water quality (Rawat et al., 2011). In this regard, different physicochemical treatment technologies, like chlorination, electrocoagulation, advanced oxidation, ozonation, membrane filtration, are implemented to treat wastewater emerging from different sources; however, these processes produce other intermediatory hazardous and noxious by-products (Das et al., 2020f). Moreover, overall capital and operational costs of these processes are much higher due to the requirement of costly materials for reactor fabrication and use of costly chemicals/catalysts, which make these processes economically infeasible. To mitigate these roadblocks of conventional technologies, researchers have focused on the application of environment-friendly techniques, i.e., biological processes for the treatment of biodegradable matters present in wastewater.

In the present scenario, microbial wastewater treatment is the most significant frontier area of environmental research. Biological wastewater treatment is categorized into three major classes, namely aerobic, anaerobic, and anoxic treatment processes. In aerobic processes, microorganisms utilize O_2 as terminal electron acceptor to oxidize organic compounds present in the wastewater for their cellular metabolism and produce CO_2. On the other hand, in anaerobic treatment, the microbiota can degrade the organic matter in the absence of O_2 and produces CO_2 and value-added methane. However, in anoxic treatment microbes utilize sulphate, nitrate, and nitrite as an electron acceptor for supporting

M. Ghangrekar, *Wastewater to Water*, https://doi.org/10.1007/978-981-19-4048-4_9

microbial growth and metabolism. Different microorganisms, like bacteria, fungi, algae, and phytoplankton, are extensively utilized in the biological wastewater treatment process, which can remove organic and inorganic pollutants effectively with the minimum production of sludge. Thus, utilization of microorganisms in wastewater treatment not only assists to solve the economic difficulties but can also facilitate recovery of value-added products from wastewaters and produce reusable quality treated water.

Generally, the wastewater treatment is classified into three different classes, such as primary, secondary, and tertiary treatment. Basically, physical forces are involved in the primary treatment of wastewater to eliminate floating and settleable organic and inorganic suspended solids prior to the secondary wastewater treatment. Afterwards, dissolved and colloidal organic matters are removed via aerobic/anaerobic microorganisms by employing secondary wastewater treatment processes, i.e., biological treatment processes. However, after primary and secondary treatment of wastewater, the secondary treated effluent still comprises of few residual organic and inorganic matters, nutrients (nitrogen and phosphorous), heavy metals, coliforms, pathogens and other contaminants (Barsanti & Gualtieri, 2014; Das et al., 2020c). Hence, secondary treated effluent requires further treatment by employing advanced physico-chemical or biological treatments termed as a tertiary treatment before being discharged into the environment or facilitating reuse of treated water. However, huge chemical requirement, high operational cost, and production of toxic sludge, while eliminating the pollutants present in wastewater, make these physico-chemical treatment processes both economically and environmentally non-sustainable. Thus, to overcome these above-mentioned bottlenecks, biological tertiary treatment utilizing algae has gained considerable attention due to its eco-friendly and economical nature (Das et al., 2020a, 2020b).

This chapter articulates the possibility of treating wastewater utilizing different microorganisms, like bacteria, algae, fungi and other protozoa, and their symbiotic interaction that promotes cellular growth with simultaneous recovery of nutrients and other value-added products. Moreover, different physico-chemical factors and growth kinetics that affect microbial growth and wastewater treatment efficiency, utilizing bacteria and algal with their drawbacks, and algae-based wastewater treatment technologies are also highlighted in this chapter. Furthermore, the future perspective of the microbial-based treatment and advanced technology used in wastewater treatment processes are also described elaborately.

9.2   Bacterial Metabolism and Use in Wastewater Treatment

Microbial communities specifically bacteria have an essential role in the removal of pollutants from wastewater. In wastewater treatment, three different types of bacteria, such as aerobic, anaerobic, and facultative are extensively utilized for the elimination of organic and inorganic pollutants from wastewater. In this respect, Wang et al. articulated that

Proteobacteria and Betaproteobacteria are predominant and are the most utilized bacterial class in wastewater treatment due to their wide-ranging application for the removal of organic compounds and nutrients from wastewater (Zhou et al., 2012). Additionally, other bacterial groups, such as Bacteroidetes, Chloroflexi and Acidobacteria, are also responsible for the bioremediation of wastewater (Wan et al., 2011; Zhou et al., 2012).

9.2.1 Bacterial Metabolism in Wastewater Treatment

The chemical processes that occur simultaneously within each bacterial cell to provide energy for their growth are called metabolism. The metabolism can be divide into two categories, namely anabolism and catabolism (Sidhu et al., 2017). Catabolism is an energy-yielding metabolism of cells; whereas, anabolism is a biosynthetic metabolism in which microbial cell synthesis occurs and it is energy consuming process (Fig. 9.1). In the catabolism process, the chemical energy stored in organic substances is released, and this energy is used in the anabolic process for the synthesis of cellular components. The net growth of cells or organisms is the result of a balance between anabolism and catabolism.

The organic contaminant removal through wastewater treatment is caused due to catabolic metabolisms in which the microbes catabolize organic pollutants. Two types of catabolism are employed for wastewater treatment as aerobic catabolism, i.e., oxidative catabolism, and anaerobic catabolism formally known as fermentative catabolism. The catabolic process in which oxidizing agents, such as O_2, nitrates, or sulphate, oxidize the organic pollutants present in wastewater through bacterial catalysed reactions is

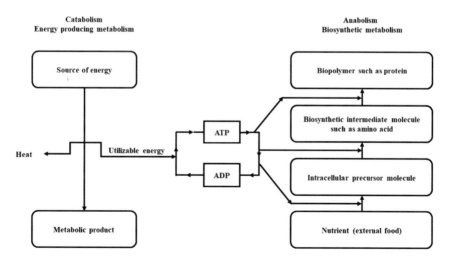

Fig. 9.1 Relation between anabolism and catabolism

known as oxidative catabolism. The chemical reaction involved in aerobic catabolism can be expressed by Eq. 9.1.

$$C_6H_{12}O_6 + 6O_2 \rightarrow 6CO_2 + 6H_2O \qquad (9.1)$$

Anaerobic catabolism is a complex multi-step pathway of serial and parallel reactions (Meegoda et al., 2018). The organic pollutants in this pathway are degraded in four steps, namely hydrolysis, acidogenesis, acetogenesis, and methanogenesis. These steps are explained in details in Chap. 12. Hydrolysis is the first step of anaerobic catabolism, where complex large molecules such as protein, carbohydrates, and lipids are converted to monomeric or dimeric compounds. The degradation of polymer to monomer is caused due to the action of exo-enzymes; the monomers are accessible to cross the cell barriers of acidogenic bacteria. After enzymatic hydrolysis, proteins are hydrolysed to the amino acid, polysaccharide to simple sugar, and lipid to long-chain fatty acid. Acidogenic bacteria produce exo-enzyme for the hydrolysis of complex polymers, and the product of enzymatic hydrolysis is used as substrates for acidogenic bacteria. The rate of hydrolysis depends on various factors, such as temperature, pH, particle size, and enzyme concentration; therefore, hydrolysis is considered as the rate-limiting step in the entire anaerobic digestion process, when organic matter is present in suspended form in the wastewater.

Acidogenesis starts after hydrolysis, here in acidogenesis, the hydrolysis product is diffused inside the acidogenic bacterial cell and fermented into short-chain fatty acid, alcohol, CO_2, and H_2 as well as other intermediate organic acids, such as propionic acid and butyric acid. Acidogenesis is considered as the most rapid reaction in the anaerobic digestion process. The end-products of acidogenesis are used as a substrate for acetogenesis and methanogenesis. Acidogenesis can be a two-directional process due to the action of acidogenic bacteria. Products, such as CO_2 and H_2, are used as the substrate for methanogenesis. However, other products such as short-chain fatty acid and alcohol may not be used as the substrate for methanogenesis and must be digested by acetogenic bacteria to acetate, the substrate that can be used by methanogenic bacteria.

Acetogenesis is thus the third step of anaerobic digestion in which the non-gaseous product of acidogenesis is further oxidized to acetate, CO_2, and H_2. The intermediate products of acidogenesis, such as propionate and butyrate, are the main substrates for acetogenic bacteria. These intermediate higher organic acids are converted to acetate as well as H_2, which are then utilized by methanogenic bacteria to produce methane. At the same time, CO_2 and H_2 can also be converted in to acetic acid and this process is known as homo-acetogenesis.

The final step of anaerobic digestion is methanogenesis, where acetate, CO_2 and H_2 are used as the substrate for methanogenic bacteria to produce CH_4. Methanogens can be classified into two groups: acetoclastic methanogens and hydrogenotrophic methanogens. Acetoclastic methanogens are acetate utilizing bacteria (AUB) that converts acetate to methane; however, hydrogenotrophic methanogens are hydrogen utilizing bacteria

(HUB) that converts CO_2 and H_2 to methane. Generally, about 70% of methane is produced from acetoclastic methanogenesis and the remaining about 30% is produced from hydrogenotrophic methanogenesis in anaerobic digestion of organic contaminants. The growth rate of AUB is low, which explains why an anaerobic reactor requires a longer start-up time as well as a higher sludge concentration in the reactor. Apart from that, the growth rate of HUB is very high, which results in maintaining low partial pressure of hydrogen that is essential for AUB, which explains the stability of an anaerobic reactor.

9.2.2   Catabolism and Anabolism

A portion of the available organic or inorganic substrate present in the wastewater is oxidized by the bio-chemical reactions, being catalysed by large protein molecules known as enzymes produced by microorganism to liberate energy. The oxidation or dehydrogenation can take place both in aerobic and anoxic conditions. Under aerobic conditions, the oxygen acts as the final electron acceptor for the oxidation. Under anoxic conditions sulphates, nitrates, nitrites, carbon dioxide and organic compounds act as an electron acceptor. Metabolic end products of the respiration are true inorganics like CO_2, water, ammonia, and H_2S.

The energy derived from the respiration is utilized by the microorganisms to synthesize new protoplasm through another set of enzymes catalysed reactions, from the remaining portion of the substrate. The heterotrophic microorganisms derive the energy required for cell synthesis exclusively through oxidation of organic matter and autotrophic microorganisms derive the energy for synthesis either from the oxidation of inorganic compounds or from photosynthesis. The energy is also required by the microorganisms for maintenance of their life activities. In absence of any suitable external substrate source, the microorganisms derive this energy through the oxidation of their own protoplasm. Such a process is known as endogenous respiration (or decay). The metabolic end products of the endogenous respiration are similar to that in primary respiration.

The metabolic processes in both aerobic and anaerobic processes are almost similar. The yield of energy in an aerobic process, using oxygen as electron acceptor, is much higher than in anaerobic condition. This is the reason why the aerobic system produces more new cells than the anaerobic systems. The most important mechanism for the removal of organic matter in biological wastewater treatment system is by bacterial metabolism. Metabolism refers to the utilization of the organic matter, either as a source of energy or as a source for the synthesis of cellular matter. When organic matter is used as an energy source, it is transformed into stable end products, a process known as catabolism. In the process of anabolism, the organic matter is transformed and incorporated into cell mass. Anabolism is an energy consuming process and it is only possible if catabolism occurs at the same time to supply the energy required for the synthesis of the cellular matter. Thus, the processes of catabolism and anabolism are interdependent and occur simultaneously.

9.2.3   Nutritional Requirements for Microbial Growth

For reproduction and proper functioning of a microorganisms, they must have (i) a source of energy, (ii) carbon for the synthesis of new cellular material, and (iii) nutrients such as N, P, K, S, Fe, Ca, Mg, etc. Energy required for the cell synthesis may be supplied by light or by chemical oxidation reaction catalysed by the bacteria. Accordingly, the microbes can be classified as stated below.

9.2.3.1   Source of Energy

Phototrophs: These are microorganisms that are using light as an energy source. These may be heterotrophic (certain sulphur reducing bacteria) or autotrophic (photosynthetic bacteria and algae).

Chemotrophs: Microorganisms that derive their energy from chemical reaction. These may be either heterotrophic, those derive energy from organic matter, like protozoa, fungi, and most bacteria or may be autotrophic like nitrifying bacteria. Accordingly, they are called as Chemoheterotrophs (those derive energy from oxidation of organic compounds) and chemoautotrophs, those obtain energy from oxidation of reduced inorganic compounds, such as ammonia, nitrite, sulphide.

9.2.3.2   Source of Carbon

Availability of carbon in the medium is necessary for supporting synthesis of new cells. The source of carbon for synthesis of new cells could be organic matter as used by heterotrophs or carbon dioxide as used by autotrophs.

9.2.3.3   Nutrient and Growth Factor Requirement

The principal inorganic nutrients required by microorganisms are N, S, P, K, Mg, Ca, Fe, Na, Cl, etc. Some of the nutrients are required in trace amount, i.e., very small amount, such as Zn, Mn, Mo, Se, Co, Ni, Cu, etc. In addition to the inorganic nutrients, organic nutrients may also be required by some microorganisms and they are known as 'growth factors'. These are compounds required by an organism as precursors or constituents of organic cell material that cannot be synthesized from other carbon sources. Requirements of these nutrients differ from organism to organism. For aerobic processes generally minimum COD:N:P ratio of 100:10:1–5 is necessary for effective treatment of the wastewater. In case of anaerobic treatment process minimum COD:N:P ratio of 350:5:1 is considered essential. The nutrient requirement is lower for anaerobic process due to slower growth rate of microorganisms as compared to aerobic process. While treating sewage external macro (N, P, K, S) and micro (trace metals) nutrients addition is not necessary; however, in case of industrial wastewater treatment, external addition of these may be required depending upon the characteristics of the raw wastewater.

9.2.4 Types of Microbial Metabolism

Aerobic: When molecular oxygen is used as terminal electron acceptor in respiratory metabolism it is referred as aerobic respiration. The organisms that exist only when there is molecular oxygen supply are called as obligately aerobic microorganisms.

Anoxic: For some respiratory microorganisms oxidized inorganic compounds, such as sulphate, nitrate and nitrite can function as an electron acceptor in absence of molecular oxygen. These organisms are called as anoxic microorganisms.

Obligately anaerobic: These are the microorganisms those generate energy by fermentation and they can exist and metabolize in absence of oxygen.

Facultative anaerobes: These microorganisms have ability to grow in absence or in presence of oxygen. These can be divided in two types: (a) *True facultative anaerobes*: those can shift from fermentative to aerobic respiratory metabolism, depending on oxygen available or not; (b) *Aerotolerant anaerobes*: these microorganisms follow strictly fermentative metabolism and are insensitive if oxygen is present in the system.

9.2.5 Bacterial Growth Kinetics

Bacterial growth kinetics describes the relationship between the specific growth rate of bacteria and the concentration of substrate that is consumed by bacteria for their cellular metabolism. Also, this type of growth kinetics predicts the rate of conversion of substrate into biomass as a product. Therefore, the growth of a bacterial culture is linked to both, the synthesis of cellular components (biomass) as well as an increase in number by multiplying individual bacterial cells to produce offspring or individual organisms. Generally, the common mechanism of bacterial cell multiplication is binary fission, in which a cell doubles all its cellular components, which increases the length of the cell and finally splits into two virtually identical sister cells (Najafpour, 2015).

Bacterial population growth follows a geometrical or an exponential pattern in which one cell splits into two cells, then 2 daughter cells in 4 cells, then 4 daughter cells in 8 cells, and so forth (2^0, 2^1, 2^2, 2^3 ... 2^n); thus, after n divisions, the number of cells will be 2^n. Although, bacterial cell growth depends on the specific time duration. The time required for a bacterial cell to grow and split into two virtually identical sister cells is known as generation time or doubling time. In this respect, Najafpour stated that under controlled environmental condition with respect to temperature, pH, and different sources of nutrients; the doubling time for any specific bacteria remains constant. Najafpour also described that the bacteria enhance their cell multiplication exponentially (Fig. 9.2a),

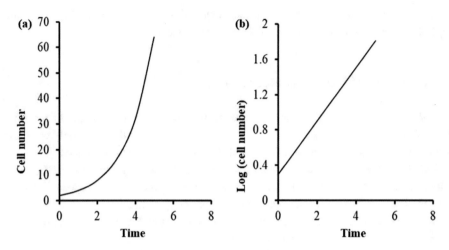

Fig. 9.2 Bacterial growth curve: **a** cell number versus time; **b** Log (cell no.) versus time

whereas the graphical representation of the logarithmic value of bacterial cell number versus time becomes a straight line (Fig. 9.2b).

In literature, there are two approaches available for defining bacterial growth rate for any specific bacterial growth medium (Najafpour, 2007). The first approach of bacterial growth kinetics is based on the number of bacteria (N) as a function of time (t); whereas, the second approach is based on the weight quantification of bacterial biomass with time (t) (Najafpour, 2007). However, in wastewater treatment, usually dry weight of bacterial biomass is preferred for the determination of bacterial growth rate.

9.2.5.1 Bacterial Growth Kinetics Based on Cell Number

The relationship between the number of bacteria (N_t) in a culture medium at a given time (t) is the number of bacteria present (N_o) at the beginning and the number of divisions (n) during this period which can be expressed by Eq. 9.2.

$$N_t = N_o \times 2^n \tag{9.2}$$

For a constant division rate (k), n can be calculated as per Eq. 9.3.

$$n = k \times t \tag{9.3}$$

The generation time (g) also known as doubling time (t_d), can be related to k as given in Eq. 9.4.

$$k = \frac{1}{g} \tag{9.4}$$

Therefore, Eq. 9.2 can also be expressed as Eq. 9.5 and solved to get Eq. 9.6.

$$N_t = N_0 \times 2^{k \times t} = N_0 \times 2^{\frac{1}{g} \times t} \tag{9.5}$$

$$k = \frac{1}{g} = \frac{ln\,N_t - ln\,N_0}{t \times ln\,2} \tag{9.6}$$

9.2.5.2 Bacterial Growth Kinetics Based on Biomass

The rate of increase in bacterial biomass in a culture medium is correlated with specific growth rate (μ) and the biomass concentration (X) that can be expressed by the differential Eq. 9.7 and upon solving Eq. 9.8 through Eq. 9.10 can be obtained.

$$\frac{dX}{dt} = \mu \times X \tag{9.7}$$

$$X_t = X_0 \times e^{\mu t} \tag{9.8}$$

$$\mu = \frac{ln\,X_t - ln\,X_0}{t} \tag{9.9}$$

$$ln\,X_t = ln\,X_0 + \mu t \tag{9.10}$$

9.2.6 Bacterial Growth Kinetics in Batch Culture

Batch culture is a dynamic process in which the bacterial and nutrient concentration changes with time (Muloiwa et al., 2020). The bacterial growth curve in batch culture follows four distinct phases, such as lag phase, log phase, stationary phase, and death phase (Fig. 9.3). Generally, in the lag phase, the bacterial cell adapts to the new environment by increasing their cell mass only and there is no net increase in cell number in this phase. Moreover, the length of the lag phase depends on the bacterial species as well as the condition of the bacterial culture medium. However, when the culture medium contains more than one carbon source, and therefore multiple lag phases are observed among bacterial species; this type of phenomenon is known as diauxic growth. In diauxic bacterial growth, bacteria first consume the easily metabolized carbon source and thereafter when this carbon source is exhausted, the bacterial metabolic pathway shifts to utilize the other carbon source. In this regard, Bren et al. cultured *Escherichia coli* in a growth medium containing glucose and lactose as carbon sources. It was observed that the bacterial species *E. coli* first utilized glucose for their metabolism and thereafter shifted its metabolism to utilize lactose after completely exhausting glucose (Bren et al., 2016).

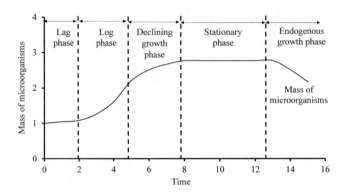

Fig. 9.3 Bacterial growth curve in batch culture

However, in case of continuously operated reactor the bacterial cells will get acclimatized to the environment over the time and then this lag phase will not exist.

The log phase (also known as the exponential growth phase) is the second phase of bacterial growth, where the bacterial growth rate is the highest. In this growth phase, bacterial cell mass as well as cell numbers increase exponentially with time and therefore this growth is known as the exponential growth phase. The bacterial growth rate in log growth phase follows a first-order rate, and the natural log of cell concentration vs. time gives a straight-line (Fig. 9.2). Since, the substrate availability will reduce after some duration in batch mode of operation, ample substrate will not be available for the bacteria to contine log growth phase and they will demonstrate decline growth phase. In the stationary phase, the growth rate of bacteria is equal to the death rate, i.e., the net growth rate is zero. Moreover, in this stationary phase, bacterial death is caused due to the depletion of an essential nutrient in the medium or due to the accumulation of toxic by-products in the growth medium. The growth in this phase can be expressed as $dX/dt = 0$. Though the bacterial net growth is zero in the stationary phase, the bacterial cells are still active and are prone to the production of secondary metabolites, like antibiotics and other value-added products.

The final phase of bacterial growth in batch culture is the death phase, in which the bacterial death rate becomes greater than the growth rate (Fig. 9.3). Bacterial death as per Eq. 9.11 is caused due to nutrient exhaustion or accumulation of toxic wastes. However, after the stationary phase, when nutrition is limited in the culture medium, bacteria oxidize their own cellular mass to survive in the environment, which is known as the endogenous growth phase or endogenous decay phase. In this phase, bacterial cells die and lyse to spill their contents into the medium environment making these as nutrients for survival of other bacteria. Therefore, the cell mass of bacteria is reduced via cellular death during the endogenous growth phase.

$$dX/dt = -K_d \cdot X \qquad (9.11)$$

where K_d is the specific death constant also known as endogenous decay coefficient.

9.2.7 Bacterial Growth Kinetics in Continuous Feed Culture

Continuous feed culture is the most applicable mode of operation for higher biomass productivity of bacteria. Generally, the continuous culture process is a steady-state process in which bacterial cells, substrate concentration and product concentration almost remain constant. In this process, the fresh nutrient medium is continuously supplied to the reactor as well as the products formed are also withdrawn continuously from the reactor. The continuous culture is of mainly two types, such as Chemostat and Turbidostat (Vandamme and Speybroeck, 1996). In chemostat, the rate of nutrient supply limits bacterial growth rate and the flow rate of culture medium. Also, the microbial culture volume is kept constant by the uninterrupted removal of culture at the same rate and the reactor volume is also kept constant in a chemostat. Therefore, the macromolecular composition of cell biomass and its functional characteristics can be maintained at a constant level. However, environmental conditions, like pH and temperature, can be varied, whilst maintaining a constant specific growth rate of microorganisms in a chemostat. On the other hand, in Turbidostat, the cell concentration in the reactor is kept constant by varying the flow rate.

Bacterial specific growth rate depends on the substrate concentration in the medium. Often a single substrate, i.e., carbon, plays a vital role in the regulation of growth rate and hence, this substrate is known as growth rate-limiting substrate. Moreover, at a condition of a single limiting substrate, the specific growth rate (μ) is generally a function of limiting substrate concentration (S), the maximum specific growth rate (μ_m), and a substrate-specific constant (K_s) (Okpokwasili & Nweke, 2006). This function can be expressed by Eq. 9.12.

$$\mu = \frac{\mu_m \times S}{K_s + S} \tag{9.12}$$

The bacterial growth rate in continuous culture depends on both the volumetric flow rate of the nutrient supply into the reactor and the dilution rate (D) as per Eq. 9.13.

$$D = \frac{F}{V} \tag{9.13}$$

where D is the dilution rate, F is flow rate of nutrient supply, and V is the reactor volume.

Therefore, the net change in cell concentration in the reactor over a time period may be expressed as per Eq. 9.14.

$$\frac{dx}{dt} = \text{(rate of growth in the reactor)} - \text{(wash-out due to dilution)}$$

$$\frac{dX}{dt} = \mu X - DX \tag{9.14}$$

Under the steady-state condition, rate of bacterial growth = rate of biomass loss, hence $\frac{dX}{dt} = 0$. Therefore, $\mu X - DX = 0$ or $\mu = D$.

Furthermore, at a higher dilution rate, the microbial biomass concentration reduces to zero, which is known as the cell washout condition. The washout of biomass occurs, when the rate of biomass removal in the chemostat outlet stream, termed as amputation rate, is greater compared to the rate of microbial cell growth. The dilution point at which this phenomenon occurs is known as critical dilution rate (D_c).

The D_c can be obtained by using Eq. 9.12 and using $\mu = D_c$ as expressed in Eq. 9.15.

$$D_c = \frac{\mu_m \times S_r}{K_s + S_r} \tag{9.15}$$

Here, S_r is the residual substrate concentration under the steady-state condition at a fixed dilution rate.

In continuous feed reactor 'food to microorganism' ratio (F/M) controls the rate of metabolism in the biological reactor. At low F/M ratio, food available is less, hence at very low F/M ratio the bacteria will exhibit endogenous growth of microorganisms (Fig. 9.4). For high F/M ratio, food available is abundant, hence the growth phase will be log growth phase. In between the growth rate will be declined growth phase. The biological reactors are typically operated at declining growth phase or endogenous growth phase with sufficient F/M ratio so that the mass of microorganisms is at least constant, and not depleting.

From operational point of view of the biological wastewater treatment system the settling capability of the sludge developed is also important. The sludge produced under log phase is having very poor settling characteristics due to ability of holding water surrounding the cells. Whereas, the sludge produced in the endogenous decay phase has better settling properties and settles well and it is more stable. Thus, the sludge with good settling characteristics will settle well in secondary sedimentation tank and will not pose any problem in sludge recycling. Whereas, the sludge developed under log growth phase will pose a problem of sludge bulking in the secondary sedimentation tank.

Fig. 9.4 Rate of metabolism in continuous feed reactors at different F/M ratio

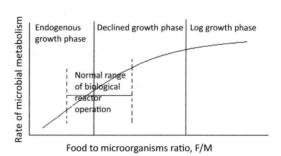

Example: 9.1 In a laboratory bacterial growth experiment, the culture was inoculated with 10^9 bacterial cells. The doubling time of bacterial cells is 3 h, what will be the bacterial cell number after 6, 24, and 36 h?

Solution

Given, $N_o = 10^9$; $t_d = 3$ h; N_t is to be determined at 6, 24, and 36 h.

Using Eq. 9.5, $N_t = N_0 \times 2^{\frac{t}{t_d}}$.

After 6 h,

$$N_6 = 10^9 \times 2^{\frac{6}{3}} = 4 \times 10^9$$

Similarly, after 24 h, $N_{24} = 10^9 \times 2^{\frac{24}{3}} = 2.56 \times 10^{11}$.

Thus after 36 h, $N_{36} = 10^9 \times 2^{\frac{36}{3}} = 4.096 \times 10^{12}$.

Hence, the bacterial cell number in that culture medium after 6, 24, and 36 h will be 4×10^9, 2.56×10^{11} and 4.096×10^{12}, respectively.

Example: 9.2 If a bacteria culture shows 40% growth in 8 h under exponential growth phase under optimum growth conditions, what will be the doubling time of the bacterial culture?

Solution

Given, 40% growth in t = 8 h; and t_d need to be determined.

40% growth $(X_t) = X_o + 40\%$ of $X_o = \frac{7X_o}{5}$.

Using Eq. 9.9, $\mu = \frac{\ln X_t - \ln X_o}{t}$.

or, $\mu = 0.042$ (approx.)

Therefore, $t_d = \frac{\ln 2}{0.042} = 16.5$ h.

Example: 9.3 The bacterial growth of a laboratory culture can be expressed by Monod's kinetics, which has maximum growth rate $(\mu_m) = 0.7$ h^{-1}, and substrate-specific constant $(K_s) = 3$ g/L. If the initial substrate concentration (S_o) in the media was 20 g/L, what will be the specific growth rate (μ) when 20%, 50%, and 80% of the substrate has been consumed?

Solution

Given, $\mu_m = 0.7$ h^{-1}; $K_s = 3$ g/L; and $S_o = 20$ g/L.

Using Eq. 9.12, $\mu = \frac{\mu_m \times S}{K_s + S}$.

Therefore, when 20% substrate is removed S = 20 − 20 × 0.2 = 16 g/L

$$\mu_{20} = \frac{0.7 \times 16}{3 + 16} = 0.59\,\text{h}^{-1}\ \text{(approx.)}$$

$$\mu_{50} = \frac{0.7 \times 10}{3 + 10} = 0.54\,\text{h}^{-1}\ \text{(approx.)}$$

$$\mu_{80} = \frac{0.7 \times 4}{3 + 4} = 0.4\,\mathrm{h}^{-1}$$

Hence, the specific growth rate (μ) when 20%, 50%, and 80% of the substrate has been consumed will be 0.59, 0.54, and 0.4 h^{-1}, respectively.

9.2.8 Principles of Biological Wastewater Treatment

Under proper environmental conditions, the soluble organic substances present in the wastewater are destroyed by biological oxidation. Part of this organic matter is oxidized, while rest is converted into biological mass in the biological reactors. The end products of the bacterial metabolisms are either gas or liquid and on the other hand, the synthesized biological mass can flocculate easily and it can be easily separated out in secondary clarifiers. Therefore, the biological treatment system usually consists of (a) a biological reactor, and (b) a sedimentation tank, to remove the produced biomass called as sludge.

9.3   Factors Affecting the Bacterial Growth

Different environmental and physiochemical parameters such as nutrients, temperature, pH, oxygen, and salinity have a significant role in the bacterial growth (Atolia et al., 2020). Temperature is an important parameter for bacterial growth, since different bacterial species grow in different temperature ranges. In this regard, Atolia et al. demonstrated that the optimal temperature range for most of the bacteria present in wastewater is 25–40 °C (Atolia et al., 2020). However, thermophilic bacteria like *Geobacillus stearothermophilus* grow best at a relatively high temperature ranging between 55 and 80 °C (Atolia et al., 2020).

Likewise, pH is another key factor for bacterial growth. Bacteria that grow at the natural pH of 7.0 are known as neutrophilic bacteria. The bacteria which grow best in an acidic medium are known as acidophilic (e.g., *Lactobacillus*). Those bacteria that grow in a basic pH medium are known as basophilic bacteria (e.g., *Vibrio cholerae*).

Bacterial species are very much sensitive to the presence and absence of oxygen. They have varying oxygen requirements for their growth. Based on oxygen requirement, bacteria can be classified as obligate aerobes, microaerophilic, facultative, aero-tolerant anaerobes, and obligate anaerobes. Obligate aerobic bacteria can grow only in the presence of oxygen (i.e., *Micrococcus luteus, Pseudomonas aeruginosa*). The bacterial species that can grow only in a very low oxygen concentration level are microaerophilic bacteria (e.g., *Campylobacter*). Facultative bacteria can grow in the presence and the absence of oxygen (e.g., *E. coli, Staphylococcus*). Some bacterial species can grow in the presence of oxygen; however, they do not use oxygen for their growth; they are tolerant to the presence of oxygen; these bacterial species are known as aero-tolerant anaerobes (e.g., *Lactobacillus*). The bacterial species that are very sensitive to oxygen and cannot grow in the presence of oxygen are known as obligate anaerobes (e.g., *Clostridium*).

Aerobic bacteria use oxygen as the final electron acceptor in their aerobic respiration process. In aerobic respiration, energy and carbon source is completely oxidized to CO_2, H_2O, and energy in the form of adenosine triphosphate (ATP). In aerobic respiration, conversion of adenosine diphosphate (ADP) to ATP takes place, which is known as oxidative phosphorylation. Anaerobic bacteria, instead of oxygen, utilize compounds like nitrates or sulphates as the final electron acceptor in their respiration process. Another anaerobic metabolism of bacteria is fermentation, in which complex organic compounds are broken down into simpler organic end products such as organic acids and alcohols, as well as gas (carbon dioxide, methane and hydrogen) by enzymatic action without the use of oxygen.

Other parameters that also affect bacterial growth are moisture, H^+ ion concentration, osmotic pressure, light, mechanical stress, etc. Water is the fundamental requirement for bacterial growth and reproduction. The water requirement varies with bacterial species. Bacteria grow best at ideal osmotic pressure, i.e., when the salt concertation of microbial cytoplasm is equal to the external environment, i.e., the osmotic pressure is optimum. The sudden transfer of bacterial culture to a hypertonic environment can cause plasmolysis (osmotic removal of water from cytoplasm leading to shrinkage of cell and death). On the other hand, if the salt concentration of the external environment is less than the cytoplasm, the osmotic gradient will result in the transfer of water into the cytoplasm, causing the microorganism to swell and burst. Moreover, some microbes are tolerant to high salt concentrations, they are known as halophiles. Some bacteria are phototropic and they require optimum light for their growth. However, most of the bacteria grow well in the absence of light. Moreover, radiation from UV ray can reduce bacterial growth. Hence, for optimum bacterial growth, maintaining an optimum physicochemical environmental condition is necessary (Stanaszek-Tomal, 2020).

9.4   Role of Enzymes in Biological Wastewater Treatment

Enzymes are versatile, high molecular weighted proteins (ranging from 10,000 to 2,000,000 Da) produced by living cells that act as efficient biocatalysts for cellular reactions (Brandelli et al., 2015). Generally, under ambient conditions, enzymes possess highly rated catalytic properties compared to other chemical catalysts (Yao et al., 2020). Enzymes degrade pollutants present in wastewater and produce biological macromolecules from complex molecules. However, the action of regulatory enzymes and their metabolic pathways are highly synchronized to yield a harmonious interplay among the many activities necessary to sustain bacterial life.

There are many advantages of bacterial enzymatic treatment over other techniques, such as floatation, coagulation, flocculation, precipitation, membrane filtration, advanced

Fig. 9.5 Primary structure of an enzyme

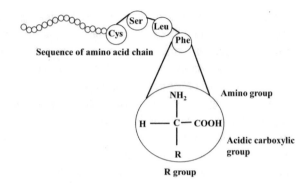

oxidation, etc., for the treatment of wastewater. The advantages are enzymatic processes require lower energy, have higher reaction kinetics, and incurs lesser operating expenses than other physicochemical treatments. Furthermore, enzymatic treatment is a cost-effective technique as the enzymes only catalyse the reaction without being consumed during the reactions and are consequently recycled after the end of the reactions (Karigar & Rao, 2011). Also, enzymes cannot compete with other microbial life forms, whereas microbes do this for their existence. Hence, the performance of enzymatic treatment is higher compared to other microbe-based biological treatments (Pandey et al., 2017).

Moreover, enzymes also work under mild reaction conditions, like ambient temperature and pH. Different micro and macro-organisms, like bacteria, fungi and plants, produce enzymes with a wide scope of uses across industries, like paper and pulp, food, textile, leather, pharmaceuticals, etc., and also remediate the pollutants emitted from these industries (Pandey et al., 2017). Their easily adaptable nature in different environments and ability to remove pollutants in mild reaction conditions encourage the researchers to use enzymes as a promising, sustainable, and renewable strategy for the treatment of wastewater.

9.4.1   Mechanism of Enzymatic Reactions

9.4.1.1   Structure of an Enzyme

Enzymes are basically proteins or glycoproteins and they consist of at least one polypeptide moiety. Enzymes are made up of different amino acids, which are associated with each other through amide or peptide bonds in a linear chain (Fig. 9.5). In an enzyme, the protein or glycoprotein part is termed as the apoenzyme; whereas, the non-protein part is called as the prosthetic group or cofactor. Also, the active form of the enzyme is called holoenzyme and it consists of apoenzyme combined with the prosthetic group. Moreover, enzymatic processes have strong reaction kinetics and specificity to bind different substrates.

Enzymes have active sites by which, they can be directly involved in catalytic reaction processes and may have other groups that are essential for their catalytic activity. During enzymatic processes, the specific active sites of enzymes have the capability to bind with their specific substrates that facilitate the conversion of the substrate into valuable products. In this catalytic process, enzymes provide a favourable condition that can reduce the activation energy of the reaction. Furthermore, enzymes could also save the time required for the transportation of substrates into cells, thus making the process more effective.

9.4.1.2   Mechanism of Enzymatic Action

The mechanism of enzymatic action depends on the nature of an enzyme and substrate interaction, where enzymes attach to the specific substrate to form an intermediate enzyme–substrate complex before the formation of the final product. This catalytic reaction occurs in the active or catalytic site, which is constituted by some amino acids, joined through several peptide chains. Moreover, these amino acids are folded with each other to form the structure of enzymes. Furthermore, in the catalytic site of the enzymes, the side chains of the amino acid residue provide enzymatic groups to bind with specific groups of the substrate. The binding induces a conformational reaction in the active site. Thus, the enzymes create a transition state complex during this catalytic reaction. However, the enzyme returns to its original state after disassociation of the products in the reaction. There are two different models proposed for the mechanism of enzyme action named as Fischer template model or lock and key model and induced fit model (Fig. 9.6).

Fischer template model: In 1894, a rigid model, the Fisher template model (formally known as lock and key model) was proposed by German chemist Emil Fischer (Lemieux & Spohr, 1994). This model elucidates the interaction among the substrate used in enzymatic reaction and an enzyme in terms of a lock and key analogy. According to the lock and key theory, both the enzyme and substrate have complementary geometric shapes and orientations with each other in which they fit easily into another as a 'key' fits into a 'lock' (Lemieux & Spohr, 1994). The active sites of the enzyme represent a 'lock' and have the ability to accept specific substrates, whereas the substrates are exemplified as 'keys'. According to the Fischer template model, the shape of both the enzymes and substrates do not affect each other, thus the substrate remains stabilized as both enzyme and substrate have already predetermined complementary geometric shape and orientation. However, the rigidity of the active or catalytic site of an enzyme is the only bottleneck of this model.

Induced fit model: To overcome the drawback of the 'lock and key' model, in 1963 scientist Koshland postulated an induced fit model for the flexibility of the catalytic site of an enzyme (Koshland Jr, 1995). According to this model, the substrate induces a conformational change in the catalytic site of the enzyme. Therefore, the substrate perfectly fits into the active site in the most convenient manner and catalyses the chemical reaction successfully. This model also proposes competitive inhibition, allosteric variation and denaturation of enzymes during inactivation (Csermely et al., 2010).

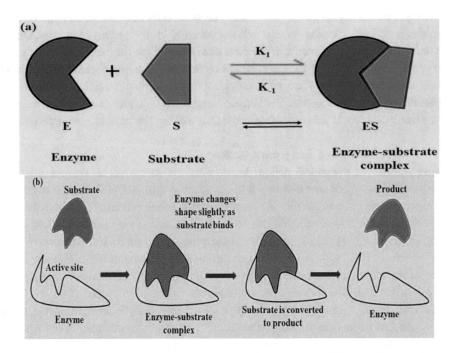

Fig. 9.6 The schematic presentation of **a** Fischer template model, and **b** Induced fit model

9.4.1.3 Mechanistic Models for Enzymatic Reaction and Kinetics

In 1902, V. C. R. Henri and in 1913, L. Michaelis and M. L. Menten developed a mathe-matical model for the single substrate-enzymatic reaction kinetics, which are often known as Michaelis–Menten kinetics or saturation kinetics. This enzymatic saturation kinetics is obtained through the reversible reaction between the enzyme (E) and substrate (S), which can be further dissociated to form an enzyme–substrate complex (ES) as listed in Eq. 9.16.

$$E + S \overset{K_1}{\rightleftharpoons} ES \rightarrow E + P$$

$$E + S \underset{K_{-1}}{\overset{K_2}{\rightleftharpoons}} ES \rightarrow E + P \qquad (9.16)$$

where K_1, K_{-1} and K_2 are specific reaction rate constants that express the relationship between the enzymatic reaction rates of the concentration of the reacting substances and P is the concentration of the product.

However, it is hypothesised that the ES complex is formed in the enzymatic reaction very quickly and the assumption of an irreversible second reaction often holds only when the product accumulation is negligible at the beginning of the reaction. Therefore, two major assumptions named as rapid-equilibrium and quasi-steady-state assumptions are established to develop the rate expression of the enzyme catalytic reaction.

According to quasi-state approximation, the rate of product formation (v) is illustrated in Eq. 9.17.

$$v = \frac{d[P]}{dt} = K_2[ES] \tag{9.17}$$

where v is the rate of product formation or substrate consumption in moles/L sec.
The rate of variation of the ES complex is given by Eq. 9.18.

$$\frac{d[ES]}{dt} = K_1[E][S] - K_{-1}[ES] - K_2[ES] \tag{9.18}$$

The enzyme itself is not consumed in this reaction, thus the conservation equation on the enzyme yield is demonstrated in Eq. 9.19.

$$[E] = [E0] - [ES] \tag{9.19}$$

In this regard, Henri, Michaelis and Menten established a rapid equilibrium assumption between the enzyme and substrate to form a $[ES]$ complex. Therefore, for the expression of $[ES]$ in terms of $[S]$, the equilibrium coefficient is required as shown in Eq. 9.20.

$$K'_m = \frac{K_{-1}}{K_1} = \frac{[E][S]}{[ES]} \quad \text{or, } [ES] = \frac{[E][S]}{[K'm]} \tag{9.20}$$

where K'_m is the equilibrium constant.
Substituting Eq. 9.19 in the Eq. 9.20, $[ES]$ is obtained as per Eq. 9.21.

$$[ES] = \frac{[E0][S] - [ES][S]}{[K'm]}$$

$$\text{or,} \quad [ES]\left[K'_m\right] + [ES][S] = [E0][S]$$

$$\text{or,} \quad [ES] = \frac{[E0][S]}{K'm + [S]} \tag{9.21}$$

Substituting Eq. 9.21 in Eq. 9.17,

$$\text{The rate of product formation } (v) = \frac{d[P]}{dt} = K_2 \frac{[E0][S]}{\left(\frac{K_{-1}}{K_1}\right) + [S]} = \frac{V_m[S]}{K'm + [S]} \tag{9.22}$$

where the dissociation constant of the ES complex (K'_m) is known as the Michaelis–Menten constant (mole m^{-3}), which reflects the affinity of an enzyme for its substrate and V_m (mol m^{-3} sec^{-1}) is the maximum forward velocity of the reaction. When K'_m is very low, the enzyme has a high affinity for the substrate. Moreover, K'_m is the concentration of substrate in which the reaction velocity of the enzyme is equivalent to half of the maximum velocity of the enzymatic reaction.

Example: 9.4 The kinetic reaction of a cellulase enzyme is demonstrated below:

$$E + S \underset{K_{-1}}{\overset{K_1}{\rightleftharpoons}} ES \overset{K_2}{\rightarrow} E + P$$

In this reaction, $K_1 = 10^7 \, M^{-1} \, s^{-1}$, $K_{-1} = 2.5 \times 10^3 \, s^{-1}$ and $K_2 = 10^2 \, s^{-1}$.

(a) Find out the value of the Michaelis Menten constant for cellulase enzyme. (b) When the enzyme concentration is 10^{-6} M, calculate the initial rate of product formation at a substrate concentration of 10^{-3} M.

Solution

Using Eq. 9.22, the rate of product formation $(v) = \frac{d[P]}{dt} = K_2 \frac{[E0][S]}{\left(\frac{K_{-1}}{K1}\right) + [S]}$.

Therefore, $K'_m = \frac{K_{-1}}{K_1} = \frac{(2.5 \times 10^3)}{10^7} = 2.5 \times 10^{-4} \, M$.

Rate of product formation (substrate consumption), $v = 10^2 \frac{10^{-6} \times 10^{-3}}{(2.5 \times 10^{-4}) + 10^{-3}}$.

$= 8 \times 10^{-5}$ M/sec.

Example: 9.5 The initial concentration of substrate [S] = 0.002 M and the value of $K'_m = 4.5 \times 10^{-5}$ M. Find out the ratio between the rate of product formation (v) and the maximum forwarded velocity of the reaction (V_{max}).

Solution

Using Eq. 9.22, $v = \frac{V_{max}[S]}{K'_m + [S]}$.

Therefore, $\frac{v}{V_{max}} = \frac{[S]}{K'_m + [S]} = 0.978$ (**Answer**).

9.4.2 Selection of Enzymes for Treatment of Wastewater

According to the International Union of Biochemists, enzymes are categorized into six functional classes based on their different kinds of catalytic reactions. The six classes of enzymes are hydrolases, oxidoreductases, transferases, isomerases, lyases and ligases. These types of enzymes have been used in biological wastewater treatment processes to remove persistent pollutants from wastewater (Boyer & Krebs, 1986). However, hydrolases and oxidoreductases are widely used enzymes in wastewater treatment due to their wide range of biocatalytic properties with most pollutants present in wastewaters (Boyer & Krebs, 1986). Moreover, laccase and peroxidase are mostly employed for eliminating organic micro contaminants from wastewater, due to their wide-ranging substrate specificities (Daâssi et al., 2016; Garg et al., 2020).

Generally, the efficiency and activity of enzymes in wastewater treatment depends on the type of enzyme and their respective reaction mechanisms. Also, specific enzymes could catalyse only specific types of pollutants present in wastewater due to their specificity with the substrate. As an example, lipases enzyme have the capability to degrade animal fat, oil and grease; whereas, protein component is decomposed by protease (Yao et al., 2020). On the other hand, carboxylesterases can also catalyse the phosphates, organophosphates and hydrolyse the ester bonds of carbamates, chlorinated compounds and other organic components (Cummins et al., 2007). Also, organophosphate compounds are hydrolysed via phosphotriesterases enzyme, whilst carbon and halogen bonds of halogenated aliphatic compounds are cleaved by haloalkane dehalogenases enzyme (Mishra et al., 2020; Romeh & Hendawi, 2014).

However, laccase and peroxidase enzymes have high specificity to select their substances, although having a wide range of reactions with the substrate (Catherine et al., 2016; Varga et al., 2019). Furthermore, researchers experimented that horseradish peroxidase and Pleurotusostreatus laccase enzyme could eliminate pharmaceuticals successfully from the wastewater. In this regard, Stadlmair et al. demonstrated that the horseradish peroxidase has the ability to transform diclofenac and sotalol completely after 4 h, and instantly it can converse acetaminophen, while the pleurotusostreatus laccase enzyme required lesser time (30 min) compared to horseradish peroxidase (Stadlmair et al., 2017).

In another investigation, Daâssi et al. (2016) examined different laccase enzymes collected from fungi for the biotransformation of Bisphenol A. Additionally, they also revealed that laccase collected from *Coriolopsis gallica* has a higher and speedier oxidation rate compared to other fungus laccases (Daâssi et al., 2016). Additionally, Meng et al. (2017) articulated that different types of lipase enzyme hydrolysed animal fat, vegetable oil, and floating grease present in food waste (Meng et al., 2017). Under the hydrolytic condition, (duration of 24 h and temperature of 40–50 °C) three types of lipases namely, Lipase-I, Lipase-II, and Lipase-III collected from *Aspergillus* sp., *Candida* and *Porcine pancreatic*, respectively, degraded animal fat, vegetable oil, and floating grease. However, Lipase-I and Lipase-II successfully hydrolysed the long-chain fatty acids with the hydrolysis efficiency of 74% and 94%, respectively, whilst comparatively low hydrolysis (41.7–46.7%) was perceived for Lipase-III enzyme (Meng et al., 2017). Therefore, different kinds of enzymes and their different sources have a significant role in the performance of wastewater treatment technologies.

9.4.3   Role of Microbial Enzymes for Wastewater Treatment

The presence of different inorganic and organic components, suspended and volatile compounds, toxic pharmaceuticals, xenobiotics, intractable compounds, and other contaminants in wastewater can pollute the aquatic environment. Furthermore, a higher concentration of chemical oxygen demand (COD) and biochemical oxygen demand

(BOD), which reduce the dissolved oxygen of water, should be removed substantially prior to the discharge of wastewater into natural water bodies (Joseph et al., 2020). For instance, Aisien et al. demonstrated that the mixture of carbohydrase, protease, and lipase enzyme removed 90% of COD and 90% of BOD from brewery wastewater (Aisien et al., 2009). With the enzymatic treatment duration of 96 h, the combined enzymes were found to reduce 50% and 100% of total suspended solids (TSS) and total hydrocarbon (THC), respectively. In addition to this, two major nutrients such as ammonium nitrogen and phosphate were also eliminated with a removal efficiency of 100% and 91.6%, respectively, from brewery wastewater (Aisien et al., 2009).

Sangave et al. developed a combined wastewater treatment technique including both enzymatic hydrolysis and aerobic oxidation process for the treatment of distillery effluent (Sangave & Pandit, 2006). Furthermore, the maximum COD removal efficiency of 54.3% was achieved at a pH of 4.8 from distillery wastewater after 72 h of aerobic oxidation (Sangave & Pandit, 2006). Another remarkable investigation on enzymatic wastewater treatment was the utilization of mixed enzymes. The protease, amylase and lipase enzyme extracted from organic kitchen wastes was effective for the removal of 93.5% of TSS, 90.8% of total dissolved solid (TDS), 94.3% of oil and grease, 75% of BOD, 97.4% of COD and 99.30% of chloride from domestic wastewater (Joseph et al., 2020).

The detoxification treatment of toxic organic and inorganic pollutants using different types of microbial enzymes are a notably eco-friendly and sustainable approach for the treatment of wastewater (Bollag et al., 1998; Gianfreda et al., 1999) (Table 9.1). Microorganisms extract energy from organic matter present in the wastewater through energy-yielding biological reactions facilitated by enzymes. In this oxidation–reduction reaction, the pollutants present in wastewater are oxidized by microbiota with the help of their cellular enzymes, thus producing less harmful compounds. Therefore, biological treatment through microbial enzyme is quite advantageous compared to other treatment technologies, like ASP, due to the action of enzymes on pollutants even when they are present in very dilute solutions (Bollag et al., 1998; Gianfreda et al., 1999). Thus, the enzymes extracted from different bacteria, fungi and higher plants have an important role in the degradation of pollutants present in the wastewaters.

In this regards, oxidoreductases enzyme can contribute to the humification of different phenolic compounds that are produced from the decomposition of lignin in soil or aqueous environment (Stevenson, 1994). Likewise, oxidoreductases can remove different toxic xenobiotic compounds, like phenolic and anilinic components, present in the wastewater via polymerization or copolymerization with humic substances (Park et al., 2006). Moreover, oxidoreductase enzymes have also been utilized for the decolourization and degradation of azo dyes from textile effluent (Williams, 1977). For instance, Yang et al. investigated that two yeasts, namely, *Candida tropicalis* and *Debaryomyces polymorphus*, and filamentous fungi *Penicillium geastrivorus* and *Umbelopsis isabellina* completely decolourized 100 mg/L of synthetic dye using manganese-dependent oxidoreductase enzyme within 16–48 h (Yang et al., 2003).

Table 9.1 Role of enzyme for the treatment of different types of wastewaters

Enzymatic source	Enzyme	Source of wastewater	Observation	References
Pleurotus ostreatus	Laccase	Olive oil mill wastewater	Removed 80% of phenolic compounds	Aggelis et al. (2003)
Gliocladium virens	Hemicellulase	Paper and pulp effluent	Decolorized 42% of pulp	Murugesan (2003)
Ipomoea batatas, Allium sativum, Sorghum bicolor and Raphanus sativus	Peroxidase	Leather industry	The maximum degradation efficiency of the phenolic compound was 93, 76, 77, and 72% by *A. sativus, S. bicolor, I. batatas*, and *sativum*, respectively	Diao et al. (2011)
Horseradish sp.	Horseradish peroxidase	Azo dye synthetic wastewater	Methyl orange decolorizing efficiency was 93.5% at pH 6.0	Bilal et al. (2018)
Porcine pancreatic	Lipase	Swine slaughterhouse	A maximum 94% of hydrolysis rate for floatable grease was achieved	Meng et al. (2017)
Industrial enzyme	Combination of lipase, protease and cellulase	Sewage sludge obtained from anaerobic digesters of municipal wastewater treatment plant	Reduction of 30–50% of total suspended solids and improved settling of solids from sewage sludge	Parmar et al. (2001)
Coriolopsis gallica., Bjerkandera adusta and Trametes versicolor	Laccase	Synthetic wastewater	Removed 91%, 85% and 72% of Bisphenol A using *C. gallica, B. adusta* and *T. versicolo*, respectively	Daâssi et al. (2016)
Trichophyton sp.	Laccase	Synthetic wastewater	Removed 90% of xenobiotic compound from synthetic wastewater	Jung et al. (2003)

(continued)

Table 9.1 (continued)

Enzymatic source	Enzyme	Source of wastewater	Observation	References
Polyporus pinsitus and *Myceliophthora thermophila*	Recombinant laccase	Synthetic wastewater	Used as a biosensor to determine phenolic compounds like 1-naphthol, o-phenylenediamine, guaiacol, o-anizidine, benzidine	Kulys and Vidziunaite (2003)
Industrial enzyme	Immobilized Papain	Synthetic wastewater	Maximum 96% of mercury was recovered at pH of 4.0	Bhattacharyya et al. (2010)
Immobilized on activated charcoal	Bromelain	Tannery effluent	Removed 98.3% of lead and 92.1% of chromium	Chatterjee et al. (2016)
Organic kitchen waste	Combination of protease, amylase, and lipase	Domestic wastewater	Removed 93.5% of TSS, 90.8% of TDS, 75% of BOD, 97.4% COD and 99.3% of chloride	Joseph et al. (2020)

Also, the oxidoreductase enzyme could remove chlorinated phenolic components, which are the most abundant recalcitrant waste materials generated due to the partial degradation of lignin found in the effluents of paper and pulp industry (Czaplicka, 2004). The white-rot fungi releases various extracellular oxidoreductase enzymes, like laccase and manganese peroxidase, from their mycelium into the environment and thus assist to degrade industrial wastewater containing different xenobiotic and phenolic compounds (Rubilar et al., 2008). In this regard, Sasaki et al. designed a pulp bio-bleaching system, where manganese peroxidase extracted from *Phanerochaete chrysosporium* bleached 88% of chlorinated phenol from the pulps (Sasaki et al., 2001).

Additionally, laccases are also a part of the oxidoreductase enzyme group that can catalyse the oxidation–reduction reaction of phenolic compounds and methoxy-phenolic acids with simultaneous reduction of sub-atomic oxygen to water. Microbiota like *Bacillus subtilis*, *Trametes versicolor* and *Pleurotus eryngii* produces intra and extracellular laccases, which can also catalyse the oxidation of different phenolic compounds, such as orthophenols, polyphenols, aminophenols, aromatic amines, aryl diamines, lignins, as well as inorganic ions like Cu, Mn and other toxic metal ions (Couto & Herrera, 2006). Generally, these toxic compounds end up in the wastewater from coal conversion, resins and plastics, wood safeguarding, oil refining, metal coating, dye and other synthetic substances, textiles, mining, pulp and paper manufacturing industries, and have an adverse effect on the aquatic ecosystem (Ullah et al., 2000). Therefore, the removal of these toxic components from water is of utmost necessity from an environmental point of view. In this regard, Camarero et al. utilized three different types of fungal laccases collected from *Trametes versicolor*, *Pycnoporus cinnabarinus*, and *Pleurotus eryngii* for the delignification of high-quality paper pulp and almost 90% of lignin was removed after a laccase-mediator treatment via H_2O_2 bleaching (Camarero et al., 2004).

Different hydrolytic enzymes such as amylase, protease, cellulase and lipase produced via hydrolytic microbes have the ability to degrade carbohydrates, fats, proteins and other cellular components present in wastewaters (Liew et al., 2020). Hydrolytic enzymes like cellulases can degrade cellulose microfibrils, generally produced at the time of cloth washing and the consumption of cotton-based constituents in water. Moreover, cellulase is also employed for the reduction of toxic dye during the recycling of paper from the paper and pulp industrial effluent. Another ubiquitous hydrolytic enzyme is lipase that catalyses the hydrolysis of triacylglycerols and produces glycerol and free-fatty acids.

In the process of sewage treatment, lipases are employed in the activated sludge process for the reduction of skimmed fats-rich liquid from the surface of aeration tanks to permit oxygen transport to maintain living conditions for the microbial biomass (Benjamin & Pandey, 1998). In this sense, lipases from *Candida rugose* was utilized for the digestion of lipid in lipid-rich wastewater (Benjamin & Pandey, 1998). Other lipase producing microorganisms like *Pseudomonas* sp., *Staphylococcus pasteurii*, *Bacillus* sp. and *Acinetobacter* sp. have also been reported to reduce hydrocarbon drastically from lipid-rich wastewaters (Behera et al., 2019; Sharma et al., 2011).

In addition to this, protease enzyme also have a significant role in the treatment of wastewater from beverages, leather, poultry, fishery, pharmaceutical and detergent industries (Beena & Geevarghese, 2010). In this regard, Singh reported that extracellular protease extracted from *Chrysosporium keratinophilum* hydrolysed up to 88% of buffalo skin from keratinic waste effluents (Singh, 2003). However, the extraction of enzymes for the treatment of wastewater is not only limited by using a single strain. Different extracellular enzymes extracted from mixed microbial consortium present in wastewater have a higher catalysing ability to degrade contaminants compared to other intracellular enzymes extracted from a single strain (Zumstein & Helbling, 2019). For instance, the crude exocellular enzyme extracted from two co-cultured yeast named *Yarrowia lipolytica* and *Candida rugosa* successfully eliminated triglycerides from palm oil mill effluent with a removal efficiency of 98.5% in 120 h (Theerachat et al., 2017). Therefore, a highly efficient enzymatic treatment is a well desired sustainable biorefinery approach for the remediation of recalcitrant pollutants from wastewater.

9.4.4 Technologies Using Enzyme for Enhancing Wastewater Treatment Efficiency

Removal of pollutants and other contaminants from wastewater through adsorption process using immobilized enzymes has a lot of advantages over free enzyme due to its recyclability, requirement of low-maintenance, easy metal and nutrient recovery, ecological and cost-effective approach. According to Bhattacharyya et al. immobilized papain enzyme on sodium alginate matrix could remove mercury (II) from aqueous solution (Bhattacharyya et al., 2014). The maximum mercury (II) removal efficiency of 98.88% was successfully obtained within 8 min at 35 °C and at a pH of 9.0 in a batch contactor. Moreover, Krajewska reported that the immobilized urease can remove different heavy metal ions, such as mercury (II), copper (II), silver (I), zinc (II) and cadmium (II), from wastewater (Krajewska, 1991).

Another noteworthy advanced technique is the utilization of genetically engineered microorganisms, whose genomic component is reformed through recombinant deoxyribonucleic acid technology to create a character-specific competent strain for the treatment of wastewater. The recombinant microorganisms can survive in extreme environmental conditions and ameliorate the degradation capability of a wide range of pollutants from wastewaters compared to non-modified strain. In this regard, Dixit et al. reviewed the removal of heavy metals using genetically modified microorganisms for the bioremediation of natural resources like the soil and aquatic environments (Dixit et al., 2015).

Researchers also reported that transgenic plants developed with specified genes from animals, microbiota and other plants could degrade different xenobiotic compounds. As an example, Abhilash et al. reported that Cytochrome P450 genes collected from human and mammalian rat inserted in *Nicotiana tabaccum, Solanum tuberosum, Oryza sativa* or *Arabidopsis thaliana* enhanced metabolization of xenobiotic compounds, like herbicides or volatile halogenated hydrocarbons from contaminated groundwater (Abhilash et al., 2009). Genetically modified microbes can also be utilized as "microbial biosensors" for the rapid and precise detection of the pollutants present in wastewaters. In this regard, recombinant *E. coli* and *Moreaxella* sp. bacteria secreted phytochelatin synthase enzyme on their cell surface and accumulated 25 times more Cd and Hg compared to the wild-type strains (Verma & Singh, 2005). However, appropriate field-scale investigations are required to be performed prior to the confirmation of the efficacy of these enzymes for the treatment of wastewater (Sayler & Ripp, 2000).

9.5 Wastewater Treatment Using Bacteria

Broadly the biological methods of wastewater treatment can be classified as aerobic, anaerobic, and anoxic (Sutton, 2006). The biological treatment method to be applied for wastewater treatment generally depends on different factors, such as nature of wastewater, concentration of organic matter present in the wastewater, energy demand required for the treatment, treatment time required, microbial concentration, and desired effluent quality. Aerobic treatment is suitable for wastewater with lower BOD and COD; however, anaerobic treatment is generally preferred, when organic matter concentration in wastewater is very high. In some cases, the combination of anoxic, aerobic, and anaerobic processes is used to reduce the sludge yield. The different types of biological reactors used in practice are summarized in Fig. 9.7.

Based on growth behaviour, a biological reactor can be designed as a suspended growth reactor and attached growth reactor. The reactor in which active biomass is sustained in the liquid bulk volume is known as a suspended growth process, whereas the reactor in which active biomass is grown onto or within a solid fixed media is known as an attached growth bioreactor. In recent years, hybrid bioreactors such as reactors involving the use of fixed or movable film media located in the suspended growth reactors to offer the combined advantages of suspended and attached growth reactors are becoming popular. However, the distinction between these biological reactors is always not clear.

9.5.1 Aerobic Treatment of Wastewater

The aerobic treatment of wastewater is a biological process in which aerobic microorganisms break down the organic and other pollutants, such as nitrogen and phosphorus, in the presence of oxygen. In this treatment process, oxygen is continuously mixed into the wastewater using mechanical aeration devices, such as air blower, paddlewheel aerator, or introducing compressed air through diffusers or using both mechanical as well as diffused air aeration. The organic matter oxidized by the microorganisms is often measured in terms of BOD, which is defined as the amount of dissolved oxygen required by the aerobic microorganisms to oxidize these organic matters. A high level of BOD indicates a higher concentration of biodegradable organic matter is present in the wastewater.

The aerobic process is suitable for low strength wastewater (COD < 1000 mg/L) (Chan et al., 2009); however, anaerobic treatments are preferred when the organic matter concentration in the wastewater is higher. Hence, aerobic treatment is generally provided after anaerobic treatment in case of treatment of high strength wastewaters. Moreover, depending upon the concentration of organic matter present in the wastewater, aerobic treatment can be used as stand-alone or as a polishing wastewater treatment after anaerobic processes to remove residual BOD from effluent of anaerobic reactor. The aerobic process produces an odour-free sludge that need to be stabilized further and dried to be used as a fertilizer. The most widely used aerobic processes for wastewater treatment are trickling filters, activated sludge process (ASP), and their modifications, which include sequencing batch reactor (SBR), moving bed biofilm reactor (MBBR) and rotating biological contactor (RBC).

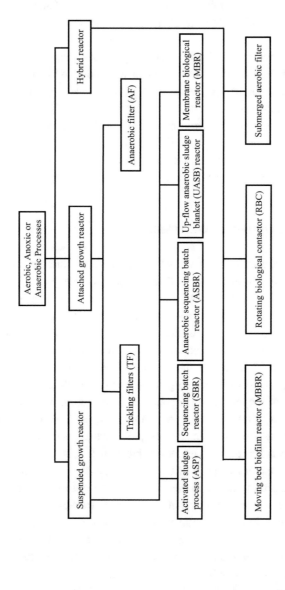

Fig. 9.7 Some of the widely used biological reactors

9.5.2   Anaerobic Treatment of Wastewater

Anaerobic treatment of wastewater is a biological process in which organic contaminants present in the wastewater are degraded with the help of a diverse group of microorganisms (bacteria and archaea) in the absence of oxygen. In anaerobic wastewater treatment, the microorganism converts the organic matter to biogas that majorly contains methane and carbon dioxide. Biogas produced in this process is used as the source of renewable energy. Anaerobic treatment is considered as a mature technology as it can be used to treat medium to high strength wastewaters, such as wastewater coming from a brewery, distillery, beverage industry, food industry, paper manufacturing industries, and petrochemical sources. Moreover, the anaerobic process produces very little excess sludge; hence, reduces the cost of sludge-handling apart from eliminating energy cost for aeration as required in aerobic processes.

Like aerobic microorganisms, anaerobic microorganisms also exhibit two types of growth, namely attached growth and suspended growth. Based on the growth of microbes, the reactor used in wastewater treatment can be classified as an attached growth reactor and a suspended growth reactor. The popularly used anaerobic reactors for wastewater treatment are an anaerobic filter, up-flow anaerobic sludge blanket (UASB) reactor, and anaerobic MBBR.

9.5.3   Anoxic Treatment

Anoxic is the state of the environment, which is free of molecular oxygen (O_2); however, bound oxygen such as nitrite (NO_2^-), nitrate (NO_3^-) are present. The electron acceptor molecules present in the environment plays a crucial role in energy generation occurring in a microbial cell. When different electron acceptors are present in the environment, microbes use that electron acceptor which is dominant, to produce the highest amount of energy. This is the reason why microbes first use O_2 instead of bound oxygen. When O_2 is exhausted in the environment, the system becomes non-aerobic and the microbes shift their respiration metabolic pathway for cellular energy production and uses this bound oxygen as electron acceptor, such as NO_2^- and NO_3^-. This shift in the respiration metabolism is known as anoxic respiration, and that environment is known as an anoxic environment. This anoxic wastewater treatment is popularly used for removal of nitrate from the aerobically treated wastewater in the denitrification reactor.

Under anoxic conditions, different nitrogen fixing anaerobes, namely *Pseudomonas aeruginosa* and *Clostridium* sp., purple non-sulphur phototrophic bacteria *Rhodopseudomonas* sp. and different denitrifying heterotrophic facultative bacteria, like *Serratia* sp., *Thiobacillus denitrificans*, *Candidatus* sp. and *Achromobacter* sp., can grow in reduced oxygen concentration and can remove nitrogen and ammonia through up-flow anammox sludge bed (UAnSB) reactor and other anaerobic up-flow reactors (Yorkor & Momoh,

2019). As an example, Reino et al. (2018) removed 82% of ammonium nitrogen from urban wastewater utilizing Candidatus Brocadia anammoxidans through UAnSB reactor with the initial nitrogen loading rate of 1.8 g N/L day. Therefore, biological secondary wastewater treatment process utilizing different aerobic, anaerobic, and facultative bacterial species have an incredible proficiency to treat wastewater in an economical way compared to the conventional aerobic wastewater treatment processes.

9.6   Role of Algae in Wastewater Remediation

9.6.1   Potential of Algae for Wastewater Treatment

Algae, a budding aspirant for renewable and sustainable energy sources have a significant role in wastewater treatment processes because of the ability to remove pollutants, nutrients (nitrogen and phosphorus), toxic metals, and pathogens from wastewater. Researchers employed photosynthetic algae for tertiary wastewater treatment due to its easy adoptable nature in the aquatic environment along with high biomass productivity compared to other photosynthetic microorganism (Das et al., 2020d; Whitton et al., 2015). Moreover, the algal biomass harvested from wastewater can be used as a feedstock for biofuel production (Acién Fernández et al., 2018). Therefore, algal-based treatment can be intended to be very economical and useful for the biological wastewater treatment. However, primary and secondary treatments are required to reduce the concentration of organic matter and other toxic pollutants, if present, in the wastewater (Fig. 9.8).

In algal-based tertiary treatment, the symbiosis of algae and bacteria ameliorates the treatment efficiency as microalgae possesses the ability to remove impurities via their

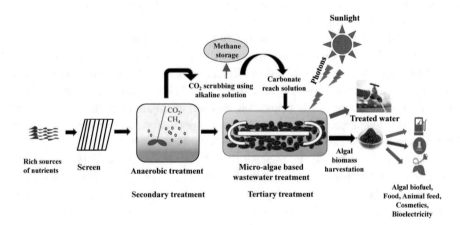

Fig. 9.8 Schematic representation of algae-based wastewater treatment along with valuable products recovery

cellular metabolism, which simultaneously provides oxygen during the photosynthesis to support the aerobic bacterial oxidation of organic compounds present in the wastewaters (Das et al., 2019b; Praveen et al., 2018). Thus, the algae-assisted tertiary treatment process provides an economical approach by reducing the energy consumption and carbon footprint of conventional wastewater treatment processes. Moreover, algal-based treatment is also able to eliminate almost 90% of the nitrogen and phosphorus present in the wastewater (Alcantaraa et al., 2019; Das et al., 2020e). In addition to this, harvested algal biomass from different pond based wastewater treatment plants could also be transformed into valuable carbon–neutral biofuels, such as bioethanol, biodiesel, and other valuable by-products, namely biogas, bio-fertilizers, cosmetics, medicines, and animal's feed (Das & Ghangrekar, 2018; Das et al., 2019a). Hence, the microalgae assisted wastewater treatment provides newer avenues for the biological wastewater treatment, which possesses significant advantages over conventional wastewater treatment processes.

9.6.2   Types of Algae Used in Wastewater Treatment

Algae, a unique and tiny (generally 1–50 μm in size) photosynthetic microorganisms, are a promising biofuel feedstock and an important member of the microbial community in aquatic ecosystems due to capability of pollutant degradation from wastewaters. During the photosynthetic process, microalgae utilize wastewater containing nutrients in presence of solar light and atmospheric CO_2 as an energy source and carbon source, respectively, for the synthesis of their cellular carbohydrate, protein and lipid (Barsanti & Gualtieri, 2014). Also, algae are known as a robust microorganism as they can survive in extreme aquatic environments, such as a wide range of saline environments, warm and cold weather, and in different illumination conditions, both in freshwater and wastewater environments.

Generally, microalgal species can be classified as eukaryotic and prokaryotic cyanobacteria (blue-green algae) based upon the utilization of energy and carbon sources for their intracellular metabolism. However, different types of unicellular green microalgal species, like *Chlorella* sp. and *Scenedesmus* sp., have been widely used in wastewater treatment due to their robustness, higher biomass growth rates and more nutrient uptake capabilities compared to other microalgal strains (Cai et al., 2013b).

Researchers also categorized the diverse group of algal species, like Chlorophyceae, Cyanophyceae, Chrysophyceae and Bacillariophyceae, which are formally known as 'Green algae', Blue-green algae', 'Golden algae' and 'Diatoms', respectively, that are majorly found in different wastewater treatment plants (Cai et al., 2013a). Mainly, the algal groups are classified based upon their different type of intra and extra-cellular structure, life span, their pigmentation and storage yields. However, tolerance of organic and inorganic pollutants, and removal of heavy metals and pathogens present in the wastewater depends on the type of algal species used. For instance, toxic metals like chromium can be successfully removed by *Oscillatoria* sp., whereas other heavy metals, like cadmium, copper and zinc, are removed by *Chlorella vulgaris* (Ting et al., 1989).

Furthermore, *Chlamydomonas* sp. and *Scenedesmus chlorelloides* have the capability to remove toxic lead and molybdenum from industrial wastewaters (Filip et al., 1979; Ting et al., 1991). Additionally, *Oscillatoria* sp., *Euglena* sp., *Chlamydomonas volvox*, *Chlorella* sp., *Navicula* sp., *Nitzschia* sp., and *Stigeoclonium* sp. have been designated as the most resistant genera for the removal of pollutants from rivers, lakes and other natural water reservoir (Bush et al., 1961; Chekroun et al., 2014). Many other *Chlorella* sp., namely *Chlorella sorokiniana*, *Chlorella kessleri*, *Chlorella reinhardtii* and other green algal species, like *Botryococcus* sp., *Phormidium* sp., *Chlamydomonas* sp., and *Scenedesmus* sp., have also been widely utilized for the removal of nutrients, heavy metals and other pollutants from wastewater (Cai et al., 2013b).

9.6.3   Favourable Growth Condition for Culturing of Algae

Different environmental and physiological parameters, such as illumination, temperature, alkalinity, carbon, nitrogen and other nutrient sources and rate of aeration can affect the growth of algae (Das et al., 2020d). Among the operational parameters, light intensity is the key parameter governing the growth of algae. During the photosynthesis process, the quality of light and photoperiods govern the algal macromolecular components, like carbohydrate, protein and lipid content, that enhances the yield of algal biomass (Grobbelaar, 2009). In this regard, Cheirsilp and Torpee reported that microalgae require both light and dark phases for their photosynthesis (Cheirsilp & Torpee, 2012). In the photochemical phase, algae generate energy-producing compounds, like adenosine triphosphate (ATP) and nicotinamide adenine dinucleotide phosphate-oxidase, in the presence of sunlight; whilst the dark phase is required for the biochemical synthesis of carbohydrate, protein and lipid components (Cheirsilp & Torpee, 2012).

Furthermore, Chisti et al. revealed that in the nocturnal period, algae can consume 25% of their biomass to sustain the activity of cells until sunrise (Chisti, 2007). However, biomass growth can be inhibited due to excess illumination, which could also enhance the temperature of the culture media. Moreover, the light availability could also be reduced by the algae themselves in case of dense cultures due to their shading effect. Therefore, biomass productivity is inversely correlated to the depth of open algal ponds or raceway ponds. Generally, a very shallow depth ranging from 0.5 to 1.2 m for the algal pond is used for wastewater treatment, so that the solar radiation can penetrate the entire depth of the pond, thus facilitating algal photosynthesis.

Temperature is also one of the crucial factors affecting the cultivation of algae in wastewater. Generally, algae can produce higher biomass at an optimum temperature and if the temperature is increased after this optimal temperature, their growth is ceased (Ras et al., 2013). In this regard, Laura and Paolo reported that usually cultured microalgal

species can generate their maximum biomass at the temperature range of 16–27 °C
(Barsanti & Gualtieri, 2014). However, comparatively a higher temperature range of
38–42 °C is required for the cultivation of *Chlorella* sp. to obtain the highest biomass productivity (Kessler, 1985). Moreover, researchers have also found that the overall biomass
productivity is adversely affected beyond the temperature range of 20–45 °C (Butterwick et al., 2005). Therefore, beyond the threshold limit of the optimum temperature, the
growth rate of algal biomass decreases linearly.

Another significant factor affecting algal growth is alkalinity, which also affects the
cultivation of microalgae. Generally, the culture media of microalgae is more alkaline
in nature and has a higher pH value due to the photosynthetic CO_2 assimilation. Also,
microalgae can enhance the pH value of culture media by up taking nitrogen as nitrate
ions and reduce ammonium ions by forming OH^- ions (Liang et al., 2013). However, the
most commonly cultured microalgal species in wastewater have shown their maximum
growth at the pH range in between 7.0 and 9.0 (Hodaifa et al., 2009).

Algal growth is also influenced by saline conditions. According to Laura and Paolo,
the optimum biomass growing conditions for the most of the algal species is at a salinity
level of 20–24 g/L (Barsanti & Gualtieri, 2014). However, researchers have investigated
that the lipid component of microalgae can be increased by increasing salinity, whereas
it can reduce the productivity of biomass (Asulabh et al., 2012).

Dissolved oxygen (DO) concentration is another crucial parameter that also governs
microalgal growth. According to Ohnishi et al. the growth of the microalgae is adversely
affected during the photorespiration and photo-inhibition process via the generation of
reactive oxygen species and biomass productivity declines linearly with an increase in DO
(Ohnishi et al., 2005; Raso et al., 2012). Therefore, beyond the optimum concentration of
oxygen, microalgal biomass growth can be inhibited (Lee, 2001). Also, the growth rate
of the microalgae depends on the availability of different nutrients like carbon, nitrogen,
and phosphorus. According to Chisti et al. the stoichiometric formula of an algal cell
is $CO_{0.48}H_{1.83}N_{0.11}P_{0.01}$, which can support the fact that nutritional requirement is the
foremost factor affecting the growth of microalgae (Chisti, 2007).

It was also investigated that the rate at which algae consume specific nutrients depends
on the diffusion rates through their cell wall and the concentration of their intra and
extra-cellular biomass. Furthermore, Redfield demonstrated that the optimum molar ratio
of C:N:P for algal cellular metabolism is 106:16:1, that is generally required for the
growth of algal biomass (Redfield, 1958). However, a higher concentration of carbon and
nitrogen can adversely affect algal biomass productivity. Usually, the ratio of the concentration of carbon and nitrogen existing in sewage ranges between 5 and 10. Hence, this
investigation proved that algae can also grow well using a sufficient amount of nutrients
present in sewage instead of freshwater (Park & Craggs, 2010). Additionally, other important growth parameters such as proper mixing of biomass, dilution rate of the biomass
concentration, depth of the open pond or closed photobioreactor, photoperiod, harvesting

frequency for the fatty acid production also has a significant impact in culturing microalgae at a lab-scale as well as in the field-scale. Therefore, suitable culture conditions should be maintained during the cultivation of microalgae for an enhanced biomass production during wastewater treatment.

9.6.4   Algal Cultivation Techniques Employed in Wastewater Treatment

According to Chen et al., the growth of algae depends on culture conditions and cultivation systems (Chen et al., 1996). Generally, algal culture conditions are classified into four major categories, such as autotrophic, mixotrophic, heterotrophic and photoheterotrophic cultivation (Chen et al., 1996). When algae utilize atmospheric CO_2 as an inorganic carbon source in the presence of sunlight during the photosynthetic process, this condition is called autotrophic condition (Huang et al., 2010). Also, an autotrophic condition diminishes the operating cost of the cultivation process due to the utilization of atmospheric CO_2; thus, autotrophic culture can be easily scaled-up for the holistic treatment of wastewater (Mata et al., 2010).

Under heterotrophic culture conditions, microalgae consume only organic carbon like glucose, fructose, sucrose, lactose, acetic acid, galactose and glycerol as both carbon and energy source in the absence of sunlight for their cellular metabolism (Liang et al., 2009). However, in the absence of light, algae synthesize more lipid compared to other culture conditions. In this regard, Xu et al. cultured green microalgae *Chlorella protothecoides* in heterotrophic condition for improving the lipid content (Xu et al., 2006). The productivity of lipid was improved by 40% compared to the autotrophic condition (Xu et al., 2006). Although, the biomass yield and chlorophyll content of algae is reduced as photosynthesis is restricted in the heterotrophic culture, which can be enhanced under photoheterotrophic culture condition (Brennan & Owende, 2010).

In the photoheterotrophic culture condition, microalgae utilize sunlight as an energy source and organic carbon as the carbon source. For instance, Chen et al. elucidated that under photoheterotrophic condition, *Chlorella protothecoides* utilized glucose, acetic acid, and glycerol as carbon sources and improved their biomass and lipid productivity compared to other cultivation conditions (Chen & Walker, 2011). In fed-batch mode, a maximum of 45.2 g/L of biomass and 24.6 g/L of lipid content was obtained after 8.2 days of cultivation (Chen & Walker, 2011). However, the utilization of organic carbon during algal cultivation makes this process uneconomical, which is a major roadblock that can be solved by using lignocellulose-based sugars (Zheng et al., 2012). Moreover, the lignocellulose-based heterotrophic culture tends to be contaminated easily by other microbes present in wastewater (Chen et al., 1996).

Another significant culture condition is the mixotrophic condition, where algae consume both inorganic and organic carbon as the carbon source for their cellular metabolism during the process of photosynthesis. As an example, Yeh and Chang reported that robust microalga *C. vulgaris* ESP-31 survive in both phototrophic, photoheterotrophic and mixotrophic conditions (Yeh & Chang, 2012). However, *C. vulgaris* ESP-31 have shown their highest lipid productivity under mixotrophic growth conditions. Under mixotrophic growth conditions, the maximum lipid productivity of 144 mg/L day was achieved, which was almost two fold higher as compared to autotrophic growth conditions (67 mg/L day) (Yeh & Chang, 2012). Therefore, it was proved that robust algal species could survive under different culture conditions that could abet in the scaling-up of algal wastewater treatment processes.

9.6.5 Algal Cell Growth Kinetics

The specific growth rate for the exponential growth phase (μ, per day) of algal biomass is determined using Eq. 9.23.

$$\mu = ln\left(W_f/W_0\right)/\Delta t \tag{9.23}$$

where W_f and W_0 are the biomass concentration at the end and in the beginning of a batch run, respectively, and Δt is the cultivation time in days between the two measurements in the exponential growth phase.

Doubling time for exponential growth phase is determined following Eq. 9.24.

$$t_d = ln2/\mu \tag{9.24}$$

where t_d is the doubling time (day).

Also, the overall biomass productivity $P_{overall}$ (g/L day) is calculated from the variation in biomass concentration (g/L) for a cultivation time (in days) according to Eq. 9.25.

$$P_{overall} = \Delta X/\Delta t \tag{9.25}$$

where ΔX and Δt are the difference between initial and final biomass concentration (g/L), and cultivation time (day), respectively (Jiang et al., 2013).

Example: 9.6 Under autotrophic conditions, green algae *Chlorella sorokiniana* was cultured in a 500 L bubble column photobioreactor with the initial concentration of 30 mg/L for the treatment of municipal wastewater. After seven days of cultivation, the final concentration of algal biomass was 2.2 g/L. Find out the specific growth rate and doubling time of *C. sorokiniana*. Also, calculate the biomass productivity of algae?

Solution

Using Eq. 9.23, the specific growth rate $(\mu) = \ln(W_f/W_0)/\Delta t$.

$= \ln (2.2/0.03)/(7 - 0) = 0.61$ g/L day.

Using Eq. 9.24, the doubling time $(t_d) = \frac{\ln 2}{\mu} = \frac{\ln 2}{0.61} = 1.13$ day.

Using Eq. 9.25, the biomass productivity $P_{overall} = \frac{\Delta X}{\Delta t} = \frac{2.2-0.03}{7-0} = 0.31$ g/L day.

9.7 Application of Algae in Wastewater Treatment

Over the past decade, researchers utilized microalgae for the treatment of wastewater due to its eco-friendly, economical, and less energy demanding nature, which renders sustainability to this process. Moreover, harvested algal biomass can be commercialized as feedstock for the production of biofuel and other valuable products recovery (Oswald et al., 1957). Therefore, utilization of algae in wastewater treatment offers a pioneering approach for the holistic treatment of wastewater with concomitant resource recovery from wastewaters, which can be deliberated as one of the successful models of circular economy (Fig. 9.8).

9.7.1 Removal of Nitrogen and Phosphorus

After primary treatment and secondary treatment via anaerobic or aerobic microbial degradation technologies, residual organic matter and inorganic components are still persistent in the secondary treated effluent (Barsanti & Gualtieri, 2014). Therefore, tertiary treatment is required for the further removal of organic compounds and nutrients, especially nitrogen and phosphorus (Table 9.2). Generally, nitrogen, phosphorus and other organic nutrients are released into the wastewater because of microbial synthesis of proteins, nucleic acid and other anthropogenic activities (Cai et al., 2013b). In the waste stream, nitrogen exists in the form of different nitrogenous ions, such as ammonium (NH_4^+), nitrite (NO_2^-), nitrate (NO_3^-) and the accumulation of these nitrogenous compounds at a higher concentration is also a serious threat for aquatic microorganism as well as human (Barsanti & Gualtieri, 2014).

Various forms of phosphorus components, namely polyphosphate, orthophosphate, and organophosphate, are also present in the wastewaters and the higher concentration of these compounds can also disrupt the food chain of aquatic animals. Furthermore, the unionized nitrite, nitrate, phosphate, and other ammonium and phosphate salts can lead to algal blooms, formally known as eutrophication in water bodies, which also adversely affects aquatic organisms (Conley et al., 2009). However, recovery of these nutrients is important for reutilizing them as well as to impart sustainability to the treatment process. Therefore, recovery and removal of nitrogen and phosphorus at an acceptable level are required prior to discharge of wastewater into the natural waterbodies.

Table 9.2   Application of different kinds of algal species in wastewater treatment

Algal species	Type of wastewater	Treatment efficiency	References
Scenedesmus obliquus and *C. vulgaris*	Secondary effluent of a domestic wastewater treatment plant	*S. obliquus* removed 55.2–83.3% of TP and 96.6–100% of TN, whereas *C. vulgaris* removed 53.3–80.3% of TP and 60.1–80.0% of TN	Ruiz-Marin et al. (2010)
Cyanophyceae sp.	Secondary effluent of a municipal wastewater treatment plant	98% of COD removal along with 54.5–72.6% of TP and 76.6–97.8% of TN were removed	Su et al. (2011)
Chlorella sp.	Effluent of a municipal wastewater treatment plant	Removed 90.8% of COD with 80.9% of TP and 89.1% of TN	Li et al. (2011)
A mixed consortium of algae	Effluent of a domestic wastewater treatment plant	Removed 99% of total coliforms via HRAP with 22 g/m^2 day of biomass production	Shelef et al. (1984)
C. proboscideum	Synthetic wastewater	Removed almost 100% of lead at 23 °C after 20 h and accumulated 60% of cadmium metal within 24 h	Soeder et al. (1978)
Immobilized *C. vulgaris*	Synthetic wastewater	Algal beads removed more than 90% of copper and almost 70% of nickel within 1 h	Mehta and Gaur (2001)
C. reinhardtii	Synthetic wastewater	Removed almost 100% of Trichlorfon at a maximum concentration of 100 mg/L of Trichlorfon	Wan et al. (2020)
Mixed culture of *Pseudokirchneriella subcapitata* and *Daphnia magna*	Synthetic wastewater	Removed 85% of the XN compound phenanthrene with the presence of fullerene nanoparticles	Baun et al. (2008)

COD: Chemical oxygen demand; TP: Total phosphorus; TN: Total nitrogen; HRAP: High-rate algal pond; XN: Xenobiotic

Generally, microalgae uptake nutrients, nitrogenous and phosphorus compounds, for their metabolism and store them in their biomass as a polyphosphate for their further necessities (Cai et al., 2013b; Powell et al., 2009). Therefore, algae can successfully remove nutrients from wastewater via consumption into their cell (Rasoul-Amini et al., 2014). In this regard, Zhai et al. cultured cyanobacterium *Arthrospira platensis* in synthetic municipal wastewater under autotrophic conditions for the removal of nitrogen and phosphorus (Zhai et al., 2017). Maximum nitrogen removal efficiency of 81.51% and phosphorus removal efficiency of 80.52% was achieved from synthetic wastewater with 262.5 mg/L of biomass concentration (Zhai et al., 2017). Wastewater containing phosphate and orthophosphate usually originate from domestic and agronomic wastes, and these compounds are successfully removed during algae-based tertiary treatment (Chatterjee & Ghangrekar, 2017). As an example, Chatterjee and Ghangrekar (2017) removed 85% and 91% of ammonium nitrogen and phosphate through high rate algal pond (HRAP) from the effluent of UASB reactor treating sewage with initial ammonia nitrogen and phosphate concentrations of 20 mg/L and 4 mg/L, respectively. Hence, the removal of an excess amount of nutrients from wastewater during algae-based treatment is a new avenue in biological wastewater treatment.

9.7.2   Removal of Chemical and Biochemical Oxygen Demand

Different inorganic and organic components, toxic surfactants, xenobiotics and recalcitrant compounds, pathogens and other detrimental contaminants present in the wastewaters can diminish the purity of water, which severely affects aquatic life and human health. Wastewater containing oxidizable organic matter can possess a huge amount of BOD and COD that can reduce the DO of water, thus directly affecting aquatic flora and fauna (Barsanti & Gualtieri, 2014). Therefore, the removal of this oxygen demand from waste streams is of utmost necessity prior to its disposal in natural water sources. Su et al. demonstrated that mixed consortia of filamentous cyanobacteria removed 98.2% and 84% of COD and total organic carbon, respectively, from the effluent collected from domestic wastewater treatment plant (Su et al., 2011). Thus, algae-based tertiary treatment can successfully reduce BOD and COD from wastewater in an economical way compared to other biological treatment processes.

9.7.3   Reduction of Coliforms and Other Pathogens

According to microbiologist Horan, a coliform population of approximately about 10^6 cells per 100 mL exist in sewage, which is not eradicated properly during primary and secondary wastewater treatment (Horan, 1989). Consequently, waterborne disease-causing pathogens like *Salmonella typhi*, *Shigella*, protozoa and other human enteric viruses, such as norovirus, enterovirus, and aichivirus are the main worries, which persist in the secondary treated sewage. Therefore, after primary and secondary treatment, harmful pathogens should be removed effectively prior to the reutilization of treated wastewater in different applications, such as landscaping, gardening, and unrestricted irrigation. Elimination of coliforms from municipal sewage is a major aspect to maintain public health issues (El-Kowrany et al., 2016).

Shelef et al. reported that microalgae removed 99% of total coliforms successfully from municipal wastewater via HRAP with 22 g/m^2 day of biomass production (Shelef et al., 1984). Furthermore, Chatterjee et al. eliminated a maximum of 3 log-scale of coliforms from the effluent of an UASB reactor using HRAP. After HRAP treatment, most probable number of coliform bacteria ranged between 600 and 800 per 100 mL; however, this range is acceptable for the reuse of the treated water for surface irrigation (Chatterjee & Ghangrekar, 2017). Likewise, Colak and Kaya revealed that microalgae have the ability to inhibit the growth of pathogens and coliforms like *Salmonella* via antibiotic production (Colak & Kaya, 1988).

For the self-sufficient biological tertiary treatment, waste stabilization pond is also a significant treatment unit, where a symbiotic relationship occurs between algae and other microbial communities to improve the removal efficiency of organic and inorganic load and other contaminants (Hosetti & Frost, 1995). Generally, waste stabilization ponds are operated for the treatment of high strength wastewater with limited hydraulic retention time using the cultivation of microalgae. Thus, a high rate sewage stabilization pond can be intended as an efficient biological wastewater treatment system for the reduction of coliforms compared to the other sewage treatment processes (Shelef et al., 1984).

9.7.4 Biosorption of Heavy Metals

Rapid industrialization and anthropogenic activities have led to the discharge of heavy metals through wastewaters. Usually, the effluent of different metal processing activities, such as mining, electroplating, tanneries, and some chemical industries, are sources of heavy metals in wastewater. These heavy metals are not easily biodegradable, thus they still persist after primary and secondary treatment in the water body and affect the aquatic environment adversely (Agarwal, 2005). Moreover, carcinogenic metals, like cadmium, mercury, and lead, present in wastewater disrupts the aquatic food chain, which also directly affects the human being by causing serious health hazards (Gupta et al., 2006). Therefore, removal of these toxic metals and recovery of heavy metals from wastewater is utmost necessary for our planet and utilization of microorganism is an economical and green approach for the biosorption of heavy metals (Azimi et al., 2017; Modestra et al., 2017; Vareda et al., 2019).

Researchers experimented that microalgae have the potential to uptake heavy metals, such as cadmium, cobalt, copper, molybdenum, zinc, chromium, selenium, and lead, from wastewater by adsorbing it or binding it to their cell surface extracellularly (Gupta et al., 2006). In this regard, Soeder et al. reported that *Coelastrum proboscideum* is a competent algal species for the accumulation of lead and cadmium. After 20 h of cultivation, *C. proboscideum* can remove almost 100% of lead from 1.0 mg/L of lead present in synthetic wastewater at 23 °C. Simultaneously, within 24 h, *C. proboscideum* can also accumulate 60% of cadmium metal from 40 mg/L of lead mixed with synthetic wastewater (Soeder et al., 1978). Similarly, few other tertiary treatments for the removal of other contaminants and hazardous materials along with their corresponding removal efficiencies are mentioned in Table 9.2.

9.7.5   Removal of Pesticides and Other Xenobiotic Compounds

Pesticides can be discharged into natural water bodies due to agricultural activities through the transportation of surface runoff or soil erosion. Water soluble insecticides and other contaminants due to anthropogenic activities have been often noticed in domestic wastewater, which causes serious health issues against aquatic organisms and mammals. Consequently, harmful pesticides and xenobiotic compounds should be removed effectively from waste streams. For instance, Wan et al. reported that *Chlamydomonas reinhardtii* could be tolerant against the pesticide Trichlorfon (TCF). The microalgae C. *reinhardtii* effectively degraded almost 100% of TCF from the synthetic wastewater at a maximum concentration of 100 mg/L of TCF (Wan et al., 2020). In another investigation, Kurade et al. removed Diazinon, one of the most used organophosphorus pesticides in farming. A maximum of 94% of insecticide Diazinon was removed from synthetic wastewater at 20 mg/L concentration of Diazinon (Kurade et al., 2016). Therefore, tertiary treatment employing algae has emerged as an economical wastewater treatment process compared to the conventional wastewater treatment processes (De Philippis et al., 2011).

9.8   Wastewater Treatment Using Other Microorganisms

Researchers have found that different kinds of fungi, mould, protozoa, prokaryote and other microorganisms also have a significant impact on wastewater treatment performance (Pandey et al., 2017). Different fungal strains, like *Aspergillus* sp., *Rhizopus* sp., *Cephalosporium eichhorniae Trametes versicolor*, and *Fusarium solani*, can degrade the oxidizable soluble pollutants to a reduced insoluble form, which simultaneously produces energy using intracellular and extracellular enzymes (Pandey et al., 2017). On the other hand, plants like *Prosopis juliflora, Ipomoea batatas*, roots of horseradish and radish also have an important role in wastewater treatment (Ely et al., 2017; Torres et al., 2016). For an instance, Garg et al. demonstrated that peroxidase enzymes collected from mesquite plants have the ability to eliminate almost 90% of phenol from wastewater within 30 min, whereas peroxidases derived from soyabean, potato and horseradish required higher residence time for a similar range of phenol reduction from wastewater (Ahirwar et al., 2017; Al-Ansari et al., 2010; Garg et al., 2020).

In the removal of radioactive and heavy metals, microbiota gains electrons from organic substrate present in wastewater with the utilization of radioactive metals as the final electron acceptor (Leung, 2004). Baldrian demonstrated that white-rot fungi absorb heavy metals from wastewater for their cellular metabolism and has the capability to accumulate those heavy metals in their mycelia (Baldrian, 2003). Also, other wood decaying fungal species and mushroom species *Deadalea quercina, Ganoderma applanatum, Schizophyllum commune* and *Stereum hirsutum* have the ability to absorb and accumulate heavy metals from wastewater (Baldrian, 2003).

In the treatment of methyl orange azo dye, Chiong et al. used 0.5 mL crude soyabean peroxidase enzyme for the degradation and elimination of recalcitrant dyes from textile effluents (Chiong et al., 2016). The maximum methyl orange dye decolourization efficiency of 81.4% was achieved from synthetic wastewater with the initial concentration of 30 mg/L using 0.5 mL crude soyabean peroxidase enzyme at 30 °C temperature for 1 h of incubation using 2 mM of H_2O_2 at pH of 5.0 (Chiong et al., 2016). Furthermore, the same group also found that the peroxidase enzyme extracted from *Luffa acutangula* removed methyl orange dye with the maximum dye removal efficiency of 75.3% with the initial concentration of 10 mg/L under 40 min of incubation at 40 °C with 2 mM of H_2O_2 at a pH of 3.0 (Chiong et al., 2016).

Theerachat et al. cultured *Candida rugosa* for the removal of triglycerides from palm oil mill effluent (Theerachat et al., 2017). In another investigation, Das et al. reported that chlorpyrifos was degraded using laccase enzyme extracted from *Trametes versicolor* effectively (Das et al., 2017). However, the degradation of pollutants for the treatment of wastewater is not only limited by using a single strain. Extracellular enzymes extracted from mixed microbial consortium present in wastewater have a higher catalysing ability to degrade contaminants compared to other intracellular enzymes extracted from a single strain (Zumstein & Helbling, 2019). For instance, the crude exocellular enzyme extracted from two co-cultured yeast named as *Yarrowia lipolytica* and *Candida rugosa* successfully eliminated triglycerides from palm oil mill effluent with a removal efficiency of 98.5% in 120 h (Theerachat et al., 2017). Thus, biological wastewater treatment via bacteria, fungi, yeasts, plants, and other microorganisms successfully removes organic and inorganic pollutants.

Questions

9.1. What is microbial generation time?
9.2. What is exponential growth? Explain in detail.
9.3. What is a bacterial growth curve? What are the growth phases in the bacterial growth curve in batch culture?

9.4. How batch and continuous cultures are different? Explain in detail.

9.5. Define and differentiate Chemostat and Turbidostat.

9.6. What are the bacterial growth parameters?

9.7. Describe the factors affecting the bacterial growth.

9.8. Define aerobic, anaerobic, and anoxic processes of wastewater treatment.

9.9. What are the anaerobic and aerobic wastewater treatment processes?

9.10. What is the relation between anabolism and catabolism?

9.11. Explain the metabolic pathways of anaerobic digestion of biomass. Explain hydrolysis, acidogenesis, acetogenesis, and methanogenesis.

9.12. In a laboratory culture the bacterial growth can be expressed by Monod's kinetics, which is growing at maximum specific growth rate with the following data:

Bacterial biomass (mg/L)	Time (day)
25	0
41	1
70	2
111	3
186	4
305	5

What will be the μ_{max}, and doubling time? [**Ans.** $\mu_{max} = 0.5$ d^{-1}; $t_d = 1.37$ day].

9.13. The initial bacterial cell number of a culture was 10^5, the cell population after 5 h was 10^8. What will be the doubling time of the bacterial culture? [**Ans.** 0.5 h].

9.14. If the generation time of a bacterial culture is 1.75 h, calculate the time required to increase the cell number from $10^{0.5}$ to 10^5. [**Ans.** 26.16 h].

9.15. If the μ_{max} of a bacterial growth culture is 0.49 h^{-1}, calculate the time required to increase the cell number from 3×10^7 to 10^{12}. [**Ans.** 56.39 h].

9.16. The initial bacterial cell number of a culture was 10^5, the cell population after 4 h was 10^{13}, at what time the bacterial population will be 10^{19}? [**Ans.** 7 h].

9.17. What is an enzyme? Explain the classification of different types of enzymes.

9.18. How do enzymes work for the degradation of the target pollutant?

9.19. What are the characteristics and drawbacks of the lock and key model?

9.20. Explain the significance of the Induced fit model.

9.21. Describe the application of various enzymes used in wastewater treatment.

9.22. What is the advantage and disadvantage of enzymatic wastewater treatment?

9.23. Write down the present technologies used in enzymatic wastewater treatment.

9.24. What is Michaelis Menten equation for enzymatic kinetics? Define Km and Vm.

9.25. The following data were received from enzymatic oxidation of toxic xenobiotic compounds by oxidoreductase enzyme at diverse concentrations of xenobiotic compounds. Calculate the rate constant (K_m) and maximum forward velocity (V_m).

Substrate concentration (S) mg/L	10	15	20	30	42	55	65	75	80	90
Rate of product formation (v) mg/L h	3	3.5	4.2	5.5	6	10	11	13	9	8

[**Ans**. Using a double-reciprocal plot of 1/v versus 1/[S] yields a linear line with X-axis intercept of $-1/K_m$ is $= -0.25$ is obtained. Therefore $K_m = 4$ and from the Y-axis intercept $1/V_m = 0.07$, then $V_m = 14$ mg/L.h]

9.26. Michaelis–Menten kinetics can define the reaction kinetics of intracellular enzymes and illustrate $[E_0] = [S_0]$. However, the quasi-steady-state hypothesis does not hold for $[E_0] = [S_0]$ in any in vitro batch reactors. Therefore, the rapid equilibrium assumption will also not hold. Moreover, the Michaelis–Menten kinetics and the quasi-steady-state approximation are still reasonable descriptions of intracellular enzyme reactions. Explain this phenomenon.

9.27. Assume the following reversible product-formation reaction in an enzymatic catalytic reaction as listed below:

$$E + S \underset{K_{-1}}{\overset{K_1}{\rightleftharpoons}} ES \underset{K_{-2}}{\overset{K_2}{\rightarrow}} E + P$$

Find out the rate expression for product-formation using the quasi-steady state assumption.

Ans: The rate expression for product-formation (v) $= \dfrac{dp}{dt} = \dfrac{\frac{Vs}{K'm}[S]-[p]\frac{Vp}{K'p}}{1+\frac{[S]}{[K'm]}+\frac{[P]}{[K'p]}}$

where, $Vs = K_2[E0]$; $V_p = K_{-1}[E0]$; $K'_m = \dfrac{K_{-1}+K_2}{K_1}$ and $K'_p = \dfrac{K_{-1}+K_2}{K_{-2}}$

9.28. Why tertiary treatment is required after primary and secondary treatment of wastewater?

9.29. Describe the different types of culture conditions employed for algal cultivation.

9.30. How do temperature and light intensity affect microalgal cultivation?

9.31. Describe the role of nutrients, which is essential for microalgal growth.

9.32. What are the different types of pond systems used in wastewater treatment? Explain them.

9.33. Mention a few photobioreactors used in algae-based wastewater treatment.

9.34. Explain the role of algae in wastewater treatment?

9.35. Under the mixotrophic growth condition, green algae *C. vulgaris* was cultured in a 1 L tubular photobioreactor for the removal of faecal coliforms from sewage. If the number of coliforms present in a reactor is 10^6 MPN per 100 mL and the death rate is 1.2 per day. Then find out the removal rate of coliform per day? [**Ans.** 93.69% per day].

References

Abhilash, P., Jamil, S., & Singh, N. (2009). Transgenic plants for enhanced biodegradation and phytoremediation of organic xenobiotics. *Biotechnology Advances, 27*(4), 474–488.

Acién Fernández, F. G., Gómez-Serrano, C., & Fernández-Sevilla, J. M. (2018). Recovery of nutrients from wastewaters using microalgae. *Frontiers in Sustainable Food Systems, 2,* 59.

Agarwal, S. K. (2005). *Water pollution.* APH publishing.

Aggelis, G., Iconomou, D., Christou, M., Bokas, D., Kotzailias, S., Christou, G., Tsagou, V., & Papanikolaou, S. (2003). Phenolic removal in a model olive oil mill wastewater using Pleurotus ostreatus in bioreactor cultures and biological evaluation of the process. *Water Research, 37*(16), 3897–3904.

Ahirwar, R., Sharma, J. G., Singh, B., Kumar, K., Nahar, P., & Kumar, S. (2017). A simple and efficient method for removal of phenolic contaminants in wastewater using covalent immobilized horseradish peroxidase. *Journal of Materials Science and Engineering, 7,* 27–38.

Aisien, F., Oyakhilomen, G., & Aisien, E. (2009). Bio-treatment of brewery effluent using enzymes. *Advanced Materials Research.* Trans Tech Publ., 774–778.

Al-Ansari, M. M., Modaressi, K., Taylor, K. E., Bewtra, J. K., & Biswas, N. (2010). Soybean peroxidase-catalyzed oxidative polymerization of phenols in coal-tar wastewater: Comparison of additives. *Environmental Engineering Science, 27*(11), 967–975.

Alcantaraa, C., de Godosa, I., & Munoza, R. (2019). Wastewater treatment and biomass generation with algae. In *Wastewater treatment residues as resources for biorefinery products and biofuels* (1st ed., pp. 229–249). Elsevier.

Asulabh, K., Supriya, G., & Ramachandra, T. (2012). Effect of salinity concentrations on growth rate and lipid concentration in Microcystis sp., Chlorococcum sp. and Chaetoceros sp. In *National Conference on Conservation and Management of Wetland Ecosystems. School of Environmental Sciences, Mahatma Gandhi University, Kottayam, Kerala.*

Atolia, E., Cesar, S., Arjes, H. A., Rajendram, M., Shi, H., Knapp, B. D., Khare, S., Aranda-Díaz, A., Lenski, R. E., & Huang, K. C. (2020). Environmental and physiological factors affecting high-throughput measurements of bacterial growth. *Mbio, 11*(5), e01378–e1420.

Baldrian, P. (2003). Interactions of heavy metals with white-rot fungi. *Enzyme and Microbial Technology, 32*(1), 78–91.

Barsanti, L., & Gualtieri, P. (2014). *Algae: Anatomy, biochemistry, and biotechnology.* CRC Press.

Baun, A., Sørensen, S. N., Rasmussen, R., Hartmann, N. B., & Koch, C. B. (2008). Toxicity and bioaccumulation of xenobiotic organic compounds in the presence of aqueous suspensions of aggregates of nano-C60. *Aquatic Toxicology, 86*(3), 379–387.

Beena, A., & Geevarghese, P. (2010). A solvent tolerant thermostable protease from a psychrotrophic isolate obtained from pasteurized milk. *Developmental Microbiology and Molecular Biology, 1,* 113–119.

Behera, A. R., Veluppal, A., & Dutta, K. (2019). Optimization of physical parameters for enhanced production of lipase from Staphylococcus hominis using response surface methodology. *Environmental Science and Pollution Research, 26*(33), 34277–34284.

Benjamin, S., & Pandey, A. (1998). Candida rugosa lipases: Molecular biology and versatility in biotechnology. *Yeast, 14*(12), 1069–1087.

Bhattacharyya, A., Dutta, S., De, P., Ray, P., & Basu, S. (2010). Removal of mercury (II) from aqueous solution using papain immobilized on alginate bead: Optimization of immobilization condition and modeling of removal study. *Bioresource Technology, 101*(24), 9421–9428.

Bhattacharyya, A., Chatterjee, S., Dutta, S., De, P., & Basu, S. (2014). Removal and recovery of lead (ii) from simulated solution using alginate immobilized papain (AIP). *Environmental Engineering & Management Journal (EEMJ), 13*(4).

Bilal, M., Rasheed, T., Iqbal, H. M., Hu, H., Wang, W., & Zhang, X. (2018). Horseradish peroxidase immobilization by copolymerization into cross-linked polyacrylamide gel and its dye degradation and detoxification potential. *International Journal of Biological Macromolecules, 113*, 983–990.

Bollag, J.-M., Dec, J., & Krishnan, S. B. (1998). *Use of plant material for the removal of pollutants by polymerization and binding to humic substances.* Springer.

Boyer, P. D., & Krebs, E. G. (1986). *The enzymes* (3rd ed.). Academic Press.

Brandelli, A., Sala, L., & Kalil, S. J. (2015). Microbial enzymes for bioconversion of poultry waste into added-value products. *Food Research International, 73*, 3–12.

Bren, A., Park, J.O., Towbin, B.D., Dekel, E., Rabinowitz, J.D., & Alon, U. (2016). Glucose becomes one of the worst carbon sources for E. coli on poor nitrogen sources due to suboptimal levels of cAMP. *Scientific Reports, 6*, 1–10.

Brennan, L., & Owende, P. (2010). Biofuels from microalgae-a review of technologies for production, processing, and extractions of biofuels and co-products. *Renewable and Sustainable Energy Reviews, 14*(2), 557–577.

Bush, A. F., Isherwood, J. D., & Rodgi, S. (1961). Dissolved solids removal from waste water by algae. *Journal of the Sanitary Engineering Division, 87*(3), 39–59.

Butterwick, C., Heaney, S., & Talling, J. (2005). Diversity in the influence of temperature on the growth rates of freshwater algae, and its ecological relevance. *Freshwater Biology, 50*(2), 291–300.

Cai, P.-J., Xiao, X., He, Y.-R., Li, W.-W., Zang, G.-L., Sheng, G.-P., Lam, M.H.-W., Yu, L., & Yu, H.-Q. (2013a). Reactive oxygen species (ROS) generated by cyanobacteria act as an electron acceptor in the biocathode of a bio-electrochemical system. *Biosensors and Bioelectronics, 39*(1), 306–310.

Cai, T., Park, S. Y., & Li, Y. (2013b). Nutrient recovery from wastewater streams by microalgae: Status and prospects. *Renewable and Sustainable Energy Reviews, 19*, 360–369.

Camarero, S., García, O., Vidal, T., Colom, J., del Río, J. C., Gutiérrez, A., Gras, J. M., Monje, R., Martínez, M. J., & Martínez, Á. T. (2004). Efficient bleaching of non-wood high-quality paper pulp using laccase-mediator system. *Enzyme and Microbial Technology, 35*(2–3), 113–120.

Catherine, H., Penninckx, M., & Frédéric, D. (2016). Product formation from phenolic compounds removal by laccases: A review. *Environmental Technology & Innovation, 5*, 250–266.

Chan, Y.J., Chong, M.F., Law, C.L., & Hassell, D. (2009). A review on anaerobic–aerobic treatment of industrial and municipal wastewater. *Chemical Engineering Journal, 155*, 1–18.

Chatterjee, P., & Ghangrekar, M. (2017). Biomass granulation in an upflow anaerobic sludge blanket reactor treating 500 m^3/day low-strength sewage and post treatment in high-rate algal pond. *Water Science and Technology, 76*(5), 1234–1242.

Chatterjee, S., Basu, S., Dutta, S., Chattaraj, R., Banerjee, D., & Sinha, S. (2016). Removal and recovery of lead (II) and chromium (VI) by bromelain immobilized on activated charcoal. *Materials Today: Proceedings, 3*(10), 3258–3268.

Cheirsilp, B., & Torpee, S. (2012). Enhanced growth and lipid production of microalgae under mixotrophic culture condition: Effect of light intensity, glucose concentration and fed-batch cultivation. *Bioresource Technology, 110*, 510–516.

Chekroun, K. B., Sánchez, E., & Baghour, M. (2014). The role of algae in bioremediation of organic pollutants. *Journal Issues ISSN, 2360*, 8803.

Chen, F., Zhang, Y., & Guo, S. (1996). Growth and phycocyanin formation of *Spirulina platensis* in photoheterotrophic culture. *Biotechnology Letters, 18*(5), 603–608.

Chen, Y.-H., & Walker, T. H. (2011). Biomass and lipid production of heterotrophic microalgae *Chlorella protothecoides* by using biodiesel-derived crude glycerol. *Biotechnology Letters, 33*(10), 1973.

Chiong, T., Lau, S. Y., Lek, Z. H., Koh, B. Y., & Danquah, M. K. (2016). Enzymatic treatment of methyl orange dye in synthetic wastewater by plant-based peroxidase enzymes. *Journal of Environmental Chemical Engineering, 4*(2), 2500–2509.

Chisti, Y. (2007). Biodiesel from microalgae. *Biotechnology Advances, 25*(3), 294–306.

Colak, O., & Kaya, Z. (1988). A study on the possibilities of biological wastewater treatment using algae. *Doga Biyolji Serisi, 12*(1), 18–29.

Conley, D. J., Paerl, H. W., Howarth, R. W., Boesch, D. F., Seitzinger, S. P., Havens, K. E., Lancelot, C., & Likens, G. E. (2009). Controlling eutrophication: Nitrogen and phosphorus. *Science, 323*(5917), 1014–1015.

Couto, S. R., & Herrera, J. L. T. (2006). Industrial and biotechnological applications of laccases: A review. *Biotechnology Advances, 24*(5), 500–513.

Csermely, P., Palotai, R., & Nussinov, R. (2010). Induced fit, conformational selection and independent dynamic segments: An extended view of binding events. *Nature Precedings, 1.*

Cummins, I., Landrum, M., Steel, P. G., & Edwards, R. (2007). Structure activity studies with xenobiotic substrates using carboxylesterases isolated from Arabidopsis thaliana. *Phytochemistry, 68*(6), 811–818.

Czaplicka, M. (2004). Sources and transformations of chlorophenols in the natural environment. *Science of the Total Environment, 322*(1–3), 21–39.

Daâssi, D., Prieto, A., Zouari-Mechichi, H., Martínez, M. J., Nasri, M., & Mechichi, T. (2016). Degradation of bisphenol A by different fungal laccases and identification of its degradation products. *International Biodeterioration & Biodegradation, 110*, 181–188.

Das, S., & Ghangrekar, M. (2018). Value added product recovery and carbon dioxide sequestration from biogas using microbial electrosynthesis. *Indian Journal of Experimental Biology, 56*, 470–478.

Das, A., Singh, J., & Yogalakshmi, K. (2017). Laccase immobilized magnetic iron nanoparticles: Fabrication and its performance evaluation in chlorpyrifos degradation. *International Biodeterioration & Biodegradation, 117*, 183–189.

Das, S., Das, S., Das, I., & Ghangrekar, M. (2019a). Application of bioelectrochemical systems for carbon dioxide sequestration and concomitant valuable recovery: A review. *Materials Science for Energy Technologies, 2*(3), 687–696.

Das, S., Das, S., & Ghangrekar, M. (2019b). Quorum-sensing mediated signals: A promising multifunctional modulators for separately enhancing algal yield and power generation in microbial fuel cell. *Bioresource Technology, 294*, 122138.

Das, I., Das, S., Dixit, R., & Ghangrekar, M. (2020a). Goethite supplemented natural clay ceramic as an alternative proton exchange membrane and its application in microbial fuel cell. *Ionics, 26*, 3061–3072.

Das, I., Das, S., & Ghangrekar, M. (2020b). Application of bimetallic low-cost CuZn as oxygen reduction cathode catalyst in lab-scale and field-scale microbial fuel cell. *Chemical Physics Letters, 751*, 137536.

Das, I., Das, S., Sharma, S., & Ghangrekar, M. (2020c). Ameliorated performance of a microbial fuel cell operated with an alkali pre-treated clayware ceramic membrane. *International Journal of Hydrogen Energy, 45*(33), 16787–16798.

Das, R., Das, S., & Bhattacharjee, C. (2020d). CO_2 Sequestration using algal biomass and its application as bio energy. *Encyclopedia of Renewable and Sustainable Materials, 3*, 372–384.

Das, S., Das, I., & Ghangrekar, M. (2020e). Role of applied potential on microbial electrosynthesis of organic compounds through carbon dioxide sequestration. *Journal of Environmental Chemical Engineering, 8*(4), 104028.

Das, S., Mishra, A., & Ghangrekar, M. (2020f). Production of hydrogen peroxide using various metal-based catalysts in electrochemical and bioelectrochemical systems: Mini review. *Journal of Hazardous, Toxic, and Radioactive Waste, 24*(3), 06020001.

De Philippis, R., Colica, G., & Micheletti, E. (2011). Exopolysaccharide-producing cyanobacteria in heavy metal removal from water: Molecular basis and practical applicability of the biosorption process. *Applied Microbiology and Biotechnology, 92*(4), 697.

Diao, M., Ouédraogo, N., Baba-Moussa, L., Savadogo, P. W., N'Guessan, A. G., Bassolé, I. H., & Dicko, M. H. (2011). Biodepollution of wastewater containing phenolic compounds from leather industry by plant peroxidases. *Biodegradation, 22*(2), 389–396.

Dixit, R., Malaviya, D., Pandiyan, K., Singh, U. B., Sahu, A., Shukla, R., Singh, B. P., Rai, J. P., Sharma, P. K., & Lade, H. (2015). Bioremediation of heavy metals from soil and aquatic environment: An overview of principles and criteria of fundamental processes. *Sustainability, 7*(2), 2189–2212.

El-Kowrany, S. I., El-Zamarany, E. A., El-Nouby, K. A., El-Mehy, D. A., Ali, E. A. A., Othman, A. A., Salah, W., & El-Ebiary, A. A. (2016). Water pollution in the Middle Nile Delta, Egypt: An environmental study. *Journal of Advanced Research, 7*(5), 781–794.

Ely, C., Lourdes Borba Magalhães, M. d., Henrique Lemos Soares, C., & Skoronski, E. (2017). Optimization of phenol removal from biorefinery effluent using horseradish peroxidase. *Journal of Environmental Engineering, 143*(12), 04017075.

Filip, D. S., Peters, V. T., Adams, E. D., & Middlebrooks, J. (1979). Residual heavy metal removal by an algae-intermittent sand filtration system. *Water Research, 13*(3), 305–313.

Garg, S., Kumar, P., Singh, S., Yadav, A., Dumée, L. F., Sharma, R. S., & Mishra, V. (2020). *Prosopis juliflora* peroxidases for phenol remediation from industrial wastewater—An innovative practice for environmental sustainability. *Environmental Technology & Innovation, 19*, 100865.

Gianfreda, L., Xu, F., & Bollag, J.-M. (1999). Laccases: A useful group of oxidoreductive enzymes. *Bioremediation Journal, 3*(1), 1–26.

Grobbelaar, J. U. (2009). Upper limits of photosynthetic productivity and problems of scaling. *Journal of Applied Phycology, 21*(5), 519–522.

Gupta, V., Rastogi, A., Saini, V., & Jain, N. (2006). Biosorption of copper (II) from aqueous solutions by *Spirogyra* species. *Journal of Colloid and Interface Science, 296*(1), 59–63.

Hodaifa, G., Martínez, M. E., & Sánchez, S. (2009). Influence of pH on the culture of Scenedesmus obliquus in olive-mill wastewater. *Biotechnology and Bioprocess Engineering, 14*(6), 854–860.

Horan, N. J. (1989). *Biological wastewater treatment systems: Theory and operation* (1st ed.). Wiley.

Hosetti, B. B., & Frost, S. (1995). A review of the sustainable value of effluents and sludges from wastewater stabilization ponds. *Ecological Engineering, 5*(4), 421–431.

Huang, G., Chen, F., Wei, D., Zhang, X., & Chen, G. (2010). Biodiesel production by microalgal biotechnology. *Applied Energy, 87*(1), 38–46.

Jiang, H.-M., Luo, S.-J., Shi, X.-S., Dai, M., & Guo, R.-B. (2013). A system combining microbial fuel cell with photobioreactor for continuous domestic wastewater treatment and bioelectricity generation. *Journal of Central South University, 20*(2), 488–494.

Joseph, A., Joji, J. G., Prince, N. M., Rajendran, R., & Nainamalai, D. (2020). Domestic wastewater treatment using garbage enzyme. In *Proceedings of the International conference on system energy and environment (ICSEE)*, Kerala, India, December 24, 2020.

Jung, H., Hyun, K., & Park, C. (2003). Production of laccase and bioremediation of pentachlorophenol by wood-degrading fungus *Trichophyton* sp. LKY-7 immobilized in Ca-alginate beads. *Journal of Korea Technical Association of the Pulp and Paper Industry, 35*(5), 80–86.

Karigar, C. S., & Rao, S. S. (2011). Role of microbial enzymes in the bioremediation of pollutants: A review. *Enzyme Research, 2011.*

Kessler, J. O. (1985). Hydrodynamic focusing of motile algal cells. *Nature, 313*(5999), 218–220.

Koshland, D. E., Jr. (1995). The key–lock theory and the induced fit theory. *Angewandte Chemie International Edition in English, 33*(23–24), 2375–2378.

Krajewska, B. (1991). Urease immobilized on chitosan membrane. Inactivation by heavy metal ions. *Journal of Chemical Technology & Biotechnology, 52*(2), 157–162.

Kulys, J., & Vidziunaite, R. (2003). Amperometric biosensors based on recombinant laccases for phenols determination. *Biosensors and Bioelectronics, 18*(2–3), 319–325.

Kurade, M. B., Kim, J. R., Govindwar, S. P., & Jeon, B.-H. (2016). Insights into microalgae mediated biodegradation of diazinon by Chlorella vulgaris: Microalgal tolerance to xenobiotic pollutants and metabolism. *Algal Research, 20,* 126–134.

Lee, Y.-K. (2001). Microalgal mass culture systems and methods: Their limitation and potential. *Journal of Applied Phycology, 13*(4), 307–315.

Lemieux, R. U., & Spohr, U. (1994). Concept for enzyme specificity 1. *Advances in Carbohydrate Chemistry and Biochemistry, 50*(1).

Leung, M. (2004). Bioremediation: Techniques for cleaning up a mess. *BioTeach Journal, 2,* 18–22.

Li, Y., Chen, Y.-F., Chen, P., Min, M., Zhou, W., Martinez, B., Zhu, J., & Ruan, R. (2011). Characterization of a microalga *Chlorella sp.* well adapted to highly concentrated municipal wastewater for nutrient removal and biodiesel production. *Bioresource Technology, 102*(8), 5138–5144.

Liang, Y., Sarkany, N., & Cui, Y. (2009). Biomass and lipid productivities of *Chlorella vulgaris* under autotrophic, heterotrophic and mixotrophic growth conditions. *Biotechnology Letters, 31*(7), 1043–1049.

Liang, Z., Liu, Y., Ge, F., Xu, Y., Tao, N., Peng, F., & Wong, M. (2013). Efficiency assessment and pH effect in removing nitrogen and phosphorus by algae-bacteria combined system of Chlorella vulgaris and Bacillus licheniformis. *Chemosphere, 92*(10), 1383–1389.

Liew, Y. X., Chan, Y. J., Manickam, S., Chong, M. F., Chong, S., Tiong, T. J., Lim, J. W., & Pan, G.-T. (2020). Enzymatic pretreatment to enhance anaerobic bioconversion of high strength wastewater to biogas: A review. *Science of the Total Environment, 713,* 136373.

Mata, T. M., Martins, A. A., & Caetano, N. S. (2010). Microalgae for biodiesel production and other applications: A review. *Renewable and Sustainable Energy Reviews, 14*(1), 217–232.

Meegoda, J. N., Li, B., Patel, K., & Wang, L. B. (2018). A review of the processes, parameters, and optimization of anaerobic digestion. *International Journal of Environmental Research and Public Health, 15*(10), 2224.

Mehta, S. K., & Gaur, J. P. (2001). Removal of Ni and Cu from single and binary metalsolutions by free and immobilized *Chlorella vulgaris*. *European Journal of Protistology, 37*(3), 261–271.

Meng, Y., Luan, F., Yuan, H., Chen, X., & Li, X. (2017). Enhancing anaerobic digestion performance of crude lipid in food waste by enzymatic pretreatment. *Bioresource Technology, 224,* 48–55.

Mishra, B., Varjani, S., Agrawal, D. C., Mandal, S. K., Ngo, H. H., Taherzadeh, M. J., Chang, J.-S., You, S., & Guo, W. (2020). Engineering biocatalytic material for the remediation of pollutants: A comprehensive review. *Environmental Technology & Innovation,* 101063.

Modestra, J.A., Velvizhi, G., Krishna, K.V., Arunasri, K., Lens, P.N.L., Nancharaiah, Y., & Venkata Mohan, S. (2017). Bioelectrochemical Systems for Heavy Metal Removal and Recovery. In: E.R. Rene, E. Sahinkaya, A. Lewis, P.N.L. Lens (Eds.), *Sustainable Heavy Metal Remediation: Principles and Processes,* Springer International Publishing. Cham, Switzerland, *1,* 165–198.

Muloiwa, M., Nyende-Byakika, S., & Dinka, M. (2020). Comparison of unstructured kinetic bacterial growth models. *South African Journal of Chemical Engineering, 33,* 141–150.

Murugesan, K. (2003). Bioremediation of paper and pulp mill effluents. *Indian Journal of Experimental Biology, 41,* 1239–1248.

Najafpour, G.D. (2007). Sterilisation. *Biochemical Engineering and Biotechnology*, Elsevier, Netherlands, 342–350.

Najafpour, G. (2015). *Biochemical engineering and biotechnology*. Elsevier.

Ohnishi, N., Allakhverdiev, S. I., Takahashi, S., Higashi, S., Watanabe, M., Nishiyama, Y., & Murata, N. (2005). Two-step mechanism of photodamage to photosystem II: Step 1 occurs at the oxygen-evolving complex and step 2 occurs at the photochemical reaction center. *Biochemistry, 44*(23), 8494–8499.

Okpokwasili, G., & Nweke, C. (2006). Microbial growth and substrate utilization kinetics. *African Journal of Biotechnology, 5*(4), 305–317.

Oswald, W., Gotaas, H., Golueke, C., Kellen, W., Gloyna, E., & Hermann, E. (1957). Algae in waste treatment. *Sewage and Industrial Wastes, 29*(4), 437–457.

Pandey, K., Singh, B., Pandey, A. K., Badruddin, I. J., Pandey, S., Mishra, V. K., & Jain, P. A. (2017). Application of microbial enzymes in industrial waste water treatment. *International Journal of Current Microbiology and Applied Sciences, 6*, 1243–1254.

Park, J., & Craggs, R. (2010). Wastewater treatment and algal production in high rate algal ponds with carbon dioxide addition. *Water Science and Technology*, 633–639.

Park, J.-W., Park, B.-K., & Kim, J.-E. (2006). Remediation of soil contaminated with 2, 4-dichlorophenol by treatment of minced shepherd's purse roots. *Archives of Environmental Contamination and Toxicology, 50*(2), 191–195.

Parmar, N., Singh, A., & Ward, O. (2001). Enzyme treatment to reduce solids and improve settling of sewage sludge. *Journal of Industrial Microbiology and Biotechnology, 26*(6), 383–386.

Powell, N., Shilton, A., Chisti, Y., & Pratt, S. (2009). Towards a luxury uptake process via microalgae–defining the polyphosphate dynamics. *Water Research, 43*(17), 4207–4213.

Praveen, P., Guo, Y., Kang, H., Lefebvre, C., & Loh, K.-C. (2018). Enhancing microalgae cultivation in anaerobic digestate through nitrification. *Chemical Engineering Journal, 354*, 905–912.

Ras, M., Steyer, J.-P., & Bernard, O. (2013). Temperature effect on microalgae: A crucial factor for outdoor production. *Reviews in Environmental Science and Bio/technology, 12*(2), 153–164.

Raso, S., Van Genugten, B., Vermuë, M., & Wijffels, R. H. (2012). Effect of oxygen concentration on the growth of *Nannochloropsis* sp. at low light intensity. *Journal of Applied Phycology, 24*(4), 863–871.

Rasoul-Amini, S., Montazeri-Najafabady, N., Shaker, S., Safari, A., Kazemi, A., Mousavi, P., Mobasher, M. A., & Ghasemi, Y. (2014). Removal of nitrogen and phosphorus from wastewater using microalgae free cells in bath culture system. *Biocatalysis and Agricultural Biotechnology, 3*(2), 126–131.

Rawat, I., Kumar, R. R., Mutanda, T., & Bux, F. (2011). Dual role of microalgae: Phycoremediation of domestic wastewater and biomass production for sustainable biofuels production. *Applied Energy, 88*(10), 3411–3424.

Redfield, A. C. (1958). The biological control of chemical factors in the environment. *American Scientist, 46*(3), 205–221.

Reino, C., Suárez-Ojeda, M. E., Pérez, J., & Carrera, J. (2018). Stable long-term operation of an upflow anammox sludge bed reactor at mainstream conditions. *Water Research, 128*, 331–340.

Romeh, A., & Hendawi, M. (2014). Bioremediation of certain organophosphorus pesticides by two biofertilizers, Paenibacillus (Bacillus) polymyxa (Prazmowski) and Azospirillum lipoferum (Beijerinck). *Journal of Agricultural Science and Technology, 16*(2), 265–276.

Rubilar, O., Diez, M. C., & Gianfreda, L. (2008). Transformation of chlorinated phenolic compounds by white rot fungi. *Critical Reviews in Environmental Science and Technology, 38*(4), 227–268.

Ruiz-Marin, A., Mendoza-Espinosa, L. G., & Stephenson, T. (2010). Growth and nutrient removal in free and immobilized green algae in batch and semi-continuous cultures treating real wastewater. *Bioresource Technology, 101*(1), 58–64.

Sangave, P. C., & Pandit, A. B. (2006). Enhancement in biodegradability of distillery wastewater using enzymatic pretreatment. *Journal of Environmental Management, 78*(1), 77–85.

Sasaki, T., Kajino, T., Li, B., Sugiyama, H., & Takahashi, H. (2001). New pulp biobleaching system involving manganese peroxidase immobilized in a silica support with controlled pore sizes. *Applied and Environmental Microbiology, 67*(5), 2208–2212.

Sayler, G. S., & Ripp, S. (2000). Field applications of genetically engineered microorganisms for bioremediation processes. *Current Opinion in Biotechnology, 11*(3), 286–289.

Sharma, D., Sharma, B., & Shukla, A. (2011). Biotechnological approach of microbial lipase: A review. *Biotechnology, 10*(1), 23–40.

Shelef, G., Sukenik, A., & Green, M. (1984). *Microalgae harvesting and processing: A literature review* (pp. 1–65). Technion Research and Development Foundation. SERI/STR-231-2396, Solar Energy Research Institute, Golder Colorado, USA. https://doi.org/10.2172/6204677

Sidhu, C., Vikram, S., & Pinnaka, A. K. (2017). Unraveling the microbial interactions and metabolic potentials in pre-and post-treated sludge from a wastewater treatment plant using metagenomic studies. *Frontiers in Microbiology, 8*, 1382.

Singh, C. J. (2003). Optimization of an extracellular protease of Chrysosporium keratinophilum and its potential in bioremediation of keratinic wastes. *Mycopathologia, 156*(3), 151–156.

Singh, A. K., Rana, H. K., Yadav, R. K., & Pandey, A. K. (2020). Dual role of microalgae: Phycoremediation coupled with biomass generation for biofuel production. In: A. K. Upadhyay, R. Singh, D. Singh (Eds.), *Restoration of wetland ecosystem: A trajectory towards a sustainable environment*. Springer. Singapore.

Soeder, C. J., Payer, H.-D., Runkel, K.-H., Beine, J., & Briele, E. (1978). Sorption and concentration of toxic minerals by mass cultures of Chlorococcales. *Internationale Vereinigung Für Theoretische Und Angewandte Limnologie: Mitteilungen, 21*(1), 575–584.

Stadlmair, L. F., Letzel, T., Drewes, J. E., & Graßmann, J. (2017). Mass spectrometry based in vitro assay investigations on the transformation of pharmaceutical compounds by oxidative enzymes. *Chemosphere, 174*, 466–477.

Stevenson, F. J. (1994). *Humus chemistry: Genesis, composition, reactions.* Wiley.

Su, Y., Mennerich, A., & Urban, B. (2011). Municipal wastewater treatment and biomass accumulation with a wastewater-born and settleable algal-bacterial culture. *Water Research, 45*(11), 3351–3358.

Sutton, P. M. (2006). Membrane bioreactors for industrial wastewater treatment: Applicability and selection of optimal system configuration. *Proceedings of the Water Environment Federation, 2006*(9), 3233–3248.

Theerachat, M., Tanapong, P., & Chulalaksananukul, W. (2017). The culture or co-culture of *Candida rugosa* and *Yarrowia lipolytica* strain rM-4A, or incubation with their crude extracellular lipase and laccase preparations, for the biodegradation of palm oil mill wastewater. *International Biodeterioration & Biodegradation, 121*, 11–18.

Ting, Y., Lawson, F., & Prince, I. (1989). Uptake of cadmium and zinc by the alga Chlorella vulgaris: Part 1 individual ion species. *Biotechnology and Bioengineering, 34*(7), 990–999.

Ting, Y., Prince, I., & Lawson, F. (1991). Uptake of cadmium and zinc by the alga Chlorella vulgaris: II Multi-ion situation. *Biotechnology and Bioengineering, 37*(5), 445–455.

Torres, J. A., Chagas, P. M. B., Silva, M. C., Dos Santos, C. D., & Corrêa, A. D. (2016). Evaluation of the protective effect of chemical additives in the oxidation of phenolic compounds catalysed by peroxidase. *Environmental Technology, 37*(10), 1288–1295.

Ullah, M. A., Bedford, C. T., & Evans, C. S. (2000). Reactions of pentachlorophenol with laccase from Coriolus versicolor. *Applied Microbiology and Biotechnology, 53*(2), 230–234.

Varga, B., Somogyi, V., Meiczinger, M., Kováts, N., & Domokos, E. (2019). Enzymatic treatment and subsequent toxicity of organic micropollutants using oxidoreductases-A review. *Journal of Cleaner Production, 221*, 306–322.

Vareda, J.P., Valente, A.J., & Durães, L. 2019. Assessment of heavy metal pollution from anthropogenic activities and remediation strategies: A review. *Journal of Environmental Management, 246*, 101–118.

Verma, N., & Singh, M. (2005). Biosensors for heavy metals. *BioMetals, 18*(2), 121–129.

Wan, C.-Y., De Wever, H., Diels, L., Thoeye, C., Liang, J.-B., & Huang, L.-N. (2011). Biodiversity and population dynamics of microorganisms in a full-scale membrane bioreactor for municipal wastewater treatment. *Water Research, 45*(3), 1129–1138.

Wan, L., Wu, Y., Ding, H., & Zhang, W. (2020). Toxicity, biodegradation, and metabolic fate of organophosphorus pesticide trichlorfon on the freshwater algae Chlamydomonas reinhardtii. *Journal of Agricultural and Food Chemistry, 68*(6), 1645–1653.

Whitton, R., Ometto, F., Pidou, M., Jarvis, P., Villa, R., & Jefferson, B. (2015). Microalgae for municipal wastewater nutrient remediation: Mechanisms, reactors and outlook for tertiary treatment. *Environmental Technology Reviews, 4*(1), 133–148.

Williams, P. P. (1977). Metabolism of synthetic organic pesticides by anaerobic microorganisms. *Residue Reviews*, 63–135.

Xu, H., Miao, X., & Wu, Q. (2006). High quality biodiesel production from a microalga *Chlorella protothecoides* by heterotrophic growth in fermenters. *Journal of Biotechnology, 126*(4), 499–507.

Yang, Q., Yang, M., Pritsch, K., Yediler, A., Hagn, A., Schloter, M., & Kettrup, A. (2003). Decolorization of synthetic dyes and production of manganese-dependent peroxidase by new fungal isolates. *Biotechnology Letters, 25*(9), 709–713.

Yao, Y., Wang, M., Liu, Y., Han, L., & Liu, X. (2020). Insights into the improvement of the enzymatic hydrolysis of bovine bone protein using lipase pretreatment. *Food Chemistry, 302*, 125199.

Yeh, K.-L., & Chang, J.-S. (2012). Effects of cultivation conditions and media composition on cell growth and lipid productivity of indigenous microalga *Chlorella vulgaris* ESP-31. *Bioresource Technology, 105*, 120–127.

Yorkor, B., & Momoh, Y. (2019). A review of anoxic wastewater treatment: An overlooked aspect in wastewater treatment in Nigeria. *American Journal of Water Resources, 7*(4), 136–145.

Zhai, J., Li, X., Li, W., Rahaman, M. H., Zhao, Y., Wei, B., & Wei, H. (2017). Optimization of biomass production and nutrients removal by Spirulina platensis from municipal wastewater. *Ecological Engineering, 108*, 83–92.

Zheng, Y., Yu, X., Zeng, J., & Chen, S. (2012). Feasibility of filamentous fungi for biofuel production using hydrolysate from dilute sulfuric acid pretreatment of wheat straw. *Biotechnology for Biofuels, 5*(1), 50.

Zhou, M., He, H., Jin, T., & Wang, H. (2012). Power generation enhancement in novel microbial carbon capture cells with immobilized *Chlorella vulgaris*. *Journal of Power Sources, 214*, 216–219.

Zumstein, M. T., & Helbling, D. E. (2019). Biotransformation of antibiotics: Exploring the activity of extracellular and intracellular enzymes derived from wastewater microbial communities. *Water Research, 155*, 115–123.

Aerobic Wastewater Treatment Systems

10

The biological processes for treatment of wastewater can be classified into aerobic, anoxic and anaerobic systems. An aerobic treatment process takes place in the presence of oxygen, wherein the aerobic microorganisms convert organic matter present in the wastewater into carbon dioxide and new cell biomass. This chapter focuses on different aerobic processes used in wastewater treatment with special emphasis on suspended and attached growth aerobic treatment systems. The system in which the microorganisms responsible for conversion of organic matter are maintained in suspension is termed as suspended growth process and the most common of this type of process is the activated sludge process (ASP). Whereas, the processes in which the microorganisms responsible for treatment are attached to an inert media material are collectively termed as attached growth processes. The most common aerobic attached growth system used for wastewater treatment is trickling filter. These reactor types used in aerobic biological wastewater treatment are described in this chapter with design details.

10.1 Activated Sludge Process

10.1.1 Overview

The activated sludge process is an aerobic biological wastewater treatment process in which the dissolved and colloidal form of non-settleable organic materials present in the wastewater are partly oxidized to carbon dioxide and remaining are converted to settleable cellular biomass, thus facilitating removal of these organic matters from wastewater. The process performance depends upon aerobic microorganisms, which are maintained in contact with the organic matter present in the wastewater entering the system. The nature and

© The Author(s), under exclusive license to Springer Nature Singapore Pte Ltd. 2022
M. Ghangrekar, *Wastewater to Water*, https://doi.org/10.1007/978-981-19-4048-4_10

the reactions performed by the microorganisms present in an ASP are determined by the biochemical environment they survive in.

The invention of the ASP took place way back in 1880s, when a group of researchers conducted research on aeration of wastewater and evaluated its effect on oxidation of organic matter. However, the importance of activated sludge and its significance in wastewater treatment was established by Ardern and Lockett (1914) and the same group developed the name 'activated sludge process' as it involved activated mass of aerobic microorganisms capable of stabilizing the organic matter present in wastewater. In situations where there is a demand for high quality effluent and low space availability, the ASP is being widely used for treating domestic wastewaters (Joshua Amarnath et al., 2015) as well as industrial wastewaters, including wastewater from pharmaceutical industry (Rana & Shah, 2014), chemical fertilizer plant (Bin Md Hafiz et al., 2021), petroleum refinery (Mirbagheri et al., 2014), removal of microplastics (Olmos et al., 2019), cheese whey (Valta et al., 2017), beverages industry (Abdel-Fatah et al., 2017), slaughterhouses (Farzadkia et al., 2016) and many other industrial effluents.

10.1.2 Process Description

An activated sludge process, as the name suggests is an aerobic biological suspended growth wastewater treatment process in which the decomposition of the organic matter present in the wastewater is accelerated by the addition of biologically active microbial sludge (activated sludge) to it. An activated sludge is a mixture of microorganism, majorly bacteria with other microorganisms, such as protozoa, fungus, etc., maintained under suspension by providing adequate mixing either by employing mechanical aeration or diffused aeration. The following units are integral parts and the essence of any continuous-flow ASP: (a) aeration tank (reactor); (b) secondary sedimentation tank (SST); and (c) sludge recirculation and excess sludge removal system (Fig. 10.1).

In the aeration tank the organic matter is stabilized by the action of active aerobic bacteria under aeration. In the secondary sedimentation tank, the biological cell mass

Fig. 10.1 Schematic diagram of a completely mixed activated sludge process

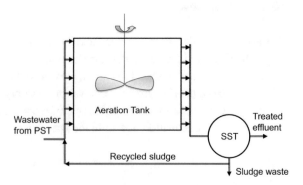

formed during the biochemical reactions in aeration tank is separated from the effluent of aeration tank. The bacterial sludge settled in sedimentation tank is partly recycled to the aeration tank and remaining sludge is wasted. The biochemical reactions, catalysed by bacteria, oxidize part of the organic matter to CO_2 as end product and liberate energy in this reaction. The remaining organic matter present in the wastewater as well as certain inorganic matters present in wastewater entering the aeration tank are converted into cell biomass through microbial anabolism. This mixed liquor from the aeration tank is sent to SST for settling of this cellular mass. A part of the settled sludge in sedimentation tank is recycled back to aeration basin for improving the treatment efficiency of the process by maintaining necessary concentration of active microbial biomass in the aeration tank. Hence, the core of the process are the microorganisms that are capable of converting the organic matter, represented as BOD present in the wastewater, into carbon dioxide and new cells with the aid of oxygen.

The major factors that affect the growth of microorganism includes food (organic matter), flow of wastewater, dissolved oxygen in aeration tank and rate of oxygen transfer, temperature, nutrients availability, pH and toxicity present in wastewater, if any (Metcalf & Eddy, 2003). The more soluble the food, i.e., organic matter, the more is the ease of consumption by microorganism. The food admitted to the aeration tank, i.e., organic matter present in the wastewater, is quantified by analysing the BOD of the influent. The influent flow of the wastewater should be set in such a way that the microorganisms get enough retention time for oxidation of the organic matter present in it. Higher flow may result in reduced treatment efficiency due to reduction in contact time.

Oxygen is prerequisite for the aerobic microorganisms to breakdown the substrate. Hence, the dissolved oxygen concentration in the aeration basin should be regularly monitored and maintained above 1.5 mg/L. As most of the microorganisms perform best at moderate mesophilic temperature. Hence, it is necessary to monitor and maintain an optimum temperature for proper operation of the system. A pH in the range of 6.0–9.0 is preferred for microorganisms to survive and metabolize.

The microorganisms present in the aeration tank are quantified in terms of volatile suspended solids (VSS) concentration present in the aeration tank. Since, the contents of the aeration tank are under mixed conditions to keep microorganisms in suspension, hence the cell mass present in the aeration tank is represented as mixed liquor volatile suspended solids (MLVSS). When the organic matter is combusted at 550 °C and above, the organic matter gets volatilized. Hence, here in this case to quantify the microbial concentration in the aeration tank, the microbial cells being organic matter, this VSS measurement, expressed as MLVSS, is used. Whereas, the total mass concentration of the sludge present in the aeration tank is expressed by mixed liquor suspended solids (MLSS). The ratio of MLVSS/MLSS for ASP sludge is generally remains in the range of 0.7–0.85.

Loading rate: The organic matter loading rate applied to the reactor, i.e., aeration tank, is quantified as kg of BOD applied per unit volume of the reactor per day, called as volumetric loading rate or organic loading rate (OLR). The loading rate on aeration tank is also expressed as specific loading rate as food to microorganisms ratio (F/M) in terms of kg of BOD applied per day per unit mass of microorganisms present in the aeration tank. This can be estimated as stated below using Eq. 10.1 and Eq. 10.2, respectively.

$$\text{Volumetric loading rate, kgBOD/m}^3 \text{ day} = \left(Q \times BOD \times 10^{-3}\right)/V \qquad (10.1)$$

$$\text{F/M, kgBOD/kgVSS day} = Q \times BOD/(V \times X_t) \qquad (10.2)$$

where, BOD is the influent BOD_5 of wastewater entering in the aeration tank, mg/L; Q is the flow rate of wastewater, m³/day; V is the volume of aeration tank, m³; X_t is the MLVSS concentration in the aeration tank, mg/L.

The F/M ratio is the main factor controlling BOD removal because organic matter oxidation is catalyzed by the microorganisms. Hence, the organic load applied per unit mass of microorganisms present in the reactor is the main responsible parameter for utilization and conversion of substrate, i.e., BOD present in the wastewater. This substrate utilization rate depends upon the capability of the microorganisms present in the aeration tank. Though the actual capability of the microbes to convert substrate (i.e., organic matter, represented by BOD) is much higher, however to avoid overloading the lower F/M values are recommended for design. Lower F/M values will give higher BOD removal. The F/M can be varied by varying MLVSS concentration in the aeration tank.

In an investigation by Travers and Lovett, the fat degradation was 90% at 4 mg/L of DO; whereas, it was about 60% at DO < 0.5 mg/L (Travers & Lovett, 1984). The investigation suggested that in presence of adequate DO low F/M ratio (0.2–0.4) favours the higher removal efficiencies of COD (96%), total suspended solids (TSS, 95%), total Kjeldahl nitrogen (TKN, 96%), phosphorus (47%), and oil-grease (94%) (Travers & Lovett, 1984).

Controlling volumetric loading rate actually ensures that per unit mass of BOD load applied the sufficient volume of wastewater is retained in the aeration tank to support necessary dissolved oxygen required for aerobic oxidation of organic matter. Though the air is continuously supplied to the aeration tank for supplying oxygen, simultaneously oxygen is utilized by the aerobic microorganism for oxidation of organic matter. Hence, DO around 2 mg/L will be generally observed in aeration tank. Thus, controlling volumetric loading rate actually helps in ensuring that sufficient oxygen is made available to support aerobic oxidation of organic matter. Hence, though the organic matter oxidation and conversion is actually carried out by the microorganisms, this volumetric loading rate is also important to maintain aerobic conditions for ensuring proper performance of ASP.

Solid retention time (SRT) or mean cell residence time (MCRT): The performance of the ASP for organic matter removal depends on the duration for which the microbial mass is retained in the system. Controlling this SRT to some minimum value ensures that the microbial cells hold in the system are matured enough to establish maximum substrate utilization rate. The retention of the sludge depends on the settling rate of the sludge in the SST. If sludge settles well in the SST then proper recirculation of the sludge in aeration tank is possible, which will help in maintaining desired SRT in the system. If the sludge developed in aeration tank has poor settling properties, it will not settle well in the SST and recirculation of the sludge will be difficult and this may reduce the SRT in the system. The SRT can be estimated using Eq. 10.3. Generally, the VSS lost in the effluent of SST are neglected as this is very small concentration as compared to the concentration of the sludge while carrying out artificial wasting from the sludge recycle line or directly from the aeration tank.

$$\text{SRT} = \frac{\text{Total kg of MLVSS present in aeration tank}}{(\text{kg of VSS wasted per day} + \text{kg of VSS lost in effluent per day})} \quad (10.3)$$

Sludge volume index: The quantity of the return sludge is expressed on volumetric basis. The sludge volume index (SVI) is the volume of the sludge in mL occupied by one gram of dry weight of suspended solids (SS), measured after 30 min of settling. Thus, SVI represents settling ability of the sludge and the sludge having better settleability will occupy lower volume per unit of its mass, hence representing lower SVI values. The SVI is measured by filling a 1 L graduated cylinder with well mixed sample of activated sludge suspension and it is allowed to settle for 30 min. The sludge settled volume (SSV) thus obtained after 30 min of settling is recorded and dividing it by MLSS concentration will give value of SVI as expressed in Eq. 10.4. For activated sludge, the SVI varies from 50 to 150 mL/g of SS. Lower SVI indicates better settling of sludge.

$$SVI\left(\frac{mL}{g}\,of\,SS\right) = \frac{Settled\;sludge\;volume(mL) \times 1000(mg/g)}{MLSS(mg)} \quad (10.4)$$

A higher value of SVI indicates poor sludge settleability, filamentous growth as well as sludge bulking and hence need to upsurge recycle sludge flow rate. A low value of SVI indicates the presence of small pin point flocs, which are having good settling property.

Quantity of return sludge: Usually MLSS concentration of about 1500 to 3000 mg/L (MLVSS 80% of MLSS) is maintained for conventional ASP and 3000 to 6000 mg/L for completely mixed ASP. Depending on the type of the activated sludge process the F/M and MLVSS concentration to be maintained in the aeration tank varies. The quantity of return sludge is to be determined based on the concentration to be maintained in aeration tank. This sludge return ratio varies for the type of ASP from 20% to even more than100%.

Sludge Bulking: The sludge which does not settle well in secondary sedimentation tank is called as bulking sludge. Because of the sludge bulking, concentration of the sludge settled at the bottom of sedimentation tank will reduce and thus, in the return sludge volume less sludge mass will be present. Due to this less mass of the returned sludge the F/M in the aeration tank will keep increasing, shifting the bacterial growth towards log growth and hence further favouring bulking sludge formation.

The sludge bulking may occur due to either: (a) the growth of filamentous microorganisms that are not allowing desirable compaction; or (b) due to the production of non-filamentous highly hydrated biomass. There are many reasons for this sludge bulking to occur. The presence of toxic substances in influent, lowering of temperature, insufficient aeration, and shock loading can also cause sludge bulking. Proper supply of air and proper design and operation of ASP to maintain near endogenous growth phase of metabolism will not produce bulking sludge. The sludge bulking can be controlled by restoring air supply to meet the oxygen demand, eliminating shock loading to the reactor, or by increasing temperature of the wastewater or by small hypochlorite dosing to the return sludge line to avoid the growth of filamentous hygroscopic microorganisms.

Mixing conditions: The aeration tank can be of plug flow type or completely mixed type. In the plug flow tank, the F/M and oxygen demand will be highest at the inlet end of the aeration tank and it will then progressively decrease as the wastewater is approaching towards outlet end of the aeration tank. In the completely mix system, the F/M and oxygen demand will be uniform throughout the tank.

Flow scheme: The wastewater addition may be done at a single point at the inlet end of the aeration tank or wastewater can be added at several points along the aeration tank. The sludge return is carried out from the underflow of the SST to the aeration tank, called as retuned sludge line. The excess sludge wasting can be done from this return sludge line or directly from the aeration tank itself. Sludge wasting from the aeration tank will have better control over the process, however higher sludge waste volume needs to be handled in this case due to lower concentration as compared to when wasting is done from underflow of SST, i.e., from sludge return line. In case of diffused air aeration, the compressed air may be applied uniformly along whole length of the tank or it may be tapered from the head of the aeration tank to its end. Thus, higher air supply is provided near inlet, which is gradually reduced to have least air supply at out let end, to match the oxygen demand.

10.1.3 Type of Aeration Provided in ASP

The aeration mechanism provided in ASP can be classified as: (a) diffused air aeration; (b) mechanical aeration; and (c) combined mechanical and diffused air aeration.

10.1.3.1   Diffused Air Aeration

In diffused air aeration, pre-filtered compressed air is blown through diffusers in the aeration tank. The tanks of these units are generally in the form of long narrow rectangular channels made by providing baffles in the rectangular or square tank. The air diffusers are placed at the bottom of tank (Fig. 10.2). The air before passing through diffusers must be passed through air filter to remove dirt. The required pressure to overcome depth of water column and friction loss is maintained by means of air compressors.

Fig. 10.2 Typical air diffusers arrangement provided in aeration tank of ASP

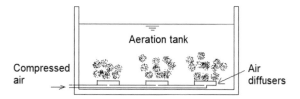

The air diffusers provided in aeration tank can be further classified in the following categories depending on the size of air bubble they introduce in the aeration tank.

(a) *Jet diffusers*: These diffusers give direct stream of air in the form of jet downward and strike against a small bowl kept just below the nozzle of the jet. The air spatters over the surface of the bowl and escapes in the form of smaller size air bubbles in wider area after impacting the bowl surface.

(b) *Porous diffusers*: These are manufactured in the form of tubes and plates from grains of crushed quartz, aluminum oxide or carbon fused to form a porous structure or porous plates with resin or perforated membrane or plastic tubes. These are tile shaped or tubular shape or dome shape. About 10–20% area of the aeration tank is covered with these porous tiles (Fig. 10.3).

The supply of air is done through pipeline laid on the floor of the tank and it is controlled by the valves. Depending upon the size of the air bubbles these porous diffusers can be classified as fine or medium bubble diffused-air aeration device. In common practice, porous dome type air diffusers of 10 to 20 cm diameter are used. These are directly fixed on the top of C.I. pipes laid at the bottom of the aeration tanks. These are economical in initial as well as from maintenance cost point of view.

Air supply: Normally air is supplied under pressure of 0.55–0.7 kg/cm^2 depending on depth of wastewater in the aeration tank and frictional losses in the air line. The quantity of air supplied varies from 1.25 to 9.50 m^3/m^3 for treatment of sewage depending on the

(a)   (b)

Fig. 10.3 Photograph of the cylindrical air diffusers placed at the floor of the aeration tank **a** close view and **b** from top of the aeration tank. (*Courtesy* MM Enviro Projects Pvt. Ltd, Nagpur, India)

BOD of the sewage to be treated and degree of treatment required. The oxygen transfer capacity of the aerators depends on the size of air bubbles introduced in the aeration tank. For example, for fine bubble aeration the oxygen transfer capabilities of aeration device is 0.7–1.4 kg O_2/kW·h; whereas, for medium bubble aeration it is 0.6–1.0 kg O_2/kW·h, and for coarse bubble diffuse aeration it is 0.3–0.9 kg O_2/kW h.

10.1.3.2 Mechanical Aeration Unit

The main objective of mechanical aeration is to splash the wastewater droplets in the air and to bring every time new surface of wastewater in contact with air to avoid stagnation of air–water contact surface, so as to enhance the oxygen mass transfer rate in wastewater. In diffuse air aeration, depending on the depth of wastewater provided in aeration tank, only about 5 to 8% of the total quantity of the air compressed is utilized for oxygen transfer and hence for oxidation of organic matter and rest of the air is provided for mixing. Hence, mechanical aeration system was developed. For this surface aerators either fixed or floating type can be used (Fig. 10.4). The rectangular aeration tanks are divided into squares and for each square section one mixer is provided, i.e., mechanical aerator. The impellers are so adjusted that when electric motors start, they suck the wastewater from the centre, with or without uptare tube support, and throw it in the form of a thin spray over the surface of the wastewater. When the wastewater is sprayed in the air more surface area of wastewater is thus brought in contact with the air and hence oxygen mass transfer will occur at accelerated rate.

10.1.4 Types of Activated Sludge Process

There are different variants of ASP based on certain operational conditions. Mixing regime is one of the parameters based on which the aeration tanks are categorised as plug flow and completely mixed. In complete mixing, the contents of the aeration tank are well stirred and uniform throughout. However, in a plug flow type, an orderly flow of mixed liquor is maintained without any mixing taking place along the path of flow

Fig. 10.4 Typical arrangement of the surface aerator supported on conical bottom tube

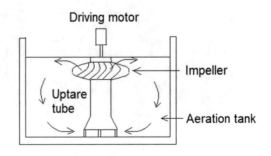

apart from the lateral mixing (Von Sperling, 2015). Thus, longitudinal mixing is considered negligible in plug flow ASP. Description of each variant of ASP is provided in this section.

Classification Based on the F/M ratio or MCRT: Based on the F/M ratio or mean cell residence time (MCRT) the variants of ASP are conventional activated sludge process and extended aeration system. A system characterised by low MCRT and higher F/M can be categorised as conventional activated sludge process or it could be completely mixed ASP with low retention time and MCRT. Whereas, the one having high MCRT and lower F/M falls under extended aeration system.

Conventional ASP: In conventional ASP the flow model adopted in aeration tank is plug flow type. Both the influent wastewater and recycled sludge enter at the head of the aeration tank and are aerated for about 5 h for sewage treatment (Fig. 10.5). The influent and recycled sludge are mixed by the action of the diffusers or mechanical aerators. Rate of aeration is constant throughout the length of the aeration tank. During the aeration period the adsorption, flocculation and oxidation of organic matter take place. The F/M ratio of 0.2–0.4 kg BOD/kg VSS day and volumetric loading rate of 0.3–0.6 kg BOD/m^3 day are used for designing this type of ASP. Lower mixed liquor suspended solids (MLSS) concentration is maintained in the aeration tank of the order of 1500–3000 mg/L and mean cell residence time of 5–15 days is maintained. The hydraulic retention time (HRT) of 4–8 h is required for sewage treatment. Higher HRT may be required for treatment of industrial wastewaters having higher influent BOD concentration. The sludge recirculation ratio for this type of ASP is generally in the range of 0.25–0.5.

Tapered aeration: Based on the physical configuration of the aeration basin, there are several variants of the continuous-flow activated sludge systems, which are tapered aeration, step aeration and completely mixed system. In plug flow type aeration tank BOD load is maximum at the inlet of aeration tank and it reduces as wastewater moves towards the effluent end. Thus, oxygen demand is maximum near the inlet of the aeration tank and it gradually get reduced while wastewater moves towards outlet end of the tank. Hence, accordingly in tapered aeration maximum air is applied at the beginning of aeration tank and it is reduced in steps towards the exit end, hence it is called as tapered aeration (Fig. 10.6).

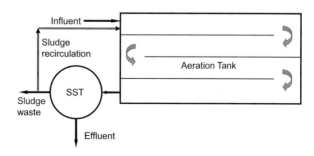

Fig. 10.5 Conventional activated sludge process following plug flow

Fig. 10.6 Schematic of tapered aeration activated sludge process

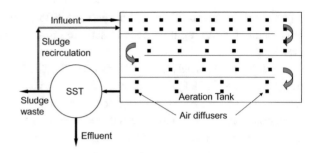

By tapered aeration the efficiency of the aeration unit will be increased and it will also result in overall economy by balancing oxygen demand and supply. The F/M ratio and volumetric loading rate of 0.2–0.4 kg BOD/kg VSS day and 0.3–0.6 kg BOD/m^3 day, respectively, are adopted in design of tapered aeration ASP. Other design recommendations are mean cell residence time of 5–15 days, MLSS of 1500–3000 mg/L, HRT of 4–8 h and sludge recirculation ratio of 0.25–0.5. Although, the design loading rates are similar to conventional ASP, tapered aeration gives better performance results.

Step aeration: If the wastewater is admitted at more than one point along the aeration channel, rather than admitting entire wastewater at the entry of aeration tank, the process is called as a step aeration (Fig. 10.7). This will reduce the load on returned sludge as the return sludge is admitted at the inlet of aeration tank. In this type of ASP, the aeration is provided uniform throughout the aeration tank. The F/M ratio and volumetric loading rate of 0.2–0.4 kg BOD/kg VSS day and 0.6–1.0 kg BOD/m^3 day, respectively, are recommended for design. Other design recommendations are mean cell residence time of 5–15 days, MLSS of 2000–3500 mg/L, HRT of 3–5 h and sludge recirculation ratio of 0.25–0.75. In step aeration the design loading rates recommended are slightly higher than conventional ASP. Due to reduction in the organic load on the return sludge, by admitting wastewater at multiple points rather than single point, it gives better performance.

Fig. 10.7 Schematic of step aeration activated sludge process

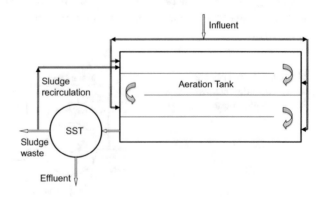

Fig. 10.8 Schematic of completely mixed activated sludge process

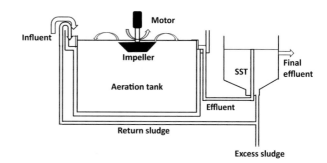

Completely mixed ASP: In completely mixed ASP, flow regime in aeration tank is completely mixed. The wastewater to be treated is distributed along with return sludge uniformly from one side of the tank and effluent is collected at another end of the tank (Fig. 10.8). The F/M ratio of 0.2–0.6 kg BOD/kg VSS day and volumetric loading rate of 0.8–2.0 kg BOD/m³ day are used for designing this completely mixed ASP. Higher MLSS is maintained in the aeration tank of the order of 3000–6000 mg/L and the mean cell residence time of 5–15 days is maintained. Generally, HRT of 3–5 h is sufficient for treatment of sewage, however higher retention time will be required for treatment of industrial wastewater having higher BOD value.

A completely mixed ASP has better capability to handle toxicity and fluctuations in organic matter concentration in the influent wastewater. This improved ability of handling organic matter fluctuations or marginal toxicity in the influent is because the incoming wastewater is getting completely mixed in the larger volume of the wastewater present in the aeration tank, hence resulting increment in organic matter concentration will be marginal and toxicity of any compound may not remain due to dilution. Hence, most of the industries prone to fluctuation in wastewater characteristic employ completely mixed activated sludge process for treatment of effluent.

Extended aeration ASP: An extended aeration ASP is one of the modifications of conventional ASP in which the endogenous respiration phase of the bacterial growth is targeted and hence it requires low organic loading rate and prolong aeration time. Since, the cells undergo endogenous respiration, the excess sludge generation in this process is low and the sludge has better dewatering capability. Thus, the excess sludge can be directly applied on the sand drying beds for drying and does not need treatment in anaerobic digester as required for excess sludge generated from conventional ASP. Unlike the conventional activated sludge process, in an extended aeration system generally the primary clarifier is not used and the secondary clarifier is designed with lower hydraulic loading rate to ensure maximum sludge retention. This type of activated sludge process is suitable for small capacity plant, such as package sewage treatment plant or industrial wastewater treatment plants of small capacity of less than 3000 m³/day. A comparative analysis of conventional and extended aeration ASP is furnished in Table 10.1.

Table 10.1 Comparison among the conventional and extended aeration activated sludge process for the treatment of domestic sewage

Parameter	Conventional ASP	Extended aeration ASP
Primary sedimentation unit	Present	Absent
F/M ratio (kg BOD/kg VSS day)	0.2–0.4	0.05–0.15
Sludge age (days)	5–15	20–30
Volume of aeration basin	Low (as HRT is 4–6 h)	High (as HRT is 12–24 h)
Area of SST	Lower	Higher
Treatment efficiency		
• BOD removal efficiency	85–95%	93–98%
• Suspended solid	85–95%	85–95%
• Ammonia	85–95%	90–95%
• Coliform	60–90%	70–95%
Oxygen requirement	Lower, due to the lower respiration by the biomass and because of the primary sedimentation of settleable organic solids	Higher, due to the oxygen requirement for the endogenous respiration of the biomass in aeration tank and also due to the non-existence of PST entry of suspended organic matter in aeration tank which demand additional oxygen
Energy requirements	Low (as the oxygen requirement is lower)	Higher (as the oxygen requirement is higher)

Lower F/M ratio of 0.05–0.15 kg BOD/kg VSS day and volumetric loading of 0.1–0.4 kg BOD/m³ day are used for designing extended aeration ASP. The MLSS concentration of the order of 3000–6000 mg/L is recommended and mean cell residence time of 20–30 days is generally adopted for design. Higher mean cell residence time is necessary to maintain endogenous growth phase of microorganisms. The HRT of 18–36 h is provided for treatment of wastewater like sewage. For this extended aeration the sludge recirculation ratio is generally in the range of 0.75–1.5.

An investigation carried out to compare the cost benefit of conventional ASP and extended aeration ASP in Tehran city revealed that an increase of MLSS concentration in aeration tank decreased the total project cost (Jafarinejad, 2017). Also, another research on treatment of wastewater from pulp and paper industry have proven that extended aeration ASP to be economical and technically optimum for treatment of low strength and medium strength wastewaters (Buyukkamaci & Koken, 2010). These variants of ASP discussed are having variation in the operational features, however the type of microbial culture developed for treatment of particular wastewater will be similar for any type of ASP chosen. Table 10.2 provides summary of the difference in features of major variants of ASPs.

Contact Stabilization: The BOD removal in ASP occurs in two phases, in the first phase absorption and in second phase actual oxidation of the absorbed organic matter occurs. Contact stabilization is developed to take advantage of the absorptive properties of activated sludge. The absorptive phase requires about 30–40 min, and during this phase most of the colloidal, finely divided suspended solids and dissolved organic matter get absorbed on the activated sludge recycled in the aeration tank. In contact stabilization, these two phases are separated out and they occur in two separate tanks (Fig. 10.9). The

Variants	Description
Completely mixed	• Uniform concentrations of pollutant throughout the reactor
	• Better overloads and toxicity handling capability
	• Suitable for industrial wastewater and for sewage
Plug flow	• Predominantly longitudinal dimension of aeration tank
	• More efficient than the complete-mixed reactor
	• Produces better settleable sludge
	• The oxygen demand decreases along the reactor
	• The DO concentration increases along the reactor
	• Lower capability to handle shock loads
Tapered aeration	• Similar to the plug flow
	• Aeration decreases along the length of the reactor
	• Optimal oxygen utilization hence better performance
Step aeration	• Wastewater is admitted at multiple points in aeration tank
	• The BOD load on the return sludge is reduced
	• More or less homogeneous oxygen demand
	• F/M is maintained within limited range
	• Flexibility in operation

Table. 10.2 Description of the main features of activated sludge processes

Fig. 10.9 Contact stabilization
activated sludge process

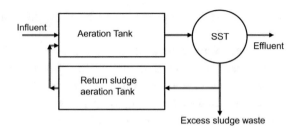

settled wastewater from primary sedimentation tank is mixed with re-aerated activated
sludge and aerated in the contact tank for the duration of 30 to 90 min. During this period
the organic matter present in the wastewater is absorbed on the sludge flocs. The sludge
with absorbed organic matter is separated from the wastewater in the SST. A portion of
the sludge is wasted to maintain requisite MLVSS concentration in the aeration tank. The
return sludge is aerated before sending it to aeration tank for 3 to 6 h in return sludge
aeration tank, where the absorbed organic matter is oxidized to produce energy and new
cells. After oxidation of absorbed organic matter this active cell mass is recycled to the
aeration tank.

The aeration tank volume requirement in contact stabilization is approximately 50%
of that of conventional ASP. It is thus possible to enhance the capacity of the existing
ASP by converting it to contact stabilization. Minor change in piping and aeration will
be required in this case. Contact stabilization is effective for treatment of wastewater
containing organic matter in colloidal form, as in case of sewage. However, its use for
the treatment of industrial wastewater may be limited when the organic matter present in
the wastewater is mostly in the dissolved form, since performance of contact stabilization
may not be that effective in this case. Existing treatment plant can be upgraded to contact
stabilization by changing the piping and providing partition in the aeration tank. This
modification will enhance the capacity of the existing plant.

Oxidation ditch: It is particular type of extended aeration activated sludge process,
where aeration tank is constructed in the ditch shape (oval shape) as indicated in the
Fig. 10.10. The oxidation ditch consists of a ring-shaped channel of 1.0–1.5 m depth of
liquid and of suitable width forming a trapezoidal or rectangular channel cross-section
for wastewater flow. An aeration rotor consisting of Kessener brush or paddle aerator are
placed across the ditch to provide aeration and wastewater circulation at velocity of about
0.3–0.6 m/s. When the wastewater will move with this velocity in the ditch it will ensure
that all the microbial mass will be kept in suspension, not allowing deposition of it, thus
ensuring contact between the microbes and organic matter present in the wastewater.

The oxidation ditch can be operated as intermittent with fill and draw cycles consisting
of (a) closing inlet valve and aerating the wastewater for duration equal to designed
detention time, (b) stopping aeration and circulation device and allowing the sludge to
settle down in the ditch itself, and (c) opening the inlet and outlet valve allowing the
incoming wastewater to displace the clarified effluent. In case of continuous flow through

Fig. 10.10 Oxidation ditch operated with continuous inflow of wastewater

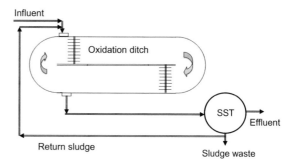

operation, where wastewater is continuously admitted and effluent leaves the ditch, it is called as 'Carrousel process'.

The mechanical aerators mounted on horizonal shaft are used to provide oxygen supply and at the same time to provide sufficient horizontal velocity for not allowing the cells to settle at the bottom of the ditch. In case of continuous flow of operation, separate sedimentation tank is used to settle the sludge and the settled sludge is re-circulated to maintain necessary MLVSS concentration in the oxidation ditch. Oxidation ditch is generally operated as extended aeration type activated sludge process and design values recommended for extended aeration ASP are used for design. The excess sludge generation in oxidation ditch is less than the conventional ASP and due to better dewatering ability of the excess sludge, it can be directly applied to the sand-beds for drying.

10.1.5 Bacterial Growth Kinetics in ASP

In aerobic biological wastewater treatment process, bacteria oxidise the organic matter, in the presence of oxygen and nutrients, to produce the end products carbon dioxide, bacterial cell mass and release ammonia. During the oxidation of organic matter, the organic matter acts as electron donor and the oxygen acts as electron acceptor. The oxidation and synthesis reaction responsible for the removal of biodegradable organic matter from the wastewater in the activated sludge process can be represented as per Eq. 10.5.

$$\underset{\text{(Organic matter)}}{\text{CHONS}} + O_2 + \text{Nutrients} \xrightarrow{Bacteria} CO_2 + H_2O + NH_4^+ + \underset{\text{(New cells)}}{C_5H_7O_2N} \qquad (10.5)$$

When the organic matter present in the wastewater is depleted, the bacteria are forced to undergo endogenous respiration. The endogenous respiration breaks down the cell protoplasm into elemental constituents, i.e., oxidation of protoplasm can be represented as per Eq. 10.6.

$$C_5H_2O_2N(113g) + 5O_2(160g) \rightarrow CO_2 + H_2O + NH_3 \qquad (10.6)$$

Thus, according to the stoichiometric relationship between the molecular mass, 160 g of oxygen is required for the stabilisation of 113 g of biodegradable cell mass. Hence, this relationship tells that for cellular sludge mass the oxygen requirement is equal to 1.42 mg O_2/mg VSS.

The total biomass present in the aeration tank is the matter of interest rather than the number of organisms for the mixed cultures used in the activated sludge process. The rate of biomass increase during the log growth phase is directly proportional to the initial biomass concentration, which is represented by the first order equation (Eq. 10.7).

$$\frac{dX}{dt} = \mu \cdot X \qquad (10.7)$$

However, taking into account the endogenous respiration occurring simultaneously, the above equation can be modified as Eq. 10.8.

$$\frac{dX}{dt} = \mu X - K_d X \qquad (10.8)$$

where, dX/dt represents the growth rate of biomass (g/m^3.day); X is the biomass concentration (g/m^3); μ is the specific growth rate (day^{-1}), which represents the mass of the cells produced per unit mass of the cells present per unit time; and K_d is the endogenous decay coefficient (day^{-1}).

As the maximum bacterial growth takes place during the exponential growth phase, μ is equal to μ_{max} when ample substrate concentration is available. However, the specific growth rate in operational ASP will never be equal to μ_{max}, and it depends on substrate concentration, which is maintained very low in the aeration tank. Hence, the growth condition for the microorganisms in the aeration tank will always be substrate limited growth condition, under which they will not be able to exhibit maximum growth rate. Monod (1949) showed experimentally that the biomass growth rate is a function of biomass concentration and limiting nutrient concentration. This Monod equation (Eq. 10.9) is the most widely used expression for representing the specific growth rate of bacteria as a function of the substrate utilization.

$$\mu = \mu_m \frac{S}{K_s + S} \qquad (10.9)$$

Substituting the value of μ in Eq. 10.8 will give the expression for time-rate of change of biomass as shown (Eq. 10.10.

$$\frac{dX}{dt} = \mu_m X \left(\frac{S}{K_s + S} \right) - K_d . X \qquad (10.10)$$

where, μ_m represents the maximum specific growth rate, day^{-1}; S is the limiting substrate concentration, mg/L; K_s is the substrate concentration corresponding to half of μ_{max}, also

Fig. 10.11 Specific growth rate versus limiting substrate concentration

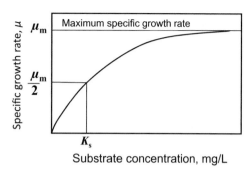

referred as half velocity constant, mg/L (Fig. 10.11). Thus, K_s is concentration of substrate S, when $\mu = \mu_m/2$, mg/L.

If all the substrate (i.e., organic matter, S) could be converted to biomass, then the substrate utilization rate will be equal to biomass growth rate. However, all the substrates cannot be converted into biomass because of energy liberating catabolic reaction, i.e., energy generation from oxidation of organic matter is must for supporting anabolic reaction (biomass synthesis) in the biochemical conversion process. Therefore, a cell yield coefficient ($Y < 1$) is used to correlate the substrate utilization rate with the biomass growth rate as per Eqs. 10.11 and 10.12.

$$-\frac{dS}{dt}Y = \frac{dX}{dt} \tag{10.11}$$

$$-\frac{dS}{dt} = \frac{1}{Y}\frac{\mu_m SX}{K_S + S} \tag{10.12}$$

where, Y is the cell yield coefficient, i.e., fraction of substrate converted to biomass, (g/m^3 of biomass)/(g/m^3 of substrate). The value of Y typically varies from 0.4 to 0.8 mg VSS/mg BOD (0.25 to 0.4 mg VSS/mg COD) in aerobic biological wastewater treatment systems.

The mean cell residence time θ_c can be expressed by the Eq. 10.13. For maintaining system under steady state condition, so as no accumulation of sludge occurs in the aeration tank, whatever the rate of sludge generation at the same rate the excess sludge should be wasted, i.e., dX/dt.

$$\theta_c = \frac{Total\ mass\ of\ VSS\ in\ reactor}{VSS\ wasted\ per\ day} = \frac{X}{dX/dt} \tag{10.13}$$

Substituting the value of dx/dt (Eq. 10.10) in Eq. 10.13 and rearranging the Eq. 10.14 shall be obtained.

$$\frac{1}{\theta_c} = \mu_m\left(\frac{S}{K_s + S}\right) - K_d \tag{10.14}$$

Solving the Eq. 10.14 for 'S', the general equation to estimate the effluent soluble BOD from a completely mixed activates sludge process can be obtained as expressed by Eq. 10.15.

$$S = \frac{K_s\,[1 + K_d \times \theta_c]}{[\mu_m \times \theta_c - K_d \times \theta_c - 1]} \qquad (10.15)$$

However, if in certain case S is significantly less than K_s, then $K_s + S$ will be nearly equal to K_s, hence this term can be substituted by K_s, which makes Eq. 10.14 simplified to Eq. 10.16 as a special case.

$$\frac{1}{\theta_c} = \mu_{max}\left(\frac{S}{K_s}\right) - K_d \qquad (10.16)$$

To summarise, in a complete-mix system under the steady state condition, since the parameters K_d, K_s, μ_m are constants, it can be concluded that the effluent BOD concentration (S) depends majorly on the MCRT (θ_c). Therefore, in the extended aeration system, due to the higher sludge age, the concentration of soluble effluent BOD obtained will be lower.

10.1.6   Process Analysis of Completely Mixed Reactor with Sludge Recycling

The design of an activated sludge process for oxidation of organic matter and thus removal form wastewater requires a computational approach that takes care of all the major aspects of the process. The approach should include the following: (a) for the given influent wastewater characteristics ensuring the desired effluent quality; (b) deciding the design mean cell residence time required; (c) determination of the biological solids production; (d) recommendation of the MLVSS and MLSS concentrations in the aeration tank; (e) determination of the aeration tank volume and hydraulic retention time required; (f) estimation of the excess sludge production so that the same amount of sludge to be wasted and recommending return sludge rate; (g) estimating the oxygen requirement; and (h) design of the secondary sedimentation tank.

10.1.6.1   Representation of the Substrate and Solids

Substrate i.e., BOD concertation present at different location in the activated sludge process is represented by 'S' and microorganism by 'X', i.e., MLVSS present in the case of aeration tank. The influent BOD concentration present in the incoming wastewater flow Q_o is represented by S_o and the effluent BOD concentration is expressed as S (Fig. 10.12). The solids in an activated sludge process can either be represented as total suspended solids or mixed liquor suspended solid (MLSS) or as a volatile suspended solids/ mixed liquor volatile suspended solid (X, MLVSS). The biomass in the aeration tank is normally

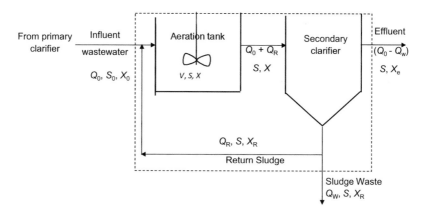

Fig. 10.12 Representation of the main variables in the analysis of activated sludge process

represented as 'X', whereas it is preferable to represent the solids in secondary clarifier in terms of suspended solids for the purpose of analysing the solid mass balance. The part of the sludge settled at the bottom of sedimentation tank, having a higher concentration of suspended solids, is recycled to the aeration tank and it is termed as return sludge (X_R). In order to maintain the equilibrium of the reactor, it is necessary to waste the excess sludge (Q_w) from the sedimentation tank. If the biomass production is not compensated with an equivalent wastage of sludge, the sludge build-up will occur in ASP and proper functioning of the process will be affected.

To elaborate further, the wastewater entering the aeration tank (influent) with a flow rate of Q_o (m^3/day) will have a BOD concentration of S_o (mg/L) and MLVSS concentration of X_o (mg/L). The treated effluent leaving the secondary sedimentation tank with a flow rate of Q_e will have a BOD concentration of S (mg/L) and VSS concentration of X_e (mg/L). The sludge leaves the sedimentation tank bottom with an underflow rate of Q_u (m^3/day), which is divided as excess sludge waste flow rate (Q_w, m^3/day), which is wasted and recirculation sludge flow rate (Q_R, m^3/day). The excess sludge moves out of the system, whereas the recirculation sludge enters the aeration tank along with influent wastewater. However, the VSS concentration in all the above three flow lines (underflow, excess sludge and recirculation sludge) will be the same and it is represented by X_R (mg/L). This is typically referred as the sludge recycle line from where the sludge wasting is generally carried out.

Kinetic models, which have been proposed to describe the activated sludge process, have been developed on the basis of steady-state conditions within the treatment system. The completely mixed reactor with sludge recycling is considered in the following discussion as a model for activated sludge process. The mass balance equations used to develop the kinetic models are based on the assumptions as described below:

- The biomass concentration in the influent is considered negligible. Though wastewater like sewage contains microorganisms responsible for biodegradation of organic matter, the concentration is lower as compared to the concentration maintained in the aeration tank, hence incoming biomass concentration in the influent is neglected. For other industrial wastewaters they are anyway free from the presence of aerobic bacteria participating in oxidation of organic matter, hence this assumption is adhering to reality.
- There is complete mixing condition that is prevailing in the aeration tank.
- The substrate concentration in the influent wastewater is considered to be constant.
- Waste stabilization is considered to occur only in the aeration tank. All biochemical oxidation reactions take place in the aeration basin only. Thus, volume of the reactor considered is the volume of the aeration tank alone. Due to completely mixed condition in aeration tank the substrate concentration in aeration tank is same as substrate concentration leaving the aeration tank.
- There is no microbial degradation of organic matter and no biomass growth that is occurring in the secondary clarifier. Hence, the substrate concentration entering the secondary clarifier and that in the final effluent are considered to be same.
- Steady state condition prevails throughout the system.
- The volume used for calculation of mean cell residence time includes volume of the aeration tank only.

10.1.6.2 Biomass Mass Balance

A mass balance for the microorganisms in the completely mixed reactor (Fig. 10.12) can be written as given in Eq. 10.17.

$$\begin{matrix} \text{Net rate of change in} & & \text{Rate at which} & & \text{Rate at which} \\ \text{biomass within the} & = & \text{biomass enters in} & - & \text{biomass leaves} \\ \text{system boundary} & & \text{the system} & & \text{the system} \end{matrix} \qquad (10.17)$$

The above mass balance statement for the microorganisms can be simplified to Eq. 10.18.

$$\text{Accumulation} = \frac{\text{Inflow of}}{\text{biomass}} + \frac{\text{Net growth}}{\text{of biomass}} - \frac{\text{Outflow of}}{\text{biomass}} \qquad (10.18)$$

It is assumed that steady state condition prevails in the system; hence, accumulation of biomass in the system will be zero, thus the rate at which the sludge is generated, at the same rate it is wasted per day from the system. Therefore the Eq. 10.18 can be rearranged as Eq. 10.19.

$$\begin{array}{c} \text{Influent} \\ \text{biomass} \end{array} + \begin{array}{c} \text{Biomass} \\ \text{production} \end{array} = \begin{array}{c} \text{Effluent} \\ \text{biomass} \end{array} + \begin{array}{c} \text{Wasted} \\ \text{biomass} \end{array} \qquad (10.19)$$

Substituting the variables declared earlier for the activated sludge process (Fig. 10.12), the Eq. 10.20 can be obtained.

$$Q_0 X_0 + V \frac{dX}{dt} = (Q_0 - Q_W) X_e + Q_W X_R \qquad (10.20)$$

where, Q_0 is the influent flow rate (m³/day); X_0 is influent biomass concentration (kg/m³); V is the volume of the aeration tank (m³); Q_W is flow rate of waste sludge (m³/day); X_e is the effluent biomass concentration (kg/m³); and X_R is the biomass concentration in the sludge wasted (kg of VSS/m³).

As per the basic assumption the biomass concentration in the influent wastewater and in the effluent from the secondary sedimentation tank is negligible, i.e., $X_0 = X_e = 0$. Therefore, after substituting these values in Eq. 10.20, it will be simplified to the Eq. 10.21.

$$V \frac{dX}{dt} = Q_W X_R \qquad (10.21)$$

Substituting Eq. 10.10 for dX/dt in Eq. 10.21, following Eq. 10.22 will be obtained.

$$V \left[\left(\frac{\mu_m S}{K_s + S} \right) X - K_d X \right] = Q_W X_R \qquad (10.22)$$

Rearranging will result in Eq. 10.23.

$$\left(\frac{\mu_m S}{K_s + S} \right) = \frac{Q_W X_R}{V X} + K_d \qquad (10.23)$$

If r_g' is considered to be the net growth of microorganisms, then from Eq. 10.22, $r_g' = Q_w X_R / V$.

Or we can write,

$$Q_w \cdot X_R / V \cdot X = r_g' / X \qquad (10.24)$$

Also,

$$r_g' = -Y.r_{su} - K_d \cdot X \qquad (10.25)$$

where, r_{su} is the substrate utilization rate, mass/unit volume time; and Y is the cell yield coefficient as explained earlier, mg cells/mg of substrate.

Substituting value of r_g' in Eq. 10.24, following Eq. 10.26 will be obtained.

$$Q_w \cdot X_R/V \cdot X = -(Y.r_{su}/X) - K_d \tag{10.26}$$

The left-hand side of the equation is the reciprocal of the mean cell residence time θ_c, hence substituting it in above equation will result in Eq. 10.27.

$$1/\theta_c = -(Y \cdot r_{su}/X) - K_d \tag{10.27}$$

Also the r_{su} can be expressed from the actual substrate utilization as per Eq. 10.28.

$$r_{su} = -Q(S_o - S)/V = (S_o - S)/\theta \tag{10.28}$$

where, θ is the hydraulic retention time (day).

From Eqs. 10.27 and 10.28 following relation given by Eq. 10.29 will be obtained.

$$1/\theta_c = [Y(S_o - S)/\theta_i X] - K_d \tag{10.29}$$

Solving for V after substituting $\theta = V/Q$ will result in following Eq. 10.30.

$$V = \frac{Q \cdot \theta_c \cdot Y(So - S)}{X(1 + K_d \cdot \theta_c)} \tag{10.30}$$

Above equation (Eq. 10.30) is used for estimating volume of the aeration tank when the kinetic coefficients are known. Estimating the volume of the aeration tank using this equation will give optimum volume of the aeration tank required once the values of the kinetic coefficients involved in the equation are known for that wastewater.

10.1.6.3   Substrate Mass Balance

Similar to the mass balance for biomass, a mass balance for the substrate in the completely mixed reactor (Fig. 10.12) using the volume of the aeration tank can be written as per Eq. 10.31.

$$
\begin{array}{lll}
\text{Net rate of change in} & \text{Rate at which} & \text{Rate at which} \\
\text{substrate inside the} \;=\; & \text{substrate enters} \;-\; & \text{substrate leaves} \\
\text{system boundary} & \text{in the system} & \text{the system}
\end{array}
\tag{10.31}
$$

When the steady state condition prevails in the system, the above mass balance for the substrate can be simplified as represented by Eq. 10.32.

$$
\begin{array}{llll}
\text{Inflow of} & \text{Consumption} & \text{Outflow of} & \text{Wasted} \\
\text{substrate} \;-\; & \text{of substrate} \;=\; & \text{substrate} \;+\; & \text{substrate}
\end{array}
\tag{10.32}
$$

Using the respective variable used in Fig. 10.12, the above Eq. 10.32 can be written as Eq. 10.33.

$$Q_0 S_0 - V\frac{dS}{dt} = (Q_0 - Q_W)S + Q_W S \tag{10.33}$$

Substituting Eq. 10.12 in Eq. 10.33, following Eq. 10.34 shall be obtained.

$$Q_0 S_0 + V\left[\frac{1}{Y}\left(\frac{\mu_m S X}{k_S + S}\right)\right] = (Q_0 - Q_W)S + Q_W S \tag{10.34}$$

After rearranging Eq. 10.34, the Eq. 10.35 can be obtained.

$$\frac{\mu_m S X}{k_S + S} = \frac{Q_0 Y}{V X}(S_0 - S) \tag{10.35}$$

Rearranging after combining with Eq. 10.23 to get the Eq. 10.36, which is used for predicting effluent BOD concentration of the completely mixed ASP. This is same as earlier Eq. 10.15 after substituting $K = \mu_m/Y$, i.e., it is maximum rate of substrate utilization per unit mass of microorganism.

$$S = \frac{K_s(1 + K_d.\theta_c)}{\theta_c(YK - K_d) - 1} \tag{10.36}$$

10.1.6.4   Estimation of Values of Other Operating Parameters

Hydraulic retention time: The HRT in the aeration tank of ASP can be estimated using Eq. 10.37.

$$\theta = \frac{V}{Q_0} \tag{10.37}$$

The usual practice is to keep the detention period of about 5 h while treating wastewater like sewage. The volume of aeration tank is also decided by considering the return sludge, which is about 25–75% of the wastewater volume as per the requirement to maintain F/M in desired range.

Mean cell residence time: The mean cell residence time representing sludge age (θ_c) is the mean duration in days up to which the sludge is being retained in the system. This is also called the solids retention time (SRT). It can be determined using simple relation of total mass of microorganism present in the aeration tank divided by the mass of microorganisms wasted from the system per day (Eq. 10.38).

$$\theta_C = \frac{Total\ mass\ of\ MLVSS\ in\ reactor}{VSS\ wasted\ per\ day} = \frac{VX}{Q_W X_R + (Q_0 - Q_W)X_e} \tag{10.38}$$

As the value of X_e is negligible as it is very less as compared to X_R, hence Eq. 10.38 will be simplified to Eq. 10.39.

$$\theta_C = \frac{VX}{Q_W X_R} \tag{10.39}$$

The SRT is higher than the HRT as a fraction of the sludge settled in secondary sedimentation tank is recycled back to the aeration tank. Thus, by maintaining higher SRT, substantial removal of organic matter can be obtained during only about 3–5 h of wastewater retention time in the aeration tank. Thus, by manipulating MCRT the BOD removal efficiency of the ASP can be manipulated. The typical MCRT values used for conventional activated sludge process and extended aeration activated sludge process are 5 - 15 days and 20–30 days, respectively.

Food to microorganism ratio: The BOD loading with regard to the microbial mass present in the system is expressed as the food to microorganism (F/M) ratio. The F/M ratio is one of the significant design and operational parameters of ASP. A balance between substrate consumption, biomass generation and sludge waste will help in achieving equilibrium in the system. The F/M ratio is responsible for the oxidation of organic matter. The type of activated sludge system can be classified by the F/M ratio adopted as stated below:

- Extended aeration, $0.05 < F/M < 0.15$ kg BOD_5/kg VSS day
- Conventional activated sludge system, $0.2 < F/M < 0.4$ kg BOD_5/kg VSS day
- Completely mixed, $0.2 < F/M < 0.6$ kg BOD_5/kg VSS day
- High-rate ASP, $0.4 < F/M < 1.5$ kg BOD_5/kg VSS day.

The F/M ratio, kg BOD_5/kg MLVSS day, is determined using Eq. 10.40, where $\theta = V/Q$.

$$\frac{F}{M} = \frac{\left[\text{ BOD of wastewater } (kg/m^3)\right]\left[\text{ Influent flow rate } (m^3/d)\right]}{\left[\text{ Reactor volume } (m^3)\right]\left[\text{ Reactor biomass } (kg \text{ of VSS } /m^3)\right]} = \frac{S_0 Q_0}{V X} = \frac{S_0}{\theta X}$$

(10.40)

Estimating quantity of excess sludge to be wasted: The excess sludge remaining in the secondary clarifier after being necessary fraction recycled to the aeration basin has to be wasted to maintain a stable concentration of MLVSS in the aeration tank. The excess sludge quantity increases with increase in F/M ratio and decreases with the increase in temperature. The excess sludge wasting can be done either from the sludge recycle line or directly from the aeration basin as mixed liquor. Although sludge wasting from sludge return line is convention and generally preferred due to lower volume of the sludge to be handled, however it is more desirable to waste the excess sludge from the aeration basin for better plant control. Sludge wasting from aeration basin is also beneficial for subsequent sludge thickening operations, as higher solid concentrations can be achieved when dilute mixed liquor is thickened rather than the concentrated sludge withdrawn from bottom of SST.

The excess sludge generation under steady state condition of operation can be estimated using Eq. 10.39 by solving for Q_w from the selected MCRT. Alternately following Eq. 10.41 shall be used for estimation of the quantity of the sludge generated per day, equal amount of it should be wasted per day to keep the system under equilibrium.

$$P_x = Y_{obs} Q_0 (S_0 - S) \times 10^{-3} \qquad (10.41)$$

where, P_x is the net mass of activated sludge produced each day, kg/day and Y_{obs} is the actual observed sludge yield $= Y/(1 + K_d \cdot \theta_c)$.

The net mass production of VSS (P_x) can also be estimated as the difference between the gross production of VSS (X_v) and biological destruction of cells (X_b) due to endogenous respiration in the system as represented by Eq. 10.42.

$$
\begin{aligned}
P_x &= \text{Gross production of } X_v - \text{Destruction of } X_b \\
&= Y \cdot Q_o \cdot (S_o - S) - K_d \cdot X_y \cdot V \qquad (10.42)
\end{aligned}
$$

where, Q_o is the influent flow (m³/day), S_o and S represents the respective influent and effluent substrate concentration (mg/L), Y is the sludge yield coefficient (g VSS produced per g BOD removed) and K_d represents the endogenous decay coefficient (day^{-1}). The typical values of Y and K_d are 0.4 to 0.7 g VSS/g BOD$_5$ removed and 0.06 to 0.10 g VSS/g VSS·day, respectively.

Sludge recycling: From the settled activated sludge in a secondary clarifier, fraction of it is returned to the aeration tank to mix with incoming wastewater to maintain necessary concentration of MLVSS in the aeration tank as per design recommendations. The return of activated sludge from the sedimentation tank to the aeration tank is essential to achieve desired degree of treatment. The return activated sludge (RAS) is important in maintaining high population of active aerobic microorganisms capable of degrading the organic matter entering in the aeration tank. The secondary clarifier operating condition as well as the RAS flow rates determines the concentration of VSS in the RAS. The RAS flow rate depends majorly on the concentration of the settled sludge in the secondary clarifier. The higher the active biomass concentration in the sludge the lower will be the RAS flow. This can be determined by sludge characteristics including sludge settleability (SVI) and thickening in the secondary clarifier, i.e., good settleability sludge results in better suspended solid concentration in the settled sludge.

The return sludge ratio can be determined from the mass balance of solids for aeration tank as per Eq. 10.43.

$$\text{Accumulation} = \text{Inflow} - \text{Outflow} + \text{Production} - \text{Consumption} \qquad (10.43)$$

Considering steady state condition, the mass accumulation in the system will be zero as the bacterial growth is balanced with the excess sludge removal, hence the difference between the production and consumption can be taken as equivalent to sludge wasted. Thus, the Eq. 10.43 gets simplified for mass balance of aeration tank to inflow of sludge to aeration tank is equal to outflow from aeration tank. Hence, the above mass balance can be rewritten based on Fig. 10.12 as per Eq. 10.44.

$$Q_0 X_Q + Q_R X_R = (Q + Q_R)X \tag{10.44}$$

where X_o, X_R and X represents the VSS concentration in inlet raw wastewater, RAS and effluent of aeration tank going to SST (same as MLVSS), respectively. Since, based on the assumption that microorganisms concentration in the influent is neglected, i.e., $X_o = 0$, hence the Eq. 10.45 will be obtained.

$$Q_R X_R = (Q + Q_R)X \tag{10.45}$$

Alternately, this Eq. 10.45 can also be understood better from writing a mass balance equation for aeration tank, where $Q_R X_R$ represents the sludge coming to the aeration tank and $(Q + Q_R)X$ represents the sludge leaving the aeration tank.

The return sludge ratio, $R = \frac{Q_R}{Q}$ can be obtained by rearranging the Eq. 10.45 as stated below in Eq. 10.46.

$$\frac{Q_R}{Q} = \frac{X}{X_R - X} \tag{10.46}$$

where, Q_R is recycle rate and Q is the influent flow rate of wastewater in aeration tank. The sludge settleability is determined by sludge volume index (SVI). If it is assumed that sedimentation of suspended solids in laboratory is similar to that in the secondary clarifier, then $X_R = (\text{VSS/SS ratio})10^6/\text{SVI}$. Values of SVI between 50 and 150 mL/g of SS indicate a good settling of the suspended solids. The X_R value may not be taken as more than 10,000 g of SS/m^3 unless separate thickeners are provided to concentrate the settled solids or secondary clarifier is designed to have a higher value of settled sludge concentration.

To summarize, the RAS flow and excess sludge waste flow in the activated sludge system have a major influence on the solid balance in the system. The balance between the solids in the reactor and secondary clarifier is determined by the RAS flow rate, whereas the excess sludge flow have an effect on the total cellular solids mass in the system.

Sludge quality: A number of design and operational parameters determine the quality of sludge generated in an activated sludge process. A very low value for MCRT (i.e., sludge age) shows a dispersion tendency of the bacteria instead of flocculent growth. Similarly, a very high value of MCRT implies floc that predominantly consist of a highly mineralised residue of endogenous respiration, with a small flocculation capacity. The type of ASP used also determines the nature of sludge developed. For instance, a plug flow type ASP produces sludge having a better settling characteristic than a completely-mix reactor.

Environmental conditions are yet another factor that affect the quality of sludge. As seen, the growth of filamentous microorganisms predominates at low dissolved oxygen concentrations as well as at low pH. Very low F/M ratio, high sulphide concentration and nutrient deficiency can also lead to growth of filamentous microorganisms. The filamentous microorganisms can disturb the settling of sludge and can cause bulking condition in the secondary clarifier, which can otherwise lead to movement of sludge out of secondary clarifier along with the effluent. The sludge quality is therefore an indirect indication of the effectiveness of the treatment process.

Estimating oxygen requirement: Oxygen is used as an electron acceptor in the energy metabolism of the aerobic heterotrophic microorganisms present in the activated sludge process. Oxygen is required in the activated sludge process for oxidation of the influent organic matter and to support endogenous respiration of the microorganisms. The aeration equipment must be capable of maintaining a dissolved oxygen concentration of about 2 mg/L in the aeration tank, while providing adequate mixing of solids and liquid phase.

In an aerobic biological wastewater treatment, the oxygen is consumed for oxidation of carbonaceous matter and also for the nitrification process if it is occurring. The oxidation of carbonaceous matter involves both the oxidation of influent organic matter as well as the endogenous respiration of the microorganisms. The carbonaceous oxygen demand is the demand of oxygen for removal of the ultimate BOD (BODu) from the system, which also correspond to the demand for substrate oxidation and endogenous respiration of microorganism. The BODu can be represented as BOD_5 multiplied by a conversion factor $(1/f)$, where f represents ratio of $BOD_5/BODu$. However, an investigation on effect of low DO on the efficiency of extended aeration process revealed that the low oxygen concentration was not limiting the development of the kinetics of process involved in the removal of organic matter (Bueno et al., 2019). Also, it was seen that a minimum duration of DO presence per day is required to prevent the nitrifying bacteria from being washed out of the system (Choubert et al., 2005).

The oxygen requirement for an activated sludge process can be estimated by knowing the ultimate BOD of the wastewater and the amount of biomass wasted from the system each day as per Eq. 10.47 (Metcalf & Eddy, 2003).

$$\text{Total O}_2 \text{ requirement } (kg/\text{ day }) = \frac{Q(S_0 - S) \times 10^{-3}}{f} - 1.42 Q_w X_R \qquad (10.47)$$

All the substrate that is organic matter present in the wastewater is not oxidized for energy. A portion of the organic carbon is oxidized to liberate energy and remaining is utilized for synthesis of new biomass, which will not demand oxygen. As it is assumed that the system is under steady state condition, there is no accumulation of biomass and the amount of biomass produced is equal to the amount of biomass wasted. Therefore, the equivalent amount of organic matter synthesized to new biomass is not oxidized in the system and exerts no oxygen demand. The oxygen requirement for oxidizing 1 unit of biomass = 1.42 units (as explained earlier in Eq. 10.6). Thus, the oxygen requirement for oxidation of biomass produced (equal to biomass wasted) as a result of substrate utilization is required to be subtracted from the theoretical oxygen requirement as given in Eq. 10.47 to get the actual oxygen requirement.

The above oxygen requirement corresponds to the oxygen demand for carbonaceous BOD removal. The oxygen demand for the oxidation of the nitrogenous matter (nitrification) is based on a stoichiometric relation with the oxidised ammonia. However, the organic nitrogen and ammonia, i.e., the total Kjeldahl nitrogen (TKN, mg N/L), generate oxygen demand for supporting nitrification process in presence of nitrifying bacteria in the activated sludge process. When nitrification has to be considered, the oxygen requirement will be given by Eq. 10.48.

$$\text{Total } O_2 \text{ required (kg/ day)} = \frac{Q(S_0 - S) \times 10^{-3}}{f} - 1.42 Q_w X_R$$
$$+ 4.57 Q_o (N_o - N) \times 10^{-3} \qquad (10.48)$$

where, N_o is the influent TKN concentration, mg/L, N is the effluent TKN concentration, mg/L and 4.57 is the amount of oxygen required for complete oxidation of TKN.

The air supply in aeration tank must be adequate to:

- Satisfy the BOD present in the wastewater,
- Satisfy the endogenous respiration of the microorganisms occurring in the system,
- Provide adequate mixing (15 to 30 kW/10^3 m^3) to keep biomass in suspension,
- Maintain minimum DO of 1 to 2 mg/L throughout the aeration tank.

Typical air requirement for conventional ASP is 30–55 m^3/kg of BOD removed. Whereas, for fine air bubble diffusers installed in aeration tank, the air requirement is 24–36 m^3/kg of BOD removed and for extended aeration ASP the air requirement is highest of the order of 75–115 m^3/kg of BOD removed. To meet the peak demand the safety factor of two should be used while designing aeration equipment.

10.1.7   Aeration and Mixing Systems

The oxygen dissolved in the aeration tank as well as the oxygen transfer efficiency plays a major role in determining the treatment efficiency of ASP. Oxygen is transferred to the mixed liquor in an aeration tank by dispersing compressed air bubbles through submerged diffusers with compressors. Alternately, using mechanical aeration the oxygen mass transfer occurs due to splashing of wastewater droplets in to atmosphere, thus increasing surface area of contact of wastewater with air and every time bringing new surface of wastewater in contact with air in aeration tank. An air diffuser is a perforated membrane, porous disc, or other device used for discharging air into the aeration basins. Air filters, blowers, air piping, etc. are the major constituents of an air diffusion system. The major factors that influence energy consumption in a diffused aeration system are the type of aeration equipment (fine bubble, medium bubble or coarse bubble diffuser), oxygen transfer efficiency, placement of diffuser, operating pressure and maintenance.

Oxygen transfer and utilization: Oxygen transfer is a two-step process. Gaseous oxygen is first dissolved in the mixed liquor by diffusers or mechanical aeration or both. The dissolved oxygen is utilized by the microorganisms in the process of metabolism of the organic matter. When the rate of oxygen utilization exceeds the rate of oxygen dissolution, the dissolved oxygen in the mixed liquor will be depleted. This condition should be therefore avoided. The dissolved oxygen should be measured periodically, especially during the period of peak loading, and the air supply should be adjusted accordingly. Typically, a minimum DO concentration of 1–2 mg/L is to be maintained in the aeration basin of an activated sludge process.

10.1.8 Secondary Sedimentation Tank for ASP

After wastewater being biologically treated in the aeration tank, the mixed liquor is passed to the SST, wherein the sludge, i.e., microbial biomass, settles down and the clarified effluent is discharged. A part of the settled sludge is recycled back to the aeration tank and the rest is wasted from bottom of the sedimentation tank. The design of secondary settling tank is important in the operation of ASP. The most commonly adopted shapes of SST is either central feeding circular tank or horizontal flow rectangular tank. The surface area of the sedimentation tank is the major aspect of its design. The area can be determined from either of the two methods as explained in Eqs. 10.49 and 10.50 satisfying other within stipulated range.

$$\text{Surface overflow rate (SOR)} = Q/A \tag{10.49}$$

where, Q is the influent flow to the sedimentation tank and A represents the surface area of the sedimentation tank.

The area of SST shall also satisfy the solid loading rate (SLR), which is the ratio of total solids applied to the surface area of tank as given by Eq. 10.50.

$$\text{SLR} = \frac{(Q + Q_R)X}{A} = \frac{Q_R \times X_R}{A} \tag{10.50}$$

where, Q is the influent flow (m³/day); Q_R is the RAS flow rate (m³/day); X is the MLSS concentration (kg of SS/m³); X_R is the returned sludge concentration expressed in kg of SS/m³; and A is the surface area of the sedimentation tank (m²).

Example: 10.1 Design a completely mixed activated sludge process to treat a primary treated wastewater flow of 2000 m³/day to meet the total effluent BOD concentration of less than or equal to 30 mg/L. The aeration tank is proposed to have mixed liquor volatile suspended solids concentration of 3500 mg/L and MCRT of 10 days. The influent wastewater to the aeration tank contains average BOD of 320 mg/L and TKN of 35 mg/L. Consider the cell yield coefficient Y having value of 0.6; endogenous coefficient K_d equal to 0.06 day^{-1}; and MLVSS/MLSS ratio of 0.8. The effluent of secondary sedimentation tank contains about 15–20 mg/L of biological solids with $\frac{2}{3}$ biodegradable fraction. The suspended solids concentration in return sludge is 1%, i.e., 10,000 mg/L. The BOD_5/BOD_u ratio is 0.68.

Solution

Wastewater flow Q = 2000 m^3/day; influent BOD (S_o) = 320 mg/L; effluent total BOD = 30 mg/L; influent TKN = 35 mg/L; VSS/SS = 0.8; MLVSS = 3500 mg/L, hence MLSS = 3500/0.8 = 4375 mg/L.

Determine the BOD removal efficiency required for the ASP

First estimate soluble BOD required in the effluent of ASP. Since the effluent contain up to 20 mg/L of biosolids with 67% biodegradable fraction, the BOD represented by these solids will be estimated as stated below:

Considering effluent solids concentration of 20 mg/L, the biodegradable solids in the effluent will be = 20 × 0.67 = 13.4 mg/L.

The ultimate BOD for these 13.4 mg/L of solids will be = 13.4 × 1.42 = 19 mg/L.

BOD$_5$ for these solids present in the effluent will be = 19 × 0.68 = 12.94 mg/L ~ 13 mg/L.

Hence, soluble effluent BOD required = 30–13 = 17 mg/L.

Therefore, the soluble BOD removal efficiency required = (320–17) × 100/320 = 94.69%

Whereas, overall BOD removal efficiency = (320–30) × 100/320 = 90.63%

Determination of aeration tank volume required

$$V = \frac{Y.Q.\theta_c(S_0 - S)}{X(1 + kd.\theta_c)}$$

$$V = \frac{0.6 \times 2000 \times 10 \times (320 - 17)}{3500(1 + 0.06 \times 10)}$$

$V = 649.29$ m^3, hence provide aeration tank volume of 650 m^3.

Calculate the mean hydraulic retention time for the aeration tank

$$\text{HRT} = \frac{V}{Q} = \frac{650}{2000} \times 24 = 7.8\,\text{h}$$

Though this HRT is greater than general guidelines, however this HRT is required for proper operation of ASP, hence it should to be provided.

Check for F/M ratio

$$\frac{F}{M} = \frac{Q S_0}{V X} = \frac{2000 \times 320}{650 \times 3500} = 0.28 \, \text{kgBOD/kg VSS.day}$$

This is within 0.2–0.6 kg BOD/kg VSS day range recommended for design, hence acceptable.

Check for volumetric loading

Volumetric loading $= Q \times S_0/V = 2000 \times 320 \times 10^{-3}/650 = 0.985$ kg BOD/m^3 day.
This is also within the range recommended for design, i.e., 0.8 to 2.0 kg BOD/m^3 day.

Estimating quantity of sludge waste

$$Y_{obs} = Y/(1 + K_d \times \theta_c) = 0.6/(1 + 0.06 \times 10) = 0.375 \, \text{mg/mg}$$

Therefore, mass of volatile waste activated sludge produced

$$P_x = Y_{obs} Q_0 (S_0 - S) \times 10^{-3} = 0.375 \times 2000(320 - 17) \times 10^{-3}$$

$= 227.25$ kg VSS/day.
Therefore, mass of sludge based on total SS $= 227.25/0.8 = 284.06$ kg SS/day.
With concentration of sludge in recycle line of SST of 10 g of SS/L, the sludge flow to be wasted
$= 284.06/10 = 28.41$ m^3/day.

Estimating sludge waste volume based on mean cell residence time

$$\theta_C = \frac{V X}{Q_W X_R} = \frac{650 \times 3500}{Q_W \times 10000 \times 0.8} = 10 \, \text{days}$$

Hence, $Q_w = 28.43$ m^3/day (when wasting is done from the recycled line of SST, which is same as estimated above).

Estimation of the return sludge ratio

The suspended solid concentration of sludge return line is 10,000 mg SS/L.
The return sludge ratio, $R = Q_R/Q$ can be obtained by writing mass balance for aeration tank as below:

$$3500(Q + Q_R) = 10000 \times 0.8 \times Q_R$$

Thus, after rearranging, $R = \frac{Q_R}{Q} = \frac{3500}{10000 \times 0.8 - 3500} = 0.78$.
Therefore, the return sludge flow $Q_R = 0.78Q = 1560$ m^3/day.

Estimation of the oxygen requirement

Oxygen demand for organic matter

$$\text{Total O}_2 \text{ requirement (kg/ day)} = \frac{Q(S_0 - S) \times 10^{-3}}{f} - 1.42 Q w X_R$$

$$\begin{aligned}
\text{kg of oxygen required} &= \left[(2000 \times (320 - 17) \times 10^{-3})/0.68\right] \\
&\quad - 1.42 \times 28.41 \times 8000 \times 10^{-3} \\
&= 568.44 \text{ kg O}_2/ \text{ day}
\end{aligned}$$

Oxygen demand for nitrification

Also, oxygen demand will be there for nitrification and as per the stoichiometry of 1 g TKN requires 4.57 g O_2 for conversion to NO_3^-.

$$\begin{aligned}
\text{Oxygen required (kg/ day)} &= 4.57 \times Q \times \text{TKN}/1000 \\
&= 4.57 \times 2000 \text{ m}^3/\text{day} \times 35 \text{ mg/L}/1000 \\
&= 319.9 \text{ kg O}_2/ \text{ day}
\end{aligned}$$

Hence total oxygen demand will be $= 568.44 + 319.9 = 888.34$ kg O_2/day.

Volume of air required, considering air contain 23% oxygen by weight and density of air 1.201 kg/m^3.

$= 888.34/(1.201 * 0.23) = 3215.94$ m^3/day.

Considering oxygen transfer efficiency of 8%, the air required $= 3215.94/0.08$.

$= 40{,}199.3$ m^3/day $= 27.92$ m^3/min.

Considering safety factor of 2, the air requirement is $= 2 \times 27.92 = 55.83$ m^3/min.

Check for air volume

Air requirement per unit volume $= 40{,}199.3/2000 = 20.09$ m^3/m^3.

Which is more than the limit of 3.75 to 15 m^3/m^3.

Air requirement per kg of $BOD_5 = 40{,}199.3/[(320-17) \times 2000 \times 10^{-3}]$.

$= 66.33$ m^3/kg of BOD_5.

This is also greater than the limit 30–55 m^3/kg of BOD_5 because of consideration of oxygen required for nitrification.

Example: 10.2 Design conventional ASP to treat soluble wastewater from bottle washing plant containing a soluble organic waste having a COD of 450 mg/L. From extensive laboratory studies for untreated wastewater the BOD_5/COD ratio was found to be 0.60. The average flow rate of effluent is 1.0 MLD, which is to be treated in ASP so that effluent SS and soluble BOD_5 should be less than 20 mg/L. Consider following conditions are applicable.

- Return sludge concentration = 8000 mg/L as SS = 6400 mg/L as VSS
- MLSS = 2500 mg/L and MLVSS/MLSS = 0.8
- Mean cell residence time θ_c = 8 days
- Y = 0.50 kg cells/ kg substrate (BOD) consumed, K_d = 0.06 day^{-1}.

Determine reactor volume, sludge wasting rate, recirculation ratio, oxygen requirement and hydraulic retention time for the reactor. Also determine specific substrate utilization rate and F/M ratio.

Solution
Soluble BOD$_5$ in the effluent is to be 20 mg/L. Also, influent COD is 450 mg/L, hence influent BOD will be $450 \times 0.6 = 270$ mg/L.

$$\text{Hence treatment efficiency required} = \frac{270 - 20}{270} \times 100 = 92.59\%$$

Estimating aeration tank volume required

$$V = \frac{Y.Q.\theta_c(S_0 - S)}{X(1 + K_d \cdot \theta_c)} = \frac{0.5 \times 10^3 \times 8(270 - 20)}{2500 \times 0.8(1 + 0.06 \times 8)} = 337.84 \, \text{m}^3$$

Estimating sludge production rate and hence sludge wasting

$$Y_{obs} = \frac{Y}{1 + K_d \cdot \theta c} = \frac{0.5}{1 + 0.06 \times 8} = 0.338 \frac{\text{kgVSS}}{\text{kgBODremoval}}$$

$$\text{Biomass production rate} = Y_{obs} \times Q \times (S_0 - S)$$

$$= 0.338 \times 1000 \times \frac{(270 - 20)}{1000} = 84.5 \frac{\text{kgVSS}}{\text{day}}$$

Sludge wasting rate from recycle line

$$\theta_c = \frac{V.X}{Qw.X_R}$$

$$Q_w = \frac{337.84 \times 2500 \times 0.8}{8 \times 6400} = 13.197 \, \text{m}^3/\text{day}$$

Sludge wasted in mass (kg)

$$\text{Wasted sludge} = \frac{8000}{1000} \times 13.197 = 105.575 \frac{\text{kgSS}}{\text{day}}$$

Estimating recycle ratio

$$Q_R \cdot X_R = (Q + Q_R)X$$

$$\frac{Q_R}{Q} = \frac{X}{X_R - X} = \frac{2000}{6400 - 2000} = 0.45$$

Hydraulic retention time $= \frac{V}{Q} = \frac{337.84}{1000} = 0.338\,day = 8.1\,h.$

Specific substrate utilization rate

$$U = \frac{S_0 - S}{\theta.X} = \frac{(270 - 20)}{0.338 \times 2000} = 0.37 \frac{kg\,BOD\,utilised}{kg\,VSS.d}$$

$$F/M = \frac{S_0}{\theta.X} = \frac{270}{0.338 \times 2000} = 0.399 \frac{kg\,BOD}{kg\,VSS.d}$$

$$\frac{F}{M} \times efficiency = 0.37 \frac{kg\,BOD\,utilised}{kg\,VSS.d} = U$$

$$\text{Oxygen requirement} = \frac{Q(S_0 - S)}{0.6} - 1.42(Px) = \frac{1000(270 - 20)}{0.6 \times 1000} - 1.42 \times 84.5$$

Thus, oxygen requirement $= 296.7$ kg O_2/day.

Considering factor of safety two provide oxygen supply capacity of 593.3 kg O_2/day.

10.2 Analysis of Gas Transfer

The two-film theory is generally applied for the analysis of the gas mass transfer in the liquid. This analysis is better than other theories to obtained reasonably accurate results. Figure 10.13, illustrates when a gas and water interface are in contact with each other. There exists a liquid film in the aqueous phase, where gradient in the concentration of gas getting soluble exists, with higher concentration at gas–water interface than the bulk liquid. Whereas, there also exists a gradient in the gas film where concentration of the particular gas of interest is varying with least concentration of that gas at gas–water interface than in bulk gas phase, due to solubility of the gas in liquid phase.

Based on the two films that are existing each in the gas and in liquid phase, it can be inferred that these two films provide resistance to the passage of gas molecules from gas phase to the liquid phase. The gas which is slightly soluble encounters primary resistance to mass transfer from liquid film. Whereas, a highly soluble gases encounter primary resistance from gaseous film for the mass transfer. The gases with intermediate solubility encounter significant resistance from both films for mass transfer.

Fig. 10.13 Gas–water interface in contact where phase transfer is occurring. P and Pi are partial pressure of solute in bulk gas and at the interface, atm, respectively. C and Ci are solute concentration in bulk water and at interfaces, mg/L, in the liquid phase, respectively

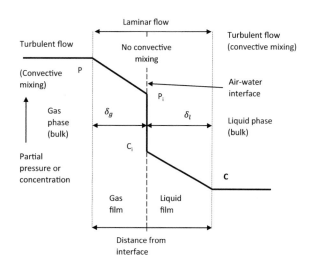

While any gas is getting solubilized in the liquid phase, the rate of gas transfer is proportional to the difference between equilibrium concentration and existing concentration of the gas in solution, i.e., difference between saturation concentration and existing concentration. This will be represented by Eq. 10.51.

$$r_m = K_g A(C_s - C) \qquad (10.51)$$

where, r_m is the rate of mass transfer, K_g is the coefficient of diffusion for gas; A is the area through which gas is diffusing, i.e., area of gas–water interface; C_s is saturation concentration of gas in solution; and C is existing concentration of gas in solution.

Under condition of mass transfer in field $r_m = V \cdot {}^{dc}/_{dt}$ and $r_c = \frac{r_m}{V}$; thus, the rate of change of concentration r_c can be expressed as per Eq. 10.52, where V is liquid volume.

$$r_c = \frac{dc}{dt} = K_g \frac{A}{V}(C_s - C) \qquad (10.52)$$

The term $K_g \left(\frac{A}{V}\right)$ is replaced by '$K_L a$' a proportionality factor related to existing exposer conditions, thus the expression 10.53 can be obtained after this replacement.

$$r_c = \frac{dc}{dt} = K_L a(C_s - C) \qquad (10.53)$$

where, r_c is rate of change of concentration, mg/L sec, and $K_L a$ represents overall mass transfer coefficient, sec^{-1}.

Integrating between the limits $C = C_0$ at $t = 0$ and $C = C_t$ at $t = t$, Eq. 10.54 can be obtained.

$$\int_c^{c_t} \frac{dc}{C_s - C} = K_L a \int_0^t dt \tag{10.54}$$

Solving, the Eq. 10.55 can be obtained

$$\frac{C_s - C_t}{C_s - C_0} = e^{-(K_L a).t} \tag{10.55}$$

where, (Cs–Ct) and (Cs–Co) represents the final and initial gas saturation deficits, respectively.

In case of already supersaturated solution of a particular gas instead of the gas getting solubilized it will be released from the liquid phase (from solution) to the gas phase. Thus, for such supersaturated solution, wherein the gas is to be degassed, then Eq. 10.55 can be modified as Eq. 10.56.

$$\frac{C_t - C_s}{C_o - C_s} = e^{-(K_L a).t} \tag{10.56}$$

10.2.1 Evaluation of O_2 Transfer Coefficient

Transfer of O_2 in clean water: For evaluating the oxygen mass transfer coefficient ($K_L a$) for the clean water, first remove the DO from known volume of water by adding sodium sulphite. After removal of DO completely, carry out reoxygenation to reach near the saturation level. During this reaeration monitor the DO concentration at several representative points in the aeration basin. The data obtained is analyzed by above equation (Eq. 10.55) to estimate $K_L a$ and equilibrium concentration $C_x{}^*$. This $C_x{}^*$ is obtained when aeration period approaches to infinity. The term C_s is substituted by $C_x{}^*$ considering this is close to the saturation concentration that could be achieved under field condition. A nonlinear regression is employed to fit the equation to the DO profile measured at each determination point. In this way, estimate $K_L a$ and $C_x{}^*$ for each point. Adjust it to standard conditions (i.e., temperature of 20 °C) and average it to get overall $K_L a$.

Evaluation of oxygen transfer coefficient in case of wastewater: In ASP, the $K_L a$ can be determined by accounting the uptake rate of O_2 by microorganisms simultaneously, while oxygen is being transferred to the aeration basin by the aerators provided. Typical O_2 concentration of 1–2 mg/L is generally maintained in aeration basin, and it is used by microorganisms rapidly. Thus, rate of change of oxygen concentration can be expressed as per Eq. 10.57

$$\frac{dc}{dt} = K_L a(C_s - C) - r_{micro} \tag{10.57}$$

where, r_{micro} represents the rate of O_2 uptake by the microorganism, which is typically 2 to 7 g/g of MLVSS day. After reaching steady state of operation the DO level will become constant, hence dc/dt will be zero, thus from the Eq. 10.57 one can get $r_{micro} = K_La (C_s - C)$. The value of r_{micro} can be determined in laboratory by using Warburg apparatus or respirometer. Once value of r_{micro} is known, using Eq. 10.58 the value of K_La can be obtained.

$$K_{La} = \frac{r_{micro}}{C_s - C} \qquad (10.58)$$

10.2.2 Factors Affecting Oxygen Transfer

(a) **Effect of temperature**: The effect of temperature is similar as that in the case of BOD reaction rate coefficient. Using van't Hoff-Arrhenius relationship (Eq. 10.59) the K_La determined at 20 °C can be converted to other working temperatures.

$$K_{La(T)} = K_{La(20°C)}\theta^{(T-20)} \qquad (10.59)$$

where, K_La $_{(T)}$ is oxygen mass-transfer coefficient at temperature T °C, h^{-1}; K_La $_{(20 °C)}$ is the oxygen mass transfer coefficient at 20 °C, h^{-1}; θ is temperature coefficient and value vary with test conditions in the range of 1.015–1.040. For diffused and mechanical aeration device typical θ value considered is 1.024.

(b) **Effect of mixing intensity and tank geometry**: The differences in mixing intensities under standard test condition than the field condition will affect the oxygen mass transfer. Also, the geometry of the aeration basin will have effect on the oxygen mass transfer, because of differences in mixing and area of gas–water interface to volume ratio. This effect is difficult to estimate on theoretical basis. While designing the ASP the aeration devices are selected based on efficiency of O_2 transfer provided by manufacturer under standard test condition. However, under field condition this oxygen transfer efficiency is expected to reduce due to deviations from standard test condition. This efficiency of oxygen transfer is strongly related to K_La value associated with given aeration unit. In most cases aeration devices are rated under standard test condition with range of operating parameters using tap water, i.e., using low TDS water. Hence, to account for these deviations the correction factor (α) is used to estimate actual K_La for aeration system under field conditions. This correction is represented by Eq. 10.60.

$$\alpha = \frac{K_La(wastewater)}{K_La(tapwater)} \qquad (10.60)$$

This value of α will vary from 0.3 to 1.0 with type of aeration device used, basin geometry adopted, degree of mixing occurring and characteristics of wastewater being aerated. For diffused air aeration system typically α will be in the range of 0.4–0.8 and for mechanical aeration it will typically vary from 0.6 to 1.0.

(c) **Effect of wastewater characteristics**: Due to difference in saturation concentration value of fresh water used during standard test condition and wastewater being aerated under field condition a correction factor 'β' (Eq. 10.61) is used to correct the test result for differences in O_2 solubility due to constituents, such as salts, particulates and surface-active substances. This correction factor β will be in the range of 0.7–0.98 and a value of 0.95 is generally used for wastewater.

$$\beta = \frac{C_s(wastewater)}{C_s(tapwater)} \qquad (10.61)$$

10.2.3 Application of Correction Factors

These correction factors are applied to predict field oxygen transfer efficiency based on the measurements made in experimental test facilities. For mechanical surface aerator Eq. 10.62 shall be used for evaluating oxygen transfer under field conditions.

$$O_f = O_T \frac{(\beta C_S - C_W)}{C_{S20}} \theta^{(T-20)}(\alpha) \qquad (10.62)$$

where, O_f is the actual O_2 mass transfer rate under field conditions, kg O_2/kW h; O_T is the standardized O_2 mass transfer rate under test conditions, i.e., at 20 °C and zero DO, kg O_2/kW h, and $O_T = K_La\ V\ C_s$. The C_S is O_2 saturation concentration for tap water at field–operating conditions, mg/L; C_W is the operating dissolved oxygen concentration in wastewater, mg/L; and C_{S20} is O_2 saturation concentration for tap water at 20 °C, mg/L.

For diffused air aeration system, the value of C_S must be corrected to account for higher than atmospheric O_2 saturation concentration achieved due to application of air at the reactor bottom. The C_s is taken as average DO attained after infinite time of aeration under experimental conditions.

Example: 10.3: Oxygen mass transfer

An activated sludge process is being designed using the diffused air aeration system. For the period of peak oxygen demand in the aeration basin following design values are adopted.

Oxygen transfer requirement = 1 kg of O_2/kg of BOD applied

Temperature of mixed liquor = 17 °C

Minimum allowable DO = 1.5 mg/L

$\alpha = 0.70$; $\beta = 0.90$

Oxygen transfer coefficient $(K_La)_{20} = 11.5$ h^{-1}

O_2 transfer efficiency = 8%

Pressure at mid depth of diffusers = 830 mm of Hg

Consider C_S at 830 mm Hg and 17 °C as 10.56 mg/L and C_S at 20 °C under normal atmospheric pressure is 9.8 mg/L

Evaluate the rate of oxygen transfer (*dc/dt*) and what will be maximum BOD volumetric loading rate applicable. Also estimate the volume of air required per kg of BOD applied.

Solution

Correcting $(K_La)_{20}$ for 17 °C.

K_La at 17°C $= (KLa)_{20} (1.024)^{17-20} = 11.5 \times (1.024)^{-3} = 10.71$ h^{-1}

Now dc/dt $= \alpha K_La(\beta C_S - C_t)$.

$= 0.70 \times 10.71(0.90 \times 10.56 - 1.5) = 60.0$ mg/L h

Considering 1 kg of oxygen utilized per kg of BOD applied (which will change if actual cell production and ultimate BOD is considered), then maximum applicable volumetric loading rate will be determined as stated below.

Maximum applicable BOD loading rate $= 60\frac{\text{mg}}{\text{L.h}} = 1.44 \frac{\text{kg BOD}}{\text{m}^3.\text{day}}$.

Thus, considering the oxygen transfer rate and keeping some margin a volumetric BOD loading rate up to 1.3 kg BOD/m^3.day can be recommended.

The efficiency of oxygen transfer is proportional to the rate of oxygen transfer. For $(K_La)_{20}$ the oxygen transfer efficiency is 8%, hence for actual field condition determining the oxygen transfer efficiency.

$$E_{\text{actual}} = 8.0\frac{\alpha.K_{La}.(\beta Cs - Ct)}{(K_{la})_{20}(Cs)_{20}}$$

$$= \frac{8.0 \times 0.7 \times 10.71(0.9 \times 10.56 - 1.5)}{11.5 \times 9.8}$$

$$= 4.26\%$$

The volume of standard air required at an actual oxygen transfer efficiency of 4.26% per kg of BOD load, (considering 1 kg O_2 utilized per kg of BOD applied) is

$$\frac{1.0 \, \text{kg of O}_2}{0.0426 \times 0.2797 \, \text{kg of O}_2/\text{m}^3} = 83.93 \, \text{m}^3 \text{ of air /kg of BOD applied}$$

One cubic meter of air at temperature and pressure of 17 °C and 760 mm, respectively, has density of 1.216 kg/m^3 and considering oxygen weight fraction in air as 23%, thus oxygen content in air is 0.2797 kg O$_2$/m^3.

10.3 Sequencing Batch Reactor

In an activated sludge process wastewater is treated in continuous or flow through basis, whereas in sequencing batch reactor (SBR) wastewater is treated in batches, i.e., sequences of operation are performed in the same tank. In SBR, the unit operation, such as primary sedimentation, unit process of biochemical oxidation of organic matter and settling of the bacterial cells produced are occurring in the same aeration tank. The five stages generally involved in this sequence are: fill cycle, reaction cycle, settle cycle, decant cycle and an ideal phase is used sometimes, which could be optional.

This process is popular because entire treatment stages occur in one reactor basin, however multiple SBRs are required to handle the wastewater continuously received at the treatment plant. Each of the SBR is operated under staggered cycle of operation to handle the inflow received, thus at any given time there will be one SBR under the fill cycle. These are generally constructed modular, hence require less space. In the treated effluent total suspended solids concentration of less than 10 mg/L can be achieved consistently through the use of effective decanters that eliminate the need for a separate clarifier. The treatment cycle can be adjusted to undergo aerobic, anaerobic/anoxic conditions in order to achieve biological organic matter and nutrient removal, including nitrification, denitrification and some phosphorus removal.

In brief, the fill cycle consists of filling the tank with wastewater to be treated. The aeration can be put on or off during this feed cycle. The reaction cycle deals with completion of biochemical reactions with the assistance of microorganisms. Continuous aeration is provided throughout this cycle. Then in settling phase, aeration is stopped and it allows the solids–liquid separation, which functions similar to a secondary sedimentation tank in a conventional ASP. This is followed by the decant phase, wherein the treated effluent is drawn from the system (Fig. 10.14). The final phase is named idle, wherein the main objective is to adjust the operational cycle of one reactor with the operational cycle of another and it is only used with multiple tanks.

During this idle phase the aeration can be put on or off. In absence of aeration, it will help in supporting denitrification of nitrified wastewater present in SBR. The idle stage is redundant in a SBR system, wherein the time is efficiently adjusted to match the filling stage of one reactor with react, settle and decant phase of other SBRs. To support denitrification in this case an anoxic compartment is maintained within the SBR

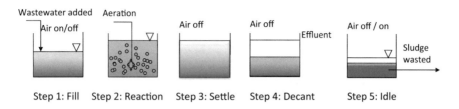

Step 1: Fill Step 2: Reaction Step 3: Settle Step 4: Decant Step 5: Idle

Fig. 10.14 Different cycles of operation involved in the sequencing batch reactor

and internal circulation of the wastewater from aerobic zone to anoxic zone is practiced to support denitrification.

Since all the wastewater receive equal amount of treatment in SBR, due to batch nature of operation, the process performance is reliable. Also, the performance of the process is not affected by the fluctuation in SVI of the sludge, as the sludge is retained in the same tank. However, since among the multiple SBRs present in the treatment plant each one is under different phase of operation, manual control of the process is not practical and automation and control is essential for successful operation of SBR to have reliable performance. The major advantage of this technology compared to the conventional activated sludge process is generally less space requirement due to modular design and no separate settling tank required, low aeration, and automated operation of these SBRs.

The decant volume varies between 25 and 30% of the total reactor volume. The settled sludge is partly removed from the system after specified intervals to maintain the required effective volume of the reactor vessel. As SBR is essentially a suspended growth process, hence the shape and type of the bio-aggregates is vital towards its functioning. Moreover, as the sludge settling is an integral stage of SBR technology, hence the shape of the formed bio-aggregates that favour such settling is essential.

Past investigations indicate that flocculant particles tend to settle less, while granular bio-aggregates enables good settling. This aggregate formation of biomass is governed by different factors both in case of aerobic as well as anaerobic SBRs. To aid in the formation of granules, presence of calcium in the influent wastewater has been proved to be beneficial. In addition to such inorganic cementing fillers, secretion and role of extra cellular polymeric substances (EPS) is also vital (Flemming et al., 2007). The EPS, as the name suggests is a cellular secretion consisting of hydrated polymers of proteins, polysaccharides and DNA (Flemming et al., 2007). The secretion of EPS is often in response to any environmental signal. However, attainment of granular configuration for the bio-aggregates is a complex phenomenon and it is yet to be understood completely. Among other things, maintaining a low F/M ratio and a high relative shear ensures granulation of sludge as smaller and denser granules are formed in such case (Beun et al., 1999). Apart from maintaining the F/M ratio, the following factors affect the size of the aerobic granular biomass formed in SBR (Liu et al., 2005):

- Substrate composition
- Organic loading rate
- Hydrodynamic shear force
- Reactor configuration
- Dissolved oxygen concentration
- Solid retention time
- Volume exchange ratio and
- Settling time.

According to Liu and co-researchers, irrespective of all other parameters, the formation of granules in the aerobic SBR is governed by the volume exchange ratio and the settling time (Liu et al., 2005). In other words, no matter what optimization is done with other operational parameters, without appropriate settling time and volume exchange ratio, granulation would not be possible. It is worth noting here that the volume exchange ratio in SBR is the ratio of the volume of the effluent withdrawn to the total volume of the reactor. Adding on to this hypothesis, Arrojo et al. (2004) suggested that discharge time is another parameter that affects the formation of granules. Combining these three assumptions, Liu et al. (2005) deduced a unified equation to choose the appropriate settling time, discharge time and volume exchange ratio as given in Eq. 10.63.

$$V_{s,min} = \frac{L}{t_s + \frac{(t_d - t_{d,min})^2}{t_d}}$$

(10.63)

where, $V_{s,min}$ is the minimum settling velocity; L is the depth of the outlet port in the SBR tank or in other words, the depth traversed in settling time t_s; t_d is the discharge time and $t_{d,min}$ is the minimum discharge time.

The design of SBR is based on the assumption that the reactor is operated in batch mode, which means flow rate Q is zero. Hence the substrate mass balance equation can be modified to represent the equation applicable for SBR (Eq. 10.64).

$$K_s \ln \frac{S_0}{S_t} + (S_0 - S_t) = X \left(\frac{\mu_m}{Y} \right) t$$

(10.64)

where, S_0 is the initial substrate concentration, S_t is the substrate concentration at time 't', the other terms bearing their usual meaning as defined previously. This equation can be modified further to represent the kinetics of nitrification by replacing S with N, the total Kjeldahl nitrogen concentration, X with X_n, the concentration of nitrifiers in the tank, and corresponding Monod coefficients for nitrification represented by μ_{mn} and Y_n. Further, in order to account for the effect of dissolved oxygen on the growth of the nitrifying bacteria, a term is added to right hand side to obtain a final equation as given by Eq. 10.65.

$$K_s \ln \frac{N_0}{N_t} + (N_0 - N_t) = X \left(\frac{\mu_{mn}}{Y_n} \right) \left(\frac{DO}{K_o + DO} \right) t \qquad (10.65)$$

where, DO is the dissolved oxygen concentration in mg/L and K_0 is the half saturation constant for dissolved oxygen in mg/L. These equations can be used for designing
the SBR using an iterative method by determining the fill and decant volume, aeration
requirements, etc.

The simplicity of SBR technology has led to the application of this technology for
treatment of different types of wastewaters, ranging from sewage to industrial wastewaters. For sewage treatment, different strategies have been implemented in SBR-based
technologies. Incorporation of polyurethane foam cubes as suspended media inside the
SBR tank with an aim of enhancing the solids retention was an effective strategy adopted
for a pilot-scale SBR having 1.2 m^3 volume (Sarti et al., 2007). The SBR technology
has also been explored for anaerobic ammonia oxidation process in different experimental cases. A case study for a period of 500 days demonstrated successful development of
DEAMOX process (a combination of partial denitrification and ANAMMOX) for removal
of 94% of total nitrogen with initial NH$_4$-N and NO$_3$-N concentrations of 64 and 69 mg
L^{-1}, respectively (Du et al., 2017).

The SBR technology has also been extensively used for treatment of landfill leachate
(Jagaba et al., 2021), poultry industry wastewater (Su et al., 2018), for removal of
pharmaceutical compounds from wastewater (Kołecka et al., 2020), agro-industry based
wastewater (Lim & Vadivelu, 2014), etc. Different variants of SBR, namely sequencing
biofilm batch reactor, sequencing granular batch reactor, cyclic activated sludge process,
intermediate cycle extended aeration system, etc., have been developed over time (Dutta &
Sarkar, 2015). For further details of these developed SBR variants, the review done by
Dutta and Sarkar (2015) can be referred.

The SBR has no secondary clarifier and more often does not require a primary clarifier.
The size of the SBR tanks is dependent on wastewater characteristics and site conditions;
moreover, the SBR system can be implemented if space is limited at the proposed site
(Chan et al., 2010). According to an investigation of dairy wastewater treatment using
SBR; 90% of COD, 80% of N, and 67% of P removal efficiencies were achieved at a
cycle time of 8 h (Schwarzenbeck et al., 2005). Another investigation revealed that the
low-strength municipal wastewater treatment resulted in 90% of COD, and 95% of N
removal in SBR (Ni et al., 2009). Therefore, SBR has significant potential to reduce
COD with simultaneous removal of nutrients like N and P.

The SBR has even shown up to 90–97% of COD and nitrogen removal efficiency while
treating sewage (Dan et al., 2021). Also, a total phosphorous removal up to 76% was
achieved in a SBR operated in Wafangdian. The same investigation reported the temperature dependence of the process, wherein a higher phosphorous removal was achieved in
winter and lower removal in summer (Li et al., 2008). Another investigation has reported
biodiesel recovery from the excess sludge generated in SBR (Hatami et al., 2021). The

technology has also proven its efficiency in treating a variety of wastewater streams including domestic wastewater, pharmaceutical wastewater (Mareai et al., 2020), swine wastewater (Dan et al., 2020), etc.

As it can be seen that the SBR technology has been implemented for different types of wastewater and target pollutants, which highlights the obvious fact that the design strategy for SBR will vary for each type. However, a few general design guidelines for the treatment of sewage can be summed up to aid the designer. The typical fill volume of SBR varies between 20 and 30%. In special cases with advanced decanting techniques, the fill volume can be assumed to be 40%. To ensure granulation in a conventional type SBR without any inert media, organic loading rate should be kept low as compared to other biological processes. The CPHEEO (2013) stipulates the F/M ratio of SBRs to be kept similar to extended aeration processes. The typical design guidelines for the SBR as per CPHEEO Part 7 (2013) is given in Table 10.3.

Example: 10.4: Design of SBR Design a SBR for a sewage treatment plant of 10 MLD capacity with the following influent characteristics as given in Table 10.4. Also determine the oxygen requirement and the decant pump rate. The MLVSS concentration in the SBR shall be 3500 mg/L. The effluent characteristics should be meeting quality norms as stated in the Table 10.4. Consider MLVSS/MLSS ratio to be 0.7. The specific sludge yield can be considered as 0.4 gVSS/gCOD. The diffused air aeration system installed provides 3% of O_2 transfer efficiency per m depth of water column under standard condition. Consider pressure at mid depth of diffusers as 830 mm of Hg. The saturation concentration of oxygen at 830 mm Hg and at water temperature of 17 °C is 10.56 mg/L and C_S at 20 °C under

Table. 10.3 Typical design parameters for SBR

Sl. No	Parameters	Units	Continuous flow and intermittent decant	Intermittent flow and Intermittent decant
1	F/M ratio	kg BOD/kg VSS day^{-1}	0.05–0.08	0.05–0.3
2	Sludge age	days	15–20	4–20
3	Sludge yield	kg solids/kg BOD	0.75–0.85	0.75–0.85
4	MLSS	mg/L	3000–4000	3500–5000
5	Cycle time	h	4–8	2.5–6
6	Settling time	h	>0.5	>0.5
7	Decant depth	m	1.5	2.5
8	Fill	–	Peak flow	Peak flow
9	Process oxygen			
9.1	BOD	kg O_2/kg-BOD	1.1	1.1
9.2	TKN	kg O_2/kg-TN	4.6	4.6

Table 10.4 Influent and desired effluent characteristics

Sl. no	Parameters	Influent characteristics, mg/L	Effluent characteristics ($\leq$, mg/L)
1	BOD	250	20
2	COD	450	100
3	TSS	300	30
4	TKN	50	2
5	TP	10	2
6	$NO_3{}^--N$	0.5	10

normal atmospheric pressure is 9.8 mg/L. Oxygen transfer coefficient $(K_La)_{20} = 11.5\ h^{-1}$ and minimum temperature of wastewater shall be considered as 17 °C. A DO of 2 mg/L is targeted in the SBR.

Solution

Flow in the reactor $= 10\ MLD = 10,000\ m^3/day = 416.67\ m^3/h = 0.116\ m^3/sec$

Assuming a peak factor of 1.5,

Peak flow rate $= 1.5 \times 10\ MLD = 15\ MLD = 625\ m^3/h = 0.174\ m^3/sec$

Assume a fill and aeration, aeration, settling and decant time of 1, 2, 0.5 and 0.5 h, respectively. The time required for the $t_a + t_s + t_d$ should be adjusted with the fill time of the other reactors. This can be adjusted by providing eight basins with operating schedule as given in Table 10.5. The aeration will be kept on during the fill cycle of the SBR.

Hence, the total cycle duration $= 2 + 0.5 + 0.5 + 1 = 4\ h$

No. of cycles per basin per day $= \frac{24}{4} = 6$.

Total aeration time per day with 3 h of aeration per cycle $= 6 \times 3 = 18\ h$

Table 10.5 Operating schedule of SBRs

Reactor no	Time (h)											
	0.5	1	1.5	2	2.5	3	3.5	4	4.5	5	5.5	6
R1	F/A	F/A	A	A	A	A	S	D	F/A	F/A	A	A
R2	S	D	F/A	F/A	A	A	A	A	S	D	F/A	F/A
R3	A	A	S	D	F/A	F/A	A	A	A	A	S	D
R4	A	A	A	A	S	D	F/A	F/A	A	A	A	A
R5	F/A	F/A	A	A	A	A	S	D	F/A	F/A	A	A
R6	S	D	F/A	F/A	A	A	A	A	S	D	F/A	F/A
R7	A	A	S	D	F/A	F/A	A	A	A	A	S	D
R8	A	A	A	A	S	D	F/A	F/A	A	A	A	A

Assuming a F/M ratio of 0.15, the volume required would be

$$\frac{F}{M} = \frac{QS_o}{VX} = 0.15$$

Or, $V = \frac{QS_o}{0.15 \times X} = \frac{10000 \times 250}{0.15 \times 3500} = 4761.9 \text{ m}^3 \approx 4762 \text{ m}^3$

Hence, volume required per basin $= \frac{4762}{8} = 595.25 \text{ m}^3$

Alternatively, number of basins receiving flow at the same time $= 2$

Hence fill volume $= \frac{416.67}{2} = 208.35 \text{ m}^3$

Assuming a volume exchange ratio of 0.3 and considering that the fill volume = the decant volume. Based on this the total volume required for each basin will be $= \frac{208.35}{0.3} = 694.5 \text{ m}^3$.

Hence choosing among higher volume, the volume per basin is 694.5 m^3.

Total volume of eight basins provided $= 694.5 \times 8 = 5555.6 \text{ m}^3$.

Hence, overall HRT $= \frac{5555.6}{416.67} = 13.33$ h.

Fill volume and depth calculations.

Peak flow rate per hour $= 625 \text{ m}^3$/h.

Fill volume to be accommodated during peak flow per basin $= 625/2 = 312.5 \text{ m}^3$.

Assuming a total depth of 5 m, the decant depth would be $= 5 \times 0.3 = 1.5$ m.

Hence, the area to be provided for the basin $= 312.5/1.5 = 208.33. \approx 210 \text{ m}^2$.

Providing a length of 15 m and the width of the tank would be 14 m.

Considering a specific sludge yield of 0.4 gVSS/gCOD, the sludge generated per day.

$= 0.4 \times (450 - 100) \times 10000 \times 10^{-3} = 1400$ kg/day

Hence, to maintain steady state this amount of sludge should be wasted from the SBRs per day.

Sludge wasting

Sludge production is 1400 kg VSS per day, i.e., in terms of SS it will be 1400/0.7 = 2000 kg/day.

If the SVI of the sludge is 100 mL/g of SS after settling, then volume of sludge to be wasted.

$= 2000 \times 100 \times 10^{-3} = 200 \text{ m}^3$/day.

Hence, the solids retention time would be $= \frac{V \times MLVSS}{Sludgewaste \times 1000} = \frac{5555.6 \times 3500}{1400 \times 1000} = 13.9$ days.

Air requirements

BOD removed per day $= (250 - 20) \times 10^{-3} \times 10000 = 2300$ kg/day.

Assuming a theoretical O_2 requirement of 1.3 kg/kg-BOD removed,

O_2 requirement $= 2300 \times 1.3 = 2990$ kg/day.

Assuming a nitrogen assimilation of 5 g/kg-BOD removed in bacterial cell mass produced, N_2 assimilation $= 2300 \times \frac{5}{1000} = 11.5$ kg.

Inlet N_2 load $= 10,000 \times \frac{50}{1000} = 500$ kg.

Outlet N_2 load $= 10,000 \times \frac{2}{1000} = 20$ kg.

N_2 removed in a day $= 500 - 11.5 - 20 = 468.5$ kg.

kg of O_2 required per day for nitrification $= 468.5 \times 4.6 = 2155.1$ kg.

Hence, total O_2 requirement $= 2990 + 2155.1 = 5145.1$ kg.

The oxygen transfer efficiency is 3% per metre depth under standard conditions, hence at mean depth of aeration $(5 + 3.5)/2$ it will be $4.25 \times 3 = 12.75\%$.

Estimating oxygen transfer efficiency under field condition.

Correcting $(K_L a)_{20}$ for 17 °C.

$K_L a$ at $17°C = (KLa)_{20} (1.024)^{17-20} = 11.5 \times (1.024)^{-3} = 10.71$ /h

Considering, $\alpha = 0.7$ and $\beta = 0.9$

Oxygen transfer under field condition, $E_{actual} = 12.75 \frac{\alpha.K_{La}.(\beta Cs - Ct)}{(K_{la})_{20}(Cs)_{20}}$.

$= \frac{12.75 \times 0.7 \times 10.71 (0.9 \times 10.56 - 2)}{11.5 \times 9.8} = 6.36\%$

Considering density of air as 1.201, weight fraction of O_2 as 23% in air, hence, the air requirement in m^3/day will be

Air requirement $= \frac{Theoretical\ requiment\ of\ O_2}{\rho_{air} \times O_2 \times X_{O_2}} = \frac{5145.1}{1.201 \times 0.23 \times 0.0636} = 292,864$ m^3/day

As aeration is done for 18 h, hence, the total air requirement per hour

$= \frac{292864}{18} = 16270.2$ m^3/h

Total air requirement per basin $= \frac{16270.2}{8} = 2033.78$ m^3/h, basin

Hence provide eight blowers along with one spare stand-by of capacity $= 2050$ m^3/h.

Decant rate

Designing decant rate at peak flow, volume of water to be decanted $=$ fill volume $= 625$ m^3.

Time for decanting $= 0.5$ h.

Decant rate $= 625/0.5 = 1250$ m^3/h.

Considering the weir loading rate of 185 m^3/m^2 day the length of weir required per basin for the decanter, since two basins will be under decant phase of operation simultaneously, will be $= 1250/(185 \times 2) = 3.378 \sim 3.4$ m. Though the decanter is designed for the peak flow, the normal practice is to operate the decanter for the average flow. Thus the inflow to the SBR may be variable, however constant outflow is maintained.

10.4 Aerated Lagoon

Aerated lagoons are one of the aerobic suspended growth biological wastewater treatment processes. In aerated lagoon the wastewater is treated on a flow through basis with or without solids recycling from secondary sedimentation tank. Oxygen is usually supplied by means of surface aerators, instead of photosynthetic oxygen yield as in case of oxidation pond. The aerators installed in aerated lagoons are usually floating aerators type, which are preferred to address the variation in water level due to variation in evaporation losses with seasons. However, for small lagoons aerators fixed on platforms or diffused air aeration units can also be used.

The mixing action of the mechanical aerators or the rising air bubbles from the diffuser is responsible to keep the biomass present in the basin under suspension. These lagoons are constructed with depth varying from 2.5 to 5 m. The depth is governed by the aerators selected depending on the depth of influence of these. The contents of an aerobic lagoon are partially mixed or completely mixed. Depending on detention time adopted, the effluent comprises about $\frac{1}{3}$ to $\frac{1}{2}$ the value of the incoming BOD in the form of cell biomass. Hence, before the effluent can be discharge, the solids must be removed by settling. If the solids are recycled back to the lagoon from settling tank (Fig. 10.15), then there is no difference between this and extended aeration ASP.

Instead of final clarification provided by settling tank for removal of bacterial cells from the effluent of aerated lagoon, a facultative stabilization pond can be provided after the aerated lagoon. This facultative lagoon will have dissolved oxygen maintaining aerobic zone in the upper part, whereas at the bottom the suspended solids settled due to insufficient mixing will undergo anaerobic decomposition. This sludge settled at bottom need to be removed from the facultative lagoon once in few years (5–10 years). No final clarifier is provided after facultative lagoon, hence the solids in the final effluent may not remain always under control.

Aerated lagoons require less space as compared to stabilization ponds and these are having lesser construction and operational cost as compared to the ASP, hence these are often preferred for treatment of biodegradable wastewaters. However, the land area required for the aerated lagoon is more than that of ASP and cost of construction is more than stabilization ponds, due to installation of aerators.

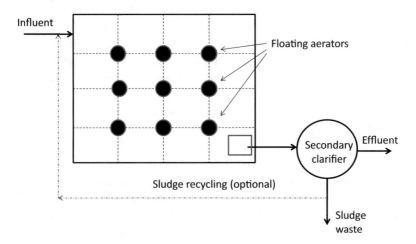

Fig. 10.15   Schematic of aerobic lagoon

The mean cell retention time should be properly selected to ensure that (a) the suspended microorganisms will easily flocculate by sedimentation and (b) that the adequate safety factor shall be provided when compared to mean cell residence time resulted during cell washout. The oxygen requirement is as per the activated sludge process. In general, the amount of oxygen required has been found to vary from 0.7 to 1.4 times the amount of BOD_5 removed. Aerated lagoons have the advantages, such as ease of operation and maintenance, equalization of wastewater, and a high capacity of heat dissipation when required. The disadvantages of aerated lagoons are large area requirement, high effluent suspended solids concentration, and sensitivity of process efficiency to variation in ambient air temperature.

Aerobic lagoons: In aerobic lagoons, power levels are great enough to maintain all the solids present in the lagoon under suspension and also to provide dissolved oxygen throughout the liquid volume. Aerobic lagoons are operated with high F/M ratio and short MCRT. These systems achieve little organic solids stabilization but convert the soluble organic matter into cellular organic material. Based on the solid handling manner, the aerobic lagoons can be classified into (i) aerobic flow through with partial mixing, and (ii) aerobic lagoon with solid recycle and nominal complete mixing.

Aerobic flow through with partial mixing: This type of aerobic lagoon is operated with sufficient energy input to meet the oxygen requirement, however the energy input is insufficient to keep all the biomass in suspension. The HRT and SRT are the same in this type of lagoon. Typically, while treating municipal wastewater retention time of 3–5 days is adequate to demonstrate more than 50% of BOD removal. As compared to activated sludge process very low MLSS concentration is maintained in the lagoon, which is around 300 mg SS/L. The effluent from this lagoon is settled in an external sedimentation facility to remove the solids prior to discharge.

Aerobic lagoons with solid recycle: This type of lagoon is same as the extended aeration activated sludge process with the exception that the aeration is carried out in an earthen basin instead of a reinforced concrete reactor basin and longer HRT than the extended aeration process is provided. The air requirement in this type of lagoon is higher than the aerobic flow through lagoons to keep all the biomass in suspension. The analysis of this type of lagoons is same as the extended aeration activated sludge process.

10.4.1   Design of Aerobic Flow-Through Type Lagoons

The design of aerobic flow-through type lagoons is similar to the aeration tanks used in activated sludge process; however, in case of aerobic flow-through type lagoons no recycling of solids is practiced. In the lagoon, the hydraulic detention time θ, is equal to the mean cell residence time, Θ_c, since no recycling of solids is practiced, thus, $\theta = \Theta_c$. A typical retention time of 3 to 6 days is adopted for aerated lagoon. Once this Θ_c is decided the soluble substrate concentration can be estimated knowing other kinetic coefficients as used in ASP design. Alternately, the BOD removal from aerated lagoon prior to settling in sedimentation tank can be modelled following a pseudo-first order kinetics (Eq. 10.66).

$$-\frac{dS}{dt} = KS \qquad (10.66)$$

where, dS/dt is the rate of BOD removal, mass/vol, time; K is the reaction rate constant and S is the substrate remaining at any time t.

Thus, writing a material mass balance for substrate as per Eq. 10.67.

$$\text{Accumulation} = \text{Inflow} - \text{outflow} - \text{Decrease due to reaction} \qquad (10.67)$$

Considering the completely mixed condition in the lagoon Eq. 10.68 shall be written for the lagoon.

$$V(dS) = QS_o dt - OS_t dt - VKS_t dt \qquad (10.68)$$

where, V is the volume of the aerated lagoon; Q is the wastewater flow rate; S_o and S_t are the substrate concentration entering and leaving the lagoon, respectively; and $KS_t dt$ equals to the dS (from Eq. 10.66).

Dividing the Eq. 10.68 with Vdt, will yield Eq. 10.69.

$$\frac{dS_t}{dt} = \frac{Q}{V}S_o - \frac{Q}{V}S_t - KS_t \qquad (10.69)$$

Under steady state conditions dS_t/dt will be zero and V/Q represents hydraulic retention time (θ), the above equation can be rearranged to give Eq. 10.70.

$$\frac{S_t}{S_0} = \frac{1}{1 + K\left(\frac{V}{Q}\right)} = \frac{1}{1 + K\theta} \quad \text{Or } K = \frac{S_0 - S_t}{S_t \theta} \qquad (10.70)$$

The Eq. 10.70 can be used for the design of aerated lagoons. As the completely mixed model is used and no sludge recycle is practiced in such lagoons, a mass balance for the microbial solids present in the lagoon at equilibrium can be obtained. Thus, performing laboratory experiment for longer duration to reach steady state and with retention time of several days the value of K can be estimated. The flow rates can be increased in steps and allowing steady state to prevail under each flow rate the influent and effluent BOD values are monitored along with concentration of biological solids. Such experiments need to be performed for about five flow rates and then plotting value of $(S_o - S_t)$ on y-axis versus $S_t\theta$ on x-axis the slope of the line can give value of K.

By using the following relation (Eq. 10.71) proposed for activated sludge process by Reynolds and Yang (1966) for completely mixed reactor, the value of Y and K_d can be estimated.

$$\frac{S_0 - S_t}{X\theta} = \frac{K_d}{Y} + \frac{1}{Y\theta_c} \tag{10.71}$$

where, X represents average mass of microbial cells in the system, termed as mixed liquor volatile suspended solids (MLVSS); S_0 and S_t represents influent and effluent BOD concentrations, respectively; θ is the hydraulic retention time; K_d represents endogenous decay rate constant, per unit time; Θ_c represents mean cell residence time and Y is the cell yield coefficient per unit of substrate utilized.

By plotting the $\frac{S_0 - S_t}{X\theta}$ on y-axis versus $\frac{1}{\theta_c}$ on x-axis will give a straight line with slope of $1/Y$ and the y intercept of K_d/Y, from which these values can be estimated.

The biomass concentration in aerated lagoon is relatively less, which is generally in the range of 50–300 mg/L and hence the performance of it will be significantly affected by the variation in temperature. The reaction rate constant K is typically equal to 0.8–2.1 per day on filtered BOD basis at 20 °C. For other temperatures it can be estimated using following relation given in Eq. 10.72.

$$K_T = K_{20}\theta^{(T-20)} \tag{10.72}$$

where, K_T is rate constant at temperature T °C; K_{20} is rate constant at 20 °C; θ is temperature coefficient having value in the range of 1.06 to 1.1 for temperatures less than 20 °C and for warmer climate, like South Asia it can be considered as 1.035.

The temperature of wastewater in the lagoon is governed by the influent wastewater temperature, heat loss due to convection, radiation and evaporation. Also, the lagoon water will gain the heat due to solar radiation. The temperature of the wastewater in a lagoon is given by the Eq. 10.73 (Mancini & Barnhart, 1968).

$$T_i - T_w = \frac{fA(T_w - T_a)}{Q} \tag{10.73}$$

where, T_i, T_w and T_a represent the temperature of influent wastewater, lagoon water, and air temperature, °C; Q is the wastewater inflow rate, m^3/day; A is the area of the lagoon, m^2; and f is the experimental factor, m/day, having value of 0.49 typically; however, value of f could be higher as 0.81 for Gulf countries.

10.4.2 Oxygen Requirements

The oxygen supply is required to be adequate to meet the demand for biochemical oxidation of carbonaceous matter and nitrification, if any, occurring in aerobic lagoons. The total oxygen required for the lagoon can be given by Eq. 10.74. In terms of five-day BOD about 1.6 times the oxygen will be required.

$$\text{Oxygen required/day} = \left[\text{solubleBOD}_u \text{ removed/day}\right]$$
$$- \left[\text{BOD}_u \text{ of solids leaving the system/day}\right]$$
$$+ \text{ Nitrification oxygen demand} \qquad (10.74)$$

Aeration power requirement: It is important to note that a minimum power level equal to or higher than 2.75–5.0 kW per 1000 m^3 lagoon volume (typical 4 kW per 1000 m^3) is required for providing proper aeration in the aerobic flow-through lagoon, otherwise facultative environment will be developed in the lagoon that will require longer retention time to meet the expected BOD removal in the lagoon.

10.4.3 Other Details

Shape and detention time: Generally, aerated lagoons are rectangular shape, with length to width ratio of 2:1. At least two lagoons should be provided for operational flexibility with plumbing arrangement so as to operate them in series or parallel. The detention time for an aerobic flow-through lagoon is maintained in the range of two to three days in favourable climates and up to five days or more in colder climates for treating municipal wastewaters that can achieve BOD removal efficiency up to 50–60%.

Proper protection of embankment should be given from inside to protect the sides from wave induced erosion. The sides and bottom should be provided with clay or polymer liner to avoid seepage of wastewater in soil. The outside surface of the dikes forming the lagoon shall also be given proper protection to avoid erosion during rains. A concrete pad shall be provided below the aerators to protect bottom from erosion.

Solids concentration: The total VSS present in an aerobic flow-through lagoon is composed of 60–80% of the total SS present in the lagoon. The SS concentration varies from 100 to 300 mg/L depending on the nature of the wastewater, the lagoon organic loading, the efficiency, and power input per unit volume.

Final effluent BOD: The total BOD in effluent of an aerobic flow-through lagoon is always observed to be high, since the presence of large amounts of solids in the final effluent adds to the final BOD value. Thus, the total BOD can be calculated by using (Eq. 10.75). Hence, a final clarifier or facultative pond shall be provided after lagoon to produce effluent with reduced concentration of biosolids.

$$\text{BOD}_5, \text{mg/L} = \left[S_t, \text{mg/L}\right] + [0.5 \text{ to } 0.8(\text{SS, mg/L})] \qquad (10.75)$$

Example: 10.5 For a wastewater with flow rate of 10,000 m^3/day and BOD$_5$ of 230 mg/L design an aerobic flow through type lagoon to serve a town using the ideal completely-mixed model. One third of the incoming BOD load is getting removed in the primary treatment. The effluent should have BOD less than or equal to 30 mg/L in normal seasons. Consider K = 1.2 per day at 20 °C. Determine retention time required, total volume and area for lagoon, oxygen requirement on the basis of 1.6 kg oxygen/kg BOD applied to the lagoon and power, considering the oxygen transfer capacity of aerator as 1 kg O_2/kW-h. Provide minimum two lagoons. Estimate BOD removal in winter when the influent wastewater temperature is 20 °C and air temperature is 12 °C.

Solution

Lagoon volume and area required
   Influent BOD to the lagoon = 230–230/3 = 153.33 mg/L.
   Considering Eq. 10.70, $K = \frac{S_o - S_t}{S_t \theta}$.
   Hence, $1.2 = \frac{153.33 - 30}{30 \times \theta}$.
   Therefore $\theta = 3.426$ days.
   Lagoon volume = 10,000 × 3.426 = 34,260 m^3.
   Let lagoon depth be 3 m, hence lagoon area = 34,260/3 = 17,130 m^2.
   Provide two lagoons, hence area for each lagoon = 17,130/2 = 8,565 m^2.
   Providing length to width ratio of 2:1; 2 W × W = 8565; hence, width of lagoon = 65.44 m and length = 130.88 m. Thus, provide two lagoons with length of 130.9 m and width of 65.5 m.
   Estimating temperature of wastewater in lagoon in winter using Eq. 10.73 and considering $f = 0.49$ and influent wastewater temperature of 20 °C.

$$T_i - T_w = \frac{f A (T_w - T_a)}{Q}$$

$20 - T_w = \frac{0.49 \times 17130 \times (T_w - 12)}{10000}$, hence T_w is equal to 16.35 °C.
   The reaction rate in winter.
   K (16.35 °C) = K (20 °C). $(1.035)^{16.35 - 20}$.
   Thus, K at 16.35 °C = 1.06 per day.
   Effluent BOD in winter.
   $\frac{S_t}{S_0} = \frac{1}{1 + K\theta}$ Hence $S_t = \frac{153.33}{1 + 1.06 \times 3.426} = 33.1$ mg/L (soluble BOD_5).
   VSS in lagoon will be $X = \frac{Y(S_0 - S)}{1 + K_d t} = \frac{(0.6)(153.33 - 33.1)}{1 + (0.06)(3.426)} = 59.84$ mg/L (during winter).
   If the effluent BOD concentration is to be kept under control even in winter, then HRT required will have to be 3.88 days instead of 3.426 days.

Oxygenation Required
   Oxygen required = 1.6 × 10,000 × 153.33 × 10^{-3} = 2453.28 kg/day = 102.22 kg/h.
   The power requirement with aerator capacity of 1 kg O_2/kW-h = 102.22 kW.
   The power intensity = 102.22/34.26 = 2.98 kW/1000 m^3.
   The provide power should be at least at the rate of 2.75 kW/1000 m^3, hence the provided power is acceptable.

10.5   Trickling Filter

10.5.1   Overview

Attached growth aerobic biological wastewater treatment processes can be of three categories, namely (1) unsubmerged attached growth, (2) submerged attached growth process and (3) moving bed attached growth process. The former two have stationary, i.e., fixed bed support over which bacterial biofilm forms, whereas in the moving bed reactor the media is moving in the reactor. The trickling filter comes under the category of unsubmerged attached growth process, wherein the wastewater is distributed from the top of the filter media and it slowly trickles down, thus giving the process the name "Trickling filter". However, as the name suggests the treatment, i.e., oxidation of organic matter present in wastewater, is not by filtration. Rather oxidation of organic matter occurs through initially absorption and subsequent oxidation while the wastewater trickles over the active biofilm formed over the filter bed. This biofilm formed comprises of active aerobic microorganisms comprising of bacteria, protozoa and fungi. Thus, the aerobic microorganisms that form biofilm on the filter media surface are responsible for the organic matter oxidation and ammonia oxidation to offer treatment to the wastewater.

10.5.2   History

Trickling filters were considered as the most efficient secondary wastewater treatment system in US during the first half of the twentieth century. Even though the technology existed in US and Great Britain during 1890's in the form of a big tank filled with boulders, the first installation as a large public treatment system was made in Madison, US, in 1912 (Metcalf & Eddy, 2003). However, later with the emergence of suspended growth activated sludge process, which was considered as most efficient treatment technology, use of trickling filter became less common. Later the extensive research on improving efficiency of trickling filter led to invention of synthetic media, which improved its treatment efficiency and commanded to resurgence of this technology.

10.5.3 Physical Description

A trickling filter mainly consist of (a) a filter media, over which the wastewater to be treated is sprayed on top to trickle over this media bed, (b) a rotary distribution system through which wastewater is sprayed uniformly over entire filter bed, and (c) an underdrainage system that carries the filtrate out of the system (Fig. 10.16).

Filter Media: The main purpose of the filter media in a trickling filter is to provide a surface for the microorganisms to grow, so that these microbes catalyse oxidation reaction of organic matter present in the wastewater, while the wastewater trickle over the filter bed. Hence, high surface area to volume ratio, higher durability, easy availability, low cost and adequate porosity to allow the air circulation and wastewater to pass through are certain prerequisite of the media. During the initial period hard stones, including quartzite, dolomite, granite, etc., were employed as filter media. However, recently use of plastic media has become the preferred choice owing to its properties including lighter weight, increased surface area for biofilm growth, and enhanced percentage of voids, which enables it to achieve improved treatment efficiency allowing higher hydraulic as well as organic loading rates. The plastic media can be built in square, round or as modules of corrugated shapes. In conventional low rate or high-rate trickling filter the media depth is kept less than 3 m, whereas for super rate trickling filter media depth of even more than 8 m is used in practice. Also, a minimum clearance of 15 cm is to be maintained between media and rotary distribution arms.

Rotary Distribution System: A distribution system consists of a central feeder pipe that supports two or more horizontal arms with number of nozzles, through which the wastewater is spread over the filter media. The rotation of the arms occurs either using

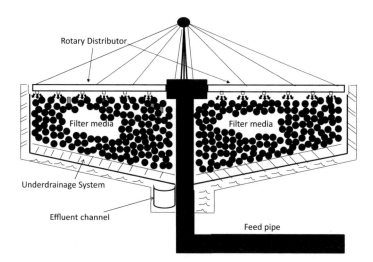

Fig. 10.16 A sectional elevation of trickling Filter

the reaction driven rotary distributers or mechanically driven by electric motors. In the former, the wastewater is circulated through only nozzles located on one side of the arms and the jet action of the discharge causes the arms to rotate about the central feed column and hence the speed of rotation depends on the wastewater flow rate.

Recirculation of wastewater is necessary during low flow rate periods to avoid issues arising due to non-uniform distribution of wastewater over the filter bed. Otherwise in absence of periodic distribution of the wastewater, the biofilm will get dry leading to the death of microorganisms. In addition to reduction in treatment efficiency, an uneven distribution can cause different patterned growth of algae as well as increased filter flies. Proper maintenance of distributor material and bearings, cleaning of nozzles and flushing distribution arms are equally important to ensure the even distribution of wastewater over the filter bed. Trickling filters that use electric motors to mechanically drive the distributors are also used in the recent designs.

Underdrain System: The underdrain collection system in a trickling filter carries the treated wastewater and solids discharged from the filter media and transport it to the subsequent sedimentation unit, wherein the solids are removed. Apart from this, the underdrain system assists in providing ventilation to the filter in order to maintain the free air circulation through natural draft to maintain aerobic environment within the filter bed. Hence, the open drains in underdrainage system are designed to flow only half full at design discharge to ensure free space for air circulation. In case if natural ventilation is inadequate, forced air pumped through the underdrain system can be used to enhance oxygen requirements or to change any filter operating conditions in later stage.

The underdrains have to be strong enough to support the overlying filter bed load and the incoming wastewater and hence are usually constructed using vitrified clay and concrete. The floor of the underdrainage system slope (1–5%) to the main drainage channel, which are designed to provide minimum velocity of 0.6 m/sec to avoid deposition of sloughed biofilm that can lead to the clogging of drains (Metcalf & Eddy, 2003). Cleaning of underdrains need to be done to prevent any septic conditions by flooding the filter after closing the filter effluent line followed by opening the valve that allow the water head in the filter to flush off the drains.

10.5.4 Types of Trickling Filters

Low-rate and high-rate trickling filter: Based on hydraulic or organic loading rates applied, the trickling filters can be classified as: (i) Low-rate trickling filter (Fig. 10.17a) and (ii) High-rate trickling filter (Fig. 10.17b). Treated wastewater recycling is not provided in case of low-rate trickling filter; whereas, treated wastewater recirculation is employed in high-rate tricking filters to improve efficiency. The recirculation helps in providing seeding to the filter bed during initial start-up and during normal operation it dilutes the organic matter concentration in the influent wastewater. Dilution is the major

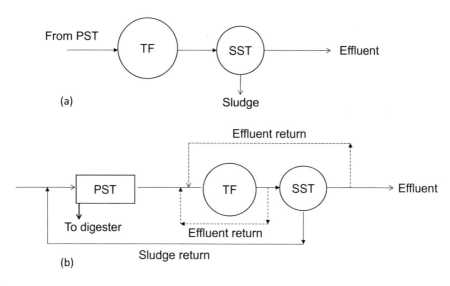

From PST

TF → SST → Effluent

(a)

Sludge

Effluent return

→ PST → TF → SST → Effluent

To digester

Effluent return

(b) Sludge return

Fig. 10.17 **a** Low-rate trickling filter and **b** high-rate trickling filter

objective behind the recirculation to match the influent BOD load with the limited oxygen resource of the filter bed.

Super-rate trickling filter: It is also called as 'Roughing filter' or 'Bio-tower'. Most super-rate trickling filters are designed using plastic media and are capable of treating organic load of more than 1.6 kg BOD/m^3 day. Also, the lower power requirement for BOD removal as compared to ASP is making this configuration popular in recent times. They are used ahead of the existing trickling filter or ASP and are generally constructed above ground. The diameter of the bio-tower can vary from 3 to 70 m. The bio-tower could also be constructed in rectangular shape. The walls of the bio-tower can be made from RCC or when modular plastic media is used the walls can be made from the plastic, since there is no hydrostatic pressure on the walls. Air blower or exhaust fan may be provided in addition to natural air draft in bio-tower to enhance oxygen resources of the system to handle higher organic loading rates (Fig. 10.18). Instead of rotary distribution arrangement used in trickling filter, a fixed wastewater distribution arrangement is provided at the top of the bio-tower to ensure that the wastewater is equally sprayed over the entire area of the bio-tower.

Increased rate of hydraulic loading rate is achieved by adopting recirculation of the wastewater in super-rate trickling filters, which is done by pumping back a portion of the settled effluent to the filter inlet. Therefore, the loading on the filter can be enhanced and as a result, a higher degree of treatment can be achieved in lesser volume of the reactor. As the power requirement per unit of BOD removed is comparatively lesser for this type of filter in comparison to ASP, these are becoming quite prevalent recently. In absence of external air circulation, such as blower or air exhaust, the super rate trickling

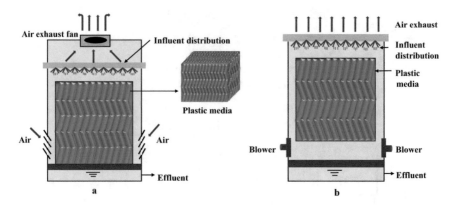

Fig. 10.18 Schematics of super-rate trickling filter with **a** air exhaust fan or **b** air blower provided to enrich the oxygen resources

filters are of shallow depth (0.9–2.5 m) to provide enough aeration required to sustain the high rate of oxidation of organic matter present in the influent wastewater. However, with artificial air draft, using air blower, they can be constructed up to 12 m depth, with 5–8 m typically adopted depth. Typically, synthetic materials such as plastic modules are used as the filter media on which biofilm is formed. Plastic media is mostly used worldwide as it provides high specific surface area, high void ratio and low density. In these plastic media, interlocking plastic surfaces are arranged to produce a honeycomb like structure, which is highly porous, clog resistant and thus permits the application of higher organic and hydraulic loading rates in these types of filters.

10.5.5 Process Description

As mentioned above, the wastewater after the primary treatment units is distributed to the trickling filter bed through the nozzles of the rotary distribution system. Even though the treatment majorly occurs with the assistance of the biological slime layer formed on the filter media, certain other mechanisms including flocculation and sedimentation in the voids of bed material also assists in the overall treatment process.

Diameter of the trickling filter depends on the mechanical rotary distributor used for spraying the wastewater. Diameter more than 12 m for single filter unit is common. Rotary arm rotates as a result of jet reaction as the wastewater exits the distributor through the nozzle to get sprayed horizontally on the filter bed; hence, external power is generally not required for rotation of the arm. However, for trickling filter of small diameter (less than 6 m) power driven rotary arm may be provided. Commercial packing media, with varying geometrical configuration and specific surface area, is available to be used in filter as bed material. These include vertical-flow random packed and cross flow media made of rock, poly-grid, plastic media or asbestos sheets. In order to avoid filter clogging, a maximum specific surface area of 100 m^2/m^3 is recommended for carbonaceous BOD removal from wastewater and up to 300 m^2/m^3 for nitrification, because of slow growth rate of nitrifiers. Overall performance of the trickling filter depends upon the hydraulic and organic loading rates applied, wastewater pH, operating temperature and availability of air through natural draft within the pores, and mean time of contact of wastewater with biofilm, etc.

The mean time of contact of wastewater with the filter surface is related to the filter depth, applied hydraulic loading rate and nature of filter packing. This contact time can be estimated as per Eq. 10.76 (Eckenfelder, 2000):

$$T = C.D/Q^n \qquad (10.76)$$

where T is mean detention time; D is depth of the filter bed; Q is the hydraulic loading m^3/m^2.day; C and n are constants related to specific surface area and configuration of the packing media used. The mean retention time in the filter bed increases considerably with formation of biofilm as compared to new filter media and it could be up to 4 times once the matured biofilm is formed on the media surface.

Process biology: As the wastewater trickles down the filter media, biological slime layer is developed on the surface of the filter media. This biofilm formed over the media surface consists of the microorganisms responsible for biodegradation of organic matter present in the wastewater, which gradually grow attached to the media and these microorganisms extract the organic matter and inorganic nutrients from the wastewater. The major population in the biofilm includes aerobic, anaerobic and facultative (predominant) bacteria, algae, fungus, protozoans, rotifers along with higher organisms including snails, flies, worms, etc.

The aerobic microorganisms present in the outermost part of the biofilm degrade the organic matter. However, the oxygen is unable to penetrate through the think biofilm layer as it thickens due to microbial growth. Usually, up to a depth of 0.1–0.2 mm the slime layer can have aerobic properties and beyond this the anaerobic zone develops, breeding anaerobic organisms. A rapid growth is observed in the microorganisms near the surface of the media; however insufficient food and limited oxygen availability lead the microorganisms in the lower portion in state of starvation. As the biological slime layer grows and become thicker, the organisms near the filter media surface die and lose their ability to cling to the surface, causing its fall or sloughing. This sloughed biomass is removed from the system via underdrain system, which is subsequently removed in the clarifier by settling (Fig. 10.17).

Nitrification: Along with BOD removal, nitrification can also be achieved in trickling filter when operated at low organic loading rates and suitable temperature. However, nitrification process gets initiated at a BOD concentration less than 20 mg/L (Henze & Harremoes, 1983). A total nitrogen removal of up to 35% can be achieved in a trickling filter. However, the nutrient removal efficiency of trickling filters greatly depends on the operating conditions, and while some sources indicate a high removal of ammonia (U.S. EPA, 2000), whereas others indicate no capacity of trickling filters for nutrients removal (UNEP, 2018). However, research on enhancing nutrient removal in trickling filter is being carried out by researchers globally.

Sludge retention: The sludge is retained in the trickling filter for very long time as compared to ASP and typically the mean cell residence time (θc) of 100 days or more can be easily achieved. Estimation of actual biomass present in the reactor is practically difficult, hence exact measurement of θc is not possible. Excess sludge generation in this process is expected to be lower due to prolong retention time of biomass, thus supporting endogenous decay within the biofilm. Sludge generation in trickling filter is typically 60 to 70% lower than that of ASP treating same wastewater. The sludge generation in high-rate trickling filter is more than the low-rate trickling filter.

Air supply: Natural air draft through the filter bed is responsible for supply of air in low-rate and high-rate trickling filter. When influent wastewater temperature is lesser than the ambient temperature there will be downward flow of air in the trickling filter bed; whereas, when the wastewater temperature is higher than the ambient temperature there will be upward flow of air. To allow air circulation, the under-drainage system should be designed to flow not more than half full while carrying maximum wastewater flow for which the trickling filter is designed.

Details of the rotary arm: It rotates with the speed of 0.5–2 revolutions per minute (rpm). The peripheral speed for two arm system will be 0.5–4 m/min. The arm length could be as low as 3 m to as high as 35 m depending on the diameter of the filter bed. This rotary arm delivers the wastewater about 15 cm above the filter bed. The velocity of wastewater moving through arm should be more than 0.3 m/sec to prevent deposition of solids. Number of ports, generally of equal diameter, are provided on this arm to deliver wastewater in horizontal direction. Minimum two arms are provided, whereas there could be four number of arms for larger dimeter trickling filter. The guidelines for design values to be used for the trickling filters are provided in the Table 10.6.

10.5.6   Advantages and Disadvantages of a Trickling Filter

The trickling filter offers following advantages:

- It can be operated under a wide range of organic and hydraulic loading rates,
- Simple biological wastewater treatment process,
- Resistant to shock loadings,
- Efficient nitrification can be achieved at low organic loading rates,
- High effluent quality in terms of BOD and suspended solids removal,
- Low power requirements, and

Table 10.6 Design values recommended for design of different types of trickling filters

Parameter	Low-rate trickling filter	High-rate trickling filter	Super-rate roughing filter
Hydraulic loading, m^3/m^2 day	1–4	10–40	40–200
Volumetric loading, kg BOD/m^3 day	0.10–0.40	0.40–1.80	1.0–6.0
Depth, m	1.5–3.0	1.0–2.0	4–12
Recirculation ratio	Nil	1–4	1–4
Power requirement, $kW/10^3$ m^3	2–4	6–10	10–20
Dosing intervals	Less than 5 min	15–60 sec	Continuous
Sloughing	Intermittent	Continuous	Continuous
Effluent quality	Fully nitrified	Nitrified only at low loading	Nitrified only at low loading

- Not land intensive treatment system, as compared to stabilization ponds.

However, the disadvantages of trickling filter are:

- High capital costs,
- Additional treatment may be required to meet more stringent discharge standards,
- Requires expert design and construction,
- Requires a constant source of wastewater flow and regular monitoring,
- Vector and odours are often problematic,
- Pre-treatment of wastewater is necessary
- Treatment of sloughed biomass settled in SST is obligatory,
- Risk of clogging is relatively high.

10.5.7 Design Equations for Trickling Filter

The design of trickling filter is based on empirical, semi-empirical and mass balance concepts. The design formulations of trickling filters have been given by several authors, out of which the NRC formula (1946), Schultz formula (1960), Eckenfelder formula (1963), etc. are a few, which are widely used. The NRC formulae are mostly applied to single-stage and multistage rock filters. The Schultz formula and Eckenfelder formula are mostly used to design the trickling filter employing plastic media.

Formula to calculate the rotational speed of rotary distributer is given in Eq. 10.77 (Metcalf & Eddy, 2003).

$$n = (1 + R)(q)(10^3 \text{mm/m})/(NA)(DR)(60\,\text{min/h}) \tag{10.77}$$

where, n is the rotational speed, rpm; q is the influent hydraulic loading rate, $\text{m}^3/\text{m}^2.\text{h}$; R is the recycle ratio; NA is the number of arms in rotatory distributor assembly; and DR is the dosing rate, mm/pass of distributor arm.

10.5.7.1 NRC Formula for Design of Trickling Filter

(a) For the first stage filter, the NRC equation (Metcalf & Eddy, 2003) is given in Eq. 10.78.

$$E_1 = 100 / \left\{ 1 + 0.4432 \sqrt{\left(\frac{L}{VF}\right)} \right\} \tag{10.78}$$

where, E_1 is the efficiency of BOD removal for first stage filter at 20 °C including recirculation and sedimentation, %; L is the BOD loading to filter, kg/day; V is the volume of filter media, m^3; F is recirculation factor, given by Eq. 10.79.

$$\text{Recirculation factor } F = 1 + R/(1 + 0.1R)^2 \tag{10.79}$$

where, R is the recirculation ratio.

(b) For a two-stage trickling filter system, the BOD removal efficiency of the second stage is given in Eq. 10.80 as follows:

$$E_2 = 100 / \left\{ 1 + \frac{0.4432}{1 - E1} \sqrt{\left(\frac{L_2}{VF}\right)} \right\} \tag{10.80}$$

where, E_2 is the efficiency of BOD removal for second stage at 20 °C including recirculation and sedimentation, %; E_1 is the efficiency of BOD removal for first stage; and L_2 is the BOD loading applied to second stage filter, kg BOD/day.

10.5.7.2   Rankine's Formula

For single stage filters: The BOD of influent to the filter (including recirculation) shall not exceed three times the BOD required for settled effluent. This can be expressed as per Eq. 10.81 or Eq. 10.82.

$$S2 + R1(S4) = 3(1 + R1)S4 \tag{10.81}$$

Or

$$S4 = S2/(3 + 2R1) \tag{10.82}$$

where, S2 is BOD of settled influent, S4 is BOD of trickling filter effluent after SST, R1 is recirculation ratio, and if E is efficiency of BOD removal, then it can be expressed as per Eq. 10.83.

$$E = (1 + R1)/(1.5 + R1) \tag{10.83}$$

The value of recirculation is given by Eq. 10.84.

$$R = (Q_T - Q)/Q \tag{10.84}$$

where, Q_T is total flow including recirculation and Q is wastewater flow to be treated.

For second stage filter: The BOD of the wastewater applied to the second stage filter including recirculation shall not exceed two times the effluent BOD. Therefore, this condition can be represented by Eq. 10.85 and Eq. 10.86.

$$S4 + R2(S6) = 2(1 + R2)S6 \tag{10.85}$$

Or

$$S6 = S4/(R2 + 3) \text{ and efficiency } = (1 + R2)/(2 + R2) \tag{10.86}$$

where, S4 is BOD of influent to second stage filter, S6 is effluent BOD of second-stage trickling filter after SST, and R2 is recirculation ratio.

10.5.7.3   Eckenfelder's Formula

One of the most common kinetic equations for filter performance while treating municipal wastewater was developed by Eckenfelder (1961). The second order reaction kinetics was considered while developing the kinetics equation. Based on this second order equation, the specific rate of substrate removal can be expressed as Eq. 10.87.

$$-\frac{1}{X}\frac{dS}{dt} = KS^2 \tag{10.87}$$

where, X is the mass of microorganisms present in the filter bed; S is the BOD concentration; and K is the reaction rate constant. Integrating within limit of time 0 to t and solving the above equation, the Eq. 10.88 will be obtained.

$$\frac{S_t}{S_0} = \frac{1}{(1 + S_0 \cdot K \cdot X.t)} \tag{10.88}$$

Substituting, $t = C\left(\frac{D^{0.67}}{Q_L^{0.5}}\right)$ and combining constant S_o, K, X and C to a single constant 'C', as for the same wastewater treatment under steady state condition these values will be fairly constant, thus the equation will become Eq. 10.89.

$$\frac{S_t}{S_0} = \frac{1}{1 + C\left(\frac{D^{0.67}}{Q_L^{0.5}}\right)} \tag{10.89}$$

where S_t is BOD$_5$ of effluent, mg/L, S_o is BOD$_5$ of the influent, mg/L, C is constant (typical value of 5.358), D is filter depth provided, m, and Q_L is hydraulic loading rate per unit are of the bed, m^3/m^2.day. This equation is popularly used for the design of the low-rate and high-rate trickling filters.

Later Eckenfelder (Eckenfelder & Ford, 1970) has also developed performance equation for the specific rate of substrate removal based on a pseudo-first-order reaction. Based on this first order reaction, the specific rate of substrate removal can be expressed as Eq. 10.90.

$$-\frac{1}{X}\frac{dS}{dt} = K \cdot S \tag{10.90}$$

Where, $\frac{1}{X}\frac{dS}{dt}$ is the specific rate of substrate utilization, $\frac{\text{Mass of substrate}}{\text{Microbial mass} \times \text{Time}}$; $\frac{dS}{dt}$ is the rate of substrate utilization, $\frac{\text{Mass}}{(\text{Volume}) \times \text{Time}}$; K is rate constant, $\frac{\text{Volume}}{(\text{Mass of microbes}) \times \text{Time}}$; and S is substrate concentration, mass/volume.

Rearranging the above Eq. (10.90) for integration, the Eq. 10.91 shall be obtained.

$$\int_{S_0}^{S_t} \frac{dS}{S} = -KX \int_0^t dt \tag{10.91}$$

where, X is average cell mass concentration, $\frac{\text{mass}}{\text{Volume}}$; S_o is substrate concentration applied over the filter bed that is resulting after recirculation; S_t is the substrate concentration after contact time, t.

Upon integrating the Eq. 10.91, following Eq. 10.92 will be obtained.

$$\frac{S_t}{S_0} = e^{-K \cdot X \cdot t} \tag{10.92}$$

The X is proportional to surface area of the media (A_s), i.e., $X \approx A_s{}^m$. Where, A_s is the specific surface area of the packing media used in the bed.

The mean contact time 't' for a filter is given by Howland (1950) (Reynolds & Richard, 1996) as given by Eq. 10.76 earlier. Substituting this in Eq. 10.92 after elimination of constant C as it will be taken into account in K, the following Eq. 10.93 will be obtained.

$$\frac{S_t}{S_0} = e^{\frac{(-K A_s^m D)}{Q_L^n}} \tag{10.93}$$

where, m is a constant specific to the media configuration and type of wastewater being treated and value of it can be experimentally determined. Similarly, the value of 'n' depends on flow characteristics through packing and usually about 0.5 to 0.67. For specific wastewater and filter media Eq. 10.93 may be simplified by combining K, A_s and m to K so as to give Eq. 10.94.

$$\frac{S_t}{S_o} = e^{\frac{-K D}{Q_L^n}} \tag{10.94}$$

For a specific media configuration used in the filter bed the value of K and n can be evaluated from pilot experiments. With temperature variation the value of K can be converted as per Eq. 10.95.

$$K_T = K_{20} \times 1.035^{(T-20)} \tag{10.95}$$

where, K_T is the rate constant at temperature T, K_{20} is rate constant at 20 °C, and T is temperature, °C.

10.5.8 Recent Developments in Trickling Filter Research

Most of the recent research have been focused mainly on the material used as media in trickling filter, as the filter media acts as the core of the treatment system. A higher biological nitrogen removal was achieved in a zeolite bed trickling filter, which is explained on the basis of coupling action of physicochemical characteristics of zeolite and biochemical reactions of microorganism (Liu et al., 2021). Investigations also have been conducted on wastewater treatment system utilizing maize cob as trickling filter media, which was able to achieve a BOD removal efficiency up to 84% (Arsalan et al., 2021). Granular activated carbon (GAC) has also been proven to be efficient material for removal of ammonia and turbidity (Forbis-Stokes et al., 2018). Also, performance evaluation of trickling filter-based wastewater treatment system utilizing cotton sticks as filter media demonstrated a COD removal efficiency up to 80% (Aslam et al., 2017).

Effects of hydraulic retention time on the performance of a pilot-scale trickling filter was also investigated showing the highest COD removal at an HRT of 48 h (Rehman et al., 2020). Temperature was also seen to influence the performance of the system to a greater extent. A temperature of 24.3 °C was reported to be an ideal in removing maximum ammonium nitrogen in a nitrifying trickling filter (Godoy-Olmos et al., 2016). Another area which most of the researchers investigated was on different applications of trickling filters. Apart from nitrification, application of trickling filter has been extended in elimination of hydrogen sulphide from biogas by a two-stage trickling filter system using effluent from anaerobic–aerobic wastewater treatment (Tanikawa et al., 2018). Also, a 75% oxidation of manganese was reported to be achieved in a trickling filter inoculated with the Mn oxidizing bacteria, *Pseudomonas putida* (McKee et al., 2016).

Example: 10.6 Determine the size of a two-stage trickling filter using the NRC formula with the following conditions. Assume both filters have same efficiency. The sewage flow rate to be treated is 4000 m^3/day having a BOD of 300 mg/L. The effluent should have BOD less than or equal to 20 mg/L. The depth of filter adopted is to be 2 m. The recirculation ratios recommended for first and second stages trickling filters are 1.5 and 0.8, respectively.

Solution

Estimation of BOD loading

Assuming 35% BOD is being removed in primary sedimentation, i.e., $= 0.35 \times 300$ mg/L $= 105$ mg/L, hence BOD remaining $= (300 - 105)$ mg/L $= 195$ mg/L.

BOD load on filter, $L_1 = QC_1 = 4000$ m^3/day $\times 195$ mg/L $\times 10^{-3} = 780$ kg BOD/day

Recirculation factor, $F1 = 1 + r_1/(1 + 0.1r_1)^2 = 1 + 1.5/(1 + 0.1 \times 1.5)^2 = 1.89$

Recirculation factor, $F2 = 1 + r_2/(1 + 0.1r_2)^2 = 1 + 0.8/(1 + 0.1 \times 0.8)^2 = 1.54$

Determining efficiency E_1 and E_2 of each filter

$$\text{Overall efficiency} = [(195 - 20)/195]100\% = 89.7\%$$

$$\text{Overall efficiency} = E_1 + E_2(1 - E_1) = 0.897$$

$$E_1 + E_2 - E_1E_2 = 0.897$$

As $E_1 = E_2$, hence $2E_1 - E_1^2 = 0.897$

$$E_1 = 0.68 = 68\%$$

Determination of the volume of first stage filter

$$E = 100 / \left[1 + 0.4432 \sqrt{\left(\frac{L}{VF} \right)} \right]$$

$$68 = 100 / \left[1 + 0.4432 \sqrt{\left(\frac{780}{1.89V} \right)} \right]$$

$V = 365$ m^3.
Calculate the diameter of filter no. 1
Area = Volume/depth = 365/2 = 182.5 m^2
$\pi d^2/4 = 182.5$ m^2
Hence, diameter d = 15.24 m

Estimate the BOD load on second filter
$L_2 = L_1 (1 - E) = 780 (1 - 0.68) = 250$ kg BOD/day

Estimating the volume of second filter

$$E = 100 / \left[1 + \frac{0.4432}{1 - E1} \sqrt{\left(\frac{L}{VF} \right)} \right]$$

$$68 = 100 / \left[1 + \frac{0.4432}{1 - 0.68} \sqrt{\left(\frac{250}{V \times 1.54} \right)} \right]$$

$V = 1394$ m^3
Calculate the diameter of second filter
Area = Volume/depth = 1394/2 = 697 m^2
$\pi d^2/4 = 697$ m^2; hence, diameter d = 29.79 m

Check BOD loading rate of each filter

(a) For first stage filter, BOD loading rate = 780 /365 = 2.13 kg BOD/m^3.day
(b) For second stage filter, BOD loading rate = = 250/1394 = 0.179 kg BOD/m^3.day

Check hydraulic loading rate of each filter

(a) For first stage filter, hydraulic loading rate; $r_1 = 1.5$

$$\text{HLR} = \frac{(1 + 1.5)4000}{182.5} = 54.79 \text{ m}^3/\text{m}^2\text{day}$$

(b) For second stage filter, hydraulic loading rate; $r_2 = 0.8$

$$HLR = \frac{(1 + 0.8)4000}{697} = 10.32 \, m^3/m^2 \, day$$

Note: The BOD loading rate and hydraulic loading rate in the first stage filter are crossing the limits given in Table 10.6, hence if these values given in table to be followed the area of first stage filter need to be increased accordingly.

Example 10.7: Design of Low-Rate Trickling Filter: Design low-rate trickling filter for secondary treatment of sewage generated from a residential academic campus with total population of 30,000 residents with rate of water supply of 170 LPCD. The BOD_5 after primary treatment is 110 mg/L and BOD_5 of final effluent should be less than or equal to 20 mg/L. Consider $C = 5.358$.

Solution
The Eq. 10.89 is used for the design.

$$\frac{St}{So} = \frac{1}{1 + C\left(\dfrac{D^{0.67}}{Q_L^{0.5}}\right)}$$

Provide depth $D = 1.5$ m.
Average sewage flow $= 30{,}000 \times 170 \times 0.80 \times 10^{-3} = 4080 \, m^3/day$

$$\frac{20}{110} = \frac{1}{1 + 5.358\left(\dfrac{1.5^{0.67}}{Q_L^{0.5}}\right)}$$

Solving the above to get flow/area, Q_L as
$Q_L = 2.441 \, m^3/m^2$ day
Plan area required for the trickling filter $= \frac{4080}{2.441} = 1671.45 \, m^2$
Hence diameter required for trickling filter $= 46.14$ m
Instead of single filter of diameter 46.14 m, two low-rate trickling filters with each having diameter of 32.65 m will be recommended with depth of filter bed of 1.5 m.

Example 10.8: Design of High-Rate Trickling Filter: Design first stage high-rate trickling filter for secondary treatment of sewage generated from a residential academic campus with total population of 30,000 residents with rate of water supply of 170 LPCD. The BOD_5 after primary treatment is 110 mg/L and BOD_5 of effluent from trickling filter should be less than or equal to 40 mg/L, since after first stage treatment in trickling filter further polishing treatment is intended to be given in high-rate algal pond. Consider recirculation ratio of 2 and filter depth of 1.8 m. Also consider C = 5.358.

Solution

The BOD of the wastewater to be treated is 110 mg/L, whereas the recirculation flow will bring BOD of 40 mg/L in the influent. Hence, the BOD of the mixture at the inlet will have to be determined.

Estimating BOD in the influent,

$Q \times 110 + 2Q \times 40 = (1 + 2) \, Q \, S_0$.

$\therefore S_0 = 63.33$ mg/L

$$\frac{40}{63.3} = \frac{1}{1 + 5.358 \left(\frac{1.80.67}{Q_L^{0.5}} \right)}$$

$Q_L = 186$ m^3/m^2-day.

Estimating the wastewater flow $= 30{,}000 \times 170 \times 0.80 \times 10^{-3} = 4080$ m^3/day.

Hence, the recycle flow $= 8160$ m^3/day.

and total inflow $= 4080 \times 3 = 12{,}240$ m^3/day.

$\therefore$ Area of trickling filter $= \frac{12240}{186} = 65.81$ m^2.

$\therefore$ Diameter of trickling filter $= 9.16$ m ~ 9.2 m.

Example 10.9: Design of Bio-Tower: Design two parallelly operated super-rate trickling filters for an average wastewater flow rate of 650 m^3/day with the initial BOD of 145 mg/L. The BOD removal efficiency of primary treatment is 35% and the final effluent BOD should be less than 20 mg/L. Consider filter depth of 4.5 m, recirculation ratio of 2, $K = 2.26$ (at 20 °C), $n = 0.5$, minimum temperature of wastewater in the region is 15 °C.

Solution

$$K_{15} = K_{20} \times 1.035^{(T-20)} = 2.26 \times 1.035^{(15-20)} = 1.903 \text{ per day}$$

BOD of the wastewater entering into the filter $= 145 - 145 \times 0.35 = 94$ mg/L.

For total wastewater flow admitted to the bio-tower, including recirculation, estimate S_0

$$Q \times 94 + 2Q \times 20 = (1 + 2) \times Q \times S_0$$

Hence, $S_0 = 44.67$ mg/L

$$\frac{S_t}{S_0} = e^{\left(\frac{-K \times D}{Q_L^n} \right)} = e^{\left(\frac{-1.903 \times 4.5}{Q_L^{0.5}} \right)}$$

$$\frac{20}{44.67} = e^{\left(\frac{-1.903 \times 4.5}{\varrho_L^{0.5}}\right)}$$

Solving, $Q_L^{0.5} = 10.66$, hence $Q_L = 113.61$ m^3/m^2 day.
Total wastewater inflow to the filter $= 3 \times 650$ m^3/day $= 1950$ m^3/day.
Hence, area required $= 1950/113.61$ m$^2 = 17.16$ m^2.
Area of each filter $= 17.16/2 = 8.58$ m^2; hence, diameter of each filter should be 3.30 m.

Thus, provide two super-rate trickling filters (Bio-towers) with the diameter of 3.3 m each.

10.6   Biological Active Filter

Biological active filter (BAF) combines the principles of biological oxidation of organic matter present in the wastewater and filtration through the employed filter media for removal of organic matter present in the wastewater. Therefore, both suspended and soluble organic matter can be efficaciously removed through BAFs. Typically, in a BAF the wastewater flows downward through the filter media that serves the dual purpose of filtration of suspended solids and also provides surface for the attachment of the bacterial biomass. In addition to this, aeration is supplied from the bottom of the BAF (Fig. 10.19), which is required for the aerobic oxidation of the organic matter present in the wastewater. The treated wastewater passing through the filter is then collected from the top of BAF after filtration. Different types of media, like granular activated carbon, anthracite, sand, bituminous coal, etc., are used as media in BAFs depending upon the degree of removal efficiency required and the characteristics of the influent wastewater. The BAFs require regular backwashing by applying water and air to prevent clogging, to maintain high filtration rate and to slough out inactive biomass tethered onto the filter media so that only active media remains attached to the filter. The BAFs can be operated at reasonable volumetric and organic loading rates even for smaller volume of reactors, thus exemplifying smaller construction footprint.

The BAFs are cost-effective and multi-barrier wastewater treatment technology, which are effective for the removal of by-products formed during tertiary treatment and emerging contaminants found at trace concentrations, thus paving the way towards water reuse/recycle and saving. However, as BAFs function on the principle of aerobic decomposition of organic matter, aeration requirement increases the capital and operating cost of these setups. As clogging is a major issue for BAFs, they are used after tertiary treatment for the removal of residual organic matter and disinfection by-products, thus partly eliminating the problem of clogging due to the presence of less concentration of organic pollutants in the influent when used after tertiary treatment. Also, the frequent requirement of backwashing can be avoided if BAFs are employed to treat secondary or tertiary

Fig. 10.19 A schematic of biologically active filter

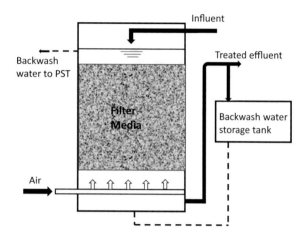

treated wastewater containing comparatively less organic load. Therefore, due to these reasons BAFs are majorly used when reuse of treated water is targeted and are employed after tertiary treatment of wastewater containing comparatively less organic matter.

10.7   Fluidised Bed Bioreactor

The phenomenon in which suspension of particles takes place by the up-flow velocity of the liquid passing through the reactor is called as fluidisation. This fluidisation has many advantages, such as proper mixing, increased specific surface area, higher mass transfer ratio, uniform mixing of particles and consistent temperature distributions inside the reactor. In most of the fluidised bed bioreactors (FBBRs), liquid–solid fluidisation is implemented (Nelson et al., 2017). A FBBR is a modern wastewater treatment technology that uses tiny fluidized media to immobilise and retain bacterial cells. The growth of biofilms on the fluidized media results in a higher biomass concentration in the FBBR ensuring better performance and stability without resulting in higher cell washout. Mostly anaerobic FBBRs are used to treat low strength wastewater; whereas, aerobic FBBRs are used to treat groundwaters polluted by hazardous wastes. In aerobic FBBRs, either the effluent has to be recirculated or it has to be passed through an oxygenation tank to pre-dissolve the oxygen required for the growth of microbes (Fig. 10.20). Activated carbon is mostly used as the packed media in an aerobic FBBR for the degradation of organic pollutants.

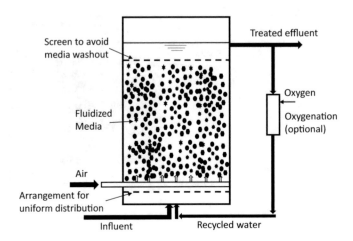

Fig. 10.20   A schematic of fluidized bed reactor

10.7.1   Different Types of FBBRs

In aerobic FBBRs, the oxygen limitation can be avoided by providing two different types: oxygenated FBBR and aerated FBBR. For the oxygenated FBBR, oxygen in liquid form is used to maximise its transfer and availability in the reactor. Ahead of FBBR treatment, wastewater can also be passed externally through an oxygenator (Fig. 10.20) so that there will be no biofilm separation from the fluidised media due to turbulence caused by gas bubbles. This recirculation of treated wastewater can assist in maintaining the fluidisation of these bioparticles. By diluting the influent wastewater, oxygen limitation in FBBR can be circumnavigated.

The bed height increases as the bed matures due to the formation of the biofilm, which can be further regulated by the periodic sloughing of the bioparticles. Moreover, in contrast to the oxygenated FBBR, an aerated FBBR as shown in Fig. 10.21, the air is directly introduced through a draft tube arranged internally to enhance oxygen transfer and mixing. Liquid circulation and gas effervescence work together to keep bioparticle fluidisation in an aerated FBBR. As a result, there is no need of recirculation of the effluent in case of aerated FBBR. Biofilm is continuously cleared from the fluidised media by bioparticle impact and attrition, as well as gas effervescence, negating the need for deliberate extended bed height control (Shieh & Li, 1989).

10.7.2   Minimum Fluidizing Velocity

When the fluid passes through a reactor packed with media and gradually increases the fluid velocity, the particles are propelled upward. Correspondingly, the minimum

Fig. 10.21 Schematics of an aerated fluidized bed bioreactor

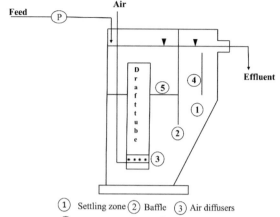

① Settling zone ② Baffle ③ Air diffusers
④ Degasification zone ⑤ Support to align the draft tube

fluidizing up-flow velocity is given by Ergun, as shown in Eqs. 10.96 and 10.97.

$$u_{mf} = 0.0055 \left(\frac{e_{mf}^3}{\left(1 - e_{mf}\right)^2} \right) \left(\frac{d^2(\rho_s - \rho)g}{\mu} \right) \tag{10.96}$$

$$u_{mf} = \frac{\mu}{d\rho} Re'_{mf} \tag{10.97}$$

where, Re'_{mf} is particle Reynolds number at minimum fluidising velocity; u_{mf} is minimum fluidising velocity; d is particle diameter; μ is dynamic viscosity; e_{mf} is voidage at minimum fluidised velocity; ρ_s is particle density; g is acceleration due to gravity; and ρ is fluid density.

Since e_{mf} is the function of surface properties, shape and size distribution of particles, typically $R\, e'_{mf}$ is provided by assuming voidages at 0.4 and 0.45 as given by Eq. 10.98 and Eq. 10.99, respectively.

$$\left(Re'_{mf} \right)_{e_{mf}=0.4} = 25.7 \left(\sqrt{(1 + 5.53 \times 10^{-5} G_a)} - 1 \right) \tag{10.98}$$

$$\left(Re'_{mf} \right)_{e_{mf}=0.45} = 23.67 \left(\sqrt{(1 + 9.39 \times 10^{-5} G_a)} - 1 \right) \tag{10.99}$$

where, G_a is the Galileo number (Archimedes number) as shown in Eq. 10.100.

$$G_a = \left(\frac{d^3 \rho(\rho_s - \rho)g}{\mu^2} \right) \tag{10.100}$$

Example: 10.10 Calculate the minimum fluidising velocity for a fluid having density of 1000 kg/m^3 and viscosity of 4 mNs/m^2 in a FBBR packed with a bed of lava rock particles of size 0.64 mm and density of 1720 kg/m^3.

Solution
From Eq. 10.100, Galileo number, $G_a = \left(\frac{d^3 \rho (\rho_s - \rho) g}{\mu^2} \right)$

$$= \left(\frac{(0.64 \times 10^{-3})^3 \times 1000 \times (1720 - 1000) \times 9.81}{\left(4 \times 10^{-3}\right)^2} \right)$$

$= 115.723$.
By assuming, $e_{mf} = 0.45$, Eq. 10.99

$$\left(Re'_{mf} \right)_{e_{mf}=0.45} = 23.67\left(\sqrt{(1 + 9.39 * 10^{-5} G_a)} - 1 \right)$$

$= 22.79$.
Thus, $u_{mf} = \frac{\mu}{d\rho} Re'_{mf} = \frac{4 \times 10^{-3}}{0.64 \times 10^{-3} \times 1000} \times 22.79 = 0.142 \text{ m/sec}$ **(Answer)**.
Therefore, the minimum fluidization velocity required is 0.142 m/sec.

Example: 10.11 At what mass flow rate (G'_{mf}) fluidisation can occur in an FBBR, when fluid is passed through a bed consisting sand particles of 0.5 mm spherical diameter and density of 2500 kg/m^3? Fluid density $= 1000 \text{ kg/m}^3$, viscosity $= 4 \text{ mNsec/m}^2$.

Solution
From Eq. 10.96, $u_{mf} = 0.0055\left(\frac{e_{mf}^3}{(1-e_{mf})^2} \right)\left(\frac{d^2 ((\rho_s - \rho) g)}{\mu} \right)$.
Since, voidage e_{mf} value is not given, consider a cube of side '2d' consisting of spherical particles of diameter 'd' of 8 in number.

$$Volume \; of \; spherical \; particles = 8\left(\frac{\pi}{6}\right)d^3$$

$$Volume \; of \; cube = (2d)^3 = 8d^3$$

Therefore, $voidage = \left(\frac{8d^3 - 8\left(\frac{\pi}{6}\right)d^3}{8d^3} \right) = 0.478$.

$u_{mf} = 0.0055\left(\frac{0.478^3}{(1-0.478)^2} \right)\left(\frac{(0.5 \times 10^{-3})^2 (2500 - 1000)(9.81)}{4 \times 10^{-3}} \right) = 2.027 \times 10^{-3}$.

Hence, the mass flow rate required for fluidization $\left(G'_{mf} \right) = \rho \times u_{mf}$
$= 1000 \times 2.027 \times 10^{-3} = 2.027 \text{ kg/m}^2 \text{ sec}$ **(Answer)**.

Questions

10.1 Describe the environmental factors that influence the growth of microorganisms used in wastewater treatment.

10.2 Describe the conditions that favour the growth of hygroscopic filamentous organisms in the activated sludge process.

10.3 Describe conventional (plug flow) activated sludge process.

10.4 Explain how solids are generated in an aeration basin, and the consequences to the operation if excess solids are not removed (wasted).

10.5 What are the major operational problems caused by hydraulic overloads in ASP?

10.6 Calculate the soluble effluent BOD concentration from a conventional activated sludge process operated with MCRT (θc) of 6 days, μ_{max} of 2.0 day^{-1} and K_s of 60 mg/L. Consider the value of K_d as 0.06 day^{-1} and cell yield coefficient as 0.60?

10.7 Estimate the oxygen consumption for the oxidation of the carbonaceous matter in the conventional activated sludge process operated with mean cell residence time θc of 10 days. The BOD removed is Q(So – S) = 100.0 kg/day; excess sludge production is 43.2 kg/day.

10.8 Design the aeration tank of an activated sludge process to treat the wastewater generated from the community. The occurrence of nitrification should be taken into consideration. Assume the required kinetic and stoichiometric parameters. The input data are as follows:

Influent flow = 9,000 m^3/day

Influent concentrations:

BOD = 341 mg/L

SS = 379 mg/L

TKN = 51 mg/L

Desired characteristics for the effluent are soluble BOD of 20 mg/L and SS of 30 mg/L.

10.9 An activated sludge process is to be designed for secondary treatment of 8000 m^3/day of wastewater. The BOD of settled wastewater after primary treatment is 180 mg/L and it is desirable to have not more than 10 mg/L of soluble BOD in the effluent. Consider $Y = 0.6$; $K_d = 0.06$ per day; MLVSS concentration in the aeration tank = 3500 mg/L and underflow concentration from the clarifier 10,000 mg/L of SS. VSS/SS = 0.80. Determine: (i) the volume of aeration tank, (ii) sludge to be wasted per day (mass and volume), (iii) the recycle ratio, and (iv) volumetric loading and F/M.

10.10 What is sludge bulking? How it can be controlled?

10.11 Define SVI. What is the best operational range of SVI? How performance of ASP will be affected by changes in SVI?

10.12 Discuss what determines the normal rotational speed of a trickling filter distributor arm.

10.13 Compare various types of media used in a trickling filter.

10.14 Describe the operational problems caused by low or high dissolved oxygen levels in trickling filters.

10.15 Describe the term "filter channelling" in trickling filter.

10.16 List ways to test and improve trickling filter ventilation.

10.17 Determine the size of a two-stage trickling filter using the NRC formula with the following conditions. Assume both filters have same efficiency. The sewage flow rate is 3785 m^3/day with BOD of 195 mg/L. The effluent should have BOD less than or equal to 20 mg/L. The depth of filter adopted is to be 2 m. The recirculation ratio adopted for first and second stage is 1.8. (Answer: Volume—1457 m^3; Area—733.5 m^2; Diameter—30.6 m).

10.18 Design two stage high-rate trickling filters for secondary treatment of sewage generated from 20,000 persons with rate of water supply 170 LPCD. The BOD$_5$ after primary treatment is 150 mg/L and BOD$_5$ of the first stage trickling filter and second stage trickling filter should be 40 mg/L and 20 mg/L, respectively. Consider C = 5.358.

10.19 Design a Bio-tower for treatment of sewage generated from the residential academic institute having population of 30,000 persons with rate of water supply 180 LPCD.

10.20 Describe sequencing batch reactor. What are the advantages and drawback of this as compared to conventional flow through ASP?

10.21 Describe how simultaneous organic matter removal and nitrification and denitrification is achieved in SBR.

10.22 Discussed what are the typical operating conditions used for SBR for treatment of sewage. How the cycle time required for the SBR is decided?

10.23 What is aerated lagoon? Describe different types of lagoons used for wastewater treatment.

10.24 What is fluidized bed reactor? Describe advantages this reactor over the stationary bed reactor.

10.25 Describe biological active filter and its application in wastewater treatment.

References

Abdel-Fatah, M. A., Sherif, H. O., & Hawash, S. I. (2017). Design parameters for waste effluent treatment unit from beverages production. *Ain Shams Engineering Journal., 8*(3), 305–310. https://doi.org/10.1016/j.asej.2016.04.008

Ardern, E., & Lockett, W. T. (1914). Experiments on the oxidation of sewage without the aid of filters. *Journal of the Society of Chemical Industry., 33*(10), 523–539. https://doi.org/10.1002/jctb.5000331005

Arrojo, B., Mosquera-Corral, A., Garrido, J. M., & Méndez, R. (2004). Aerobic granulation with industrial wastewater in sequencing batch reactors. *Water Research, 38*(14–15), 3389–3399. https://doi.org/10.1016/j.watres.2004.05.002

Arsalan, M., Khan, Z. M., Sultan, M., Ali, I., Shakoor, A., Mahmood, M. H., Ahmad, M., Shamshiri, R. R., Imran, M. A., & Khalid, M. U. (2021). Experimental investigation of a wastewater treatment system utilizing maize cob as trickling filter media. *Fresenius Environmental Bulletin., 30*(1), 148–157.

Aslam, M. M. A., Khan, Z. M., Sultan, M., Niaz, Y., Mahmood, M. H., Shoaib, M., Shakoor, A., & Ahmad, M. (2017). Performance evaluation of trickling filter-based wastewater treatment system utilizing cotton sticks as filter media. *Polish Journal of Environmental Studies, 26*(5), 1955–1962. https://doi.org/10.15244/pjoes/69443

Beun, J. J., Hendriks, A., van Loosdrecht, M. C. M., Morgenroth, E., Wilderer, P. A., & Heijnen, J. J. (1999). Aerobic granulation in a sequencing batch reactor. *Water Research, 33*(10), 2283–2290. https://doi.org/10.1016/S0043-1354(98)00463-1

de Bueno, R. F., Piveli, R.P., & Campos, F. (2019). Extended aeration activated sludge process operated under low dissolved oxygen concentration: Kinetic behavior of nitrifying heterotrophic and autotrophic bacteria. *Engenharia Sanitaria e Ambiental, 24*(4), 939–947. https://doi.org/10.1590/s1413-41522019134260

Buyukkamaci, N., & Koken, E. (2010). Economic evaluation of alternative wastewater treatment plant options for pulp and paper industry. *Science of the Total Environment, 408*(24), 6070–6078. https://doi.org/10.1016/j.scitotenv.2010.08.045

Chan, Y. J., Chong, M. F., & Law, C. L. (2010). Biological treatment of anaerobically digested palm oil mill effluent (POME) using a Lab-Scale Sequencing Batch Reactor (SBR). *Journal of Environmental Management, 91*(8), 1738–1746.

Choubert, J. M., Racault, Y., Grasmick, A., Beck, C., & Heduit, A. (2005). Maximum nitrification rate in activated sludge processes at low temperature: Key parameters, optimal value. *European Water Management Online., 7*(3), 1–13.

Dan, N. H., Rene, E. R., & Le Luu, T. (2020). Removal of nutrients from anaerobically digested swine wastewater using an intermittent cycle extended aeration system. *Frontiers in Microbiology., 3*(1), 1–9. https://doi.org/10.3389/fmicb.2020.576438

Dan, N. H., Phe, T. T. M., Thanh, B. X., Hoinkis, J., & Le Luu, T. (2021). The application of intermittent cycle extended aeration systems (ICEAS) in wastewater treatment. *Journal of Water Process Engineering, 40*, 101909. https://doi.org/10.1016/j.jwpe.2020.101909

Du, R., Cao, S., Li, B., Wang, S., & Peng, Y. (2017). Simultaneous domestic wastewater and nitrate sewage treatment by DEnitrifying AMmonium OXidation (DEAMOX) in sequencing batch reactor. *Chemosphere, 174*, 399–407. https://doi.org/10.1016/J.CHEMOSPHERE.2017.02.013

Dutta, A., & Sarkar, S. (2015). Sequencing batch reactor for wastewater treatment: Recent advances. *Current Pollution Reports, 1*(31), 177–190. https://doi.org/10.1007/S40726-015-0016-Y

Eckenfelder, W. W. (1961). Trickling filtration design and performance. *Journal of Sanitary Engineering Division, ASCE, 87*(4), 33.

Eckenfelder, W. W., & Barnhart, E. (1963). Performance of a high-rate trickling filter using selected media. *Journal of Water Pollution Control Federation, 35*, 535.

Eckenfelder, W. W., & Ford, D. L. (1970). *Water pollution control experimental procedures for process design* (pp. 173–183). Jenkins Publishing Co.

Eckenfelder, W. W. (2000). *Industrial water pollution control*. McGraw-Hill International.

Farzadkia, M., Vanani, A.F., Golbaz, S., Sajadi, H.S., & Bazrafshan, E. (2016). Characterization and evaluation of treatability of wastewater generated in Khuzestan livestock slaughterhouses and assessing of their wastewater treatment systems. *Global Nest Journal, 18*(1), 108–118. https://doi.org/10.30955/gnj.001716

Flemming, H. C., Neu, T. R., & Wozniak, D. J. (2007). The EPS matrix: The "House of Biofilm Cells." *Journal of Bacteriology, 189*(22), 7945–7947. https://doi.org/10.1128/JB.00858-07

Forbis-Stokes, A. A., Rocha-Melogno, L., & Deshusses, M. A. (2018). Nitrifying trickling filters and denitrifying bioreactors for nitrogen management of high-strength anaerobic digestion effluent. *Chemosphere, 204*, 119–129. https://doi.org/10.1016/j.chemosphere.2018.03.137

Godoy-Olmos, S., Martínez-Llorens, S., Tomás-Vidal, A., & Jover-Cerdá, M. (2016). Influence of filter medium type, temperature and ammonia production on nitrifying trickling filters performance. *Journal of Environmental Chemical Engineering., 4*(1), 328–340. https://doi.org/10.1016/j.jece.2015.11.023

Hatami, B., Ebrahimi, A., Ehrampoush, M. H., Salmani, M. H., Dalvand, A., Pirmoradi, N., Angelidaki, I., Fotidis, I. A., & Mokhtari, M. (2021). Recovery of intermittent cycle extended aeration system sludge through conversion into biodiesel by in-situ transesterification. *Renewable Energy, 163*, 56–65. https://doi.org/10.1016/j.renene.2020.08.116

Henze, M., & Harremoes, P. (1983). Anaerobic treatment of wastewater in fixed film reactors: A literature review. *Water Science and Technology., 15*(9), 1–101. https://doi.org/10.2166/wst.1983.0161

Jafarinejad, S. (2017). Cost estimation and economical evaluation of three configurations of activated sludge process for a wastewater treatment plant (WWTP) using simulation. *Applied Water Science., 7*(3), 2513–2521. https://doi.org/10.1007/s13201-016-0446-8

Jagaba, A. H., Kutty, S. R. M., Lawal, I. M., Abubakar, S., Hassan, I., Zubairu, I., Umaru, I., Abdurrasheed, A. S., Adam, A. A., Ghaleb, A. A. S., Almahbashi, N. M. Y., Al-dhawi, B. N. S., & Noor, A. (2021). Sequencing batch reactor technology for landfill leachate treatment: A state-of-the-art review. *Journal of Environmental Management., 282*, 111946. https://doi.org/10.1016/J.JENVMAN.2021.111946

Joshua Amarnath, D., Thamilamudhan, R., & Rajan, S. (2015). Comparative study on wastewater treatment using activated sludge process and extended aeration sludge process. *Journal of Chemical and Pharmaceutical Research., 7*(1), 798–802.

Kołecka, K., Gajewska, M., Cytawa, S., Stepnowski, P., & Caban, M. (2020). Is sequential batch reactor an efficient technology to protect recipient against non-steroidal anti-inflammatory drugs and paracetamol in treated wastewater? *Bioresource Technology., 318*, 124068. https://doi.org/10.1016/J.BIORTECH.2020.124068

Li, H., Chen, Y., Gu, G., & Liu, Y. (2008). Phosphorous removal in intermittent cycle extended aeration system wastewater treatment plant: Effect of temperature, in: 2nd international conference on bioinformatics and biomedical engineering. *ICBBE, 4*(1), 1–12. https://doi.org/10.1109/ICBBE.2008.1075

Lim, J. X., & Vadivelu, V. M. (2014). Treatment of agro based industrial wastewater in sequencing batch reactor: Performance evaluation and growth kinetics of aerobic biomass. *Journal of Environmental Management, 146*, 217–225. https://doi.org/10.1016/J.JENVMAN.2014.07.023

Liu, Y., Wang, Z. W., & Tay, J. H. (2005). A unified theory for upscaling aerobic granular sludge sequencing batch reactors. *Biotechnology Advances, 23*(5), 335–344. https://doi.org/10.1016/J.BIOTECHADV.2005.04.001

Liu, L., Li, N., Tao, C., Zhao, Y., Gao, J., Huang, Z., Zhang, J., Gao, J., Zhang, J., & Cai, M. (2021). Nitrogen removal performance and bacterial communities in zeolite trickling filter under different influent C/N ratios. *Environmental Science and Pollution Research, 28*(4), 15909–15922. https://doi.org/10.1007/s11356-020-11776-y

Mancini, J. L, & Barnhart E. L. (1968). Industrial wastewater treatment in aerated lagoon. In E. F. Gloyna, & W.W. Ekenfelder Jr (Eds.), *Advances in water quality improvement*. University of Texas Press

Mareai, B. M., Fayed, M., Aly, S. A., & Elbarki, W. I. (2020). Performance comparison of phenol removal in pharmaceutical wastewater by activated sludge and extended aeration augmented with activated carbon. *Alexandria Engineering Journal, 59*(6), 5187–5196. https://doi.org/10.1016/j. aej.2020.09.048

McKee, K. P., Vance, C. C., & Karthikeyan, R. (2016). Biological manganese oxidation by Pseudomonas putida in trickling filters. *Journal of Environmental Science and Health: Part A Toxic/hazardous Substances and Environmental Engineering., 51*(7), 523–535. https://doi.org/ 10.1080/10934529.2016.1141618

Md Hafiz, M. F. U. B, Mohamed Kutty, S. R. B., & Hakmi, S. N. B. S. I. (2021). Impact of treating ammonia-nitrogen contamination from chemical fertilizer plant using extended aeration activated sludge system. In: *Lecture notes in civil engineering*, vol 132. Springer. https://doi.org/10.1007/ 978-981-33-6311-3_19

Metcalf, W., & Eddy, C. (2003). *Metcalf and eddy wastewater engineering: Treatment and reuse. Wastewater engineering: Treatment and reuse.* McGraw Hill.

Mirbagheri, S. A., Ebrahimi, M., & Mohammadi, M. (2014). Optimization method for the treatment of Tehran petroleum refinery wastewater using activated sludge contact stabilization process. *Desalination and Water Treatment, 52*(3), 156–163. https://doi.org/10.1080/19443994.2013. 794105

Nelson, M. J., Nakhla, G., & Zhu, J. (2017). Fluidized-bed bioreactor applications for biological wastewater treatment: A review of research and developments. *Engineering, 3*(3), 330–342.

Ni, B.-J., Xie, W.-M., Liu, S.-G., Yu, H.-Q., Wang, Y.-Z., Wang, G., & Dai, X.-L. (2009). Granulation of activated sludge in a pilot-scale sequencing batch reactor for the treatment of low-strength municipal wastewater. *Water Research, 43*(3), 751–761.

Olmos, S., López-Castellanos, J., & Bayo, J. (2019). Are advanced wastewater treatment technologies a solution for total removal of microplastics in treated effluents? *WIT Transactions on Ecology and the Environment, 229*(3), 109–116. https://doi.org/10.2495/WRM190111

Rana, K., & Shah, M. (2014). Extended aeration activated sludge process of pharmaceutical wastewater. *International Journal of Advanced Engineering Research and Science, 1*(2), 89–92.

Rehman, A., Ayub, N., Naz, I., Perveen, I., & Ahme, S. (2020). Effects of hydraulic retention time (HRT) on the performance of a pilot-scale trickling filter system treating low-strength domestic wastewater. *Polish Journal of Environmental Studies, 29*(4), 249–259. https://doi.org/10.15244/ pjoes/98998

Reynolds, T. D., & Yang, J. T. (1966). Model of the completely mixed activated sludge process. In *Proceedings of the 21st industrial waste conference*. Purdue University, Engineering Extension Series No. 121, p. 696.

Reynolds, T. D., & Richards, P. A. (1996). *Unit operations and processes in environmental engineering.* PWS Publishing company.

Sarti, A., Garcia, M. L., Zaiat, M., & Foresti, E. (2007). Domestic sewage treatment in a pilot-scale anaerobic sequencing batch biofilm reactor (ASBBR). *Resources, Conservation and Recycling, 51*(1), 237–247. https://doi.org/10.1016/J.RESCONREC.2006.09.008

Schwarzenbeck, N., Borges, J., & Wilderer, P. (2005). Treatment of dairy effluents in an aerobic granular sludge sequencing batch reactor. *Applied Microbiology and Biotechnology, 66*(6), 711–718.

Shieh, W. K., & Li, C. T. (1989). Performance and kinetics of aerated fluidized bed biofilm reactor. *Journal of Environmental Engineering., 115*(1), 65–79.

Su, J. J., Huang, J. F., Wang, Y. L., & Hong, Y. Y. (2018). Treatment of duck house wastewater by a pilot-scale sequencing batch reactor system for sustainable duck production. *Poultry Science., 97*(11), 3870–3877. https://doi.org/10.3382/PS/PEY251

Tanikawa, D., Fujise, R., Kondo, Y., Fujihira, T., & Seo, S. (2018). Elimination of hydrogen sulfide from biogas by a two-stage trickling filter system using effluent from anaerobic–aerobic wastewater treatment. *International Biodeterioration and Biodegradation., 130*, 98–101. https://doi.org/10.1016/j.ibiod.2018.04.007

Travers, S., & Lovett, D. (1984). Activated sludge treatment of abattoir wastewater—II: Influence of dissolved oxygen concentration. *Water Research, 18*(4), 435–439.

U.S. Epa. (2000). *United States environmental protection agency.* Wastewater Technology Fact Sheet: Wetlands: Subsurface Flow (EPA 832-F-00-023).

UNEP. (2018). *United Nations environmental program.* Technical University of Denmark (DTU), Copenhagen, Denmark.

Valta, K., Damala, P., Angeli, E., Antonopoulou, G., Malamis, D., & Haralambous, K. J. (2017). Current treatment technologies of cheese whey and wastewater by Greek cheese manufacturing units and potential valorisation opportunities. *Waste and Biomass Valorization, 8*(4), 1649–1663. https://doi.org/10.1007/s12649-017-9862-8

Von Sperling, M. (2015). *Activated sludge and aerobic biofilm reactors.* IWA Publishing. https://doi.org/10.2166/9781780402123

Hybrid Aerobic Wastewater Treatment Systems

11

Suspended and attached growth aerobic biologic wastewater treatment processes used in practice have been discussed in the previous chapter. To overcome drawback of these systems and to improve reliability of the wastewater treatment further, instead of using either suspended or stationary attached growth process, the reactor can be configured in such a way that it will offer hybrid system. Such reactor types popularly used for the wastewater treatment are submerged aerobic filter, moving bed biofilm reactor, rotating biological contactor and membrane bioreactors. Even in recent time, the sponge is also being encouraged as a media support and hanging sponge biofilm reactors are also being employed for the treatment of the sewage. These hybrid systems are reported to give satisfactory and reliable wastewater treatment. The sewage treated from this reactor can be reused after suitable tertiary treatment. These hybrid aerobic wastewater treatment technologies are discussed in this chapter.

11.1 Submerged Aerobic Filter

11.1.1 Description of Submerged Aerobic Filter

Activated sludge process provided with some packing material in it are being used for the treatment of sewage in past two to three decades. These packing materials may be fixed or may be movable in the rector. Accordingly, when the media is stationary the reactor is named as submerged fixed film reactor or submerged aerobic filter; whereas, when the media is movable it is referred as either fluidized bed reactor or moving bed biofilm reactor. Different types of synthetic plastic media are being used in these types of reactors. Due to introduction of the media, over which the biofilm grows, more biomass can be retained in the reactor; thus, enhancing the mean cell residence time and this will result in improving performance of the reactor. Also, it will result in reducing the volume of the aeration tank as compared to the volume required for the suspended growth process alone. Apart from removal of organic matter, these types of reactors can also be configured for nitrification and by providing some portion of the media as anoxic zone denitrification can also be achieved, thus supporting nitrogen removal.

The submerged attached growth process, also referred as submerged aerobic filter (SAF) or biological aerobic filter (BAF), can be classified as down-flow packed bed reactor or up-flow packed bed reactor depending on the direction of the wastewater flow in the reactor (Fig. 11.1). Oxygen is supplied from the bottom of the reactor through diffused air aeration system. The type and size of the packing media used is an important parameter determining performance of the reactor. Generally, the synthetic plastic packing material offering specific surface area of 90–500 m^2/m^3 is used in the SAF. However, the media with specific surface area ranging from 90 to 200 m^2/m^3 is typically used in practice.

Location and placement of the packing material is important in SAF. For wastewater with higher BOD concentration packing material with lower specific surface area shall be used; whereas, when nitrification is only being targeted higher specific surface area material can be used. If simultaneous denitrification is also to be achieved along the direction of wastewater flow in the reactor some anoxic zone is maintained to support it, accordingly the aeration arrangement should be made. The secondary clarifier is optional for this type of reactor and the biomass sloughed in the reactor and the suspended solids in the influent are trapped in the reactor itself when granular media is used and during periodic cleaning, once in a day, these solids are removed from the reactor and taken for digestion (Mendoza-Espinosa & Stephenson, 1999). Backwashing arrangement of the filter is also necessary, particularly while using high specific surface area material. This backwashing arrangement could be just like rapid sand filter. When a filter bed with coarser pore size is used, generally lamella clarifier is used for removal of sloughed biomass from the effluent of the filter to produce final effluent with less suspended solids.

Low retention time required (about 2 h) for treatment of dilute wastewater, like sewage, reduces the volume required for the reactor. Also, no sludge recirculation is involved in this reactor, which also reduces the operating cost of the treatment plant. In addition, the performance of the system is not dependent on the settling property of the sludge, as in case of suspended growth ASP, as here in case of SAF the biomass is retained in the form of biofilm grown on the filter media. However, higher capital cost due to involvement of media, more complex instrumentation and control and difficulty in scaling for large capacity sewage treatment are disadvantages of this technology.

11.1.2 Design Recommendation for SAF

Hydraulic loading in the range of 2–5 m^3/m^2 h can be adopted for designing the SAF. Nearly 50% or more of the empty volume of the reactor is filled with the media, since the

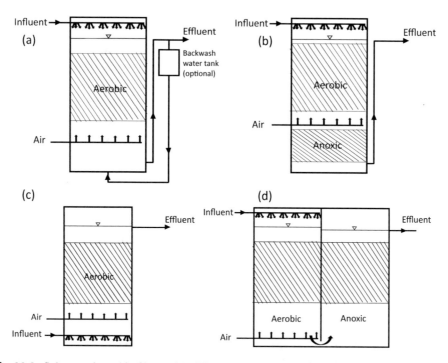

Fig. 11.1 Submerged aerobic filters with different arrangements of aeration and media location. **a** Downflow SAF for removal of organic matter, if more space is provided below the aeration where anoxic suspended biomass can be retained then some denitrification can be achieved. **b** Downflow SBR with top aerobic zone and below anoxic zone and media placed in both. The media below air diffusers will support denitrification. **c** Up-flow SAF, which may support nitrification at low OLR, however will not support denitrification due to complete aerobic operation. **d** Separate aerobic and anoxic section of SAF with media placement in each to support simultaneous organic matter removal and nitrification and denitrification

performance is majorly dependent on the biofilm area offered by the reactor. The organic loading rate (OLR) in the range of 1.5–3 kg BOD/m^3 day can be used for the designing for treatment of sewage. Loading up to 5 kg BOD/m^3 day has also been successfully used in practice to support organic matter removal and nitrification. The depth of the reactor can be in the range of 2–4 m, with 3 m typical. If properly designed and operated this SAF is capable to produce treated effluent with BOD and SS less than 10 mg/L and simultaneous nitrogen removal can also be achieved within shorter hydraulic retention time (HRT) (Stensel & Reiber, 1983).

11.1.3 Performance of SAF

When granular media is used the suspended particles present in the wastewater are stuck by the filter grains that offer a surface for biofilm growth to support aerobic degradation (Gilles et al., 1990). In this case the SAF removes carbonaceous organic matter through the mechanism of solid filtration, absorption, and aerobic oxidation (Stensel et al., 1988). Removal of COD for domestic wastewater up to 84% is reported in the experimental study carried out by González Martínez and Duque-Luciano (1992) at OLR of 20 g COD/m^2 day and 290 min of HRT. Paffoni et al. (1990) achieved 5 mg/L of effluent concentration of BOD with 86% removal efficiency for SS and BOD and 75% removal efficiency for NH$_4$-N with an ammonia loading rate of 0.50 kg N/m^2 day from a wastewater at operating temperature of 13 °C and HRT of 1 h. Whereas, while treating alcohol distillery wastewater with influent COD in the range of 3000–3500 mg/L, the SAF can achieve a COD removal efficiency of 82% at an HRT of 23 h and OLR of 3.3 kg COD/m^3 day (Reis & Sant'Anna, 1985).

Moore et al. (2001) suggested media selection is an important parameter for the design and operation of SAF to achieve higher organic matter removal efficiency from domestic wastewater. They found that by using foamed clay media of 1.5–4.5 mm size, 88% of COD removal was achieved at HRT of 81 min and OLR of 8.1 kg COD/m^3 day[1]. Morgan-Sagastume and Noyola (2008) also carried out performance investigation of SAF using volcanic scoria stones for treatment of a mixture of sewage with sugar, while the SAF was operated at OLR between 0.45 and 9.4 kg COD/m^3 day. A total COD removal efficiency of 80% was obtained for OLR up to 3 kg COD/m^3 day, which dropped down to 54% at OLR of 9.4 kg COD/m^3 day.

While treating hospital wastewater in SAF under variable OLRs of 0.25–1.25 kg COD/m^3 day, the effluent COD, BOD$_5$ and Phosphate concentration of 96–116 mg/L, 22–28 mg/L, and 3.06–3.27 mg/L was observed with a removal efficiency of 48%, 46%, and 29%, respectively, under HRT of 21 h (Khan et al., 2021). In another investigation by Priya and Philip (2015) for pharmaceutical wastewater treatment in SAF, with an influent COD range of 1200–1600 mg/L and SAF operated under an OLR of 3.09 ± 0.05 kg COD/m^3 day at HRT of 12 h, a COD removal efficiency of 92% was reported. For treatment of industrial wastewater containing 1,000 mg/L of phenol, 400 mg/L of total nitrogen, 30,000 mg/L salinity, and 2,800 mg/L of COD in a SAF, treated effluent recirculation of 600% was adopted (Ramos et al., 2007). A COD removal efficiency of 95.75% and total nitrogen removal of 83% was reported in this case.

Ferraz et al. (2014) reported treatment of landfill leachate at an HRT of 24 h in SAF. The COD and BOD removal efficiencies of 80% and 98%, respectively, at an OLR of 0.34 kg COD/m^3 day were reported. In another investigation on treatment of landfill leachate by Galvez et al. (2008), while treating influent average COD concentration of 18,683 mg/L using aerated and non-aerated submerged biofilters, the aerated submerged biofilter proved to be an effective system demonstrating maximum efficiencies of 66.7%,

91.2%, and 21.7% for COD, BOD_5, and TS, respectively, when operated under OLR of 25.1 kg COD/m^3 day at 15.9 h of HRT. In another investigation on treatment of an industrial hazardous waste landfill leachate using SAF with influent COD concentration of 3,628 mg/L, the COD removal efficiency of 76% under an OLR of 2.65 kg COD/m^3 day at an HRT of 1.37 day was reported (Smith, 1995).

In an attempt to remove amoxicillin from wastewater having an influent concentration ranging from 0.01 to 10 mg/L and soluble COD concentration of 1000 mg/L, Baghapour et al. (2015) employed SAF. The maximum soluble COD removal efficiency of 45.7% and amoxicillin removal efficiency of 50.7% was reported under an HRT of 12 h. Another effort on amoxicillin degradation by bacterial consortium in a SAF demonstrated 50.8% and 45.3% removal efficiencies of amoxicillin and soluble COD for an influent concentration of 10 mg/L and 1000 mg/L, respectively at 12 h of HRT (Baghapour et al., 2014). Baghapour et al. (2013) also experimented removal of atrazine using SAF and the maximum efficiencies of 97.9 and 98.9% under an HRT of 24 h were reported for atrazine and soluble COD with an influent concentration of 0.01–10 mg/L and 1000 mg/L, respectively.

Thus, the literature supports that SAF can be successfully employed for the treatment of domestic sewage, municipal wastewaters and industrial wastewaters. Successful application of SAF is also reported for treatment of landfill leachate and hospital wastewater containing pharmaceuticals. For obtaining satisfactory performance, lower OLR of around 3 kg COD/m^3 day is recommended. Short HRT of 2–3 h is adequate for treatment of sewage; whereas, while treating industrial effluents having higher concentration of organic matter HRT more than 12 h is required; and for treatment of complex organic matters, like pharmaceuticals and leachates, an HRT of more than a day may be essential to get satisfactory performance results.

11.2 Moving Bed Biofilm Reactor

11.2.1 General Description of Moving Bed Biofilm Reactor

Norwegian company, Kaldnes Miljoteknologies developed moving bed biofilm reactor (MBBR). This is mainly attached growth biological process, where media is not stationary and it moves freely in the reactor to improve substrate removal kinetics. Small cylindrical shaped polyethylene carrier elements (sp. density around 0.96 g/cm^3) are added in aerated or non-aerated reactors, depending on aerobic or anaerobic mode of operation, to support biofilm growth (Fig. 11.2). Cylinders of 10 mm ø and 7 mm thick with a cross walls inside and made with polyethylene are popularly used as media in MBBR.

The biofilm carriers are retained in the reactor by the use of a perforated plate (5 × 25 mm slots) or any similar screen at the effluent pipe of the reactor. Thus, this media having larger size cannot escape the reactor along with the effluent. Air agitation is used to continuously circulate the media and to keep it moving so as to establish optimum contact

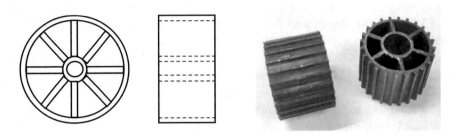

Fig. 11.2 Typical polyethylene media used in MBBR: **a** schematic and **b** photograph of the typical media used

with substrate present in the wastewater and bacteria attached to the media (Fig. 11.3). In case of anaerobic operation mixer is used to promote movement of the media. The media may fill about 25–50% of the empty reactor volume. The typical specific surface area of the media is about 200–500 m^2/m^3 of bulk media volume.

Provision of media inside the reactor in MBBR offers advantage that no sludge recycling from secondary sedimentation tank is required, hence minimizing operating cost by eliminating sludge recirculating pump operation. Also, since the media is moving there is no chance of blocking the media, which otherwise require back washing as in case of fixed bed stationary media. A final clarifier is used to settle sloughed biomass. Another advantage is use of more efficient fine bubble aeration equipment is not required, which would require periodic drainage of aeration tank and removal of packing for cleaning of diffusers.

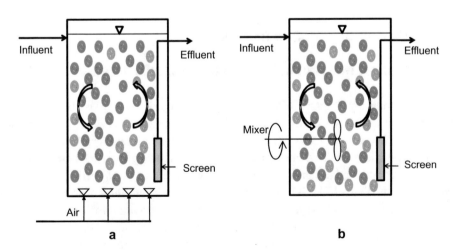

Fig. 11.3 Typical MBBR used with suspended packing materials: **a** aerobic and **b** anaerobic/anoxic with internal mixer

Since last decade, MBBRs are finding increasing application for post treatment of anaerobically treated industrial effluents and also as a secondary treatment system for treatment of sewage. These reactors can be used for removal of organic matter and also for achieving nitrification and denitrification for nitrogen removal. Employing single stage MBBR for primary treated low strength sewage treatment may meet the effluent standards for organic matter. Whereas, to achieve nitrification and denitrification along with organic matter removal multistage MBBRs are used with intermediate settler or settler provided at the end for both the reactors. As per the need the first MBBR will be anoxic to achieve carbonaceous organic matter removal and denitrification and this will be followed by aerobic MBBR mainly for nitrification and remaining organic matter removal. The nitrified effluent of aerobic MBBR is recycled back to anoxic MBBR for denitrification. For satisfactory nitrogen removal generally more than 100% recycling of nitrified effluent from second stage aerobic MBBR is required.

The performance of MBBR is affected by the applied OLR, temperature of the wastewater and adequacy of oxygen supplied in case of aerobic MBBR. For achieving satisfactory performance of denitrification adequate COD to nitrogen ratio need to be ensured. The denitrification process consumes about 3.7 g COD per g of NO_3^-N getting reduced and produces about 3.57 g of alkalinity per g of NO_3^-N and 0.45 g VSS. For supporting nitrification ensuring adequate alkalinity of about 7.14 mg as $CaCO_3$ per mg of N is necessary. The oxygen mass transfer efficiency prescribed by the manufacturer of diffusers per meter depth of MBBR gets reduced under field condition and this should be kept in mind while evaluating the oxygen requirement.

11.2.2 Design Recommendations for MBBR

The typical HRT used for MBBR while treating sewage is in the range of 3–5 h. The media with specific surface area of 200–250 m^2/m^3 is generally used in practice. The volumetric loading rate (OLR) normally adopted while designing MBBR for sewage treatment is in the range of 1.0–2.0 kg BOD/m^3 day. Considering the recirculation of the nitrified effluent from aerobic MBBR to anoxic MBBR an HRT in the range of 1–2 h is provided in the later. For designing secondary clarifier for MBBR, the surface overflow rate of 12–20 m^3/m^2 day is recommended. For designing MBBR for a particular industrial wastewater, the loading rate to be adopted may be determined from the laboratory experiments using the same media as proposed for the full-scale reactor to ensure satisfactory performance. The oxygen requirement shall be estimated as estimated in ASP design.

11.2.3   Performance of MBBR

Municipal and industrial wastewater treatment technology based on biofilm growth was developed in order to minimise several limitations of conventional wastewater treatment methods (Biswas et al., 2014). Andreottola et al. (2000) conducted an experiment on treatment of municipal wastewater containing influent total COD concentration of 231 mg/L and 32 mg/L of NH_4-N using aerobic MBBR technology. A 76% of total COD and 90% of NH_4-N removal efficiency was reported when this MBBR was operated under OLR of 0.61 kg COD/m^3 day at an HRT of 6.7 h. For treatment of laundry wastewater containing 727–944 mg/L of COD, 335–542 mg/L of BOD_5, 24–43 mg/L of anionic surfactants, and 27.8–59.6 mg/L of non-ionic surfactants in an aerobic MBBR, double stage treatment was adopted (Bering et al., 2018). The removal efficiencies of 95–98%, 89–94%, and 85–96% for BOD_5, COD, and a sum of anionic and non-ionic surfactants, respectively, were reported for operation at OLR of 1.5 kg COD/m^3 day at 7 h of HRT.

Effluents from hospitals contribute a lot of pharmaceuticals to municipal wastewater. Removal efficiencies of COD and NH_4-N from hospital wastewater up to 81% and > 99% for an influent concentration of 274 ± 190 mg/L and 54.0 ± 15.8 mg/L, respectively, were reported in the experimental investigation done by Casas et al. (2015) using three aerobic MBBRs in series at 6 h of HRT in each MBBR and overall OLR of 1.1 kg COD/m^3 day. Khan et al. (2020) used aerobic MBBR followed by ozonation for removal of ibuprofen and ofloxacin drugs from pre-treated manganese oxide wastewater. The removal efficiencies for both the drugs were more than 90% when MBBR was operated under an HRT of 16–20 h.

Wastewater containing terephthalic acid (TPA) if discharged untreated can degrade the quality of receiving water body rigorously. In an attempt to remove TPA, Liu et al. (2019) carried out an experiment using aerobic MBBR inoculated with a bacterial sludge of *Delftia* sp. WL-3 as degrader strain. A COD and TPA removal efficiency of 68 and 76% were obtained for an influent concentration of 1000 mg/L and 700 mg/L, respectively, under an OLR of 2.5 kg COD/m^3 day under 24 h of HRT. In an attempt to treat palm oil mill effluent having influent COD concentration of 1500 ± 19 mg/L and NH_3-N concentration of 115 ± 66 mg/L, Bakar et al. (2020) used aerobic MBBR filled with black plastic media (BPM) and Hexafilter (HEX) as a biofilm carrier for three media filling fractions (MFFs) of 25, 50, and 70%. A better COD removal of 58.6 and 92.8% of NH_4-N removal was observed for both BPM and HEX at MFF of 50% when operated under OLR of 0.5 kg COD/m^3 day and HRT of 72 h.

Zkeri et al. (2021) investigated treatment of dairy wastewater having influent COD, NH_4-N, TKN, and PO_4-P concentration of 2499 ± 812, 89.1 ± 34.8, 120 ± 10, and 20.7 ± 9.6 mg/L, respectively, using two-stage MBBR comprised of a methanogenic MBBR and an aerobic MBBR. A COD, NH_4-N, TKN and PO_4-P removal efficiencies of $93 \pm 4\%$, $97 \pm 3\%$, $99 \pm 1\%$ and $49 \pm 15\%$, respectively, were observed under OLRs of 2.5 ± 0.81 kg COD/m^3 day and 2.14 ± 0.7 kg COD/m^3 day with HRT of 24 h and 28.8 h for a methanogenic MBBR and an aerobic MBBR, respectively. In another investigation, Santos et al. (2020) treated dairy wastewater having four different influent COD concentrations of 600, 800, 1100, and 1200 mg/L, using aerobic MBBR. The removal efficiencies of 98%, 98%, 80% and 80% under HRTs of 7 h, 6 h, 7 h, and 7 h, respectively, were reported.

In an effort to treat tannery wastewater with the influent concentration of total COD, soluble COD, TN, and NH_4-N of 1656–2010, 1150–1326, 314–346, and 210–229 mg/L, respectively, Sodhi et al. (2021) evaluated performance of anoxic-aerobic MBBRs. The removal efficiencies of 93.5%, 94.8%, 88.7%, and 95.2% for total COD, soluble COD, TN, NH_4-N, respectively, were obtained at an OLR of 1.64 kg COD/m^3 day and an HRT of 14.7 h for anoxic MBBR and under HRT of 14 h for aerobic MBBR. For treatment of petrochemical wastewater containing $1,322 \pm 289$ mg/L of COD, 110 ± 23 mg/L of BOD_5, 680 ± 118 mg/L of TOC, and 385 ± 70 mg/L of TSS in an aerobic MBBR, the activated sludge from a high TDS wastewater was used as a halo-tolerant bacterial strains source (Ahmadi et al., 2019). A COD removal efficiency of 80% was reported for operation under an OLR of 3.05 ± 0.15 kg COD/m^3 day and HRT of 12 h.

A comparative study for the treatment of aquaculture systems wastewater containing 9.75 mg/L of NH_4-N in an aerobic MBBR using novel sponge biocarriers (SB) and K5 plastic carriers was conducted by Shitu et al. (2020). The removal efficiency of NH_4-N as $91.65 \pm 1.3\%$ was obtained for SB; whereas, for K5 plastic carriers removal efficiency of $86.67 \pm 2.4\%$ was achieved under 6 h of HRT. Thus, the MBBR is coming up as a promising technology for treatment of different wastewaters. Literature supports successful applications of MBBR for treatment of low strength as well as high strength wastewaters from different industries, like laundry, dairy, petrochemical, etc. Satisfactorily removal of organic matter, nitrogen and various other pollutants could be achieved in MBBR operated under optimum HRTs and OLRs.

Example: 11.1 Design an aerobic moving bed biofilm reactor for treatment of sewage having average flow rate of 1300 m^3/day consisting of average BOD concentration of 180 mg/L. Estimate the volume of media required and air requirement for supporting the BOD removal. The effluent should have maximum soluble BOD of 20 mg/L. The efficiency of oxygen transfer specified by the manufacturer of diffuser is 3% per meter of water column and consider oxygen saturation concentration at mid depth of MBBR equal to 10.5 mg/L and saturation concentration of DO at atmospheric pressure at 20 °C as 9.1 mg/L. The targeted DO in the MBBR is 2.0 mg/L. Oxygen transfer coefficient $(K_La)_{20} = 11.0$ h^{-1}, and minimum temperature of sewage in winter shall be considered as 17 °C.

Solution
Provide the organic loading rate of 1.5 kg BOD/m^3 day, the volume of MBBR required will be:

Volume = Flow × BOD/OLR = $1300 \times 0.180/1.5 = 156$ m^3.

Provide water depth of 5 m, hence area required will be 31.2 m^2.

Hence, diameter of circular MBBR $= (31.2 \times 4/\pi)^{1/2} = 6.3$ m.

Provide the MBBR media as 30% of total reactor volume, hence media packing volume required

$$= 156/3 = 52 \text{ m}^3$$

Estimating oxygen requirement

The kg of BOD removal $= (180 - 20) \times 10^{-3} \times 1300 = 208$ kg BOD per day.

Hence, ultimate BOD removed $= 208/0.68 = 305.88$ kg BOD per day.

Since, endogenous decay of biomass will also be supported in the biofilm formed on the media, hence for safety considering the oxygen required will be 305.88 kg/day.

The efficiency of oxygen transfer specified by the manufacturer of diffuser is 3% per meter of water column, which is proportional to the rate of oxygen transfer under field conditions. For $(K_La)_{20}$ the oxygen transfer efficiency is 3%, hence for actual field condition the oxygen transfer efficiency will be given as below.

Correcting $(K_La)_{20}$ for 17 °C

$$K_La \text{ at } 17\,^\circ C = (KLa)_{20}(1.024)^{17-20} = 11.0 \times (1.024)^{-3} = 10.24 \text{ h}^{-1}$$

$$Eactual = 3.0 \frac{\alpha . K_{La}.(\beta Cs - Ct)}{(K_{la})_{20}(Cs)_{20}}$$

$$= \frac{3.0 \times 0.7 \times 10.24(0.9 \times 10.5 - 2.0)}{11.0 \times 9.1}$$

$$= 1.6\%$$

Thus, with per meter of liquid height oxygen transfer efficiency will be 1.6%, hence for 5 m of water column depth in MBBR the oxygen transfer will be $1.6 \times 5 = 8\%$

Hence air required will be $= \frac{305.88}{1.201 \times 0.23 \times 0.08}$

$$= 13842 \text{ m}^3/\text{day} = 9.61 \text{ m}^3/\text{min}$$

11.3   Rotating Biological Contactor

11.3.1 Process Description

A rotating biological contactor (RBC) is an attached growth hybrid process providing the advantages of both biological fixed film and partial mixing reactor that is well suited for secondary and/or advanced biological treatment. The RBC process consists of a large disc with 1–3 m diameter and spaced 20–30 mm apart on a central shaft with radial and concentric passages slowly rotating with the speed of 0.75–1.5 RPM in a tank (Fig. 11.4). During the rotation, generally about 40% of the media surface area is under submergence in the wastewater. The rotation and subsequent exposure to oxygen from atmospheric air allows microorganisms to oxidize substrate and multiply to form a thin layer of biofilm over the disc surface.

This large, active population of microorganisms cause the biological degradation of organic pollutants. Excess biomass shears off at a steady rate from the discs and it is carried through the system to a secondary clarifier for removal of these biosolids from the treated effluent. Media rotation can be provided by either mechanical drives or air-motivated rotation. The main advantages of an aerobic RBC are: (i) short HRT, thus allowing a smaller footprint, (ii) high specific surface area, (iii) high biomass holdup in reactor, (iv) insensitivity to toxic substrate, (v) less accumulation of sloughed bio-film and partial mixing, (vi) low energy consumption (1 kW per 1000 population equivalent, PE), and (vii) operational simplicity (Chowdhury et al., 2010).

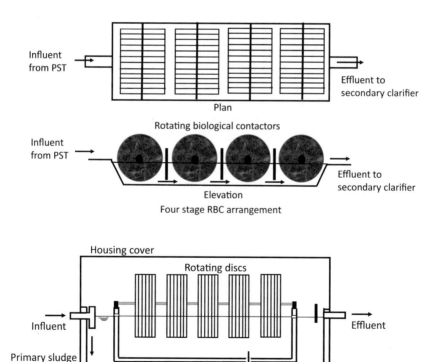

Fig. 11.4 Schematics showing configuration of multistage rotating biological contactor

11.3.2 Factors Affecting Performance of RBC

Like all biological units, alkalinity, pH, nutrients, temperature, oxygen, biomass population balance and concentrations of pollutants in the influent wastewater impact efficiency of RBCs. A large biofilm surface area is supported on the discs of RBC and performance rely on the mass transfer of oxygen and substrates from the bulk liquid to the biofilm. The complexity in the physical and hydrodynamic characteristics requires that the design of the RBC be based on fundamental information derived from performance of pilot plants and field-scale installations. Other than these regular factors, efficiency of RBC is also affected by disc rotational speed, loading rate, disk submergence, media composition, and staging.

There are many design parameters that regulate the performance of RBCs; these are organic and hydraulic loading rates, HRT, rotational speed, type of media used, operating temperature, DO level, wastewater and biofilm characteristics, and sludge recirculation (Cortez et al., 2008). In an investigation, to identify the effect of hydraulic loading rates, six experiments with different HRTs were performed, and the final result confirmed the BOD removal of 80–56% when increasing hydraulic loading rates from 38 to 76 L/day m^2 (Oğuz & Oğuz, 1993).

Disk Rotational Speed: The rotation of disc leads to bulk fluid mixing, convection through media/biofilm pores, compound diffusion to the film and subsequent product exchange with the reactor and surroundings (Rittmann & McCarty, 2001). Rotation of media creates a head difference leading to convective air/water exchange. Disc rotation mostly impacts the oxygen transfer, which takes place through the air–liquid interface in the RBC tank due to the turbulence caused by discs rotation and through the liquid film adhering to the discs during the aeration phase when discs come out of liquid submergence during rotation. Increasing tip speed increases the total oxygen transfer rate in a pseudo-linear fashion (Rittmann et al., 1983). However, the energy usage for the driving motor increases exponentially with increasing rotational speed. For minimal operating cost the lowest rpm should be selected and rotor speeds of 0.7–2.0 rpm are commonly adopted in practice (Mba et al., 1999). However, some high-rate systems are known to exceed this speed.

Loading rate: For RBC, the design organic loading rates used for carbonaceous oxidation range from 0.5 to 1.0 kg BOD_5/m^3 day, and for nitrification it range from 0.16 to 0.24 kg BOD_5/m^3 day (Wang et al., 2009). The organic loading in the first stage cells should be reduced to prevent oxygen from becoming limiting, by placing more first stage cells in parallels, or by step feeding the influent wastewater to the first two cells. Because the performance depends on the amount of biomass on the media, a design loading based on media surface area is a better approach. The maximum soluble organic matter loading limitations on the first stage recommended by the RBC manufactures is 0.0244 kg BOD/m^2 day using standard density media (Wang et al., 2009). The overall soluble BOD_5 loading rates vary from 0.0049 to 0.0146 kg/m^2 day (Wang et al., 2009).

For nitrification, the recommended ammonia nitrogen loading rates vary from 0.0012 to 0.0024 kg/m^2 day depending on the effluent ammonia nitrogen concentration requirements (Wang et al., 2009). Also, the stages where nitrification takes place should not have a BOD$_5$ concentration higher than 30 mg/L (15 mg/L soluble BOD$_5$) (Wang et al., 2009). Design hydraulic loading rates in the range of 0.03–0.16 m^3/m^2 day of media surface area are used for secondary treatment, and in the range of 0.01–0.08 m^3/m^2 day of media surface area are used for treatment with nitrification (Wang et al., 2009).

For successful treatment, the loading rates must be within the oxygen transfer capability of the system. When oxygen demand exceeds the oxygen transfer capability due to overloading, it can lead to poor performance. Under overloaded conditions, anaerobic conditions develop deep in the attached film. Sulphates are reduced to H$_2$S, which diffuses to the outer layer of the biofilm, where oxygen is available. *Beggiatoa*, a filamentous bacterium which can oxidize the reduced sulphur compounds, forms a biofilm that does not slough under normal rotational sheer conditions.

Disk submergence: Partially submersed discs are used for nitrification, while completely submersed discs are used for denitrification, however, it can lead to longer start-up times (Teixeira & Oliveira, 2001). Disk submergence typically impacts the physical properties, composition and activity of biofilm growing on the discs. About 1–4 mm thickness of biofilm will develop on the surface of the media. Partially submerged discs usually will support thicker and denser biofilm compared to fully submerged ones. It is recommended that about 40% of the media be submerged in the tank (Wang et al., 2009).

Media composition: RBC media can be present as discs, mesh plates, saddlesor rings in a packed bed reactor, which resembles a partially submerged, rotating, moving bed biofilm reactor (Sirianuntapiboon & Chumlaong, 2013). The RBC media commonly has a specific surface area of 150–250 m^2/m^3 for biofilm growth, which supports high removal rates at low HRTs. Lower density media is normally applied at the front-end of the RBC stack, which typically receive high organic loading rates (Cortez et al., 2008). Support media should be insoluble and non-degradable, have high mechanical and biological stability, and be a cost effective (Leenen et al., 1996). The media physicochemical composition and architecture both impact on the microbial biofilm and the removal rate of substrates (Stephenson et al., 2013).

Using media with high specific unit surface area will increase the treatment efficiency. However, the use of high specific surface area media in the first stages of treatment may result in clogging owing to the smaller clearances and, if used, requires a very low organic loading, which is offsetting the advantage of using high specific surface area media. The arrangement of media in a series of stages has been shown to increase treatment efficiency significantly. High density polyethylene is used for the disk media to offer surface area of 120 m^2/m^3. A single RBC unit can be 3.5 m in diameter and 7.5 m long with 1.5 m depth of water. Thus, single unit contains about 9300 m^2 disk surface area. The RBC units having higher surface area per unit, up to 13,900 m^2, are also available.

Staging of RBC units: Staging is compartmentalizing of the RBC discs to form a series of independent cells. The number of stages depends on the treatment goals, with two to four stages for removal of organic matter and six or more stages for nitrogen removal. Staging promotes a variety of conditions, where different organisms can flourish in varying degrees in different stages, depending primarily on the soluble organic matter available in the bulk liquid. For domestic sewage treatment, performance increases with liquid to surface area ratio up to 0.0049 m^3/m^2, with no improvement in further increase in area (Eckenfelder, 2000). Significant improvement in two to four stages can be obtained for sewage treatment; however, beyond fourth stage there is no advantage. For treatment of high strength industrial wastewaters more than four stages may be desirable. In this case the first stage may be enlarged and intermediate settler may be provided to avoid anaerobic condition in RBC basin followed by subsequent contactor stages. For designing an RBC for treatment of industrial wastewaters usually a pilot plant performance evaluation is recommended to ascertain the design parameters for satisfactory performance.

Enclosure to RBC: Fiberglass reinforced plastic cover is usually provided over each shaft of RBC. The enclosure is provided to satisfy following:

- Protect plastic discs from deterioration due to ultra-violate light.
- Protect the process form low temperature.
- Control the build-up of algae in the process.
- Protect equipment from any possible damage.

11.3.3 Process Design

Different models have been proposed for designing the RBC, which are described below.

US EPA model: The US EPA model, Eq. 11.1, can be applied to an RBC system; however, this model does not include parameters on microbial kinetics, substrate limitation or changes to influent/temperature.

$$Le/Lo = \exp[-K_p(V/695Q)^{0.5}] \tag{11.1}$$

where, V is media volume, ft^3; Q is wastewater design flow excluding recycle flow, MGD; L_e is reactor effluent BOD$_5$, mg/L; L_o is reactor influent BOD$_5$, mg/L; and K_p is performance measurement parameter. An average K_p value of 0.30 at 20 °C can be used for design.

Modified US EPA model: Another similar formula is also frequently used for design of RBC systems as given in Eq. 11.2.

$$Le/Lo = \exp[-K_t(Ac/695Q)^{0.5}] \tag{11.2}$$

where K_t is treatability function related to surface area; and A_c is media surface area, ft^2. An average K_t value of 0.066 at 20 °C has been used by many professional engineers for RBC design.

The following are the boundary conditions and design equations for RBC systems in evaluation of coefficients K_p and K_t. For wastewater temperatures above 13 °C, the $(K_p)_T = (K_p)_{20}$ and $(K_t)_T = (K_t)_{20}$ shall be considered. Whereas, for wastewater temperatures below 13 °C, for any other temperature the relation is given by Eq. 11.3 and Eq. 11.4.

$$(K_p)_T = (K_p)_{20}(1.018)^{T-20} \tag{11.3}$$

$$(K_t)_T = (K_t)_{20}(1.018)^{T-20} \tag{11.4}$$

where, $(K_p)_T$ and $(K_p)_{20}$ are the performance measurement parameters at wastewater temperature T °C and 20 °C, respectively; and $(K_t)_T$ and $(K_t)_{20}$ are the treatability parameters at wastewater temperature T °C and 20 °C, respectively.

Grady's Model: A second order model is used to estimate the surface area required for RBC as stated below in Eq. 11.5. This second-order model (US EPA, 1985) can be used for prediction of effluent BOD, which is later converted to SI units by Grady et al. (1999).

$$Ln = \frac{-1 + \sqrt{1 + (4)(0.00974)\left(\frac{As}{Q}\right)Sn} - 1}{(2)(0.00974)(As/Q)} = \frac{-1 + \sqrt{1 + 0.039\left(\frac{As}{Q}\right)Sn} - 1}{0.0195\left(\frac{As}{Q}\right)} \tag{11.5}$$

where, L_n is soluble organic matter concentration in stage n, mg/L, A_s is disc surface area in stage n, m^2 and Q is flow rate in m^3/day.

Since only soluble BOD is used in the model for estimating effluent BOD after secondary clarifier, appropriate BOD$_5$/BODu should be used (0.5–0.75 for sewage). For first stage, organic loading rate should be kept equal to or less than 12 to 15 g BOD$_5$/m^2 day to determine the first stage disk area and effluent BOD$_5$ concentration from above equation.

Clark's model: Removal rate of organic matter in RBC can be determined from influent/effluent conditions and microbial growth rate as per the model illustrated in Eq. 11.6.

$$r_a = \left(\frac{\mu_{max}}{X_a}\right)/Y_a \tag{11.6}$$

where, r_a is substrate removal rate (mg/L^2.sec), μ_{max} is maximum specific growth rate (sec^{-1}), X_a is concentration of attached biomass (mg/L), Y_a is yield coefficient for attached biomass.

Kincannon and Stover model: The model represented in Eq. 11.7 integrates substrate removal rate over disc area of RBC.

$$r_a = \left(\frac{K_C}{U_{max}}\right) \cdot \left(\frac{A_s}{QL_n}\right) + \left(\frac{1}{\mu_{max}}\right) \tag{11.7}$$

where, K_C is the half saturation constant (mg/L), U_{max} is maximum substrate removal rate (mg/m^2 sec), L_n is soluble organic matter concentration in stage n (mg/L), A_s is disc surface area on stage n (m^2) and Q is flow rate (m^3/day), μ_{max} is maximum specific growth rate (sec^{-1}).

Kinetic model: The kinetic model for the RBC can be expressed in terms of the surface area of the disk as presented in the Eq. 11.8.

$$Q \times (S_o - S)/A = K.S \tag{11.8}$$

where, Q is flow rate; A is surface area of disk; S_o is the influent substrate concentration; S is the effluent substrate concentration; and K is reaction rate.

For wastewater with variable influent organic matter concentration, the kinetic equation can be modified as per Eq. 11.9.

$$Q \times (S_o - S)/A = K.S/S_o \tag{11.9}$$

The maximum BOD removal rate under given operating condition, rpm adopted and oxygen contact, etc. will depend on influent BOD concentration of the wastewater. For multiple contactors operated in series, the performance can be defined as presented by Eq. 11.10.

$$S/So = \left(\frac{1}{1 + K\left(\frac{A}{Q}\right)}\right)^n \tag{11.10}$$

where, n is the number of stages.

In case of multistage RBC, the loading for each stage must be checked due to O$_2$ limitations in some system. Loading will be limited due to oxygen limitation, and increase in effluent BOD will be observed beyond certain applied loading rate sharply.

11.3.4 Process Operation

While operating an RBC the following checks should be maintained:

i. Check peak organic loads: It should be less than twice the daily average, otherwise improve pre-treatment or expand plant.
ii. Check peak hydraulic loads: It should be less than thrice the daily average, otherwise use flow equalization before RBC.
iii. Desired range of pH is 6.5–8.5 for secondary treatment and 8.0–8.5 for nitrification. Exposure to pH below 5.0 or above 10.0 can cause sloughing of biomass from disks.
iv. Temperatures less than 13 °C will reduce efficiency of RBC.

The use of RBCs for biological wastewater treatment to remove organic matter and ammonia has been well established for the last three decades. Application has largely been at the lower end of the wastewater treatment scale, usually for up to 2000 P.E. (Griffin & Findlay, 2000). The process can be adapted to remove nutrients, both nitrogen and phosphorus, as with other biological processes. Control of disc immersion can be used to stimulate denitrification. The simplicity of operation, adaptability, less land use and maintenance requirement, and high volumetric activity are some of the advantages of RBC.

Example: 11.2 Designed a staged RBC treatment system for wastewater with raw and treated wastewater characteristics given in Table 11.1.

Solution
Determine number of shafts required in first stage.
Consider BOD loading of 15 g BOD_5/m^2 day in first stage.
BOD_5 loading $= 4000 \times 90 = 360,000$ g/day.
Hence, disk area required $= (360,000$ g/d)/(15 g/m^2 day) $= 24,000$ m^2.
Using surface area per shaft $= 9,300$ m^2/shaft.
Number of shafts required $= 24,000/9300 = 2.6 \approx$ say 3 parallel RBCs.
Number of trains and number of stages required.
Provide three trains with 3 stages each train as shown in Fig. 11.5.
Flow rate /train $= 4000/3 = 1333.3$ m^3/day.

Table 11.1 Characteristics of the untreated wastewater and required for the effluent

Parameter	Unit	Primary effluent	Target effluent
Flow rate	m^3/day	4000	
BODu	g/m^3	140	20
BOD$_5$	g/m^3	90	10
TSS	g/m^3	70	20

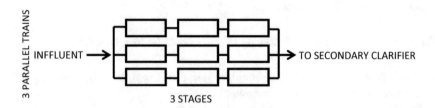

Fig. 11.5   Arrangement of stages proposed in three parallel trains

Calculation of soluble BOD in each stage.
a. Stage 1

$$S1 = \frac{-1 + \sqrt{1 + (4)(0.00974)\left(\frac{As}{Q}\right)So}}{(2)(0.00974)(As/Q)} = \frac{-1 + \sqrt{1 + 0.039\left(\frac{As}{Q}\right)So}}{0.0195\left(\frac{As}{Q}\right)}$$

$So = 90 \text{ g/m}^3$
$\left(\frac{As}{Q}\right) = 9300/1333.3 = 6.97 \text{ day/m}$

$$S1 = \frac{-1 + \sqrt{1 + 24.6}}{0.136} = 29.8 \text{ g/m}^3 \text{ (mg/L)}$$

Similarly, $S_2 = 14.82 \text{ g/m}^3$ and $S_3 = 9.1 \text{ g/m}^3 < 10 \text{ g/m}^3$ satisfying the requirement. Hence three stages are adequate to meet the required effluent standard.
Determination of organic and hydraulic loadings.

a. First stage organic loading

$$OLR1 = \frac{4000\left(\frac{m^3}{d}\right) * 90(g\frac{BOD_5}{m^3})}{(3)(9300 \text{ m}^2)} = 12.9 \text{ g BOD}_5/\text{m}^2 \text{ day}$$

This is within the 10 to 15 g BOD_5/m^2 day hence fine.

b. Overall organic loading in terms of ultimate BOD

$$OLRover = \frac{4000 * 140}{3 * 9300} = 21.1 \text{ g BODu/m}^2 \text{ day}$$

c. Hydraulic loading

$$\frac{\text{HLR}}{\text{per}}\text{shaft} = \frac{4000}{3*9300} = 0.143\,\text{m}^3/\text{m}^2\text{day}$$

Estimating volume of each RBC tank

Consider the tank volume requirement as $0.0049\,\text{m}^3/\text{m}^2$ of the disc area.

Hence, for $9300\,\text{m}^2$ disc area of each shaft the volume $= 9300 \times 0.0049 = 45.6\,\text{m}^3$.

Providing liquid depth of 1.5 m, the area required is $30.38\,\text{m}^2$.

For rectangular cross section, the width of 4.0 m and length of 7.6 m is recommended.

11.4 Hanging Sponge Reactor

11.4.1 Process Description

The downflow hanging sponge (DHS) reactor is a modification of trickling filter system equipped with sponge as media, developed as a low-cost aerobic treatment system (Tandukar et al., 2007). The highlight of the DHS reactor is that it can be operated without aeration or with low aeration requirements, as oxygen is naturally getting dissolved in wastewater in this system. In addition, the sponge media supports a large amount of biomass as well as high microbial diversity on the surface and inner section of the sponge media. Biomass is retained both inside and outside the sponge media. Therefore, the nitrification–denitrification reactions proceed owing to the unique properties of the sponge media, i.e., media may provide aerobic environment near the media surface (the nitrification reaction portion) and an anaerobic environment deeper inside the media (the denitrification reaction portion). The high microbial diversity in this ecosystem, with an extremely long food chain, reduces the production of excess sludge (Onodera et al., 2014).

Other than loading rate and pH, which is common to all biological treatment processes, performance of DHS is also impacted by physical parameters, like contact time, sponge bulk volume, sponge pore size. Higher retention time, higher sponge volume and lower pore size are the best suited conditions for maximum removal efficiency in a DHS reactor. The DHS reactor can be operated in multiple modes:

- Without effluent recirculation
- With recirculation of effluent
- Anoxic compartments to ensure denitrification.

The first prototype of a DHS was developed in 1995 and it is referred to as the first-generation DHS (DHS G1 or cube-type DHS). Sponge is first cut into small cubes (1.5 cm per side), which then are connected to each other diagonally in series with the help of nylon string (Machdar et al., 1997). The second-generation DHS (DHS G2 or curtain-type DHS) was developed to overcome a few drawbacks of DHS G1. The modified version, DHS G2 is constructed with long triangular polyurethane sponges that are glued to both sides of a polyvinyl sheet (Okubo et al., 2015). The basic design concept of the third-generation DHS (DHS G3) was like that of a TF. However, instead of the solid packing materials used in TFs, small sponge pieces with an outer supporting material were used. Construction of a reactor with this type of DHS is easier, as it involves simply packing the sponge media randomly (Okubo et al., 2016).

Differing from DHS G3, the fourth-generation DHS (DHS G4) was developed to create an absolute gap between sponge units, which provides a space for the contact of air and wastewater running through the reactor. The reactor is constructed with long sponge strips that are placed inside a net-like cylindrical plastic cover to provide rigidity (Tandukar et al., 2006). The fifth-generation DHS (DHS G5) is a direct improvement to the design of DHS G2. Several DHS curtains are lined up to construct a DHS module. A single-curtain unit is constructed by gluing a sponge sheet with an undulating, or wavy, surface to both sides of a thin plastic sheet (Tandukar et al., 2007). The sixth-generation DHS (DHS G6) is based on the concept of the DHS G3. The only difference is the type of sponge material used. For DHS G6, the sponge is hardened by introducing epoxy resin into the polyurethane network structure. This approach removes the need for the supporting plastic ring because the sponge media itself is reasonably rigid (Onodera et al., 2014).

A modification of the DHS reactor was used by Chatterjee et al. (2016, 2018) for treatment of UASB effluent. The hybrid reactor could achieve chemical oxygen demand (COD) removal of $99 \pm 1\%$ with an effluent COD concentration of around 7 ± 3 mg/L at an organic loading rate (OLR) of 2 kg COD/m^3 day and a total nitrogen removal of $89 \pm 8\%$ by combination of nitrification–denitrification processes at a nitrogen loading rate of 0.2 kg TN/m^3 day. Innovative reactor design with moving rope bed supporting oxic and anoxic microenvironments favoured high nitrogen removal rate (Chatterjee et al., 2016).

The DHS provides a unique environment for use as microbial habitat. Water quality drastically changes along with the height of the reactor. In addition, the redox potential changes from the outer layer to the inner part of the sponge medium. For this reason, both aerobic and anaerobic microorganisms are detected in DHS sludge, and different community structures are found along the height of the reactor.

11.4.2 Performance of DHS Reactor

The DHS rector wastewater treatment technology was developed by Machdar et al. (1997). In this first ever experiment of DHS reactor was used for treatment of effluent generated from UASB reactor treating municipal wastewater. The influent to DHS reactor had total-COD, total-BOD and TKN as 144 mg/L, 68 mg/L, and 56 mg/L, respectively, and up to 72% of total-COD, 97% of total-BOD, and 89% of TKN removal efficiency was achieved at an OLR of 2.66 kg COD/m^3 day and HRT of 1.3 h in DHS reactor. In another investigation by Machdar et al. (2000) on post treatment of UASB reactor effluent in DHS reactor with an influent concentration of total-COD of 161 mg/L, total-BOD of 51 mg/L, and TKN of 51 mg/L the removal efficiencies of 60%, 92%, and 57%, respectively, were reported under HRT of 2 h and at an OLR of 1.932 kg COD/m^3 day in DHS reactor.

In an attempt to treat low strength sewage using DHS rector having influent BOD in the range of 20–50 mg/L, Yoochatchaval et al. (2014) achieved removal efficiency of 85% at an HRT of 1.5 h and an OLR of 1.34 kg COD/m^3 day. In another investigation by Tandukar et al. (2005) for municipal sewage treatment using combination of UASB reactor and DHS, the wastewater with a total-BOD of 78 mg/L, total-COD of 195 mg/L, and TKN of 40 mg/L was applied to DHS reactor, which was operated under an OLR of 2.34 kg COD/m^3 day at 2 h of HRT. Removal efficiencies of 88%, 76% and 30% were reported respectively for total-BOD, total-COD and TKN. Tawfik et al. (2010) also experimented on treatment of UASB reactor effluent in DHS reactor and the maximum efficiency of 88% was reported under an HRT of 2.6 h when total-COD of 307 mg/L was admitted to DHS reactor. In another investigation on treatment of septic tank effluent by Machdar et al. (2018) using DHS reactor, while treating influent total-BOD concentration of 130 mg/L, the maximum removal efficiency of 80% was observed at HRT of 4 h and OLR of 2.49 kg COD/m^3 day.

In an investigation on treatment of agricultural drainage water by using DHS system with influent total-COD and BOD$_5$ concentration of 441.5 ± 41.5 mg/L and 212.9 ± 32.9 mg/L, the maximum removal efficiencies of $89.3 \pm 5.4\%$ and 94.4 ± 3.3, respectively, were demonstrated, when DHS reactor was operated under OLR of 2.01 kg COD/m^3 day at 5.26 h of HRT (Fleifle et al., 2013). The natural rubber processing wastewater was post treated in DHS reactor after first stage treatment in anaerobic baffled reactor. The DHS reactor demonstrated $64.2 \pm 7.5\%$ total-COD removal efficiency for an influent concentration of 197 ± 6 mg/L at 4.8 h of HRT under an OLR of 0.97 ± 0.03 kg COD/m^3 day (Watari et al., 2017).

Thus, the literature supports that DHS reactor can be successfully employed for the post treatment of UASB reactor effluent to offer complete treatment to municipal wastewaters. Successful application of DHS is also reported in literature for post treatment of effluent generated from UASB reactor or other anaerobic reactors for the treatment of industrial wastewaters. For post treatment of municipal wastewaters in DHS reactor an HRT of 2 h is found to be adequate; whereas, while post treating industrial effluents with higher organic matter concentrations an HRT of about 4 h is found to be necessary.

11.5   Membrane Bio-reactor

11.5.1 Process Description

Membrane bio-reactor or membrane biological reactor (MBR) consist of an aeration tank working as a bio-reactor, like ASP, with suspended active aerobic microbial sludge for oxidation of organic matter present in the incoming wastewater. However, the secondary clarifier used in ASP is replaced in MBR with typically microfiltration membrane having pore size of 0.1–0.4 μm. This membrane module provided could be flat sheet membrane or hollow fibre membrane, typically installed in aeration tank itself. Thus, it will not allow biomass to escape from the aeration tank through membrane and to appear in the effluent and retaining biomass inside the aeration tank of MBR itself. Thus, instead of secondary clarifier participating in recycling the necessary biomass to aeration tank, here in MBR the membrane is not allowing the microbial sludge mass to escape and hence sludge recirculation is not required as it is retained in the reactor.

The MBR could be operated as aerobic MBR or anaerobic MBR. Due to involvement of microfiltration membrane the effluent quality of MBR is superior in terms of BOD and suspended solids as compared to conventional ASP for the treatment of domestic as well as industrial effluents. Also, reliable performance of MBR can be obtained, since performance is independent of SVI of the sludge developed because biomass is retained in the reactor by membrane and sludge recirculation is not involved.

Generally, the MLSS concentration maintained in MBR is higher than conventional ASP and hence higher volumetric loading rate can be applicable, reducing required volume of the reactor. Due to retention of biomass by membranes, higher SRT is provided in the reactor, resulting in lesser sludge production. If operated at lower DO, even simultaneous nitrification and denitrification can be achieved in MBR. However, due to involvement of membrane, higher capital and maintenance cost is involved in this because of fouling of the membrane requiring replacement, which is the major drawback of this system. Also, energy requirement for MBR is higher than the conventional ASP.

The MBR can be designed as membrane module housed in the main aeration tank or it can be designed as the side stream membrane module, where the mixed liquor from the aeration tank is transferred to separate membrane module situated outside the bioreactor and the sludge retained by membrane is recycle back to the aeration tank (Fig. 11.6). Similar arrangement can be made even in case of anaerobic MBR, where membrane module is housed outside the bioreactor and sludge retained by the membrane module is recycled back to the bioreactor after separation of the permeate by membrane, which forms the effluent.

As shown in Fig. 11.6a, the membrane cassettes or module, i.e., membrane assembly comprising of membranes, support structure for membrane and inlet and outlet pipes with an overall casing of the module, is inserted in the main bioreactor. A vacuum of around 50 kPa (0.5 Bar or 0.5 atm) is applied to the membrane to draw the permeate from the

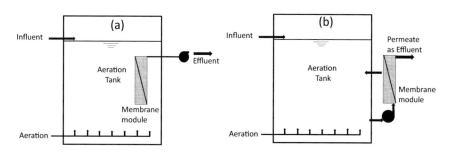

Fig. 11.6 Types of membrane bioreactors: **a** MBR with submerged membrane module in bioreactor and **b** MBR with external membrane module and recycling the sludge retained by membrane to bioreactor

mixed liquor, while retaining biosolids in the bioreactor. To minimize biofouling of the membrane an arrangement for air score along the membrane surface is made by applying the compressed air, so that solids attached to the membrane surface shall be removed. This air requirement for membrane scouring is additional quantity than required for oxidation of organic matter. The membrane cartridge can also be taken out of the reactor for periodic external thorough cleaning with chemical and water.

11.5.2 Process Design

The MLSS concentration up to 15,000–20,000 mg/L can be used in MBR, which is much higher than the conventional ASP. However, recommended range of MLSS in design is 8,000–10,000 mg/L to make the process more cost effective. Depending on the type of wastewater treated and SRT used the MLVSS/MLSS of the sludge for aerobic MBR is in the range of 0.7–0.8. The typical value of SRT adopted in designing MBR is in the range of 5–20 days. The permeate flux rate through the membrane is specified by the manufacturer of the membrane under applied particular negative pressure. This flux rate can get reduced while maintaining higher MLSS concentration in the aeration tank. The flux rate of permeate through membrane is generally in the range of 600–1200 L/m^2 day.

The OLR of 1–3 kg BODu/m^3 day is used for designing MBR for treatment of sewage. For treatment of industrial wastewater, suitable OLR can be evaluated based on the performance evaluation of laboratory experiments. For sewage treatment HRT of 4–6 h is adequate, again this required HRT for industrial wastewater need to be determined based on applicable OLR. Though, F/M ratio in the range of 0.1–0.4 kg BOD/kg VSS day is generally used for MBR, even higher F/M values have been used in practice. For treatment of sewage adopting design values in this range, the MBR is capable to produce treated effluent with BOD less than 10 mg/L, COD around 30 mg/L and effluent turbidity can also be maintained below 5 NTU (Nephelometric turbidity unit).

11.5.3 Control of Membrane Fouling

In MBR, when the effluent is withdrawn through the membrane applying suction, the biomass present in the mixed liquor gets adhered to the exterior of the membrane. Some of the microorganisms may colonize on the membrane surface and the extra polymeric substances excreted by microbes adhere to the membrane surface. Also, the finer particles present in the wastewater may enter into pores of the membrane and blocking it. This membrane fouling thus lead to increased pressure loss and higher pressures need to be applied to maintain desired membrane water flux. Hence, strategies are adopted to control this membrane fouling in MBR.

An air scour wash is adopted to generate agitation within the membrane module; thus, membrane agitate among each other leading to mechanical scoring of the attached biomass on the surface and also due to scouring effect of the rising air bubbles. The membrane can also be intermittently backwashed by reversing the direction of flow to remove solids adhered in pores and on surface. This can be done for 30–45 min every day and mild dose of chlorine (5 mg/L) can be introduced in this back-wash water for removal of microbes that have formed biofilms on membrane surface.

Once in three to six months the membrane module shall be taken out of the aeration tank and cleaned with hypochlorite or citric acid solution by submerging in a separate external tank. Upon this cleaning the membranes are back-washed with clean water and then only restored in the MBR. Once in a week in situ cleaning can be done without taking the membrane module out of aeration tank by flushing with hypochlorite solution with concentration of 100 mg/L for about 45 min, followed by treated water back-washing for about 15 min. Even by adopting the strategy of air-scour wash or chemical wash, over the period of operation membrane fouling is bound to happen as these methods are not very effective in completely restoring the membrane water flux to original value.

11.5.4 Performance of MBR

11.5.4.1 Domestic Wastewater

Membrane technology is considered a critical component of sophisticated wastewater treatment and reuse schemes and recently it is being used across the world. The MBR is a potential biomass retention technique that combines activated sludge treatment with membrane filtration (Melin et al., 2006). Holler and Trösch (2001) investigated urban wastewater treatment using aerobic jet-loop MBR configuration with integrated membrane microfiltration operated at high OLRs of 6–13 kg COD/m^3 day. More than 95% removal of COD was reported with influent COD concentration of 400–870 mg/L at 1.3–1.9 h of HRT. In an investigation carried out by Lin et al. (2011) for treatment of primarily treated municipal wastewater having an average influent COD of 425 mg/L, using a flat sheet membrane submerged anaerobic MBR, the removal efficiency of 90% was reported at OLR of 1.0 kg COD/m^3 day at 10 h of HRT.

An aerobic hollow-fibre hybrid MBR (HMBR) by introducing Kaldnes K3 biofilm carriers in conventional MBRs for municipal wastewater treatment has proved to be efficient for organic matter and nutrients removal (Liu et al., 2010). Removal efficiencies of COD, TN and TP from HMBR were 94.2%, 51% and 80.5% for an average influent concentration of 467 mg/L, 40.4 mg/L and 7.3 mg/L, respectively, at a constant HRT of 10 h under an OLR of 1.12 kg COD/m^3 day. Bohdziewicz et al. (2008) reported that landfill leachate dilution with synthetic wastewater in anaerobic submerged MBR with the capillary ultra-filtration gives improved COD removal over 95% for influent COD of 2800–5000 mg/L with leachate addition of 10% and 20% (v/v) at an HRT of 2 days and OLR of 2.5 kg COD/m^3 day. Wen et al. (2004) investigated the treatment of hospital wastewater using submerged aerobic hollow-fibre MBR. The COD and NH_4^+–N removal efficiencies of 80% and 93% were reported with the average effluent concentration of <25 mg/L and <1.5 mg/L, respectively, when operated under OLR of 0.4 kg COD/m^3 day and HRT of 7.2 h.

11.5.4.2   Performance of MBR for Industrial Wastewater Treatment

In an investigation conducted by Artiga et al. (2008) using an aerobic hollow-fibre hybrid MBR, for the treatment of saline wastewater from a fish canning factory, operated at OLR of 4 kg COD/m^3 day and initial COD concentration of 18,000 mg/L, the COD removal efficiency of 92% was reported at an HRT of 5 days. The MBR system shows high removal efficiency of oily pollutants (Scholz & Fuchs, 2000) and 99.99% of fuel oil and lubricating oil can be removed from the aerobic reactor followed by external tubular cross-flow ultrafiltration membrane unit operated at an HRT of 13.3 h. Removal efficiencies for COD and TOC of the order of 93–98% and 95–98%, respectively, under the oil loading rates of 3–5 kg/m^3 day were reported. Pendashteh et al. (2012) investigated the treatment of hypersaline oily wastewater using an aerobic tubular membrane sequencing batch reactor for synthetic and real wastewater. Under an OLR of 1.124 kg COD/m^3 day and an HRT of 48 h, removal efficiencies of 97.5%, 97.2%, and 98.9% for COD, TOC and oil and grease from synthetic wastewater and 86.2%, 90.8%, and 90% removal efficiencies, respectively, from real wastewater were reported.

Kornboonraksa and Lee (2009) investigated the performance of hollow-fibre MBR for piggery wastewater treatment under variation in organic and nitrogen concentrations adopting intermittent 60 min of on and off aeration cycle for alternate nitrification and denitrification. Using influent concentration of COD as 2,050 mg/L, BOD of 1,198 mg/L and NH_3–N of 248 mg/L the removal efficiencies of 92.0%, 92.7% and 69.5%, respectively, were reported at an HRT of 2.8 day under OLR of 0.77 kg COD/m^3 day. The performance of heavy metal removal from submerged aerobic hollow-fibre MBR was evaluated by Katsou et al. (2011) for primary treated municipal wastewater spiked with multi-metal solution enriched with Cu(II), Pb(II), Ni(II) and Zn(II) with respective influent concentration of 5.7–6.2, 6.4–13.1, 9.9–12.2 and 6.5–9.9 mg/L. The average removal efficiencies reported for these metals were 80%, 98%, 50% and 77% for Cu(II), Pb(II), Ni(II) and Zn(II), respectively, under an HRT of 10.3 h and OLR of 1.2 kg COD/m^3 day.

Decolouration of azo dye from wastewater can be achieved by using MBR technology. Spagni et al. (2012) reported decolouration of the reactive orange 16 as a model of an azo dye using submerged anaerobic flat-sheet MBR. More than 99% decolouration was achieved under a steady HRT of 2.5 days, when operated under an OLR of 2.7 kg COD/m^3 day and 94% of COD removal was obtained with an influent COD concentration of 6,750 mg/L. Brik et al. (2006) studied the performance of an activated sludge process connected to an external tubular cross-flow ultrafiltration membrane unit for the treatment of textile wastewater. The COD removal efficiency varied between 60 and 95% for an average influent concentration of 3,348 mg/L and colour removal was reported to be above 87%, when this MBR was operated under OLR of 2.1 kg COD/m^3 day at an HRT of 2.9 days. Dhaouadi and Marrot (2008) investigated olive mill wastewater treatment in an external ceramic MBR followed by aerobic bioreactor. Under OLR of 0.32 kg COD/m^3 day and 4.75 h of HRT, the COD removal efficiency of 81% was achieved for influent COD concentration of 1,500 mg/L and for influent phenol concentration of 5,410 mg/L, greater than 92% of phenol removal was reported.

The anaerobic MBRs are also adopted for the treatment of industrial wastewaters. Ng et al. (2015) investigated pharmaceutical wastewater treatment using bioaugmented anaerobic hollow-fibre MBR operated under an OLR of 13.0 ± 0.6 kg COD/m^3 day. The total COD and TDS of influent were 16,249 ± 714 mg/L and 29,450 ± 2209 mg/L, respectively. Under an HRT of 30.6 h, COD removal efficiency of 60.3 ± 2.8% was reported. In another experiment conducted by Svojitka et al. (2017) for pharmaceutical wastewater treatment using an anaerobic reactor with external cross-flow ceramic tubes membrane unit, an addition of 25 g/L methanol as COD demonstrated improved COD removal of up to 97%, while operated under an OLR of 4.5–14.2 kg COD/m^3 day at HRT of 3–3.5 days. In an attempt to treat a brewery wastewater, Chen et al. (2016) used an advanced anaerobic membrane reactor having a hollow-fibre membrane configuration. A COD removal efficiency of 98% was reported at influent COD concentration of 11,100 ± 2700 mg/L under an OLR of 10 kg COD/m^3 day. A total nitrogen and phosphorous removal of 54 ± 13% and 28 ± 11% was reported with an influent concentration of 180 ± 100 mg/L and 55.0 ± 35.0 mg/L, respectively, at an HRT of 44 h.

Highly concentrated food wastewater treatment using an anaerobic reactor followed by cross-flow flat-plate MBR module was reported by He et al. (2005). The COD and SS removal efficiencies in the range of 81–94% and greater than 99% were reported for influent total COD concentration of 2,000–15,000 mg/L and SS concentration of 600–1,000 mg/L, respectively, when MBR was operated under OLR of 4.5 kg COD/m^3 day with an HRT of 60 h.

Thus, the literature encourages the application of MBR system for treatment of the municipal wastewaters and various industrial effluents. Most of the high strength wastewaters have been successfully treated by aerobic and anaerobic MBRs. This has been confirmed by the use of MBR for wastewater treatment in brewery industry, landfill leachate, fish canning factory, pharmaceutical wastewater, food industry, and textile industry. By controlling the HRT and OLR to an optimum condition, the best performance of MBR can be achieved. Also, by adopting in situ air score wash and application of back wash water, with periodic external cleaning of the membrane, the membrane fouling can be kept under control for getting long term operational duration of the membrane modules and to keep the cost of membrane module replacement under control.

Questions

11.1 Explain the different types of submerged aerobic filters used for wastewater treatment.

11.2 Write a short note on treatment of various wastewaters in the submerged aerobic filter and discuss possible OLR and HRT that can be used for obtaining appreciable treatment of wastewater.

11.3 For simultaneous organic matter removal and nitrogen removal suggest appropriate configuration schematic for SAF and describe how these pollutants will be removed.

11.4 Describe moving bed biofilm reactor. What are the advantages this will offer over conventional activated sludge process?

11.5 Discuss the performance of MBBR while treating sewage and industrial wastewaters.

11.6 Discuss the applicability of RBC system for biological nitrification and denitrification.

11.7 A three-stage RBC system is to be designed based on the following environmental and design conditions, RBC influent BOD_5 concentration is 135 mg/L and desired RBC effluent BOD_5 concentration is 30 mg/L. The design risk anticipated is 20% for the treatment of wastewater having average daily flow of 1 MGD and minimum winter temperature (T) of 10 °C. Consider treatability parameter $(K_t) = 0.066$ at 20 °C. The design flow shall be peak flow of the month, i.e., 1.45 times average day flow. Determine the following: effluent BOD concentration, total surface area of the rotating media, and surface area of each stage.

11.8 Describe various configuration of media used in downflow hanging sponge reactor.

11.9 How is MBR different than conventional activated sludge process?

11.10 How the fouling of the membrane can be controlled in MBR?

11.11 Write short note on application of MBR for treatment of industrial wastewaters.

11.12 Discuss the design guidelines to be adopted for MBR for treatment of sewage.

Solutions

Q. 11.7: Design risk = 20%, therefore, design effluent BOD_5 concentration = 30 − 0.2 * 30 = 24 mg/L. $(K_t)_T = (K_t)_{20}(1.018)^{T-20}$, $(K_t)_{10} = 0.055$; $Le/Lo =$

$exp[-Kt(Ac/695Q)^{0.5}]$; therefore, $Ac = 993,855\text{ft}^2$. Therefore, area of first stage $=$ 350,000 ft^2, 2nd stage $=$ 350,000 ft^2 and third stage $=$ 300,000 ft^2.

References

Ahmadi, M., Ahmadmoazzam, M., Saeedi, R., Abtahi, M., Ghafari, S., & Jorfi, S. (2019). Biological treatment of a saline and recalcitrant petrochemical wastewater by using a newly isolated halotolerant bacterial consortium in MBBR. *Desalination and Water Treatment, 167*, 84–95.

Andreottola, G., Foladori, P., Ragazzi, M., & Tatàno, F. (2000). Experimental comparison between MBBR and activated sludge system for the treatment of municipal wastewater. *Water Science and Technology, 41*(4–5), 375–382.

Artiga, P., García-Toriello, G., Méndez, R., & Garrido, J. M. (2008). Use of a hybrid membrane bioreactor for the treatment of saline wastewater from a fish canning factory. *Desalination, 221*(1–3), 518–525.

Baghapour, M. A., Nasseri, S., & Derakhshan, Z. (2013). Atrazine removal from aqueous solutions using submerged biological aerated filter. *Journal of Environmental Health Science and Engineering, 11*(1), 1–9.

Baghapour, M. A., Shirdarreh, M. R., & Faramarzian, M. (2014). Degradation of amoxicillin by bacterial consortium in a submerged biological aerated filter: Volumetric removal modeling. *Journal of Health Sciences & Surveillance System, 2*(1), 15–25.

Baghapour, M. A., Shirdarreh, M. R., & Faramarzian, M. (2015). Amoxicillin removal from aqueous solutions using submerged biological aerated filter. *Desalination and Water Treatment, 54*(3), 790–801.

Bakar, S. N. H. A., Hasan, H. A., Mohammad, A. W., Abdullah, S. R. S., Ngteni, R., & Yusof, K. M. M. (2020). Performance of a laboratory-scale moving bed biofilm reactor (MBBR) and its microbial diversity in palm oil mill effluent (POME) treatment. *Process Safety and Environmental Protection, 142*, 325–335.

Bering, S., Mazur, J., Tarnowski, K., Janus, M., Mozia, S., & Morawski, A. W. (2018). The application of moving bed bio-reactor (MBBR) in commercial laundry wastewater treatment. *Science of the Total Environment, 627*, 1638–1643.

Biswas, K., Taylor, M. W., & Turner, S. J. (2014). Successional development of biofilms in moving bed biofilm reactor (MBBR) systems treating municipal wastewater. *Applied Microbiology and Biotechnology, 98*(3), 1429–1440.

Bohdziewicz, J., Neczaj, E., & Kwarciak, A. (2008). Landfill leachate treatment by means of anaerobic membrane bioreactor. *Desalination, 221*(1–3), 559–565.

Brik, M., Schoeberl, P., Chamam, B., Braun, R., & Fuchs, W. (2006). Advanced treatment of textile wastewater towards reuse using a membrane bioreactor. *Process Biochemistry, 41*(8), 1751–1757.

Casas, M. E., Chhetri, R. K., Ooi, G., Hansen, K. M., Litty, K., Christensson, M., & Bester, K. (2015). Biodegradation of pharmaceuticals in hospital wastewater by staged Moving Bed Biofilm Reactors (MBBR). *Water Research, 83*, 293–302.

Chatterjee, P., Ghangrekar, M. M., & Rao, S. (2016). Organic matter and nitrogen removal in a hybrid upflow anaerobic sludge blanket—Moving bed biofilm and rope bed biofilm reactor. *Journal of Environmental Chemical Engineering, 4*, 3240–3245.

Chatterjee, P., Ghangrekar, M. M., & Rao, S. (2018). Sludge granulation in an UASB–moving bed biofilm hybrid reactor for efficient organic matter removal and nitrogen removal in biofilm reactor. *Environmental Technology (United Kingdom), 39*, 298–307.

Chen, F., Zeng, S., Luo, Z., Ma, J., Zhu, Q., & Zhang, S. (2020). A novel MBBR–MFC integrated system for high-strength pulp/paper wastewater treatment and bioelectricity generation. *Separation Science and Technology, 55*(14), 2490–2499.

Chen, H., Chang, S., Guo, Q., Hong, Y., & Wu, P. (2016). Brewery wastewater treatment using an anaerobic membrane bioreactor. *Biochemical Engineering Journal, 105*, 321–331.

Chowdhury, P., Viraraghavan, T., & Srinivasan, A. (2010). Biological treatment processes for fish processing wastewater—A review. *Bioresource Technology, 101*, 439–449.

Cortez, S., Teixeira, P., Oliveira, R., & Mota, M. (2008). Rotating biological contactors: A review on main factors affecting performance. *Reviews in Environmental Science and Bio/technology, 7*, 155–172.

Dhaouadi, H., & Marrot, B. (2008). Olive mill wastewater treatment in a membrane bioreactor: Process feasibility and performances. *Chemical Engineering Journal, 145*(2), 225–231.

Eckenfelder, W. W. (2000). *Industrial water pollution contro* (3rd ed.). McGraw-Hill International Editions, Environmental Engineering series, Boston, USA.

Ferraz, F. D. M., Povinelli, J., Pozzi, E., Vieira, E. M., & Trofino, J. C. (2014). Co-treatment of landfill leachate and domestic wastewater using a submerged aerobic biofilter. *Journal of Environmental Management, 141*, 9–15.

Fleifle, A., Tawfik, A., Saavedra, O. C., & Elzeir, M. (2013). Treatment of agricultural drainage water via downflow hanging sponge system for reuse in agriculture. *Water Science and Technology: Water Supply, 13*(2), 403–412.

Gálvez, A., Zamorano, M., Hontoria, E., & Ramos, A. (2006). Treatment of landfill leachate with aerated and non-aerated submerged biofilters. *Journal of Environmental Science and Health, Part A, 41*(6), 1129–1144.

Gilles, P., Rogalla, F., Gousailles, M., & Paffoni, C. (1990). Aerated biofilters for nitrification and effluent polishing. *Water Science and Technology, 22*(7/8), 181–189.

González-Martínez, S., & Duque-Luciano, J. (1992). Aerobic submerged biofilm reactors for wastewater treatment. *Water Research, 26*(6), 825–833.

Grady, C. P. L., Jr., Daigger, G. T., & Lim, H. C. (1999). *Biological wastewater treatment.* Marcel Dekker Inc.

Griffin, P., & Findlay, G. E. (2000). Process and engineering improvements to rotating biological contactor design. *Water Science and Technology, 41*, 137–144.

He, Y., Xu, P., Li, C., & Zhang, B. (2005). High-concentration food wastewater treatment by an anaerobic membrane bioreactor. *Water Research, 39*(17), 4110–4118.

Holler, S., & Trösch, W. (2001). Treatment of urban wastewater in a membrane bioreactor at high organic loading rates. *Journal of Biotechnology, 92*(2), 95–101.

Husain Khan, A., Abdul Aziz, H., Khan, N. A., Ahmed, S., Mehtab, M. S., Vambol, S., & Islam, S. (2020). Pharmaceuticals of emerging concern in hospital wastewater: Removal of Ibuprofen and Ofloxacin drugs using MBBR method. *International Journal of Environmental Analytical Chemistry.* https://doi.org/10.1080/03067319.2020.1855333

Katsou, E., Malamis, S., & Loizidou, M. (2011). Performance of a membrane bioreactor used for the treatment of wastewater contaminated with heavy metals. *Bioresource Technology, 102*(6), 4325–4332.

Khan, N. A., Bokhari, A., Mubashir, M., Klemeš, J. J., El Morabet, R., Khan, R. A., & Show, P. L. (2021). Treatment of Hospital wastewater with submerged aerobic fixed film reactor coupled with tube-settler. *Chemosphere*, 131838.

Kornboonraksa, T., & Lee, S. H. (2009). Factors affecting the performance of membrane bioreactor for piggery wastewater treatment. *Bioresource Technology, 100*(12), 2926–2932.

Leenen, E. J. T. M., Dos Santos, V. A. P., Grolle, K. C. F., Tramper, J., & Wijffels, R. (1996). Characteristics of and selection criteria for support materials for cell immobilization in wastewater treatment. *Water Research, 30*, 2985–2996.

Lin, H., Chen, J., Wang, F., Ding, L., & Hong, H. (2011). Feasibility evaluation of submerged anaerobic membrane bioreactor for municipal secondary wastewater treatment. *Desalination, 280*(1–3), 120–126.

Liu, J., Zhou, J., Xu, N., He, A., Xin, F., Ma, J., & Dong, W. (2019). Performance evaluation of a lab-scale moving bed biofilm reactor (MBBR) using polyethylene as support material in the treatment of wastewater contaminated with terephthalic acid. *Chemosphere, 227*, 117–123.

Liu, Q., Wang, X. C., Liu, Y., Yuan, H., & Du, Y. (2010). Performance of a hybrid membrane bioreactor in municipal wastewater treatment. *Desalination, 258*(1–3), 143–147.

Machdar, I., Harada, H., Ohashi, A., Sekiguchi, Y., Okui, H., & Ueki, K. (1997). A novel and cost-effective sewage treatment system consisting of UASB pre-treatment and aerobic post-treatment units for developing countries. *Water Science and Technology, 36*, 189–197.

Machdar, I., Muhammad, S., Onodera, T., & Syutsubo, K. (2018). A pilot-scale study on a down-flow hanging sponge reactor for septic tank sludge treatment. *Environ. Eng. Res., 23*(2), 195–204.

Machdar, I., Sekiguchi, Y., Sumino, H., Ohashi, A., & Harada, H. (2000). Combination of a UASB reactor and a curtain type DHS (downflow hanging sponge) reactor as a cost-effective sewage treatment system for developing countries. *Water Science and Technology, 42*(3–4), 83–88.

Mba, D., Bannister, R. H., & Findlay, G. E. (1999). Mechanical redesign of the rotating biological contactor. *Water Research, 33*, 3679–3688.

Melin, T., Jefferson, B., Bixio, D., Thoeye, C., De Wilde, W., De Koning, J., & Wintgens, T. (2006). Membrane bioreactor technology for wastewater treatment and reuse. *Desalination, 187*(1–3), 271–282.

Mendoza-Espinosa, L., & Stephenson, T. O. M. (1999). A review of biological aerated filters (BAFs) for wastewater treatment. *Environmental Engineering Science, 16*(3), 201–216.

Moore, R., Quarmby, J., & Stephenson, T. (2001). The effects of media size on the performance of biological aerated filters. *Water Research, 35*(10), 2514–2522.

Morgan-Sagastume, J. M., & Noyola, A. (2008). Evaluation of an aerobic submerged filter packed with volcanic scoria. *Bioresource Technology, 99*(7), 2528–2536.

Ng, K. K., Shi, X., & Ng, H. Y. (2015). Evaluation of system performance and microbial communities of a bioaugmented anaerobic membrane bioreactor treating pharmaceutical wastewater. *Water Research, 81*, 311–324.

Oğuz, M., & Oğuz, M. (1993). Characterization of Ankara meat packing plant wastewater and treatment with a rotating biological contactor. *International Journal of Environmental Studies, 44*(1), 39–44.

Okubo, T., Kubota, K., Yamaguchi, T., Uemura, S., & Harada, H. (2016). Development of a new non-aeration-based sewage treatment technology: Performance evaluation of a full-scale down-flow hanging sponge reactor employing third-generation sponge carriers. *Water Research, 102*, 138–146.

Okubo, T., Onodera, T., Uemura, S., Yamaguchi, T., Ohashi, A., & Harada, H. (2015). On-site evaluation of the performance of a full-scale down-flow hanging sponge reactor as a post-treatment process of an up-flow anaerobic sludge blanket reactor for treating sewage in India. *Bioresource Technology, 194*, 156–164.

Onodera, T., Tandukar, M., Sugiyana, D., Uemura, S., Ohashi, A., & Harada, H. (2014). Development of a sixth-generation down-flow hanging sponge (DHS) reactor using rigid sponge media for post-treatment of UASB treating municipal sewage. *Bioresource Technology, 152*, 93–100.

Paffoni, C., Gousailles, M., Rogalla, F., & Gilles, P. (1990). Aerated biofilters for nitrification and effluent polishing. *Water Science and Technology, 22*(7–8), 181–189.

Pendashteh, A. R., Abdullah, L. C., & Fakhru'l. (2012). Evaluation of membrane bioreactor for hypersaline oily wastewater treatment. *Process Safety and Environmental Protection, 90*(1), 45–55.

Priya, V. S., & Philip, L. (2015). Treatment of volatile organic compounds in pharmaceutical wastewater using submerged aerated biological filter. *Chemical Engineering Journal, 266*, 309–319.

Ramos, A. F., Gomez, M. A., Hontoria, E., & Gonzalez-Lopez, J. (2007). Biological nitrogen and phenol removal from saline industrial wastewater by submerged fixed-film reactor. *Journal of Hazardous Materials, 142*(1–2), 175–183.

Reis, L. C., & Sant'Anna, G. L., Jr. (1985). Aerobic treatment of concentrated wastewater in a submerged bed reactor. *Water Research, 19*(11), 1341–1345.

Rittmann, B. E., & McCarty, P. L. (2001). *Environmental biotechnology: Principles and applications.* McGraw-Hill Education, New York. USA

Rittmann, B. E., Suozzo, R., & Romero, B. R. (1983). Temperature effects on oxygen transfer to rotating biological contactors. *Journal of Water Pollution Control Federation, 55*, 270–277.

Santos, A. D., Martins, R. C., Quinta-Ferreira, R. M., & Castro, L. M. (2020). Moving bed biofilm reactor (MBBR) for dairy wastewater treatment. *Energy Reports, 6*, 340–344.

Scholz, W., & Fuchs, W. (2000). Treatment of oil contaminated wastewater in a membrane bioreactor. *Water Research, 34*(14), 3621–3629.

Shitu, A., Zhu, S., Qi, W., Tadda, M. A., Liu, D., & Ye, Z. (2020). Performance of novel sponge biocarrier in MBBR treating recirculating aquaculture systems wastewater: Microbial community and kinetic study. *Journal of Environmental Management, 275*, 111264.

Sirianuntapiboon, S., & Chumlaong, S. (2013). Effect of Ni^{2+} and Pb^{2+} on the efficiency of packed cage rotating biological contactor system. *Journal of Environmental Chemical Engineering, 1*, 233–240.

Smith, D. P. (1995). Submerged filter biotreatment of hazardous leachate in aerobic, anaerobic, and anaerobic/aerobic systems. *Hazardous Waste and Hazardous Materials, 12*(2), 167–183.

Sodhi, V., Singh, C., Cheema, P. P. S., Sharma, R., Bansal, A., & Jha, M. K. (2021). Simultaneous sludge minimization, pollutant and nitrogen removal using integrated MBBR configuration for tannery wastewater treatment. *Bioresource Technology*, 125748.

Spagni, A., Casu, S., & Grilli, S. (2012). Decolourisation of textile wastewater in a submerged anaerobic membrane bioreactor. *Bioresource Technology, 117*, 180–185.

Stensel, H. D., & Reiber, S. (1983). Industrial wastewater treatment with a new biological fixed-film system. *Environmental Progress, 2*(2), 110–115.

Stensel, H. D., Brenner, R. C., Lee, K. M., Melcer, H., & Rakness, K. (1988). Biological aerated filter evaluation. *Journal of Environmental Engineering, 114*(3), 655–671.

Stephenson, T., Reid, E., Avery, L. M., & Jefferson, B. (2013). Media surface properties and the development of nitrifying biofilms in mixed cultures for wastewater treatment. *Process Safety and Environment Protection, 91*, 321–324.

Svojitka, J., Dvořák, L., Studer, M., Straub, J. O., Frömelt, H., & Wintgens, T. (2017). Performance of an anaerobic membrane bioreactor for pharmaceutical wastewater treatment. *Bioresource Technology, 229*, 180–189.

Tandukar, M., Ohashi, A., & Harada, H. (2007). Performance comparison of a pilot-scale UASB and DHS system and activated sludge process for the treatment of municipal wastewater. *Water Research, 41*, 2697–2705.

Tandukar, M., Uemura, S., Ohashi, A., & Harada, H. (2006). Combining UASB and the "fourth generation" down-flow hanging sponge reactor for municipal wastewater treatment. *Water Science and Technology, 53*, 209–218.

Tandukar, M., Uemura, S., Machdar, I., Ohashi, A., & Harada, H. (2005). A low-cost municipal sewage treatment system witha combination of UASB and the "fourth-generation" downflow hanging sponge reactors. *Water Science and Technology, 52*(1–2), 323–329.

Tawfik, A., Ohashi, A., & Harada, H. (2010). Effect of sponge volume on the performance of down-flow hanging sponge system treating UASB reactor effluent. *Bioprocess and Biosystems Engineering, 33*, 779–785.

Teixeira, P., & Oliveira, R. (2001). Denitrification in a closed rotating biological contactor: Effect of disk submergence. *Process Biochemistry, 37*, 345–349.

Wang, L. K., Wu, Z., & Shammas, N. K. (2009). Rotating Biological Contactors. In L. K. Wang, N. C. Pereira, & Y.-T. Hung (Eds.), *Biological treatment processes* (pp. 435–458). Humana Press.

Watari, T., Cuong Mai, T., Tanikawa, D., Hirakata, Y., Hatamoto, M., & Syutsubo, K., ... Yamaguchi, T. (2017). Development of downflow hanging sponge (DHS) reactor as post treatment of existing combined anaerobic tank treating natural rubber processing wastewater. *Water Science and Technology, 75*(1), 57–68

Wen, X., Ding, H., Huang, X., & Liu, R. (2004). Treatment of hospital wastewater using a submerged membrane bioreactor. *Process Biochemistry, 39*(11), 1427–1431.

Yoochatchaval, W., Onodera, T., Sumino, H., Yamaguchi, T., Mizuochi, M., Okadera, T., & Syutsubo, K. (2014). Development of a down-flow hanging sponge reactor for the treatment of low strength sewage. *Water Science and Technology, 70*(4), 656–663.

Zkeri, E., Iliopoulou, A., Katsara, A., Korda, A., Aloupi, M., Gatidou, G., & Stasinakis, A. S. (2021). Comparing the use of a two-stage MBBR system with a methanogenic MBBR coupled with a microalgae reactor for medium-strength dairy wastewater treatment. *Bioresource Technology, 323*, 124629.

Anaerobic Process for Wastewater Treatment **12**

12.1 Introduction

Biological wastewater treatment processes in the absence of molecular oxygen, where for the respiratory reaction the terminal electron acceptors are carbon dioxide, organics, nitrates and sulphate or fermentative reduction reactions undergoes are generally referred to as anaerobic processes. These processes are similar to those that occur naturally in stomach of ruminant animals, marshes, organic sediments from lakes and rivers, and sanitary landfills. The main gaseous by-products are methane (CH_4), carbon dioxide (CO_2), and trace gases, such as hydrogen sulphide (H_2S), hydrogen (H_2), ammonia (NH_3), and a liquid or semi-liquid by-product known as digestate. The digestate consists of microorganisms, nutrients (nitrogen, phosphorus, etc.), metals precipitate, undegraded organic matter and inert materials.

In principle, majority of biodegradable organic compounds can be degraded by an anaerobic process, which offers more efficient and economical solution for wastewater remediation. Anaerobic digesters have been largely used in the treatment of solid wastes, including agricultural wastes, animal excrements, sludge from sewage treatment plants and urban wastes, and it is estimated that millions of anaerobic digesters have been built all over the world with this purpose. Since last three decades, anaerobic digestion has also been largely used in the treatment of effluents from agricultural, food and beverage industries, both in developed and developing countries. High-rate anaerobic digestion processes are also being used for municipal wastewater treatment in many countries.

12.2   Advantages and Limitations of Anaerobic Process

In aerobic systems, only about 40 to 50% of biological stabilization of organic matter occurs, with its subsequent conversion into CO_2. About 50–60% of the organic matter is incorporated into biomass and leaves the system as an excess biological sludge. The organic material, not converted into carbon dioxide or into biomass, leaves the reactor as non-degraded material (5–10%) depending upon the complexity of the organic matter present in the wastewater that is being treated. In anaerobic process, most of the biodegradable organic matter present in the waste is converted into biogas (about 70–90%), which is removed from the liquid phase and leaves the reactor in a gaseous form. Only a small portion of the organic material is converted into microbial biomass (about 5–15%). Besides being small in quantity, the excess sludge is usually more concentrated having better dewatering characteristics, thus eliminating the need for further treatment before dewatering this sludge generated from anaerobic process. Hence, the excess sludge generated can be directly dewatered on sand drying beds or in other mechanical dewatering systems. The material that could not be converted into biogas or into biomass leaves the reactor as nondegraded material (10–30%).

Aerobic processes require external energy input for aeration; whereas this energy can be saved in anaerobic digestion as no aeration is required. However, depending on the climatic condition of that location heating of anaerobic digester might be required during winter. This energy required for heating in anaerobic treatment can be obtained from the produced biogas, where its production and recovery are economically viable. Moreover, the by-product of aerobic processes (biological sludge mass) can also be treated by anaerobic processes; just as the aerobic processes can be used for post-treatment of the effluent generated from anaerobic processes.

For sewage treatment, aerobic processes can ensure a better effluent quality than anaerobic processes in terms of carbonaceous and nitrogenous pollutants removal. However, for treatment of high strength wastewaters, it may be economical to have anaerobic treatment first that is followed by aerobic polishing treatment. For low strength wastewaters, such as municipal wastewaters, low biogas yield makes anaerobic treatment less cost effective than aerobic treatment. However, in tropical countries with high ambient temperatures, anaerobic pre-treatment of low strength wastewaters can offer a net positive energy gain due to the relatively low external energy required to bring the system to the more efficient mesophilic or thermophilic temperature ranges. Anaerobic pre-treatment of municipal wastewater is currently in use in many countries, e.g., Brazil, Columbia, Ghana, India etc. In some instances, the objective of anaerobic pre-treatment for low strength wastewaters may be simply to convert complex organic molecules into simpler organic acids, and consequently reduce the oxygen (and energy) demand for a subsequent aerobic polishing treatment, thereby encouraging the use of low energy and ecological systems for this purpose.

In addition to the foremost advantage of methane recovery, the loading on anaerobic reactor is not limited to the oxygen resource of the system, hence very high volumetric loading rates in terms of COD can be applied to anaerobic reactors. The loading typically adopted in practice for designing anaerobic reactor is up to 15 kg COD/m^3 day, though successful operation of anaerobic reactor is reported up to 30 kg COD/m^3 day. This volumetric loading rate is much higher than the aerobic counter part of the anaerobic reactor, thus leading to saving in the volume of the reactor required. Also, the anaerobic reactors can be constructed taller as compared to aerobic reactor, where depth in most cases is limited due to exposing more surface area of the wastewater to air for enhancing oxygen transfer. Since such limitation is not present in anaerobic reactor constructing taller reactor can reduce the footprint of the reactor, leading to saving in area required.

Since anaerobic bacteria have slower growth rate as compared to aerobic process, the nutrient (N and P) requirement for anaerobic process is less. This is particularly advantageous for the treatment of industrial wastewaters where external supplementation of nutrients is required, because selecting anaerobic process for treatment of such wastewaters will lead to the saving in the cost of chemicals for external nutrients supplementation. However, for treatment of wastewater containing higher concentration of nutrients, the removal of these nutrients in anaerobic process will be less, as compared to aerobic process, and it will be a drawback of anaerobic process. Also, non-feed conditions to the anaerobic reactor for 2–3 months will not very adversely affect the sludge present in anaerobic reactor and upon resuming the feed, the reactor may take-up full load within a week. Whereas such non-feed condition in aerobic process for a week will lead to complete destruction of biomass due to endogenous decay.

Another weakness of anaerobic process is the long start-up time required when the reactor is to be commissioned fresh. This is due to slow growth rate of anaerobic bacteria, particularly the methanogens. Typically, 2–4 months duration will be required for anaerobic reactor to reach to full load condition. Whereas the aerobic process can be commissioned within couple of weeks due to faster growth rate of the aerobic bacteria. Also, if any process upset occurs during normal operation of the reactor, the anaerobic reactor will take longer time to restore normal performance as compared to aerobic process. Since, anaerobic bacteria, particularly methanogens are more sensitive to the environmental conditions, such as operating temperature, pH and volatile fatty acids accumulation in the reactor, a skilled operator is generally considered essential for anaerobic reactor, to keep rapid fluctuation of these parameters under control to ensure stable performance of the reactor.

12.3 Principles of Anaerobic Digestion

Anaerobic digestion can be considered as an ecosystem where several groups of microorganisms work interactively in the conversion of complex organic matter into final

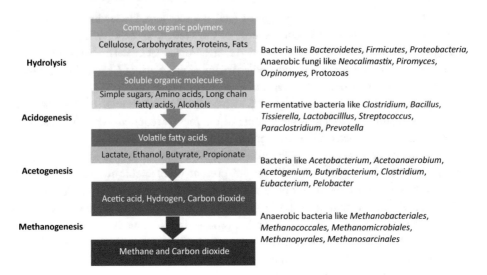

Fig. 12.1 Metabolic pathways and microbial groups involved in anaerobic digestion

products, such as methane, carbon dioxide, hydrogen sulphide, water and ammonia, besides new bacterial cells. It can be subdivided into various metabolic pathways, with the participation of several microbial groups, each with a different physiological behaviour as discussed in the following sections (Fig. 12.1) (Lettinga et al., 1996).

12.3.1 Hydrolysis

Hydrolytic bacteria are strict anaerobes with a doubling time of around 2–3 h. Complex particulate materials (polymers) are hydrolysed into simpler dissolved materials (smaller molecules), by the action of exoenzymes, which can penetrate through the cell membranes of fermentative bacteria. Factors like temperature, residence time, substrate composition and size, and operational pH impact the rate of hydrolysis.

12.3.2 Acidogenesis

Hydrolysis degradation products are further fermented into volatile fatty acids (VFAs), lactic acid, and pyruvic acid in this step. Acidogenesis is carried out by a large and diverse group of fermentative bacteria with a doubling time of 2–3 h. Usual species involved in acidogenesis belong to the *Clostridia* group, which comprises anaerobic species that form spores and are able to survive in very adverse environments, and the family *Bacteroidaceaea*, organisms commonly found in digestive tracts, that participate in the degradation of sugars and amino acids.

12.3.3 Acetogenesis

Products generated in the acidogenic phase are further oxidized to acetic acid, hydrogen and carbon dioxide in this phase by obligate hydrogen producing acetogens (OHPA). Low hydrogen partial pressure favours these reactions as stated in Eqs. 12.1 and 12.2.

$$CH_3CH_2COOH + 2H_2O \rightarrow CH_3COOH + CO_2 + 3H_2 \tag{12.1}$$

$$CH_3CH_2CH_2COOH + 2H_2O \rightarrow 2CH_3COOH + 2H_2 \tag{12.2}$$

Gibbs free energy changes for reactions presented in Eqs. 12.1 and 12.2 are + 76.1 kJ/mole and +48.1 kJ/mole, respectively. A negative value of free energy change is necessary for any reaction to proceed without input of external energy. This theory apparently suggests that hydrogen producing acetogenic bacteria cannot obtain energy for growth from these reactions. However, the value of free energy change in the actual environment surrounding the bacteria, is different from that of Gibbs free energy and depends on the concentrations of substrates and products and the energy change can only be negative at low partial pressure of hydrogen. The bottleneck here is that though a low partial pressure of hydrogen is essential for OHPA to produce acetate from different substrate, they themselves produce hydrogen. Based on thermodynamics associated with this reactions Harper and Pohland (1986) indicated that propionic acid oxidation to acetate becomes favourable only at hydrogen partial pressure below 10^{-4} atm; while the butyric acid oxidation becomes favourable at 10^{-3} atm H_2 or below.

Hydrogen utilizing methanogenic archaea can receive hydrogen as a substrate and as such serve a thermodynamically favourable conditions for hydrogen producing acetogenic bacteria in anaerobic reactors. The interrelationship between these two groups of microorganisms is called interspecies hydrogen transfer, which also exists between acidogenic bacteria and methanogenic archaea. Acidogenic bacteria produce more hydrogen and acetate than propionate or lactate and obtain more energy under low hydrogen partial pressure, which is kept by methanogenic archaea. The interspecies hydrogen transfer is favourable but not essential for acidogenic bacteria; while it is indispensable for hydrogen producing acetogenic bacteria.

12.3.4 Methanogenesis

The final phase in the overall anaerobic degradation process of organic compounds into methane and carbon dioxide is performed by the strictly anaerobic methanogenic archaea. The doubling time of these archaea is 2–10 days. There are two distinct pathways for methanogenesis depending on the substrate used, that is acetic acid (acetotrophs) and hydrogen (lithotrophs) (Eqs. 12.3 and 12.4).

$$CH_3COOH \rightarrow CO_2 + CH_4 \qquad\qquad (12.3)$$

$$4H_2 + CO_2 \rightarrow CH_4 + 2H_2O \qquad\qquad (12.4)$$

12.3.5 Microorganisms Involved in Methanogenesis

Particulate materials present in wastewater are converted into dissolved materials by the action of exoenzymes excreted by the hydrolytic fermentative bacteria during hydrolysis. Acidogenesis is carried out by a large and diverse group of fermentative bacteria. Acetogenic bacteria are part of an intermediate metabolic group that produces substrate for methanogenic microorganisms. In view of their affinity for substrate and extent of methane production, methanogenic microorganisms are divided into two main groups, one that forms methane from acetic acid or methanol (acetoclastic methanogens), and the other that produces methane from hydrogen and carbon dioxide (hydrogenotrophic methanogens or lithotrophs). Although, only a few of the methanogenic species (*Methanosarcina* and *Methanosaeta*) can form methane from acetate, these are usually the microorganisms prevailing in anaerobic digestion accounting for about 60 to 70% of total methane production. Unlike the acetoclastic microorganisms, practically all the well-known methanogenic species (*Methanobacterium*, *Methanospirillum*, *Methanobrevibacter*) can produce methane from hydrogen and carbon dioxide.

Among the methanogenic species, the predominant organism is *Methanosaeta* (formerly known as *Methanothrix*), which can form very long filaments of size 200–300 μm, followed by cocci shaped *Methanosarcina*, and some filamentous *Methanobacterials* and cocci shaped *Methanococcals* (Hulshoff Pol et al., 2004). The SEM images as shown in Fig. 12.2b, c depicts the structure and morphology of anaerobic sludge granules. Each granule acts as a functional unit, comprising of different microbial communities necessary for the degradation of contaminants in wastewater thereby producing extracellular polymeric substances and biomass (Calderón et al., 2013). A study conducted by (Chatterjee et al., 2018) demonstrated that spherical granules consisting of loosely intertwined filamentous bacteria resembling acetoclastic *Methanosaeta* (or *Methanothrix*) and hydrogen utilizing *Methanobacterials* at the surface of the granules and less populous spherical

Methanosarcina attached and entrapped in this mesh as illustrated in Fig. 12.2d. These filamentous *Methanosaeta* cells are methanogens as observed and reported by Li and Sung, (2015) and they are most predominant in the core of the granules and present in lower numbers in the outermost layer (Liu et al., 2003). *Methanosaeta* population dominates at low organic loading rates and have high affinity for acetate as substrate (Hulshoff Pol et al., 2004).

When these organisms grow without attachment to a solid support particle, a loosely intertwined structure of filaments, with very poor settling characteristics is obtained (Hulshoff Pol et al., 2004). Group of *Methanosarcina* bacteria secretes extracellular polymeric substances (EPS) that forms a nucleus around which both *Methanosarcina* and *Methanosaeta* can attach themselves giving rise to a granular structure (Liu et al., 2003). The structured aggregates further grow due to cellular multiplication of the entrapped bacteria and become denser and acquire more or less spherical shape by the hydrodynamic shear force caused by the up-flowing liquid and biogas (Liu et al., 2003). Porous structure confirms gas production and movement within the granules (Najafpour et al., 2006). With increasing organic loading rate, growth of *Methanosarcina* increases and they secrete high

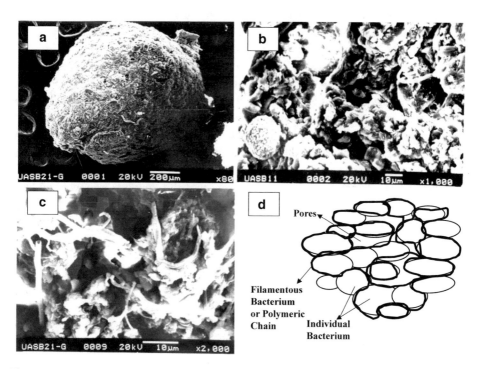

Fig. 12.2   a SEM images of whole anaerobic sludge granules, **b** and **c** micrograph of interior of the granule, and **d** schematic representation of granular sludge

quantity of EPS to form larger clumps, and then *Methanosaeta* (*Methanothrix*) fills those clumps, thus leading to a dense granular structure (Liu et al., 2003).

The *Methanosaeta* species are unable to use hydrogen in combination with CO_2 and these are *non-hydrogen-oxidizing acetotrophs* (NHOA). In contrast, *Methanosarcina* can utilize H_2/CO_2 as well as acetate, carbon monoxide, methanol, and methylamines as growth substrates. Due to their ability to use both H_2/CO_2 and acetate, these archaea are classified as *Hydrogen oxidizing acetotrophs* (HOA). *Hydrogen-oxidizing methanogens* (HOM) do not cleave acetate but utilize H_2/CO_2 and format as substrates (Harper & Pohland, 1986). The HOA are unique in their capability to utilize multiple (one and/or two carbon) substrates. This ability affords a higher potential for survival when competing with sulphate reducing bacteria (SRB) and nitrate reducing bacteria (NRB) for hydrogen and acetate. At hydrogen partial pressure $>10^{-4}$ atm, HOA use H_2/CO_2 in favour of acetate, whereas acetate cleavage by NHOA is unaffected by hydrogen.

The NHOA have a much higher affinity for acetate than the HOA. Hence, NHOA may outcompete HOA at acetate concentrations below 50 mg/L; while above 250 mg/L of acetate, the HOA are more competitive (Speece, 1983). As a result of this comparative kinetics, *Methanosaeta* (NHOA) may be found in reactors operated with lower organic loading rates. Whereas *Methanosarcina* are more predominant in low retention time reactors operated under higher organic loading rates, such as in the lower reaches of plug flow anaerobic filters and in two phase reactor system.

Oxidation of reduced organic products to bicarbonate and acetate also occurs due to NRB and SRB. Higher organic matter conversion rates can be availed through SRB than through methanogenesis. Moreover, SRB and NRB are not limited to one-and two-carbon substrates, as in the case of methanogens. However, from process engineering perspective, such an approach has disadvantages, including the loss of energy available from methane and the production of hydrogen sulphide or ammonia. Since, sulphide and ammonia are much more soluble than methane, their dissolved components can contribute significantly to effluent COD (Harper & Pohland, 1986). However, this approach may hold possibilities for reducing propionic acid and hydrogen, as well as acetic acid in an organically overloaded stressed anaerobic reactor, to more rapidly re-establish the equilibrium with the existing hydrogen removal system.

12.3.6 Estimation of Methane Production

Estimation based on chemical composition of waste: The Buswell stoichiometric equation (Eq. 12.5) shall be used for estimation of the production of methane from a given chemical composition of the organic matter present in wastewater.

$$C_n H_a O_b N_d + \left(n - \frac{a}{4} - \frac{b}{2} + \frac{3d}{4} \right) H_2 O \rightarrow C H_4$$

$$+ \left(n - \frac{a}{8} - \frac{b}{4} + \frac{3d}{8} \right) C O_2 + (d) N H_3 \qquad (12.5)$$

where, $C_n H_a O_b N_d$ represents the chemical formula of the biodegradable organic compound subjected to the anaerobic degradation process. Assumptions of this equation are the maximum stoichiometrically possible methane production is considered, and production of bacterial biomass is neglected and not accounted.

Considering degraded COD: Another method of evaluating the production of methane is from the estimation of the COD degradation according to Eq. 12.6:

$$CH_4 + 2O_2 \rightarrow CO_2 + 2H_2O \qquad (12.6)$$

One mole of methane requires two moles of oxygen for its complete oxidation to carbon dioxide and water. Therefore, every 16 g of CH_4 produced and lost to the atmosphere corresponds to the removal of 64 g of COD from the waste. Under normal temperature and pressure conditions, this corresponds to 350 mL of CH_4 for each gram of degraded COD. Theoretical production of methane can be calculated from Eq. 12.7:

$$V_{CH_4} = \frac{COD_{CH_4}}{K(t)} \qquad (12.7)$$

where, V_{CH_4} is the volume of methane produced (L) and COD_{CH_4} is COD removed from the reactor (g COD). The $K(t)$ is temperature correction factor (g COD/L), value of it can be obtained using Eq. 12.8.

$$K(t) = \frac{P \times K}{R \times (273 + T)} \qquad (12.8)$$

where, P is atmospheric pressure (1 atm); K is COD corresponding to one mole of CH_4 (64 g COD/mole); R is gas constant (0.08206 atm L/mole K); and T is temperature (°C).

However, not all the organic matter present in the influent is utilized by the methanogens. The influent wastewater to anaerobic reactor may also consists of sulphates, which are reduced to sulphides and SRB consume about 0.67 kg of COD per kg of sulphate being reduced (Eq. 12.9).

$$S^{2-} + 2O_2 \rightarrow SO_4^{2-} \qquad (12.9)$$

One mole of S^{2-} requires two moles of oxygen for its oxidation to sulphate. Therefore, every 96 g of SO_4^{2-} present in the waste consume 64 g of COD. Also, some portion of biogas will remain in soluble form in the reactor effluent due to high partial pressure of biogas inside the anaerobic reactor. Typically, about 15–16 mg/L of methane will be lost along with the effluent of the anaerobic reactor.

Example: 12.1 Calculate the energy output from anaerobic treatment of 4 MLD sewage having COD of 500 mg/L and sulphate of 80 mg/L. The COD removal efficiency of the reactor is 75% and sulphate removal is 80%.

Solution

Methane production in litres $= 1.28 \times T$ (K) per kg of COD removed

$= 1.28 \times (273 + 30)$ per kg of COD removed

$= 387.84$ L per kg of COD removed $= 0.387$ m^3/kg COD removed

The COD removal efficiency of the reactor is 75%.

Hence, kg of COD removed $=$ Flow $\times$ COD $\times 0.75$.

$= 4000 \times 500 \times 0.75 \times 10^{-3}$ kg/day

$= 1500$ kg/day

The total COD removed in the reactor $= 1500$ kg/day.

However, not all the organic matter present in the influent is utilized by methanogens. The influent also consists of sulphates, which are reduced to sulphides and consume about 0.67 kg of COD per kg of sulphate.

$SO_4{}^{2-} \rightarrow S^{2-}$

Sulphate removal of reactor is 80%, hence the total sulphate reduction.

$= 0.8 \times 4000 \times 80 \times 10^{-3} = 256$ kg/day

COD consumed in sulphate reduction $= 256 \times 0.67 = 171.52$ kg/day.

Hence COD available for methane production $= 1500 - 171.52 = 1328.48$ kg/day.

Also, some portion of biogas will remain in soluble form in the reactor effluent due to high partial pressure of biogas inside the reactor. Typically, about 16 mg/L of methane will be lost along with the effluent.

Methane that can be collected $= 1328.48 \times 0.387 - 4000 \times 16 \times 10^{-3} = 450.122$ m^3/day.

Also, the biogas collection efficiency of the collection domes will be about 85 to 90%, hence actually methane collected at 85% of collection efficiency will be $= 382.60$ m^3/day $= 382.60 \times 0.656 = 251$ kg/day.

Since, density of methane at STP $= 0.656$ kg/m^3 and per kg of methane contains 13,284 kcal.

Total energy obtained $= 251 \times 13,284 = 3334,284$ kcal $= 3334,284 \times 1.163$ W-h $= 3,877.8$ kW-h

Assuming 30% conversion efficiency of this energy to electrical power,

The electrical power output will be $= 3,877.8 \times 0.3 = 1163$ kW-h.

Example: 12.2 Consider the treatment of 500 m^3/day of wastewater at a temperature of 26 °C. Chemical composition analysis of the wastewater revealed that it consists of three wastewater streams each containing individually either sucrose, formic acid or acetic acid mixed in the volumetric ratio of 2.5:1:1.5 and concentration of each was 380 mg/L, 430 mg/L and 980 mg/L, respectively. Determine: (a) the total concentration of COD in the wastewater, (b) the maximum theoretical methane production, assuming yield coefficients for acidogenic and methanogenic organisms as 0.15 g $COD_{cel}/g\ COD_{remov}$ and 0.03 g $COD_{cel}/g\ COD_{remov}$, respectively.

Solution

(a) *Estimating COD concentration*

Volumetric load of sucrose containing wastewater $= \frac{2.5}{5} \times 500 = 250\ m^3/day$.

Volumetric load of formic acid containing wastewater $= \frac{1}{5} \times 500 = 100\ m^3/day$.

Volumetric load of acetic acid containing wastewater $= \frac{1.5}{5} \times 500 = 150\ m^3/day$.

Estimation of COD for the sucrose

$$C_{12}H_{22}O_{11} + 12O_2 \Rightarrow 12CO_2 + 11H_2O$$

Since, 342 g of sucrose requires 384 g of oxygen, therefore 380 mg/L of sucrose corresponds to $(384 \times 380/342 =)$ 427 mg COD/L, hence COD load due to the sucrose will be:

$= 250\ m^3/day \times 0.427$ kg $COD/m^3 = 106.8$ kg COD/day

Estimating COD of the formic acid wastewater stream

$$CH_2O_2 + 0.5O_2 \Rightarrow CO_2 + H_2O$$

Since, 46 g of formic acid requires 16 g of oxygen, therefore 430 mg/L corresponds to $(16 \times 430/46=)$ 150 mg COD/L, hence COD load due to the formic acid.

$= 100\ m^3/day \times 0.150$ kg $COD/m^3 = 15.0$ kg COD/day

Estimation of COD in the acetic acid wastewater stream

$$C_2H_4O_2 + 2O_2 \Rightarrow 2CO_2 + 2H_2O$$

Since, 60 g of acetic acid requires 64 g of oxygen, therefore 980 mg/L corresponds to 1045 mg COD/L, hence COD load due to the acetic acid.

$= 150\ m^3/day \times 1.045$ kg $COD/m^3 = 156.8$ kg COD/day

Therefore, final concentration of the waste in terms of COD

$=$ Total load/total flow $= (106.8 + 15.0 + 156.8$ kg COD/day$)/500\ m^3$/day

$= 0.557$ kg $COD/m^3 = 557$ mg COD/L

(b) The maximum theoretical methane production occurs when the removal efficiency of COD is 100%, and there is no sulphate reduction in the system, since it is not present in this wastewater.

COD load removed in the treatment system is $= 106.8 + 15.0 + 156.8 = 278.6$ kg COD/day.

Organic matter converted into acidogenic biomass

$= COD_{acid} = Y_{acid} \times 278.6 = 0.15 \times 278.6 = 41.2$ kg COD/day

Organic matter converted into methanogenic biomass

$= COD_{methan} = Y_{methan} \times (278.6 - 41.2) = 0.03 \times 237.4 = 7.1$ kg COD/day

Hence, COD available for conversion into methane (COD_{CH4})

$=$ Total COD load $-$ COD load converted into biomass

$= 278.6 - 41.2 - 7.1 = 230.3$ kg COD/day

$K(t) = \frac{P \times K}{R \times (273+T)} = (1 \text{ atm} \times 64 \text{ g COD/mole})/[0.0821 \text{ atm·L/mole·K} \times (273 + 26\,°C)]$

$= 2.61$ g COD/L

Therefore, the theoretical production of methane $= V_{CH_4}$

$= \frac{COD_{CH_4}}{K(t)} = (230.3 \text{ kg COD/d})/(2.61 \text{ kg COD/m}^3) = 88.24 \text{ m}^3\text{/day}$

Alternately, per kg COD removed the methane production in litre is $1.28\ T$, where T is in Kelvin. In this case per kg of COD removed at 26 °C, the methane production will be 382.72 L, i.e., 0.383 m^3/kg COD removed. Thus, for 230.3 kg COD removed per day, the methane production will be $230.3 \times 0.383 = 88.21$ m^3/day.

12.4   Anaerobic Reactor Types

Anaerobic reactors can be classified into two large groups (Fig. 12.3). The separation between conventional and high-rate systems are based on solids retention mechanisms in the system, hydraulic retention time and volumetric loads adopted. Conventional systems with low solids retention, high HRT and low volumetric loading rates are evolving towards high-rate systems. The most used anaerobic reactors are septic tanks, sludge digesters, anaerobic pond, anaerobic filter and up-flow anaerobic sludge blanket (UASB) reactor. These reactor types are discussed in more detail in the following subsections.

12.4.1 Anaerobic Sludge Digesters

Conventional anaerobic digesters are mainly used for the stabilization of primary and secondary sludges, originating from sewage treatment plants, and for the treatment of waste with a high concentration of suspended solids. They usually consist of covered circular or egg-shaped tanks of reinforced concrete or stainless steel. The bottom walls are usually inclined to form hopper shape, to favour the sedimentation and removal of the most concentrated digested solids. The covering of the digester can be fixed or floating (movable). Typically, sludge digesters are of three types, without mixing (low-rate), with

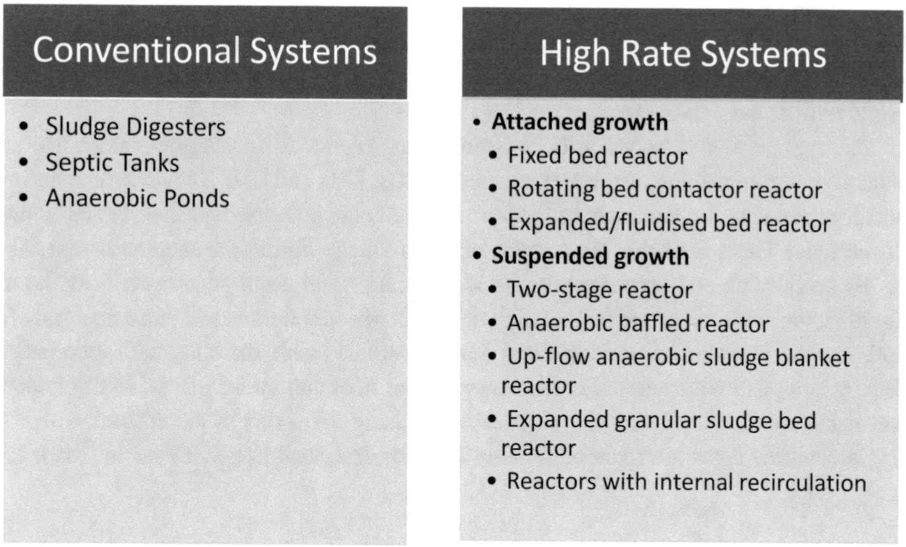

Conventional Systems	High Rate Systems

• Sludge Digesters
• Septic Tanks
• Anaerobic Ponds
	• **Attached growth**
	  • Fixed bed reactor
	  • Rotating bed contactor reactor
	  • Expanded/fluidised bed reactor
	• **Suspended growth**
	  • Two-stage reactor
	  • Anaerobic baffled reactor
	  • Up-flow anaerobic sludge blanket reactor
	  • Expanded granular sludge bed reactor
	  • Reactors with internal recirculation

Fig. 12.3 Classification of anaerobic reactors

mixing (high-rate) and two stage digesters. The supernatant and stabilized sludge are periodically removed from the digester.

Because of absence of mixing, there is sludge stratification in low-rate digesters, no more than 50% of the digester volume is used in the digestion process. Hence, large reactor volume is being required to achieve good sludge stabilization. In view of these limitations, low-rate digesters are mainly used in small sewage treatment plants. The single stage high-rate digester incorporates supplemental heating and mixing mechanisms, besides being operated at uniform feeding rates and with the previous thickening of the raw sludge, to guarantee more uniform conditions in the whole digester. As a result, the tank volume can be reduced, and the stability of the process is improved. The two-stage digester consists of incorporation of a second tank, operating in series with a high-rate primary digester. The second tank is meant for the storage and thickening of the digested sludge, leading to the formation of a clarified supernatant.

12.4.2 Septic Tanks

In developing countries, either the entire sewage treatment system is based on centralized facilities or where this centralized system is not available, almost every household owns or shares a septic tank. The septic tank is a unit that carries out the multiple functions of sedimentation and removal of floatable materials, besides acting as a low-rate digester. The settleable solids present in the influent sewage settle at the bottom of the tank and form a sludge zone. The oils, grease and other lighter materials present in the influent wastewater float on the surface of the tank, forming a scum layer. The sewage, free from the settled and floated material, flows between the sludge and scum layers and leaves the septic tank at the opposite end.

The organic matter settled at the bottom of the septic tank undergoes facultative and anaerobic decomposition, and it is converted to CO_2, CH_4 and H_2S. The anaerobic decomposition provides a continuous reduction of the sludge volume deposited at the bottom of the tank. There is always an accumulation of sludge during the months of operation of the septic tank and consequently the sludge and scum accumulation reduces the net liquid storage volume of the tank, which demands periodic removal of these materials. To optimize the retention of settleable and floatable solids inside the tank, the septic tank is usually equipped with internal baffles close to the inlet and outlet points. Multiple compartments are also used with the purpose of reducing the solids in the effluent, although single-chamber tanks are more commonly used, as described in detail later in Chap. 19.

12.4.3 Anaerobic Pond or Anaerobic Lagoons

Anaerobic ponds provide alternative for sewage treatment in warm-climate regions, and they are usually combined with facultative ponds. They are also frequently used for the treatment of wastewaters with a high concentration of organic matter, such as those from slaughterhouses, distilleries, etc. Anaerobic ponds are operated at a low organic loading rates and have high hydraulic retention time. In their typical configuration, the operation of the anaerobic ponds is very similar to that of septic tanks and uses the same basic removal mechanisms. One major difference with septic tanks is that there is no need for the systematic removal of the sludge deposited at the bottom of the anaerobic ponds that frequently, and cleaning/desludging is expected to be required at intervals of a few years. This is described in more detailed in next chapter.

12.4.4 Anaerobic Filters

The more commonly known example of anaerobic reactors with an attached bacterial growth, in a fixed bed, is the anaerobic filter. The wastewater in the anaerobic filter system passes the reactor usually with vertical flow, either up-flow or downflow. This anaerobic filter is characterized by the presence of a stationary packing material, in which the biological solids can attach to or be kept within the interstices. A variety of materials have been examined as supporting matrices including polyvinyl chloride sheets, potters' clay, red drain tile clay, needle-punched polyester and glass. The anaerobic filter is a fixed film reactor in which anaerobic microorganisms are allowed to grow on the surface of the filter media to form biofilm. As wastewater flow through the filter, the microorganisms

Fig. 12.4 Schematic
representation of anaerobic
filter

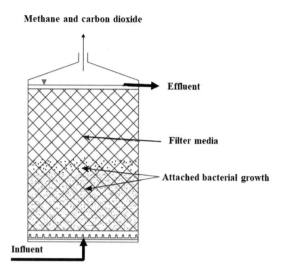

attached to filter media traps the organic contaminants to degrade the organic matter. This
reactor contains one or more filtration chambers in series. Depending on the design configuration, the influent flow could be either downwards or upwards; however, to prevent
the biofilm washout, upward flow of influent is preferred (Fig. 12.4).

The mass of microorganisms attached to the support material or kept in their interstices
degrades the substrate contained in the wastewater flow and, although the biomass is
released occasionally, the average solid retention time (SRT) in the reactor is usually
very high of about 100 days. The long SRT, associated with the short HRT, provide the
anaerobic filter with a great potential for application to the treatment of low-concentration
wastewaters like sewage. The main disadvantage of anaerobic filters is the accumulation of
biomass at the bottom of up-flow reactors, where it can lead to blockage or the formation
of hydraulic short circuits. Hence, for treatment of wastewater containing organic matter
in suspension, downflow anaerobic filter shall be preferred than up-flow filter to facilitate
easy cleaning in case of filter media blockage.

The performance of an anaerobic filter much depends on OLR, HRT, and wastewater source. In an investigation pertaining to the treatment of dairy wastewater, more than
90% COD removal was achieved at an OLR of 5–6 kg COD/m^3 day (Omil et al., 2003).
However, at an OLR of 23.25 kg COD/m^3 day, the COD removal was found to be 64%
(Goyal et al., 1996). Moreover, COD removal of 85–90%, as well as 100% degradation of phenol, were achieved after 8 h of HRT from industrial oil refinery wastewater
(Jou & Huang, 2003). The major limitations of anaerobic filters are clogging of filter
media as this process is mainly suitable for the remediation of dissolved waste, and this
is a comparatively costly process. The system volume is relatively large due to inert
packaging material. However, by using plastic media, which offers high surface area with
very reduced material volume, this drawback of larger reactor volume requirement can

be overcome. Anaerobic filters with high-density polyethylene media are widely used for treatment of municipal sewage as well as industrial wastewaters.

12.4.5 Up-Flow Anaerobic Sludge Blanket Reactor

Up-flow anaerobic sludge blanket (UASB) reactor was developed by Lettinga and co-workers, being initially largely applied in the Netherlands. The UASB reactors are the most widely used high-rate anaerobic wastewater treatment process and several full-scale reactors have been in operation world-wide (Buntner et al., 2013). Most of the successful applications of UASB reactors are to treat high strength industrial wastewaters (Lim & Kim, 2014). Municipal sewage treatment using UASB reactors is restricted to tropical regions, where temperature of the raw sewage allows fast hydrolysis of suspended organic solids (Zhang et al., 2013).

The UASB reactor can be divided in three parts, sludge bed, sludge blanket and gas–liquid-solid (GLS) separator provided at the top of reactor (Fig. 12.5). The sludge bed consists of high concentration of active anaerobic bacteria (40–100 g VSS/L) and it occupies about 40–60% of reactor volume. Majority of organic matter degradation (>95%) takes place in this zone. The sludge consists of biologically formed granules or thick flocculent active anaerobic sludge. Biochemical conversion of organic matter present in the wastewater occurs as the wastewater meets the granules and/or thick flocculent sludge. The biogas produced causes internal mixing in the sludge bed portion of the reactor. Some of the biogas produced within the sludge bed gets attached to the biological granules. The free biogas and the sludge particles with the attached biogas rise to the top of the reactor due to buoyancy.

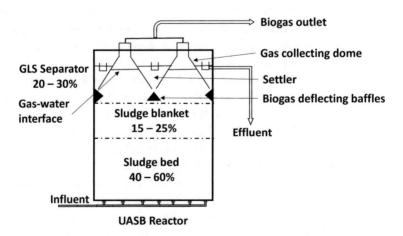

Fig. 12.5 A schematic of the UASB reactor

Above the sludge bed and below GLS separator, thin concentration of sludge is maintained, which is called as sludge blanket. This zone occupies 15–25% of reactor volume. Maintaining sludge blanket zone is important to dilute and further treat the wastewater stream that has bypassed the sludge bed portion following the rising biogas bubbles. The GLS separator occupies about 20 to 30% of the reactor volume. The sludge particles that raise to the liquid surface strike the bottom of the biogas deflecting baffles or the turbulence created at gas–water interface inside the gas collection dome causing the attached biogas bubbles from the cell mass to be released. The degassed granules typically drop back to the surface of the sludge bed. The free biogas and biogas released from the granules is captured in the gas collection domes located at the top of the reactor. Liquid containing some residual solids and disintegrated biological granules pass into a settling chamber, which is formed outside the gas collection dome by prohibiting entry of biogas to support quiescent condition, where the residual solids are separated from the liquid by gravity settling. The separated solids fall back through the sloping deflectors to the top of the sludge bed. The provision of deflector below the entry of wastewater in the settling zone does not allow biogas to enter in this zone, thus maintaining conditions supportive for gravity settling of sludge without causing any biogas induced mixing.

One of the fundamental principles of the UASB reactor is its ability to develop a high-activity biomass. This biomass can be in the form of flocs or granules (1 to 5 mm aggregate size). Anaerobic granules are self-immobilized aggregates that act as enriching agents in anaerobic reactors, like UASB reactor, which gives distinct engineering advantages for reactor operation and are especially important, however difficult to obtain while treating low strength wastewater like sewage (Najafpour et al., 2006). The granules are dense microbial consortia packed with several synergistic bacterial species mostly hydrolytic bacteria, acidogenic bacteria and syntrophic association of acetogenic bacteria and methanogenic archaea and typically contain millions of organisms per gram of biomass (Li & Sung, 2015). Granular sludge plays an important role in treating wastewater due to their advantages of denser and stronger aggregate structure, better settleability, better solids retention improving solid-effluent separation, higher biomass concentration and greater resistance to shock loadings (Xing et al., 2015). This property is particularly important while dealing with low-strength wastewater, where adequate biomass retention can be possible even at higher up-flow velocity (1.75 m/h) under peak flow conditions (Bhunia & Ghangrekar, 2007).

Biomass granulation depends on several factors starting from the source of inoculums, the type of organic matter, water hardness, presence of nutrients, applied hydraulic loading rate, pH, temperature, presence of metal ions, hydraulic retention time (HRT), liquid up-flow velocity and even on the design of the reactor (Chen et al., 2015) and applied organic loading rates (OLRs) (Ghangrekar et al., 1996). Formation of granules from non-granular inoculum occurs due to microbial self-immobilization and subsequently, aggregate formation and growth (Li & Sung, 2015). Effectiveness of different metal ions like magnesium

(Chen et al., 2015) to improve sludge granulation has been studied previously. Development of granular biomass in UASB reactor using synthetic or natural polymer addition has also been evaluated by the researchers (Bhunia & Ghangrekar, 2008).

Use of internal packing as an alternative for granular sludge growth has been investigated (Najafpour et al., 2006). Packing medium in a UASB reactor is expected to accelerate granulation and increase solids retention by dampening short circuiting, improving gas/liquid/solid separation, and providing surface for biomass attachment (Najafpour et al., 2006). The properties of anaerobic granules are important factors that affect process performance and stability; hence it is very important to develop granular biomass in UASB reactor with appropriate characteristics to support maximum sludge hold-up and high substrate conversion rate even at low organic matter concentration in the feed wastewater.

The second fundamental component of the UASB reactor is the presence of a GLS separator, which is in the upper part of the reactor. The main purpose of this device is the separation of the biogas contained in the liquid mixture, so that a zone favouring sedimentation is created at the upper part of the reactor. Presence of this internal settler also eliminates the requirement of external sedimentation tank as required for other biological reactors involved in secondary treatment. Thus, effluent suspended solids can also be kept under control due to presence of this GLS separator. Solid retention time (SRT) of more than 35–40 days is essential for effective operation of anaerobic reactors. By retaining cell mass in the reactor, this GLS separator also helps to maintain such high SRT in the UASB reactor.

The UASB reactor can effectively treat high-strength wastewater with the COD of 7000 ± 800 mg/L as well as tyrosine concentration of 20–200 mg/L. This UASB reactor resulted in about 70–75% COD removal as well as 95% tyrosine reduction (Radjenovic et al., 2007). Moreover, an investigation demonstrated 80–85% COD removal efficiency for the treatment of agricultural-based pulp mill wastewater at 20 h of HRT, when the OLR was 7.5 kg COD/m^3 day (Chinnaraj & Rao, 2006). Whereas, while treating sewage the UASB reactor generally gives COD removal efficiency ranging from 60 to 70%.

12.4.6 Anaerobic Fluidised Bed Bioreactor

An anaerobic fluidized bed reactor (FBBR) is an attached growth anaerobic process in which the operating principle is fundamentally same as that of expanded bed bioreactor (Fig. 12.6). Here the reactor is packed with support media of inert material such as silica sand, activated carbon or with any other plastic material. The bed gets fluidised by the up-flow velocity of the liquid, which provide particles with a larger surface area for the growth of biomass (Sokol, 2001). The up-flow velocity of the fluid is sufficient enough to suspend the media, however not large enough to carry them out of the reactor. This up-flow velocity agitates the media so that quick mixing takes place in the reactor.

Fig. 12.6 Schematics of an anaerobic fluidized bed bioreactor

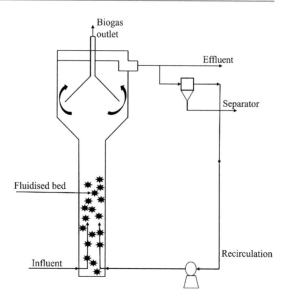

This fluidization can be controlled by varying its up-flow velocity of the liquid and this expansion can be from 30 to 100%.

In general, the depth of the bed of a FBBR is in the range of 3–4 m. The size of media i.e. granular activated carbon is about 0.5–0.7 mm with a specific surface area of 900 m²/g (Chen et al., 1988). Generally, 20–40% of the empty reactor volume will be filled with the media. The up-flow velocity is maintained in the range of 20–40 m/h and the recirculation of the effluent is done to maintain upward velocity within the specified retention time. The hydraulic retention time can be ranged from 15 to 60 min. The volumetric loading for anaerobic FBBR can be as high as 20–30 kg COD/m³ day for medium and high strength wastewater for an anticipated removal efficiency of 70–90%. With due course of time, the biofilm size increases, and makes the packing media lighter, where it reaches the top and subsequently removed. Hence, arrangement to control media washout should be there for example by providing internal settler as shown in Fig. 12.6.

12.4.7 Anaerobic Baffled Reactor

This anaerobic baffled reactor (ABR) consists of 3–5 chambers and wastewater is allowed to flow upward direction in each compartment or up-flow and downflow mode in alternate compartments (Fig. 12.7). This reactor can give reliable treatment efficiency particularly for the treatment of low strength wastewater containing particulate organic matter. Experience of operation of anaerobic baffled reactor operation has shown that the compartment having downward flow of the wastewater contributes very less for organic matter removal.

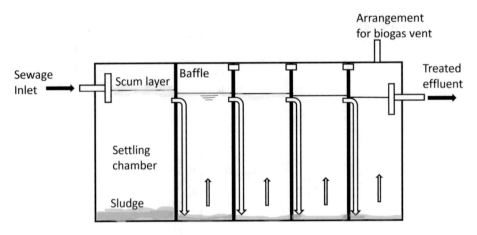

Fig. 12.7 Schematic of anaerobic baffle reactor

Hence, up-flow mode of ABR operation is preferred by transporting the wastewater by down-streamer pipe and maintaining up-flow mode in each compartment.

In the last compartment, GLS separator arrangement, as used in UASB reactor, can be provided to improve the effluent quality by controlling the suspended solids concentration in the effluent. Presence of such GLS will improve sludge retention in the ABR, favouring improved performance of organic matter removal. This ABRs are also being used for the treatment of the septic tank sludge. This ABR can also be used as improved septic tank, instead of conventional septic tank for onsite sewage treatment. While treating sewage if properly maintained, with periodic desludging, the ABR can give more than 80% BOD removal. Whereas if desludging is not done for many years, this efficiency may fall down to 50–60% for BOD removal and even the effluent will have high suspended solids.

12.4.8 Anaerobic Sequencing Batch Reactor

The anaerobic sequencing batch reactor (AnSBR) was developed as a modification of anaerobic contact and anaerobic activated sludge processes. It was invented by Richard R. Dague in 1993. In the operation of AnSBR, four major sequences of cycles are involved; feed, react, settle, and decant. The process starts with the feeding of the reactor, then goes to the reaction phase, where microbes are mixed well. After digestion, mixing is stopped to favour the settling. The final step is decanting in which supernatant is drained off from the reactor as treated effluent (Fig. 12.8) (Shao et al., 2008).

The AnSBR possesses some unique characters that enable the reactor to achieve higher organic matter removal. After immediate feeding to a low substrate at the end of the reaction step, the high substrate concentration causes the variable F/M ratio in the AnSBR.

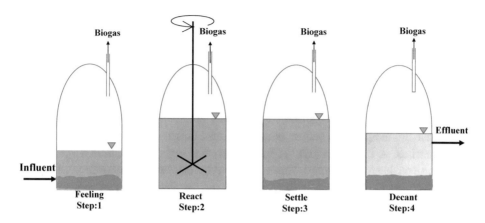

Fig. 12.8 Schematic representation of an anaerobic sequencing batch reactor

High F/M ratio just after feeding results in higher substrate removal and maximum organic matter conversion into biogas as predicted by Monod's kinetics. However, as the reaction step stops and settling starts, there is a low F/M ratio, and therefore, there is low biogas production that is a favourable condition for biomass flocculation and settling. Due to this internal settling phenomenon, an external clarifier is eliminated as required in other anaerobic processes.

The AnSBR process has been extensively explored in lab-scale as well as in field-scale for the removal of organic pollutants from different high strength wastewaters, like meat processing plant effluent, landfill leachate, domestic wastewater, dairy wastewater and brewery wastewater (Bodík et al., 2002; Timur & Özturk, 1999). The geometry of the AnSBR can affect organic matter removal. The pilot-scale AnSBR for domestic sewage treatment with a height to diameter (H/D) ratio of 3 resulted in 43% of total COD and 67% of TSS removal; whereas total COD and TSS removal was reported to be 39 and 64%, when H/D ratio was 1.5 (Sarti et al., 2007). Pilot-scale AnSBR for brewery wastewater treatment operated at OLR of 1.5–5 kg COD/m^3 day and HRT of 24 h resulted in more than 90% of COD removal (Shao et al., 2008). Therefore, AnSBR technology can be considered as an effective technology for removal of COD, TSS and other pollutants from high strength wastewaters.

12.5   Factors Affecting Anaerobic Digestion

Several environment factors can affect anaerobic digestion, either by enhancing or inhibiting parameters such as specific growth rate, decay rate, gas production, substrate utilization rate, start-up duration, and response to changes in influent characteristics. These parameters affecting the performance of the anaerobic digestion are described in the following sections.

12.5.1 Temperature

As in all biological processes, anaerobic processes are affected by operating temperature. The higher the temperature, higher is the microbial activity until an optimum temperature is reached. A further increase of the temperature beyond its optimum value results in decrease in bacterial activity. Anaerobic process can take place over a wide range of temperatures (4–60 °C); however, most of the anaerobic reactors are operated in the mesophilic range (30–40 °C). The effect of increasing temperature on biochemical reaction rate in the range of 4–25 °C is profound.

Although anaerobic digesters have been reported to operate at substantially lower temperatures, such as 20 °C, anaerobic growth under these temperature conditions is slow, requiring prolonged start-up time and difficulties in operation. In situations where the ambient and wastewater temperatures are low, start-up will be benefited if initiated at temperature of approximately 35 °C by providing heating arrangement to the reactor. At temperature of less than 25 °C, the digestion rate decreases sharply and conventional anaerobic digesters in operation at ambient temperatures in cooler climates may require detention time of as much as 12 weeks for the treatment of sewage sludge. Increase in microbial reaction rates at the elevated temperatures of thermophilic processes (50–60 °C), and hence decrease in SRT may prove advantageous under some circumstances. Thermophilic digestion is most practical where wastewater stream to be treated is discharged at an high temperature and the digester is present on site.

In psychrophilic (below 20 °C), mesophilic (25–40 °C), or thermophilic temperature ranges (50–60 °C), uniformity of temperature over the entire reactor contents is of paramount importance to anaerobic digestion. Temperature lowering of even a few degrees can result in a marked upset in microbial metabolism and rapid alterations in reactions in the reactor and may necessitate several days for the recovery. Seasonal changes in temperature will be accommodated with little effect on the bacterial population if gradual alterations of approximately 1 °C/day in internal system temperature are introduced. Such gradual slow changes permit microbial adaptation to the changed temperature. In temperature-controlled anaerobic systems, no alteration more than 1 °C should occur as reactor vessel contents pass over or through heating elements. A consistent temperature throughout the reactor can be provided by adequate mixing of the reactor by paddle, gas sparging, or flow overheat exchangers.

12.5.2 Hydrogen Ion Concentration (pH)

Methanogenic microorganisms are vulnerable to the minute changes in the pH values. Optimum operating pH range of 6.6–7.6 is considered favourable for the methane producing archaea, which cannot tolerate the pH fluctuations. The non-methanogenic bacteria do not exhibit such strong sensitivity for environmental conditions and are able to function in a wide range of pH form 5.0–8.5. The pH maintained inside the reactor, due to the anaerobic process results from the interaction of the carbon dioxide-bicarbonate buffering system and volatile acids-ammonia formed by the process.

It is necessary to prevent the accumulation of acids to a level, which may become inhibitory to the methanogenic archaea. For this, it is important that there should be sufficient buffering capacity present in the reactor, which may prevent the reactor from souring due to accumulation of VFAs. Although, the carbonates and bicarbonates of sodium and calcium are required to be added to the digesters to provide buffering action, lime is mostly used for this purpose. While there is no lag period prior to system recovery after pH restoration from high (basic) pH values, whereas considerable time is required when a low pH below 6.5 period exceeded 3 days (Lin Chou et al., 1979). Reactor configurations like fluidized bed reactor have an inherent buffering capacity promoting increased tolerance of influent pH fluctuations.

12.5.3 Physical Parameters

The efficient application of mixing, fluid flow, gas sparging, paddles or impellers best utilizes the reactor volume available, minimises the inhibitory effects of local build-up of VFAs and other digestion products, allows even dispersion of suspended materials, such as granules or flocs, to promote maximum biomass/wastewater contact. Provision of adequate mixing permits rapid and uniform distribution of fresh influent throughout the reactor volume and also allows maintenance of a constant temperature over the contents of the vessel, thus reducing convection currents and thermophoresis. The biogas produced in the reactor is also responsible for providing mixing and rate of this biogas production is directly proportional to the volumetric organic loading rate applied to the reactor.

In UASB reactor mixing also partly controls the size of flocculated or granulated bioactive particles. Inactive debris can also accumulate at the bottom of these type of reactors, if insufficient agitation is applied, and build-up of inorganic solids can decrease effective system volume. At low shear rates large flocs maintain their basic structure but tend to aggregate, while shear stresses encountered at higher shear rates are generally large enough to cause complete floc disintegration. Efficiency of operation in fixed-film reactors, such as the anaerobic filter, and expanded and fluidized beds depends greatly on good hydraulic distribution and maximal substrate availability to the digester microorganisms. As fluid velocity increases at higher flow rates, biofilm accumulation decreases and detachment can occur, a condition which eventually culminates in process failure.

The maximum volume of biofilm attached to the media surface in a turbulent flow regime could be limited by fluid shear stress. The increase of shear removal rate was found to be proportional to interfacial fluid shear stress. At high film volumes, the critical shear stress in fixed biofilm systems is that value at which all the cells are removed from media surface; the proportion of the initial bacterial population remaining after a specified period of shear is indicated by the adhesion number. The physical forces resulting from reactor operation vary under different conditions and with different digester configurations. These forces are the consequence primarily of externally applied energies and as such can be controlled by proper attention to operational parameters and careful system design.

12.5.4 Nutrients (Nitrogen and Phosphorus)

For supporting the growth of anaerobic microorganisms present in the reactor is it necessary to ensure that sufficient concentration of nitrogen and phosphorous is available in the wastewater to be treated. If this N and P is not sufficient in wastewater then external supplementation of this is necessary for supporting microbial growth and activity. Optimum N/P ratio for anaerobic processes can be 7. The theoretical minimum COD/N—ratio is 350/7. A value around 400/7 for COD/N is considered reasonable for high-rate anaerobic processes operated in sludge loading rate (SLR) of 0.8–1.2 kg COD/kg VSS d). For low-rate processes, operated at SLR < 0.5 kg COD/kg VSS d, the COD/N-ratio has been observed to be increased dramatically to values of 1000/7 or more (Van den Berg and Lentz, 1979).

12.5.5 Trace Metals

In addition to fundamental requirements for macronutrients, such as nitrogen and phosphorous, the inability of a great number of anaerobes to synthesise some essential vitamin or amino acid necessitates the supplementation of the bacterial medium with certain trace metals, such as cobalt, molybdenum, selenium, tungsten, and nickel, which are probably necessary for the activity of several enzyme systems. Microbial nutritional requirements may also be interdependent in the presence of low concentrations of other cations, such as potassium. For example, sodium may be able to satisfy some of the potassium requirement. In addition, nutrients not obligatory for growth, such as calcium, may however be needed for process stability. If the wastewater to be treated by anaerobic digestion is not rich in micronutrients, supplementation, either chemically or by the addition of another waste stream may be necessary. The design and optimisation of the growth environment, however, should include consideration of the technical and economic constraints imposed by the nature and quantity of the waste to be converted.

The metal nutrients can be classified based on their concentration in the cells into major cations (K, Mg, Ca), and micronutrients (Mn, Fe, Co, Cu, Mo, Ni, Se, W). The Fe is utilized in the transport system of the methanogenic archaea for the conversion of CO_2 to CH_4, and functions both as an electron acceptor and donor. However, Mn acts as an electron acceptor in anaerobic respiration processes, while Zn takes part in the functioning of enzymes involved in methanogenesis. All the trace elements are responsible for the functioning of one or other enzymes.

12.5.6 Inhibitory Substances

Inhibition of the anaerobic digestion process can be mediated to varying degrees by toxic materials present in the system. These inhibitory substances may be components of the influent wastewater, or by-products of the metabolic activities of the digester bacteria. Inhibitory toxic compounds include sulphides, consequential in the processing of wastes from such sources as molasses fermentation, petroleum refining and tanning industries, and volatile acids, microbial products, which can accumulate and exceed the reactor buffering capacity. Sulphide concentration exceeding 100 mg/L in the reactor will start noticeable reduction in the methane production from the anaerobic reactor. Inhibition may also arise as the consequence of the levels of ammonia, alkali and the alkaline earth metals, and heavy metals in the system. The latter have been considered the most common and major factors governing reactor failure.

Volatile acids inhibition: Failure of the methanogenic population growth due to environmental disruptions such as shock loadings, nutrient depletion or infiltration of inhibitory substances leads to a decrease in pH of the reactor. A reduced pH shifts the equilibrium of the fatty acids to their undissociated form. The undissociated nature of these acids allow them to penetrate the bacterial cell membrane more efficiently than their ionized counterparts, and once assimilated, induce an intracellular decrease in pH and hence a decrease in microbial metabolic rate. The VFA concentration in the anaerobic reactor should be maintained below 500 mg/L (measured as acetate) at any point of time and preferably below 200 mg/L for optimum performance.

Ammonia–Nitrogen Inhibition: Although ammonia is an important buffer in anaerobic processes, high ammonia concentration can be a major cause of operational failure. Shock loadings of high ammonia concentration can cause rapid production of VFAs and the buffering capacity of the system may not be able to compensate for the decrease in pH. Further depression of alkalinity and reduction of pH may result in the reactor operational failure. Free ammoniacal concentration of 100 mg/L will dent the methane production of the anaerobic reactor and inhibition of the methanogenesis occurs at higher concentrations of ammonia.

Heavy Metals Inhibition: The most common agents of inhibition and failure of anaerobic reactors are heavy metals. Heavy metals in the soluble state are of more significance to reactor toxicity than the insoluble forms. The metals like copper, chromium, nickel, lead can induce toxicity in the reactor when present in higher concentration, and acceptable concentration in the wastewater to be treated differs from metal to metal. The concentration at which a metal inhibits methanogenesis by 50%, is defined as C50. There has been extensive research on heavy metal inhibition, however, different researchers report different values of inhibition.

Although heavy metal toxicity in UASB reactors has been studied abundantly, reported values of toxic concentrations vary considerably between different authors (Alrawashdeh et al., 2020; Fang & Hui, 1994; Lin & Chen, 1997). A study by Alrawashdeh et al. (2020) reports that the safe concentration of Fe, Zn, Cr, Pb, and Ni can improve the anaerobic digestion process (in terms of increasing biogas and methane production and increasing TCOD, SCOD, volatile solids, and polyphenols removal) if they are lower than 2.9, 0.335, 1.211, 0.297, and 0.082 mg/L, respectively. These values can be considered as the limiting concentration above which inhibition starts.

A different study by Fang and Hui (1994) reports that methanogenic activity is reduced by 50% when each gram of biomass was in contact individually with 105 mg of zinc, 120 mg of nickel, 180 mg of copper, 310 mg of chromium, or >400 mg of cadmium. It should be noted here that while the study by Alrawashdeh et al. (2020) reports the liquid concentration, Fang and Hui (1994) reports the toxicity per gram of biomass. Yet another study by Lin and Chen (1997) reports a much lower C50 value per gram of biomass as 1.6 mg Cd, 1.0 mg Cr, 0.9 mg Cu, 2.3 mg Ni, 4.0 mg Pb and 1.6 mg Zn. The study by Lin and Chen (1997) was conducted in batch bottles, indicating a higher capacity of the reactors to tolerate heavy metals when operated in continuous mode.

12.6   Design of Anaerobic Reactors

Hydraulic Retention Time: The hydraulic detention time refers to the average time of residence of the liquid inside the reactor, calculated by using the Eq. 12.10.

$$\theta = V/Q \qquad (12.10)$$

where, θ is hydraulic detention time (h), V is volume of the reactor (m^3), Q is the average influent flowrate (m^3/day).

Hydraulic Loading Rate: The hydraulic loading rate refers to the volume of wastewater applied daily per unit area of filter packing medium for anaerobic filters (Eq. 12.11).

$$HLR = Q/A \qquad (12.11)$$

where, *HLR* is the hydraulic loading rate (m³/m² day), Q is average influent flowrate (m³/day), and A is the surface area of the packing medium (m²).

Organic Loading Rate: The volumetric organic loading rate or simply organic loading rate (OLR) refers to the load of organic matter applied daily per unit volume of the reactor and it is expressed as per Eq. 12.12.

$$L_v = \frac{Q \times S_0}{V}$$
(12.12)

where, L_v is the OLR (kg COD/m³ day); Q is the average influent flowrate (m³/day), S_0 is the influent COD concentration (kg COD/m³), and V is total volume of the reactor (m³).

Solid Loading Rate: Solids or sludge loading rate (SLR) refers to the amount (mass) of organic matter applied daily to the reactor per unit mass of biomass present in the reactor as given in Eq. 12.13. This represents the specific loading rate of organic matter applied in the reactor per unit mass of microorganisms present in the reactor, thus expression for this loading rate is similar to the F/M used in case of activated sludge process.

$$L_s = \frac{Q \times S_0}{M}$$
(12.13)

where, L_s is the specific or sludge loading rate (kg COD/kg VSS·day), Q is average influent flowrate (m³/day), S_0 is influent substrate concentration (kg COD/m³), and M is the mass of microorganisms present in the reactor (kg VSS/m³).

Up-flow Velocity: The up-flow velocity of the liquid is calculated from the relation between the influent flowrate and the cross-sectional area of the reactor as given in Eq. 12.14.

$$v = \frac{Q}{A}$$
(12.14)

where, v is the up-flow velocity (m/h); Q is flow rate (m³/h); and A is the area of the cross section of the reactor, in this case the surface area (m²). Alternatively, the up-flow velocity can also be expressed as the ratio of the height of the reactor and the HRT as per Eq. 12.15.

$$v = \frac{Q \times H}{V} = \frac{H}{\theta}$$
(12.15)

where, H is the height of the reactor (m) and θ is HRT (h).

Sludge production: Estimation of the mass production of sludge in anaerobic reactors can be done using the Eq. 12.16.

$$P_s = Y \times COD_{rem} \tag{12.16}$$

where, P_s is the production of biosolids in the system (kg VSS/day); Y is the sludge yield or production coefficient (kg VSS/kg COD$_{rem}$); COD$_{rem}$ is COD load removed from the system (kg COD). The values of Y reported for the anaerobic treatment of domestic sewage are in the range of 0.06 to 0.15 kg VSS/kg COD$_{rem}$.

Biogas production: The biogas production can be evaluated from the estimated influent COD load to the reactor that is converted into biogas. In a simplified manner, the portion of COD that can be converted into methane can be determined using Eq. 12.17.

$$COD_{CH_4} = Q \times (S_0 - S) - Y_{obs} \times Q \times (S_0 - S) \tag{12.17}$$

where, COD_{CH_4} is the COD load converted into methane (kg COD_{CH_4}/day); Q is average influent flow (m^3/day); S_0 and S are influent and effluent COD concentrations, respectively (kg COD/m^3); Y_{obs} is coefficient of solids production in the system, in terms of COD (0.08 to 0.20 kg COD$_{sludge}$/kg COD$_{rem}$).

12.7   Design of UASB Reactor

For sewage treatment, the design of UASB reactor or anaerobic filter at higher loading rates is not possible due to limitations of up-flow velocity, and maximum OLR of about 2–3 kg COD/m^3 day can be adopted for design. Similarly, for high strength wastewater, such as distillery, satisfying minimum velocity criteria and maximum HRT limit is difficult. Therefore, categorization of wastewater based on COD concentration is necessary for generalizing the design procedure of UASB reactor to meet the recommended operating conditions to the maximum extent. Thus, the COD concentration of the wastewater is suitably divided in four categories. It has been proposed to adopt loading conditions as recommended in the Table 12.1, for design of UASB reactor depending on the average COD concentration of the raw wastewater. Thus, selecting suitable OLR the volume of the reactor required shall be estimated. Based on the recommendation of HRT and up-flow velocity the height and area required for the reactor can be worked out.

Based on the suitable value of OLR, for given COD concentration of the wastewater, the volume of reactor required is estimated using Eq. 12.12. For the suitable SLR values for that COD concentration range (Table 12.1), the volume of sludge required should be worked out considering the average concentration of VSS between 25 and 35 g/L for medium and high strength wastewaters, and 15 to 25 g/L for low strength wastewater. This volume of sludge should be less than 50 to 60% of the reactor volume estimated based on OLR, to avoid overloading of the reactor with respect to SLR. If the volume is not meeting the requirements, the OLR shall be reduced to increase the total reactor volume. The HRT should generally not be allowed to be less than 6 h for any type of

Table. 12.1 Recommended design criteria for UASB reactor based on the COD values of the wastewater to be treated

Category	COD (mg/L)	OLR (kg COD/m³ day)	SLR (kg COD/kg VSS day)	HRT (h)	Liquid up-flow velocity (m/h)
Low strength	<750	1–3	0.1–0.4	6–18	0.25–0.7
Medium strength	750–3000	2–5	0.2–0.5	6–24	0.25–0.7
High strength	3000–10,000	5–10	0.2–0.6	12–36	0.15–0.7
Very high strength	>10,000	5–15	0.2–1	36–240	0.05–0.3

Source Modified from Ghangrekar et al., (2003)

wastewaters. For very high strength of the wastewater, COD greater than 10,000 mg/L, the HRT may be allowed to exceed even 200 h as per the requirement.

Higher up-flow velocities, favours better selective process for the sludge and improve mixing in the reactor. However, at very high up-flow velocity, greater than 1.0–1.5 m/h, the inoculum may get washed out during start-up or during normal operation granules may get disintegrated, and the resulting fragments can easily wash out of the reactor. The maximum liquid up-flow velocity allowed in design should not exceed 1.2–1.5 m/h at peak flow. Up-flow velocities in the range of 0.25–0.8 m/h are favourable for granular biomass growth and accumulation during operation of the reactor and maximum up-flow velocity up to 1.5 m/h at peak flow conditions for short duration can be used in design.

Taller reactor can be designed to reduce the plan area, cost of GLS device and influent distribution arrangement. However, the height of the reactor provided should be sufficient to have enough sludge bed depth to avoid channelling and to keep liquid up-flow velocity within maximum permissible limits. In order to minimise channelling the minimum depth of sludge bed should be about 1.5 m. For this reason, the minimum height of the reactor should be restricted to 4.0 m, to conveniently accommodate sludge bed, sludge blanket and GLS separator. The maximum height of the reactor can be about 8 m. The height of the reactor adopted in practice is usually between 4.5 m and 8 m and 6 m is used as the typical height for the UASB reactors.

After finalizing volume and height, the plan area can be worked out and suitable dimensions of the reactor can be adopted. Generally, the maximum diameter or side length of single reactor should be kept less than 20 m. Restricting length or diameter will help in appropriate design of the influent distribution system if pump feeding is adopted. Very large side length or diameter will pose difficulty in uniform distribution of the wastewater to entire plan area of the reactor. Before finalizing the dimensions of the reactor, it is necessary to consider the dimensions required for GLS separator, because to accommodate the GLS separator meeting all requirements, it may be necessary to alter height and plan area of the reactor.

12.7.1 Design of Gas–Liquid-Solid Separator for UASB Reactor

In order to achieve highest possible sludge retention under operational conditions, it is necessary to equip a GLS separator device at the top of a UASB reactor. The main objective of this design is to facilitate the sludge retention without help of any external energy and control device. The guidelines for shapes and design of GLS separator are given by Lettinga and Hulshoff (1991). This GLS separator should be designed to meet the requirements, such as provision of enough gas–water interface area inside the gas dome, providing sufficient settling area outside the biogas collection dome to control surface overflow rate; and sufficient aperture opening at bottom from where the wastewater is entering in the settling compartment to avoid turbulence due to high inlet velocity of liquid in the settler, and thus to allow proper return of sludge solids back to the reactor. Attention is required to be paid to the geometry of the unit and its hydraulics, to ensure proper working of the GLS separator.

The shape of the GLS device that can be considered in design is presented in Fig. 12.9. The gas–water interface inside the biogas collection dome is considered at the depth Δh from top of the dome. In the beginning, the height of GLS separator can be considered as 25% of the total reactor height. The angle of inclination of the dome wall with horizontal is generally kept between 45° and 70°. Thus, sludge particles settled on this slope in the settling zone will easily slide back by gravity and get settled in the sludge bed. For estimating the number of domes required, initially the angle of dome with horizontal can be assumed as 45°, and base width of dome (Wb) can be estimated using relation $2(h + \Delta h)/\tan\theta$. The Δh generally provided is in range of 0.3 to 0.5 m. Even higher value of Δh is provided in many of the working UASB reactors treating wastewaters which are liable for higher foam formation in the gas collection dome. The number of domes required for given diameter (or length for rectangular reactor) can be estimated by dividing reactor length or diameter by W_B, and rounding this number. Where, $W_B = Wb + Wa$, and Wa (the width of aperture opening at entry of settling compartment) can be in the range of 0.2–0.3 m. After deciding the number of domes, the flow rate shared by each dome should be estimated in proportion to the base area occupied by each dome, including proportional aperture width, to the total area of the reactor.

The area of aperture (Ap) required for the wastewater to enter in settling zone of GLS can be computed based on the maximum inlet velocity of liquid to be allowed. This area can be estimated as proportional flow rate per settling zone in case of rectangular reactor (or central settling zone in case of circular) divided by maximum velocity to be allowed. The maximum inlet velocity of 3 m/h is considered safe for medium and high strength wastewater and for low strength wastewater lower aperture velocity should be preferred.

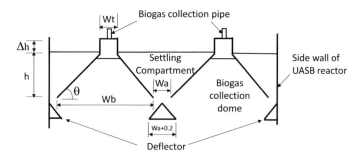

Fig. 12.9 Details of the gas–liquid-solid (GLS) separator

The width of aperture (*Wa*) is to be computed as aperture area divided by width (or in case of circular reactor by diameter) of the reactor. It is recommended to use minimum aperture width of 0.2 m and if the width required is greater than 0.5 m, then increase the number of domes.

The gas production expected in the reactor can be estimated based on the OLR selected for the design and expected COD removal efficiency in the range of 70 to 90%. The methane production can be estimated as 0.35 m^3/kg COD removed at ambient temperature and methane content of 65–70% in biogas. From this methane production the biogas collection per dome is estimated in proportion with percentage of area covered by the dome. The biogas loading at gas–water interface can be estimated as gas collection per dome divided by area. The loading of biogas at gas–water interface should be kept less than 80 m^3 biogas/m^2 day, i.e., about 3 m/h (Ghangrekar et al., 2003). Generally, top width (*Wt*) of 0.3–0.6 m shall be adopted in design with maximum of 1.0 m.

When even with maximum top width, if biogas loading is greater than 3.0 m/h, reduce the height of GLS separation device to 20% and repeat the earlier steps of GLS separator design, with fresh number of domes. Even with reduction in height of GLS separator if these checks are not satisfying, provide additional layer of gas collector dome. When two or more layer of gas collectors are used the height of each layer can be 15–20% of the overall reactor height, with minimum height of each layer as 1.2 m and maximum up to 1.5–2.0 m. For treatment of wastewater with COD greater than 10,000 mg/L, generally it is essential to provide multilayer dome (2–3) to keep biogas loading under control at gas water interface (Fig. 12.10). Excessive gas loading at gas water interface will lead to sever foam formation and this turbulence will push even the sludge particles in the gas collection pipe along with foam and eventually leading to blockage of gas collection pipe. In the event of blockage of gas collection pipe, the whole biogas will be pushed to the settling compartment and that will lead to the sludge washout from the UASB reactor and hence such situation will lead to performance failure due to reduction in solids retention time.

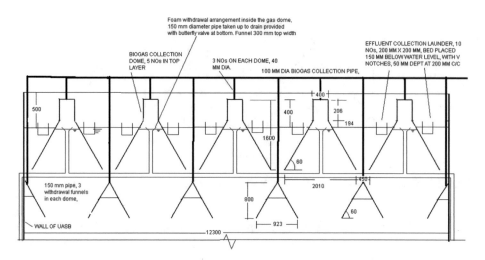

Fig. 12.10 Typical two-layer dome structure used in GLS separator of the UASB reactor

The width of the water surface (*Ws*) available for settling of solids between the two adjacent gas domes, at top of the reactor, can be estimated from difference of total length of the reactor minus total top width of the domes and dividing it with number of biogas collection domes. The corresponding surface overflow rate is estimated as hydraulic flow rate per dome divided by area of settling compartment. It is recommended that the surface overflow rate for effective settling of solids back to the reactor should be less than 20 m³/m² day at average flow and should be less than 36 m³/m² day under peak flow conditions. When it is exceeding the limits recommended, it is necessary to reduce the height of the reactor; thus, for same volume of the reactor more plan area will be available. However, the minimum height of the reactor should be restricted to preferably 4.5 m.

Deflector baffles of sufficient overlap (0.1–0.2 m) should be provided below the aperture between the gas collectors (Fig. 12.9) from where wastewater enters in settling compartment to avoid entry of biogas. The diameter of the gas exhaust pipes should be sufficient to guarantee easy removal of the biogas from the gas collection dome, particularly in case of foaming.

12.7.2 Effluent Collection System

The effluent has to leave the UASB reactor via number of launders distributed over entire area, in between the domes, discharging to main launder provided at periphery of the reactor. The effluent launders are designed in such a way that the weir loading (m³/m day) should not exceed the design criteria of secondary settling tank (i.e.,185 m³/m day). The width of the individual effluent collection launders may be minimum 0.20 m to facilitate cleaning and maintenance. The depth of the launder is estimated using open channel flow design. Additional depth of 0.10–0.15 m shall be provided to facilitate free fall of treated wastewater in the launder. On both sides of the launder 'V' notches shall be provided at 15–20 cm centre to centre. When effluent launders are provided with scum baffles, the 'V' notches will be protected from clogging as the baffles retain the floating materials. A scum layer may form at the top of reactor and sludge accumulation can occur in the launder; hence, periodical cleaning of launders and removal of scum should be carried out.

12.7.3 Design of Feed Inlet System

It is important to establish optimum contact between the sludge available inside the UASB reactor and organic matter present in the wastewater, and to avoid channelling of the wastewater through the sludge bed. Hence, proper design of inlet distribution system is necessary. Depending on topography, pumping arrangement, and likelihood blocking of inlet pipes, one could provide either (i) gravity feed from top (preferred for wastewater with high suspended solids, Fig. 12.11), or (ii) pumped feed from bottom through manifold and laterals (preferred in case of soluble industrial wastewaters).

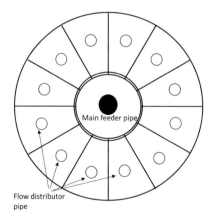

Fig. 12.11 Gravity fed inlet distribution system, showing typical arrangement made for dividing the flow in twelve parts, **a** schematics and **b** actual photograph of flow splitter box installed on the top of UASB reactor for distributing the flow at 12 different locations at the UASB reactor bottom (Courtesy IIT Kharagpur, India). The sewage is pumped through the central main feeder pipe over which deflector is provided to break jet action. From this central chamber through the V-notches provided at the periphery of this central chamber the inflow is getting distributed in twelve equal parts. This is taken to the reactor bottom and distributed at different location as indicated later in Fig. 12.14

The numbers of feed inlet points required in UASB reactor are recommended by Lettinga and Hulshoff (1991) for different concentration of the sludge present in sludge bed of the reactor and applicable organic loading rates. In general, the area to be served by each feed inlet point should be between 1 and 3 m^2. Lower area per inlet point (about 1 m^2) is provided for reactor designed to operate at lower OLR of about 1 kg COD/m^3 day, and higher area (2–3 m^2) per inlet point can be provided in the reactor designed for OLR greater than 2 kg COD/m^3 day.

Apart from the number of feed inlet points, the minimum and maximum outflow velocity through the feed inlet nozzles should also be given due consideration while designing. This outflow velocity through nozzles can be kept between 0.5 and 4.0 m/s. However, nozzle with diameter less than 20 mm shall be avoided due to possibility of clogging. The equation of 'condition for maximum power transfer through series of nozzles' shall be used for working out diameter of manifold (main feeder pipe) in case of pumped feeding. Number of laterals are attached to this main feeder line and nozzles are located at specified spacing on these laterals (Fig. 12.12). The clogging of the nozzles may represent serious problem resulting in uneven distribution of the wastewater over reactor bottom, particularly when treating partially soluble wastewater. Hence, arrangement should be provided for facilitating cleaning or flushing of the influent distribution system. The nozzles are provided on the laterals in downward directions with 30° to 45° angle with vertical axis of pipe, to avoid blockage in case the lateral is touching the floor of the reactor.

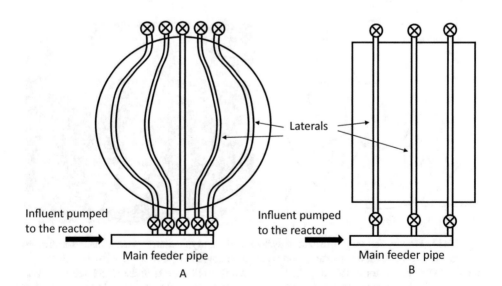

Fig. 12.12 Influent laterals and main feeder line arrangement for **a** circular and **b** rectangular UASB reactor

12.7.4 Other Requirements

It is necessary to keep provision for removal of excess sludge from the reactor at periodic time interval. Although, the excess sludge should be wasted from about middle height of the reactor during regular sludge wasting, it is also necessary to make arrangement at bottom of the reactor for withdrawal of sludge. In addition, 5 to 6 numbers of valves should be provided over reactor height to facilitate sampling of the sludge and wastewater for analysis. For treating high strength wastewater, it is recommended to apply effluent recycle, in order to dilute COD concentration of influent and to improve contact between sludge and wastewater. For treating wastewater with COD concentration greater than 4– 5 g/L, it is recommended to apply dilution during start-up, for proper granulation of biomass inside UASB reactor.

Other auxiliary equipment required to be installed are for addition of essential nutrients and alkalinity for control of pH of the influent as per the requirements. In addition, arrangement for measurement of influent pH, temperature, influent flow rate, and gas production rate should be made. It is necessary to provide pressure gauge on the gas collection dome and in biogas collection line fire arrester and foam arrester (Fig. 12.13) shall be provided as a safety measures. In the event of non-utilizing biogas, for some reason, an arrangement of the flare stack must be there to flare the biogas generated.

Example: 12.3 Design a UASB reactor for treatment of sewage having average influent flow (Q_{av}) of 3,000 m^3/day. The maximum hourly influent flow (Q_{max-h}) is 5,400 m^3/day

Fig. 12.13 Typical schematics of the foam arrester used in the biogas line before flare stack

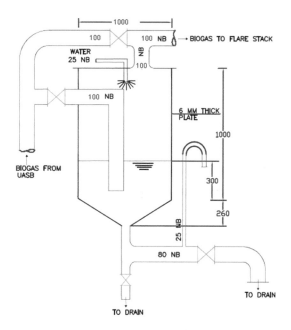

(225 m^3/h). The average influent COD (S$_o$) is 600 mg/L with average influent BOD of 350 mg/L. The least temperature of sewage reaches to 23 °C (average of the coldest month). Sludge yield coefficient (Y) shall be considered as 0.12 kg TSS/kg COD$_{rem}$. The solids yield coefficient, in terms of COD (Y$_{obs}$) can be considered as 0.17 kg COD$_{sludge}$/kg COD$_{rem}$. Expected concentration of the discharge sludge, C, is 4%. Consider the sludge density: $\gamma =$ 1,020 kg/m^3.

Solution

Calculation of the average influent COD load (L$_o$)

$\quad$ L$_o$ = S$_o$ × Q$_{av}$ = 0.600 kg/m^3 × 3,000 m^3/day = 1,800 kg COD/day

Calculation of reactor volume

Adopt a value for the hydraulic detention time, θ = 8.0 h

Hence, the total volume of the reactor (V)

V = Q$_{av}$ × θ = 125 m^3/h × 8 h = 1,000 m^3

Adopt the number of reactor modules, N = 2.

Although there is no limitation to the volume of the reactor, it is recommended that the reactor volume shall not exceed 1,500 m^3, due to construction and operational limitations, particularly with respect to influent distribution and GLS separator design. In the case of small systems for the treatment of domestic sewage, the adoption of modular reactors presents advantages. It is typical to use modules with volumes of about 400 to 500 m^3.

Hence, volume of each module (V$_u$)

V$_u$ = V/N = 1,000 m^3/2 = 500 m^3

Estimating reactor area

Adopt a height of the reactor H = 4.5 m

Hence, the area of each module (A)

A = V$_u$/H = 500 m^3/4.5 m = 111.1 m^2

Adopt rectangular reactors of 7.45 m × 15.00 m (A = 111.8 m^2)

Verification of the corrected area, volume, and detention time

Corrected total area: A$_t$ = N × A = 2 × 111.8 m^2 = 223.6 m^2

Corrected total volume: V$_t$ = A$_t$ × H = 223.6 m^2 × 4.5 m = 1,006 m^3

Corrected hydraulic detention time: θ = V$_t$/Q$_{av}$ = 1,006 m^3/ (125 m^3/h) = 8.0 h

Verification of the loading rates applied

Volumetric hydraulic load = Q/V = (3,000 m^3/day)/1,006 m^3 = 2.98 m^3/m^3·day

Volumetric organic load: L$_v$ = Q$_{av}$ × S$_o$/V = (3,000 m^3/day × 0.600 kg COD/m^3)/1006 m^3

= 1.79 kg COD/m^3·day

Considering sludge bed occupies half of the reactor volume (i.e., 503/2 = 251.5 m^3) with a sludge concentration of 20 g VSS/L.

Sludge loading rate = (3,000 m^3/day × 0.600 kg COD/m^3)/(251.5 × 20)

= 0.36 kg COD/kg VSS day

Hence, to keep SLR under control this reactor should be operated while under full flow condition with sludge bed volume of 50 to 60% of the total reactor volume.

Verification of the up-flow velocity

- for Q_{av}: $v = Q_{av}/A = (125 \text{ m}^3/\text{h})/223.6 \text{ m}^2 = 0.56 \text{ m/h}$

This is also equal to reactor height/HRT = 4.5/8 = 0.56 m/h

- for Q_{max-h}: $v = (225 \text{ m}^3/\text{h})/223.6 \text{ m}^2 = 1.01 \text{ m/h}$

Up-flow velocity should be in the range of 0.5 to 0.7 m/h at average flow (Q_{av}) and preferably less than 1.2 to 1.5 m/h at peak flow (Q_{peak}).

Design of inlet distribution system

Adopting an influence area of 2.25 m^2 per distribution point, number of feed inlet tubes (Fig. 12.14) required:

$N_d = A/A_d = 223 \text{ m}^2/2.25 \text{ m}^2 = 99$ distributors

Due to the necessary symmetry of the reactor, let us adopt 100 distributor points, as follows:

- along the length of each reactor (15.00 m): 10 tubes
- along the width of each reactor (7.45 m): 5 tubes

Thus, each UASB reactor module will have 50 (10 × 5) influent distribution pipes, each serving an influence area equivalent to: $A_d = 223.6 \text{ m}^2/100 = 2.24 \text{ m}^2$.

Estimation of biogas production

Biogas production, considering 70% COD removal

Fig. 12.14 Inlet gravity feed distribution system **a** photograph outside, **b** photograph inside a UASB reactor. (*Courtesy* IIT Kharagpur, India)

$$COD_{CH_4} = Q \times (S_0 - S) - Y_{obs} \times Q \times (S_0 - S)$$

$= 3{,}000$ m^3/day $\times (0.600 - 0.180$ kg COD/m$^3) - (0.17$ kg COD$_{sludge}$/kg COD$_{rem}$ $\times$ $3{,}000$ m^3/day $\times (0.600 - 0.180$ kg COD/m$^3) = 1260 - 214.2 = 1045.8$ kg COD/day

$K(t) = \frac{P \times K}{R \times (273 + T)}$ (1 atm $\times 64$ g COD/mol)/ [(0.08206 atm·l/mol·K $\times (273 + 23$ °C)]
$= 2.63$ kg COD/m^3

$V_{CH_4} = \frac{COD_{CH_4}}{K(t)} = (1045.8$ kg COD/day)/ $(2.63$ kg COD/m$^3) = 397.64$ m^3/day

Considering a methane content of 65% in the biogas

Biogas total volume $= V_{CH_4}/0.65 = (397.64$ m^3/day)/0.65 $= 611.75$ m^3/day

Design of GLS separator

Provide max liquid velocity at aperture, i.e., at inlet of the settler $= 3$ m/h

Area of opening at inlet of settler $= 3000/(3 \times 24) = 40.56$ m^2

Hence, for each reactor the area of opening at inlet of settler $= 40.56/2 = 20.28$ m^2

Provide width of opening at inlet of settler $= 0.3$ m, and width of the deflector beam below as 0.5 m

Hence, length of opening required will be $20.28/0.3 = 67.6$ m

Since dome are laid along the width, the number of domes required $= 67.6/7.45 = 9$

Thus, there will be two 0.15 m gaps between the reactor wall and dome and eight numbers of 0.3 m gaps between the domes

Thus, base width of each dome $= (15 - (9 \times 0.3))/9 = 1.367$ m

Provide a height of 1.2 m for each dome up to water level in the reactor and 0.30 m top width of dome at gas–water interface (Fig. 12.15).

Hence, angle of inclination $= \tan^{-1} \frac{1.2}{(1.367 - 0.30)/2} = 66°$

Check for surface overflow rate in settler

The area available for settling at the top of the reactor $= (7.45 \times 15) - (9 \times 0.3 \times 7.45)$
$= 91.635$ m^2

This calculation is based on the provision of GLS domes with MS plates, which will have minimum thickness of 5 mm to 10 mm including corrosion protection cover. If the

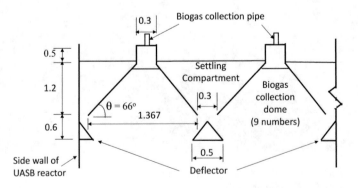

Fig. 12.15 Schematic of the GLS separator provided (not to the scale)

GLS is proposed to be constructed with RCC then wall thickness of GLS need to be considered in this calculation.

Hence, surface overflow rate at average flow in the settling compartment of each reactor $= 3000/(2 \times 91.635) = 16$ m^3/m^2 day (It is acceptable, as less than 20 m^3/m^2 day)
SOR at peak flow $= 5,400/(2 \times 91.635) = 29.46$ m^3/m^2 day
This is also acceptable as it is less than 36 m^3/m^2 day
Checking for gas loading rate
The biogas release rate should be under control to overcome a possible scum layer, but high enough to quickly release the gas from the sludge, not allowing the sludge to be dragged and consequently accumulated in the gas exit piping. A maximum rate of 3.0 m^3 biogas/m^2·h is recommended.

Area of gas–water interface inside the dome in each reactor $= 9 \times 0.30 \times 7.45 = 20.115$ m^2

Total biogas production each reactor $= 611.75/2 = 305.875$ m^3/day
Therefore, biogas loading at gas–water interface inside the dome $= \frac{305.875}{20.115 \times 24} = 0.63$ m^3/m^2 h
This is less than 3 m^3/m^2.h, hence it is acceptable
Design of effluent collection launders
Provide 10 effluent collection channels with a width of 0.3 m and depth of 0.3 in each reactor, provided with 90° V notch at 0.15 m c/c. The discharge to be collected in each effluent collection channel will be $= 1500/10 = 150$ m^3/day $= 0.00174$ m^3/sec.

Provide 50 mm drop in the bed of effluent collection channel on one side of the reactor where common effluent collection launder is provided. Thus, the slope of the channel is 1 in 149 m. Using the Manning's equation, the depth 'd' of the wastewater in each channel will be estimated.
$$Q = A \times v = 0.3 \times d \times \frac{1}{0.013} \times \left(\frac{0.3d}{2d+0.3}\right)^{\frac{2}{3}} \left(\frac{1}{149}\right)^{1/2} = 0.00174 \text{ m}^3/\text{sec}$$
Hence, depth of wastewater in each channel $= 0.016$ m, hence the total depth of 0.3 m provided for launder is acceptable. Using similar procedure, the common effluent launder, which will collect the discharge from these 10 launders, for each reactor can be designed for the flow rate of 1500 m^3 day for each reactor and dimensions of this common effluent launder shall be finalized.

12.8   Operation of Anaerobic Reactors

12.8.1 Start-Up of Anaerobic Reactors

The reduction of the period necessary for the start-up and improved operational control of the anaerobic processes are important factors to increase the efficiency and the competitiveness of the high-rate anaerobic systems. The start-up of the anaerobic reactors and their operation has been considered by technicians as a barrier, possibly due to bad experiences linked to the use of unsuitable operational strategies. Therefore, systematized operational procedures are very much important, mainly during the start-up of high-rate systems, notably in the case of UASB reactors, where self-immobilization of biomass is targeted to develop granular biomass in the reactor.

The start-up can be basically achieved in three different manners:

- By using seed sludge adapted to the wastewater to be treated: In this case the start-up of the reactor occurs fast, in a satisfactory way, as there is no need for acclimatization of the sludge, hence this duration required for adaptation will be saved leading to faster commissioning of the reactor.
- By using seed sludge not adapted to the wastewater to be treated: In this case, the start-up of the system goes through an acclimatization period, including a microbial selection phase. Hence, longer start-up time will be required here in this case and scrupulous control of the reactor operation is necessary particularly to avoid organic overloading to the reactor, so that appropriate microbial species can be developed in the reactor.
- With no use of seed sludge: This is considered the most unfavourable form to start-up the anaerobic reactor, however it is feasible only for the domestic sewage, which contains necessary microorganisms, though at much lower concentration than required. As the concentration of microorganisms in the wastewater is very less, the time required for the retention and selection of a large microbial mass can be very long time consuming and may take about 4 to 6 months for the reactor to reach full load condition.

As stated, the excess sludge from the existing anaerobic reactor treating the similar wastewater shall be treated as a preferred inoculum. Even in this case, if the amount of sludge available is lesser than the total sludge required for full load condition, then start-up duration will be required by initially controlling the inflow to the reactor to keep the OLR and SLR under control. Once the reactor performance stabilized at low flow rate, then flow rate can be increased in steps to reach full flow. When, sludge source for treatment of similar wastewater is not available then the anaerobic sludge from sewage treatment plant or septic tank after proper screening, to remove debris, shall be use as inoculum. In absence of even such source, pond bottom sediments or waste activated sludge can also work as inoculum, however longer start-up time will be required in this case.

12.8.2 Procedure Preceding the Start-Up of a Reactor

1. *Characterization of the seed sludge*: Once the use of seed sludge is defined for the start-up of the reactor, analyses should be carried out for its qualitative and quantitative characterization, such as total solids (TS), volatile solids (VS), and specific methanogenic activity (SMA). Besides the parameters referred to above, a visual and olfactory characterization of the sludge should be carried out.
2. *Characterization of the raw wastewater*: To establish the start-up routine of the anaerobic reactor, a qualitative and quantitative characterization campaign of the influent raw wastewater should be carried out.
3. *Estimation of the seed sludge volume necessary for the start-up of the reactor*: Based on the results of the characterization analyses of the sludge and the influent sewage/industrial wastewater, the seed sludge volume necessary for the start-up of the reactor can be estimated.

Typically, while starting up the reactor the available inoculum is filled in the reactor. Depending on the mass of VSS present in the reactor and the methanogenic activity the organic loading rate is controlled to the reactor during initial days. The organic loading rate of 1–3 kg COD/m^3 day and SLR of 0.1–0.3 kg COD/kg VSS day are recommended to be adopted during start-up (Ghangrekar et al., 1996). By controlling this loading, it will be ensured that the VFA accumulation in the reactor will not be excessive, to provide ample opportunity for the methanogens to grow in the sludge.

Once at this lower loading rates stable reactor performance is demonstrated, typically giving COD conversion in the range of 70–80%, depending on the type of wastewater being treated, the loading to the reactor is increased in steps to reach to the full load conditions. This step increase in loading is time consuming process due to slow growth rate of methanogens to come up to the expected fraction in the sludge hold-up within the reactor. Hence, it might take up to two to four months of time for the reactor to reach full load condition, which of course depends upon the type of inoculum used, strength of the wastewater, alkalinity generating capacity of the wastewater, alkalinity control exercised during start-up, design OLR and SLR of the reactor under full load conditions and operating temperature.

Example: 12.4 Estimate the amount of sludge necessary for the inoculation of a UASB reactors, if influent flowrate: $Q_{av} = 3,000$ m^3/day, sewage COD concentration: $S_o = 600$ mg COD/L (adopted as an average of the characterization campaign), concentration of volatile solids in the seed sludge: $C = 30,000$ mg VS/L (3%) (adopted as an average of the samples analysed), density of the seed sludge: $\gamma = 1,020$ kg/m^3, volume of the reactor: $V = 1,003.5$ m^3, sludge loading rate adopted during the start-up of the reactor: $L_s = 0.3$ kg COD/kg VS·day.

Solution

Applied organic load, $L_o = Q_{av} \times S_o = 3,000 \, m^3/d \times 0.600 \, kgCOD/m^3 = 1,800 \, kg \, COD/day$

Necessary seed sludge mass, $M_s = L_o/L_s = (1,800 \, kg \, COD/day)/(0.3 \, kg \, COD/kg \, VS \cdot day)$
$= 6,000 \, kg \, VS$

Resulting seed sludge volume, $V_s = P_s/(\gamma \times C_s) = (6,000 \, kg \, VS)/(1,020 \, kg/m^3 \times 0.03)$
$= 196 \, m^3$

As the necessary seed sludge volume is relatively high ($196 \, m^3$), equivalent to approximately 32 tank trucks, the possibility of not applying the total organic load can be evaluated, diverting (by-passing) part of the influent sewage to the post treatment unit of the treatment plant during the initial days of the reactor start-up.

Possible alternatives for inoculation of the reactor can be evaluated:

- For application of 100% of the influent flowrate, considering a sludge with a concentration of volatile solids equal to 3%, a seed sludge volume equal to approximately $200 \, m^3$ is necessary.
- For application of 50% of the influent flow, considering a sludge with a concentration of volatile solids equal to 5%, a seed sludge volume equal to approximately $60 \, m^3$ is necessary.

12.8.3 Procedure During the Start-Up of an Anaerobic Reactor

Inoculation of the reactor: The inoculation can be done with the reactor either full or empty, although the inoculation is preferable with the reactor half full, to reduce sludge losses during the transfer process. For this second situation, the following procedures can be adopted:

- Transfer the seed sludge to the reactor, ensuring that it is discharged at the bottom of the reactor. Avoid turbulence and excessive contact with air.
- After filling the inoculum add the wastewater so as to make a water level of about 0.5–1 m above the inoculum sludge settled level so as to minimize contact with air. Leave the sludge at rest for an approximate period of 12–24 h, allowing its gradual adaptation to local temperature.
- Fill the reactor with wastewater and leave the reactor unfed for next 24 h period. At the end of this period, and prior to beginning the next feeding, collect supernatant samples from the reactor and analyse temperature, pH, alkalinity, volatile acids and COD. Should these parameters be within acceptable ranges, continue the feeding process. Acceptable values: pH between 6.8 and 7.4 and volatile acids below 200 mg/L

(as acetic acid). For wastewaters having lesser alkalinity or alkalinity liberation capacity, external supplementation alkalinity is necessary during start-up to keep pH under control. This alkalinity should be preferably added in calcium bicarbonate form.

- Operate the reactor under control flow rate for another 24 h duration. At the end of this period, collect new samples for analyses and proceed as previously stated.
- If the parameters analysed are within the established ranges, feed the reactor continuously under control flow and loading rate condition. In accordance with the amount of seed sludge used and the flow percentage to be applied need to be decided depending on the applicable SLR values.
- Implement and perform a routine monitoring of the treatment process performance.
- Increase the influent flow gradually in steps, initially after 2–3 weeks, in accordance to the system response and allowing it to stabilize at each loading rate. This interval can be either increased or reduced, depending on the results obtained.

12.9 Monitoring of the Treatment Process

For the monitoring of the treatment process, the sample collection and the routine physical–chemical parameters to be analysed should be defined during the start-up period. An example of a monitoring programme for start-up of UASB reactors is presented in Table 12.2.

12.10 Troubleshooting of the Process

Set of information that can help to detect and correct operational problems in anaerobic reactors is presented in Table 12.3.

Questions

12.1 Explain the reactions involved in the anaerobic treatment of wastewater while the organic matter is finally removed from the wastewater.

12.2 Describe different types of microorganisms and their role in anaerobic digestion of organic matter to final end product.

12.3 Describe advantages of anaerobic treatment and describe drawback of anaerobic process for treatment of low strength wastewaters.

12.4 Discuss the factors that can affect the anaerobic reactor performance adversely.

12.5 Describe different types of reactors used for anaerobic treatment of wastewaters.

12.6 What is high-rate anaerobic process? Name different high-rate anaerobic reactors.

12.7 With the help of schematic explain UASB reactor and its working.

Table. 12.2   Monitoring program of UASB reactor

	Parameter	Unit	Monitoring points and frequency				
			Inlet	Sludge bed	Effluent	Gas outlet	Sludge outlet
Treatment efficiency	Settleable Solids	mg/L	Daily	–	Daily	–	–
	TSS	mg/L	Alternate days	–	Alternate days	–	–
	COD	mg/L	Alternate days	–	Alternate days	–	–
	BOD	mg/L	Weekly	–	Weekly	–	–
	Biogas production	m^3/day	–	–	–	Daily	–
Operational stability	Temperature	°C	Daily	Daily	–	–	–
	pH	–	Daily	Daily	–	–	–
	Alkalinity	mg/L	Alternate days	–	Alternate days	–	–
	VFA	mg/L	Alternate days	–	Alternate days	–	–
	Biogas composition	$\%CH_4$	–	–	–	Weekly	–
Sludge quality	TS	mg/L	Analysed at various points along the height of the bed and sludge blanket (3 to 6 points), to obtain the profile and the mass of solids inside the reactor				Monthly
	VS	mg/L					Monthly
	SMA	g COD/g VS day					Alternate week
	SVI	mL/g					Monthly

12.8   What is GLS separator? What are the design guidelines for GLS separator?

12.9   Describe advantages of sludge granulation in UASB reactor.

12.10   Why post treatment is necessary for anaerobic reactor effluents? What post treatment you will recommend after treatment of sewage in UASB reactor?

12.11   Describe organic loading rates used for design of UASB reactor. What is the role of up-flow velocity for proper functioning of UASB reactor and how it is controlled?

12.12   Design a UASB reactor for treatment of 5 MLD of sewage having COD of 500 mg/L and BOD of 260 mg/L. Make suitable assumptions for the design.

12.13   Describe feed distribution arrangement used in UASB reactor. What is relation of OLR with number of feed inlet points in the reactor?

12.14   Write a short note on start-up of the anaerobic reactors.

Table. 12.3 Performance troubleshooting of UASB reactor or any other anaerobic reactor

Observation	Probable cause	Verify	Solution
Distribution lateral does not receive sewage	Blockage	Physical verification	Unblock
High level of settleable solids in the effluent	Excessive hydraulic load	Flow	Reduce flow
	Excessive solids in the reactor	Sludge mass	Waste the excess sludge
Gas production lower than normal	Biogas leakage	Gas collection	Eliminate leakage
	Defective gas meter	Gas meter	Repair or replace
	Reduced flow	Influent flow rate	Unblock feeder pipes
	Toxic material in the influent	SMA	Identify and act on the toxic source
	Excessive organic load	SMA	Reduce organic load
Sludge production higher than normal	Overloaded sludge,	Sludge stability	Reduce applied load
	Coarse and/or inorganic solids entering the reactor	Pre-treatment Operation (screen, grit chamber)	Re-establish operation of the pre-treatment units
Sludge production lower than normal	Small flow	Influent flow rate	Unblock sewers, clean screen and feeder lines
	Deficient sludge retention	Settleable solids in the effluent	Repair separator, if required, or clean the separator for removal of any sludge settled on the slopping base
Floating sludge	Excessive hydraulic load, reduced temperature	Organic and hydraulic loads	Reduce organic and hydraulic loading, heating, control flow
Reduced efficiency in the removal of organic matter	Excessive load	Loading rate and pH	Reduce organic loading rate, supplement alkalinity if required to stabilize pH

12.15 What are the different sources of inoculum sludge that can be used for commissioning of the anaerobic reactor?

12.16 When it is necessary to have multilevel gas collection domes in UASB reactor? With a schematic describe the multilevel GLS separator used in UASB reactor.

12.17 What auxiliary equipment are required on biogas collection line of anaerobic reactor?

12.18 What is the role of provision of foam trap in the biogas collection line, before flare stack, in anaerobic reactor?

12.19 What are the possible substances that can cause toxicity and lead to reduction of performance of anaerobic reactor?

12.20 When gravity feeding and pumped feeding is used in UASB reactor?

References

Alrawashdeh, K. A., Bkoor, Gul, E., Yang, Q., Yang, H., Bartocci, P., & Fantozzi, F. (2020). Effect of heavy metals in the performance of anaerobic digestion of olive mill waste. *Processes, 8*(9), 1629.

Bhunia, P., & Ghangrekar, M. M. (2008). Effects of cationic polymer on performance of UASB reactors treating low strength wastewater. *Bioresource Technology, 99*, 350–358.

Bhunia, P., & Ghangrekar, M. M. (2007). Required minimum granule size in UASB reactor and characteristics variation with size. *Bioresource Technology, 98*, 994–999.

Bodík, I., Herdová, B., & Drtil, M. (2002). The use of up flow anaerobic filter and AnSBR for wastewater treatment at ambient temperature. *Water Research, 36*(4), 1084–1088.

Buntner, D., Sánchez, A., & Garrido, J. M. (2013). Feasibility of combined UASB and MBR system in dairy wastewater treatment at ambient temperatures. *Chemical Engineering Journal, 230*, 475–481.

Calderón, K., González-Martínez, A., Gómez-Silván, C., Osorio, F., Rodelas, B., & González-López, J. (2013). Archaeal diversity in biofilm technologies applied to treat urban and industrial wastewater: Recent advances and future prospects. *International Journal of Molecular Sciences, 14*, 18572–18598. https://doi.org/10.3390/ijms140918572

Chatterjee, P., Ghangrekar, M. M., & Rao, S. (2018). Sludge granulation in an UASB–moving bed biofilm hybrid reactor for efficient organic matter removal and nitrogen removal in biofilm reactor. *Environmental Technology (united Kingdom), 39*, 298–307.

Chen, H., He, L.-L., Liu, A.-N., Guo, Q., Zhang, Z.-Z., & Jin, R.-C. (2015). Start-up of granule-based denitrifying reactors with multiple magnesium supplementation strategies. *Journal of Environmental Management, 155*, 204–211.

Chen, S. J., Li, C. T., & Shieh, W. K. (1988). Anaerobic fluidized bed treatment of a tannery wastewater. *Chemical Engineering Research and Design, 66*, 518–523.

Chinnaraj, S., & Rao, G. V. (2006). Implementation of an UASB anaerobic digester at bagasse-based pulp and paper industry. *Biomass and Bioenergy, 30*(3), 273–277.

Fang, H. H. P., & Hui, H. H. (1994). Effect of heavy metals on the methanogenic activity of starch-degrading granules. *Biotechnology Letters, 16*, 1091–1096.

Ghangrekar, M. M., Asolekar, S. R., Ranganathan, K. R., & Joshi, S. G. (1996). Experience with UASB reactor start-up under different operating conditions. *Water Science and Technology, 34*, 421–428.

Ghangrekar, M. M., Kahalekar, U. J., & Takalkar, S. V. (2003). Design of upelow anaerobic sludge blanket reactor for treatment of organic wastewaters. *Indian Journal of Environmental Health, 45*, 121–132.

Goyal, S., Seth, R., & Handa, B. (1996). Diphasic fixed-film biomethanation of distillery spentwash. *Bioresource Technology, 56*(2–3), 239–244.

Harper, S. R., & Pohland, F. G. (1986). Recent development in hydrogen management during anaerobic biological wastewater treatment. *Biotechnology and Bioengineering, 28*, 585–602.

Hulshoff Pol, L.W., de Castro Lopes, S.I., Lettinga, G., & Lens, P.N.L. (2004). Anaerobic sludge granulation. *Water Research, 38*, 1376–1389.

Jou, C.-J.G., & Huang, G.-C. (2003). A pilot study for oil refinery wastewater treatment using a fixed-film bioreactor. *Advances in Environmental Research, 7*(2), 463–469.

Van den Berg, L., & Lentz, C. P. (1979). Comparison between up- and downflow anaerobic fixed film reactors of varying surface-to-volume ratios for the treatment of bean blanching waste. In *Proceedings of the 34th Purdue industrial wastes conferences,* Purdue University, pp. 319–325.

Li, X., & Sung, S. (2015). Development of the combined nitritation–anammox process in an upflow anaerobic sludge blanket (UASB) reactor with anammox granules. *Chemical Engineering Journal, 281*, 837–843.

Lim, S. J., & Kim, T.-H. (2014). Applicability and trends of anaerobic granular sludge treatment processes. *Biomass and Bioenergy, 60*, 189–202.

Lin, C.-Y., & Chen, C.-C. (1997). Toxicity-resistance of sludge biogranules to heavy metals. *Biotechnology Letters, 19*, 557–560.

Lin Chou, W., Speece, R. E., & Siddiqi, R.H. (1979). Acclimation and degradation of petrochemical wastewater components by methane fermentation. In *Biotechnol. Bioeng. Symp.*, Symp. No.8, pp. 391–414.

Liu, Y., Xu, H.-L., Yang, S.-F., & Tay, J.-H. (2003). Mechanisms and models for anaerobic granulation in upflow anaerobic sludge blanket reactor. *Water Research, 37*, 661–673.

Najafpour, G. D., Zinatizadeh, A. A. L., Mohamed, A. R., Hasnain Isa, M., & Nasrollahzadeh, H. (2006). High-rate anaerobic digestion of palm oil mill effluent in an upflow anaerobic sludge-fixed film bioreactor. *Process Biochemistry, 41*, 370–379.

Omil, F., Garrido, J. M., Arrojo, B., & Méndez, R. (2003). Anaerobic filter reactor performance for the treatment of complex dairy wastewater at industrial scale. *Water Research, 37*(17), 4099–4108.

Radjenovic, J., Petrovic, M., & Barceló, D. (2007). Analysis of pharmaceuticals in wastewater and removal using a membrane bioreactor. *Analytical and Bioanalytical Chemistry, 387*(4), 1365–1377.

Sarti, A., Fernandes, B. S., Zaiat, M., & Foresti, E. (2007). Anaerobic sequencing batch reactors in pilot-scale for domestic sewage treatment. *Desalination, 216*(1–3), 174–182.

Shao, X., Peng, D., Teng, Z., & Ju, X. (2008). Treatment of brewery wastewater using anaerobic sequencing batch reactor (ASBR). *Bioresource Technology, 99*(8), 3182–3186.

Speece, R. E. (1983). Anaerobic biotechnology for industrial wastewater treatment. *Environmental Science and Technology, 17*, 416a–427a.

Sokol, W. (2001). Operating Parameters for a gas-liquid-solid fluidized bed bioreactor with a low-density biomass support. *Biochemical Engineering Journal, 8*, 203–212.

Timur, H., & Özturk, I. (1999). Anaerobic sequencing batch reactor treatment of landfill leachate. *Water Research, 33*(15), 3225–3230.

Xing, B.-S., Guo, Q., Yang, G.-F., Zhang, Z.-Z., Li, P., Guo, L.-X., & Jin, R.-C. (2015). The properties of anaerobic ammonium oxidation (anammox) granules: Roles of ambient temperature, salinity and calcium concentration. *Separation and Purification Technology, 147*, 311–318.

Zhang, L., Hendrickx, T. L. G., Kampman, C., Temmink, H., & Zeeman, G. (2013). Co-digestion to support low temperature anaerobic pretreatment of municipal sewage in a UASB–digester. *Bioresource Technology, 148*, 560–566.

Ponds and Wetlands for Treatment of Wastewater

13

13.1   Introduction

Treatment of wastewater through different types of aerobic technologies, such as activated sludge process, sequential batch reactor, aerobic filter, etc., require energy in the form of electricity for aeration, wastewater pumping and sludge or wastewater recycling, etc. These reactors also demand substantial cost for their operation and maintenance. Hence, nature-based technologies are developed for the treatment of wastewater without employing sophisticated machines; thus, these treatment systems can be operated without dependency on electricity and continuous attention is not (or less) required from the operators for these nature-based treatment systems so as to reduce the operational cost.

In the natural environment, the organic matter present in the wastewater is degraded by different microorganisms and plant species adopting the biochemical process that occurs within the cells during their cellular metabolisms. The biochemical oxidation of organic matter in the natural systems is facilitated through healthy coordination of symbiotic relations between the intra and inter-species organisms. However, the natural systems are often disturbed by numerous physical and environmental factors, such as a change in weather, temperature, water-level, etc., that constrain the rate of biological degradation of organic matters. Thus, a selected land area is engineered by mimicking the natural-environment that provides enough volume for holding the influent wastewater, where microorganisms and plant species are allowed to grow, thus to achieve the biochemical oxidation of organic matters and removal of other pollutants present in the wastewater. Also, the presence of plant species in the natural systems helps in removal of nutrients (N and P) and trace metals from the wastewaters. The nature-based engineered treatment systems are categorized as pond system and constructed wetland depending on the design and types of microorganisms that are grown with/without selective water-loving plants.

M. Ghangrekar, *Wastewater to Water*, https://doi.org/10.1007/978-981-19-4048-4_13

13.2   Pond Systems

Pond systems are artificially made on land generally with shallow depth and clay bottom and side embankment made with soil, open to air and light that can accommodate the volume of the wastewater to be treated for desired duration. The symbiotic growth of algae and bacteria occurs in the pond that facilitates the degradation of organic matters. In the pond systems the wastewater is treated by maintaining a high retention time, from few days to few weeks. The wastewater treatment performance of the pond depends on the design and the mode of microbial respiration; hence, accordingly the ponds are categorized in to three types, i.e., (i) aerobic pond or oxidation pond, (ii) facultative waste stabilization pond, and (iii) anaerobic lagoons.

13.2.1 Oxidation Pond

Oxidation ponds are having a large surface area and shallow depth. These are designed to treat wastewater by employing the naturally grown microorganisms, such as bacteria and algae (Fig. 13.1). In the presence of sunlight, algae start to accomplish photosynthesis in the presence of water and carbon dioxide, while releasing substantial amount of oxygen. This photosynthetic oxygen released results in increasing the dissolve oxygen (DO) in the wastewater that is getting treated in the pond and supports oxidation of organic matter by aerobic bacteria. The aerobic bacteria oxidize the organic compounds present in the wastewater and release simplified nutrients and carbon dioxide, which are utilized by algae for synthesis of algal cells.

The soluble nitrogen and phosphorus compounds present in the wastewater that are not suitably removed by bacterial metabolism, prominently consumed by algae for its cellular growth. Thus, establishing an algae-bacteria symbiotic relationship in the oxidation pond. There are many forms of algae growing in wastewater treatment ponds. The two most

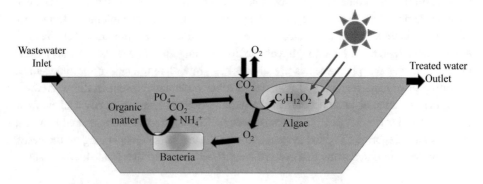

Fig. 13.1 Shematic of oxidation pond system

common types are green and blue-green algae. Green algae, which give the green colour to the ponds, predominate when pond conditions and wastewater treatment performance are good. Blue-green algae, such as *Spirulina*, *Merismopedia* and *Anacystis* are filamentous and indicate poorer pond operating conditions, such as high organic loading, low DO, low nutrients, and warm water conditions. They often form unsightly and odorous mats. The algal species like *Chlorella*, *Scenedesmus*, *Euglena*, *Chlamydomonas*, *Actinastrum*, *Pediastrum*, *Hydrodictyon*, and *Ankistrodesmus* are typically found to be commonly present in the oxidation ponds.

The bacterial sludge produced from the aerobic oxidation of organic matter in the oxidation pond gets settled down at the pond bottom, which is required to be periodically removed through mechanical means to restore the volume of the pond for holding the wastewater. Similarly, the suspended algal biomass in the oxidation pond also settled down upon death of the cells or forming aggregation with bacterial cells. Oxidation ponds are kept with a depth of 0.15–0.45 m, when only nitrogen removal is a target, otherwise a depth of 0.5–1.2 m is generally preferred for the treatment of wastewater like sewage. The main idea of maintaining a low-depth in the oxidation pond, is to allow the penetration of sun light through the entire depth of the pond that could lead to photosynthesis and release of oxygen, thus an aerobic environment can be maintained in the entire pond.

An oxidation pond is called as a 'maturation pond', when the depth is kept below 0.5 m and secondary treated wastewater with a very low-concentration of organic matter is applied in the pond to offer tertiary treatment. In the maturation pond, the growth of algae is higher than the bacteria. Due to low oxygen demand of organic matter in these ponds, these maturation ponds produce surplus oxygen during photosynthesis that is released in atmosphere during day time.

13.2.2 Facultative Stabilization Ponds

All the ponds present in natural conditions are generally facultative in nature, where both aerobic and anoxic respiration as well as anaerobic digestion is established through different types of microorganisms. These are also named as 'Facultative waste stabilization ponds' or simply 'Waste stabilization ponds'. In facultative pond also the syntrophic association of algae and bacterial is present at the upper layer. Since in this case, the depth of the pond is more than the depth provided for oxidation pond, three distinct zones exist in the pond (Fig. 13.2). In the top zone, an aerobic zone is created due to the photosynthetic action of algae that combined with aerobic bacteria to facilitate the aerobic oxidation of organic matters present in the wastewater. The bottom zone is typically operated under anaerobic condition, since no photosynthetic activity is taking place at this depth. In this bottom zone, anaerobic bacteria digest the suspended organic matter present in the wastewater which gets settled along with dead algal and bacterial biomass in this zone. The middle zone, between the top aerobic and bottom anaerobic zone, is developed as a

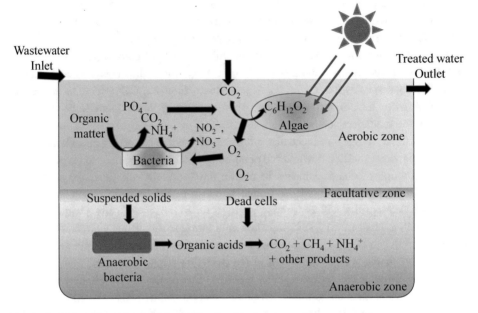

Fig. 13.2 Schematic of facultative stabilization pond

facultative zone, where the dissolve oxygen available is less since maximum amount of oxygen is consumed by the top-aerobic zone for the oxidation of organic matters. Hence, in the facultative zone both aerobic and facultative bacteria are present and play the key role for the degradation of organic matters.

Presence of aerobic and anaerobic treatment processes is very crucial for the treatment of some wastewaters, such as dyes, which are degraded better when both anaerobic and aerobic treatment is provided (Verma et al., 2017). The nuisance associated with the anaerobic reaction is eliminated due to the presence of top aerobic zone. Maintenance of aerobic zone in the top of the facultative pond is highly indispensable, and it depends on solar radiation, wastewater characteristics, BOD loading and operational temperature. If properly operated with controlled organic loading and under favourable environmental conditions, the performance of these ponds can be comparable with conventional aerobic wastewater treatment processes.

13.2.3 Anaerobic Lagoons

In anaerobic lagoons, also named as anaerobic ponds, the entire depth is under anaerobic condition except an extremely shallow top layer of the pond, if exposed to atmosphere. Normally these ponds are used in series and these are followed by facultative or aerobic

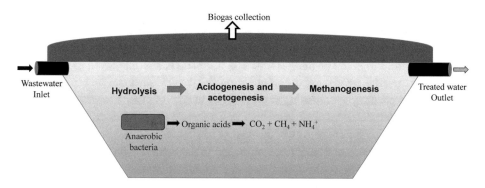

Fig. 13.3 Schematic of anaerobic lagoon

pond for complete treatment of wastewater. Depth of these ponds is in the range of 2.5–6 m. They are generally used for the treatment of high strength industrial wastewaters, like distillery effluent, and sometimes for municipal wastewater and sludges. Depending upon the strength of the wastewater, longer retention time of up to 50 days or more is maintained in the anaerobic ponds. Anaerobic lagoons are covered these days by polymer bladder for biogas recovery and eliminating smell problem and also to eliminate the greenhouse gas emission in atmosphere (Fig. 13.3). The effluent from the anaerobic lagoons is further treated with aerobic ponds to stabilize the physiochemical properties, such as pH, organic matter concentration, and pathogens, before final discharge of the treated effluent in the water bodies.

13.2.4 Other Types of Ponds

13.2.4.1 Fish Pond

Fish pond can be a part of maturation pond or altogether separate pond, in which fish are reared. Sometimes, fish are also reared in the end compartment of primary oxidation pond. The phytoplankton rapidly grow on the pond surface in the presence of sun light, while consuming the oxidation products of the bacteria and algae symbiosis. Fish can directly eat the phytoplankton biomass developed in the aerobic ponds. Different types of fish are cultured in the fish pond receiving wastewater, e.g., *Cyprinus carpio, Catla catlax, Cirrhina mrigala, Labeo rohita, Hypophthalmichthys molitrix, Aristichthys nobilis* including genus *Tilapia*.

The effect of these fish, grown in wastewater, on human health is matter of concern due to the presence of pathogens, and the accumulation of other contaminates, such as heavy metals in the fish tissue (bioaccumulation). Hence, after the growth of fish in wastewater fed aquaculture, these are incubated in a clean-fresh water ponds for at least two weeks, which facilitates in eliminating the residual unpleasant odours and pathogens and make the fish acceptable for human consumption. Hence, culturing fish in wastewater and sell-ing these fishes can be a good source of revenue generation. Though there is a risk of causing health issues in humans; however, many countries are using treated wastewater for culturing the fish in the ponds systems. The fish pond is also employed to monitor the treated water quality by observing the effect on fish health. Fish can be cultured in the last cell of the ponds system treating wastewater or in case of other treatment technology, the treated wastewater is used to grow fish in a separate pond. This experiment of fish growth in the treated water can certify the quality of treated water in terms of harmfulness to aquatic life, and thus, can safely be discharged to the natural water bodies.

13.2.4.2 Aquatic Plant Ponds

The aquatic plant ponds are the types of ponds, similar to oxidation pond, where instead of algae aquatic plant species, e.g., water hyacinths, duckweeds, etc., are grown to remove contaminates present in the wastewater. The main purpose for the aquatic pond is that the aquatic plants float on the surface and these are capable of absorbing heavy metals from the wastewater through meshy root structure, while also consuming nitrogen and phos-phorus from the wastewater. Thus, the aquatic plant ponds are preferred for treatment of the secondary treated wastewater. The growth of these aquatic plants on water sur-face develops floating root systems, providing a high surface area, that provides dissolve oxygen to the water bodies and also facilitates the enrichment of the aerobic bacteria on the root system, thus providing a healthy plant-microbes interaction in the aquatic plant ponds. Therefore, in the aquatic plant ponds the removal of contaminants, such as heavy metal and some level of organic matter oxidation, occur via the plants-aerobic bacteria symbiosis.

The excess biomass of aquatic plants produced in the pond is easy to harvest and this biomass can be used as the source for production of bioenergy, such as biodiesel, bioethanol, combustible gas or as animal feed stock depending on quality of biomass and pollutants present in the wastewater. The advantages of using macrophyte in aquatic plant ponds is that they can be easily harvested from the pond compared to the harvesting of the algal biomass in case of oxidation ponds. Also, presence of top cover of this macrophyte reduces the algal biomass concentration in the pond and can be effective in controlling algae concentration in the effluent. However, due to lower dissolved oxygen present in this aquatic pond, higher organic matter concentration cannot be handled by these ponds, hence these are only suitable for tertiary treatment of the wastewater.

13.2.4.3 High-Rate Algal Ponds

The high-rate algal pond (HRAP), apart from effective in nutrient removal, is potentially an effective disinfection mechanism within the requirements of imparting sustainability to sewage treatment. The nutrient removal mechanisms are also active in the HRAP, specifically those involved in the removal of phosphate. These ponds are not generally designed for optimum purification efficiency; however, they are designed for maximum algal production. The harvested algae from these ponds can be utilized for a variety of uses, principally being high quality algal protein.

The ponds are shallow lagoons of 0.20–0.50 m deep with a wastewater retention period of 1–3 days. The whole pond is kept aerobic by maintaining a high algal concentration and using some form of mechanical mixing. Mixing is normally carried out for short periods at night to prevent the formation of a sludge layer. Mixing may also be required for short periods during the day to prevent a rise in pH in the surface water due to photosynthesis. The pond is commissioned in the same way as a facultative pond, except that continuous loading should not be permitted until an algal bloom has developed. The organic loading rate applied on the HRAP depends on solar radiation, and the average loading throughout the year could be 100–200 kg BOD/ha.day. Sewage with high organic matter concentration inhibits the photosynthetic action due to high ammonia concentrations, which results in the pond becoming anaerobic.

High-rate algal ponds are designed to promote the symbiosis between the microalgae and aerobic bacteria, each utilizing the major metabolic products of the other. Microalgae grow profusely releasing oxygen from water by photosynthesis. This oxygen is made available to bacteria to oxidize most of the soluble and biodegradable organic matter remaining in the wastewater. The HRAPs are shallower than oxidation ponds and operate at shorter hydraulic retention time. At the rapid growth of algae, the pH can raise to above 9.0, since at peak algal activity carbonate and bicarbonate ions react to provide more carbon dioxide for the algae, leaving an excess of hydroxyl ions. A pH above 9.0 for 24 h ensures a 100% killing of E. coli and presumably most of the pathogenic bacteria present in sewage.

Among open pond systems, high rate algal pond (HRAP), the most reminiscent aerobic pond is fabricated based on the principles of raceway pond system, where paddle wheels (Fig. 13.4) are usually employed for proper mixing, which prevents the precipitation of algal biomass (Whitton et al., 2015). Additionally, to decrease the head loss, this type of pond is designed with a lower length to width ratio of channels along with fewer bends. In this regard, Acien et al. reported that the ratio of length and width of channels ranging between 10 to 20 and the water channel depth ranging from 0.2–0.4 m is well accepted to allow maximum sunlight to penetrate for higher growth of algae in HRAP (Acién et al., 2017). Lower water depth with proper mixing could be recommended to operate HRAP for the enhancement of light penetration, stability, and the growth of biomass.

13.2.4.4 Primary and Secondary Ponds

Ponds receiving untreated wastewaters are referred to as raw or primary waste stabilization ponds. Those receiving primary treated or biologically treated wastewaters for further treatment are called as secondary waste stabilization ponds. The maturation pond is a secondary pond receiving already treated sewage either from the ponds or other biological wastewater treatment process, like UASB reactor or ASP. The detention time of 5–7 days is provided in maturation ponds, with the main purpose of achieving natural bacterial

Fig. 13.4 Photograph of
HRAP treating effluent of
UASB reactor employed for
sewage treatment (Courtesy:
IIT Kharagpur, India)

die-off to desired levels. In warm climate these maturation ponds often constitute an economical alternative for chlorination. They are lightly loaded in terms of organic loading
and the oxygen generated by photosynthesis is generally more than the oxygen demand.

13.3   Other Features of Ponds

13.3.1 Modes of Operation of Ponds

A waste stabilization pond system is the assembly of number of individual pond cells, and
their mode of wastewater feeding is followed in to basic pathways, i.e., either in series
or in parallel. In case of series mode, the wastewater is allowed to flow through each cell
(pond) in sequence (e.g., 1st cell to 2nd cell and then to 3rd (finishing cell) as shown
in Fig. 13.5a. The series mode of operation of waste stabilization ponds facilitates the
highest quality of treatment for wastewater and minimizes the algae in the final effluent.
In parallel mode of operation of ponds, the influent wastewater is divided between two or
more primary ponds (Fig. 13.5b). Parallel operation is adopted to evenly distribute high
organic loading so that the high-concentrated wastewater is mixed with low concentrated
wastewater present in each pond. Thus, the concentration of organic matter and nutrients is
reduced in each pond to keep it under acceptable limits that helps in improving biological
oxidation of organic matters; hence, a better treated water quality can be obtained.

The arrangement of ponds is kept in series or parallel and the wastewater is allowed to
flow in the ponds in a particular sequence by maintaining a desired retention time in each
pond. The flow diagram is illustrated in Fig. 13.6. Thus, ponds are operated as primary
and secondary ponds depending on the organic matter concentration in the wastewater
received. For example, the pond systems where raw wastewater with high organic load is
fed are operated as primary waste stabilizing pond; whereas the secondary pond, receiving
secondary treated wastewater from the primary pond with less concentration of organic
matter, is used for the further treatment, hence operated as secondary waste stabilization
pond. A maturation pond shall be used as last pond with detention time of 5–7 days to kill

(a) Ponds in Series

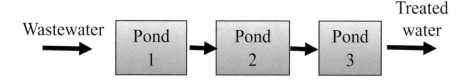

(b) Ponds in Parallel connection

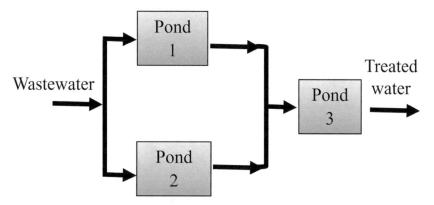

Fig. 13.5 Series and parallel arrangements for operation of ponds

the pathogenic bacteria and to get the *Escherichia coli* count to an acceptable limit. These pathogens are killed in the presence of hydroxyl radicals, generated in the water during the photosynthesis of the algal cells in presence of sunlight. The pond systems provide a low-cost alternative to the chlorination or other types of costly technologies used for the disinfection of municipal wastewaters.

13.3.2 Components of a Pond System

The stabilizing pond is designed with different check points at different location of the pond for the precise operation (Fig. 13.7). These checkpoints are described below.

i. *Headworks and screening*: Screen is used to remove plastics, rags and large objects before wastewater enters the waste stabilization pond.

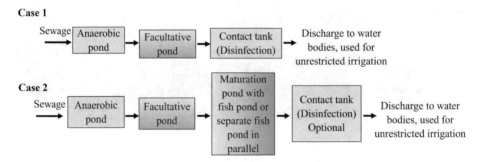

Fig. 13.6 Schematics of the flow chart of waste stabilization ponds

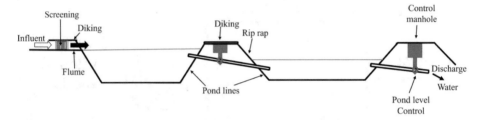

Fig. 13.7 Schematics of pond operational parts

ii. *Flow meter and weirs*: These devices are used to measure influent or effluent discharge of wastewater.

iii. *Dikes*: These are the earthen pond sides, which give the pond its shape and depth.

iv. *Rip rap*: Stone pitching is placed at the operating water levels of the waste stabilization pond in order to prevent erosion of the inner slope of earthen embankment that could occur from wind actions.

v. *Pond Liner*: Generally, a liner with clay or synthetic polymer is provided at the bottom and side wall of the pond before filling with wastewater, which will avoid wastewater from seeping into the soil. In absence of such liner and where at the pond site the soil is permeable, the wastewater from the pond will meet the groundwater due to seepage and it will lead to severe groundwater pollution. Hence, provision of clay or polymer liner is utmost important for the ponds.

vi. *Control structures*: Provision of influent flow control, effluent discharge control and pond water level control is necessary. The wastewater is filled in the pond to a certain level marked in meter/feet at the different depths in the pond. The filling of wastewater in the pond is controlled through manholes, located at the centre of embankment, using boards or valves by inserting or removing (in case of boards) or by turning left or right (in case of valves) that cause the rise or lowering of the wastewater level in the pond (Fig. 13.7). A flume is made up of concrete structure like an open channel for transferring wastewater to the pond.

13.3.3 Operation and Maintenance of Waste Stabilization Pond

13.3.3.1 Significance of Soil Type for Pond-Based Treatment

Soil is categorized as sand, silt or clay that directly determines the permeability. For example, the particle size of sand is largest, thus the sandy soils are very permeable. Therefore, in case of sandy soil the wastewater will infiltrate at a faster rate compared to the silt and clay soil, which are having smaller particle sizes, hence prevents seepage of wastewater to the groundwater. Thus, the silt and clay types of soils are better option for construction of pond-based wastewater treatment system. If the soil is permeable, clay lining or polymer lining should be given to the pond bottom and inner face of soil embankment to avoid wastewater seepage. The sludge is deposited at the bottom of the pond over the years and it occupy the working volume of the pond, thus this needs to be monitored on a regular basis. The excess sludge should be periodically removed from the pond to restore pond working volume.

13.3.3.2 Monitoring Groundwater Movement

Groundwater flows from a higher elevation to a lower elevation and the flow rate is very slow of few inches to a few feet per day depending on the permeability of the soil and also on the location, where the ground water is traveling through. It is highly essential to monitor the seepage of wastewater, hence transport of contaminants or organic matter from pond to the groundwater, because it will pollute groundwater and make it unsuitable for use. Natural bacterial degradation of organic matter from groundwater will not occur, since microbial activity in soil can be found only up to a depth of only few meters below the ground level (maximum up to 10 m). Pollutant once entered in groundwater will remain in it for longer duration, hence extreme care is required while employing pond system for not allowing wastewater seepage through ponds. The water samples from wells located at hydraulically up gradient and down gradient to the pond are used to monitor the possibility of groundwater contamination by comparing the components present in both the water samples. The groundwater sample analysis shall be done once in a year for this monitoring.

13.3.3.3 Algal Growth and Control in Stabilization Ponds

The growth of algae in waste stabilization pond is essential to maintain the aerobic environment, however excess proliferation of algae causes algal bloom that develops 'pea soup' and heavily pollute the treated water and thus increase the total suspended solids (TSS) at the outlet. The growth of algae is found higher during summer, when higher intensity of sunlight that stays for longer period during the day is available. In addition, the growth of algae can be induced with higher influent nutrient (N and P) concentration and organic matter concentration. Therefore, removal or inhibition of excess algal biomass in the ponds is a matter of concern.

Few important steps are required to be followed for controlling the algal growth in ponds. The volume of pond and the wastewater level should be maintained in such a way that there is no overload in the pond. The detention time for wastewater in stabilization pond should be kept high, such that maximum removal of organic matter should occur during this time period so that a very low concentrations of nutrients and organic matter remain in the last stage of pond, which could inhibit the algae growth and also reduce the excess algal biomass in final outlet. Along with this, a few strategies can be adopted for controlling the excess algae growth in the pond, for example application of barley straw, alum (settle down the algal cells), or the last polishing pond can be an aquatic macrophyte pond with floating plant covering pond surface or covered with polyethylene in order to avoid the penetration of sunlight that will inhibit the growth of microalgae.

13.3.4 Advantages and Disadvantages of Pond Systems

The operation and maintenance of waste stabilization ponds is easy as compared to other wastewater treatment systems. The advantages and disadvantages of the ponds are described in Table 13.1.

13.4   Factors Affecting Performance of Ponds

13.4.1 Seasonal Changes and Environmental Factors

The environmental factors, such as light, temperature and wind direction and velocity, can directly affect the treatment efficiency of waste stabilization ponds and these environmental factors are modulated with the changes in seasons. Wastewater treatment efficiency of the pond becomes lesser during the winter due to the fall in the temperature that reduces the microbial activity in the pond. The days become shorter in winter; thus, the availability of sunlight is also decreased that limits the photosynthesis of algae and reduce dissolved oxygen that is required for oxidation of organic matter by aerobic bacteria. During spring, the temperature and the sunlight become sufficient in the pond that facilitate the growth of algae and bacteria.

Summer is the best time for the growth of algae and bacteria; however, the excess growth of algae could cause serious problems, like formation of algal bloom in pond, hence proper operational strategy is pivotal. During summer, the rate of evaporation is highest due to higher solar radiation and heat. Thus, net evaporation (evaporation minus

Table 13.1 Advantages and disadvantages in pond treatment systems

Advantages	Disadvantages
• Low construction cost	• Large land requirements
• Low operational and maintenance cost	• Chances of seepage of wastewater from the pond bottom and side walls to groundwater, hence in this case risk of groundwater pollution
• Low energy or electricity usage	• Climatic conditions, such as sunlight, temperature, wind, etc., affect wastewater treatment performance
• Can accept surge loadings	• Suspended solids (algae bloom) removal is a big problem
• Low chemical usage	• Possible odour problems for the nearby environments
• Fewer mechanical problems	• Animal problems (muskrats, turtles, etc.)
• Easy operation and no requirement of skilled operators	• Vegetation problems (rooted weeds, duckweed, algae), excess vegetation in the pond can prevent the penetration of sunlight required for photosynthesis and reduce the dissolve oxygen
• No continuous sludge handling	• Though sludge removal is periodic, it is labour intensive and costly

rainfall) must be taken into account during the design of facultative and maturation ponds. Similarly, during the monsoon seasons, in the countries like India, the rainfall can also influence the wastewater treatment efficiency of the ponds. The growth of algae and bacteria is also affected due to the fluctuations in the temperature and availability of sun light during the monsoon, and also the excess storm water entry in the pond can dilute the organic matter and nutrients present in the pond, affecting performance of it.

13.4.2 Effect of Photosynthesis Activity

Photosynthesis is a biochemical reaction in algae and plants, where chlorophyll acts as the reaction centre that produce carbohydrate and oxygen by using carbon dioxide and water in the presence of sunlight. This carbohydrate is used in algae and plants for production of energy molecule like adenosine triphosphate (ATP) and cell growth and maintenance. Photosynthesis assists in increasing DO in the wastewater present in the pond that is utilized by aerobic bacteria for the oxidation of organic matter present in the wastewater. During day time, the intensity of solar radiation is highest and the algae in the pond perform photosynthesis and consume maximum carbon dioxide while releasing oxygen.

However, at night bacteria and algae both consume oxygen and release carbon dioxide that react with water and form carbonic acid (H_2CO_3), thus this will lower the pH of wastewater in the pond. Thus, the pH of the pond water drops at night and opposite is true during day time. During long sunny days of summer, the pH of the pond can rise above 9.0 in the presence of algae. This fluctuation of pH in the pond also affects the efficiency of metabolic activity of bacterial cells, hence affects the overall treatment efficiency of the pond.

13.4.3 Other Factors to Be Considered for Designing Ponds

Wastewater treatment performance of pond depends on environmental and biological factors as discussed earlier, thus some basic principles are required to be considered for designing pond systems (Arceivala & Asolekar, 2007):

i.   Fluctuation in physio-chemical and biological characteristics of inlet wastewater,
ii. Environmental factors, such as solar radiation, sky clearance, temperature, wind and their variation,
iii. Algal growth pattern and their fluctuation during diurnal and seasonal variation,
iv. Bacterial growth pattern and decay rates,
v.   Hydraulic transport pattern of wastewater in the pond,
vi. Evaporation and seepage from the pond,
vii. Solids settlement, liquefaction, gasification, upward diffusion, and sludge accumulation,
viii. Gas transfer at air-wastewater interface.

Generally, about 6% of the visible light energy received will be converted to the algal energy. The calorific value of the algal biomass can be considered as 6 kcal/g. The visible light radiation which can penetrate the water surface will be limited to 4000–7000 Å (Arceivala & Asolekar, 2007). Based on the data obtained from the weather department on the sky clearance factor, the average radiation received can be estimated using Eq. 13.1.

$$\text{Average radiation} = \text{Min.radiation} + [(\text{max.radiation} - \text{min.radiation}) \times \text{Sky clearance factor}] \qquad (13.1)$$

Example: 13.1 Considering the energy capture efficiency of 6%, determine theoretical algal production per hectare of pond area located at 20 °N, where max. and min. radiation received are 182 and 120 cal/cm^2.day. If the BOD removal efficiency of 85% is expected from this pond determine the BOD_u loading that can be applied on the pond. Consider sky clearance factor of 75%.

Solution

Average radiation received $= 120 + [(182 - 120) \times 0.75] = 166.5$ cal/cm^2.day

Thus, per hectare the radiation received will be $= 166.5 \times 10^8$ cal/ha.day.

Thus, average algal production $= 166.5 \times 10^8 \times 0.06/(6000 \times 10^3) = 166.5$ kg algae/ha.day.

On weight basis, the oxygen production of the algae is about 1.3 times the algal biomass produced.

Hence, for ensuring the aerobic condition the ultimate BOD applied to the pond should not exceed the oxygen production due to photosynthesis, hence the ultimate BOD that can be applied to the pond for 85% efficiency will be:

$= 166.5 \times 1.3/0.85 = 254.65$ kg BOD/ha.day **(Answer)**

13.5 Terminologies Used for Defining Performance of Pond Systems

13.5.1 Hydraulic Loading Rate

The hydraulic loading rate is the inflow of wastewater per unit area of the pond considering that the wastewater is evenly spread over an entire area of the pond. Thus, hydraulic loading rate is calculated by dividing the influent flow rate by total area of the pond and expressed as m^3/acre.day or m^3/ha.day.

13.5.2 Organic Loading Rate

In case of ponds, the rate at which the BOD load is applied to the pond is expressed as mass of BOD ultimate applied per unit area of the pond per day, since performance of the oxidation pond is majorly correlated with photosynthetic activity that is related to the surface area of pond. This organic loading rate on the ponds is expressed as kg BOD applied/ha.day. This loading for oxidation pond needs to be lower than the photosynthetic oxygen yield to ensure sufficient oxygen yield is available for maintaining aerobic condition in the pond. For facultative waste stabilization ponds the BOD load applied to the pond is more than photosynthetic oxygen yield as part of the organic matter undergo anaerobic digestion in these ponds.

13.5.3 Nitrogen and Phosphorous Removal

Nitrogen removal in the pond occurs firstly following uptake of the nitrogen for synthesis of algal cells and subsequent washout of the algae along with effluent, secondly the loss through seepage, if there is any, and third responsible reaction is denitrification. Denitrification is the principal mechanism responsible for major removal, while loss due to washout of algae in the effluent may account for only 6–9% nitrogen contained by weight in algae. Denitrification is supported more in ponds operated under warmer climates. Nitrogen removal in ponds can be enhanced by adopting lighter organic loading on the ponds, by providing increased area of interface between pond water and the sediment settled at the base of pond and providing ponds cells in series rather than the single pond with equivalent large area.

The phosphorous content in the pond is majorly in the form of inorganic phosphate, though in small fraction organic form of phosphorous is also present. The phosphorous is majorly removed from the pond through precipitation as calcium hydroxyapatite (Ca_{10} $(PO_4)_6$ $(OH)_2$). Precipitation of phosphate is dependent on operating pH of the pond. Phosphorus removal is promoted by algal growth, which raises pH value to above 8.2. At this alkaline pH, phosphate will get precipitated and in warmer climate more phosphorous removal shall be obtained. Also, when ponds are operated in series the P removal is more in the later stage of the ponds, where pH value will be typically higher and anaerobicity will be absent. In addition to precipitation, removal of phosphorous also occur from the pond along with the algal cell washout through effluent. However, algae contain only about 1% phosphorus by weight; hence, removal through washout of cell along with effluent would only account for removal of about 1 mg/L of phosphorous.

13.6   Design Guidelines for Oxidation Ponds

Depth of Pond: The depth of the oxidation pond is kept in the range between 1.0 m and 1.5 m so that sun light can easily penetrate till the bottom of pond that encourages the algal growth and maintain dissolve oxygen throughout the pond. When these ponds are used for sewage treatment the primary objective is organic matter removal and a depth of 1.0–1.2 m is used. Shallow ponds experience higher temperature variation than deeper ponds, hence an optimum pond depth is necessary. When the depth, worked out based on the area required and detention time, is greater than 1.5 m, the depth of the pond should be restricted to 1.5 m and area of the pond should be increased to satisfy necessary detention time.

Surface area of Pond: The photosynthesis and production of dissolve oxygen in pond is directly proportional to the active surface area that receives the sun light. The dissolve oxygen available in the pond should be sufficient for satisfying the BOD of the wastewater being treated in the pond. This photosynthetic oxygen yield depends upon latitude of the location of the pond, its elevation above the mean sea level, sky clearance and season. The recommended loading rate on the pond are provided in Table 13.2 (IS:5611: 1987). These loading are inclusive of the BOD associated with suspended fraction of organic matter present in the sewage.

Table 13.2 Expected photosynthetic oxygen yield, indicating BOD loading rate that can be applied to the pond per unit area at different latitudes

Latitude (^{0}N)	Applicable BOD loading rate (kg/ha.day)
8	325
12	300
16	275
20	250
24	225
28	200
32	175
36	150

This photosynthetic oxygen yield is estimated at sea level and considering that for 75% of the time in a year sky is clear. For ponds located at higher elevations, the applicable BOD loading rate should be divided by a factor $(1 + 0.003EL)$, where EL is the elevation of the pond above mean sea level in hundred meters. At a location where sky is clear for less than 75% days in a year then for every 10% decrease in sky clearance a 3% increase in pond area should be adopted. The individual pond area shall not be kept more than 0.5 ha and hence for higher area required, multiple ponds should be provided that are operated in series. Ponds operated in series with three to five cells give very satisfactory results. Beyond five ponds in series have no further benefits, hence in such cases when multiple ponds are required 3–5 ponds can be kept in series and such parallel series can be provided as per the number of ponds required. The size of the pond required will be half when plug flow pattern is maintained, which is possible in series operation, rather than completely mixed conditions. Thus, providing ponds in series is preferred. A 25% more area will be required than that calculated to account for side embankments of the ponds.

Substrate removal rate: Substrate removal rate (Kp) depends on the temperature. The extent of rate organic matter oxidation in the pond, i.e., BOD removal rate, Kp per unit time, can be determined from evaluating performance of existing pond or from laboratory scale pond unit by monitoring the influent and effluent substrate concentrations, S_o and S, respectively, the flow rate Q, the pond volume V, and the operating temperature. The slope of plot of $(S_o-S)Q/V$ versus S will provide the value of Kp. Based on the pond data analysed for various locations, such as from USA, Canada, Thailand, India and elsewhere, for domestic sewage after taking actual dispersion conditions and loadings into account, Arceivala (1980) gave the Eq. 13.2 for estimating the value of Kp.

$$Kp(20\,^\circ C) = 0.132\log(BOD_u \text{ load }) - 0.169 \tag{13.2}$$

where, BOD_u load is expressed as kg of ultimate BOD applied to the pond per hectare per day. In practice, depending on the pond BOD_u load varying from 100 to 300 kg BOD_u/ha.day the values of Kp at 20 °C are found to be varying from 0.10 to 0.15 per day in most cases. For aerobic ponds with lower depth the Kp value could be up to 0.20 per day at higher temperature of 25 °C (Arceivala & Asolekar, 2007). The equation for calculating Kp for any other temperature at any location is presented in Eq. 13.3.

$$Kp\,(T\,°C) = Kp\,(20\,°C)\,(1.035)^{(T-20)} \tag{13.3}$$

Usually, two to three ponds are provided in series rather than the single pond. Since in nature the ponds will have arbitrary flow through pattern, rather than ideal mixing; hence, Wehner and Wilhelm (1958) equation (Eq. 13.4) is used for determining performance of such ponds having hydraulic regime between completely mixed and plug flow.

$$\frac{S}{S_o} = \frac{4ae^{1/2d}}{(1+a)^2 e^{a/2d} - (1-a)^2 e^{-a/2d}} \tag{13.4}$$

where, S and S_o are the effluent and influent substrate concentration, respectively; $a = \sqrt{1 + 4ktd}$; d is dispersion number (=D/UL); D is axial dispersion, m²/h; U is fluid velocity, m/h; L is the length of the reactor, k is first order reaction rate constant, h⁻¹; and t is time in hours.

Thirumurthi (1969) developed the graphical representation of Eq. 13.4 (Fig. 13.8), where the removal efficiency is plotted against the dimensionless number $Kp.t$ for dispersion number varying from zero for ideal plug flow reactor to infinity for a completely mixed reactor. For most of the stabilization ponds the dispersion number is within the range of 0.1–2.0.

Detention time (t): It can be varied depending on the organic loading rate, photosynthetic oxygen yield at that location and the treatment efficiency of the pond required. Basically, the value of '*t*' is optimized, such that it is adequate enough for stabilizing the organic load to an acceptable level to produce effluent with acceptable BOD value.

Sulphide production: The sulphide production in an oxidation pond can be estimated using Eq. 13.5 (Arceivala & Asolekar, 2007).

$$S^{2-}\left(\frac{mg}{l}\right) = (0.0001058 \times BOD_5 - 0.001655 \times t + 0.0553) \times SO_4^{2-} \tag{13.5}$$

where, BOD_5 is a five day BOD loading rate in kg/ha.day, t is detention time in days, and $SO_4{}^{2-}$ is expressed in mg/L.

It should be noted that the sulphide ion concentration should be within 4 mg/L, otherwise beyond this concentration it can inhibit the algal growth in the pond.

Coliform removal: Coliform count in the effluent from the pond is an important parameter and it should be below 1000/100 mL if the treated water is to discharge for irrigation purpose. The first order reaction rate equation for coliform removal is presented in Eq. 13.6 (Arceivala & Asolekar, 2007):

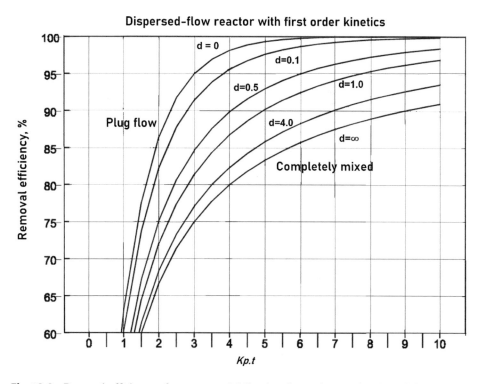

Fig. 13.8. Removal efficiency of a compound following first-order reaction (e.g., BOD), for the main hydraulic model given by Eq. 13.4

$$\frac{dN}{dt} = K_d \cdot N \tag{13.6}$$

where, N represents the number of coliform organisms at any given time (t), and K_d represents the death rate per unit time (1–1.2 per day at 20 °C).

Sludge accumulation: The excess biomass or sludge is settled at the bottom of the pond at the rate 0.05–0.08 m^3/capita.year, and occupy the working volume of the pond. Therefore, the desludging, i.e., removing the sludge deposited at the bottom of the pond, should be carried out every 6–10 years to restore the capacity. In case of ponds operated in series, the maximum deposition of the sludge will occur in the primary pond, i.e., first pond in the series.

Pre-treatment: Medium screens and grit removal devices should be provided before the ponds to avoid entry of the floating matters and inert grit matter deposition in the pond. However, in practice only screen is provided and provision of grit chamber is eliminated to avoid removal of grit from the grit chamber every day, which will require additional manpower attention. Hence, when grit chamber is not provided both primary

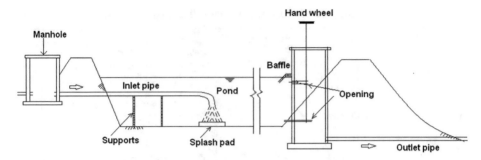

Fig. 13.9 Sectional elevation of waste stabilization pond

and secondary treatment is occurring in the pond. In this case, grit constitutes only about 5–10% of the accumulated sludge volume in the pond.

Other requirements: If influent is getting discharged in pond through pipe, then a bell mouth at its end shall be provided, which is discharging the wastewater near the inlet end at centre of width of the pond (Fig. 13.9). Precaution is necessary to handle the overflow situation in pond during seasonal changes and it is generally controlled through multiple valve draw-off lines provided at different depths of the pond. The soil type at site is important and especially in case of sandy soil, which is more pervious, proper sealing is required to prevent the seepage of wastewater from the pond. As discussed earlier either clay or polymer lining shall be provided to eliminate seepage.

Example: 13.2 Design an oxidation pond for following plug flow conditions to treat the sewage generated from 40,000 persons with per capita water demand of 150 LPCD. The sewage has average BOD concentration of 150 mg/L and the pond should generate effluent with BOD less than or equal to 30 mg/L. The proposed pond is located at a place where the latitude is 20° N and elevation is 260 m above MSL. The sky clearance factor at that place is 65%. The mean monthly temperature of the wastewater is 32 °C maximum and 18 °C minimum. Consider the pond removal rate constant has a value of 0.20 per day at 20 °C. For the pond consider dispersion number of 0.2 considering operation under plug flow condition and for BOD removal of 80% the value of $Kp.t$ is 2.0, as obtained from Wehner and Wilhelm equation plotted versus substrate remaining for various dispersion factor (Fig. 13.8).

Solution

Oxygen production by photosynthesis at 20^0 N $= 250$ kg/ha.day.

Correcting it for elevation of 260 m, the elevation correction $= 1 + 0.003 \times 2.6 = 1.0078$.

Hence, applicable BOD loading rate $= 250/1.0078 = 248$ kg BOD/ha.day.

Correction for sky clearance, since it is only 65% clear, hence to increase the area required for the pond by 3%, the loading applied will be corrected to

$$248/1.03 = 240.84 \text{ kg BOD/ha.day}$$

Estimation of the BOD load of the sewage and pond area.
Considering 80% of the water supply will result in sewage generation.
BOD load generated per day $= 40{,}000 \times 150 \times 0.8 \times 150 \times 10^{-6} = 720 \text{ kg BOD/day}$.
Effluent BOD is 30 mg/L, hence the BOD load in the effluent

$$40000 \times 150 \times 0.8 \times 30 \times 10^{-6} = 144 \text{ kg BOD/day}$$

Hence BOD load to be removed in the pond $= 720 - 144 = 576 \text{ kg BOD/day}$.
Ultimate BOD load $= 576/0.68 = 847.06 \text{ kg BOD}_u$ per day.
Hence, pond area required $= 847.06/240.84 = 3.52$ ha.
Considering 25% of the extra area required for side embankments, the total area required for the pond $= 3.52 \times 1.25 = 4.396 \text{ ha} \sim 4.4$ ha.
Estimating retention time and pond volume required.
Substrate removal rate, K_p at 20 °C is 0.20 day^{-1}, hence rate constant at 18 °C will be:
$K_{18} = K_{20} (1.035)^{(T-20)} = 0.20 \times 1.035^{-2} = 0.187$ per day.
As the oxidation pond is to be designed for dispersion number of 0.2 for which at 80% removal the value of $Kp.t$ given is 2.0.
Hence, detention time, $t = K_p t/K_p = 2.0/0.187 = 10.695$ days.
Thus, volume of pond $=$ flow of wastewater $\times$ detention time.
$= 40{,}000 \times 150 \times 0.8 \times 10^{-3} \times 10.695 = 51{,}336 \text{ m}^3$.
Hence, depth of pond $= 51{,}336/(3.52 \times 10^4) = 1.46$ m.
For total area of 4.4 ha provide 8 number of ponds with four operated in series and two such series in parallel, with each pond having area of 0.55 ha. Thus, dimension of each pond will be length $= 110$ m, breadth $= 50$ m, and total depth of the pond with free board $= 1.46 + 0.7 = 2.16$ m.
The ponds each of area of 0.55 ha can also be arranged as $4 + 2 + 2$, thus four ponds working as a primary pond and then in second and third stage provided with two ponds each. Thus, two primary ponds will deliver treated wastewater in a secondary pond and this secondary pond will deliver effluent in a third stage secondary pond.

13.7   Design of Facultative Stabilization Pond

In design of facultative stabilization ponds the oxygen resources of the pond are equated to the part (nearly half) of the applied organic loading. The principal source of oxygen in the pond water is photosynthesis and that is dependent on solar energy received at that location. The solar energy again is related to geographical, meteorological and astronomical phenomenon, and it varies principally with time in year and the altitude of the place. The photosynthetic oxygen yield at different latitude is given earlier in Table 13.1.

Yield of photosynthetic oxygen can be estimated directly if the amount of solar energy in calorie/m^2.day and the efficiency of conversion of light energy to fix energy in the form of algal cells are known.

The settleable organic solids present in the influent wastewater will settle at the bottom of the pond and also bacterial cells and algal biomass present in the upper aerobic zone of the pond will get settled upon end of the life or due to aggregation of the biomass improving its settling. This settled fraction of organic matter undergoes anaerobic decomposition in the facultative stabilization ponds. Thus, while designing the facultative ponds part of the organic matter (about 50%) is considered to undergo anaerobic decomposition and the photosynthetic oxygen yield is equated to the remaining organic matter to support aerobic oxidation. The organic matter loading in kg of BOD per hectare per day that is applied on the pond can be estimated using Eq. 13.7, which is obtained by expressing total BOD load applied (i.e., flow × BOD) to the pond divided by the area of the pond in hectares.

$$Lo = 10(d/t)BOD_u \tag{13.7}$$

where, Lo is the organic loading in kg BOD_u/ha.day; d is the depth of pond in m; t represents detention time in days; and BOD_u is ultimate soluble BOD expressed in mg/L.

In this case also the corrections for elevation and sky clearance need to be applied to estimate correct value of photosynthetic oxygen yield at that location of the pond. Thus, the organic loading shall be modified for elevations above MSL by dividing with a factor $(1 + 0.003\ EL)$. Where EL is elevation of pond site above MSL in hundred meters. In addition, for every 10% decrease in the sky clearance factor below 75%, the pond area shall be increased by 3%.

Example: 13.3 Design a facultative stabilization pond for treatment sewage generated from 40,000 persons with per capita water demand of 150 LPCD. The average BOD of the raw sewage is 150 mg/L and the treated effluen have BOD less than or equal to 30 mg/L. The facultative pond is located at a place having latitude 20° N and 260 m above mean sea level and at this location the sky is clear for 65% days in a year.

Solution

$$\text{Sewage flow} = 40000 \times 150 \times 0.8 \times 10^{-3} = 4800\ \text{m}^3/\text{day}$$

At 20° N, oxygen production by photosynthesis = 250 kg/ha.day.
Since the place is 260 m above mean sea level, the oxygen yield

$$= 250/(1 + 0.003 \times 2.6) = 248.06\ \text{kg/ha.day}$$

Applying sky clearance correction for 10% decrease in sky clearance below 75%, hence the applicable BOD load

$$= 248/1.03 = 240.84 \text{ kg BOD/ha.day}$$

Now, $BOD_5 = BOD_u$ $(1 - e^{-k.t})$ and considering, $k = 0.23$ day^{-1}.
$BOD_u = BOD_5/(1 - e^{-k.t}) = 150/(1 - e^{-0.23*5}) = 219.5$ mg/L.
The applicable BOD_u loading on the pond can be estimated as per Eq. 13.7.
Hence, $L_o = 10$ (d/t) $BOD_u = 10$ (d/t) $\times 219.5$
Assuming 50% of this BOD_u load is non settleable, and it will undergo aerobic decomposition in the top layer of the pond.
The oxygen requirement $= 10$ (d/t) $\times 219.5 \times 0.5$
Equating this to photosynthetic oxygen yield of 240.84 kg/ha.day.
$240.84 = 10$ (d/t) $\times 219.5 \times 0.5$
Therefore, d/t $= 0.219$.
Provide d $= 1.5$ m, hence t $= 1.5/0.219 = 6.835$ days.
Now, volume $=$ flow $\times$ detention time $=$ depth $\times$ surface area.
Therefore, surface area required $= 4800 \times 6.835 /1.5 = 21,872$ $m^2 = 2.18$ ha.
Provide four cells of the facultative stabilization ponds with each having area of 5,468 m^2 with two ponds operated in series and two parallel series.
Considering the area requirement for the side embankments, 25% more area will have to be provided for these ponds.

Example: 13.4 An algal raceway pond, located at a place where the latitude is 20° N and the yield of photosynthetic oxygen is 250 kg/ha.day. The temperature of the influent wastewater is 28 °C. The rate of wastewater flow is 0.8 MLD. The influent BOD is 180 mg/L and the effluent BOD required is less than 30 mg/L, thus targeting 85% of treatment efficiency. Determine the removal rate of the substrate. Also find out the area and depth of the pond. Consider the raceway pond was designed for plug flow conditions with $K_p t = 2.5$ and $K_p = 0.152$ per day.

Solution
Using, $K_p(T°C) = K_p(20°C) \times 1.035^{(T-20)}$.
    Therefore, substrate removal rate (K_p) at 28 °C $= 0.20$/day.
    As the pond was designed for plug flow conditions,
    Therefore, detention time $(t) = \frac{K_p t}{K_p} = 2.5/0.2 = 12.5$ day.
    The flow rate of wastewater $(Q) = 0.8$ MLD $= 800$ m^3/day.
    Therefore, the volume of raceway pond $= t \times Q = 800 \times 12.5 = 10,000$ m^3.
    As the yield of photosynthetic oxygen produced by algae during photosynthesis $= 250$ kg/ha.day at 20° N latitude.

Therefore, the maximum BOD loading on the raceway pond $= 250/0.85 =$ 294.11 kg/ha.day (as the efficiency of BOD removal for the pond is 85%).

Influent ultimate BOD $=$ Influent $BOD_5 \times (1/0.68) \times$ Flow rate

$$= 1.47 \times 180 \times 800 = 211.76 \text{ kg BOD/day}$$

Hence, the area of the raceway pond $= 211.76/294.11 = 0.72$ ha (7200 m^2).

Depth of the pond $=$ Pond volume/area of the pond $= 10,000/7200 = 1.39$ m.

Hence, including free board a total depth of 2.0 m shall be provided.

13.8 Constructed Wetland Systems

Natural wetlands are the lands, saturated or submerged with water, either surface or ground water, saline or fresh water, where selective water-loving plants grow. For example, marshes, the edge of a lake or ocean, the delta at the mouth of a river, and low-lying areas that are frequently flooding with water up to a certain height that promotes the growth of plants are considered as natural wetlands. In a wetland, plants grow and perform photosynthesis on their green leaves using water, carbon dioxide and sunlight, which inevitably release oxygen as a by-product. The oxygen produced in leaves is diffused towards the root tips of the plant and finally released to the water surrounded by the root systems, where it encourages the growth of aerobic microorganisms coupled with stabilization of organic matters available in the wetland. Thus, an aerobic-anaerobic zone is established in the wetland spreading from the plant root zone (rich in oxygen) towards the bottom of the wetland (poor or absence of oxygen), which accommodates the growth of numerous microorganisms that eventually involved in the degradation of organic matters and nitrification–denitrification process. Notably, wetland stabilizes organic matters in a cost-effective and natural manner; thus, based on these principles, artificial wetland systems are built for the treatment of wastewaters, which are referred as constructed wetlands (CWs).

The CWs are built using soil, gravels, sands, selective plants and sources of active microorganisms (generally cow dung or activated sludge). In CWs, the aerobic zone around the root system promotes the growth of aerobic bacteria and these bacteria participate in the oxidation of complex organic matters in to simpler forms that can be easily taken up by plants through root systems; thus, a healthy plant-microbes symbiotic interaction is established in the CW. The bottom zone of the CW, where dissolve oxygen gradually becomes low or negligible, an anaerobic zone is developed where anaerobic bacteria are enriched and the organic matter present in wastewater reaching this zone undergo anaerobic digestion.

In CW systems, wastewater admitted passes through the wetland bed either vertically or horizontally through the bed media (combination of soil, coarse sand and gravels) and reaches to the outlet end, where it is collected by means of water level control arrangement. As the wastewater flows from inlet to outlet through the bed media in CW, initially it passes through plant roots zone, where most of the organic matter is oxidized in the presence of aerobic bacteria and the nutrients and trace metals present in the wastewater are being removed due to uptake by the plants. The remaining organic matter fraction present in the wastewater reaches at the bottom of CW, where it is degraded by the anoxic and anaerobic bacteria, and a high-quality treated water is possible to be produced at the outlet of the CW.

The treated effluent from the CW can be used for irrigation or discharged in to the water bodies. Application of CW for treatment of wastewater is an economical and nature-based remedy; hence, it has been in practice from last 2 to 3 decades. The CWs are generally used for treatment of sanitary sewage to reduce the organic load, nitrogen and phosphorus. More importantly the treatment of wastewater in CW is achieved without any consumption of electricity in very less operating cost as compared to conventional treatment processes. Also, the operation and maintenance of CW does not require skilled manpower or technicians, making it further economical.

13.8.1 Classification of Constructed Wetlands

The CWs are classified into two basic categories, either surface flow constructed wetland (SF-CW) or sub-surface flow constructed wetlands (SSF-CW) based on the water level maintained in the CW (Fig. 13.10). Based on the water flow pattern in CW, these are further categorized as horizontal flow constructed wetlands, vertical flow constructed wetlands and up-flow constructed wetlands (Fig. 13.11). The selection of the plant species for application in CWs is solely dependent upon the design of CW, water level and the water flow pattern maintained in the CWs.

The flow pattern of wastewater in CWs has certain impact on the treatment efficiency and other factors, e.g., in case of SSF-CW, it prevents the mosquito breeding and odour problem compared to SF-CW. However, the construction cost for SSF-CW is higher compared to SF-CW, and also there is more chances of clogging in case of SSF-CW compared to SF-CW during continuous operation, which again add on extra cost for replacement and maintenance of the bed media used in SSF-CW. To overcome such issues in CWs, hybrid CWs systems are developed with combination of both surface and subsurface flow wetlands, with vertical, horizontal and up-flow pattern in CWs, depending upon the requirements. For example, in a combination of SF-CW and SSF-CW, the SF-CW is employed first that is fed with wastewater having higher organic matter concentration and the treated effluent from SF-CW with reduced organic loading is then fed to SSF-CW that can minimize the clogging in the filter media of the SSF-CW and also achieves a high-quality treated water at the final outlet.

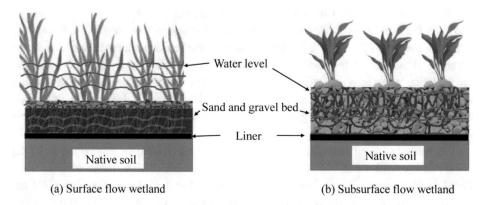

(a) Surface flow wetland   (b) Subsurface flow wetland

Fig. 13.10 **a** Surface flow wetland: water level is above the ground surface; vegetation is rooted and emerges above the water surface. In this the water flow is primarily above ground. **b** Subsurface flow wetland: water level is below ground; water flow is through a sand or gravel penetrating up to the bottom of the bed

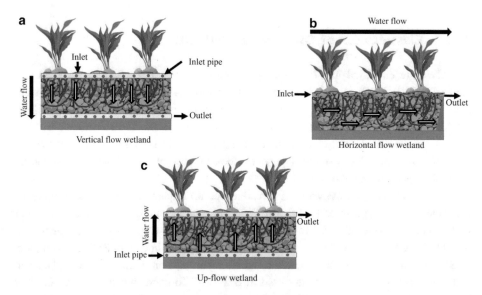

Fig. 13.11 Schematics of subsurface flow wetland with three different water flow patterns: **a** vertical water flow **b** horizontal water flow **C** up-flow constructed wetland

13.8.1.1 Surface Flow Constructed Wetlands

Surface flow constructed wetlands, also known as free surface water constructed wetlands (FSW-CWs), holds submerged and floating plants in open water area. In this type of wetlands wastewater flows in long and narrow channels, safeguarding near plug-flow regime, which is regulated by less velocity of flow, low water depth, and availability of deep roots

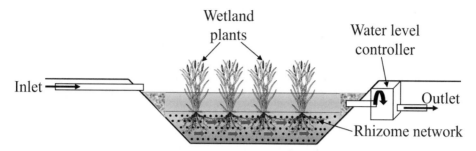

Fig. 13.12 Surface flow constructed wetland

and stalks of wetland plants (Fig. 13.12). Treatment of wastewater is carried out by physical operations like filtration, sedimentation, UV contact from solar radiation, chemical processes such as adsorption and precipitation and biological processes like uptake from root zone, microbial degradation, and nutrient transformations during the movement of wastewater through the CW.

Sedimentation and filtration in SF-CW support the removal of solids and organic matter, which are effectively degraded further via microbial breakdown. Nitrogen removal is governed by various factors, like amount of dissolved oxygen available, temperature of wastewater, organic load at inlet, and seasonal variations. Nitrification takes place in top layer of aerobic zones by aerobic microorganisms followed by denitrification of nitrate at bottom layer of anoxic zones by numerous microorganisms, thus supporting ammonia removal. Some of the nitrogen removal also occurs through uptake from the plant roots.

The SF-CWs provide phosphorus removal, although at a sluggish pace, through a combination of physical, chemical, and biological processes. Phosphorus enters CW as both soluble form (PO_4) and particulate (attached to suspended material like small soil particles and clays). The dissolve form of phosphorus is spontaneously consumed by plants and microorganisms present in the CW for synthesis of proteins and nucleic acids, like deoxyribonucleic acid (DNA) and ribonucleic acid (RNA) in their cells. The soluble phosphate is also reacted with aluminium, iron oxides and hydroxides present in the wastewater and precipitate to form aluminium, iron, and calcium phosphates, respectively. The particulate form of phosphorus is deposited in CWs through sedimentation; leaves and stems of plants allow the solid particles to settle out by slowing the water flow in the CW. Thus, the plant biomass (dead plants, leaves, root debris) and soil (non-decomposable parts of dead algae, bacteria, fungi, and invertebrates) are required to remove periodically from the bottom to improve the phosphorus removal in CWs.

In case of SF-CW, the wetland bed is composed of soil, coarse sand, and gravels. The depth of the bed is maintained up to 0.6–0.8 m and a bed slope of 1–3% is provided. The porosity of the bed depends on the size of the gravels and the other packaging materials used, which is approximately in the range of 0.3–0.42. To prevent the ground water contamination through the leakage of wastewater from the wetland bed, the peripheral outer surface of the bed is properly lined with clay materials or in some cases lined with plastic medium. The inlet and outlet points of the SF-CW are designed with water level controller, proper piping systems and channels for distribution of wastewater on the bed (Fig. 13.12). The number of plants for the CW are generally kept to 3–5 plants per m^2, which are grown up to a height of 3–4 m tall. The SF-CWs are mostly used for the treatment of municipal wastewater, as well as for storm water runoff and mine drainage water.

13.8.1.2 Subsurface Flow Constructed Wetlands

In SSF-CW, water level is kept below ground. Thus, the applied wastewater flows through a sand or gravel or brick bats bed rooted with plants, penetrate till the bottom of the bed and collected at the outlet point. The SSF-CW is further divided into three category such as horizontal, vertical and up-flow vertical SSF, depending upon the flow patterns of wastewater in the CW.

Horizontal subsurface flow constructed wetlands: In horizontal subsurface flow (HSSF) CWs, wastewater admitted to the wetland gently flows through the bed media in a horizontal manner, where the vegetation is planted, to reach to the outlet (Fig. 13.13). When Common reed (*Phragmites australis*) is planted in CW, HSSF-CWs are also known as "Reed Beds" and "Reed Bed Treatment System" (RBTS). To avoid seepage and to guarantee controlled outflow, the bottom and sides of HSSF-CWs are usually sealed with a linear either made of clay or in some case plastic mats are used. As the wastewater flows from inlet to the outlet through wetland bed, which is working also as a filter media, it passes through aerobic (roots and rhizomes of the plants), anoxic and anaerobic zones (bottom of the wetland bed). At the bottom of the HSSF-CWs an anaerobic environment exists due to lack of oxygen supply from plants roots. In this oxygen depleted zone organic matters are decomposed in the presence of anaerobic microorganisms, which also support the denitrification process.

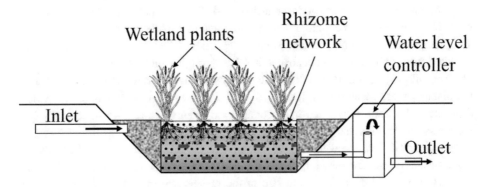

Fig. 13.13 Horizontal subsurface flow constructed wetland

The anaerobic respiration of the organic matter occurs at the bottom zone of the bed media that releases heat through the reaction (exothermic reaction). This is an advantage for maintaining a favourable temperature in the wetland bed and supports the other microbial activities present in the HSSF-CW against the cold temperatures during night and winter seasons. Microbial activities directly correlate with the temperature and therefore maintaining a suitable temperature in the CW treating wastewater is also an important factor for achieving a better treatment efficiency. Thus, compared to the FSW systems, HSSF-CWs can be operated effectively at low temperature conditions, such as during the night and it can also be adopted to seasonal variations. The HSSF-CWs is also suitable for cold regions of the climatic zones where the ice covers the surface, in such cases, though the surface of the CW is covered with ice, however the temperature produced in the anaerobic decomposition inside the filter bed is enough to keep the minimum microbial activity for the treatment of wastewater. In addition, the plants standing on the CW and litter deposited on the surface of the wetland act an insulation layer against the freezing temperature during cold and snow fall and thus, advantageous for the microbial activities in the media bed of CW (Brix, 1997).

In HSSF-CWs, oxygen release from different rhizomes is actually not adequate enough to sustain the oxygen demand for the aerobic degradation and nitrification of the organic matters present in wastewater (Brix, 1990; Brix & Schierup, 1990). Introduction of external aeration in HSSF-CWs enhances oxygen transport capacity, for this properly designed diffuser tubing and air blowers are sometimes provided to improve the removal efficiency (Vymazal, 2011). The effect of external aeration on BOD removal in aerated HSSF-CWs and non-aerated HSSF-CWs was evaluated, where the size of each HSSF-CW was 5 m^2 per population equivalent (m^2 PE^{-1}), operated with an organic loading of 8 g/m^2.day. The BOD values in the final effluents of non-aerated and aerated conditions in HSSF-CWs were observed to be 28 mg/L and 15 mg/L, respectively. The higher BOD removal in case of aerated HSSF-CWs compared to non-aerated HSSF-CWs is ascertained to the availability of sufficient O_2 in the bed through external aeration (Vymazal, 2011).

Similarly, the TSS removal efficiency for non-aerated and aerated conditions in HSSF-CWs was also determined with an initial loading of 7 g/m^2.day, and the TSS values in the effluent of non-aerated and aerated conditions in HSSF-CWs of 25 mg/L and 20 mg/L, respectively, were observed. The higher TSS removal efficiency for aerated HSSF than non-aerated HSSF is attributed to the higher growth of protozoa that is involved in the grazing of suspended bacteria in the wetland bed, whereas these protozoa cannot survive under anaerobic conditions (Vymazal, 2011). Thus, in comparison to non-aerated, aerated HSSF-CWs contribute enhanced removal efficiencies of BOD_5, total suspended solids (TSS) and total Kjeldahl nitrogen (TKN), although there was no influence on total P or faecal coliform removal.

Moreover, phosphorus removal is also possible in the media bed of HSSF-CWs through the adsorption in the clay materials and assimilation via synthesis of proteins and nucleic acids in microorganisms and plant species present in the CW. Similarly, suspended particles are screened out or removed due to sedimentation in the micro-pockets of the media bed and the roots of the plants. Although HSSF-CWs are frequently employed for removal of organic matter as secondary treatment of municipal wastewater, a wide range of additional applications have been documented that include removal of solids contents, biological contaminations including pathogenic microorganisms, and heavy metals from the wastewaters.

Vertical subsurface flow constructed wetlands: Vertical subsurface flow (VSSF) CWs are made up of graded gravel covered with sand and seeded with macrophytes (Fig. 13.14). Wastewater applied from top passes through the media and treated wastewater is collected at the bottom by a drainage network. In contrast to HSSF-CWs, the VSSF-CWs are filled with wastewater intermittently with considerable volume of wastewater on an irregular basis, saturating the surface (Cooper, 1999; Vymazal, 2007, 2010). Thus, during the dry condition in VSSF-CWs, the bed allows air to replenish the pores; thus, increases oxygen transport into the filter media, which is resulting in enhancing oxidation of organic matter and nitrification for the next batch of wastewater applied (Cooper, 1999). The remaining organic matters and nitrogen-content in the wastewater undergo anaerobic decomposition in the anoxic environment at the bottom of VSSF-CW, which also supports denitrification. However, it is observed that, the denitrification process is not adequately supported in the VSSF-CW. Hence, the effluent from the wetland is again recycled back to the sedimentation tank, attached prior to VSSF-CW, in order to promote the dentification in the sedimentation tank (Arias et al., 2005).

Vertical flow systems require less land than HSSF-CWs, typically 1–3 m^2 PE^{-1} (population equivalent) (Vymazal, 2010). In the early research, the VSSF systems were divided into stages, with 2–4 beds in the first stage being supplied with wastewater in rotation. These VF systems referred to as first-generation VF-CWs systems. The VF-CWs with only one bed have recently been introduced and termed as second-generation VF-CWs (or space-saving vertical flow beds). The bed matrix in case of second-generation VF-CWs is constructed with a deep sand layer (approximately 800 mm) compared to the

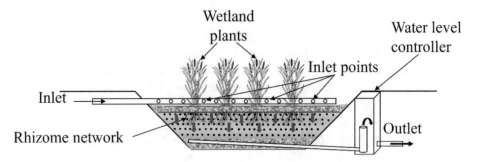

Fig. 13.14 Vertical flow subsurface constructed wetland

100–150 mm sand bed designed for first-generation VF-CWs, thus the thick sand bed is not getting disturbed during the weed removal or the burrowing of small mammals in the bed. The interval between the influent applications to the filter bed is scrupulously regulated in case of second-generation VF-CWs with a single-chamber compared to the first-generation VF-CWs with multiple chambers, for providing a proper resting period in the filter bed. These VF-CWs are widely employed to treat on-site household wastewaters, sewage from small societies, industrial wastewaters and storm water runoff (Vymazal, 2010).

Up-flow vertical subsurface flow constructed wetlands: In up-flow vertical CWs, the wastewater enters from the bottom of the wetland, flows in upward direction through the filter media and exits at the top level of the wetland (Fig. 13.15). In recent times "tidal flow" CWs are used to treat the wastewater that works on the principle of "fill, react, drain and rest". In tidal flow CWs, the wastewater enters through aeration pipes at the bed's bottom, then flows upward till it reaches the top level of bed surface. When the surface has been completely saturated, the pump is turned off, and the wastewater comes in contact with the microorganisms grown on the surface of the bed media having the plant root systems. After a fixed time period, the wastewater is drained out as a result, voids in the filter media are filled with air, and the treatment cycle is completed after the complete drainage of wastewater from the CW. Thus, when the wastewater enters from the bottom of the filter bed, initially, the wastewater is treated via aerobic oxidation and gradually an anaerobic environment is developed due to the consumption of oxygen coupled with oxidation of organic matter.

Thus, before starting a new cycle, air is allowed to fill in the filter bed that develops an aerobic environment to combat the oxygen transfer limitation that occurs due to the up-flow mode of operation in an up-flow vertical CWs. Clogging of the filter substrate

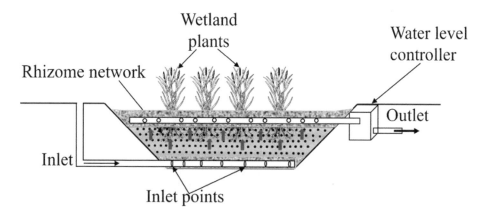

Fig. 13.15 Up-flow vertical flow constructed wetland

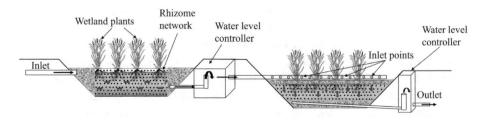

Fig. 13.16 Hybrid horizontal and vertical flow constructed wetlands

is a key issue in VF-CWs. Hence, selection of bed material and estimation of the appropriate hydraulic loading rate are very crucial key factors to achieve uniform spreading of wastewater throughout the wetland surface and a better treatment efficiency.

13.8.1.3 Hybrid Constructed Wetlands

Single CW is capable of removing organic matters, nitrogen and other pollutants present in wastewater; however, in terms of effective removal of pollutants, each type of CW is only efficient for the removal of either organic carbon or nitrogen (Vymazal, 2007). For, example, in HF-CW systems, nitrification does not occur completely due to the poor oxygen-transfer capacity and on the other hand in VF-CW systems, only nitrification occurs smoothly, however no denitrification is supported (Fig. 13.16). Therefore, the advantages of the horizonal flow and vertical flow systems can be combined to achieve both nitrification and denitrification (Cooper, 1999). Different combination sequences are used in a hybrid CW system, e.g., VF-HF, HF-VF, free water surface-HF, multistage VF and HF hybrid (VF-VF-HF, VF-HF-VF, HF-HF-VF), which have been practiced for treatment of both sewage and industrial wastewater (Vymazal, 2013). The design is divided into two phases, with many parallel VF-CWs followed by two or three HF-CWs in succession. The large size HF-CW can also be provided in the beginning of the hybrid treatment systems to reduce organics and suspended particles loading. Whereas, a small sized periodically loaded VF-CW is provided for further removal of organics and suspended particles, as well as to achieve nitrification (Fig. 13.16). If necessary, this can be followed by third stage of HF-CW to facilitate denitrification.

13.8.2 Plants Used in Constructed Wetlands

While selecting plants species for CWs following points should be kept in mind: (a) the plant should be able to withstand large organic and nutritional loadings; (b) the selected plant offers denser root systems and rhizomes to supply substrate for associated bacteria and oxygenation of regions close to roots and rhizomes; and (c) plants should have a lot of aboveground biomass for insulation during winter seasons, as well as nutrient removal through harvesting. In addition, one of the most noticeable characteristics of wetlands is the presence of macrophytes, particularly in case of free water surface CW, which distinguish CW from unplanted soil filters or aerobic lagoons. The macrophytes that develop in CWs have characteristics that make them an important part of the treatment process. The plant species which are generally used in CWs are discussed below.

Phragmites spp.: *Phragmites australis* (Common reed) is the most widely employed plant for CWs. It is a perennial, flood-tolerant plant with a large rhizome network that can reach depths of 0.6–1.0 m. These plants stems are stiff with hollow internodes, with shoot heights ranging from less than 0.5–8 m tall. The maximum aboveground biomass of *P. australis* varies greatly depending on geography, temperature, pH, water level, eutrophication, and their interactions. *Phragmites australis* is found in Europe, Canada, Australia, most of Asia (excluding India and Nepal). In India and Nepal, *Phragmites karka* species is used in CWs.

Typha spp.: *Typha* spp. (Cattails, Typhaceae) are rhizomatous perennial plants with upright stems and no joints. The plants may grow up to 3 m tall and have a large branching horizontal rhizome network. The backs of the leaves are plane or slightly curved, and the basal portions are porous. It grows in lakes, rivers, shallow bays, ponds, irrigation ditches and in fresh as well as brackish water swamps. It is widely used in CWs for treatment of both municipal and domestic sewage. *Typha angustifolia L.* (Narrow-leaved cattail), *Typha latifolia L.* (Broadleaved cattail, Common cattail), *Typha glauca* (blue cattail), etc. are the most significant *Typha* species planted in wetlands. The most widespread species is *T. latifolia*, which is distributed across the entire temperate northern hemisphere. *T. angustifolia* is another widespread variety, however it does not extend as far north. *Typha domingensis* has a more southern American distribution, and also in Australia. *Typha orientalis* is widespread species found in Asia, Australia, and New Zealand. *Typha laxmannii*, *Typha minima*, and *Typha shuttleworthii* are largely restricted to Asia and southern Europe.

Scirpus (Schoenoplectus) spp.: *Scirpus* (Cyperaceae) species are perennial plants that grow in clusters or huge colonies. The stem of these plants is roughly triangular or gently rounded and lightly sloped, reaching a height of up to 3 m, in certain species even higher height can be achieved. Roots of the *Scirpus* species extend to a depth of 70–80 cm, increasing root-zone oxygenation and microbial nitrification. *Scirpus* is most commonly seen in Australia, New Zealand, India and North America.

Phalaris arundinacea: *Phalaris arundinacea L.* (Poaceae) (Reed canary grass) is a perennial grass that grows up to a height of 1–3 m. It forms thick stands with conspicuous systems of strong roots and rhizomes that penetrate to a depth of 30–40 cm, enabling for aggressive vegetative propagation. Because of the increased nutrient availability, it grows aboveground. Reed canary grass blooms in a variety of wetland environments, including damp meadows and river banks, and widely grown under cold and damp conditions. This grass has a wide distribution in Europe, Asia, northern Africa and North America.

Iris spp.: *Iris pseudacorus L.* (*Iridaceae,* Yellow flag) is a beautiful annual shrub with a strong rhizome that grows up to 1.5 m tall, grows in ponds, lakes, slowly moving streams and rivers, and moist fields. The stem is erect, curved to flat, and branching. It is spread all over Europe, the Middle East, and North Africa.

13.8.3 Salient Features of Wetland Systems

International Union for Conservation of Nature and Natural Resources (IUCN) has defined natural systems for wastewater treatment as "actions to protect, sustainably manage, and restore natural or modified ecosystems that address societal challenges effectively and adaptively, simultaneously providing human well-being and biodiversity benefits" (Cohen-Shacham et al., 2016). The CWs act as "living solutions continuously supported by nature, which are designed to address various societal challenges in a resource-efficient and adaptable manner and to provide simultaneously economic, social, and environmental benefits" (Pineda-Martos & Calheiros, 2021). Natural treatment systems are the methodologies that imitate natural processes in urban landscapes, with less use of mechanical equipment, energy and chemicals. A few silent features of CWs are stated below.

  i.   Low-cost and sustainable solution for wastewater treatment in terms of construction, operations and maintenance costs.
  ii. Applicable for the treatment of domestic wastewater, agricultural runoff, storm water runoff and for some of the industrial effluents including mining drainage.
  iii. No energy required for operations of CWs or very less energy in case of pumping requirement.
  iv. It can be operated for long term without any requirement of frequent maintenance.
  v.   Treated effluent from CWs can be reused for different applications.
  vi. Preserve groundwater and surface water levels.
  vii. Facilitate the conservation and environmental protection to the habitat for plants, animals and insects.
  viii. Accommodate volume for storage of water including the storm water.
  ix. Does not release odour and contribute to natural aesthetics.

13.8.4 Design of Constructed Wetland

The area required for the bed of CW can be determined using Kickuth's equation (UN-HABITAT, 2008) as given in Eq. 13.8.

$$A = \frac{Q(ln C_i - ln C_o)}{K_{BOD}} \tag{13.8}$$

Table 13.3 Variation of rate constant, K_{BOD} with temperature for HF-CW and VF-CWs

Temperature, °C	Rate constant, K_{BOD} (m/d)	
	HF-CW	VF-CW
10	0.100	0.130
15	0.130	0.185
20	0.175	0.230
25	0.230	0.310

Source UN-HABITAT (2008)

where, A is CW bed surface area, m^2; Q is average flow of wastewater, m^3/day; C_i is BOD$_5$ concentration in the influent wastewater, mg/L; C_o is BOD$_5$ concentration in the outlet, i.e., in effluent, mg/L; K_{BOD} is rate constant, m/day. The values of rate constant, K_{BOD} varies with temperature and type of CWs. Typical values for K_{BOD} are given in Table 13.3. It is recommended to use an average depth of 40 cm for HF-CW and 70 cm in case of VF-CW. Length to width ratio for HF-CW bed varies from 1:1–3:1 and for VF-CW bed it varies from 1:1–2:1.

Example: 13.5 Design a HF-CW for treatment of sewage generated from a population of 300 persons with per capita wastewater generation of 85 LPCD. The effluent shall have BOD$_5$ concentration of less than or equal to 30 mg/L. In laboratory analysis the BOD$_5$ of influent was found out to be 425 mg/L and 25% of BOD$_5$ is removed during the primary treatment. Consider K$_{BOD}$ as 0.2 m/day.

Solution

Effluent BOD$_5$ concentration (C$_o$) = 30 mg/L.

   Rate constant, K$_{BOD}$ = 0.2 m/day.

   Average flow of wastewater, Q = 300 × 85/1000 = 25.5 m^3/day.

   Since, 25% of BOD$_5$ is removed during primary treatment, hence BOD$_5$ concentration at HF-CW inlet (C_i) = 425 – 0.25 × 425 = 319 mg/L.

   Estimating surface area required for HF-CW.

   Surface area = $A = \frac{Q(lnC_i - lnC_o)}{K_{BOD}} = \frac{25.5(\ln 319 - \ln 30)}{0.2} = 301.41$ m^2.

   Provide a depth of 0.4 m and consider length to width ratio as 2:1,

$$L \times B = 2B \times B = 2B^2 = 301.41$$

B = 12.28 ≈ 12.3 m; hence, L = 12.3 × 2 = 24.6 m.

   Provide HF-CW bed of 24.6 m length, 12.3 m width and 0.4 m depth.

Example: 13.6 Design the VF-CW for data provided in Example 13.5. Consider K$_{BOD}$ as 0.25 m/day.

Solution

Effluent BOD_5 concentration (C_o) = 30 mg/L.

Rate constant, K_{BOD} = 0.25 m/day.

Average flow of wastewater, Q = $300 \times 85/1000 = 25.5$ m^3/day.

Since, 25% of BOD_5 is removed during primary treatment, hence BOD_5 concentration at VF-CW inlet (C_i) = $425 - 0.25 \times 425 = 319$ mg/L.

Estimating surface area required for VF-CW

$$\text{Surface area} = \text{Surface area} = A = \frac{Q(lnC_i - lnC_o)}{K_{BOD}} = \frac{25.5(\ln 319 - \ln 30)}{0.25} = 241.13\,m^2$$

Provide a depth of 0.7 m and length to width ratio as 1.5:1,

$$L \times B = 1.5B \times B = 1.5B^2 = 241.13$$

B = $12.68 \approx 12.7$ m; hence, L = $12.7 \times 1.5 = 19.05$ m ~ 19.1 m.

Provide VF-CW bed with length of 19.1 m and width of 12.7 m with depth of 0.7 m.

Questions

13.1 Describe different types of pond systems and constructed wetlands used for treatment of wastewaters.

13.2 A high-rate algal pond is to be designed with 80% efficiency for the treatment of municipal wastewater, where the yield of photosynthetic oxygen is 225 kg/ha.day. The temperature of the influent wastewater is 28 °C and the wastewater flow is 3 MLD along with an influent BOD concentration of 220 mg/L. Estimate the minimum area and depth required for the pond. Consider that this raceway pond is designed for plug flow condition. [**Ans.** Minimum area is 3.66 Ha and depth is 1.02 m]

13.3 A facultative stabilization pond having depth of 1.3 m removes substrate at the rate of 0.23 per day from 2.5 MLD of municipal sewage. If the photosynthetic oxygen yield is 250 kg/ha.day and BOD_5 of the sewage at 20 °C is 180 mg/L. Determine the area of the pond. Consider 50% of organic compound is non-settleable and aerobically oxidised in the upper level of the pond.[**Ans.** Area = 1.32 ha]

13.4 Describe oxidation pond, maturation pond, facultative stabilization pond and anaerobic lagoons used for treatment of wastewater.

13.5 How syntrophic association of algae and bacteria play a role in treatment of wastewater in the algal ponds?

13.6 Write short notes on aquatic plant-based ponds used for wastewater treatment. Which type of wastewaters can be effectively treated in such ponds?

13.7 Describe in details the application of high-rate algal ponds for sewage treatment. What are the benefit and drawbacks this pond can offer over other variant of the ponds?

13.8 What are the different components involved in a pond system employed for wastewater treatment?

13.9 How the algal concentration in the final effluent of the ponds can be kept under control?

13.10 What are the operating factors which will affect the wastewater treatment performance of the ponds?

13.11 Discuss the recommendations for designing the oxidation pond.

13.12 As compared to conventional wastewater treatment what are the advantages and drawbacks of the ponds system?

13.13 Discuss classification of constructed wetlands. What are the advantages of these over the conventional wastewater treatment?

13.14 Describe different types of constructed wetlands that can be used for wastewater treatment.

13.15 Describe constructed wetlands. Also, illustrate the salient features of the constructed wetlands.

13.16 Discuss different plant species that are used in constructed wetlands in south Asia.

References

Acién, F. G., Molina, E., Reis, A., Torzillo, G., Zittelli, G. C., Sepúlveda, C., & Masojídek, J. (2017). Photobioreactors for the production of microalgae. In *Microalgae-based biofuels and bioproducts*. Elsevier, pp. 1–4.

Arceivala, S.J. (1980). Discussion of "Design Criteria for Aerobic Aerated Lagoons." *Journal of Environmental Engineering Division 106*, 250–252.

Arceivala, S. J., & Asolekar, D. S. R. (2007). *Wastewater Treatment for Pollution Control and Reuse* (3rd ed.). Tata McGraw-Hill.

Arias, C. A., Brix, H., & Marti, E. (2005). Recycling of treated effluents enhances removal of total nitrogen in vertical flow constructed wetlands. *Journal of Environmental Science and Health Part A Toxic/Hazardous Substances & Environmental Engineering, 40* (6–7), 1431–1443.

Brix, H. (1997). Do macrophytes play a role in constructed treatment wetlands? *Water Science and Technology, 35*, 11–17.

Brix, H. (1990). Gas exchange through the soil-atmosphere interphase and through dead culms of phragmites australis in a constructed reed bed receiving domestic sewage. *Water Research, 24*, 259–266.

Brix, H., & Schierup, H.-H. (1990). Soil oxygenation in constructed reed beds: The role of macrophyte and soil-atmosphere interface oxygen transport. In P. F. Cooper & B. C. Findlater (Eds.), *Constructed Wetlands in Water Pollution Control* (pp. 53–66). Pergamon Press.

Cohen-Shacham, E., Walters, G., Janzen, C., & Maginnis, S. (2016). Nature-based solutions to address global societal challenges. IUCN, Gland, Switzerland, pp. 97.

Cooper, P. (1999). A review of the design and performance of vertical-flow and hybrid reed bed treatment systems. *Water Science and Technology, 40*(3), 1–9.

Pineda-Martos, R., & Calheiros, C. S. C. (2021). Nature-based solutions in cities—contribution of the Portuguese national association of green roofs to urban circularity. *Circular Economy and Sustainability, 1*, 1–17.

Thirumurthi, D. (1969). Design Principles of Waste Stabilization Ponds. *Journal of Sanitary Engineering Division 95*, 311–332.

UN-HABITAT. (2008). Constructed Wetlands Manual. UN-HABITAT Water for Asian Cities Programme, Nepal, Kathmandu.

Verma, A. K., Nath, D., Bhunia, P., & Dash, R. R. (2017). Application of ultrasonication and hybrid bioreactor for treatment of synthetic textile wastewater. *Journal of Hazardous, Toxic, and Radioactive Waste, 21*(2), 1–8.

Vymazal, J. (2013). The use of hybrid constructed wetlands for wastewater treatment with special attention to nitrogen removal: A review of a recent development. *Water Research, 47*(14), 4795–4811.

Vymazal, J. (2011). Constructed wetlands for wastewater treatment: Five decades of experience. *Environmental Science and Technology, 45*(1), 61–69.

Vymazal, J. (2010). Constructed wetlands for wastewater treatment. *Water, 2*(3), 530–549.

Vymazal, J. (2007). Removal of nutrients in various types of constructed wetlands. *Science of the Total Environment, 380*(1–3), 48–65.

Wehner, J.F., & Wilhelm, R.H. (1958). Boundary conditions of flow reactor. *Chemical Engineering Science 6*, 89.

Whitton, R., Ometto, F., Pidou, M., Jarvis, P., Villa, R., & Jefferson, B. (2015). Microalgae for municipal wastewater nutrient remediation: Mechanisms, reactors and outlook for tertiary treatment. *Environmental Technology Reviews, 4*(1), 133–148.

Biological Processes for Nutrient Removal

14

Domestic wastewater typically contains nitrogen and phosphorus in various forms. Nitrogen is usually detected as ammonia, nitrite, nitrate or as combined with organic molecules; whereas, phosphorus is found in either organic form (attached with carbon-based molecules) or inorganic form in dissolved or suspended state. The concentration of nitrogen and phosphorus in domestic wastewater is generally in a range of 10–50 mg of N/L as total nitrogen and 5–15 mg of P/L as total phosphorus, respectively. However, in industrial wastewaters, such as petrochemical, fertilizer, food industries, tannery etc., the concentration of nitrogen and phosphorus can be significantly high. The discharge of untreated or partially treated wastewater with nutrients (nitrogen and phosphorus) causes build-up of nutrients in the receiving water bodies. This develops a favourable condition for algal growth, which further deteriorates the water quality via eutrophication.

This human induced accelerated eutrophication of water bodies due to wastewater discharge severely affects the water quality and the local aquatic ecosystem (Anderson et al., 2008). Additionally, ammonia is oxidized to nitrite and then to nitrate causing a reduction in dissolved oxygen in the water bodies, thus affecting the ecosystem and the local fauna. To prevent this, the wastewater must be free from nutrients prior to its discharge to the water bodies. The biological nitrogen removal via nitrification–denitrification is the most commonly used technique for the removal of nitrogen from wastewater during the secondary treatment. More recently anaerobic ammonia oxidation (Anammox) was developed, in which ammonia is oxidized under anoxic conditions using nitrate as electron acceptor. These biological processes of nutrient removal are described in this chapter in details.

14.1   Nitrification

Nitrification is a two-step process in which, ammonia is oxidized to nitrite in first step and this nitrite is further oxidized to nitrate in the second step. Nitrification is commonly achieved in the secondary aerobic biological wastewater treatment process along with the degradation of carbonaceous matter. Similar to the removal of carbonaceous organic matter, nitrification also can be performed in suspended as well as attached growth biological processes. This can be accomplished in suspended growth biological treatments, such as activated sludge process, sequential batch reactor, membrane bioreactor, or attached growth biological treatments, such as moving bed biofilm reactor, rotating biological contactors, trickling filter, constructed wetland, etc.

However, the effectivity of simultaneous carbonaceous organic matter oxidation and nitrification depends on the BOD_5/TKN ratio. When nitrification is supported along with organic matter oxidation for BOD_5/TKN ratio greater than 4–5, it is said to be simultaneous carbonaceous organic matter oxidation and nitrification. If BOD_5/TKN ratio of less than 3, the system is termed as separate nitrification process (Parker et al., 1975). The suspended growth system consists of aeration tank, secondary settling tank, and sludge recycling system, which is also called as single-sludge system (Fig. 14.1a). However, in case if wastewater has high organic matter concentration or presence of toxic compounds, a two-sludge system is recommended that comprising of two sets of aeration tanks and secondary sedimentations tanks (Fig. 14.1b), meant for separate organic matter oxidation and nitrification. The first aeration tank is operated at a shorter sludge retention time and it is designed for organic matter removal; whereas, the following aeration tank is mainly adopted for nitrification, which is operated with high hydraulic retention time and sludge retention time.

Since the nitrifying microorganisms have slow growth rate as compared to heterotrophic microorganisms, the sludge retention time of nitrifying tank is kept higher. In attached growth aerobic processes, the heterotrophic microorganisms dominate the attached growth surface area of biofilm grown over media and the growth of nitrifying microorganisms is suppressed. Hence, for combined (simultaneous BOD removal and nitrification) attached growth systems, BOD loading rates should be kept lower than that used for system with only BOD removal. A separate system can be a viable option for accomplishing nitrification in the attached growth process, which is provided after removal of organic matter in either attached growth or suspended growth treatment.

Although autotrophic microorganisms are responsible for nitrification, two stages of nitrification, i.e., ammonia to nitrite conversion and nitrite to nitrate conversion involves different group of microorganisms. Both groups of microorganisms are autotrophic; however, they are distinctly different from each other. The prime microorganisms responsible for ammonia to nitrite and nitrite to nitrate conversion are *Nitrosomonas* and *Nitrobacter*, respectively. *Nitrosococus*, *Nitrosospira*, *Nitrosolobus*, *Nitrosorobrio* are some of the other

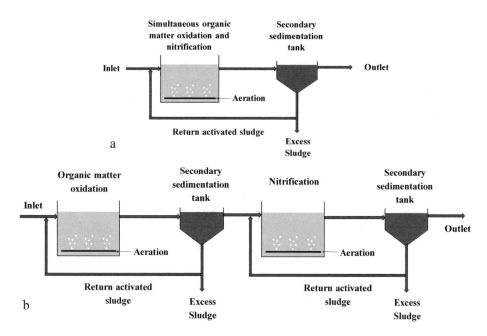

Fig. 14.1 Biological nitrification process using **a** single-sludge system, **b** two-sludge system

autotrophic microorganisms capable of oxidizing ammonia to nitrite. Similarly, *Nitrococcus*, *Nitrospina*, and *Nitroeystis* are other gene of microorganisms capable of oxidizing nitrite to nitrate.

The stoichiometric equations of two stages of nitrification are as follows (Eq. 14.1 through Eq. 14.3):

The equation of oxidation of ammonia to nitrite:

$$2NH_4^+ + 3O_2 \rightarrow 2NO_2^- + 4H^+ + 2H_2O \tag{14.1}$$

The equation of oxidation of nitrite to nitrate:

$$2NO_2^- + O_2 \rightarrow 2NO_3^- \tag{14.2}$$

Combined nitrification reaction:

$$NH_4^+ + 2O_2 \rightarrow NO_3^- + 2H^+ + H_2O \tag{14.3}$$

Based on ammonia oxidation to nitrite (Eq. 14.1), 3.43 g O_2 is required per g of ammonia oxidation; whereas, for nitrite oxidation 1.14 g O_2 is required per g of nitrite (Eq. 14.2). Hence, the oxygen requirement for complete nitrification (ammonia to nitrate conversion) is 4.57 g O_2/g N (Eq. 14.3). Apart from oxygen requirement, alkalinity is also essential to progress the nitrification reaction (Eq. 14.4). For each gram of nitrogen 7.14 gram of alkalinity as $CaCO_3$ is essential.

$$NH_4^+ + 2HCO_3^- + 2O_2 \rightarrow NO_3^- + 2CO_2 + 3H_2O \tag{14.4}$$

If $C_5H_7O_2N$ is the chemical representation of bacterial cell, a part of ammonia nitrogen is integrated into microbial cell (Eq. 14.5).

$$4CO_2 + HCO_3^- + NH_4^+ + H_2O \rightarrow C_5H_7O_2N + 5O_2 \tag{14.5}$$

Due to the assimilation of some part of ammonia into the microbial cell, the practical oxygen requirement for nitrification is less than 4.57 g O_2/g N as the microbial cell synthesis is not considered in Eq. 14.3. For designing nitrification process, the constituents of wastewater, such as nitrogen concentration, available alkalinity, available oxygen, carbonaceous organic matter concentration, and presence of inhibitory substances to nitrifying microorganisms, cause a vital role. Additionally, phosphorus, CO_2, and trace metals are also required for supporting the growth of nitrifying microorganisms.

Based on Monod kinetics the growth of nitrifying microorganisms can be modelled as per Eq. 14.6.

$$\mu_N = \mu_{mN} * \frac{N}{K_N + N} \tag{14.6}$$

where, μ_N is specific growth rate of nitrifying microorganisms; μ_{mN} is maximum growth rate of nitrifying microorganisms (per day); K_N represents half saturation constant; and N is the concentration of NH_3-N.

At steady state conditions, biomass produced and quantity of biomass leaving the system is equal; hence the solid retention time (Θ_c) can be estimated using the Eq. 14.7.

$$\theta_c = \frac{1}{\mu_N - d_N} \tag{14.7}$$

In this equation, d_N represents endogenous decay coefficient for nitrifying microorganisms (per day), which is very less for the nitrifying microorganisms compared to heterotrophic microorganisms. The value of Θ_c obtained from this equation is a theoretical value for steady state conditions. To compensate for change in unknown factors, such as pH, dissolved oxygen, ammonia concentration, temperature, and other inhibitory substances, a factor of safety of about 1.5 or greater is applied and the designed value of Θ_c is increased accordingly.

14.2 Factors Affecting Nitrification

The factors that affect the growth of nitrifying microorganisms and the efficiency of nitrification process are described below.

Inhibiting compounds: Nitrifying microorganisms are sensitive to a wide range of organic and inorganic compounds, in presence of which the rate of nitrification will be reduced or in presence of any toxicity, the nitrifying microorganisms might be destroyed. Certain compounds, such as amines, proteins, phenols, alcohols, benzenes, etc., are known to have a detrimental impact on the growth of nitrifying microorganisms. Hence in such cases, two-sludge system is recommended, in which the concentration of inhibitory substances is reduced in the first aeration tank.

The pH: The rate of nitrification is highly dependent on the pH of wastewater. Optimum range of pH for nitrification is 7.5–8.0 and it is significantly affected in the wastewater pH below 6.8. In order to elevate the pH of wastewater, external alkalinity is induced in the form of either sodium bicarbonate, lime, soda ash, etc. The quantity of alkalinity required to be externally added depends on the available alkalinity in the wastewater and the amount of ammonia nitrogen present in the wastewater.

Dissolved oxygen concentration: Nitrifying microorganisms are aerobic microorganisms and a minimum dissolved oxygen concentration of 1 mg/L is essential for nitrification to proceed. Below this level, the oxygen becomes a limiting factor, thus reducing the rate of nitrification. Under anaerobic condition the nitrification is totally suppressed. Hence, in anaerobic treatments, such as UASB reactor or anaerobic filters, the nitrification cannot occur and unless the nitrogen present in the influent wastewater is in the form of nitrate, or converted to nitrate, the biological nitrogen removal cannot be achieved. However, in anaerobic reactor denitrification can be supported and if influent has nitrogen in the form of nitrate, then denitrification can occur to support biological nitrogen removal if sufficient carbon source is present in the wastewater.

Ammonia concentration: High concentration of free ammonia and nitrous acid acts as inhibitors to the nitrifying microorganisms. It is noted that the free ammonia concentration of ~150 mg/L will be inhibitory for its oxidation (ammonia oxidation to nitrite); whereas, ammonia concentration of ~18 mg/L will be inhibitory for nitrite oxidizing bacteria converting it to nitrate (Zhang et al., 2018).

14.3 Denitrification

For the reuse of treated effluent, only nitrification of wastewater might not be adequate, since regulatory norms at most places is in the form of total nitrogen. The presence of nitrate can cause eutrophication. Additionally, if treated effluent is used for ground water recharge, high concentrations build-up of nitrate is not advisable; since, nitrate concentration more than 45 mg/L will make this ground water unsuitable for drinking.

In such cases, nitrate removal is essential prior to wastewater discharge, which can be accomplished via biological denitrification process.

Denitrification is the process of reduction of nitrate to nitric oxide, nitrous oxide, and finally to nitrogen gas using microorganisms. In the biological processes, nitrate formed via nitrification can be eliminated through two mechanisms; one via assimilation and other via dissimilating reduction. The microbial cell synthesis requires ammonia, which is obtained from reduction of nitrate to ammonia and it is progressed, if ammonia nitrogen is not available in wastewater. This mechanism takes places in assimilation reduction of nitrate. While in the dissimilation reduction, nitrate is used as a terminal electron acceptor for anoxic oxidation of other electron donors available in the wastewater.

Denitrifying microorganisms use either dissolved oxygen or nitrate as the oxygen source for cell metabolism. Hence for effective denitrification to occur, the dissolved oxygen level should be close to zero and preferably less than 0.2 mg/L. A wide range of microorganisms both heterotrophic and autotrophic have the ability to demonstrate denitrification of wastewater via reduction of nitrate to nitrogen gas. Microorganisms, such as *Bacillus, Pseudomonas, Rhodopseudomonas, Vibrio, Chromobacterium, Spirillum*, etc., can undertake denitrification reaction in wastewater. *Pseudomonas* are most common denitrifying microorganisms, which are commonly detected in wastewaters. Moreover, under anaerobic conditions, *Nitrosomonas europaea* (nitrifying microorganism) can oxidize ammonia via use of nitrate with the release of nitrogen gas.

The nitrate reduction involves four steps reduction from nitrate to nitrite, nitric oxide, nitrous oxide, and nitrogen gas successively as illustrated in Eq. 14.8.

$$\underset{\text{(nitrate)}}{NO_3^{\,-}} \rightarrow \underset{\text{(nitrate)}}{NO_2^{\,-}} \rightarrow \underset{\text{(nitric oxide)}}{NO} \rightarrow \underset{\text{(nitrous oxide)}}{N_2O} \rightarrow \underset{\text{(nitrogen gas)}}{N_2}$$

$$(14.8)$$

For biological denitrification to occur, a sufficient carbon source must be available for growth of heterotrophic denitrifying microorganisms. Theoretically, 4.2 g COD is required per g of N for total nitrogen removal with glucose as carbon source (Issacs & Henze, 1995). However, for practical purpose depending on the biodegradability of the carbon source, this ratio can be between 4 to 15 g COD/g N (Peng et al., 2007). In order to meet this requirement of carbon, external addition of methanol as a carbon source can be implemented (Eq. 14.9); else some fraction of untreated wastewater can be admitted to the denitrification reactor as electron donor (Eq. 14.10). If the carbon source used is from outside of the treatment plant (e.g., methanol), it is termed as external source. Likewise, if the carbon source is from within the treatment plant (e.g., influent wastewater, biological sludge), it is termed as internal carbon source (UP EPA 2013).

$$5CH_3OH + 6NO_3^- \rightarrow 3N_2 + 5CO_2 + 7H_2O + 6OH^- \qquad (14.9)$$

$$C_{10}H_{19}O_3N + 10NO_3 \rightarrow 5N_2 + 10CO_2 + 3H_2O + NH_3 + 10OH^- \qquad (14.10)$$

When wastewater is added as a carbon source, the non-nitrified part of nitrogen present in the wastewater remains unaffected during denitrification and this fraction of nitrogen will appear in the effluent. Due to this methanol or acetate (pure carbon source without nitrogen) addition can be a suitable option. The chemical representation of the biodegradable organic matter of wastewater is $C_{10}H_{19}O_3N$. These equations (Eqs. 14.9 and 14.10) are representative of denitrification process using methanol and wastewater as carbon source that is oxidized and nitrate is reduced. In the denitrification process, 3.57 g of alkalinity (as $CaCO_3$) per g of nitrate is produced.

14.4   Factors Affecting Denitrification

The operating conditions of the reactor and parameters affecting the growth of denitrifying microorganisms will affect the performance of denitrification. These factors are discussed below in details.

Dissolve oxygen concentration: The denitrification can effectively occur when dissolved oxygen concentration in the reactor is maintained less than 0.2 mg/L. As nitrate is the prime electron acceptor in denitrification, the presence of oxygen will lead to use of oxygen as a terminal electron acceptor, subsequently suppressing the denitrification reaction.

Availability of carbon source: In the denitrification sufficient amount of electron donor in terms of carbonaceous matter should be available in the wastewater. Generally, 4.2 g of COD to TKN ratio is essential for denitrification. Although methanol is suitable (due to absence of nitrogen content) as a carbon source, it increases the cost of treatment; hence wastewater itself is considered as an ideal source of carbon for denitrification. The total quantity of methanol can be estimated using Eq. 14.11, in absence of the precise data.

$$Methanol_{dose} = 2.47 N_{nitrate} + 1.53 N_{nitrite} + 0.87 DO_{removed} \qquad (14.11)$$

In this Eq. 14.11, $N_{nitrate}$ and $N_{nitrite}$ resembles the nitrate nitrogen removed and nitrite nitrogen removed, respectively. The added methanol also serves the purpose of reducing the dissolved oxygen in denitrification tank, if there is any. The quantity of methanol required for DO removal is termed as $DO_{removed}$.

The pH: As alkalinity is produced per g of nitrate, the pH of wastewater during denitrification stage is high. The pH of 6.5–8.0 is appropriate for biological denitrification to occur using microorganisms; while at pH lower than 6.5, the rate of denitrification is reduced.

Fig. 14.2 The activated sludge floc capable of simultaneous nitrification and denitrification with aerobic exterior and anoxic interior

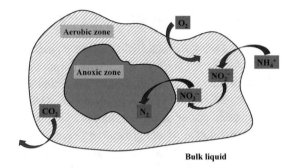

14.5 Systems Used for Nitrification and Denitrification

14.5.1 Simultaneous Nitrification and Denitrification

The dissolved oxygen level of bulk wastewater in the aerobic system does not reflect the same dissolved oxygen concentration within the bio-solids/sludge flocs. In the low dissolved oxygen concentrations, the outside of sludge floc is exposed to aerobic conditions, whereas within the flocs anoxic conditions are observed. Due to the higher availability of non-oxidized matter on the exterior of sludge floc, the rate of dissolved oxygen consumption will be faster. Due to utilization of *DO* at the outer layer, the dissolved oxygen cannot permeate inside the sludge flocs; as a result, anoxic conditions are developed within the floc. This corroborates possible nitrification on the flocs exterior surface and denitrification inside the flocs (Fig. 14.2). Similarly, in aerobic tanks with dissolved oxygen level of less than 0.5 mg/L, these two zones aerobic and anaerobic, can be observed depending on the mixing conditions. The anoxic conditions are observed as the distance from mechanical aerators increases. Hence, nitrification and denitrification can be achieved in a single aerobic tank itself. However, the rate of both processes is slower than that obtained in a separate system mainly due to effect of dissolved oxygen.

14.5.2 Pre-anoxic Single-Sludge System

As the name suggests, in the pre-anoxic single-sludge process, the influent wastewater is fed to anoxic tank followed by aerobic tank, before admitting the effluent into the secondary sedimentation tank/secondary clarifier. In aeration tank, apart from the degradation of organic matter, ammonia is oxidized to nitrate. The nitrified effluent from the aerobic tank is recycled back into the anoxic tank along with the sludge settled in the clarifier (Fig. 14.3). The non-oxidized organic matter in the inlet wastewater serves as a carbon source for denitrification reaction in the anoxic tank. During denitrification alkalinity is produced; whereas in nitrification alkalinity is required and it is consumed. Denitrification

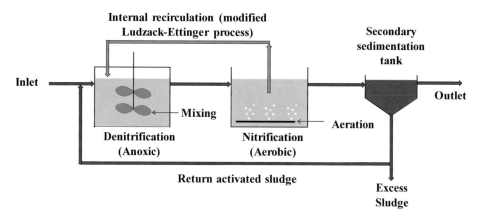

Fig. 14.3 Pre-anoxic single-sludge system with internal recirculation

in the anoxic tank ensures that the alkalinity is produced prior to nitrification, thus bene-fiting the overall process by reducing dependency on external addition of alkalinity. When using return activated sludge alone (without recycling of nitrifying effluent), the denitrifi-cation rate is less and it is dependent on the rate of return activated sludge recycling into anoxic tank.

The efficiency of denitrification can only be improved via provision of internal recir-culation from aerobic tank to anoxic tank and the process is also known as modified Ludzack-Ettinger process (US EPA, 1993) (Fig. 14.3). The internal recirculation allows entry of nitrified effluent into anoxic tank, thus allowing denitrification for removal of nitrogen. It is observed that addition of internal recirculation in pre-anoxic single-sludge system increases the nitrogen removal rate than standalone return activated sludge recircu-lation. For treating domestic wastewater ratio of internal recirculation to inflow wastewater is kept between 2 and 4. Another modification to this type of system is via use of step feed strategy, in which wastewater is fed in the anoxic system at several inlets. In this manner, inflow wastewater also acts as a carbon source for denitrification. The subsequent aerobic zone is critical for complete nitrification of the effluent. The effluent recycle flow rate from the aerobic zone will be dependent on the desired effluent nitrogen concentration. In general, the efficient anoxic zone and high recirculation of nitrified effluent to anoxic zone could achieve 70–80% of nitrogen removal using pre-anoxic single-sludge system.

14.5.3 Post-anoxic Single and Two Sludge System

In the post-anoxic single sludge system, aerobic reactor is used first and it is followed by anoxic reactor, in which the nitrified effluent from aerobic treatment is admitted (Fig. 14.4). In such systems, the external carbon source in the form of methanol is

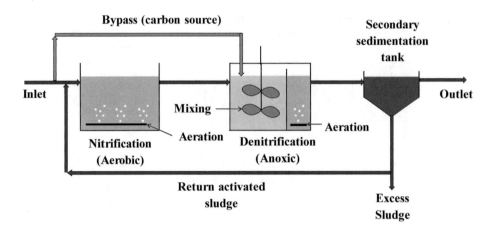

Fig. 14.4 Post-anoxic single sludge system

required to be added to the anoxic reactor for achieving denitrification. However, sufficient hydraulic retention time and sludge retention time is required in the anoxic reactor for ensuring degradation of methanol and denitrification of the wastewater. The denitrification reaction releases nitrogen and CO_2 from the wastewater. These gases adhere to sludge flocs, which reduces its settleability. Hence short aeration of about 15–60 min is provided subsequent to anoxic treatment so that the gases attached to the sludge flocs can be released due to this air agitation, thus aiding in better solid removal in the secondary sedimentation tank. A similar settleability to that of secondary biological sludge can be achieved following aeration.

The use of methanol can cause difficulty in treatment plants operated under fluctuating nitrogen loading, as methanol dose needs to be optimized depending upon the nitrate content. Excessive methanol addition than required will result in unnecessary higher BOD in the treated effluent; whereas less quantity of methanol will cause discharge of nitrate laden wastewater from the anoxic tank. Post-anoxic two-sludge system is another variation, which includes two sedimentation tanks one each after nitrification tank and denitrification tank to facilitate recycling of sludge rich in bacteria responsible for nitrification and denitrification, respectively.

The pre-anoxic system has advantages over post-anoxic system of nitrogen removal as described below.

- During the nitrate reduction, a small quantity of oxygen is released, which caters to some quantity of oxygen required for oxidation of incoming organic matter present in the wastewater.
- External methanol addition is not required as incoming wastewater serves as the carbon source required for denitrification.

- Aeration is not required prior to sludge settling for releasing nitrogen gas attached to sludge.
- However, in the pre-anoxic system, continuous recycling of nitrified effluent is essential to achieve successful denitrification.

14.5.4 Closed Loop Systems

Closed loop system, such as oxidation ditch, is a type of extended aeration process used for the treatment of wastewater in small to medium size plants. Such systems are operated with a longer hydraulic retention time of 20–30 h and sludge retention time (>20 days). Oxidation ditch is capable of achieving ~90% nitrification, provided the temperature is above 10 °C. With re-positioning of aerators and controlling the rate of aeration, the aerobic zone inside the oxidation ditch can be shortened (Fig. 14.5). The aerobic zone exists immediately near the vicinity of the aerators and the dissolved oxygen concentration continuously reduces due to microbial oxygen uptake as wastewater moves away from the aerator. The portion in the ditch where oxygen concentration is depleted, anoxic zone is created serving ideal atmosphere for the denitrifying microorganisms. The aerators installed in the oxidation ditch not only aerates the wastewater but also controls the velocity of wastewater flow in the channel. In such manner, some portion can serve as aerobic zone; whereas the wastewater on the other side of baffle wall will undergo anoxic condition, in which denitrification can take place.

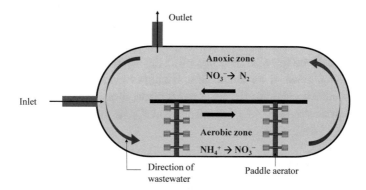

Fig. 14.5 Closed loop system for nitrification and denitrification

14.6 Design Guidelines for Biological Nitrification and Denitrification

In the case of biological nitrification and denitrification, majority of the biological kinetic equations described in the Chap. 10 are applicable. However, nitrification and denitrification being sophisticated process, some correcting factors are applied to the basic equations.

Correction for maximum specific growth rate ($\mu_{N\,max}$) of nitrifying and denitrifying microorganisms: The process of biological nitrogen removal is susceptible to the external factors such as temperature, dissolved oxygen concentration, pH, etc. Hence based on the practical conditions, the corrections are applied to the kinetic coefficient as follows (Eq. 14.12 through 14.17).

$$\mu_{N\,max\,actual} = F_T \times F_{DO} \times F_{pH} \times \mu_{N\,max} \tag{14.12}$$

$$F_T = \theta_T^{T-20} \tag{14.13}$$

$$F_{DO} = \frac{DO}{K_{DO} + DO} \tag{14.14}$$

$$\text{For pH} < 7.0,\ F_{pH} = 0.0004017 \times e^{1.0946 \times pH} \tag{14.15}$$

$$\text{For pH of 7.2,}\ F_{pH} = 2.35^{pH-7.2} \tag{14.16}$$

$$\text{For pH of 7.2} < \text{pH} < 9.5,\ F_{pH} = \frac{1.13(9.5 - pH)}{9.8 - pH} \tag{14.17}$$

where, DO is dissolved oxygen concentration in the aeration tank; K_{DO} is dissolved oxygen half saturation constant for ammonia oxidizing or nitrite oxidizing microorganisms; F_T is temperature correction factor; θ_T is temperature correction coefficient for ammonia oxidizing or nitrite oxidizing microorganisms; and F_{pH} is the pH correction factor.

For the case of denitrification,

$$\mu_{DN\,max\,actual} = F_T \times F_{DO} \times F_{NO_3} \times \mu_{DN\,max} \tag{14.18}$$

In this equation the except F_{NO_3} the other corrections factors are same as that mentioned in Eqs. 14.13 and 14.14. The Correction factor for nitrate is calculated by using Eq. 14.19.

$$F_{NO_3} = \frac{F_{NO_3}}{F_{NO_3} + K_{NO_3}} \tag{14.19}$$

where, K_{NO_3} denotes nitrate half saturation constant for denitrifying microorganisms.

Specific growth rate: It is dependent on the value of $\mu_{N\,max}$ (after applying the corrections), which is given by Eq. 14.20.

$$\mu_N = \frac{\mu_{N\,max} \times N}{K_N + N} - K_{DN} \tag{14.20}$$

where, $\mu_{N\,max}$ is maximum growth rate for nitrifying microorganisms (per day); μ_N is specific growth rate for nitrifying microorganism (per day); K_N represents half saturation concentration for nitrification (mg/L); N is TKN concentration in the wastewater; and K_{DN} is decay coefficient for nitrifying microorganisms (per day).

Effluent TKN concentration: Considering the substrate removal rate, maximum specific nitrifying microorganism growth rate, solid retention time, and the decay coefficient, the actual effluent TKN concentration will vary. This effluent TKN concentration can be obtained using Eq. 14.21.

$$N = \frac{K_N \times (1 + K_{DN} \times \theta_{CN})}{\theta_{CN} * (\mu_{N\,max} - K_{DN}) - 1} \tag{14.21}$$

where, θ_{CN} is solid retention time for nitrifying microorganisms.

Fraction of nitrifying microorganisms in the aerobic reactor: In the aerobic system, heterotrophic and autotrophic microorganisms are present together. However, for the ammonia oxidation, only ammonia oxidizing microorganisms are actively involved. Hence, the quantification of specific quantity of microorganisms is essential for designing nitrification system, which is estimated based on the biomass growth yield of both microorganisms.

The biomass growth of heterotrophic microorganisms (P_{XC}) is given by Eq. 14.22.

$$P_{XC} = Y_c(C_o - C) \tag{14.22}$$

The biomass growth of autotrophic microorganisms (P_{XN}) is given by Eq. 14.23.

$$P_{XN} = Y_N(N_o - N) \tag{14.23}$$

where, Y_c and Y_N represents observed biomass yield for heterotrophic and autotrophic microorganisms in the aeration tank, respectively. The C_o and C denotes initial and final concentration of COD; whereas, N_o and N denotes initial and final TKN concentration in the wastewater, respectively.

Based on this, the fraction of autotrophic microorganisms present in the biomass (MLVSS) can be estimated using Eq. 14.24.

$$F_N = \frac{Y_N \times (N_o - N)}{Y_N \times (N_o - N) + Y_C \times (C_o - C)} \tag{14.24}$$

14.7   Anammox Process

Ammonia oxidation and reduction to nitrogen for its removal under aerobic and anaerobic conditions, respectively, was discovered in nineteenth century. The latest modification to ammonia removal in the form of Anammox was invented in 1990s (Mulder et al., 1995), which led to the discovery of another ammonia conversion pathway. The Anammox process provides biological nitrogen removal without the need of external carbon source, whereas external carbon source is essential in conventional denitrification process of nitrogen removal (Fig. 14.6).

Ammonia removal using Anammox process comprises of two steps; first being oxidation of NH_4 to NO_2^- (Eq. 14.25) and subsequently N_2 formation from oxidization of NH_4 with NO_2^- (Eq. 14.26). The first step is called as partial nitrification; while second step is the Anammox process is ammonia oxidation. During the initial days after invention, the two steps were applied in two separate reactors; however, with recent advances in understanding both the processes can be applied in a single reactor.

$$NH_4^+ + 1.5O_2 \rightarrow NO_2^- + H_2O + 2H^+ \qquad (14.25)$$

$$NH_4^+ + NO_2^- \rightarrow N_2 + 2H_2O \qquad (14.26)$$

In the conventional nitrification process, ammonia is converted to nitrite and nitrate successively under aerobic conditions. Whereas, in Anammox process the ammonia oxidation is restricted to nitrite only; thus, additional oxygen demand in terms of aeration

Fig. 14.6 Anammox process combining partial nitrification (nitritation) and ammonia oxidation

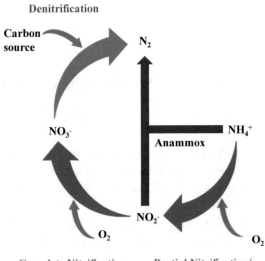

energy for oxidation of nitrite to nitrate is eliminated (Francis et al., 2007). Anammox process can reduce 25% of oxygen requirement during the nitrification stage, while eliminating the external carbon source requirement as required in denitrification stage.

The Anammox bacteria are autotrophic, which can oxidize the ammonia using nitrite as electron acceptor under anoxic conditions. These bacteria are slow growing and have 10 folds lower maximum specific growth rate compared to that of nitrifying microorganisms. Hence, a sludge retention time of minimum 20 days is required for proper growth of Anammox bacteria. Anammox bacteria, produce extracellular polymeric substances, which cause better agglomeration and granule formation. This causes a high settling rate of sludge flocs, thus allowing easy settling in the secondary clarifiers.

Bacteria, namely *Brocadia, Kuenenia, Scalindua, Anammoxoglobus, Jettenia,* etc., are microbial genera capable of demonstrating Anammox reaction pathway. The Anammox bacteria exist in some of the natural systems under oxygen depleted areas, such as marine ecosystem, freshwater, seawater ecosystem, and terrestrial ecosystem. Although, Anammox process is an attractive option for nitrogen removal, the bacteria capable of undertaking Anammox reaction are slow growing with doubling rate of 10–14 days at 30 °C. Also, these species are highly susceptible to changing environment, thus making it difficult to operate the reactor as compared to conventional ammonia oxidizers and nitrate reducers. Factors affecting the Anammox process are described below in details.

Free ammonia: The ammonium concentration of up to 1 g/L can be handled by Anammox bacteria; however, the free ammonia is identified as the process inhibitor. The free ammonia concentration of less than 20–25 mg/L is required for the stable operation of Anammox process. During the start-up of Anammox process, Anammox bacteria require even low concentration of free ammonia, which relates that through acclimatization the free ammonia toleration ability could be developed in Anammox bacteria.

Dissolved oxygen concentration: The dissolved oxygen of about 0.6–0.8 mg/L is beneficial for Anammox bacteria. These bacteria die if exposed to excessive dissolved oxygen levels. Moreover, at higher dissolved oxygen concentration the nitrite reduction will be inhibited, thus affecting the overall Anammox process.

pH: The pH range of 6.7–8.3 is well suited for the functioning of Anammox process. As the pH reduces free nitrous acid concentration increases; whereas under alkaline pH, free ammonia concentration increases. Hence, pH of near neutral is preferred so as to avoid inhibition due to free nitrous acid or free ammonia.

Temperature: Temperature affects the microbial growth rate and temperature of 30–40 °C is considered as the optimum range for Anammox process. Although Anammox bacteria have high tolerance for temperature variation, at temperatures of <15 and >40 °C the nitrogen removal ability is minimized (Van Hulle et al., 2010).

Salinity: Salinity causes built-up of osmotic pressure and it is harmful to microorganisms at high concentrations. Since Anammox bacteria are naturally detected in the marine environment, it is still a promising option for treating wastewater with high salinity. High salinity causes a reduction in performance of Anammox system; however, the performance can still be improved via acclimatization of bacteria. Moreover, the lab experiments proved that the effect of salinity is dependent on the type of salt and the Anammox species (Jin et al., 2012).

The Anammox process can be implemented in either single reactor system or two reactor system in the wastewater treatment plant. Single reactor system comprises of simultaneous partial nitrification and the successive Anammox process in a single reactor. Ammonia oxidizing bacteria grow in outer layer of biofilm, which consumes oxygen for partial nitrification. Whereas, in the deeper layers, anoxic conditions are observed that creates ideal conditions for Anammox bacteria. The control over dissolved oxygen concentration is a critical operational parameter in this configuration. In the two-reactor system, both processes (partial nitrification and Anammox) occur in separate reactors, where individual optimal parameters are maintained. In suspended growth aerobic processes, via integration of intermittent aeration, the dissolved oxygen levels can be controlled to 0.3 mg/L. Low dissolved oxygen concentration in aerobic suspended growth process provides ideal condition for the growth of Anammox bacteria; while limiting the growth of nitrifying microorganisms.

Anammox process offers following benefits over conventional nitrification–denitrification process for nitrogen removal from wastewater.

- The requirement of oxygen is reduced due to partial nitrification (ammonium to nitrite).
- Anammox bacteria do not require addition of external carbon source for nitrogen removal.
- This process minimizes release of CO_2 emissions during biological nitrogen removal since use of carbon source is eliminated.
- Reduced sludge generation is another advantage of Anammox process.

Although, Anammox process is considered as an attractive option than conventional nitrification–denitrification process for nitrogen removal, it is associated with technical challenges as described below, which are required to be resolved for field applications.

- Longer start-up duration required for Anammox process compared to conventional nitrification–denitrification process, due to slow growth rate of Anammox bacteria.
- Anammox bacteria are susceptible to high nitrite concentration and high oxygen levels, hence precise process control is essential.
- Skilled operators are required to operate and maintain the process.

Example: 14.1 Estimate the maximum specific growth rate for nitrifying bacteria when the reactor is operated as follows: temperature = 15 °C, dissolved oxygen concentration = 2.5 mg/L, operating pH = 7.2, $\mu_{N\,max} = 0.9$/day, $K_{DN} = 0.15$/day, $K_{DO} = 0.45$ mg/L and temperature coefficient = 1.072.

Solution

The actual value of $\mu_{N\,max\,actual}$ is dependent on the field operating parameters, such as pH, dissolved oxygen, and temperature.

Determination of the temperature correction factor.

$F_T = \theta_T^{T-20}$ where (θ_T) at 20 °C = 1.072.

$F_T = 1.072^{15-20} = 0.706$.

Determination of the dissolved oxygen correction factor

$$F_{DO} = \frac{DO}{K_{DO} + DO}$$

$$F_{DO} = \frac{2.5}{0.45 + 2.5} = 0.847$$

Determination of the pH correction factor.
The pH correction factors are pH dependent,

For pH < 7.0, $F_{pH} = 0.0004017 * e^{1.0946*pH}$.

For pH of 7.2, $F_{pH} = 2.35^{pH-7.2}$.

For pH of 7.2 < pH < 9.5, $F_{pH} = \frac{1.13(9.5 - pH)}{9.8 - pH}$.

$F_{pH} = 2.35^{7.2-7.2} = 1$.

The $\mu_{N\,max\,actual}$ is estimated as:

$$\mu_{N\,max\,actual} = F_T \times F_{DO} \times F_{pH} \times \mu_{N\,max}$$

$$\mu_{N\,max\,actual} = 0.706 \times 0.847 \times 1 \times 0.9 = 0.538 \text{ per day}$$

Example: 14.2 Design an activated sludge process for the treatment of wastewater with a flow of 4 MLD having influent BOD_5 and TKN as 140 and 30 mg/L, respectively. The desired concentration of BOD_5 and nitrate is 30 and 1 mg/L, respectively. The MLVSS content in the aeration tank is 3500 mg/L. The values of other parameters are as follows: $\mu_{max} = 0.5$/day, $\mu_{N\,max} = 0.45$/day, $K_S = 55$ mg/L, $K_N = 0.4$ mg/L, dissolved oxygen = 2 mg/L, $K_{DN} = 0.04$/d, $Y_N = 0.16$, $Y_{DN} = 0.2$, $Y_C = 0.6$, $K_D = 0.07$/day.

Solution

The nitrification and organic matter oxidation are being achieved in a single aerobic reactor. Hence the hydraulic retention time for nitrification as well as organic matter degradation should be computed and the maximum out of these two shall be considered for designing.

Calculation of solid retention time for nitrifying microorganisms

$$\mu_N = \frac{\mu_{N\,max} \times N}{K_N + N} \times \frac{DO}{K_N + DO} - K_{DN}$$

$$\mu_N = \frac{0.45 \times 1}{0.4+1} \times \frac{2}{0.4+2} - 0.04; \text{ thus, } \mu_N = 0.227\,per\,day$$

$$\theta_{CN} = \frac{1}{\mu_N}$$

$$\theta_{CN} = \frac{1}{0.227}; \text{ hence } \theta_{CN} = 4.4\,\text{day}$$

Increasing the θ_{CN} by 2 times as factor of safety, hence $\theta_{CN} = 8.8$ days.

Calculation of N value in the effluent of aerobic process

$$N = \frac{K_N \times (1 + K_{DN} \times \theta_{CN})}{\theta_{CN} \times (\mu_{N\,max} - K_{DN}) - 1}$$

$$N = \frac{0.4 \times (1 + 0.04 \times 8.8)}{8.8 \times (0.45 - 0.04) - 1} \quad \text{Thus, } N = 0.207\,\text{mg/L}$$

Calculation of fraction of nitrifying microorganisms (F_N) in the aerobic reactor

$$F_N = \frac{Y_N \times (N_o - N)}{Y_N \times (N_o - N) + Y_C \times (C_o - C)}$$

$$F_N = \frac{0.16 \times (30 - 0.207)}{0.16 \times (30 - 0.207) + 0.6 \times (140 - 30)} \quad \text{Therefore } F_N = 0.067$$

This means that out of the total biomass in the aerobic reactor, 6.7% is nitrifying microorganisms.

Since MLVSS is 3500 mg/L; hence, nitrifying microorganisms (MLVSS$_N$)

$= 3500 \times 0.067 = 235$ mg/L

Determination of hydraulic retention time for nitrification

Calculation of substrate utilization rate of nitrifying microorganisms (U_N)

$$U_N = \frac{\mu_{N\,max} \times N}{Y_N \times (K_N + N)}$$

$$U_N = \frac{0.45 \times 0.207}{0.16 \times (0.4 + 0.207)} \quad \text{hence, } \quad U_N = 0.96\,\text{mgN/mg VSS day}$$

$$\theta_N = \frac{N_o - N}{U_N \times MLVSS_N}$$

$\theta_N = \frac{30 - 0.207}{0.96 \times 235}$ hence, $\theta_N = 0.132\,\text{day} = 3.17\,\text{h}$ (for nitrification)

Calculation of BOD$_5$ value in the effluent of aerobic process

$$S = \frac{K_S \times (1 + K_D \times \theta_{CN})}{\theta_{CN} \times (\mu_{\max} - K_D) - 1}$$

In this case, solid retention time for nitrifying microorganisms (θ_{CN}) and heterotrophic microorganisms (θ_C) is equal as it is a combined process.

$$S = \frac{55 \times (1 + 0.04 \times 8.8)}{8.8 \times (0.5 - 0.07) - 1} \quad S = 26.71\,\text{mg/L}$$

The value of effluent BOD$_5$ concentration (26.71 mg/L) is less than the desired value of 30 mg/L. Hence, the computed value of (θ_{CN}) for nitrifying microorganisms is suitable for the heterotrophic microorganisms as well.

Calculation of fraction of heterotrophic microorganisms in the aerobic reactor.
Nitrifying microorganisms = 6.7%, hence heterotrophic microorganisms' fraction is 93.3%

MLVSS$_C$ = 3500 – 235 = 3265 mg/L

Determination of hydraulic retention time for the oxidation of carbonaceous organic matter.

Calculation of substrate utilization rate of heterotrophic microorganisms (U)

$$U = \frac{\mu_{\max} \times S}{Y_C \times (K_S + S)}$$

$$U = \frac{0.5 \times 26.71}{0.6 \times (55 + 26.71)} \quad \text{Therefore } U = 0.273\,\text{mg N/mg VSS day}$$

$$\theta_C = \frac{S_o - S}{U \times MLVSS_C}$$

$\theta_C = \frac{140 - 26.71}{0.273 \times 3265}$ hence, $\theta_C = 0.127\,day = 3.05\,\text{h}$ (for carbonaceous matter removal).

Considering hydraulic retention time maximum of θ_N and θ_C. Hence $\theta_C = \theta_N = 3.17\,\text{h}$.

Volume of aerobic reactor and loading rates

$$\theta = \frac{V}{Q}$$

$$V = \frac{3.17 \times 4 \times 10^3}{24} \qquad V = 529 \, \text{m}^3$$

$$\frac{F}{M} = \frac{Q \times S_o}{MLVSS_C \times V}$$

$$\frac{F}{M} = \frac{4 \times 10^3 \times 140 \times 10^{-3}}{3265 \times 10^{-3} \times 529} \quad \frac{F}{M} = 0.325 \text{ kg BOD/kg VSS day}$$

$$Organic\,loading\,rate = \frac{Q \times S_o}{V}$$

$$Organic\,loading\,rate = \frac{4 \times 10^3 \times 140 \times 10^{-3}}{529} = 1.05 \text{ kg BOD}_5/\text{m}^3.\text{day}$$

Air requirement

$$Sludge\,production\,(P_{XVSS}) = \frac{Y_C \times Q \times (S_o - S)}{1 + K_D \times \theta_{CN}}$$

For heterotrophic bacteria

$$P_{XVSS} = \frac{0.6 \times 4 \times 10^3 \times \left((140 - 26.71) \times 10^{-3}\right)}{1 + 0.07 \times 8.8} = 168.25 \text{ kg/day}$$

For nitrifying bacterial sludge

$$P_{XVSS} = \frac{0.16 \times 4 \times 10^3 \times \left((30 - 0.207) \times 10^{-3}\right)}{1 + 0.04 \times 8.8} = 14.1 \text{ kg/day}$$

Hence, total excess sludge VSS produced, which is wasted per day from the aeration tank

$$= 168.25 + 14.1 = 182.35 \text{ kg/day}$$

$$Oxygen\,requirement = 1.47 \times Q(S_o - S) + 4.57 \times Q(N_o - N) - 1.42 P_{XVSS}$$

$$\begin{aligned} Oxygen\,requirement &= 1.47 \times 4 \times 10^3 \times \left((140 - 26.71) \times 10^{-3}\right) \\ &\quad + 4.57 \times 4 \times 10^3 \times \left((30 - 0.207) \times 10^{-3}\right) - 1.42 \times 182.35 \\ &= 951.83 \text{ kg O}_2/\text{day} \end{aligned}$$

Estimating volume of air required considering air contains 23% oxygen by weight, air density as 1.2 kg/m^3, and oxygen transfer efficiency as 8%.

$$Volume\,of\,air = \frac{951.83}{0.23 \times 1.2 \times 0.08} = 43108.24 \, \text{m}^3/ \text{day} \left(29.94 \, \text{m}^3/\text{min}\right)$$

Example: 14.3 In a biological nutrient removal process, methanol is used as a carbon source for the denitrification of nitrified effluent. Estimate the daily methanol requirement for the treatment plant with wastewater flow of 300 m^3/day, if the nitrate, and nitrite concentration in the anoxic tank is 22 and 3.5 mg/L, respectively, along with dissolved oxygen concentration of 0.75 mg/L.

Solution
According to the Eq. 14.11,

$$Methanol_{dose} = 2.47 N_{nitrate} + 1.53 N_{nitrite} + 0.87 DO_{removed}$$

$$= (2.47 \times 22\,mg/L) + (1.53 \times 3.5\,mg/L) + (0.87 \times 0.75\,mg/L)$$

$Methanol_{dose} = 60.35$ mg/L $= 0.06035$ kg/m^3.
Daily amount of methanol $= (0.06035\,kg/m^3 \times 300\,m^3/day)$.
$= 18.105$ kg methanol per day.

14.8   Biological Phosphorus Removal

14.8.1 Process Description

Phosphorus is mainly released in the municipal wastewaters from domestic sources, industrial wastewater (food processing and fertilizer industrial wastewater), and agricultural runoff. Domestic wastewater typically contains phosphorus in the form of orthophosphate along with other compounds such as phosphorylated proteins, nucleic acids, etc. (Yeoman et al., 1988). The phosphorus can be removed from wastewater using physico-chemical, biological or combination of both treatments. The physico-chemical phosphorus removal techniques are effective; however, these methods require chemical addition and often additional processing unit (solid separation) as a follow-up treatment. Adsorption, ion exchange and chemical precipitation (using calcium, aluminium, iron) are the examples of physico-chemical phosphorus removal techniques (Ramasahayam et al., 2014).

The aerobic secondary biological treatment process is designed mainly for the oxidation of carbonaceous organic matter and ammonia nitrogen from wastewater. As elaborated in Chap. 10, in activated sludge process, influent wastewater is treated in the aeration tank in presence of dissolved oxygen and the aerobic microorganisms. Some amount of phosphorus is essential for the growth of microorganisms, which causes minor amount of phosphorus removal from wastewater during the secondary treatment. For the biological phosphorus removal enhanced biological phosphorus removal (EBPR) is a cost-effective method and is considered as an alternative to the chemical precipitation. In the EBPR process, polyphosphate accumulating organisms convert the phosphorus from wastewater into sludge in the form of intracellular polyphosphate. The biomass sludge generated from conventional activated sludge process contains only ~1.5–2% phosphorus (of the dry weight), which can be enhanced using EBPR process. Hence, in conventional aerobic wastewater treatment plants, only about 10–20% of the phosphorus removal can be accomplished.

The growth of microorganisms capable of EBPR is favourable in successive anaerobic followed by aerobic conditions. As a result, the phosphorus removal in EBPR can be enhanced to 80–90%. The biological phosphorus removal process is dependent on the concentration of phosphorus accumulating organisms in the activated sludge process. These organisms can accumulate greater amount of poly-phosphate inside their cell, which results in higher phosphate removal. The dried biomass generated from EBPR process contains 4.5–5% phosphorus.

Phosphorus accumulating organisms are aerobic microorganisms and they cannot multiply under the anaerobic conditions. However, under harsh anaerobic conditions, phosphorus accumulating organisms have the ability to use volatile fatty acids. During the process, these organisms degrade polyphosphate (stored in the cell), thus causing release of phosphorus in the surrounding wastewater. This causes increase in the dissolved phosphorus concentration during the anaerobic treatment. In the subsequent aerobic zone, the phosphorus accumulating organisms are able to grow and consume the available dissolved phosphorus and store it as intracellular polyphosphate. The phosphorus accumulating organism are selective and excessive phosphorus accumulation occurs in the aerobic zone. This means more storage of phosphorus than equivalent phosphorus released under the anaerobic conditions. In this manner, when the biomass is removed from the reactor as excess biomass after aerobic cycle, the removal of phosphorus occurs in the EBPR process.

The sludge with enriched phosphorus accumulating organisms can only be obtained through a sequence of anaerobic followed by aerobic treatment. The EBPR configuration consist of anaerobic zone in which the influent wastewater is admitted and it is mixed with returned activated sludge from the secondary sedimentation tank. The effluent from the anaerobic zone along with suspended biomass is transferred in the aerobic zone prior to separation of sludge and treated effluent using secondary sedimentation tank. To achieve simultaneous nitrogen and phosphorus removal, the process can be slightly modified via incorporation of anoxic zone, which is termed as A^2O process (anaerobic-anoxic-aerobic). The nitrate formed in the aerobic tank is recirculated in the anoxic tank for denitrification. This ensures that minimum amount of nitrate reaches the anaerobic zone along with returned activate sludge (Fig. 14.7).

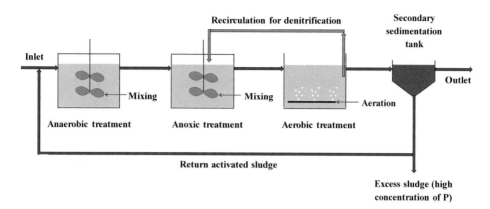

Fig. 14.7 Biological nutrient (phosphorus and nitrogen) removal process using anaerobic followed by aerobic treatment

14.8.2 Factors Affecting Enhanced Biological Phosphorus Removal Process

Influent wastewater characteristics: The phosphorus accumulating organisms mainly require volatile fatty acids source to release phosphorus that is stored in the cell. The volatile fatty acids can be in-situ synthesized in the anaerobic zone or can be externally supplemented. The BOD: total phosphorus ratio of about 20:1 to 25:1 is sufficient for the phosphorus accumulating organisms to survive in the anaerobic zone (Jeyanayagam, 2005). Lower ratio of BOD: total phosphorus than this might create unfavourable conditions for the fermentation as a result of which, complete phosphorus might not be released in the wastewater.

Hydraulic retention time: In the anaerobic zone, shorter the hydraulic retention time, the higher is the volatile fatty acid concentrations. Thus typically, hydraulic retention time of 1–2 h is sufficient for fermentation and release of phosphorus in the wastewater. However, very short retention time in the anaerobic zone might affect the denitrification of wastewater. Therefore, the trade-off between the contact time for denitrification and phosphorus release needs to be worked out.

Dissolved oxygen concentration: Sufficient amount of dissolved oxygen is required in the aeration tank so as to create favourable atmosphere for the growth of phosphorus accumulation organisms. Dissolved oxygen concentration of 4.5–5 mg/L promotes growth of glucose accumulating organisms; whereas that of 2.5–3 mg/L creates conducive atmosphere for the growth of phosphorus accumulating organisms (Izadi et al., 2020). Additionally, the returned activated sludge should be without any dissolved oxygen so as to promote volatile fatty acid formation in the anaerobic treatment.

Questions

14.1. Why it is necessary to remove nutrient from wastewaters?

14.2. Describe biological nitrification and denitrification process.

14.3. How ANAMMOX process is different from biological nitrification and denitrification?

14.4. How the phosphors removal occurs in enhanced biological phosphorus removal process?

14.5   What should be the treatment train for the removal of nutrients from municipal wastewater with an aim of removing nutrients via biological pathway?

14.6. What is the effect of BOD to TP ratio on the phosphorus removal?

14.7   What is the differentiation between single sludge and two sludge system during nitrification–denitrification? Under what circumstances single sludge system is suitable?

14.8. Why pre-anoxic system is preferred over post-anoxic treatment for biological nitrogen removal?

14.9. What is the importance of anaerobic zone in the enhanced biological phosphorus removal process?

14.10. Determine the actual maximum specific growth rate for denitrification with the operating conditions as follows: temperature $= 11\ °C$, dissolved oxygen concentration $= 0.1$ mg/L, nitrate concentration in the anoxic tank $= 7$ mg/L, temperature coefficient for denitrifying microorganisms $= 1.072$, $K_{DO} = 0.1$ mg/L, $\mu_{N\,max} = 1.1$/day (**Answer**: 0.607/day).

References

Anderson, D. M., Burkholder, J. M., Cochlan, W. P., Glibert, P. M., Gobler, C. J., Heil, C. A., Kudela, R. M., Parsons, M. L., Rensel, J. E. J., Townsend, D. W., Trainer, V. L., & Vargo, G. A. (2008). Harmful algal blooms and eutrophication: Examining linkages from selected coastal regions of the United States. *Harmful Algae, 8*, 39–53. https://doi.org/10.1016/j.hal.2008.08.017

Francis, C. A., Beman, J. M., Kuypers, M. M. M. (2007). New processes and players in the nitrogen cycle: the microbial ecology of anaerobic and archaeal ammonia oxidation. *ISME Journal, 11*(1), 19–27. https://doi.org/10.1038/ismej.2007.8

Issacs, S. H., & Henze, H. (1995). Controlled carbon source addition to an alternating nitrification-denitrification wastewater treatment process including biological p removal. *War Resolution, 29*(1), 77–89. https://doi.org/10.1016/0043-1354(94)E0119-Q

Izadi, P., Izadi, P., & Eldyasti, A. (2020). Design, operation and technology configurations for enhanced biological phosphorus removal (EBPR) process: A review. *Reviews in Environmental Science & Biotechnology, 19*, 561–593. https://doi.org/10.1007/s11157-020-09538-w

Jeyanayagam, S. (2005). True confessions of the biological nutrient removal process. *Florida Water Resource Journal, 1*, 37–46.

Jin, R. C., Yang, G. F., Yu, J. J., & Zheng, P. (2012). The inhibition of the Anammox process: A review. *Chemical Engineering Journal, 197*, 67–79. https://doi.org/10.1016/J.CEJ.2012.05.014

Mulder, A., Graff, A., Roberston, L., & Kuenen, J. (1995). Anaerobic ammonium oxidation discovered in a denitrifying fluidized bed reactor. *FEMS Microbiology Ecology, 16*, 177–184. https://doi.org/10.1016/0168-6496(94)00081-7

Parker D., S., Stone, R. W., Stenquist, R. J., & Culp, G. (1975). *Process design manual for nitrogen control*. US Environmental Protection Agency, Washington, DC, Office of Technology Transfer.

Peng, Y., Yong, M., & Wang, S. (2007). Denitrification potential enhancement by addition of external carbon sourcesin a pre-denitrification process. *Journal of Environmental Sciences, 19*, 284–289. https://doi.org/10.1016/S1001-0742(07)60046-1

Ramasahayam, S., Guzman, L., Gunawan, G., & Viswanathan, T. (2014). A comprehensive review of phosphorus removal technologies and processes. *Journal of Macromolecular Science Part A, 51*(6), 536–545. https://doi.org/10.1080/10601325.2014.906271

United States Environmental Protection Agency. (1993). *Manual nitrogen control, EPA/625/R-93/010*. Office of research and development.

United States Environmental Protection Agency, Wastewater Treatment Fact Sheet: External Carbon Sources for Nitrogen Removal, EPA 832-F-13-016, Office of Wastewater Management, August 2013.

Van Hulle, S. W. H., Vandeweyer, H. J. P., Meesschaert, B. D., Vanrolleghem, P. A., Dejans, P., & Dumoulin, A. (2010). Engineering aspects and practical application of autotrophic nitrogen removal from nitrogen rich streams. *Chemical Engineering Journal, 162*, 1–20. https://doi.org/10.1016/J.CEJ.2010.05.037

Yeoman, S., Stephenson, T., Lester, J., & Perry, R. (1988). The removal of phosphorus during wastewater treatment: A review. *Environmental Pollution, 49*, 183–233. https://doi.org/10.1016/0269-7491(88)90209-6

Zhang, F., Yang, H., Wang, J., Liu, Z., & Guan, Q. (2018). Effect of free ammonia inhibition on NOB activity in high nitrifying performance of sludge. *RSC Advances, 8*, 31987. https://doi.org/10.1039/C8RA06198J

Sludge Management

15

Wastewater treatment operations and processes are primarily aimed at the removal of impurities from the bulk liquid. These impurities, when concentrated into solid form are referred to as residue or sludge, which includes organic and inorganic solids, algae, bacteria, virus, colloids and chemicals used for precipitating dissolved salts along with ionic species present in wastewater. These are removed by various solid–liquid separation processes employed in wastewater treatment systems. The semisolid residuals produced during biological wastewater treatment processes are termed as *biosolids or biomass*. Sludge management is an integral part of wastewater treatment systems and it involves the engineering tasks of sludge collection, pumping, treatment, disposal and reuse.

The objective of sludge management, from an engineering perspective, is usually to minimize the mass and volume of solids that must ultimately be disposed-off, after properly recovering recyclable materials, reducing biodegradable fraction present in it and reducing the water content of the sludge. The sludge treatment processes and disposal facilities usually deal with the raw sludge, which may be 1–5% of the total wastewater volume handled by the wastewater treatment plant; however, it accounts for 40–60% of capital cost and about 50% of the operating cost of wastewater treatment plant, and 90% of operational problems at treatment plants (Henry & Heinke, 1996). Therefore, the ultimate disposal of sludge has been one of the challenging and cost intensive tasks in the field of wastewater engineering.

15.1 Importance of Sludge Treatment and Disposal

Sludge is made of materials separated from raw wastewater or produced as a result of chemical coagulation and biochemical oxidation in the wastewater treatment processes. It typically contains large volumes of water as much as 98% and usually appears to be in

© The Author(s), under exclusive license to Springer Nature Singapore Pte Ltd. 2022   619
M. Ghangrekar, *Wastewater to Water*, https://doi.org/10.1007/978-981-19-4048-4_15

the form of a semisolid liquid slurry. However, the solid content of resulting sludge may be varied from 0.12 to 12%, depending upon the types of unit operations and processes employed in wastewater treatment plants. The sludge containing high moisture content becomes bulkier for its storage and handling. Hence, the sludge treatment processes are concerned with removing large fraction of water from the sludge. The sludge is putrescible due to its high organic contents and it could be hazardous or toxic to human as it may contain inorganic contaminants. The sludge generated from sewage treatment plant contains pathogenic organisms, which may contribute to the outbreak of water borne diseases and thus it offers a threat to public health. Therefore, handling and disposal of sludge is a complex process. Thus, the objectives of sludge treatment and disposal are:

- To reduce the organic content of the sludge to make it stable, inoffensive and more mineralized materials.
- To reduce the volume of sludge by minimizing water content in it.
- To reduce the presence of pathogens, present in it.
- To enhance the fertile value of sludge for agricultural use.
- To prevent soil pollution and pollution of groundwater when it is disposed-off.
- To dispose it off in a safe and aesthetically acceptable manner.

15.2   Sources and Characteristics of Sludge

Sewage sludge consists of clumpy form of suspended and colloidal solids scattered in water (Kiely, 2007). The sludge includes moisture, dry solids, volatile solids, and inorganic solids with varying physical characteristics, such as density, colour, texture, fluidity, plasticity, specific gravity, shear strength, energy content and particle size. The chemical characteristics include metals, nutrients and salts present in the dry solids. Sludge behaves as a liquid when moisture content is above 90%, and below 90%, it behaves as a non-Newtonian fluid exhibiting plasticity.

The type of sludge produced depends on the type of solids separation processes, size of the treatment plant and characteristics of wastewater. Sewage sludge consists of organic and inorganic solids removed from raw sewage in primary clarifier, and organic cellular biomass generated in secondary clarifier. The inorganic solid fraction in sewage sludge is found to have a specific gravity of about 2.5, while the organic matter has specific gravity ranging from 1.01 to 1.6. However, industrial wastewaters produce very different types of sludge that are organic or inorganic with or without presence of toxic materials, such as heavy metals.

15.2.1 Primary or Raw Sludge

During treatment of sewage, the sludge collected from primary sedimentation tank contains organic solids and finer inorganic particles those escaped the grit chamber. It is grey in colour and has an offensive faecal odour with about or more than 95% of water content. This sludge contains more than 70% of organic solids and it is difficult to dewater. Due to presence of high moisture content, it cannot be easily dried on sand drying beds and it necessitates digestion to improve dewaterability before drying. The primary sludge contains 5–9% dry solids (Hammer, 2000). This primary sludge can be readily digested either aerobically or anaerobically.

15.2.2 Secondary Sludges

The secondary sludge contains flocculated biosolids, suspended organic matter biosorbed on to floc and inert matters that are wasted from biological treatment processes. This sludge can get digested with or without addition of primary sludge. The organic content (volatile solids) of a typical secondary sludge varies from 60 to 80%. Depending upon the type of biological wastewater treatment process employed the characteristics of the sludge will vary, e.g., the sludge generated from conventional ASP operated with low solid retention time will be difficult to dewater and requires treatment before dewatering, whereas the sludge generated from extended aeration type ASP can be easily be dewatered. In case secondary process involving chemical oxidation, the sludge contains large amounts of metal salts. The different sources and characteristics of the sludge generated from wastewater treatment are described below.

Activated sludge: The sludge from a properly operating activated sludge process appears to be golden brown with inoffensive earthy odour. When it undergoes septic condition in the absence of aeration it becomes dark in colour. The sludge with lighter colour is produced due to under aeration and tends to settle slowly. Waste activated sludge from activated sludge process contains large concentration of flocculated microbial mass biosorbed on inert suspended and colloidal solids (Hammer, 2000). The organic content of activated sludge varies from 59 to 88%. Waste activated sludge typically contains 0.8–2.5% dry solids. Thus, moisture content of this sludge could be around 98–99%. However, when mixed with primary sludge the mixed sludge contains about 2.0% of dry solids.

Trickling filter sludge: The sludge generated from trickling filter has brown colour and appears to be flocculent biomass (humus), which settles rapidly as it is a biofilm sloughed out of the filter matrix. It decomposes very slowly and does not concentrate to high density without digestion. Trickling filter sludge contains typically 1–3% of dry solids.

Chemical precipitation sludge: Sludge from chemical precipitation mainly consists of metals salts such as hydrates of iron and aluminium forming gelatinous precipitates. It is dark in colour usually or reddish-brown if iron content is more or greyish-brown colour for lime sludge. Depending on the type of wastewater being treated the organic fraction in this sludge will vary and even if organic fraction is present, it will undergo reduced rate of decomposition.

15.2.3 Digested Sludge

Aerobically digested sludge: It is a well digested flocculated mass of slurry under aerobic condition with colour varying from brown to dark. It has no offensive odour and it can be dewatered on sand drying beds. However, only about 50% of the volatile solids are converted to gaseous end products making it expensive for dewatering and hence aerobic digestion has limited application for waste activated sludge.

Anaerobically digested sludge: Anaerobically digested sludge is not offensive in odour and dark brown or black in colour. The moisture content of primary and secondary digested sludge may be typically about 94–96%. Sludge from up-flow anaerobic sludge blanket reactor contains solids ranging from 3 to 6% of dry solids and a per capita digested sludge production ranges from 12 to 18 g SS/capita day (Andreoli et al., 2007).

Compost: The composted solids usually appear to be similar colour as that of a digested sludge, however the colour may vary depending upon the type of bulking agents used in composting process.

15.2.4 Tertiary Sludges

Tertiary treatment process involves the removal of nutrients like nitrogen and phosphorous from wastewater. The characteristics of the sludges from tertiary treatments depend on the nature of process employed, such as biological removal nitrogen, chemical precipitation of phosphate, etc. When chemical phosphorous removal is accomplished after activated sludge process, the resulting chemical sludge when combined with biological sludge makes it more difficult to treat. Denitrification process for the removal of nitrogen results in biological sludge with similar characteristics as that of sludge generated from the anaerobic biological reactor.

15.3 Quantity and Mass-Volume Relationship of Sludge

15.3.1 Quantity Estimation

The quantity and nature of sludge produced in a biological wastewater treatment process depends on its influent wastewater characteristics, type of unit operations and processes used, organic loading rate adopted and removal efficiency of the biological process employed. The typical quantity of solids contribution in domestic wastewater is 90 g/capita day of suspended solids with concentration ranging from 100 to 300 mg/L.

For primary sedimentation tank, the quantity of sludge produced, on a dry basis, can be estimated from percent removal of suspended solids and its concentration present in the influent wastewater. However, the quantity of sludge produced by aerobic biological processes depends on the organic loading rate and sludge age adopted. The biological solids generated from attached growth aerobic processes usually ranged between 0.4 and 0.7 g VSS/g of BOD_5 applied with varying organic loading rates. The biological solids production (net VSS yield) for a suspended growth process varies with F/M ratio employed and is less than 0.3 g VSS/g of BOD_5 for extended aeration and above 0.45 and up to 0.8 g VSS/g of BOD_5 removed for conventional activated sludge process. The biomass produced in case of anaerobic process is about one-fifth of that for aerobic process. The combined daily primary and secondary sludge production can be estimated by using Eq. 15.1 (Gray, 1999).

$$W_{ds} = Q[K_1 S + (1 - K_2)Y \cdot b \cdot K_3] \, \text{kg/day} \tag{15.1}$$

where, W_{ds} is the sum of dry mass of primary and secondary sludges, kg/day, Q is wastewater flow rate, m^3/day; S is suspended solids concentration in primary sludge, kg/m^3; K_1 is the fraction of suspended solids removal in primary settling tank and K_2 is fraction of total BOD_5 removed in primary treatment; b is the total BOD_5 present in wastewater, kg/m^3; K_3 is the fraction of BOD_5 removed during biological treatment process; and Y is the sludge yield coefficient, kg VSS/kg BOD_5 removed.

Wastewater treatment plant sludges constitute primarily water and less solids; thus, the volume of sludge is estimated from the solids content and the specific gravity of the sludge. The wet volume of sludge decreases considerably as the concentration of solids increases two-fold (Hammer, 2000). However, direct correlation of concentration and volume does not occur in reality due to change in density of the sludge with change in concentration. The exact density of sludge depends on the solids content and it is calculated by adopting mass balance as illustrated in Eq. 15.2.

$$\rho_{sl} . V_{sl} = \rho_w . V_w + \rho_{ds} . V_{ds} \tag{15.2}$$

where, $\rho_{sl} . V_{sl}$ represents the mass of the sludge as a product of density and volume of liquid sludge; $\rho_w . V_w$ represents the mass of water as a product of density (1000 kg/m^3) and volume of water; and $\rho_{ds} . V_{ds}$ represents the mass of solids with product of density of dry solids in sludge and volume of solids.

Example: 15.1

A conventional activated sludge process treats 2000 m^3/day of municipal wastewater having BOD of 250 mg/L and suspended solids concentration of 200 mg/L. The activated sludge is concentrated using a floatation thickener and it is blended with primary sludge in a holding tank. Estimate the quantity and solids contents of the primary, secondary and mixed sludges. Assuming 65% suspended solids and 30% BOD are reduced in the primary settling tank. Primary sludge has water content 95%, solids content of activated sludge and sludge from flotation thickener are 1.5% and 3.5%, respectively. Consider sludge yield coefficient, $Y = 0.5$/day and BOD removal in the aeration basin of 90% (K_3).

Solution

$$\text{Weight of primary sludge} = Q \times K_1 \times S$$
$$= 2000 \times 0.65 \times 200 \times 10^{-3} = 260 \, \text{kg/day}$$

$$\text{Volume of primary sludge,} \, W_p = 260/[(1 - 0.95) \times 1000]$$
$$= 5.2 \, \text{m}^3/\text{day}$$

$$\text{Weight of secondary sludge,} \, W_s = Q[(1 - K_2) \times Y \times b \times K_3]$$
$$= 2000 \times 10^{-3}[(1 - 0.3) \times 0.5 \times 250 \times 0.9]$$
$$= 157.5 \, \text{kg/day}$$

$$\text{Volume of secondary sludge} = 157.5/(15)$$
$$= 10.5 \, \text{m}^3/\text{day}$$

Assuming 100% solids capture after floatation thickening,

$$\text{Volume secondary thickened sludge} = 157.5/(35)$$
$$= 4.5 \, \text{m}^3/\text{day}$$

The mixed sludge has the following characteristics:

$$\text{Weight of mixed sludges} = 260 + 157.5$$
$$= 417.5 \, \text{kg/day}$$

$$\text{Volume of the mixed sludges} = 5.2 + 4.5$$
$$= 9.7 \, \text{m}^3/\text{day}$$

$$\text{Solids content of the mixed sludges}(P_s) = (100 \times 417.5)(1000 \times 9.7)$$
$$= \mathbf{4.30\%}$$

15.3.2 Mass-Volume Relationship

The sludge volume mainly depends on the water content and only slightly on the physical characteristics of the solids matter present. A sludge having 5% of solids is having 95% of water in it. When the sludge comprises of fixed solids and volatile solids, the specific gravity of this mixed sludge can be determined using Eq. 15.3.

$$\frac{W_s}{S_s \rho_w} = \frac{W_f}{S_f \rho_w} + \frac{W_v}{S_v \rho_w} \tag{15.3}$$

where, W_s is weight of solids; S_s is specific gravity of solids; ρ_w is density of water; W_f weight of fixed solids; S_f specific gravity of fixed solids; W_v is the weight of volatile solids; and S_v is the specific gravity of volatile solids.

For a sludge containing 5% solids, out of which 1/3 are fixed solids with specific gravity of 2.5 and 2/3 are volatile solids with specific gravity of 1.02, then the specific gravity of this sludge can be estimated as stated below using Eq. 15.3.

$$\frac{1}{S_s} = \frac{0.33}{2.5} + \frac{0.67}{1.02} = 0.789$$

Hence, $S_s = 1/0.789 = 1.267$.

Now, considering the specific gravity of the water to be 1.0, the specific gravity of the sludge S_{sl} with 5% solids present in it can be estimated as stated below.

$$\frac{1}{S_{sl}} = \frac{0.05}{1.267} + \frac{0.95}{1.0} = 0.9895$$

Hence, $S_{sl} = 1/0.9895 = 1.011$.

The volume of the sludge can be estimated using the Eq. 15.4.

$$V = \frac{W_s}{\rho_w S_{sl} P_s} \tag{15.4}$$

where, V is the volume of the sludge, m^3; W_s is the mass of dry solids, kg; ρ_w is the density of water, 1000 kg/m^3; S_{sl} is the specific gravity of the wet sludge; and P_s is the percentage of solids expressed in fraction.

The volume of the sludge varies approximately with the inverse of the percent of solids matter contained in the sludge as given by the relation in Eq. 15.5.

$$\frac{P_1}{P_2} = \frac{V_2}{V_1} \tag{15.5}$$

where, V_1 and V_2 represents sludge volume and P_1 and P_2 represents percentage of solids.

15.4 Sludge Treatment

Sludge with moisture content more than 95% has 70% water in free form and the remaining is bound to the sludge (Gray, 1999; McGhee, 2014). Free water, refers to the water that surrounded the sewage sludge solids, and it is removed by sedimentation allowing thickening of sludge. Floc water or interstitial water is that part of the water, which is trapped in the interstices of floc particles and it can be removed by mechanical dewatering. Capillary water is removed through compaction (dewatering). Bound water, that is chemically bound water within the bacterial cells, can be removed by destruction of cells utilizing digestion process, and chemical or thermal processes. Bound water comprises of intercellular (adsorbed on surface) and intracellular (absorbed) water. These mechanisms are shown in Fig. 15.1. The relative proportions of all forms of water and solids in waste-activated sludge are shown in Table 15.1. As the capillary and bound water content increases, the sludge becomes more difficult to dewater.

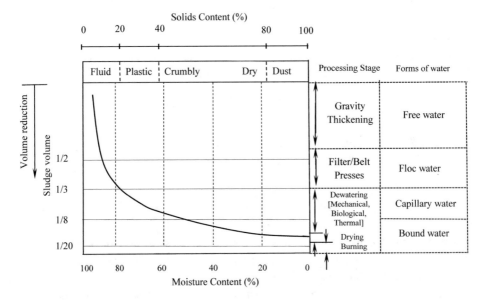

Fig. 15.1 Mechanism of sludge dewatering and variation of moisture content

Table 15.1 Relative proportion of water in waste-activated sludge

Constituents	Percentage by weight	Removal method(s)
Free water	70	Gravity settling
Floc water	25	Mechanical dewatering
Capillary water	2	Compaction (dewatering)
Bound water	2	Chemical, biological, thermal
Solids	1	All separation processes

The processes selected for sludge treatment depends primarily on nature and characteristics of the sludge and on the final disposal option adopted (Eckenfelder, 2000). The sludge generated from a wastewater treatment plant undergoes usually series of treatment steps, such as thickening, stabilization, conditioning, dewatering, volume reduction and final disposal. Typical activated sludge can effectively be concentrated by flotation than by gravity thickening and further the solids can be considerably reduced by incineration as the final disposal option. Thickening is the process of separating as much water as possible by gravity or flotation. Stabilization converts the organic solids into more mineralized form so that it can be used as soil conditioner without causing any adverse consequences through biochemical oxidation processes. Conditioning referred to as treating the sludge with chemical or heat so that the separation of water is rapid. Dewatering is the process of separating water under the conditions of vacuum, pressure or solar drying. Volume reduction of the sludge is the process in which the volume of the sludge is decreased by converting the solids into stable-inert solids by wet oxidation and incineration. The different unit operations and processes that can be adopted for treatment of the sludge are illustrated in the Fig. 15.2.

Sludge processing and disposal are most challenging part of a wastewater treatment processes and are often neglected due to economic constraints and land availability. The primary sludge produced in a municipal wastewater treatment plant consists of organic and inorganic solids that are readily settled out in an hour and thus it is usually capable of thickening by physical processes. Secondary sludges are less dense with smaller particle size than primary sludge and this can be thickened or dewatered along with primary sludge. However, the solids content of the primary and secondary sludges or mixed sludge varies considerably based on the nature and characteristics of solids, type of treatment process used and its removal efficiency.

The principal methods of sludge treatment are thickening (concentration), conditioning, dewatering, drying, digestion, composting, and incineration (volume reduction). Thickening, conditioning, dewatering and drying are usually employed for removing moisture content. Digestion, composting or incineration are commonly used for stabilizing organic

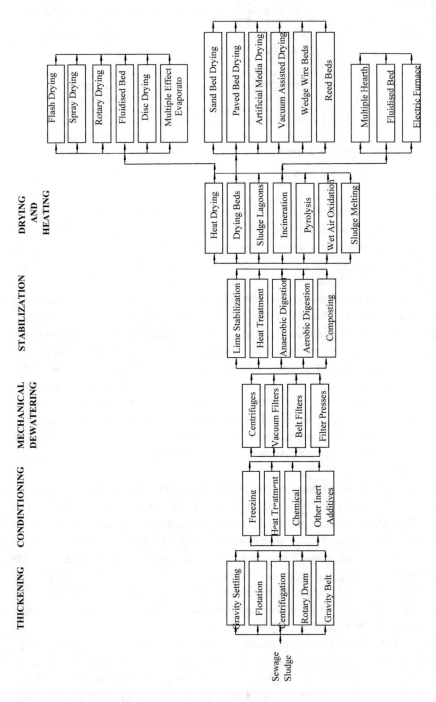

Fig. 15.2 Major unit operation and process alternatives employed for sludge management

solids present in the sludge. The biological stabilization of sludge is usually accomplished by either aerobic or anaerobic digestion to achieve mineralization of organic solids. The anaerobic digestion has widespread acceptance due to recovery of methane gas. Aerobic digestion has high energy cost associated with supplementing molecular oxygen during the process.

Thickening is the process of concentrating sludge by gravity or mechanical means. It will not produce solids content greater than 5% and the sludge still acts like a liquid. The process of thickening allows to remove some water from sludge. Dewatering is the process by which more water is removed from sludge that increases solids content as high as 15–40% of dry solids. Centrifuge can attain solids concentrations as high as 40%. Dewatered sludge behaves like a solid and it is difficult to pump but it can be managed by using belt conveyers. Dewatering is accomplished by mechanical and thermal processes.

Conditioning refers to chemical or thermal treatment to improve the efficiency of thickening or dewatering process. Chemicals used for conditioning sludge are inorganic chemicals or organic polyelectrolytes or both. Thermal conditioning can be accomplished by heating or freezing (freeze and thaw processing) the sludge. Stabilization processes treat the sludge generated from primary and secondary treatment process, converting them to a stable and inert product for beneficial use or disposal. Stabilisation also reduce pathogens and odour from sludge. Commonly used stabilisation processes are anaerobic digestion, aerobic digestion, composting, alkaline stabilisation and combustion. Alkaline stabilisation involves addition of chemicals to raise pH to about 12.0, at which pathogens or microorganisms are inactivated.

Composting and digestion processes reduce volatile content of sludge. These are not necessary if the sludge is to be incinerated. Drying is the process of removing water from sludge by evaporation by thermal means. This can be accomplished either by natural or by artificial drying processes. Figure 15.3 shows the possible sludge processing train for sludge treatment that depends largely on the method of ultimate disposal and use.

15.5   Preliminary Operations Used in Sludge Handling

Raw or digested sludge may contain large and rigid materials such as plastic and rags. These materials are liable to clog or wrap around rotating elements of machines used in sludge processing facilities. They are removed in the preliminary operations, which are primarily aimed at removing materials from sludge, but are necessary to provide relatively constant and homogenous feed to the sludge processing units. Following preliminary operations are used as per the need in a comprehensive sludge handling facility.

De-gritting: Grit is necessary to be removed before further processing of sludge in the wastewater treatment plant with inadequate grit removal facility ahead of the primary settling tank. In such situation grit matter will settle along with primary sludge. The most effective method of grit removal from sludge is achieved by cyclonic de-gritter, which

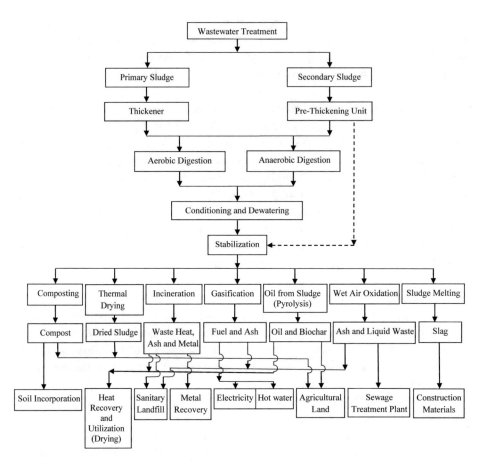

Fig. 15.3 Typical processing routes for sludge treatment

separate the grit by the application of centrifugal forces in the sludge stream. Its efficiency is affected by pressure and sludge concentration. Grit separation is very effective for dilute sludge stream with solids content ranging from 1 to 2% on dry solids basis.

Crushing: It is the process in which large materials in sludge are cut or crushed into small particles so as to prevent clogging and wear and tear of pumps, solid-bowl centrifuge and belt-filter press.

Screening: Materials, which cause nuisance, is to be removed from sludge and it is done by passing the sludge stream through a screen with 5 mm openings followed by a fine screen with 2 mm of diameter perforations that might result in achieving sufficient compaction and dewatering.

Mixing: Sludge generated in primary, secondary and tertiary settling tank of wastewater treatment consists of settleable sewage solids, biological solids and chemical sludge. These solids form the sludge by blending to produce uniform mixture, which is fed to the downstream operation and process used in sludge treatment facility. Sludge mixing can be accomplished in several ways. The common practices adopted are mixing in primary sedimentation tank, in pipes, in digester with completely-mixed system and in a separate mixing tank. In smaller treatment plant, mixing is accomplished in primary sedimentation tank; whereas in larger installations, the sludges are separately thickened and mixed to obtaining maximum efficiency.

Storage: Storage should be provided for holding the sludge when there are fluctuations in the rate of biosolids production from wastewater treatment process. The storage also allows to accumulate the sludge when downstream processes not operating for various reasons. Storage is most significant in maintaining uniform feed rate to the operation/process, such as mechanical dewatering, lime stabilization, heat drying and heat reduction. Short term storage is accomplished in settling tanks or thickening tanks; however, long term storage can be accomplished during the stabilization process in aerobic and anaerobic digesters with long detention times or in a separate, specially designed tank.

15.6 Sludge Thickening

Thickening is the first step, and regarded as pre-processing of sludge that is employed in the sludge processing for volume reduction thereby increasing the solids content of the sludge. A typical waste activated sludge of 0.8% dry solids can be thickened to about 4% solids content, resulting in about considerable reduction in sludge volume, whereas about three-fold decrease in the volume of raw primary sludge can be achieved by increasing the solids content from 1.0–2.5 to 6.0–8.0% of solids (Metcalf & Eddy, 2003). The reduction in sludge volume by increasing solids concentration in a thickening process facilitates the subsequent sludge treatment process, such as digestion, dewatering, drying and incineration in the following ways:

- Increasing solids loading on the digester that lowers the size of digesters and equipment,
- Increasing feed solids concentration in the downstream processes,
- Cost of handling and transport of sludge is minimized,
- Reduction in chemicals required for conditioning the sludge,
- In case of digester heating and consumption of auxiliary fuel for heat drying and incineration, the fuel consumption is reduced.

Thickening is generally accomplished by physical processes such as gravity settling, co-settling, dissolved air flotation, solid bowl centrifugation, gravity belt and rotary drum.

15.6.1 Co-settling Thickening in Primary Settling Tank

The free water in sludge is readily separated from solids in settling tanks with or without addition of chemicals, such as ferric chloride and polyelectrolyte. Depending upon the overflow rate applied, addition of chemicals can enhance the settling and underflow concentration of solids to about 5%. It is commonly observed that a primary sedimentation basin offers a suspended solids removal efficiency of about 60–70% in 2 h of retention time at an average hydraulic loading rate of 40 m^3/m^2 day. Primary sedimentation tank is capable of concentrating solids from 4 to 12% dry solids while treating wastewater, if thickening is combined. The thickening and settling of primary or mixed sludge accomplished simultaneously in primary sedimentation tank is termed as co-settling thickening.

The co-settling thickening is not a conventional practice in primary sedimentation tank and it is employed only when the provision is made in an activated sludge process to handle excess activated sludge with the influent of primary sedimentation tank for settling and consolidation with primary sludge. The co-settling thickening is possible in the sedimentation basins which handle light flocculent solids of 98–99.5% water content with influent solids concentration ranging from 1500 to 10,000 mg/L (Metcalf & Eddy, 2003). In order to achieve thickening, a sludge blanket must be created to consolidate the solids without being affecting the normal discharge of clarified water. Long solids retention time is not usually provided in the primary settling tank to achieve the desired levels of solids concentration for underflow; since, long retention time for sludge can lead to decomposing the solids when activated sludge is added with the influent.

15.6.2 Gravity Thickening

Gravity thickening is the most commonly adopted technique for sludge volume reduction. It is usually employed for efficient thickening of primary and chemical sludges, which thicken well under the action of gravity (Table 15.2). Gravity thickening is traditionally accomplished in a deeper circular settling tank equipped with slowly rotating rake mechanism that breaks the bridge between solid particles, thereby increasing settling and compaction of solids and encouraging the water to rise up to the top as supernatant for withdrawal. Figure 15.4 shows schematic diagram of circular gravity sludge thickener. The dilute sludge is fed through a central column of circular tank in which the solids are allowed to settle and consolidate over a maximum period of 24 h (Andreoli et al., 2007).

The thickened sludge is withdrawn from the bottom of the tank and pumped to dewater equipment or digester as required for further processing. Vertical pickets or deep trusses attached to collector arms stir the sludge gently and releases the free water trapped in the sludge, so as to concentrate it. Supernatant overflows the outlet weir, which is drawn off and re-circulated to inlet of the primary sedimentation tank. The picket and fence

Table 15.2 Typical design criteria recommended for gravity thickener

Type of sludge	Solid concentration, %		Hydraulic loading m^3/m^2 day	Solids loading kg/m^3 day
	Feed sludge	Thickened sludge		
Primary sludge	2–7	5–10	25–35	90–150
Waste-activated sludge	0.8–1.5	2–3	2–6	20–40
Combined primary and waste-activated sludge	0.5–2	4–6	4–11	25–70
Trickling filter sludge	1–4	3–6	2–6	40–50
Chemical sludge	3–4	12–16	20–35	25–60

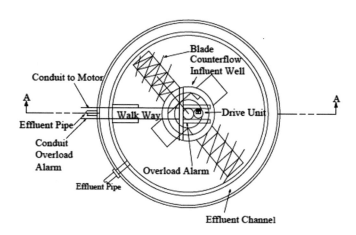

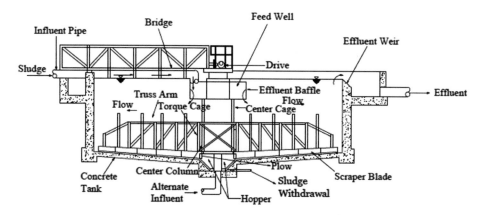

Fig. 15.4 Schematic diagram of circular gravity sludge thickener

thickeners are traditionally operated as batch process utilizing two numbers of tanks. A continuous consolidation thickening is very recent development and becoming a standard practice.

Typical solid loading criteria based on the operation of the existing plants are listed in the Table 15.2. A proper control of hydraulic loading needs to be maintained, because high hydraulic loading results in solids washout and a low overflow may cause septic condition followed by evolution of foul odour and sludge floatation. In order to alleviate such operational problems, provision must be made for addition of dilution water or clarified effluent and occasional chlorine addition. Addition of dilution water at the rate of 25–30 m^3/m^2 day maintains the aerobic condition and flushing of soluble organic and inorganic compounds from sludge. This will result in saving of conditioning chemical required for dewatering.

The addition of dilution water contributes more supernatant with additional volume, and it is recycled to the inlet of the primary sedimentation tank. Therefore, the hydraulic loading due to the recycling of the supernatant must be taken into account in the design of primary sedimentation tank. The sludge thickening facility is designed based on laboratory and pilot-scale operation and testing as there are wide variations in the thickening characteristics of wastewater solids. The underflow concentration governs the height of the sludge blanket in the consolidation zone of the tank. The sludge blanket height is maintained in the range of 0.5–2.0 m and a shallow depth is recommended during warmer climates to avoid anaerobic activity.

Example: 15.2
A sewage sludge contains 3% dry solids having a specific gravity (S_{ds}) of 1.25. Estimate the volume occupied by 100 kg of sludge with this 3% dry solids before digestion. This raw sludge contains 30% of fixed solids with specific gravity of 2.5 and remaining volatile solids with specific gravity of 1.02. After digestion 65% of the volatile solids got destructed and the dry solids concentration was 10%. Determine volume of the sludge after digestion.

Solution
Estimation of the average specific gravity of the solids in the raw sludge before digestion

$$\frac{1}{S_s} = \frac{0.30}{2.5} + \frac{0.70}{1.02} = 0.12 + 0.686 = 0.806$$

Hence, $S_s = 1/0.806 = 1.24$.
Thus, the specific gravity of the raw sludge containing 3% dry solids will be:

$$\frac{1}{S_s} = \frac{0.03}{1.24} + \frac{0.97}{1.0} = 0.0242 + 0.97 = 0.9942$$

Hence, $S_s = 1/0.9942 = 1.0058$.
Therefore, the volume of the 100 kg of dry solids will be

$$V = \frac{M_s}{\rho_w \, S_{sl} \, P_s} = \frac{100}{1000 \times 1.0058 \times 0.03} = 3.314 \, \text{m}^3$$

Estimating volatile solids after digestion

Volatile solids in the raw sludge $= 100 \times 0.7 = 70$ kg.

Hence, volatile solids destructed during digestion $= 70 \times 0.65 = 45.5$ kg.

Volatile solids left after digestion $= 70 - 45.5 = 24.5$ kg.

Fixed solids in the raw sludge $= 100 \times 0.3 = 30$ kg.

Hence, total solids left after digestion $= 30 + 24.5 = 54.5$ kg.

Hence percentage of volatile solids after digestion in the total solids
$= 24.5 \times 100/54.5 = 44.95\%$

Average specific gravity of the solids in digested sludge

$$\frac{1}{S_s} = \frac{0.5505}{2.5} + \frac{0.4495}{1.02} = 0.2202 + 0.4407 = 0.6609$$

Hence, $S_s = 1/0.6609 = 1.5131$.

Estimating the specific gravity of the digested sludge with 10% solids in it.

$$\frac{1}{S_{ds}} = \frac{0.1}{1.5131} + \frac{0.90}{1.0} = 0.0661 + 0.9 = 0.966$$

Hence, $S_{ds} = 1/0.966 = 1.0351$.

Therefore, volume of the digested sludge $= \frac{54.5}{1000 \times 1.0351 \times 0.1} = 0.527 \, \text{m}^3$.

Volume reduction of the sludge $= (3.314 - 0.527)/3.314 = 84.1\%$

Thus, more than sixfold volume reduction of the sludge occurred after digestion.

15.6.3 Floatation Thickening

Floatation is the process of transforming suspended, colloidal, and emulsified substances into floating substances (Hess et al., 1953). Very fine particle that is difficult to settle from suspension can be flocculated and buoyed to the liquid surface by utilizing the lifting power of minute air bubbles, which attach themselves to these finer suspended solids. The sludge flocs so agglomerated can readily be removed from the liquid surface by skimming. The scum concentrated at the water surface is continuously collected and drained off. The solids content of the scum that floats is increased two-fold by flotation method. The water is squeezed out of the scum while concentrating the particles. The sludge float is stable and odour free. The floatation of sludge can be achieved by applying air under pressure or maintaining vacuum condition.

Floatation thickening is mostly preferred for the gelatinous sludges from suspended growth process, such as waste-activated sludge or nitrification process, having a density very close to that of water and thus these sludges are readily buoyed to the liquid surface. This technique makes the waste-activated sludge to become concentrated to about 3–5%

with a solid recovery of 85% without using any floatation chemical aids. It is an expensive technique as it requires equipment, power to operate and skilled supervision. However, it is advantageous for effective removal of oil and grease, finer solids and odour control. Flotation efficiency can be increased by the addition alum and polyelectrolyte. Using floatation aid, such as polymer, the waste activated sludge with around 10,000 mg/L of solids can be concentrated to 4–6% in the floatation device. Alum increases the sludge quantity during the process, whereas polyelectrolyte enhances the solids capture to as high as 98%.

15.6.3.1 Dissolved Air Flotation

The dissolved air flotation works on the principle that when the pressure is released, the bubbles of air attach themselves to solids with less or comparable specific gravity than that of water, and these solids rise to the surface of water. In dissolved air flotation, a small quantity of supernatant effluent from the flotation unit is pressurized to about 3–4 kg/cm^2 in presence of sufficient air within it to form an air-saturated pressurized liquid. This super saturated liquid is then released in the floatation unit through which the sludge is getting passed at atmospheric pressure. In the floatation unit the air is released in the form of tiny rising bubbles throughout the tank from this mixture of pressurized recycle water and the sludge. This tiny rising air bubbles entrap the suspended, colloidal and some emulsified particles and lead to floating of these to the liquid surface. The floated sludge by these air bubbles is usually continuously skimmed off and removed from the top of the tank by sludge scrapper (Fig. 15.5). The heavier settleable solids are removed from the floor by means of racking mechanism.

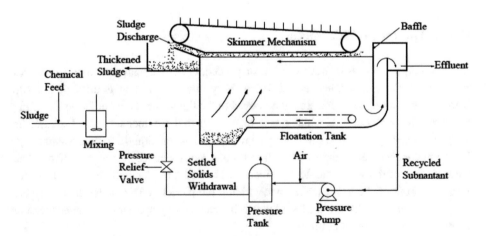

Fig. 15.5 Schematic diagram of dissolved air floatation

The primary operational parameters of dissolved air floatation are recycle-ratio, feed solids concentration, solid loading ratio, hydraulic loading ratio, air–solid ratio and SVI of sludge. Typical value of design criteria adopted are stated below.

Pressure	280–480 kPa
Pressurized effluent cycle	20–100%
Solid loading rate	40–90 kg/m^2 day
A/S ratio	0.005–0.06 kg/kg

It is reported that lowering concentration of the feed sludge by dilution increases the floated solids concentration. The nature of influent sludge affects significantly the performance of floatation thickening. A well flocculated sludge may achieve thickening to about 4–5%; whereas, a filamentous bulking activated sludge may not achieve 2% solids content. The supernatant or other extracted liquid from underflow will contain finer suspended particles and fairly high organic matter concentration that needs to be returned back to the wastewater treatment processes.

15.6.3.2 Vacuum Floatation

In a typical vacuum flotation unit, the waste sludge is first aerated at the rate of 0.0092 m^3 air per m^3 of sludge for a short period of 30 s, followed by a short period de-aeration at atmospheric pressure to remove large bubble in it. The sludge, saturated with dissolved air is then passed to a vacuum tank, which is closed and maintained under vacuum of about 230 mm of Hg. This vacuum developed in the tank gives rise to bubbles, which cause the sludge solids to float on the surface because of imparted buoyancy due to attachment with fine air bubbles erupting out of whole sludge volume during vacuum application.

A major distinction between the dissolved air floatation and vacuum flotation is based on its mechanisms of dissolved-air and dispersed-air bubble formation (Nemerow, 1971). Dispersed air flotation produces air bubbles as a result of diffusion of air or mechanical shear of propeller or homogenous blending of air and liquid stream. Dissolved air floatation produces tiny air bubbles from liquid, which is supersaturated with air under pressure. Dissolved-air floatation produces smaller bubble size than that of dispersed-air bubbles, generally not more than 80 μm. However, the sizes of dispersed-air bubbles grow as high as 1000 μm.

15.6.3.3 Centrifugal Thickening

Centrifuges are commonly employed for thickening and dewatering of sludge. The application of centrifugation is based on the principle of separation of sludge solids under centrifugal force, which is about 2000 times that of gravity. The basic type of centrifuge used for water and wastewater is the solid-bowl scroll centrifuge. It consists of a rotating long cylindrical bowl mounted horizontally and tapered at one end and an interior rotating screw conveyer. The feed sludge, entering from the centre of screw conveyer, passes

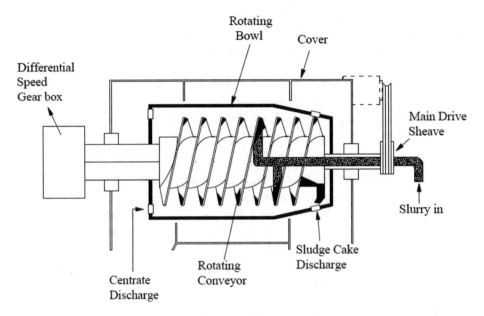

Fig. 15.6 Schematic diagram of a solid bowl centrifuge

through the discharge nozzles, and is held against the bowl wall by centrifugal force. The internal helical scroll (conveyer) rotates inside the bowl and pushes the thickened sludge forward. The sludge is compacted by centrifugal force along the wall towards the conical area and pushed out (Fig. 15.6).

As the feed rate to centrifuge increases the solid recovery from the centrifuge and retention time of sludge decreases. Feed rate are usually limited to 3–14 m³/day kW to achieve a desired solids recovery with more than 5% solids. Moreover, a higher solid concentration and addition of polyelectrolyte result in an increased solid recovery. Increased removal of fine particles along with chemical addition can lower the cake dryness. Centrifuge thickening process requires high maintenance and power costs, and requires skilled supervision. Because of limited space availability in sewage or effluent treatment plants, the centrifugal thickening process could be an attractive option for dewatering the sludge.

The performance of the centrifuge is largely affected by the type of machine and process operational variables. The significant process variables are the following: (i) solid bowl speed and conveyer speed; (ii) feed sludge characteristics; (iii) hydraulic loading rate applied; (iv) depth and volume of the liquid pool in the bowl; (v) chemical addition and dosage; (vi) operating temperature; and (vii) differential speed of screw conveyer. Since, the performance of the centrifuge widely varies with above factors, it needs to be tested on pilot-scale prior to its full-scale application.

15.6.3.4   Gravity Belt Thickening

Application of belt press for sludge dewatering is an advanced thickening technology. The principle of gravity drainage in sludge dewatering using belt press is found to be effective even with solids contents less than 2%. However, the porosity of belt used for sludge thickening is different than that used for sludge dewatering. The equipment consists of a gravity belt, which moves around the rollers driven by variable speed motor. The sludge, conditioned with polymer, is applied evenly across the width of the moving belt which moves forwards, the water is drained out from the sludge that is concentrated at the discharge end of the thickener. The plow blades placed along the direction of the motion of the belt facilitate the water to drain through the belt by making ridges and furrow. The thickened sludge is removed and the belt moves forwards to the wash cycle. The belt thickening is widely employed for thickening the waste activated sludge and aerobically or anaerobically digested sludge.

Design criteria adopted for gravity belt thickener are hydraulic loading rate and solids loading rate that are typically 750 L/m min and 550 kg/m h, respectively. The equipment is designed to achieve solids content of around 7% and solids recovery typically range from 92 to 98%. The desirable solids content is achieved by changing belt speed. Polymer dosing range from 2 to 7 kg of dry polymer per tonne of dry solids that can offer a solids content of 5–15%.

15.6.3.5   Rotary Drum Thickening

The rotary drum thickener comprises of two cylindrical drums placed horizontally and stacked in concentric manner. The impervious outer one is conditioning drum and another is rotating cylindrical screens. The dilute sludge is conditioned with polymer in the first drum and it is then passed to the cylindrical screens, where the flocculated solids are separated from water. The inner drum slowly rotates and the vacuum maintained in the cake formation section inside the inner drum forms the sludge cake on the filter medium attached on the periphery of this drum. Further as the drum moves, the filtrate drops through the screen wall mounted on the circumference of the drum; whereas the thickened sludge is removed at the end of the drum rotation using scrapper (Fig. 15.7). Rotary-drum unit coupled with belt combines the performance for both thickening and dewatering in a single system. The rotary drum thickener can act as a pre-thickening stage prior to belt-press dewatering of waste activated sludge.

15.7   Sludge Stabilization

Stabilization is aimed at reducing nuisance potential of the sludge. The specific requirements of sludge stabilization are to reduce pathogenic organisms, eliminate foul odour and control possible decomposition of organic residuals in the sludge. The presence of organic fraction in the sludge flourishes microorganisms on it. This will subsequently

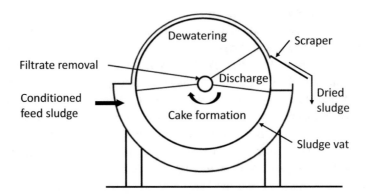

Fig. 15.7 Rotary vacuum drum dryer

cause the microbial degradation of volatile solids present in the sludge and release of foul odours. The nuisance related to the putrefaction of volatile solids can be eliminated by the addition of chemicals, which will prevent the biodegradation of organic solids and render them unsuitable for survival of microorganism. The stabilization is done by using chemicals, such as lime and chlorine, that prevents the decay of organic materials and enables the mechanism of disinfection.

The complex organic compounds in the sludge are biochemically digested so that they become non-putrescible and more dewaterable, thereby achieving strength and volume reduction. In case of the incineration of sludge, no stabilization is required after dewatering; however, the sludge necessarily required to be stabilized to control the odour and pathogens if it is applied on land. The extent of volatile solids destruction is the measure of effectiveness of a stabilization process. However, the most common indicator parameter of stabilization process is the order of magnitude of pathogen reduction. Various methods adopted for sludge stabilization are aerobic and anaerobic digestion, chemical oxidation or lime stabilization and thermal condition (drying process). Anaerobic digestion process is the most widely preferred stabilization process as it does not require much energy and chemicals. The anaerobic digestion is usually employed for sludge volume reduction, production of methane gas and improving dewaterability of sludge.

15.7.1 Lime Stabilization

In order to eliminate the nuisance conditions in organic fraction of the sludge, such as survival of microorganism, putrefaction and evolution of offensive gases, an alkaline material is used to render the sludge unfavourable for the survival of microorganism. Lime stabilization is a most common method in which sufficient quantity of lime is added to raise the pH to about 12.0. The alkaline pH in the sludge inhibits or halts microbial activity substantially. The sludge remains stable against the biodegradation as long as the pH is maintained at this level. This process can also inactivate or suppress virus and other pathogens present in the sludge.

Lime stabilization process alters the chemical composition of the sludge through variety of chemical reactions. If calcium, phosphorous, carbon dioxide and organic acids are present in the sludge, the reaction results in the formation of $CaCO_3$, $Ca_3(PO_4)$ and calcium salts of acids. If biological activity is not sufficiently inhibited by lime stabilization, then the release of carbon dioxide and organic acids will take place due to bacterial decomposition. These compounds will readily react with lime to lower the pH sufficient enough to continue biochemical reactions releasing ammonia and other volatile gases. Therefore, excess lime is added in the stabilization process. Lime is added in the form of $Ca(OH)_2$ and CaO or fly ash or cement kiln dust can be added before or after dewatering. Addition of lime prior to sludge dewatering is referred to as lime pre-treatment; whereas, addition after dewatering is termed as lime post-treatment.

Lime pre-treatment: Lime pre-treatment is used for direct disposal of liquid sludge on land or achieving the combined sludge conditioning and stabilization prior to dewatering. Direct disposal of sludge pre-treated with lime is limited to small treatment plants, while considering the large quantity of liquid sludge to be transported to the disposal sites. A pressure type filter press is used to dewater the sludge after lime pre-treatment; however, centrifuge or belt filter press are not commonly used as they are subjected to wear and scale formation. The quantity of lime required is more than that is used for application after dewatering. More dose and contact time are necessary for efficient pathogen removal. By addition of lime the sludge is typically maintained at pH of around 12.0 for a contact time of 2 h. The bacterial reduction is more efficient after lime stabilization, while comparing it with the results obtained from digestion processes. Lime pre-treatment stabilization is an outdated technique because of the production of large quantity of lime sludge, which is 25–40% more than its initial weight. Amount of lime added typically ranges between 120 and 200 kg/tonne of dry solids.

Lime post-treatment: Addition of lime is exothermic process, which results in substantial rise in temperature. Quick lime is commonly used in lime post-treatment as it can inactivate worm eggs with rising temperature above 50 °C during the reaction of quick lime. The advantages of lime post treatment are: (i) lime is added in dry state and (ii) scaling and associated maintenance problems of sludge dewatering equipment are eliminated. The advances in lime stabilization based on the use of cement kiln dust and fly ash are modifications on dry lime stabilization process. Pasteurization under carefully controlled conditions with adequate mixing at temperature of 70 °C for 30 min can inactivate pathogens.

15.7.2 Heat Stabilization

Heat stabilization is also called heat disinfection. Stabilization of sludge, by application of heat beyond 200 °C under pressure, is gaining much attention in sludge processing. The heat treatment technique also serves the purpose of sludge thickening, conditioning, and dewatering at high temperatures and pressures. Heat treatment process destroys microbial cells in the sludge resulting stabilization and sterilization of sludge. High degree of pathogens killing is achieved in the stabilized sludge under high pressure and temperature.

15.7.3 Anaerobic Digestion

Anaerobic digestion is a biochemical oxidation process of sludge treatment and it is employed for reducing the organic content of the sludge in wastewater treatment plants. It is a traditional unit process adopted for the treatment of sludge generated from industrial, agricultural and municipal wastewater treatment plants world-wide. It is accomplished in the absence of molecular oxygen under which anaerobic bacteria convert the organic waste into stable, inert gaseous end products such as methane (CH_4), carbon dioxide (CO_2), hydrogen sulphide (H_2S) and ammonia (NH_3). The objective of anaerobic digestion of sludge is to biochemically convert bulky high organic solids containing raw sludge to a relatively inert material with less obnoxious odour and good dewaterability.

The process is carried out in an air tight digester to which primary sludge or a combination of primary and secondary sludge is introduced continuously or intermittently. The sludge undergoes digestion process under suitable environmental conditions for long period of time. The stabilized sludge is withdrawn continuously or intermittently from the tank for further processing, such as dewatering and drying. The digester gases produced during sludge digestion process are collectively termed as biogas, which could be used as fuel for digester heating or utilized for electricity production or flared off. The digested sludge is relatively stabilized or mineralized and has good fertilizer value. The supernatant withdrawn as effluent having high organic content, present in both soluble and suspended form, is returned back to the primary sedimentation tank. The sludge digestion ensures substantial reduction in the pathogens present in the sludge. Sludge digestion proves to be a dominant process for stabilization of sludge, as it addresses the energy conservation and recovery of value-added products, while managing the pollution potential of high strength organic wastes.

15.7.3.1 Fundamentals of Anaerobic Digestion Process

The biochemical transformation process of organic matter consists of three successive stages of metabolic processes that occur simultaneously in digesting sludge. These three stages are: (i) liquefaction, (ii) fermentation, and (iii) gasification. In the first stage, complex organic compounds such as carbohydrate, protein, lipids and polysaccharides are hydrolysed by extracellular enzymes excreted by hydrolytic bacteria to simpler soluble products. The hydrolytic bacteria are normally heterogeneous group of facultative and obligate anaerobic bacteria. In the second stage, the products of hydrolysis, long-chain fatty acids, amino acid, sugars, alcohol and triglycerides are fermented to short-chain fatty acids, alcohol, ammonia, hydrogen and carbon dioxide. The fermentation process results in the formation of simple organic compounds and hydrogen in a process, which is called as acidogenesis. The short-chain fatty acids so produced are mainly volatile fatty acids and alcohol. These short-chain fatty acids are converted to acetate, hydrogen and carbon dioxide during acetogenesis the third step. While these products of hydrolysis are converted to organic acids, alcohol and new bacterial cells, commonly referred to as acid fermentation, negligible change in the BOD or COD is observed.

In the fourth stage, methane is produced from two pathways through which hydrogen and acetate are utilized separately by hydrogenotrophic methanogens and acetoclastic methanogens, respectively. This stage is generally referred to as methane fermentation. The principal component of volatile fatty acids during acid fermentation phase is acetic acid, which is the precursor for methane formation. Methane producing heterotrophs are obligate anaerobes and very sensitive to environmental conditions, such as temperature and pH. Moreover, methanogens have slow growth rate than the acid formers in the second stage and are very specific in food supply requirements. On the other hand, acidogens are relatively tolerant to changes in pH and temperature and have much higher growth rate than methanogens (as described in details in Chap. 12).

Liquefaction of suspended organic solids and final step of methane fermentation are generally the rate-limiting steps in the overall anaerobic digestion process of sludge. The stability of digestion process depends on the dynamic balance between acidogenesis and methanogenesis. Transient increase in organic loading (shock loading) or sharp change in operating temperature will lead to building up of organic acids. Production of excess amount of organic acids beyond the assimilative capability of the methane forming bacteria suppresses the methanogenic activity by lowering the pH from the optimum value. This imbalance will eventually result in a souring of digester, in which the methane ceases to produce, unless the organic loading is lowered to recover the methanogenic activity.

15.7.3.2 Factors Affecting the Digestion Process

In anaerobic digestion, complex interactions of several varieties of bacteria and archaea involved in the process must be in equilibrium in order for the digester to remain stable. The major factors that influence the digestion are pH, temperature, C/N ratio, retention time, bacterial competition, nutrient content, the presence of toxic compounds and solids content.

Operating pH: The pH inside the digester is a determinant factor for the health of digester, which changes in response to biological conversions during the different processes of anaerobic digestion. A stable pH indicates system equilibrium and digester stability. The anaerobic digestion process operates in a pH range of 6.6–8.0 and an optimum performance is obtained with value of pH close to 7.0. Acid forming bacteria works in a relatively wide pH range, while acetate-splitting methane forming archaea work in a narrow pH range of 6.7–7.4. Excess production of volatile fatty acids depresses the pH and it is necessary to supplement sufficient bicarbonate alkalinity to neutralize these excess acids produced.

Solid and hydraulic retention time: Anaerobic digester is designed based on the solid retention time (SRT) and hydraulic retention time (HRT) to allow the significant destruction of volatile suspended solids. The SRT and HRT are the mean time for solids and liquid to be held in the reactor for digestion process, respectively. Variations in SRT values can vary the extent of the metabolic reactions (hydrolysis, fermentation and methanogenesis) in the digester. The operation below the minimum SRT will result in failure of the digestion process.

Temperature: Temperature has a significant effect on the rate of digestion, especially on the rate of hydrolysis and methane formation. Anaerobic digesters are commonly designed either to operate under the mesophilic temperature range, between 25 and 40 °C, or in the thermophilic temperature range of 45–65 °C. Most of the acid forming microorganisms grows under mesophilic conditions; however, for methanogens, a higher temperature is favourable. The optimum mesophilic temperature is around 36 °C and the optimum thermophilic temperature is 55 °C (Metcalf & Eddy, 2003). In cold countries, where digester temperature falls below 20 °C and substantial changes in temperature occurs it can affect the process performance. In such cold climates, the digesters are required to be heated externally to bring the temperature to mesophilic range, that is typically above 25 °C.

15.7.3.3 Advantages and Disadvantages of Anaerobic Sludge Digestion

The anaerobic digestion of the sludge offers following advantages:

- Reduction of pollution potential of organic sludge waste,
- Reduction in greenhouse gas emission,
- Reduction in volume of sludge and improving dewatering characteristics that makes easy to dewater,
- Elimination of pathogens,
- Production and recovery of biogas as alternate energy source,
- Enhanced fertilizer value of the digested sludge,
- Low capital and operating cost,
- Low nutrients requirements, hence less nutrient removal during digestion.

However, the anaerobic digestion of the sludge has following drawbacks:

- It is difficult to maintain process performance at lower temperature, hence might require external heating,
- Accumulation of heavy metal and recalcitrant compounds in the digested sludge,
- Very sensitive to temperature and other environmental factors, hence affecting its performance,
- Long solid retention time required for digestion.

15.7.4 Types of Anaerobic Digesters

Anaerobic digesters are used for the sludge digestion and volume reduction of organic sludge under controlled conditions. Reactor used for sludge digestion consists of an air tight circular hopper bottom tank made of concrete or steel with proper corrosion protection coating and it is covered with fixed or floating type of roof. Two types of sludge digestion are commonly employed in practice, which are: (i) standard-rate or low-rate digester and (ii) high-rate digesters.

15.7.4.1 Standard-Rate or Low-Rate Digesters

In a standard-rate digestion process, the digester content is not subjected to continuous mixing or heating and it is allowed to stratify into zones, so that acidification, methane fermentation and sludge thickening are accomplished in single tank. Conventional or low-rate digesters are operated under intermittent sludge feeding, intermittent mixing and intermittent sludge withdrawal (Reynolds, 1982). However, if necessary, the rate of fermentation is increased by heating the digester that will result in shorter retention time for sludge for digestion. The sludge digestion time ranges usually between 30 and 60 days in this low-rate digester. Figure 15.8 shows a schematic diagram of a standard rate anaerobic digester.

In the absence of proper mixing, the digester contents stratify with scum layer on top, supernatant, underlying active digesting sludge as intermediate zones and a concentrated digested sludge, which occupies at the bottom layer. The volume utilized for active digestion is approximately one-third of the total volume and remaining volume consists of scum layer, supernatant and digested sludge. This results in larger volume requirement for low-rate digester. The recommended digestion period for different operational temperatures for low-rate digester is provided in Table 15.3.

The basic functions of a low-rate digester with floating roof are stabilization of volatile solids, gravity thickening and storage of digested sludge. When the digester is first put in operation, it is seeded with the digester sludge obtained from another operating digester. The raw sludge from primary settling tanks is pumped into the zone of active digestion of the digester on intermittent basis and it is passed through sludge heating, if necessary. The digested sludge is accumulated at the bottom and it is normally withdrawn once

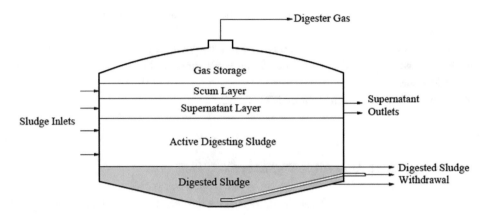

Fig. 15.8 Schematic diagram of a standard-rate (low-rate) anaerobic digester

Table 15.3 Digestion time required for the low-rate digester at different operating temperatures

Temperature, °C	10	15	20	25	30	35	40
Detention period, days	75	55	41	33	27	26	25

every two weeks from the bottom of digester, if it is appropriate to deliver the sludge to the subsequent treatment and disposal facilities. During rainy days, the sludge is not withdrawn and it is stored in the digester till the weather favours to apply on sand drying beds. The supernatant is withdrawn periodically and returned back to the inlet of the primary sedimentation tank. The digester has conical (hopper) bottom, which facilitates the easy sludge withdrawal under hydrostatic pressure, while floating cover suitably accommodates the volume changes due to the addition of raw sludge and withdrawal of digested sludge, supernatant and biogas from the digester.

The low-rate digester encounters the problem of acidification due to absence of mixing. The dimensions of conventional low-rate digester are restricted in the range of 5–38 m in diameter and 4.5–9 m side water depth with a free board of 0.6–0.7 m. The design criteria adopted for the low-rate digester is provided in the Table 15.4. The digester can be constructed partially above and partly below the ground level and should be situated above the ground water table.

The low-rate digester can also be of fixed cover type, which is generally suitable for a population not more than 10,000 persons. The sludge feeding and withdrawals of digested sludge and biogas are restricted in the operation of fixed cover digester. The biogas inside the digester gets compressed during addition of fresh sludge and thus an increase in the gas pressure occurs, which should be restricted to maximum allowable pressure of 0.205 m of water column. Similarly, the biogas inside the digester gets expanded on

Table. 15.4 Design criteria for anaerobic digesters

Parameter	Low-rate digestion (intermittent mixing)	High-rate digestion (heating and mixing)
Volatile solids loading, kg VSS/m³ day	0.5–1.5	2.0–6.0
Hydraulic retention time, days	30–80	10–20
Solid retention time, days	30–80	10–20
Gas production, m³/kg VSS added	0.55	0.65
Solids content in feed sludge, % (dry basis)	2–6	4–6
Solids content in digested sludge, % (dry basis)	4–7	4–6
Volatile solids reduction, %	35–50	50–65
Methane content, %	65	65
Diameter, m	5–38	5–38
Height, m	6–9	6–9

withdrawal of supernatant or digested sludge and thus the negative pressure develops inside the digester, which should not exceed 76 mm of water column.

Fixed cover digester has more problem of grease floating and drying. To overcome these operational problems the digester can be provided with floating cover as shown in Fig. 15.9. For efficient performance, the size of reticulating pump used to provide mixing should have a capacity to pump one tank volume in 30 min. The digester contents are mixed for one hour in a day. While considering the simplicity of operation and economic reason, smaller treatment plants processing wastewater less than 3,500 m³/day generally employ standard-rate digester.

15.7.4.2 High-Rate Digesters

High-rate anaerobic digester is operated on continuous mixing and heating (if necessary), continuous or intermittent sludge feeding and sludge withdrawal. High-rate digesters are efficient and require relatively less digesting volume than low-rate digesters. The digester contents are mixed to improve effective contact between the fresh sludge and active microorganisms. For operation in colder climate the sludge is heated to enhance the rate of microbial metabolism, thus hastening the digestion process. The mixing is continuous and hence the entire digester volume remains active suspension, which is 50% in case of low-rate digesters. This mixing provided will allow higher organic loading rates and thus shorten retention time. In the operation of the single-stage digester, it becomes necessary to stop the mixing and allow the sludge to stratify and settle at the bottom, prior to the withdrawal of digested sludge and supernatant (Fig. 15.10a). Then the fresh sludge

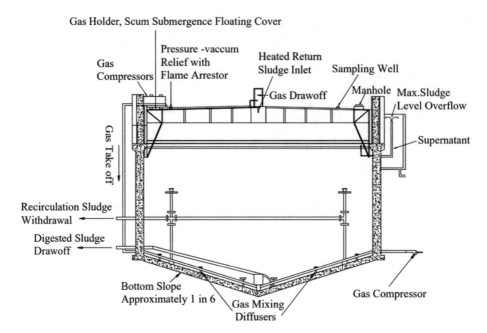

Fig. 15.9 Schematic of cross section of a typical low-rate anaerobic digester with floating dome type

feeding and mixing are restored. Alternatively, when fresh sludge is added, the digested sludge may be displaced to a holding tank, from where the supernatant is separated. High-rate digester is evolved as a result of continuing efforts to improve the performance of standard-rate digester. Two stage digestion (Fig. 15.10b) is an extension of high-rate digestion in which the functions of fermentation and solids-liquid separation are separately accomplished in two tanks.

15.7.4.3 Two-Stage Digesters

Although the volatile solids contents are reduced in the high-rate digester, the volume of the digested sludge remains unchanged as there is no thickening and dewatering that occur in it. Therefore, a single-stage digester followed by a second-stage digester similar to a standard-rate digester (no heating and mixing provided) is adopted for alternative dewatering system. The contents in the first-stage high-rate unit are intimately mixed and if necessary, the sludge is heated to enhance the fermentation rate. The mixing of digester contents enables better microbial contact with solids and more uniform distribution of temperature in the entire volume of digester. Gas production in the second-stage digester is little compared to the first-stage digester; however, the influent of second-stage is super-saturated with gas, which is released in the second stage digester. Therefore, the provision for gas collection and recovery is necessary with second-stage digester.

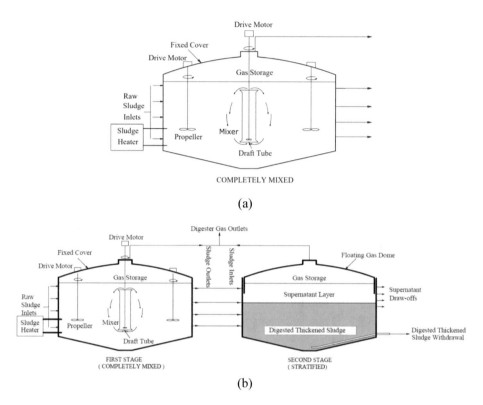

Fig. 15.10 Schematic of high-rate anaerobic digesters **a** single-stage process, and **b** two-stage digestion process

The primary functions of first stage are the liquefaction of organic solids, digestion (fermentation) of soluble organics and gasification. The first stage is usually designed as high-rate digester equipped with fixed or floating cover and provision for continuous mixing (Fig. 15.10b). The second stage is primarily for digested sludge storage, gas storage and solid/liquid separation (sludge and supernatant separation). The second stage is usually conventional standard-rate digester with floating type cover, equipped with intermittent mixing and it is often unheated. The organic loading applied to the first stage is much more than second stage since raw sludge if fed to the first stage. The first stage digester of a high-rate system is characterized by a completely mixed reactor without recycle and hence it has equal solids retention time and hydraulic retention time. The key operating parameters that affect VSS reduction are digestion temperature and biological solids retention time.

The organic content in the supernatant from the digester is still quite high. The suspended solids and TKN are as high as 11,000 mg/L and 1000 mg/L, respectively (Davis and Cornwell, 2010). Hence, the supernatant is re-circulated to the inlet of the primary sedimentation tank. The sludge removed from the digester contains lower volatile solids content than raw feed sludge as about 2/3 of the volatile solids are getting converted into biogas during digestion. The digested sludge is taken for further processing, such as conditioning, dewatering and drying for ultimate disposal. Two-stage digestion may be beneficial for some plants, while conventional low-rate unit may be found suitable for other cases. The choice of the digester type is based on the factors such as capacity of treatment plant, flexibility of sludge handling processes, required storage capacity, ultimate solids disposal techniques, and local climatic conditions.

Proper mixing of high-rate digester contents is most significant consideration in achieving desired digester performance. Various systems employed for mixing of digester contents are: (i) biogas recirculation, (ii) mechanical stirring, and (iii) recirculating digester contents. Figure 15.11 shows different types of mixing systems adopted in high-rate anaerobic digesters. The advantages and disadvantages of various systems are listed in Table 15.5.

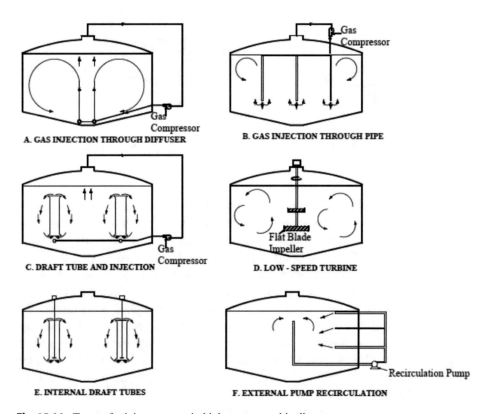

Fig. 15.11 Types of mixing systems in high-rate anaerobic digesters

Table. 15.5 Advantages and disadvantageous of various mixing systems

Mixing system	Advantages	Disadvantageous
Biogas recirculation	Less maintenance and less hindrance to cleaning, efficient against building up of scum, good mixing	Corrosion of gas piping and equipment, can result in only local mixing hence incomplete mixing, foaming and scum formation, potential gas seal problem
Mechanical stirring	Mixing efficiency is high, breaking up of scum layer	Wear and tear of impeller and shafts, bearing failure, gas leaks
Recirculating pumps	Good vertical mixing, easy maintenance, better control on mixing	Corrosion, bearing and gear box failure, impeller wear, plugging of pumps by rags, etc.

15.7.5 Start-up and Operation of Digesters

Operation of anaerobic digester encounters with several challenges in achieving stable process performance, principally because of the inherent characteristics of methanogenesis making the digester operation complicated. The mass of active viable methanogenic archaea is low in raw sludge compared to acid forming bacteria. In order to overcome this problem, the digester is inoculated with seed sludge from another operational digester to maintain the active biomass during start-up. During the start-up one-tenth of designed sludge feed rate is applied in the beginning to ensure that the digester is loaded with low organic matter. The buffering capacity of the digester is very important in the stable operation of digester; thus, lime is added to supplement the alkalinity for maintaining the pH near 7.0. Incorrect dosage of lime (calcium hydroxide) may lead to sharp change in pH to alkaline condition, which is also detrimental to the growth of bacteria. Nitrogen and phosphorous are vital nutrients required for bacterial growth and these are required at minimum concentration of 2.5 and 0.5% of dry VSS content of the sludge. The nutrient requirement for anaerobic digestion in terms of COD is COD: N: P equal to 350: 5: 1.

Once the digestion process becomes stable at low loading of sludge, the sludge feeding rate is increased gradually by small increment in steps to reach the designed loading rate. The digester operation is monitored by plotting a response curve of daily gas production with respect to fresh sludge feed rate, percentage of CO_2 in the digester gas and volatile fatty acids concentration in the digested sludge. Properly operating digester has typical composition of digester gas as 65–72% of methane, 25–30% of CO_2 and other trace gases, such as NH_3, H_2, H_2S, etc. Any change in the gas composition, rate of gas production, and increase in volatile acids indicates failure of the digestion process that needs immediate attention to change in the feed rate pattern of the digester.

During digestion of mixed primary and secondary sludge from a domestic sewage treatment a typical biogas production of about 0.5 m^3/kg VSS or 0.375 m^3/kg TSS added to the digester can be observed. In a conventional sludge digestion process, typically, 50–60% of organic solids are digested to produce biogas and nearly 10% of these solids are converted to biomass. The heat content (gross calorific value) of biogas is approximately 22,400 kJ/m^3. The fuel equivalent of 100 m^3 of biogas produced by the anaerobic digestion is about 140 kWh. After digestion, the solids content of anaerobic digester sludge typically ranges from 10 to 15% for primary sludge and 6 to 10% for mixed sludge from primary and secondary clarifier.

Ultrasonic disintegration is a novel technology developed to disintegrate the sludge bio-solids using ultrasonic frequency sonication before feeding to the anaerobic sludge to the digester (Tiehm et al., 2001). The disintegration results in a more dispersed and homogeneous flocs of activated sludge intensifying the biogas production and improving the quality of digested sludge. Laboratory scale studies reported that a digester with SRT of 15 days under mesophilic condition showed an increase of 20% biogas production and 10% increase of methane production when compared to sludge solids without the ultrasonic disintegration (Tomczak-Wandzell et al., 2011). The enhanced biodegradability resulted in the ultrasonic disintegration was attributed to the transfer of organic sludge solids into aqueous phase. The advantages of increase of sludge digester efficiency, increase of biogas production, and methane content in biogas, and reduced volume of digested sludge by 12% were demonstrated in the experiment (Tomczak-Wandzell et al., 2011). Full-scale studies in Singapore reported that ultrasound pre-treatment with raw sludge increased the daily biogas production by 45% (Xie et al., 2007).

15.7.6 Design Considerations

The basic design criterion of standard-rate digester is to provide sufficient capacity to have an adequate solids retention time in the digester. The SRT commonly adopted is in the range of 25–30 days. Therefore, the total digester volume is a function of volume of raw sludge added per day (sludge loading rate), volume of digested sludge produced per day (digested sludge accumulation rate), volume of biogas produced, volume of supernatant and volume of the digested sludge stored. Volume of biogas produced and volume of supernatant are neglected since the gas volume is relatively insignificant and supernatant is removed from digester as it is produced. Thus, the change in volume of digesting sludge is assumed to be parabolic function with respect to the digestion time. Hence the average volume of digesting sludge is initial volume minus $\frac{2}{3}$rd of the difference between the initial and final volume as stated in Eq. 15.6 (Fair et al., 1968). The total volume of the sludge present in digester and volume of the digester required are given by Eq. 15.7 and Eq. 15.8, respectively.

Average volume of digesting sludge per day, m^3/day

$$V_d = V_1 - \frac{2}{3}(V_1 - V_2) \tag{15.6}$$

The total volume of sludge in digester, m³,

$$V_s = V_d \cdot t_d + V_2 \cdot t_s \tag{15.7}$$

Total digester volume, m³,

$$V_T = 2 \cdot V_S \tag{15.8}$$

where, V_1 is the volume of raw sludge added per day, m³/day; V_2 is the volume of digested sludge produced per day, m³/day; V_s is total sludge volume, m³ [i.e., digesting + digested]; V_d is average digesting sludge volume per day, m³/day; t_d is digestion period, day; t_s is digested sludge storage period, day; and V_T is total digester volume, m³.

The digesters can also be designed based on typical mean cell residence time (Eq. 15.9) and organic loading rates (kg VSS/m³ day).

The mean cell residence time,

$$\theta_c = \frac{X}{\Delta X} \tag{15.9}$$

where, X is the kg of dry solids in the digester and ΔX is kg of dry solids produced per day in the digested sludge, which is equal to the sludge wasted per day.

The SRT in digester is equal to HRT because of no sludge recycling. The number of cells in the sludge fed to the digester is negligible as compared to the cells present in the digester sludge and digested sludge. Though θ_c is considered equal to θ_H, however while wasting the digested sludge the cells which are going out of digester may be more thus affecting the mean cell residenace time in reality. Since, θ_c is critical parameter in performance of the digester it should not fall below some critical value. These minimum θ_c value below which digestion will not occur depends on temperature as given in Table 15.6 (McCarty, 1964).

Actual value of θ_c used in design is recommended to be 2.5 times higher since it is a critical parameter affecting performance of the digester. Therefore, volume required for high-rate digesters shall be estimated using Eq. 15.10.

$$V = Q \times \theta_c = Q \times \theta_H \tag{15.10}$$

where, Q is digested sludge volume per day, m³/day; θ_c is designed mean cell residence time, days; θ_H is hydraulic retention time, days.

Table. 15.6 Suggested values of θ_c at different temperature

Temperature, °C	18	24	30	35	40
θ_c, days	11	8	6	4	4

Example: 15.3

Design a low-rate digester of the required capacity to treat primary and waste activated sludge generated from a sewage treatment plant serving a population equivalent of 25,000. The per capita sludge contribution is 0.12 kg per day on dry basis and the raw sludge has solids content of 3% and a volatile solids content of 80% of DS. A specific gravity of sludge is 1.015. The sludge digestion will be destroying 65% of volatile solids and a solids content of 7% dry solids with a specific gravity of 1.03 is expected to be produced. Consider a digestion time and storage time of sludge as 25 days and 45 days, respectively.

Solution

Estimation of raw sludge production

Raw sludge production, $W_r = 25,000 \times 0.12 = 3,000$ kg/day.

Volume of raw sludge admitted per day, $V_1 = \frac{W_r}{P_s \times S_s \times \rho_w} = \frac{3000}{0.03 \times 1.015 \times 1000} = 98.52$ m^3.

Estimation of digested sludge volume

Volatile solids in the sludge $= 3,000 \times 0.8 = 2,400$ kg/day.

Fixed solid in the sludge $= 3,000 \times 0.2 = 600$ kg/day.

Volatile solids destroyed $= 2,400 \times 0.65 = 1,560$ kg/day.

Undigested volatile solids fraction $= 2,400 - 1,560 = 840$ kg/day.

Total sludge accumulated in the digester, $W_d =$ Undigested VS + Fixed solids $= 840 + 600 = 1440$ kg/day.

Volume of digested sludge produced, $V_2 = \frac{W_d}{P_s \times S_d \times \rho_w} = \frac{1440}{0.07 \times 1.03 \times 1000} = 19.97$ m^3.

Average volume of digesting sludge, $V_d = V_1 - \frac{2}{3}(V_1 - V_2)$ $= 98.52 - 0.67(98.52 - 19.97) = 46.15$ m^3/day.

Total volume of sludge in the digester $= V_d \cdot t_d + V_2 \cdot t_s$ $= (46.15 \times 25) + (19.97 \times 45) = 2052.4$ m^3.

Hence, the volume of the digester, $V_T = 2 \times 2052.4 = \mathbf{4104.8\ m^3}$.

Provide floating cover circular digestion tank of 29.5 m diameter with 6 m height.

Example: 15.4

Design a high-rate digester for the required capacity to treat primary and waste activated sludge generated from a sewage treatment plant serving a population equivalent of 25,000 persons. The per capita sludge contribution is 0.12 kg per day on dry basis and the raw sludge has solids content of 3% and a volatile solids content of 80% DS, a specific gravity of 1.015. Consider an operating temperature of digester as 30 °C.

Solution

Estimation of raw sludge production

Raw sludge production, $W_r = 25,000 \times 0.12 = 3,000$ kg/day.

Volume of raw sludge admitted per day, $V_1 = \frac{W_r}{P_s \times S_s \times \rho_w} = \frac{3000}{0.03 \times 1.015 \times 1000} = 98.52$ m³.

Estimation of digested sludge volume

From Table 15.6, for operating temperature of 30 °C, the minimum digestion time required is 6 days. Thus, considering the design mean cell residence time of $2.5 \times 6 = 15$ days.

Since, for completely mixed reactor without sludge recycling the hydraulic retention time and mean cell residence time will be same.

The volume of the digester required, $V = Q \times \theta_H = 98.52 \times 15 = 1477.8$ m³.

Providing cylindrical digester with liquid depth of 6 m plus 1.5 m of gas collection zone, thus making total depth of 7.5 m, the diameter required for the digester is 17.7 m. Heating, mixing and biogas collection arrangement need to be provided to this digester.

15.7.7 Aerobic Digestion

Sludge can also be stabilized by prolonged aeration that results in biological destruction of volatile solids. Aerobic digestion of biological sludges is most common method of sludge digestion, which is an extension of the principle of aerobic metabolism as applied in activated sludge process. Aerobic digestion is usually applied to digest the waste activated sludge from aerobic treatment plant operating without primary clarifier. It is found successful in the digestion of dilute suspensions, such as activated sludge, which comprises of wide range of aerobic microbial species. It is most commonly applied to the stabilization of excess sludge from extended aeration systems.

15.7.7.1 Fundamentals of Aerobic Digestion

Aerobic digestion is endogenous respiration process in which the microorganisms are forced to metabolize their cell mass in the absence of external food source. A portion of the organic fraction removed from cell mass is utilized for synthesis of new biomass. The remaining organic matter is transformed through energy metabolism, referred to as respiration, and oxidized to carbon dioxide, water, ammonia and soluble inert material to provide energy for both synthesis and life maintenance functions of bacteria as summarized in Eq. 15.11.

$$\underset{\text{cell}}{C_5H_7NO_2} + H_2O \xrightarrow{\text{Aerobic bacteria}} 5CO_2 + NH_3 + 2H_2O + \underset{\text{biomass}}{Cell} + Heat \qquad (15.11)$$

The aerobic digestion results in more mineralized material, in which the remaining organic solids are principally cell wall or other cell fragments, which are not readily subjected to biodegradation. An extended period of aerobic digestion can further convert ammonia to nitrates. Aerobic process is accomplished through many different pathways involving various trophic groups of microorganisms, which can be employed in the bio-chemical oxidation of complex organic matter. This is the reason why aerobic systems have lesser chance to be affected by toxic materials. The volatile solids reduction in aer-obic digestion is typically in the range of 30–50%, while that for anaerobic systems it is typically 65%. The digestion process also reduces the pathogens and offensive odours.

15.7.7.2 Operation of Aerobic Digester

The application of aerobic digestion is limited to treat excess sludge from (i) waste acti-vated sludge, (ii) secondary sludge (ASP or TF) plus primary sludge, and (iii) waste sludge of ASP designed without PST, i.e., extended aeration ASP. The factors that affect the aerobic digestion are temperature, rate of degradation, solid loading rate, oxygen require-ments, solid retention time, and sludge characteristics. The SRT required for operation at 20 °C is typically more than 20 days; however, for practical operations SRTs are kept in between 40 and 60 days (U.S. EPA, 1979). The oxygen requirement for complete oxida-tion of organic matter is 2 kg of O_2/kg of VSS. Typical design criteria adopted for aerobic digestion are listed in the Table 15.7.

Aerobic digesters are single or multiple open tanks equipped with diffused or mechan-ical type aerator. It is operated by continuous feeding of raw sludge and intermittent withdrawals of digested sludge and supernatant. The raw sludge is aerated in the tank for a typical digestion period of 2–3 weeks. The stabilized sludge is allowed to settle by gravity after discontinuing the aeration and the gravity thickened sludge is separated for disposal. Many a times, the aerobic digester is followed by a settling tank, which thickens the sludge before being disposed off on land in slurry form. The digester sludge has high specific gravity of up to 1.72 and hence more thickened sludge with solids content of 3% is obtained in the underflow (Kelly, 2005). The supernatant water typically has sus-pended solids of around 300 mg/L and BOD_5 of the order of 500 mg/L. The supernatant is decanted and recycled back to the inlet of the primary sedimentation tank.

Table. 15.7 Typical values of design parameters for aerobic digestion

Parameters	Values
Suspended solids content	1%
Solid retention time, θ_c at 20 °C	40 days
Solid retention time, θ_c at 10 °C	60 days
Air required (diffused aeration) Activated sludge only Activated sludge plus primary sludge	18–36 L/min m^3 54–66 L/min m^3
Solids loading rate	1.6−4.8 kg VSS/m^3 day*

* *Source* Adopted from Metcalf and Eddy (2003)

Aerobic digestion is more robust and rugged than anaerobic and will operate at low temperature and low pH of even 5.5. However, the performance is temperature dependant as increased microbial activity shall be obtained at higher temperatures. The sludge is stabilized to highly fertile humus substances, which can be used as soil conditioner. However, in contrast to anaerobic process, aerobic digestion is energy intensive. Mechanical equipment used for aerobic digestion process require high capital investments and are conventionally employed for large municipal sewage treatment plants. In addition to this, aerobically digested sludge has poor dewatering characteristics and it requires mechanical equipment for dewatering, which adds on to the financial burden on sludge treatment facility.

15.7.7.3 Advantages and Disadvantages of Aerobic Digestion of Sludge

The aerobic digestion of the sludge can offer following advantages.

- Less susceptible to reaction upsets than anaerobic systems
- Easy operation and process control
- Not as sensitive to environmental factors as in case of anaerobic digestion
- Emission of non-explosive gases (CO_2 and NH_3) and no emission of toxic and explosives gases, such as H_2S and CH_4
- Low capital cost
- Low odour nuisance
- Better quality of supernatant than in anaerobic digestion
- Sludge is stabilized to highly fertile humus substance.

However, the disadvantages of the aerobic digestions are:

- No energy content with digester gas (since it is CO_2)
- Digested sludge is still difficult to dewater
- High operation and maintenance costs
- Reduced performance during cold climates.

15.7.8 Composting

Composting is a low-cost and environment friendly alternative technique for stabilization of wastewater treatment plant sludges. In composting an artificial conducive environment is created for natural aerobic decomposition of organic matter. Composting has not been a conventional treatment method for stabilization of sludge in large sewage treatment plant. However, the recent advancement in sludge characteristics studies and composting techniques have resulted in the application of composting process as an appropriate method for sludge processing and its acceptance in developed nations lead to invest more on installation of full-scale composting facility with sewage treatment plant. Objectives of the composting are digestion of putrescible organic matter, destruction of pathogenic organisms, and reduction of mass and volume of sludge by removing water content. When compared with anaerobic digestion, composting is a more robust process, as it is operated at a wide pH range of 6.5–9.5.

15.7.8.1 Fundamentals of Composting

Composting process involves the breakdown of complex organic matter with formation of intermediate product, such as humic acid, to form completely stabilized end product. The process of complete degradation of organic solids, such as protein, lipids, and fats, is irreversible and the final end product obtained is fully stabilized compost. The microorganisms involved in the degradation process are of three categories: bacteria, actinomycetes and fungi. The decomposition of complex organics and cellulose is attributed to the action of varying levels of fungi and actinomycetes in mesophilic and thermophilic temperature range.

Composting can be either aerobic or anaerobic biological stabilization process. The aerobic composting is widely employed, which produces major end products, such as compost, carbon dioxide, water and heat (Eq. 15.12). Other than composted biomass, the anaerobic composting yields methane, hydrogen sulphide, carbon dioxide and significantly less heat. In both the cases the compost is one of the products, which is highly fertile humus substance beneficial to use in agricultural sector. Anaerobic composting is likely to emit odour. Based on the moisture content of sludge (35–50% of dry solids), composting can also be classified as dry process; whereas, aerobic and anaerobic digestion are categorized as wet processes in which solids content is less than 3% for aerobic digestion and 3–8% for anaerobic digestion (Andreoli et al., 2007; Kiely, 2007; Vesilind, 2003).

$$\text{Organic solids} + O_2 \xrightarrow{\text{Aerobic microorganisms}} CO_2 + NH_3 + H_2O + \text{New cells} + SO_4 + \text{Heat} \tag{15.12}$$

Composting is accomplished through wide range of temperatures ranging from psychrophilic (10–20 °C), mesophilic (20–40 °C) and thermophilic (50–60 °C). The rate of degradation and stabilization of organic matter is higher at thermophilic range in which the dominant microorganisms are thermophilic bacteria, actinomycetes and thermophilic fungi. Due to the exothermic characteristics of the compost process, the heat produced in the compost pile decreases the moisture content of the compost material and inactivates pathogenic microorganisms. This stage is followed by a cooling stage (curing) that characterized by reduction of microbial activity, shift of thermophilic bacteria to mesophilic bacteria and fungi, further loss of water by evaporation from compost materials and volume reduction. Besides, stability of pH and completion of formation of humic acids are achieved during cooling stage.

Sewage sludge has relatively uniform characteristics and solids particle size that makes it less complex for composting than composting of municipal solid waste. The time period required to stabilize the organic matter during aerated static pile composting or windrows composting is about 28 days for composting followed by a curing period of 30 days or longer (Metcalf & Eddy, 2003).

15.7.8.2 Composting Operations

Composting operations involves following steps:

i. Dewatered sludge, by chemical conditioning, using lime or ferric chloride or poly-electrolytes and partially digested in a digestion process, is mixed with an organic amendment or a bulking agent to decrease the bulk density.
ii. Sludge is aerated either by the addition of air or by mechanical turning or by both for a minimum of five times during composting period.
iii. Recovery of the bulking agents at the end of curing period.
iv. Storage of compost, allowing further stabilization and cooling.
v. Screening the compost by separating inert material like plastic, metals etc. and size reduction by grinding.

Dewatered sludge cake is commonly too wet and dense mass with solids content of 20–30%. If it is not mixed with another substance, the sludge cake will form a dense mass with wet anaerobic interior and a dried outer crust. Bulk density of the dewatered sludge cake is to be reduced for successful composting. Addition of organic amendments or bulking agents reduces the bulk density, absorbs excess water and provides adequate porosity for aeration inside the pile. Therefore, the composting sludge is given a pre-treatment by adding solid materials. Common choices of organic amendments are saw dust, straw, peat, rice husks manure, refuse, garbage and yard trimmings. Bulking agents widely used are wood chips, pelleted refuse, shredded tyres and peanut shells.

The bulking agents increases the voids for better penetration of air to provide adequate aeration and the sludge trapped in the inter spaces of bulking agents undergoes aerobic decomposition. The bulking agents will also facilitate to dry out the blended mixture of sludge and agent. The organic amendments in the compost mixture decomposes during composting and maintain the C:N ratio of the mixture with an increase in the organic content of the compost. Composting is biological process, which is affected by many direct and indirect factors that contribute to the overall mechanism of composting. The most important operational factors are discussed below.

Aeration: Air supply is necessary for compost pile to obtain the optimum results. Aeration is accomplished by natural convection and diffusion of air through regular turning-over using equipment or by supplying air by forced aeration. The oxygen concentration available in entrapped air should be at least 50% of the fresh air oxygen concentration to take care of remaining composting material and it is found to be critical concentration when it is lowered to 15%, below which the process shifts to anaerobic. Oxygen is not only required for the oxidation of various organic compounds but also for drying and cooling of compost material. Consumption of oxygen during composting is directly proportional to microbial metabolic activity and it is maximum at temperatures between 30 and 57 °C.

Temperature: Microbial respiration during composting process releases heat, which maintains high temperature in the compost pile. Heat generated in the compost pile is dissipated slowly and that is confined in the compost pile, since the compost material is good insulator for heat. Heat loss from compost pile is a function of the temperature gradient with outside temperature and the rate of metabolic activity depends on the rate of oxygen supply. Compost maintains maximum temperature until all the volatile solids have been decomposed. Factors that govern the composting temperature are moisture content, air temperature, rainfall, type of microorganisms and size and shape of pile.

In an aerated pile, it takes 1–2 days to reach 60 °C; whereas, in an unaerated windrow system the oxygen supply is limited and hence the similar temperature is attained after 5 days. The best results are obtained in composting, when the temperature is maintained between 50 and 56 °C for first few days and 54 and 60 °C for remaining days of active composting period. Most microorganisms can't survive at elevated temperatures beyond 60 °C. Fungi, which degrade the cellulose and lignin cannot withstand this temperature.

Moisture content: Moisture content is related to the aeration of compost pile. The biological activity (rate of decomposition) gets reduced or stopped at low moisture content, especially less than 20%, though the compost is stable physically but not biologically. Moisture content more than 60% reduces the porosity of compost mixture, resulting in poor air ventilation and leading to anaerobic activity causing foul smell. Hence, composting process is effective when the sludge is in solid state and the moisture content ranges between 40 and 60%.

C:N ratio: Optimum composting process necessitates a minimum level of organic carbon and nutrient requirements. Microorganisms involving composting process utilize carbon for growth and nitrogen for protein synthesis. For a good microbial growth in an aerobic system, the optimum value for carbon to nitrogen ratio shall be about 12. The carbon present within the bulking agent is not readily available, therefore the optimum C:N ratio for composting may be adjusted in the range of 25–30. The ratio, above 30 inhibits the composting process, and the ratio far below 20 leads to incomplete composting. The C:N ratio of sewage sludge is in the range of 5–20. The supplemental carbon can be provided by the amendment and mixing of bulking agents that improves energy balance and C:N ratio of the mixture (Vesilind, 2003).

The approximate C:N ratio of sewage sludge (activated sludge) is 6:1 and for digested sludge is 16:1 (Gray, 1999). By the addition of bulking agents, the C:N ratio is increased. High C:N ratio will slow down the decomposition until the carbon gets oxidized. Low C:N ratio will also slow down the decomposition process and initiate the loss of nitrogen through ammonia volatilization at high pH. Therefore, the final compost product must have a C:N ratio less than 12 otherwise the soil microbes immediately utilize the available nitrogen, which will limit the plant uptake for its growth and impair the purpose of serving the compost as a soil conditioner (Gray, 1999).

pH: For aerobic composting the optimum operating pH range is between 6.5 and 9.5 for successful composting process. However, pH of 7.0–7.4 should remain for optimum aerobic decomposition. There is a sharp decline of pH at the beginning due to the production of organic acids by acid producing bacteria. Higher pH along with high temperature can result in nitrogen loss.

15.7.8.3 Composting Systems

Two methods of composting generally employed are open system and closed system. Open system does not require a reactor for operation and composting is done in turned piles or windrows and aerated static pile with or without air supply. In closed system specially designed reactor is used, which is equipped with forced aeration. Commonly adopted closed system is called in-vessel composting.

Windrows: In windrows system, oxygen is supplied by natural diffusion and periodic turning. It is more popular than aerated static piles and it is efficient in space utilization. Windrows are constructed on concrete base with 1–2 m width and 3–5 m length at regular intervals so that they can be periodically turned over by mechanical means with ease at sewage treatment plants. The rows are regularly mixed and turned mechanically during composting period. It should be ensured that windrows are turned for a minimum of 5 times during composting period, and the sludge cake is maintained at 55 °C in the centre of the pile for an adequate period of composting, which is about 21–30 days. Following the active composting period, the curing period extends from 30 days to 9 months to improve the quality of finished compost.

In windrow composting, aerobic conditions are not uniformly maintained. Anaerobic condition might develop within windrows and periodic turning results in the release of offensive odour. However, the microbial activity accomplished within the pile may be aerobic, facultative and anaerobic depending upon how often pile is turned. Turning the compost material promotes the gas exchange, which supply oxygen to the pile. Oxygen also enters the pile by diffusion and natural ventilation caused by convection currents, which is induced by escape of various gases through the pile. Periodic turning at 6–7 days interval results in a cyclic variation of oxygen concentration within the pile and the biological oxidation is never maintained at maximum efficiency due to the fluctuations in the oxygen availability for a composting mass. Turning operation leads to the release of spores-containing dust, which is hazardous to the plant operator.

Aerated static pile: Aerated static pile system is not turned for aeration, it remains static and the oxygen is supplied by natural ventilation or more commonly by forced aeration. The amount of oxygen supplied to the pile can be controlled and so as with moisture content and temperature of the compost. Static pile requires less area than windrow and less manpower. The compost is not handled manually till the process is completed and hence the risk of operators suffering from pulmonary disease could be eliminated.

Aerated static pile system requires equipment and structures for forced aeration or promoting natural ventilation. Air can either be drawn or blown through the pile to maintain sufficient oxygen and control the temperature. Porosity of compost mixture is maintained by wood chips, which absorb the water content initially. A layer of wood chips is prepared over a network of perforated pipes, over which mixture of sludge cake and wood chips are placed to a depth of 2–2.4 m. On the top of the pile, a layer of finished compost is placed as an insulator. Air is supplied by a blower, intermittently through the pile to control the excessive cooling and that continues for 3–4 weeks. The blower exhaust is circulated through finished compost to deodorize it. The compost is allowed to cure and dry for 4–6 weeks. The compost is highly stabilized with low moisture content, when air is blown through the compost to enhance evaporation.

In-vessel composting: In-vessel composting is closed system composting that is accomplished in a specially constructed reactors or vessels. The compost mixture is fed into the reactor under controlled conditions with forced air supply. The system is designed to reduce odour problem and composting time by controlling temperature and oxygen concentration. Small plants are operated as batch system of vessels in series or as continuous flow through system. Rotating drum is adopted for larger ones. In the operation, caustic soda is added to control the pH and the compost is recycled to seed the incoming mixture with active population of microorganisms. The advantages include small area requirement, good compost quality, reduced odour problem and less labour costs and digestion time. Recent advancement has resulted in two-fold reduction in ammonia emission, while operating an anaerobic vessel composting compared to aerobic composting. It is also reported that ammonia emission is 10 times less in anaerobic in-vessel composting system than aerobic composting system (Rijs, 1992).

Straw composting: It is an open system of composting sewage sludge with straw as bulking agent. It is cheap in terms of capital and operational costs. The straw decomposes with sludge and screening of finished compost is eliminated. Careful management of the system is essential, otherwise transportation cost makes it expensive. Maintaining adequate C:N ratio is rather difficult for mixture of sewage and straw. Straw has a C:N ratio of 48 and an optimum C:N ratio of 12 is good for microbial metabolism. The necessary ratio to mix sludge and straw on dry basis is 4:1 for dewatered digested sludge with 20% dry solids. Outdoor composting that requires high degree of operational care should have covered forced aeration, which results in successful composting.

Co-composting with municipal solid waste: Co-composting of sludge generated from wastewater treatment plants and municipal solid waste or yard waste is an alternative with integrated solid waste disposal facilities. Advantages of this include improved C:N ratio of mixture, sludge dewatering, less metal content in composted material than that of composted sludge alone. This system requires careful management as it becomes expensive due to transport of sewage sludge or municipal solid waste.

15.7.8.4   Compost Quality

The composting process typically achieves the volatile solids reduction of about 50% and moisture content less than 40%. Fully composted sludge appears to be dark brown with pleasant earthy smell. Compost containing low nutrients cannot be a substitute of inorganic fertilizer; however, it is an excellent soil conditioner. The nutrient value of final compost product can be improved during composting process by blending sewage sludge and bulking agents or amendments. Co-composting with municipal solid waste is a possible alternative to improve the compost quality. Compost is similar to soil humus and when mixed with soil, it enhances the water retention capacity of soil. Presence of metals in compost is a problem that restrict land disposal. Since composting process involves partial aerobic thermophilic decomposition that will result in considerable reduction in number of pathogens; however, it does not comply with the standards for pathogens reduction. The problem of odour from composting process is due to the emission of ammonia, hydrogen sulphide and sulphide compounds. These problems can be eliminated by adopting bio filtration technologies, which are commonly employed in industrial air pollution control.

15.8   Sludge Conditioning

Digested sewage sludge from a digester contains more than 90% water with solid content typically ranging from 2 to 6%. Using some pre-treatment to increase the solids content and changing characteristics of the sludge solids to remove the water associated with sludge, the volume of the sludge can be reduced. The purpose of sludge conditioning is to improve dewatering process by: (i) enhancing filtration characteristics of sludge; (ii) increasing degree of flocculation of sludge solids particles; and (iii) preventing the small particles clogging the filter cloths in filter presses. Sludge conditioning is chemical or thermal process employed on raw or digested sludge to enhance the efficiency of thickening and dewatering. Sludge conditioning also results in some level of sludge stabilization to reduce its nuisance potential upon disposal. Most common conditioning processes include chemical conditioning, thermal conditioning, freezing, elutriation, and addition of inorganic bulking agents.

The intercellular form of bound water is retained in the biological sludge solids due to chemical bonding. This bond may be broken by the addition of chemicals, which causes destabilization of electrically charged sludge solids particles. Mechanical dewatering processes remove free water and intercellular water from sludge; whereas the intracellular water is removed by heating and freezing that break the walls of the microbial cell.

15.8.1 Chemical Conditioning

The objective of chemical conditioning is to reduce zeta potential by adding chemicals, which can change the surface charge of particles so as to make it more efficient for dewatering in mechanical processes. The higher the zeta potential the more tightly and stable the water is bound to the sludge solid particles. The destabilization of the sludge solids is accomplished by the reduction of zeta potential using chemical coagulants. The mechanism of chemical conditioning involves the addition of coagulants to neutralize the surface charge and formation of a polymeric interparticle bridging of individual sludge solids particles that results in a clump of solids particles with rigidity and porosity permitting water to easily drained out of it. Chemical conditioning can be effective in reducing the moisture content of the sludge typically from 99% to about 70%. In order to achieve the moisture content to that extent, mechanical systems, such as centrifugation, belt-filter press and pressure filter press are employed. Most commonly used chemicals in chemical conditioning are inorganic chemicals and organic polyelectrolytes.

Chemical conditioning using inorganic chemicals: Inorganic chemicals used for conditioning are lime, ferric chloride and combination of lime and ferric chloride. Lime is used for conditioning or stabilization; however, the doses are different. Much higher dose and contact time are required for lime stabilization to raise the pH to 12.0. Doses of lime range from 1 to 12% of dry mass of sludge solids. Sludge from activated sludge process requires highest dose for conditioning, while primary sludge requires the lowest dose. Chemically conditioned sludges are not readily dewatered by air drying; however, a combination with mechanical processes can be used. The main disadvantage of using inorganic chemicals is that every addition of chemical as conditioning agent results in the production of extra sludge.

Chemical conditioning using organic polymer: Polymers or organic polyelectrolytes are widely used in conditioning of primary and digested sludges. Organic polymers are characterized with long chain and water soluble synthetic organic chemicals. They are typically cationic polyacrylamide capable of destabilizing the ionic charge of sludge solids with dosages typically ranging from 0.01 to 0.65% of sludge solids on dry basis. They are easy to handle, require less storage space, less dosage, very effective and produce relatively less sludge.

High molecular weight polymers provide floc strength to resist the shear forces offered by mechanical systems, which is a serious cause of concern for operational troubles and costs in dewatering process. The required dose of polymer may change depending on variation in the sludge composition. Therefore, the dose is optimized by performing test on a pilot plant or bench tests. The factors that influence the selection of type and dosage of conditioning agents are pH, alkalinity, phosphate concentration, solids concentration, type of mixing devices and dewatering technology used. The dosage of polymer will also depend on its molecular structure, ionic strength and activity level.

15.8.2 Thermal Conditioning

Heat treatment has been traditionally used for conditioning and stabilization of sludge. Heat treatment for sludge conditioning results in the destruction of bacterial cells with high degree of removal of moisture. The sludge is mixed with air and heated to about 200 °C for retention time of 25–30 min at pressures ranging from 1000 to 1800 kPa (McGhee, 2014; Metcalf & Eddy, 2003). On heating, the sludge undergoes changes such as break down of gel structure and coagulation of solids particles. Consequently, the sludge is stabilized and sterilized. The bound water is removed from sludge solids with an improvement in dewatering characteristics of the sludge. The supernatant so released from the heat treatment unit contains high BOD that requires pre-treatment and it is re-circulated to the inlet of primary clarifier.

The thermal conditioning process offers following advantages:

- No requirement for chemical conditioning
- Thermally conditioned sludge dewaters better than chemically conditioned sludge
- High degree of stabilization and pathogen removal
- The process is not affected by the changes in the composition of sludge
- Utilisation of residual heat of volatile solids for pre-heating the sludge.

However, the disadvantages of the thermal conditioning of the sludge are:

- Highly contaminated gaseous and liquid wastes are generated from the process,
- High capital and operating costs,
- Skilled supervision and maintenance required,
- Emission of odours gases,
- Scale formation and hence requirement of acid washing of reactor vessel.

The volatile solids in the sludge contains residual heat value of 27 kJ/g. Utilization of waste heat obtained from a source for pre-heating can increase solids concentration to 6% with conditioned sludge at 60 °C in pilot testing centrifuge (Garelli et al., 1992). Preheating, which saves the cost of producing heat, is found to be effective and beneficial method of conditioning to enhance the sludge dewatering characteristics.

15.8.3 Freeze–Thaw Conditioning

Freezing of wastewater sludge during cold climate enhances dewatering characteristics. During freezing, the free water starts to freeze, then form a crystalline structure. As it grows in larger size, the floc particles are pushed out of the ice front and the floc (interstitial) water is extracted by diffusion and added to frozen water (Vesilind & Martel, 1990).

The outcome of freezing process is a condition favourable for converting jellylike sludge into granular material that readily and easily drains out when it melts. The freezing process is more effective when the sludge freezes throughout its depth, which is a function of differential temperature below freezing surface and duration of freezing period. The operational variables for optimized freeze/thaw process are initial solids concentration in the sludge, rate of freezing and duration of freezing condition.

The following methods are also employed for achieving conditioning of sludge under special circumstances.

Elutriation: Washing of sludge with fresh water or effluent is termed as elutriation. It is done by gentle mixing of water with sludge for about 10 min either mechanically or by diffused aeration and then allowing the sludge to settle. It reduces the chemical requirement for conditioning or removes ammonia compounds, which interfere with coagulants making the sludge difficult to dewater. Anaerobically digested sludge contains high amount of alkalinity ranging from 3000 to 5000 mg/L, which is very high for lime conditioning to occur, therefore the sludge is washed with fresh water or water of low alkalinity to bring the alkalinity required for dosage of ferric chloride and lime.

Addition of inorganic bulking agents: Materials like fly ash, newspaper pulp and pulverized coal are mixed with sludge to obtain relatively dry sludge with solids content 40–50% after mechanical dewatering, however the volume reduction of the sludge is minimal as it already contains bulk volume of additives.

15.9 Sludge Dewatering

Dewatering is a physical operation that is employed to reduce moisture content of the sludge and thereby increasing solids content. Primary sludge of 2% solids concentration can be readily concentrated by thickening to about twice the solids content (4%). Digested sludge (10–18% DS) is dewatered to 15–35% solids on dry basis. Dewatering is accomplished either by mechanical dewatering process or air-drying process. Digested sludge is dewatered mechanically in large plants; whereas, in small treatment plants it is applied on sand drying beds. Mechanical dewatering without prior thickening is not a feasible method of sludge processing for activated sludge with high water content in the range of 98–99%. Aerobically digested sludge is not immediately amenable to mechanical dewatering process, unless it is further stabilized anaerobically. Handling of anaerobically digested sludge in dewatering equipment can increase odour problem. The selection of dewatering technology is based on the type of sludge, characteristics of dewatered sludge, space availability, downstream sludge processing used, if any, and final disposal option.

15.9.1 Mechanical Dewatering

Mechanical dewatering operation involves the water separation mechanisms, such as pressure filtration, vacuum filtration, squeezing, capillary action, and centrifugal compaction. The mechanical devices commonly used are: (1) centrifuge, (2) belt filter press, (3) plate and frame filter press, and (4) vacuum filter. Belt filter press and plate and frame filter press are of pressure filter types. Enclosures are fabricated to control the odour nuisance associated with mechanical dewatering devices, which are adopted based on the type of sludge.

15.9.1.1 Centrifugation

Centrifuges are generally employed for both thickening and dewatering processes of sludge in sludge handling facility. It is commonly used in industries for thickening and separation of liquids and solids. However, it is used in municipal wastewater treatment plants to dewater primary, digested and activated sludges, but not as common as belt filters press. Dewatering performance and plant economics are the criteria for the final choice of dewatering device to be selected as either centrifuge or filter press.

The basic types of centrifuges are: (i) solid-bowl centrifuge and its modifications and (ii) basket centrifuges. Imperforate basket centrifuge is not commonly used in new installations for dewatering sewage sludge. The sludge centrifuge works on the principle of centrifugal force, which compacts the sludge against the bowl wall and the internal scroll or conveyor supply the sludge (Fig. 15.6). The compacted sludge is conveyed due to the differential speed of bowl and conveyor, along the bowl wall towards the conical area to discharge (Eckenfelder, 2000).

Sludge is introduced at constant flow rate to a rotating bowl. The dilute liquid (called centrate) is produced and separated from the sludge due to centrifugal action of rotating bowl and the sludge is thus compacted to a dense cake. The rotating helical scroll or screw conveyor moves forward and discharges the compacted solids (concentrated solids) into a hopper or onto a conveyor belt. The centrate, from the centrifuge contains high organics and fine suspended solids, is recirculated back to the inlet of the primary clarifier. The sludge cake obtained from centrifuges typically contains moisture content ranging from 60 to 80%.

Centrifuge concentrates the sludge to a solids content ranging from 15 to 30% dry solids. The solid capture is in the range of 55–80% without chemical conditioning, while solid recovery of 80–95% is achieved with the addition of polymer or ferric chloride by centrifugation. The conditioning chemicals are added either to sludge feed line or to the sludge in the centrifugal bowl. Dosage of polymer for conditioning the sludge prior to the centrifugation ranges from 1 to 1.8 g/kg solids on dry basis. High solid centrifuge is a modified form of solid-bowl centrifuge, which provides higher solid recovery efficiency due to increased residence time and improved compaction. Proper ventilation should be provided for centrifuge units to control odour nuisance. Odour from the centrifuge can also be controlled by the addition of lime with untreated sludge.

Sludge dewatering using centrifuge has the following advantages:

- Recovery of solids is high,
- It is capable of achieving comparatively dry form of sludge cake,
- Low initial cost,
- Not much problem of odour as centrifuges are enclosed in casing,
- Less area requirement for installation compared with other dewatering devices

However, the disadvantages of sludge centrifuge are:

- Higher power cost,
- Requires skilled personnel and high maintenance cost of scroll due to wear,
- Pilot tests are required to be run for plant design,
- Larger motor demands adequate source of electricity supply source,
- It requires special type of foundation and sound proofing.

15.9.1.2   Belt Filter Press

Belt filter press is the most popular sludge dewatering machine, classed under pressure filtration dewatering techniques. It is operated on continuous mode under the principle of chemical conditioning, gravity drainage and application of pressure mechanically. A belt press consists of two continuous porous belts, between which the sludge is applied, that passes over a series of rollers, which squeeze the water from the sludge. The porous belts are supported on an open framework. Dewatering occurs in three phases through distinct mechanisms. Initially, the chemically conditioned sludge is fed evenly at a uniform rate onto a moving porous belt, which permits gravity drainage; or sometimes vacuum condition is established to increase the water removal at this stage. The belt, at this gravity zone, allows almost half of the free water to drain out by gravity through the porous belt to a collection tray from which the filtrate if pumped back to the inlet of the primary clarifier.

Then the lower belt moves with dewatering sludge, which gradually gets compressed between upper and lower belts as they come together in the wedge zone. Upper belt is brought down towards the moving sludge and gradually compress the sludge between the belts to squeeze excess water, while moving the compression zone. As the paired belts move around the series of rollers, the high pressure and shear force developed due to their relative rotational speed with each other, contribute to the discharge of free water from the sludge cake sandwiched in the belts (Fig. 15.12). Further, the pressure gradually increases as the belts move through the series of smaller and closer rollers in this stage. In some cases, a partial vacuum is applied towards the end of this zone for effective dewatering.

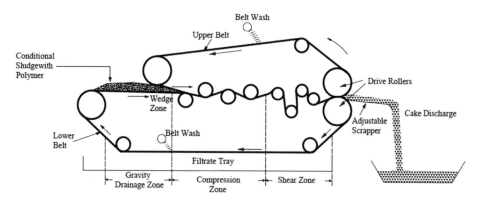

Fig. 15.12   Belt filter presses

The belts then separate the sludge cake, which is removed by scraping the belts with a blade held against the belts or by passing it around smaller radius rollers.

The operating variables to control the process performance are polymer dosage and filter solids loading rates. The key factors that affect the filtration performance are sludge characteristics, method of conditioning process, pressure applied, system configuration, and porosity, speed and width of belt. The sludge cake obtained is expected to have a solids content ranging from 15 to 45%, with typically 25% on dry solid basis. Raw and digested sludge with or without thickening can be dewatered by belt filter press. For primary sludge with 2–10% solids content is concentrated by dewatering to about 25–45% with solids loading rate of 350–690 kg/h per metre width of belt filer press and 1–6 g/kg polymer dosages. However, waste activated sludge and aerobically digested sludge are less amenable to dewatering by belts filters.

The width and length of the belts typically adopted are 1–3 m and 3–9 m, respectively. The life spans of the belts typically range from 1200 to 2000 h, which is subjected to vary with damages on belts fabric. The efficiency of the belts depends on the pressure developed between the rollers due to its clearance and the retention time, which is determined by the length and speed of the belts. Belt tracking and belt life are the major operational problems of the belts. The belt tracking is considered as the ability to maintain control of the desired path of a conveyor belt once it is aligned and installed. With new system design, vertical belts and spring-loaded rollers, belt filter press can offer the production of sludge cake with 35% solids on dry basis. The belt filter press offers the following advantages:

- Less capital and operating costs,
- Capable of achieving dry sludge cake in high pressure machines,
- Easy to maintain,
- Less complexity in operation,

- Low energy requirement,
- Easy to shut down the machine.

The disadvantages of the belt filter press are:

- Requires preliminary treatment in feed sludge as the sand or grit can damage the belt,
- Very sensitive to varying feed characteristics,
- High odour problem,
- Machines are manually operated.

15.9.1.3 Plate and Frame Filter Press

The plate filter press is operated as batch process. Fixed volume and variable volume plate filter presses are the two types employed for larger installation. Fixed volume filter press comprised of a series of rectangular plates held vertically, face to face, on a metallic frame and recess on both sides (Fig. 15.13). The face of each plate is fitted with filter cloth. The conditioned sludge is pumped into the filter press plates under some pressure until the chambers are completely filled. The frames are pressed together after feeding the sludge either electromechanically or hydraulically between a fixed and movable head so as to form a sealed chamber that can withstand the pressure applied during the filter press. The pressure is maintained till the sludge cake is formed and the flow of filtrate through the filter and the plate outlet is stopped. The plates are then separated, and sludge cake is dislodged. The filtrate usually recycled back to the influent of the sewage treatment plant in primary clarifier.

The dewatered sludge cake obtained from the press has often solids content as high as 50%; however, typically it varies from 30 to 45%. The pressure filter is primarily used and effective for raw, mixed primary and secondary, digested and thermally conditioned sludges. The improved performance of the plate filter is dictated by degree of conditioning of the sludge either with lime or ferric chloride or polymer prior to the pressure filtration process. The applied pressure ranges from 700 kN/m^2 and the filtration cycle takes about 3–5 h to complete (Gray, 1999). The factors governing the rate of filtration are pump pressure, choice of filter fabric, condition of filter fabric, filterability of solids in the sludge. Filter cloths are required to be washed periodically between the cycles using a high-pressure water jet. The frequency of washing, typically range from 20 to 25 cycles, depends on the type of chemical conditioner and conditions of cloth.

In variable volume plate filter, a rubber diaphragm is placed behind the filter cloth that expands and exerts the squeeze pressure to sludge cake thereby reducing the cake volume during compression stage. It is designed to dewater with an initial pressure of 700 kN/m^2 followed by maintaining a constant compression of 1700 kN/m^2.

Plate and frame filter press offers advantages such as low chemical consumption, high solids concentration in the filtered cake, low suspended solids in the filtrate, and lower

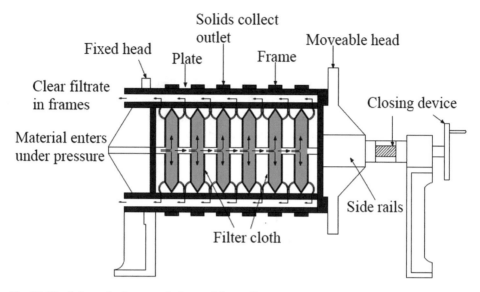

Fig. 15.13 Schematic diagram of plate and frame filter press

sludge volume after filtration due to high solids content and lower moisture. However, it has disadvantages like high labour cost due to involvement of skilled labour for operation and maintenance, limited filter cloth life requiring replacement, high equipment cost involved, necessity of batch mode of operation, and larger floor area requirement.

15.9.1.4 Vacuum Filtration

Vacuum filter is largely used in industrial wastewater treatment plants and not very common for use in newly installed sewage treatment plants for sludge dewatering. In vacuum filtration, the water is sucked through a filter cloth under vacuum that is carried on a slowly revolving drum immersed partially in the sludge. The type of vacuum filter widely used is the rotary drum filter (Fig. 15.7) having travelling media and precoat media filter (Parker et al., 2008).

In vacuum filtration, chemically conditioned sludge is fed to a vat or a basin in which a cylindrical drum filter is suspended above and immersed partially in the sludge (Metcalf & Eddy, 1990). The filtration medium is made of either cloth of natural or synthetic fibres, or wire-mesh fibres. As the drum revolves slowly, the water is drawn from the sludge through the filter medium inside the drum due to application of a vacuum inside the drum. The solids are captured on the circumference of the drum and the filtrate water is withdrawn from the drum. The drum continues to rotate and allows the sludge to remove by a scraper mechanism that involves a blade pointing towards the moving sludge on the drum.

The solids content of the cake may range from 7 to 40%. The filtration is usually aided with chemical conditioning using lime and ferric chloride. Polymer addition is not very effective in vacuum filtration process. The design variables are filter medium, vacuum level, cycle time, cake discharge mechanism and sludge conditioning chemicals. Solid recovery of 96–99% is achieved in a precoat rotating filter. Dewaterability of a sludge using filtration is governed by specific resistance. Although coagulants are used to reduce the specific resistance of sludge, it may not be a good choice from economic point of view and may dictate alternate dewatering methods.

15.9.2 Air Drying Processes

Air drying processes involve the methods that remove moisture from sludge by natural evaporation and gravity induced drainage. Air drying processes are accomplished either in open-air or under covered roof of glass. Digested sludge from digesters contains typically more than 90% of water and it is required to be further dewatered to reduce the volume of sludge. Reduction of moisture content by drying up the sludge on an open sand bed is most common and suitable in tropical countries like India. Whereas, in European countries the dewatering is generally done by mechanical dewatering process, such as high-speed centrifuges, pressure filters and vacuum filters. Most widely used air drying methods are sludge drying beds, lagooning and reed beds.

15.9.2.1 Sludge Drying Beds
Open-air drying of anaerobically digested sludge is the most popular and oldest method of sludge dewatering using drying beds. Drying beds are considered to be the cheapest method of dewatering the sludge. Sludge is placed on the layer of clean sand bed of specified depth and it is allowed to dry. Simplicity in operation and easy maintenance make these beds popular for employing in small treatment plants. They are suitable for plants serving small and medium communities with an average flow rate of less than 8000 m³/day due to large land area requirement and possible odour and flies nuisance. Five types of sand drying beds that are generally used include: (i) conventional sand bed (ii) paved (iii) artificial medium (iv) vacuum assisted, and (v) solar (Metcalf & Eddy, 2003).

15.9.2.2 Construction and Operation
Sludge drying beds are open beds of land constructed in shallow tank with a layer of clean sand having specified depth (about 0.3 m), supported on layers of graded gravel medium. Sand beds larger in size are partitioned by masonry or concrete walls and each cell is independently operated with distribution of sludge through a header pipe from digester. Figure 15.14 shows schematic constructional details of sludge drying bed. Sludge placed on the sand bed gets dewatered by percolation of water through the sludge mass and

supporting sand and gravel bed, after which air drying commences evaporation of water remaining in the sludge. On drying, sludge cake shrinks, producing more cracks, which speeds up the evaporation of water from the increased sludge surface exposed to air.

Dewatering occurs mainly due to rapid drainage, typically in 2 days and progressively decreases until the sludge solidifies. Seepage collected in the laterals of underdrainage lines is returned to the primary clarifier. Climatic factors that affect dewatering by evaporation are wind, humidity, and solar radiation. Further, rainfall dilutes the solids concentrations. Sand drying bed can be constructed with or without mechanical removal of dried sludge and roof cover for protecting from rain.

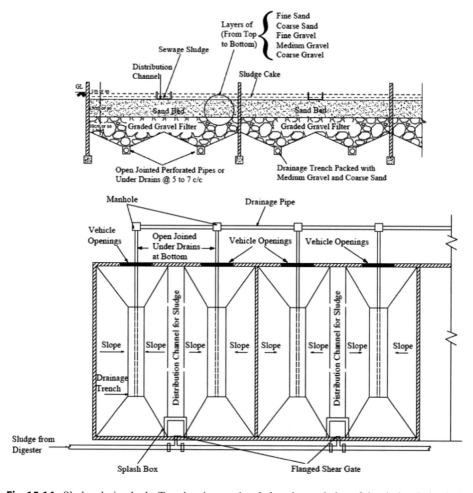

Fig. 15.14 Sludge drying beds. Top showing sectional elevation and plan of the sludge drying beds at the bottom

The performance of sand drying beds depends mainly on the factors, such as type of sludge, solids content, dewatering characteristics of sludge, sludge loading rate, drying period, porosity of sand, atmospheric temperature, rainfall, humidity, wind velocity, degree of stabilization of sludge, grease content, depth and frequency of sludge application. Depending on the rate of gravity drainage and evaporation, the drying time may extend from several days to few weeks. Sludge is not usually conditioned by adding coagulants prior to the application of sludge on the sand drying beds. However, addition of aluminium chloro-hydroxide and polyelectrolytes may improve dewatering characteristics and reduce the exposure time. Moreover, secondary sludge does not get dewatered that easily and hence conditioning is necessary before applying them on drying beds.

The sand drying beds offer major advantages such as low cost, occasional attention required for operation, and high solids content in dried cake. However, the disadvantages associated with this are challenges in drying process and fluctuation in characteristics due to the effect of climate, labour intensive operation for the removal of dried sludge, problem of insects and odour, and large land area required for the drying beds, which occupies about 40% of the plant area.

15.9.2.3 Design Criteria for Sand Drying Beds

Area of beds: The drying area depends on the volume of the sludge, climatic conditions, location, cycle time required to retain the sludge on sand bed for dewatering, drying and removing dried sludge and time required for making the sand bed ready for next batch of sludge application. Cycle time for dewatering and drying process depends on the depth of sludge applied on the bed. The cycle time also depends on the general factors governing the performance of sand beds as discussed above. Typically, the cycle time varies up to 2 weeks for warm climate and 3–6 weeks in unfavourable conditions. The land area required for drying bed range from 0.1 to 0.15 m^2/capita with dry solids loading of 120–150 kg/m^2 of area per year for anaerobically digested primary sludge (Metcalf & Eddy, 2003). For digested mixed sludge, area requirement is 0.16 to 0.23 m^2/capita with dry solids loading of 60–120 kg/m^2 of area per year. Land area required for digested primary sludge is 50–60% less than that required for digested mixed sludge from primary and secondary clarifiers.

Underdrains: It is a lateral drainage lines made of perforated plastic pipes or vitrified clay pipes laid with open joints, slopped at a minimum of 1% and spaced 2.5–6 m centre to centre. Drainage from beds, if not disposed of directly, shall be recycled back to the plant wet well for treatment.

Gravel: Lateral drainage lines are properly placed and supported in a centrally located trench, which are covered with graded gravel or crushed aggregates around the underdrain pipes in layers up to 30 cm, with minimum of 15 cm above underdrain pipes. At least 8 cm of the top layer is made of gravel with 3–6 mm size.

Sand: Clean sand with an effective size of 0.5–0.75 mm and uniformity coefficient not more than 4 is used. The depth of sand bed may vary from 20 to 30 cm (CPHEEO, 2013).

Dimensions: Drying beds are commonly of 6–8 m wide and 30–45 m long. With the bed slope of 0.5%, the length of travel of sludge on the bed from the pipe header at the inlet should not be more than 30 m for a single discharge point of wet sludge. For larger sized sand beds, multiple discharge points should be provided to restrict longer wet sludge travel. Free board of at least 0.4 m is should be provided.

Sludge inlet: Piping to the sludge beds is designed for minimum velocity of 0.75 m/sec and a minimum diameter of 200 mm to supply sludge to the beds. The pipe should discharge at a minimum height of 0.3 m above the sand bed. Splash plates are placed at the sludge discharge points to spread the sludge uniformly over the beds and to prevent erosion of the sand bed.

Removal of sludge: With favourable drying condition the sludge becomes fit for removal within 2 weeks. Dried sludge cake from the beds is removed mechanically or manually when moisture content is reduced to less than 70%. Mechanical equipment employed for shovelling and lifting the sludge cake and then loading it in a truck result in excessive loss of sand and disturbance of gravel media. As the moisture content reduces to 40% the sludge cake becomes suitable for grinding. Some sand gets removed with sludge cake on which it clings. The loss of sand bed when decreased to about 10 cm shall be maintained periodically by placing fresh, clean sand to its original depth (about 30 cm). After removing dried sludge cake, the sand bed is prepared by levelling for next cycle of sludge application.

Example: 15.5

Design a sludge drying bed for digested sludge from a primary clarifier and excess sludge from activated sludge process of a sewage treatment plant serving for a population equivalent of 1,00,000. Consider a total solid remaining in digested mixed sludge on per capita basis as 60 g/day. The digested sludge has a solids content of 6% of DS with specific gravity of 1.03.

Solution

For a total solids in a digested mixed sludge per capita of 60 g/day, the daily digested sludge production

$= 1,00,000 \times 60 \times 10^{-3} = 6000$ kg/day.

Adopt a dry solids loading of 120 kg/m^2 year.

Therefore, area of bed required $= \frac{6000 \times 365}{120} = $ **18,250 m^2**.

Thus, the area required for drying bed per capita $= 18,250/10000 = 0.18$ m^2.

This is well within the range of 0.16−0.23 m^2/capita.

Number of beds required with bed size of 8 m width and 30 m length $= \frac{18250}{8 \times 30} = $ **76**.

Assume 2 months of rainy season in a year and 3 weeks for drying and for preparation of bed.

No. of cycles per year $= \frac{(12-2)4}{3} = 13.33 \sim 13$.

The digested sludge has a solids content of 6% of DS with specific gravity with 1.03.

$V_{sl} = \frac{W_{ds}}{P_s S_{sl} \rho_w} = \frac{6000}{0.06 \times 1.03 \times 1000} = 97.09$ m^3/day.

Therefore, the depth of application of sludge on the bed $= \frac{97.09 \times 365}{76 \times 8 \times 30 \times 13} = 0.149 \sim \textbf{15.0 cm}$.

Alternately, considering the depth of application of sludge as 0.3 m, the area required for the bed shall be worked out as illustrated below.

Considering 8 weeks rainy period when sludge cannot be applied on the sand drying beds, hence no. of weeks available for application of sludge $= 52 - 8 = 44$.

Considering 3 weeks cycle time, the number of cycles per year $= 44/3 = 14.66$, say 14.

Hence, area required for the bed $= \frac{97.09 \times 365}{14 \times 0.3} = 8437.58$ m^2.

For each bed of size 8 m $\times$ 30 m, number of sand beds required $= \frac{8437.58}{8 \times 30} = 35$.

Thus, provide 35 sand beds of size 8 $\times$ 30 m, with sludge application depth of 0.3 m on the bed.

15.9.3 Thermal Drying

Thermal sludge drying is a method of sludge volume reduction that involves application of heat to evaporate the water thereby reducing the moisture content of the sludge. Thermal drying or heat drying processes require pre-treatments, which include thickening, anaerobic digestion, sludge conditioning, and dewatering to obtain solids content of around 25%. This drying process can result in a granular form of cake with solids content as high as 95% on dry solids basis. Advantages of heat drying include volume reduction of the sludge, reduced transportation costs and pathogen removal. The conditions favourable for effective heat transfer are:

- Large surface area of exposure between the sludge and thermal carrier
- High heat content of thermal carrier
- Long contact time between the sludge and the thermal carrier
- Velocity and turbulence of drying air.

Heat drying is classified as direct, indirect, combination of direct–indirect and infrared depending upon the heating carrier (medium) and transfer mechanisms used. Thermal carriers commonly used as source of heat energy are condensing steam, hot air, combustion gases, superheated water, thermal oil and infrared radiation for heat drying. The auxiliary fuels, such as coal, gas and oil are burnt to heat the ambient air acting as heating medium to provide the latent heat of evaporation. The biogas generated from anaerobic digester is generally used as heat source for drying process. The mechanisms of heat transfer that drive the drying process are conduction, convection and radiation.

Convection: Convection is a direct drying process in which the wet sludge is exposed to direct contact with the heat transfer medium, such as hot air or gas. Under equilibrium condition, the heat transfer rate is directly proportional to the area of wetted sludge surface exposed and the temperature difference between the gas and the interface of the sludge-gas.

Conduction: Conduction is an indirect drying process in which a solid medium separates the wet sludge and the heat transfer medium is usually a stream or hot air. Heat transfer occurs when heat from the heating medium transfer through solid medium to the wet sludge.

Radiation: Infrared drying system comprising of infrared lamps or gas fired heating system supply radiant heat energy that transfers to the wet sludge to evaporate water.

15.9.3.1   Direct Dryers

In direct drying, the sludge is exposed to direct contact with the heating medium, such as hot air or gas. The water vapour coming from the sludge leaves with drying medium from the dryers. The drying medium not only dries the sludge but also conveys the sludge through the dryer. Flash dryer, rotary dryer and fluidized bed dryer are classified under direct (convection) dryers that are conventionally employed for drying municipal wastewater sludge. Fluidized bed drying for municipal sludge is more common in new installations. However, rotary dryers are seldom employed for municipal sludge and it is generally used in industries.

Flash dryers: Flash drying involves pulverizing the wet sludge along with a recycled part of already dried sludge in a cage mill. The mixture may typically have moisture approximately 40–55%. Hot gases from furnace or heat exchanger create enough turbulence to remain in contact with dispersed solids particles in a duct so as to accomplish mass transfer of water from sludge to the hot gases. Drying process occurs in the duct through which hot gases and sludge are forced up. Dried solids and moisture-laden hot gas are separated in a cyclone separator to produce dried solids having moisture content of around 10%. A portion of the dried solids is recycled back to the mixture in which the dried solids are mixed with incoming wet sludge to feed the dryer. Dried solids so produced are transferred for incineration or packaging as fertilizer.

Rotary dryer: Rotary dryers are commonly used for drying raw, secondary and digested sludges generated from wastewater treatment plant. A rotary dryer consists of a revolving cylindrical steel drum at 5–8 rpm mounted on bearings with its axis slightly inclined with horizontal. The dryer is fed with mixture of dewatered sludge and previously dried solids exposing it to a hot gas stream at a temperature of around 600 °C for a period of 30–60 min. The solids mixture and hot gases are conveyed through the dryer progressively, while rotating the mixer in the interior. The resulting product is of solids content of approximately 95% and it is excellent in the use as a fertilizer or as a soil conditioner. This drying process is required to install air pollution control devices.

Fluidized bed dryer: A fluidized dryer consists of an enclosed chamber similar to that of a rotary dryer, in which a bed of sand is kept in fluidized state by an up-flow hot air to provide evaporation. A sand bed acts as a heat exchanger to transfer the heat supplied by a steam boiler for drying the sludge solids. Fluidized bed dryer maintains a uniform temperature of approximately 120 °C through an intimate contact between sand particles in bed and recycled low oxygen content fluidizing air. The advantages include lesser footprint, production of good quality fertilizer and automatic operation. Major disadvantage of it is large electrical power consumption.

15.9.3.2 Indirect Dryer

In indirect drying, the sludge is not exposed directly with a heating medium. Disc dryer is an indirect dryer in which the separation wall between the sludge and heating medium is at a temperature range from 160 to 260 °C. The sludge, water vapour and heating medium are released from dryer in two different routes. Commonly used indirect dryers are either horizontal or vertical configuration. Horizontal dryer is a hollow core with jacketed shell through which the heated medium such as steam or oil is recirculated. The dryer is fed by dewatered sludge, which moves horizontally but it revolves in a spiral manner through the dryer. The sludge particles are brought to contact with heated metallic surface of the dryer. The water vapour released from drying operation is drawn off by an induced-draft fan in the off-gas duct. The dried sludge with a solids content in the range of 70–95% can be achieved as desired for ultimate use or disposal.

15.10 Thermal Volume Reduction

15.10.1 Incineration

Incineration of sewage sludge is a technological solution adopted for both sludge treatment and disposal in most of developed nations. Incineration process of sludge solids involves drying and combustion that result in reduction of water and volatile organic matter leading to the reduction in volume of sludge solids to a great extent by converting the solids into inert ash. Therefore, the transportation of large quantity of sludge and its disposal on land are substantially reduced as the ash can be disposed of easily on soil, unless otherwise it has metal contamination in it. In case of metal containing ash, either metal recovery or disposal on landfill need to be adopted.

Sludge solids comprise of high amount of combustible volatile matter and less of fixed carbon content. When sludge is dried and burnt, it produces considerable amount of heat due to exothermic oxidation reaction. Therefore, incineration of sludge is a process of thermal oxidation in which all the volatile organic solids are converted to form oxidized end products, such as carbon dioxide, water and ash along with release of heat energy. Raw sewage sludge is a potential source of energy than digested sludge as it contains more organic fraction. The energy content of raw sludge ranges from 15,000 to 24,000 kJ/kg of dry solids; however, for digested sludge it ranges from 10,000 to 14,000 kJ/kg dry solids. Sludge with solids content less than 30% will require auxiliary fuel to burn it, because the heat energy released during combustion is not enough to that required for the evaporation of the water present in it.

The oxygen required for the complete combustion can be determined stoichiometrically. However, to ensure the complete combustion, about 50% of excess amount of air is supplied than the theoretical requirement. A proper material balance must be made across the system to include the effect of inorganic material and moisture content in sludge and air for heat requirement on combustion process. The heat requirement can be lowered, primarily, by reducing the moisture content, which determines the additional fuel required to accomplish combustion.

The total heating value can be determined based on the percentage composition of the sludge solids as per Eq. 15.13 (WEF, 1998).

$$Q = 14{,}544 \times C + 62{,}208(H - O/8) + 4050 \times S \qquad (15.13)$$

where, Q is total heat value, kJ/kg dry solids; C, H, O and S are the fraction of carbon, hydrogen, oxygen and sulphur, respectively. The fuel value of raw primary sludge is found to be higher as it generally contains grease and other skimming materials, which improve its thermal content.

The major advantages of incineration of sludge are: maximum extent of volume reduction of sludge occurs, it works as a potential source of heat energy if feed sludge has less moisture content, and complete destruction of pathogens and toxic biorefractory organic compounds occur during incineration. Such toxic compounds, if present in the sludge, they may remain in the sludge during other drying methods. However, the disadvantages associated with incineration are high capital and operating cost, requirement of skilled personnel for operation and maintenance, requires air pollution control devices due to the emission of toxic gases of incomplete combustion and ash, and difficulty to dispose the residue ash, as it may be hazardous in nature.

15.10.2 Multiple Hearth Incineration

Multiple hearth incineration is widely employed for converting sewage sludge into ash and gases in large treatment plants as well as in small facilities, where land is inadequate for sludge disposal or at chemical plants for recalcination of lime sludge. The incinerator consists of cylindrical steel shell in which a series of hearths are arranged vertically in a stack and a central rotating hollow shaft is connected with cooling fan and drive assembly.

Sludge cake enters at the top hearth and it is slowly raked towards the centre, where it is dropped to the next hearth, from where it is pushed towards the periphery of the shell, and then drops to the next lower hearth, where it is raked towards the centre. In the upper hearths (drying zone), maintained at the temperature of 550 °C, vaporization of moisture and cooling of exhaust gases occur. In the middle hearth (combustion zone), sludge solids are burnt at 1000 °C. The bottom zone cools the ash to 350 °C and removes the ash.

The hollow central shaft is cooled by forced air supplied by cooling fan. The preheated air from the shaft is returned to the lowest hearth where it is further heated by hot ash and passes through the combustion zone then cools as it heats to dry the incoming sludge cake. The air after two circulations in furnace is discharged to the wet scrubber to remove fly ash and other volatile gases. The operation of incinerator will result in the emission of particulate matter, which is trapped with air pollution control devices. Cyclonic water jet scrubber and perforated plate type scrubber are used to control the particulate emission to meet the air quality standard. These devices are also able to meet the stack emission standards of nitrogen oxide and odour.

Feed sludge with solids content not less than 20% is desirable for incineration and can be burned with additional fuel at 20–30% dry solids. This problem can generally be overcome by blending the sludge with refuse to enhance the combustibility of the sludge. However, auxiliary fuel is required to start the process and may be necessary for regular operation. Sludge with 30–50% of dry solids is burnt without fuel. Solids content more than 50% may lead to creating excess furnace temperature, which is beyond refractory and metallurgical limits. Combustible sludge solids contain grease, carbohydrate and proteins and on its combustion with excess air the resulting end products are CO_2, SO_2 and water vapour. The sludge volume is decreased to less than 15% of its original volume. The main advantages of multiple hearth incineration are high sludge loading rate, low fuel consumption, stability in operation, easy process control and operation and reduced air pollution emissions. However, the major problems associated with the operations are with ineffective sludge dewatering over sizing of the furnace is required, that can increase the heat requirement by consuming more fuel.

15.10.3 Fluidized Bed Incinerator

In fluidized bed incinerator, a cylindrical steel shell containing a bed of preheated sand is fluidized by upward moving air into which the sludge solids are injected. The air is injected into the incinerator at a pressure range from 25 to 35 kN/m^2 to fluidize the sand bed. Sludge solids introduced into bed are mixed rapidly by turbulent condition created in the bed due to upward moving hot gases. The temperature of the sand is maintained between 760 and 820 °C resulting in rapid drying and burning of the sludge. Combustion gases along with ash move upwards and get released through the outlet at the top of the furnace. The ash is removed from exhaust gases by venture scrubber. A cyclone separator is used to separate the ash and scrubbed water. The parameters controlling the process to achieve complete oxidation of organic matter are sludge feed rate and air flow to reactor. Continuous operation of the process, even though with short duration of shutdown, does not require the use of auxiliary fuel after the start up. Fluidized bed system is reliable; however, the complexity necessitates skilled supervision and trained personnel for operation. This is normally used either in medium or in large sewage treatment plant for sludge incineration.

With the development of successful technologies to control the particulate and other gaseous end products, the incineration has become an integral part of sludge treatment facility in developed countries. In last few decades, Germany, The Netherlands and Japan have been employing fluidized bed incineration to treat about 20% of the sewage sludge produced.

15.10.4 Electrical Furnace

Electrical furnace involves an infrared based heating system for heating the sludge on a moving woven wire mesh to attain the ignition point. Sludge with low moisture content and high fuel value is favourable for operation of the system. It is less expensive than other types of incinerators and it is suitable for small installations as it operates intermittently.

15.11 Pyrolysis

Many organic substances are thermally unstable and when subjected to extreme heat in oxygen free condition are broken up through combined thermal cracking and condensation reactions resulting into gaseous, liquid and solids fractions as its end products. Pyrolysis is the process in which organic matter is chemically decomposed at elevated temperature in the absence of oxygen under extremely dry condition. The process typically accomplishes at temperatures above 450 °C under atmospheric pressure, but can be operated at vacuum and positive pressures. The process is irreversible involving the changes in both physical state and chemical composition. Pyrolysis is widely employed to convert organic substances into a solid inert residue containing ash and carbon along with small quantities of liquid and gases. At elevated temperatures, pyrolysis process yields the inert residue as carbon that is referred to as carbonization.

Ideally pyrolysis does not involve reaction with water, oxygen and other chemical substances. However, it is practically impossible to maintain an absolute oxygen free environment. The main disadvantages of pyrolysis are: the product stream is more complex than for many of the alternative treatments, and the product gases cannot be vented directly in the cabin without further treatment because of the high CO concentrations.

Sludge treatment using pyrolysis process produces: (i) a gaseous mixture comprised of H_2, CH_4, CO, CO_2 and others; (ii) a char of almost pure form of carbon; and (iii) liquid stream containing oil and tar. Thus, the three domains of area focused on pyrolysis of sewage sludge are oil production, gasification and biochar production from sludge.

15.11.1 Production Oil and Biochar from Sludge (OFS Process)

In the oil production process from sewage sludge, the organic solids are preheated. The dried sludge is then converted to liquid fuel during heat treatment. To accomplish this, dried sludge with more than 25% DS is heated to about 450 °C in the absence of oxygen that results in vaporization of nearly half the sludge. Tar produced, during thermal decomposition of sludge, contains aliphatic hydrocarbons (C_nH_m). The vapours so produced during the process are in contact with the tar residue of the sludge, and this tar captures the aliphatic hydrocarbons in gaseous phase (Kiely, 2007). The end products of the process are oil, char, non-condensable gas (CO_2, CO, H_2, N_2, C_nH_m and CH_4) and water. These by-products (tar or bio-oil and syngas) are combustible at about 880 °C to produce energy for drying. Dried raw sludge typically yields 30% of oil and 50% of char.

Use of biochar in agriculture sector: Indiscriminate disposal of sludge poses hazardous effect on soil and water environment as well as threat to public health. Preparation of biochar from sewage sludge has gained attention of researchers to explore the potential application in agricultural field as it improves the soil quality and reduces the dispersal and uptake of heavy metals. Biochar has proved to be excellent in soil amendment substance in agricultural use.

Utilization of pyrolysis products as energy resource: All pyrolysis products, such as bio-oils and gases, have potential to be secondary energy resources. The bio-oil is similar in characteristics as that of diesel and could be a substitute for engine fuels. However, biochar has high energy content as it contains higher carbon content. The pyrolysis oil has heating value of 25 MJ/kg and biochar has 31 MJ/kg, which is higher than bio-oils produced from wood and agriculture residue (20 MJ/kg) and comparable to commercially available coals (Ben et al., 2019).

Application of biochar in water treatment process: Properties of biochar include large specific surface area, enhanced surface functional groups and porous structure. Therefore, it can be a good low-cost and effective adsorbent. The ability of biochar in removing heavy metals, refractory organic pollutants and nutrients, such as N and P, has been reported (Racek et al., 2020). Production of biochar represents a sustainable solution involving circular economy, carbon sequestration, contaminant immobilization, greenhouse gas reduction, soil fertilization and improvement of water retention capacity.

15.11.2 Gasification

Gasification involves converting dried sludge with more than 80% DS to combustible gas, which in turn is utilized to produce electricity or hot water. It is normally accomplished in a reactor in which dried sludge is heated to 500 °C for distillation followed by carbonation at 800 °C. The solids yield the gases without combustion to take place as the air flow rate is limited. The gas is cleaned of tar and oils at temperatures more than 1200 °C. Ash is also an end product of this process.

15.12 Other Alternatives for Sludge Treatment

15.12.1 Wet Air Oxidation

The wet air oxidation does not require sludge dewatering as a pre-treatment and it is adequate if the sludge is thickened to 5% DS. In wet air oxidation organic solids are oxidized under an aerobic condition and high pressure and temperature. It involves liquid phase reaction of organic matter between water and oxygen at typically about 170 °C and a pressure of about 10 MPa (Kiely, 2007). The process accomplishes the oxidation of large part of organic sludge into inorganic sludge, which can be readily dewatered to 40–50%. The condition is favourable for production of clear liquid stream of low BOD concentration than that from a thermal sludge conditioning process. Volatile solids reduction of about 80–90% can be achieved with a volume reduction of more than 96% after dewatering (McGhee, 2014; Sincero & Sincero, 1999).

The wet air oxidation of the sludge can offer following advantages.

- A variety of organic compounds can be treated simultaneously in single step process,
- Wastes are treated in liquid phase and thus the problem associated with air pollution is reduced,
- The process is less energy intensive than incineration.

However, the major disadvantages of the process are as follows:

- Wet air oxidation doesn't perform a complete mineralization of organic waste,
- The process is limited to waste containing organic and inorganic compounds and it doesn't destroy some halogenated aromatics and some pesticides.
- Corrosion is a severe problem with wet air oxidation technology, which requires control of pH.

15.12.2 Sludge Melting

Sludge melting is a process of sludge volume reduction in which the sludge is heated to more than 1200 °C to evaporate water and thermally decompose under controlled air ratio and melt the organic fraction. The melting process results in the conversion of molten inorganic material into a beneficial product as slag (Kiely, 2007). The process of melting involves several steps such as, dewatering as pre-treatment, melting, heat recovery, waste gas purification and slag production. Sludge melting is generally accomplished in methods that are either melting after drying and melting after carbonation. Carbonizing process is a controlled combustion process under air ratio less than 1. The characteristic of slag, which is obtained after melting, depends on the rate of cooling. Slow rate of cooling produces a high-strength slag. In incineration, the volume reduction of organic sludges is about one-nineteenth of their original volume, whereas it is about one-thirtieth by the melting process (Kiely, 2007).

15.12.3 Sludge Lagoons

A lagoon is an earthen basin into which raw or digested sludge is pumped and stored for long period. The sludge undergoes stabilization producing obnoxious odours. Stabilised sludge accumulates at bottom. Supernatant is withdrawn and then returned to the treatment plant. Dried sludge from the lagoons is removed periodically.

15.12.4 Ultrasound Waves for Sludge Dewatering

Recent approach to achieve sludge volume reduction is the application of ultrasound waves for improving dewatering capabilities of the sludge. When an ultrasonic wave is applied to a liquid medium, minute bubbles are formed during the cavitation process. These cavitation bubbles implode under extreme forces of temperature and pressure. When sewage sludge is subjected to ultrasound waves, the sludge flocs are disintegrated as a result of powerful hydro-mechanical shear force generated by bubble implosion. Disintegration rate of sludge particles increases with increased duration of ultrasound treatment due to increased concentration of extracellular polymeric substances (EPS). This prolonged contacts with ultrasound waves lead to increasing bacterial cell wall collapse and the extraction of intracellular water, proteins and nutrients into the sludge. Moreover, as the intensity of ultrasound increases, the particle size distribution and surface charge of the sludge get altered. Extracellular polymeric substances released along with alteration of size distribution and surface charge of the sludge particles will improve the dewatering characteristics of the sewage sludge.

Recent trend of research shows the potential application of ultrasound to reduce the sludge volume by about 50%. About half of the cell mass comprises of water therefore the application of ultrasound could be a pre-treatment to thermal sludge drying process to reduce energy consumption. Moreover, it has the advantage that the sludge treated with ultrasound is expected to be free from pathogens. The other advantages of ultrasound treatment are: no sludge production during the treatment; easy operation; increased biogas production from the sludge after ultrasound treatment; improvement in sludge dewaterability; no secondary pollution problem; and effective in degradation of complex materials. However, the disadvantages associated with this are not much reported and successful operational feedback of full-scale installation is not yet available. However, additional energy demand for operation is a drawback and exact economic analysis and adaptability are required to be explored further.

15.12.5 Value Added Product Recovery

The sludge generated from wastewater treatment plants can be utilized as a feed stock to explore value added product recovery. Some of the options available that need to be exploited further are stated below.

i. Ash from incineration process contains metals that can be recovered.
ii. The slag produced from sludge melting process is combined with other material to manufacture bricks or can be used to construct road pavements.
iii. Application of sludge as a soil amendment to agriculture land can provide essential plant nutrients. Macro nutrient like potassium (K) is not supplemented by sludge in significant quantity; however, soils which are deficient in Zn may be amended by sludge.
iv. Utilization of sewage sludge in digester would produce biogas that can be utilized for in-plant energy requirement and organic sludge produced can be used as manure for agricultural purpose after amending it with organic additives to maintain C:N ratio.
v. Biochar recovered from sludge contains solid carbonaceous material, which has applications in energy and agricultural sector due to its high heating and fertilizer value.

15.13 Sludge Disposal

The sludge generated from wastewater treatment plants is usually disposed off after providing suitable treatment by adopting any of the following ways.

15.13.1 Ocean Dumping

This was most common disposal route in the past for island countries and coastal urban areas. However, dumping of sludge into sea is banned by almost all the countries since 1998. It is reported that the main impact on marine environments is organic matter enrichment leading to changes in species richness, relative abundance and biomass of macro-invertebrate species.

15.13.2 Incineration

It is commonly operated as a sludge treatment technology rather than a sludge disposal method. Dewatered sludge of about 25% dry solids can be incinerated for complete elimination of organic matters, water and pathogens, leaving only inorganic ash residues.

15.13.3 Land Spreading

Sewage sludge is beneficial for land application as a soil amendment in agricultural sector. Intensive agricultural practices make the soil deficient in soil nutrients that can be supplemented by the addition of sewage sludge, which comprised of all essential nutrients. The factors which influence the land spreading of sludge are: climate and rainfall pattern, existing soil nutrients level, chemical composition of sludge, and potential contamination for cropland and ground water. The sludge generated from municipal wastewater treatment plant, if contains high concentration of any metals, may not be suitable for soil application, due to hazardous nature. Soil amended with sawdust and digested sludge enhances C:N ratio. The application rate for land spreading of sewage sludge on to the land is limited to 10 tonnes DS/ha year and it is estimated based on the levels of either N or P of the crop grown on a particular type of soil (Wagner, 1992a, 1992b).

15.13.4 Land Revegetation

Land, which is overgrazed and hence expended in its soil nutrients may be brought back into agricultural practices quickly by the application of sewage sludge. The condition for application rate of sludge on land revegetation is restricted by the levels of metals present in sludge than nutrients. The application rate may vary from 2 to 200 tonne DS/ha year.

15.13.5 Land Reclamation of Abandoned Mining Sites

Revegetation of mining sites by the application of sewage sludge has been practiced for decades. Restoration of abandoned mines by utilizing large quantity of sewage sludge is a wide spread practice around the world now. Common problem encountered with restoration are low pH, low field capacity and presence of high level of heavy metals. The exceedingly large application rate of sludge can contribute to the potential problem, such as leaching of nitrogen to ground water, transport of metal through surface run off and transmission of pathogens.

15.13.6 Land Spreading in Forestry

Though the practice of land application on the forest land has a beneficial end use, forest lands are always beyond the limit of the economic travel distance of wastewater treatment plants. Besides, the uniform application of sludge is difficult as there is an interference of trees from a technical stand point, presence of pathogens and nitrate leaching are other problems of prime concern for promoting this application.

15.13.7 Landfilling

Landfill as a disposal option has remained a preferred solution for sewage sludge disposal. It is the most preferred method if sanitary landfill is available regardless of stabilization of sludge, presence of grease, etc. Sludge is deposited on the landfill and proper compaction is carried out using tractor or roller and then a soil cover layer of 30 cm is placed on the top. The present solid waste regulation requires monitoring of ground water contamination and protection of ground water quality that are mandatory to ensure the environment protection. However, the ash produced from incinerated sludge may be directly landfilled.

Disposal of sewage sludge on sanitary landfills is a cause of serious concern due to following facts:

a. Contamination of ground water or nearby water bodies, if any, with leachate containing metals and organic compounds,
b. Risk of transmission of pathogens,
c. Enhanced methane production and release in atmosphere due to decomposition of organic matter,
d. Odour problem, fly nuisance and other public health issues in nearby area.

Therefore, the two major possible disposal routes for the sewage sludge are incineration and application on agricultural land. If economy of the scale supports, then biochar and bio-oil recovery could also offer a sustainable option.

Questions

15.1. Describe the relative distribution of the forms of water in sludge and state their removal mechanisms.
15.2. Why is it necessary to treat sludge before final disposal?
15.3. State purpose of sludge stabilization.
15.4. What is sludge digestion? What are the two basic types of sludge digesters?
15.5. Write short notes on the principle of anaerobic digestion.
15.6. Compare aerobic digestion and composting technique adopted for sludge stabilization.
15.7. Name and describe the most common methods available for volume reduction of sludge.
15.8. With the help of a neat sketch explain two stage digestion process.
15.9. Describe the most common methods adopted for ultimate disposal of sludge.
15.10. Write notes on: (i) sludge conditioning; (ii) sludge dewatering; (iii) methods of composting; (iv) sludge drying.
15.11. Explain various methods of heat treatment of sewage sludge.
15.12. Explain the sludge thickening using filter press and centrifuge.
15.13. Briefly explain the advantages and disadvantages of sludge drying beds.
15.14. Describe sludge thickening by air floatation.
15.15. Determine the liquid volume before and after digestion and percentage reduction for 700 kg (dry mass) of primary sludge having the following characteristics.

Characteristics	Primary	Digested
(i) Solids, %	3	10
(ii) Volatile solids (%)	63	35
(iii) Specific gravity of fixed solids	2.5	2.5
(iv) Specific gravity of volatile solids	1.2	1.2

[**Answer**: Volume of primary sludge $= 23.10$ m^3; Digested sludge volume 5.0 m^3 and Volume reduced to 78.36%]

15.16 A low-rate digester is to be designed for waste sludge generated from an activated sludge process treating sewage generated from a community having 25,000 persons. The fresh sludge has 0.11 kg dry solids/capita day (VS $= 70\%$ ds). The dry solids are 5% of the sludge and specific gravity is 1.01. During digestion 65% of VS are destroyed and fixed solids remained unchanged. The digested sludge has 7% dry solids and a wet specific gravity is 1.03. Operating temperature of digester is proposed to be maintained at 35 °C and sludge storage time of 45 days to be provided. Determine the digester volume required. Assume digestion time of 23 days. [**Answer**: 3339 m^3]

15.17 A sewage sludge contains 6% of dry solids having a specific gravity (S_{ds}) of 1.25. Calculate the volume occupied by 1 tonne of sludge with 6% of dry solids before dewatering and after dewatering to 35% of dry solids. Consider suitable other data as desired.

15.18 Design suitable sludge drying bed for the 1200 m^3/day digested sludge generated from the sewage treatment plant.

15.19 Describe the scope of value-added product recovery from the sludge generated from sewage treatment plant.

15.20 Discuss the application of wet air oxidation process for sludge treatment.

15.21 Explain process of pyrolysis adopted for the sludge generated from effluent treatment plant.

15.22 How biochar can be recovered from the sludge generated from the sewage treatment plant? What are the different application potentials for this biochar?

15.23 Write a short note on the final disposal of the sludge generated from the wastewater treatment plants.

15.24 How ultra sound pre-treatment is effective in improving dewatering characteristic of the sludge?

15.25 Write short note on production oil and biochar from sludge.

References

Andreoli, C. V., von Sperling, M., & Fernandes, F. (2007). *Sludge treatment and disposal. Biological wastewater treatment series* (Vol. 6). IWA Publishing.

Ben, H., Wu, F., Wu, Z., Han, G., Jiang, W., & Ragaukas, A. J. (2019). *A comparative characterization of pyrolysis oil from softwood barks.* Open source article, http://creativecommons.org/licenses/by/4.0, Licensee MDPI, Basel, Switzerland.

CPHEEO. (1994). *Manual on sewerage and sewage treatment systems.* Ministry of Urban Development.

CPHEEO. (2013). *Manual on sewerage and sewage treatment systems.* Ministry of Urban Development.

Davis, M. L., & Cornwell, D. A. (2010). *Introduction to environmental engineering* (4th ed.). Tata McGraw Hill.

Eckenfelder, W. W. (2000). *Industrial water pollution control* (3rd ed.). McGraw-Hill.

Fair, G. M., Geyer, J. Ch., & Okun, D. A. (1968). *Water and wastewater engineering* (Vol. 2). Wiley.

Garelli, B. A., Swartz, B. J., & Dring, R. W. (1992). Improved centrifuge dewatering by steam and carbon dioxide injection. *Water Environment & Technology, 4*(6) (Marcel Dekker).

Gray, N. F. (1999). *Water technology, introduction for environmental engineers and scientist* (2nd ed.). IWA Publishing.

690 15 Sludge Management

Hammer, M. J. (2000). *Water and wastewater technology*. Prentice-Hall of India.

Henry, J. G., & Heinke, G. W. (1996). *Environmental science and engineering* (2nd ed.). PHI Learning Private Ltd.

Hess, R. W., et al. (1953). 1952 industrial waste forum. *Sewage Industrial Wastes, 25*, 709.

Kiely, G. (2007). *Environmental engineering*. Tata McGraw Hill.

Kelly, B. C. O. (2005). Mechanical properties of dewatered sewage sludge. *Waste Management, 25*(1), 47–52.

McCarty, P. L. (1964). Anaerobic wastewater treatment fundamentals. *Public Works Magazine, 95*(9–10), 107–123.

McGhee, T. (2014). *Water supply and sewerage* (6th ed.). McGraw Hill Education.

Metcalf and Eddy, Inc. (1990). *Wastewater engineering: Treatment, disposal, reuse*. Revised by G. Tchobanoglous (2nd ed.). Tata McGraw-Hill.

Metcalf and Eddy, Inc. (2003). *Wastewater engineering: Treatment, disposal, reuse*. Revised by G. Tchobanoglous (4th ed.). Tata McGraw-Hill.

Nemerow, N. L. (1971). *Liquid waste of industry: Theories, practices, and treatment*. Addison-Wesley Publishing Company.

Parker, R., Morris, N., Fair, F. N., & Bhatia, S. C. (2008). *Wastewater engineering*. CBS Publishers and Distributors.

Peavy, H. S., Rowe, D. R., & Tchobanoglous, G. (1985). *Environmental engineering*. McGraw Hill International Editions.

Qasim, S. R. (1985). *Wastewater treatment plants: Planning*. CBS Publishing Japan Ltd.

Racek, J., Sevcik, J., Chorazy, T., Kucerik, J., & Hlavinek, P. (2020). Biochar-recovery, material from pyrolysis of sewage sludge: A review. *Waste and Biomass Valorization, 11*, 3677–3709.

Rao, M. N., & Datta, A. K. (1987). *Wastewater treatment* (2nd ed.). Oxford IBH Publishing Co., Ltd.

Reynolds, T. D. (1982). *Unit operations and processes in environmental engineering*. PWS-Kent Publishing CO.

Rijs, G. B. J. (1992). Advanced techniques for minimizing the volume of sewage sludge. In *Proceeding of Conference on the Future Direction of Municipal Sludge (Biosolods) Management*, Portland, Oregon, Water Environment Federation, Virginia, USA.

Sincero, A. P., & Sincero, G. A. (1999). *Environmental engineering*. Prentice-Hall of India.

Tiehm, A., Nickel, K., Zellhorn, M., & Neis, U. (2001). Ultrasonic waste activated sludge disintegration for improving anaerobic stabilization. *Water Resources, 35*(8), 2003–2009.

Tomczak-Wandzell, R., Ofverstrom, S., Dauknys, R., & Medrzycka, K. (2011). *Effect of disintegration pre-treatment of sewage sludge for enhanced anaerobic digestion. In Proceeding of the 8th International Conference*. Vilnius Gediminas Technical University.

U.S. Environmental Protection Agency. (1979). *Process design manual for sludge treatment and disposal*. EPA-625/4-78-012.

Vesilind, P. A. (2003). *Wastewater treatment plant design*. IWA Publishing.

Vesilind, P. A., & Martel, J. (1990). Freezing of water and wastewater sludges, America Society of Civil Engineers. *Journal of Environmental Engineering Division, 116*(5), 854–862.

Wagner, E. O. (1992a). Seeking a common vision for sludge. In *Proceedings of Sludge 2000 Conference*, Cambridge, UK.

Wagner, G. (1992b). The benefits of biosolids, a farmer's perspective. In *Proceeding of Conference on the Future Direction of Municipal Solids (Biosolids) Management*, Portland, Oregon, Water Environment Federation, Virginia, USA.

WEF. (1998). *Design of wastewater treatment plants* (4th ed.). Manual of Practice no. 8 (Vol. 3, Chaps. 17–24). Water Environment Federation.

Xie, R., Zing, Y., Ghani, Y. A., Ooi, K. E., & Ng, S. W. (2007). Full-scale demonstration of an ultra-sonic disintegration technology in enhancing anaerobic digestion of mixed primary and thickened secondary sewage sludge. *Journal of Environmental Engineering and Science, 6,* 533–541.

Tertiary Wastewater Treatment Systems

16

16.1 Introduction

Tertiary wastewater treatments are follow-up treatments after secondary processes. Although easily biodegradable organic compounds, nutrients and suspended solids are removed from wastewater in primary and secondary treatments, the additional treatments are required for elimination of colloids, turbidity, and other dissolved ions, pathogens, and other contaminants including complex refractory organic matters. Due to importance given on the reuse of treated wastewater and the stringent discharge norms, the need of wastewater treatment has become even more rigorous in terms of quality of the treated effluent. In order to satisfy the discharge norms or to meet the specific reuse guidelines, the existing secondary treatments have to be revamped and/or additional tertiary treatments have to be installed. In this regard, this chapter includes tertiary wastewater treatment systems such as pressure filters, adsorption, ion exchange, desalination, electrodialysis, and membrane filtration processes. Now a days, the tertiary wastewater treatment systems are necessary due to following factors:

- Removal of inorganic pollutants from secondary treated wastewater, e.g., heavy metals, dissolved solids, etc.
- Enhanced removal of organic pollutants from secondary treated wastewater so that the treated wastewater meets the reuse norms.
- Removal of suspended solids and turbidity for effective disinfection of secondary treated wastewaters.

16.2   Pressure Filters

Filtration in water and wastewater treatment can be defined as process of separating suspended solids via passage of wastewater through a porous media. In filtration where solids are accumulated in the filter depth via colliding and adhering with the filter media is termed as depth filtration. In this the particle size of filtered solids is much lower than the pore size of the filter bed. The pressure filter consists of top layer of sand/coal and underlying layer of coarser material that supports the filter media. The depth filtrations technique is most common in water treatment; however, it has gained equal popularity in wastewater treatment as well. Filtration in wastewater treatment is primarily focused on the removal of suspended solids from the secondary treatment effluents for effective disinfection downstream or to facilitate any other reuse. It has been widely used for the elimination of colloidal particles, which cannot be removed via secondary sedimentation and it is a common pre-treatment for disinfection and advanced oxidation processes with an aim of reducing the turbidity. The prime mechanism associated with the removal of finer suspended solids in the filtration process is straining. Interception and adhesion can be other removal mechanisms; however, their magnitude is small as compared to straining.

Pressure filters are preferred over gravity filters in the wastewater treatment plants. In pressure filters, wastewater passes through the layer of filter media under a pressure greater than atmospheric pressure. These filters are operated in a closed vessel under pressurized conditions attained via pumping. The applied pressure helps in overcoming the frictional resistance offered by the filter media. Due to this, pressure sand filters are operated at a higher terminal head loss than gravity filters, thus allowing lesser number of backwash cycles. The typical pressure filter comprises of sand as filter media, which is supported on gravel and pebbles of larger size. The sand bed is actively involved in the filtration, thus removing the suspended solids. The top distributer uniformly distributes the influent wastewater over the entire surface area, whereas a under drainage system collects the filtered effluent from the system. Typical cross-sectional view of a vertical pressure filter and actual photograph of installed pressure filter with various valve arrangements is showed in Fig. 16.1.

During the operation of pressure filter, the head loss gradually increases with time due to accumulation of suspended solids in the sand bed, as a result, the treated effluent flow rate is reduced. Similar to gravity filters, backwashing is required to clean the filter media in pressure filters. Along with water backwashing, air scouring can also be implemented prior to backwashing to loosen the intercepted impurities (solids) between the filter media. However, these filters suffer disadvantages during the backwashing operations as the process cannot be directly observed by the plant operators. Therefore, pressure filters sometimes are fitted with an open tank type arrangement prior to inlet, which allows the physical inspection of wash water during backwashing passing through this open tank. Such open box can allow the operator to visually verify the washing out of filter media and the backwashing flow rate can be immediately reduced to prevent the washing of

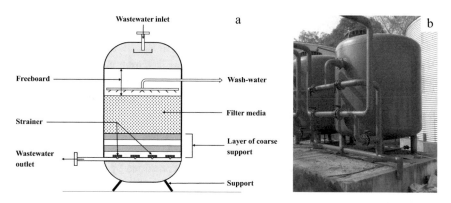

Fig. 16.1 **a** Sectional view of vertical pressure filter, and **b** photograph of actual pressure sand filter installed at IIT Kharagpur, India, for tertiary treatment of secondary treated sewage

filter media, and also to minimize excessive use of filter backwash water by stopping the backwash operation after getting clean water.

A freeboard equivalent to a minimum of 50% of media depth is provided in the pressure sand filtration vessel for allowing expansion of sand during backwashing operation. The grain size of the sand used is between 0.5 and 1.5 mm with uniformity coefficient of ~1.5–1.8, and depth of ~0.5–2.0 m is generally adopted. The rate of filtration through pressure filters is generally in between 15 and 20 m³/m².h.

Example: 16.1 Design a pressure sand filter for treatment of secondary treated wastewater flow of 100 m³/day to be operated under a filtration rate of 15 m³/m².h.

Solution

Wastewater flow, $Q = 100$ m³/day $= 4.167$ m³/h

Required surface area for pressure sand filter $= \dfrac{4.167 \frac{m^3}{h}}{15 \frac{m^3}{m^2 h}} = 0.278$ m² $= \frac{\Pi}{4} d^2$

Hence, diameter $= 0.59 \sim 0.6$ m

Considering depth of sand as 1 m, depth of gravel and underdrainage 0.5 m, and 1 m clearance above sand bed.

Overall height of pressure sand filter $= 2.5$ m

Considering backwashing rate of 25 m³/m².h for 10 min. Hence the quantity of filtered wash water required for backwashing $= \frac{25}{60} \frac{m^3}{m^2 \min} \times 10$ min $= 4.17$ m³

16.3   Adsorption

Adsorption can be defined as a process of molecule transfer from one phase to other, i.e., liquid to solid, gas to solid, etc. This technique is commonly used in water and wastewater treatment for the removal of odour, and other soluble refractory compounds. Adsorption in the context of wastewater treatment is completely different from water treatment due to presence of considerably high content of organic pollutants and colloidal matters in wastewaters. A majority of readily biodegradable organic matter from wastewater gets removed during aerobic and/or anaerobic secondary treatment processes. However, biorefractory organic compounds cannot be consumed by microorganisms in the secondary treatment processes, due to which they remain in wastewater even after secondary treatments. Adsorption is a technique used for the removal of such bio-refractory organic compounds from wastewaters. Majority of these compounds include chlorinated compounds, hydrocarbons, heavy metals, pesticides, fungicides, surfactants, etc. These contaminants available in the solution are concentrated on the surface of a solid, which is termed as adsorbent. The dissolved contaminant that is intended to be removed from the wastewater is termed adsorbate or sorbate.

16.3.1 Types of Adsorptions

The process of adsorption can be categorised into three types viz. physical adsorption, chemical adsorption, and exchange. Physical adsorption (physisorption) is solely a result of weak attraction force, also called as van der Waals forces of attraction, and there is no chemical bonding between the adsorbent and the adsorbate. In this, the target pollutant or adsorbate can freely move on the surface of adsorbent. Moreover, physical adsorption is multi-layered adsorption and can be easily reversible, i.e., reduction of adsorbate concentration in the solution can cause desorption. Chemical adsorption, also termed as chemisorption is due to strong force of attraction compared to van der Waals force leading to chemical bonding between adsorbent and adsorbate. The adsorbed impurities are fixed in adsorbed position and this type of adsorption is monolayer adsorption. For desorption in the chemisorption, a mere concentration gradient is not sufficient, instead high temperature is required to remove the adsorbed impurities.

Both processes (physical and chemical adsorption) are exothermic; however, the released heat is more in chemisorption. For the application in water and wastewater treatment, physical adsorption is preferred over chemical owing to its ability of ease in regeneration. A comparison between physisorption and chemisorption is provided in Table 16.1. Ion exchange on the other hand, is governed by the electrical attraction between adsorbate and the surface of the adsorbent. The ions with higher surface charge (trivalent ions) are attracted more dominantly towards active sites on adsorbent than monovalent ions. Moreover, the attraction is inversely proportional to the particle size of ion.

Table 16.1 Comparison between physisorption and chemisorption

Parameters	Physisorption	Chemisorption
Type of bonding	Van der Waals forces of attraction	Chemical bonds at surface
Reaction type	Reversible	Generally irreversible
Surfaces for adsorption	Multilayer	Monolayer
Position of pollutant adsorption	Adsorbate is free to move over the adsorbent surface	Adsorbate once adsorbed are fixed in the position
Heat released during adsorption	4–40 kJ/mol	>200 kJ/mol

Ion exchange is one of the best examples of an exchange type of adsorption, which is elaborated in details in Sect. 16.4.

Adsorption being a surface phenomenon is dependent on the surface area of the adsorbent. Activated carbon (granular or powdered) is by far the most widely used adsorbent in water and wastewater treatment for the removal of adsorbates (target contaminants). Moreover, other natural adsorbents, activated alumina, synthetic polymer-based adsorbents as well as waste-derived activated carbon or biochar have also been used in adsorption experiments. Activated carbon is synthesized from wood or coal by heating it to temperatures up to 700 °C. The char produced from this stage is further activated at higher temperatures (800–900 °C) using CO_2 and steam. This activation results in the development of porous structure with numerous macro- to microspores causing very high surface area in relation to its weight. The precise surface properties, pore size distribution, and the functional groups of activated carbon are a function of raw material and the process of synthesis.

Regeneration of the adsorbent is the process of recovering the adsorptive capacity of the exhausted adsorbent material. It can be achieved via chemical addition, steam application, usage of solvents, and biological processes. The exhausted adsorbent on regeneration loses approximately 4–10% of adsorption capacity, which demands the requirement of a new adsorbent after a specific number of regeneration cycles. The loss in adsorption capacity depends on the method of regeneration and type of adsorbate. Hence, the economic use of adsorbent depends on the effectiveness of regeneration and adsorptive capacity of the adsorbent following regeneration. For granular activated carbon same process as used for synthesis is used for regeneration as well.

16.3.2 Progression of Adsorption

The adsorption is progressed in four stages: bulk transfer, film transport, pore transport, and adsorption (Fig. 16.2). Bulk transport involves the transfer of dissolved impurities from the bulk liquid to the boundary layer of fixed film around the adsorbent. The impurities are transported near the adsorbent via advection and dispersion. In the film transport,

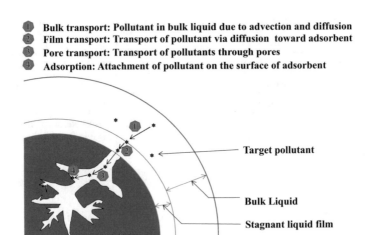

Fig. 16.2 Process of adsorption of pollutant on the adsorbent

the pollutants are transferred via diffusion through the stationary liquid film around the pores of adsorbent. Subsequently, pore transport involves a flow of pollutants inside the pores of adsorbent due to diffusion. While the final stage of adsorption involves the attachment of pollutants on the available site of the adsorbent surface. Attachment of pollutants can be on the surface of the adsorbent or inside the macro and micropores.

Amongst these steps, the rate limiting step is the one, which is the slowest of other processes and the rate of adsorption process is governed by the corresponding rate limiting step. Intensive turbulence causes minimization of liquid film surrounding the adsorbent, which increases the film diffusion. On the contrary, pore diffusion is dependent on the molecular size of the adsorbate and pores of adsorbent. This implies that pore diffusion can be sluggish if the molecular size of adsorbate is more than the pore size. In physical adsorption, diffusion transport is the rate limiting stage as the process of adsorption is rapid. On the other hand, in chemical adsorption, the process of adsorption itself is the rate limiting step.

The feasibility of specific adsorbent can be assessed by performing lab-scale experiments and plotting a graph of q_e (mass of adsorbate adsorbed per unit mass of adsorbent) versus C_e (concentration of adsorbate in the solution). The graphical representation of the experimental results can predict if the adsorption is favourable, linear, or unfavourable (Fig. 16.3). The adsorption is termed as favourable when a higher quantity of adsorbate is adsorbed on the surface of the adsorbent even at a lower adsorbate concentration. The adsorption is termed linear if the amount of adsorbate adsorbed is directly proportional to the concentration of adsorbate in the solution. Whereas, in case of unfavourable adsorption, the adsorbed mass on the adsorbent is lower even at a higher adsorbate concentration.

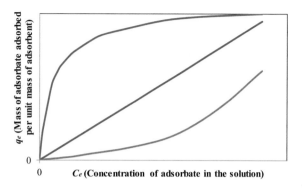

Fig. 16.3 Graphical representation for nature of adsorption. Blue line represents favourable adsorption, red line represents linear adsorption, and green line represents unfavourable adsorption

16.3.3 Adsorption Isotherms

An adsorption isotherm is the equation corresponding to the amount of pollutant adsorbed on the surface of the adsorbent (q_e) and the equilibrium concentration of pollutant remaining in the solution (C_e) at a constant temperature. The relation between q_e and C_e is established graphically and can be fitted into one or more isotherm models. In the development of adsorption isotherm, the q_e can be determined by Eq. 16.1. Freundlich and Langmuir isotherm are two of the most commonly used isotherms in practice.

$$q_e = \frac{(C_o - C_e)V}{M} \tag{16.1}$$

where C_o is the initial adsorbate concentration in the solution before start of adsorption process; V is the volume of liquid; and M is the mass of adsorbent.

16.3.3.1 Freundlich Isotherm
Freundlich isotherm is the mathematical representation of adsorption between a liquid or gaseous fluid and solid adsorbent. The Eq. 16.2, was derived by Freundlich in the year 1912 to represent the mass of adsorbed adsorbate per gram of the solid at equilibrium (q_e) and the concentration of adsorbate in the solution at equilibrium (C_e). The K_f and n are Freundlich capacity factor and Freundlich intensity parameter, respectively.

$$q_e = K_f \times C_e^{\frac{1}{n}} \tag{16.2}$$

Converting the Eq. 16.2 into a linearized form (Eq. 16.3) and by plotting a graph of $\ln q_e$ against $\ln C_e$, the Freundlich equilibrium constants and the Freundlich isotherm can be obtained. If the value of n is 1 then adsorption is linear and all sites have equal affinity for adsorption.

$$\ln q_e = \ln K_f + \frac{1}{n} \times \ln C_e \tag{16.3}$$

16.3.3.2 Langmuir Isotherm

Langmuir isotherm is an expression (Eq. 16.4) of equilibrium between adsorbent and adsorbate, in which the adsorption is limited to only one layer or monolayer. The Langmuir isotherm is based on the assumptions that: (1) adsorption is localized on a fixed number of active sites and all active sites have the same energy, (2) adsorption is monolayer, and (3) adsorption is reversible.

$$q_e = \frac{a \times b \times C_e}{1 + b \times C_e} \qquad (16.4)$$

where, a and b denote the maximum mass of adsorbate adsorbed on the unit quantity of adsorbent and an empirical constant, respectively. These constants and the Langmuir isotherm can be obtained by plotting $1/q_e$ against $1/C_e$.

16.3.3.3 BET Isotherm

The isotherm was formulated by Stephen Brunauer, Paul Emmett, and Edward Teller in the year 1938. This type of adsorption is based on multilayer adsorption and it is a complex extension of Langmuir isotherm, which is given by Eq. 16.5.

$$q_e = \frac{a \times b \times C_e}{(C_s - C_e) \times \left[1 + (b - 1) \times \frac{C_e}{C_s}\right]} \qquad (16.5)$$

where, C_s is the saturation concentration of adsorbate in the solution. The BET isotherm assumes that Langmuir isotherm is applied to each layer of adsorbate accumulated at the surface of adsorbent.

16.3.3.4 Temkin Isotherm

The effect of indirect adsorbate interaction on the adsorption is considered in the Temkin isotherm, which is expressed mathematically by Eq. 16.6. It is based on assumption that the heat of sorption of molecules reduces with increase in surface coverage.

$$q_e = \frac{RT}{b} \ln K_T + \frac{RT}{b} \ln C_e \qquad (16.6)$$

where, R universal gas constant, T is temperature in Kelvin scale, K_T represents Temkin isotherm constant (L/g) reflecting maximum binding energy, and b is heat of sorption in J/mol.

16.3.4 Factors Affecting Adsorption

The surface area of adsorbent: Adsorption being a surface phenomenon is directly affected by the available surface area of adsorbent. With an increase in the surface area of adsorbent, there is an increase in the rate of adsorption as well as the amount of adsorption on adsorbent. The activated carbon is therefore considered as one of the best adsorbents owing to its high surface area even up to 2000 m^2 g^{-1}.

pH of solution: At the acidic and alkaline pH, the concentration of H^+ and OH^- ions, respectively, will be very high in the solution. Depending on the nature of adsorbate, either H^+ or OH^- ions will compete for the adsorption sites. Mainly, the adsorption of a majority of the organic pollutants in water and wastewater treatment is increased under the acidic pH.

Characteristics of adsorbate: Based on the characteristics of adsorbate, it can be classified into hydrophilic or hydrophobic. The hydrophobic adsorbates have less affinity towards water, which makes them comparatively easy to remove via adsorption. On the other hand, hydrophilic adsorbates have liking towards water, thus making them difficult to be easily removed via adsorption. The size of adsorbate molecules also affects its ability to fit within the pores of adsorbent.

16.3.5 Materials Used as Adsorbent

Activated carbon: Activated carbon is the most widely used adsorbent in water and wastewater treatment, apart from its use for the removal of gaseous pollutants. Granular activated carbon and powdered activated carbon are the two classes of activated carbon classified solely on the basis of particle size. The diameter of granular and powdered activated carbons is greater than 0.1 mm and less than 0.074 mm, respectively. The surface area of powdered activated carbon is higher compared to the same quantity of granular activated carbon, which makes powdered activated carbon a superior adsorbent.

The powdered activated carbon can be added to the contacting basin following secondary biological treatment of wastewater. However, due to the finer particle size, coagulation and settling or filtration is recommended for the separation of powdered activated carbon from the treated effluent. To avoid the excessive cost of treatment posed by requirement of additional units, granular activated carbon-based columnar systems are used for the field-scale applications. Typical characteristics of powdered and granular activated carbon are represented in Table 16.2.

Activated alumina: Activated alumina is synthesized by dehydration and/or activation of aluminium trihydrate. Based on the heating temperature, the original material undergoes complex transformations. Other additives such as alkali metal oxides can also be added during synthesis of activated alumina to improve its adsorptive properties (Serbezov et al., 2011). It has been used for the adsorptive removal of fluorides from wastewater.

Table 16.2 Properties of granular and powdered activated carbon

Parameters	Granular activated carbon	Powdered activated carbon
Surface area (m^2/g)	700–1300	800–1800
Particle diameter (mm)	Greater than 0.1 (generally 0.5–3 mm)	Less than 0.074
Bulk density (kg/m^3)	400–500	360–740
Ash content (%)	Less than 8	Less than 6
Mode of operation in water and wastewater treatment	Operation in column or continuous mode	Operation in batch mode
Practical applications	In the form of continuous filters as tertiary treatment	Added into contact basin following secondary treatment
Pollutant removal efficiency	Less than powdered activated carbon	Higher than granular activate carbon

Synthetic polymers: Polymers are long repeating chain molecules bound by covalent bonds. Polymers and polymer-based composites gain adsorptive properties through blending, surface functionalization, and crosslinking. Better chemical stability, stability in harsh environmental conditions, and high adsorption capacity for various pollutants are some of the benefits of synthetic polymer-based adsorbents in wastewater treatment. However, a high cost of synthesis is one of the major drawbacks of using synthetic polymers.

16.3.6 Modes of Operation

Batch mode: Powdered activated carbon is preferred as an adsorbent in the batch investigations and it is performed to assess the suitability of adsorbent for the adsorptive removal of target pollutant, i.e., adsorbate. The best fit reaction kinetics (reaction order), the applicable isotherm, maximum adsorption capacity (q_m), and time required to achieve the same can be identified using batch experiments. Preferably the duration of adsorption is decided based on the equilibrium conditions of adsorbate in the effluent.

Continuous mode: For the field-scale operations, granular activated carbon filters, in either up-flow or downflow configuration are used. The downflow beds are more common in wastewater treatment as there is a lesser chance of entrapping solids at the bottom of the filter bed, which then could be difficult to remove during backwashing. The arrangement inside the filter is such that a higher quantity of solution will be in contact with the adsorbent. In a downflow carbon filter, water/wastewater is applied from the top and it is distributed through headers to pass through filter layers. Typically, the carbon filter comprises of granular activated carbon, coarse gravel and/or pebbles, and strainer plates.

The area of carbon filter bed, which actively takes part in adsorption is termed as mass transfer zone. The adsorbate concentration is reduced to its minimum concentration, once the solution is filtered through the depth of adsorption column equals to the height of the mass transfer zone (HMTZ). In this case, there is no possibility of adsorption below the mass transfer zone. As the top layer of activated carbon column is saturated with adsorbate (exhausted), the mass transfer zone moves down successively. The process is continued till the breakthrough is achieved (Fig. 16.4). The breakthrough is said to have been achieved when the effluent concentration of adsorbate reaches 5% of the initial concentration; whereas, exhaustion of activated carbon bed is achieved when effluent concentration of adsorbate reaches 95% of inlet concentration. At complete exhaustion of carbon, the influent and effluent concentrations of adsorbate will be equal.

Mass transfer zone is a function of the characteristics of activated carbon and the applied hydraulic loading rate. The height of mass transfer zone will exceed the overall depth of activated carbon filter under a higher hydraulic loading rates and in such circumstances, adsorbates cannot be completely removed in the column. To utilize the bottom portion of activate carbon; multiple columns in series are used in actual practice. The multiple columnar arrangements ensure that switching is possible once one of the columns

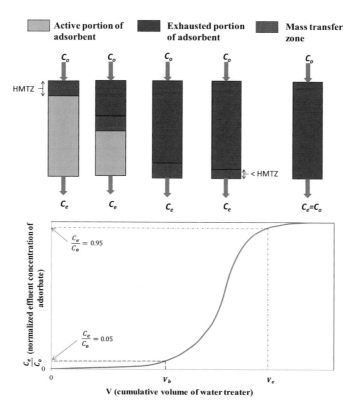

Fig. 16.4 Breakthrough curve of activated carbon column with movement of mass transfer zone with cumulative volume of polluted water being treated

is exhausted. This multiple column arrangement is beneficial to avoid deterioration in the effluent quality in case any one of the columns achieves a breakthrough. The height of mass transfer zone can be calculated by using Eq. 16.7.

$$Height\ of\ mass\ transfer\ zone = Z \times \left[\frac{V_e - V_b}{V_e - 0.5 \times (V_e - V_b)} \right] \tag{16.7}$$

In this equation, Z is the total height of activate carbon column/filter; V_e is the throughput liquid volume at exhaustion; and V_b is the total throughput liquid volume at breakthrough.

The adsorption capacity in the field-scale adsorption column will be less than the theoretical adsorption capacity determined from the lab-scale analysis. The field-scale breakthrough adsorption capacity of a single activated carbon column is assumed to be 25–50% of the theoretical breakthrough adsorption capacity. The typical GAC design specifications are given in Table 16.3. The time for breakthrough (t_b) can be computed using the Eq. 16.8.

$$t_b = \frac{X/m \times M_{ac}}{Q\left(C_o - C_b/2\right)} \tag{16.8}$$

where, X/m is the field scale adsorption capacity of the activated carbon; M_{ac} is the mass of activated carbon in the column; Q is a flowrate; C_o is concentration of adsorbate in the influent; and C_b is concentration of adsorbate at the breakthrough moment.

Table 16.3 Typical design guidelines for field-scale granular activated carbon filter

Design parameter	Typical value
Density of granular activated carbon	350–550 kg/m^3
Hydraulic loading rate	5–15 m/h
Depth of granular activated carbon bed	1.8–4.8 m
Depth to diameter ratio	1.5: 1 to 4: 1
Bed volume	10–50 m^3
Cross sectional area	5–30 m^2
Effective contact time	1–2 min
Specific throughput	50–200 m^3/kg
Inlet suspended solid concentration	<5 mg/L

16.4 Ion Exchange

Ion exchange is a process of displacing dissolved ions from the solution to an insoluble exchange material by other ions, which are non-objectionable in the treated wastewater. In water treatment for the removal of hardness, ion exchange is the most common technique used for the displacement of Ca and Mg ions (hardness causing) from water with Na. Whereas, in wastewater treatment, it is used for the removal of various forms of nitrogen, phosphate, sulphate, heavy metals, and other dissolved solids. In simple words, ion exchange means the replacement of one ion with other, in which the displacing ion is the pollutant in solution and displaced ion is the part of solid matrix or ion exchange material (resin).

The exchange materials contain a high number of ionizing groups bonded on the solid surface via electrostatic force. Ion exchange materials are broadly classified into two categories, i.e., natural and synthetic ion exchange materials. Zeolite is the naturally occurring ion exchange material used for application in water treatment, mainly for the removal of hardness and ammonium ions. Synthetic ion exchange materials are synthesized via copolymerization of styrene and divinylbenzene. The styrene serves the purpose of base matrix, whereas divinylbenzene for cross-linking of polymers.

Exchange capacity, particle size, and stability are the three important factors of exchange material, which affect the performance of ion exchange. Exchange capacity is the measure of exchangeable ions that can be exchanged on ion exchange material. It is expressed as eq/L or eq/kg. The higher the exchange capacity, the better is ion exchange material due to requirement of fewer numbers of regeneration cycles. Synthetic materials, such as resins, have an exchange capacity of 2–10 eq/kg compared to 0.05–0.1 eq/kg of zeolite. Synthetic materials can have a defined and uniform exchange capacity compared with natural materials. Due to this reason, synthetic ion exchange resins are widely used in water and wastewater treatment. Apart from exchange capacity, particle size is a deciding factor for the rate of exchange; while stability is associated with long-term use of the ion exchange material.

Synthetic ion exchange materials also termed as resins can be cationic or anionic resins exchanging cations and anions, respectively. Sodium cation exchanger is used for the removal of hardness (Eqs. 16.9 and 16.10), which is regenerated using sodium chloride (Eqs. 16.11 and 16.12). Similarly, cation exchangers are used for exchanging other cations (Eq. 16.13), which are regenerated using mineral acids such as sulphuric acid (Eq. 16.14).

$$R - Na_2 + Ca(HCO_3)_2 \rightarrow R - Ca + 2NaHCO_3 \tag{16.9}$$

$$R - Na_2 + Mg(SO_4) \rightarrow R - Mg + Na_2SO_4 \tag{16.10}$$

R-Na$_2$ is the sodium cation exchanger resin; R-Ca and R-Mg are hardness components attached to the cationic resin following ion exchange.

$$R - Ca + 2NaCl \rightarrow R - Na_2 + CaCl_2 \qquad (16.11)$$

$$R - Mg + 2NaCl \rightarrow R - Na_2 + MgCl_2 \qquad (16.12)$$

$$R - H_2 + CaSO_4 \rightarrow R - Ca + H_2SO_4 \qquad (16.13)$$

R-H_2 is hydrogen cation exchanger resin; R-Ca is hardness components attached to the cationic resin following ion exchange.

$$R - Ca + H_2SO_4 \rightarrow R - H_2 + CaSO_4 \qquad (16.14)$$

The anion exchangers remove anions (Cl^-, NO_3^-, SO_4^{2-}) using OH^- as anionic exchangers (Eq. 16.15).

$$R - 2NH_3OH + HCl \rightarrow R - 2NH_4Cl + 2H_2O \qquad (16.15)$$

R-$2NH_3OH$ is a weak basic anionic exchanger material; R-$2NH_4Cl$ is exchanged pollutant fraction on resin.

The cationic resins are further classified as strong acid exchangers and weak acid exchangers as described below.

Strong acid cation exchanger: Strong word is related to the ease with which the functional group can be released from the resin. In this type of cation exchange resin, sulfonate group acts as an exchange site. The functional group in strong acid exchangers is derived from strong acids, thus allowing hydrogen to be dissociated and exchanged with target cation in a wide pH range (generally complete pH range of 1–14).

Weak acid cation exchanger: For the weak acid exchangers, the functional group is derived from weak acids, such as carboxylic acids, and such resins work in the alkaline pH regime. Weak acid cation exchangers have pK_a value in a range of 4–5, hence they work in a pH range greater than 6.0–7.0.

Similar to cationic exchangers, anionic exchangers are also classified into strong base and weak base anion exchangers.

Strong base anion exchanger: The strong base anion exchangers have a pK_a value in a range of 0–1, thus making it operational in pH up to 13.0. These exchangers can be regenerated using sodium hydroxide. In water treatment, they are commonly used for elimination of nitrate, arsenic, etc.

Weak base anion exchanger: In weak base anionic exchangers, tertiary amine group is the functional group and they are effective in solution pH of <6.0. These exchangers can be regenerated using sodium carbonate.

The ion exchange process can be performed either in batch or continuous mode of operation. In a batch process, ion exchange material is added to the solution and the mixture is stirred until the exchange reaction is completed. The used exchange material is settled and separated from the solution. In a continuous mode of operation, column is packed with exchange material, and the water/wastewater to be treated is continuously passed through the column. Similar to activated carbon, the spent ion exchange material is regenerated before reuse. The affinities of ions towards ion exchange material are dependent on following factors:

- Ions with high valence are preferred compared with low valence ions. For example, the affinity for cationic ions will be in an order $Fe^{3+} > Ca^{2+} > Na^+$ and that for anions will be $PO_4^{3-} > SO_4^{2-} > Cl^-$.
- The exchange reaction is preferred for ions with higher atomic number for ions having same valence, e.g., $K^+ > Na^+$; $Ca^{2+} > Mg^{2+}$.
- For very high ionic concentration in the solution, the ion exchange process can be reversed. This principle is used for the regeneration of ion exchange resins.

16.4.1 Applications of Ion Exchange Process

Removal of heavy metals: Ion exchange can even be used for the removal of trace quantities of heavy metals from the solution. Resins having specific selectivity towards desired heavy metal(s) can be manufactured in synthetic resins. Thus, the resin with high exchange capacity and selectivity causes improvement in the exchange process and the overall economy. Heavy metals, such as lead, mercury, cadmium, nickel, vanadium, chromium, copper, etc., have been successfully removed using ion exchange process (Dąbrowski et al., 2004).

Removal of nitrogen: In the wastewater stream nitrogen can be available in NH_4^+, NO_2^-, and NO_3^- forms, which can be exchanged via the use of cationic and/or anionic exchange resins. Zeolite is one of the natural materials that can be used for the removal of ammonia from wastewater (Koon & Kaufman, 1975). Wastewater contains sulphate ions and sulphate ions have a greater affinity towards resins compared with nitrate. If resin with exchanged nitrate is operated beyond a breakthrough point, then there is a possibility of sulphate ions (available in solution) displacing exchanged nitrate onto the resin. This will again release the nitrate in the solution and the phenomenon is termed as nitrate dumping. To prevent this synthetic nitrate selective resins are beneficial.

Removal of other dissolved solids (cationic and/or anionic): For the removal of dissolved solids from wastewater, both cationic and anionic exchange resins are used in series. The wastewater is initially passed through cationic exchange resins, in which the cationic ions are exchanged with hydrogen. The effluent from cationic exchanger is passed through anionic exchanger, where anions in the solution are replaced by hydroxide ions.

16.4.2 Limitations of Ion Exchange Process

- **Formation of precipitates**: The reaction between the component in the wastewater and the chemical used for exchanger can lead to formation of precipitates inside the ion exchange column. For example, if water contains high calcium ion and resins with sulphate is used as an exchange material, the reaction will form calcium sulphate and the precipitates will be deposited inside the column.
- **Presence of turbidity**: The presence of turbidity can interfere with the exchange process thereby reducing the overall performance.
- **Microbial contamination**: Ion exchange is not a disinfection process and due to presence of dissolved organic matter in the wastewater, bacterial growth can be induced over the resins. This can reduce the available sites for possible exchange and can reduce the overall efficiency.
- **Operational cost**: Ion exchange is a costly process and continuous maintenance is required for proper functioning of the process. Additionally, the process is dependent on requirement of higher amount of chemicals for the regeneration, which further increases the cost of operation.

Example: 16.2 Powdered activated carbon was used as adsorbent for the removal of fluoride with an initial concentration of 900 mg/L from the wastewater. The samples were filtered and the fluoride concentration was measured after 30 h of contact time. Determine the best fit isotherm (from Langmuir and Freundlich) and calculate the relevant constants for the data given below.

Mass of PAC (mg)	450	375	300	200	75
Effluent Fl concentration (mg/L)	78	100	221	302	589

Solution

The values of $\frac{1}{q_e}$, $\frac{1}{C_e}$ and ln (q_e), ln (C_e) are derived from the given experimental data for plotting Langmuir and Freundlich isotherm, which is presented below in Table 16.4.

Converting Langmuir isotherm into linearized form, $\frac{1}{q_e} = \frac{1}{ab}\left(\frac{1}{C_e}\right) + \frac{1}{a}$.

Converting Freundlich isotherm into linearized form, $\ln q_e = \ln K_f + \frac{1}{n} \ln C_e$.

Figure 16.5 indicate that Freundlich isotherm fits better for this experimental dataset.

$\ln (q_e) = 0.3748(\ln C_e) - 1.0434$

From this the values of $n = 2.66$ and $K_f = 0.352$

Thus, Freundlich isotherm equation for this experimental data will be

Table 16.4 Calculating values for plotting isotherms

Mass of PAC (M) (mg)	Initial concentration (C_o) (mg/L)	Effluent Fl concentration (C_e) (mg/L)	$qe = (\frac{C_0 - C_e}{M})$	$\frac{1}{q_e}$	$\frac{1}{C_e}$	$\ln(q_e)$	$\ln(C_e)$
450	900	78	1.827	0.547	0.0128	0.602	4.357
375	900	100	2.133	0.469	0.0100	0.758	4.605
300	900	221	2.263	0.442	0.0045	0.817	5.398
200	900	302	2.990	0.334	0.0033	1.095	5.710
75	900	589	4.147	0.241	0.0017	1.422	6.378

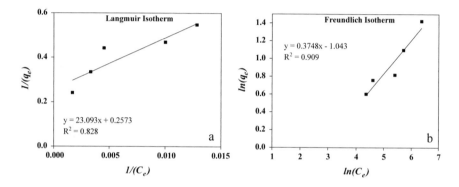

Fig. 16.5 **a** Langmuir and **b** Freundlich isotherm plot for determination of best fit isotherm

$$q_e = K_f \times C_e^{\frac{1}{n}} = 0.352 \times C_e^{\frac{1}{2.66}}$$

Example: 16.3 Activated carbon is used for treatment of wastewater with a flow rate of 100 m^3/day. The phenol concentration was reduced from 3 mg/L to 0.5 mg/L following adsorption. The lab experiments proved that Freundlich isotherm was applicable for adsorption with Freundlich intensity parameter $(1/n)$ as 0.6 and Freundlich capacity factor (K_f) as 200 (mg/g)(L/mg). If the cost of activated carbon is Rs. 200/kg, compute the annual cost of activated carbon required for the treatment of wastewater without considering the regeneration.

Solution

Flow rate $= 100$ m^3/day $= 69.44$ L/min

$q_e = \frac{(C_0 - C_e) \times V}{M}$ from this,

$\frac{M}{V} = \frac{(C_0 - C_e)}{q_e}$ substituting the given values in the equation, we get,

$\frac{M}{V} = \frac{(3 - 0.5)}{q_e}$, using Freundlich isotherm Eq. 16.2, $q_e = 200 \times 0.5^{0.6}$

$\frac{M}{V} = \frac{(3-0.5)}{200 \times 0.5^{0.6}} = 0.0151$ g/L

Annual cost of activate carbon, $= \frac{0.0151 \times 69.44 \times 1440 \times 365 \times 200}{1000} = $ Rs. 110,222.94/year

Example: 16.4 Determine the theoretical quantity of cationic and anionic exchanger used for the treatment of 10 m^3/day of industrial effluent. Assume the exchange capacities of cationic and anionic exchange resin as 40,000 and 20,000 g CaCO$_3$/m^3.cycle, respectively. Assuming the frequency of regeneration cycle of 1 per day, compute the quantities of regeneration chemicals. Consider H$_2$SO$_4$ (90% purity) and NaOH (93% purity) is used for regeneration. The characteristics of the effluent are given below:

Ions	Ca^{2+}	Mg^{2+}	Na^+	NO_3^-	Cl^-	SO_4^{2-}
Concentration (mg/L)	140	55	75	60	200	150

Solution

The total ion exchange capacity of ion exchange resin is expressed in terms of CaCO$_3$. Hence the given concentrations of cationic and anionic ions are converted as CaCO$_3$.

Sample calculation

Concentration of Ca^{2+} = 140 mg/L

Equivalent weight of Ca = 20, CaCO$_3$ = 50

Thus, Ca^{2+} concentration as CaCO$_3$ = $\frac{140 \frac{mg}{L} \times 50 (eq\ wt\ of\ CaCO_3)}{20 (eq\ wt\ of\ Ca)}$.

= 350 mg/L as CaCO$_3$

Similarly, concentrations of all cations and anions are computed and expressed as CaCO$_3$, these results are presented in Tables 16.5 and 16.6, respectively.

Total quantity of cations exchanged per cycle = 738.46 g/m^3 × 10 m^3/day

= 7384.6 g CaCO$_3$/day

Volume of cation exchange resin required $\frac{7384.6\ g\ CaCO_3/day}{40000\ g\ CaCO_3/m^3} = 0.185$ m^3

Total quantity of anions exchanged per cycle = 486.33 g/m^3 × 10 m^3/day

= 4863.3 g CaCO$_3$/day

Table 16.5 Estimation of total cations present in the wastewater

Cation	Concentration (mg/L)	Equivalent weight	mg/L as CaCO$_3$
Ca^{2+}	140	20.0	350.00
Mg^{2+}	55	12.2	225.41
Na^+	75	23.0	163.04
Total			**738.46**

Table 16.6 Estimation of total anions present in the wastewater

Cation	Concentration (mg/L)	Equivalent weight	mg/L as CaCO3
NO_3^-	60	62.0	48.39
Cl^-	200	35.5	281.69
SO_4^{2-}	150	48.0	156.25
Total			**486.33**

Volume of anion exchange resin required $= \frac{4863.3\,g\,CaCO_3/day}{20000g\,CaCO_3/m^3} = 0.244\ m^3$

H_2SO_4 with 90% purity is used for regeneration of cationic exchange resin

Quantity of $H_2SO_4 = \frac{49.1\,eq.wt.of\,H_2SO_4 \times 7384.6g\,\frac{CaCO_3}{day}}{0.9\,purity \times 50\,eq.wt.of\,CaCO_3 \times 1000\frac{g}{kg}} = 8.06$ kg H_2SO_4/cycle

NaOH with 93% purity is used for regeneration of anionic exchange resin

Quantity of NaOH $= \frac{40eq.wt.of\,NaOH \times 4863.3g\,\frac{CaCO_3}{day}}{0.93\,purity \times 50eq.wt.of\,CaCO_3 \times 1000\frac{g}{kg}} = 4.2$ kg NaOH/cycle

Example: 16.5 Design an ion exchange column for the hardness removal for treating wastewater with a flow of 25,000 m³/day. The initial hardness of the wastewater is 250 mg/L as $CaCO_3$ and desired effluent hardness is 75 mg/L. The resin used has an exchange capacity of 90 kg $CaCO_3$/m³ and it is regenerated once in a day. Consider that after treatment with ion exchange no hardness is left in the effluent. Assume the applicable flow rate as 0.4 m³/m².min per unit plan area of the column.

Solution

Developing mass balance to quantify the actual flow of wastewater to be bypassed (Fig. 16.6):

$Q =$ Total flow $= 25,000$ m³/day

$C_o =$ Inlet hardness concentration $= 250$ mg/L

$C_b =$ Concentration of hardness in bypass line $= 250$ mg/L

$Q_b =$ Flow rate of bypass stream

$C_i =$ Concentration of hardness after ion exchange $= 0$ mg/L

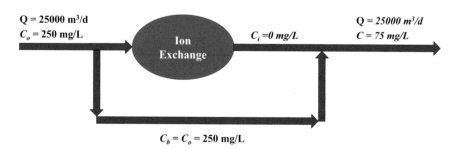

Fig. 16.6 Mass balance of wastewater flow

Q_i = Flow rate passing through ion exchange column
C = Desired concentration of hardness = 75 mg/L

$$(Q_b \times C_b) + (Q_i \times C_i) = (Q_b + Q_i) \times C$$

$$\frac{\left(Q_b \frac{m^3}{d} \times 250 \frac{mg}{L}\right) + \left(Q_i \frac{m^3}{d} \times 0 \frac{mg}{L}\right)}{(Q_b + Q_i)} = 75 \frac{mg}{L}$$

Since, $Q_b + Q_i = Q$.
$$\frac{\left(Q_b \frac{m^3}{d} \times 250 \frac{mg}{L}\right)}{Q} = 75 \frac{mg}{L}; Q_b = \frac{75}{250} \times Q.$$
$Q_b = 0.3 \times Q$ and $Q_i = 0.7 \times Q$.

Determination of total hardness to be removed

$Q_i = 17{,}500$ m³/day, thus, $Q_i = 12.153$ m³/min and $C_o = 0.25$ kg/m³

Total hardness to be removed = $17{,}500 \frac{m^3}{day} \times 0.25 \frac{kg}{m^3} = 4375$ kg CaCO₃/day

Determination of volume of resin with regeneration cycle of 1 per day

Volume of resin = $\frac{4375 kg \frac{CaCO_3}{d}}{90 kg \frac{CaCO_3}{m^3}} = 48.62$ m³

Surface area of ion exchange column with a flow rate of 0.4 m³/m² min

Plan area of column required = $\frac{12.153 \frac{m^3}{min}}{0.4} = 30.38$ m²

Provide 10 columns of 1.97 m Ø having surface area of 3.04 m². Considering volume of resin as 75% in a column so as to leave extra volume for loosening of resins during backwashing, hence each column volume will be = $\frac{48.62}{10 \times 0.75} = 6.49$ m³

Depth of column = $\frac{6.49}{3.04} = 2.13$ m

Regeneration of resin

Considering NaCl is use for regeneration,

Quantity of NaCl = $4375 kg \frac{CaCO_3}{d} \times \frac{58.5 \frac{g}{eq} \text{ as } NaCl}{50 \frac{g}{eq} \text{ as } CaCO_3} = 5118.75$ kg NaCl

Twice the quantity of NaCl is considered for actual use, which will be 10,238 kg NaCl per day.

Considering 4% NaCl solution for regeneration, with solution density as 1200 kg/m³,

Volume of regeneration = $\frac{10238 kg NaCl}{0.04 \times 1200 \frac{kg}{m^3}} = 213.68$ m³/cycle

16.5   Electrodialysis

Electrodialysis is a tertiary treatment unit designed for the removal of dissolved solids from the secondary treated wastewater. It is a separation process in which, the potential difference is generated between the two adjacent electrodes leading to separation of ions from the solution, which is wastewater in wastewater treatment unit. It uses selective permeable membranes that allow migration of only selective ions. A typical electrodialysis

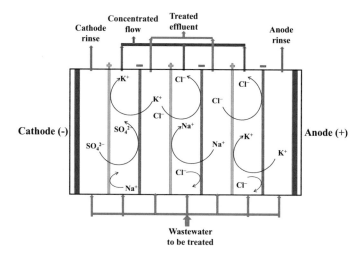

Fig. 16.7 Working of electrodialysis unit (grey colour-anion exchange membrane and yellow colour-cation exchange membrane)

unit consists of series of alternatively placed cation permeable and anion permeable membranes (Fig. 16.7). The cation permeable membranes are negatively charged, which allows passage of only cations, such as Na^+, K^+, Mg^{+2}, etc., and repels the anions. Similarly, the anion permeable membranes are positively charged, which allows the passage of only anions, such as Cl^-, F^-, SO_4^{-2}, etc., and repels the passage of cations. Owing to the alternate placement of cation and anion permeable membranes, concentrated or reject flows and treated effluents are generated from the alternate chambers (Xu & Huang, 2008).

A typical electrodialysis cell has hundreds of such cells comprising of spacers (space between the two adjacent membranes) for passage of wastewater through it. The wastewater to be treated is passed through the spacers and under the action of electrical potential, the cations migrate towards cathode; whereas, the anions migrate towards the anode. The ions from the influent wastewater passing through every alternate spacer (except adjacent to cathode and anode) are migrated through adjacent cation and anion permeable membranes. As a result, ions diminished treated effluent is generated. At the same time, concentrated effluent also termed as brine is generated in the next spacer, which is enriched with cations and anions.

The membrane fouling and membrane scaling are the common drawbacks of electrodialysis. Electrodialysis is the tertiary treatment technique and it is preceded by secondary sedimentation unit and sand filtration to remove the suspended solids and colloidal matters from wastewater. Even then, fine solids can pass through these pre-treatments and can be deposited on the membrane surface. Moreover, the presence of high concentration of multivalent ions in the wastewater near the saturation level can cause its precipitation on the membrane surface. The membrane fouling on the other hand is caused due to precipitation of organic anions such as humates or natural organic matter on the anion exchange membranes. Organic anions of larger size than the pore size of anion exchange membrane cannot easily pass through the membranes. This causes increase in the electrical resistance (Oztekin & Altin, 2016).

The membrane fouling and scaling can be prevented via electrodialysis reversal, which is same as electrodialysis except charge reversal. If the polarity is reversed the anions and cations can pass through the anion and cation permeable membranes respectively, back to the bulk wastewater. Thus, the treated effluent stream becomes concentrated stream. During electrodialysis reversal, the treated effluent is not collected for this short period. In this manner, the membrane fouling and precipitation can be minimized.

One Faraday of electricity is equal to the electricity required for migration of one gram of equivalent substrate from one electrode to other. Based on the Faraday's law, the current required for the electrodialysis can be computed by Eq. 16.16.

$$i = \frac{F \times Q_p \times N \times E}{n \times E_c} \tag{16.16}$$

where, i is current in ampere; F is Faraday's constant; Q_p is flowrate of treated effluent in L/sec; N is normality of feed wastewater in g eq./L; E is expected salt removal efficiency; n is number of cells in pair; and E_c is the current efficiency.

The ratio of current density (CD) and wastewater normality (N) is known as CD/N ratio, which indicates the presence of charge in solution to carry the electrical current. Higher value of CD/N means wastewater has sufficiency of salts to carry the electrical current. The typical range of CD/N ratio is between 500 and 800.

Example: 16.6 Estimate the power consumption per litre of wastewater treated and area requirement for an electrodialysis unit designed for the treatment of the secondary treated wastewater having a flow of 100 m^3/day. The details of the electrodialysis unit are as follows: influent TDS = 0.03 g.eq/L, expected TDS removal efficiency = 70%, current efficiency = 80%, resistance = 10 Ω, number of pairs per electrodialysis cell = 50, CD/N = 600 (mA/cm^2)/(g.eq/L).

Solution

Design flow rate = 100 m^3/day = 1.16 L/sec

Calculation of power requirement

$i = \frac{F \times Q_p \times N \times E}{n \times E_c} = \frac{96485 \times 1.16 \times 0.03 \times 0.7}{50 \times 0.8} = 58.76$ A

Power = $i^2 \times R = 58.76^2 \times 10 = 34{,}528$ W = 34.53 kW

Power per litre of wastewater = $\frac{34528}{100 \times 1000} = 0.345$ W/L of wastewater

Calculation of membrane surface area requirement per cell in pair

$\frac{CD}{N} = 600$ $CD = 600 \times 0.03 = 18$ mA/cm^2

Membrane area required for each cell in pair (A) $= \frac{i}{CD} = \frac{58.76 \times 1000}{18} = 3265 \text{ cm}^2$

Considering square configuration, 58 cm × 58 cm membrane shall be used per pair of cells. Such 50 pairs of cells will be provided in the electrodialysis cell.

16.6   Solar Still

Solar still is a low-cost technique for the removal of dissolved solids from wastewater. Solar still utilizes the solar energy for the evaporation of wastewater, which is condensed in an attached unit leaving behind the dissolved salts. It is dependent only on the solar energy and does not require any supplemented energy for its operation. The process is thus considered as a low-cost alternative that can be ideal choice in rural areas. Other desalination techniques such as electrodialysis and reverse osmosis are energy and cost intensive, which requires huge capital cost and skilled labours for its operation and regular maintenance.

On the other hand, solar still requires no external energy and it is free from moving parts; thus, the operation and maintenance cost of solar still is very less. However, the main drawbacks of the solar still is the requirement of high surface area. Moreover, the yield of treated effluent is very less compared with other desalination techniques. The yield of treated effluent using solar still is dependent on the environmental factors such as solar radiation, surrounding temperature, wind velocity and operating conditions like dissolved salts concentration, water depth (Abujazar et al., 2016; Hoque et al., 2019). Due to this, solar still can be considered as an effective point-of use treatment.

16.7   Membrane Filtration

The growing demand of high-quality drinking water or recycled wastewater with reduced pollutant content promoted the application of advanced membrane technologies as tertiary treatment unit. In general, the terminology 'membrane', used in water and wastewater treatment regime, is the material which allows certain components to pass through it selectively from its source by acting as physical barrier to prevent contaminants/foreign particles to cross it. As it is perm-selective in nature, the constituents passing though it are called permeate and those rejected form retentate (Fig. 16.8). The selectivity of different membranes generally depends on their pore sizes.

Fig. 16.8 Schematic of typical
function of membrane

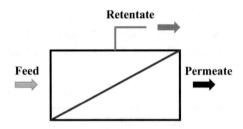

16.7.1 Classification of Membranes as Per Pore Size

According to the molecular weight of the solute that the membrane can reject and the pore size, there are four types of membrane generally used in the water and wastewater treatment systems. In other way, the effective pore diameter (in meters) and the equivalent mass of the smallest solute molecule in Daltons (Da), 90% of which is retained by the membrane, also known as molecular weight cut-off (MWCO), are the defining factors in classification of four different types of membrane.

Microfiltration (MF): The membranes with a range of 0.1 micron to 1 micron of pore size called microfiltration membranes (Fig. 16.9). The pore sizes are measured using porometer, which works with the principle of measuring the gas flow rate through membrane after filling the pores up with fluids of known surface tension.

Ultrafiltration (UF): The pore size in the range of 0.01–0.1 micron with a MWCO range of 1,000–300,000 Da represent ultrafiltration membranes (Fig. 16.9). Other than its major usage in the industrial sectors, it is also used in protein separation processes. Due to the formation of the cake layer, rejection of actual solute sizes by MF and UF are almost same regardless of their pore sizes. Because in actual filtration process, the cake layer acts as a dynamic membrane which sometimes cause higher rejection efficiency of

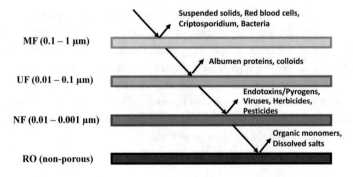

Fig. 16.9 An approximate monograph of particles that is being rejected by the membranes based on their pore sizes

UF than MF for tiny solutes with less than 10,000 Da. The MF and UF membranes run at lower range of operating pressure (< 6 bar).

Nanofiltration (NF): The membranes with a range of 1–10 nm of pore sizes and MWCO ranging from 200–1,000 Da are called nanofiltration membranes (Fig. 16.9). It is highly efficient in removing divalent (Ca^{2+}, Mg^{2+} etc.) and trivalent (Fe^{3+}, etc.) ions than monovalent ions (K^+, Cl^- etc.) because of its looser skin layer structure and pore size. Generally, the nanofiltration membrane has a very high water permeability, but the ideal permeability of solutes might be near zero or some higher value, depending on the solute and application. For example, it may not allow any pesticide molecule to pass through the membrane but may allow Ca ions to some extent. Transmembrane pressures are generally in the range of 5–15 bar for these nanofiltration membrane operations.

Reverse osmosis (RO): The pore sizes of the RO membranes with less than 1 nm are impossible to determine using porometers and therefore sometimes refer as non-porous (Fig. 16.9). However, microscopical analysis is performed to determine the same or simply the MWCO is measured to classify RO membranes. In case of RO membrane, because of the high pressure induced chemical potential in the feed side, the solvent material (water in general) diffuses and dissolves into the membrane.

On the other hand, because of lower chemical potential, water desorption happens from the permeate side, which eventually showcases the major role of solubility of feed solution in the RO membrane material. This indeed is the crucial reason for higher rejection of RO membranes in addition to its sieving mechanism. The same phenomenon supports the fact that lithium ions (6.9 Da) are rejected more than ethanol (46.1 Da) by RO membranes. Because lithium ions are very unstable when dissolve in membrane materials due to the enhanced charge repulsion by the lower electrical permittivity of membrane material. However, ethanol does not have net charge and thus does not experience high repulsive force when dissolved in membrane material. This is the reason why small organic molecules typically showcased lower rejection efficiency by RO membranes. On the other hand, due to stronger destabilization of divalent ions, they have higher rejection efficiencies than monovalent ions in RO membranes. The NF and RO membranes typically run at higher operating pressure (>8 bar) compare to UF or MF.

All the cases above, the pressure is used to drive the solute through the membranes. However, there are filtration processes where the membrane not only retains the pollutants and allow the solvent to cross through it, but rather used to: (1) selectively extract the constituents (extractive) like in electrodialysis or membrane extraction (ME) processes, or (2) introducing a component in molecular form (diffusive) like pervaporation (PV) or gas transfer (GT).

Electrodialysis: When the separation is achieved by virtue of difference in ionic sizes, charge, and charge densities of solute ions by using different ion exchange membranes.

Pervaporation (PV): It has almost same mechanism as RO; however, the volatile solute partially vaporized in the membrane by partially vacuuming the permeate in the PV.

Gas transfer (GT): The gas is transferred under a partial pressure gradient into or out of water in molecular form in this technology.

Membrane extraction (ME): Here the constituents are removed by virtue of a concentration gradient between retentate and permeate side of the membrane.

Most of the commercial membranes used in water and wastewater treatment are either pressure-driven or electrodialysis ones, which are used to extract the target ions like nitrate or the ions associated with the hardness and salinity of target water source. The contaminant removal process emerges a very fundamental constraint in any membrane processes, which is membrane fouling. This is because the rejected contaminants tend to accumulate over the upstream side of the membrane and inculcate various physico-chemical phenomena, which lead to a reduced water flow through the membrane (i.e., water flux) at any given transmembrane pressure (TMP). Provided the fact that the fouling is the limiting factor in membrane technology, most of the researches on membrane material and processes involve the characterization and mitigation of fouling of the membrane.

16.7.2 Membrane Materials

In general, there are two types of materials being used in membrane preparation, i.e., polymer and ceramic. There are certain researches where metallic membranes are as well used for some specific purposes. As a thumb rule, any membrane comprises of a very thin surface layer with comparatively smaller pore sizes, to provide enough perm-selectivity for the target contaminant removal. This thin surface layer is supported on top of a thicker and more open support layer to provide mechanical stability. Also, a classical nature of membrane is always anisotropic in structure, which means it should have symmetry towards the plane orthogonal to its surface. The desired membrane should have higher surface porosity and higher total porous cross-sectional area with narrow pore size distribution for higher throughput with very selective degree of rejection. The membrane should also have to be mechanically strong with enough structural integrity to resist thermal and chemical extremities to offer fouling resistance.

The most popular polymer materials used in fabrication of membranes are poly-ethylsulphone (PES), polyvinylidene difluoride (PVDF), polyethylene (PE), polypropylene (PP) and polyamide. With specific manufacturing techniques, these materials can showcase desirable chemical and physical properties. However, to further improve the physical and chemical stabilities, different surface modification techniques are being followed like grafting, chemical oxidation, plasma treatment, etc.

16.7.3 Membrane Configurations

The geometry of the membrane and the way it is oriented and mounted in relation to the water flow is very important to demonstrate successful process performance. Ideally, the configuration is expected to provide the following:

i. A high degree of turbulence to promote the mass transfer in the upstream/feed side,
ii. A higher available membrane area to bulk volume ratio of the module,
iii. A low operational cost per unit area of the membrane,
iv. A low energy requirement per unit volume of the product water,
v. Proper cleaning mechanism, and
vi. Provide options for retrofitting and modularization.

Moreover, some of the listed characteristics are mutually exclusive in nature. Per se, if the turbulence is promoted, it results in higher energy expense or in order to go for direct mechanical cleaning, it should have low surface area to volume ratio.

There are few general structural configurations being employed in the field of membrane processes based on its geometrical shapes (either a planner or cylindrical):

a. *Plate and frame type/flat sheet (FS) type*—It is mainly useful in applications with higher membrane fouling potential. However, the structure of the module is little bit complex due to the space in between the membrane plates and loss of pressure due to zigzag flow pattern inside the module. Nowadays, it is mostly used in leachate filtration, electrodialysis, etc. Once added with additional mechanical agitation unit, the module has high application potential for digestate and livestock wastewater treatment, frac/flowback water, fruit juice, etc.
b. *Hollow fibre (HF)*—This type of module is generally comprised of tens of bundled hollow fibres cemented at both of the ends, which are then placed inside a tubular module, as shown in Fig. 16.10. It is popularly used for the tertiary filtration of secondary effluent from wastewater treatment plants, surface water treatment and product recovery, etc.
c. *Spiral wound (SW)*—This type of membrane modules became famous in late 1970's by designing as more compact FS type modules with low fouling potential. Because of its higher specific surface area, it reduces the operating cost. It is commonly used for membranes with significantly tight pore sizes like NF and RO.
d. *Tubular*—It is traditionally used in pressure filtration units with feed water recirculation through its lumen.

The packing density is crucial for HF membrane, as higher the packing density lower the interstitial gap and which can cause clogging of membrane. Furthermore, the membrane module should be strong enough to not to break or buckle when the backflushing will be

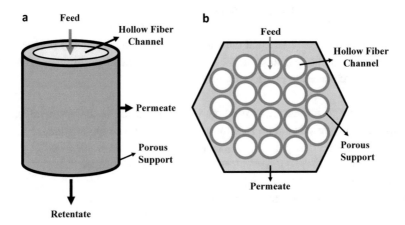

Fig. 16.10 Schematic of (**a**) a single hollow fibre and (**b**) hollow fibre bundle in module

done for cleaning. Table 16.7 is showcasing different types of membrane configuration used with its possible applications.

Table 16.7 Membranes and their applications

Membrane	Module configurations and/or operating method	Driving force	Classification	General applications
Flat Sheet (FS)	Plate and frame type (PF)	Pressure driven	MF/UF	ED, WWT
	Spiral wound type (SW)	Pressure driven	UF/NF/RO	PR, DS
Hollow Fibre (HF)	Contained in pressure vessels	Pressure driven	MF/UF/RO	PR, WT
	Immersed module without the pressure vessels	Vacuum driven	MF/UF	WT
Tubular (TB)	Pressure filtration type	Pressure driven	MF/UF	PR, WWT

WWT: Wastewater treatment, ED: Electrodialysis, DS: Desalination, PR: Product recovery, WT: Water treatment

16.7.4 Classification Based on Filtration Modes

Depending upon the direction of flow on the surface of the membrane, the filtration modes can be divided into crossflow and dead-end filtration modes as shown in Fig. 16.11.

- *Crossflow filtration*: In case of crossflow filtration, the movement of feed is parallel to the filter medium in order to generate enough shear stress for scouring the surface (Fig. 16.11a). Although some extra energy is required for it, the thickness of the cake layer can be easily controlled. Therefore, this type of filtration is effective when the foulant (suspended solids etc.) concentration is pretty high.
- *Dead-end filtration*: In case of dead-end filtration, there is no crossflow rather the movement of feed is towards the filter medium (Fig. 16.11b). Backwashing has to be performed to remove the accumulated solids with time or the filter medium has to be replaced after a stipulated life span. Most of the tertiary filtration treatments or pre-treatment of seawater RO involve dead-end filtration mode.

The filtration methods can also be divided depending upon where the rejection of particles is taking place.

- In case of surface filtration, the contaminants are rejected by the membrane filter surface itself and do not enter to the membrane medium (Fig. 16.12a). Most of the membrane filtrations fall into this category.
- In case of depth filtration, the contaminants can intrude the filter medium and captured through collision (Fig. 16.12b). Sand filter, multi-media filter, air filters, some of the ceramic filters, etc. fall into this category.

In case of HF membrane, there are two different kinds of filtration modes depending on the permeate flow direction as shown in Fig. 16.13.

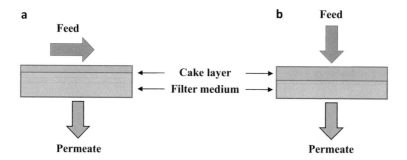

Fig. 16.11 Different types of filtration modes—(**a**) crossflow and (**b**) dead-end mode

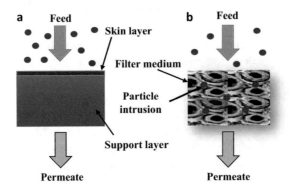

Fig. 16.12 **a** Surface filtration by skin layer and **b** depth filtration by filter medium

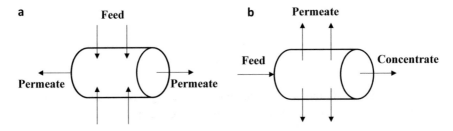

Fig. 16.13 **a** Outside-in and **b** inside-out filtration modes in hollow fibre membrane

- Outside-in mode—The outside-in filtration mode can withstand very high suspended solids concentration and most of the commercial HF modules are of this type only (Fig. 16.13a). There is provision of air scouring, which prevents the accumulation of solids on the membrane surface. The moment the TMP exceeds the threshold limit or the pre-determined cycle time is reached, the accumulated solids are removed by performing back flushing in a periodic manner.
- Inside-out mode—This type of HF membranes can retain uniform hydrodynamics inside the lumen side of the membranes; however, this mode will fail to create enough turbulence to mitigate the formation of cake layer due to its low water velocity and small inner diameter (Fig. 16.13b).

16.7.5 Concentration Polarization

During the permeation of the fluids through membrane, solutes are being rejected and generate a high solute concentrated layer near the surface of the membrane. Because of

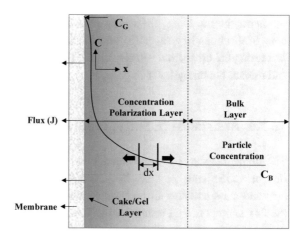

Fig. 16.14 Schematic diagram of solute concentration profile and concentration polarization phenomenon in the solution

the higher concentration of the solute particles in this layer, few solute particles tend to diffuse back to the bulk solution if not already being fixed in the cake layer. The settlement of concentration profile depends on the equilibrium between the convective transport of particles towards the membrane surface and back transport of diffused particles towards the bulk solution (Fig. 16.14). This phenomenon is called the concentration polarization and considered to be the fundamental limiting factor for membrane performance.

The water flux at any point of operation can be calculated by balancing the diffusive back transport of the solute and convective transport of particles towards membrane as shown in Eq. 16.17.

$$JC = -D_{eff}\frac{dC}{dx} \qquad (16.17)$$

where, J is the water flux at steady state in m/sec, C is the particle concentration in mg/L, x is the distance from membrane surface in m, D_{eff} is the effective diffusion coefficient in m^2/sec.

The negative sign in the Eq. 16.17 shows the reflection of negative concentration gradient along the x-axis. On the other hand, the effective diffusion co-efficient (D_{eff}) adds the combined effect of thermodynamic diffusion, shear induced diffusion, and all the other hydrodynamic forces that travels the particles away from the surface of the membrane.

Using the boundary conditions at the steady state, i.e., when x = 0, C = C_G and when x = δ, C = C_B with δ is the thickness of boundary layer in m and C_G and C_B are the concentration of particle/solute in the cake layer and in the bulk solution, respectively, the Eq. 16.18 can be derived.

$$J_{ss} = -\frac{D_{eff}}{\delta}\ln\frac{C_B}{C_G} \qquad (16.18)$$

As per the Eq. 16.18, the steady state flux (J_{ss}) increases with the decrement in thickness of the boundary layer (δ) and increment in effective diffusion co-efficient (D_{eff}). Increasing the cross-flow velocity on the surface of the membrane can help in higher D_{eff} and thinner boundary layer in general.

16.7.6 Resistance in Series Model

The simple resistance in series model can describe the relationship between the TMP and flux of a membrane module. In fact, this model in univocal for other applications like heat and mass transfer, air and water flow physics, electric current flow, etc. It says that the flux is proportional to the driving force (TMP in this case) and inversely proportional to all the resistances as illustrated in the Eq. 16.19.

$$J = \frac{\Delta P_T}{\mu(R_m + R_c + R_f)} \tag{16.19}$$

where, J is the water flux in m/sec or kg/m^2.sec, ΔP_T is the trans-membrane pressure in Pa or kg/m.sec^2, μ is the viscosity of permeate in kg/m.sec or cP or Pa.sec, 1.00×10^{-3} Pa.sec for water at 20 °C, R_m is the membrane resistance in per m (or m^2/kg), R_f is the irreversible fouling resistance in per m (or m^2/kg), R_c is the cake resistance in per m (or m^2/kg).

All the three resistances mentioned in the Eq. 16.19 can be operationally defined and measured as follows. Assuming R_f and R_c are zero, the *value of R_m* can be calculated using Eq. 16.19 by filtering only the clean water before the new membrane is utilized for treating wastewater provided that J, ΔP_T and μ are known. The sum of R_f, R_c, and R_m can be measured by the operational data obtained from the filtration of wastewater. Subtracting R_m from this total resistance gives the $R_f + R_c$. After the membrane is cleaned to get rid of the cake layer either or both by chemical or water jet application, the filtration is performed to obtain R_f. Thus, R_c can also be calculated. However, R_f is often considered to be negligible and included in the R_m because of its irreversible nature.

This model is very much useful and universally accepted to understand the resistances incur during the filtration process and the troubleshooting related to the improvement of flux. However, it is to be well understood that the value of each of the resistances is obtained experimentally by a certain method and experimental set-up. Hence, when repeated for another experimental set-up, additional care should be given to measure the same. For example, the values can even vary depending upon how precisely and thoroughly the cake layer is removed in each case. Furthermore, the internal pressure is related to the dimensions of the fibres, flux rate, etc. Therefore, for fibres with different dimensions and flux rates, the value of R_m obtained can be drastically varied with the measured R_m for even the same type of fibres.

16.7.7 Compaction of Cake Layer

The formation of cake layer acts as a filtration layer over the membrane surface. It causes the TMP to increase at constant flux mode. This change in pressure can be calculated by using Eq. 16.20.

$$\Delta P = \frac{5\mu S^2 (1-\varepsilon)^2 J}{\varepsilon^3} \Delta l \qquad (16.20)$$

where, J is the water flux based on cake surface area in m/sec; ΔP is the pressure drop in cake layer in Pa, N/m^2 or kg/m.sec^2; S is the specific surface area or total surface area per unit bulk volume in per m (or /m), ε is the porosity of cake layer; μ is the viscosity of liquid permeating through cake layer, kg/m.sec; and l is the depth of cake layer, m.

As per the Eq. 16.20, the value of ΔP increases with the increament in S and decreament in ε at constant flux (J). Furthermore, ΔP defines the force acting on squeezing per unit area of cake layer, which is why higher the ΔP stronger the chance of squeezing of cake layer over the membrane surface. However, actually the flux rate is the primarily governed by the TMP for squeezing of the cake layer. As it is witnessed that despite running at much higher operating pressure (5–80 bar) in RO compared to UF (0.1–1 bar) membrane, the cake layer does not squeezed in case of RO membrane much strongly than UF membrane due to lower flux rate in general in case of RO.

In case of constant flux mode, the rise in TMP value is slow in the beginning of the filtration cycle; however, after a certain point of time, it raises exponentially. It is known as 'sudden TMP rise or jump' because of its abrupt nature. It is realized to be the reason of self-accelerating nature of compaction of gel/cake layer at constant flux mode. The compaction of cake layer starts from the very bottom of the cake layer due to higher compression force because of cumulative nature of dynamic pressure loss in the cake layer. This dynamic pressure loss acts as squeezing force for the cake layer and once it reaches a threshold limit, the filtration resistance increases and cake layers start collapsing. Consequence to this, in order to compensate the flux loss, the TMP must rise. Furthermore, the increment of TMP causes the sub-layers at the bottom of the cake layer to be more compact and which in turn cause further increment in TMP values.

16.7.8 Back Transport

The solute particles experience various forces acting on them from different directions during the filtration process. The drag forces with its axial (also known as tangential) and the lateral (also known as permeation) components of fluid flow carry solute (particulate species) along the streamlines. Simultaneously, the solute particles are subjected to a number of other forces that helps them to cross the streamlines. Those are inertial

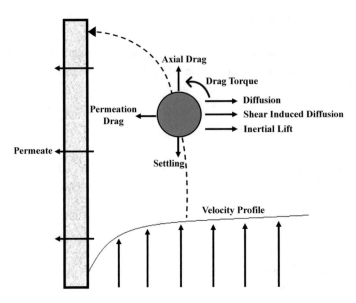

Fig. 16.15 Different forces and torques acting on a charged spherical particle suspended in a viscous fluid undergoing laminar flow in the proximity of a flat porous surface

lift, charge repulsion, van der Waals attractions, thermodynamic (or Brownian) diffusion, shear induced diffusion, etc. as shown in Fig. 16.15. In general, the sedimentation velocity is negligible in comparison to the convective flows and that is why it is being neglegted in calculations. Most importantly the effective particle deposition velocity can be determined from the difference in values between the overall back transport velocity and the permeation velocity (or flux).

For submerged membrane system, it is hard to estimate the actual back transport velocity as it is difficult to define the hydrodynamics of two-phase flow over the membrane surface. However, in case of closed slit channels, it is worth to witness how the particle back transport phenomenon looks like. It is though well assured that the underlying principle behind the particle back transport should be identical for both the systems except the additional irregular shear stress imposed by the air bubbles in case of the submerged membrane system.

The Brownian diffusion induced back transport velocity decreases with the increase in particle size. On the other hand, inertial migration and shear induced diffusion improve with the size of the particles. At the steady-state condition, the accumulation of particles near the surface of the membrane is balanced by the diffusive back transport velocity to the bulk. The equilibrium back transport velocity, V_D can thus be expressed by Eq. 16.21.

$$V_D \propto (D_B^2 y_w)^{\frac{1}{3}} \ln \frac{C_W}{C_B} \tag{16.21}$$

where, V_D is the diffusion induced back transport velocity in m/sec; D_B is the Brownian diffusivity of particles, m²/sec; γ_w is the shear rate at the membrane surface, per sec (or /sec); C_w is the particle concentration at the membrane surface, mg/L; C_B is the particle concentration in bulk liquid, mg/L.

The diffusivity, DB, is calculated by Stokes–Einstein equation (Eq. 16.22) as

$$D_B = \frac{K_B T}{6\pi\mu r_{\mathrm{p}}} \tag{16.22}$$

where, K_B is the Boltzmann constant, J/K; T is the absolute temperature, K; μ is the fluid viscosity, kg/m.sec; and r_p is the particle radius, m.

The Brownian diffusion induced particle back transport is in general thermodynamically motivated and it is only effective when there is a concentration gradient. In this case, for larger particles, the resulting particle back transport is weaker because of smaller Brownian diffusivity. However, shear diffusion induced particle back transport is kinematically motivated and it is only effective when the fluid movement caused a velocity gradient. The resulting particle back transport in this case is way stronger for the larger particles because of the higher chances of collision and the larger inertial moment of them.

Because of the particle back transport phenomenon a stratification of cake layer happens with the larger particles being at the bottom and the smaller ones at the top. This is more prominent in constant pressure mode, when the flux reduces over time gradually (Fig. 16.16). Because of declining flux condition, at the beginning of filtration cycle the larger particles with higher back transport velocities can deposit due to their convective flow velocities towards the membrane can exceed the particle back transport velocities. On the other hand, in case of constant flux mode, the particle size distribution does not change as the convective flow that pulls particle towards the membrane is constant in nature.

There is a trick to reduce the cake layer resistance by manipulating the TMP at the very beginning of the operational cycle. By allowing higher flux temporarily at the beginning

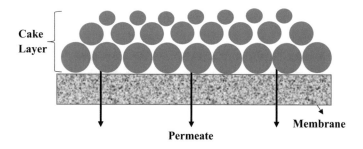

Fig. 16.16 Schematic of cake layer formed under constant pressure mode

of filtration cycle, the deposition of larger particles can be induced to form faster cake layer over the membrane surface. This cake layer can act as a barrier for the deposition of finer particles at a lower flux rate later on. Addition to that, because of the presence of larger particles at the bottom of the cake layer, it is less prone to collapse and can delay the cake layer compaction and respective loss of water flux.

16.7.9 Operational Modes and Their Effects on Membrane Fouling

There are majorly two types of operational modes for membrane technology depending on the control parameter that is required to be kept constant, i.e., flux or TMP. In case of membranes driven by vacuum pressure, they run under constant flux mode and their TMP is constantly monitored to track the status of membrane fouling. All the immersed membranes used in membrane bioreactor (MBR) are of this type. On the other hand, membranes which are driven by positive pressure, run in constant pressure mode and flux is monitored to track the fouling of the membrane. Mostly tubular and plate and frame type membranes are falling into this category with a few exceptions.

The constant flux mode has an inherent benefit in perspective to the membrane fouling. As excessively high flux can expedite the cake layer formation and make them more compact, hence it is more feasible to go for constant flux mode for long term operation. However, in case of constant TMP mode, due to higher initial flux, the formation period of cake layer is also very fast and causes serious decline in water flux due to more compact cake layer formation.

In Fig. 16.17, the comparative diagram is illustrated for constant flux, constant pressure, and modified constant pressure modes. The Fig. 16.17a, i.e., in case of constant flux, the flux has to be kept low and maintained in the same level, whereas the TMP is allowed to go up. The Fig. 16.17b, i.e., in case of constant pressure mode, the initial TMP has to be kept higher than required, which results in high initial flux compared to the constant flux mode. However, the flux decreases fast under this mode due to faster accumulation of cake layer. The Fig. 16.17c, i.e., in case of modified constant pressure mode, initial TMP has to be kept just enough to obtain the target flux. If the flux declines, the TMP is raised slightly to obtain the target flux. Overall, this mode imitates the constant flux mode.

16.7.9.1 Membrane Fouling Roadmap Under Constant Flux Mode
The membrane fouling occurs during the interaction of the complex feed while filtration is taking place. Under constant flux mode, fouling occurs in three main stages.

(a) *Stage 1—Initial adsorption*: Because of initial adsorption of macromolecules (e.g., extra-cellular polymeric substances (EPS) and soluble microbial products or SMP) on the membrane surface, a short-term rise in TMP value occurs even at zero flux.

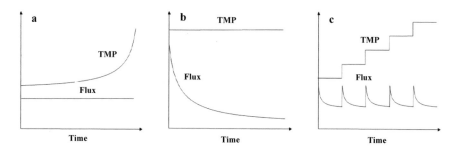

Fig. 16.17 Conceptual diagram of different modes of membrane operation: **a** constant flux mode, **b** constant TMP mode, and **c** semi-constant TMP mode

However, the adsorptive fouling can be minimized by more hydrophilic membrane materials as it prevents macromolecules to come in direct contact with membrane.

(b) *Stage 2—Slow TMP rise*: Due to continuous deposition of EPS and SMP followed by gradual compaction of the deposit layer, a long-term rise in TMP value takes place.

(c) *Stage 3—Sudden TMP rise (or jump)*: As cake layer compaction takes place, the TMP must rise to compensate the loss due to permeability under this constant flux mode, which again in turn accelerates further the cake layer compaction and cause exponential rise in TMP value (Fig. 16.17a).

16.7.9.2 Membrane Fouling Roadmap Under Constant Pressure Mode

(a) *Stage 1—Rapid cake layer formation and compaction*: Because of high TMP and flux, quick formation of cake layer takes place and consequently the flux starts decreasing rapidly with further formation and compaction of cake layer (Fig. 16.17b).

(b) *Stage 2—Slow cake layer growth with continuous cake layer compaction*: With the time of operation, the growth of cake layer increases and thus the flux keeps on decreasing until meeting the critical flux.

(c) *Stage 3—Pseudo steady state*: At one point of time, the flux reaches an extreme value (also known as critical flux), when the particle back transport velocity is equal to the particle deposition velocity and thus the particle deposition becomes negligible. Therefore, the cake layer almost stops growing and membrane permeability becomes stable.

16.7.9.3 Physical and Chemical Cleaning for Fouling Control

The major impact on the flux during the filtration is the cleaning procedure involved with it, which is either physical or chemical. The physical cleaning is majorly achieved by the backflushing, i.e., by reversing the flow or simply ceasing the permeation whilst still continuing the scouring of membrane by air bubbles. Chemical cleaning is majorly done with the help of different mineral or organic acids, such as sodium hypochlorite, caustic soda, etc., either in situ (also known as cleaning in place, CIP) or ex situ mode. Furthermore, during backflushing, sometimes small quantity of chemical cleaning agent is also added for chemically enhanced backflush. Although physical cleaning is less time consuming and less harmful to the membrane, it is less effective than the chemical cleaning process. However, the permeability of original virgin membrane can never be restored and the remaining residual resistance known as irrecoverable fouling determines the life of the membrane.

Questions

16.1   What is the difference between ion exchange process and adsorption?

16.2   Why physisorption is preferred in wastewater treatment compared to chemisorption?

16.3   Explain the adsorption process in detail.

16.4   What is the working principle of ion exchange process?

16.5   Write a short note on electrodialysis.

16.6   Determine the best fit isotherm from Langmuir and Freundlich isotherm and calculate its constants for following experimental data.

Dose of adsorbent (ug/L)	0	0.9	1.7	4	7	10
Equilibrium pollutant concentration (ug/L)	20	13	10	6	4	3

(**Answer**: Best fit isotherm-Freundlich isotherm: $K_f = 0.545$, $n = 0.966$)

16.7   Describe the relationship between the TMP and flux of a membrane module using simple resistance in series model.

16.8   Define back transport phenomenon. What are the factors at which the particle deposition velocity depends on?

16.9   State different operational modes of membrane filtration process and their effect on membrane fouling roadmap.

16.10 A microfiltration membrane operating with pure water as feed produces a flux of 0.05 kg/m^2.sec when operated with a TMP of 140 kPa at 20 °C.

(a)   What is the resistance due to membrane? Specify the units.

(b)   If operated with a contaminated water sample at a pressure difference of 180 kPa, a flux of 0.02 kg/m^2.sec is measured at steady-state. What is the resistance due to the cake build-up? Specify the units.

(c)   Also, estimate the pressure drop required to achieve a flux of 0.035 kg/m^2.sec.

[**Answer**: (a) Taking $\mu = 1 \times 10^{-3}$ Pa.sec for water at 20°C, $R_m = 2.8 \times 10^9$ m^2/kg; (b) $R_c = 6.2 \times 10^9$ m^2/kg; (c) $\Delta P_T = 315$ kPa.]

References

Abujazar, M., Fatihah, S., Rakmi, A., & Shahrom, M. (2016). The effects of design parameters on productivity performance of a solar still for seawater desalination: A review. *Desalination, 385,* 178–193. https://doi.org/10.1016/j.desal.2016.02.025

Dąbrowski, A., Hubicki, Z., Podkościelny, P., Robens, E. (2004). Selective removal of the heavy metal ions from waters and industrial wastewaters by ion-exchange method. *Chemosphere, 56*(2), 91–106. https://doi.org/10.1016/j.chemosphere.2004.03.006

Hoque, A., Abir, A., & Shourov, K. (2019). Solar still for saline water desalination for low-income coastal areas. *Applied Water Science, 9,* 104. https://doi.org/10.1007/s13201-019-0986-9

Koon, J. H., & Kaufman, W. J. (1975). Ammonia removal from municipal wastewaters by ion exchange. *Journal of Water Pollution Control Federation, 47,* 448–465.

Oztekin, E., & Altin, S. (2016). Wastewater treatment by electrodialysis system and Fouling problems. *Journal of Science and Technology, 6*(1), 91–99.

Serbezov, A., Moore, J. D., & Wu, Y. (2011). Adsorption equilibrium of water vapor on selexsorb-CDX commercial activated alumina adsorbent. *Journal of Chemical & Engineering Data, 56,* 1762–1769. https://doi.org/10.1021/je100473f

Xu, T., & Huang, C. (2008). Electrodialysis-based separation technologies: A critical review. *AIChE Journal, 54*(12), 3147–3159. https://doi.org/10.1002/aic.11643

Advanced Oxidation Processes

17

17.1 Introduction

In the recent decades, the growing population and changes in the human lifestyle has led to the increase in the concentration of trace organic chemicals, such as pesticides, pharmaceuticals, endocrine disrupting compounds (EDCs) and personal care products (PCPs) in the aquatic environment (Wang & Xu, 2012). Thus, it has increased the polluted water bodies due to presence of these compounds and increased the stress on freshwater resources. The decreasing availability of freshwater resources requires proper management to maintain future sustainability. The established wastewater treatment technologies, such as adsorption, coagulation, ultrafiltration and biological treatments, are only changing the phase of these refractory organic contaminants and do not fully degrade these compounds (Priyadarshini et al., 2020). Therefore, as an alternative, advanced oxidation processes (AOPs) have gained the wide attention of researchers in the recent past due to their higher removal efficiency and complete degradation of biorefractory contaminants into harmless end products.

The AOP is a versatile technology, where reactive radicals are generated for the removal of organic or inorganic contaminants present in the wastewater. These generated reactive radicals contain unpaired electrons in their valence shell, which can very effectively react with the contaminants and simultaneously oxidize them to carbon dioxide (CO_2) and water (H_2O) or other simpler end products those may be biodegradable. In 1894, Fenton observed the degradation of tartaric and racemic acids in the presence of ferrous (Fe^{2+}) salt and hydrogen peroxide (H_2O_2) (Fenton, 1894). Later in 1932, Haber and Wesis investigated the catalytic decomposition of H_2O_2 in presence of Fe^{2+} salt, which resulted in the production of hydroxyl radicals ($OH^·$) (Haber & Weiss, 1932).

Subsequently, in 1987, Glaze et al. defined AOP as a technology, where free radicals, such as $OH^·$, superoxide ($O_2^{·-}$), hydroperoxyl ($HO_2^·$), and sulphate anions ($SO_4^{·-}$), were generated for wastewater treatment by using ozone (O_3), H_2O_2 and ultraviolet (UV) radiation (Glaze et al., 1987). In 1975, Koubek for the first time investigated the formation of OH˙ radicals through photochemical decomposition of H_2O_2 for the destruction of organic contaminants present in the aqueous stream (Koubek, 1975). Later researchers have attempted to explain the role of OH˙ radical in the decolourization of dyes, remediation of wastewater contaminated with heavy metals and homolytic bond dissociation of naphthalene and benzene rings (Dutta et al., 2005).

The major benefits of using AOPs for wastewater treatment are: (i) AOPs can efficiently mineralize the refractory organic contaminants without transferring them into another phase; (ii) AOPs, such as UV, O_3 and cavitation, can be used for the disinfection of water; (iii) heavy metals can also be effectively removed from wastewater; and (iv) AOPs produce less or no sludge during the treatment, which makes AOPs economically advantageous over conventional wastewater treatment technologies, such as activated sludge process. Despite of these benefits, high operating costs, requirement of external chemical reagents and separation of residual oxidants from wastewater are the major bottlenecks of AOPs. Hence, presently researchers are directing efforts to minimize the operating cost and chemical requirements to overcome these limitations, which would help to establish a sustainable field-scale applications of AOPs.

17.2   Classification of Advanced Oxidation Processes

The AOP technology is broadly classified based on the sources used for the production of reactive radical species, such as UV, ozonation, electrochemical, cavitation and Fenton processes (Table 17.1). For example, in the Fenton process, Fenton's reagents (Fe^{2+} and H_2O_2) are used to generate reactive radicals. Similarly, in the case of UV-based AOPs, UV light or the integration of UV light with various radical promoters are used as a source to produce radicals in the system. Furthermore, the emerging AOP technologies are identified based on the articles published in the recent times. The subsequent sections discuss the fundamentals, applications, benefits and drawbacks of both conventional and emerging AOP technologies. Furthermore, the bench-scale and large-scale implementations of each conventional and emerging technologies are also discussed in this chapter.

17.3   Fenton-Based AOPs

The Fenton-based AOP is the oldest and most extensively investigated AOP, in which Fe^{2+} and H_2O_2, known as Fenton's reagents, are utilized to generate reactive radicals for the degradation of organic and inorganic toxic refractory pollutants present in the wastewater. Generation of reactive radicals in Fenton process follows two typical pathways. In the first pathway, oxidation of Fe^{2+} to Fe^{3+} ion occurs to disintegrate H_2O_2 into OH˙ radicals

Table 17.1 Classification of AOPs based on the source used for the production of reactive radicals

Advanced oxidation processes	Source of reactive radicals
UV-based methods	UV irradiation
Electrochemical oxidation	Electricity
Heterogeneous photocatalysis	TiO_2/UV $TiO_2/UV/H_2O_2$
Sonochemical oxidation	Ultrasounds (US)
Ozone based methods	O_3 O_3/UV O_3/H_2O_2 $O_3/UV/H_2O_2$
Fenton based methods	H_2O_2/Fe^{2+} (Fenton process) $H_2O_2/Fe^{2+}/UV$ (Photo Fenton process) Fe^{2+}/US (Sono-Fenton process)

UV—Ultra violet, TiO_2—Titanium dioxide, H_2O_2—Hydrogen peroxide, US—Ultrasound, Fe^{2+}—Ferrous salt, O_3—Ozone

under acidic condition (Eq. 17.1). Whereas, in the second pathway, the generated Fe^{3+} ions are reduced by excess H_2O_2 to produce again Fe^{2+} ions and $HO_2 \cdot$ radicals (Eq. 17.2). However, the excess production of Fe^{2+} ions can scavenge the produced $OH \cdot$ and $HO_2 \cdot$ radicals (Eqs. 17.3 and 17.4) and the produced $HO_2 \cdot$ radicals can further react with Fe^{3+} ions for Fe^{2+} ions regeneration (Eq. 17.5).

$$Fe^{2+} + H_2O_2 \rightarrow Fe^{3+} + OH^- + OH^\bullet \tag{17.1}$$

$$Fe^{3+} + H_2O_2 \rightarrow Fe^{2+} + HO_2^\bullet + H^+ \tag{17.2}$$

$$Fe^{2+} + OH^\bullet \rightarrow Fe^{3+} + OH^- \tag{17.3}$$

$$Fe^{2+} + HO_2^\bullet \rightarrow Fe^{3+} + HO_2^- \tag{17.4}$$

$$Fe^{3+} + HO_2^\bullet \rightarrow Fe^{2+} + O_2 + H^+ \tag{17.5}$$

17.3.1 Major Benefits and Drawbacks of Fenton-Based AOPs

The Fenton process is generally carried out under ambient temperature and the Fenton's reagents are readily available, easy to store and handle, which are some of the added

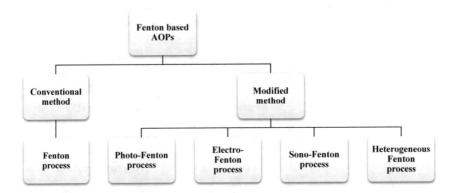

Fig. 17.1 Classification of Fenton processes

advantages of this process. However, two major bottlenecks associate with this technique are: (i) the production of high amount of iron sludge and (ii) lower treatment efficiencies in alkaline pH. Thus, to overcome these shortcomings of the traditional Fenton process, many modifications are suggested by researchers, which are detailed in the subsequent sections (Fig. 17.1).

17.3.2 Photo-Fenton Process

The combination of Fenton process with UV or visible light radiation is known as photo-Fenton process. The main mechanisms of photo-Fenton process are the photo reduction of Fe^{3+} into Fe^{2+} in the presence of photon source and direct photolysis of H_2O_2 to generate $OH^{•}$ radicals, which could improve the rate of degradation and also reduces the use of iron and ferric sludge formation in the system (Eqs. 17.6 and 17.7). However, direct photolysis of H_2O_2 generates lesser $OH^{•}$ radicals in the existence of iron complexes, which generally absorb maximum light radiations. This process offers better performance and higher efficiency under acidic pH, because $Fe(OH)^{2+}$ complexes are more photoactive and soluble at pH of 3.0.

$$[Fe(OH)]^{2+} + light \rightarrow Fe^{2+} + OH^{•} \tag{17.6}$$

$$H_2O_2 + light \rightarrow 2OH^{•} \tag{17.7}$$

The efficiency and feasibility of photo-Fenton process have been successfully reported in many research articles on treating non-biodegradable toxic organic matters, such as phenol (Bauer & Fallmann, 1997), herbicides (Sun & Pignatello, 1993), surfactants (Amat et al., 2004) and ethylene glycol (McGinnis, 2000). In some cases, instead of UV light

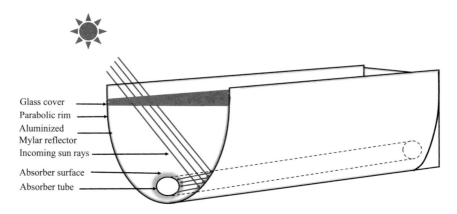

Fig. 17.2 Schematic view of a compound parabolic collector (Modified from Yogev & Yakir, 1999)

irradiation, sunlight has been used for the activation of $Fe(OH)^{2+}$ complex, which effectively minimize the treatment cost. However, the design of an efficient photoreactor for large-scale utilization of natural sunlight is the main challenging issue that limits the practical applications of solar driven Fenton process. The compound parabolic collector (CPC) is one of the most popular photoreactor widely used for photochemical applications. The CPC is a non-imaging optical concentrator, where incident solar radiations, after reflection from the reflector gets concentrated on the absorber surface (Fig. 17.2).

The energy accumulated on the absorber surface of CPC can be estimated by using Eq. 17.8 (Cabrera Reina et al., 2020).

$$Q_{UV,n} = \frac{t_n.UV.S_{CPC}}{V_t} \qquad (17.8)$$

where, $Q_{UV,n}$ is the total energy accumulated per unit volume at time n (J/L), S_{CPC} is the illuminated surface area of CPC (m^2), t_n is the treatment time (sec), V_t is the volume of treated water (L) and UV is the mean yearly UV radiation at a particular location (W/m^2).

The amortization cost of photo-Fenton treatment process can be determined by using Eq. 17.9 (Sánchez Pérez et al., 2013).

$$AC_f = \frac{I_f}{V_t.D.L} \qquad (17.9)$$

where, AC_f is the unit amortization cost in euro (€)/m^3, D is the number of days the treatment plant is expected to work in a year, L is the life span of the treatment plant in years, V_t is the volume of treated water (m^3/day) and I_f is the total investment costs of photocatalytic system, which can be assessed as per Eq. 17.10.

$$I_f = C_{base}\left(\frac{S_{CPC}}{S_{base}}\right)^{0.6}$$ (17.10)

where, C_{base} is the cost of treatment plant, S_{base} is the area of treatment plant (m^2) and S_{CPC} is the illuminated surface area of CPC (m^2).

The operating cost (OC, €/m^3) can be determined by applying Eq. 17.11.

$$OC = C_{staff} + 0.02AC_f + C_E + C_{reactant}$$ (17.11)

where, C_{staff} is the unit cost of personnel (€/m^3), $C_{reactant}$ is the unit cost of reactants (€/m^3), maintenance cost is considered as 2% of AC_f and C_E is the unit electric cost (€/m^3), which can be determined using Eq. 17.12.

$$C_E = \frac{P_{electricity}.W_{pump}.t_{sun}}{V_t}$$ (17.12)

where, $P_{electricity}$ is the electricity cost (€/kWh), W_{pump} is the pump power (kW), t_{sun} is the average daily sun time available at the location (h) and V_t is the volume of treated water (m^3).

Finally, total cost (TC) can be obtained using Eq. 17.13.

$$TC = AC_f + OC$$ (17.13)

The quantum efficiency of photo reduction of Fe^{3+} into Fe^{2+} can be determined as per Eq. 17.14 (Pignatello et al., 2006).

$$\{\phi(Fe^{2+})\} = \frac{d[Fe^{2+}]}{dt}\frac{1}{P_a}$$ (17.14)

where, $d[Fe^{2+}]/dt$ is the rate of Fe^{2+} formation (M sec^{-1}) and P_a is the photon flux absorbed by Fe^{3+} complex over entire wavelength range (Einstein L^{-1} sec^{-1}), which is obtained as per Eq. 17.15 (Pignatello et al., 2006).

$$P_a = \sum P_{a,\lambda} = \left(\frac{RP_e}{V.N_A}\right).\sum\left[\frac{S_{e,\lambda}(1 - 10^{-A_\lambda})}{E_{ph,\lambda}}\right]$$ (17.15)

where, $P_{a,\lambda}$ is the photon flux absorbed at wavelength λ (Einstein L^{-1} sec^{-1}), V is the volume of the solution (L), N_A is the Avogadro's number, RP_e is the radiant power emitted by the light (W), $E_{ph,\lambda}$ is the energy of a photon of wavelength λ (J), A_λ is the average absorbance of the solution at wavelength λ during irradiation and $S_{e,\lambda}$ is the spectral distribution of radiant power emitted by the light (usually obtained from the provider).

The radiant power (RP_e) can be determined using ferrioxalate actinometry test. The RP_e can be estimated by using Eq. 17.16 (Bossmann et al., 2001).

$$RP_e = \frac{nFe^{2+}}{t} \left\{ \sum \left[\frac{S_{e,\lambda}\phi_{ac,\lambda}\left(1 - 10^{-A_{ac,\lambda}}\right)}{E_{ph,\lambda}} \right] \right\}^{-1} \qquad (17.16)$$

where, nFe^{2+} is the number of Fe^{2+} ions formed during irradiation time t (sec), $A_{ac,\lambda}$ is the absorbance of the ferrioxalate actinometric solution at wavelength λ and $\phi_{ac,\lambda}$ is the quantum yield of the actinometric solution at wavelength λ.

Major benefits and drawbacks of photo-Fenton process: Photo-Fenton process is very advantageous as compared to the classical Fenton process, because the use of light source that accelerate the reduction of Fe^{3+} to Fe^{2+}. This reduces the production of iron sludge and also harvests additional $OH^·$ radicals through the decomposition of H_2O_2. However, high operational costs, low utilization rate of light energy and high costs for the design of photo reactors limit the commercialization of this technology.

17.3.3 Heterogeneous Fenton Process

Heterogeneous Fenton process has been developed to overcome the demerits of classical Fenton process, which is limited to a narrow working pH range and also generates large quantity of iron sludge as waste. In this process, solid catalysts containing catalytic active components are used as Fenton catalysts instead of Fe^{2+} ions. Heterogeneous catalysts, such as iron minerals including pyrite, ferrihydrite, magnetite, hematite, goethite and zero-valent iron, soils including clay and laterite, etc., are utilized as source of iron in the heterogeneous Fenton process (Nidheesh, 2015). Moreover, industrial waste containing iron, such as blast furnace dust, fly ash, pyrite ash, etc., have also been used as Fenton catalyst for the degradation of toxic contaminants (Zhang et al., 2019). In the recent decades, to prevent the loss of iron compounds from the aqueous solution along with treated effluent, researchers have used support materials, like activated carbon, silica, clay, zeolite, etc., to immobilize the iron compounds and minimize leaching in the wastewater that is being treated (Nidheesh, 2015).

The decomposition of organic pollutants in heterogeneous Fenton process potentially follows complex series of reactions as presented in Eq. 17.17 through Eq. 17.20 (Araujo et al., 2011).

$$Fe^{3+} + H_2O_2 \rightarrow Fe^{2+} + HO_2^· + H^+ \qquad (17.17)$$

$$Fe^{2+} + H_2O_2 \rightarrow Fe^{3+} + OH^- + OH^• \qquad (17.18)$$

$$HO_2^{\cdot} \leftrightarrow H^+ + O_2^{\cdot-} \qquad (17.19)$$

$$Fe^{3+} + HO_2 \rightarrow Fe^{2+} + H^+ + O_2 \qquad (17.20)$$

Major benefits and drawbacks of heterogeneous Fenton process: Heterogeneous Fenton process has a series of advantages, such as low iron sludge production, less leaching of iron, wide working pH range and prolong stability of catalysts. However, the main drawbacks of this process are the high cost for the synthesis of catalysts and complicated synthesis routes that limit the development of heterogeneous Fenton process on a large scale (Ganiyu et al., 2018).

17.3.4 Sono-Fenton Process

In this process, ultrasound (frequency range = 20 kHz to 2 MHz) coupled with Fenton process is used to generate high concentration of reactive radicals through two distinct types of mechanisms, such as physical (or direct) and chemical (or indirect) mechanism (Riesz et al., 1985). The direct or physical mechanism involves the rapid formation of microbubbles by the application of ultrasonic wave, which grows and then collapse by releasing large magnitude of energy with extremely high temperature (temperature = 2000–5000 K) and pressure (about 6×10^4 kPa) (Fig. 17.3) (Oturan & Aaron, 2014). Under these extreme conditions, sonolysis (action of breakdown or decomposing a substance using ultrasound) of H_2O_2 occurs and that produces $OH^{\cdot}$ radicals in the system (Eq. 17.21). In indirect or chemical mechanism, the Fe-OOH^{2+} can be effectively dissociated into Fe^{2+} and $HO_2^{\cdot}$ under the ultrasonic irradiation, which further reacts with Fe^{2+} ions and produces $OH^{\cdot}$ radicals that can improve the removal rate of organic contaminants present in the wastewater as shown in Eq. 17.22 through Eq. 17.26 (Bagal & Gogate, 2014).

$$H_2O_2 \rightarrow 2OH^{\cdot} \qquad (17.21)$$

$$Fe^{2+} + H_2O_2 \rightarrow Fe^{3+} + HO^- + OH^{\cdot} \qquad (17.22)$$

$$H_2O_2 + Fe^{3+} \rightarrow Fe\text{-}OOH^{2+} + H^+ \qquad (17.23)$$

$$Fe\text{-}OOH^{2+} + US \rightarrow Fe^{2+} + HO_2^{\cdot} \qquad (17.24)$$

$$Fe^{2+} + HO_2^{\cdot} \rightarrow Fe^{2+} + H^+ + O_2 \qquad (17.25)$$

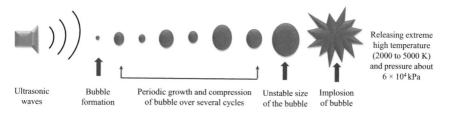

Ultrasonic waves   Bubble formation   Periodic growth and compression of bubble over several cycles   Unstable size of the bubble   Implosion of bubble   Releasing extreme high temperature (2000 to 5000 K) and pressure about 6×10^4 kPa

Fig. 17.3 Schematic representation of growth and collapse of bubble in sono-Fenton process

$$Fe^{2+} + H_2O_2 \rightarrow Fe^{3+} + HO^- + OH^{\cdot} \qquad (17.26)$$

The efficiency of sono-Fenton process can be future enhanced by the application of UV or visible light irradiation, in this case it is known as sono-photo-Fenton process. Researchers have reported that the combination of sonochemistry with different AOPs, such as Fenton's reagents, ozonation, photocatalysis and electrochemical oxidation, has significantly increased the efficacy of wastewater remediation.

Major benefits and drawbacks sono-Fenton process: The sono-Fenton process produces variety of reactive species like $H^{\cdot}$, $OH^{\cdot}$ and $HO_2{}^{\cdot -}$, which is an added advantage of this technology. Furthermore, in sono-Fenton process, ultrasonic waves can improve the mass transfer rate and disaggregation of catalyst particles (Chowdhury & Viraraghavan, 2009). However, the notable drawbacks of this technology are the high cost of operation and uncertain distribution of cavity formation in the reactor that hinder its practical applications (Pliego et al., 2015).

Example: 17.1 In a decomposition reaction H_2O_2 (aq) $\rightarrow H_2O$ (l) $+$ ½ O_2 (g), the concentration of H_2O_2 is 0.26 M after 20 s, and the rate of reaction is 10×10^{-4} M/sec. What will be $[H_2O_2]$ formation after 40 sec?

Solution

The rate of the reaction can be defined as the change in concentration over time, recognizing it as decomposition reaction the initial rate for the loss of H_2O_2 will be negative at t = 20 sec, when $[H_2O_2]$ = 0.26 M.

$$Rate = -10 \times 10^{-4} M/sec = \frac{\Delta[H_2O_2]}{\Delta t}$$

$$Rate = -10 \times 10^{-4} M/sec = \frac{[H_2O_2] - 0.26\ M}{(40 - 20)}$$

Hence, H_2O_2 concentration after 40 sec = 0.24 M.

Example: 17.2 In Experiment no. 1, the product X, with an initial concentration $[X]_0$ of 1.512 M, was found to be 1.496 M after 30 sec. In Experiment no. 2, with an initial concentration $[X]_0$ of 2.584 M, the product $[X]$ was found to be 2.552 M at time $t = 1$ min. What is the order of this reaction?

Solution
First, calculate the rate of reaction of each experiment. Then by using method of initial rate, order of the reaction can be obtained.

Experiment no	$[X]_0$ (M)	$[X]$ (M) at t (sec)	Rate (M/sec)
1	1.512	1.496 at 30 sec	$\frac{1.496-1.512}{30-0} = -5.3 \times 10^{-4}$
2	2.584	2.552 at 60 sec	$\frac{2.552-2.584}{60-0} = -5.3 \times 10^{-4}$

Since the rates are same for both experiments, the reaction must be zero order.

$$\frac{\text{Rate 1}}{\text{Rate 2}} = \frac{-5.3 \times 10^{-4}}{-5.3 \times 10^{-4}} = 1 = \frac{k[X]_1^m}{k[X]_2^m} = \frac{[1.512]^m}{[2.584]^m} = (0.5851)^m,$$

Which can only be true for m = 0; hence the reaction is zero order.

17.3.5 Applications of Fenton-Based AOPs

The Fenton-based AOPs have been used widely to treat a variety of organic contaminants, such as dyes, pharmaceuticals, surfactants and agrochemicals, etc. In an investigation, photo-Fenton process was used for the removal of Reactive Red 198 dye with initial concentration, C_o, of 50 mg/L from the aqueous solution and the result of the investigation displayed 99% removal of the dye within 120 min by employing Fe^{2+} ions of 10 mg/L and H_2O_2 concentration of 75 mg/L at pH of 3.0 (Dehghani et al., 2015). The degradation of Carbofuran (C_o of 20 mg/L) was investigated by employing sono-Fenton process and almost 99% degradation of Carbofuran was obtained within 30 min under optimum conditions of H_2O_2 dose of 200 mg/L and Fe^{2+} concentration of 10 mg/L (Ma & Sung, 2010). In another investigation, pyrite was used as a heterogeneous Fenton catalyst for the degradation of Diclofenac (C_o of 100 mg/L) and almost 100% degradation of Diclofenac was achieved by adopting 1 g/L of pyrite and 0.3 M of H_2O_2 at pH of 5.1 (Oral & Kantar, 2019). Similarly, other prominent research articles are also published on Fenton-based AOPs for wastewater treatment, these performance results are summarized in Table 17.2.

Table 17.2 Organic contaminants removal by Fenton processes

Process	Target pollutant	C_o (mg/L)	Type of catalyst	Experimental conditions	Source of energy	Efficiency (%)	References
Heterogeneous Fenton process	p-Nitrophenol	20.9	Ferrihydrite	Catalyst dosage = 0.2 g/L, $[H_2O_2]$ = 0.45 mM, Time = 5 h	–	94.5 TOC	Wu et al., (2013)
	Methylene blue	10	Niobium (Nb)-doped hematite	Catalyst dosage = 10 mg/L, $[H_2O_2]$ = 8 mol/L, Time = 120 min	–	75	Oliveira et al., (2015)
	Phenol	47	Natural Fe clay	Catalyst dosage = 1 g/L, $[H_2O_2]$ = 7 mM, Time = 6 h	–	70 TOC	Djeffal et al., (2014)
Photo-Fenton process	Flutamide	5	$FeSO_4.7H_2O$	Catalyst dosage = 5 mg/L, $[H_2O_2]$ = 150 mg/L, Time = 120 min	Solar	58	Della-Flora et al., (2020)

(continued)

Table 17.2 (continued)

Process	Target pollutant	C_o (mg/L)	Type of catalyst	Experimental conditions	Source of energy	Efficiency (%)	References
	2-chlorophenol	1500	FeCl$_3\cdot$6H$_2$O	Catalyst dosage = 28 mg/L, [H$_2$O$_2$] = 10,316 mg/L Time = 150 min	UV	95 TOC	Poulopoulos et al., (2008)
	Phenol	50	Fe-zeolites	Catalyst dosage = 2 g/L, [H$_2$O$_2$] = 1 g/L, Time = 90 min	Solar light	90 TOC	Gonzalez-Olmos et al., (2012)
Sono-Fenton process	Ibuprofen	20	FeSO$_4$.7H$_2$O	Catalyst dosage = 10 mg/L, [H$_2$O$_2$] = 6.4 mM, US frequency = 20 kHz, Time = 180 min	-	50 TOC	Adityosulindro et al., (2017)

(continued)

Table 17.2 (continued)

Process	Target pollutant	C_O (mg/L)	Type of catalyst	Experimental conditions	Source of energy	Efficiency (%)	References
	Malachite green	200	$FeSO_4 \cdot 7H_2O$	Catalyst dosage = 0.5 mM, $[H_2O_2]$ = 10 mM, US frequency = 25 kHz, Time = 60 min	-	79.19 COD	Melo et al., (2020)
	2, 4, 6 Trinitrotoluene	10	$FeSO_4 \cdot 7H_2O$	Catalyst dosage = 2 mM, $[H_2O_2]$ = 40 mM, US frequency = 25 kHz, Time = 20 min	-	100	Hashemi and Sagharlo (2020)

$FeSO_4 \cdot 7H_2O$—Ferrous sulphate heptahydrate, $FeCl_3 \cdot 6H_2O$—Ferric chloride hexahydrate, H_2O_2—Hydrogen peroxide, US—Ultra sound, UV—Ultra violet, TOC—Total organic carbon, COD—Chemical oxygen demand

17.3.6 Scaling-Up of Different Fenton-Based AOPs

The lab-scale investigations on Fenton-based AOPs have been well documented in the literature. However, a very limited number of research articles have explained the practical applications of Fenton-based AOPs. For example, a solar photo-Fenton-based AOP having a capacity 250 L was operated with a flow rate of 600 L/h for the removal of Ofloxacin from secondary treated domestic wastewater. It was observed that 92.7% removal of Ofloxacin was achieved after 180 min by using Fe^{2+} and H_2O_2 concentration of 5 mg/L and 75 mg/L, respectively (Michael et al., 2012). Moreover, based on the pilot-scale experiments and the preliminary economic analysis for a full-scale treatment unit having a flow rate of 150 m^3/day the cost per unit of treatment was estimated to be 0.85 €/m^3 (Michael et al., 2012).

Similarly, a pilot-scale unit having the total capacity of 10 L was operated for the treatment of real petroleum refinery wastewater through solar Fenton process. This setup demonstrated 79.6% of COD and 73.2% of TOC removal within 180 min under optimal operating conditions with dosing of 694.7 mg/L of H_2O_2 and 67.3 mg/L of Fe^{2+} at pH of 3.2 (Pourehie & Saien, 2020). Furthermore, the total operating cost was estimated as 10.4 dollars ($)/$m^3$, which was lesser than the total cost estimated ($17.4/$m^3$) for the pilot plant operated with solar/Fe^{2+}/$S_2O_8{}^{2-}$ to treat petrochemical wastewater (Babaei & Ghanbari, 2016). Outcome reported in other research articles published on the pilot-scale demonstration of Fenton processes are presented in Table 17.3.

17.4   UV-Based AOPs

In last two decades, UV photolysis and UV-based AOPs have received immense attention of the researchers for the removal of organic refractory pollutant from wastewaters. In these processes, sole UV irradiation or the integration of UV light with other radical promoters, such as H_2O_2, Cl_2, O_3, $SO_4{}^{\cdot-}$, etc., are used to degrade the organic pollutants present in the wastewater. The UV fluence usually greater than 200 mJ/cm^2 is applied for the removal of toxic pollutants in UV-based AOPs (USEPA, 2006). However, in addition to the removal of organic contaminants, UV can also be used for the microbial disinfection of both drinking water and wastewater.

Generally, the most common sources of UV light are low pressure (LP) or medium pressure (MP) mercury lamps with mono or polychromatic spectra, respectively. For UV/H_2O_2 and UV/O_3 treatment processes, LP-UV lamps with a wavelength close to 254 nm are generally used for the oxidation process. Table 17.4 represents the specifications of various types of LP-UV lamps that are typically available in market. Nowadays, UV light emitting diode (LED) has been used for inactivating the microorganisms present in aqueous solutions (Song et al., 2016). Moreover, UV LEDs possess many advantages

Table 17.3 Performance results of the large-scale investigations on Fenton processes

Process	Target pollutant	C_o (mg/L)	Experimental conditions	Efficiency (%)	Reference
US/Solar photo-Fenton process	Bisphenol A	$TOC_0 = 80$	Volume = 10 L, Flow rate = 7 L/min, time = 300 min, US = 400 kHz, Fe^{2+} = 5 mg/L	75 TOC	Papoutsakis et al., (2015)
Solar photo-Fenton process	Municipal wastewater treatment plant (WWTP) effluent	$COD_0 = 61.9$	Volume = 16 L, Flow rate = 900 L/m².day, Fe^{2+} = 0.1 mM, H_2O_2 = 0.88 mM, time = 20 min	60 COD	Arzate et al., (2020)
Photo-Fenton process	Municipal WWTP effluent	$TOC_0 = 7.5$	Volume = 37 L, Flow rate = 14 m³/h, H_2O_2 = 50 mg/L, Fe^{2+} = 4 mg/L	88 TOC	De la Cruz et al., (2013)
Solar photo-Fenton process	Phenol	94.11	Volume = 190 L, Flow rate = 100 L/h, Fe^{3+} = 0.27 mmol/L, H_2O_2 = 21.30 mmol/L	100	Rodríguez et al., (2005)
Two stage Fenton process	6-Nitro-1-diazo-2-naphthol-4-sulfonic acid	1775 ± 56	Volume = 2 m³, Flow rate = 16 m³/h, H_2O_2 = 4.9 g/L, pH = 2.5	99.3	Gu et al., (2012)

TOC_0—Initial total inorganic carbon, COD_0—Initial chemical oxygen demand, US—Ultra sound, Fe^{2+}—Ferrous salt, Fe^{3+}—Ferric salt, H_2O_2—Hydrogen peroxide.

compared to conventional LP and MP mercury lamps, such as: (i) compact size, (ii) elimination of use of mercury, hence, eco-friendly, (iii) short start-up phase, and (iv) unique peak emission wavelengths (Autin et al., 2013). In the subsequent sections, the current status of implementations, mechanisms, applications, major benefits and drawbacks of various UV-based AOPs are briefly discussed.

Table 17.4 Specifications of commercially available LP-UV lamps

Manufactures	Product description	UV model	Electrical power (W)	UV output (W)	Diameter (mm)	Length (mm)
LightTech	High output	GHO893T5L	95	30	15	900
		GHO436T5L	48	13	15	360
Osram	Puritec HNS	ZMP 4021026	25	6.9	26	438
		ZMP 4021031	55	18	26	895
Philips	TUV T5	927970204099	40	15	19	1600
	TUV TL mini	927971204099	11	2.6	19	328

17.4.1 UV/H$_2$O$_2$

In this process, photolysis of H_2O_2 molecule undergoes homolytic cleavage to produce $OH^\cdot$ radicals in the presence of UV radiations with wavelength ranging from 200 to 300 nm (Eq. 17.27) (Oturan & Aaron, 2014). The produced $OH^\cdot$ radicals possess high redox potential (E_0) of 1.8–2.7 V versus normal hydrogen electrode (NHE), which makes them able to react with the organic contaminants and simultaneously break the aromatic rings of the organic contaminants and mineralize them into CO_2 and H_2O. The redox potential (E_0) is described as the tendency of a chemical species to either gain or lose electrons through ionization. The increment of initial H_2O_2 concentration enhances the production of $OH^\cdot$ radicals up to a certain optimum dose. However beyond the optimum dose, H_2O_2 acts as a $OH^\cdot$ radicals scavenger and produces less reactive $HO_2^\cdot$ radicals, which could significantly reduce the degradation efficacy of the UV/H_2O_2 process as shown in Eqs. 17.28 and 17.29 (Wang & Xu, 2012).

$$H_2O_2 + UV \rightarrow 2OH^\cdot \tag{17.27}$$

$$OH^\cdot + H_2O_2 \rightarrow H_2O + HO_2^\cdot \tag{17.28}$$

$$OH^\cdot + HO_2^- \rightarrow HO_2^\cdot + OH^- \tag{17.29}$$

Major benefits and drawbacks of UV/H$_2$O$_2$ process: A large amount of bench and field-scale investigations pertaining to UV/H_2O_2 process have been well documented for decontamination of groundwater and degradation of refractory organic pollutants, like benzene, trichloroethylene and tetrachloroethylene (Eckenfelder, 1997). The UV/H_2O_2 process has many advantages, such as: (i) simple operation, (ii) can work for disinfection, (iii) no sludge production, and (iv) does not use any special type of equipment for its operation. However, certain drawbacks of this process are: (i) UV light penetration can be affected when treating highly turbid wastewater, and (ii) the presence of compounds, such as nitrate, can interfere with the absorbance of UV light.

17.4.2 UV/Cl$_2$

The UV/Cl$_2$ process is an emerging technology, where initially Cl$_2$ reacts with H$_2$O to form HOCl and Cl$^-$ by disproportionation reaction (Eq. 17.30) and further HOCl dissociates and forms ClO$^-$ ions (Eq. 17.31). However, the produced HOCl and ClO$^-$ are homolytically cleaved by UV radiations and forms radical species, like OH$^{\cdot}$, Cl$^{\cdot}$, Cl$_2^{\cdot-}$, which then oxidize the biorefractory pollutants present in aqueous solution as shown in Eq. 17.32 through Eq. 17.35 (Lee et al., 2021). The Cl$^{\cdot}$ and Cl$_2^{\cdot-}$ are more selective oxidants that usually prefer to react with electro-rich-moieties like conjugated double bonds and amines (Xiang et al., 2020). Moreover, Cl$^{\cdot}$ and Cl$_2^{\cdot-}$ have E_0 of 2.4 V and 2.0 V versus NHE, respectively (Mártire et al., 2001). In this process, generally hypochlorite and chlorine dioxide are used as precursor to harvest OH$^{\cdot}$ and reactive chlorine species (RCS).

$$Cl_{2(aq)} + H_2O \rightarrow HOCl + Cl^- + H^+ \tag{17.30}$$

$$HOCl \leftrightarrow H^+ + ClO^- \tag{17.31}$$

$$HOCl + UV \rightarrow OH^{\cdot} + Cl^{\cdot} \left(pH < 7.6 \text{ at } 25^{\circ}C \right) \tag{17.32}$$

$$ClO^- + UV \rightarrow O^{\cdot-} + Cl^{\cdot} \left(pH > 7.6 \text{ at } 25^{\circ}C \right) \tag{17.33}$$

$$O^{\cdot-} + H_2O \rightarrow OH^{\cdot} + OH^- \tag{17.34}$$

$$Cl^{\cdot} + Cl^- \rightarrow Cl_2^- \tag{17.35}$$

Major benefits and drawbacks UV/Cl$_2$ process: The UV/Cl$_2$ process offers advantages, such as: (i) does not require skilled labour to operate the system, (ii) easy installation and (iii) removes pathogens, taste and odour from wastewater. However, major disadvantage of this process is the formation of disinfection by-products (DBPs), such as trihalomethanes, haloacetontriles, halonitromethanes and haloacetamide. These DBPs are highly toxic and their prolonged occurrence in the environment can cause bladder cancer, miscarriages and ulcerative colitis in human. Therefore, the presence of DBPs requires further polishing treatment to make treated water suitable for discharge or reuse.

17.4.3 UV/O₃

The UV/O$_3$ is another promising AOP technology, where UV irradiation (λ < 300 nm) results in the photolytic cleavage of O$_3$ in water, which forms OH$^{\cdot}$ radicals as shown in Eqs. 17.36 and 17.37. In this process, O$_3$ reacts with the organic contaminants present in wastewater either directly by electrophilic substitution or indirectly by radical chain reactions (Wang & Xu, 2012). However, it has been reported in literature that the reactivity of O$_3$ towards organic pollutants is very slow as compared to OH$^{\cdot}$ radicals, which indicates that the reaction of OH$^{\cdot}$ radicals with organic pollutants is an important mechanism in the UV/O$_3$ process (Garoma & Gurol, 2004). Other investigations have demonstrated that the combination of other techniques, such as H$_2$O$_2$, ultrasound, TiO$_2$ with UV/O$_3$ process also improved the production of OH$^{\cdot}$ radicals, which further intensified the rate of degradation of organic contaminants and efficacy of the system (Cuerda-Correa et al., 2019).

$$O_3 + H_2O + UV \rightarrow 2H_2O_2 + O_2 \qquad (17.36)$$

$$2O_3 + H_2O_2 \rightarrow 2OH^{\cdot} + 3O_2 \qquad (17.37)$$

Major benefits and drawbacks UV/O₃ process: The UV/O$_3$ process has advantages, such as: (i) it does not produce sludge as waste, while removing colour and organic pollutants from wastewater, (ii) photolysis of O$_3$ generates H$_2$O$_2$ in the system, which enhances the disintegration rate of pollutants, and (iii) installation of this system is simple and requires small space. Apart from these advantages, the major drawbacks of this system are: (i) formation of by-products, which may be more toxic than the original target contaminant, (ii) the use of both UV lamps and O$_3$ generator requires a considerable amount of electrical energy that will escalate the operating cost of the treatment process and (iii) this process typically requires an air permit and off-gas treatment system for O$_3$ emissions and O$_3$ destruction, respectively, which could also increase the operational and capital cost of the system.

17.4.4 UV/SO₄$^{\cdot -}$

In recent years, UV/SO$_4$$^{\cdot -}$ technique has gained much attention of scientists in the field of wastewater treatment. In this process, peroxy-mono-sulphate (PMS; HSO$_5$$^-$) and persulphate (PS; S$_2O_8$$^{2-}$) salts are used as oxidants to generate SO$_4$$^{\cdot -}$ radicals. The formation of SO$_4$$^{\cdot -}$ radicals in the presence of UV radiations typically follows two pathways. In the first pathway homolytic dissociation of O–O bond occurs in the presence of UV radiation as shown in Eqs. 17.38 and 17.39. The second pathway explains UV radiational dissociation of the water molecule to produce electrons, which further reacts with PMS and PS to produce $SO_4{}^{\cdot-}$ radicals as shown in Eq. 17.40 through Eq. 17.42.

$$S_2O_8^{2-} + UV \rightarrow 2SO_4^- \tag{17.38}$$

$$HSO_5^- + UV \rightarrow SO_4^- + OH^\cdot \tag{17.39}$$

$$H_2O + UV \rightarrow H^\cdot + OH^\cdot \tag{17.40}$$

$$S_2O_8^{2-} + H^\cdot \rightarrow SO_4^- + SO_4^{2-} + H^+ \tag{17.41}$$

$$HSO_5^- + H^\cdot \rightarrow SO_4^- + H_2O \tag{17.42}$$

Major benefits and drawbacks UV/SO$_4$$^{\cdot-}$: The UV/SO$_4$$^{\cdot-}$ has emerged as a promising alternative to OH$^\cdot$-based AOPs because SO$_4$$^{\cdot-}$ is a strong oxidizing species having E_0 of 2.5 V to 3.1 V versus NHE, which is higher than the E_0 of OH$^\cdot$ radical (1.8–2.7 V vs. NHE). Secondly, the half-life of SO$_4$$^{\cdot-}$ radical is 30–40 μsec, which is higher than the half-life of OH$^\cdot$ radicals (20 nsec). Furthermore, SO$_4$$^{\cdot-}$ radicals demonstrate robust behavior over a wide pH range of 2.0–8.0. The molar absorption coefficient (ε) of $S_2O_8^{2-}$ is 22 M^{-1} cm^{-1}, which is slightly lesser than ε of H_2O_2 (18 M^{-1} cm^{-1}) at $\lambda = 254$ nm. However, the release of sulphate ions as end product might require further polishing treatment to the effluent generated, which may make the process more expensive.

The term molar absorption coefficient 'ε' represents the ratio of absorbance to the molar concentration of a substance with a path length of ℓ cm at a particular wavelength. It is an intrinsic property of the substance that is dependent upon their structure and chemical composition. The ε value of a substance is determined by using the Beer-Lambert law, which is given in Eq. 17.43.

$$A = \varepsilon c \ell \tag{17.43}$$

where, A is the absorbance of the substance (M^{-1} cm^{-1}), c is the molar concentration of the substance (mol), ε is molar absorption coefficient and ℓ is the path length (cm).

Rearranging the Beer-Lambert equation in order to determine the molar absorption coefficient, Eq. 17.44 is obtained.

$$\varepsilon = A/c\ell \tag{17.44}$$

17.4.5 Design of UV-Based AOPs Reactor

From the engineering point of view, the design of any reactor for wastewater treatment requires preliminary bench-scale evaluation, which could help the designer to decide the possible alternatives for dealing with the scaling-up issue. Moreover, it will also identify the necessary operating parameters that affect the performance of the process (Table 17.5). Therefore, to identify the most relevant operating parameters, bench-scale experiments are conducted using a reactor, which is similar to the reactor used for the field-scale treatment. This is important because most of the operating parameters are affected by the configuration of reactor, and if values selected for these parameters are changed, the performance outcome will also be changed. In addition, design of the reactor not only depends on the geometry of the reactor, but also depends upon the targeted effluent characteristics, optical path, quartz sleeve and type of the UV lamp used, which are quite relevant parameters that require optimization before scaling-up.

17.4.5.1 The UV-Based AOPs Reactor Design Process

For design of UV/H_2O_2 reactor the following steps can be adopted.

(i) The first step is to estimate the reactor volume (Eq. 17.45), which depends upon the spatial retention time and pollutant removal efficiency (Eq. 17.48).

Table 17.5 Set of parameters that can affect the performance of UV-based AOPs

Parameters	Relevance
Turbidity	The presence of suspended solids can obstruct the penetration of UV light in to the wastewater to be treated
Carbonate and bicarbonate	These chemicals scavenge the free radicals, which can affect the process performance
Dissolved organic carbon	It will affect the intensity of UV radiation and it is also responsible for consumption of H_2O_2 and $OH^\cdot$ radicals
Soluble iron and manganese	These metals can be oxidized by H_2O_2 and precipitate on to the quartz sleeve, thereby reducing the transmission of UV radiation
UV light transmittance	Effluent with small transmittance will require high UV radiation intensity for the degradation process
Optical path on the reactor	It is the distance that UV radiation needs to penetrate in the water or wastewater being treated in order to effectively react with radial promoters to produce reactive radicals. Generally, optical path more than 1 cm is not recommended, since above this value UV-light intensity will decrease exponentially

UV—Ultra violet, $OH^\cdot$—Hydroxyl radical, H_2O_2—Hydrogen peroxide

$$V = Q \times t \tag{17.45}$$

where, V is the volume of the reactor (m^3), Q is the water/wastewater flow rate in the reactor (m^3/sec) and t is the spatial retention time (sec), which can be determined as per Eq. 17.46 or Eq. 17.47.

$$[C] = \frac{[C_0]}{k't + 1} \text{ (For continuous stirred tank reactor, CSTR)} \tag{17.46}$$

$$[C] = [C_0].e^{-k'.t} \text{ (For plug - flow reactor)} \tag{17.47}$$

where, C and C_0 are the outlet and inlet pollutant concentration (mg/L), respectively, and k' is the pseudo first order degradation rate constant (sec^{-1}).

$$Removal efficiency(\%) = 100 \times \left(1 - \frac{[C]}{[C_0]}\right) \tag{17.48}$$

For the desired pollutant removal efficiency (Eq. 17.48), the ratio of $\frac{[C]}{[C_0]}$ value can be obtained, which will allow to determine the spatial retention time (Eq. 17.46) for CSTR and (Eq. 17.47) for plug-flow reactor and also the volume of the reactor can be obtained using Eq. 17.45.

(ii) Knowing the quartz sleeve diameter, reactor volume and optical path, it is possible to

determine the equivalent length and UV exposed area of the reactor using Fig. 17.4 as a reference. From the Fig. 17.4, the diameter of UV reactor (dR) and the volume of the reactor (V) can be obtained per Eqs. 17.49 and 17.50.

$$d_R = d_{qs} + 2OP \tag{17.49}$$

$$V = S_{annular} L_{eq} \tag{17.50}$$

where, d_{qs} is the diameter of quartz sleeve (m); L_{eq} is the equivalent length of the reactor (m); OP is the optical path (m); and $S_{annular}$ is the area of the annular region (m^2), which can be assessed as per Eqs. 17.51 and 17.52.

$$S_{annular} = S_{reactor} - S_{qs} \tag{17.51}$$

$$S_{annular} = \frac{\pi d_R^2}{4} - \frac{\pi d_{qs}^2}{4} \tag{17.52}$$

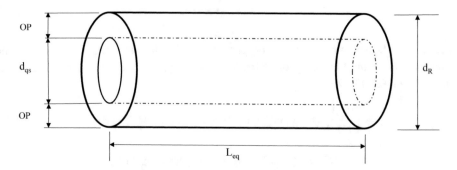

Fig. 17.4 Schematic representation of total length of an UV reactor

Now, by replacing Eq. 17.49 in Eq. 17.52, the Eq. 17.53 is obtained, which is then substituted in Eq. 17.50 to obtain Eq. 17.54.

$$S_{annular} = \frac{\pi}{4}\left(4 \times (d_{qs} \cdot OP + OP^2)\right) \tag{17.53}$$

$$L_{eq} = \frac{V}{\pi(d_{qs} \cdot OP + OP^2)} \tag{17.54}$$

It is assumed that the UV exposed area is the internal reactor surface, which can be determined using Eq. 17.55 (Mierzwa et al., 2018).

$$S_{exp} = L_{eq} \cdot \pi \cdot \left(d_{qs} + 2 \cdot OP\right) \tag{17.55}$$

where, S_{exp} is the UV exposed area in the reactor (m^2); L_{eq} is the equivalent length of the reactor (m); d_{qs} is the diameter of quartz sleeve (m); d_R is the diameter of UV reactor (m); and OP is the optical path on the UV reactor (m).

(iii) The third step is to calculate the applied UV power as per Eq. 17.56. For this calculation, it is necessary to account for lumen depreciation over the time of operation, which is generally 15–20%. Thus, considering the decrease in UV light efficiency the effective UV irradiance can be estimated using Eq. 17.57 (Mierzwa et al., 2018).

$$Applied_{UV_{power}} = \frac{I_{effective} \times S_{exp}}{\frac{\%T_{qs}}{100} \times \frac{\%T_{effluent}}{100} \times \frac{(100\% - \%UV_{el})}{100}}$$

$$= \frac{100^3 \times I_{effective} \times S_{exp}}{\%T_{qs} \times \%T_{effluent} \times (100\% - \%UV_{el})} \tag{17.56}$$

$$I_{effective} = \frac{D}{t} \tag{17.57}$$

where, $I_{effective}$ is the effective UV irradiance (W/m^2); T_{qs} is the quartz sleeve transmittance (%); $T_{effluent}$ is the transmittance of the effluent (%); UV_{el} is the UV lamp efficiency lost (%); and D is the effective UV dosage (mJ/cm^2).

(iv) Total number of UV lamps ($N_{UV\ lamps}$) and number of quartz sleeves (N_{qs}) required for the treatment unit can be obtained from Eq. 17.58. For this calculation, the specifications of UV lamp are required.

$$N_{UVlamps} = N_{qs} = \frac{Applied_{UV_{power}}}{Lamp_{UV_{output}}} \tag{17.58}$$

where, $Lamp_{UV_{output}}$ is the power of UV lamp (W) and $Applied_{UV_{power}}$ is the UV power applied to the system (W).

By using the total number of quartz sleeve and the net UV lamp length ($L_{UV\text{-}net}$), it is possible to obtain the reactor's total volume (V_{total}) using Eq. 17.59.

$$V_{total} = V + \frac{N_{qs} . \pi . (d_{qs})^2 L_{UV-net}}{4} \tag{17.59}$$

Now, by using V_{total}, the UV-reactor diameter (d_R) can be obtained using Eq. 17.60.

$$d_R = \sqrt{\frac{4V_{total}}{\pi L_{UV-net}}} \tag{17.60}$$

where, V_{total} is reactor total volume (m^3) and $L_{UV\text{-}net}$ is the length of UV lamp (m).

For a plug flow reactor, it is recommended to use L_{UV-net}/d_R ratio equal to or greater than 3, however for lower L_{UV-net}/d_R ratios, it is important to use more number of reactors operated in parallel or more adequate L_{UV-net}/d_R ratio can be provided by changing the model of UV lamp.

17.4.5.2 Estimation of Total Cost Per Order of Wastewater Treatment

The total cost ($Cost_{total}$) is estimated to evaluate the feasibility of various UV-based processes, which includes the electrical energy cost ($Cost_{UV}$) and oxidant cost ($Cost_{ox}$). The $Cost_{UV}$ and $Cost_{total}$ can be estimated as per the Eq. 17.61 through Eq. 17.66.

$$Cost_{UV}(\$/m^3) = EE_{UV}(kWh/m^3) \times electricity\ cost(\$/kWh) \tag{17.61}$$

$$Cost_{UV}\left(\$/m^3\right) = \frac{P \times t \times 1000}{V \times 60 \times \log\left(\frac{C_0}{C_t}\right)} \times electricity\ cost \tag{17.62}$$

$$\log\frac{C_t}{C_o} = -0.4343k't \tag{17.63}$$

By replacing Eq. 17.63 in Eq. 17.62, the Eq. 17.64 is obtained.

$$Cost_{UV}\left(\$/m^3\right) = \frac{38.4 \times P}{Vk'} \times electricity cost \tag{17.64}$$

$$Cost_{OX}\left(\$/m^3\right) = \frac{total\ oxidant\ consumed \times oxidant\ cost}{V} \tag{17.65}$$

$$Cost_{total}\left(\$/m^3\right) = Cost_{UV} + Cost_{OX} \tag{17.66}$$

where, EE_{UV} is the electrical energy, i.e., the electrical energy in kW-h required to degrade the target pollutant by one order of magnitude in 1 m^3 of water; P is the lamp power output (kW); t is the irradiation time (h); V is the reactor volume (L); k' is the pseudo first order degradation rate constant (min^{-1}) and C_t and C_o are the initial and final concentration of target compound (mg/L), respectively.

Example: 17.3 A textile industry treats its effluents to eliminate the residual of a toxic contaminants utilized in the dyeing of cloths. The effluent is pre-treated by physico-chemical process; however, the residual concentration of the toxic contaminants is 20 times higher than that of the limit set by the environmental protection agency. The designer responsible for the treatment of the effluent conducted some bench-scale experiments using UV/H$_2$O$_2$ method in a batch reactor and obtained the pseudo-first-order degradation rate constant for the pollutant as 9×10^{-2} sec^{-1} by employing an UV dosage of 200 mJ/cm^2, optical path of 0.75 cm, diameter of quartz sleeve of 3 cm, quartz sleeve transmittance of 90% and H$_2$O$_2$ dosage of 10 g/m^3. The cost of electricity and H$_2$O$_2$ are \$0.1 per kWh and \$1.5 per kg, respectively. The characteristics of the effluent and specifications of the UV lamp are displayed in Table 17.6. Based on the displayed data, design a continuous plug-flow reactor to obtain a removal efficiency higher than 96%.

Solution
For ensuring removal efficiency of greater than 96%, let us design the system for removal efficiency of 99%.

Flow $(Q) = 6/(60 \times 60) = 1.66 \times 10^{-3}$ m^3/sec.
UV dosage $(D) = 200 \times 10,000/1000 = 2000$ J/m^2.
Optical length $(OP) = 0.75/100 = 7.5 \times 10^{-3}$ m.

Table 17.6 Characteristics of the pre-treated effluent and specifications of UV-lamp

Parameter	Value	Units
Flow	6	m^3/h
pH	6.6	Unitless
Transmittance	90	% for an optical path of 0.75 cm
Turbidity	<1	NTU
Colour	<15	Colour units
Bicarbonate	35	mg/L
UV lamp specifications (Philips-927970204099)		
Electric power	40	Watt
UV power (254 nm)	15	Watt
Length of UV-lamp	1.6	m
Diameter of UV-lamp	19	mm

UV—Ultra violet, NTU—Nephelometric turbidity unit

Diameter of quartz sleeve $(d_{qs}) = 3 \times 10^{-2}$ m
Quartz sleeve transmittance $(\%T_{qs}) = 90\%$
Initially, it is necessary to calculate $\frac{[C]}{[C_0]}$ ratio by using Eq. 17.48.

$$\frac{[C]}{[C_0]} = 1 - \frac{Removal\,Efficiency(\%)}{100} = 1 - \frac{99}{100} = 0.01$$

With this $\frac{[C]}{[C_0]}$ ratio, spatial retention time can be estimated by using Eq. 17.47.

$$t = -\frac{\ln([C]/[C_0])}{k\prime} = -\frac{\ln(0.01)}{9 \times 10^{-2}} = 51.17\ sec$$

Now, using spatial retention time, the reactor volume can be obtained by using Eq. 17.45.

$$V = Q.t = 1.66 \times 10^{-3} \times 51.17 = 0.085\ m^3$$

The reactor equivalent length is obtained by using Eq. 17.54.

$$L_{eq} = \frac{V}{\pi(d_{qs}.OP + OP^2)} = \frac{0.085}{3.1416 \times \left(3 \times 10^{-2} \times 7.5 \times 10^{-3} + \left(7.5 \times 10^{-3}\right)^2\right)}$$

$$= 96.2\ m$$

Now, by using L_{eq}, UV exposed area can be calculated by using Eq. 17.55.

$$S_{exp} = L_{eq}.\pi.(d_{qs} + 2.OP)$$
$$= 96.2 \times 3.1416 \times (3 \times 10^{-2} + 2 \times 7.5 \times 10^{-3}) = 13.6 \text{ m}^2$$

The applied UV power is obtained by using Eq. 17.56. Let us assume that the UV lamp efficiency is lost by 20%. The effective irradiance is obtained using Eq. 17.57.

$$I_{effective} = \frac{D}{t} = \frac{2000}{51.17} = 39.09 \text{ W/m}^2$$

$$Applied_{UV_{power}} = \frac{100^3 \times I_{effective}S_{exp}}{\%T_{qs} \times \%T_{effluent} \times (100\% - \%UV_{el})}$$
$$= \frac{100^3 \times 39.09 \times 13.6}{90 \times 90 \times (100 - 20)} = 820.41 \text{ W}$$

The total number of UV lamp and quartz sleeve, can be estimated as per Eq. 17.58.

$$N_{UVlamps} = N_{qs} = \frac{Applied_{UV_{power}}}{Lamp_{UV_{output}}} = \frac{820.41}{15} = 54.7$$

The number UV lamps required for a single reactor is about 55, which might increase the temperature of the reactor. Therefore, it is recommended to use $\frac{55}{8} = 6.88 \sim 7$ reactors in parallel with 8 UV lamps each in a single reactor.

The total reactor volume obtained by Eq. 17.59.

$$V_{total} = V + \frac{N_{qs}.\pi.(d_{qs})^2 L_{UV-net}}{4} = (0.085/7) + \frac{8 \times 3.1416 \times (3 \times 10^{-2})^2 \times 1.6}{4}$$

$$= 0.021 \text{ m}^3$$

The diameter of the UV reactor (d_R) required can be worked as per Eq. 17.60.

$$d_R = \sqrt{\frac{4V_{total}}{\pi L_{UV-net}}} = \sqrt{\frac{4 \times 0.021}{3.1416 \times 1.6}} = 0.129 \text{ m}$$

Now, the ratio of $\frac{L_{UV-net}}{d_R} = \frac{1.6}{0.129} = 12.4$, which is higher than the recommended value of 3. Therefore, the resulting reactor configuration seems much suitable for the treatment of wastewater. This is because the distribution of UV light will be more uniform, which will provide more effective irradiance and also enhance the oxidation process (Fig. 17.5).

Evaluation of total cost per order of wastewater treatment.

Here, H_2O_2 price = $ 1.5 per kg, electricity cost = $ 0.1 per kWh, UV lamp power $(P) = 0.015$ kW, H_2O_2 dosage = $Q \times 10$ g/m^3 = 6 m^3/h $\times$ 10 g/m^3 = 60 g/h.

Total H_2O_2 consumption = 60/3600 $\times$ 51.17 = 0.85 g.

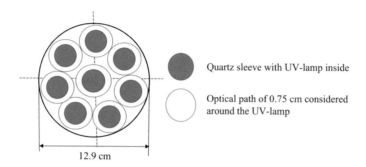

Fig. 17.5 Final arrangement for UV-lamps and quartz sleeves inside the reactor

Treatment process	UV/H$_2$O$_2$
k$'$ (h^{-1})	324
EE$_{UV}$ (kWh/m^3)	1.185
Cost$_{UV}$ ($/m^3)	0.117
Cost$_{ox}$ ($/m^3)	0.015
Cost$_{total}$ ($/m^3)	0.132

Table 17.7 Total cost analysis of UV/H$_2$O$_2$ treatment system

The total cost of the treatment system is measured as per Eqs. 17.61 and 17.66, which is shown in Table 17.7. $k' = 0.09$ per second $= 324$ per hour

$$Cost_{UV}(\$/m^3) = \frac{38.4 \times P}{Vk'} \times electricity\ cost = \frac{38.4 \times 0.015 \times 7 \times 8}{0.085 \times 324} \times 0.1 = 0.117$$

$$Cost_{OX}(\$/m^3) = \frac{total\ oxidant\ consumped}{V} \times oxidant\ cost = \frac{0.00085 \times 1.5}{0.085} = 0.015$$

$$Cost_{total}(\$/m^3) = Cost_{UV} + Cost_{OX} = 0.117 + 0.015 = 0.132$$

17.4.6 Applications of Various UV-Based AOPs

Application of various UV-based AOPs for treating a wide variety of toxic organic contaminants have been reported by many researchers. In an investigation, the degradation of Tyrosol (C_o of 0.5 mM), a phenolic compound generally found in olive mill wastewater, was observed through UV/S$_2$O$_8$$^{2-}$ process. The result of the investigation depicted that by using 640 mJ/cm^2 UV fluence and S$_2$O$_8$$^{2-}$ dosage of 2 mM and pH of 6.8, almost complete degradation of Tyrosol was achieved (Kilic et al., 2019). In another investigation, the degradation of Sucralose, an artificial sweetener was examined by O_3 and UV/O_3 system. This research demonstrated about 89.8% mineralization of Sucralose (C_o of 50 mg/L) through the process of UV/O_3 within 120 min by applying UV intensity of 33.4 W/m^2 and O_3 dosage of 19.4 mg/L at neutral pH. While under the same reaction conditions, only 39.1% mineralization of Sucralose was achieved in sole O_3 process (Xu et al., 2018). Similarly, in another investigation, the degradation of Amitriptyline hydrochloride, a tricyclic antidepressant, was examined by UV/Cl_2 process. The result showed that almost 90% degradation of Amitriptyline hydrochloride was achieved within 7 min by using an UV-C lamp and Cl_2 dosage of 10 μM at neutral pH (Javier Benitez et al., 2017). Performance of other investigations carried out by using various UV-based AOPs are summarized in Table 17.8.

17.4.7 Scaling-Up of UV-Based AOPs

The UV-based AOPs have been extensively investigated at lab-scale for the removal of organic pollutants; however, till date very few investigations have been undertaken for the commercial applicability of this technology. For example, an UV/$SO_4{}^{\cdot-}$ field-scale treatment unit having a flow rate of 12,000 m^3/day was scaled-up for the tertiary treatment of wastewater in Estiviel WWTP located in the city of Toledo, Spain (Rodríguez-Chueca et al., 2018). Moreover, the operating cost of UV/$S_2O_8{}^{2-}$ and UV/$HSO_5{}^-$ was compared with the UV/H_2O_2 process to determine the best option for the routine treatment in the WWTP. The result of the field tests achieved 55% removal of micropollutants within 18 sec through UV/H_2O_2 process by using 0.5 mM of H_2O_2. While under same operating conditions, 48% and 10% removal efficiency was achieved through UV/$HSO_5{}^-$ and UV/$S_2O_8{}^{2-}$ processes, respectively (Rodríguez-Chueca et al., 2018). Additionally, UV/$HSO_5{}^-$ (for 0.5 mM; 48%, 0.585 €/m^3) and UV/$S_2O_8{}^{2-}$ (for 0.5 mM; 10%, 0.090 €/m^3) treatments required higher operating costs because of high cost of oxidants in comparison to H_2O_2 (for 0.5 mM; 55%, 0.035 €/m^3) (Rodríguez-Chueca et al., 2018). The performance of other field- and pilot-scale investigations on UV-based AOPs for wastewater treatment are summarized in Table 17.9.

17.5   Ozone-Based AOPs

Ozone (O_3) has been traditionally utilized as an oxidant and disinfectant for the treatment of wastewater. In this process, abatement of organic pollutants occurs through two different pathways: (a) direct reaction by O_3 molecules and (b) indirect reaction by OH$^{\cdot}$ radical (Fig. 17.6). In direct reaction, O_3 molecule attacks the organic pollutants through a series

Table 17.8 Organic contaminants removal by various UV-based AOPs

Process	Target pollutant	C_o (mg/L)	Experimental conditions	Efficiency (%)	References
UV/HSO$_5^-$	Di-(2-ethylhexyl) phthalate	10	HSO$_5^-$ dosage = 100 mg/L, UV intensity = 15.3 mW/cm^2, Time = 70 sec	90	Huang et al., (2017)
UV/HSO$_5^-$	Ciprofloxacin	16.6	HSO$_5^-$ dosage = 1 mM, UV fluence rate = 9.3×10^{-7} Einstein/sec, Time = 60 sec	97	Mahdi-Ahmed and Chiron (2014)
UV/S$_2$O$_8^{2-}$	Chloramphenicol	10.01	S$_2$O$_8^{2-}$ dosage = 0.5 mM, UV fluence rate = 2.43 mW/cm^2, Time = 60 sec	100	Ghauch et al., (2017)
UV/H$_2$O$_2$	Estrone	0.125	H$_2$O$_2$ dosage = 0.5 mol/L, UV fluence rate = 0.069 mW/cm^2, Time = 2 min	97	Zhang and Li (2014)
UV/H$_2$O$_2$	Methyl orange	10	H$_2$O$_2$ dosage = 300 mg/L, UV fluence rate = 6.30×10^{-5} mol/cm^2.sec, Time = 120 sec	72 COD	Navarro et al., (2019)
UV/H$_2$O$_2$	2,4-Dichlorophenoxyacetic acid	100	H$_2$O$_2$ dosage = 2.5 mol, UV fluence rate = 700 mJ/cm^2, Time = 6 min	97	Adak et al., (2019)
UV/O$_3$	p-Nitroaniline	10	O$_3$ dosage = 0.9 g/h, UV fluence rate = 254 mW/m^3, time = 40 min	81	Rajabizadeh et al., (2020)

(continued)

of reactions, such as electrophilic substitution reaction, nucleophilic reaction, oxidation–reduction reaction and cycloaddition reaction. Furthermore, in direct ozonation, O$_3$ with E_0 of 2.07 V (vs. NHE) very selectively attacks some classes of organic compounds containing activated aromatic rings, double bonds and amines as shown in Eq. 17.67. However, the formation of OH$^·$ radical through the indirect ozonation process usually occurs through radical chain reactions (Eqs. 17.68–17.72).

$$O_3 + HO_2^- \rightarrow O_3^{·-} + HO_2^·$$

(17.67)

Table 17.8 (continued)

Process	Target pollutant	C_o (mg/L)	Experimental conditions	Efficiency (%)	References
UV/O$_3$	Caffeine	5	O$_3$ Dosage $= 1$ g/h, UV fluence rate $= 0.7$ mW/cm^2, Time $= 22.5$ min	95	Souza and Féris (2015)
UV/O$_3$	Ketoprofen	25.4	O$_3$ Dosage $= 4.93 \times 10^{-6}$ M, UV fluence rate $= 5.7 \times 10^{-6}$ Einstein/sec, Time $= 3$ min	100	Illés et al., (2014)
UV/Cl$_2$	Ciprofloxacin	0.34	Cl$_2$ Dosage $= 0.1$ mM, UV-C lamp, Time $= 2$ min	99.7	Yang, Li, et al. (2019)a, Yang, Zhou, et al., 2019)
UV/Cl$_2$	Clofibric acid	10	Cl$_2$ Dosage $= 1$ mM, UV fluence rate $= 0.1$ mW/cm^2, Time $= 120$ min	85.8	Lu et al., (2018)
UV/Cl$_2$	Phenacetin	1.8	Cl$_2$ Dosage $= 300$ μM, UV fluence rate $= 0.059$ mW/cm^2, Time $= 20$ min	75.4	Zhu et al., (2018)

UV—Ultra violet, HSO$_5{}^-$—Peroxymonosulfate, S$_2$O$_8{}^{2-}$—Persulfate, H$_2$O$_2$—Hydrogen peroxide, O$_3$—Ozone, Cl$_2$—Chlorine, COD—Chemical oxygen demand

$$O_3 + OH^- \rightarrow HO_4^- \qquad (17.68)$$

$$HO_4^- \leftrightarrow HO_2^\cdot + O_2^{\cdot-} \qquad (17.69)$$

$$O_2^{\cdot-} + O_3 \rightarrow O_2 + O_3^{\cdot-} \qquad (17.70)$$

$$O_3^{\cdot-} \rightarrow O_2 + O^{\cdot-} \qquad (17.71)$$

$$O^{\cdot-} + H_2O \rightarrow OH^\cdot + OH^- \qquad (17.72)$$

Ozonation is a pH dependent process because at low pH (pH < 4.0) direct ozonation dominates the reaction process; whereas, under alkaline conditions (pH > 9.0) indirect ozonation prevails. The rate of degradation of organic pollutants increases under basic conditions because basic pH favours the decomposition of O$_3$ molecules and generates a

Table 17.9 Performance of pilot-scale UV-based AOPs for wastewater treatment

Process	Target pollutant	C_o (mg/L)	Experimental conditions	Efficiency (%)	References
UV/H_2O_2	Sulfolane	12.6	Volume = 5.6 L, Flow rate = 46 L/min, time = 500 min, H_2O_2 dosage = 40 mg/L	95	Yu et al., (2020)
UV/H_2O_2	Acid orange 10	26.9	Volume = 100 L, Flow rate = 9.32 m^3/day, time = 500 min, H_2O_2 dosage = 40 mg/L	99	Shu and Chang (2005)
UV/H_2O_2	Natural organic matter	20–26 COD_0	Volume = 7.6 L, Flow rate = 2000 L/h, time = 120 sec, H_2O_2 dosage = 4 mM, UV fluence rate = 480 $\mu W/cm^2$	–	Audenaert et al., (2011)
UV/O_3	Post-treatment of biologically treated wastewater	–	Volume = 38 m^3, Flow rate = 4 L/min, time = 28 h, O_3 dosage = 0.045 g O_3/g DOC, UV fluence rate = 6.6 J/cm^2	100	Krakkó et al., (2021)
UV/O_3	Polycyclic aromatic hydrocarbons	–	Volume = 1.3 m^3, Flow rate = 1800 m^3/day, time = 8 h, O_3 dosage = 239.1 ± 1.2 g/h	75	Lin et al., (2014)

UV—Ultra violet, H_2O_2—Hydrogen peroxide, O_3—Ozone, COD_0—Initial chemical oxygen demand, DOC—Dissolved organic carbon

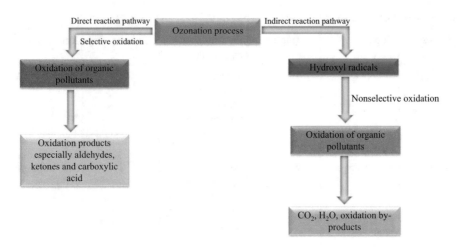

Fig. 17.6 Reaction pathway of ozonation process

high quantity of reactive radicals as compared to acidic conditions. Moreover, O_3 has a ε of 3300 M^{-1} cm^{-1}, which is significantly higher than that of ε of H_2O_2 (18 M^{-1} cm^{-1}) at $\lambda = 254$ nm (Miklos et al., 2018).

The O_3 is generated on-site by using a device called ozone generator, which converts O_2 from various sources such as dry air, ambient air or concentrated O_2 into O_3. The ozone generators produce O_3 by disintegrating the stable O_2 molecules into two oxygen atoms (O), which recombine with other O_2 molecules to form O_3. The O_3 is then used for the degradation of bio-refractory organic and inorganic contaminants and water disinfection. The requirement of O_3 dose for wastewater treatment and the output of an ozone generator can be determined as per Eq. 17.73 (Rakness, 2011).

$$\text{Flow rate (Lpm)} \times O_3 \text{ concentration} \left(\frac{g}{m^3}\right) \times \frac{1m^3}{1000\,L} \times 60\frac{min}{h} = O_3 \text{ required (g/h)}$$

$$(17.73)$$

where, Lpm is the litres per minute.

17.5.1 Major Benefits and Drawbacks of Ozonation

The primary advantage of ozonation process is its ability to eliminate impurities as well as odour and colour from wastewater. However, this process has certain drawbacks like high operational cost due to poor utilization of gaseous O_3 and generation of toxic DBP due to selective interaction of O_3 with pollutants that leads to the incomplete oxidation of some of the refractory compounds. Therefore, to deal with the problems of poor utilization of

O_3 and incomplete mineralization of organic pollutants, in the recent past O_3 has been combined with other catalysts (e.g., TiO_2, Cu-TiO_2, Al_2O_3 and MgO), H_2O_2 and UV to intensify the rate of degradation.

17.5.2 O_3/H_2O_2

The integration of O_3 with H_2O_2 is commonly known as Peroxone process, where initially H_2O_2 dissociates into water and hydroperoxide ions (HO_2^-) that react rapidly with O_3 to form HO_5^- as shown in Eqs. 17.74 and 17.75. Subsequently, HO_5^- decomposes into $HO_2^{\cdot}$ and $O_3^{\cdot-}$ (Eq. 17.76) and finally the decaying of $O_3^{\cdot-}$ gives rise to $OH^{\cdot}$ radicals following Eqs. 17.71 and 17.72. The residual concentration of H_2O_2 in the solution might need to be eliminated before discharging the effluent in to the natural water bodies.

$$H_2O_2 \leftrightarrow H^+ + HO_2^- \tag{17.74}$$

$$HO_2^- + O_3 \leftrightarrow HO_5^- \tag{17.75}$$

$$HO_5^- \rightarrow HO_2^{\cdot} + O_3^{\cdot-} \tag{17.76}$$

Major benefits and drawbacks O_3/H_2O_2 process: The O_3/H_2O_2 process offers significant advantages over the conventional AOPs. This process shows higher germicidal property than individual O_3 and chlorination process and also has a high potential of generating $OH^{\cdot}$ radicals in comparison to O_3 and H_2O_2/UV process. However, apart from these significant advantages, O_3/H_2O_2 process possesses drawbacks, such as potential of bromate formation, requirement of post treatment to remove residual H_2O_2 and requirement of off-gas treatment system for ozone destruction, which could enhance the operational cost.

17.5.3 Catalytic Ozonation

Researchers have demonstrated that catalytic ozonation process has gained attention as an effective alternative to the conventional ozonation process. In this process, the addition of either homogeneous or heterogeneous catalysts can improve the decomposition of O_3 to generate $OH^{\cdot}$ radical. The homogenous catalysts, such as metal or transition metal ions (e.g., Zn^{2+}, Mn^{2+}, Co^{2+}, Cu^{2+} and Fe^{3+}), are generally used to decompose O_3 molecules to form reactive radicals as presented in Eq. 17.77 through Eq. 17.80.

$$Mn^{2+} + O_3 + H^+ \rightarrow Mn^{3+} + OH^{\cdot} + O_2 \tag{17.77}$$

$$O_3 + OH^{\cdot} \rightarrow O_2 + HO_2^{\cdot} \tag{17.78}$$

$$Mn^{3+} + HO_2^{\cdot -} + OH^- \rightarrow Mn^{2+} + H_2O + O_2 \tag{17.79}$$

$$Mn^{2+} + OH^{\cdot} \rightarrow Mn^{3+} + OH^- \tag{17.80}$$

The metal oxide (e.g., TiO_2, MnO_2, Al_2O_3 and MgO) and metal/metal oxide supports (e.g., Re-CeO_2, Cu-TiO_2, Cu-Al_2O_3,) are most widely used as heterogeneous catalysts to decompose the O_3 molecules. The interaction of heterogeneous catalysts with O_3 and organic contaminants occurs through three typical pathways. In the first pathway, O_3 gets absorbed on to the catalyst surface and subsequently decomposed to form reactive radicals, which then react with the organic contaminants. In the second pathway, the organic contaminants absorbed on the surface of catalysts are attacked by O_3 and other reactive species present in the wastewater. Similarly in the third pathway, both organic contaminants and O_3 are absorbed on the catalysts surface and then they react with each other. However, heterogeneous catalysts are usually preferred compared to homogenous catalysts, since heterogeneous catalysts could be easily recovered even after several treatment cycles.

Major benefits and drawbacks of catalytic ozonation process: Compared to traditional ozonation process, catalytic ozonation process is more economical as it does not require UV or pH adjustment and also due to its effectiveness over a wide pH range. However, the treatment of residuals, regeneration of catalysts, and complex surface reactions between O_3 and catalysts are the major challenges associated with this technology. Moreover, the addition of metal ions might result in the generation of secondary pollution, which might limit their successful field-scale implementation.

17.5.4 Applications of Catalytic Ozonation Process

The catalytic ozonation process has shown excellent performance in both water and wastewater treatment. The catalytic ozonation of toxic organic contaminants, including nitrobenzene, phenol, pharmaceuticals and pesticides has been extensively investigated. Liu et al. investigated the degradation and mineralization of dimethyl phthalate by heterogeneous catalytic ozonation process using Cu–Fe–O nanoparticles as catalyst. The result of the investigation showed complete degradation and 50.1% mineralization of dimethyl phthalate within 20 min through Cu–Fe–O assisted ozonation process; whereas, 93.9% degradation and 18.6% mineralization of dimethyl phthalate was obtained within 120 min by sole O_3 process (Liu et al., 2019). The results indicated that Cu–Fe–O nanoparticles improved the degradation and mineralization of dimethyl phthalate, which was mainly due to the amount of OH˙ radicals generated in the catalytic ozonation process were much higher than that of O_3 at pH of 5.7.

Similarly in another investigation, the catalytic ozonation of 1-Amino-4-bromoanthraquinone-2-sulfonic acid (BBA; C_o of 50 mg/L) was observed by using Mn-CeO$_x$/Al$_2$O$_3$ as a catalyst (Wu et al., 2018). Outcome of this investigation demonstrated almost complete mineralization of BBA within 120 min by employing 1 g/L of Mn-CeO$_x$/Al$_2$O$_3$ and 20 mg/L of O_3 concentration, while under the same operating conditions 32.7% removal of BBA was achieved in absence of a catalyst. Other researchers have also reported the successful application of catalytic oxidation for the removal of organic pollutants, which are summarized in Table 17.10.

17.5.5 Scaling-Up of Various Ozone-Based AOPs

Ozone has been demonstrated to be a highly effective for the disinfection and mineralization of organic contaminants at both laboratory-scale and field-scale demonstrations. Recently, a field-scale wastewater treatment plant based on catalytic ozonation process having an effluent flow rate of 100 m^3/h was scaled-up to treat the biologically treated coking wastewater. The field-scale system demonstrated COD removal of 45.6%, which was obtained within 90 min by using 180 g/L of Mn$_x$Ce$_{1-x}$O$_2$/γ-Al$_2$O$_3$ catalyst and ozone concentration of 80 mg/L (He et al., 2020). Similarly, another field-scale wastewater treatment plant having volume of 350,000 m^3/day was operated to remove pathogens and organic contaminants from domestic and industrial wastewaters produced in the city of Melbourne, Australia. The results of the field tests demonstrated that more than 98% and 69% of organic pollutants and pathogens, respectively, were removed at ozone dose of 1 mg of O_3 mg^{-1} dissolved organic carbon (Blackbeard et al., 2016).

Example: 17.4 How much O_3 production is essential to dose 2 g/m^3 into 22 Lpm of wastewater flow?

Solution
Flow rate $= 22$ Lpm, O_3 dosage $= 2$ g/m^3.
The production of O_3 is measured as per Eq. 17.73.

$$22\frac{L}{min} \times 2\frac{g}{m^3} \times \frac{1\ m^3}{1000\ L} \times 60\frac{min}{h} = 2.64\frac{g}{h}$$

Therefore, 2.64 g/h of O_3 dose is required to the water to have 2 g/m^3 of ozone.

Example: 17.5 The O_3 concentration exiting an ozone generator is 120 g/m^3 at 6 Lpm of O_2 flow. What is the output of the ozone generator in g of O_3/h?

Table 17.10 Organic contaminants removal by various ozonation process

Process	Target pollutant	C_o (mg/L)	Type of catalyst	Operating conditions	Time (min)	Efficiency (%)	References
Ozonation	Ofloxacin	20	–	O_3 dosage = 15 mg/L	60	55 TOC	Chen and Wang (2021)
	Methylene Blue	400	–	O_3 dosage = 24 g/m^3	12	100	Turhan et al., (2012)
	Brilliant Green	13.31	–	O_3 dosage = 1.7 mg/sec	60	80 TOC	Khuntia et al., (2015)
Heterogeneous catalytic ozonation	Diclofenac	15	Fe-MCM-48	O_3 dosage = 100 mg/L, catalyst dosage = 0.15 g/L	60	50 TOC	Li et al. (2018)
	Phenol	200	Fe–Mn/granular activated carbon	O_3 dosage = 22 mg/L, catalyst dosage = 1 g/L	80	100 COD	Xiong et al., (2019)

(continued)

Table 17.10 (continued)

Process	Target pollutant	C_O (mg/L)	Type of catalyst	Operating conditions	Time (min)	Efficiency (%)	References
	Acid Orange 7	100	$CuAl_2O_4$	O_3 dosage = 10.06 mg/min, catalyst dosage = 0.5 g/L	120	96	Xu et al., (2019)
	Atrazine	2	g-C_3N_4	O_3 dosage = 5 mg/min, catalyst dosage = 0.5 g/L	15	92.91	Yuan et al., (2019)
Homogenous catalytic ozonation	Phenol	254	Ce^{3+} catalyst	O_3 dosage = 130 mg/L, catalyst dosage = 50 mM	120	71 TOC	Matheswaran et al., (2007)
	Reactive Red 2	276.9	Fe^{2+} catalyst	O_3 dosage = 2 mg/L, catalyst dosage = 0.9 mM	60	100 COD	Zhang et al., (2013)
	m-Dinitrobenzene	168.1	Mn^{2+} catalyst	O_3 dosage = 12 mg/L, catalyst dosage = 1 g/L	120	80 COD	Trapido et al., (2005)

Fe-MCM-48—Iron-Mobil Composition of Matter No 48, Fe–Mn—Iron-Manganese, $CuAl_2O_4$—Copper aluminate, g-C_3N_4—Graphitic carbon nitride, Ce^{3+}—Cerium, Fe^{2+}—Ferrous salt, Mn^{2+}—Manganous ion, O_3—Ozone, TOC—Total organic carbon, COD—Chemical oxygen demand

Solution

Flow rate of $O_2 = 6$ Lpm, O_3 concentration $= 120 \, g/m^3$.

The output of an ozone generator can be measured as per Eq. 17.73.

$$6 \frac{L}{min} \times 120 \frac{g}{m^3} \times \frac{1 \, m^3}{1000 \, L} = 0.72 \frac{g}{min}$$

Therefore, the ozone generator will produce $0.72 \times 60 = 43.2$ g/h of O_3 at 6 Lpm of O_2 flow.

17.6   Emerging AOP Technologies

The UV, O_3 and Fenton-based AOPs are already well established and operated at field-scale for drinking water treatment and water reuse purposes. However, in recent years, other emerging AOPs, such as gamma radiation, cavitation, non-thermal plasma and electrochemical-based AOPs, have been gaining attraction among researchers. In the following section the major mechanisms, advantages, disadvantages and principles of radical generation by the emerging technologies are briefly discussed.

17.6.1 Gamma Radiation-Based AOP

Gamma radiation-based AOP has been utilised to remove persistent pollutants and to disinfect treated water and sludge. In this process, gamma radiation is used to cause radiolysis of water molecules to produce reactive species including hydrate electrons (e_{aq}^-), hydronium ion (H_3O^+) and $OH^{\cdot}$, etc. (Eqs. 17.81 and 17.82). The Co-60 and Cs-137 radioisotopes are most commonly used as sources to generate high intensity of ionizing radiation for wastewater treatment and disinfection. The generated reactive species further reacts with the organic pollutants and convert them into CO_2 and H_2O through the oxidation process. However, the generation of reactive species depends on the solution pH. Under acidic pH (pH < 4.0), e_{aq}^- reacts with H^+ ion to form hydrogen radical ($H^{\cdot}$) through diffusion-controlled reaction, whereas in strong basic medium, $H^{\cdot}$ is converted to e_{aq}^-.

$$H_2O + \gamma \rightarrow OH^-(0.29), \, H^{\cdot}(0.06), \, e_{aq}^-(0.28), \, H_2(0.047), \, H_2O_2(0.07), \, H_3O^+(0.27)$$
$$(17.81)$$

$$H_2O_2 + \gamma \rightarrow 2OH^{\cdot}(0.29 + 0.29 = 0.58) \qquad (17.82)$$

The values written in the brackets represent the radiochemical yield (G-values, μmol/J) of the respective species, which quantifies the number of free radicals, molecules or atoms produced or consumed per 100 eV of absorbed energy and it is given by Eq. 17.83 (Rivas-Ortiz et al., 2017).

$$G = \frac{6.023 \times 10^{23} C}{D \times 6.24 \times 10^{16}} \qquad (17.83)$$

where, C is the concentration of reactive radicals produced (mg/L) and D is the gamma radiation absorbed dose (Gray (Gy)).

17.6.1.1 Major Benefits and Drawbacks

The major advantages of this technology are: (i) unlike UV radiation, gamma rays are highly penetrating that makes the technology more efficient in treating multi-component waste streams without further addition of any chemical reagents, (ii) less generation of by-products compared to other AOPs, and (iii) high process reaction rate and efficiency due to production of variety of reactive species. However, the process has two obvious drawbacks, such as: (i) this system requires gamma irradiators or accelerators for safe irradiation that in turn enhances the capital costs, and (ii) it requires skilled operators and protective measures to handle gamma ray.

17.6.1.2 Scaling-Up of Gamma-Based Technology

The application of gamma radiation-based AOP has been extensively investigated at a lab-scale (Table 17.11); however, limited literature is available addressing the commercial applicability of this technology. A field-scale treatment unit based on gamma radiation technology was scaled up in India at Sludge Hygienization Research Radiator (SHRI), Vadodara to disinfect the sewage sludge having a flow rate of 110 m³/day. It was observed that at a radiation dose of 3 kGy with Co-60 gamma source, more than 99.9% of disease-causing bacteria were destructed from the sewage sludge. Similarly, other countries, including Germany, Italy, Canada and United States have successfully adopted gamma radiation-based treatment systems to disinfect the sludge prior to the agricultural land application (Swinwood et al., 1994).

17.6.2 Cavitation-Based AOP

In this process, radicals are generated by the application of ultrasonic waves or by creating high velocity and pressure variations in the aqueous medium, which result in the formation of microbubbles or cavities in the cavitation system. The generated microbubbles undergo series of compression and rarefaction cycles in the liquid medium, which grows and then collapses by releasing large magnitude of energy in the form of high temperature and pressure. Based on the source of formation of microbubbles in the cavitation chamber, the cavitation process is classified into two types, such as (a) acoustic or ultrasonication

Table 17.11 Organic contaminants removal by different emerging AOP technologies

Process	Pollutant	C_o (mg/L)	Experimental conditions	Efficiency (%)	References
Hydrodynamic cavitation	Rhodamine B	20	Pressure = 3 bar, time = 150 min	91.11	Li et al., (2020)
Hydrodynamic cavitation	Dicofol	50	Pressure = 7 bar, time = 60 min	85	Panda and Manickam (2019)
Hydrodynamic cavitation	Pentachlorophenol	10	Pressure = 5.77 bar, time = 60 min	100	Choi et al., (2020)
Hydrodynamic cavitation	Naproxen	10	Pressure = 4 bar, time = 40 min	100	Thanekar et al., (2020)
Acoustic cavitation	Tylosin	10	US frequency = 40 kHz, time = 40 min	76.5	Yousef Tizhoosh et al., (2020)
Acoustic cavitation	Ampicillin	10.5	US frequency = 375 kHz, time = 60 min	80	Montoya-Rodríguez et al., (2020)
Acoustic cavitation	Methylene blue	5	US frequency = 35 kHz, time = 50 min	98	Sun et al., (2020)
Acoustic cavitation	Benzenesulfonic acid	15.81	US frequency = 350 kHz, time = 275 min	50 TOC	Thomas et al., (2020)
Electro-Fenton process	Ofloxacin	36.1	Current density = 9.37 mA/cm², Boron doped diamond anode, time = 480 min	100	Yang et al., (2020)
Electro-Fenton process	Imatinib	34.5	Current density = 16.66 mA/cm², dimensionally stable anode, time = 480 min	100	Yang, Zhou, et al. (2019))

(continued)

(US) and (b) hydrodynamic cavitation (HC). In acoustic cavitation, the microbubbles are formed by using US frequency ranging from 20 kHz to 2 MHz; while in HC, microbubbles are generated by the application of high velocity or pressure variations in the stream of flowing liquid as shown in Eq. 17.84 through Eq. 17.88.

$$H_2O + US/HC \rightarrow OH^. + H^. \tag{17.84}$$

Table 17.11 (continued)

Process	Pollutant	C_o (mg/L)	Experimental conditions	Efficiency (%)	References
Anodic oxidation	Acid orange 7	100	Current density = 20 mA/cm², Fe-doped PbO₂ anode, time = 60 min	87.15 COD	Xia et al., (2020)
Anodic oxidation	Sulfadiazine	50	Current density = 10 mA/cm², anode = Titanium suboxide mesh, time = 60 min	100	Teng et al., (2020)
Gamma radiation	Reactive Blue-19	50	Radiation absorbed dose = 12 kGy	60	Arshad et al., (2020)
Gamma radiation	p-Nitrophenol	50	Radiation absorbed dose = 9 kGy	100	Yu et al., (2010)
Gamma radiation	Sulfadiazine	25	Radiation absorbed dose = 1 kGy	100	Rivas-Ortiz et al., (2017)
Gamma radiation	Indigo carmine	150	Radiation absorbed dose = 5 kGy	96	Zaouak et al., (2018)

US—Ultra sound, Fe-doped PbO₂—Iron doped lead dioxide, TOC—Total organic carbon, COD—Chemical oxygen demand

$$OH^. + \text{organic pollutants} \ \rightarrow CO_2 + H_2O \qquad (17.85)$$

$$OH^. + OH^. \rightarrow H_2O_2 \qquad (17.86)$$

$$H^. + O_2 \rightarrow O_2^{.-} + H^+ \qquad (17.87)$$

$$O_2^{.-} + \text{Organic pollutants} \ \rightarrow CO_2 + H_2O \qquad (17.88)$$

The intensity of cavitation depends upon various operating parameters, such as inlet pressure, oxidant dosage, temperature, vapour pressure, surface tension and viscosity of the liquid. The cavitation is most often defined by the cavitation number (C_v), which is a dimensionless number utilized for measuring the intensity of cavitation process through the Eq. 17.89.

$$C_V = \frac{P_2 - P_V}{\frac{1}{2} \times \rho V_0^2} \tag{17.89}$$

where, P_2 is the recovered pressure downstream of the constriction (Pascal), P_v is the vapour pressure of the liquid (Pascal), ρ is the density of water (kg/m^3) and V_0 is the velocity at the throat of the cavitating constriction (m/sec).

With the rise in inlet pressure the velocity of the liquid also increases, which therefore decreases the cavitation number. Hence, smaller is the cavitation number greater will be the number of cavitation bubbles generated in the system. However, beyond the optimum inlet pressure, the generation of reactive radical reduces due to the formation of cavity cloud, which ultimately decreases the collapse intensity of the individual cavity.

The field-scale implementation of this process majorly depends on the economics of degradation, which can be estimated on the basis of factors like cavitation yield, the time required for degradation and energy consumption. In the case of HC, pump is used for the inception of the microbubbles in a cavitation reactor, which is the most expensive component of the system that can enhance the operating cost during scaling-up. The power required and energy consumption for pumps can be estimated by using Eqs. 17.90 and 17.91.

$$Pump\ power\ (kW) = \frac{\Delta P Q}{\eta} \tag{17.90}$$

where, ΔP is the pressure drop (Pascal), Q is the volumetric flow rate (m^3/sec) and η is the efficiency.

$$Energy\ consumption\left(\frac{kWh}{m^3}\right) = \frac{Pump\ power\ (kW) \times treatment\ time\ (h)}{treatment\ volume\ (m^3)} \tag{17.91}$$

The energy efficiency of HC process is also obtained based on the cavitation yield, which is defined as the ratio of the amount of contaminants degraded to the energy consumed by the system and it can be expressed as per Eq. 17.92 (Panda & Manickam, 2019).

$$Cavitation\ yield\left(\frac{mg}{J}\right) = \frac{Amount\ of\ contaminant\ degraded\left(\frac{mg}{L}\right)}{\frac{P_m \times t}{V}} \tag{17.92}$$

where, P_m is the pump power (W), t is the operation time (sec) and V is the volume of the wastewater (L).

The synergetic effect (f) is an effect seen when two or more AOPs combined to create an effect that is greater than the addition of their separate effects (Eq. 17.93) and it can be estimated based on the kinetic rate constants (K), COD and TOC of the combined and individual processes. If value of f is greater than 1, that means it indicate the percentage removal of contaminant by the combined system is greater than that sum of the individual systems due to synergy between them.

$$f = \frac{K_{hybrid\ process}}{K_{process1} + K_{process2}} \tag{17.93}$$

17.6.2.1 Major Benefits and Drawbacks of Cavitation Process

The cavitation process possesses distinct advantages as compared to other commercially available wastewater treatment methods, such as: (i) limited use of oxidants, like O_3, H_2O_2 etc., (ii) works efficiently in both acidic and basic medium, (iii) disinfects the wastewater efficiently, and (iv) eliminates the mass transfer limitation by creating microbubbles and turbulence in the solution. However, despite of these advantages, cavitation process has certain limitations. (i) In acoustic cavitation, the large volume of wastewater can absorb the ultrasonic waves generated by the sonic transducers, thus lowering the energy received from cavitation per unit volume of wastewater being treated. Therefore, for treatment of large volume of wastewater greater number of sonic transducers are required, which will increment the cost of operation and energy consumption. (ii) The violent nature of cavitation can cause erosion in the inner surface of the cavitation chamber, which raises the maintenance costs.

17.6.2.2 Scaling-Up of Cavitation-Based Technology

The effectiveness of cavitation processes has been demonstrated by many researchers at both bench-scale and pilot-scale (Table 17.11). To the best of our knowledge, till now there is no such large-scale treatment plant based on acoustic cavitation process that exist; however, only few large-scale treatment plants were noticed so far for HC process. A wastewater treatment plant based on HC system having flow rate of 54–163 KLD was scaled-up at March Air Force Base, Riverside, California to remove methyl tert-butyl ether from wastewater and almost 80% reduction of methyl tert-butyl ether was achieved within few hours of treatment (Sievers, 2011). Similarly, in another investigation, a large-scale treatment unit having a capacity of 200 L was operated for the disinfection of wastewater through HC process. The result of the investigation demonstrated that at a discharge pressure of 30 pounds per square, 56% of pathogens elimination was achieved within 5 min of treatment (Save et al., 1997).

Example: 17.6 In an experiment, a combined HC/H_2O_2 system was used for the treatment of hospital wastewater. After 90 min of treatment, 70.12% TOC removal was observed by HC/H_2O_2 process, while 40.18% and 10.86% removal efficiencies were observed by individual HC and H_2O_2 process, respectively. Estimate the synergistic effect of the combined system.

Solution
The synergetic effect (f) of the combined HC/H_2O_2 system can be obtained by the using Eq. 17.93.

$$f_{HC+H_2O_2} = \frac{[\%TOCremoval]_{HC+H_2O_2}}{\%TOCremoval_{HC} + \%TOCremoval_{H_2O_2}}$$

$$= \frac{70.12}{40.18 + 10.86} = 1.37$$

The value of f obtained is greater than 1, which indicates the efficiency of combined HC/H_2O_2 system is greater than that sum of the individual systems, thus indicating synergy.

Example: 17.7 The lab-scale degradation of a dye was evaluated by HC process. It was observed that after 60 min of treatment, TOC reduced from 27.24 to 22.35 mg/L during hydrodynamic cavitation with power input of 1100 J/sec. The volume of the solution present in the lab-scale HC reactor was 6 L. Estimate the energy efficiency of the treatment system.

Solution
Total mg of TOC reduced during 60 min of treatment

$$= (\text{Initial TOC} - \text{Final TOC})(\text{mg/L}) \times \text{ Volume of solution in the reactor (6L)}$$

$$= (27.24 - 22.35) \times 6 = 29.34 \text{ mg}$$

Energy input to the treatment system in 60 min = (Power input to the HC) (J/sec) × Time (sec)

$$= 1100 \times 3600 = 3.96 \times 10^6 \text{ J}$$

Energy efficiency = mg of TOC reduced/Energy input (J)
$$= 29.34/(3.96 \times 10^6) = 7.40 \times 10^{-6} \text{ mg/J (Answer)}$$

17.6.3 Electrochemical Oxidation-Based AOPs

Electrochemical oxidation technology is based on the in-situ production of $OH^{\cdot}$ radicals via the processes of direct or indirect oxidation. In direct oxidation or anodic oxidation, $OH^{\cdot}$ radicals are directly formed on the surface of anode by the oxidation of H_2O and at the same time H_2O_2 is produced on the cathode surface through cathodic reduction of O_2 as shown in Eq. 17.94 and Eq. 17.95, respectively. This process is also known as anodic oxidation because $OH^{\cdot}$ radicals are electrochemically generated at anode surface.

$$H_2O \rightarrow OH^{\cdot} + H^+ + e^- \tag{17.94}$$

$$O_2 + 2H^+ + 2e^- \rightarrow H_2O_2 \tag{17.95}$$

In indirect oxidation process, $OH^{\cdot}$ radicals are generated electrochemically with an addition of different chemicals, like Fenton's reagent, Cl_2, hypochlorite and peroxodisulphate, into the aqueous medium. Among these chemicals, Fenton's reagent is most commonly used for the indirect oxidation of organic matter and other refractory contaminants present in the wastewater. The in-situ electrochemical generation of $OH^{\cdot}$ radicals and Fenton's reagent using externally added Fe^{2+} ions is known as electro-Fenton process (Eq. 17.1).

The performance of electrochemical oxidation process typically assessed through three main parameters, such as mineralization current efficiency (*MCE*), specific energy consumption and electron efficiency (η_c). The *MCE* (in %) is defined as the charge required for the degradation of single organic pollutant at a given electrolysis time and it can be estimated by using Eq. 17.96 (Ridruejo et al., 2018).

$$\%MCE = \frac{nFV(TOC_0 - TOC_t)}{4.32 \times 10^7 mIt} \times 100 \tag{17.96}$$

where, $TOC_0 - TOC_t$ is the experimental decay of TOC (mg/L), n is the theoretical number of electrons exchanged during mineralization of organic pollutant, m is the number of carbon atoms present in the organic pollutant, I is the current applied (A), V is the volume of solution (L), F is the Faraday's constant (96,485 C mol^{-1}), 4.32×10^7 is a conversion factor ($= 3600$ sec/h $\times 12,000$ mg mol^{-1}) and t is the electrolysis time (h).

The specific energy consumption (*SEC*) per unit mass of TOC (*SEC*$_{TOC}$, in kWh/g TOC) and per unit volume of wastewater (*SEC*$_v$, in kWh/m^3) being treated can be estimated from Eq. 17.97 and Eq. 17.98, respectively.

$$(SEC)_{TOC} = \frac{E_{cell} \times I \times t}{V_s \times (TOC_0 - TOC_t)} \tag{17.97}$$

$$(SEC)_V = \frac{E_{cell} \times I \times t}{V_s} \tag{17.98}$$

where, E_{cell} is the average cell potential (V), $TOC_0 - TOC_t$ is the experimental decay of TOC (mg/L), I is the applied current (A), t is the electrolysis time (h) and V_s is the volume of solution (L).

The electron efficiency (η_c) represents the efficiency of electrons that are used by the electrode to remove the organic contaminants during electrolysis, which can be determined as per Eq. 17.99 (Pacheco et al., 2007).

$$Electron\,efficiency\,(\eta_c) = \frac{32}{12} \times \frac{n}{4m} \times \frac{dTOC}{dCOD} \tag{17.99}$$

where, n is the number of electrons exchanged from anode during mineralization of organic pollutant, m is the number of carbon atoms present in the organic pollutant and $\frac{dTOC}{dCOD}$ is the ratio of destruction of TOC (mg/L) to the rate of COD reduction (mg/L). This expression is not valid for multicomponent systems. It is assumed that if $\eta_c = 1$, the degradation of organic pollutant results in the complete mineralization into end products (CO_2, H_2O and O_2), whereas if $\eta_c < 1$, other intermediates by-products are formed as end products.

The energy efficiency of the electrochemical oxidation process depends upon the electrochemical performance of electrodes and other operational conditions, like current density (mA/cm^2), concentration of pollutants and electrolysis time. The electrical energy (EE) is utilized to estimate the energy efficiency of the treatment system. The EE of the electrochemical treatment technology can be determined as per Eq. 17.100 (Meng, 2018).

$$EE\left(kWh/m^3\right) = \frac{U \times J \times A \times t}{V \times log\left(\frac{C_o}{C_t}\right)} \tag{17.100}$$

where, U is the applied voltage (V), J is the current density (mA/cm^2), A is the surface area of electrode (cm^2), V is the volume of the reactor (m^3), t is the reaction time (h) and C_o and C_t are initial and final concentration of target compound, respectively (mg/L).

However, the above Eq. 17.100 can be written as Eq. 17.101 by using Eq. 17.63.

$$EE\left(kWh/m^3\right) = \frac{6.39 \times 10^{-4} \times U \times J \times A}{V \times k'} \tag{17.101}$$

where, 6.39×10^{-4} is a conversion factor ($= 1\,h/3600\,sec \times 0.4343$), for conversion of natural log to logarithm with base 10, i.e., 1/2.303, hence the conversion 0.4343 is used, U is the applied voltage (V), J is the current density (mA/cm^2), A is the surface area of electrode (cm^2), V is the volume of the reactor (m^3) and k' is the pseudo first order degradation rate constant (sec^{-1}).

Example: 17.8 An electrochemical oxidation process was used for the removal of tetra-caine ($C_{15}H_{24}N_2O_2$) from wastewater. Result of the investigation demonstrated that with 0.561 mM of tetracaine, applied current of 0.100 A and electrolysis time of 120 min, the decay of TOC and COD of 53 mg/L and 207.23 mg/L, respectively, occurred. The volume of treated solution and average cell potential used were 0.150 L and 2.7 V, respectively. Faraday's constant is 96,485 C mol^{-1}. Determine the MCE, SEC_{TOC}, SEC_V and η_c of the electrochemical oxidation process.

Solution

The number of carbon atoms (m) and theoretical number of electrons (n) exchanged during the mineralization of tetracaine molecule are taken as 15 and 74, respectively, according to the following reaction (Eq. 17.102):

$$C_{15}H_{24}N_2O_2 + 28H_2O \rightarrow 15CO_2 + 2NH_4^+ + 72H^+ + 74e^- \tag{17.102}$$

$$
\begin{aligned}
\%MCE &= \frac{nFV(TOC_0 - TOC_t)}{4.32 \times 10^7 mIt} \times 100 \\
&= \frac{74 \times 96,485 \times 0.15(53)}{4.32 \times 10^7 \times 15 \times 0.100 \times 2} \times 100 = 43.79\%
\end{aligned}
$$

$$(SEC)_{TOC} = \frac{E_{cell} \times I \times t}{V_s \times (TOC_0 - TOC_t)} = \frac{2.7 \times 0.100 \times 2}{0.150 \times \frac{53}{10^3} \times 10^3} = 0.067 \frac{kWh}{g \, TOC}$$

$$(SEC)_V = \frac{E_{cell} \times I \times t}{V_s} = \frac{2.7 \times 0.100 \times 2}{\frac{0.150}{10^3} \times 10^3} = 3.6 \frac{kWh}{m^3}$$

$$Electronefficiency(\eta_c) = \frac{32}{12} \times \frac{n}{4m} \times \frac{dTOC}{dCOD} = \frac{32}{12} \times \frac{74}{4 \times 15} \times \frac{53}{207.23} = 0.84$$

Here the value of η_c is less than 1 that means the degradation of tetracaine will form some intermediate by-products rather than complete mineralization to CO_2, H_2O and O_2.

Example: 17.9 In a laboratory test, a 99% degradation of 20 mg/L Ofloxacin was achieved within 90 min by using electrochemical oxidation process. The test was conducted in a cylindrical glass beaker having capacity of 250 mL, electrode surface area of 10 cm^2, average voltage of 6.2 V and current density of 30 mA/cm^2. Determine the electrical energy (EE) required to reduce one order concentration of the Ofloxacin.

Solution

$V = 250$ mL $= 0.00025$ m^3; $A = 10$ cm^2; $t = 90$ min $= 1.5$ h; $U = 6.2$ V; $J = 30$ mA/cm^2
$= 0.03$ A/cm^2; $C_0 = 20$ mg/L

$$\text{Degradation efficiency } (\%) = \frac{(C_0 - C_t) \times 100}{C_0}$$

$$99\% = \frac{(20 - C_t) \times 100}{20} \text{ and by solving we get the value of } C_t = 0.2 \text{ mg/L}$$

$$EE\left(kWh/m^3\right) = \frac{U \times J \times A \times t}{V \times \log\left(\frac{C_0}{C_t}\right)} = \frac{6.2 \times 0.03 \times 10 \times 1.5}{0.00025 \times \log\left(\frac{20}{0.2}\right) \times 1000} = 5.58$$

Thus, 5.58 kWh/m^3 of electrical energy is required to destroy one order of the contaminants during electrolysis.

17.6.3.1 Major Benefits and Drawbacks of Electrochemical Oxidation-Based AOPs

The use of electrochemical oxidation in the treatment of wastewater has shown many advantages over the conventional AOPs, such as shorter reaction time, no sludge production and the in-situ electrogeneration of reagents entails cost saving as well as omits the risk associated to their transport, storage and dosing. However, the high operating costs due to current consumption, high expense of electrode materials and electrode fouling pose the challenge, which need to be resolved for field-scale application of this technology.

17.6.3.2 Scaling-Up of Electrochemical Oxidation-Based AOPs

The electrochemical oxidation process has been experimentally demonstrated effective for the degradation of organic pollutants at the lab-scale (Table 17.11). However, there is no adequate information available regarding their commercial applicability. To date, there is only one article elucidating the large-scale application of electrochemical oxidation technique for the disinfection of real swimming pool water (Naji et al., 2018). The performance of electrochemical oxidation process was evaluated by employing niobium boron-doped diamond (Nb/BDD) and titanium coated with platinum (Ti/Pt) electrodes. The results of the investigation showed that the real swimming pool water initially contained 10^6 CFU (colony-forming units)/100 mL and after imposing a current of 1.5 and 3.0 A, and using Nb/BDD and Ti/Pt electrodes, the effluent was almost completely disinfected (Naji et al., 2018).

17.7 Factors Affecting Efficiency of Advanced Oxidation Processes

The performance of AOPs depends upon various operational parameters, such as pH, light intensity, inorganic ions, and concentration of target pollutant and oxidant. These parameters directly affect the production of reactive radicals in the system.

17.7.1 Effect of Carbonate and Bicarbonate Ions

The carbonate ions present in the wastewater can react with the free radicals such as $OH^{\cdot}$ and $SO_4^{\cdot-}$, and transform them into $CO_3^{\cdot-}$ (Eqs. 17.103 and 17.104).

$$OH^{\cdot} + CO_3^{2-} \rightarrow CO_3^{\cdot} + OH^- \tag{17.103}$$

$$SO_4^{\cdot-} + CO_3^{2-} \rightarrow CO_3^{\cdot-} + SO_4^{2-} \tag{17.104}$$

The $CO_3{}^{\cdot-}$ possesses lower E_0 (1.59 V vs. NHE) and higher selectivity compared to $OH^{\cdot}$ radical. In some processes, such as O_3 and UV/H_2O_2, the presence of carbonate ions increases the production of carbonate radicals compared to $OH^{\cdot}$ radicals (Wang & Wang, 2021). The generated $CO_3{}^{\cdot-}$ in the system enhances the removal of organic contaminants up to a certain level. However, continuous rise of $CO_3{}^{\cdot-}$ concentration can scavenge the $OH^{\cdot}$ and $SO_4{}^{\cdot-}$ radicals and also change the solution pH, which can change the existing form of carbonate ions (Black & Hayon, 1970). Similarly, the continuous increase in the concentration of bicarbonate ions can also act as an $OH^{\cdot}$ and $SO_4{}^{\cdot-}$ scavenger and produce less reactive $HCO_3{}^{\cdot-}$ radicals, which could significantly reduce the efficiency of AOPs (Eqs. 17.105 and 17.106). These generated reactive species are usually detected by electron paramagnetic resonance (EPR), transient absorption spectrum, electron spin resonance (ESR) and high-performance liquid chromatography (HPLC).

$$OH^{\cdot} + HCO_3^- \rightarrow HCO_3^{\cdot-} + OH^- \tag{17.105}$$

$$SO_4^{\cdot-} + HCO_3^- \rightarrow HCO_3^{\cdot-} + SO_4^{2-} \tag{17.106}$$

17.7.2 Effect of Phosphate Ions

The phosphate ions also react with $OH^{\cdot}$ and $SO_4{}^{\cdot-}$ radicals and generate new type of reactive species as shown in Eq. 17.107 through Eq. 17.110. The existence of radicals varies with the variation in solution pH. Generally, phosphate ions are formed when the solution is having pH higher than 12.0 (Eqs. 17.107 and 17.108), whereas hydrogen phosphate ions are formed for pH lower than 11.0 (Eqs. 17.109 and 17.110) (Black & Hayon, 1970). The presence of phosphate ions in water can scavenge the $OH^{\cdot}$ and $SO_4{}^{\cdot-}$ radicals and also forms different intermediate products compared to $OH^{\cdot}$ radicals, which could reduce or improve the removal efficiency of the target contaminants (Wang and Wang, 2018).

$$OH^{\cdot-} + PO_4^{3-} \rightarrow PO_4^{\cdot2-} + OH^- \tag{17.107}$$

$$SO_4^{\cdot-} + PO_4^{3-} \rightarrow PO_4^{\cdot2-} + SO_4^{2-} \tag{17.108}$$

$$OH^{\cdot} + HPO_4^{2-} \rightarrow HPO_4^{\cdot-} + OH^- \tag{17.109}$$

$$SO_4^{\cdot} + HPO_4^{2-} \rightarrow HPO_4^{\cdot-} + SO_4^{2-} \tag{17.110}$$

17.7.3 Effect of Target Pollutant Concentration on Efficiency of AOPs

The target pollutant concentration also affects the degradation efficiency of AOPs and
the degradation efficiency is inversely proportional to the target pollutant concentration.
This is because as the concentration of target pollutant rises the requirement of reactive
radicals also increases. Since the catalyst loading and irradiation conditions are constant,
the availability of reactive species generated in the system remains constant. Therefore,
the insufficient number of reactive species are available to degrade the greater number
of target pollutant molecules that can significantly lower the removal efficiency of AOPs.
For example, in a demonstration, the impact of target contaminant concentration for photocatalytic degradation of Carbamazepine was observed by using TiO$_2$ as catalyst. With
catalytic load of 1.5 g/L, process time of 90 min and initial concentrations from 5 to
20 mg/L, the removal of Carbamazepine was 92.8%, 68.4%, 54.0%, and 41.2% for the
initial concentration of 5, 10, 15 and 20 mg/L, respectively (Carabin et al., 2015). In
another demonstration, the removal of Norfloxacin was observed by using gamma-ray
radiation and it was found that an increase in Norfloxacin concentration from 3.4 to
16.1 mg/L resulted in the reduction of removal efficiency from 94.0 to 79.1% (Sayed
et al., 2016).

17.7.4 Effect of pH

The pH is another most crucial operating parameter that affects the performance of AOPs.
It has been noticed that pH can affect the physicochemical properties of catalysts, produces new reactive species and change the concentration of reactive radicals in various
AOPs. For example, in sulphate-based AOP, the change of pH can affect the formation of
reactive species in two pathways. The first pathway is the conversion of SO$_4$$^{\cdot-}$ radicals
into OH$^{\cdot}$ radicals (Eq. 17.111) and the second pathway is the base induced activation of
sulphate ions to produce new reactive species such as O$_2$$^{\cdot-}$ and singlet oxygen (^{1}O$_2$) that
takes part in the removal of pollutants (Eq. 17.112 through Eq. 17.116) (Furman et al.,
2010).

$$SO_4^{\cdot-} + OH^- \rightarrow OH^{\cdot} + SO_4^{2-} \tag{17.111}$$

$$S_2O_8^{2-} + 2H_2O \rightarrow HO_2^- + 2SO_4^{2-} + 3H^+ \tag{17.112}$$

$$HO_2^- + S_2O_8^{2-} \rightarrow SO_4^{\cdot-} + SO_4^{2-} + H^+ + O_2^{\cdot-} \tag{17.113}$$

$$HSO_5^- \rightarrow SO_5^{2-} + H^+ \tag{17.114}$$

$$SO_5^{2-} + H_2O \rightarrow H_2O_2 + SO_4^{2-} \qquad (17.115)$$

$$HSO_5^- + H_2O_2 \rightarrow H^+ + SO_4^{2-} + H_2O +^1 O_2 \qquad (17.116)$$

Similarly, in UV/H_2O_2 method under acidic pH conditions, $OH^·$ radicals are the main reactive species (Eq. 17.117); however, the rising of pH can cause H_2O_2 dissociation to form HO_2^-, which can scavenge the $OH^·$ radicals (Eq. 17.118).

$$H_2O + UV \rightarrow e_{aq}^- + H^· + OH^· + H^+ \qquad (17.117)$$

$$HO_2^- + OH^· \rightarrow H_2O + O_2^{·-} \qquad (17.118)$$

17.7.5 Effect of Oxidant Concentration

The oxidants, such as H_2O_2, $S_2O_8^{2-}$, HSO_5^- etc., are used as sources to generate reactive radicals in various AOPs. Many researchers have reported that the oxidant concentration affects the performance of AOPs. For example, in an investigation the influence of H_2O_2 concentration on the performance of UV/H_2O_2/TiO_2 process was examined by employing Amoxicillin, Ampicillin and Cloxacillin antibiotics as target pollutants. The results revealed that the COD removal efficiencies of 14.8%, 26.3%, 23.1%, 16.3%, and 12.3% were observed at 50, 100, 150, 200, and 300 mg/L of H_2O_2 concentration, respectively, with a reaction time of 300 min. However, while increasing H_2O_2 concentration from 150 to 300 mg/L, COD removal efficiency decreased due to the scavenging effect of H_2O_2 at higher concentration, which thereby affects the degradation efficacy of the system (Elmolla & Chaudhuri, 2010).

Similarly, in another investigation, removal of Clofibric acid was observed by using UV/$S_2O_8^{2-}$ process. The outcome of the investigation showed the rate constant increased from 0.00684 to 0.0176 min^{-1} with raising $S_2O_8^{2-}$ concentration from 0.25 to 0.75 mM. However, the removal rate decreased from 0.0176 min^{-1} to 0.0158 min^{-1} with increasing the dosage from 0.75 mM to 1.0 mM due to the scavenging effect of excess $S_2O_8^{2-}$ and recombination of $SO_4^{·-}$ radicals (Eq. 17.119 through Eq. 17.121) (Lu et al., 2018).

$$SO_4^{·-} + S_2O_8^{2-} \rightarrow SO_4^{2-} + S_2O_8^{·-} \qquad (17.119)$$

$$OH^· + S_2O_8^{2-} \rightarrow S_2O_8^{·-} + OH^- \qquad (17.120)$$

$$SO_4^{·-} + SO_4^{·-} \rightarrow S_2O_8 \qquad (17.121)$$

17.7.6 Effect of Light Intensity

The influence of intensity of light is another most significant parameter that affects the kinetics of removal of target contaminants. In general, activation of semiconductor photocatalysts requires light irradiation for transfer of electrons from valence band (VB) to conduction band (CB), which results in the generation of electrons and holes (Eq. 17.122). The generated electrons (e^-) and holes (h^+) react with atmospheric oxygen and water molecules to produce $O_2{}^{\cdot-}$ and $OH^{\cdot}$ radicals as shown in Eqs. 17.123 and 17.124 (Fig. 17.7). The impact of light intensity on the performance of UV/TiO$_2$ method was examined by using three pharmaceuticals, namely Metronidazole, Atenolol and Chlorpromazine. The result demonstrated that the degradation efficiency increased from 53.36% to 61.38% for Atenolol, 82.59–87.17% for Metronidazole and 55.36–77.71% for Chlorpromazine, when the intensity of light was raised from 12.2 to 47.2 W/m^2 (Khatee et al., 2013).

$$TiO_2 + UV \rightarrow h^+ + e^- \tag{17.122}$$

$$H_2O + h^+ \rightarrow OH^{\cdot} + H^+ \tag{17.123}$$

$$O_2 + e^- \rightarrow O_2^{\cdot-} \tag{17.124}$$

Questions

17.1 Design a continuous plug flow UV/SO$_4{}^{\cdot-}$ reactor based on the data given below:
$Q = 5$ m^3/h; $k' = 8.8 \times 10^{-2}$ sec^{-1}; UV dosage = 210 mJ/cm^2; OP = 0.5 cm;

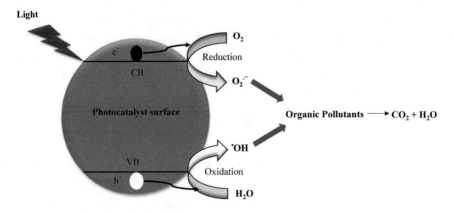

Fig. 17.7 Activation of photocatalyst through light irradiation

$SO_4{}^{\cdot-}$ dose $= 10$ g/m^3; Quartz sleeve diameter $= 0.03$ m; Electrical power $= 40$ W;

UV output $= 15$ W; Diameter $= 19$ mm; $L_{UV\text{-net}} = 0.8$ m;

Assume the other suitable data as necessary.

[Answer: t $= 52.93$ sec, V $= 0.074$ m^3, Leq $\doteq 134.6$ m, Sexp $= 16.9$ m^2, Ieffective $= 39.67$ W/m^2, Applied UVpower $= 1034.6$ W, Neq $= 68$, Vtotal $= 0.112$ m^3, dR $= 0.42$ m, LUV-net/dR $= 1.9$].

17.2 Briefly discuss the factors affecting the performance of various UV-based AOPs.

17.3 Discuss the mechanism of OH$^{\cdot}$ radical production through Fenton process.

17.4 Explain the advantages and disadvantages of UV/Cl$_2$ process as a water disinfectant.

17.5 A 0.2 M solution in a test tube with a path length of 1 cm has an absorbance of 0.07 at 560 nm. Determine the molar absorption coefficient of the solution and what will be the absorbance if the concentration of the solution changes to 0.5 M? [**Answer** $\varepsilon = 0.35$, A $= 0.175$ M^{-1} cm^{-1}]

17.6 How is UV dosage expressed mathematically?

17.7 How pH affects the efficiency of the ozone based advanced oxidation process?

17.8 Describe utility and effectiveness of ozonation and catalytic ozonation for removal of emerging contaminants.

17.9 What do you mean by cavitation yield? Briefly discuss the mechanism of cavity formation in acoustic and hydrodynamic cavitation process.

17.10 Explain the purpose and principle of CPC with neat sketch. Also describe the design aspects of solar photo Fenton process.

17.11 What is G-value of a reactive species? How is it calculated?

17.12 Define molar absorption coefficient. How do you determine the molar absorption coefficient of a chemical species?

17.13 Describe how presence of carbonate and bicarbonate will affect the performance of advanced oxidation process.

17.14 Write a short note on classification of advanced oxidation processes.

17.15 Describe advantages and drawback of Fenton based AOPs.

17.16 Discuss application of sono-Fenton process for removal of emerging contaminants, stating advantages that it can offer along with present drawbacks.

17.17 Describe different types of UV based AOPs that can be used for removal of emerging contaminants. Also state advantages and drawbacks of each of them.

17.18 Describe different types of ozone based AOPs that have been developed for removal of emerging contaminants. Also state advantages and drawbacks of each of them.

17.19 Write short note on Gamma radiation based AOP

17.20 How the emerging contaminants are getting removed in electrochemical oxidation based AOPs?

References

Adak, A., Das, I., Mondal, B., Koner, S., Datta, P., & Blaney, L. (2019). Degradation of 2,4-dichlorophenoxyacetic acid by UV 253.7 and UV-H_2O_2: Reaction kinetics and effects of interfering substances. *Emerging Contaminants, 5*, 53–60. https://doi.org/10.1016/j.emcon.2019.02.004

Adityosulindro, S., Barthe, L., González-Labrada, K., Jáuregui Haza, U. J., Delmas, H., & Julcour, C. (2017). Sonolysis and sono-Fenton oxidation for removal of ibuprofen in (waste)water. *Ultrasonics Sonochemistry, 39*, 889–896. https://doi.org/10.1016/j.ultsonch.2017.06.008

Amat, A. M., Arques, A., Miranda, M. A., & Seguí, S. (2004). Photo-fenton reaction for the abatement of commercial surfactants in a solar pilot plant. *Solar Energy, 77*, 559–566. https://doi.org/10.1016/j.solener.2004.03.028

Araujo, F. V. F., Yokoyama, L., Teixeira, L. A. C., & Campos, J. C. (2011). Heterogeneous Fenton process using the mineral hematite for the discolouration of a reactive dye solution. *Brazilian Journal of Chemical Engineering, 28*, 605–616. https://doi.org/10.1590/S0104-66322011000400006

Arshad, R., Bokhari, T. H., Khosa, K. K., Bhatti, I. A., Munir, M., Iqbal, M., Iqbal, D. N., Khan, M. I., Iqbal, M., & Nazir, A. (2020). Gamma radiation induced degradation of anthraquinone Reactive Blue-19 dye using hydrogen peroxide as oxidizing agent. *Radiation Physics and Chemistry, 168*, 108637. https://doi.org/10.1016/j.radphyschem.2019.108637

Arzate, S., Campos-Mañas, M. C., Miralles-Cuevas, S., Agüera, A., García Sánchez, J. L., & Sánchez Pérez, J. A. (2020). Removal of contaminants of emerging concern by continuous flow solar photo-Fenton process at neutral pH in open reactors. *Journal of Environmental Management, 261*, 110265. https://doi.org/10.1016/j.jenvman.2020.110265

Audenaert, W. T. M., Vermeersch, Y., Van Hulle, S. W. H., Dejans, P., Dumoulin, A., & Nopens, I. (2011). Application of a mechanistic UV/hydrogen peroxide model at full-scale: Sensitivity analysis, calibration and performance evaluation. *Chemical Engineering Journal, 171*, 113–126. https://doi.org/10.1016/j.cej.2011.03.071

Autin, O., Romelot, C., Rust, L., Hart, J., Jarvis, P., MacAdam, J., Parsons, S. A., & Jefferson, B. (2013). Evaluation of a UV-light emitting diodes unit for the removal of micropollutants in water for low energy advanced oxidation processes. *Chemosphere, 92*, 745–751. https://doi.org/10.1016/j.chemosphere.2013.04.028

Babaei, A. A., & Ghanbari, F. (2016). COD removal from petrochemical wastewater by UV/hydrogen peroxide, UV/persulfate and UV/percarbonate: Biodegradability improvement and cost evaluation. *Journal of Water Reuse and Desalination, 6*, 484–494. https://doi.org/10.2166/wrd.2016.188

Bagal, M. V., & Gogate, P. R. (2014). Wastewater treatment using hybrid treatment schemes based on cavitation and Fenton chemistry: A review. *Ultrasonics Sonochemistry, 21*, 1–14. https://doi.org/10.1016/j.ultsonch.2013.07.009

Bauer, R., & Fallmann, H. (1997). The Photo-Fenton oxidation—A cheap and efficient wastewater treatment method. *Research on Chemical Intermediates, 23*, 341–354. https://doi.org/10.1163/156856797X00565

Black, E. D., & Hayon, E. (1970). Pulse radiolysis of phosphate anions $H_2PO_4^-$, HPO_4^{2-}, PO_4^{3-}, and $P_2O_7^{4-}$ in aqueous solutions. *Journal of Physical Chemistry, 74*, 3199–3203. https://doi.org/10.1021/j100711a007

Blackbeard, J., Lloyd, J., Magyar, M., Mieog, J., Linden, K. G., & Lester, Y. (2016). Demonstrating organic contaminant removal in an ozone-based water reuse process at full scale. *Environmental Science: Water Research & Technology, 2*, 213–222. https://doi.org/10.1039/c5ew00186b

Bossmann, S. H., Oliveros, E., Göb, S., Kantor, M., Göppert, A., Braun, A. M., Lei, L., & Yue, P. L. (2001). Oxidative degradation of polyvinyl alcohol by the photochemically enhanced Fenton reaction. Evidence for the formation of super-macromolecules. *Progress in Reaction Kinetics and Mechanism, 26*, 113–137. https://doi.org/10.3184/007967401103165208

Cabrera Reina, A., Miralles-Cuevas, S., Cornejo, L., Pomares, L., Polo, J., Oller, I., & Malato, S. (2020). The influence of location on solar photo-Fenton: Process performance, photoreactor scaling-up and treatment cost. *Renewable Energy, 145*, 1890–1900. https://doi.org/10.1016/j.renene.2019.07.113

Carabin, A., Drogui, P., & Robert, D. (2015). Photo-degradation of carbamazepine using TiO_2 suspended photocatalysts. *Journal of the Taiwan Institute of Chemical Engineers, 54*, 109–117. https://doi.org/10.1016/j.jtice.2015.03.006

Chen, H., & Wang, J. (2021). Degradation and mineralization of ofloxacin by ozonation and peroxone (O_3/H_2O_2) process. *Chemosphere, 269*, 128775. https://doi.org/10.1016/j.chemosphere.2020.128775

Choi, J., Cui, M., Lee, Y., Kim, J., Son, Y., Lim, J., Ma, J., & Khim, J. (2020). Application of persulfate with hydrodynamic cavitation and ferrous in the decomposition of pentachlorophenol. *Ultrasonics Sonochemistry, 66*, 105106. https://doi.org/10.1016/j.ultsonch.2020.105106

Chowdhury, P., & Viraraghavan, T. (2009). Sonochemical degradation of chlorinated organic compounds, phenolic compounds and organic dyes—A review. *Science of the Total Environment, 407*, 2474–2492. https://doi.org/10.1016/j.scitotenv.2008.12.031

Cuerda-Correa, E. M., Alexandre-Franco, M. F., & Fernández-González, C. (2019). Advanced oxidation processes for the removal of antibiotics from water. An overview. *Water, 12*, 102. https://doi.org/10.3390/w12010102

De la Cruz, N., Esquius, L., Grandjean, D., Magnet, A., Tungler, A., de Alencastro, L. F., & Pulgarín, C. (2013). Degradation of emergent contaminants by UV, UV/H_2O_2 and neutral photo-Fenton at pilot scale in a domestic wastewater treatment plant. *Water Research, 47*, 5836–5845. https://doi.org/10.1016/j.watres.2013.07.005

Dehghani, M., Taghizadeh, M. M., Gholami, T., Ghadami, M., Keshtgar, L., Elhameyan, Z., Javaheri, M. R., Shamsedini, N., Jamshidi, F., Shahsavani, S., & Ghanbarian, M. (2015). Optimization of the parameters influencing the photo-Fenton process for the decolorization of reactive red 198 (RR198). *Jundishapur Journal of Health Sciences, 7*(2), e28243. https://doi.org/10.5812/jjhs.7(2)2015.28243

Della-Flora, A., Wilde, M. L., Thue, P. S., Lima, D., Lima, E. C., & Sirtori, C. (2020). Combination of solar photo-Fenton and adsorption process for removal of the anticancer drug Flutamide and its transformation products from hospital wastewater. *Journal of Hazardous Materials, 396*, 122699. https://doi.org/10.1016/j.jhazmat.2020.122699

Djeffal, L., Abderrahmane, S., Benzina, M., Fourmentin, M., Siffert, S., & Fourmentin, S. (2014). Efficient degradation of phenol using natural clay as heterogeneous Fenton-like catalyst. *Environmental Science and Pollution Research, 21*, 3331–3338. https://doi.org/10.1007/s11356-013-2278-5

Dutta, P. K., Pehkonen, S. O., Sharma, V. K., & Ray, A. K. (2005). Photocatalytic oxidation of arsenic(III): Evidence of hydroxyl radicals. *Environmental Science and Technology, 39*, 1827–1834. https://doi.org/10.1021/es0489238

Eckenfelder, W. (1997). *Chemical oxidation: technologies for the nineties*. CRC Press.

Elmolla, E. S., & Chaudhuri, M. (2010). Photocatalytic degradation of amoxicillin, ampicillin and cloxacillin antibiotics in aqueous solution using UV/TiO_2 and UV/H_2O_2/TiO_2 photocatalysis. *Desalination, 252*, 46–52. https://doi.org/10.1016/j.desal.2009.11.003

Fenton, H. J. H. (1894). Oxidation of tartaric acid in presence of iron. J. Chem. Soc. *Trans., 65*, 899–910. https://doi.org/10.1039/CT8946500899

Furman, O. S., Teel, A. L., & Watts, R. J. (2010). Mechanism of base activation of persulfate. *Environmental Science and Technology, 44*, 6423–6428. https://doi.org/10.1021/es1013714

Ganiyu, S. O., Zhou, M., & Martínez-Huitle, C. A. (2018). Heterogeneous electro-Fenton and photoelectro-Fenton processes: A critical review of fundamental principles and application for water/wastewater treatment. *Applied Catalysis B: Environmental, 235*, 103–129. https://doi.org/10.1016/j.apcatb.2018.04.044

Garoma, T., & Gurol, M. D. (2004). Degradation of tert-butyl alcohol in dilute aqueous solution by an O_3/UV process. *Environmental Science and Technology, 38*, 5246–5252. https://doi.org/10.1021/es0353210

Ghauch, A., Baalbaki, A., Amasha, M., El Asmar, R., & Tantawi, O. (2017). Contribution of persulfate in UV-254 nm activated systems for complete degradation of chloramphenicol antibiotic in water. *Chemical Engineering Journal, 317*, 1012–1025. https://doi.org/10.1016/j.cej.2017.02.133

Glaze, W. H., Kang, J.-W., & Chapin, D. H. (1987). The chemistry of water treatment processes involving ozone, hydrogen peroxide and ultraviolet radiation. *Ozone Science and Engineering, 9*, 335–352. https://doi.org/10.1080/01919518708552148

Gonzalez-Olmos, R., Martin, M. J., Georgi, A., Kopinke, F. D., Oller, I., & Malato, S. (2012). Fe-zeolites as heterogeneous catalysts in solar Fenton-like reactions at neutral pH. *Applied Catalysis B: Environmental, 125*, 51–58. https://doi.org/10.1016/j.apcatb.2012.05.022

Gu, L., Nie, J.-Y., Zhu, N., Wang, L., Yuan, H.-P., & Shou, Z. (2012). Enhanced Fenton's degradation of real naphthalene dye intermediate wastewater containing 6-nitro-1-diazo-2-naphthol-4-sulfonic acid: A pilot scale study. *Chemical Engineering Journal, 189–190*, 108–116. https://doi.org/10.1016/j.cej.2012.02.038

Haber, F., & Weiss, J. (1932). The catalytic decomposition of hydrogen peroxide by mixed catalysts. *Journal of Physical Colloid Chemistry, 53*, 1070–1091. https://doi.org/10.1021/j150472a009

Hashemi, M., Sagharlo, N. G. (2020). Optimization and evaluation of the efficiency of sono-Fenton and photo-Fenton processes in the removal of 2, 4, 6 trinitrotoluene (TNT) from aqueous solutions. *Journal of Advances in Environmental Health Research, 8*, 38–45. https://doi.org/10.22102/jaehr.2020.196050.1139

He, C., Wang, J., Wang, C., Zhang, C., Hou, P., & Xu, X. (2020). Catalytic ozonation of bio-treated coking wastewater in continuous pilot- and full-scale system: Efficiency, catalyst deactivation and in-situ regeneration. *Water Research, 183*, 116090. https://doi.org/10.1016/j.watres.2020.116090

Huang, J., Li, X., Ma, M., & Li, D. (2017). Removal of di-(2-ethylhexyl) phthalate from aqueous solution by UV/peroxymonosulfate: Influencing factors and reaction pathways. *Chemical Engineering Journal, 314*, 182–191. https://doi.org/10.1016/j.cej.2016.12.095

Illés, E., Szabó, E., Takács, E., Wojnárovits, L., Dombi, A., & Gajda-Schrantz, K. (2014). Ketoprofen removal by O_3 and O_3/UV processes: Kinetics, transformation products and ecotoxicity. *Science of the Total Environment, 472*, 178–184. https://doi.org/10.1016/j.scitotenv.2013.10.119

Javier Benitez, F., Real, F. J., Acero, J. L., & Casas, F. (2017). Assessment of the UV/Cl_2 advanced oxidation process for the degradation of the emerging contaminants amitriptyline hydrochloride, methyl salicylate and 2-phenoxyethanol in water systems. *Environmental Technology (united Kingdom), 38*, 2508–2516. https://doi.org/10.1080/09593330.2016.1269836

Khataee, A. R., Fathinia, M., & Joo, S. W. (2013). Simultaneous monitoring of photocatalysis of three pharmaceuticals by immobilized TiO_2 nanoparticles: Chemometric assessment, intermediates identification and ecotoxicological evaluation. *Spectrochimica Acta, Part a: Molecular and Biomolecular Spectroscopy, 112*, 33–45. https://doi.org/10.1016/j.saa.2013.04.028

Khuntia, S., Majumder, S. K., & Ghosh, P. (2015). A pilot plant study of the degradation of Brilliant Green dye using ozone microbubbles: Mechanism and kinetics of reaction. *Environmental Technology (united Kingdom), 36*, 336–347. https://doi.org/10.1080/09593330.2014.946971

Kilic, M. Y., Abdelraheem, W. H., He, X., Kestioglu, K., & Dionysiou, D. D. (2019). Photochemical treatment of tyrosol, a model phenolic compound present in olive mill wastewater, by hydroxyl and sulfate radical-based advanced oxidation processes (AOPs). *Journal of Hazardous Materials,* *367,* 734–742. https://doi.org/10.1016/j.jhazmat.2018.06.062

Koubek, E. (1975). Photochemically induced oxidation of refractory organics with hydrogen peroxide. *Industrial and Engineering Chemistry Process Design and Development, 14,* 348–350. https://doi.org/10.1021/i260055a025

Krakkó, D., Illés, Á., Licul-Kucera, V., Dávid, B., Dobosy, P., Pogonyi, A., Demeter, A., Mihucz, V. G., Dóbé, S., & Záray, G. (2021). Application of (V)UV/O$_3$ technology for post-treatment of biologically treated wastewater: A pilot-scale study. *Chemosphere, 275,* 130080. https://doi.org/10.1016/j.chemosphere.2021.130080

Lee, J.-Y., Lee, Y.-M., Kim, T.-K., Choi, K., & Zoh, K.-D. (2021). Degradation of cyclophosphamide during UV/chlorine reaction: Kinetics, byproducts, and their toxicity. *Chemosphere,* *268,* 128817. https://doi.org/10.1016/j.chemosphere.2020.128817

Li, G., Yi, L., Wang, J., & Song, Y. (2020). Hydrodynamic cavitation degradation of Rhodamine B assisted by Fe^{3+}-doped TiO$_2$: Mechanisms, geometric and operation parameters. *Ultrasonics Sonochemistry, 60,* 104806. https://doi.org/10.1016/j.ultsonch.2019.104806

Li, X., Chen, W., Tang, Y., & Li, L. (2018). Relationship between the structure of Fe-MCM-48 and its activity in catalytic ozonation for diclofenac mineralization. *Chemosphere, 206,* 615–621. https://doi.org/10.1016/j.chemosphere.2018.05.066

Lin, C., Zhang, W., Yuan, M., Feng, C., Ren, Y., & Wei, C. (2014). Degradation of polycyclic aromatic hydrocarbons in a coking wastewater treatment plant residual by an O$_3$/ultraviolet fluidized bed reactor. *Environmental Science and Pollution Research, 21,* 10329–10338. https://doi.org/10.1007/s11356-014-3034-1

Liu, Y., Wu, D., Peng, S., Feng, Y., & Liu, Z. (2019). Enhanced mineralization of dimethyl phthalate by heterogeneous ozonation over nanostructured Cu-Fe-O surfaces: Synergistic effect and radical chain reactions. *Separation and Purification Technology, 209,* 588–597. https://doi.org/10.1016/j.seppur.2018.07.016

Lu, X., Shao, Y., Gao, N., Chen, J., Deng, H., Chu, W., An, N., & Peng, F. (2018). Investigation of clofibric acid removal by UV/persulfate and UV/chlorine processes: Kinetics and formation of disinfection byproducts during subsequent chlor(am)ination. *Chemical Engineering Journal,* *331,* 364–371. https://doi.org/10.1016/j.cej.2017.08.117

Ma, Y. S., & Sung, C. F. (2010). Investigation of carbofuran decomposition by a combination of ultrasound and Fenton process. *Sustainable Environment Research, 20,* 213–219.

Mahdi-Ahmed, M., & Chiron, S. (2014). Ciprofloxacin oxidation by UV-C activated peroxymonosulfate in wastewater. *Journal of Hazardous Materials, 265,* 41–46. https://doi.org/10.1016/j.jhazmat.2013.11.034

Mártire, D. O., Rosso, J. A., Bertolotti, S., Carrillo Le Roux, G., Braun, A. M., & Gonzalez, M. C. (2001). Kinetic study of the reactions of chlorine atoms and Cl$_2^{\bullet-}$ radical anions in aqueous solutions. II. Toluene, benzoic acid, and chlorobenzene. *Journal of Physical Chemistry A, 105,* 5385–5392. https://doi.org/10.1021/jp004630z

Matheswaran, M., Balaji, S., Chung, S. J., & Moon, I. S. (2007). Studies on cerium oxidation in catalytic ozonation process: A novel approach for organic mineralization. *Catalysis Communications, 8,* 1497–1501. https://doi.org/10.1016/j.catcom.2006.12.017

McGinnis, B. D. (2000). Degradation of ethylene glycol in photo Fenton systems. *Water Research,* *34,* 2346–2354. https://doi.org/10.1016/S0043-1354(99)00387-5

Melo, J. M. O., Duarte, J. L. S., Ferro, A. B., Meili, L., & Zanta, C. L. P. S. (2020). Comparing Electrochemical and Fenton-Based Processes for Aquaculture Biocide Degradation. *Water, Air, and Soil Pollution, 231,* 79. https://doi.org/10.1007/s11270-020-4454-9

Meng, X. (2018). The development of electrochemical advanced oxidation processes for wastewater treatment : From two-dimensional to three-dimensional. Dissertation. Georgia Institute of Technology. Atlanta, USA.

Michael, I., Hapeshi, E., Michael, C., Varela, A. R., Kyriakou, S., Manaia, C. M., & Fatta-Kassinos, D. (2012). Solar photo-Fenton process on the abatement of antibiotics at a pilot scale: Degradation kinetics, ecotoxicity and phytotoxicity assessment and removal of antibiotic resistant enterococci. *Water Research, 46,* 5621–5634. https://doi.org/10.1016/j.watres.2012.07.049

Mierzwa, J. C., Rodrigues, R., Teixeira, A. C. S. C. (2018). UV-hydrogen peroxide processes. Chapter 2. In S. C. Ameta & R. Ameta (Eds.), *Advanced oxidation processes for wastewater treatment: Emerging Green chemical technology* (pp. 13–48). Academic Press, Elsevier. https://doi.org/10.1016/B978-0-12-810499-6.00002-4

Miklos, D. B., Remy, C., Jekel, M., Linden, K. G., Drewes, J. E., & Hübner, U. (2018). Evaluation of advanced oxidation processes for water and wastewater treatment—A critical review. *Water Research, 139,* 118–131. https://doi.org/10.1016/j.watres.2018.03.042

Montoya-Rodríguez, D. M., Serna-Galvis, E. A., Ferraro, F., & Torres-Palma, R. A. (2020). Degradation of the emerging concern pollutant ampicillin in aqueous media by sonochemical advanced oxidation processes—Parameters effect, removal of antimicrobial activity and pollutant treatment in hydrolyzed urine. *Journal of Environmental Management, 261,* 110224. https://doi.org/10.1016/j.jenvman.2020.110224

Naji, T., Dirany, A., Carabin, A., & Drogui, P. (2018). Large-scale disinfection of real swimming pool water by electro-oxidation. *Environmental Chemistry Letters, 16,* 545–551. https://doi.org/10.1007/s10311-017-0687-2

Navarro, P., Pellicer, J. A., & Gómez-López, V. M. (2019). Degradation of azo dye by an UV/H_2O_2 advanced oxidation process using an amalgam lamp. *Water Environment Journal, 33,* 476–483. https://doi.org/10.1111/wej.12418

Nidheesh, P. V. (2015). Heterogeneous Fenton catalysts for the abatement of organic pollutants from aqueous solution: A review. *RSC Advances, 5,* 40552–40577. https://doi.org/10.1039/c5ra02023a

Oliveira, H. S., Almeida, L. D., De Freitas, V. A. A., Moura, F. C. C., Souza, P. P., & Oliveira, L. C. A. (2015). Nb-doped hematite: Highly active catalyst for the oxidation of organic dyes in water. *Catalysis Today, 240,* 176–181. https://doi.org/10.1016/j.cattod.2014.07.016

Oral, O., & Kantar, C. (2019). Diclofenac removal by pyrite-Fenton process: Performance in batch and fixed-bed continuous flow systems. *Science of the Total Environment, 664,* 817–823. https://doi.org/10.1016/j.scitotenv.2019.02.084

Oturan, M. A., & Aaron, J. J. (2014). Advanced oxidation processes in water/wastewater treatment: Principles and applications. A review. *Critical Reviews in Environment Science and Technology, 44,* 2577–2641. https://doi.org/10.1080/10643389.2013.829765

Pacheco, M. J., Morão, A., Lopes, A., Ciríaco, L., & Gonçalves, I. (2007). Degradation of phenols using boron-doped diamond electrodes: A method for quantifying the extent of combustion. *Electrochimica Acta, 53,* 629–636. https://doi.org/10.1016/j.electacta.2007.07.024

Panda, D., & Manickam, S. (2019). Hydrodynamic cavitation assisted degradation of persistent endocrine-disrupting organochlorine pesticide Dicofol: Optimization of operating parameters and investigations on the mechanism of intensification. *Ultrasonics Sonochemistry, 51,* 526–532. https://doi.org/10.1016/j.ultsonch.2018.04.003

Papoutsakis, S., Miralles-Cuevas, S., Gondrexon, N., Baup, S., Malato, S., & Pulgarin, C. (2015). Coupling between high-frequency ultrasound and solar photo-Fenton at pilot scale for the treatment of organic contaminants: An initial approach. *Ultrasonics Sonochemistry, 22,* 527–534. https://doi.org/10.1016/j.ultsonch.2014.05.003

Pignatello, J. J., Oliveros, E., & MacKay, A. (2006). Advanced oxidation processes for organic contaminant destruction based on the fenton reaction and related chemistry. *Critical Reviews in Environment Science and Technology, 36*, 1–84. https://doi.org/10.1080/10643380500326564

Pliego, G., Zazo, J. A., Garcia-Muñoz, P., Munoz, M., Casas, J. A., & Rodriguez, J. J. (2015). Trends in the intensification of the Fenton process for wastewater treatment: An overview. *Critical Reviews in Environment Science and Technology, 45*, 2611–2692. https://doi.org/10.1080/106 43389.2015.1025646

Poulopoulos, S. G., Nikolaki, M., Karampetsos, D., & Philippopoulos, C. J. (2008). Photochemical treatment of 2-chlorophenol aqueous solutions using ultraviolet radiation, hydrogen peroxide and photo-Fenton reaction. *Journal of Hazardous Materials, 153*, 582–587. https://doi.org/10.1016/ j.jhazmat.2007.09.002

Pourehie, O., & Saien, J. (2020). Homogeneous solar Fenton and alternative processes in a pilot-scale rotatable reactor for the treatment of petroleum refinery wastewater. *Process Safety and Environment Protection, 135*, 236–243. https://doi.org/10.1016/j.psep.2020.01.006

Priyadarshini, M., Das, I., & Ghangrekar, M. M. (2020). Application of metal organic framework in wastewater treatment and detection of pollutants: Review. *Journal of the Indian Chemical Society, 97*, 507–512.

Rajabizadeh, K., Yazdanpanah, G., Dowlatshahi, S., & Malakootian, M. (2020). Photooxidation process efficiency (UV/O$_3$) for P-nitroaniline removal from aqueous solutions. *Ozone Science and Engineering, 42*, 420–427. https://doi.org/10.1080/01919512.2019.1679614

Rakness, K. L. (2011). *Ozone in drinking water treatment: Process design, operation and optimization*, 1st edn. American Water Works Association, Denver, USA.

Ridruejo, C., Centellas, F., Cabot, P. L., Sirés, I., & Brillas, E. (2018). Electrochemical Fenton-based treatment of tetracaine in synthetic and urban wastewater using active and non-active anodes. *Water Research, 128*, 71–81. https://doi.org/10.1016/j.watres.2017.10.048

Riesz, P., Berdahl, D., & Christman, C. L. (1985). Free radical generation by ultrasound in aqueous and nonaqueous solutions. *Environmental Health Perspectives, 64*, 233–252. https://doi.org/10. 1289/ehp.8564233

Rivas-Ortiz, I. B., Cruz-González, G., Lastre-Acosta, A. M., Manduca-Artiles, M., Rapado-Paneque, M., Chávez-Ardanza, A., Teixeira, A. C. S. C., & Jáuregui-Haza, U. J. (2017). Optimization of radiolytic degradation of sulfadiazine by combining Fenton and gamma irradiation processes. *Journal of Radioanalytical and Nuclear Chemistry, 314*, 2597–2607. https://doi.org/10.1007/s10 967-017-5629-8

Rodríguez-Chueca, J., Laski, E., García-Cañibano, C., Martín de Vidales, M. J., Encinas, Á., Kuch, B., & Marugán, J. (2018). Micropollutants removal by full-scale UV-C/sulfate radical based advanced oxidation processes. *Science of the Total Environment, 630*, 1216–1225. https://doi. org/10.1016/j.scitotenv.2018.02.279

Rodríguez, M., Malato, S., Pulgarin, C., Contreras, S., Curcó, D., Giménez, J., & Esplugas, S. (2005). Optimizing the solar photo-Fenton process in the treatment of contaminated water. Determination of intrinsic kinetic constants for scale-up. *Solar Energy, 79*, 360–368. https://doi.org/10. 1016/j.solener.2005.02.024

Sánchez Pérez, J. A., Román Sánchez, I. M., Carra, I., Cabrera Reina, A., Casas López, J. L., & Malato, S. (2013). Economic evaluation of a combined photo-Fenton/MBR process using pesticides as model pollutant. Factors affecting costs. *Journal of Hazardous Materials, 244–245*, 195–203. https://doi.org/10.1016/j.jhazmat.2012.11.015

Save, S. S., Pandit, A. B., & Joshi, J. B. (1997). Use of hydrodynamic cavitation for large scale microbial cell disruption. *Food and Bioproducts Processing, 75*, 41–49. https://doi.org/10.1205/ 096030897531351

Sayed, M., Khan, J. A., Shah, L. A., Shah, N. S., Khan, H. M., Rehman, F., Khan, A. R., & Khan, A. M. (2016). Degradation of quinolone antibiotic, norfloxacin, in aqueous solution using gamma-ray irradiation. *Environmental Science and Pollution Research, 23*, 13155–13168. https://doi.org/10.1007/s11356-016-6475-x

Shu, H. Y., & Chang, M. C. (2005). Pilot scale annular plug flow photoreactor by UV/H_2O_2 for the decolorization of azo dye wastewater. *Journal of Hazardous Materials, 125*, 244–251. https://doi.org/10.1016/j.jhazmat.2005.05.038

Sievers, M. (2011). Advanced oxidation processes. *Treatise Water Science, 4*, 377–408. https://doi.org/10.1016/B978-0-444-53199-5.00093-2

Song, K., Mohseni, M., & Taghipour, F. (2016). Application of ultraviolet light-emitting diodes (UV-LEDs) for water disinfection: A review. *Water Research, 94*, 341–349. https://doi.org/10.1016/j.watres.2016.03.003

Souza, F. S., & Féris, L. A. (2015). Degradation of caffeine by advanced oxidative processes: O_3 and O_3/UV. *Ozone Science and Engineering, 37*, 379–384. https://doi.org/10.1080/01919512.2015.1016572

Sun, M., Yao, Y., Ding, W., & Anandan, S. (2020). N/Ti^{3+} co-doping biphasic TiO_2/Bi_2WO_6 heterojunctions: Hydrothermal fabrication and sonophotocatalytic degradation of organic pollutants. *Journal of Alloys and Compounds, 820*, 153172. https://doi.org/10.1016/j.jallcom.2019.153172

Sun, Y., & Pignatello, J. J. (1993). Photochemical reactions involved in the total mineralization of 2,4-D by iron(3+)/hydrogen peroxide/UV. *Environmental Science and Technology, 27*, 304–310. https://doi.org/10.1021/es00039a010

Swinwood, J. F., Waite, T. D., Kruger, P., & Rao, S. M. (1994). Radiation technologies for waste treatment: A global perspective. *IAEA Bulletin, 36*, 11–15.

Teng, J., Liu, G., Liang, J., & You, S. (2020). Electrochemical oxidation of sulfadiazine with titanium suboxide mesh anode. *Electrochimica Acta, 331*, 135441. https://doi.org/10.1016/j.electacta.2019.135441

Thanekar, P., Garg, S., & Gogate, P. R. (2020). Hybrid treatment strategies based on hydrodynamic cavitation, advanced oxidation processes, and aerobic oxidation for efficient removal of naproxen. *Industrial and Engineering Chemistry Research, 59*, 4058–4070. https://doi.org/10.1021/acs.iecr.9b01395

Thomas, S., Rayaroth, M. P., Menacherry, S. P. M., Aravind, U. K., & Aravindakumar, C. T. (2020). Sonochemical degradation of benzenesulfonic acid in aqueous medium. *Chemosphere, 252*, 126485. https://doi.org/10.1016/j.chemosphere.2020.126485

Trapido, M., Veressinina, Y., Munter, R., & Kallas, J. (2005). Catalytic ozonation of m-dinitrobenzene. *Ozone Science and Engineering, 27*, 359–363. https://doi.org/10.1080/01919510500250630

Turhan, K., Durukan, I., Ozturkcan, S. A., & Turgut, Z. (2012). Decolorization of textile basic dye in aqueous solution by ozone. *Dyes and Pigments, 92*, 897–901. https://doi.org/10.1016/j.dyepig.2011.07.012

USEPA. (2006). National primary drinking water regulations: Long term 2 enhanced surface water treatment rule (40 CFR Parts 9, 141, and 142). *Federal Register, 71*, 654–786.

Wang, J., & Wang, S. (2021). Effect of inorganic anions on the performance of advanced oxidation processes for degradation of organic contaminants. *Chemical Engineering Journal, 411*, 128392. https://doi.org/10.1016/j.cej.2020.128392

Wang, J. L., & Xu, L. J. (2012). Advanced oxidation processes for wastewater treatment: Formation of hydroxyl radical and application. *Critical Reviews in Environment Science and Technology, 42*, 251–325. https://doi.org/10.1080/10643389.2010.507698

Wang, S., & Wang, J. (2018). Radiation-induced degradation of sulfamethoxazole in the presence of various inorganic anions. *Chemical Engineering Journal, 351*, 688–696. https://doi.org/10.1016/j.cej.2018.06.137

Wu, Y., Chen, R., Liu, H., Wei, Y., & Wu, D. (2013). Feasibility and mechanism of p-nitrophenol decomposition in aqueous dispersions of ferrihydrite and H_2O_2 under irradiation. *Reaction Kinetics, Mechanisms and Catalysis, 110*, 87–99. https://doi.org/10.1007/s11144-013-0571-4

Wu, Z., Zhang, G., Zhang, R., & Yang, F. (2018). Insights into mechanism of catalytic ozonation over practicable mesoporous $Mn-CeO_x/\gamma-Al_2O_3$ catalysts. *Industrial and Engineering Chemistry Research, 57*, 1943–1953. https://doi.org/10.1021/acs.iecr.7b04516

Xia, Y., Wang, G., Guo, L., Dai, Q., & Ma, X. (2020). Electrochemical oxidation of Acid Orange 7 azo dye using a PbO_2 electrode: Parameter optimization, reaction mechanism and toxicity evaluation. *Chemosphere, 241*, 125010. https://doi.org/10.1016/j.chemosphere.2019.125010

Xiang, H., Shao, Y., Gao, N., Lu, X., An, N., & Chu, W. (2020). Removal of β-cyclocitral by UV/persulfate and UV/chlorine process: Degradation kinetics and DBPs formation. *Chemical Engineering Journal, 382*, 122659. https://doi.org/10.1016/j.cej.2019.122659

Xiong, W., Chen, N., Feng, C., Liu, Y., Ma, N., Deng, J., Xing, L., & Gao, Y. (2019). Ozonation catalyzed by iron- and/or manganese-supported granular activated carbons for the treatment of phenol. *Environmental Science and Pollution Research, 26*, 21022–21033. https://doi.org/10.1007/s11356-019-05304-w

Xu, Y., Lin, Z., Zheng, Y., Dacquin, J. P., Royer, S., & Zhang, H. (2019). Mechanism and kinetics of catalytic ozonation for elimination of organic compounds with spinel-type $CuAl_2O_4$ and its precursor. *Science of the Total Environment, 651*, 2585–2596. https://doi.org/10.1016/j.scitotenv.2018.10.005

Xu, Y., Wu, Y., Zhang, W., Fan, X., Wang, Y., & Zhang, H. (2018). Performance of artificial sweetener sucralose mineralization via UV/O_3 process: Kinetics, toxicity and intermediates. *Chemical Engineering Journal, 353*, 626–634. https://doi.org/10.1016/j.cej.2018.07.090

Yang, H., Li, Y., Chen, Y., Ye, G., & Sun, X. (2019). Comparison of ciprofloxacin degradation in reclaimed water by UV/chlorine and UV/persulfate advanced oxidation processes. *Water Environment Research, 91*, 1576–1588. https://doi.org/10.1002/wer.1144

Yang, W., Zhou, M., Oturan, N., Bechelany, M., Cretin, M., & Oturan, M. A. (2020). Highly efficient and stable $Fe^{II}Fe^{III}$ LDH carbon felt cathode for removal of pharmaceutical ofloxacin at neutral pH. *Journal of Hazardous Materials, 393*, 122513. https://doi.org/10.1016/j.jhazmat.2020.122513

Yang, W., Zhou, M., Oturan, N., Li, Y., & Oturan, M. A. (2019). Electrocatalytic destruction of pharmaceutical imatinib by electro-Fenton process with graphene-based cathode. *Electrochimica Acta, 305*, 285–294. https://doi.org/10.1016/j.electacta.2019.03.067

Yogev, A., & Yakir, D. (1999). Bioreactor and system for improved productivity of photosynthetic algae. *United States Patent US, 5*(958), 761.

Yousef Tizhoosh, N., Khataee, A., Hassandoost, R., Darvishi Cheshmeh Soltani, R., & Doustkhah, E. (2020). Ultrasound-engineered synthesis of $WS_2@CeO_2$ heterostructure for sonocatalytic degradation of tylosin. *Ultrasonics Sonochemistry, 67*, 105114. https://doi.org/10.1016/j.ultsonch.2020.105114

Yu, L., Iranmanesh, S., Keir, I., & Achari, G. (2020). A field pilot study on treating groundwater contaminated with sulfolane using UV/H_2O_2. *Water, 12*, 1200. https://doi.org/10.3390/w12041200

Yu, S., Hu, J., & Wang, J. (2010). Gamma radiation-induced degradation of p-nitrophenol (PNP) in the presence of hydrogen peroxide (H_2O_2) in aqueous solution. *Journal of Hazardous Materials, 177*, 1061–1067. https://doi.org/10.1016/j.jhazmat.2010.01.028

Yuan, X., Xie, R., Zhang, Q., Sun, L., Long, X., & Xia, D. (2019). Oxygen functionalized graphitic carbon nitride as an efficient metal-free ozonation catalyst for atrazine removal: Performance and mechanism. *Separation and Purification Technology, 211*, 823–831. https://doi.org/10.1016/j.sep pur.2018.10.052

Zaouak, A., Noomen, A., & Jelassi, H. (2018). Gamma-radiation induced decolorization and degradation on aqueous solutions of Indigo Carmine dye. *Journal of Radioanalytical and Nuclear Chemistry, 317*, 37–44. https://doi.org/10.1007/s10967-018-5835-z

Zhang, A., & Li, Y. (2014). Removal of phenolic endocrine disrupting compounds from waste activated sludge using UV, H_2O_2, and UV/H_2O_2 oxidation processes: Effects of reaction conditions and sludge matrix. *Science of the Total Environment, 493*, 307–323. https://doi.org/10.1016/j.sci totenv.2014.05.149

Zhang, M. H., Dong, H., Zhao, L., Wang, D. X., Meng, D. (2019). A review on Fenton process for organic wastewater treatment based on optimization perspective. *Science of the Total Environment, 670*, 110–121. https://doi.org/10.1016/j.scitotenv.2019.03.180

Zhang, X., Dong, W., & Yang, W. (2013). Decolorization efficiency and kinetics of typical reactive azo dye RR2 in the homogeneous Fe(II) catalyzed ozonation process. *Chemical Engineering Journal, 233*, 14–23. https://doi.org/10.1016/j.cej.2013.07.098

Zhu, Y., Wu, M., Gao, N., Chu, W., Li, K., & Chen, S. (2018). Degradation of phenacetin by the UV/chlorine advanced oxidation process: Kinetics, pathways, and toxicity evaluation. *Chemical Engineering Journal, 335*, 520–529. https://doi.org/10.1016/j.cej.2017.10.070

Disinfection of Wastewater

18

Domestic wastewater contains pathogens and microorganisms of virulence, which are detrimental to the human health. A report by the world health organisation suggests that globally, at least two billion people use water contaminated with faecal matter for potable purpose (Fact sheet, WHO). Water borne diseases, such as cholera, diarrhoea, amoebic dysentery, are epidemic menaces that have lurched in different parts of the world till the last decade. Even today, in under developed countries, a major fraction of population is deprived of safe and pathogen free water. One of the major reasons for such pathogenic contamination is the discharge of untreated or partially treated sewage in water bodies and open defecation. The malpractice of open defecation can be eradicated by providing sanitary toilet facilities to general population by government and non-governmental social initiatives. A classic example of the same would the construction and/or aid for construction of toilet under the 'Swachh Bharat' initiative launched by the Indian government in 2014. However, to control the pathogen content of treated wastewater, disinfection strategies for the sewage is important. More importantly, the reuse of treated sewage is being increasingly adopted, which demands proper disinfection of sewage to ensure safe usage.

18.1 Introduction

As the concept of disinfection started with water supply systems, hence the modes or agents used as disinfectants are same for both wastewater and water treatment plants. Although, the strategies adopted are similar, the disinfection dose vary drastically owing to the high pathogen contents of sewage. Moreover, the characteristics of wastewaters are more complex than fresh water, thus the methodology, dose–response patterns and the disinfection models are different from those applied for the water treatment plants. The pathogen count of the sewage is expressed in terms of most probable number (MPN) of

© The Author(s), under exclusive license to Springer Nature Singapore Pte Ltd. 2022
M. Ghangrekar, *Wastewater to Water*, https://doi.org/10.1007/978-981-19-4048-4_18

viable cells per 100 mL and colony-forming units (*cfu*) per 100 mL. In a typical raw sewage, the *cfu*/100 mL for *Escherichia coli*, which is an indicator of faecal coliforms, is in the order of 10^5–10^{10} (Chahal et al., 2016).

The discharge quality norms of treated sewage vary according to intended mode of disposal or usage and also country of implementation. For instance, the Indian norms for the non-potable contact usage, such as car washing, fire-fighting as well as construction water, require complete removal of pathogens to ensure nil pathogen count (CPHEEO, 2013). For other purposes such as discharge over land the limit is 100 MPN/100 mL of treated sewage. Although the standards vary from country to country, the need for disinfection of treated sewage is recognized globally. The disinfection strategies range from chemical to physical perturbative forces. This chapter provides a detailed description of different strategies adopted for disinfection along with solved design problems for different type of disinfection strategy. This chapter is dedicated towards understanding the current disinfection practices for sewage. Established processes, such as chlorination, ozonation, UV radiation and combined processes of UV/H_2O_2 and O_3/H_2O_2 are also discussed as prospective future processes in context of present case studies done on these integrated technologies.

18.1.1   Need for Disinfection

The discharge of sewage in inland fresh water bodies should be preceded by decontamination of this treated sewage by following different disinfection strategies. Present practice in developing as well as developed countries however confines to reducing the pathogen to a certain threshold value prior to discharge. The threshold value is often governed by two factors: (a) the first being the idea that the treated sewage would not immediately come in contact with humans, and (b) the second being subsequently since the water is not directly put to contact usage over the period of time, natural death of pathogens will occur because of not getting conducive environment for their survival when released in environment. Hence, it is considered as targeting complete disinfection would be a redundant cost in the treatment chain. However, with the advanced understanding of the diseases caused by the microbiota and the paradigm shift in fate of treated sewage towards reuse for non-potable contact usage in different countries, considerable attention is being given to disinfection of sewage.

Disinfection either by chemical reactive species or by physical forces require additional cost and infrastructure owing to the high pathogens load in the sewage. Hence, the pretreatment options for disinfection are often chosen in purview of reducing the organic contaminant load to an acceptable operating level for the application of the disinfectant. For example, removal of dissolved organic matter via bacteriological metabolism can reduce the chlorine dose required during chlorination as the chlorine would react both with the cellular matrix as well as the dissolved organic compounds indiscriminately. Similarly, the presence of suspended and dissolved organic matter in high concentration also affects the performance of the UV-C irradiation by absorbing the radiation and thereby increasing the operating cost as UV dose has to be increased in such case.

A common practice for disinfection of secondary treated sewage is to provide tertiary treatment as multiple stage physical barriers in the form of settlers, flocculators and multi-grade filters as pre-treatment to the disinfection units. Such pre-treatment options reduce the organic matter content thereby reducing load on disinfection units. The provision of multi-grade filter as pre-treatment also reduces the protozoa as well as to some extent the helminth eggs content of the sewage. However, such strategies are only successful to the extent that would only enable easy functioning of the downstream disinfection units and cannot be considered as substitute to the same. Even with such preventive barriers in place, the pathogen load varies between 10^3 and 10^5 levels of *cfu*/mL for the influent going into the disinfection unit. This makes us understand the dire need of disinfection prior to utilizing this water for human contact reuse purposes.

18.1.2 Present Disinfection Practices

The disinfection strategies that are recognized and practiced on a commercial scale range from addition of chemicals to mechanical means for disinfecting secondary treated sewage. In addition, few experimental trials of electromagnetic waves other than UV irradiation, pasteurization have also been performed, which however requires more investigation prior to commercialization. The prevalent strategies which are commonly adopted are described below.

18.1.2.1 Chemical Agents

The chemical species for disinfection include halogens (chlorine and iodine), potassium permanganate, ozone, surfactants and in some cases metals, such as Ag. Among the wide range of chemical disinfectants available, chlorination and ozonation are considered as most effective strategies. Application of permanganate to water causes aesthetic issues as it imparts pink colour to the water. Application of immobilized Ag nanoparticles have been advocated in the past research investigation; however, the usage is limited owing to risk of contamination to higher life forms in case of leaching. Additionally, Ag metal is precious and increases the cost of the disinfection many folds.

Iodine has a greater chemical stability, less reactive with nitrogenous species and a wider applicable pH range as compared to chlorine. In spite of these advantages, iodine is less preferred to chlorine, owing to the less reactivity of iodine, which begets higher dosage. Also, the effect of higher concentrations of iodine on higher life forms is yet to be established by conducting thorough scientific investigations. Unlike iodine, bromine was extensively used as a disinfectant since early 1930s in the countries like USA (Fact Sheet, Lenntech). However, the usage of bromine as disinfectant in wastewater is completely stopped owing to its high potential of forming carcinogenic disinfection-by-products (DBPs).

18.1.2.2 Physical Forces and Radiation

The physical forces including heat, UV radiation and ultrasound wave are used for disinfection. In case of chemical disinfectants, the major mode of disinfection is achieved by lysis of the cell membrane by the reactive species generated from the applied chemical disinfectants. Unlike chemical oxidation, the mechanism of disinfection by physical forces varies from one to another. The application of heat has colloidal effect on the cytoplasmic protein, thus terminating the vital life processes of the microorganism, while the UV radiation alters the DNA structure. The UV rays stimulate the formation of dimers of thymine, one of the bases of the nucleic acid of the DNA, which in turn prevents replication of the microbe. In case of RNA, the UV radiation affects the uracil, which is another pyrimidine like thymine. In both cases, the formation of dimer in the corresponding pyrimidines of DNA and RNA prevent replication of DNA/RNA and eventually leads to inactivation of the bacteria.

Other than UV irradiation, the effect of other electromagnetic radiations, such as gamma rays, also have been explored for disinfection. However, the commercial application of such radiation is yet to be realised. Presently the prevalent physical disinfecting mechanisms include exposure of the secondary treated sewage to UV radiation after sufficient pre-treatment to remove solids and to a major extent turbidity to avoid shading effect. In certain cases, ultrasound vibration enabled were reported as disinfectant. The ultrasound vibrations require very high energy for effective disinfection; hence, it is often used as a complementary unit for chemical species mediated disinfection to activate the reactive species.

As mentioned, the disinfection of sewage can be realised by adding chemical species to water that can dissociate or react in water to form highly reactive species or by introducing physical agitating forces, such as UV-C radiation or ultrasonic vibration. In case of chemical reactive species addition, there might be a residual concentration of the reactive species in the effluent of the disinfection unit, which is often beneficial for combatting re-contamination during storage, transportation and reuse. In fact, similar to water treatment schemes it is recommended to adjust the dosage of the chemical species in order to produce a residual concentration for ensuring protection again re-contamination.

In case of chlorine or other reactive species present in dissolved form, such residual action can be achieved by drawing the break-point curves or experimental curves to determine the initial dose required for producing a desired residual concentration. However, in case of ozone, administering a residual concentration is difficult owing to the fugitive nature of ozone and difficulty in monitoring, which requires expensive online ozone sensors. In case of UV and US, which can be implemented as physical perturbations for the inactivation/destruction of pathogen, such residual effect cannot be registered. Hence, in certain cases, where the strategy is to integrate two or more combinations of disinfectant

Fig. 18.1 Comparison of the physical and chemical agents for disinfection (DBP-disinfection by-products)

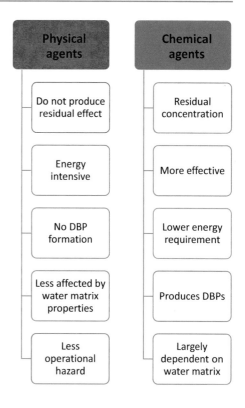

techniques together to achieve a surer result and reliability, both in terms of real time disinfecting ability and to ensure a shield against future contamination. The benefits and drawbacks of different processes are highlighted in Fig. 18.1.

18.1.3   Selection of Disinfectant and Influencing Parameters

The selection criteria for disinfectant are governed by: (1) availability of the disinfectant material, (2) affordability in terms of chemical cost/energy consumption, (3) should not be corrosive to moving parts and other material used in the plant, (4) should concomitantly deodorise the wastewater and should not leave any residual offensive odour, (5) can be transported easily, (6) should be effective in a wide range of ambient temperature and pH range of secondary treated sewage, and (7) should be effective against wide range of microorganisms. The criteria are self-explanatory and can be followed while selecting any chemical or physical perturbative force as a disinfectant. However, the checklist given above is a generalised criteria and the type of disinfectant to be administered might also be decided by the designer.

In addition to these criteria, there are a set of universal parameters that influence the performance of a disinfecting technique and can also aid in selecting a disinfectant and determining its dose. The influencing parameters are: retention time; dose to be administered; purity of the disinfectant in case of chemical disinfectants; intensity of physical agents, such as UV and US; and organic matter content in the influent of the disinfection unit. In addition to these parameters, the different technology/technique specific influencing factors will be elaborated in the respective subsections. A general overview of these parameters is a pre-requisite for the practicing engineer to design disinfection units for sewage treatment.

18.1.3.1 Influence of Retention Time

The effect of contact time of disinfectant can be summarized by Chick's law, which states that the extent to which disinfection/inactivation is achieved is directly proportional to the concentration N_t at a given time t (Chick, 1908). The law can be summarized in a differential equation form by establishing relationship of the change in the concentration of the organism dN_t in a time period dt with the initial concentration (Eq. 18.1).

$$\frac{dN_t}{dt} = -kN_t \tag{18.1}$$

This indicates that the Chick's law considers the reaction rate to be only dependent on the concentration of organisms at a given time t and follows a first order rate kinetics. However, in reality, the residence time is a more complex parameter, which can be dependent on temperature, presence of unwanted reactive moieties and also on the purity of the disinfectant. In case of physical forces, such as UV, the naturally occurring organic and inorganic suspended and dissolved content that diminishes UV-C radiation will prolong the time t.

18.1.3.2 Disinfectant Concentration

The concentration of the disinfectant (C) has a profound effect on the inactivation rate constant. The relationship was established in 1908 by Watson, while working in England for disinfection of wastewater (Watson, 1908). The relationship of k with C can thus be written as per Eq. 18.2.

$$k = k'C^n \tag{18.2}$$

where, k' is the die-off constant and n is the coefficient related to dilution. Combining the two Eq. 18.1 with Eq. 18.2, we can get the modified equation for the relationship between N_t, t and C, which can be integrated to derive the Eq. 18.3.

$$N_t = N_0.e^{-k'C^n t} \tag{18.3}$$

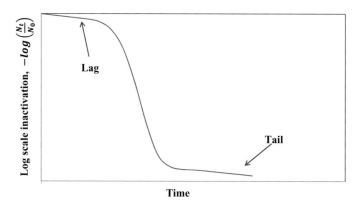

Fig. 18.2 Lag and tailing effect owing to presence of dissolved organic matter and other competing ions

The Eq. 18.3 is rearranged to a linearized form for expressing C in terms of t and other terms (Eq. 18.4). A log plot of concentration versus time can aid in determining the value of k' from the slope and intercept value of the graph for the linear form of the Eq. 18.4.

$$\ln C = -\frac{1}{n}\ln t + \frac{1}{n}\ln\left[\frac{1}{k'}\left(-\ln\frac{N_t}{N_0}\right)\right] \qquad (18.4)$$

However, in reality, the C versus t graph does not follow this pattern and it is found to be a double shouldered graph with a lag or a tail phase owing to the influence of competing ions and dissolved organic matter (Fig. 18.2).

18.1.3.3 Effect of Temperature

Similar to other chemical reactions, the effect of temperature on the disinfection mechanism can be understood by the van't Hoff-Arrhenius equation (Eq. 18.5). The relationship also takes into account the dependency on time for achieving a certain order of pathogen killing. The relationship indicates that a higher temperature can results in a rapid killing or in other words within a short interval of time the pathogens can be destructed.

$$\ln\frac{t_1}{t_2} = \frac{E_a(T_2 - T_1)}{RT_2T_1} \qquad (18.5)$$

where, R is the universal gas constant having a value of 1.99 cal/mol K and E_a is the activation energy in cal/mol K, t_1 and t_2 are the time required for achieving specified log scale kill at temperatures T_1 and T_2, respectively.

Among other factors that determines the selection, the properties of a disinfecting agent, the types of microorganisms, and the quality of the water to be disinfected are also

Table 18.1 N_t values for different disinfectant concentration and contact time

Time, mins	Disinfectant concentration (mg/L)		
	2	4	6
	Coliform count in MPN/100 mL		
5	65,111	54,111	24,331
10	36,433	19,324	9,765
15	17,111	6718	2789
30	4533	546	91

coming into picture. Chemical disinfectants, such as ozone and chlorine, release reactive oxygen species (ROS) and reactive chlorine species (RCS), that are scavenged by carbonate and other interfering chemical species present in the water matrix. In addition to the inorganic dissolved species, the dissolved organic matter (DOM) also interferes with the ROS and the RCS released by the O_3 and Cl_2 disinfectants, respectively. The subsequent sections will provide detailed description for the popular disinfection techniques.

Example: 18.1 Determine the constant value for the Chick and Watson equation for a given value of dataset in Table 18.1. The average concentration of coliforms in raw sewage is 2.512×10^6 MPN/100 mL. Using these values predict whether the disinfectant will be able achieve the discharge standard of 300 MPN/100 mL for a given retention time of 25 min. Also, predict the value of the constants n and k' for the estimated N/N_0 value corresponding to effluent standard.

Solution

The data provided is used to find the $-\ln\left(\frac{N_t}{N_0}\right)$ values for the different contact times and corresponding disinfectant concentration. For instance, for a retention time of 10 min and a disinfectant concentration of 4 mg/L, a MPN value of 19,324 per 100 mL in the effluent is reported, hence the $-\ln (N/N_0)$ value is estimated as:

$$-\ln\left(\frac{N_t}{N_0}\right) = -\ln\left(\frac{19324}{2.512 \times 10^6}\right) = 4.867$$

Similarly, the deduced values for other concentrations of the same and different retention times are presented in Table 18.2.

Subsequently, a plot for $-\ln\left(\frac{N_t}{N_0}\right)$ versus contact time was made as given in Fig. 18.3. The target coliform count in outlet is 300 MPN/100 mL. Hence, the $-\ln\left(\frac{N_t}{N_0}\right)$ value will be:

$$-\ln\left(\frac{N_t}{N_0}\right) = -\ln\left(\frac{300}{2.512 \times 10^6}\right) = 9.032$$

Time, mins	Disinfectant concentration (mg/L)		
	2	4	6
	$-\ln(N/N_0)$ values		
5	3.653	3.838	4.637
10	4.233	4.867	5.550
15	4.989	5.924	6.803
30	6.317	8.434	10.226

Table 18.2 Values for $-\ln\left(\frac{N_t}{N_0}\right)$ at different disinfectant concentration and contact time

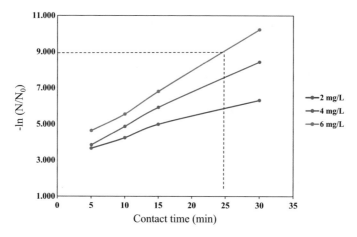

Fig. 18.3 Plot of $-\ln(N/N_0)$ versus contact time in min

Fig. 18.4 ln (C) versus ln (t) graph

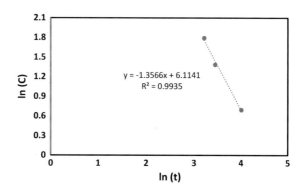

The corresponding retention time from the plot for this $-\ln\left(\frac{N_t}{N_0}\right)$ value for the three different disinfectant concentration is estimated from the plot and values are presented in Table 18.3.

Table 18.3 Cl_2 dose and corresponding retention time in minutes	C (mg/L)	t, min
	2	54.98
	4	31.43
	6	24.84

Hence, it can be seen from the values presented in Table 18.3 that a disinfectant concentration of 6 mg/L would be required for achieving 300 MPN/100 mL within a retention time of 25 min. The Chick-Watson equation constants, the concentration versus retention time was plotted as shown in Fig. 18.4 using the values from Table 18.3.

Form the slope of the straight-line fit, $-\frac{1}{n} = -1.3566$.

Hence, $n = 0.73714$.

Further, solving for k', $\ln C = -\frac{1}{n} \ln t + \frac{1}{n} \ln \left[\frac{1}{k'} \left(-\ln \frac{N_t}{N_0} \right) \right]$.

For a given retention time of 25 min and a disinfectant concentration of 6 mg/L, the final coliform is ca. 300 MPN/100 mL, the $\ln \frac{N_t}{N_0}$ value being -9.302. Therefore, putting the values in the above equation we get:

$$\ln 6 = -1.3566 \times \ln 25 + 1.3566 \times \ln \left[\frac{1}{k'} (9.032) \right]$$

Solving,

$$\frac{1}{1.3566} \ln 6 = \ln \left[\frac{9.032}{25 \times k'} \right]$$

Taking antilog on both sides,

$$6^{\frac{1}{1.3566}} = \frac{9.302}{25 \times k'}$$

Hence, $k' = 0.0965$.

Thus, the final equation is $\ln \frac{N_t}{N_0} = -0.0965 C^{0.737}$.

18.2   Chlorination

Chlorination is perhaps the oldest disinfection strategy adopted for disinfection of water and eventually wastewater. The first use of chlorine as a disinfectant was possibly reported in 1897 when it was used as in a water supply system in Maidstone, Kent, UK to curb the effects of a typhoid outbreak (Chamberlin, 1948). Following this, continuous application of hypochlorous acid (produced by mixing ferric chloride coagulant with calcium hypochlorite) was conducted at Middelkerke, Belgium as a disinfectant in water supply systems in 1902. Subsequently, in the years 1903, 1905 and 1908, the usage of chlorine in the form of sodium hypochlorite and bleaching powder became a common practice for controlling water borne pandemics and for general decontamination of river and other sources of water. Over the turn of the century, chlorination was adopted as the default strategy of disinfection by the industrialized countries followed by the developing countries in the later decades (Humans, 1991).

The first usage of chlorine for disinfection of wastewater was recorded in 1914 in Altoona and Milwaukee, USA (Chamberlin, 1948). The advantage of chlorine is that it easily dissolves in water (solubility of chlorine gas at 15.5 °C is 7 g/L). Chlorine can be administered as Cl_2 gas or as hypochlorite of sodium or calcium or as chlorine dioxide. The gaseous form can be liberated from unconfined liquid chlorine, which is an amber coloured liquid with a specific gravity of 1.47. However, liquid chlorine readily vaporises to gaseous form to release around 450 L of gas per litre of liquid chlorine. Although, the liquid chlorine is a cheaper option for chlorine-based disinfection, the logistic hazard associated with the same and the risk of leakage and contamination of operating personnel makes liquid Cl_2 as a less popular choice as a chlorination agent.

Among the other two forms, sodium hypochlorite is a light and heat sensitive liquid that has to be stored in a cool-dark place or black coloured corrosion resistant opaque containers to avoid loss of strength. Sodium hypochlorite is costly as compared to liquid chlorine and calcium hypochlorite. In comparison to NaOCl, the $Ca(OCl)_2$, is available in both solid and wet form. Commercial $Ca(OCl)_2$ is less costly than NaOCl and is sold with a purity ranging between 65 and 70%. Bleaching powder, a common form of chlorine-based disinfectant has $Ca(OCl)_2$ as its main ingredient. It is also known as calcium oxychloride and it is prepared by passing chlorine gas over dry slaked lime $(Ca(OH)_2)$ (Eq. 18.6). The bleaching powder liberates chlorine when dissolved in water and it is commonly used to bleach cloths and as a disinfectant for public swimming pools and in water treatment plants.

$$2Ca(OH)_2 + 2Cl_2 \rightarrow Ca(OCl)_2 + CaCl_2 + 2H_2O \qquad (18.6)$$

Although, the main ingredient of bleaching powder is calcium hypochlorite, presence of other impurities, such as lower hydrated salts of $CaCl_2$, unreacted $Ca(OH)_2$, and the basic chloride $CaCl_2 \cdot Ca(OH)_2 \cdot H_2O$, reduces the available chlorine (Bunn et al., 1935). This basic chloride of calcium prevents the deliquescence of the $CaCl_2$ and hence provides a consistency to the commercial bleaching powder. However, the $CaCl_2$ is an unwanted side product that reduces the available chlorine content in the powder to a maximum of 35–40% even in the most controlled operating conditions (Bunn et al., 1935). The available chlorine for bleaching powder is thus usually in the range of 25–35% based on the manufacturer's recommendation. Nevertheless, it is recommended to test the strength of the bleaching powder prior to administering bulk usage to avoid incomplete sterilization owing to compromised or reduced strength due to environmental exposure of the granules or powder.

A different chlorination technique is electro-chlorination by generating chlorine ions in situ from a chloride salt using an electrochemical setup. In an electro-chlorinator concept, an electrolysis cell is fed with brine water to produce NaOCl and H_2 as the end product. This is a cleaner technology as compared to procurement of Cl_2 as it reduces the risk of logistics and storage issues. However, this process is energy intensive and requires additional supply of distilled or demineralised water for preparing brine solution.

18.2.1 Strategies for Chlorination

As mentioned previously, chlorination can be done by dosing liquid chlorine, chlorine dioxide or hypochlorite of sodium or calcium that liberate chlorine when dissolved in water. The chlorine gas dissolves in water to form HOCl and HCl (Eq. 18.7). The formed HOCl further ionizes to H^+ and OCl^- (Eq. 18.8)

$$Cl_2 + H_2O \leftrightarrow HOCl + H^+ + Cl^- \tag{18.7}$$

$$HOCl \leftrightarrow H^+ + OCl^- \tag{18.8}$$

It should be noted that the disinfection efficiency of HOCl is 40 times more than OCl^- and hence for disinfection, it is desirable that the dissolved chlorine prevails in the HOCl form. It is thus important to ascertain the percentage/fraction of HOCl present in the solution at a particular temperature and pH. The fraction of HOCl present is given as per Eq. 18.9.

$$\frac{[HOCl]}{[HOCl] + [OCl^-]} = \frac{1}{1 + [OCl^-]/[HOCl]} \tag{18.9}$$

Further, the dissociation of HOCl is governed by the ionization constant of HOCl dissociation, (K_i) which can be evaluate from Eq. 18.10. Combining these two expressions, we get the fraction of HOCl present in the solution as a function of ionization constant and pH (Eq. 18.11). The ionization constant K_i of acid is expressed as the acid dissociation constant K_a, and the corresponding values of the pK_a can be utilized to find the fraction of HOCl at any given temperature and pH. Table 18.4 gives the acid dissociation constants expressed as pK_a (Morris, 1966), while the relationship between pK_a is given in Eq. 18.12.

$$K_i = \frac{[H^+][OCl^-]}{[HOCl]} \tag{18.10}$$

$$\frac{[HOCl]}{[HOCl] + [OCl^-]} = \frac{1}{1 + K_i/[H^+]} = \frac{1}{1 + K_i 10^{pH}} = \frac{1}{1 + K_a 10^{pH}} \tag{18.11}$$

$$pK_a = -\log_{10}(K_a) \tag{18.12}$$

When hypochlorite salts are added to water, the respective hydroxides and hypochlorous acid are formed (Eqs. 18.13 and 18.14). It should be noted that the available chlorine in a compound as well as the purity of the compound needs to be determined prior to ascertaining weight of compound to be dosed to the water for effective disinfection. Hence, in case of commercial products, such as the bleaching powder that contains 35–40% available chlorine, the required weight of the compound can be found out by using Eq. 18.15.

$$Ca(OCl)_2 + 2H_2O \rightarrow HOCl + Ca(OH)_2 \tag{18.13}$$

$$NaOCl + H_2O \rightarrow HOCl + NaOH \tag{18.14}$$

$$Cl_2\,avialable = \frac{weight\,of\,Cl_2\,in\,the\,compound}{weight\,of\,the\,compound} \times purity\,of\,the\,compound \tag{18.15}$$

The dosage of chlorine required for disinfection is affected by the water matrix parameters as mentioned earlier. For instance, when the chlorine-based disinfectant is added to wastewater, the chlorine first reacts with reactive reducing species present in water and then with the ammonium compounds (refer Sect. 18.2.2 for details). The oxidation of both live and dissolved organic matter starts after these species are consumed. Hence, the dose required for a wastewater has to be determined by first conducting the breakpoint chlorination test, which is explained in detail in the next section.

Example: 18.2 Determine the fraction of HOCl present in a wastewater at 25 °C and at a pH of 7.5. If the initial dose of HOCl is 10 mg/L, what is the absolute concentration of

Table 18.4 Acid dissociation constant values of HOCl at different temperatures

Temperature in C°	Corresponding pK_a values
0	7.825
5	7.754
10	7.690
15	7.633
20	7.582
25	7.537
30	7.497
35	7.463

Source Data adopted from Morris (1966)

HOCl and OCl$^-$ considering a NaOCl to HOCl conversion ratio of 85%. The pK_a value at 25 °C is 7.537.

Solution

At 25 °C, the pK_a value is 7.537. Hence, the corresponding K_a value (Eq. 18.12) is = $10^{-7.537} = 2.904 \times 10^{-8}$ mole/L

$$K_a \times 10^{pH} = 2.904 \times 10^{-8} \text{mole/L} \times 10^{7.5} = 0.917$$

$$\frac{[\text{HOCl}]}{[\text{HOCl}] + [\text{OCl}^-]} = \frac{1}{1 + K_a 10^{pH}} = \frac{1}{1 + 0.917} = 0.521$$

Thus, the fraction of HOCl present in the wastewater is 0.521 and that for OCl$^-$ is 0.479.

The stoichiometric equation for dissociation of NaOCl in water is given by Eq. 18.14. It can be seen from the equation that, 1 mol of NaOCl produces 1 mol HOCl.

NaOCl + H$_2$O → HOCl + NaOH.

However, the conversion efficiency of NaOCl to HOCl is 85%. Hence only $\frac{85}{100} \times 10 = 8.5$ mg/L of NaOCl gets converted to HOCl.

Hence, molar concentration of NaOCl equivalent to 8.5 mg/L = $\frac{8.5 \times 10^{-3}}{74.44} = 0.1142$ mM.
Corresponding quantity of HOCl moles liberated = 0.114 mM. The concentration of HOCl liberated is = $0.1142 \times 52.46 = 5.99$ mg/L.

Since the fraction of HOCl at equilibrium is 0.521, hence the concentration of HOCl at equilibrium is = $5.99 \times 0.521 = 3.12$ mg/L.

Similarly, the equilibrium concentration of OCl$^-$ is = $5.99 \times 0.479 \times 51.46/52.46 = 2.815$ mg/L.

Example: 18.3 A company is selling bleaching powder with a label of 20% available chlorine. For disinfection of wastewater, desired Cl$_2$ concentration is 4 mg/L. Calculate the bleaching powder required for a disinfection of secondary treated sewage with a daily flow of 300 MLD. What is the required dose of bleaching powder if the available chlorine is 15% by weight? What is the additional cost implication considering cost of Rs. 20/kg-bleaching powder?

Solution

Daily Cl$_2$ requirement = $300 \times 10^6 \times 4 \times 10^{-6} = 1200$ kg.

Considering 20% available chlorine, bleaching powder required per day = $\frac{1200}{0.2}$ = 6000 kg.

Cost of the bleaching powder required per day = 6000×20 =Rs.1,20,000.

Considering 15% available chlorine, cost of bleaching powder = $\frac{1200}{0.15} \times 20$ = Rs.1,60,000.

Hence, additional cost = Rs. 1,60,000–Rs. 1,20,000 = Rs. 40,000 per day (**Answer**).

18.2.2 Determination of Chlorine Dose Using Chlorination Model

The concept of $C_R t$ is used popularly in disinfection process to express the dose and retention time relationship as well as to control the disinfection process. This concept stems from the fact that to achieve a certain level of disinfection (say 3-log scale inactivation), both disinfectant dose as well as retention time are important. The concept of $C_R t$, which is the product of the residual Cl_2 concentration and the contact time t, takes into account both these deciding factors to achieve the desired level of disinfection. Modelling of chlorination has been demonstrated by different past investigations. The most accepted among them is the model developed by Collins in 1970's, which was later modified by Selleck and subsequently by White in 1978 and 1999, respectively (Selleck et al., 1978; White, 1999). The final modified model is given by Eq. 18.16.

$$\frac{N}{N_0} = 1 \text{ for } C_R t < b$$

$$\frac{N}{N_0} = \left[(C_R t)/b \right]^{-n} \text{ for } C_R t > b \qquad (18.16)$$

where, C_R is chlorine demand remaining at time t; t represents contact time; n is the slope of inactivation curve; and b is the value of intercept of horizontal axis when $N/N_0 = 1$ or $\log N/N_0 = 0$.

The values of n and b are reported in literature as 2.8 and 3.0 for faecal coliform in secondary treated effluent (Robert et al., 1980; White, 1999). These are however indicative values and site-specific disinfection experiments should be conducted to determine these values.

Example: 18.4 Estimate the dose required for treating secondary effluent having a coliform count of 4.5×10^7 MPN/ 100 mL to achieve a final MPN concentration of 100 MPN/ 100 mL for a retention time of 45 min using the data given in Example 18.3. Also estimate the cost of chlorination in this case assuming 20% available chlorine for bleaching powder. The typical values of b and n are given as 3 and 2.8. The initial chlorine demand is 3 mg/L and the demand due to decay during chlorination is 1.5 mg/L.

Solution

Using the equation, required chlorine demand,

$$C_R = \frac{b}{t} \left(\frac{N_0}{N} \right)^{\frac{1}{n}} = \frac{3}{45} \left(\frac{4.5 \times 10^7}{100} \right)^{\frac{1}{2.8}} = 6.98 \approx 7.00 \text{ mg/L}$$

Total Cl_2 required $= 7 + 3 + 1.5 = 11.5$ mg/L $= 11.5$ g/m^3.

For a daily flow of 300 MLD, the required Cl_2 will be $= 300 \times 10^3 \times 11.5 \times 10^{-3} =$ 3450 kg.

For 20% available Cl_2, cost $= \frac{3450}{0.2} \times 20 = $ Rs.3,45,000/- (**Answer**).

18.2.3 Formation of Chloramines and Its Effect

The HOCl reacts with the residual ammonia in the secondary treated sewage to form chloramines. The reaction of HOCl with the chloramines occurs in three subsequent steps forming monochloramine, dichloramine and trichloramine compounds (Eq. 18.17 through Eq. 18.19). At each step of the reaction, one hydrogen atom of ammonia or the chloramine is replaced by one chlorine atom. The formation of monochloramine is more probable at normal pH range of sewage and when the chlorine: ammonia weight ratio is between 1:3 and 1:5 (Committee, 1980). At lower pH and higher chlorine to ammonia concentration, the formation of dichloramine and trichloramine occurs.

$$NH_3 + HOCl \rightarrow NH_2Cl + H_2O \qquad (18.17)$$

$$NH_2Cl + HOCl \rightarrow NHCl_2 + H_2O \qquad (18.18)$$

$$NHCl_2 + HOCl \rightarrow NCl_3 + H_2O \qquad (18.19)$$

The chloramines are collectively referred to as combined chlorine and these compounds too have a secondary disinfecting effect other than the free chlorine in the form of HOCl and OCl^-. The sum of the combined chlorine and the free chlorine is known as the total chlorine. To measure the residual total and free chlorine, iodometric method is adopted for water or wastewater. The principle of the test is that chlorine will liberate free iodine from KI solution at pH of 8.0 or less. The liberated iodine is titrated with a standard solution of acidified 0.1 N sodium thiosulphate ($Na_2S_2O_3$) solution with starch as an indicator. The pH is maintained at 3.0–4.0 to ensure that the combined forms of chlorine also react with the thiosulphate. Also, at pH of 7.0, the thiosulphate is oxidised to sulphate, which is not desirable.

To prepare the sodium thiosulphate solution, 25 g of $Na_2S_2O_3$ is added to freshly boiled distilled water. The 0.1 N $Na_2S_2O_3$ solution is further standardized by iodate or dichromate method as detailed in Standard Methods (APHA, 2005). The starch indicator is prepared by adding 5 g starch to 10–15 mL of cold water with thorough mixing with mortar to make a paste followed by making up the final volume to 1 L with hot water. The supernatant is used as indicator by discarding the settled part. The starch solution is preserved with 25 g salicylic acid and 4 g zinc chloride. The sample volume is such selected that it does not require more than 20 mL and no less than 0.2 mL of 0.1 N $Na_2S_2O_3$ titrant to achieve the starch-iodide endpoint.

Typically, for a chlorine dose range of $1-10$ mg/L, a sample volume of 500 mL is used and proportionately smaller volume should be used for higher concentrations. The sample is taken in an Erlenmeyer flask and mixed with acetic acid for maintaining pH of 3.0–4.0 during titration and 1 g KI is added and mixed with a stirring rod. Titration is conducted until the yellow colour of the iodine is discharged. At this point 1 mL starch indicator solution is added and further titrated till the colour of the solution becomes blue (end point A). Blank is recorded by conducting titration with distilled water (end point B) and the total residual chlorine concentration is calculated by using Eq. 18.20. Care must be taken during the titration to avoid direct contact with sunlight.

$$\text{mg/L of } Cl_2 = \frac{(A - B) \times N \times 35450}{\text{mL sample}} \tag{18.20}$$

As mentioned previously, the addition of any chlorine-liberating chemical in water is followed by successive reactions that define the dosage as well as the retention time. Although, the usual practice is to adopt a retention time of half an hour for effective chlorination, however, it is recommended to study the disinfecting potential of a given dose of the chlorine liberating compound if the treated water is intended for non-potable contact applications, such as construction water, gardening, toilet flushing, etc. For understanding the phenomena of chlorination, a close look at the break-point curve is necessary (Fig. 18.5). As mentioned earlier, the addition of chlorine to wastewater or treated wastewater is followed by the rapid reaction of the highly reactive HOCl and OCl$^-$ with the reducing species and the ammonium compounds. This phase is followed by the reaction of the chlorine with the dissolved organic matter to form organo-chloro compounds commonly known as the organic DBPs. The combination of these fugitive reactions is represented as the immediate demand in the breakpoint curve (till point P, phase P-1).

Fig. 18.5 Breakpoint chlorination curve

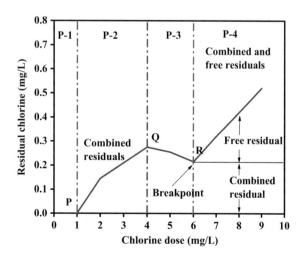

Once this immediate chlorine demand is met, the chlorine reacts with ammonia to form chloramines. This is phase P-2, wherein the residual chlorine value is owing to the combined chlorine residual concentration. However, after certain threshold value designated by point 'Q', the value of the combined residuals reduces due to the destruction of the formed chloramines to release N_2 and N_2O along with the reduction of the chlorine to chloride ion. This is represented by phase P-3 of the breakpoint curve and some amount of nitrogen trichloride is also formed in this stage. As the chloramines are destroyed, the chlorine residual concentration also goes down till the point 'R' in the curve, which is known as the breakpoint. The concentration of chloramines is minimum at the breakpoint and further chlorine addition leads to the increase in the total chlorine residual concentration owing to the increase in the free chorine concentration (P-4). The chlorine dose that has to be administered to achieve a desired free residual concentration is termed as the chlorine demand.

When chlorination is performed in the final treatment stage prior to discharge or reuse of treated water to control the pathogen content of the sewage then it is termed as post-chlorination. Usually, post-chlorination is done by adding hypochlorite or other sources of chlorine directly to the secondary treated effluent or after rendering some polishing treatment to the secondary treated sewage for removal of suspended solids. Although, by definition, post-chlorination is the terminal stage for disinfecting sewage, it is recommended that post-chlorination should be followed by an adsorptive stage, usually in the form of a carbon filter. Pre-chlorination is also performed for raw sewage for controlling odour, fly problems and for removal of grease. In addition to pre- and post-chlorination, chlorination can also be performed both at the beginning as well as the end of the treatment in which case, it is termed as split-chlorination.

18.2.4 Drawbacks of Chlorination

Chlorination is a proven technique and the same is used prevalently in sewage and water treatment plants for disinfection, the chlorine-based disinfection has different operational, environmental and commercial limitations that begets research for alternatives for disinfection. The major limitation of chlorination is the cost of the chemical when hypochlorite or its commercial form bleaching powder is used. Typically, for Indian scenarios, the cost of bleaching powder ranges from 20 to 25 Indian Rupees per kg (INR/kg), the cost of hypochlorite being Rs. 500/L for standard 4% available chlorine solutions. In addition, the energy cost of mixing and the infrastructural cost of installing the dosing system for larger sized sewage treatment plants (STPs) poses additional burden on the cost of sewage treatment.

When liquefied chlorine gas cylinders are used for directly purging highly soluble chlorine gas in the secondary treated sewage, the chemical cost of delivering 1 kg of chlorine vapour is around 2–5 INR. However, the handling of chlorine gas cylinders can be hazardous and requires highly trained operators and the dosing system is prone to leakage and contamination. Moreover, the installation cost of the chlorine dosing system in case of chlorine gas requires additional cost of using stainless steel appurtenances and sensors for arrangement owing to highly corrosive nature of chlorine gas. As an alternative approach, electro-chlorination has been adopted in different STPs, which relies on onsite generation of chlorine radicals using an electrochemical setup and brine solution. Although, the setup is effective, but this again requires a source of distilled water for preparation of brine solution and it is highly energy intensive process. The maintenance cost of replacement of electrodes is an additional cost that has to be borne in case of electro-chlorination.

In addition to the cost component, the formation of toxic and carcinogenic DBPs, during the chlorination of treated water is a major environmental burden that has to be accounted. With advanced understanding of emerging contaminants and the interactive pathways of the oxidising species with the constituents of the water matrix that generates such DBPs, there is a growing concern over the usage of chlorine-based disinfection for treated sewage. It has been claimed in different investigations in the past that among the different chemical species mediated disinfection and advanced oxidation, reactive chlorine species are responsible for producing more toxic DBPs. Such DBPs can be carcinogenic, mutagenic, cause chromosomal aberrations, and sperm abnormalities. Different classes of compounds namely, haloacetic acids, haloforms, nitrosamines, haloacetonitriles, haloketones and chlorophenols are produced as DBPs during advanced oxidation of wastewater with chlorine.

Formation of such compounds of halogen is dependent on the characteristics of the influent to the chlorination tank. For example, the formation of a prevalent nitrosamine, nitrosodimethyl amine (NDMA) is governed by the alkaline pH (pH above 10.0) and presence of monochloramine and dimethylamine (Mitch & Sedlak, 2001). Similarly, the formation of haloacetic acids and haloketones are governed by the dissolved organic matter (DOM) content in the sewage (Ike et al., 2019). As a preventive measure, filtration and flocculation of secondary treated sewage using dual media filter beds and bio-polymer-based flocculants should be employed. Further as a preventive measure, activated carbon filter is to be provided as a post-treatment to simultaneously adsorb the formed DBPs as well as for dechlorination.

18.2.5 Dechlorination

Dechlorination refers to the removal of residual chlorine from treated water if the water has to be discharged to the environment. Dechlorination can be done by dosing the chlorinated water with sulphur dioxide, activated carbon, sodium sulphite and for laboratory purposes, sodium thiosulphate. Use of sodium or calcium sulphites, meta-sulphites and bi-sulphites is an effective approach that reduces the chlorine to HCl, although at the expense of increase ion content downstream owing to formation of sodium sulphate (Eq. 18.21) and sodium hydrogen sulphate (Eqs. 18.22 and 18.23).

$$Na_2SO_3 + Cl_2 + H_2O \rightarrow Na_2SO_4 + 2HCl \qquad (18.21)$$

$$NaHSO_3 + Cl_2 + H_2O \rightarrow NaHSO_4 + 2HCl \qquad (18.22)$$

$$Na_2S_2O_5 + Cl_2 + 3H_2O \rightarrow 2NaHSO_4 + 4HCl \qquad (18.23)$$

The reaction of chloramines with the sulphites also produces sodium sulphate and sodium hydrogen sulphate salt along with ammonium ion (Eq. 18.24 through Eq. 18.26).

$$Na_2SO_3 + NH_2Cl + H_2O \rightarrow Na_2SO_4 + Cl^- + NH_4^+ \qquad (18.24)$$

$$NaHSO_3 + NH_2Cl + H_2O \rightarrow NaHSO_4 + Cl^- + NH_4^+ \qquad (18.25)$$

$$Na_2S_2O_5 + 2NH_2Cl + 3H_2O \rightarrow Na_2SO_4 + H_2SO_4 + 2Cl^- + 2NH_4^+ \qquad (18.26)$$

Use of granular activated carbon (GAC) beds are general prevalent in larger wastewater and water treatment plants for dechlorination. The removal of free residual chlorine can be achieved effectively with such setup. However, chloramines require higher bed contact time in GAC filters. For dechlorination, keeping a bed contact time of 15–20 min and a loading rate of 3000–4000 L/m^2 day is the usual practice. Dechlorination using GAC is an effective, yet expensive technique that is applicable when simultaneous residual organics as well as residual chlorine removal is desired. The removal of both free and combined residual occurs via reaction of the chlorine species with the carbon to produce CO_2 while reducing the chlorine to HCl in case of free Cl_2 (Eq. 18.27) and NH_4Cl in case of combined Cl_2 (Eq. 18.28).

$$C + 2Cl_2 + 2H_2O \rightarrow 4HCl + CO_2 \qquad (18.27)$$

$$C + 2NH_2Cl + H_2O \rightarrow NH_4^+ + 2Cl^- + CO_2 \qquad (18.28)$$

One of the most effective methods for dechlorination is the usage of SO_2 gas. Similar to Cl_2 gas, SO_2 comes in liquefied cylinders and can be handled to produce gas using similar setups as the chlorination units for gaseous chlorine. The SO_2 reacts with water to form sulphurous acid that further dissociates to form HSO_3^- that further reduces chlorine to chloride ion and produces sulphate ion (Eqs. 18.29 and 18.30).

$$SO_2 + HOCl + H_2O \rightarrow Cl^- + SO_4^{2-} + 3H^+ \qquad (18.29)$$

$$SO_2 + NH_2Cl + 2H_2O \rightarrow Cl^- + SO_4^{2-} + NH_4^+ + 3H^+ \tag{18.30}$$

As the stoichiometric ratio of SO_2 to chlorine is 0.9, hence in real life applications, for every 1 mg/L of residual chlorine concentration, SO_2 requirement ranges between 1 and 1.2 mg/L. The reaction is instantaneous and does not require additional reactor vessel unlike other de-chlorinating species, such as sulphites or activated carbon. Care must be taken to ensure appropriate measurement of residual chlorine as in case of overestimation, excess SO_2 dosing can exert oxygen demand in the treated effluent.

Example: 18.5 Estimate the quantity of sodium sulphite required to dechlorinate a treated effluent having a total residual chlorine concentration of 2.75 mg/L and a daily flow rate of 100 m^3/day. Assume the purity of sodium sulphite as 75%. The fraction of combined residual chlorine in the wastewater is 0.64.

Solution
Sodium sulphite reacts with free chlorine and combined chlorine to form HCl and NH_4Cl, respectively. Assuming that the combined chlorine is in monochloramine form, the quantity of monochloramine in the treated water is $2.75 \times 0.64 = 1.76$ mg/L. Free chlorine concentration is $2.75 - 1.76 = 0.99$ mg/L.
As per Eq. 18.24, one gram-equivalent of monochloramine reacts with one gram-equivalent of sodium sulphite (63.025 g). One gram-equivalent of monochloramine has 35.5 g chlorine. Thus, 35.5 g of chlorine reacts with 63.025 g of Na_2SO_3. Hence, with a purity of 75%, Na_2SO_3 required $= 1.76 \times 100 \times \frac{63.025}{35.5} \times \frac{1}{0.75 \times 1000} = 0.416$ kg/day.
As per Eq. 18.21 one gram-equivalent weight of chlorine (35.5 g) reacts with one gram-equivalent of Na_2SO_3 (63.025 g). Hence quantity of sodium sulphite required to quench free chlorine is:

$$= 0.99 \times \frac{63.025}{35.5} \times 100 \times \frac{1}{0.75 \times 1000} = 0.234 \, \text{kg/day}$$

Hence, total sodium sulphite required every day $= 0.234 + 0.416 = 0.65$ kg/day (**Answer**).

18.3   Ozonation

18.3.1 Application Areas

The historical development of ozone utilization for disinfection started in the early 1900s in European countries. It has been used since for controlling odour, taste and disinfection in water treatment plants. The usage of ozone for wastewater was promoted as early as 1973 (Rosen, 1973). This was concomitant to the discovery of trihalomethanes occurring in natural water as a by-product of chlorination, thus promoting ozonation as an alternative disinfectant in water treatment plants (Lovelock et al., 1973; Murray & Riley, 1973). The STP for the community of Indian town in the state of Florida, USA first installed ozone for the disinfection of secondary treated sewage. As early as 1989, around 45 STPs in USA implemented ozonation as disinfection systems. However, ozone generation required energy intensive processes, which limited the usage of ozone in wastewater.

Interest in ozone grew once more in the past two decade as an oxidant for removal of refractory organics occurring in trace quantities in sewage. This led to intense research on ozone-based techniques as advanced oxidation process (AOP) for implementation in STPs. However, ozone mediated disinfection/oxidation also produces DBPs/OBPs, which are extensively researched till now for finding a clear solution to such problems. The DBPs formation will be taken up in the next section and the present section sketches the principles, mechanisms, current practices and efficacy of ozone-based disinfection for secondary treated sewage.

Ozone is a gas at room temperature that undergoes dissociation to produce oxygen gas. Ozone is commonly produced by electric discharge ozone generator in WTPs and STPs. A typical ozone generation unit thus comprises of five components: the oxygen storage/concentrator, an electric arc generator for ionising the oxygen for ozone production, ozone contact tank, off-gas ozone destruction unit and ambient ozone sensor (Fig. 18.6). When pure oxygen is used as feed gas for ozone generator, the ozone yield is very high and hence the ozone purging time is less. In case of oxygen concentrator, air is used as a pre-cursor, which is dried and then passed through the ozone generation chamber.

In an electric arc type ozone generator, an electric discharge is given to the air, following which the oxygen molecules dissociate to produce two oxygen atoms that recombine

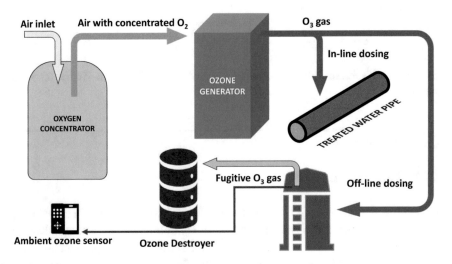

Fig. 18.6 Ozonation arrangement for in-line and off-line dosing

with other oxygen molecules to form ozone. Generation of ozone can also be realised by exciting the oxygen atoms by ultraviolet radiation similar to the phenomena that occurs in stratosphere. In this case the oxygen atom is excited by absorbing the photons to further dissociate as nascent oxygen, which recombines with another oxygen molecule to form ozone.

For ozone dosing, both offline as well as inline dosing systems can be opted. The offline system comprises of a batch reactor, wherein the diffusers be used to purge the wastewater with ozone gas. Online ozone dosing can be in counter-current direction in which case the water flows from top to bottom and the ozone is purged from below. This is an effective technique as it allows maximum exposure of the ozone molecule with the target molecules. The fugitive gas from the batch reactor is passed through the ozone destroyer prior to atmospheric discharge.

Other forms of ozone generation include generation of ozone at high pressure and voltage via electrolysis of water (Tanner et al., 2004). The occurrence of ozone in water is short-lived and it dissociates to form $\cdot OH$ and $HO_2 \cdot$ The oxidation of pollutants occurs by O_3 molecule (Eq. 18.31), $HO_2 \cdot$ (Eq. 18.32) and $\cdot OH$ (Eq. 18.33).

$$X + O_3 \rightarrow X_{oxidised} \tag{18.31}$$

$$O_3 \rightarrow HO_2^{\cdot} + X \rightarrow X_{oxidised} \tag{18.32}$$

$$O_3 \rightarrow {}^{\cdot}OH + X \rightarrow X_{oxidised} \tag{18.33}$$

The oxidation potential of ozone molecule is 2.07 eV, which makes it a suitable alternative to Cl_2 based disinfectants. Disinfecting action of ozone is realised by four distinct mechanisms: (i) direct lysis of cell wall leading to leakage of cytoplasmic content outside the cell; (ii) reaction of cell matrix with reactive oxygen species released from ozone; (iii) breakage of carbon–nitrogen bonds leading to depolymerization; and (iv) damage to the constituents of the nucleic acids (purines and pyrimidines). Several past investigations have explored oxidative treatment using ozone with the purpose of reusing the treated water. Secondary treated wastewater was disinfected and oxidatively treated with ozone-based standalone and integrated systems by varying initial ozone dose (5, 10 and 15 mg/L) and retention time (2, 5 and 10 min) (Tripathi et al., 2011).

The implementation of ozone as a disinfectant depends on the contact time of ozone and the transfer efficiency of the gas to the liquid phase. Ozone gas has a moderate solubility (12 mg/L at 20 °C). For improved transfer efficiency of ozone, diffusers are to be used in specially designed chambers. For achieving adequate disinfection with ozone (i.e., in case ozone is used as a standalone disinfection technique), two chambers of ozone contact tank need to be designed. The ozone can be transferred in the chamber in cross-current flow keeping the ozone diffuser at the bottom of the first chamber, while ensuring the flow of secondary treated wastewater is in down flow direction.

Similar to chlorination, the ozonation efficiency is also affected by the characteristics of the wastewater matrix. The biodegradable organic matter exerts an ozone demand, thus increasing the ozone dose requirement. The dissociation of ozone is adversely affected by the presence of humic substances as these also exert an ozone demand. The introduction of ozone leads to the oxidation of inorganic species, such as Fe and NO_3^-. Modelling the dose of ozone follows a similar approach like ascertaining the chlorine demand. The required input dosage D, can be computed by Eq. 18.34.

$$D = U\left(\frac{100}{TE}\right) \tag{18.34}$$

where, U is the dose actually administered to wastewater and TE is the ozone transfer efficiency varying typically between 80 and 90%. The actual dose can be determined from the disinfection model similar to that of chlorination. The relationship between U and the N, bacterial concentration can be expressed by Eq. 18.35.

$$N/N_0 = \left(\frac{U}{q}\right)^{-n}$$

Or, in the log linear version

$$\log N/N_0 = -n \log U + n \log q \tag{18.35}$$

where, q is the initial ozone demand and n is the slope of the log N/N_0 versus log U curve.

Ozonation is favoured at alkaline pH as the dissociation of ozone molecules is better in this pH regime. The presence of carbonate or hydroxyl alkalinity is thus favourable for ozonation and does not increase the ozone demand. The dosage of ozone varies according to type of wastewater and has to be determined conducting pilot studies by using Eqs. 18.34 and 18.35. The ozone dosage required for different types of wastewaters considering a contact time of 15 min varies between 3 and 30 mg/L for achieving about 6-log removal of bacteria. For instance, the effluent from sand filter requires an ozone dose of 12–20 mg/L for 6-log scale removal, while that for permeate of microfiltration the dosage ranges between 4 and 8 mg/L. Similarly, for unfiltered secondary treated water, the dose can be higher (16–30 mg/L). Although, the initial ozone demand and the required dosage should be determined for the real wastewater on case-to-case basis.

Example: 18.6 Determine the constants for the Eq. 18.34 using the data from the sewage treatment plant as furnished in Table 18.5. Determining the ozone dose (D) that has to be administered for achieving a MPN count of 100 MPN/100 mL for a wastewater with an initial MPN count of 3.166×10^7/100 mL. Assume suitable ozone transfer efficiency.

Further, consider that the secondary biological treatment followed by filtration is rendering 3 log scale removal of faecal coliforms (Table 18.18.5).

Solution

The N/N_0 values for the different transferred ozone doses are computed as presented in Table 18.6 and $\log(N/N_0)$ are plotted against the logarithmic values of ozone dose (Fig. 18.7). For example, the log (N/N_0) value for initial MPN of 2,445,000 and final effluent of 465 MPN/100 mL is:

$$\log\left(\frac{N}{N_0}\right) = \log\left(\frac{465}{2445000}\right) = -3.721$$

The equation as obtained from the regression line (Fig. 18.7) is

$$Y = -3.5913X + 0.059$$

Which corresponds to the Eq. 18.35,

$$\log\left(N/N_0\right) = -n\log U + n\log q$$

The slope $-n = -3.5913$, hence $n = 3.5913$.
The intercept is, $n\log q = 0.059$, solving, $\log q = 0.01643$

Table 18.5 Disinfection data from treatment plant

Sl. no	Initial MPN/100 mL	Ozone dose transferred (U)	Final MPN/100 mL
1	1,560,000	2	1,10,000
2	2,333,000	4	24,000
3	3,121,000	6	9100
4	1,687,000	8	1110
5	2,445,000	10	465

Table 18.6 Estimation of $\log(N/N_0)$ for different ozone doses

Sl. no	Initial MPN/100 mL	Ozone dose transferred (U)	Log U	Final MPN/100 mL	$\log(N/N_0)$
1	1,560,000	2	0.301	1,10,000	−1.152
2	2,333,000	4	0.602	24,000	−1.988
3	3,121,000	6	0.778	9100	−2.535
4	1,687,000	8	0.903	1110	−3.182
5	2,445,000	10	1.000	465	−3.721

Fig. 18.7 Log U versus Log N/N_0 plot

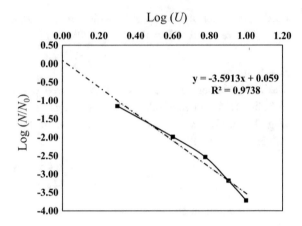

$$q = 10^{0.01643} = 1.0385$$

Hence, the equation $N/N_0 = \left(\frac{U}{q}\right)^{-n} = \left(\frac{U}{1.0385}\right)^{-3.5913}$.

Initial concentration of faecal coliform in raw sewage $= 3.166 \times 10^7$ MPN/100 mL.

Estimating faecal coliform concentration after secondary biological treatment and filtration.

Initial log scale value for faecal coliform $= \log(3.166 \times 10^7) = 7.5$

Log scale after filtration $= 7.5 - 3 = 4.5$

Hence, coliform concentration is $10^{4.5} = 3.162 \times 10^4$ MPN/100 mL.

The dose required for a final effluent with 100 MPN/100 mL will be

$$U = 1.0385 \times \left(100/3.162 \times 10^4\right)^{-\frac{1}{3.5913}} = 5.16 \text{ mg/L}$$

The dose D that has to be administered for an ozone transfer efficiency of 80% is

$$D = \frac{5.16}{80} \times 100 = 6.45 \text{ mg/L (Answer)}$$

18.3.2 Energy Requirement and Energy Per Order

The energy requirement for ozonation is documented as 0.35 kWh/m^3 for domestic wastewater (Katsoyiannis et al., 2011). In a different investigation, an energy requirement of 12–15 kWh/kg of O_3 was estimated (Ried et al., 2009). The ozone requirement and in turn the energy consumption however increases when industrial wastewater is considered, which has more complex organic compound loading (Rice, 2010). For instance, a paper and pulp industry process wastewater with an initial COD of 156 mg/L, required

600 mg/L ozone dose for 61% COD removal (Mainardis et al., 2020). The high ozone dose requirement was owing to the refractory nature of the organic matter present in the effluent. In a textile wastewater recycling plant, a COD removal of 67% was obtained with an ozone dose of 40 mg/L and a contact time of 30 min.

As the operating conditions, retention time and the dosage vary for different ozone installations in industrial effluent treatment plants and STPs, the energy requirement of ozone is often expressed as the electrical energy per order (EE/O) figure of merit. This term was introduced by Bolton and co-researchers in the year 1996. According to the definition of electrical energy per order, it is the energy required in kWh for removal of 1-log scale (i.e., 90%) of a target pollutant in 1 m^3 of contaminated water (Bolton et al., 1996; Wardenier et al., 2019). The generalised formula can be expressed as given in Eq. 18.36.

$$EE\big/_O = \frac{P.t.1000}{V.60\log\left(\frac{C_0}{C_t}\right)} \tag{18.36}$$

where, P is the power in kW, t is the contact time in minutes, V is the volume of wastewater treated in time t in litres, C_0 and C_t are the initial and final concentration of the target pollutant, respectively.

Example: 18.7 Calculate the energy per order for ozonation when nearly 90% of coliforms removal is achieved in 10 min from a secondary treated sewage with volume of 20,000 L. The power rating for ozonator is given as 1 kW and the initial and final count of *cfu*/100 mL is 31,000 and 3,300, respectively.

Solution

$\log\left(\frac{C_0}{C_t}\right) = \log\left(\frac{31000}{3300}\right) = 0.973$.

Therefore, using Eq. 18.36, $EE\big/_O = \frac{1\times10\times1000}{20000\times60\times0.973} = \frac{1}{2\times60\times0.973} = 0.0085\,kWh/m^3$ (**Answer**).

18.3.3 Drawbacks of Ozonation

The formation of DBPs in case of ozonation does not include the organo-chloro compounds as formed during chlorination. This reduces the risk of environmental contamination during discharge of water. However, oxygenated organic DBPs, such as aldehydes, ketones and organic acids and inorganic DBPs, such as bromate and nitrate, are formed during ozonation. In addition, nitrosamines such as NDMA are also formed during ozonation. Removal of DBPs formed during ozonation can be realised by implementing a granular activated carbon filter as a post-treatment step of ozonation.

Ozonation facilities require careful handling, as the off-gas generated during ozonation is highly toxic for human health. Ozone at higher concentrations can cause damage to the respiratory tracts, aggravate lung diseases, such as asthma emphysema and chronic bronchitis. Continued exposure also makes the lungs susceptible to exposure. Adverse effects also include scratchy throat, shortness of breath, chest pain while taking deep breath (Fact sheet, US EPA). Direct exposure to higher concentrations of off-gas may also result in death of affected person. In order to safeguard operating personnel working with ozonation systems, ambient ozone monitoring units must be installed in the facility to monitor the ozone concentration below the acceptable ambient ozone standard. For Indian scenario, for the exposure of 1 h and 8 h the acceptable concentration of ozone in air is 180 and 100 $\mu g/m^3$, respectively. In addition, proper arrangement for ozone destruction unit for the off-gas and regular maintenance of the ozonation equipment has to be taken up.

18.4 Ultraviolet Radiation

18.4.1 Application Areas

Utilization of UV radiation for disinfection is a time-tested technology that is well documented in scientific investigations and extensively applied in field-scale setups (Darby et al., 1993; Kang et al., 2004; Lindenauer & Darby, 1994). The first application of UV radiation for the disinfection of water started in the early 1900s and for wastewater in the 1990s with advancement in electronics, superior design and novel materials available offering better options. The UV radiations emitted in three wavelength ranges are classified as UV-A, UV-B and UV-C. The UV-C type radiation that has a wavelength range of 200–280 nm is absorbed by DOM and is capable of penetrating the live bacterial cells to cause damage to the DNA structure of cells, thus deactivating the bacterial cell (Amin et al., 2009; Gray, 2013). The UV-B, having a wavelength range of 280–320 nm is also capable of damaging the DNA of the cell although to a milder extent than the UV-C radiation (Kim et al., 2013). The xenobiotic compounds can either be oxidised by the ROS generated, when photocatalyst releases electron from the valence band by absorbing UV-A radiation, that has the longest wavelength range (315–400 nm) or can be directly lysed by absorbing the UV-C radiation mentioned earlier.

The UV light system used for disinfection will fall under three types: (a) low pressure low-intensity lights, (b) low pressure high-intensity lamps, and (c) medium-pressure high-intensity lamps. The low-pressure low-intensity UV system use mercury-argon lamps to emit monochromatic UV light with wavelength of 254 nm, which is close to the wavelength at which the maximum microbial inactivation occurs (260 nm). The lamps operate at 35–45 °C with a Hg pressure of 0.007 mm. The Hg vapour has to be maintained at a temperature that ensures the Hg does not solidify back to liquid. The typical life of the

UV lamps can vary between 9,000 and 13,000 h and the life of the quartz encasing is 4–8 years.

The low-pressure high-intensity UV systems utilize mercury-indium amalgam for producing UV-C rays and operate at a pressure of 0.001–0.01 mm Hg. The intensity outputs of these lamps are around 2–5 times that of the low-pressure low-intensity lamps. The operating temperature regime of these lamps are 90–150 °C. The medium-pressure high-intensity UV system is used for disinfection in special cases for high flows, such as storm water or for modular plants, which have low space availability thus leading to accommodation of fewer lamps. The total UV-C output of these lamps are approximately 50 – 100 times of the low-pressure low-intensity UV lamps and the operating temperature is 600–800 °C. These medium-pressure high-intensity UV lamps operate at high energy and can be modulated to different power settings without altering the spectral distribution of the lamp.

UV disinfection units can either be open channel or closed channel systems. In case of open channel UV system, the flow is divided into equal number of channels, each channel comprising of two to three horizontal or vertical lamp modules. The architecture of the lamp module allows water to flow around quartz encased UV lamps kept in close vicinity (usually 75 mm) to enable adequate exposure of the flow to UV radiation. The UV lamps are placed inside the module symmetrically in even numbers of 2, 4, 6, 8 or 16. The flow is controlled by gate and float arrangement in the chambers. In case of closed channel flows, low-pressure high-intensity and medium-pressure high-intensity UV lamps are used. The lamps can be placed perpendicular to the flow or can be parallel to the direction of flow. The UV dose (D) is expressed in mJ/cm^2, and can be obtained by multiplying the intensity (I, in mW/cm^2) of the UV lamps with exposure time (t, in sec) (Eq. 18.37). The concept of UV dose is thus similar to the $C_R t$ concept of chlorination.

$$D = I \times t \tag{18.37}$$

The performance of UV disinfection is affected by the placement of the lamps and the flow regime. In open channel flows, it is difficult to maintain a uniform and laminar flow to ensure minimum turbulence. Reduced turbulence is required for ensuring equivalent UV exposure of the entire flow. Other than the asymmetrical placement of the UV lamps, the design of the open channel basin and the closed channel inlet and outlet arrangements also impact the flow regime. In addendum to the hydraulic impacts, the effect of high turbidity, and flocculated particles that disperse and shield the micro-organisms also negatively affect the UV performance. Especially, the presence of dyes, humic substances and metals alter transmittance of the UV rays. The UV dose can be determined by the collimated-beam bioassay test. In this test a low-pressure low-intensity UV lamp is used to generate a fixed UV dose for disinfecting a known sample with fixed sample depth. The intensity is estimated by the Beer's law stated in the form as given in Eq. 18.38.

$$I_{avg} = I_o \times \frac{\left(1 - 10^{-\alpha d}\right)}{(\alpha d)} \qquad (18.38)$$

where, I_{avg} is the average UV intensity, mW/cm^2, I_0 is the average incident intensity on the sample surface in mW/cm^2, α is the adsorption coefficient in arbitrary unit (a.u.) /cm and d is the depth of the liquid sample.

The value of α is dependent on the effluent characteristics and varies between 0.5 and 0.8 per cm for primary treated effluent and 0.2 and 0.4 per cm for filtered secondary treated wastewater. The dose corresponding to the average intensity could be estimated by using the Eq. 18.37. The results from the collimated-beam study for different UV dosage conducted in triplicate should be used to determine the dose required for a desired level of bacterial inhibition. In order to derive a reliable result that takes into account the unaccounted variability in the replicate test conditions, statistical significance of the result must be tested by suitable hypothesis test.

Although, the above discussion indicates that the UV dose determination is a case specific activity, as a general guideline, for membrane filtration effluent, sand filtration effluent and reclamation water, a minimum dose of 50, 80 and 100 mJ/cm^2, respectively, is recommended. The recommended average transmittance is 55, 65 and 90% for sand filtration, microfiltration and reverse osmosis effluent, respectively. The transmittance can also be used to calculate the absorbance (α) by considering the Eq. 18.39.

$$T(\%) = 10^{-(\alpha)} \times 100 \qquad (18.39)$$

18.4.2 Drawbacks of UV

The UV disinfection technology does not generate any DBPs unlike its chemical counterparts such as ozone and chlorine. However, the drawback of the technology lies in the fact that the energy requirement and the biofouling of the quartz sleeves reduces the disinfecting action. Also, no residual effect is imparted to treated water once it leaves the UV system, hence any downstream contamination after UV module will remain as it is. Conventional Hg-based systems generate heating problems and has to be carefully operated to avoid accidental damage during long operational periods. The formation of filamentous bacterial or fungal growth is more pronounced on the surface of the lamps in the open channel modules. This leads to attenuation of the UV ray transmittance, thus leading to a shielding effect for the virulent and pathogenic organisms in the effluent. Periodic cleaning of the modules with the peracetic acid or hypochlorite and covering the channels to avoid entry of visible light can reduce the problem of biofouling.

Energy expense of the UV systems are decrementing with the advent of newer technologies. The advent of UV-LEDs has changed the overall design perspective of UV disinfection systems. The advantage of UV-LED over the low-pressure Hg lamps is that the UV-LED configuration can be tailored unlike the reactors with low-pressure Hg lamps that needs to be architected to accommodate the tubular shaped low-pressure Hg lamps (Martín-Sómer et al., 2017). The UV-LEDs also have longer life span and it is projected to incur less energy expense once the wall plug efficiency of these LEDs surpass the conventional low-pressure Hg lamps. It is possible to control the wavelength of the emitted UV radiation in case of UV-LEDs unlike low-pressure Hg lamp.

Comparison studies between low-pressure Hg lamps and UV-LEDs demonstrate that while the disinfection efficiency of both is comparable, the reactivation of UV-LED irradiated bacterial cells are less (Li et al., 2017). The effectiveness of fungal spore inactivation and reduced extent of reactivation was observed for UV-LED combination of 260/280 nm as compared to low-pressure low intensity Hg lamps (Wan et al., 2020). However, further validation and standardisation of the UV-LEDs is required for replacing the low-pressure Hg lamps with UV-LEDs (Song et al., 2016). Also, further research is required to improve the wall plug efficiency or to achieve equivalent disinfection with lower UV-LED intensity.

18.5 Hydrogen Peroxide

18.5.1 Application of Hydrogen Peroxide

Hydrogen peroxide was first synthesized in the year 1818 by Louis Jacques Thénard by reacting barium peroxide with nitric acid. The synthesis procedure was further refined by using hydrochloric acid followed by sulphuric acid to precipitate barium chloride. Presently, H_2O_2 is produced industrially by anthraquinone process. The H_2O_2 has a weak O–O single bond, which dissociates readily to yield water as a final product. The reaction proceeds by production of ROS in the form of two ·OH (Eq. 18.40) or as the hydroperoxyl radical and a proton (Eq. 18.41). The oxidising potential of ·OH is 2.8 eV, which is higher than H_2O_2 (1.8 eV). This suggests that when H_2O_2 dissociates, the ROS produced has more oxidising power than the chemical itself.

$$H_2O_2 \rightarrow 2^{\cdot}OH \tag{18.40}$$

$$H_2O_2 \rightarrow HO_2^{\cdot} + H^+ \tag{18.41}$$

There are different strategies for rapid dissociation of H_2O_2, among which addition of Fe catalyst is most prevalent. This was established by H.J.H Fenton and hence it is known as the Fenton's reaction, while the combination of H_2O_2 and Fe is known as the Fenton's reagent. In presence of Fe^{+2}, the H_2O_2 is dissociated into two ·OH radicals, by oxidising Fe^{+2} to Fe^{+3} (Eq. 18.42). The ferric ion thus generated is further reduced to Fe^{+2} while dissociating the H_2O_2 to $HO_2^{\cdot}$ and H^+ (Eq. 18.43). Although the Fenton's process is a promising approach towards the utilization of H_2O_2 as an oxidant/disinfectant; however, the formation of $Fe(OH)_3$ sludge poses an operational constraint. Hence, as a modification to prevent the loss of Fe^{+3} ions as ferric hydroxide sludge, electro-Fenton was implemented (Petrucci et al., 2016).

$$H_2O_2 + Fe^{+2} \rightarrow \ \dot{}OH + OH^- + Fe^{+3} \tag{18.42}$$

$$H_2O_2 + Fe^{+3} \rightarrow Fe^{+2} + \ \dot{}OOH + H^+ \tag{18.43}$$

The Fenton's reaction is suitable for laboratory grade or analytical experiments and it is yet to be realised as a common disinfection strategy. In wastewater, the dissociation of H_2O_2 is facilitated by hybrid advanced oxidation processes, such as ozone/H_2O_2 (peroxone process) or UV-C/H_2O_2 process. Another alternative peroxide is the per-acetic acid, which is represented by the stoichiometric formula $C_2H_4O_3$ and consists of a mixture of acetic acid and H_2O_2. Application of peracetic acid for disinfection of treated water prior to discharge has been implemented at pilot-scale at different locations (Lefevre et al., 1992; Wagner et al., 1992). The dosing system for per-acetic acid or H_2O_2 is similar to the hypochlorite dosing system with certain modifications to avoid excess light exposure of the peroxide. The UV-C/H_2O_2 process has been implemented in pilot-scale studies for removal of trace contaminants from wastewater and for disinfection of wastewater (Burns et al., 2007; Krystynik et al., 2018).

Dose of peroxide-UV processes vary with source characteristics of wastewater as well as target log-scale inactivation. For a total inactivation of total coliforms, a per-acetic acid dose of 4 mg/L and UV dose of 192 mJ/cm^2 was required (Caretti & Lubello, 2003). Using per-acetic acid (PAA) individually, a 4-log scale inactivation was achieved with 8 mg/L dose and 30 min of contact time. For both cases the average count of total coliform was estimated as 3×10^5 cfu/100 mL (Caretti & Lubello, 2003). In addition to disinfection, the UV/H_2O_2 advanced oxidation process is also implemented for the photo-Fenton type oxidation of trace contaminants. The UV/H_2O_2 system is capable of removing trace contaminants that are present in municipal wastewater (Miklos et al., 2018).

18.5.2   Drawbacks of Hydrogen Peroxide Process

The hydrogen peroxide can be safely used unlike chlorine, which produces DBPs. The residual H_2O_2 breaks down to release O_2 and water. This is advantageous as there is no formation of carcinogenic DBPs and it has no known toxic effect to aquatic flora and fauna unlike chlorine-based disinfection. However, the reactivity of H_2O_2 largely depends on the dissociation of the same to produce hydroxyl radicals. The dissociation has to be catalysed by chemical species or by physical perturbations.

As discussed previously, Fenton's reaction delivers promising results but contaminates the water due to dosing or leaching of Fe species. Peroxone process, that involves the formation of ROS in the presence of O_3/H_2O_2 in equimolar ratio is an effective process. The energy consumption is lower than single stage ozonation process. However, the installation of such hybrid process is yet to be standardised as the trade-off between energy and performance enhancement is not established as on date. Moreover, this process involves external addition of H_2O_2 or in situ electro-synthesis, either of which is an additional cost.

18.6 Summary

The advent of different new advanced oxidation processes and other energetic physical forces is opening new areas of disinfection strategies. Disinfection is an important prerequisite prior to usage of the treated water for different contact usage and also in some cases for safe discharge where the fresh water intake for water supply is in the vicinity. This chapter described the established techniques in details and introduced a few upcoming technologies to the reader. The intention is to develop a basic understanding of disinfection in context of wastewater so that the reader as a practicing engineer or a designer is capable of implementing the concepts in future projects.

Questions

18.1. Explain necessity of disinfection prior to encourage reuse of treated sewage for various purposes.
18.2. What are the different options available for disinfection of secondary treated sewage?
18.3. What type of pre-treatment is essential for effective disinfection of secondary treated sewage?
18.4. Describes the parameters that will affect the performance of disinfection.
18.5. Describe how the dose of chlorine required is estimated.
18.6. Write a note on use of bleaching powder as a disinfectant.
18.7. Determine the fraction of HOCl present in a wastewater at 25 °C and at a pH of 7.5. If the dose of NaOCl is 6 mg/L, what is the absolute concentration of HOCl and OCl$^-$ considering a NaOCl to HOCl conversion ratio of 85%. The pK_a value at 25 °C is 7.537.

   (**Answer**: Fraction of HOCl present 0.521 and that for OCl$^-$ is 0.479; HOCl = 1.87 mg/L; OCl = 1.69 mg/L)

18.8. Describe formation of chloramines during disinfection using chlorine.
18.9. What is necessity of dichlorination? How it is done in wastewater treatment plant?
18.10. Describe advantages and drawbacks of the ozonation used for disinfection of wastewater.

18.11. Describe disinfection with ozone. What dose of ozone is typically required?

18.12. What are the types of UV lamps that are used in wastewater treatment?

18.13. Compare conventional Hg lamp-based UV with UV-LED.

18.14. What are the advantages and drawbacks of application of UV for disinfection?

18.15. Discuss use of hydrogen peroxide as a disinfectant for secondary treated sewage.

18.16. What are the different disinfectants that can be used for disinfection of sewage? What is the target effluent quality expected?

References

Amin, M. M., Hashemi, H., Bina, B., Attar, H. M., Farrokhzadeh, H., & Ghasemian, M. (2009). Pilot-scale studies of combined clarification, filtration, and ultraviolet radiation systems for disinfection of secondary municipal wastewater effluent. *Desalination, 260*, 70–78. https://doi.org/10.1016/j.desal.2010.04.065

APHA. (2005). *Standard methods for the examination of water and wastewater* (21st ed.). American Public Health Association/American Water Works Association/Water Environment Federation.

Bolton, J. R., Bircher, K. G., Tumas, W., Tolman, C. A. (1996). Figures-of-merit for the technical development and application of advanced oxidation processes. *Journal of Advanced Oxidation Technologies, 1*, 13–17. https://doi.org/10.1515/jaots-1996-0104

Burns, N., Hunter, G., Jackman, A., Hulsey, B., Coughenour, J., Walz, T. (2007). The return of ozone and the hydroxyl radical to wastewater disinfection. *29*, 303–306. https://doi.org/10.1080/01919510701463206

Caretti, C., & Lubello, C. (2003). Wastewater disinfection with PAA and UV combined treatment: A pilot plant study. *Water Research, 37*, 2365–2371. https://doi.org/10.1016/S0043-1354(03)00025-3

Chahal, C., van den Akker, B., Young, F., Franco, C., Blackbeard, J., & Monis, P. (2016). Pathogen and particle associations in wastewater: Significance and implications for treatment and disinfection processes. *Advances in Applied Microbiology, 97*, 63–119. https://doi.org/10.1016/bs.aambs.2016.08.001

Chamberlin, N. (1948). Chlorination of sewage. *Sewage Works Journal, 20*(2), 304–318. Retrieved August 18, 2021, from http://www.jstor.org/stable/25030833

Chick, H. (1908). An investigation of the laws of disinfection. *Journal of Hygiene, 8*(1), 92–108. https://doi.org/10.1017/2Fs0022172400006987

CPHEEO (2013). Manual on sewerage and sewage treatment systems: Part A engineering, ministry of urban development. Government of India.

Darby, J. L., Snider, K. E., & Tchobanoglous, G. (1993). Ultraviolet disinfection for wastewater reclamation and reuse subject to restrictive standards. *Water Environment Research, 65*, 169–180. https://doi.org/10.2175/wer.65.2.10

Fact Sheet, US EPA, n.d. Health Effects of Ozone Pollution | US EPA [WWW Document]. https://www.epa.gov/ground-level-ozone-pollution/health-effects-ozone-pollution. Accessed 18 Aug 21.

Fact sheet WHO, n.d. Drinking-water [WWW Document]. https://www.who.int/news-room/fact-sheets/detail/drinking-water. Accessed 17 Aug 21.

Gray, N. F. (2013). Ultraviolet disinfection. In: *Microbiology of waterborne diseases: Microbiological aspects and risks*, 2nd edn (pp. 617–630). Elsevier Ltd. https://doi.org/10.1016/B978-0-12-415846-7.00034-2

Humans, I.W.G. on the E. of C.R. to, (1991). Chlorinated drinking-water. IARC monographs on the evaluation of carcinogenic risks to humans/World Health Organization. *International Agency for Research on Cancer, 52*, 45–141

Ike, I. A., Karanfil, T., Cho, J., & Hur, J. (2019). Oxidation by products from the degradation of dissolved organic matter by advanced oxidation processes: A critical review. *Water Research, 164*, 114929. https://doi.org/10.1016/j.watres.2019.114929

Kang, S. J., Allbaugh, T. A., Reynhout, J. W., Erickson, T. L., Olmstead, K. P., Thomas, L., & Thomas, P. (2004). Selection of an ultaviolet disinfection system for a municipal wastewater treatment plant. *Water Science and Technology, 50*, 163–169. https://doi.org/10.2166/wst.2004.0445

Katsoyiannis, I. A., Canonica, S., & von Gunten, U. (2011). Efficiency and energy requirements for the transformation of organic micropollutants by ozone, O_3/H_2O_2 and UV/H_2O_2. *Water Research, 45*, 3811–3822. https://doi.org/10.1016/J.WATRES.2011.04.038

Kim, S., Ghafoor, K., Lee, J., Feng, M., Hong, J., Lee, D. U., & Park, J. (2013). Bacterial inactivation in water, DNA strand breaking, and membrane damage induced by ultraviolet-assisted titanium dioxide photocatalysis. *Water Research, 47*, 4403–4411. https://doi.org/10.1016/j.watres.2013.05.009

Krystynik, P., Masin, P., & Kluson, P. (2018). Pilot scale application of $UV-C/H_2O_2$ for removal of chlorinated ethenes from contaminated groundwater. *Journal of Water Supply: Research and Technology-Aqua, 67*, 414–422. https://doi.org/10.2166/AQUA.2018.144

Lefevre, F., Audic, J. M., & Ferrand, F. (1992). Peracetic acid disinfection of secondary effluents discharged off coastal seawater. *Water Science and Technology, 25*, 155–164. https://doi.org/10.2166/WST.1992.0347

Li, G. Q., Wang, W. L., Huo, Z. Y., Lu, Y., & Hu, H. Y. (2017). Comparison of UV-LED and low pressure UV for water disinfection: Photoreactivation and dark repair of *Escherichia coli*. *Water Research, 126*, 134–143. https://doi.org/10.1016/J.WATRES.2017.09.030

Lindenauer, K. G., & Darby, J. L. (1994). Ultraviolet disinfection of wastewater: Effect of dose on subsequent photoreactivation. *Water Research, 28*, 805–817. https://doi.org/10.1016/0043-1354(94)90087-6

Lovelock, J.E., Maggs, R.J., & Wade, R.J. (1973). Halogenated hydrocarbons in and over the Atlantic. *Nature, 241*(5386), 194–196. https://doi.org/10.1038/241194a0

Mainardis, M., Buttazzoni, M., de Bortoli, N., Mion, M., & Goi, D. (2020). Evaluation of ozonation applicability to pulp and paper streams for a sustainable wastewater treatment. *Journal of Cleaner Production, 258*, 120781. https://doi.org/10.1016/J.JCLEPRO.2020.120781

Martín-Sómer, M., Pablos, C., van Grieken, R., & Marugán, J. (2017). Influence of light distribution on the performance of photocatalytic reactors: LED vs mercury lamps. *Applied Catalysis B: Environmental, 215*, 1–7. https://doi.org/10.1016/J.APCATB.2017.05.048

Miklos, D. B., Hartl, R., Michel, P., Linden, K. G., Drewes, J. E., & Hübner, U. (2018). UV/H_2O_2 process stability and pilot-scale validation for trace organic chemical removal from wastewater treatment plant effluents. *Water Research, 136*, 169–179. https://doi.org/10.1016/J.WATRES.2018.02.044

Mitch, W. A., & Sedlak, D. L. (2001). Formation of *N*-Nitrosodimethylamine (NDMA) from dimethylamine during chlorination. *Environmental Science and Technology, 36*, 588–595. https://doi.org/10.1021/ES010684Q

Morris, J. C. (1966). The acid ionization constant of HOCl from 5 to 35°. *Journal of Physical Chemistry, 70*, 3798–3805. https://doi.org/10.1021/J100884A007

Murray, A. J., & Riley, J. P. (1973). Occurrence of some chlorinated aliphatic hydrocarbons in the environment. Nature, 242(5392), 37–38. https://doi.org/10.1038/242037a0

Petrucci, E., da Pozzo, A., & di Palma, L. (2016). On the ability to electrogenerate hydrogen peroxide and to regenerate ferrous ions of three selected carbon-based cathodes for electro-Fenton processes. *Chemical Engineering Journal, 283*, 750–758. https://doi.org/10.1016/j.cej.2015.08.030

Rice, R. G. (2010). Applications of ozone for industrial wastewater treatment—A review. *The Journal of the International Ozone Association, 18*(6), 477–515. https://doi.org/10.1080/01919512.1997.10382859

Ried, A., Mielcke, J., & Wieland, A. (2009). The potential use of ozone in municipal wastewater. *The Journal of the International Ozone Association, 31*(6), 415–421. https://doi.org/10.1080/01919510903199111

Roberts, P. V., Aieta, E. M., Berg, J. D., & Chow, B. M. (1980). *Chlorine dioxide for wastewater disinfection: A feasibility evaluation, technical report 21*. Stanford University, Stanford, USA.

Rosen, H. M. (1973). Use of ozone and oxygen in advanced wastewater treatment on JSTOR. *Water Pollution Control Federartion, 45*, 2521–2536.

Safe Drinking Water Committee and National Research Council (1980). *The chemistry of disinfectants in water: Reactions and products in drinking water and health*, vol. 2. National Academies Press (US), Washington DC, USA.

Selleck, R. E., Saunier, B. M., & Collins, H. F. (1978). Kinetics of bacterial deactivation with chlorine. *Journal of Environmental Engineering, American Society of Civil Engineers, 104*, 1197–1212.

Fact Sheet, Lenntech, n.d. Bromine as a disinfectant [WWW Document]. https://www.lenntech.com/processes/disinfection/chemical/disinfectants-bromine.htm. Accessed 01 Nov 21.

Song, K., Mohseni, M., & Taghipour, F. (2016). Application of ultraviolet light-emitting diodes (UV-LEDs) for water disinfection: A review. *Water Research, 94*, 341–349. https://doi.org/10.1016/J.WATRES.2016.03.003

Tanner, B. D., Kuwahara, S., Gerba, C. P., & Reynolds, K. A. (2004). Evaluation of electrochemically generated ozone for the disinfection of water and wastewater. *Water Science and Technology, 50*, 19–25. https://doi.org/10.2166/wst.2004.0007

Tripathi, S., Pathak, V., Tripathi, D. M., & Tripathi, B. D. (2011). Application of ozone based treatments of secondary effluents. *Bioresource Technology, 102*, 2481–2486. https://doi.org/10.1016/j.biortech.2010.11.028

Wagner, M., Brumelis, D., & Gehr, R. (1992). Disinfection of wastewater by hydrogen peroxide or peracetic acid: development of procedures for measurement of residual disinfectant and application to a physicochemically treated municipal effluent. *Water Environment Research, 74*, 33–50.

Wan, Q., Wen, G., Cao, R., Xu, X., Zhao, H., Li, K., Wang, J., & Huang, T. (2020). Comparison of UV-LEDs and LPUV on inactivation and subsequent reactivation of waterborne fungal spores. *Water Research, 173*, 115553. https://doi.org/10.1016/J.WATRES.2020.115553

Wardenier, N., Liu, Z., Nikiforov, A., van Hulle, S. W. H., & Leys, C. (2019). Micropollutant elimination by O_3, UV and plasma-based AOPs: An evaluation of treatment and energy costs. *Chemosphere, 234*, 715–724. https://doi.org/10.1016/J.CHEMOSPHERE.2019.06.033

Watson, H. E. (1908). A note on the variation of the rate of disinfection with change in the concentration of the disinfectant. *The Journal of Hygiene, 8*(4), 536–542. https://doi.org/10.1017/s0022172400015928

White, G. C. (1999). *Handbook of chlorination and alternative disinfectants*, 4th edn. A Wiley-Interscience Publication, Wiley, New York, USA.

Onsite Sanitation Systems

19

19.1 Introduction

Adequate management of sanitation facilities and offering sustainable solution for sanitary wastewater management is a serious challenge in developing countries. Many rural and urban areas in developing countries are lacking sustainable sanitation infrastructure. There is enormous scope for improvement in this sector. There are several historical evidences of sewage collection and sanitation facilities; however, the modern days flush toilet concept was first time developed by Thomas Crapper in the year 1860 (Antoniou et al., 2016). Research in the domain of sanitation increased considerably after 1990 and almost more than 90% of the research articles were published in the last three decades (Zhou et al., 2018). The majority of the research articles published in the field of sanitation are from developed countries, especially from the United States of America or the United Kingdom; however developing countries are the one who are facing severe challenges in sanitation infrastructure (Zhou et al., 2018). A significant deviation in the accessibility of improved sanitation facilities was observed based on the socioeconomic status of the citizens in developing and underdeveloped countries and most prominently in the African continent (Armah et al., 2018).

South American nations are also seriously lacking in the construction of sustainable sanitation infrastructure for their citizens (Mukherjee et al., 2019). In India, adequate sanitation infrastructure was lacking till last decade and local people were forced to practice open defecation (Chaudhuri & Roy, 2017). Almost 60 million people staying in urban India were not having an access to healthy sanitation facilities. Riverbanks were used for sanitation practices in different parts of India and almost two-third of the sanitary wastewater was released to the nearby river or open land without getting sufficient treatment or minimal treatment (Wankhade, 2015). The scenario of rural India was also similar and open defecation or unhygienic sanitation practices were very common (Mukherjee et al.,

2019). However, the country as a whole has done remarkable progress in last 5 years to improve sanitation infrastructure and to provide access to sanitation for every individual.

The unhealthy sanitation practices cause pathogenic contamination in the nearby freshwater sources and pose severe health hazards (Nweke & Sanders, 2009). The pandemic causing COVID-19 viruses SARS-CoV-2 can also sustain and spread through untreated sewage, which is causing serious threats for human civilization (Bhowmick et al., 2020). Unavailability of resources, appropriate technologies, and primitive traditional behaviour of the residents are some major reasons behind this situation to resists improvement in sanitation practices (Jain et al., 2020).

Sustainable sanitation is a prime need for developing countries. In September 2015, world leaders of the United Nations (UN) set a goal to "ensure the availability and sustainable management of water and sanitation for all" (Zhou et al., 2018). Prime Minister of India has also announced Swachh Bharat Mission (Clean India Mission) in 2014 to obsolete open defection and ensure a toilet for each household. According to statistical analysis, it was observed that for the last 30 years, groundwater faecal coliform concentration (2002–2017, ~2.5% per year) and acute diarrheal cases (1990–2016, ~3% per year) were considerably decreased with an increase in toilet construction (1990–2017, ~2.6% per year) with simultaneous improvement in sanitation infrastructure (Mukherjee et al., 2019). Noticeable improvements in groundwater quality and human health have been perceived since the year 2014, after the announcement of the Clean India (Swachh Bharat) Mission.

China also proposed a "toilet revolution" in 2015 for rural areas (Zhou et al., 2018). After considerable efforts, significant numbers of toilets have been constructed and the sanitation scenario is getting improved since then (Zhou et al., 2018). Although toilet structures are abundantly constructed, the construction of adequate treatment facilities for toilet waste is being overlooked in many places. Construction of toilet structure may obsolete the open defecation; however, this is concentrating human waste in the pit of the toilet and without compulsory treatment of human waste, pollution problem remains unchanged. Additionally, the absence of affordable and adequate black water treatment options to handle the waste generated from the toilets causes faecal coliform contamination in groundwater and probably even lead to contamination of potable water source, if located nearby the toilet pits (Jangam & Pujari, 2019).

19.2   Classification of Onsite Sanitation Solutions

Two primary options for the treatment of sanitary wastewater are adopted in urban areas for the management of faeces, such as offsite sanitation systems and onsite sanitation systems. Offsite sanitation system is the arrangement where the sanitary wastewater is conveyed from each household using a water carriage system to a single point for centralized wastewater treatment and disposal to nearby water bodies or irrigation. Whereas, onsite sanitation systems (OSS) are alternative options, where faecal waste is collected in a containment system and may or may not be treated properly before disposal. India and other developing countries do not have adequate sanitary sewerage infrastructure and significantly depend on OSS. Different primitive types of OSSs are popular in rural and urban India, such as aqua privy, pit privy, borehole latrine, dug well latrines, etc. The performance of these latrines is unsatisfactory and in most of the scenarios these practices are not adequately managed and these are majorly responsible for contamination of water bodies (Nakagiri et al., 2016).

Other sanitation facilities, such as septic tank systems, are also installed in rural and urban households, where significant care for effluent disposal is required (Butler & Payne, 1995). A septic tank is a specially designed sedimentation tank where a longer detention time is maintained for the efficient digestion of settled sludge. A conventional type of septic tank is a combined sedimentation cum digestion tank, which is only able to remove chemical oxygen demand (COD) of about 30–40% from sewage and the remaining organic matters along with pathogenic microorganisms are left to the environment to pollute the water bodies (Almomani, 2016). Hence, provision of proper soak pits arrangement is necessary to further dispose of the effluent generated from the septic tank to avoid pollution of groundwater.

19.3   Advantages and Disadvantages of Onsite Sanitation

Centralized wastewater treatment is the most popularly adopted wastewater treatment in the twentieth century. A centralized treatment unit is a solo treatment option situated away from the city (usually situated downstream riverbank of the city) for treating wastewater collected from a large area or a city (Massoud et al., 2009). According to statistical data of the Central Pollution Control Board of India, the overall capacity of a sewage treatment plant in Class I and Class II cities in India is 78.7% less than the required capacity and considering under construction sewage treatment plants this deficiency reduces to 72.7% (CPCB, 2021). Additionally, there are considerable numbers of non-compiled and under-construction or under-maintenance sewage treatment plants, which provides only partial treatment and releases water to the river or other waterbodies. According to a report published by United Nations Educational, Scientific and Cultural Organization (UNESCO), around 80% of the total wastewater generated in the world and over 95% of the wastewater generated in the under-developed countries is released without any treatment (UN Water, 2017). Management of this huge amount of sewage is difficult to accomplish in centralized treatment systems because of lack of space for future expansion.

Wastewater from each household and other similar wastewater generation sources is collected and conveyed by drains, i.e., underground pipes, and intermediate pumping is required to convey it to the centralized treatment unit. This requires enormous piping and drainage network and excessive energy consumption for intermediate pumping stations that are required in the collection system. Adequate sludge management is also absent in most of these kinds of treatment units. Reuse of treated water in this centralized treatment system is also uneconomical and not adopted as similar massive piping network and pumping is required for conveying the treated water back to each household back to the city, which is typically at the higher elevation relative to treatment plant in most cases.

Decentralized wastewater treatment plants are suitable for small locality or single household-based wastewater treatment units, where wastewater collection and conveyance can be achieved effortlessly. The decentralized treatment system is also advantageous because the reuse of treated water is easy in this system and does not require a huge distribution network for water reuse for gardening, car washing, toilet flushing and several other non-potable purposes (Libralato et al., 2012; Massoud et al., 2009; Suriyachan et al., 2012). Due to the small size of these treatment units, easy installation and maintaining necessary precision is possible to improve their performance. Suitable choice of treatment units can be adopted or modified depending upon the wastewater characteristics and local site conditions without disrupting regular life.

In an unsewered locality, sanitary wastewater or black water released from the toilet is not treated with a centralized municipal wastewater treatment plant because of the difficulty in transporting it. Majorly, separate centralized black water treatment units or decentralized septic tank units are constructed in rural and urban India at present. However, because of the very high cost of collection and transportation of sanitary waste from door to door, centralized treatments are now mostly avoided. On the other hand, partial treatment of wastewater without meeting the necessary quality requirement is a serious disadvantage of the septic tank and this needs further disposal of treated effluent through soak pits. However, because of aquifer conditions and limited availability of space, direct discharge of this partially treated wastewater from septic tank to stormwater drain is noticed, which creates pollution issues. Therefore, the only use of septic tanks without proper soak pits is not safe and healthy for mankind, and modification based on modern technologies—like anaerobic digestion, incineration, chemical or bio-electrochemical process, etc., are called for eliminating pollution of downstream water bodies or land.

19.4 Pit Toilets

Pit toilets are primitive low-cost onsite sanitary wastewater management technology, where faecal sludge is stored and treated because of the interactions between soil microorganisms and faecal microorganisms in the pit. The treatment efficiency of pit toilets is very poor and insufficient; however, these types of toilets are still observed in rural and urban India and Africa or other developing and underdeveloped countries. Twin-pit toilet system is a modification of the primitive pit toilets and this toilet is consisting of a superstructure (toilet) and treatment units (two chambers). This special toilet structure was installed throughout India under the scheme Clean India Mission. The two underground chambers

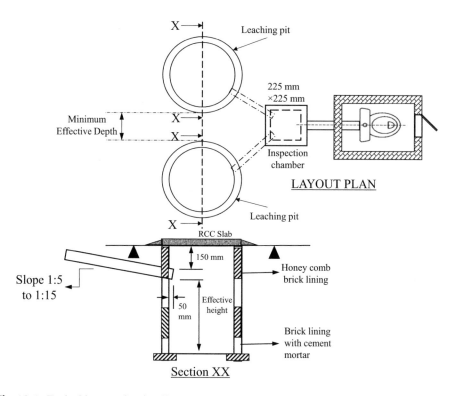

Fig. 19.1 Typical layout of a pit toilet

(pits) are provided to hold faecal sludge and only one pit is functional at a time, while the other is allowed to rest as the liquid leaches out of the pit in this system while solids dry inside the pit (Fig. 19.1).

According to the design specifications, two pits are constructed outside the toilet structure and at least at a considerable distance (minimum 1 m) apart from each other. A single pipe from the toilet leads to a small inspection chamber, from which two separate pipes lead to the two underground pits. The pits are lined or unlined based on the water table of the surroundings and constructed by honeycomb brick missionary. Each pit is designed to hold at least 1 year accumulation of faecal sludge. Wastewater is discharged to one pit until it is filled with faecal sludge and after that the discharge is then diverted to the second pit. Just before the second chamber is filled with faecal sludge, the anaerobically stabilized organic contents of the first pit are withdrawn. The capacity of a twin pit is estimated by calculating the sludge accumulation rate. Sludge accumulation rate is a function of a wide range of variables, including water table, pit age, water and excreta loading rates, microbial conditions in the pit, temperature and local soil conditions, and the type of material used for cleansing while using toilet.

Twin pit toilets are designed based on IS: 12314-1987 code in India. In Africa or other tropical countries, similar design standard may be followed because of similar temperature conditions (IS: 12314-1987, 2002). However, the water usage in toilet may vary in different countries and proper estimation of water usage is necessary for efficient design of pit toilets. For the countries situated outside tropical region and where the ambient temperature is significantly lower, the microbial activity may be significantly lower and pit toilet may not be suitable in those areas. The pit toilet is typically designed based on solid accumulation rate, long term infiltration rate of the liquid fraction of the soil below the toilet, hydraulic loading of the toilet and minimum period requirement for effective pathogen destruction and optimal emptying frequency. These are major factors which need to be considered for designing the twin pit toilet (Fig. 19.1). Lining or sealing of the bottom of the pits is not proposed, when the ground water level lies 2 m or more below the bottom most level of the pit, otherwise sealing of the bottom of the pit is suggested.

Where the distance between the bottom of the pit and the maximum groundwater level throughout the year is more than 2 m, the average solid accumulation is observed to be around 0.04 m^3 per capita per annum. If the water level rises till the bottom of the pit or more than that, then sand and soil envelop is required to be provided around the pit up to 2 m from the top of the highest water table level or till the top of the ground surface, respectively to avoid the contamination of the groundwater source. Additionally, because of the less percolation of liquid, excess accumulation of sludge will occur. According to IS: 12314-1987, the solids accumulation rate for a desludging interval of 2 year is estimated as 0.095 m^3/capita annum; however, this value changes for the desludging interval of 3 years to 0.095 m^3/capita annum. For desludging frequency of once in 4, 5 and 6 years the solids accumulation rate shall be considered as 0.051, 0.041 and 0.035 m^3/capita annum, respectively.

Where the water table is higher and rises till the bottom of the pit or more, the complete lining of pit toilet may be required which restrict the percolation of urine and causes the requirement of regular removal of liquid waste from the pit. Under these circumstances, installation of pit toilets is disadvantageous and installation of advanced technologies is encouraged. In addition, it is recommended to install a twin pit in the localities where manual handling is common practice; however, a single pit can also be installed in a locality where vacuum desludging is possible. Additionally, a minimum of 2 years of emptying frequency of the sludge is proposed. Minimum 150 mm freeboard of each pit is suggested however more freeboard can also be provided based on the requirement. Few other assumptions are also considered during the design of twin pit toilets according to IS: 12314-1987 as stated below.

i.   1.5 L of water is used for cleansing by each person per use of toilet.
ii. Each person uses the toilet twice a day.
iii. 2 L water is used for each flushing of the toilet.

iv. Average volume of urine and excreta production per day per capita is approximately 1.5 L.

v. At least 2 years of cleaning interval should be considered.

In addition, the water usage for cleaning after the usage of urinal by each member per day may be calculated by considering 6–8 times toilet usage per person per day and 0.5 L water usage for cleaning, thus this value will be around 3–4 L per day.

Example: 19.1 Design of twin pit toilet for 15 users, where (a) water table is sufficiently lower (≥ 2 m) than ground level throughout the year, and (b) water table is around 0.5 m below ground level during some period of the year. Consider local soil condition as sandy loam.

Solution
(a) Wastewater flow per capita per day (L per capita per day, LPCD) = Number of times faeces passed per day (two times on an average) × (total volume of water required for flushing and cleansing) + Average volume of urine and excreta production per day per member + water needed for flushing urine = $2 \times (2 + 1.5) + 1.5 + 4 = 12.5$ LPCD.

Total wastewater flow per day considering 15 users and 12.5 L for latrine floor washing

$$= 15 \times 12.5 + 12.5 = 200 \text{ L}$$

Assuming 1.5 m internal diameter having 75 mm inside lining surrounded by brick and effective depth of 1.5 m (1.8 m total depth including free board of 300 mm).

The infiltrative surface area is = $\pi \times$ external diameter × effective depth

$$= \pi \times (1.5 + 0.15) \times 1.5 = 7.77 \text{ m}^2$$

According to IS: 12314-1987; the infiltrative loading rate of sand is 50 L/m^2-day; for sandy loam it is 30 L/m^2-day; for porous silty loams or porous clay loams it is 20 L/m^2-day and for compact silty loams, silty clay loams or clay it is 10 L/m^2-day.

Hence, the infiltration area required is = $(200/30) = 6.66$ m$^2 \leq$ infiltrative surface area provided, which is 7.77 m^2 and more than the required.

The volume of the pit required based on the solid accumulation rate of 0.04 m^3 per capita per annum = $0.04 \times 15 \times 2 = 1.20$ m^3.

The proposed effective volume of the pit = $\pi \times 0.75^2 \times 1.5 = 2.64$ m$^3 \geq$ Required volume (1.20 m^3).

(b) For ground water table 0.5 m below the ground level during some period of the year the lining is required. Estimating the solid accumulation rate with the rate of 0.095 m^3 per capita per day considering desludging period of 2 years.

The volume of the pit required in this condition is = $0.095 \times 2 \times 15$ m$^3 = 2.85$ m$^3 >$ 2.64 m^3.

Considering 300 mm free board, the dia. of each pit required will be 1.6 m and total depth of 1.8 m, where effective depth of each pit is 1.5 m.

19.5   Septic Tank

Septic tank is a kind of anaerobic digestor, in which human excreta gets a longer detention time for anaerobic biodegradation. Typically, 24–48 h of hydraulic retention time is provided during designing a septic tank. Septic tanks usually have rectangular or circular shapes and are fabricated with concrete, masonry, or fiberglass materials. Baffles are provided in front of the inlet and outlet of the septic tank to restrict floating matters and oil and grease materials to escape from the septic tank. Heavy solid materials are settled at the bottom of the septic tank and biodegradable organics are getting anaerobically digested. Due to the anaerobic digestion of settled organic solids, biogas will be produced in the septic tank, which may cause hindrance for the settling of solids. Ventilation facility is provided to release this biogas by providing a ventilation pipe extended till at least 2 m higher than the nearest highest structure within 20 m vicinities. A significant hourly flow variation is observed in septic tanks based on the toilet usage. Sudden flow surge causes turbulence in the settled solids and may cause washout of the settled solids.

In practice, the septic tanks are generally constructed by local masons according to the area availability and the economic condition of the user and not constructed according to the guidelines; hence, the treatment efficiency of the system further reduces. This practice also affects the emptying frequency and quality of the treated effluent and septage recovered from these unscientifically designed septic tanks. Most often bigger pits are constructed than required or the bottom of the tanks is not lined, to increase the interval of the sludge removal.

Septic tanks are modified by increasing the number of chambers or hybridizing with other physicochemical or biological processes to intensify the treatment efficiency; however, this modification sometimes increases the complexity of the system. Single-chambered septic tanks and two- or three-chambered septic tanks are considered as the conventional design of the septic tank. However, improved hybridized system designs, such as two-chambered septic tanks with filter or anaerobic baffled reactors with filter are also constructed.

19.5.1 Single-Chambered Septic Tank

Single-chambered septic tanks are similar to the single pit toilet with complete lining, where a simple modification of design is observed. The biodegradable solid portion of the human excreta coming from the toilets gets settled at the bottom of the single-chamber septic tank and anaerobic digestion of the settled sludge takes place. This type of OSS

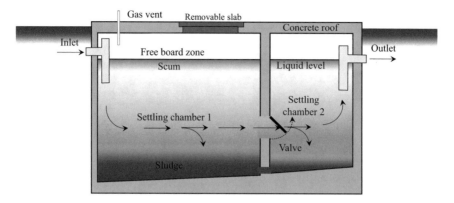

Fig. 19.2 Schematic of two chambered septic tank

requires frequent emptying as the rate of digestion of solids is comparatively low. A higher probability of the washout of the settled sludge during a high inflow of water was observed in this type of septic tank. This conventional type of single-chambered septic tank is not advisable as an OSS, because of its low wastewater treatment efficiency and requirement of high maintenance.

19.5.2 Two-Chambered Septic Tank

To overcome the bottlenecks of the single-chambered septic tank two-chambered or multi-chambered septic tanks are scientifically designed. In a typical two-chambered septic tank, two chambers are provided in such an orientation that the first chamber should be at least twice the size of the second chamber. Maximum solids settle down in the first chamber and the partition between the chambers prevents scum and solids from escaping with the effluent (Fig. 19.2). This setup reduces the possibility of unwanted sludge washout during a higher effluent discharge period. A T-shaped outlet pipe further reduces the amount of scum and solids that are discharged. Generally, these septic tanks have to be emptied every 3 years.

19.5.3 Two-Chambered Septic Tank with Filter

This type of system is a modification of two-chambered septic tank systems, where an additional filtration chamber is placed after two settling chambers (Fig. 19.3). The incorporation of an additional filter chamber improves treatment efficiency. As wastewater flows through the filter, particles are trapped and organic matter is degraded by the active biomass that is attached to the surface of the filter material. Commonly gravel, crushed

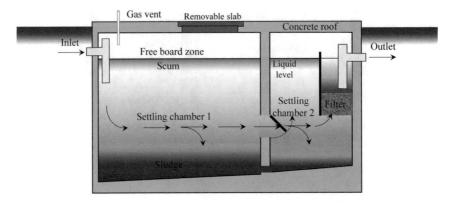

Fig. 19.3 Schematic of two chambered septic tank with filter

rocks, cinder, or specially manufactured plastic pieces are used as filter material. Typical filter material sizes range from 12 to 55 mm in diameter. Ideally, these materials have a specific surface area between 90 and 300 m^2 per m^3 of filter volume, which provides a large surface area for the bacterial mass to form biofilm and participate in removal of organic matter. Increased contact between the organic matter and active biomass effectively degrades the organic matter in this modification of the septic tank. Suspended solids and biochemical oxygen demand (BOD) removal of this type of septic tank may be as high as 85–90%, however lower nitrogen removal typically around 10–15% is observed (Darby & Leverenz, 2004). Extra care for maintaining filter media is necessary in this setup.

19.5.4 Anaerobic Baffled Reactor with Filter

An anaerobic baffled reactor with filter based onsite sanitation treatment system is further improvement of the septic tank technology (Fig. 19.4). Here more than one baffles (typically 2 to 5) are incorporated to control the movement of the sewage to flow upward from the bottom to top in each chamber until it enters into the next chamber and to improve the substrate and biomass contact (Barber & Stuckey, 1999). Additionally, a filtration unit is also installed after anaerobic baffled reactors to enhance the treatment efficiency (Luthra et al., 2017). An increasing number of baffles increases the hydraulic retention time for improved settling of sludge and digestion of sludge. This advanced setup is capable enough to manage a considerably higher organic loading rate (Barber & Stuckey, 1999).

A typical anaerobic baffled reactor can demonstrate maximum BOD removal up to 90% (Tilley et al., 2014). Majority of the solids get settled in the first compartment before the first baffle and comparatively lesser volume of solids settles in the succeeding chambers.

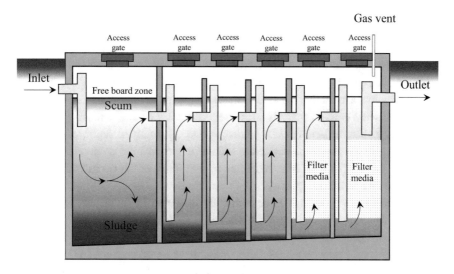

Fig. 19.4 Schematic of anaerobic baffled reactors with filter based septic tank

Filtration arrangement in the last compartment ensures the high treatment efficiency of the module and provides a tertiary treatment to the effluent. This advanced design is capable enough to handle a sludge inflow of minimum of 2 m³/day to maximum 200 m³/day with a corresponding hydraulic retention time of 24–72 h and suitable for installation in a small household scale to a community toilet or managing septic tank sludge of the small and medium community scale (Tilley et al., 2014).

Up-flow velocity of 0.6 m/h or less is suitable for efficient removal of solids with three to six numbers of compartments separated by baffles (Tilley et al., 2014). Accessibility to each chamber is essential for sludge removal and a sludge removal interval of one to 3 years is recommended for this (Tilley et al., 2014). Because of anaerobic digestion, biogas generation will occur; however, biogas collection is not sustainable for small scale reactors because of insufficient amount of biogas generated. However, proper vent arrangement is essential to release odorous and harmful gasses liberated. Sometimes during start-up of the reactor inoculum addition may be required to reduce the start-up time and sludge settled in the existing septic tank shall be used for this purpose.

19.5.5 Design of Septic Tank

According to the Central Public Health and Environmental Engineering Organisation (CPHEEO) manual of Sewerage and Sewage Treatment System, India (2013) only about 30% of the settled solids are anaerobically digested in a septic tank. Digestion rate remains unchanged after frequent desludging, which is essential for satisfactory effluent quality.

Additionally, when sludge is not removed from the septic tank for a longer period, i.e., more than the design period, a substantial portion of solids washout with the effluent will occur because of the reduction in the effective liquid depth in the septic tank due to accumulated sludge. Therefore, for efficient suspended solids removal, septic tank should have sufficient capacity with proper inlet and outlet arrangements. Additionally, it should be designed in such a fashion that the sludge should settle at the bottom and scum accumulate at the surface, while enough space is left in between, for the sewage to flow through without disturbing either the scum or the settled sludge. Typically, a septic tank is designed and desludging frequency is maintained to such an extent that the accumulated sludge and scum occupy only about half, or a maximum of two-third, of the overall tank capacity, at the end of the design storage period.

The conventional septic tank is scientifically designed in India based on IS: 2470–1985 (IS: 2470-1985, 2001), where the sewage flow is estimated based on the number of fixture units. Other tropical countries where similar temperature range exists, similar design of septic tanks may be adopted. According to standard convention (IS: 2470-1985) peak discharge coming from a household constituted with five persons are considered as a fixture unit. During design of a septic tank if sewage is coming from one or two fixture units then it should be considered as all the units are discharging simultaneously. However, for three or four fixture units it was considered that two or three fixture units, respectively, are likely to discharge simultaneously. For each fixture unit discharging simultaneously, 9 L/min peak discharge is considered based on IS: 2470-1985.

The hydraulic retention time of a typical septic tank is observed as 24–48 h; however, because of the significant flow fluctuation the instantaneous hydraulic retention time may vary significantly. For larger septic tanks, half yearly or yearly sludge removal is desirable. However, for the smaller septic tanks serving individual households, the septic tank may be advisable to be cleaned at least once in 2 years if the tank is not overloaded within this requisite time frame. The required surface area of the tank is calculated as 0.92 m^2 for every 10 L per minute of peak flow rate at an ambient temperature of 25 °C. This surface area favours the sedimentation of solids having 0.05 mm size and specific gravity of 1.2. A minimum depth for settled sludge should be considered at least 0.25–0.3 m during septic tank design.

For design of a septic tank, it is assumed that 70 g/day per capita sludge enters into the septic tank during regular operation. The capacity of digestion zone required for digesting this sludge is 0.033 m^3 per capita considering about 62 days of digestion time and the volume of the digested sludge is normally 0.00021 m^3 per capita per day. Considering this estimation, the sludge storage per capita per year can be approximated as (0.00021 m^3 × 365) or 0.076 m^3. Additionally, about 25–50 mm depth of seed volume should be considered, and care should be taken while withdrawing the sludge to leave this volume of sludge to act as seed at the bottom of the septic tank so that the efficiency of the sludge digestion should not be reduced after cleaning.

Example: 19.2 Determine the dimension of a septic tank for handling sewage generated from 15 users.

Solution

As the number of users is 15, the fixture unit simultaneously discharging is 2 and the peak discharge of the septic tank can be considered as 18 L/min.

The required surface area of the septic tank is $(0.92 \times 18/10) = 1.656$ m^2.

Assuming the required depth of settling zone as 0.3 m the total volume of the settling zone is $= (0.3 \times 1.656)$ m$^3 = 0.4968$ m^3.

The volume of the digestion zone is $= (0.033 \times 15)$ m$^3 = 0.495$ m^3.

Considering the sludge cleaning frequency of 2 years the volume of the sludge storage zone is $= (0.00021$ m$^3 \times 365 \times 15 \times 2)$ m$^3 = 2.299$ m^3.

Considering 0.3 m of free board, the volume of the free board zone

$$= (0.3 \times 1.656) \text{ m}^3 = 0.4968 \text{ m}^3.$$

Hence, the total volume of the septic tank for 15 users $= (0.4968 + 0.495 + 2.299 + 0.4968)$ m^3

$$= 3.7876 \text{ m}^3$$

The total height of the septic tank is $= [0.3 + (0.495/1.656) + (2.299/1.656) + 0.3]$ m

$$= [0.3 + 0.3 + 1.4 + 0.3] \text{ m} = 2.3 \text{ m}$$

Considering the length and width ratio of 3, the length of the septic tank is 2.25 m and width of the septic tank is 0.75 m.

19.6   Soak Pits

A soak pit, or leach pit, is a covered chamber; typically rectangular, square, or circular with porous-wall to allow the septic tank effluent to slowly soak into the ground. Sometimes the effluent of centralized wastewater treatment units or other decentralized primary treated effluents is also soaked through soak pits. As the water percolates through the soil, the small particulates are filtered and adsorbed by the soil matrix and dissolved organics are digested by microbes present in the soil. Typically the depth of the soak pit is around 1.5–4 m however in practice the depth of the soak pit is maintained more than 2 m above the groundwater table (Tilley et al., 2014). In addition, soak pits should be constructed at a considerable distance from a high traffic zone to avoid soil compaction and a safe distance of 30 m should be always maintained between the soak pit and any drinking water sources to avoid microbial contamination (Tilley et al., 2014). Soak pit is generally filled with porous materials such as coarse rock or gravel to side walls and walls are semi lined or kept unlined for efficient percolation of the water. A fine layer of sand or small gravel is spread at the bottom for uniform dispersion of the water. For the areas having high groundwater tables or flood-prone zones, a soak pit is not a suitable option and a high possibility of groundwater contamination exists.

This subsurface disposal field can serve as a further treatment system and for disposal of treated wastewater, which has undergone some reduction in SS and grease content. Many of the natural soils are suitable for such systems. The design of this system is based on the long-term capacity of the soil to percolate the water and it is decided upon conducting the standard percolation test. The subsidence rate of water in the test bore hole of 100 mm diameter with the test depth of proposed disposal field of minimum 500 mm is recorded. After removing the loose soil 50 mm of fine gravel or coarse sand is placed at the bottom of test bore. The hole is then filled with the water to the depth of 300 mm, and the depth is maintained overnight (at least 4 h) by adding water. The water depth is then adjusted to 150 mm above gravel and the drop in water level is recorded at 30 min interval for next 4 h. If no water is hold in the bore in the morning, it is filled with the water (150 mm above gravel) and the drop in water level is recorded. The drop recorded in the last 30 min is considered for the determination of percolation rate. The flow rate which can be applied per unit area as a function of percolation rate is estimated as per Eq. 19.1.

$$Q = 204/\left(t^{0.5}\right)$$

(19.1)

where, Q is flow (L/m^2 day), t is time in min, required for the water to fall 25 mm. Instead of Eq. 19.1 the CPHEEO (2013) manual recommend $Q = 130/(t^{0.5})$ for Indian conditions.

If the subsidence rate is over 0.5 mm/min, then a septic tank and the disposal system will work satisfactorily. The disposal field is constructed by using short length open joined pipe (100 mm dia.) or perforated plastic or fibre pipe. The length of individual pipe is kept up to 30 m and pipe is laid with the slope of 0.017–0.33%. The pipe is placed in a ditch (300–900 mm width) and minimum 500 mm depth of this ditch can be excavated to the depth of permeable stratum. The ditch is backfilled with gravel for a depth of 300–400 mm and over which pipes are placed. An additional 50 mm gravel cover is given to the pipe before the soil backfilling material is placed. The total length of pipe depends on the trench width, since the product of this, i.e., the plan area of the trench, should be equal to the area required to be provided as per the percolation test. Laterals (pipes) are placed about 2 m c/c.

At a location where the parent soil permeability is poor, mounds may be constructed above the surface of the ground by bringing pervious soil material. The area of mounds depends on evapotranspiration rate for disposal of the water. The remainder may be able to percolate below ground level. Grass cover and proper shaping of the mound are important to ensure that rainfall will run-off and that evapotranspiration of wastewater will be maximized. Sand filters or buried filters or intermittent sand filters can be provided when soils are relatively impermeable. Loading rate on intermittent sand filters treating septic tank effluent typically ranged from 0.16 to 0.20 m^3/m^2 day.

Example: 19.3 Determine the size of the septic tank and percolation field for an isolated residential complex away from sewerage having 200 residents. After performing the percolation test on site, the percolation rate has been observed to be 5 mm/min. The rate of water supply is 150 L/capita day.

Solution
Total wastewater volume generated per day $= 150 \times 0.8 \times 200 \times 10^{-3} = 24$ m^3.

For hydraulic retention time of 24 h, the volume of septic tank $= 24$ m^3 + volume for sludge accumulation @ 0.073 m^3/capita year $= 24 + 0.073 \times 200 = 38.6$ m^3.

(Alternately the exact area and volume requirement for the septic tank can be worked out using peak discharge of 480 L/min for 200 persons as per the procedure explained in Example 19.2).

The percolation rate, i.e., the time to fall water to a depth of 25 mm is $= 25/5 = 5$ min.

Hydraulic loading applicable $= Q = 130/(t^{0.5}) = 130/(5^{0.5}) = 58.14$ L/m^2 day.

Therefore, total trench area $= 24/0.058 = 413$ m^2.

If width of trench is 900 mm, length of trench $= 413/0.9 = 458.89 \sim 460$ m.

Provide 16 laterals of 28.7 m length each, placed at 2 m c/c.

The area dedicated to the field would be $= 32 \times 28.7 = 918.4$ m^2.

19.7 Biodigestors

Anaerobic bio-digestor is a wastewater treatment technology where biodegradable sludge is anaerobically reduced to methane, carbon dioxide, ammonia, etc., and a small amount of new biomass forms, which will also eventually get digested. Bio-digestion is a very effective measure to sustainably manage biodegradable wastes. Human waste is an ideal substrate for bio-digestion and a significant reduction in biodegradable organics and sludge mass along with efficient energy recovery in terms of methane is possible (Colón et al., 2015). Biodigesters are suitable for both small-scale household level and community based sanitary wastewater treatment (McGill, 2013). In India, Defence Research and Development Organisation has improved this biodigester technology and made this technology more efficient for application in a community or household toilets and moving vehicles (Kamboj et al., 2012). Indian railway adopted this technology for sustainable treatment of human waste under Swachha Bharat Mission.

Bio-digesters promoted by the Swachha Bharat Mission is designed to achieve around 80–90% treatment of black water from individual and cluster of few households or institutional buildings, where there is no separate sewerage network. This technology has two key features. Primarily, a specific anaerobic microbial consortium is adopted in this technology for the early decomposition of faecal matters. Secondly, an especially designed fermentation tank is used for sustainable management of the human excreta. The specific microbial consortium has been engineered by acclimatizing, enriching and bio-augmenting cold-active bacteria collected from Antarctica and other low-temperature areas. These specific microbes are composed of four groups of bacteria having hydrolytic, acidogenic, acetogenic and methanogenic nature with high efficiency of bio-degradation. A multi-chambered fermentation tank fabricated with metal or fibreglass-reinforced polymer (FRP) is designed strategically to immobilize large numbers of engineered microbes. A typical multi-chambered biodigester tank designed by DRDO is comprised of an initial screening arrangement by using steel rods at 30° angle. After screening multi-chambered digestion tanks are installed separated by polyvinyl chloride materials. Baffle arrangement is provided to restrict inoculum washout and finally for disinfection of treated water, chlorination is adopted. The hydraulic retention time of biodigester is typically maintained at around 2 days.

19.8   Bio-toilet (Aerobic)

Biochemical oxidative stabilization of sludge takes place during aerobic digestion in the aerobic Bio-Toilet system. In this technology, biodegradable solids or sludge is aerobically degraded in closed chambers. Because of the lack of the substrate, endogenous degradation of biomass takes place and the biomass is aerobically oxidized to CO_2, H_2O, NH_4^+, NO_2^-, and NO_3^-. For efficient oxidation, air or oxygen is supplied by surface aerators and mixing arrangements are also provided in several instances. Rapid degradation of sludge is the major advantage of this technology, which reduces the reactor volume. The degradation efficiency of this technology is comparable to anaerobic treatment technology and the foul smell does not generate from this process (Shammas & Wang, 2007). Aerobic bio-toilets designed by Stone India are portable bio-toilets constituted with a multi-chambered bio-tank for the storage of human waste (Fig. 19.5).

The bio-toilet reactors are typically fabricated with bricks-mortar, FRP or steel made bio-digester tank and superstructure. Multiple strains of aerobic bacteria are separately dosed into the bio-tank to efficiently degrade the waste matter following aerobic digestion. The movement of human waste is slowed down by using multiple chambers. When human waste flows from one chamber to another, it follows a longer pathway in the bio-tank, such that the multi-strain microbes present in the digestor tank can get sufficient span to digest the waste efficiently and convert it to stabilized end product. Treated effluent of the toilet is disinfected in the final chamber and released for reuse. A 1.2 m (length) ×

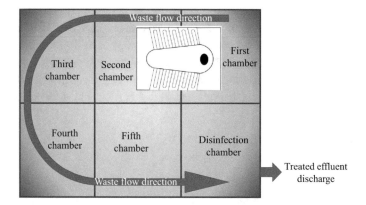

Fig. 19.5   Schematic of the top view of the aerobic Bio-toilet

1.2 m (width) × 1.2 m (height) size aerobic Bio-toilet is typically designed and attached with Indian pan or water closet for managing the waste generated from a maximum of 30 defecations per day with no limit on urination.

19.9   Other Innovative Technologies

Other than conventional sanitary wastewater treatment technologies, different dry and wet decentralized wastewater treatment technologies are also investigated by the researchers (Katukiza et al., 2012; Lourenço & Nunes, 2020; Varigala et al., 2020). Although most of the innovative technologies are in lab-scale or pilot-scale; hence, significant future research is still necessary to increase the feasibility of these technologies.

19.9.1   Incineration Toilet

Incineration Toilet is a strategically designed dry toilet, where human faeces is burned directly by using energy sources, like gasses, fuels, or by electrical means (Kufrin et al., 1972). This type of technology is highly efficient to manage human waste along with other variety of waste materials, like tissue papers or plastic wrappers, etc., within a very short retention time. This technology is also suitable in cold weather and locations having water scarcity. However, high energy consumption during the vaporization of moisture and combustion of the waste is a major drawback of this technology. Additionally, the exhaust gasses generated from incineration toilets are also causing air pollution.

These types of toilets have specific applicability like the apartments having difficulty in waste plumbing, houses without adequate drainage access or within the moving vehicles and marine vessels. This technology is presently patented and commercialized by different international organizations ("Incinolet electric incinerating toilet," 2021), however still there lies a scope for further improvement in this technology that is being offered at present. Research on different aspects of this technology is still going on to make this technology more sustainable, such as RTI International has designed an incineration toilet under the funding of Bill & Melinda Gates Foundation by electrochemical disinfection and drying of human waste and harvest energy by burning the dried waste (Morrison et al., 2014).

19.9.2 Chemical Toilet

Human waste is stabilized and managed by chemical means in such toilets. In a typical chemical toilet, human excreta is stored in a holding tank and chemical dosing is given to minimize organic loading, microbial contamination and odours (Appiah-Effah et al., 2020). These toilets are usually, but not always, self-contained and movable. A chemical toilet is designed as a relatively small compact tank, which needs to be emptied frequently. It is not connected to a hole in the ground (like a pit latrine), nor to a septic tank, nor is it plumbed into a municipal system leading to a sewage treatment plant. During the emptying of the system, the contents are usually pumped into a sanitary sewer or directly to a treatment plant. Aircraft lavatories and passenger train toilets were often designed as chemical toilets in the past but now converted to vacuum toilets.

19.9.3 Composting Toilet

Composting toilet is a special type of dry toilet, where biological composting process is used for conversion of human excreta into compost like material. Composting is carried out by utilizing microorganisms under control aerobic condition. However, complete destruction of pathogens does not occur in this process (Redlinger et al., 2001). This technology can efficiently function in the remote areas having water shortage (Anand & Apul, 2014). The toilet structures of composting toilets are made up of plastic, ceramic, or fiberglass materials. A composting toilet has typically two basic components, such as the toilet compartment and the composting tank. Often a fan and vent pipes are also installed to remove any odour problem from the toilet. A drainage arrangement is typically provided to eliminate excess leachate and access doors are provided to empty compost chamber periodically.

Composting process requires minimal or almost no use of water for conveyance of human excreta. Composting compartment is a waste collector, where the waste is collected and digested aerobically. Composting systems are sometimes modified by using earthworms, which is popularly known as vermicomposting toilet (Hill & Baldwin, 2012; Yadav et al., 2010). Sawdust, leaves, and food waste are often used as bulking agents or amendments to improve the quality of the compost, adjust carbon to nitrogen ratio (20–30), and increase porosity of the compost. Sometimes urine separation technologies are also used in the composting toilet to reduce the quantity of the waste and avoid unwanted moisture from the toilet (Lienert & Larsen, 2010).

19.9.4 Bioelectric Toilet

Bioelectric toilet technology is the application of bio-electrochemical technology, like microbial fuel cell (MFC) for the onsite treatment of human waste generated from a toilet. A typical MFC is a two-chambered (anodic and cathodic) setup, separated by a proton exchange membrane (PEM). Anaerobic oxidation of biodegradable organic matter occurs in the anodic chamber through the catalytic action of electrogenic bacteria and electrons and protons are generated. Protons migrate through the proton exchange membrane towards cathode and electrons reduces anode and flows through external circuit towards cathode and harvests electricity. The treatment of the sanitary wastewater by using MFC was first time conducted in Ghana, however only preliminary research based on nitrification was carried out and detailed research was not conducted (Castro et al., 2014). Treatment of separated urine was successfully carried out by stacking MFCs made up of ceramic in Bristol BioEnergy Centre and efficient electricity recovery was achieved (Ieropoulos et al., 2016).

For managing complete toilet waste by using MFC, Bioelectric Toilet was developed (Fig. 19.6). According to this technology, chemical energy, stored into a bio-convertible human waste and sludge is bio-electrochemically converted to electricity by employing in situ electrogenic microbiota present within the waste environment. By installing Bioelectric Toilet, apart from the direct electricity recovery, improved treatment to the black water can also be attained within the short retention period. Odour problem associated with the septic tank is minimized due to anaerobic oxidation of ammonia to nitrogen and sulphide to elemental sulphur in Bioelectric Toilet. Thus, degradation of organic matter, along with bioelectricity recovery can be simultaneously achieved in bioelectric toilets. Tertiary treatment, such as disinfection of effluent can also be achieved by using the chlorinated solution as catholyte or during post treatment of effluent generated from Bioelectric Toilet (Ghadge et al., 2015). Other polishing treatments like filtration may also be adopted to improve the quality of the final treated effluent. In addition, the Bioelectric Toilet system could be capable of generating electricity from human waste for onsite use in rural areas. Obsoleting odour problems and curbing environmental pollution without the need for external energy can also be achieved by adopting this technology.

First Bioelectric Toilet (Das, 2020) was designed and installed at a field-scale on the campus of IIT Kharagpur, which was consisting of a toilet superstructure and a toilet waste treatment unit having hexagonal shape and 1.5 m^3 internal volume (Fig. 19.6a). The hexagonal treatment unit consisted of one hexagonal central chamber and six peripheral

Fig. 19.6 **a** Bioelectric toilet established at IIT Kharagpur; **b** Top view of the bioelectric toilet-based wastewater treatment unit; **c** Illumination of toilet at night; **d** an alternative low-cost MFC based toilet waste treatment unit

chambers (Fig. 19.6b). Wastewater coming from the toilet enters into the central chamber, which primarily acts as a settling chamber as well as anodic chamber and peripheral chambers are designed as air cathode stacked-MFC modules. Hypochlorite dosing is being done in the final chamber, which acts as catholyte and this final chamber works as the cathodic chamber corresponding to the central anodic chamber. For each peripheral MFCs, membrane electrode assemblies consisting of clayware ceramic membrane separator (for charge separation) and carbon felt electrodes are being used with suitable cathode catalysts to enhance oxygen reduction reaction in this system.

More than 95% COD removal efficiency was observed from this Bioelectric Toilet and the BOD of the treated effluent was observed to be less than 30 mg/L till an organic loading rate (OLR) is kept below 0.5 kg COD/m^3 day (which corresponds to the toilet used by 10 persons). Also, by using hypochlorite dosing in the last compartment, which is also working as catholyte, disinfection of the wastewater is being achieved producing treated water safe for human contact reuse, which can be reused in toilet for flushing. The treated effluent quality is sufficient to be reused for toilet flushing or gardening purposes and by using treated water for toilet flushing almost 70–80% of freshwater requirement per use of toilet can be reduced. About 1 L of water used for cleaning per use of the toilet can compensate the water loss due to evaporation. Thus, the 300 L overhead tank once filled with fresh water will serve the purpose of making water available for toilet flushing and cleaning, as the water flushed will be treated onsite and pumped to the overhead tank once a day using solar powered pump. Thus, application of such toilet will reduce water demand per use of the toilet to about a litre (jut for cleaning) and still be able to suffice the water requirement for maintaining cleanliness in the toilet.

The sludge generation in this system is very low and desludging in a 6 months interval is recommended. The sludge after drying is suitable for land disposal and can be used as soil amendment agent or manure. No externally cultured specific inoculum dosing or air supply is required for this system. In addition, onsite electricity recovery is facilitated and by storing the electricity harvested from the toilet, onsite illumination of toilet and premises can be achieved (Fig. 19.6c). A low-cost portable Bioelectric Toilet was further demonstrated at IIT Kharagpur by using large clayware cylinders as MFCs (Fig. 19.6d). The individual MFC units were having volume of 25 L and these are suitable enough to carry from one place to another place.

19.9.5 Vacuum Toilets

Vacuum toilets are specially designed toilets, where high suction pressure is employed to successfully remove the excreta without consuming freshwater or consuming a minimal volume of freshwater of about 0.5–1.5 L (Beach et al., 2020). Vacuum toilets provide a similar level of comfort and aesthetic look as traditional flush toilets and they help to reduce cost by utilizing a nominal amount of flush water. Because of the very high organic content of the waste generated from vacuum toilets, compost treatment can be used. Vacuum toilet systems are applicable both in large and small buildings, trains, ships and aeroplanes. Water saving, hygienic, portable, and odour free nature make this technology advantageous. However, relatively high investment cost, high electrical energy required to generate adequate suction pressure are few disadvantages of this technology. Bulky materials like tissue or sanitary napkins may cause clogging of these types of toilets and might create operational nuisance.

19.10 Summary

The sanitation sector gained much attention in the past decade and world leaders are presently focusing to develop sustainable sanitation infrastructures throughout the world. Significant research efforts are being given to gain knowledge of sustainable sanitation alternatives and to spread the sustainable sanitation infrastructure throughout the world. Lack of sanitary network and requirement of enormous funding to spread the sanitary infrastructure throughout the world is a major challenge and decentralized sanitation is found to be an attractive alternative in this scenario. Different decentralized sanitation alternatives are being proposed in the recent decade and excellent treatment efficiency was demonstrated. However, high cost of installation and necessity of trained manpower is a bottleneck of most of the technologies. Reluctance for spending higher cost on modern age sanitation alternatives is another major reason behind the installation of primitive sewage treatment options, which is inviting risk of contamination of freshwater sources. Additionally, installing costly technologies by the low-income group of people is not feasible and affordable solution. Hence, there is a need to develop further affordable yet technologically sound solutions for onsite treatment of the toilet waste to provide access to safe sanitation for every individual.

Questions

19.1. Estimate the wastewater generation from a toilet which is used by 10 persons.
19.2. The human excreta generated per capita per day is 70 g in terms of COD load and water usage per capita per day is 20 L in a cold region. Suppose the volume of a toilet waste storage tank having 4 m^3 effective internal volume was situated in that region and was connected to a toilet which was used by 10 persons. Assume, in a specific day if 50% of the tank was filled with wastewater having a COD of 500 mg/L in completely mixed condition. Calculate the increase in COD of the tank after one day in completely mixed condition. Ignore the biodegradation due to cold environment and water loss due to evaporation on that specific day. (**Answer**: 272.7 mg/L)
19.3. Design a twin pit toilet for a household having 10 members situated in (a) dry region and (b) nearby a sea shore.

19.4. Describe the biodigester developed by DRDO.

19.5. Discuss design guidelines for septic tank accounting for the flow surges it is likely to receive.

19.6. With a schematic discuss the working of anaerobic baffled reactor. What will be typical performance expected from this while treating domestic sewage?

19.7. Discuss necessity of soak pits for handling effluent from septic tank.

19.8. Why percolation rate is important while designing a soak pit?

19.9. Describe how percolation test is performed.

19.10. Write a note on Bioelectric toilet.

19.11. Write a note on vacuum toilet.

19.12. Describe composting toilet.

19.13. Describe aerobic Bio-toilet and its operation for onsite treatment of wastewater generated from toilet.

19.14. Write a note on sustainable onsite sanitation.

19.15. With the recent developments what are the options available for providing onsite sanitation solution?

References

Almomani, F. (2016). Field study comparing the effect of hydraulic mixing on septic tank performance and sludge accumulation. *Environmental Technology, 37*, 521–534. https://doi.org/10.1080/09593330.2015.1074623

Anand, C. K., & Apul, D. S. (2014). Composting toilets as a sustainable alternative to urban sanitation—A review. *Waste Management, 34*, 329–343. https://doi.org/10.1016/j.wasman.2013.10.006

Antoniou, G. P., De Feo, G., Fardin, F., Tamburrino, A., Khan, S., Tie, F., Reklaityte, I., Kanetaki, E., Zheng, X. Y., Mays, L. W., & Angelakis, A. N. (2016). Evolution of toilets worldwide through the millennia. *Sustainability (switzerland), 8*(8), 779. https://doi.org/10.3390/su8080779

Appiah-Effah, E., Duku, G. A., Dwumfour-Asare, B., Manu, I., & Nyarko, K. B. (2020). Toilet chemical additives and their effect on faecal sludge characteristics. *Heliyon, 6*, e04998. https://doi.org/10.1016/j.heliyon.2020.e04998

Armah, F. A., Ekumah, B., Yawson, D. O., Odoi, J. O., Afitiri, A. R., & Nyieku, F. E. (2018). Access to improved water and sanitation in sub-Saharan Africa in a quarter century. *Heliyon, 4*. https://doi.org/10.1016/j.heliyon.2018.e00931

Barber, W. P., & Stuckey, D. C. (1999). The use of the anaerobic baffled reactor (ABR) for wastewater treatment: A review. *Water Research, 33*, 1559–1578. https://doi.org/10.1016/S0043-1354(98)00371-6

Beach, D., Gee, K., Yu, J., Sommerfeldt, S., Murray, G., Boodaghians, R., & Kim, D. J. C. (2020). Multiple rinse injections to reduce sound in vacuum toilets. US010683652B1.

Bhowmick, G. D., Dhar, D., Nath, D., Ghangrekar, M. M., Banerjee, R., Das, S., & Chatterjee, J. (2020). Coronavirus disease 2019 (COVID-19) outbreak: Some serious consequences with urban and rural water cycle. *npj Clean Water, 3*, 32. https://doi.org/10.1038/s41545-020-0079-1

Butler, D., & Payne, J. (1995). Septic tanks: Problems and practice. *Building and Environment, 30*, 419–425. https://doi.org/10.1007/BF02249053

Castro, C. J., Goodwill, J. E., Rogers, B., Henderson, M., & Butler, C. S. (2014). Deployment of the microbial fuel cell latrine in Ghana for decentralized sanitation. *Journal of Water Sanitation and Hygiene for Development, 4,* 663–671. https://doi.org/10.2166/washdev.2014.020

Chaudhuri, S., & Roy, M. (2017). Rural-urban spatial inequality in water and sanitation facilities in India: A cross-sectional study from household to national level. *Applied Geography, 85,* 27–38. https://doi.org/10.1016/j.apgeog.2017.05.003

Colón, J., Forbis-Stokes, A. A., & Deshusses, M. A. (2015). Anaerobic digestion of undiluted simulant human excreta for sanitation and energy recovery in less-developed countries. *Energy for Sustainable Development, 29,* 57–64. https://doi.org/10.1016/j.esd.2015.09.005

CPCB. (2021). Status of STPs [WWW Document]. CPCB. Retrieved December 21, 2021 from https://cpcb.nic.in/status-of-stps/

CPHEEO. (2013). Manual of sewerage and sewage treatment system, Ministry of Urban Development, Government of India.

Darby, J. L., & Leverenz, H. (2004). *Technical completion report of virus, phosphorus, and nitrogen removal in onsite wastewater treatment processes.* University of California.

Das, I. (2020). *Bioelectric toilet: For onsite treatment of blackwater to facilitate reuse of treated water and electricity generation for onsite applications.* Ph.D. thesis, Indian Institute of Technology Kharagpur, India.

Ghadge, A. N., Jadhav, D. A., Pradhan, H., & Ghangrekar, M. M. (2015). Enhancing waste activated sludge digestion and power production using hypochlorite as catholyte in clayware microbial fuel cell. *Bioresource Technology, 182,* 225–231. https://doi.org/10.1016/j.biortech.2015.02.004

Hill, G. B., & Baldwin, S. A. (2012). Vermicomposting toilets, an alternative to latrine style microbial composting toilets, prove far superior in mass reduction, pathogen destruction, compost quality, and operational cost. *Waste Management, 32,* 1811–1820. https://doi.org/10.1016/j.wasman.2012.04.023

Ieropoulos, I. A., Stinchcombe, A., Gajda, I., Forbes, S., Merino-Jimenez, I., Pasternak, G., Sanchez-Herranz, D., & Greenman, J. (2016). Pee power urinal-microbial fuel cell technology field trials in the context of sanitation. *Environmental Science: Water Research & Technology, 2,* 336–343. https://doi.org/10.1039/c5ew00270b

Incinolet Electric Incinerating Toilet [WWW Document]. (2021). https://incinolet.com/

IS: 12314-1987. (2002). Indian Standard Code of practice for sanitation with leaching pits for rural communities. Bureau of Indian Standards, New Delhi, India.

IS: 2470-1985. (2001). Indian Standard Code of practice for installation of septic tanks. Bureau of Indian Standards, New Delhi, India.

Jain, A., Wagner, A., Snell-Rood, C., & Ray, I. (2020). Understanding open defecation in the age of Swachh Bharat Abhiyan: Agency, accountability, and anger in rural Bihar. *International Journal of Environmental Research and Public Health, 17*(4), 1384. https://doi.org/10.3390/ijerph170 41384

Jangam, C., & Pujari, P. (2019). Impact of on-site sanitation systems on groundwater sources in a coastal aquifer in Chennai, India. *Environmental Science and Pollution Research, 26,* 2079–2088. https://doi.org/10.1007/s11356-017-0511-3

Kamboj, D. V., Singh, L., Gangwar, V. K., & Vijayaraghavan, R. (2012). Self-sustained bio-digester for onboard degradation of human waste. WO2012042526A1. Indian Patent filed by Director General, Defence Research & Development Organisation, India.

Katukiza, A. Y., Ronteltap, M., Niwagaba, C. B., Foppen, J. W. A., Kansiime, F., & Lens, P. N. L. (2012). Sustainable sanitation technology options for urban slums. *Biotechnology Advances, 30,* 964–978. https://doi.org/10.1016/j.biotechadv.2012.02.007

Kufrin, F. W., Virnoche, P. R., Allen, D. J., Gokey, P. E., & Rose, F. A. (1972). Disposal of human waste by incineration. US3694825A. US Patent Application filed by Polar Ware Co., USA.

Libralato, G., Volpi Ghirardini, A., & Avezzù, F. (2012). To centralise or to decentralise: An overview of the most recent trends in wastewater treatment management. *Journal of Environmental Management, 94*, 61–68. https://doi.org/10.1016/j.jenvman.2011.07.010

Lienert, J., & Larsen, T. A. (2010). High acceptance of urine source separation in seven European countries: A review. *Environmental Science and Technology, 44*, 556–566. https://doi.org/10.1021/es9028765

Lourenço, N., & Nunes, L. M. (2020). Review of dry and wet decentralized sanitation technologies for rural areas: Applicability, challenges and opportunities. *Environmental Management, 65*, 642–664. https://doi.org/10.1007/s00267-020-01268-7

Luthra, B., Bhatnagar, A., Matto, M., & Bhonde, U. (2017). Septage management: A practitioner's guide. Centre for Science and Environment.

Massoud, M. A., Tarhini, A., & Nasr, J. A. (2009). Decentralized approaches to wastewater treatment and management: Applicability in developing countries. *Journal of Environmental Management, 90*, 652–659. https://doi.org/10.1016/j.jenvman.2008.07.001

McGill, G. (2013). Community-appropriate biodigesters for Cambodia. *Water: Journal of the Australian Water Association, 40*, 74–80.

Morrison, L., Hossain, A., Elledge, M., Stoner, B., & Piascik, J. (2014). User-centered guidance for engineering and design of decentralized sanitation technologies. In *RTI Press research brief*. RTI Press. https://doi.org/10.3768/rtipress.2018.rb.0017.1806

Mukherjee, A., Duttagupta, S., Chattopadhyay, S., Bhanja, S. N., Bhattacharya, A., Chakraborty, S., Sarkar, S., Ghosh, T., Bhattacharya, J., & Sahu, S. (2019). Impact of sanitation and socioeconomy on groundwater fecal pollution and human health towards achieving sustainable development goals across India from ground-observations and satellite-derived nightlight. *Science and Reports, 9*, 15193. https://doi.org/10.1038/s41598-019-50875-w

Nakagiri, A., Niwagaba, C. B., Nyenje, P. M., Kulabako, R. N., Tumuhairwe, J. B., & Kansiime, F. (2016). Are pit latrines in urban areas of Sub-Saharan Africa performing? A review of usage, filling, insects and odour nuisances. *BMC Public Health, 16*, 1–16. https://doi.org/10.1186/s12889-016-2772-z

Nweke, O. C., & Sanders, W. H. (2009). Modern environmental health hazards: A public health issue of increasing significance in Africa. *Environmental Health Perspectives, 117*, 863–870. https://doi.org/10.1289/ehp.0800126

Redlinger, T., Graham, J., Corella-Barud, V., & Avitia, R. (2001). Survival of fecal coliforms in dry-composting toilets. *Applied and Environment Microbiology, 67*, 4036–4040. https://doi.org/10.1128/AEM.67.9.4036-4040.2001

Shammas, N. K., & Wang, L. K. (2007). Aerobic digestion. In L. K. Wang, N. K. Shammas, & Y.-T. Hung (Eds.), *Biosolids treatment processes* (pp. 14–15). Humana Press. https://doi.org/10.1007/978-1-59259-996-7_6

Suriyachan, C., Nitivattananon, V., & Amin, N. T. M. N. (2012). Potential of decentralized wastewater management for urban development: Case of Bangkok. *Habitat International, 36*, 85–92. https://doi.org/10.1016/j.habitatint.2011.06.001

Tilley, E., Ulrich, L., Lüthi, C., Reymond, P., & Zurbrügg, C. (2014). *Compendium of sanitation systems and technologies* (2nd Revised ed.). Swiss Federal Institute of Aquatic Science and technology (Eawag). ISBN 978-3-906484-57-0.

Varigala, S. K., Hegarty-Craver, M., Krishnaswamy, S., Madhavan, P., Basil, M., Rosario, P., Raj, A., Barani, V., Cid, C. A., Grego, S., & Luettgen, M. (2020). Field testing of an onsite sanitation system on apartment building blackwater using biological treatment and electrochemical disinfection. *Environmental Science: Water Research & Technology, 6*, 1400–1411. https://doi.org/10.1039/c9ew01106d

Wankhade, K. (2015). Urban sanitation in India: Key shifts in the national policy frame. *Environment and Urbanization, 27*, 555–572. https://doi.org/10.1177/0956247814567058

Water, U. N. (2017). *The United Nations world water development report: Wastewater: The untapped resource.* UNESCO.

Yadav, K. D., Tare, V., & Ahammed, M. M. (2010). Vermicomposting of source-separated human faeces for nutrient recycling. *Waste Management, 30*, 50–56. https://doi.org/10.1016/j.wasman.2009.09.034

Zhou, X., Li, Z., Zheng, T., Yan, Y., Li, P., Odey, E. A., Mang, H. P., & Uddin, S. M. N. (2018). Review of global sanitation development. *Environment International, 120*, 246–261. https://doi.org/10.1016/j.envint.2018.07.047

Emerging Technologies for Treatment of Wastewaters

20

20.1 Bio-electrochemical Processes

20.1.1 Introduction

Bio-electrochemical processes (BEPs) are innovative and emerging group of processes that not only treat wastewaters but also produce valuable by-products. Due to this major benefit of concomitant valuable recovery during wastewater treatment, which is not the case for traditional wastewater treatment technologies, the BEPs have gained a lot of attention of the researchers in the recent past. The BEPs work on the principle of electro-chemistry in conjugation with the principles of biological systems. This multidisciplinary approach opens up new avenues of research in the field of BEPs employed for the treatment of wastewater.

There are different types of BEPs employed for the remediation of wastewaters; however, the first concept of the application of microbes for the generation of electricity from wastewater was demonstrated by Potter (1911). This concept was further worked upon by researchers in the late twentieth century and then the concept of microbial fuel cells (MFCs) was formulated. Later numerous other derivatives of MFCs were conceptualized for recovering different types of value-added products with simultaneous wastewater treatment. The major benefit of employing BEPs for wastewater treatment is the generation of additional revenue from the valuables recovered while treating wastewater. Hence in this way, the operating cost of these systems can be partly recovered; thus, rendering them economically beneficial in comparison to other traditional wastewater treatment technologies, like activated sludge process.

© The Author(s), under exclusive license to Springer Nature Singapore Pte Ltd. 2022
M. Ghangrekar, *Wastewater to Water*, https://doi.org/10.1007/978-981-19-4048-4_20

20.1.2 Types of Bio-electrochemical Processes and Performance Evaluation

20.1.2.1 Microbial Fuel Cells

There are different variations of BEPs commonly employed by researchers globally to treat wastewater and concomitantly recover valuables in the process. However, MFC is the most prevalent type of BEPs, which exemplifies wastewater treatment with simultaneous bioelectricity recovery (Logan, 2008). In a typical MFC there are two separate chambers for housing anode and cathode, namely anaerobic anodic chamber and aerobic cathodic chamber, which are separated by a proton exchange membrane (PEM). The PEM allows only protons to migrate from the anodic chamber to the cathodic chamber of a MFC and also prevents oxygen and substrate crossover between these two chambers (Fig. 20.1). The recovery of bioelectricity in MFCs can be majorly attributed to the electrogens residing in the anodic chamber, where the organic matter present in the wastewater is oxidized by them to produce electrons. These electrons reduce the anode and are passed to the cathode through an external circuit. On the other hand, the oxidation of organic matter in the anodic chamber produces protons, which are transmitted to the cathodic chamber through the PEM.

In the aerobic cathodic chamber, oxygen is supplied, where it reacts with protons and electrons in an oxygen reduction reaction (ORR) to produce water (Eq. 20.1) and thus completing the circuit to generate usable electricity. If specific metallic cathode catalysts are used for ORR to support two electron oxygen reduction pathway, hydrogen peroxide can be generated instead of water (Eq. 20.2), which can also be used as a disinfectant for the removal of pathogens from the anodic effluent (Das et al., 2020e; Gupta et al., 2020).

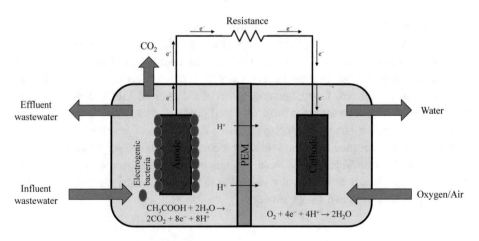

Fig. 20.1 Schematic of a microbial fuel cell

$$O_2 + 4H^+ + 4e^- \rightarrow 2H_2O \tag{20.1}$$

$$O_2 + 2H^+ + 2e^- \rightarrow H_2O_2 \tag{20.2}$$

This ORR occurring in the cathodic chamber of an MFC, using plain carbon as electrode material, is a sluggish reaction; hence, restricts the power generation of an MFC. Therefore, researchers have employed different low-cost cathode catalysts with higher catalytic activity to improve the power generation of MFCs (Bhowmick et al., 2020). Moreover, the activity of electrogenic microorganisms present in the anodic chamber and efficient transfer of electrons from the electrogens to the anode also govern the power generation of an MFC. Thus, to facilitate proficient conduction of electrons from the electrogens to the anode, at times electron mediators are used as anode catalysts, which significantly improves the power generation of MFCs (Das & Ghangrekar, 2019). These are some of the prevalent techniques employed by the researchers to overcome the bottleneck of lower electricity yield of MFCs, which in the future could pave the way towards the successful field-scale implementation of MFCs.

The microbial culture used in anodic chamber and electrode and membrane material used in MFCs are imperative in dictating the efficacy of these systems. Generally, mixed anaerobic culture, e.g., septic tank sludge, is preferred as inoculum in MFCs, with some pre-treatment, due to the robust nature of mixed cultures and the difficulties associated with the maintenance of aseptic condition in field-scale setups required for efficient growth and sustenance of pure cultures. As a separator, generally polymer-based membranes are employed in lab-scale MFCs owing to their superior proton conductivity (Ghorai et al., 2020). However, the major drawback of polymer-based membranes is extremely high cost, hence lately ceramic-based membranes have gained a lot of attention due to their low-cost (Das et al., 2020c). Moreover, in field-scale MFCs, the use of polymer membranes become challenging due to their inherent high cost and inability to sustain higher water heads, which is commonly encountered in field-scale applications. Therefore, researchers are currently opting for ceramic-based membranes for scaling-up investigation pertaining to MFCs due the above-mentioned advantages.

Carbon-based materials like carbon felt, graphite felt, carbon cloth, graphite rod and sheet are preferred as the base electrode materials in MFCs due to their cost-effectiveness, superior catalytic activity, conductivity and specific surface area (Bhowmick et al., 2019). However, these base materials are generally coated with other highly active catalysts to further improve the performance of these materials. Moreover, the cost of these materials also governs the economic sustainability of field-scale implementations of MFCs. Researchers in this field are toiling hard to develop low-cost, however highly efficient electrode and membrane materials, that will not only improve the performance of a field-scale MFC but also would drastically reduce the fabrication cost of the same; which would be a significant step toward still unattained commercialization of this inventive technology.

In this regard, novel low-cost cathode catalyst CuZn was developed and applied in both lab and field-scale MFCs and increment in power density at both the scales was observed (Das et al., 2020b). The lab- and field-scale MFCs operated with CuZn as cathode catalyst demonstrated four- and 64-times higher power density with respect to the control MFC operated using bare carbon felt as cathode without any catalysts. Also, this novel catalyst was 300-folds cheaper than Pt-based cathode catalyst that is commonly employed in MFCs. Moreover, the field-scale MFC operated with CuZn as cathode catalyst also exhibited 38% higher COD removal efficiency ($68 \pm 8\%$) from actual sewage in comparison to the control MFC ($49 \pm 10\%$) operated without any cathode catalysts (Das et al., 2020b). Thus, such novel, low-cost and efficient cathode catalysts could be envisaged to be used in field-scale MFCs and further expedite the commercialization of this technology.

Typically, lab-scale MFCs demonstrate around 70–90% of COD removal efficiency from wastewater, which vary depending on the type of wastewater used as the anolyte, bacterial culture and electrocatalysts used in the MFC setup (Zhuang et al., 2012). However, the COD removal efficiency reduces to around 60–70% for field-scale MFCs as other operational parameters comes into the play, when MFCs are scaled-up and electrochemical losses add up, diminishing the performance (Du et al., 2007). Moreover, MFCs have been used to treat a variety of wastewaters starting from industrial effluents to domestic wastewater and promising results have been reported in terms of the treatment of these different types of wastewaters through MFCs (Chae et al., 2009; Pant et al., 2010). Thus, MFCs can be efficiently used to treat a variety of wastewaters, exemplifying its wide range applicability.

Anode catalysts or mediator are also commonly employed in MFCs to mediate the electrons generated by the electrogens, thus reducing the loss of electrons in parasitic side reactions and in turn increasing the power generation of MFCs (Das & Ghangrekar, 2019; Varanasi et al., 2016; Wang et al., 2014). In this context, the application of WO_3 as electrode mediator/anode catalysts exhibited a coulombic efficiency (CE, explained latter) of 20.2%, which was more than twice than the CE obtained for the control MFC (10.0%) operated with carbon felt as anode without any catalysts (Das & Ghangrekar, 2019).

Generally, polymer-based membranes are used in lab-scale MFCs; however, due to their high cost and inability to handle high water head, they are not usually employed in field-scale MFCs. Instead of polymer-based membranes, researchers have developed low-cost clayware ceramic membranes, blended with montmorillonite as cation exchanger, that are equally effective as the polymer-based membranes and these are extremely low-cost (Das et al., 2020a; Ghadge & Ghangrekar, 2015). Therefore, research should be focused on developing more such low-cost, however efficient material, for the application in field-scale MFCs, which is the way forward toward the successful commercialization of this technology.

20.1.2.2 Performance Evaluation of MFCs

There are many parameters that are used to determine the performance efficiency of MFCs and some of these parameters measure the efficiency of an MFC to produce bioelectricity; whereas, some are used to measure the degree of treatment offered to the wastewater. The combination of these two factors is represented by Coulombic efficiency (CE), which is the measure of the ability of a MFC to convert moles of carbon present in wastewater into coulombs of electricity (Logan, 2008). The amount of current generated can be estimated by measuring operating voltage (OV) across an external resistance and then by applying Ohm's law (Eq. 20.3) the current generation can be determined.

$$V = IR \tag{20.3}$$

where, V is the average OV of the MFC (V), I is the current generated (A) and R is the external resistance connected to the MFC (Ω).

If the voltage generated by an MFC is measured across an infinite resistance, i.e., open circuit mode of operation, i.e., when electrodes are not electrically connected to each other for allowing current to flow, then this measured voltage is termed as an open circuit voltage (OCV) and in such case the current generation is nil. The OCV of an MFC gives an idea about how much maximum voltage can be generated by an MFC under certain condition; whereas, OV gives definite idea about the sustainable voltage generation ability of an MFC, when an external load is connected to it. It should be always kept in mind that OV tends to be equal to OCV, when external resistance is very high or tending towards infinity and OCV should always be higher than OV for normal conditions. From the average OV of a MFC, the sustainable power density (SPD) of a MFC can be estimated as per Eqs. 20.4 and 20.5.

$$SPD_v = \frac{V^2}{R \times v} \tag{20.4}$$

$$SPD_A = \frac{V^2}{R \times A} \tag{20.5}$$

where, SPD_v and SPD_A are the sustainable power density normalized to the volume of the anodic chamber (v) and projected surface area of the anode (A), respectively. Generally, SPD_v and SPD_A are reported in W/m^3 and mW/m^2, respectively, by normalizing to the anodic chamber volume or projected surface area of anode, as anodic chamber is the primary working chamber of an MFC majorly responsible for power generation.

The SPD gives an idea about the average power generation from an MFC; however, it fails to elucidate the maximum power generating capability of an MFC. Hence, to determine the maximum power generation that is possible by an MFC in a certain set of conditions, polarization test is carried out. In a typical polarization test, initially a very high external resistance is connected to the MFC and the stable OV at this external resistance is noted. The applied external resistance is then reduced slowly in steps and at each external resistance the corresponding OV is noted. After noting down these values, the current generated at every external resistance is determined as per Eq. 20.3 and corresponding current density (CD) is also noted as per Eq. 20.6 or Eq. 20.7.

$$CD_v = \frac{I}{v} \tag{20.6}$$

$$CD_A = \frac{I}{A} \tag{20.7}$$

where, CD_v and CD_A are the current density normalized to the volume of the anodic chamber (v) and projected (or specific, used in in-depth electrochemical analysis) surface area of the anode (A), respectively.

Similarly, power density at different external resistances is also determined as per Eqs. 20.4 and 20.5 and the external resistance which corresponds to the highest power density is noted as the internal resistance of the MFC. This maximum power density calculated from polarization demonstrates the maximum power that can be harvested from an MFC under the test condition for a unit volume of the anodic chamber or unit area of the anode. Moreover, maximum power that can be harvested from an MFC is when the internal resistance is equal to the external resistance, which can be estimated as per Eq. 20.8.

$$P = \frac{OCV^2 \times R_{ex}}{(R_{in} + R_{ex})^2} \tag{20.8}$$

where, P is the power that is harvested, R_{in} and R_{ex} are internal and external resistances of the MFC. In Eq. 20.8, when R_{in} is equal to R_{ex}, maximum power that can be produced by an MFC shall be obtained.

The degree of wastewater treatment offered by an MFC is measured by chemical oxygen demand (COD) removal efficiency during a feed cycle as per Eq. 20.9.

$$COD_\eta = \frac{(COD_I - COD_f) \times 100}{COD_i}\% \tag{20.9}$$

where, COD_i and COD_f are the respective influent and effluent COD of a feed cycle and COD_η is the COD removal efficiency of MFC.

The overall coulombs capturing efficiency of an MFC in terms of recovering current from wastewater is measured by CE as per Eq. 20.10 when MFC is operated under batch mode and for continuous mode of operation the CE is estimated using Eq. 20.11.

$$CE = \frac{M \times \int_0^t I dt}{F \times b \times v \times \Delta COD} \times 100\% \tag{20.10}$$

$$CE = \frac{M \times I}{F \times b \times q \times \Delta COD} \times 100\% \tag{20.11}$$

where, I is current produced (A); M is molecular weight of oxygen; b is the number of electrons exchanged per mole of oxygen $= 4$; F is Faraday's constant $= 96,485$ C/mol; ΔCOD is the difference in the influent and effluent COD concentration (g/L) after time t in sec and q is wastewater flow rate (m^3/sec).

The CE which is measure of the efficiency of a MFC to harvest current from wastewater, decreases drastically when MFCs are scaled-up, i.e., when the volume of the anodic chamber is increased (Logan, 2010). This is due to the unavoidable losses that magnify and different fresh parameters that come into play, whenever MFCs are scaled-up. Generally, for lab-scale MFCs the CE in the range of 10–40% is reported, which reduces to around 5% for the field-scale MFCs (Cheng & Logan, 2011). However, these wide range of CE depends on multiple factors like shape, size and configuration of the MFC, type of electrode and membrane material used, characteristics of the influent wastewater and inoculum used (Gajda et al., 2018). Currently, the CE obtained in field-scale MFCs are quite inferior for the real-life applications and commercialization of this technology; hence, researchers are working hard to develop novel materials to ameliorate the CE of field-scale MFCs so that this innovative technology can be effortlessly commercialized.

Another important parameter termed as normalized energy recovery (NER) is recently coined and it is also used to elucidate the efficiency of an MFC to harvest bioelectricity from wastewater. The NER explains the potential of a MFC to generate bioelectricity from unit volume of wastewater being treated or unit mass of organic matter oxidized from wastewater as per Eqs. 20.12 and 20.13, respectively (Ge et al., 2013).

$$NER_w = \frac{I \times V \times t}{v_w} = \frac{I \times V}{Q} \tag{20.12}$$

$$NER_c = \frac{I \times V}{Q \times \Delta COD} = \frac{I \times V \times t}{v_w \times \Delta COD} \tag{20.13}$$

where, NER_w and NER_c are the NER normalized to the unit volume of wastewater treated (W-h/m^3) and per unit of COD removed (W-h/kg of COD), respectively, and v_w is the volume of the wastewater (m^3) treated during batch mode of operation in time t (h) and Q is flow rate of wastewater for continuous mode of operation (m^3/h). Generally, the average NER with acetate as the carbon source is 0.40 kW-h/kg of COD and that with glucose it is 0.12 kW-h/kg of COD. The average NER for an MFC operated with domestic wastewater is around 0.17 kW-h/kg of COD and for industrial wastewaters the same can be as low as 0.04 kW-h/kg of COD depending on complexity of organic matter present (Ge et al., 2013).

Example: 20.1 An MFC is fabricated with the total working volume of 120 mL divided equally among the two chambers, and with the cathode and anode surface area of 8 cm^2 each. The average OV across 100 Ω of external resistance and OCV of the MFC are 305 and 528 mV, respectively. Calculate the sustainable power density and current density both normalized to the volume of the anodic chamber and area of the anode.

Solution

$A = 8 \text{ cm}^2 = 8 \times 10^{-4} \text{ m}^2$, $v = 120/2 \text{ mL} = 60 \times 10^{-6} \text{ m}^3$, $V = 305 \text{ mV} = 0.305 \text{ V}$, $R = 100 \, \Omega$

$$SPD_v = \frac{V^2}{R \times v} = \frac{0.305^2}{100 \times (60 \times 10^{-6})} = 15.50 \text{ W/m}^3$$

$$SPD_A = \frac{V^2}{R \times A} = \frac{0.305^2}{100 \times (8 \times 10^{-4})} = 0.581 \text{W/m}^2 = 581 \text{ mW/m}^2$$

$$I = \frac{V}{R} = \frac{0.305}{100} = 0.003 \text{A} = 3 \text{ mA}$$

$$CD_v = \frac{I}{v} = \frac{0.003}{60 \times 10^{-6}} = 50 \text{ A/m}^3$$

$$CD_A = \frac{I}{A} = \frac{0.003}{8 \times 10^{-4}} = 3.75 \text{ A/m}^2$$

Example: 20.2 A lab-scale MFC was operated under a fed batch mode with the retention time of 3 days using synthetic wastewater having an average COD of 3220 mg/L, fed to the anodic chamber. Average effluent COD after batch operation was found to be 871 mg/L, and the average OV (across 150 Ω of external resistance) and OCV were found to be 347 and 569 mV, respectively. The volume of the anodic chamber of the MFC was 85 mL. Determine the CE, COD removal efficiency, NER_w and NER_c of the MFC. Also determine the power density required to be produced by this MFC from the same wastewater without compromising with the treated water quality to attain CE of 40%.

Solution

$M = 32 \text{ g/mol}$, $OV = 0.347 \text{ V}$, $R = 150 \, \Omega$, $t = 3 \text{ days} = 3 \times 24 \times 3600 = 259{,}200 \text{ s}$, $F = 96{,}485 \text{ C/mol}$, $b = 4$, $v = 85 \text{ mL} = 85 \times 10^{-6} \text{ m}^3$, $COD_i = 3220 \text{ mg/L}$, $COD_f = 871 \text{ mg/L}$, $\Delta COD = (3220 - 871) \text{ mg/L} = 2349 \text{ mg/L} = 2349 \text{ g/m}^3$.

$I = OV/R = 0.347/150 = 0.0023 \text{ A} = 2.3 \text{ mA}$

$$COD_\eta = \frac{(COD_i - COD_f) \times 100}{COD_i} \% = \frac{(3220 - 871) \times 100}{3220} \% = 72.95\%$$

$$CE = \frac{M \times \int_0^t I dt}{F \times b \times v \times \Delta COD} \times 100\%$$
$$= \frac{32 \times 0.0023 \times 259200}{96485 \times 4 \times (85 \times 10^{-6}) \times 2349} \times 100\% = 24.77\%$$

$$NER_w = \frac{I \times V \times t}{v} = \frac{0.0023 \times 0.347 \times 3 \times 24}{\left(85 \times 10^{-6}\right)} = 676.36 \text{ Wh/m}^3 = 0.68 \text{ kWh/m}^3$$

$$NER_c = \frac{I \times V \times t}{v \times \Delta COD} = \frac{0.0023 \times 0.347 \times 3 \times 24}{\left(85 \times 10^{-6}\right) \times 2349 \times 10^{-3}} = 287.8 \frac{\text{Wh}}{\text{kg of COD}}$$

$$= 0.288 \text{ kWh/kg of COD}.$$

Now let the OV across 150 Ω of resistance be 'x' V, which is required to attain CE of 40%. Therefore, current I $=$ x/150. Hence by substituting the values in Eq. 20.10, and solving, the value of x can be obtained.

$$40 = \frac{32 \times \frac{x}{150} \times 259200}{96485 \times 4 \times \left(85 \times 10^{-6}\right) \times 2349} \times 100\%$$

Thus, x $= 0.558$ V $= 558$ mV.

Hence, power density required would be $= 0.558^2/(150 \times 85 \times 10^{-6}) = 24.42$ W/m^3.

Example: 20.3 Determine the maximum power density that can be harvested from an MFC having anode surface area of 16 cm^2 operated with an external resistance of 150 Ω. The OV across the external resistance was noted to be 570 mV and OCV was 735 mV. From polarization, the internal resistance of this MFC was determined to be 100 Ω.

Solution

OV $= 570$ mV $= 0.57$ V, OCV $= 735$ mV $= 0.735$ V, $R_{in} = 100$ Ω, $R_{ex} = 150$ Ω, A $=$ 16 cm^2 $= 16 \times 10^{-4}$ m^2.

As per Eq. 20.8, maximum power can be harvested if $R_{ex} = R_{in}$. Hence, substituting the values,

$$P = \frac{OCV^2 \times R_{ex}}{\left(R_{in} + R_{ex}\right)^2} = \frac{0.735^2 \times 100}{\left(100 + 100\right)^2} = 0.00135 \text{ W} = 1.35 \text{ mW}$$

Hence, maximum power density that could be obtained $= 1.35/(16 \times 10^{-4}) =$ 843.75 mW/m^2.

20.1.2.3 Other Variants of BEPs

Different variants of BEPs like microbial carbon-capture cell (MCC), microbial desalination cell (MDC) and microbial electrolysis cell (MEC) are developed to simultaneously recover other valuables while treating wastewater (Fig. 20.2). Other than MFC, the MEC is the mostly employed BEP for valuable recovery with wastewater treatment in which an external potential is imposed on the cathode to produce different valuables and wastewater treatment concomitantly occurs in the anodic chamber (Fig. 20.3). The type of valuable recovered through MEC is governed by the cathode material and catholyte used in the setup, e.g., if catalyst that follow the two-electron pathway is used in the cathodic chamber, hydrogen peroxide can be produced (Das et al., 2020d). For instance, by the application of −0.8 V (vs. standard hydrogen electrode, SHE) of cathodic potential and the application of Ni-Pd as cathode catalysts, 233 ± 16 mg/L day of hydrogen peroxide was produced in the cathodic chamber of a MEC (Gupta et al., 2020).

   Furthermore, this hydrogen peroxide produced in cathodic chamber was used for killing of pathogens and removal of residual recalcitrant organic matters present in the anodic effluent of MEC, which resulted in the 5 log-scale pathogens reduction and further 70% removal of organic matter (in terms of COD) in the cathodic chamber of MEC

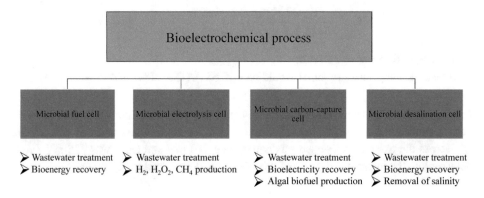

Fig. 20.2 Different types of bio-electrochemical processes developed for wastewater treatment with concomitant valuables recovery

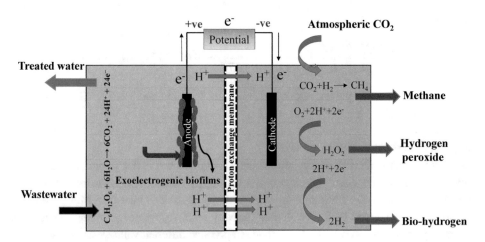

Fig. 20.3 Schematic of a microbial electrolysis cell

(Gupta et al., 2020). Therefore, such MECs can exemplify holistic treatment of wastewater by employing both chambers and thus would reduce the need of extra reactors for the comprehensive treatment of wastewater. Moreover, other metallic catalysts, such as Au, Pd, Ni, etc., have also been used in MEC for the production of hydrogen peroxide employing two electron pathway (Eq. 20.2) and demonstrated promising results (Das et al., 2020e). Additionally, other valuables, like hydrogen, acetate, etc., have also been reported to be produced through MEC, thus exemplifying its versatility to produce a wide range of products with concomitant wastewater treatment (Chen et al., 2012; Logan et al., 2008).

The other variant of BEP, which is quite similar to an MFC is MCC, where algae is cultured in the cathodic chamber utilizing the CO_2 liberated during the anodic oxidation of organic matter present in the wastewater (Das et al., 2019). The oxygen liberated by the algal cells acts as an external electron acceptor in the cathodic chamber, thus reducing the operating cost as required for providing external aeration (Fig. 20.4). An MCC generates bioelectricity with concomitant wastewater remediation for removal of organic matter in anodic chamber and the algal cells grown in the cathodic chamber can be harvested to further produce valuables from them (Ghangrekar et al., 2021; Neethu et al., 2020). Also, anodic effluent if sent to cathodic chamber for further treatment, the organic matter and nutrient removal from wastewater can occur along with 2 to 3 log-scale reduction of pathogens. Even trace metals present in the wastewater can also be removed due to uptake by algae.

In this regard, MCC operated with the pure culture of *Chlorella vulgaris* in the cathodic chamber of MCC produced 5.6 W/m^3 of power, when operated with sewage as a catholyte

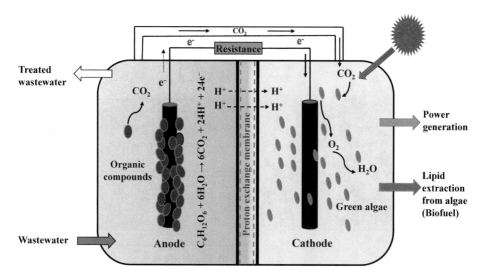

Fig. 20.4 Schematic of a microbial carbon-capture cell

(Wang et al., 2010). Moreover, 94 and 85% of CE and carbon-capture efficiency, respectively, was reported in the same investigation. In another investigation, 86.69% of COD, 70.52% of ammonium nitrogen and 69.24% of phosphorus was removed from municipal wastewater through a novel air lift type MCC, which also generated 972.5 mW/m^3 of power (Hu et al., 2015). Moreover, the lipid extracted from the harvested algae, cultivated in the cathodic chamber of MCC, was also used to produce bio-diesel. Therefore, through the application of MCC, comprehensive wastewater treatment can be achieved with the simultaneous production of valuables and bioenergy.

Another important version of BEP is MDC, where organic matter removal from wastewater, desalination of brackish water and bioelectricity generation all occur in a single reactor (Fig. 20.5). Therefore, this version of BEP can be used in coastal areas, where seawater can be desalinated and simultaneously power can also be recovered in the process (Pradhan & Ghangrekar, 2015). In this context, an investigation demonstrated 34% of desalination efficiency from 20 g/L of saline solution and 78% of COD removal efficiency from domestic wastewater with the concurrent power generation of 931 ± 29 mW/m^2 (Qu et al., 2012). In another research, 90% desalination of 35 g/L of salt solution was exhibited with 2 W/m^3 of power generation through MDC (Cao et al., 2009). A 1 L capacity MDC also demonstrated 70% desalination efficiency, when operated with artificial sea water with concurrent power generation of 28.9 W/m^3 (Jacobson et al., 2011). Therefore, these investigations prove that MDC can be successfully used as a multipurpose tool for the desalination of brackish water, organic matter, nutrient and trace metal removal from wastewaters and bioelectricity generation, thus exemplifying its superiority in comparisons to other traditional technologies.

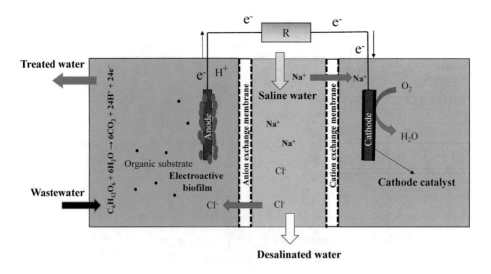

Fig. 20.5 Schematic of a microbial desalination cell

These major benefits of BEPs can render environmental sustainability to these innovative technologies, which already possessed a lot of potential of replacing the traditional wastewater treatment technologies. Though these BEPs possess the above-mentioned advantages, the major bottlenecks of these technologies are high fabrication cost and low yield of valuables, which are presently restricting a proper field-scale applications of these innovative technologies. The cost of electrode and membrane material are the major contributor in the elevated fabrication cost of BEPs; however, recently numerous low-cost electrodes and membrane materials have been developed to be employed in BEPs to reduce the fabrication cost of BEPs (Das et al., 2020a, b).

Moreover, the yield of valuables recovered through BEPs also needs to be improved considerably so that enough revenue can be generated to compensate the cost of wastewater treatment. Therefore, presently the researchers are working hard on developing innovative technologies to improve the yield of BEPs and use of low-cost materials to overcome these major bottlenecks of the technology, which would lead to the successful field-scale application of BEPs. Moreover, the electrochemical parameters, like charge transfer resistance and overpotential losses, reducing the yield of BEPs need to be diminished significantly prior to the effective real-life field-scale applications of BEPs.

20.2 Electrocoagulation

20.2.1 Introduction

The technology of electrocoagulation (EC) was first proposed during the late eighteenth century (Moreno-Casillas et al., 2007). In 1889, a treatment plant was built for the treatment of sewage by electrolyzing it with seawater as it contains salts for coagulation. In EC the destabilization of organic and inorganic substances is achieved through the production of in-situ electro-generated coagulants (Moreno-Casillas et al., 2007); whereas, in conventional coagulation external coagulants in the form of chemicals are added to the wastewater. Nonetheless, in early twentieth century, A.E. Dietrich first patented the EC to treat bilge water from ships. In 1909, first time sacrificial aluminium and iron electrodes were used to treat wastewater as a result, J.T. Harris filed a patent for this investigation (Vik et al., 1984).

Further, in the year 1940, an electronic coagulator was discovered by Matteson and co-workers, which produced electro-generated coagulants from aluminium electrodes to form aluminium hydroxide (Matteson et al., 1995). The produced hydroxides, coagulate and flocculate the organic and inorganic substances on to their surface; thereby, treating water. Similarly, in 1956, aluminium was replaced by iron to treat river water through this process (Moreno-Casillas et al., 2007). Since then, EC has been widely developed to treat contaminated water arising from different sources, such as municipal, hospital and textile industries due to the improvement in operational cost and desired efficacy of inorganic and organic pollutant removal. Moreover, EC has been explored to remove COD, turbidity, suspended solids, colloids and emerging organic contaminants of anthropogenic in nature. In this regard, this section focuses on the mechanism and influencing factors of EC for inorganic and organic pollutant removal. Moreover, recent application of EC as well as cost analysis has been also discussed.

20.2.2   Colloidal Stability and Destabilization

The microscopic particles that usually range between 1 nm and 2 μm are termed as colloids, owing to their very small size these particles have very small mass to surface area ratio. Hence, due to larger surface area compared to the size and mass, gravitational force acting on any colloidal particle is generally negligible; thus, the surface phenomenon is predominant in case of colloidal suspension. Moreover, these particles are resistant to agglomeration due to the fact that colloids carry similar negative charge on their surface; as a result, they repel each other and remain in suspension. Thereby, these particles are often termed as stable colloids and do not form flocs for their separation from contaminated water. In order to make colloids participate during coagulation and flocculation these particles must be destabilized by neutralizing the charge present on their surface, which can be provided by attaching the counter charge particles on the surface of stable colloids forming an electric double layer (Fig. 8.41, Chap. 8). To diffuse this electric double layer coagulants, such as aluminium sulphate [$Al_2(SO_4)_3$] or ferric sulphate ($FeCl_3$), are mixed to the suspension to neutralize negative charge of the colloids.

The double layer consists of inner and outer layer, where the inner layer is often known as the 'Stern layer' and in this region counter ions are tightly bound to the surface of particle; whereas, the outer layer possesses free moving mixed ions due to diffusion from colloidal surface. The interface of this inner and outer layer is usually known as the shear surface, which is the limit of stern layer. At the surface of colloid, maximum potential occurs, which is also known as the 'Nernst potential' and it decreases across the stern layer due to the presence of oppositely charge ions resulting in the zeta potential, which is measured at shear plane.

The zeta potential is the measure of extent of repulsion between colloids with same charge on its surface and it is also the major reason influencing colloidal stability. The particles with higher zeta potential tend to be more repulsive; thus, these particles are highly stable in a colloidal suspension. Generally, particles having zeta potential higher than +30 mV or lesser than −30 mV are considered as stable colloids. Additionally, the stability of two colloids in proximity with each other is also governed by the attractive Van der Waals forces arising from permanent or induced dipoles within the dispersed particles and repulsive electrostatic forces resulting from the overlapping of diffuse, stern and electric double layer (Fig. 20.6).

Fig. 20.6 Forces acting on two colloids (Conceptualized from Tahreen et al., 2020)

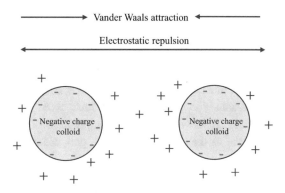

Fig. 20.7 Force field between colloids of similar charges (Conceptualized from Peavy et al., 1985)

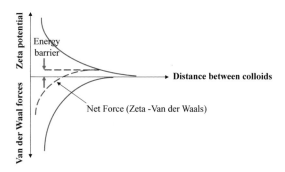

The magnitude of Van der Waals force decays exponentially with distance and the rate of decay of this force is also rapid as compared to the electrostatic forces. Moreover, for a shorter distance between the colloids, Van der Waals force is the strongest attractive force and vice-versa. Therefore, the reason to add the coagulant is to overcome the energy barrier, which is maximum net repulsive force; thereby, making the Van der Waals force attractive between colloids and ensuring destabilization of organic and inorganic substances (Fig. 20.7). This criteria of colloid stability and destabilization is also known as Derjaguin Landua Verwey Overbeek (DLVO) theory. Hence, the main aim of adding coagulants is to overcome the energy barrier for facilitating contact between two or more particles; thereby, forming flocs for the separation of colloidal particles through sedimentation and filtration.

20.2.3 Mechanism of Colloidal Destabilization

The destabilization of colloidal substances in polluted water can be achieved by following the mechanisms as discussed below.

Electric double layer compression: The stability and the amount of repulsion between the colloids are governed by the thickness of double layer. The magnitude of repulsive force is directly proportional to the thickness of double layer. As a result, decreasing the thickness of double layer would decrease the repulsion force between the colloids; thereby, making them attracted towards each other. Therefore, to decrease the thickness, coagulants with positive charge, such as Al^{+3} or Fe^{+3}, are added to the colloidal suspension. Moreover, when the compression of the electric double layer occurs, the zeta potential at the shear surface also gets reduced; thus, favouring attraction between the colloids.

Charge neutralization: The addition of counter ions to the solution will neutralize the charge on the surface of colloidal particles by adsorption as a result, the Van der Waal forces become predominant by overcoming the energy barrier of net maximum repulsive force; thus, enhancing attraction of particles.

Inter-particle bridging: When aluminium or iron coagulants are dissociated in the solution, metal hydroxide complexes can be formed and some of these complexes have high molecular weight and long chain, especially when polymers are used as coagulants. These polymeric molecules have the ability to attach stable colloidal particles due to its highly reactive surface and branching; hence, several other polymer-colloidal groups are formed, which are enmeshed to each other. Thus, polymer act as a bridge between two stable colloids to make them attached on its surface, resulting in a settable mass.

Particle entrapment: When metallic salts of divalent or trivalent ions are added into the solution, metal complexes in the form of metal hydroxides $[M(OH)_n]$ flocs are formed during the coagulation, which are gelatinous in nature, possesses high specific surface area and weight, and settles under the action of gravity. Hence, as the flocs settle, they entrap and adsorb the other dispersed colloidal particles present in their settling path due to its sticky surface. This phenomenon is also known as sweep coagulation as the metal complexes sweeps down the colloidal particles while settling. This has been discussed in details in Chap. 8 (Fig. 8.42).

20.2.4 Coagulation and Flocculation

Coagulation is the process of chemical addition in the form of trivalent metallic salts $[Al_2(SO_4)_3$ or $FeCl_3]$ to the solution for the destabilization of the colloidal substances; thereby, ensuring agglomeration of two or more colloids. Whereas, flocculation is the slow mixing of the destabilized colloids in a solution, as a result two or more colloids bind together to form flocs. Thus, in follow-up clarifier, the flocs can be either removed by sedimentation and if required remaining flocs after clarification shall be removed by filtration to produce very low turbidity treated water. Thus, during conventional water treatment system coagulation and flocculation are applied prior to sedimentation and filtration.

20.2.5 Electrocoagulation

The electrocoagulation process is an emerging technology, wherein on the application of external current between the electrodes, in-situ production of electro-generated metal coagulants for the remediation of organic and inorganic substances takes place. In an EC setup (Fig. 20.8), the metal electrodes are installed as the pair of an anode and a cathode and a DC power supply is used to provide electric current between the electrodes. When electric current is applied, anodic oxidation and cathodic reduction between the electrodes takes place; as a result, anode produce positive metal ion (M^+), whereas cathode produce hydroxyl ions (OH^-) as well as liberate the hydrogen gas (H_2) (Eq. 20.14 through Eq. 20.18).

Anodic reactions:

$$M(s) \rightarrow M^{n+}(aq) + ne^- \tag{20.14}$$

$$2H_2O\ (l) \rightarrow 4H^+(aq) + O_2(g) + 4e^- \tag{20.15}$$

where, l stands for liquid, aq for aqueous phase and g for gas.

Cathodic reactions:

Fig. 20.8 The mechanism of EC for the inorganic and organic pollutant removal (modified from Ahmad et al., 2020)

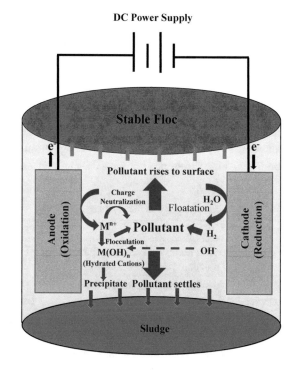

$$\text{Mn}^{n+}(\text{aq}) + \text{ne}^- \rightarrow \text{M(s)} \qquad (20.16)$$

$$2\text{H}_2\text{O(l)} + 2\text{e}^- \rightarrow \text{H}_2(\text{g}) + 2\text{OH}^- \qquad (20.17)$$

Overall reaction:

$$\text{Mn}^{n+} + n\text{QHH}^- \rightarrow \text{M(OH)}_n(\text{s}) \qquad (20.18)$$

where, s stands for solid.

The metal ions react with hydroxyl ion to form metal hydroxide complexes [M(OH)ₙ], which can adsorb and entrap the organic and inorganic colloids onto their surface due to its gelatinous nature and high specific surface area (Ghernaout, 2019).

Hence, as the organic and inorganic substances get attached to the surface of metal hydroxides, the specific density of flocs also increases; thus, ensuring sedimentation of heavy flocs. On the other hand, the produced hydrogen gas would also buoyant the dispersed particles to the surface of water; thereby, some particles are removed through floatation. The amount of metal ions generated into the solution is given by Faraday's Law (Eq. 20.19)

$$m = \frac{ItM_w}{zF} \qquad (20.19)$$

where, m is the mass of metal ions produced (g), I is the applied current (A), t is the electrolysis time (s), M_w is the molecular weight of the metal (g/mol), F is Faraday's constant (96,485 C/mol), z is the number of electrons involved in the reaction.

Thus, the addition of external chemical in the form of trivalent metallic salts in chemical coagulation to provide positive metal ions for counteracting stable colloids and in-situ production of positive metal ions by application of electric current to the sacrificial metal electrode is the basic difference between the conventional coagulation and EC. The advantages of EC over conventional coagulation are described in Table 20.1 to further understand the difference between EC and conventional coagulation.

Table 20.1 Advantages and disadvantages of EC

Advantages	Disadvantages
EC is effective in removing secondary inorganic and organic pollutant present in wastewater	Periodic replacement of anode is required due to its sacrificial nature
Lower sludge volume is produced as compared to conventional coagulation	EC can be expensive in places of high electrical cost
Efficient in removing small colloidal particles, turbidity and colour from wastewater	High conductivity wastewater is required for efficient performance
Employing solar EC can be economical than conventional methods of power supply	Electrode passivation is one of the major problems associated with EC

20.2.6 Operating Factors Affecting the Efficiency Electrocoagulation

The inorganic and organic pollutant removal efficiency of EC is influenced by various operating parameters because the amount of electro-generated metal ions, metal complex flocs, and bubble formation in the EC setup is dependent on electrode arrangement, current density (CD), electrolysis time, electrode gap, initial pH of wastewater, electrolyte concentration, initial inorganic and organic pollutant concentrations, etc. The role of different operating factors affecting the inorganic and organic pollutant removal efficiency of EC has been discussed here.

20.2.6.1 Current Density

Current density is one of the most important influencing factors in an EC process, which is defined as the ratio of current applied to the effective surface area of electrode. The CD is usually measured in mA cm^{-2} or A m^{-2}. The inorganic and organic pollutant removal efficiency is directly proportional to CD up to optimum value of CD. However, increasing CD above optimum value would impart negative effect on the performance of EC process due to over dosage of metal ions in the wastewater, which can reverse the charge of colloidal substances, thus making them stable rather than de-stabilizing them (Tahreen et al., 2020). The amount of metal ions generated is controlled by the CD, as it determines the release rate of electron from the metal electrode.

The operational cost of an EC process is dependent on CD, hence it is important to optimize the CD along with several other parameters, such as electrolysis time, pH and electrolyte concentration. In an investigation, the influence of CD on COD removal was explored using copper electrodes for the treatment of automobile wastewater. When CD was changed from 5 to 30 A m^{-2}, the COD removal efficiency was increased from 54.5 to 89.7% due to the increase in dissolution rate of electrode material (Priya & Jeyanthi, 2019). Similarly, the sodium dodecyl sulphate surfactant removal was increased from 59 to 81.6%, when CD was increased from 0.15 to 0.5 mA cm^{-2} using iron electrodes (Yüksel et al., 2009). Thus, an increase in CD can increase the inorganic and organic pollutant removal efficiency up to the optimum point.

20.2.6.2 Electrolysis Time

The inorganic and organic pollutant removal efficiency in an EC setup increases with increase in electrolysis time at constant CD, which can be attributed to the formation of metal hydroxide flocs and bubbles for the adsorption and floatation of organic and inorganic substances from wastewater. However, after a specific time the inorganic and organic pollutant removal rate becomes constant, indicating its optimum value due to occupied active sites of the metal flocs (Zaied et al., 2020). For instance, the COD removal efficiency for the treatment of licorice processing wastewater was increased from 10 to 88.3% as the electrolysis time was increased from 30 to 90 min (Abbasi et al., 2020). Hence, similar to CD, electrolysis time also needs to be optimized for the efficient treatment of wastewater through EC, otherwise only extra energy is consumed without affecting much the inorganic and organic pollutants removal efficiency.

20.2.6.3 Inter Electrode Distance

The electrostatic field between the electrodes is controlled by inter electrode distance, which consequently affects the formation of metal flocs. When the gap between the anode and cathode is lesser as compared to its optimal value, the produced metal hydroxides can be degraded by collision due to strong electrostatic attraction between the electrodes, as a result inorganic and organic pollutant removal efficiency is lower (Khandegar & Saroha, 2013). On the other hand, increasing the distance can also defer the generation of metal complexes due to reduced electrostatic forces as the metal ions and hydroxyl ions are destroyed in the path by covering the larger distance from anode and cathode before forming complexes; thus, resulting in lower removal efficiency of organic and inorganic substances. In an exploration, when inter electrode gap of iron electrode was increased from 1.5 to 2.4 cm, Ciprofloxacin (CIP) removal was decreased from 86 to 75.9% with the initial CIP concentration of 85 mg L^{-1} (Yoosefian et al., 2017). Hence, it is vital to optimize inter electrode distance for cost-effective remediation of organic and inorganic substances through EC process.

20.2.6.4 Initial pH of Solution

Initial pH of contaminated water is also one of the important parameters affecting the EC process. In an EC process, the maximum removal efficiency of inorganic and organic pollutants can be achieved at an optimum pH value for different electrode material. This can be attributed to the solubility of metal ions generated by different materials used as anode. As the solubility increases the tendency of ions to generate flocs decreases; thereby, a lesser removal of inorganic and organic pollutant is achieved and vice-versa (Nidheesh, 2018). Additionally, in an EC process, the solution pH increases, which is one of the advantages of applying EC to treat acid mine drainage as it can bring the final pH near to neutral, which is safe for disposal from environmental aspect (Sivaranjani et al., 2020). Moreover, in an alkaline medium, EC displays a buffer capacity to the solution.

The influence of pH for treating high strength wastewater was explored using aluminium electrodes in EC. At the optimum CD of 80 A m^{-2} and operating time of 25 min, about 32–66.5% colour and 10.5–18.4% COD removal was achieved for the pH range of 3.0–11.0 with initial colour and COD of 790 Pt–Co scale and 34,400 mg L^{-1}, respectively (Deshpande et al., 2010). However, the optimum removal efficiency of colour and COD was 70.3 and 24%, respectively, which was observed at near neutral pH. This can be attributed to the low solubility of Al^{+3} species at near neutral pH of 7.2; as a result, more organic and inorganic substances got adsorbed onto the surface of these species and eventually precipitated in the form of aluminium complexes (Deshpande et al., 2010). The precipitated metal hydroxide complexes are then removed through filtration or sedimentation and higher removal efficiency was obtained at near neutral pH.

20.2.6.5 Electrolyte Concentration

In an EC process, the inorganic and organic pollutant removal efficiency is also influenced by the conductivity of the wastewater being treated. The increase in concentration of the electrolyte, such as sodium chloride (NaCl) in a solution, would increase the conductivity of solution; hence, the internal resistance between the working electrode decreases and more numbers of metal ions can be generated for coagulation with the lower consumption of electricity (Nandi & Patel, 2017). For instance, when the concentration of NaCl was increased from 0.5 to 1 g L^{-1}, Imidacloprid pesticide removal from wastewater was increased from 76.5 to 95% using iron electrodes (Ghalwa & Nader Farhat, 2015). Similarly, the brilliant green dye removal efficiency was increased from about 97.27 to 99.67% as NaCl concentration was increased from 0.1 to 0.3 g L^{-1} (Nandi & Patel, 2017). Hence, the removal efficiency increases with the increase in electrolyte concentration. Which can also be attributed to the elimination of passive layer on the surface of anode with oxidation due to the presence of chloride ion by the addition of NaCl, otherwise the passive layer would reduce the formation of metal ions in the solution.

20.2.6.6 Initial Concentration of Inorganic and Organic Pollutants

The basic step of EC for organic and inorganic matter removal is the adsorption of these substances on to the specific surface of the metal hydroxides complexes. Hence, it is essential to understand the effect of inorganic and organic pollutant concentration on removal efficiency. The treatment efficiency is inversely proportional to the concentration of organic and inorganic substances as at constant CD only specific amount of coagulants are produced, which can only adsorb specific concentration of organic and inorganic substances; thereby, at higher concentration only limited number of produced flocs are available for the adsorption of organic and inorganic substances (Ahmad et al., 2020).

In an exploration, influence of initial pollutant concentration was investigated for the removal of CIP from hospital wastewater using aluminium electrodes. When CIP concentration was increased from 32.5 to 77.5 mg L^{-1}, the removal efficiency was decreased from about 83.52 to 67.31% during 17.5 min of electrolysis time (Ahmadzadeh et al., 2017). Similarly, when the Amoxicillin (AMX) concentration was increased from 10 to 100 mg L^{-1}, the AMX removal was decreased from 98.82 to 74.3% (Balarak et al., 2019). Hence, for a particular CD, the concentration of inorganic and organic pollutant can also be optimized to achieve efficacious treatment of wastewater.

20.2.6.7 Electrode Arrangement

The arrangement of electrode is also another important factor as it controls the current consumption of the EC process. Electrodes such as aluminium or iron can either be simply arranged vertically with a single pair of anode and cathode in an EC setup (Fig. 20.9). Other than this, multiple number of electrodes can be connected forming a complex arrangement such as monopolar parallel (MP-P), monopolar series (MP-S) and bipolar series (BP-S), which is based on different connection of anode and cathode to DC power

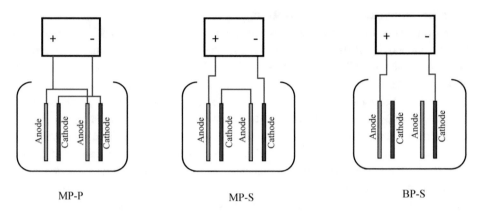

Fig. 20.9 Different modes of electrode arrangement in electrocoagulation set-up

supply (Fig. 20.9). Out of the different electrode arrangement MP-P arrangement is more cost effective as it provides higher current efficiency with low operating cost due to lower potential difference (Hakizimana et al., 2017).

However, it is difficult to conclude which is better in removing a particular inorganic or organic pollutant considering only cost, as in some cases BP-S has also shown higher removal efficiency. For instance, effect of mode of connection with iron electrodes was observed for the removal of natural organic matter (NOM) from contaminated water. After 20 min of electrolysis time, about 99% of NOM removal was achieved with BP-S connection, which is about 19% more as compared to MP-S connection (about 80% removal) with initial NOM concentration of 50 mg L^{-1} due to high voltage (Alimohammadi et al., 2017). Thus, ratio of effectiveness and cost for same degree of removal should be further explored for each configuration of electrode to explore which is best for EC process.

20.2.7 Application of Electrocoagulation in Wastewater Treatment

Over the years, EC has been broadly investigated for treating contaminated water streams from various sources, such as municipal wastewater, urban runoff, and effluents from textile, tannery, laundry, restaurants, distillery and oily wastewater (Tahreen et al., 2020). Moreover, EC has also been explored to treat emerging contaminants, such as dyes, phenols, pharmaceuticals, surfactants and pesticides present in contaminated water. Nutrient removal in the form of struvite and remediation of heavy metals from industrial wastewaters have also been achieved using EC (Ahmad & Gupta, 2019). In most of the cases more than 85% of inorganic and organic pollutant removal efficiency was achieved through EC (Table 20.2); thus, making EC a reliable and efficacious technology for the treatment of wastewater as compared to conventional treatment technologies. An overview of different applications of EC is provided in this section.

Table 20.2 Application of EC for the remediation of inorganic and organic pollutants from contaminated water

Source	Pollutant	Process parameters	Observations	References
Ground water	Chromium (Cr^{+3})	Anode and cathode: Iron; pH: 8.0; Temperature: 25 °C; Process time: 5 min; Current density: 7.94 mA cm^{-2}	100% of Cr^{+3} removal with initial concentration of 1 mg L^{-1}	Hamdan and El-Naas (2014)
Synthetic wastewater	Manganese (Mn)	Anode and cathode: Aluminium; pH: 9.0; Electrode gap: 2 cm; Process time: 195 min; Voltage: 10 V	92% of Mn removal with initial concentration of 360 mg L^{-1}	Omranpour Shahreza et al. (2020)
Synthetic wastewater	COD	Anode and cathode: Aluminium; pH: 4.1; Electrode gap: 1 cm; Process time: 90 min; Voltage: 15 V	51% of COD removal with initial concentration of 10 mg L^{-1}	Nguyen et al. (2020)
Distillery spent wash	COD and colour	Anode and cathode: Aluminium; pH: 8.7; Initial COD: 5803 mg L^{-1}; Process time: 400 min; Current density: 10.75 A cm^{-2}	94.88% of COD and 78.65% of colour removal was achieved	(Wagh et al., 2020)
Vegetable wastewater	Dicofol (DFL) pesticide	Anode and cathode: Aluminium; pH: 5.0; Stirring speed: 700 rpm; Process time: 120 min; Voltage: 15 V	95.52% of DFL removal was achieved	(Rao et al., 2017)
Textile wastewater	Colour	Anode and cathode: iron; NaCl electrolyte: 1 g L^{-1}; Process time: 7 min; Current Density: 10 mA cm^{-2}	80.64% of colour removal efficiency was achieved	Malinović et al. (2017)
Synthetic wastewater	Indigo carmine dye	Anode and cathode: magnesium; pH: 7.0; Electrode gap: 0.5 cm; Process time: 240 min; Current Density: 50 A m^{-2}	86% of dye removal with influent concentration of 100 mg L^{-1}	Donneys-Victoria et al. (2019)

20.2.7.1 Heavy Metals Removal

The process of EC has been successfully demonstrated for the remediation of heavy metals, such as zinc, arsenic, chromium, cadmium, lead, mercury, boron and several more metals from contaminated water. These metallic substances can be discharged through various sources, such as mining and industries, and are toxic and non-biodegradable;

thereby, possess serious threat to the environment. Golder et al. (2007) investigated the performance of EC using mild steel electrodes for the removal of chromium (Cr^{+3}) from wastewater with bipolar and monopolar electrode arrangement (Golder et al., 2007). The observations revealed that more than 99% removal of Cr^{+3} was achieved using bipolar configuration at 50 min of electrolysis time and 1 A of current, which was about 18.4% more as compared to monopolar configuration (81.5%) with initial Cr^{+3} concentration of 1700 mg L^{-1} for both the configuration (Golder et al., 2007). Similarly, Mansoorian et al. (2014), explored iron electrodes for the removal of Pb and Zn from battery industry wastewater. At CD of 10 mA cm^{-2} and electrolysis time of 40 min, about 98% of Pb removal with initial concentration of 9 mg L^{-1} and 97.7% of Zn removal with initial concentration of 3 mg L^{-1} was reported (Mansoorian et al., 2014). Thus, these observations revealed that EC is efficient in removing heavy metals from wastewater with capability to demonstrate more than 95% removal efficiency.

20.2.7.2 Chemical Oxygen Demand Removal

The effluent from many industries and municipal wastewater consists of high concentration of COD; thus, the basic aim of treating this wastewater is to bring down the COD within the permissible limit before treated water is being discharged into water body. In this regard, EC has been widely explored to remove COD from wastewater arising from pharmaceutical, petroleum, hospital, institutional, paper mill and several other sources. For instance, EC was applied to treat mineral processing wastewater with iron electrodes and at CD of 19.23 mA cm^{-2}, pH of 7.1 and electrolysis time of 70 min, and about 82.8% of COD removal with initial COD of 424.3 mg L^{-1} was reported (Jing et al., 2020).

Moreover, EC was found to be effective for the recovery of minerals such as sphalerite (65.33%) and galena (36.06%) from wastewater; thereby, proclaiming the reuse of treated wastewater in mining industry (Jing et al., 2020). On the other hand, differential floatation with chitosan has shown only about 20 and 10% recovery of galena and sphalerite, respectively, at pH of 7.0 with the chitosan concentration of 0.67 mg L^{-1} and flotation time of 3 min (Huang et al., 2013). Hence at neutral pH of 7.0, EC demonstrated better recovery of both sphalerite and galena than differential floatation with chitosan (Huang et al., 2013). Further, Priya and co-worker explored the application of EC for the removal of organic matter from automobile wash water with copper anode and aluminium cathode. The demonstration revealed that at optimum CD of 25 A m^{-2}, electrolysis time of 40 min and pH of 6.0, about 95.1% of COD removal was obtained with the initial COD of 320 mg L^{-1} (Priya & Jeyanthi, 2019). Thus, EC is advantageous for the treatment of wastewater having high organic content and can bring the concentration of organic substances below to safe limit.

20.2.7.3 Emerging Contaminants

Apart from treating actual effluent, EC has been also investigated to treat specific substances of recalcitrant nature. The emerging contaminants are also known as xenobiotics and include phenols, pesticides, pharmaceuticals, surfactants, as well as synthetic dyes. Xenobiotics are not effectively removed through conventional wastewater treatment technologies due to their low biodegradability and recalcitrant nature. Moreover, conventional technologies, such as activated sludge process and trickling filters, are designed to remove biodegradable organic matter containing carbon and nitrogen. Hence, EC plays a vital role in the remediation of these highly toxic and bioaccumulated substances from contaminated water (Ahmad et al., 2020).

The inorganic and organic pollutant removal mechanism in EC is a complex process, which include numerous mechanism, such as oxidation, charge neutralization, adsorption, precipitation and floatation (Oden and Sari-Erkan, 2018). In an exploration, EC was applied for the treatment of azo dye with stainless steel (SS) and aluminium electrodes. With the CD of 22 mA cm^{-2} and treatment time of 60 min, about 99 and 80% colour removal was achieved with SS and aluminium electrodes, respectively (Arslan-Alaton et al., 2008). The greater removal of colour in case of SS electrode is attributed to the direct anodic oxidation at the SS electrode and reduction of dye at cathode along with the adsorption of organic and inorganic substances on the Fe(OH)$_3$ metal complexes (Arslan-Alaton et al., 2008). Moreover, between pH of 7.0 and 9.0, rapid oxidation of iron anode occurs, which is also one of the mechanism for enhanced organic matter removal in EC (Sasson et al., 2009).

The EC process was explored to remove three different pharmaceuticals, namely Diclofenac (DCF), Amoxicillin (AMX) and Carbamazepine (CBZ) using aluminium anode and stainless-steel cathode. At optimum operating condition with CD of 0.5 mA cm^{-2}, electrolysis time of 38 h, about 90% of DCF, 77% of AMX and 70% of CBZ removal was achieved with initial concentration of each pharmaceutical of 10 mg L^{-1} (Ensano et al., 2017). Similarly, remediation of Malathion (MTH) pesticide from contaminated water using EC was also explored. With a CD of 10 mA cm^{-2}, initial pH of 6.0 and electrolysis time of 10 min, more than 90% of MTH was removed with initial pesticide concentration of 40 mg L^{-1} (Behloul et al., 2013). Further, about 81.6% of sodium dodecyl sulphate (SDS) surfactant was removed using EC with iron electrode with CD of 0.5 mA cm^{-2} and 10 min of electrolysis time (Yüksel et al., 2009). Thus, these investigations revealed that EC is also efficient for the remediation of emerging contaminants from wastewater with more than 80% removal.

20.2.8 Cost Analysis of Electrocoagulation

The economic analysis of any treatment system plays an important role pertaining to desired organic and inorganic contaminant removal from wastewater. The cost of an EC setup is influenced by various factors, such as CD, electrolysis time, conductivity, inter electrode gap and arrangement of electrodes as the energy consumption, degradation of electrodes and amount of sludge produced are mainly dependent on these factors. There-fore, these influencing factors must be optimized in order to achieve higher treatment at lower cost. Nonetheless, the total operating cost of EC is incorporated by cost of energy, cost of metal electrodes, cost of extra chemicals, if required to increase conductivity, and sludge handling cost. These cost components can be correlated with operating cost as per Eq. 20.20.

$$\text{Operating cost } (\text{Cost m}^{-3}) = \alpha C_{\text{electriciat}} + \beta C_{\text{electrodes}} + \gamma C_{\text{chemical}}$$
$$+ \text{ Sludge handling cost} \qquad (20.20)$$

where, constants α, β and γ are per unit cost of electricity ($\$ \text{ kWh}^{-1}$), metal electrodes ($\$ \text{ kg}^{-1}$), and chemicals ($\$ \text{ kg}^{-1}$), respectively, and $C_{electricity}$, $C_{electrodes}$ and $C_{chemical}$ are the consumption of electrical energy in kWh m^{-3}, electrodes in kg m^{-3} and consumption of chemicals in kg m^{-3} of treated contaminated water, respectively. For instance, comparison of iron and aluminium was explored for the treatment of printing ink wastewater using EC and with aluminium electrode the cost of treating per kg of COD was between 0.13 and 0.36 €, which is costlier as compared to iron electrodes (0.07–0.33 €) (Papadopoulos et al., 2019). Similarly, Bayramoglu et al. investigated aluminium and iron electrodes for treating textile wastewater and observed that the per kg cost of COD removal is about 0.086 and 0.260 €, while using iron and aluminium electrodes, respectively (Bayramoglu et al., 2004). Thus, it can be concluded that the utilization of iron electrode in EC is cost effective as compared to aluminium electrode.

Example: 20.4 A batch mode EC was used to treat textile industry wastewater with alu-minium electrodes at optimum applied current of 1.5 A and electrolysis time of 80 min. Estimate the amount of aluminium generated (in milligrams) to destabilize the organic substances.

Solution
The mass of metal generated is given by Eq. 20.19.

$$m = \frac{ItM_w}{zF}$$

Given: I = 1.5 A, t = 80 min or 540 sec, F = 96,485 C/mol, and molecular weight of Al (M_w) = 26.98 g mol^{-1}.
Also, anodic reaction of aluminium is given by $Al(s) \rightarrow Al^{3+} (aq) + 3e^-$
Hence, number of electrons involved in the reaction (z) = 3.
Thus, $m = \frac{1.5 \times 540 \times 26.98}{3 \times 96485} = 0.075$ g = 75 mg.
Therefore, about 75 mg of aluminium mass was generated at the optimum applied current.

Example: 20.5 Calculate the current density of an EC reactor, which consists of pair of copper electrodes with area of 15 cm (height) $\times$ 4 cm (width) for the treatment of *E. coli* from municipal wastewater at an applied current of 1 A. The unsubmerged height of electrode is 3 cm from the top of the reactor.

Solution

The effective height of electrode = Total height − unsubmerged height = 15 cm − 3 cm = 12 cm.

Total effective area of one electrode = 12 cm $\times$ 4 cm $\times$ 2 = 96 cm^2 (neglecting thickness).

Hence,

$$Current\ density = \frac{Current}{effective\ area\ of\ electrodes} = \frac{1 \times 1000\ \text{mA}}{96\ \text{cm}^2} = 10.42\ \text{mA cm}^{-2}.$$

Thus, the CD of 10.42 mA cm^{-2} is required for the treatment of *E. coli* from municipal wastewater.

20.3   Electro-oxidation

Electro-oxidation (EO) is an innovative and versatile emerging technology employed for the remediation of wastewater containing complex organic substances. The process of EO works on the in-situ generation of hydroxide radicals at the anode surface via oxidation of water molecule, which can mineralize the organic substances into carbon dioxide, water and other inorganic molecules. Moreover, at the same time hydrogen peroxide is also produced, which can also facilitate the oxidation of complex inorganic substances; thereby, rendering treatment to wastewater. The in-depth mechanism of EO has been discussed in Sect. 16.6.3 of this book.

20.3.1 Factors Influencing Electro-oxidation of Inorganic and Organic Pollutants

The efficiency of EO in terms of degradation of inorganic and organic pollutants present in the contaminated water is influenced by different factors, such as initial pH, current density, electrolysis time, by-product formation, type of the electrode used, initial concentration of organic and inorganic substances and addition of supporting electrolyte. For example, adding electrolyte, such as sodium chloride, can increase the conductivity of wastewater, which in turn minimizes the energy consumption for the same amount of removal efficiency as compared to wastewater treated without the addition of electrolyte. In this regard, this section provides a discussion on factors influencing efficiency of EO in terms of inorganic and organic pollutant removal. Refer to parameters of EC presented in Sect. 20.2.6 for detail discussion of few other factors like current density, initial concentration of inorganic and organic pollutant, electrolysis time, electrolyte concentration and electrodes arrangement as these factors are common for both EO and EC.

20.3.1.1 Electrode Materials

The electrode material is one of the important parameters as the selectivity and efficiency of EO is highly dependent on the electrode material used. The electrode material must possess high physical and chemical stability, high electrical conductivity, high surface area, corrosion resistance and low cost to life ratio (Anglada et al., 2009). In the recent past, most of the researchers have applied mixed metal oxide (MMO) electrodes due to their commercial availability and corrosion resistance nature. The MMO are also known as dimensionally stable electrodes (DSA) in which the base material such as titanium is coated with metal oxide of tin or lead providing stability to the electrodes. In addition, boron doped diamond electrodes (BDD) have also been used in EO due to its higher oxidative potential as compared to MMO (El-Ghenymy et al., 2014).

In this regard, Lebik-Elhadi et al. (2018) demonstrated the application of BDD electrodes for the degradation of Thiamethoxam pesticide and reported that with the CD of 16 mA cm^{-2}, electrolysis time of 6 min and 0.1 M Na$_2$SO$_4$, about 76% of Thiamethoxam removal was achieved with initial Thiamethoxam concentration of 2 mg L^{-1} (Lebik-Elhadi et al., 2018). Similarly, Steter et al. (2016) explored the application of DSA with RuO$_2$ active layers electrodes for the degradation of methyl paraben (MP) and reported that after 6 h of electrolysis time and CD of 66.7 mA cm^{-2} about 43% of MP removal was achieved with initial MP concentration of 158 mg L^{-1} (Steter et al., 2016). Thus, choosing proper electrode material is essential as it can influence the removal efficiency of the inorganic and organic pollutant present in the wastewater.

20.3.1.2 By-Product Formation

The EO process is highly efficient in treating wastewater; however, due to high anodic potential, inorganic and organic toxic by-products are formed in an EO setup, which can pose serious risk to human life such as birth defects and bladder cancer (Jasper et al., 2017). Thus, it is important that an ideal EO setup is one which completely degrades and mineralizes the organic substances along with the minimization of toxic by-products formation. For example, the EO treatment of saline water with BDD electrodes produced by-products, such as haloacetonitriles, haloacetic acids (HAAs), haloketones, 1, 2-dichloroethane, trihalomethanes (THMs) and other organic chlorine compounds (Lin et al., 2020). Similarly, during the EO of pesticides and pharmaceuticals, perchlorate, chlorate ions and chloramines were also formed (Garcia-Segura et al., 2015). Hence, the US-EPA has set up the maximum concentration limit for combined THM and HAA of 80 μg L^{-1} in drinking water for four regulated THM as well 60 μg L^{-1} for five regulated HAAs (Lin et al., 2020). Due the risk associated with the by-products formed during EO of wastewater, toxicity analysis and final concentration of these by-products should be critically analysed before releasing the treated wastewater in the environment.

20.3.1.3 Solution pH

The pH of the electrolyte is also one of the major parameters affecting the efficiency of EO for wastewater treatment. Sufficient hydroxyl radicals are produced at acidic pH, which is beneficial to oxidize the complex substances due to higher oxidative potential and lower side reactions (Saha et al., 2020). However, at basic pH of electrolyte, chances of occurrence of side reactions are higher; as a result, limited number of hydroxyl radicals would be available for the EO of organic and inorganic substances. Thus, acidic pH is favoured due to the higher removal efficiency of EO setup at acidic pH as compared to basic pH (Xia et al., 2020). In an exploration the effect of pH for the removal of Trypan blue dye was investigated with an increase in pH from 3.0 to 7.0 and then to 9.0, the dye removal decreased from 80 to 61.6% and finally to 58%, respectively, with electrolysis time of 2 h and initial dye concentration of 500 mg L^{-1} (Ghime and Ghosh, 2020). Hence, at acidic pH of electrolyte, maximum pollutant removal was achieved; however, near neutral pH has also demonstrated to be effective for EO of organic and inorganic substances compared to basic pH.

20.3.1.4 Electrode Modification

One of the major drawbacks of the electrodes used in EO is the sluggish electron transfer rate; thereby, researchers have doped different cost efficient nano-materials with noble electrodes in order to develop electrodes with higher selectivity, stability and low-cost along with enhanced electron transfer rate. Moreover, by applying different synthesis procedure, properties such as decrement in activation energy, increase in specific surface area and higher number of active sites can be achieved (Ahmad et al., 2021). For instance, BDD electrodes were modified with tin oxide and fluorine for the degradation of chlorinated polyfluorinated ether sulfonate and a maximum pollutant degradation of 95.6% was achieved with modified electrodes, which is about 15% more than unmodified BDD electrodes (Zhuo et al., 2020). Hence, electrode modification can aid in higher removal of organic and inorganic substances compared to conventional electrodes.

20.3.2 Applications of Electro-oxidation

The EO process has received considerable interest of the researchers for the oxidation of wide variety of refractory organic substances commonly found in the wastewaters. In recent years, EO has been applied to treat emerging contaminants, such as dyes, phenols, pesticides and others. Moreover, EO is also used for the treatment of wastewaters arising from olive mill, pulp bleaching, tannery, paper mill and many other industrial effluents along with the disinfection of water (Särkkä et al., 2015). In a demonstration, EO was used for the removal of sulfamethoxazole (SMX) (initial concentration of 30 mg L^{-1}) with BDD electrodes from synthetic wastewater and the result revealed that at current density of 30 mA cm^{-2}, pH of 7.0 and electrolysis time of 3 h, more than 99% of SMX was removed (Hai et al., 2020).

The EO of phenol from petrochemical wastewater was investigated under the applied current density of 3 mA cm^{-2} and electrolysis time of 15 min. More than 99% removal of phenol was reported with initial phenol concentration of 6.8 mg L^{-1} (Abou-Taleb et al., 2020). Similarly, Ramirez et al. explored the removal efficiency of methyl orange dye in an EO setup with BDD electrodes operated under the flow rate of 12 L min^{-1} and electrolysis time of 138 min. About 94% of decolorization of dye was achieved with initial dye concentration of 100 mg L^{-1} (Ramírez et al., 2013). Performance of EO for removal of inorganic and organic pollutants from wastewaters is summarized in Table 20.3.

Table 20.3 Application of EO for removal of inorganic and organic pollutants from wastewater

Source	Pollutant	Electrodes	Observations	References
Olive mill wastewater	Phenol and colour	Ti/IrO$_2$	100% removal of phenol and colour was attained at CD of 50 mA cm^{-2}	Chatzisymeon et al. (2009)
Synthetic urine	Uric acid	Sb–Sn–Ta–Ir/Ti	95.35% removal of uric acid with CD of 7.46 mA cm^{-2} and electrolysis time of 42.79 min	Singla et al. (2019)
Synthetic wastewater	Amoxicillin (AMX) and TOC	Ti/RuO$_2$	60% AMX and 48% of TOC removal in 60 and 240 min, respectively, with CD of 5.88 mA cm^{-2}	Kaur et al. (2019)
Synthetic wastewater	COD and TOC	Pb/PbO$_2$	79.64% of COD and 73.25% of TOC removal with CD of 44.69 mA cm^{-2} and 52.68 min of electrolysis time	Leili et al. (2020)
Municipal wastewater	Sodium dodecyl sulphate (SDS)	SnZrK	75.2% of SDS removal was achieved under CD of 0.6 mA cm^{-2} and electrolysis time of 6 h	Saha et al. (2020)
Biologically treated wastewater	Coliform bacteria	BDD	4 log reduction with CD of 120 mA cm^{-2} and electrolysis time of 1.4 min	Schmalz et al. (2009)

Note CD—Current density

20.4 Nano-technology

20.4.1 Introduction

The idea of nanotechnology was proposed by physicist Richard Feynman on 29 December, 1959 at an American Physical Society meeting in California (Feynman, 1960). However, over 14 years later, the actual term nanotechnology was coined by Norio Taniguchi in a lecture on the development of dimensional accuracy (Ali et al., 2020). Nanotechnology is a new class of technology, which represents the application of nano-materials in physics, chemistry, medical, material science, environmental engineering, etc. Nanomaterials are predominantly elucidated as materials having any dimension smaller than 100 nm and contain materials with significant alteration in chemical, physical, and biological properties as compared to the bulk materials. Moreover, at this scale of dimension, materials tend to develop novel size-dependent properties and because of the presence of a smaller number of atoms due to their small size, nanomaterials are distinct from their larger counterparts (Kunduru et al., 2017). Currently, nanotechnology is explored to enhance the efficiency of existing wastewater treatment technologies along with developing the novel techniques for the remediation of organic and inorganic substances, such as heavy metals, phenols, pesticides, and pharmaceuticals, present in contaminated water.

Advancements in nanotechnology have proved that it can be highly efficacious in providing solutions for water crisis. For instance, nanomaterials are used in adsorption due to high specific surface area and easy reuse. Similarly, due to high photocatalytic activity and high stability under UV spectra, it is widely used in photocatalysis. Nanomaterials, such as polymer nanoparticles, metal nanoparticles, zeolites, biopolymers, iron nanoparticles, carbon-based nanomaterials, nanoscale semiconductor photocatalyst, etc., have been explored for the removal of organic and inorganic substances from contaminated water (Qu et al., 2013). The following section focuses on the application of nanotechnology in wastewater treatment. Moreover, the eco-toxic effect of nanomaterials on the environment and regeneration of nanoparticles have also been elucidated.

20.4.2 Applications of Nanotechnology in Wastewater Treatment

The remediation of contaminated water using nanotechnology to facilitate reuse is still lagging behind on commercial-scale applications. However, researchers have successfully demonstrated the lab-scale applications of nanotechnology for the remediation of organic and inorganic substances from wastewater (Fig. 20.10). In environmental engineering applications, nanotechnology has shown promising potential for the treatment of emerging pollutants, such as pesticides, phenols, pharmaceuticals and surfactants, which

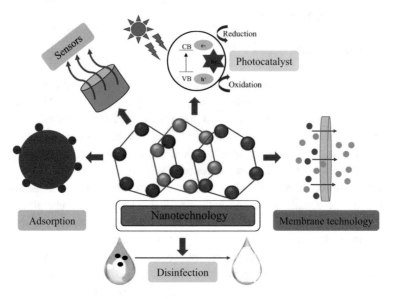

Fig. 20.10 Applications of nanotechnology for the treatment of wastewater

is one of the major advantages of incorporation of nanotechnology in wastewater treatment systems compared to conventional treatment methods (Karthigadevi et al., 2021). This section focuses on the application of nanotechnology for the remediation of organic and inorganic substances from contaminated water to make it safe for discharge in water bodies, which are likely to be used as source of drinking water downstream.

20.4.2.1 Application in Adsorption

Adsorption is a surface phenomenon wherein contaminated particles, often known as adsorbate, are adsorbed on a solid surface called as adsorbents. Generally, the adsorbents are the materials that are selected on the basis of high specific surface area. However, the adsorption efficiency of conventional adsorbents, such as activated carbon, is restricted due to lack of active sites or surface area, selectivity and adsorption kinetics (Qu et al., 2013). In this regard, nano-adsorbents, which are applicable for the removal of organic and inorganic substances from contaminated water possess the potential to provide large adsorption sites for the interaction with organic and inorganic substances by offering high specific surface area, improved catalytic potential surface chemistry and tunable pore size. Thus, the improvements in these properties make nano-adsorption an ideal for the removal of organic and inorganic substances from water and wastewaters. Metal-based, carbon-based, polymeric, magnetic or nonmagnetic oxide composites and zeolites are commonly used as nano-adsorbents for the reclamation of contaminated water (Kunduru et al., 2017).

Metal based nano-adsorbents: Metal oxides adsorbents, such as titanium dioxide, iron oxide, zinc and aluminium oxide, are efficient and low-cost materials for the removal of heavy metals, like chromium, arsenic, nickel, etc., from water and wastewater. These metal-based adsorbents have shown potential to surpass activated carbon-based adsorbents in terms of inorganic and organic pollutants removal efficiency. The major controlling mechanism for sorption is the complexation between dissolved metal in contaminated water and oxygen in metal oxides. Moreover, sorption generally occurs in two steps; firstly, rapid adsorption of organic and inorganic substances, such as metal ions, on external surface of the material; secondly, rate limiting intra particle diffusion along the pore walls of the absorbents (Trivedi & Axe, 2000).

In comparison with bulk metal oxides, nanoscale adsorbents have faster kinetics and high adsorption capacity due to smaller particle size, high specific surface area, and shorter intra particle diffusion distance. For instance, when the particle size of nano-magnetite was reduced to 11 nm, from 300 nm, about 100 folds increase in arsenic adsorption capacity was observed (Yean et al., 2006). This can be attributed to 'nano-scale effect' of the adsorbents, wherein new adsorption sites or vacancies are created due to the change in surface structure of magnetite at nanoscale for the adsorption of metal ions. The adsorption capacity at equilibrium (q_e) can be estimated using mass balance formula as presented in Eq. 20.21.

$$q_e = \frac{(C_o - Ce)V_w}{M_a} \tag{20.21}$$

where, C_o and C_e are in initial and final concentration of pollutant (mg/L), V_w is the volume of contaminated water (L) and M_a is weight of dry absorbent used (g).

Additionally, with the application of moderate pressure, metal oxide nanoparticles can be transformed into porous pellets without significantly affecting the specific surface area. The pore size and pore volume can be controlled by consolidation pressure (Lucas et al., 2001). Thus, the nanometal oxide can be applied in the form of porous pellets and fine powder, which are generally used in industrial application of inorganic and organic pollutants removal from wastewaters.

Carbon based nano-adsorbents: Carbon based nanotubes (CNTs) are mostly used as adsorbents, which have demonstrated efficacious removal efficiency as compared to activated carbon for the removal of organic substances as well as heavy metals from contaminated water. Depending on their preparation, CNTs can be classified as single walled and multi walled nanotubes. The external surface area of CNTs provides distributed hydrophobic sites for the adsorption of organic substances. Moreover, CNT contains grooves and interstitial spaces on its surface, which are highly active adsorption sites for the removal of organic substances.

Further, owing to their enhanced sorption sites and larger pores in aggregates, CNTs have much higher adsorption capacity as compared to activated carbon for the removal of bulky organic molecules, such as pharmaceuticals. In addition to this, various contaminant-CNT interactions, which include hydrophobic effect, hydrogen bonding, covalent bonding, π–π interactions and electrostatic interactions with wide range of organic substances, make CNT a better option for the removal of low molecular weight polar organic compounds as these compounds are not efficiently removed using activated carbon due to its low adsorption capacity.

On the other hand, CNTs oxidized with nitric acid, hydrogen peroxide and potassium permanganate have high adsorption capacity along with enhanced rate kinetics for the remediation of heavy metals, such as cadmium, from aqueous solution. This can be attributed to the presence of functional groups, such as phenol, hydroxyl and carboxyl, on the surface of CNTs, which provide chemical bonding and electrostatic attraction between the target substance and adsorbents; thereby, significantly enhancing the adsorption capacity. However, high cost of CNTs is one of the major limitations for the commercialization of this technology. Nonetheless, various investigations also revealed that heavy metals, such as lead, cadmium, mercury, zinc, cobalt and copper, are efficiently removed using CNTs from contaminated water (Table 20.4). Thus, it can be concluded that CNTs along with activated carbon can be used as a pre-treatment or post polishing step to enhance the removal of recalcitrant organic and inorganic substances from wastewater.

20.4.2.2 Photocatalysis Applications

Photocatalytic oxidation is a nanomaterial based emerging technology for the removal of micro pollutants and microbial pathogens from contaminated water. Photocatalysis can be used as a pre-treatment step for the removal of organic and inorganic substances with recalcitrant nature to enhance their biodegradability, which can be further mineralized through biological processes. In addition, this technology can also be used as a post polishing step to remove xenobiotics from water and wastewater. Usually, titanium dioxide (TiO_2) is used as a photocatalyst due to its easy availability, safe and low-cost. Generally, UV light source with range between 200 and 390 nm is required for the irradiation of TiO_2, wherein pair of electron–hole ($e^- - h^+$) is photo-excited; as a result, these pair move into valance (V) and conduction band (C). Hence, depending on redox potential of substrate, charge separation for an efficient photocatalytic functioning takes place (Eq. 20.22 through 20.28).

$$TiO_2 + hv \rightarrow e^-_{(C)} + h^+_{(V)} \tag{20.22}$$

$$O_2 + e^-_{(C)} \rightarrow O_2^- \tag{20.23}$$

$$h^+(v) + H_2O \rightarrow OH^. + H^+ \tag{20.24}$$

$$OH^. + OH^. \rightarrow H_2O_2 \tag{20.25}$$

$$O_2^- + H_2O_2 \rightarrow OH^. + OH^- + O_2 \tag{20.26}$$

Table 20.4 Application of CNT-based composites for the adsorption of heavy metals from wastewater

Pollutant	Adsorbent	Operating parameters	Adsorption capacity (mg g^{-1})	References
Copper (II)	Acid chitosan	Adsorbent mass = 0.05 g; pH = 7.0; initial concentration = 800 mg L^{-1}; contact time = 5 h; temperature = 25 °C	158.7	Mubarak et al. (2014)
Cobalt (II)	Sodium alginate-hydroxyapatite	Adsorbent mass = 0.1 g; pH = 6.8; initial concentration = 400 mg L^{-1}; contact time = 9 h; temperature = 21 °C	1111.1	Karkeh-abadi et al. (2016)
Cadmium (II)	Aluminium oxide	Adsorbent mass = 1 g L^{-1}; pH = 7.0; initial concentration = 1 mg L^{-1}; contact time = 4 h; temperature = 25 °C; agitation speed = 150 rpm	27.2	Liang et al. (2015)
Mercury (II)	Sulphur integrated CNT	Adsorbent mass = 0.02 g; pH = 6.0; initial concentration = 100 mg L^{-1}; Contact time = 1.75 h; agitation speed = 110 rpm	151.5	Gupta et al. (2014)
Lead (II)	Iron oxide	Adsorbent mass = 0.10 g; pH = 6.0; initial concentration = 20 mg L^{-1}; contact time = 40 min; temperature = 25 °C	21.5	Elmi et al. (2017)

$$O_2^- + H^+ \rightarrow OOH^{\cdot} \tag{20.27}$$

$$OH^{\cdot} + \text{Organic matter} + O_2 \rightarrow CO_2 + H_2O \tag{20.28}$$

However, the energy band gap of TiO_2 (E_g) tackles broadening (ΔE_g) within the nano range; as a result, more organic substrates might participate in redox processes in accordance with necessary criteria of photocatalysis. The energy band gap can be defined as the lowest energy required to add and remove an electron from the valence band to the conduction band of an electron system, where an electron can participate in conduction (Sham & Schlüter, 1985). Materials with larger band gaps, such as non-metals, have poor electrical conductivity due to high energy requirement of an electron to excite it to conduction band. However, semiconductors have narrow band gaps and require lesser energy for the movement of an electron. On the other hand, broadening is the phenomenon of increasing apparent band gap (Eq. 20.29) due to heavy doping of a semiconductor, as a result shifting of an absorption edge to higher energy levels can be obtained, which is helpful for obtaining different optical properties of same material (Gahlawat et al., 2019).

$$\text{Apparent band gap} = \text{Actual band gap}\left(E_g\right) + \text{Broadening}\left(\Delta E_g\right) \qquad (20.29)$$

$$E_{(C)} < E^0_{A/A-} \text{ and } E_{(V)} > E^0_{D/D^+}$$

where, $E_{(C)}$ and $E_{(V)}$ are the energy edges of conduction and valance bond, respectively; and $E^0{}_{A/A}-$ and $E^0{}_{D/D}{}^+$ are the reduction and oxidation standard potential of an electron acceptor (A) and donor (D). In addition to this, for the production of $OH^\cdot$ for photocatalysis, the nanomaterial must have band gap coupled to redox potential of $H_2O/OH^\cdot$ ($OH^- \rightarrow OH^\cdot + e^-$; $E^0 = -2.8$ V vs. SHE), such as TiO_2 has a band gap of 3.2 eV. Thus, TiO_2 nanomaterials have suitable properties for its application as photocatalyst for the degradation of organic and inorganic substances. Also, theoretically, E_g can be estimated using the Tauc expression (Eq. 20.30).

$$(ah\nu) = C\left(h\nu - E_g\right)^n \qquad (20.30)$$

where, a is the absorption coefficient, h is plank's constant, ν is the light frequency, C is band tailing constant, and n is transition mode power factor with values of 0.5, 2, 1.5 and 3 for allowed direct, indirect, forbidden direct and indirect mode, respectively.

Other materials, such as zinc oxide (ZnO), iron oxide (Fe_2O_3), silver phosphate (Ag_3PO_4) and cadmium sulphide (CdS), have also been explored as a photocatalytic material due to their light inducing redox reaction properties. However, the photocatalytic degradation of these materials is limited due to the amount of light intensity being absorbed and recombination rate of charge carriers. Therefore, doping of these materials (ZnO or Fe_2O_3) with some noble metals, such as gold (Au) and silver (Ag), have been tried to create photo induced electrons and hole for enhancing photocatalytic efficiency. For instance, Ahmad and co-workers explored the potential application of Au-ZnO and compared it with ZnO for the removal of Rhodamine B dye and revealed 21% more degradation of dye with Au-ZnO nanoparticles (95%) as compared to bare ZnO (74%) due to the modification of energy-band gap after the addition of Au particles (Ahmad et al., 2021).

In another investigation, TiO_2 was infused with grapheme oxide (GO-TiO_2) for improving photocatalytic degradation of Methylene blue dye (MB) and at optimum condition with pH of 10.0, photocatalyst dose of 0.2 g L^{-1} and 4 h of treatment time, more than 99% removal of MB was reported with initial concentration of 5 mg L^{-1} (Kurniawan et al., 2020). Moreover, the removal efficiency was 30% more under same operating condition with GO-TiO_2 compared to bare TiO_2 catalyst due to efficient electron transfer efficiency and high conductivity of grapheme oxide. Similarly, Hussien (2021) synthesized the nanostructured manganese doped silver phosphate (0.15-Mn–Ag$_3$PO$_4$) for visible photo degradation of Streptomycin antibiotic (SMN) and about 98% removal of SMN was observed with the initial concentration of 50 mg L^{-1} and 30 min of treatment time. However, under similar operating conditions about 30% degradation of SMN was achieved without doping Mn with Ag$_3$PO$_4$ (Hussien, 2021). Thus, to overcome the drawbacks, such as limited utilization of light intensity and reaction selectivity, these nanoparticles can be doped with other noble metals.

20.4.2.3 Applications in Membrane Processes

Membrane is a thin layered porous material, which provides physical barrier for organic and inorganic substances, such as metals, salts, viruses, and bacteria. Membranes are also one of the key components in wastewater treatment to produce reusable quality treated water, as they separate substances based on the size of molecules and pore present in the membranes. Based on their composition, porosity, and mode of application, membranes can be of several types, such as nanocomposite, nanofiltration, nanofiber, biological, thin-film composite membranes, which are reliable and provide appreciable wastewater treatment efficiency. However, there are few limitations of membrane technology, such as consumption of high energy, as it is a pressure driven process, fouling of membranes due to which working life of membranes gets reduced along with reduction in its inorganic and organic pollutant removal efficiency and membrane selectivity.

Application of nanomaterials in membrane process has been explored to enhance functionality of the membranes by inclusion of the nanomaterials in polymeric membranes to improve fouling resistance, membrane permeability, strength, and thermal stability. For instance, addition of alumina, silica and zeolite has shown to alter membrane surface hydrophilicity, which might be helpful in increasing the adsorption of organic and inorganic substances on the surface of membrane; thereby, enhancing their separation from water (Qu et al., 2013). Similarly, Ag nanoparticles when infused with polysulphone membrane can improve the antibacterial properties of the membrane, thereby enhancing bacterial removal efficiency with one log reduction of *Escherichia coli* from contaminated water than bare polysulphone membrane (Andrade et al., 2015). Moreover, increment in pure water flux was also noticed after the addition of Ag particles. The pure water flux can be determined using the Eq. (20.31).

$$J = \frac{V}{A\Delta t} \tag{20.31}$$

where, J is the water flux (L/m^2 h), V is the volume of permeated water (L), A is the membrane area (m^2) and Δt is the permeation time (h).

In recent progress, photocatalytic based nanoparticles were also mixed with polymer membranes to simultaneously perform photo-degradation as well as physical separation, wherein TiO$_2$ nanoparticles were grafted on metallic filters through dip coating. However, application of photocatalysis in membranes can be performed only with inert material as polymeric material might get degraded during oxidation process. Additionally, thin film nanocomposite membranes (TFN), which are usually prepared to enhance the interfacial properties, such as permeability of membranes by the doping of nanomaterials, are a new class of membranes that have been applied in treatment technologies such as reverse osmosis. In this regard, nano-zeolite is most commonly used as dopant in TFN membranes, which has shown great potential in improving permeability of membranes due to high negative charge required for the neutralization of opposite charge substances and thicker polyamide active layer. Moreover, nano-zeolite was also used as antimicrobial carrier agent and as a result, fouling resistance gets enhanced (Lind et al., 2009).

Further, bio-inspired membranes, such as aquaporin-based membrane, are known to possess high flux and selectivity and have been commercialized for its application in desalination of sea water by Aquaporin Inside, Denmark (Tang et al., 2013). Thus, the application of nanomaterial in membrane technology would enhance the properties of the membranes, thereby improving performance. However, adaptation of these processes at a commercial-scale would depend on the operational cost, high cost of nanomaterial used during its synthesis and the risk involved towards aquatic life due to the nano-toxicity of these materials. In this regard, low-cost material such as low purity nanomaterials should be explored, which can minimize the membrane synthesis cost as well as provide better stability without compromising the efficiency.

20.4.3 Ecotoxicological Effects of Nanotechnology

The advancement in nanotechnology has led to the release of engineered nanoparticles in different sections of the environment. As a result, these particles can pose potential risk to the health of both human as well as aquatic organisms. Nanoparticles released from different industrial and domestic wastes usually end up in wastewater treatment systems through effluents from their production process, wastes from processes using nano-materials and unscientific disposal of nano-wastes. About 60% of these materials are used in pharmaceuticals and other industrial applications (Kurwadkar et al., 2015); hence, the probability of these particles being detected in wastewater is high (Fig. 20.11). The sources through which nanoparticles might end up in wastewater are commercial products containing nanomaterials, such as titanium dioxide and silver, as well as waste streams of production processes of different industries, such as photography, textiles, electronics, etc. However, dermal exposure and inhalation of nanoparticles in the working environment is the major pathway through which these particles enter in to the human body.

The physical, chemical, and biological changes can alter the properties of these nanoparticles in terms of shape, size, and surface functionality; consequently, the behaviour and fate of nanoparticles also get altered. For instance, silver nanoparticles

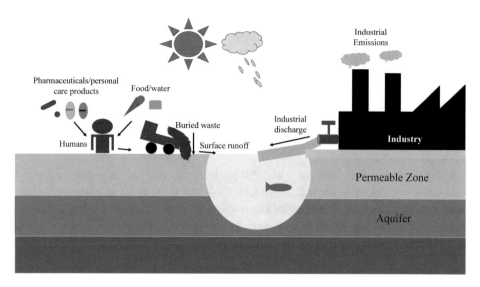

Fig. 20.11 Pathways through which nano particles enter in the environment (conceptualized from Kurwadkar et al., 2015)

after sulfidation followed by oxidation changed the surface reactivity, degree of aggregation, charge and potential to release toxic silver ions in the environment (Abbas et al., 2020). Similarly, aluminium oxide exposure to living organisms can oxidize the microbial cells and block the respiration, thereby causing death of the organisms (Arul Prakash et al., 2011). The nanoparticles due to their small size can also penetrate into the cell wall causing the disruption of cells. On the other hand, these particles might react with network of immune cells, which are located within epithelial surfaces and these particles can act as an allergic agent, stimulating the immune system causing allergic inflammation in later stages of life. Similarly, these particles can affect the lungs, which can lead to asthma or lung cancer. Thus, human and other living organism might get expose to the transformed nanoparticles of high toxicity, as a result posing severe adverse effects on these organisms.

Additionally, the major concern of nanoparticles among the researchers is due to the trophic level transfer and bioaccumulation of these particles in food web through the direct ecological exposure and intake of contaminated food. For instance, *Cyprinus carpio* fish on long term exposure for silver nanoparticles in a water tank has shown highest bioaccumulation at 0.09 mg/L (Kakakhel et al., 2021). Moreover, the highest bioaccumulation was observed in the liver with adsorption of 272.09 mg in 20 days of exposure (Kakakhel et al., 2021). The muscles have shown the minimum bioaccumulation of nanoparticles. The detrimental effect on *Cyprinus carpio* was alternation in gills and intestine with necrosis, cell lysis and degeneration at different concentration of nanoparticles. Nonetheless, the bioaccumulated nanoparticles are potentially toxic to higher tropic level consumers.

Hence, to obtain advantages of nanotechnology in the field of environmental engineering, it is important to develop further understanding of adverse effects of release of these nanoparticles in environment and monitor the toxic effects of nanoparticles on human and aquatic life. Thus, strategies should be made to minimize these limitations by adopting sustainable nano-wastes disposal techniques.

20.4.4 Recovery of Nanoparticles from Treated Effluent

The regeneration and recovery of nanoparticles is one of the key aspects of nanotechnology-based water and wastewater treatment systems as nanomaterials impart nanotoxicity as well as it influences the operational cost of the technology. However, detection of the nanoparticles is very complicated, costly and sensitive; hence, only limited methods are available for its detection in complex water matrixes. Moreover, during the recovery of these particles it is essential to know the recovery cost, otherwise the net treatment cost of the systems might get exorbitantly incremented, which would significantly influence the economics of nanotechnology-based treatment systems. Nonetheless, by applying separate systems these spent particles can be recovered for proper waste management through its application as construction material in bricks and in wastewater treatment systems.

Ghanbari and co-workers applied the electrocoagulation process for the removal of TiO_2 nanoparticles using iron electrodes from the aqueous solution. The produced sludge consisted of TiO_2 nanoparticles and iron species and it demonstrated improved catalytic activity of electro-generated sludge (Ghanbari et al., 2020). Hence, this electro-generated sludge was directly used in adsorption process for the removal of Ciprofloxacin antibiotic (CIP) from contaminated water and the results revealed that with 0.1 mg/L of electro-generated sludge and 1 h of reaction time, about 88% of CIP removal was achieved (Ghanbari et al., 2020). Thus, similar other techniques can be applied for the recovery of nanoparticles and it can be utilized in emerging wastewater treatment technologies. On the other hand, membrane filtration is considerably efficacious for the separation of these particles compared to other processes. Additionally, to separate magnetic nanocomposites, low field magnetic separation can be applied as an energy and cost-efficient technology for the recovery of these particles. Thus, recovery of nanomaterial from environment will not only minimize the toxicological effects but also it can be used as a cost reduction method due to its application in other emerging treatment systems such as adsorption, electrochemical oxidation, etc.

Example: 20.6 Calculate the adsorption capacity of a silver nanoparticle decorated activated carbon for the treatment of 5 L of wastewater containing acid blue dye with initial concentration of 10 mg/L. The removal efficiency of adsorption column is 85% using 2 g of adsorbent.

Solution

Removal efficiency,

$$\eta = \frac{(C_o - C_e)}{C_o} \text{ where, } C_o \text{ and } C_e \text{ are initial and final concentration of dye}$$

$$85/100 = (10 - C_e)/10$$

$$C_e = 1.5 \text{ mgL}^{-1}.$$

Now using,

$$q_e = \frac{(C_o - Ce)V_w}{M_a}$$

$V_w =$ Volume of contaminated water = 5 L and $M_a =$ Mass of adsorbent = 2 g. Hence, adsorption capacity $q_e = (10 - 1.5) \times 5/2 = 21.25$ mg/g

Example: 20.7 A polymeric membrane infused with zinc oxide nanoparticles with surface area of 0.5 m^2 was used for the treatment of 25 L of oily wastewater. Calculate the pure water flux with 30 min of permeate time?

Solution

A = 0.5 m^2; V = 25 L; Δt = 30 min = 0.5 h.
   Using equation, $J = \frac{V}{A\Delta t}$.

$$J = 25/(0.5 \times 0.5) = 100 \text{ L/m}^2\text{h} \quad (\textbf{Answer}).$$

20.5   Plasma Technology

20.5.1 Introduction

The development of efficacious and innovative emerging technology for wastewater treatment is still a challenge due to environmental compatibility and need for fast removal of pollutants during the treatment. In this regard, plasma technology have gained much attention of the researchers due to its eco-friendly nature as no secondary pollutants are produced in this process, fast removal of contaminants, and mild reaction condition of temperature and pressure are required to operate these reactors (Du & Lin, 2020). In 1928 Irving Langmuir first discovered the plasma technology, which is based on the fact that when energy is supplied to the matter it changes its state from solid to liquid and further liquid to gas. Further, when more energy is supplied to gas, it gets ionized into plasma state (Trivedi & Ashwin, 2013). Plasma consists of electrons, ions of both positive and negative nature, free radicals along with neutral atoms; thus, it is considered to be fourth state of the matter. The heat and gas molecules dissociate to the atomic state due to the electrical resistivity in the system and owing to the increment in temperature and loss of electrons, the ionization of molecules takes place; as a result, chemically active species such as $OH^{\cdot}$ radicals are produced. The $OH^{\cdot}$ radicals are strong and vital oxidants that dominate the plasma water and wastewater treatment technology.

Moreover, during low temperature plasma discharge, due to re-dissociation and re-combination of short lived free radicals, numerous other stable and strong oxidizing molecules, such as H_2O_2 and O_3 are also produced, which also participates in the removal of organic matter present in wastewater (Du & Lin, 2020). In the past few years, researchers have successfully demonstrated the application of plasma technology for the treatment of dyestuff, pharmaceuticals, phenols and many other organic pollutants. Thus, this section of chapter will provide a discussion on various types of plasma technology used in the field of wastewater treatment, type of discharges used in plasma treatment, factors influencing treatment efficiency and different application of this technology in wastewater treatment.

20.5.2 Classification of Plasma Technology

Plasma is ionized gas, which consists of electrons, ions and free radicals. Hence, depending upon the thermal equilibrium, wherein the temperature of electrons, ions and free radicals species is same, plasma is termed as thermal plasma (Annemie et al., 2002). On the other hand, when temperatures of all species are not same, and temperature of electrons is much higher compared to bulk gas molecules, the plasma is termed as non-thermal plasma. Thus, thermal plasma, such as torches, radio frequency and arc discharges requires high energy; whereas, non-thermal plasma, such as di-electric barrier, corona, gliding arc and glow discharge requires less energy (Jiang et al., 2014). Thermal plasma is produced by applying alternate or direct current (AC or DC) at high pressure of more than 10 kPa and temperature range of 2,000–20,000 K (Fig. 20.12) (Samal, 2017). However, non-thermal plasma is produced by applying electric potential through pulse AC or DC between two electrodes placed in a gas at low pressure. Moreover, depending on the type of gas used, the electrons can have temperature range of 10,000–100,000 K and temperature of the gas can be as low as room temperature (George et al., 2021).

20.5.3 Type of Discharges

The main goal during the plasma treatment of contaminated water is to achieve the efficient removal of the organic pollutant at lower treatment cost, which is possible by

Fig. 20.12 Thermal and
non-thermal plasma electron
temperature range versus
electron density

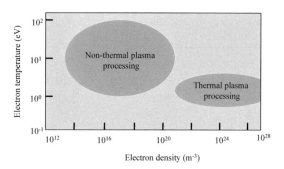

developing energy efficient plasma reactor. Thus, these reactors designed are classified on the basis of type of discharge used for the production of reactive species for oxidation of pollutants, which is influenced by type of discharge applied. Generally, for the formation of oxidative species, the discharges follow few mechanisms, such as ionization, dissociation and vibration or rotation and as a result OH⁺ radicals can be produced in the plasma setup (Eq. 20.32 through Eq. 20.38) (Joshi et al., 1995).

Ionization

$$H_2O + e^- \rightarrow 2e + H_2O^+ \tag{20.32}$$

$$H_2O^+ + H_2O \rightarrow OH^. + H_3O^+ \tag{20.33}$$

Dissociation

$$H_2O + e^- \rightarrow OH^. + H^. + e \tag{20.34}$$

Vibration

$$H_2O + e^- \rightarrow H_2O^* + e \tag{20.35}$$

$$H_2O^* + H_2O \rightarrow H_2O + H^. + OH^. \tag{20.36}$$

$$H_2O^* + H_2O \rightarrow H_2 + O^. + H_2O \tag{20.37}$$

$$H_2O^* + H_2O \rightarrow 2H^. + O^. + H_2O \tag{20.38}$$

where H_2O^* represents excited state of water molecule due to vibration.

Fig. 20.13 Pulsed discharge
plasma reactor

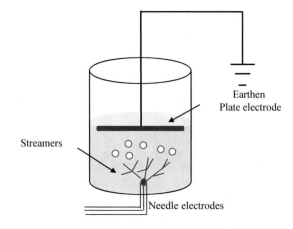

20.5.3.1 Pulsed Corona

The plasma reactor operated by this type of discharge requires pulsed electric generator, which can help to produce sharp pulse of high voltage with the duration of few nano or micro second. The discharge is also known as streamer and spark discharge. The basic configuration of pulsed setup requires two asymmetrical electrodes, one of small curvature like plate electrode and other of high curvature such as wires, rings or needle (Fig. 20.13). As a result, uniform electric field is produced on the plate electrode, which might induce a potential gradient; thereby, the discharge inception voltage gets reduced. The advantage of this type of discharge is that produced oxidative radicals in the wastewater can directly react with the pollutants.

20.5.3.2 DC Pulse-Less Corona

In the pulse-less system, high electron flux can be generated in the liquid and gases with the application of DC electric discharge; as a result, reactive species for the oxidation of refractory organic matter can be produced. Compared to pulsed system, pulse-less system can produce continuous radical species; however, due to continuous operation this system consumes more electrical energy. A capillary point shaped electrode reactor using oxygen as discharge gas can produce pulse-less corona, wherein mass transfer of reactive species for the oxidation of organic molecules is enhanced due to the capillary spraying of oxygen (Fig. 20.14).

20.5.3.3 Di-electric Barrier

In this process, one or more di-electric barrier in the form of alumina, glass, or others are placed in the discharge gap along with the application of time varying voltage between the electrodes. This configuration helps in the restriction of spark formation; as a result, fast corrosion of the electrodes can be prevented, as well as the discharge can be distributed uniformly over the electrode area. Beside the conventional parallel plate reactor,

Sampling pipe

Vent

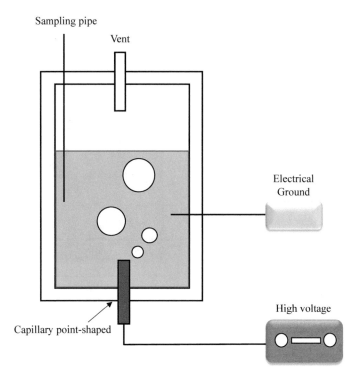

Electrical
Ground

High voltage

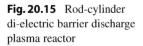

Capillary point-shaped

Fig. 20.14 Pulse-less corona discharge plasma set-up

Fig. 20.15 Rod-cylinder di-electric barrier discharge plasma reactor

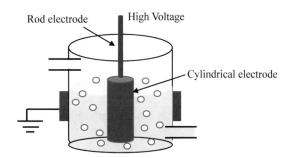

Rod electrode   High Voltage

Cylindrical electrode

rod-cylinder reactor is also used to generate this type of plasma (Fig. 20.15). Moreover, di-electric barrier is advantageous to produce industrial level O_3 for the treatment of industrial wastewater by using oxygen as discharge gas.

20.5.3.4 Gliding Arc Discharge

The gliding arc process is as an emerging process for the production of oxidative species as it can produce both thermal as well as non-thermal plasma in the same reactor; thereby,

Fig. 20.16 Gliding arc
discharge plasma set-up

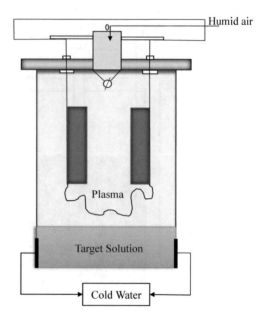

showing dual nature of discharge. Compared to pulse process, gliding arc requires high electrical power, which can help in increasing the yield of short-lived reactive species, which is advantageous for enhancing the removal rate of organic molecules. In this process, high voltage is applied between the divergent electrodes and when the electric field reaches to about 3 kV mm^{-1} in air, arc discharge can be produced (Fig. 20.16).

In case of gliding arc discharge, the discharge plasma is easily produced in gas phase compared to the discharge in liquid phase; and thus, less energy is required in case of the discharge in gas phase. Moreover, high velocity of plasma produced in gliding arc discharge blows toward the liquid surface, which can provide gas and liquid phase mixing due to magnetic stirring induced by the turbulence. Thus, short lived reactive species can react with target pollutants, while stable species can dissolve in liquid; thus, this is an added advantage of the arc discharge produced in gaseous phase.

20.5.4 Factors Affecting Performance of Plasma Technology

The wastewater treatment through plasma technology is dependent on the electrical discharge mechanism adopted to produce species of plasma, which own their specific chemical and physical characteristics. Hence, there are few factors that might behave differently with the application of electric discharges. Thus, it is essential to understand the behaviour of factors such as conductivity, gas input, voltage, temperature, and solution pH on the characteristics of produced plasma and these factors must be optimized for efficient target pollutant removal.

20.5.4.1 Conductivity

Conductivity of the solution is one of the major influencing parameters affecting the discharge properties; thereby, influencing production of plasma species. When an electrical connection between the electrodes inside the solution takes place, a conductive plasma channel is produced leading to the formation of spark discharge. Whereas, the steamers are produced for the short duration of time, when air has suffered an electrical breakdown around a high voltage conductor and gets ionized, as a result the electric charge leak into the air forming finger like discharges. Additionally, when strong electric field is provided between the electrodes, certain chain reaction occurs in the solution. As a result, electrons collide with the atoms to ionize them; thereby, creating reactive species by ionizing more atoms.

The increase of conductivity in pulsed liquid process of discharge, can activate the spark streamer mode from spark discharge mode, and finally from spark to steamer corona discharge mode. Contrary, the decrease of solution conductivity can increase the streamer length, which is the length of the plume produced during discharge process and when the streamer reaches the counter electrode, a spark is induced in the system. This induced spark can provide a more reactive environment for the production of reactive species like OH˙ radicals compared to steamer corona discharge; thereby, increasing the efficiency of plasma process for organic pollutant removal.

Higher solution conductivity however can have limitations, such as high discharge current, higher plasma temperature, and higher UV irradiation, which can lead to lower rate of production of reactive species. Jiang and co-workers have demonstrated the influence of conductivity on degradation of methyl orange dye during the treatment of synthetic wastewater, and with the increase of solution conductivity from 1.5 to 5 mS cm^{-1}, the dye degradation was decreased from 85 to 75% with the initial methyl orange concentration of 60 mg L^{-1} and treatment time of 20 min (Jiang et al., 2012). Hence, when pulsed electrical discharge technology is applied for wastewater treatment, lower conductivity is desirable.

20.5.4.2 Discharge Gas

In a pulse discharge plasma process, increment of ozone (O_3) and other oxygen based reactive species is carried out by introducing oxygen gas into the discharge system to boost the formation of the OH˙ radicals and degradation rate of organic pollutants. Oxygen has demonstrated promising results when used as a highly efficient feeding gas followed by argon, and at last nitrogen in terms of pollutant removal efficiency through the production of reactive species. When nitrogen is used as feeding gas, the electric discharge can dissociate molecules of nitrogen into reactive nitrogen radicals. In the presence of oxygen, the radicals such as N˙, H˙, O˙ and OH˙ either combine with each other or with other molecules, such as N_2, H_2O, etc., as well as with products like NO_2, HNO_2 and HNO_3. Formation of these products will change pH and conductivity of the solution, which can influence the pollutant removal rate (Du & Lin, 2020; Pinart et al., 1996).

In an investigation by Zhang et al. (2007), about 100% removal of 4-Chlorophenol (4-CP) was achieved with treatment time of 7 and 25 min using oxygen and nitrogen as working gas, respectively, with initial 4-CP concentration of 60 mg L^{-1}; thereby, indicating faster removal in case of oxygen as compared to nitrogen. This can be attributed to the formation of O_3 and hydrogen peroxide (H_2O_2) in case of oxygen as working gas; however, a part of nitrogen is utilized to produce nitrite and nitrate ions, which did not take any part in the degradation of 4-CP (Zhang et al., 2007). As a result, during the degradation with nitrogen, a prolonged time was required for same degree of 4-CP removal (Zhang et al., 2007).

Further, when argon was used as discharge gas, the metastable particles, and atoms of argon formed in the gas phase gets dissolved in water. Argon as a working gas generates the reactive species such as $H^{\cdot}$, $O^{\cdot}$ and $OH^{\cdot}$ without producing O_3; as a result, the degradation rate under argon was not effectively higher compared to oxygen as working gas. However, using argon is advantageous as it is less expensive than other gases, such as helium, and argon also has better energy transfer efficiency than helium. Thus, using oxygen can produce ozone, a highly oxidizing species along with other highly reactive radicals, which would be beneficial for improved removal rate over argon as it does not produce ozone.

20.5.4.3 Electrodes

The main objective of electrode is to provide the large electron discharge zone for efficient removal of pollutants. In plasma technology, basically two conductive electrodes of asymmetric shapes and high curvature are installed for the formation of plasma species. The addition of multiple electrodes in a system is vital as it enhances the treatment performance of the plasma process (Zhang et al., 2008). However, power input, flow rate and reactor configuration can influence the selection of number of electrodes. Moreover, one dimension electrode such as wire can generate more uniform and larger discharge zone compared to zero discharge zone formed due to the application of pin or needle.

In addition, an optimum inter electrode gap should be used as with a narrower gap, the discharge interception voltage can diminish due to an increase in the electric field; on the other hand, too large gap can reduce the production of reactive species due to small discharge current and weaken the electric field (Sano et al., 2002). Thus, an optimum gap can result in an effective plasma formation for the degradation of targeted compounds. Further, electrode materials also have significant effect on plasma process due to the release of ions to form reactive radicals from the metal surface into the wastewater to be treated. For example, in a plasma process, when platinum is used as high-voltage electrode, the degradation efficiency of organic compounds get enhanced due to the increase in production of highly oxidative $OH^{\cdot}$ radicals (Mededovic & Locke, 2007).

20.5.4.4 Solution pH

The pH of the wastewater also affects the chemical properties of plasma during electric discharge. The intensity at which OH˙ radicals are produced by pulsed discharge is influenced by pH. Moreover, the production of H_2O_2 is also pH dependent, which would help in the oxidation of organic molecules. The acid–base equilibrium also affects the potential degradation of organic molecules; for instance, about 40% of phenol was degraded at basic pH of 10.2, which is about 15% more in comparison to the degradation observed at acidic pH of 4.6 after 40 min of treatment time with the initial phenol concentration of 50 mg L^{-1} (Li et al., 2007). This can be attributed to the fact that at basic pH of 10.2; owing to the enhancement in electron liability of π bond in the aromatic ring, phenol is more reactive towards the generated OH˙ compared to acidic pH (Li et al., 2007). Thus, better removal of phenol is obtained at basic pH. Similarly, Gao and co-workers investigated the influence of pH for the decolouration of alizarin red S dye and reported that when the pH of the solution was increased from 3.65 to 10.99, the decolouration rate was also increased from 33.33 to 73.47%, suggesting that the elevated pH is favourable for the removal of organic molecules present in wastewater (Gao et al., 2008).

20.5.4.5 Energy Input

The rate of organic matter degradation via plasma treatment is dependent on the range of specific energy input, which according to the need can be changed either by changing the frequency of the discharge process or voltage. Additionally, higher density of reactive species, such as OH˙, O˙, O_3 and H_2O_2, is generated at higher electron density due to the subsequent chemical reaction between neutral gaseous molecules and electrons. Further, higher energy input also favours enhanced pollutant removal due to the more intense physical effect between the reactive species and organic molecules. In this regard, Lukes and co-workers obtained a relation between photon flux ($J_{190-280}$) (Quantum pulse^{-1}) and pulse mean power of discharge (P_p, kW) in pulsed corona with needle and plate electrode for the spectral region of 190–280 nm (Eq. 20.39) (Lukes et al., 2008).

$$J_{190-280} = 44.33 P_p^{2.11} \tag{20.39}$$

Generally, one can achieve either highest energy yield or best removal efficiency; however, not both (Jiang et al., 2012). Thus, depending upon the requirement for wastewater treatment, it is usually essential to focus on both energy yield and degradation efficiency.

20.5.5 Application of Plasma Technology in Wastewater Treatment

In recent years, plasma process has received great attention of the researchers for the degradation of a wide variety of organic pollutants. Owing to the low biodegradability and wider existence in the environment, pollutants such as, phenols, dyes and pharmaceuticals have been greatly explored for their remediation from the contaminated water via plasma technology. Phenol is a raw material, which is used in industries for the production of rubbers, dyes and plastics; as a result, it is the most commonly detected pollutant in wastewaters from these industries. Nonetheless, plasma process has been successfully demonstrated for the degradation of phenol and the addition of catalyst in the plasma process has shown improvement in the degradation of phenol.

Wang and co-workers demonstrated that kinetic oxidation constant (k) of phenol is about 1.7 folds more in case of plasma-TiO_2 process ($k = 5.2 \times 10^{-2}$ min^{-1}) compared to bare plasma ($k = 3.1 \times 10^{-2}$ min^{-1}) due to the increase in catalytic activity by the addition of TiO_2 to produce more $OH^{\cdot}$ radicals (Wang et al., 2008). Moreover, about 66.1% of total organic carbon (TOC) removal was achieved with catalyst induced plasma, which is about 11% more compared to plasma alone (Wang et al., 2008). Thus, it can be concluded that plasma along with catalyst can increase the removal rate of pollutants.

Similarly, due to the application of chemical dyes in textile industries, its content in wastewater has also increased rapidly, posing potential threat to the aquatic ecosystem. During the treatment of acid orange 7 (AO7) dye from the dielectric barrier discharge (DBD) thermal plasma process, about 60% of the dye discoloration was achieved in 2.5 min and 20 kV of applied voltage with initial dye concentration of 10 mg L^{-1} (Iervolino et al., 2020). Moreover, when H_2O_2 of 20 mM was added in the plasma reactor, the discoloration of dye was increased to 80% from 60%; thus, ensuring the improvement in removal of dye with the addition of H_2O_2.

The mineralization rate of organic dye is also influenced by the plasma treatment time, when sliding arc discharge plasma was used to mineralize AO7, the dye degradation increased from 58.9 to 99% with change in treatment time from 25 to 125 min due to the production of reactive species to oxidize the organics present in wastewater with the increase in treatment time (Du et al., 2008). Further, for the remediation of pharmaceutical amoxicillin (AMX), DBD plasma was applied with ZnO/α-Fe_2O_3 composite as nano-catalyst and with the treatment time of 18 min, initial AMX concentration of 16 mg L^{-1} and voltage of 18 kV, the degradation rate of AMX was increased from 75 to 99.3% with the addition of the nano-catalyst composite to DBD (Ansari et al., 2020). Similarly, the performance results of plasma technology reported by other investigators for treatment of contaminated water are summarized in Table 20.5.

Questions

20.1 List the factors that affect the COD removal and CE of microbial fuel cells.

20.2 What are the different variants of bio-electrochemical systems employed for wastewater treatment? Also mention the different valuables recovered through these processes while simultaneously treating wastewater.

20.3 What are the major roadblocks toward the commercialization and successful field-scale implementation of bio-electrochemical processes? How can these roadblocks be circumnavigated?

Table 20.5 Application of plasma for the removal of organic pollutants from wastewater

Pollutant	Discharge	Initial concentration (mg L^{-1})	Observations	Removal efficiency (%)	References
Phenol	Di-electric barrier	47	Treatment time of 3 h and gas discharge rate of 1.8 L h^{-1}	97	Marotta et al. (2012)
Bisphenol-A (BPA)	Gliding arc	28	Treatment time of 30 min and applied voltage of 10 kV	100	Abdelmalek et al. (2008)
Chlorophenol	Di-electric barrier	20	Treatment time of 19 min and applied voltage of 20 kV	94.8	Dojčinović et al. (2008)
Ciprofloxacin (CIP)	Surface discharge	50	Treatment time of 20 min and applied voltage of 17 kV	33.8	Li et al. (2020)
Tetracycline (TC)	Di-electric barrier with Mn/γ-Al$_2$O$_3$ catalysts	NA	Treatment time of 5 min and addition of catalyst improve removal	67.3–99.3	Wang et al. (2019)
Acid orange Dye	Di-electric barrier/Per sulphate	5	Discharge power of 36 W and 50 min of treatment time	60	Shang et al. (2017)

NA—Not available

20.4 Calculate the maximum attainable power density that can be produced from a MFC with the anode surface area and internal resistance of 20 cm^2 and 110 Ω, respectively. The OV (across 175 Ω of external resistance) and OCV of the MFC were 617 and 778 mV, respectively. (**Answer**: OCV = 0.778 V, R$_{ex}$ = R$_{in}$ = 110 Ω, A = 20 cm^2 = 20 × 10^{-4} m^2, Power = 0.00138 W = 1.38 mW, Power density = 690 mW/m^2)

20.5 Calculate the CE and COD removal efficiency of a MFC with anodic chamber volume of 150 mL generating an OV of 375 mV across 110 Ω of external resistance. The initial and final COD of the anolyte after 4 days of batch operation was 3540 and 770 mg/L, respectively. (**Answer**: V = 0.375 V, R = 110 Ω, I = 0.0034 A, t = 4 × 24 × 3600 = 345,600 s, v = 150 mL = 150 × 10^{-6} m^3, COD$_i$ = 3540 mg/L, COD$_f$ = 770 mg/L, COD removal efficacy = 78.45%, CE = 23.45%)

20.6 Explain the mechanism of colloidal de-stabilization.

20.7 What are the advantages and disadvantages of electrocoagulation over chemical coagulation?

20.8 Describe applications of electrocoagulation for wastewater treatment.

20.9 What are current density and its unit? How does it influence the performance of EC?

20.10 What is nanotechnology and explain its role in the environmental engineering?

20.11 Explain the different pathways through which nanoparticles enter into the environment and its ecotoxicological effects on human and aquatic life?

20.12 How nanotechnology is applicable in the process of adsorption and why it is essential to recover nanoparticles?

20.13 What is the role of oxygen as discharge gas in plasma technology?

20.14 Explain di-electric barrier and gliding arc discharge?

20.15 Calculate the photon flux produced in a plasma reactor, when pulse mean power of 300 kW is required to operate the system in spectral region of 190–280 nm? (**Answer**: The photon flux is given by $J_{190-280} = 44.33 P_p^{2.11}$; $At P_p = 300 kW$; $J_{190-280} = 44.33 \times 300^{2.11} = 7.47 \times 10^6 Quantumpulse^{-1}$)

20.16 Explain the mechanism of electro-oxidation.

20.17 Write a short note on by-product formation in electro-oxidation setup along with its environmental impact.

20.18 What are the advantages of electrode modification in electro-oxidation process?

References

Abbas, Q., Yousaf, B., Ullah, H., Ali, M. U., Ok, Y. S., & Rinklebe, J. (2020). Environmental transformation and nano-toxicity of engineered nano-particles (ENPs) in aquatic and terrestrial organisms. *Critical Reviews in Environment Science and Technology, 50*, 2523–2581. https://doi.org/10.1080/10643389.2019.1705721

Abbasi, S., Mirghorayshi, M., Zinadini, S., & Zinatizadeh, A. A. (2020). A novel single continuous electrocoagulation process for treatment of licorice processing wastewater: Optimization of operating factors using RSM. *Process Safety and Environment Protection, 134*, 323–332. https://doi.org/10.1016/j.psep.2019.12.005

Abdelmalek, F., Torres, R. A., Combet, E., Petrier, C., Pulgarin, C., & Addou, A. (2008). Gliding Arc Discharge (GAD) assisted catalytic degradation of bisphenol A in solution with ferrous ions. *Separation and Purification Technology, 63*, 30–37. https://doi.org/10.1016/j.seppur.2008.03.036

Ahmad, A., & Gupta, A. (2019). Urine a source for struvite: Nutrient recovery—A review. *Journal of the Indian Chemical Society, 96*, 507–514.

Ahmad, A., Das, S., & Ghangrekar, M. M. (2020). Removal of xenobiotics from wastewater by electrocoagulation: A mini-review. *Journal of the Indian Chemical Society, 97*, 493–500.

Ahmad, M., Rehman, W., Khan, M. M., Qureshi, M. T., Gul, A., Haq, S., Ullah, R., Rab, A., & Menaa, F. (2021). Phytogenic fabrication of ZnO and gold decorated ZnO nanoparticles for photocatalytic degradation of Rhodamine B. *Journal of Environmental Chemical Engineering, 9*, 104725. https://doi.org/10.1016/j.jece.2020.104725

Ahmadzadeh, S., Asadipour, A., Pournamdari, M., Behnam, B., Rahimi, H. R., & Dolatabadi, M. (2017). Removal of ciprofloxacin from hospital wastewater using electrocoagulation technique by aluminum electrode: Optimization and modelling through response surface methodology. *Process Safety and Environment Protection, 109*, 538–547. https://doi.org/10.1016/j.psep.2017. 04.026

Ali, J. A., Kalhury, A. M., Sabir, A. N., Ahmed, R. N., Ali, N. H., & Abdullah, A. D. (2020). A state-of-the-art review of the application of nanotechnology in the oil and gas industry with a focus on drilling engineering. *Journal of Petroleum Science and Engineering, 191*, 107118. https://doi.org/10.1016/j.petrol.2020.107118

Alimohammadi, M., Askari, M., Dehghani, M. H., Dalvand, A., Saeedi, R., Yetilmezsoy, K., Heibati, B., & McKay, G. (2017). Elimination of natural organic matter by electrocoagulation using bipolar and monopolar arrangements of iron and aluminum electrodes. *International Journal of Environmental Science and Technology, 14*, 2125–2134. https://doi.org/10.1007/s13762-017-1402-3

Andrade, P. F., de Faria, A. F., Oliveira, S. R., Arruda, M. A. Z., Gonçalves, M. do C. (2015). Improved antibacterial activity of nanofiltration polysulfone membranes modified with silver nanoparticles. *Water Research, 81*, 333–342.https://doi.org/10.1016/j.watres.2015.05.006

Annemie, B., Neyts, E., Gijbels, R., & Mullen, J. Van der. (2002). Gas discharge plasmas and their applications. *Spectrochimica Acta Part B, 57*, 609–658.

Ansari, M., Hossein Mahvi, A., Hossein Salmani, M., Sharifian, M., Fallahzadeh, H., & Hassan Ehrampoush, M. (2020). Dielectric barrier discharge plasma combined with nano catalyst for aqueous amoxicillin removal: Performance modeling, kinetics and optimization study, energy yield, degradation pathway, and toxicity. *Separation and Purification Technology, 251*. https://doi.org/10.1016/j.seppur.2020.117270

Arul Prakash, F., Dushendra Babu, G. J., Lavanya, M., Vidhya, K. S., & Devasena, T. (2011). Toxicity studies of aluminium oxide nanoparticles in cell lines. *International Journal of Nanotechnology and Applications, 5*, 99–107.

Balarak, D., Chandrika, K., & Attaolahi, M. (2019). Assessment of effective operational parameters on removal of amoxicillin from synthetic wastewater using electrocoagulation process. *Journal of Pharmaceutical Research International, 29*, 1–8. https://doi.org/10.9734/jpri/2019/v29i130227

Bayramoglu, M., Kobya, M., Can, O. T., & Sozbir, M. (2004). Operating cost analysis of electrocoagulation of textile dye wastewater. *Separation and Purification Technology, 37*, 117–125. https://doi.org/10.1016/j.seppur.2003.09.002

Behloul, M., Grib, H., Drouiche, N., Abdi, N., Lounici, H., & Mameri, N. (2013). Removal of malathion pesticide from polluted solutions by electrocoagulation: Modeling of experimental results using response surface methodology. *Separation Science and Technology, 48*, 664–672. https://doi.org/10.1080/01496395.2012.707734

Bhowmick, G., Das, S., Adhikary, K., Ghangrekar, M., & Mitra, A. (2019). Using rhodium as a cathode catalyst for enhancing performance of microbial fuel cell. *International Journal of Hydrogen Energy, 44*(39), 22218–22222.

Bhowmick, G. D., Das, S., Adhikary, K., Ghangrekar, M. M., & Mitra, A. (2020). Bismuth-impregnated ruthenium with activated carbon as photocathode catalyst to proliferate the efficacy of a microbial fuel cell. *Journal of Hazardous, Toxic, and Radioactive Waste, 25*(1), 04020066.

Cao, X., Huang, X., Liang, P., Xiao, K., Zhou, Y., Zhang, X., & Logan, B. E. (2009). A new method for water desalination using microbial desalination cells. *Environmental Science & Technology, 43*(18), 7148–7152.

Chae, K.-J., Choi, M.-J., Lee, J.-W., Kim, K.-Y., & Kim, I. S. (2009). Effect of different substrates on the performance, bacterial diversity, and bacterial viability in microbial fuel cells. *Bioresource Technology, 100*(14), 3518–3525.

Chen, S., Liu, G., Zhang, R., Qin, B., Luo, Y., & Hou, Y. (2012). Improved performance of the microbial electrolysis desalination and chemical-production cell using the stack structure. *Bioresource
   Technology, 116*, 507–511.
Cheng, S., & Logan, B. E. (2011). Increasing power generation for scaling up single-chamber air
   cathode microbial fuel cells. *Bioresource Technology, 102*(6), 4468–4473.
Das, S., & Ghangrekar, M. (2019). Tungsten oxide as electrocatalyst for improved power generation
   and wastewater treatment in microbial fuel cell. *Environmental Technology, 41*(19), 2546–2553.
Das, S., Das, S., Das, I., & Ghangrekar, M. (2019). Application of bioelectrochemical systems for
   carbon dioxide sequestration and concomitant valuable recovery: A review. *Materials Science for
   Energy Technologies, 2*(3), 687–696.
Das, I., Das, S., Dixit, R., & Ghangrekar, M. (2020a). Goethite supplemented natural clay ceramic
   as an alternative proton exchange membrane and its application in microbial fuel cell. *Ionics, 26*,
   3061–3072.
Das, I., Das, S., & Ghangrekar, M. (2020b). Application of bimetallic low-cost CuZn as oxygen
   reduction cathode catalyst in lab-scale and field-scale microbial fuel cell. *Chemical Physics
   Letters, 751*, 137536.
Das, I., Das, S., Sharma, S., & Ghangrekar, M. (2020c). Ameliorated performance of a microbial
   fuel cell operated with an alkali pre-treated clayware ceramic membrane. *International Journal
   of Hydrogen Energy, 45*(33), 16787–16798.
Das, S., Mishra, A., & Ghangrekar, M. (2020d). Concomitant production of bioelectricity and hydrogen peroxide leading to the holistic treatment of wastewater in microbial fuel cell. *Chemical
   Physics Letters, 759*, 137986.
Das, S., Mishra, A., & Ghangrekar, M. (2020e). Production of hydrogen peroxide using various
   metal-based catalysts in electrochemical and bioelectrochemical systems: Mini review. *Journal
   of Hazardous, Toxic, and Radioactive Waste, 24*(3), 06020001.
Deshpande, A. M., Satyanarayan, S., & Ramakant, S. (2010). Treatment of high-strength pharma-
   ceutical wastewater by electrocoagulation combined with anaerobic process. *Water Science and
   Technology, 61*, 463–472. https://doi.org/10.2166/wst.2010.831
Dojčinović, B. P., Manojlović, D., Roglić, G. M., Obradović, B. M., Kuraica, M. M., & Purić, J.
   (2008). Plasma assisted degradation of phenol solutions. *Vacuum, 83*, 234–237. https://doi.org/
   10.1016/j.vacuum.2008.04.003
Donneys-Victoria, D., Bermúdez-Rubio, D., Torralba-Ramírez, B., Marriaga-Cabrales, N., &
   Machuca-Martínez, F. (2019). Removal of indigo carmine dye by electrocoagulation using magnesium anodes with polarity change. *Environmental Science and Pollution Research, 26*, 7164–
   7176. https://doi.org/10.1007/s11356-019-04160-y
Du, Z., & Lin, X. (2020). Research progress in application of low temperature plasma technology for
   wastewater treatment. *IOP Conference Series: Earth and Environmental Science, 512*, 012031.
   https://doi.org/10.1088/1755-1315/512/1/012031
Du, Z., Li, H., & Gu, T. (2007). A state of the art review on microbial fuel cells: A promising
   technology for wastewater treatment and bioenergy. *Biotechnology Advances, 25*(5), 464–482.
Du, C. M., Shi, T. H., Sun, Y. W., & Zhuang, X. F. (2008). Decolorization of Acid Orange 7 solution
   by gas-liquid gliding arc discharge plasma. *Journal of Hazardous Materials, 154*, 1192–1197.
   https://doi.org/10.1016/j.jhazmat.2007.11.032
Elmi, F., Hosseini, T., Taleshi, M. S., & Taleshi, F. (2017). Kinetic and thermodynamic investigation
   into the lead adsorption process from wastewater through magnetic nanocomposite Fe3O4/CNT.
   Nanotechnology for Environmental Engineering, 2, 1–13. https://doi.org/10.1007/s41204-017-
   0023-x

Ensano, B. M. B., Borea, L., Naddeo, V., Belgiorno, V., de Luna, M. D. G., & Ballesteros, F. C. (2017). Removal of pharmaceuticals from wastewater by intermittent electrocoagulation. *Water (Switzerland), 9*, 1–15. https://doi.org/10.3390/w9020085

Feynman, R. P. (1960). There's plenty of room at the bottom. An invitation to open up a new field of physics. *Engineering and Science, 23*(5), 22.

Gahlawat, S., Singh, J., Yadav, A. K., & Ingole, P. P. (2019). Exploring Burstein-Moss type effects in nickel doped hematite dendrite nanostructures for enhanced photo-electrochemical water splitting. *Physical Chemistry Chemical Physics: PCCP, 21*, 20463–20477. https://doi.org/10.1039/c9cp04132j

Gajda, I., Greenman, J., & Ieropoulos, I. A. (2018). Recent advancements in real-world microbial fuel cell applications. *Current Opinion in Electrochemistry, 11*, 78–83.

Gao, J., Yu, J., Lu, Q., He, X., Yang, W., Li, Y., Pu, L., & Yang, Z. (2008). Decoloration of alizarin red S in aqueous solution by glow discharge electrolysis. *Dyes and Pigments, 76*, 47–52. https://doi.org/10.1016/j.dyepig.2006.08.033

Ge, Z., Li, J., Xiao, L., Tong, Y., & He, Z. (2013). Recovery of electrical energy in microbial fuel cells: Brief review. *Environmental Science & Technology Letters, 1*(2), 137–141.

George, A., Shen, B., Craven, M., Wang, Y., Kang, D., Wu, C., & Tu, X. (2021). A review of non-thermal plasma technology: A novel solution for CO2 conversion and utilization. *Renewable and Sustainable Energy Reviews, 135*, 109702. https://doi.org/10.1016/j.rser.2020.109702

Ghadge, A. N., & Ghangrekar, M. M. (2015). Development of low cost ceramic separator using mineral cation exchanger to enhance performance of microbial fuel cells. *Electrochimica Acta, 166*, 320–328.

Ghalwa MA, & Nader Farhat, B. (2015). Removal of imidacloprid pesticide by electrocoagulation process using iron and aluminum electrodes. *Journal of Environmental Analytical Chemistry, 02*. https://doi.org/10.4172/2380-2391.1000154

Ghanbari, F., Zirrahi, F., Olfati, D., Gohari, F., & Hassani, A. (2020). TiO2 nanoparticles removal by electrocoagulation using iron electrodes: Catalytic activity of electrochemical sludge for the degradation of emerging pollutant. *Journal of Molecular Liquids, 310*, 113217. https://doi.org/10.1016/j.molliq.2020.113217

Ghangrekar, M. M., Das, S., & Das, S. (2021). Microbial electrochemical technologies for CO2 sequestration. In *Biomass, biofuels, biochemicals* (pp. 413–443). Elsevier.

Ghernaout, D. (2019). Electrocoagulation process: A mechanistic review at the dawn of its modeling. *Journal of Environmental Science and Allied Research, 2*, 22–38. https://doi.org/10.29199/2637-7063/esar-201019

Ghorai, A., Roy, S., Das, S., Komber, H., Ghangrekar, M. M., Voit, B., & Banerjee, S. (2020). Chemically stable sulfonated polytriazoles containing trifluoromethyl and phosphine oxide moieties for proton exchange membranes. *ACS Applied Polymer Materials, 2*(7), 2967–2979.

Golder, A. K., Samanta, A. N., & Ray, S. (2007). Removal of Cr3+ by electrocoagulation with multiple electrodes: Bipolar and monopolar configurations. *Journal of Hazardous Materials, 141*, 653–661. https://doi.org/10.1016/j.jhazmat.2006.07.025

Gupta, A., Vidyarthi, S. R., & Sankararamakrishnan, N. (2014). Enhanced sorption of mercury from compact fluorescent bulbs and contaminated water streams using functionalized multiwalled carbon nanotubes. *Journal of Hazardous Materials, 274*, 132–144. https://doi.org/10.1016/j.jhazmat.2014.03.020

Gupta, A., Das, S., & Ghangrekar, M. (2020). Optimal cathodic imposed potential and appropriate catalyst for the synthesis of hydrogen peroxide in microbial electrolysis cell. *Chemical Physics Letters, 754*, 137690.

Hakizimana, J. N., Gourich, B., Chafi, M., Stiriba, Y., Vial, C., Drogui, P., & Naja, J. (2017). Electrocoagulation process in water treatment: A review of electrocoagulation modeling approaches. *Desalination, 404*, 1–21. https://doi.org/10.1016/j.desal.2016.10.011

Hamdan, S. S., & El-Naas, M. H. (2014). Characterization of the removal of Chromium(VI) from groundwater by electrocoagulation. *Journal of Industrial and Engineering Chemistry, 20*, 2775–2781. https://doi.org/10.1016/j.jiec.2013.11.006

Hu, X., Liu, B., Zhou, J., Jin, R., Qiao, S., & Liu, G. (2015). CO2 fixation, lipid production, and power generation by a novel air-lift-type microbial carbon capture cell system. *Environmental Science & Technology, 49*(17), 10710–10717.

Hussien, M. S. A. (2021). Facile synthesis of nanostructured Mn-Doped Ag3PO4 for visible photodegradation of emerging pharmaceutical contaminants: streptomycin photodegradation. *Journal of Inorganic and Organometallic Polymers and Materials, 31*, 945–959. https://doi.org/10.1007/s10904-020-01831-z

Iervolino, G., Vaiano, V., & Palma, V. (2020). Enhanced azo dye removal in aqueous solution by H2O2 assisted non-thermal plasma technology. *Environmental Technology and Innovation, 19*, 100969. https://doi.org/10.1016/j.eti.2020.100969

Jacobson, K. S., Drew, D. M., & He, Z. (2011). Use of a liter-scale microbial desalination cell as a platform to study bioelectrochemical desalination with salt solution or artificial seawater. *Environmental Science and Technology, 45*(10), 4652–4657.

Jiang, B., Zheng, J., Liu, Q., & Wu, M. (2012). Degradation of azo dye using non-thermal plasma advanced oxidation process in a circulatory airtight reactor system. *Chemical Engineering Journal, 204–205*, 32–39. https://doi.org/10.1016/j.cej.2012.07.088

Jiang, B., Zheng, J., Qiu, S., Wu, M., Zhang, Q., Yan, Z., & Xue, Q. (2014). Review on electrical discharge plasma technology for wastewater remediation. *Chemical Engineering Journal, 236*, 348–368. https://doi.org/10.1016/j.cej.2013.09.090

Jing, G., Ren, S., Gao, Y., Sun, W., & Gao, Z. (2020). Electrocoagulation: A promising method to treat and reuse mineral processing wastewater with high COD. *Water (Switzerland), 12*(2), 595. https://doi.org/10.3390/w12020595

Joshi, A. A., Locke, B. R., Arce, P., & Finney, W. C. (1995). Formation of hydroxyl radicals, hydrogen peroxide and aqueous electrons by pulsed streamer corona discharge in aqueous solution. *Journal of Hazardous Materials, 41*, 3–30. https://doi.org/10.1016/0304-3894(94)00099-3

Kakakhel, M. A., Wu, F., Sajjad, W., Zhang, Q., Khan, I., Ullah, K., & Wang, W. (2021). Long-term exposure to high-concentration silver nanoparticles induced toxicity, fatality, bioaccumulation, and histological alteration in fish (Cyprinus carpio). *Environmental Sciences Europe, 33*, 14. https://doi.org/10.1186/s12302-021-00453-7

Karkeh-abadi, F., Saber-Samandari, S., & Saber-Samandari, S. (2016). The impact of functionalized CNT in the network of sodium alginate-based nanocomposite beads on the removal of Co(II) ions from aqueous solutions. *Journal of Hazardous Materials, 312*, 224–233. https://doi.org/10.1016/j.jhazmat.2016.03.074

Karthigadevi, G., Manikandan, S., Karmegam, N., Subbaiya, R., Chozhavendhan, S., Ravindran, B., Chang, S. W., & Awasthi, M. K. (2021). Chemico-nanotreatment methods for the removal of persistent organic pollutants and xenobiotics in water—A review. *Bioresource Technology, 324*, 124678. https://doi.org/10.1016/j.biortech.2021.124678

Khandegar, V., & Saroha, A. K. (2013). Electrocoagulation for the treatment of textile industry effluent—A review. *Journal of Environmental Management, 128*, 949–963. https://doi.org/10.1016/j.jenvman.2013.06.043

Kunduru, K. R., Nazarkovsky, M., Farah, S., Pawar, R. P., Basu, A., & Domb, A. J. (2017). Nanotechnology for water purification: Applications of nanotechnology methods in wastewater treatment. *Water Purification, 2017*, 33–74. https://doi.org/10.1016/b978-0-12-804300-4.00002-2

Kurniawan, T. A., Mengting, Z., Fu, D., Yeap, S. K., Othman, M. H. D., Avtar, R., & Ouyang, T. (2020). Functionalizing TiO2 with graphene oxide for enhancing photocatalytic degradation of methylene blue (MB) in contaminated wastewater. *Journal of Environmental Management, 270,* 110871. https://doi.org/10.1016/j.jenvman.2020.110871

Kurwadkar, S., Pugh, K., Gupta, A., & Ingole, S. (2015). Nanoparticles in the environment: Occurrence, distribution, and risks. *Journal of Hazardous, Toxic, and Radioactive Waste, 19,* 04014039. https://doi.org/10.1061/(asce)hz.2153-5515.0000258

Li, J., Sato, M., & Ohshima, T. (2007). Degradation of phenol in water using a gas-liquid phase pulsed discharge plasma reactor. *Thin Solid Films, 515,* 4283–4288. https://doi.org/10.1016/j.tsf.2006.02.070

Li, H., Li, T., He, S., Zhou, J., Wang, T., & Zhu, L. (2020). Efficient degradation of antibiotics by non-thermal discharge plasma: Highlight the impacts of molecular structures and degradation pathways. *Chemical Engineering Journal, 395.* https://doi.org/10.1016/j.cej.2020.125091

Liang, J., Liu, J., Yuan, X., Dong, H., Zeng, G., Wu, H., Wang, H., Liu, J., Hua, S., Zhang, S., Yu, Z., He, X., & He, Y. (2015). Facile synthesis of alumina-decorated multi-walled carbon nanotubes for simultaneous adsorption of cadmium ion and trichloroethylene. *Chemical Engineering Journal, 273,* 101–110. https://doi.org/10.1016/j.cej.2015.03.069

Logan, B. E. (2008). *Microbial fuel cells.* Wiley.

Logan, B. E. (2010). Scaling up microbial fuel cells and other bioelectrochemical systems. *Applied Microbiology and Biotechnology, 85*(6), 1665–1671.

Logan, B. E., Call, D., Cheng, S., Hamelers, H. V. M., Sleutels, T. H. J. A., Jeremiasse, A. W., & Rozendal, R. A. (2008). Microbial electrolysis cells for high yield hydrogen gas production from organic matter. *Environmental Science & Technology, 42*(23), 8630–8640.

Lucas, E., Decker, S., Khaleel, A., Seitz, A., Fultz, S., Ponce, A., Li, W., Carnes, C., & Klabunde, K. J. (2001). Nanocrystalline metal oxides as unique chemical reagents/sorbents. *Chemistry—A European Journal, 7,* 2505–2510. https://doi.org/10.1002/1521-3765(20010618)7:12%3c2505::AID-CHEM25050%3e3.0.CO;2-R

Lukes, P., Clupek, M., Babicky, V., & Sunka, P. (2008). Ultraviolet radiation from the pulsed corona discharge in water. *Plasma Sources Science and Technology, 17,* 024012. https://doi.org/10.1088/0963-0252/17/2/024012

Malinović, B. N., Pavlović, M. G., & Djuričić, T. (2017). Electrocoagulation of textile wastewater containing a mixture of organic dyes by iron electrode. *Jouranl of Electrochemical Science and Engineering, 7,* 103. https://doi.org/10.5599/jese.366

Mansoorian, H. J., Mahvi, A. H., & Jafari, A. J. (2014). Removal of lead and zinc from battery industry wastewater using electrocoagulation process: Influence of direct and alternating current by using iron and stainless steel rod electrodes. *Separation and Purification Technology, 135,* 165–175. https://doi.org/10.1016/j.seppur.2014.08.012

Marotta, E., Ceriani, E., Schiorlin, M., Ceretta, C., & Paradisi, C. (2012). Comparison of the rates of phenol advanced oxidation in deionized and tap water within a dielectric barrier discharge reactor. *Water Research, 46,* 6239–6246. https://doi.org/10.1016/j.watres.2012.08.022

Matteson, M. J., Dobson, R. L., Glenn, R. W., Kukunoor, N. S., Waits, W. H., & Clayfield, E. J. (1995). Electrocoagulation and separation of aqueous suspensions of ultrafine particles. *Colloids Surfaces A: Physicochemical and Engineering Aspects, 104,* 101–109. https://doi.org/10.1016/0927-7757(95)03259-G

Mededovic, S., & Locke, B. R. (2007). The role of platinum as the high voltage electrode in the enhancement of Fenton's reaction in liquid phase electrical discharge. *Applied Catalysis B: Environmental, 72,* 342–350. https://doi.org/10.1016/j.apcatb.2006.11.014

Moreno-Casillas, H. A., Cocke, D. L., Gomes, J. A. G., Morkovsky, P., Parga, J. R., & Peterson, E. (2007). Electrocoagulation mechanism for COD removal. *Separation and Purification Technology, 56*, 204–211. https://doi.org/10.1016/j.seppur.2007.01.031

Mubarak, N. M., Sahu, J. N., Abdullah, E. C., & Jayakumar, N. S. (2014). Removal of heavy metals from wastewater using carbon nanotubes. *Separation and Purification Reviews, 43*, 311–338. https://doi.org/10.1080/15422119.2013.821996

Nandi, B. K., & Patel, S. (2017). Effects of operational parameters on the removal of brilliant green dye from aqueous solutions by electrocoagulation. *Arabian Journal of Chemistry, 10*, S2961–S2968. https://doi.org/10.1016/j.arabjc.2013.11.032

Neethu, B., Tholia, V., & Ghangrekar, M. (2020). Optimizing performance of a microbial carbon-capture cell using Box-Behnken design. *Process Biochemistry, 95*, 99–107.

Nguyen, Q. H., Watari, T., Yamaguchi, T., Takimoto, Y., Niihara, K., Wiff, J. P., & Nakayama, T. (2020). COD removal from artificial wastewater by electrocoagulation using aluminum electrodes. *International Journal of Electrochemical Science, 15*, 39–51. https://doi.org/10.20964/2020.01.42

Nidheesh, P. V. (2018). Removal of organic pollutants by peroxicoagulation. *Environmental Chemistry Letters, 16*, 1283–1292. https://doi.org/10.1007/s10311-018-0752-5

Omranpour Shahreza, S., Mokhtarian, N., & Behnam, S. (2020). Optimization of Mn removal from aqueous solutions through electrocoagulation. *Environmental Technology, 41*(7), 890–900. https://doi.org/10.1080/09593330.2018.1514071

Pant, D., Van Bogaert, G., Diels, L., & Vanbroekhoven, K. (2010). A review of the substrates used in microbial fuel cells (MFCs) for sustainable energy production. *Bioresource Technology, 101*(6), 1533–1543.

Papadopoulos, K. P., Argyriou, R., Economou, C. N., Charalampous, N., Dailianis, S., Tatoulis, T. I., Tekerlekopoulou, A. G., & Vayenas, D. V. (2019). Treatment of printing ink wastewater using electrocoagulation. *Journal of Environmental Management, 237*, 442–448. https://doi.org/10.1016/j.jenvman.2019.02.080

Peavy, H. S., Rowe, D. R., & Tchobanoglous, G. (1985). Environmental Engineering.

Pinart, J., Smirdec, M., Pinart, M. E., Aaron, J. J., Benmansour, Z., Goldman, M., & Goldman, A. (1996). Quantitative study of the formation of inorganic chemical species following corona discharge—I. Production of HNO2 and HNO3 in a composition-controlled, humid atmosphere. *Atmospheric Environment, 30*, 129–132. https://doi.org/10.1016/1352-2310(95)00231-M

Potter, M. C. (1911). Electrical effects accompanying the decomposition of organic compounds. *Proceedings of the Royal Society of London. Series B, Containing Papers of a Biological Character, 84*(571), 260–276.

Pradhan, H., & Ghangrekar, M. (2015). Organic matter and dissolved salts removal in a microbial desalination cell with different orientation of ion exchange membranes. *Desalination and Water Treatment, 54*(6), 1568–1576.

Priya, M., & Jeyanthi, J. (2019). Removal of COD, oil and grease from automobile wash water effluent using electrocoagulation technique. *Microchemical Journal, 150*, 104070. https://doi.org/10.1016/j.microc.2019.104070

Qu, Y., Feng, Y., Wang, X., Liu, J., Lv, J., He, W., & Logan, B. E. (2012). Simultaneous water desalination and electricity generation in a microbial desalination cell with electrolyte recirculation for pH control. *Bioresource Technology, 106*, 89–94.

Qu, X., Alvarez, P. J. J., & Li, Q. (2013). Applications of nanotechnology in water and wastewater treatment. *Water Research, 47*, 3931–3946. https://doi.org/10.1016/j.watres.2012.09.058

Rao, S. S., Srikanth, M., Neelima, P., & Student, M. T. (2017). Optimisation parameters for dicofol pesticide removal by electro-coagulation. *International Advanced Research Journal in Science, Engineering, Technology, 4*, 258–261. https://doi.org/10.17148/IARJSET.2017.4935

Samal, S. (2017). Thermal plasma technology: The prospective future in material processing. *Journal of Cleaner Production, 142*, 3131–3150. https://doi.org/10.1016/j.jclepro.2016.10.154

Sano, N., Kawashima, T., Fujikawa, J., Fujimoto, T., Kitai, T., Kanki, T., & Toyoda, A. (2002). Decomposition of organic compounds in water by direct contact of gas corona discharge: Influence of discharge conditions. *Industrial and Engineering Chemistry Research, 41*, 5906–5911. https://doi.org/10.1021/ie0203328

Sham, L., & Schlüter, M. (1985). Density-functional theory of the band gap. *Physical Review B, 32*(6), 3883–3889.

Shang, K., Wang, X., Li, J., Wang, H., Lu, N., Jiang, N., & Wu, Y. (2017). Synergetic degradation of Acid Orange 7 (AO7) dye by DBD plasma and persulfate. *Chemical Engineering Journal, 311*, 378–384. https://doi.org/10.1016/j.cej.2016.11.103

Sivaranjani, Gafoor, A., Ali, N., Kumar, S., Ramalakshmi, Begum, S., & Rahman, Z. (2020). Applicability and new trends of different electrode materials and its combinations in electro coagulation process: A brief review. *Materials Today: Proceedings, 37*, 377–382.https://doi.org/10.1016/j.matpr.2020.05.379

Tahreen, A., Jami, M. S., & Ali, F. (2020). Role of electrocoagulation in wastewater treatment: A developmental review. *Journal of Water Process Engineering, 37*, 101440. https://doi.org/10.1016/j.jwpe.2020.101440

Tang, C. Y., Zhao, Y., Wang, R., Hélix-Nielsen, C., & Fane, A. G. (2013). Desalination by biomimetic aquaporin membranes: Review of status and prospects. *Desalination, 308*, 34–40. https://doi.org/10.1016/j.desal.2012.07.007

Trivedi, R., & Ashwin, T. (2013). Application of plasma technology for coated textile. *IJSRD—International Journal for Scientific Research & Development, 1*(Issue 3), 653–655. ISSN (online).

Trivedi, P., & Axe, L. (2000). Modeling Cd and Zn sorption to hydrous metal oxides. *Environmental Science and Technology, 34*, 2215–2223. https://doi.org/10.1021/es991110c

Varanasi, J. L., Nayak, A. K., Sohn, Y., Pradhan, D., & Das, D. (2016). Improvement of power generation of microbial fuel cell by integrating tungsten oxide electrocatalyst with pure or mixed culture biocatalysts. *Electrochimica Acta, 199*, 154–163.

Vik, E. A., Carlson, D. A., Eikum, A. S., & Gjessing, E. T. (1984). Electrocoagulation of potable water. *Water Research, 18*, 1355–1360. https://doi.org/10.1016/0043-1354(84)90003-4

Wagh, M. P., Nemade, P. D., & Jadhav, P. (2020). Continuous electro coagulation process for the distillery spent wash using Al electrodes. *Techno-Societal, 2018*, 41–49. https://doi.org/10.1007/978-3-030-16962-6_5

Wang, H., Li, J., Quan, X., & Wu, Y. (2008). Enhanced generation of oxidative species and phenol degradation in a discharge plasma system coupled with TiO2 photocatalysis. *Applied Catalysis B: Environmental, 83*, 72–77. https://doi.org/10.1016/j.apcatb.2008.02.004

Wang, X., Feng, Y., Liu, J., Lee, H., Li, C., Li, N., & Ren, N. (2010). Sequestration of CO2 discharged from anode by algal cathode in microbial carbon capture cells (MCCs). *Biosensors and Bioelectronics, 25*(12), 2639–2643.

Wang, Y., Li, B., Cui, D., Xiang, X., & Li, W. (2014). Nano-molybdenum carbide/carbon nanotubes composite as bifunctional anode catalyst for high-performance Escherichia coli-based microbial fuel cell. *Biosensors and Bioelectronics, 51*, 349–355.

Wang, B., Wang, C., Yao, S., Peng, Y., & Xu, Y. (2019). Plasma-catalytic degradation of tetracycline hydrochloride over Mn/γ-Al2O3 catalysts in a dielectric barrier discharge reactor. *Plasma Science and Technology, 21*, 65503. https://doi.org/10.1088/2058-6272/ab079c

Yean, S., Cong, L., Yavuz, C. T., Mayo, J. T., Yu, W. W., Falkner, J. C., Kan, A. T., Colvin, V. L., & Tomson, M. B. (2006). Erratum: "Effect of magnetite particle size on adsorption and desorption of arsenite and arsenate". [Journal of Materials Research, Vol. 20(3255) (2005)]. *Journal of Materials Research, 21*, 1862. https://doi.org/10.1557/jmr.2005.0403e

Yoosefian, M., Ahmadzadeh, S., Aghasi, M., & Dolatabadi, M. (2017). Optimization of electrocoagulation process for efficient removal of ciprofloxacin antibiotic using iron electrode; kinetic and isotherm studies of adsorption. *Journal of Molecular Liquids, 225*, 544–553. https://doi.org/10.1016/j.molliq.2016.11.093

Yüksel, E., Şengil, I. A., & Özacar, M. (2009). The removal of sodium dodecyl sulfate in synthetic wastewater by peroxi-electrocoagulation method. *Chemical Engineering Journal, 152*, 347–353. https://doi.org/10.1016/j.cej.2009.04.058

Zaied, B. K., Rashid, M., Nasrullah, M., Zularisam, A. W., Pant, D., & Singh, L. (2020). A comprehensive review on contaminants removal from pharmaceutical wastewater by electrocoagulation process. *Science of the Total Environment, 726*, 138095. https://doi.org/10.1016/j.scitotenv.2020.138095

Zhang, Y., Zhou, M., Hao, X., & Lei, L. (2007). Degradation mechanisms of 4-chlorophenol in a novel gas-liquid hybrid discharge reactor by pulsed high voltage system with oxygen or nitrogen bubbling. *Chemosphere, 67*, 702–711. https://doi.org/10.1016/j.chemosphere.2006.10.065

Zhang, Y., Zheng, J., Qu, X., & Chen, H. (2008). Design of a novel non-equilibrium plasma-based water treatment reactor. *Chemosphere, 70*, 1518–1524. https://doi.org/10.1016/j.chemosphere.2007.09.013

Zhuang, L., Zheng, Y., Zhou, S., Yuan, Y., Yuan, H., & Chen, Y. (2012). Scalable microbial fuel cell (MFC) stack for continuous real wastewater treatment. *Bioresource Technology, 106*, 82–88.

Life Cycle Costing of Wastewater Treatment

21

Estimating cost of any engineering project is of paramount importance from the stake-holder's perspective. Unlike other process plants, the costing and estimation of sewage treatment plant construction and operation is estimated from a point of view as the treatment of sewage is the cost that we have to pay towards polluting the environment. However, with growing awareness regarding the water scarcity, this concept is changing and the life cycle costing is being modified to take into account the additional cost of further tertiary treatment as well as the infrastructural cost. With this focus of using treated water for different reuse purposes, costing exercises for STPs using different secondary wastewater treatment technologies are demonstrated in this chapter. The chapter also includes detailed calculations for assessing the present value of the different consumables and annual expenditures.

21.1 Introduction

Life cycle costing (LCC) provides a pathway for assessing the cost incurred for any project or asset throughout its life time by bringing all the estimated future expenditures to present valuation. As per IS code 13174–1994 Part 1, life cycle method of costing enables one to arrive at an equitable and economic assessment of the different design alternatives. It takes into account all the possible cost elements over the total period of project/operating life and expresses in monetary units applicable to a defined base date, usually in the financial year during which the cost estimate is being carried out. Typically, the LCC method can be done by two different methods namely, the present worth (PW) method and annualised cost (AC) method. In the PW method, all the future costs are converted to present worth by taking into consideration the discount rates and the inflation rates.

The AC method annualizes or amortizes the present expenditure as well as the future expenditures over the entire span of the project life cycle period. The judgement of the suitability of the method for any particular project depends upon the experience of the designer/estimator. However, as a rule of thumb, for projects with shorter span duration, or projects wherein rate of obsolescence (the rate at which a product loses its innovative edge or demand) is high, PW method can be adopted. The PW method is also adopted when the policy makers are less concerned about the future actual costs. In contrast, the AC method can be adopted when there is sufficient supply of present funds and the management is interested to understand the future annual cost expenditures and in turn wish to assess the future savings.

While the life cycle costing can assume different methods, for convenience, this chapter would adhere to the guidelines provided by the IS:13174–1994 (Parts 1 and 2) and the terminologies relevant to the analysis would be defined in same line. The reader may look up for further definitions for a more detailed understanding the different aspects of LCC in IS:13174.

Amortization: The process of converting the initial capital costs and the future expenditures to annual cost figures to be incurred over the life time of the asset. This would enable to remove the effect of the one-time present or future investment and its corresponding interest during the calculation of the present worth. All these costs are converted to annual cost over the life span of the project.

Amortization factor: The estimated fraction/decimal that needs to be multiplied with the initial investment figure to calculate the amortized/annualized payments required to be made every year for the entire period of operation to pay off the debt/cost of the initial one-time investment.

Present value analysis: It is estimation of the discounted present value of the future expenditures and the summation of such estimated discounted values after taking into account the effect of inflation or deflation as well as the earning potential of the money during service life period.

Baseline: The common point in time is considered as baseline, during which all expenditures are expected to start and to which all individual present values will be estimated for different future costs.

Capitalized cost: The present worth of the cost borne for maintaining and/or operating an asset/equipment throughout its design life.

Discount rate: The rate of interest based upon the investor's perception of the future value of money to calculate the present discounted rate of cost/benefit for the different elements of the project.

Nominal discount rate: Same as discount rate without considering the allowances to be made to the interest rate according to the variation in the purchasing power of the money (for e.g., due to inflation).

Real discount rate: The discount rate that is estimated by considering the variations due to the future value of money due to inflation.

Inflation: A continual rise in the level of different prices owing to the decrease in the value of money in circulation and credit relative to the available goods. In other words, it is the decrease in the purchasing power of the money.

21.2 Life Cycle Costing Sample Calculation

In order to move further with the LCC approaches, the initial assumptions are to be defined clearly. Among these assumptions, apart from those which require engineering experience and technical knowledge, few of them are financial assumptions that need discussion. Hence, this section is dedicated to the assertion of the different values to be used in the LCC process.

21.2.1 Consideration of Inflation and Discount Rates

The two most important assumptions that guides the LCC are the discount and the inflation rates. Adopting pragmatic values can enable the management to predict realistic values that help in decision making. However, inappropriate assumptions often lead to unrealistic figures which might make the selection process more tedious. For assessment of discount rates, current economic practices may be consulted. It is to be noted that the discount rate is not to be confused with the interest rates that is applicable in case of private investments and differs to a certain degree as it denotes the rate at which the monetary unit in a society loses its value on annual basis. The exact assessment of discount rate is a difficult job and requires some intensive research to get an accurate result. For convenience, in our calculations, a discount rate of 8–10% is considered. However, in real practice, economic experts might be consulted while assessing value of megaprojects. For assertion of interest rate, economic data for the last twenty to thirty years may be considered. Such data is available at a global as well as national level with world bank database. An average inflation rate estimation from such historical dataset can give an apt idea about the inflation rate to be adopted. For calculations in this chapter, an inflation rate of 4–5% is considered.

21.2.2 Determination of Present Value

Following the different described terminologies, the formulae common to both the PW and the AC method are described in this section. The subsequent section defines the method specific formulae for the PW and AC method. The most basic formula is the present worth factor (*PWF*) calculation for the discount rate i (%) and number of years t into future when the expenditure/cost is supposed to incur (Eqs. 21.1–21.4).

$$PWF_t = b^t \qquad (21.1)$$

$$\text{Where, } b = \frac{1}{(1 + \frac{i}{100})} \qquad (21.2)$$

$$\text{Hence, } PW_t = FV \times PWF_t \qquad (21.3)$$

$$\text{and, } PW = \sum PW_t \qquad (21.4)$$

The above approach requires the multiplication of the *PWF* with the future values (FV) of the items to estimate the individual present worth and subsequent present value analysis by summing these estimated present worth values.

For a uniform future annual expenditure, a uniform present worth factor (*UPWF*) can be evaluated to estimate the total present value as per Eq. 21.5.

$$UPWF = \frac{b(1 - b^t)}{(1 - b)} \qquad (21.5)$$

The total present worth for *t* years would hence be calculated directly by multiplying the future value by the *UPWF* factor (Eq. 21.6). Unlike the individual *PV* addition, the *UPWF* directly estimates the *PW*

$$PW = FV \times UPWF \qquad (21.6)$$

In case, the inflation rate *r* (%) is considered, the *PWF'* are calculated using a modified b_r (Eqs. 21.7) and 21.8 for PWF'.

$$b_r = \frac{1 + \frac{r}{100}}{1 + \frac{i}{100}} \qquad (21.7)$$

$$\text{Hence, } PWF' = b_r^t \qquad (21.8)$$

The individual modified present values (*PV'*) are estimated as per Eq. 21.9.

$$PV_t' = FV \times PWF_t' \qquad (21.9)$$

Correspondingly, the adjusted present worth (*PW'*) is estimated as the summation of the *PV'*

$$PW' = \sum PV_t' \qquad (21.10)$$

Similarly, a modified uniform present worth factor (*UPWF'*) and corresponding *PW'* are estimated as per Eq. 21.11 and Eq. 21.12, respectively.

$$UPWF' = \frac{b_r(1 - b_r^t)}{(1 - b_r)} \tag{21.11}$$

$$and\ PW' = FV \times UPFW' \tag{21.12}$$

Example: 21.1 Determine the present worth for an expenditure assuming a discount rate of 8% for a period of 4 years and a discount rate of 10% for a period of next 6 years using the PW method. The future costs for the expenditure can be assumed as 10,000 Indian Rupees per year (INR/year) for next ten years. Also calculate the present worth using the UPW factor for a uniform discount rate of 9%. Assume year 2020 as the baseline.

Solution

The present values for the individual future expenditures are estimated as given in Table 21.1 using Eqs. 21.1–21.4.

For example, for year 2022, b = $1/(1 + 0.08)$ = 0.926.

Hence, b^t = 0.926^2 = 0.857.

Therefore, the PV = FW $\times PWF_t'$ = $10,000 \times 0.8573$ = Rs. 8573.

Similarly estimating present value for all the years as presented in Table 21.1, after summation of present value for each year the present worth of Rs. 62,868 is obtained.

For assessment of b for a uniform discount rate of 9%

For finding the $UPWF$, first equivalent b is evaluated by using Eq. 21.2 and from the b value obtained the $UPWF$ is estimated by using Eq. 21.5.

Hence value of $b = \frac{1}{1+0.01 \times 9} = \frac{1}{1.09} = 0.917$

Subsequently, the value of $UPWF = \frac{0.917 \times (1 - 0.917^{10})}{1 - 0.917} = 6.4$

Table 21.1 Present values for the individual future expenditures

Year	t	i (%)	FV (INR)	b	b^t	Present value (INR)
2021	1	8	10,000	0.926	0.926	9,259.26
2022	2	8	10,000	0.926	0.857	8,573.39
2023	3	8	10,000	0.926	0.794	7,938.32
2024	4	8	10,000	0.926	0.735	7,350.30
2025	5	10	10,000	0.909	0.621	6,209.21
2026	6	10	10,000	0.909	0.564	5,644.74
2027	7	10	10,000	0.909	0.513	5,131.58
2028	8	10	10,000	0.909	0.467	4,665.07
2029	9	10	10,000	0.909	0.424	4,240.98
2030	10	10	10,000	0.909	0.386	3,855.43
Present worth, INR						62,868.28

Table 21.2 Present value for each individual future expenditure

Year	t	i (%)	r (%)	FV' (INR)	b_r	$b_r{}^t$	PV' (INR)
2021	1	8	4	10,000	0.963	0.963	9,629.63
2022	2	8	4	10,000	0.963	0.927	9,272.98
2023	3	8	4	10,000	0.963	0.893	8,929.53
2024	4	8	4	10,000	0.963	0.860	8,598.81
2025	5	10	6	10,000	0.964	0.831	8,309.33
2026	6	10	6	10,000	0.964	0.801	8,007.17
2027	7	10	6	10,000	0.964	0.772	7,716.00
2028	8	10	6	10,000	0.964	0.744	7,435.42
2029	9	10	6	10,000	0.964	0.717	7,165.04
2030	10	10	6	10,000	0.964	0.690	6,904.49
			Present worth, INR				81,968.40

Consequently, the present worth of uniform annual investment of 10,000 INR for a tenure of 10 years at a discount rate of 9% $= 6.4 \times 10,000 = 64,000$ INR.

Example: 21.2 Determine the present worth of the investment of 10,000 INR/year. A discount rate of 8% for a period of first 4 years and a discount rate of 10% for a period of next 6 years shall be considered; also, an inflation rate of 4% for first four years and 6% for the next six years shall be considered. What would be the *PW* if a uniform inflation rate of 7% and uniform discount rate of 9% is considered for the entire period of 10 years?

Solution
The present value for each individual future expenditure is estimated as given in Table 21.2 using Eqs. 21.7 and 21.9.

For finding the $UPWF'$, the equivalent b_r is evaluated by using Eq. 21.7 and using the value of b_r obtained, the $UPWF'$ is estimated by using Eq. 21.11.

Hence value of $b_r = \frac{1+0.01\times7}{1+0.01\times9} = \frac{1.07}{1.09} = 0.982$

Subsequently, the value of $UPWF = \frac{0.982\times(1-0.982^{10})}{1-0.982} = 9.061$

Consequently, the present worth of annual investment of 10,000 INR for a tenure of 10 years at a discount rate of 9% and uniform inflation rate of 7% $= 9.061 \times 10000 = 90,615$ INR.

Example: 21.3 A membrane module in a 500 m³/day capacity sewage treatment plant needs to be changed every 4 years for a design period of 20 years. The cost of this future expenditure is Rs. 3,000,000. Assuming the inflation rate to be 4.5% and the discount rate to be 8%, estimate the total present worth of this future expenditure for the entire design period.

Solution

Estimation of b_r as per Eq. 21.7.

$$b_r = \frac{1 + \frac{r}{100}}{1 + \frac{i}{100}} = \frac{1 + \frac{4.5}{100}}{1 + \frac{8}{100}} = 0.9676$$

Using this b_r, the value of corresponding PWF_t is estimated for the tth year using Eq. 21.8. For instance, the second membrane replacement will be on 8th year. Hence estimating the PWF_8 would be as shown below:

$$PWF_t = b_r^t$$

$$Or,\ PWF_8 = b_r^8 = 0.9676^8 = 0.7684$$

Further, using this PWF_t, the corresponding individual present worth is estimated for the replacement activity at the end of eight year.

$$PW_t = PWF_t \times 3{,}000{,}000$$

$$PW_8 = PWF_8 \times 3{,}000{,}000 = 0.7684 \times 3{,}000{,}000 = 2{,}305{,}200$$

Accordingly, the corresponding present worth for each replacement year is calculated and the total present worth is estimated as the summation of the present worth of the future expenditures occurring in 4th, 8th, 12th and 16th year (Table 21.3).

Example: 21.4 For the above membrane module, a suction pump of 3 HP capacity is operational for 20 h every day. Estimate the annual electricity consumption, and corresponding annual electricity cost for an electricity tariff of Rs. 10 per kW-h of electricity spent. What is the present worth of this expenditure if same inflation and discount rate (as stated in Example 21.3) is applicable?

Table 21.3 Present worth estimation for membrane module replacement

Year	b_r^t	Expenditure	PW_t, Rs.
4	0.8766	3,000,000	2,629,800.00
8	0.7684	3,000,000	2,305,200.00
12	0.6735	3,000,000	2,020,500.00
16	0.5904	3,000,000	1,771,200.00
$PW = \Sigma\ PW_t =$			8,726,700.00

Solution

Capacity of the pump $= 3$ HP $= \frac{3}{0.7457} = 2.2371$ kW.

Daily energy consumption $=$ power $\times$ time operated $= 2.2371 \times 20 = 44.7420$ kW-h.

Hence, annual energy consumption $= 44.7420 \times 365 = 16330.83$ kW-h.

Hence, annual cost of energy consumption for an energy tariff of Rs. 10/kW-h

$= 16330.83 \times 10 =$ Rs.163,308.30

The total present worth can be estimated either by taking the summation of the individual present worth of each expenditure in corresponding year or it can be directly estimated by using the uniform present worth factor (*UPWF*). Both approaches have been demonstrated here for better understanding of the reader.

Option—1: Estimation of total present worth by finding individual present worth for corresponding year

The present worth factors (b_r^t) are estimated for each year, from 1 to 20 using the estimated value of b_r (0.9676) as obtained in Example 21.3. The corresponding present worth for each year is found separately by multiplying the *PWFs* (b_r^t) with the annual cost of electricity expenditure (Rs.163,308.30). For example, for the 8th year, the b_r^8 is estimated as 0.7684 and the corresponding PW_8 is estimated as

$$PW_8 = b_r^8 \times 163,308.30 = 0.7684 \times 163,308.30 = \text{Rs.}125,486.10$$

Similarly, the individual present worth for each corresponding year is estimated and values are presented in Table 21.4.

Hence the present worth of annual energy charges spent over 20 years is $\approx$ Rs. 2,353,175.

Option—2: Estimation of total present worth by using the UPWF

As the expenditure is occurring every year, The *UPWF* factor is estimated for the entire period of 20 years using Eq. 21.11 by first estimating the value of b_r. The value of b_r is already estimated in Example 21.3 as 0.9676. Hence using this value further for the estimation of *UPWF*.

$$UPWF' = \frac{b_r(1 - b_r^t)}{(1 - b_r)} = \frac{0.9676 \times (1 - 0.9676^{20})}{1 - 0.9676} = 14.4092$$

The total present worth can be directly estimated by using this *UPWF* value

$$PW = UPWF' \times 163,308.30 = 14.4092 \times 163,308.30$$

$$= \text{Rs.}2,353,141.96 \approx \text{Rs.}2,353,142$$

Table 21.4 Present worth estimation for cost of electricity

Year	b_r^t	Electricity cost	PW_t
1	0.9676	163,308.30	158,017.11
2	0.9362	163,308.30	152,889.23
3	0.9059	163,308.30	147,940.99
4	0.8766	163,308.30	143,156.06
5	0.8482	163,308.30	138,518.10
6	0.8207	163,308.30	134,027.12
7	0.7941	163,308.30	129,683.12
8	0.7684	163,308.30	125,486.10
9	0.7435	163,308.30	121,419.72
10	0.7194	163,308.30	117,483.99
11	0.6961	163,308.30	113,678.91
12	0.6735	163,308.30	109,988.14
13	0.6517	163,308.30	106,428.02
14	0.6306	163,308.30	102,982.21
15	0.6102	163,308.30	99,650.72
16	0.5904	163,308.30	96,417.22
17	0.5713	163,308.30	93,298.03
18	0.5527	163,308.30	90,260.50
19	0.5348	163,308.30	87,337.28
20	0.5175	163,308.30	84,512.05
$PW = \Sigma\ PW_t =$			2,353,174.62

It can be observed that the present worth values estimated from both methods are similar. However, the *UPWF* method of present worth estimation is recommended for different regular and continuous future expenditures as it is facile, faster and user friendly.

21.3 Cost Comparison of STP Using Different Secondary Treatment Technologies

21.3.1 Description of Sewage Treatment Plant

The model STP considered in this case study has a sewage treatment capacity of 1 million litres per day (MLD). The plant uses a combination of physical operations, secondary biological treatment and tertiary advanced oxidation and filtration for treating sewage to produce a non-potable reuse quality treated water. As given in the schematic below

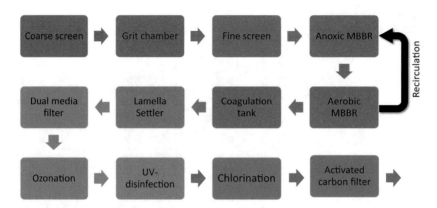

Fig. 21.1 Chosen treatment scheme for life cycle costing with two stage MBBR

(Fig. 21.1), the raw sewage is pumped from the collecting sump to the coarse screen, subsequent to which a grit chamber and a fine screen unit are provided for the removal of floating matter and grit particles from the sewage. The effluent from the primary physical operations enters an anoxic moving bed biofilm reactor (MBBR) followed by an aerobic MBBR. The effluent from the aerobic MBBR is recirculated to the upstream anoxic MBBR for denitrification to remove nitrogen using the concept of Ludzack-Ettinger process.

The effluent from the aerobic MBBR is sent to the coagulation unit, wherein provision of chemical dosing arrangement is made. The effluent from the coagulation tank is passed through the baffled flocculator and then to the lamella settler for settling of the suspended solids. The effluent from lamella settler is fed to the dual media filter (DMF) feed tank, from which the water is pumped under pressure flow condition in the DMF. The DMF effluent is stored in the ozone contact tank, wherein ozonation is achieved using ozone generator of three oxygen concentrators with a combined ozone generation capacity of 150 g/h. The ozonised water is further treated by UV disinfection and stored in chlorine contact tank, wherein liquid chlorine in the form of hypochlorite is added. The chlorinated effluent is further pumped through an activated carbon filter to remove the residual pollutants and disinfection by-products, if any. The flow sequence for the 1000 m^3/day capacity STP is given in Fig. 21.1.

The LCC for the described plant is being carried out by finding the present worth of different expenditures. The future expenditures are classified under the different heads of annual maintenance cost, manpower salaries, consumables, and future costs of one-time nature. The future costs represent the investment required for replacement of a particular component of a treatment plant at end of its life span. Corresponding one-time present worth for replacement of these components is taken into account in this one-time future cost. For ease of understanding, the assumed discount and the inflation rate are fixed at 8% and 5%, respectively. Corresponding to these figures, the *PWF* and *UPWF* are calculated for twenty consecutive years (Table 21.5).

Table 21.5 *PWF* values for corresponding years for 1–20 years

Year	PWF	Year	PWF	Year	PWF
1	0.9722	8	0.7981	15	0.6551
2	0.9452	9	0.7759	16	0.6369
3	0.9189	10	0.7543	17	0.6192
4	0.8934	11	0.7334	18	0.6020
5	0.8685	12	0.7130	19	0.5853
6	0.8444	13	0.6931	20	0.5690
7	0.8209	14	0.6739		

21.3.2 Service Life of Different Components

The service life of different components and recurring expenditures in terms of consumables have been ascertained in this section. Overall, the design life of the plant is assumed to be 20 years, while the service life of different components considered based on the manufacturer specification is given in Table 21.6. This service life is in addition to the incidental breakdown and repair of these equipment that need to be catered to on a regular basis and it is considered under the head of annual maintenance and repair.

21.3.3 Annual Maintenance, Repair and Consumable Cost Assumptions

The monthly minor repair is considered as Rs. 8000 for plant maintenance including cost of minor repair and incidental spares. The lump sum cost of painting is considered as Rs. 1,50,000 after every two years. Furthermore, the third-party lab analysis cost for ascertaining treatment performance of the plant is taken up as Rs. 12,000 per month. Since the chosen technology is two stage MBBR, hence loss of media also has to be taken into account. The reactor volume required for an assumed HRT of 2 h and 4 h for the anoxic and the aerobic MBBR, respectively, for an hourly flow of 41.67 m³ would be:

$$V_{AN} = Q \times HRT_{AN} = 41.67 \times 2 = 83.33\,\text{m}^3$$

$$V_{AERO} = Q \times HRT_{AERO} = 41.67 \times 4 = 166.68\,\text{m}^3$$

Hence, assuming a media volume of 25% in the anoxic and aerobic MBBR, the total media volume required would be:

$$V_{media} = \frac{25}{100} \times (V_{AN} + V_{AERO}) = 62.5\,\text{m}^3$$

Table 21.6 Serviceable life considered for different components of 1.0 MLD STP

Sl. No	Item	Service life (years)
1	Raw sewage pump	10
2	Flocculator	10
3	Air Blower	10
4	MBBR recirculation pump	10
5	Sludge recirculation pump	10
6	DMF feed pump	10
7	ACF feed pump	10
8	Mixer for poly dosing tank	10
9	Basket centrifuge	10
10	Poly dosing pump	10
11	HOCl dosing pump	10
13	Oxygen concentrator (3) + ozone generator (1)	10
12	UV module along with power management system	10
13	Area lighting	10
14	Online Monitoring System	10
15	UV-C module lamps	1.23
16	UV-C module quartz tube	3
17	DMF sand and gravel	4
18	ACF media	2
19	MBBR media	5
20	Paint sustenance	2

For a yearly wastage of 10% of the total media volume, the media volume to be replaced every year is 6.25 m^3. Hence, the typical media replacement cost including 5% carrying cost for a rate of 10,000 INR/m^3 of media would be

$$Cost\ of\ media\ replacement = 6.25 \times 10000 \times 1.05 = Rs.\ 65{,}625$$

For estimation of monthly chlorine dosing cost, the estimated Cl_2 requirement with an initial chlorine demand of 4 mg/L (or 4 g/m^3) would be:

$Cl_2 requirement = 1000 \times 4 = 4000\ g = 4.0\ kg/day.$

Hence, for a 30% available chlorine content and a rate of Rs. 30/kg the daily cost of bleaching powder required $= \frac{4}{0.30} \times 30 = Rs.\ 400$ per day.

Hence, the monthly cost of bleaching powder would be Rs. $400 \times 30 = Rs.\ 12{,}000$

21.3.4 Assessment of Future One-Time Costs

The majority future one-time costs in this case study relates to the replacement of equipment including blowers, pumps, dosing units, ozonation units and UV-C modules. For assessment of the present worth of each equipment, the amortization factor is estimated and multiplied with the future cost. The one-time cost of replacement of the different equipment is as given in Table 21.7. For example, in case of raw sewage pump, the future cost is Rs. 210,000. The final future cost including transportation charges would hence be:

$$210,000 + 210,000 \times \frac{5}{100} = \text{Rs.}\, 220,500$$

The *PWF* is estimated corresponding to the time of replacement (investment) i.e., $t = 10$ years, and discount rate of 8% and inflation rate of 5%.

$$b_r = \frac{1 + 0.01 \times 5}{1 + 0.01 \times 8} = \frac{1.05}{1.08} = 0.9722$$

For 10th year expenditure the present worth would be Rs. $220,500 \times 0.9772^{10}$

Table 21.7 Cost of equipment for one-time future replacement

Description	Present cost of equipment (Rs.)	Cost including a transportation cost of 5% (Rs.)
Raw sewage pump	210,000.00	220,500.00
Flocculator	76,700.00	80,535.00
Air blower	564,210.00	592,420.50
MBBR recirculation pump	115,000.00	120,750.00
Sludge recirculation pump	115,000.00	120,750.00
DMF feed pump	158,400.00	166,320.00
ACF feed pump	158,400.00	166,320.00
Mixer for poly dosing tank	29,500.00	30,975.00
Basket centrifuge	159,300.00	167,265.00
Poly dosing pump	61,360.00	64,428.00
HOCl dosing pump	92,040.00	96,642.00
Oxygen concentrator (3) + ozone generator (1)	324,500.00	340,725.00
UV	300,000.00	315,000.00
Online monitoring system	873,200.00	916,860.00

Table 21.8 Present worth of investments for the one-time replacement

Description	Landed cost including transportation (Rs.)	$PW_t = PWF_t \times$ Landed cost (Rs.)
Raw sewage pump	220,500.00	166,323.15
Flocculator	80,535.00	60,747.55
Air blower	592,420.50	446,862.78
MBBR recirculation pump	120,750.00	91,081.73
Sludge recirculation pump	120,750.00	91,081.73
DMF feed pump	166,320.00	125,455.18
ACF feed pump	166,320.00	125,455.18
Mixer for poly dosing tank	30,975.00	23,364.44
Basket centrifuge	167,265.00	126,167.99
Poly dosing pump	64,428.00	48,598.04
HOCl dosing pump	96,642.00	72,897.06
Oxygen concentrator + ozone generator	340,725.00	257,008.87
UV	315,000.00	237,604.50
Online monitoring system	916,860.00	691,587.50
Total PW for the one-time expense		2,564,235.68

$$= 220500 \times 0.7543 = \text{Rs. } 166,323.15$$

Similar to the above example, the present worth (PW_t) for each equipment is estimated, which is presented in Table 21.8.

21.3.5 Manpower and Electricity Cost Calculations

For manpower, two operators are considered in semi-skilled category, and one is considered in skilled category in addition to one security guard for night watch with a monthly salary of Rs. 12,000. The monthly remuneration for semi-skilled and skilled personal is considered as Rs. 15,000, and Rs. 21,000, respectively. Hence the yearly manpower cost (MP_1) would be:

$$MP_1 = (15,000 \times 2 + 21000 + 12000) \times 12 = 63000 \times 12 = Rs.\,756,000$$

The cost of electricity tariff is considered as Rs. 10 per kWh. The electrical load considering operation at full load condition of the equipment is presented in Table 21.9. Using this energy consumption, the energy cost per annum would be:

$$= 446123.16 \times 10 = Rs.\,4,461,231.60$$

Corresponding to the annual estimated expenditures, the present worth of the total cost incurred towards manpower and electricity component is estimated by multiplying the annual expenditures with the *UPWF* value. The *UPWF* is estimated by using Eq. 21.11.

Table 21.9 Equipment load data and corresponding energy consumption

Description	power required (HP)	Hours operated	Energy (HP-hour)
Raw sewage pump	10	24	240
Flocculator	0.5	12	6
Air blower	40	18	720
MBBR recirculation pump	7.5	12	90
Sludge recirculation pump	7.5	1	7.5
DMF feed pump	12.5	20	250
ACF feed pump	12.5	20	250
Mixer for poly-electrolyte dosing tank	1	8	8
Basket centrifuge	2	0.285	0.571
Poly-electrolyte dosing pump	0.25	12	3
HOCl dosing pump	0.25	16	4
Oxygen concentrator + Ozone generator	2	20	40
UV	0.75	20	15
Online monitoring system	0.25	20	5
Total energy consumed in HP-h			1639.07
Total energy consumed in kWh			1222.26
Total energy consumed in kW-h per annum			446,123.16

$$UPWF' = \frac{b_r(1 - b_r^t)}{(1 - b_r)} = \frac{0.9722 \times (1 - 0.9722^{20})}{1 - 0.9722} = 15.0726$$

$$PW_{electricity} = 4,461,231.60 \times 15.0726 = Rs.\,67,242,360.06$$

$$PW_{manpower} = 756,000 \times 15.0726 = Rs.\,11,394,885.60$$

For periodic expenditures such as UV lamp, DMF and ACF media replacement, the PWF and the corresponding PV are estimated. For example, in case of UV light, the cycle is 1.23 years. The number of replacements for a plant design period of twenty years would be 16. The total cost of UV lamp is estimated as below considering the cost of individual lamp as Rs. 4,500. Thus, the corresponding expenditure for 24 lamps with a carrying cost of 5% will be:

$$Cost = Rs.\,4500 \times 24 = Rs.\,108,000$$

Transportation cost @ 5% of total cost $= Rs.108,000 \times \frac{5}{100} = Rs.\,5,400$
Total procurement cost $= Rs.\,108,000 + Rs.\,5,400 = Rs.\,113,400.$
The present value is thus estimated as future procurement cost $\times PWF$.
For example, for cycle 5, the present worth factor would be estimated as:

$$b_r = \frac{1 + \frac{5}{100}}{1 + \frac{8}{100}} = 0.9722$$

Hence, for 5th cycle corresponding to 6.1645 years (i.e., $t = 6.1645$), the b_r^t will be:

$$b_r^t = b_r^{6.1645} = 0.9722^{6.1645} = 0.8405$$

Thus, for cycle 5th the present value will be $= b_r^t \times 113,400$

$$= 0.8405 \times 113,400 = Rs.\,95,312.70$$

The present value of the future procurement costs would be as given in Table 21.10.
The DMF media is to be replaced after every 4 years for which the total sand requirement will be 7200 kg.
Total cost of sand required in one cycle @ Rs. 8.16/kg of sand (including 2% carrying cost) $= 7200 \times 8.16 = Rs.\,58,752$ per replacement after 4 years.
Total gravel requirement is 4800 kg. Hence, total cost of gravel required in one cycle @ Rs. 8.16/kg of gravel (including 2% carrying cost) $= 4800 \times 5.10 = Rs.\,24,480$ per replacement after 4 years. Hence, PV of future sand replacement using corresponding PWF and total PW is estimated for the two items as illustrated in Table 21.11.
For the replacement of ACF media and painting a cycle of 2 years is considered. The cost of commercial activated carbon is Rs. 46 per kg. The total AC requirement is 900 kg.

Table 21.10 UV lamp estimated present values

No. of cycles	Year (t)	$b_r{}^t$	PV (Rs.)
1	1.2329	0.9658	1,09,521.72
2	2.4658	0.9328	1,05,779.52
3	3.6987	0.9010	1,02,173.40
4	4.9316	0.8702	98,680.68
5	6.1645	0.8405	95,312.70
6	7.3974	0.8118	92,058.12
7	8.6303	0.7840	88,905.60
8	9.8632	0.7572	85,866.48
9	11.0961	0.7314	82,940.76
10	12.3290	0.7064	80,105.76
11	13.5619	0.6822	77,361.48
12	14.7948	0.6589	74,719.26
13	16.0277	0.6364	72,167.76
14	17.2606	0.6147	69,706.98
15	18.4935	0.5937	67,325.58
16	19.7264	0.5734	65,023.56
	$PW = \Sigma$ Present value $=$		1,367,649.36

Table 21.11 Present values for DMF sand and gravel replacement

Year of replacement	PWF	DMF sand (Rs.)	DMF gravel (Rs.)
4	0.8934	52,489.04	21,870.43
8	0.7981	46,889.97	19,537.49
12	0.7130	41,890.18	17,454.24
16	0.6369	37,419.15	15,591.31
$PW = \Sigma PV =$		1,78,688.34	74,453.47

Hence the cost of AC is $= 46 \times 900 =$ Rs. 41,400. Considering a transportation charge of 5% of the total cost, the final landed cost of AC $= 41,400 \times 1.05 =$ Rs. 43,470. For painting expenditure, a lump sum cost of Rs. 150,000 is considered.

The estimated PW_t for corresponding years of expenditure for the given PWF obtained from Table 21.5 is as given in Table 21.12.

The cost of quartz tube is considered as Rs. 2500 per tube and cost for 24 quartz tubes is estimated. The cost of one future replacement including 5% transportation charges is hence estimated as:

Table 21.12 Present values for ACF replacement and painting

Year of replacement	PWF	ACF media (Rs.)	Painting (Rs.)
2	0.9452	41,087.84	1,41,780.00
4	0.8934	38,836.10	1,34,010.00
6	0.8444	36,706.07	1,26,660.00
8	0.7981	34,693.41	1,19,715.00
10	0.7543	32,789.42	1,13,145.00
12	0.7130	30,994.11	1,06,950.00
14	0.6739	29,294.43	1,01,085.00
16	0.6369	27,686.04	95,535.00
18	0.6020	26,168.94	90,300.00
$PW = \Sigma\ PW_t =$		298,256.36	1,029,180.00

$$2500 \times 24 \times 1.05 = Rs.\ 63,000$$

Correspondingly, the present values are estimated using the respective *PWF*s as presented in Table 21.13.

The calculation for general maintenance, chlorine dosing and MBBR media replacement is of annual nature. For these expenditures, a $UPWF = 15.0759$ is estimated, as explained earlier, and the corresponding total present worth of expenditure is estimated as described previously in Eq. 21.12.

For chlorine dosing, the estimated present worth is $12,000 \times 12 \times 15.0726 =$ Rs. 2,170,454.40

For MBBR media, the estimated present worth is $65,625 \times 15.0726 = Rs.\ 989,139.38$

For general maintenance including technical troubleshooting (@ 8000 per month, the estimated present worth is, $8000 \times 12 \times 15.0726 = Rs.\ 1,446,969.60$.

For lab analysis, the estimated present worth is $12,000 \times 12 \times 15.0726 =$ Rs. 2,170,454.40

Table 21.13 Present values for Quartz tube replacement

Year of replacement	PWF	Present values (Rs.)
3	0.9189	57,890.70
6	0.8444	53,197.20
9	0.7759	48,881.70
12	0.7130	44,919.00
15	0.6551	41,271.30
18	0.6020	37,926.00
$PW = \Sigma\ PW_t =$		284,085.90

21.3.6   Estimation of Total Present Worth of 1 MLD Capacity STP

The total present worth that will be incurred owing to the capital investment and annual expenditure as well as periodic expenditures is estimated. The capital cost for the STP is estimated at Rs. 18,000,000. The salvage value at the end of the operation period is considered as 10% of the capital expenditure. Hence the future cost of the salvage value is = Rs. 18,000,000 × 10/100 = Rs. 1,800,000. This salvage value has to be subtracted from the total expenditure. Using the values obtained above, the total capitalized operating cost of the plant and correspondingly, the cost of wastewater treatment per kilo-litre is estimated as presented in Table 21.14. Thus, with these calculations the estimated life cycle cost for 1 m^3 of sewage to produce reusable quality treated water suitable for any non-potable use is Rs. 15.21. Thus, this will be unit life cycle cost of the sewage treatment estimated for the plant with present economic assumptions for a capacity of 1 MLD and using anaerobic and aerobic MBBRs as secondary biological wastewater treatment to facilitate removal of organic matter and nitrogen.

21.4   Case Studies Considering Different Secondary Treatment Technologies for STP

In continuation to the above cost estimation, three alternate technologies are further selected as secondary biological processes to demonstrate a comparative cost analysis. Such comparative LCC is pre-requisite for any engineering project for rendering an idea to the management pertaining to the affordability and suitability of the different alternatives. In this particular case, extended aeration activated sludge process (EA-ASP), aerobic sequencing batch reactor (SBR) and aerobic membrane bio-reactor (MBR) technologies are chosen to compare the cost with the already described MBBR technology. The associated capital cost of the chosen technologies is described herein to arrive at the final capitalized operating cost (Table 21.15).

21.4.1   Estimation of Life Cycle Costing for MBR as Secondary Treatment

In addition to the capital cost, technology such as aerobic MBR requires consideration of additional operating cost expenditures in the form of suction pump for membrane filtration and periodic membrane replacement cost. Typical membrane module cost for 1 MLD capacity STP would be around Rs. 6,000,000 with a replacement period of 4 years. The required suction pump capacity is considered as 4 HP. Since the MBR has effluent with negligible suspended solids, hence, the cost of mixer for poly-electrolyte dosing tank and poly-electrolyte dosing pump for the one-time replacement would not be applicable.

Table 21.14 Estimation of present worth of sewage treatment plant and cost of treatment with MBBR as a secondary treatment

Sl. No	Item description	Present worth (Rs.)
1	Man power cost	11,394,885.60
2	Electricity cost	67,242,360.06
3	General maintenance	1,446,969.60
4	ACF media replacement	298,256.36
5	DMF sand replacement	178,688.34
6	DMF gravel replacement	74,453.47
7	UV lamp replacement	1,367,649.36
8	UV quartz replacement	284,085.90
9	Cost of pump replacement considering a pump and motor design life of 10 years	2,564,235.68
10	Cost of painting after every two years	1,029,180.00
11	Bleaching powder cost	2,170,929.60
12	MBBR media replacement cost	989,139.38
13	Lab analysis cost	2,170,454.40
14	Contractor's profit @ 15% for Sl. 1,3–13	3,595,339.15
15	Total operating cost (Sum Sl. 1–14)	94,806,626.91
16	Capital cost	18,000,000.00
17	Salvage value	1,800,000.00
18	Total cost (15 + 16 − 17)	111,006,626.91
19	Total water treated in 20 years, m^3	7,300,000
20	Cost of water treatment Rs. per m^3	15.21

Table 21.15 Capital cost for different chosen technologies

Sl. No	Chosen technology	Capital cost for 1 MLD STP (Rs.)
1	Aerobic MBR	40,000,000
2	EA-ASP	21,200,000
3	Aerobic SBR	25,000,000

Note The costs considered are provided by MM Enviro Projects Pvt. Ltd., Nagpur, India

Instead, a suction pump of 4 HP as mentioned above is apt for this treatment alternative. Hence, the corresponding changes in the equipment cost is presented in Table 21.16.

The cost of the 4 HP suction pump is considered as Rs. 105,000 including 5% transportation cost. The present worth for all equipment is estimated by multiplying the PWF_{10}

Table 21.16 Present worth of investments for the one-time replacement for employing MBR as a secondary treatment process

Description	Future landed cost (Rs.)	$PW = PWF_{10} \times$ landed cost (Rs.)
Raw sewage pump	220,500.00	166,323.15
Air blower	592,420.50	446,862.78
Centrifuge feed pump	105,000.00	79,201.50
Basket centrifuge	167,265.00	126,167.99
DMF feed pump	166,320.00	125,455.18
ACF feed pump	166,320.00	125,455.18
Suction pump for membrane	105,000.00	79,201.50
HOCl dosing pump	96,642.00	72,897.06
Oxygen concentrator (3) + Ozone generator (1)	340,725.00	257,008.87
UV	315,000.00	237,604.50
Online monitoring system	916,860.00	691,587.50
Total PW for the one-time expense		2,407,765.21

with the equipment cost. Finally, the total present worth of equipment replacement is estimated as the summation of the present worth for different equipment.

Apart from the altercation in the equipment replacement expenditure, the energy requirement calculations are also altered in this case. The modified energy consumption and annual energy expenditure is given in Table 21.17.

Hence, the corresponding annual energy cost for Rs. 10 per kW-h would be

$$= 451275.27 \times 10 = Rs.\,4{,}512{,}752.70$$

The corresponding total present worth of energy expenditure for twenty years would be

$$4{,}512{,}752.70 \times 15.0726 = Rs.\,68{,}018{,}916.35$$

Using the modified present worth of electricity cost and the equipment replacement, the capitalized cost of operation and the corresponding cost of wastewater treatment per kilo litre is estimated for the MBR as a secondary treatment alternative (Table 21.18). The energy and the maintenance expenditure in case of MBR as secondary treatment process is high as compared to the other chosen alternatives as shown in the subsequent calculations. Apart from the operating cost, the cost of installation for the MBR is also higher than the other secondary biological treatment process alternatives. In this case, the capital cost is considered as Rs. 40,000,000. The corresponding salvage value is estimated as 10% of the capital cost, i.e., Rs. 4,000,000.

Table 21.17 Equipment load data and corresponding energy consumption for MBR as a secondary treatment process

Description	Power required (HP)	Hours operated	Energy expenditure (HP-hour)
Influent sewage pump	10	24	240
Air blower	40	18	720
Centrifuge feed pump	5	12	60
Basket centrifuge	2	1	2
DMF feed pump	12.5	20	250
ACF feed pump	12.5	20	250
Suction pump for membrane	4	18	72.0
HOCl dosing pump	0.25	16	4
ozone concentrator (3) + ozone generator (1)	2	20	40
UV	0.75	20	15
Online monitoring system	0.25	20	5
Total energy consumed in HP-h			1658.00
Total energy consumed in kW-h per day			1236.37
Total energy consumed in kW-h per annum			451,275.27

21.4.2 Estimation of Life Cycle Costing for EA-ASP as Secondary Treatment

In case of EA-ASP, a capital cost of Rs. 21,200,000.00 is considered including an anoxic MBBR downstream. Unlike the treatment sequence for first treatment option with recirculation of nitrified effluent to the upstream anoxic MBBR (Fig. 21.1), in this case, the nitrified effluent from the extended aeration ASP is applied to a downstream anoxic MBBR with a detention time of 2 h. In this case, sludge from the EA-ASP is proposed to be used as the carbon source for denitrification. Also, the time for aeration is 23 h as compared to previous two cases wherein, the time of operation is 18 h. In this case also, the sludge generation is lower (owing to higher solids retention time) and the sludge that is generated has good settleability. Hence the poly-electrolyte dosing arrangement is not required here. Instead of recirculation pump as considered in case of two-stage MBBR, here a sludge feed pump to feed the settled sludge from the lamella clarifier to the 2nd stage anoxic MBBR is considered (Fig. 21.2).

The media replacement cost for the downstream anoxic MBBR is to be estimated. The media volume is typically 25% of the reactor volume. For estimating the loss of media (since 10% of total media volume is reduced every year), the reactor volume has to be

Table 21.18 Estimated cost of treatment for 1 MLD capacity STP using MBR as secondary treatment process

Sl. No	Item description	Present worth (Rs.)
1	Man power cost	11,394,885.60
2	Electricity cost	68,018,916.35
3	General maintenance	1,446,969.60
4	ACF media replacement	298,256.36
5	DMF sand replacement	178,688.34
6	DMF gravel replacement	74,453.47
7	UV lamp replacement	1,367,649.36
8	UV quartz replacement	284,085.90
9	Cost of pump replacement considering a pump and motor design life of 10 years	24,07,765.21
10	Cost of painting every two years	1,029,180.00
11	Bleaching powder cost	2,170,454.40
12	Membrane replacement cost	18,248,400.00
13	Lab analysis cost	2,170,454.40
14	Contractor's profit @ 15% for Sl. 1, 3–13	6,160,686.40
15	Total operating cost (Sum Sl. 1–14)	115,250,845.38
16	Capital cost	40,000,000.00
17	Salvage value	4,000,000.00
18	Total cost (15 + 16 − 17)	151,250,845.38
19	Total water treated in 20 years	7,300,000
20	Cost of water treatment Rs. per kL	20.72

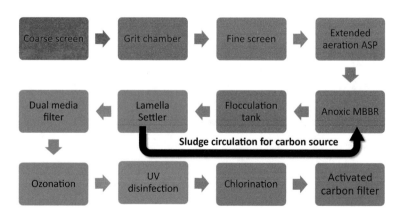

Fig. 21.2 Chosen treatment scheme for life cycle costing with EA-ASP

determined. The flow of the designed STP is 1000 m³/day i.e., 41.667 m³/h. Hence, with a hydraulic retention time of 2 h, the volume of the anoxic MBBR is:

$MBBR_{vol} = 41.667 \times 2 = 83.334$ m³.

Hence, the total media volume is $MBBR_{vol} \times \frac{25}{100} = 83.334 \times \frac{25}{100} = 20.8335$ m³.

Hence, loss of media every year $20.8335 \times \frac{10}{100} = 2.0834$ m³.

Hence annual cost of media replacement considering landed cost of media as Rs. 10,500

$$2.0834 \times 10500 = Rs.\,21,875.18$$

Hence, the total cost of media replacement using the $UPWF_{20}$ would be:

$$Rs.21,875.18 \times 15.0726 = Rs.\,329,715.84$$

The equipment details, its landed cost while replacement and the corresponding present worth of each equipment is given in Table 21.19.

Further the changes in the energy cost for the different operational hours of blower and altercations in equipment requirement is considered in Table 21.20 for estimating the energy cost.

Hence, considering energy cost for Rs. 10 per kWh the annual energy cost would be

$$= 479582.04 \times 10 = Rs.\,4,795,820.40$$

Table 21.19 Present worth of investments for the one-time replacement for employing EA-ASP as a secondary treatment process

Description	Future landed cost (Rs.)	$PW = PWF_{10} \times$ landed cost (Rs.)
Raw sewage pump	220,500.00	166,323.15
Air blower	592,420.50	446,862.78
Sludge recirculation pump	120,750.00	91,081.73
DMF feed pump	166,320.00	125,455.18
ACF feed pump	166,320.00	125,455.18
Basket centrifuge	167,265.00	126,167.99
HOCl dosing pump	96,642.00	72,897.06
Oxygen concentrator (3) + Ozone generator (1)	340,725.00	257,008.87
UV	315,000.00	237,604.50
Online monitoring system	916,860.00	691,587.50
Total PW for the one-time expense		2,340,443.94

Table 21.20 Equipment load data and corresponding energy consumption for EA-ASP as a secondary treatment process

Description	power required (HP)	Hours operated	energy expenditure (HP-h)
Raw sewage pump	10	24	240
Air blower	40	23	920
Sludge feed pump	3	12	36
Basket centrifuge	2	1	2
DMF feed pump	12.5	20	250
ACF feed pump	12.5	20	250
HOCl dosing pump	0.25	16	4
Oxygen concentrator (3) + Ozone generator (1)	2	20	40
UV	0.75	20	15
Online monitoring system	0.25	20	5
Total energy consumed in HP-h			1762.00
Total energy consumed in kW-h			1313.92
Annual energy consumption			479,582.04

The corresponding total present worth of electricity expenditure for twenty years would be

$$4,795,820.40 \times 15.0726 = \text{Rs.}\, 72,285,482.71$$

Hence, with the above altercations in the cost, the operating cost and the capitalized operating cost while using extended aeration as a secondary treatment option is demonstrated in Table 21.21. The salvage value is considered as 10% of the capital cost, i.e., Rs. 2,120,000. Thus, for EA as secondary treatment the unit life cycle cost of sewage treatment is estimated to be Rs. 16.15. This cost is Rs. 15.21 for using MBBR as secondary treatment and Rs. 20.72 while using MBR as a secondary treatment technology.

21.4.3 Estimation of Life Cycle Costing for SBR as Secondary Treatment

As a fourth option, SBR technology is chosen as a secondary treatment. The SBR operation yields sludge with low SVI and hence the requirement of poly-electrolyte dosing

Table 21.21 Estimated cost of treatment for extended aeration secondary process

Sl. No	Item description	Present worth (Rs.)
1	Man power cost	11,394,885.60
2	Electricity cost	72,285,482.71
3	General maintenance	1,446,969.60
4	ACF media replacement	298,256.36
5	DMF sand replacement	178,688.34
6	DMF gravel replacement	74,453.47
7	UV lamp replacement	1,367,649.36
8	UV quartz replacement	284,085.90
9	Cost of pump replacement considering a pump and motor design life of 10 years	2,340,443.94
10	Cost of painting every two years	1,029,180.00
11	Bleaching powder cost	2,170,929.60
12	Lab analysis cost	2,170,454.40
13	Media replacement cost for Anoxic MBBR	329,715.84
14	Contractor's profit @ 15% for Sl. 1, 3–13	3,462,856.86
15	Total operating cost (Sum Sl. 1–14)	98,834,051.98
16	Capital cost	21,200,000.00
17	Salvage value	2,120,000.00
18	Total cost (15 + 16 − 17)	117,914,051.98
19	Total water treated in 20 years	7,300,000
20	Cost of water treatment Rs. per kL	16.15

is redundant in this case as well. In case of SBR, for facilitating denitrification, a recirculation arrangement is provided. In addition, a motor arrangement for the decanter is considered. The decanter and the motor are of high-quality SS and 20 years of operational time is considered for this. However, the motor operation contributes to the energy consumption calculations and it is considered in that section. The capital cost with SBR as secondary treatment process is considered as Rs. 25,000,000.00, while the salvage value is estimated as 10% of the capital expenditure as Rs. 2,500,000.00. The equipment list and corresponding present worth of future replacement is as given in Table 21.22. The present worth for each equipment is estimated by multiplying the PWF_{10} (= 0.7543) with the landed cost.

The alterations in the energy consumption for the different operational hours of blower and altercations in equipment requirement as considered is presented in Table 21.23 for estimating the energy cost.

Hence, the corresponding annual energy cost @ Rs. 10 per kW-h would be:

Table 21.22 Present worth of investments for the one-time replacement for employing SBR as a secondary treatment process

Description	Future landed cost (Rs.)	$PW = PWF_{10} \times$ landed cost (Rs.)
Raw sewage pump	220,500.00	166,323.15
Air blower	592,420.50	446,862.78
SBR recirculation pump	120,750.00	91,081.73
DMF feed pump	166,320.00	125,455.18
ACF feed pump	166,320.00	125,455.18
Basket centrifuge	167,265.00	126,167.99
HOCl dosing pump	96,642.00	72,897.06
Oxygen concentrator (3) + Ozone generator (1)	340,725.00	257,008.87
UV	315,000.00	237,604.50
Online monitoring system	916,860.00	691,587.50
Total PW for the one-time expense		2,340,443.93

$$= 486658.73 \times 10 = Rs.\,4,866,587.34$$

The corresponding total present worth of electricity expenditure for twenty years would be

$$4,866,587.34 \times 15.0726 = Rs.\,73,352,124.34$$

Hence, with the above altercations in the cost, the operating cost and the capitalized operating cost, while using SBR as a secondary treatment option, is demonstrated in Table 21.24.

Thus, while using SBR as secondary treatment technology the life cycle cost of per unit of sewage treatment is estimated to be Rs. 16.72. Whereas, this cost was Rs. 15.21 for using MBBR as secondary treatment, Rs. 20.72 while using MBR as a secondary treatment technology and Rs. 16.15 while using extended aeration ASP as secondary treatment choice. The above comparative estimation demonstrates that the cost of treatment varies owing to the selection of the technology. In the present comparison, the technologies are compared from economic aspect without giving any weightage to the efficiency of the process. For a real field project, the final decision making should also incorporate the performance efficiency of the chosen systems to make a more pragmatic decision.

Table 21.23 Equipment load data and corresponding energy consumption for SBR as a secondary treatment process

Description	Power required (HP)	Hours operated	Energy expenditure (HP-hour)
Influent sewage pump	10	24	240
Air blower	40	20	800
SBR internal recirculation pump (100% recirculation)	7.5	20	150
DMF feed pump	12.5	20	250
ACF feed pump	12.5	20	250
Basket centrifuge	2	1	2
HOCl dosing pump	0.25	16	4
Motor for decanting arrangement	2	16	32
Oxygen concentrator (3) + Ozone generator (1)	2	20	40
UV	0.75	20	15
Online monitoring system	0.25	20	5
Total energy consumed in HP-h			1788.00
Total energy consumed in kW-h			1333.31
Annual energy consumption, kW-h			486,658.73

21.5   Summary

The cost analysis undertaken in this chapter aims to introduce the first-time reader to the different aspects of LCC. The chapter describes the different terminologies relevant to LCC and further demonstrates the estimation of present values, present worth and amortization factors with suitable examples. The chapter also describes the costing for a hypothetical STP and compares the changes in cost of treatment with different secondary treatment combinations as a part of a comparative analysis. The values assumed in the chapter have been judiciously selected to be present market relevant to render a fair comparison and provide lucid idea to the reader about the cost of sewage treatment per kilo litre.

Table 21.24 Estimated cost of treatment for 1 MLD capacity STP comprising of aerobic SBR as secondary treatment process

Sl. No	Item description	Present worth (Rs.)
1	Man power cost	11,394,885.60
2	Electricity cost	73,352,124.34
3	General maintenance	1,446,969.60
4	ACF media replacement	298,256.36
5	DMF sand replacement	178,688.34
6	DMF gravel replacement	74,453.47
7	UV lamp replacement	1,367,649.36
8	UV quartz replacement	284,085.90
9	Cost of pump replacement considering a pump and motor design life of 10 years	2,340,443.93
10	Cost of painting after every two years	1,029,180.00
11	Bleaching powder cost	2,170,454.40
12	Lab analysis cost	2,170,454.40
13	Contractor's profit @ 15% for Sl. 1, 3–12	3,413,328.20
14	Total operating cost (Sum Sl. 1–14)	99,520,973.90
15	Capital cost	25,000,000.00
16	Salvage value	2,500,000.00
17	Total cost (14 + 15 − 16)	122,020,973.90
18	Total water treated in 20 years	7,300,000
19	Cost of water treatment Rs. per kL	16.72

Questions

21.1. Define the following terms:
   a. Amortization factor
   b. Present worth
   c. Baseline
   d. Capitalized cost
   e. Discount rate
   f. Inflation rate.

21.2. Deduce the main difference between present worth and annualized cost method.

21.3. Estimate the present worth of expenditures to be incurred at different time intervals as given in Table 21.25. Use the inflation and the discount rates as 6 and 10%.

21.4. Ascertain the different amortization factors for the expenditures as displayed in Table 21.26 using the inflation and discount rates as 6 and 10%.

Table 21.25 Values of expenditures

Sl. No	Time with respect to baseline year	Future value of expenditure (Rs.)
1	5th year	100,000
2	10th year	115,450
3	15th year	120,510
4	18th year	200,000

Table 21.26 Values of future expenditures

Sl. No	Time with respect to baseline year	Future value of expenditure (Rs.)
1	4th year	100,000
2	12th year	150,000
3	15th year	180,000
4	21st year	240,000

21.5. Discuss the cost of sewage treatment associated with extended aeration ASP, MBR and SBR.

21.6. Write short note on economic analysis of the sewage treatment plant.

21.7. Based on the economic analysis performed on the existing sewage treatment plant, identify the scopes to reduce the cost of sewage treatment, which will make the sewage treatment plant operation sustainable and affordable.

21.8. Define the terminologies: (i) Present worth; (ii) Amortization; (iii) Life cycle costing.

21.9. Describe the importance of life cycle costing in taking the final decision on the selection of suitable technology for treatment of sewage.

21.10. Describe the terminology what is meant by 'present worth' and 'annualised cost' for a sewage treatment plant.

References

IS 13174 1991. Life cycle costing, part—1: Terminology. Bureau of Indian Standards, New Delhi, India.

IS 13174 1991. Life cycle costing, part—2: Methodology. Bureau of Indian Standards, New Delhi, India.

Appendix 1.1 Effluent Discharge Standards

The impacts of the discharge of urban wastewater into rivers, lakes, estuaries and the sea are a matter of great concern in most countries. An important point in this scenario is the establishment of an adequate legislation for the protection of the quality of water resources, this being a crucial point in the environmental and public health development of all countries. Most developed nations have already surpassed the basic stages of water pollution problems, and are currently fine-tuning the control of micropollutants, the impacts of pollutants in sensitive areas or the pollution caused by drainage of stormwater. However, developing nations are under constant pressure, from one side observing or attempting to follow the international trends of frequently lowering the limit concentrations of the standards, and from the other side being unable to reverse the continuous trend of environmental degradation. The increase in the sanitary infrastructure can barely cope with the net population growth in many countries. The implementation of sanitation and sewage treatment depends largely on political will and even when this is present, financial constraints are the final barrier to undermine the necessary steps towards environmental restoration and public health maintenance. Time passes, and the distance between desirable and achievable, between laws and reality, continues to enlarge.

Table A1.1 through Table A1.3 presents effluent discharge standards of various developed and developing countries. In most developed countries, compliance occurs for most of the time, and the main concern relates to the occasional episodes of non-compliance, at which most of the current effort is concentrated. However, in most developing nations the concentrations of pollutants discharged into the water bodies are still very high, and efforts are directed towards reducing the permissible values of discharge standards, and eventually achieving compliance. An adequate legislation for the protection of public health and the quality of water resources is an essential tool in the environmental development of all countries. The transfer of written codes from paper into really practicable standards, which are used not merely for enforcement but mainly as an integral part of the public health and environmental protection policy, has been a challenge for most countries.

See Tables A1.2, A1.3.

M. Ghangrekar, *Wastewater to Water*, https://doi.org/10.1007/978-981-19-4048-4

Table A1.1 Effluent discharge standard of Asian countries

Countries ▶	India (Ref. CPHEEO, 2012)	Taiwan (Ref. Revisions to Articles 2 promulgated by EPA Order Huan-Shu-Shui-Tzu No., 2014)	China (Ref. China Water Risk, 1996)	Malaysia (Ref. Malaysia's Environmental Law, 1974)	Singapore (Ref. Singapore's Environmental Law, 2000)	Japan (Ref. Ministry of the Environment, 2015)	Thailand (Ref. Kayode et al., 2018)
Characteristics of the effluent ▶							
Temp. (°C)	Shall not exceed 5 °C above the receiving water temperature	<40	–	40	45	–	40
pH	5.5–9.0	6.0–9.0	6.0–9.0	6.0–9.0	6.0–9.0	5.8–8.6	5.5–9.0
Fluorides (mg/L)	2.0	15.0	10.0	–	–	–	–
Fluorine and its compounds (mg/L)	–	–	–	–	–	Non-coastal areas: 8.0 Coastal areas: 15.0	–
Nitrate nitrogen (mg/L)	10.0	50.0	–	–	–	–	–
Nitrate (as NO$_3$) (mg/L)	–	–	–	–	20.0	–	–
Ammonia nitrogen (mg/L)	50	10.0	–	–	–	–	–
Ammonium Nitrogen (mg/L)	–	–	15.0	–	–	–	–
TKN (mg/L)	100	–	–	–	–	–	–
Free Ammonia as NH$_3$ (mg/L)	5.0	–	–	–	–	–	–

(continued)

Table A1.1 (continued)

Countries ▶	India (Ref. CPHEEO, 2012)	Taiwan (Ref. Revisions to Articles 2 promulgated by EPA Order Huan-Shu-Shui-Tzu No., 2014)	China (Ref. China Water Risk, 1996)	Malaysia (Ref. Malaysia's Environmental Law, 1974)	Singapore (Ref. Singapore's Environmental Law, 2000)	Japan (Ref. Ministry of the Environment, 2015)	Thailand (Ref. Kayode et al., 2018)
Ammonia, Ammonium compounds, Nitrate and Nitrite compounds (Total of NH_3-N multiplied by 0.4, NO_2-N and NO_3-N)	–	–	–	–	–	100	–
BOD_5 (mg/L)	30	30.0	20.0	20.0	20.0	160 (Daily Average 120)	20–60
COD (mg/L)	250	100.0	60.0	50.0	60.0	160 (Daily Average 120)	120–400
Suspended Solids (mg/L)	100	30.0	20.0	50.0	30	200 (Daily Average 150)	50
Total dissolved solids (mg/L)	–	–	–	–	1000	–	–
Total nitrogen (mg/L)	–	15.0	–	–	–	120 (Daily Average 60)	–
Total phosphorus (mg/L)	5.0	2.0	–	–	–	16.0 (Daily Average 8.0)	–
Phosphorus (as an element) (mg/L)	–	–	0.1	–	–	–	–
Orthophosphates (mg/L)	–	4.0	–	–	–	–	–

(continued)

Table A1.1 (continued)

Countries ▶	India (Ref. CPHEEO, 2012)	Taiwan (Ref. Revisions to Articles 2 promulgated by EPA Order Huan-Shu-Shui-Tzu No., 2014)	China (Ref. China Water Risk, 1996)	Malaysia (Ref. Malaysia's Environmental Law, 1974)	Singapore (Ref. Singapore's Environmental Law, 2000)	Japan (Ref. Ministry of the Environment, 2015)	Thailand (Ref. Kayode et al., 2018)
Phosphates (as P) (mg/L)	–	–	0.5	–	–	–	–
Phosphate (as PO4) (mg/L)	–	–	–	–	2.0	–	–
Organic phosphorus compounds (parathion, methyl parathion, methyl demeton and EPN only) (mg/L)	–	–	–	–	–	1.0	–
Phenols (mg/L)	–	1.0	0.5	0.001	Nil	5.0	1.0
Anionic surfactants (mg/L)	–	10.0	5.0	–	–	–	–
Cyanide (mg/L)	0.2	1.0	–	0.05	–	1.0	–
Oil and grease (mg/L)	10	10.0	–	Not detectable	1 (Total)	–	5–15
Soluble iron (mg/L)	–	10.0	–	–	–	10.0	–
Iron (Fe) (mg/L)	3.0	–	–	1.0	1.0	–	–
Soluble manganese (mg/L)	–	10.0	–	–	–	10.0	–
Total Manganese (Mn) (mg/L)	2.0	–	2.0	0.20	0.5	–	5.0
Cadmium (mg/L)	2.0	0.03	–	0.01	0.003	0.03	0.03
Lead (mg/L)	0.1	1.0	–	0.10	0.1	0.1	0.2

(continued)

Table A1.1 (continued)

Countries ▶	India (Ref. CPHEEO, 2012)	Taiwan (Ref. Revisions to Articles 2 promulgated by EPA Order Huan-Shu-Shui-Tzu No., 2014)	China (Ref. China Water Risk, 1996)	Malaysia (Ref. Malaysia's Environmental Law, 1974)	Singapore (Ref. Singapore's Environmental Law, 2000)	Japan (Ref. Ministry of the Environment, 2015)	Thailand (Ref. Kayode et al., 2018)
Total chromium (mg/L)	2.0	2.0	–	–	–	2.0	–
Hexavalent chromium (mg/L)	0.1	0.5	–	0.05	0.05	0.5	0.25
Trivalent chromium (mg/L)	–	–	–	0.20	0.05	–	–
Methyl mercury (mg/L)	–	0.0000002	–	–	–	–	–
Total mercury (mg/L)	0.01	0.005	–	0.005	0.001	0.005	–
Alkyl mercury compounds (mg/L)	–	–	–	–	–	Not detectable	–
Copper (mg/L)	3.0	3.0	0.5	0.20	0.1	3.0	2.0
Zinc (mg/L)	5.0	5.0	2.0	1.0	0.5	2.0	5.0
Silver (mg/L)	0.5	0.5	–	–	0.1	–	–
Nickel (mg/L)	3.0	1.0	–	0.20	0.1	–	1.0
Selenium (mg/L)	0.05	0.5	–	–	0.01	0.1	–
Arsenic (mg/L)	0.2	0.5	–	0.05	0.01	0.1	–
Boron (mg/L)	–	1.0	–	1.0	0.5	–	–
Sulphides (mg/L)	2.0	1.0	1.0	0.5	0.2	Non-coastal areas: 10.0 Coastal areas: 230	1.0

(continued)

Table A1.1 (continued)

Countries ▶	India (Ref. CPHEEO, 2012)	Taiwan (Ref. Revisions to Articles 2 promulgated by EPA Order Huan-Shu-Shui-Tzu No., 2014)	China (Ref. China Water Risk, 1996)	Malaysia (Ref. Malaysia's Environmental Law, 1974)	Singapore (Ref. Singapore's Environmental Law, 2000)	Japan (Ref. Ministry of the Environment, 2015)	Thailand (Ref. Kayode et al., 2018)
Sulphate (as SO$_4$) (mg/L)	–	–	–	–	200	–	–
Formaldehyde (mg/L)	–	3.0	1.0	–	–	–	–
Petroleum hydrocarbons (mg/L)	–	–	5.0	–	–	–	–
Vegetable and animal oils (mg/L)	–	–	10.0	–	–	–	–
Total cyanides (CN$^-$) (mg/L)	–	–	0.5	–	0.1	–	–
Aniline (mg/L)	–	–	1.0	–	–	–	–
Nitrobenzene (mg/L)	–	–	2.0	–	–	–	–
Benzene (mg/L)	–	–	0.1	–	–	0.1	–
Methylbenzene (mg/L)	–	–	0.1	–	–	–	–
Ethylbenzene (mg/L)	–	–	0.4	–	–	–	–
Total Organic Carbon (TOC) (mg/L)	–	–	20.0	–	–	–	–
Tin (mg/L)	–	–	–	0.20	5.0	–	–
Free chlorine (mg/L)	–	–	–	1.0	1.0	–	1.0
Chloride (as chloride ion) (mg/L)	–	–	–	–	250	–	–

(continued)

Table A1.1 (continued)

Countries ▶	India (Ref. CPHEEO, 2012)	Taiwan (Ref. Revisions to Articles 2 promulgated by EPA Order Huan-Shu-Shui-Tzu No., 2014)	China (Ref. China Water Risk, 1996)	Malaysia (Ref. Malaysia's Environmental Law, 1974)	Singapore (Ref. Singapore's Environmental Law, 2000)	Japan (Ref. Ministry of the Environment, 2015)	Thailand (Ref. Kayode et al., 2018)
Detergents (linear alkylate sulphonate as methylene blue active substances) (mg/L)	–	–	–	–	5.0	–	–
Barium (mg/L)	–	–	–	–	1.0	–	–
Beryllium (mg/L)	–	–	–	–	0.5	–	–
Metals in total (mg/L)	–	–	–	–	0.5	–	–
Calcium (as Ca) (mg/L)	–	–	–	–	150	–	–
Magnesium (as Mg) (mg/L)	–	–	–	–	150	–	–
N-hexane Extracts (mineral oil) (mg/L)	–	–	–	–	–	5.0	–
N-hexane extracts (animal and vegetable fats) (mg/L)	–	–	–	–	–	30.0	–
Coliform groups	–	–	–	–	–	Daily Average 3000 per cm^3	–
Faecal coliforms (MPN/100 mL for discharge)	Desirable—1000 Max. Permissible—10,000	–	–	–	–	–	–
PCBs (mg/L)	–	–	–	–	–	0.003	–

(continued)

Table A1.1 (continued)

Countries ▶	India (Ref. CPHEEO, 2012)	Taiwan (Ref. Revisions to Articles 2 promulgated by EPA Order Huan-Shu-Shui-Tzu No., 2014)	China (Ref. China Water Risk, 1996)	Malaysia (Ref. Malaysia's Environmental Law, 1974)	Singapore (Ref. Singapore's Environmental Law, 2000)	Japan (Ref. Ministry of the Environment, 2015)	Thailand (Ref. Kayode et al., 2018)
Trichloroethylene (mg/L)	–	–	–	–	–	0.1	–
Tetrachloroethylene (mg/L)	–	–	–	–	–	0.1	–
Dichloromethane (mg/L)	–	–	–	–	–	0.2	–
Carbon tetrachloride (mg/L)	–	–	–	–	–	0.02	–
1, 2-dichloro ethane (mg/L)	–	–	–	–	–	0.04	–
1, 1-dichloro ethylene (mg/L)	–	–	–	–	–	1.0	–
cis-1, 2-dichloro ethylene (mg/L)	–	–	–	–	–	0.4	–
1, 1, 1-trichloro ethane (mg/L)	–	–	–	–	–	3.0	–
1, 1, 2-trichloro ethane (mg/L)	–	–	–	–	–	0.06	–
1, 3-dichloropropene (mg/L)	–	–	–	–	–	0.02	–
Thiram (mg/L)	–	–	–	–	–	0.06	–
Simazine (mg/L)	–	–	–	–	–	0.03	–
Thiobencarb (mg/L)	–	–	–	–	–	0.2	–

(continued)

Table A1.1 (continued)

Countries ▶	India (Ref. CPHEEO, 2012)	Taiwan (Ref. Revisions to Articles 2 promulgated by EPA Order Huan-Shu-Shui-Tzu No., 2014)	China (Ref. China Water Risk, 1996)	Malaysia (Ref. Malaysia's Environmental Law, 1974)	Singapore (Ref. Singapore's Environmental Law, 2000)	Japan (Ref. Ministry of the Environment, 2015)	Thailand (Ref. Kayode et al., 2018)
1,4-dioxane (mg/L)	–	–	–	–	–	0.5	–
Alpha emitters (micro curie/L)	10^{-7}	–	–	–	–	–	–
Beta emitters (micro curie/L)	10^{-6}	–	–	–	–	–	–
Vanadium as V (mg/L)	0.2	–	–	–	–	–	–

Table A1.2 Effluent discharge standard of selected European countries (Ref. Environment Agency; Schellenberg et al., 2020; Preisner et al., 2020)

Country	PE Treated	pH	Temp. (°C)	SS (mg/L)	COD (mg/L)	BOD$_5$ (mg/L)	TN (mg/L)	NH$_4$-N (mg/L)	TP (mg/L)
EU Urban Wastewater Treatment Directive (UWWTD)	<2000	–	–	35 (or 90% reduction)	125 (or 70–90% reduction)	25 (or 70–90% reduction)	–	–	–
	2000–10,000	–	–				–	–	–
	10,000–100,000	–	–				15	–	2
	>100,000	–	–				10	–	1
UK	–	–	–	–	125	25	15	–	2
Ireland	≤10	–	–	30	–	20	5	20	–
	>2000	UWWTD apply as a minimum, but may be more stringent to comply with Water Framework Directive (WFD)							
France	<20	–	–	30	–	35	–	–	–
	20–2000	6.0–8.5	<25	50% reduction	60% reduction	35, 60% reduction	–	–	–
	>2000	UWWTD apply as a minimum, but may be more stringent to comply with Water Framework Directive (WFD)							
Romania	>2000	UWWTD apply as a minimum, but may be more stringent to comply with Water Framework Directive (WFD)							
Germany	<1000	–	–	–	150	40	–	–	–
	<5000	–	–	–	110	25	–	–	–
	<20,000	–	–	–	90	20	–	10	–
	<100,000	–	–	–	90	20	18	10	2
	>100,000	–	–	–	75	15	13	10	1
Sweden	>2000	–	–	–	–	12.45	15	–	0.5
	2000–100,000	–	–	–	–	12.45	15	–	0.5
	> 100,000	–	–	–	–	12.45	10	–	0.5
Denmark	–	–	–	–	75	10	8	–	0.4
HELCOM signatory countries	300–2000	–	–	–		25	35	–	2
	2000–10,000	–	–	–		15	30	–	1

(continued)

Table A1.2 (continued)

Country	PE Treated	pH	Temp. (°C)	SS (mg/L)	COD (mg/L)	BOD$_5$ (mg/L)	TN (mg/L)	NH$_4$-N (mg/L)	TP (mg/L)
	10,000–100,000	–	–	–	–	15	15	–	0.5
	>100,000	–	–	–	–	15	10	–	0.5

PE—Population equivalent

Table A1.3 Effluent discharge standard of non-EU countries/regions (Ref. Preisner et al., 2020)

Country/region	WWTP category	COD (mg/L)	BOD$_5$ (mg/L)	NH$_4^+$-N, NH$_3$-N (mg/L)	NO$_2^-$-N, NO$_3^-$-N (mg/L)	TN (mg/L)	PO$_4^{3-}$-P (mg/L)	TP (mg/L)
Belarus	<500 PE	125	35	–	–	–	–	–
	501–2000 PE	120	30	20	–	–	–	–
	2001–10,000 PE	100	25	15	–	–	–	–
	10,001–100,000 PE	80	20	–	–	20	–	4.5
	> 100,000 PE	70	15	–	–	15	–	2
Switzerland	200–10,000 PE	60	20	2 (sum of NH$_3$-N and NH$_4$-N)	0.3 (NO$_2^-$-N)	–	0.8	–
	>10,000 PE	45	15	2 (sum of NH$_3$-N and NH$_4$-N)	0.3 (NO$_2^-$-N)	–	0.8	–
USA	–	–	30	–	–	3–5 (areas sensitive to eutrophication)	–	1.0–0.1 (areas sensitive to eutrophication)
BC, Canada	Streams, rivers and estuaries	–	45 (10 if dilution ratio <40:1)	–	–	–	0.5 (MDF > 50 m^3/d)	1.0 (MDF > 50 m^3/d)

(continued)

Table A1.3 (continued)

Country/region	WWTP category	COD (mg/L)	BOD5 (mg/L)	NH4+-N, NH3-N (mg/L)	NO2−-N, NO3−-N (mg/L)	TN (mg/L)	PO43−-P (mg/L)	TP (mg/L)
	Lakes	–	45	–	–	–	0.5 (MDF > 50 m3/d)	1.0 (MDF > 50 m3/d)
	Open marine water	–	130 (MDF > 10 m3/d)	–	–	–	–	–
	Coastal waters	–	45 (MDF > 10 m3/d)	–	–	–	–	–
Dubai	Harbor area	100	50	2 (NH4+-N)	40 (NO3−-N)	10 (TKN)	2	–
	Open Sea	–	30	5 (NH3-N)	–	–	0.1	–

MDF—maximum daily flow, TKN—as a sum of organic nitrogen (N_{org}) and NH_4^+-N

References

CPHEEO (2012). *Manual on sewerage and sewage treatment, part A: engineering final draft, central public health and environmental engineering organisation.* Ministry of Urban Development, New Delhi.

China Water Risk (1996). National standard of the People's Republic of China integrated wastewater discharge standard GB 8978.

Environment Agency. Guidance waste water treatment works: Treatment monitoring and compliance limit (www.gov.uk).

Kayode, O. F., Luethi, C., & Rene, E. R. (2018). Management recommendations for improving decentralized wastewater treatment by the food and beverage industries in Nigeria. *Environments, 5*(3), 41.

Malaysia's Environmental Law (2000). Environmental quality act 1974, the Malaysia environmental quality (sewage and industrial effluents) regulations, 1979, 1999.

Ministry of the Environment (2015). Government of Japan; Uniform National Effluent Standards (Last update: October 21, 2015).

Preisner, M., Neverova-Dziopak, E., & Kowalewski, Z. (2020). An analytical review of different approaches to wastewater discharge standards with particular emphasis on nutrients. *Environmental Management, 66*(4), 694–708.

Revisions to Articles 2 promulgated by EPA Order Huan-Shu-Shui-Tzu No. 1030005842 on January 22, 2014.

Singapore's Environmental Law, Code of Practice on Pollution Control (2000). Appendix 9: Allowable limits for trade effluent discharged in to a public sewer/watercourse/controlled watercourse.

Schellenberg, T., Subramanian, V., Ganeshan, G., Tompkins, D., & Pradeep, R. (2020). Wastewater discharge standards in the evolving context of urban sustainability–The case of India. *Frontiers in Environmental Science, 8*, 30.

Appendix A: Periodic Table of the Elements

M. Ghangrekar, *Wastewater to Water*, https://doi.org/10.1007/978-981-19-4048-4

Legend

Atomic number	Symbol	Name	Atomic mass	Chemical group block
1	H	Hydrogen	1.008	Nonmetal

Periodic Table of the Elements

Group	Z	Symbol	Name	Atomic mass	Block
IA	1	H	Hydrogen	1.008	Nonmetal
0	2	He	Helium	4.002	Noble gas
IA	3	Li	Lithium	6.941	Alkali Metal
IIA	4	Be	Beryllium	9.012	Alkaline Earth Metal
IIIA	5	B	Boron	10.811	Metalloid
IVA	6	C	Carbon	12.011	Nonmetal
VA	7	N	Nitrogen	14.007	Nonmetal
VIA	8	O	Oxygen	15.999	Nonmetal
VIIA	9	F	Fluorine	18.998	Halogen
0	10	Ne	Neon	20.179	Noble gas
IA	11	Na	Sodium	22.989	Alkali Metal
IIA	12	Mg	Magnesium	24.305	Alkaline Earth Metal
IIIA	13	Al	Aluminum	26.982	Post-Transition Metal
IVA	14	Si	Silicon	28.085	Metalloid
VA	15	P	Phosphorus	30.973	Nonmetal
VIA	16	S	Sulphur	32.066	Nonmetal
VIIA	17	Cl	Chlorine	35.453	Halogen
0	18	Ar	Argon	39.948	Noble gas
IA	19	K	Potassium	39.098	Alkali Metal
IIA	20	Ca	Calcium	40.078	Alkaline Earth Metal
IIIB	21	Sc	Scandium	44.956	Transition Metal
IVB	22	Ti	Titanium	47.88	Transition metal
VB	23	V	Vanadium	50.941	Transition metal
VIB	24	Cr	Chromium	51.996	Transition metal
VIIB	25	Mn	Manganese	54.938	Transition metal
VIII	26	Fe	Iron	55.847	Transition metal
VIII	27	Co	Cobalt	58.933	Transition metal
VIII	28	Ni	Nickel	58.69	Transition metal
IB	29	Cu	Copper	63.546	Transition metal
IIB	30	Zn	Zinc	65.39	Transition metal
IIIA	31	Ga	Gallium	69.723	Post-Transition metal
IVA	32	Ge	Germanium	72.61	Metalloid
VA	33	As	Arsenic	74.922	Metalloid
VIA	34	Se	Selenium	78.96	Nonmetal
VIIA	35	Br	Bromine	79.904	Halogen
0	36	Kr	Krypton	83.80	Noble gas
IA	37	Rb	Rubidium	85.468	Alkali Metal
IIA	38	Sr	Strontium	87.62	Alkaline Earth Metal
IIIB	39	Y	Yttrium	88.906	Transition Metal
IVB	40	Zr	Zirconium	91.224	Transition metal
VB	41	Nb	Niobium	92.906	Transition metal
VIB	42	Mo	Molybdenum	95.94	Transition metal
VIIB	43	Tc	Technetium	98.907	Transition metal
VIII	44	Ru	Ruthenium	101.07	Transition metal
VIII	45	Rh	Rhodium	102.905	Transition metal
VIII	46	Pd	Palladium	106.42	Transition metal
IB	47	Ag	Silver	107.868	Transition metal
IIB	48	Cd	Cadmium	112.411	Transition metal
IIIA	49	In	Indium	114.82	Post-Transition Metal
IVA	50	Sn	Tin	118.710	Post-Transition Metal
VA	51	Sb	Antimony	121.75	Metalloid
VIA	52	Te	Tellurium	127.60	Metalloid
VIIA	53	I	Iodine	126.904	Halogen
0	54	Xe	Xenon	131.29	Noble gas
IA	55	Cs	Cesium	132.905	Alkali Metal
IIA	56	Ba	Barium	137.327	Alkaline Earth Metal
IIIB	57	*La	Lanthanum	138.906	Lanthanide
IVB	72	Hf	Hafnium	178.49	Transition metal
VB	73	Ta	Tantalum	180.948	Transition metal
VIB	74	W	Tungsten	183.85	Transition metal
VIIB	75	Re	Rhenium	186.207	Transition metal
VIII	76	Os	Osmium	190.2	Transition metal
VIII	77	Ir	Iridium	192.22	Transition metal
VIII	78	Pt	Platinum	195.08	Transition metal
IB	79	Au	Gold	196.966	Transition metal
IIB	80	Hg	Mercury	200.59	Transition metal
IIIA	81	Tl	Thallium	204.383	Post-Transition Metal
IVA	82	Pb	Lead	207.2	Post-Transition Metal
VA	83	Bi	Bismuth	208.980	Post-Transition Metal
VIA	84	Po	Polonium	208.982	Metalloid
VIIA	85	At	Astatine		Halogen
0	86	Rn	Radon	222.018	Noble gas
IA	87	Fr	Francium	223.019	Alkali Metal
IIA	88	Ra	Radium	226.025	Alkaline Earth Metal
IIIB	89	**Ac	Actinium	227.028	Actinide
IVB	104	Rf	Rutherfordium	261.11	Transition metal
VB	105	Db	Dubnium	268.126	Transition metal
VIB	106	Sg	Seaborgium	269.128	Transition metal
VIIB	107	Bh	Bohrium	270.133	Transition metal
VIII	108	Hs	Hassium	269.134	Transition metal
VIII	109	Mt	Meitnerium	277.154	Transition metal
VIII	110	Ds	Darmstadtium	282.166	Transition metal
IB	111	Rg	Roentgenium	282.169	Transition metal
IIB	112	Cn	Copernicium	286.179	Transition metal
IIIA	113	Nh	Nihonium	286.182	Post-Transition metal
IVA	114	Fl	Flerovium	290.192	Post-Transition Metal
VA	115	Mc	Moscovium	290.196	Post-Transition Metal
VIA	116	Lv	Livermorium	293.205	Post-Transition Metal
VIIA	117	Ts	Tennessine	294.211	Halogen
0	118	Og	Oganesson	296.216	Noble gas

Lanthanides (*)

Z	Symbol	Name	Atomic mass	Block
58	Ce	Cerium	140.115	Lanthanide
59	Pr	Praseodymium	140.908	Lanthanide
60	Nd	Neodymium	144.24	Lanthanide
61	Pm	Promethium	144.913	Lanthanide
62	Sm	Samarium	150.36	Lanthanide
63	Eu	Europium	151.965	Lanthanide
64	Gd	Gadolinium	157.25	Lanthanide
65	Tb	Terbium	158.926	Lanthanide
66	Dy	Dysprosium	162.50	Lanthanide
67	Ho	Holmium	164.930	Lanthanide
68	Er	Erbium	167.26	Lanthanide
69	Tm	Thulium	168.934	Lanthanide
70	Yb	Ytterbium	173.04	Lanthanide
71	Lu	Lutetium	174.967	Lanthanide

Actinides (**)

Z	Symbol	Name	Atomic mass	Block
90	Th	Thorium	232.038	Actinide
91	Pa	Protactinium	231.036	Actinide
92	U	Uranium	238.029	Actinide
93	Np	Neptunium	237.048	Actinide
94	Pu	Plutonium	244.064	Actinide
95	Am	Americium	243.061	Actinide
96	Cm	Curium	247.070	Actinide
97	Bk	Berkelium	247.070	Actinide
98	Cf	Californium	251.079	Actinide
99	Es	Einsteinium	252.083	Actinide
100	Fm	Fermium	257.095	Actinide
101	Md	Mendelevium	258.10	Actinide
102	No	Nobelium	259.101	Actinide
103	Lr	Lawrencium	260.105	Actinide

Appendix B: Dissolved Oxygen Saturation Concentration Values in Fresh and Saline Water Exposed to Atmosphere Under a Pressure of 760 mm Hg Containing 20.9% Oxygen

Temperature, °C	Dissolved oxygen concentration, mg/L, at different chloride concentration, mg/L		
	0	5000	10,000
0	14.62	13.79	12.97
2	13.84	13.05	12.28
4	13.13	12.41	11.69
6	12.48	11.79	11.12
8	11.87	11.24	10.61
10	11.33	10.73	10.13
12	10.83	10.28	9.72
14	10.37	9.85	9.32
16	9.95	9.46	8.96
18	9.54	9.07	8.62
20	9.17	8.73	8.30
21	8.99	8.57	8.14
22	8.83	8.42	7.99
23	8.68	8.27	7.85
24	8.53	8.12	7.71
25	8.38	7.96	7.56
26	8.22	7.81	7.42
27	8.07	7.67	7.28
28	7.92	7.53	7.14
29	7.77	7.39	7.00
30	7.63	7.25	6.86

© The Editor(s) (if applicable) and The Author(s), under exclusive license to Springer Nature Singapore Pte Ltd. 2022
M. Ghangrekar, *Wastewater to Water*, https://doi.org/10.1007/978-981-19-4048-4

Under any other pressure, P (mm of Hg), the solubility of oxygen, DO_p mg/L, shall be calculated from the corresponding value from the table using following relation:

$$DO_p = DO \frac{P - p}{760 - p}$$

where, DO is the solubility of oxygen at 760 mm of Hg atmospheric pressure and p is the pressure (mm of Hg) of saturated water vapour at that temperature of water. For elevation below 1000 m MSL and temperature less than 25 °C, the p can be ignored and the equation get simplified to $DO_p = DO \frac{P}{760}$.

(*Source* Whipple and Whipple, 1911. Journal of American Chemical Society, 33, 362.)

Appendix C: Physical Properties of Water

Temperature, °C	Density, kg/m^3	Dynamic viscosity $\times 10^3$, N-s/m^2	Kinematic viscosity $\times 10^6$, m^2/s
0	999.87	1.781	1.785
5	999.99	1.518	1.519
10	999.73	1.307	1.306
15	999.13	1.139	1.139
20	998.23	1.002	1.003
22	997.86	0.955	0.957
24	997.38	0.911	0.913
26	996.86	0.871	0.874
28	996.31	0.833	0.836
30	995.68	0.798	0.801
32	995.09	0.765	0.769
34	994.43	0.734	0.738
36	993.73	0.705	0.709
38	993.00	0.678	0.683
40	992.25	0.653	0.658

Note For example, the dynamic viscosity of water at 10 °C will be 1.307×10^{-3}

© The Editor(s) (if applicable) and The Author(s), under exclusive license
to Springer Nature Singapore Pte Ltd. 2022
M. Ghangrekar, *Wastewater to Water*, https://doi.org/10.1007/978-981-19-4048-4

Printed in the United States
by Baker & Taylor Publisher Services